EMBRYONIC STEM CELLS: THE HORMONAL REGULATION OF PLURIPOTENCY AND EMBRYOGENESIS

Edited by **Craig S. Atwood**

Embryonic Stem Cells: The Hormonal Regulation of Pluripotency and Embryogenesis
Edited by Craig S. Atwood

Published by InTech

Janeza Trdine 9, 51000 Rijeka, Croatia

Publishing Process Manager Katarina Lovrecic

Technical Editor Teodora Smiljanic

Cover Designer InTech Design Team

Additional hard copies can be obtained from orders@intechopen.com

Embryonic Stem Cells: The Hormonal Regulation of Pluripotency and Embryogenesis,
Edited by Craig S. Atwood
p. cm.
ISBN 978-953-307-196-1

For EU product safety concerns:
IN TECH d.o.o., Prolaz Marije Krucifikse Kozulić 3, 51000 Rijeka, Croatia,
info@intechopen.com or visit our website at intechopen.com.

Meet the editor

Craig S. Atwood, Ph.D. is an Associate Professor of Medicine at the University of Wisconsin and a Health Science Specialist with the Geriatric Research, Education and Clinical Center at the William S. Middleton Memorial Veterans Administration Hospital in Madison. Dr. Atwood completed his Ph.D. in Biochemistry from the University of Western Australia in Perth, Australia prior to post-doctoral fellowships at the National Cancer Institute, NIH, Bethesda, and Massachusetts General Hospital, Charlestown. He held faculty positions at Harvard Medical School and Case Western Reserve University prior to his current appointment where he directs the research program of the Laboratory for Endocrinology, Aging and Disease. Dr. Atwood has broad research interests related to the endocrinology of embryogenesis, adulthood and senescence as elaborated upon in 'The Reproductive-Cell Cycle Theory of Aging'. He has published over 200 scientific articles, has served on numerous review boards and is an Editor of 20 scientific journals including the Journal of Biological Chemistry. In 2006 he received the Zenith Fellows Award from the Alzheimer's Association in recognition of his research.

Contents

Preface XIII

Part 1 Pluripotency - Embryonic Stem Cells 1

Chapter 1 **Embryonic Stem Cells and the Capture of Pluripotency 3**
Timothy J. Davies and Paul J. Fairchild

Chapter 2 **Molecular Mechanisms of Pluripotency in Murine Embryonic Stem Cells 27**
Elisa A. Casanova, Kurt Bürki and Paolo Cinelli

Chapter 3 **The LIF/STAT3 Pathway in ES Cell Self-renewal 61**
Hiroshi Koide and Takashi Yokota

Chapter 4 **Effects of Recombinant Leukemia Inhibitory Factor (LIF) on Functional Status of Mouse Embryonic Stem Cells 79**
Ludmila M. Mezhevikina, Raushaniya R. Petrova, Marina P. Borisova, Olga Yu. Seraya and E.E. Fesenko

Chapter 5 **Maintaining Embryonic Stem Cells and Induced Pluripotent Stem Cells 91**
Qiang-Sun Zheng, Dong-Bo Ou, Xiong-Tao Liu and Jing-Jing Guo

Chapter 6 **The Function of Glycan Structures for the Maintenance and Differentiation of Embryonic Stem Cells 101**
Shoko Nishihara

Chapter 7 **The Function of E-cadherin in ES Cell Pluripotency 125**
Kate Hawkins and Christopher M Ward

Chapter 8 **Proteomic Analysis of Mouse ES Cells 143**
Akira Kurisaki, Yasuhiro Seki and Atsushi Intoh

Chapter 9 **Molecular Biomarkers of Embryonic Stem Cells 159**
Takumi Miura, Akihiro Umezawa and Hidenori Akutsu

Chapter 10 **Target Identification of MicroRNAs Expressed Highly and Regulated by Activin A in Human Embryonic Stem Cells 173**
Steven Shoei-Lung Li

Chapter 11 **Smads – the Intracellular Hubs of Signalling in Regulation of Pluripotency and Differentiation of Stem Cells 191**
Dmitry A. Ovchinnikov and Ernst J. Wolvetang

Chapter 12 **Determination and Regulation of 'Stemness' by MicroRNAs 217**
Kye-Seong Kim and Jaesang Kim

Part 2 Hormonal Signals that Regulate Early Embryogenesis 233

Chapter 13 **Dynamic Changes in Gene Expression during Early Trophoblast Differentiation from Human Embryonic Stem Cells Treated with BMP4 235**
Toshihiko Ezashi, Bhanu Prakash V.L. Telugu and R. Michael Roberts

Chapter 14 **Human Embryonic Stem Cells: A Model System for Delineating the Molecular Basis of Human Embryogenesis and Aging-related Diseases 251**
Craig S. Atwood and Sivan Vadakkadath Meethal

Chapter 15 **Directed Differentiation of Mesendoderm Derivatives from Embryonic Stem Cells 271**
Miriam Gordillo, Ritu Kumar, Harma K. Turbendian and Todd Evans

Chapter 16 **BNP is a Novel Regulator of Embryonic Stem Cell Proliferation 303**
Abdelalim, EM and Tooyama, I

Chapter 17 **Non-classical Signalling Mechanisms in Stem Cells 317**
Alice Pébay, Hitesh Peshavariya, Raymond C.B. Wong and Gregory J Dusting

Chapter 18 **Telomerase and Oxidative Stress in Embryonic Stem Cells 337**
Gabriele Saretzki

Part 3 Hormonal Signals that Regulate the Differentiation of Ectodermal Cells - Neurogenesis 363

Chapter 19 **Embryonic Neural Stem Cell Differentiation to Aldynoglia Induced by Olfactory Bulb Ensheathing Cell-Conditioned Medium 365**
Daniel Ortuño-Sahagún, Argelia E. Rojas-Mayorquín, Antoni Camins and Mercè Pallàs

Chapter 20 **Dopaminergic Neurons Derived from Human Embryonic Stem Cell Derived Neural Progenitors: Biological Relevance and Application 383**
Young, A. and Stice, S.L.

Chapter 21 **Dopaminergic Differentiation Potential of Neural Precursor Cells Derived from Embryonic Stem Cells 413**
Gilda Guerrero-Flores and Luis Covarrubias

Chapter 22 **A Rational Approach to Inducing Neuronal Differentiation in Embryonic Stem Cells 429**
Tadahide Furuno and Mamoru Nakanishi

Part 4 Hormonal Signals that Regulate the Differentiation of Mesodermal Cells - Cardiogenesis, Angiogenesis and Osteogenesis 445

Chapter 23 **ESC Cardiac Differentiation and Applications 447**
Nasser Aghami and Hananeh Fonoudi

Chapter 24 **Chemicals Regulating Cardiomyocyte Differentiation 471**
Hideki Uosaki and Jun K Yamashita

Chapter 25 **Embryonic Stem Cells as a Model System to Elucidate Early Events in Cardiac Specification and Determination 487**
Gabriella Minchiotti, Cristina D'Aniello, Roberto Ronca, Laura Gualandi and Patrizia Dell'Era

Chapter 26 **Differentiation, Characterization and Applications of Human Embryonic Stem Cell –Derived Cardiomyocytes 505**
Mari Pekkanen-Mattila, Erja Kerkelä and Katriina Aalto-Setälä

Chapter 27 **Recent Advances in Controlling Cardiacmyoctye Differentiation from Embryonic Stem Cells 523**
Dong-Bo Ou, Qiang-Sun Zheng, Yong He, Di Zeng and Ji-Wei Teng

Chapter 28 **Oxytocin as an Inducer of Cardiomyogenesis 537**
Jolanta Gutkowska and Marek Jankowski

Chapter 29 **Impact of Reactive Oxygen and Reactive Nitrogen Species for Stem Cell Mobilization, Function and Cardiovascular Differentiation 557**
Heinrich Sauer and Maria Wartenberg

Chapter 30 **ES Cell Differentiation as a Model to Study Cell Biological Regulation of Vascular Development 581**
Kiyomi Tsuji-Tamura, Hiroshi Sakamoto and Minetaro Ogawa

Chapter 31 **Direct Differentiation of Human Embryonic Stem Cells toward Osteoblasts and Chondrocytes through an Intermediate Mesenchyme Progenitor Lineage 607**
Basem M. Abdallah, Linda Harkness, Amer Mahmood and Moustapha Kassem

Chapter 32 **From Pluripotent Stem Cells to Lineage-Specific Chondrocytes: Essential Signalling and Cellular Intermediates 621**
Naoki Nakayama and Katsutsugu Umeda

Chapter 33 **Derivation of Mesenchymal Stem Cells from Human Embryonic Stem Cells 649**
Shelley E. Brown and Paul H. Krebsbach

Part 5 Hormonal Signals that Regulate the Differentiation of Endodermal Cells - Thymogenesis 661

Chapter 34 **Generation of Thyrocytes from Embryonic Stem Cells 663**
Reigh-Yi Lin

Preface

The human zygote possesses the ability to differentiate into all tissues of the body. The recombination of genetic material within the ovum and its cytoplasmic contents allows for germ cells to form a zygote that during its initial divisions results in cells capable of differentiating into any cell type in the body. This pluripotent state of embryonic stem cells (ESCs) has been harnessed for various applications, such as the generation of knockout or transgenic animals, drug testing, toxicology and as a potential cell source for the derivation of functionally mature cell populations for regenerative medicine applications.

Human pluripotent stem cells (hPSCs, which include both embryonic and induced pluripotent stem cells [hESCs and hiPSCs]) provide a unique model system to study early mammalian development, the topic of this book. In the first section of this book "Pluripotency - Embryonic Stem Cells", Davies and Fairchild describe the spectrum of stem cell types and a historical perspective on the derivation of ESC and recent advances in the maintenance of ESC pluripotency. Casanova et al., Koide and Yokota, Mezhevikina et al. and Zheng et al. follow up by reviewing the hormonal signaling pathways (LIF/Stat3 signaling, Wnt/β-catenin signaling, BMP signaling, and FGF signaling), downstream transcription factors, and biomaterials that have been identified as important in regulating pluripotency in mouse ESC. With respect to this last point, it is becoming well recognized that cell surface proteins are important regulators of ESC fate. Nishihara review the role of glycoproteins (proteoglycans) and glycolipids for the maintenance of self-renewal and pluripotency, and differentiation, of ESC. Heparan sulfate and chondroitin sulfate regulate the extrinsic signaling by BMP, Wnt and FGF that inhibits or is required for differentiation. Hawkins and Ward discuss the specific requirement of E-cadherin in ESC pluripotency, its role in protein cell surface localization, signaling, transcript expression and cell survival. Kurisaki et al. next describe proteomic analyses that they have used to determine the nuclear protein machinery required for the maintenance of pluripotent stem cells. These workers identify TIF1β, and the Oct4-centered interactome, as important in regulating pluripotency. They also describe proteomics approaches for identifying cell-surface markers of PCS, a subject that is comprehensively covered in the chapter by Miura et al. In their chapter, the authors describe traditional transcriptional factor (transcription factors Oct4, Sox2, and Nanog) biomarkers of PCS as well as cell surface antigen biomarkers such as CD9 and the need for new biomarkers that function in the regulation of pluripotency.

The following 3 chapters examine hESC pluripotency. Ovchinnikov and Wolvetang examine the role of Smad signaling as the intracellular hub of TGFβ/activin family member signaling in the regulation of pluripotency and differentiation of stem cells. Li next looks at how activin A regulates miRNAs expression in hESC, while Kim and Kim describe three clusters of miRNAs, miR-302, miR-209, and miR-371 that represent major regulators of pluripotent stem cells. These molecules appear essential for controlling 'stemness' including pluripotency, proliferation, and differentiation. Such molecules may be important (and safer) in the reprogramming of primary somatic cells into ESC-like cells with self-renewal and pluripotency.

In the next section of the book - 'Hormonal Signals that Regulate Early Embryogenesis' we move from signals that regulate pluripotency to physiological signals that regulate early embryonic cell proliferation and differentiation from the zygote to the blastocyst. Ezashi et al. describe the hormonal factors, including BMP4, that regulate early trophoblast differentiation from hESC, and the utility of ESC as a model system to understanding how trophoblasts differentiate into the differentiated syncytiotrophoblast and extravillous trophoblast. This process is crucial for the production of growth factors required for ESC proliferation and differentiation as described by Atwood and Vadakkadath Meethal. These authors describe how ESC can be used as a model system to understand the physiological signals that direct blastulation and neurulation (ectodermal differentiation), and focus on the roles of the trophoblastic/corpus luteal-secreted (pregnancy) hormones human chorionic gonadotropin (hCG) and progesterone in the regulation of these processes. They also describe the potential for using hESC, embryoid bodies (EBs) and neuroectodermal rosettes to gain insights into how reproductive endocrine dyscrasia associated with menopause/andropause drives aberrant cell cycle signalling mechanisms leading to age-related diseases including neurodegeneration associated with cognitive decline. Gordillo et al. next provide a comparative description of the hormonal and transcriptional regulation of mesoderm and endoderm differentiation of tissues (from mesendoderm) among various species. Abdelalim and Tooyama report on novel functions of brain natriuretic peptide, which via the regulation of GABAAR and the transcription factor Ets-1 has a role in the regulation of proliferation and survival of murine ESC, but not pluripotency. The next chapter by Pébay et al. moves away from the classical paracrine or juxtacrine cell interactions within the early embryo to review oxidant signaling through various protein kinase pathways, and intercellular communication through gap junctions in the regulation of ESC proliferation, differentiation and apoptosis. In a related paper, Saretzki highlights the importance of oxygen conditions, mitochondria, telomerase expression and their interplay in the maintenance of self-renewal, pluripotency and stem cell differentiation. High telomerase expression, as well as a special mitochondrial state, are both characteristic properties of ESCs in addition to pluripotency. The low and varying oxygen tension of the uterus might elicit some of the early signals for ESC differentiation.

From early embryonic differentiation we move into a review of the factors and conditions that promote the differentiation of ESC into various cell lineages. The first section deals with the differentiation of ectodermal cell types ('Hormonal Signals that Regulate the Differentiation of Ectodermal Cells - Neurogenesis'). Ortuño-Sahagún et al. first describe the induction of aldynoglia, a special group of macroglial cells, using olfactory bulb ensheathing cell-conditioned medium and involving Wnt, BMP and IGF-1 signaling. Young and Stice provide a very detailed description of the growth factors and signaling pathways involved in the differentiation of ESC into dopaminergic neurons. Guerrero-Flores and Covarrubias review the physiological signals that lead to the development of midbrain dopaminergic neurons as well as the growth factors and co-culturing techniques required for the induction of midbrain dopaminergic neurons from ESC. As the authors note, 'At least two major pathways appear to regulate midbrain dopaminergic neuron specification, the "Shh-Foxa2" and "Wnt1-Lmx1a/b" pathways'. These authors suggest that genetic manipulation of "master genes" is an attractive means of inducing specific stem cell differentiation (avoids identification of specific growth factors and timing issues), and this issue is addressed in the next chapter by Furuno and Nakanishi who report that NeuroD2 overexpression induces ESC differentiation into a neuronal phenotype independent of EB formation.

The next section describes advances in our understanding of mesodermal cell differentiation - 'Hormonal Signals that Regulate the Differentiation of Mesodermal Cells - Cardiogenesis, Angiogenesis and Osteogenesis'. Aghami and Fonoudi, Uosaki and Yamashita, Minchiotti et al. Pekkanen-Mattila et al. and Ou et al. describe the various chemical signals (growth factors and small molecules) and techniques (co-culture and hanging drop, forced aggregation, suspension culture, microprinting technique, manual) that have evolved for ESC differentiation into cardiac lineages. Uosaki and Yamashita elaborate on a culturing system for the generation of cardiomyocytes, endothelial cells, and pericyte from mESCs. Inducers of cardiogenesis include activin, BMPs, VEGF, FGF, DKK1, SHH, Wnt3A, cresent and cripto. A fascinating report by Gutkowska and Jankowski reviews oxytocin induction of cardiomyoctye differentiation from ESC. Oxytocin, historically known for its neuromodulatory, uterine contraction and milk let-down properties (myoepithelial cell contraction), also has well known affects on blood pressure and endothelial cell growth and angiogenesis. These authors describe the mechanism of oxytocin-induced stem cell differentiation, and the expression of oxytocin receptor in oocytes suggests oxytocin signaling as a physiological cardiomyogenic factor. In another important review, Sauer and Wartenberg draw attention to the role of reactive oxygen species (ROS) and reactive nitrogen species in the regulation of stem cell mobilization, function and cardiovascular differentiation. Indeed, the low oxygen tension of the uterus during early embryogenesis may prevent differentiation and promote pluripotency. ROS, NADPH oxidases and NO synthases have been demonstrated to be involved in various signaling pathways and the mediation of cardiomyogenesis, vasculogenesis, angiogenesis and neurogenesis – the so-called 'free radical theory of development' whereby metabolic gradients influence development. Tsuji-Tamura et al. end this section on mesodermal tissue types by reviewing the factors and conditions that regulate endothelial cell differentiation. Many of the chapters in this section also expound on the practical applications (drug discovery, toxicology and regenerative medicine) of identifying the physiological or other signals that can direct differentiation of hESC towards specific cardiac and vascular lineages.

The following chapters examine mesodermal osteogenesis and chondrogenesis. Abdallah et al. review different strategies for directing the differentiation of hESC into osteogenic and chondrogenic lineages via an intermediate mesenchyme progenitor lineage. Among them, withdrawal of feeder cells, addition of PDGF, co-culturing hESC with primary bone derived cells, OP9 cells or periodontal ligament fibroblasts, as well as blocking TGF-β signaling during hEB formation and treatment with activin B all promote differentiation of ESC into mesenchymal stem cells (MSCs). Nakayama and Umeda review the TGFβ (1/3) and BMP (2/4) molecules that regulate cartilage formation from the outgrowth area of EBs. Brown and Krebsbach end this section on MSCs by providing an overview of their derivation from ESC and their application for tissue engineering.

In the final section of the book on endodermal cell differentiation ('Hormonal Signals that Regulate the Differentiation of Endodermal Cells – Thymogenesis'), Lin reviews the factors that induce differentiation (activin A, TSH) and successful maturation into thyroglobulin-expressing (insulin and insulin-like growth factor-1) thyrocytes from mESC.

It is hoped that the research and reviews described here will help to update the ESC research community on recent advances in our understanding of pluripotency, and the hormonal and other factors involved in the differentiation of ESC into tissue types derived from the ectoderm, mesoderm and endoderm.

Craig S. Atwood

Geriatric Research, Education and Clinical Center
Veterans Administration Hospital
Department of Medicine
University of Wisconsin
Madison, USA

Part 1

Pluripotency - Embryonic Stem Cells

1

Embryonic Stem Cells and the Capture of Pluripotency

Timothy J. Davies and Paul J. Fairchild
University of Oxford, Sir William Dunn School of Pathology, South Parks Road, Oxford OX1 3RE, U.K.

1. Introduction

Stem cells have the ability to produce daughter cells, which may either differentiate into specialized cell types or remain uncommitted over repeated mitotic divisions, thereby maintaining the stem cell population. As such, stem cells offer great promise for research and medicine. They have potential uses in cell-based therapies such as blood or organ replacement, genetic engineering, drug and toxic substance screening and for studies in the fields of reproductive development and cancer. A number of different types of stem cell have been described which possess varying degrees of developmental potential. These may either be naturally-occurring *in vivo* (adult, fetal or primordial germ cells) or *in vitro* derived stem cell lines (embryonal carcinomas, embryonic germ cells, embryonic stem cells, extra-embryonic stem cells, trophoblast stem cells, epiblast stem cells or induced pluripotent stem cells). Here, we concentrate on the most extensively-researched type, embryonic stem cells (ESC), and how they are derived in mice. In order to understand why mouse ESC are regarded as the 'gold standard' for pluripotency to which other stem cell types are compared, we shall begin by outlining the origin, developmental potential and logistical suitability of the various forms of stem cell described so far.

2. The spectrum of stem cell types

2.1 Adult stem cells

The organs and tissues of adults harbour stem cells, many of which are multipotent and are, therefore, able to follow multiple differentiation pathways within a particular primary germ layer lineage. Adult stem cells have been derived from tissues such as the brain, blood, muscle, skin, pancreas, liver and from bone marrow, which contains both haematopoietic and mesenchymal stem cells. These somatic stem cells typically reside in small numbers in specific niches within the tissue or organ to which they contribute. They have a limited capacity for proliferation and may lie quiescent over many years until stimulated to divide in order to replace or repair injured tissue. Autologous somatic stem cells offer the potential for patient-specific treatment without the concern of transplant rejection. However, adult stem cells are difficult to isolate and propagate in cell culture, with a limited lifespan. The differentiation repertoire of any particular type of adult stem cell is restricted to a relatively narrow range of cell types, and most somatic stem cell research has centered on haematopoietic stem cells, isolated from bone marrow and blood.

2.2 Fetal stem cells

Further types of stem cell that arise at an early stage during the course of natural reproductive development, include fetal and umbilical stem cells. Whilst umbilical cord blood contains a population of multipotential haematopoietic stem cells (HSC), fetal stem cells are capable of differentiation towards a wider range of cell types. Cord blood-forming stem cells differ from adult haematopoietic stem cells in gene expression, developmental potential, self-renewal, and regulation (Kim et al., 2007). At three to four weeks post partum, fetal stem cells take on adult HSC properties and become quiescent. Fetal HSCs display greater regenerative capacity on adoptive transfer to irradiated hosts than do adult HSCs. Pluripotent fetal stem cells can be isolated from amniotic fluid and placenta, and their capacity to differentiate into a wider range of cell types may offer an alternative to bone marrow transplantation because they elicit a weaker immune response from the host.

2.3 Germline stem cells

Also present in the juvenile and adult are germline stem (GS) cells that reside within a specific stem cell niche and are consistently proliferating asymmetrically to produce the progenitor cells for sperm and egg. Their developmental potential is even more restrictive within the body. Embryonic germ (EG) cells, however, are mouse cell culture lines derived from isolates of primordial germ cells from the genital ridges of post-implantation conceptuses (day E8.5 - E12.5). These stem cell cultures are pluripotent, capable of generating tissues from each of the three primary germ-layers (ectoderm, endoderm and mesoderm). They are non-tumorigenic when transplanted to immunologically matched host but have a limited proliferative capacity *in vitro* (70 - 80 cell divisions). They have been reported to produce chimeras when introduced into a host embryo and permitted to continue development within the maternal environment. Although EG cells have been derived in a number of species (mouse, pig, cow, human and chicken) germline transmission to the offspring of chimeras has only been reported for the mouse (Laborsky et al., 1994; Stewart et al., 1994) and, on one occasion, in the chicken (Chang et al., 1997).

2.4 Embryonic stem cells

During mammalian pre-implantation development, the period extending from fertilization to uterine implantation, the mouse zygote, following the first four cleavage divisions, gives rise to a rounded mass of compacted cells called a morula. Over the course of the next few divisions, two morphologically-distinct tissues are generated which form a blastocyst. The blastocyst comprises a monolayer spheroid of outer trophectoderm surrounding an inner cell mass (ICM) which becomes asymmetrically located within the blastocoelic cavity. By injecting single cells from early (3.5d) and mature (4.5d) mouse ICM into genetically dissimilar host blastocysts and mapping their contributions to the post-gastrulation embryo, Papaioanou and Gardner demonstrated that the mature ICM consisted of two populations of cells, hypoblast (primitive endoderm) and epiblast (primitive ectoderm). These elegant experiments revealed that it was the epiblast tissue which contributed most to the embryo proper, with hypoblast and trophoblast restricted to extra-embryonic regions of the resulting chimeras (Papaioanou & Gardner, 1975).

ESC are an *in vitro* cultured cell line derived from the epiblast of the early embryo. They represent the earliest developmental stage from which stem cells have been obtained. Perhaps reflecting their more primitive origin, these small, highly proliferative adherent cells can generate cell and tissue types from all three primary germ layers (ectoderm, endoderm and mesoderm) of the embryo, both *in vitro* and when used to generate teratomas *in vivo*. Furthermore, when introduced into a host embryo and replaced in the maternal environment, murine ESC can populate the gonads of the resulting chimeric fetus and transmit its genome to subsequent offspring.

Since their derivation in mouse was first described in 1981 (Evans & Kaufman, 1981; Martin, 1981), ESC have engendered much excitement and expectation within the scientific community and much debate amongst the public. Their arrival opened up the possibility of a vehicle for genetic manipulation and delivery to a host, making cell replacement therapy a more achievable goal. As such, murine ESC could be used to produce novel models for human diseases. Mouse ESC can be grown and expanded through many generations in culture, their genomes can be readily altered, and they can be reproducibly differentiated along various developmental pathways *in vitro*. The capacity of ESC when placed *in vivo* to differentiate into every cell type or organ within the body also permitted cell lineage and genetic investigations via the use of gene knockout mice. Thus they provide valuable insights into the earliest stages of embryonic development.

Characteristic of ESC are their morphology, growth pattern and expression of a range of genes now known to be associated with pluripotency. Mouse ESC (mESC) are small cells with a high nucleus to cytoplasm ratio. On feeder cell layers they grow as distinct domed colonies with no discernable intercellular space between neighboring cells. They have a population doubling time of ~12 hours, with almost no gap phases (G1 and G2) in their cell cycle, such that individual colonies display discernable daily growth. If maintained properly, mESC colonies should not show high incidence of spontaneous differentiation between passages and should retain a normal karyotype. Though routinely grown on feeder layers of mitotically inactivated fibroblasts in the presence of fetal bovine serum (FBS) containing medium to which the growth factor, leukemia inhibitory factor (LIF) is often added, mouse ESC can also be cultured under feeder-free conditions. With feeder-free culture on gelatin-coated or specially-treated tissue culture plastic, mESC have a tendency to form monolayers of cells with prominent nucleoli. When grown as a cell suspension in the absence of LIF and feeders, either in hanging drops, on a surface which resists attachment or in a matrix such as methylcellulose; ESC will aggregate together to form rounded, partially-organized structures known as embryoid bodies (EB) (Doetschman et al., 1985; Kurosawa, 2007). The EB provides an environment for ESC to coordinate and differentiate in a fashion similar to embryogenesis. Initially consisting of an outer rind of endoderm surrounding a core of primitive ectoderm after only just a few days' culture, EB can be grown either clonally from a single cell, or formed from aggregates of up to a thousand or more cells. They provide a useful tool for studying progenitor cell interactions and a means of assessing gene targeted mutagenesis effects. With use of differing growth factors and attachment matrices, EB can be encouraged to differentiate to a wide variety of tissue types such as cardiomyocytes (Wobus et al., 1991), skeletal muscle (Miller-Hance et al., 1993) or neuronal (Bain et al., 1995), pancreatic (Skoudy, 2004) and haematopoeitic (Schmitt et al., 1991; Fairchild et al., 2000) cell types. The opinion voiced by many who work on ESCs, that they are like babies - expensive to look after and requiring constant attention - may be something of an exaggeration, but ESC nevertheless require daily changes of medium ("feeding") and passaging to freshly-prepared feeders on average every 3 days (Brook, 2006). A distinct advantage of ESC is that they can be stored frozen under liquid nitrogen for many years without loss of viability.

The pluripotent status of ESC is evident by their expression of a range of cell surface, protein and transcriptional markers. Stage specific embryonic antigen-1 (SSEA-1) is expressed on the cell surface of mouse ESC as well as murine early epiblast and EC cells, and it becomes down-regulated as these cell types differentiate. Interestingly, undifferentiated human ESCs don't express SSEA-1; instead SSEA-3 and SSEA-4 are indicative of pluripotency. Alkaline phosphatase activity also correlates with the undifferentiated state in these cell types, as it does in the mouse. Foremost amongst pluripotency markers are the transcription factors, Octamer binding protein 4 (Oct4), SRY-box containing gene 2 (Sox2) and Nanog which are required to maintain the pluipotent state. Other genes that are highly expressed in uncommitted ESC but which are down regulated at the transcriptional level upon differentiation, include Rex1, Dppa5, utf1, Rex2 and Rif1 (Surani et al., 2010). Early differentiation marker genes, such as Brachyury (mesoderm), nestin (neuroectoderm) and GATA 4 (endoderm) are absent from ESC until differentiation is induced by removal of LIF.

2.5 Other stem cell lines derived from the mouse conceptus

The mouse pre-implantation embryo can also give rise to other types of stem cell. Each of the three primary germ layers of the blastocyst (trophoblast, primitive endoderm and primitive ectoderm) can yield stem cells, though only the epiblast is capable of giving rise to pluripotent stem cells. Trophoblast stem cells (TS) can be derived from the mural trophectoderm layer of mouse blastocysts cultured in the presence of fibroblast growth factor (FGF)-4 and heparin on feeders, or in feeder conditioned medium (Tanaka et al., 1998). FGF-4 is produced by the ICM and binds to FGF-4 receptors on trophoblast cells. The transcription factors Cdx-2 and Eomesodermin (Eome) have been found necessary for maintenance of this cell type. Extra-embryonic endoderm (XEN) stem cells arise if no growth factors (such as LIF or FGF) are added to serum-containing culture medium on feeders (Kunath et al., 2005). These are characterized by expression of GATA-4 and GATA-6 transcription factors. Both TS and XEN cells are restricted to their parental lineages when used to make chimeras. It is not just the mouse blastocyst which can give ESC; lines have been derived from as early in development as day 2.5 (Tesar, 2005; Chung et al., 2006). These cells were found to be equivalent to ESCs derived from blastocyst and demonstrate that the window of development from which ESC can be derived extends from as early as the 4-cell stage until the late blastocyst. That the implantation period represents the closure of this window was revealed with the discovery by two independent investigations, of mouse epiblast stem cells (EpiSC) (Tesar et al., 2007; Brons et al., 2007). EpiSC are stable pluripotent cells derived from the post-implantation epiblast (E5.5-E7.75), requiring FGF-2 for self-renewal. Like ESC they can be expanded almost indefinitely in culture as undifferentiated cells yet still retain their pluripotency. In culture, they can be readily differentiated into a variety of tissue types, perhaps representing a more committed developmental stage than mESC.

EpiSC can be derived from a variety of mouse strains including those considered non-permissive for ESC derivation, such as (B6 x CBA)F_1 and NOD. However their generation from a later stage in embryonic development is reflected by their inability to populate ICM and give rise to chimeras when injected into host blastocysts. Mouse EpiSC are considered to more closely resemble human ESC than mouse ESC. Both human ESC and mouse EpiSC grow as flat epithelial colonies and require FGF and Activin/Nodal signaling to maintain their pluripotency. EpiSC represent the stem cell type which is developmentally closest to gastrulation, and so should provide valuable insight into the events pertaining to somatic and germ cell lineage determination.

2.6 Induced pluripotent stem cells

August 2006 saw the publication of potentially one of the most important findings in the field of stem cell biology. Takahashi and Yamanaka revealed that forced expression of just four reprogramming factors could transform fully differentiated somatic (mouse skin fibroblast) cells into embryonic-like stem cells, which they termed induced pluripotent stem cells (iPSC) (Takahashi & Yamanaka, 2006). Although this first iPSC line failed to generate viable chimeras, subsequent iPSC lines produced by retroviral integration of the four classical reprogramming factors, Oct3/4, Sox2, Klf4 and c-Myc, gave chimeras with germ-line transmission (Okita et al., 2007) and live mice by means of the tetraploid complementation assay (Zhao et al., 2009). Importantly other groups were able to reproduce these findings and, where feasible, extend them to human tissues (Takahashi et al., 2007; Wernig et al., 2007; Yu et al., 2007; Lowry et al., 2008). iPSCs could be readily differentiated into a wide variety of somatic tissue types. However, concerns arose that retroviral and lentiviral vectors used in reprogramming adult tissues would permanently integrate into the cells' genome and that the oncogenic nature of some of the reprogramming factors, especially c-Myc, were responsible for tumours found in chimeras made from iPSC. The rapidly-expanding field of iPSC research soon led to the development of alternative reprogramming strategies and refinements to existing protocols. It was found, for instance, that the number of reprogramming factors could be reduced from four and that others (Nanog, LIN28) could be substituted. Although retroviral vectors remain the most common and efficient method of reprogramming, the possibility that random integration of transgenes into the genome might lead to tumorigenesis, prompted the development of vector-free systems. These include non-integrating adenovirus-based vectors, plasmids, modified insect-specific baculovirus and 'gene-free' systems such as charged protein transfection (Cho et al., 2010) and small molecules, which mimic the activity of the reprogramming factors and promote cell survival. However, the initial enthusiasm that iPSCs represented 'an ethical version of ESC' that could be used for patient-specific treatment of disease or injury, has been tempered by the concerns that iPSC may have a greater propensity to form tumours than ESC (Gutierrez-Aranda et al., 2010). As iPSC came under increasing scrutiny it emerged that they retain an epigenetic memory of the cell type from which they were derived, and that they preferentially differentiated towards a specific lineage linked to their cell of origin. These transcriptional, epigenetic and differentiation differences linked to the cell of origin are seemingly, mostly but not entirely, erased with continuous passaging. Furthermore, evidence is emerging that iPSC are not as indistinguishable from ESC as first thought. Gene expression mapping techniques indicate that they have a novel gene expression profile that is different from that of ESC, but similar amongst iPSC lines (Chin et al., 2009). Additionally, there are more than 500 genes which are differentially expressed between low and high human iPSC passage number, with iPSC becoming more like human ESC at higher passage, though still distinct. It has become evident that the potential for regenerative medicine that has been opened up by the discovery of iPSC elevates the continued study of ESC to an even greater degree of importance.

3. A historical perspective on the derivation of ESC

3.1 Establishing cultures conditions

The advent of mouse ESC was preceded by pioneering research into the establishment of cell culture lines from testicular teratomas (Stevens, 1967) and teratocarcinomas (Kleinsmith & Pierce, 1964). Teratocarcinomas are a subset of germ cell tumours that contain a disorganized array of many somatic and extra-embryonic cells, together with foci of stem cells. Their isolation led to embryonal carcinoma (EC) cell lines becoming the first self-renewing pluripotent cell line to be characterized. EC cells are thought to arise from transformed germ cells in the testis or ovary, they are now regarded as the malignant counterpart of ESC. Classic experiments conducted by Kleinsmith and Pierce in which transplantation of a single EC cell to a new recipient mouse was sufficient to regenerate a new tumour, suggested the pluripotent nature of EC cells. Human EC cell lines were then later obtained from testicular tumours (Andrews et al., 1984). Although of immense value as a research tool, EC cells differ from ESC in several aspects. EC cells have a limited capacity for differentiation, with many lines becoming nullipotent at higher passage number. In addition they are often karyotypically abnormal, particularly the human EC lines. These disparities between ESC and EC may reflect differences in adaptation to culture. The development and refining of culture techniques required for EC isolation and expansion proved to be fundamental to the derivation of mouse ESC.

Studies by Edwards and Paul in Glasgow during the 1960s, into the derivation of cell lines from *in vitro* cultured rabbit pre-implantation embryos, helped pave the way for the discovery of ESC in mice. These early investigations by Edwards and Paul resulted in two long lasting cell lines being isolated from the ICMs of 6d rabbit blastocysts explanted onto a collagen-coated culture surface. The cell lines possessed good proliferation rates, and stability in their secretion of enzymes, morphology and chromosomal complement. One was a fibroblastic type cell line, the other epithelial. Although developmentally-earlier rabbit embryos failed to yield cell lines (Cole et al., 1965; Cole et al., 1966), these investigations focused interest on the mature ICM as a source of progenitor cells. They established techniques and procedures that were to be of fundamental value in revealing the role played by the epiblast in fetal development, such as micro-injection of embryo cells into a host blastocyst to generate chimeras. In 1975, Sherman performed similar experiments with the mouse. Pooled intact blastocysts outgrown *in vitro* gave four cell lines which could be maintained in culture for over a year. However they became chromosomally abnormal and were not able to generate tumours when injected into syngeneic hosts. Sherman had employed the same highly rich medium (NCTC-109) with which Cole and Paul (1965) had achieved limited success. Whilst Cole and Paul reported observing proliferation of 5 - 20% of blastocysts cultured *in vitro*, Sherman obtained 90% hatching and attachment of blastocysts in culture (Sherman, 1975). The improvement may have been attributed to differences between mouse strains (see below) or to the careful selection of the heat-inactivated Fetal Bovine Serum (FBS) used. Currently, the batch testing of FBS for ESC/embryo culture work is standard practice in every laboratory, since commercial supplies can vary extensively in their suitability. Heat-inactivation may have destroyed both contaminants, such as mycoplasma, and heat-labile growth factors that encourage differentiation, such as FGF. FBS likely consists of a milieu of growth factors and biomolecules which probably varies significantly from batch to batch. However bone morphogenetic proteins (BMPs) have now been identified as one such component that has an important role in maintaining ESC in their undifferentiated state (Ying et al., 2003; Qi et al., 2004). BMP, in particular BMP4, acts via the SMAD pathway to induce Inhibition of Differentiation (ID) genes, which suppress neural differentiation of ESC in the mouse. Why then, with critical components (medium and FBS) for long-term blastocyst culture apparently in place, was the attainment of undifferentiated pluripotent, self-renewing cell lines still so elusive?

It is now understood that the BMP contained within the FBS acts in conjunction with another growth factor produced by fibroblast feeder cells, or which can be added as a medium supplement, to maintain self-renewal of mouse ESC. That the fibroblast feeder cell layer produced both an attachment surface and secreted factor(s) conducive to stem cell growth, was known from the studies on EC cell lines. This secreted factor has now been identified as Leukemia Inhibitory Factor (LIF) (Smith et al., 1988; Williams et al., 1988). LIF is an interleukin-6 (IL-6) family multifunctional cytokine that is essential for maintaining self-renewal of undifferentiated mouse ESC, but is not required for such in either the human or rat. Adding recombinant LIF to mouse ESC culture medium removes the need for feeder cells; though many laboratories now routinely culture ESC on feeders with LIF supplemented medium. LIF, a factor also secreted by trophoblast cells, binds to a cell surface complex composed of the LIF receptor LIFRβ, and the transmembrane glycoprotein, gp130. This subsequently actives the transcription factor STAT3 (Signal Transducers and Activators of Transcription-3) that is essential and sufficient for suppression of mESC differentiation (Smith et al., 1988; Niwa et al., 1998; Matsuda et al., 1999). Whilst BMP induces ID genes to block the neural differentiation pathway, LIF stimulates STAT3 to constrain BMP from inducing mesodermal and endodermal differentiation. Recently, a chemically defined medium supplement (Knockout Serum Replacement; KSR) has been formulated which directly substitutes for FBS in media and requires only LIF to be added in order to maintain mESC in the pluripotent condition, either on feeders or under feeder-free conditions. KSR has been identified in our laboratory as an essential component in the efficient derivation of mESC lines, previously regarded as non-permissive.

One further point of interest when considering reasons for Sherman's improved mouse embryo cell line generation, is that blastocysts were cultivated in groups, rather than being plated out individually. Four hundred blastocysts were pooled together for his initial successful experiment, and cohorts of 25 blastocysts for subsequent experiments. Mouse embryos cultured in micro-drops under oil, in the absence of feeders or LIF, often give better morphological development to blastocyst stage when a number of embryos are pooled together, than when cultured individually (T.D and P.F. unpublished observations). During natural pregnancies, LIF is produced maternally by the endometrium and has an important role in trophoblast giant cell differentiation, which is a vital early step in implantation and invasion of the uterus. Addition of LIF to culture medium has been shown to enhance *in vitro* pre-implantation embryo development in several different species, including mice (Kauma & Matt, 1995), cows (Maquant-Le et al., 1993) and sheep (Fry et al., 1992), though the situation is not clear in the human (Jurisicove et al., 1995; Chen et al., 1999). Human ESC in culture do not require LIF, instead FGF-2 must be added to maintain the undifferentiated state. Mouse (and human) pre-implantation embryos produce both LIF and LIR receptor (LIF-R) mRNA transcripts, and the proportion of embryos expressing transcripts increases at the morula and blastocyst stages (Chen et al., 1999). Pooling a number of embryos together may have the effect of enhancing their collective responsiveness to LIF, in a paracrine or autocrine manner, and may result in more robust epiblast development. This conjecture is supported by a study in which mouse embryos, cultured from zygote to blastocyst in medium supplemented with or without LIF, were used for ESC derivation. LIF supplementation had beneficial effects, increasing blastocyst total cell number through increased proliferative activity, especially of the ICM, and enhanced derivation of ESC (Rungsiwiwut et al., 2008). That mESC are able to promote their own growth has been indicated by the description of a Stem cell Autocrine Factor (SAF) secreted by the ESC themselves which supports their clonal propagation in serum-free medium and which is augmented by up-regulated autocrine Nodal signaling during early colony formation (Ogawa et al., 2006). Finally, Sherman reported fibroblasts often dominated his cultures, and these would have been a source of secreted LIF.

3.2 Derivation of ESC

In 1981 two laboratories working independently, brought together the factors crucial for generation of undifferentiated proliferating mouse stem cells. These factors included:

- Culture conditions conducive to good embryo development
- Expansion of the epiblast/stem cell progenitor population
- Removal of the trophoblast
- Separation of the epiblast from the hypoblast
- Establishment of primary colonies in the presence of BMP4 and LIF
- Propagation of primary colonies in the presence of LIF

Evans and Kaufman explanted batches of six whole blastocysts onto a tissue culture plastic surface and allowed them to attach and outgrow *in vitro* for four days. Under these culture conditions, outgrown blastocysts have a 'fried egg' appearance. The trophectoderm firmly attaches and spreads out resembling the flattened egg-white, whilst the ICM elongates to form an egg-cylinder (the yolk). Endodermal cells tend to migrate away from the base of the largely ectodermal egg-cylinder structure. At this stage Evans and Kaufman picked off the epiblast, enzymatically dissociated it and dispersed the resulting cell suspension into plates containing a layer of STO fibroblast feeder cells that had previously been mitotically inactivated. The medium used was Dulbecco's modified minimal essential medium supplemented with 10% FBS and 10% newborn calf serum. Colonies of pluripotent epiblast cells became visible by five days and these could be picked and passaged to fresh feeder plates following trypinisation. In addition to employing conducive culture conditions in a protocol that successfully removes epiblast from the influences of trophectoderm, Evans and Kaufman attempted to maximize the number of stem cell progenitors in the embryo by using delayed-implanting blastocysts. Delay of implantation is a natural occurrence in many species and can be induced in mice either by lactation or by maternal ovariectomy during the early pre-implantation period. Delayed embryos develop normally *in vivo* until the expanded blastocyst stage, shedding their zona pellucida as usual but then become quiescent before giant cell transformation and fail to implant into the uterus. They reach a maximal total cell number of about 130 (half of them epiblast) and can be stimulated to exit diapause and resume normal development, by explanting to culture or by giving maternal hormone injections. By utilizing facultative delay, Evans and Kaufman hoped to maximize the number of epiblast cells available at the time of plating. Using blastocysts from 129SvE mouse strain, over a period, they achieved up to 30% derivation success, establishing some 15 novel ESC lines. The four lines from their initial experiment described above, were all karyotypically normal, with an even sex ratio. These cells were termed "EK" cells by Evans and Kaufman, but they were essentially the first true ESC derived.

Interestingly, the second laboratory to derive pluripotent cell lines at that time, used a slightly different approach. Martin (1981) did not use delayed embryos, nor did she outgrow the blastocyst in order to pick the epiblast. Instead she isolated ICMs by use of immunosurgery to destroy the outer trophectoderm layer. ICMs from about thirty (ICR X SWR/J)F_1 blastocysts produced by superovulation and cultured overnight prior to immunosurgery, were plated onto mitotically-inactivated STO feeders in Dulbecco's modified Eagle's medium (DMEM) supplemented with 10% FBS. The DMEM was supplemented with conditioned medium from EC cells cultured on fibroblasts, which had then been dialyzed and lyophilized and was effectively concentrated five fold at working dilution. Four colonies emerged within a week, and these were dissociated and re-plated onto feeder cells in conditioned medium, giving rise to embryonic stem cell lines which could be passaged on feeders without conditioned medium after the fifth passage. Isolated ICMs similarly cultured but with no conditioned medium, failed to yield ESC. The concentrated conditioned medium produced from EC cells cultured on STO feeders, would have been a rich source of LIF (and BMP4) sufficient to inhibit differentiation of the epiblast and to promote self-renewal. EC conditioned medium may contain other ESC self-renewal factors that are independent of the LIF/Stat3 pathway (Kawazoe et al., 2009). The four lines isolated, which Martin termed 'embryonic stem cells', showed some karyotypic abnormalities but expressed SSEA-1 and like the cell lines of Evans and Kaufman, were shown to be pluripotent both *in vitro* and *in vivo*. Martin was also able to demonstrate that ESC could be derived from other strains eg (C3H x C57BL/6)F_1 mice, albeit with a similar low efficiency (12%). That feeder cells can produce enough LIF has been demonstrated by Stewart et al. who showed that LIF null fibroblasts could not support self-renewal of mESC (Stewart et al., 1992). Why then were Evans and Kaufman able to derive ESC lines on feeders without additional LIF when Martin had found it an absolute requirement?

3.3 Significance of mouse strain on permissiveness of ESC derivation

It may have been serendipitous that Kaufman & Evans chose the 129 strain of mouse from which to attempt to derive ESC, or the success of their experiments may have been related to the propensity of this strain to produce teratocarcinomas from which EC cells had previously been derived (Stevens, 1958). However, other strains of mouse have since proven to be much more refractory to the derivation of ESC. Lines have been obtained from inbred stocks of C57 and BALB/c, but most emanate from 129 strain, though, genetically, this is not the ideal strain and requires time-consuming and expensive backcross breeding programs to transfer any genetic modifications to a more useful background. Despite these drawbacks, 129 are often still favoured over the commonly used C57BL/6 mouse because it has better germline competence (Seong et al., 2004). Until recently, strains such as SVB, CBA and in particular, NOD have been regarded as non-permissive for the derivation of ESC. Even when the occasional lines could be generated from these strains, they often fared badly at producing chimeras and were not able to show germline transmission (Roach et al., 1995; Chen et al., 2005). The CBA/Ca strain was originally selected for its low incidence of mammary tumours, and it seems that genetic background has a profound effect on propensity to generate ESC lines. It is probably for this reason that Martin was not able to generate ESC lines from (ICR x SWR/J)F_1 and (C3H x C57BL/6)F_1 mice by culturing on feeders alone, whilst Evans and Kaufman succeeded using the 129 strain. When she added conditioned medium equivalent to a five fold concentration of LIF, ESC lines were obtained.

Stage	Authors	Technique	Attachment substrate	BMP4	LIF	Mouse strain	Efficiency of derivation
Expand Epiblast	MJE & MHK	Delayed-implantation (2.5d for 4-6 days)	In vivo	-	Endogenous, pooled embryos (x6)	MJE & MHK 129 SvE	4x colonies from 6-12x blastocysts . (30%)
	GM	Overnight culture (3.5d to 4.5d)	None	FBS	Endogenous, pooled embryos (x30)		
Remove trophoblast	MJE & MHK	TB attaches and outgrows.	TC plastic + trophectoderm	FBS	Endogenous, pooled embryos (x6)		
	GM	Immunosurgery	-	-	-		
Isolate Epiblast from Hypoblast	MJE & MHK	Whole blastocyst cultured (4 days). Epiblast picked.	TC plastic + trophectoderm	FBS(10%)			
	GM	Cultured whole ICMs(<7 days). Epiblast picked.	Feeders cells	FBS(10%)	Feeder cells +con. med.	GM ICR x SWR/J C3H x C57BL6	4/30(13%) 1/8(12%)
Establish Primary colonies	MJE & MHK	Dissociated picked epiblast	Feeder cells	FBS(10%) + NCS(10%)	Feeder cells		
	GM	Dissociated picked epiblast	Feeders cells	FBS(10%)	Feeder cells + con. med.		
Expand Primary colonies	MJE & MHK	Dissociated colonies and passaged	Feeder cells	FBS(10%) + NCS(10%)	Feeder cells		
	GM	Dissociated colonies and passaged	Feeders cells	FBS(10%)	Feeder cells + con. med.		

Table 1. Comparison of important stages in the initial description of mESC derivation.

At this time, Evans & Kaufman iterated a number of criteria for a cell line to be regarded as a true ESC line. These included:

- The ESC must remain undifferentiated through repeated passages in cell culture
- The cells must maintain their normal karyotype
- They should retain their pluripotential capacity to differentiate both *in vitro* and *in vivo*
- An additional demonstration of true ESC status which can be tested for in mouse but not human, is that ESC should be capable of giving rise to chimeras when introduced into a host embryo and transferred to the oviducts/uterus of a pseudopregnant recipient
- That chimeric offspring display the capacity for germ line transmission to their descendents is compelling evidence of stem cell status

4. Benchmark assays for true mouse ESC status

4.1 Germline transmission

Obtaining germline transmission (GLT) is an important test of ESC pluripotency status, since it proves that the stem cell is capable of generating functional germ cells as well as somatic cell types. It involves injection of a small number (10 - 15) of ESC into the blastocoelic cavity of a genetically dissimilar host blastocyst, usually differing in coat colour. Successfully injected ESC will become incorporated into the ICM of the blastocyst (Figure 1) and when the host embryo is transferred to the oviducts or uterus of a pseudopregnant recipient mouse, the ESC can contribute fully to the normal embryological development following implantation. Contribution to the chimera is always quite variable, even for injected ESC from the same line, and some lines routinely give more chimeric off-spring than others. If the ESC are included in the formation of the gonads, germline transmission may be observed.

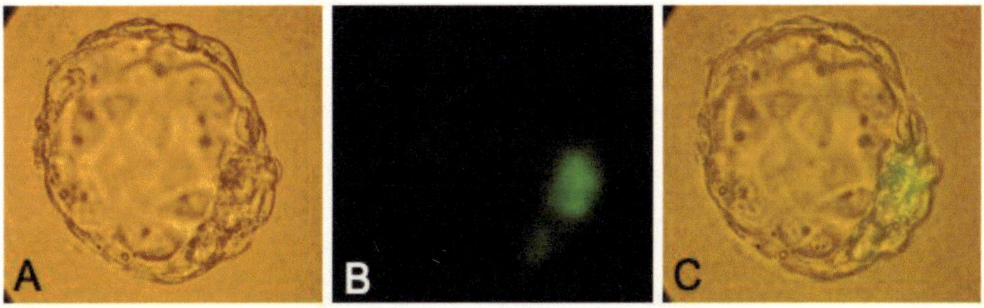

Fig. 1. Embryonic stem cells transgenic for GFP injected into GFP-negative blastocyst and cultured for 1 day become incorporated into host ICM. A) Bright-field phase contrast of an injected embryo. B) dark-field fluorescence showing the introduced ESC and C) combination of low light plus fluorescence, revealing the location of the cells within the ICM.

4.1 Tetraploid complementation

Tetraploid complementation is a very stringent test of the ESC ability to form an entire embryo. In this assay, a tetraploid host embryo is formed by fusing together two cell blastomeres and culturing until morula formation. The resulting tetraploid morula cells are then dissociated and aggregated together with a similar number of ESCs. The cell aggregate re-organizes to form a blastocyst which can be transferred to a pseudopregnant recipient, where implantation will occur. Alternatively, the ESC may be injected into the blastocoelic cavity of a tetraploid blastocyst. The tetraploid host embryo is only able to contribute to extra-embryonic tissues, so that all embryonic lineages must be ESC derived. Live offspring with germ-line transmission represent powerful evidence of the ESC pluripotentiality. The strain-to-strain genetic differences inherent in mice that make ESC derivation more refractory in some strains are not fully understood, but may include chromosomal aberrations such as small changes in DNA methylation affecting gene expression. It has been recognized that, for blastocyst injection experiments, certain combinations of ESC and host embryos are preferable. 129 ESC are often injected into C57BL/6 hosts, whilst Balb/c blastocysts are often used as recipients for C57BL/6 ESC. However, for morula aggregation experiments using 129 ESC, CD1 or ICR are the strains of choice. The reasons why some combinations work better than others may involve strain compatibility issues or relative rates of proliferation of ESC to ICM. Arai et al. have observed that NOD ESC, which typically had a very low GLT (1%), showed improved incorporation into the testes of chimeras when injected into NOD host blastocysts compared to genetically-dissimilar C57BL/6 recipient embryos (Arai et al., 2004). This may have been because their NOD ESC were significantly slower growing than the C57BL/6 ICM cells and so were out-competed at some stage in embryonic development.

4.2 Generation of chimeras by ESC injection

Strain variations are not the only factors affecting mouse chimera contribution, and subsequent germ-line transmission. Not least amongst these factors is the routine day-to-day culturing of stem cells prior to blastocyst injection. Mouse ESC are usually grown in humidified incubators at 37°C under 5% CO_2. Chromosomal aberrations accumulate with time in culture, and each passage represents an opportunity for trauma. To minimize cumulative DNA damage, passage number for blastocyst injection should be kept low. Protocols for optimal culturing of stem cells vary from laboratory to laboratory, and different cell lines may have differing requirements. All protocols agree that to avoid increased spontaneous differentiation, ESC should not be allowed to approach confluency and that their medium should not be allowed to become exhausted before feeding (as evidenced by discoloration of the medium indicating a pH change due to metabolic by-products). In addition to the quality of reagents (FBS, LIF, medium, water source, etc.) and the requirement to use fresh feeders, as we shall discuss later, plating cell density is important. At passage, ESC should be dissociated to single cells/small clusters by gentle pipetting with a 1ml 'Gilson' tip or silliconised Pasteur pipette to avoid subsequent differentiation that would occur if larger clumps persisted. However ESC are gregarious and so should not be sub-divided too much on plating to fresh feeders (p2 - p4). For routine passaging every 2 to 3 days, depending on the rate of proliferation, the range 1:3 to 1:10 is typically observed when sub-dividing. The best way to determine when to sub-divide ESC, is to observe colony morphology by phase-contrast microscopy. The colonies should not coalesce, nor lose their smoothly domed appearance by the accumulation of small rounded cells at their surface. A typical figure for cell density in a flask ready for passaging is between 1-2 x 10^5 cells per cm^2. ESC intended for blastocyst injection to make chimeras should be split 1:1 to 1:3 the day before injection so that the cells are actively proliferating when harvested.

4.3 Homologous recombination in embryonic stem cells

Within three years of the initial description of ESC, Bradley et al. had demonstrated germ line transmission of the ESC genome from three lines, with a rate of chimera formation of more than 50% of live born pups. Seven of these were functional germ line transmitters (Bradley et al., 1984). A few years after that, germ line transmission of genetically altered mESC was achieved (Gossler et al., 1986; Robertson et al., 1986) and with the establishment of homologous recombination techniques (Smithies et al., 1985), gene targeting of specific alleles has since enabled the production of more than 6000 gene knockout mice (Koller et al., 1989; Thomas & Capecchi, 1990). These early endeavors have come to fruition with the setting up of the International Mouse Phenotyping Consortium (IMPC). This is an international co-operation that aims to generate germ line transmission of targeted knockout mutations inESC for each of the 20,000 plus mouse genes on identical genetic backgrounds. The knockouts will be tested and phenotyped and a freely available database established which should provide an invaluable resource for mouse modeling of human diseases.

5. Advances in the derivation of ESC

5.1 Fibroblast feeder cells and conditioned medium

In addition to the huge influence that mouse strain has on the generation of ESC lines, other factors are also of importance. Foremost amongst these factors are the fibroblast feeder cells onto which the blastocyst/ICM is plated. A fibroblast feeder layer may not be an absolute requirement for the derivation of mESCs, but they have been found to improve growth and passaging of primary colonies, a critical stage in obtaining a cell line. Mouse primary embryonic fibroblast (MEF or PEF) feeder cells are best used at low passage (p2 - p4 for ESC derivation) and seeded the day before explanting the blastocyst. They are made mitotically inactive either by gamma irradiation or mitomycin C treatment so as not to overgrow the flask. The fibroblast seeding density should be between 5-10 x 10^4 cells per cm^2. The first ESC lines were isolated on STO fibroblasts (as described above), however mouse ESC can be derived and maintained on a range of fibroblasts. Rabbit spleen-derived fibroblasts (RSF) which express high levels of LIF and Wnt3A, have recently been shown to support mESC self-renewal in an exogenous LIF-free culture system. STO (fibroblasts from SIM mice that are thioguanine- and ouabain-resistant) are a continuous cell line utilized for EC culture. They have the advantage over MEF of not having a limited life span before they lose their proliferation enhancing abilities. However, STO have the disadvantage of requiring optimal culture conditions to prevent aberrant growth which may result in too much differentiating or too little proliferating growth factors. For some strains of mice, MEF have been shown to permit better ESC derivation than STO (Brook & Gardner, 1997). MEF are derived from E13.5 mouse fetus which has been decapitated and eviscerated prior to mincing and plating onto tissue culture plastic in medium containing 10% FBS. Fibroblasts may have their genomes transformed to include drug resistance genes which allow their survival along with ESC during selection experiments. Schoonjans et al. found that medium conditioned by rabbit fibroblasts transduced with the rabbit LIF gene permitted more efficient derivation and maintenance of mESC from ten inbred mouse strains, than adding recombinant mouse or rabbit LIF to unconditioned medium. In each case the ESC were derived on MEF (Schoonjans et al., 2003). Which factor(s) in the conditioned medium secreted by the rabbit fibroblasts were responsible are not yet known. However other IL-6 family cytokines are known to activate the gp130 signaling process that is triggered by LIF. These include related members ciliary neurotrophic factor and oncostatin M and the combination of interleukin-6 (IL-6) plus soluble interleukin-6 receptor (sIL-6R), which have been shown to support ESC derivation (Nichols et al., 1994).

5.2 Efficiency of mESC derivation

Schoonjans et al. (2003) achieved derivation rates of between 5% and 66% efficiency across the ten different inbred strains, including the CBA/CaOla mouse which was previously thought non-permissive. It is difficult to compare efficiency of derivation between publications because of inconsistencies in the criteria used to assess success rates. The efficiency rate described by Schoonjans et al. is based on number of cultured blastocysts that gave rise to ESC lines. Other reports are based on number of lines generated from pooled blastocysts (McWhir et al., 1996) or from ICMs/epiblasts isolated after blastocyst culture period (Roach et al., 1995). Another inconsistency is that a number of separate clonal lines can be raised from primary colonies originating from the same embryo, perhaps from different epiblast progenitor cells, whereas on other occasions several primary colonies from a single embryo may be mixed and regarded as one polyclonal cell line (Robertson, 1987).

5.3 Genetic modification

A number of different strategies and refinements have been used to improve the efficiency of ESC derivation. The approach employed by WcWhir et al. (1996) was to derive ESC lines from hybrid embryos genetically modified for drug resistance linked to expression of a pluripotency marker gene. Non-permissive CBA strain mice were mated with the permissive C57BL/6J strain transgenic for a neomycin-resistance gene linked to Oct3/4 promoter, such that hybrid embryos possessing different levels of CBA genetic background were generated. Under G418 selection conditions, ESC lines could be generated from embryos containing as much as 87.5% CBA genetic background, although at a lower efficiency than embryos containing just 25% CBA (10.5% compared to 22.8%). The influence of strain difference on the capture of pluripotency was also highlighted by the work of Brook and Gardner who achieved very high success rates from (NOD x 129)F_1 x 129 hybrid embryos (88%) which could not be replicated with two strains of pure NOD genetic background (Brook et al., 2003).

5.4 Serum replacement

The defined serum-free substitute for FBS, KSR has been designed for use with an osmolarity optimized DMEM formulation, Knockout D-MEM. The KSR overcomes the problems of variability between FBS batches and is free from unknown differentiation-inducing factors. Using KSR as a direct substitute for FBS, Cheng et al. observed improved establishment of ESC lines from C57BL/6J blastocysts on feeders with LIF (10^3 U/ml) at 19% - 36% efficiency, with stable karyotype and GLT (Cheng et al., 2004). The authors observed that cells cultured in KSR medium are more readily dissociated following trypsinization, which is regarded as a critical step in early culture of ESC. Fibroblast feeder cells lay down a thicker basement matrix in KSR medium, which can often be removed intact during gentle trypsinization thereby removing more fibroblasts and enriching the remaining ESC population. ESC colonies adopt a more prominent domed morphology when KSR is substituted for FBS, which may provide a cell-to-cell contact environment more suitable to ESC proliferation (Figure 2).

Whilst KSR alleviates the uncertainties associated with FBS, it seems that a low concentration of serum may be beneficial. For an ESC line (MGZ5), normally resistant to clonal growth in medium supplemented with 10% KSR and LIF (10^3 U/ml), addition of a small amount (0.3%) of FBS to the KSR medium permitted ESC propagation from single cells in low-density culture (<100 cells/cm^2) (Ogawa et al., 2006). Bryja et al. described using KSR for generating mESC from C57BL/6J blastocysts, also in conjunction with feeders and LIF (10^3 U/ml), with a success rate of 50% - 75%. In these experiments KSR was alternated between pulses of FBS used following trypsinization steps to dissociate the ICM and primary colonies. These authors also reported that blastocysts explanted on feeders in KSR retained more ESC-like properties, such as expression of Oct-4 and decreased Erk kinase activation, than embryos similarly cultured in FBS (Bryja et al., 2006). KSR medium has recently been shown to be more effective than serum-containing medium for obtaining iPSC from mouse embryonic and adult fibroblasts (Okada et al., 2010).

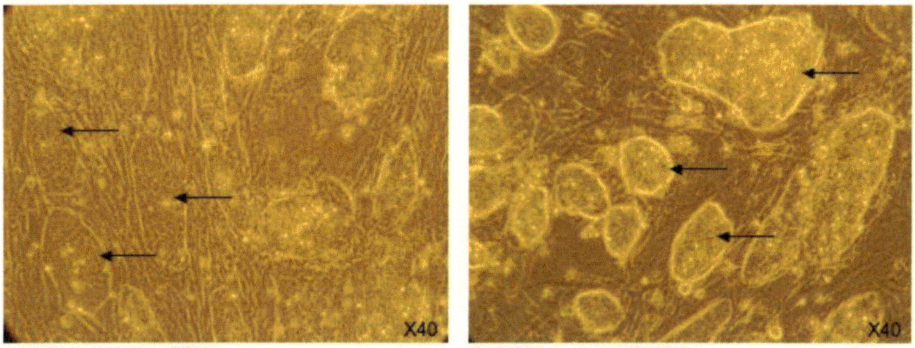

Fig. 2. Comparison of colony morphology between C57BL/6 ESC (EST29) cultured on feeders in either 15% FBS- or 15% KSR-containing medium with LIF. The arrows indicate individual ESC colonies which appear much flatter when FBS is used.

5.5 Growth factors

A number of external factors are known to contribute to stem cell self-renewal in the mouse. The LIF/gp130/JAK/Stat3, BMP4/Smad/Id and Wnt/GSK3β/β-catenin pathways combine to promote self-renewal without differentiation by activation of nuclear transcription factors such as Nanog, KLf4 and Oct3/4. Other growth factors such as SAF are also likely to be involved. However LIF plays a significant role in the induction of pluripotency. Insufficient activation of JAK/Stat3 restricts the acquisition of ESC status (Yang et al., 2010). Although it has become the standard to use LIF at 10^3 U/ml for both ESC maintenance and derivation, the range reported varies between 400 - 10^4 U/ml. A number of successful attempts of derivation from non-permissive strains have featured high levels of LIF (Baharvand et al., 2004; Yang et al., 2009). It is interesting to speculate, that, had Martin not used concentrated conditioned medium, she may not have been amongst the first to describe the derivation of ESC. Results from our own laboratory have indicated that by using elevated levels of LIF (5000 U/ml; ESGRO; Millipore UK) in conjunction with KSR medium, highly efficient rates of ESC derivation can be achieved from C57BL/6, Balb/k, CBA/Ca and even NOD mice (T.J. Davies & P.J. Fairchild, manuscript in preparation).

5.6 Small molecules

Small molecules which target specific signaling pathways have been shown to be useful chemical tools in manipulating cell developmental fate and function, by such means as replacing transcription factors. The POU transcription factor Oct-4 is a specific marker for pluripotent cells and is expressed by ICM and ESCs. Oct-4 expression by epiblast *in vivo* is lost just prior to differentiation following implantation and *in vitro* when explanted for ESC derivation. Hence it was considered that if Oct-4 could be maintained, then the number of ESC progenitor cells in the outgrowing epiblast might be increased leading to more efficient derivation. In 2003, Buehr and Smith reported that, by combining immunosurgical isolation of delayed-implanting ICM with culture in the presence of a small molecule that inhibits

Oct-4 loss, they could obtain CBA ESC at 25% efficiency (Buehr & Smith, 2003). The small molecule was PD98059, an inhibitor of the Erk activating enzyme MEK-1 that was known to be able to enhance self-renewal in established ESC lines. Cell based screening of synthetic chemical libraries revealed other candidate small molecule inhibitors that proved to be successful in mESC derivation from CBA and Balb/c strains (Lodge et al., 2005; Umehara, 2007). However these reagents, MEK-1 inhibitors PD98059 and U0126 and the GSK-3 inhibitor BIO, still required the presence of feeders and LIF. A small molecule, pluripotin (otherwise known as SC-1), was identified through feeder-, serum- and LIF-free screening which acted as a dual inhibitor of RasGAP and ERK1. Inhibition of RasGAP promotes ESC self-renewal by enhancing the phosphoinositide-3 kinase (PI3K) signaling pathway, whereas ERK1 inhibition blocks ESC differentiation. Pluripotin could be used to maintain undifferentiated ESC without LIF and when used in combination with LIF (2×10^3 U/ml), novel ESC lines were successfully derived from the normally refractory Balb/c (63%), CD-1 (80%) and NOD-scid (57%) strains (Yang et al., 2009). These derivations were conducted under an alternating FBS - KSR regime in the presence of 2x LIF (2×10^3 U/ml). However, the NOD-scid ESC required continuous presence of pluripotin and one out of the four lines displayed chromosomal abnormalities.

5.7 ESC from NOD mice

The Non-Obese Diabetic (NOD) strain is an important mouse model for human type 1 diabetes. More than 20 insulin dependent diabetes (Idd) loci are involved in the disease. It would be very desirable to have NOD ESC lines in which to perform gene knockout of potential Idd gene candidates or allele-shuffling strategies in which deletion of potential susceptibility genes could be followed by their replacement with identical genes from a diabetes-resistant strain. Unfortunately, the NOD mouse has proven to be the most refractory of all strains from which to produce ESC; and the pre-implantation NOD embryo is notoriously fragile *in vitro*. Until recently only one NOD ESC line had been generated but this had very low GLT (1%), was incompatible with feeder-free culture and prone to spontaneous differentiation on MEF (Nagafuchi et al., 1999). Interestingly, the line had been isolated using relatively high levels of LIF (10^4 U/ml) and required continuous culture in 3×10^3 U/ml LIF to prevent differentiation. The GLT of this line was later improved by batch testing of FBS and using NOD blastocysts as recipient embryos for injection of the NOD ESC (Arai et al., 2004). Even strategies which had proven successful in deriving ESCs from other non-permissive strains, such as CBA/Ca, were not able to succeed with NOD. Brook and Gardner had previously shown that by removing the inhibitory influence of the hypoblast (primitive endoderm) and trophoblast from 5 day mouse blastocysts by means of micro-dissection combined with brief enzymatic digestion; isolated epiblasts gave improved efficiency of ESC derivation when compared to whole embryos (Brook & Gardner, 1997). When epiblasts from implantation-delayed embryos were used, the efficiency of derivation of 129 ESC was increased even further, from 52% to 100%. Efficient rates of derivation using delayed isolated epiblasts cultured on MEF, were also demonstrated for other strains, including PO (56%) and CBA/Ca (21%). When LIF (10^3 U/ml) was also included 56% of the CBA/Ca embryos yielded ESC lines. Would these techniques, when applied to NOD embryos, result in novel ESC lines?

The pluripotency-suppressing nature of the NOD genome could not be overcome using these techniques (Brook et al., 2003). Brook et al. were also able to demonstrate that the ICR mouse strain, from which the NOD strain was originally derived, was equally as refractory as the NOD. Researchers sought other avenues to explore and NOD EpiSC were produced, as described above (Brons et al., 2007). The generation of NOD EpiSC from 6.75d post-implantation epiblast required FGF-2 plus Activin A/Nodal signaling, rather than LIF/BMP4. However, these pluripotent cells, in common with other EpiSC, were not able to contribute to chimeras and were, therefore, unable to transmit their genome. Using two of the reprogramming factors reported by Takahashi and Yamanaka (2006) to be responsible for the production of iPSC, Hanna et al. infected NOD ICM with constitutive lentiviruses encoding Klf4 or c-Myc (Takahashi & Yamanaka, 2006; Hanna et al., 2009). Under routine conditions for mouse ESC derivation (DMEM + FBS + LIF on feeders), the ICM produced NOD ESC with a normal karyotype, which expressed pluripotency markers and gave chimeras with GLT. Unfortunately, ectopic expression of integrated c-Myc transgene lead to tumour formation in the chimeras. When small molecule inhibitors that are known to substitute for Klf4 and c-Myc in the generation of iPS cells, were applied to NOD ICM outgrowths then genetically unmodified NOD ESC were produced. The reagents used were Kenpaullone, a GSK3β and CDK1/ cyclin B inhibitor, the glycogen synthase kinase 3 inhibitor, CHIR99021, and the ERK-cascade inhibitor PD184352. Their use in various combinations generated NOD ESC which remained stable but which required the continuous presence of the inhibitors to prevent differentiation of the ESC. Applying the same principle of using a cocktails of either two or three small molecules ('2i' or '3i') to prevent differentiation and promote ESC expansion, Ying et al. applied a serum- and feeder-free system to generate ESC from both permissive 129 and non-permissive CBA and MF1 strains (Ying et al., 2008). The important question that arose from these studies was whether this approach would finally break the NOD barrier?

5.8 The ground state of pluripotency.

In 2009, Nichols et al. cultured NOD embryos from eight-cell to blastocyst in KSOM-2 medium in the presence of the mitogen-activated protein kinase kinase (MEK) inhibitor PD0325901(1μM) and the glucogen synthase kinase-3 (GSK3) inhibitor CHIR99021(3μM) – collectively called '2i'. The blastocysts were then expanded by culturing for a day in N2B27 medium with 2i and LIF, at an undisclosed concentration, prior to removing the zona pellucida and immunosurgery to destroy the trophoblast. Isolated ICM were then plated into gelatinized wells in N2B27 medium with 2i and LIF for 5-7 days. The ICM outgrowths were then trypsinized and replated under the same feeder-free conditions until primary colonies were given which could then be expanded to give NOD ESC lines. Thirty NOD embryos yielded sixteen ESC lines. Of the eleven lines tested, seven were karyotypically normal and four of these generated chimeras, three with GLT (Nichols et al., 2009). By removing exogenous signaling factors other than LIF and attachment on gelatin, coupled with inhibition of differentiation via MEK/ERK inhibition in combination with the promotion of proliferation and self-renewal via GSK3-inhibition, true NOD ESC were at last obtained. These cells were stable and pluripotent both *in vitro* and *in vivo*, and offspring of germ-line competent chimeras could be induced to develop diabetes. Furthermore, the ESC could be genetically modified and still retain the capacity for germ-line transmission, demonstrating that they represent a functional model for the human disease. The authors speculate that pluripotency is the basal state for epiblast/ESC and is self-maintaining in the absence of commitment cues. If perturbed by exposure to extrinsic factors, such FGF or serum, differentiation occurs. Although LIF signaling is known to sustain self-renewal in mESC, it has been demonstrated not to be absolutely required for ESC generation under certain conditions (Ying et al., 2008). However its addition greatly improves efficiency. Activation of the Jak/Stat3 pathway (by LIF) is a limiting process in induction of pluripotency. The current picture is of a metastable pluripotent condition where the ESC exists in one of two states: as either in the naïve (ground state) or in the primed (committed) state (Nichols & Smith, 2009). The capture of pluripotency is, therefore, a balance between differentiation and self-renewal, increased LIF signaling tips the scales towards naïve ESC status.

5.9 Pluripotent stem cells from other species

Whilst ESC-like lines from some fifteen different species, including other mammals, chick and teleost fish, have been reported that meet some of the criteria as outlined by Evans and Kaufman and restated above, until recently only the mouse fulfilled all. In particular, no other species consistently gives as high levels of chimerism and none support germline transmission at high passage, as does the mouse. Nevertheless, the derivation, in 1995, of ES-like cells from the Rhesus monkey (Thomson et al., 1995) demonstrated that lines could indeed be made from primates. This raised the expectation that human ESC lines could also be made and with it the prospect of real advances in understanding the genetic and molecular processes of cell differentiation and proliferation that are central to conditions such as birth defects and cancer. In 1998 human ESC (Thomson et al., 1998) were isolated for the first time from surplus pre-implantation embryos produced *in vitro* by IVF and donated with consent. Traditionally, culture conditions for human ES have differed from those of mice; mice requiring LIF/BMP4 and/or feeders, whilst human require FGF-2/Nodal-Activin signaling. There are now many human ESC lines established. The advent of 2i/3i small molecule induction of pluripotency has opened up the door to production of ESC from species other than mouse (rat, vole, and rabbit). The first rat ESC derived using 3i culture conditions did not yield GLT (Buehr et al., 2008), however Li et al. derived rat ESC from explanted ICM in 2i culture conditions with an efficiency of 30-60%. Two of the nine rate ESC lines generated produced GLT (Li et al., 2008). Interestingly, none of the three rat iPSC lines so far described have been shown to be germline competent (Jacob et al., 2010). Although pluripotent stem cells have been derived from mouse, rat, monkey, pig and rabbit by reprogramming differentiated cells back to the ground state, only mouse and rat iPSC have shown GLT, although human iPSC can not be tested for obvious ethical reasons. The continued study of mouse ESC therefore remains the gold standard by which other pluripotent stem cell types should be measured.

6. Acknowledgements

Work in the authors' laboratory is supported by project grant G0802538 from the Medical Research Council U.K. T.D and P.F are members of the Oxford Stem Cell Institute.

7. References

Andrews, P.W.; Damjanov, I.; Simon, D.; Banting, G.S.; Carlin, C.; Dracopoli, N.C. & Føgh, J. (1984). Pluripotent embryonal carcinoma clones derived from the human teratocarcinoma cell line Tera-2. Differentiation in vivo and in vitro. *Lab. Invest.*, vol. 50, No. 2, pp. 147-162.

Arai, S.; Minjares, C.; Nagafuchi, S. & Miyazaki, T. (2004). Improved experimental procedures for achieving efficient germ line transmission of Non-Obese Diabetic (NOD)-derived embryonic stem cells. *Experimental Diabetes Research*, Vol. 5, pp. 219-226.

Baharvand, H. & Matthaei, K.I., (2004). Culture condition difference for establishment of new embryonic stem cell lines from the C57BL/6 and BALB/c mouse strains. *In Vitro Cell Dev Biol Anim*, Vol. 40, No.s 3-4, pp. 76-81.

Bain, G.; Kitchens, D.; Yao, M.; Huettner, J. E. & Gottlieb, D. I. (1995). Embryonic stem cells express neuronal properties in vitro. *Developmental Biology*, Vol. 168, pp. 342—357.

Bradley, A.; Evans, M.; Kaufman, M.H. & Robertson, E. (1984). Formation of germ-line chimaeras from embryo-derived teratocarcinoma cell lines. *Nature*, Vol. 309(5965), pp. 255-6.

Brons,I.G.; Smithers, L.E.; Trotter, M.W.; Rugg-Gunn, P.; Sun, B.; Chuva De Sousa Lopes, S.M.; Howlett, S.K.; Clarkson, A.; Ahrlund-Richter, L.; Pedersen, R.A. & Vallier, L., (2007). Derivation of pluripotent epiblast stem cells from mammalian embryos. *Nature* Vol. 448, pp. 191-195.

Brook, F.A. & Gardner, R.L. (1997). The origin and efficient derivation of embryonic stem cells in the mouse. *Proc Natl Acad Sci USA*, Vol. 94, pp. 5709-5712.

Brook, F.A.; Evans, E.P.; Lord, C.J.; Lyons, P.A.; Rainbow, D.B.; Howlett, S.K.; Wicker, L.S.; Todd, J.A. & Gardner, R.L., (2003). The derivation of highly germline-competent embryonic stem cells containing NOD-derived genome. *Diabetes*, Vol. 52(1), pp. 205-8.

Brook, F.A., (2006). Chapter 2: Procedures for deriving ES cell lines from the mouse, pp. 7-40, ISBN 0-19-855001-4, in *Embryonic Stem Cells: Practical Approach*, Oxford University Press.

Bryja, V.; Bonilla, S.; Cajanek, L.; Parish, C.L.; Schwartz, C.M.; Luo, Y.; Rao, M.S. & Arenas, E. , (2006). An efficient method for the derivation of mouse embryonic stem cells. *Stem Cells*, Vol. 24, pp. 844-849.

Buehr, M. & Smith, A., (2003). Genesis of embryonic stem cells. *Philos. Trans. R. Soc. Lond. B Bio.l Sci.*, Vol. 358(1436), pp. 1397-402.

Buehr, M.; Meek, S.; Blair, K.; Yang, J.; Ure, J.; Silva, J.; McLay, R.; Hall, J.; Ying, Q-L. & Smith, A., (2008). Capture of authentic Embryonic Stem Cells from Rat blastocysts. *Cell* Vol. 135(7), pp. 1287-1298.

Chang, I.K.; Jeong, D.K.; Hong, Y.H.; Park, T.S.; Moon, Y.K.; Ohno, T. & Han, J.Y. (1997). Production of germ line chimerical chickens by transfer of cultured primordial germ cells. *Cell Biol. Int*. Vol. 21, pp. 495-499.

Chen, H.; Shew, J.; Ho, H.; Hsu, W. & Yang, Y., (1999). Expression of leukemia inhibitory factor and its receptor in preimplantation embryos. *Fertility and sterility*, Vol. 72(4), pp. 713-719.

Chen, J.; Reifsnyder, P.C.; Scheuplein, F.; Schott, W.H.; Mileikovsky, M.; Soodeen-Karamath, S.; Nagy, A.; Dosch, M.H.; Ellis, J.; Koch-Nolte, F. & Leiter, E.H., (2005). "Agouti NOD": identification of a CBA-derived Idd locus on Chromosome 7 and its use for chimera production with NOD embryonic stem cells. *Mamm Genome*, Vol. 10, pp. 775-83.

Cheng, J.; Dutra, A.; Takesono, A.; Garrett-Beal, L. & Schwartzberg, P.L., (2004). Improved generation of C57BL/6J mouse embryonic stem cells in a defined serum-free media. *Genesis* Vol. 39, pp. 100-104.

Chin, M.H.; Mason, M.J.; Lowry, W.E. et al, (2009). Induced pluripotent stem cells and embryonic stem cells are distinguished by gene expression signatures. *Cell Stem Cell*, Vol. 5, pp. 111-123.

Cho, H.J.; Lee, C.S.; Kwon, Y.W.; Paek, J.S.; Lee, S.H.; Hur, J.; Lee, E.J.; Roh, T.Y.; Chu, I.S.; Leem, S.H.; Kim, Y.; Kang, H.J.; Park, Y.B. & Kim, H.S., (2010). Induction of pluripotent stem cells from adult somatic cells by protein-based reprogramming without genetic manipulation. *Blood*, Vol. 116(3), pp. 386-95.

Chung, Y.; Klimanskaya, I.; Becker, S.; Marh, J.; Lu, S.J.; Johnson, J.; Meisner, L. & Lanza, R., (2006). Embryonic and extra-embryonic stem cell lines derived from single mouse blastomeres. *Nature*, Vol. 439, pp. 216-219.

Cole, R.J.; Edwards, R.G. & Paul, J., (1965). Cytodifferentiation in cell colonies and cell strains derived from cleaving ova and blastocysts of the rabbit. *Exp.Cell Res*. Vol. 37, pp. 501-504.

Cole, R.J.; Edwards, R.G. & Paul, J., (1966). Cytodifferentiation and embryogenesis in cell colonies and tissue cultures derived from ova and blastocysts of the rabbit. *Dev. Biol.* Vol. 13, pp. 385-407.

Doetschman, T. C.; Eistetter, H.; Katz, M.; Schmidt, W. & Kemler, R. (1985). The in vitro development of blastocyst-derived embryonic stem cell lines: formation of visceral yolk sac, blood islands and myocardium. *Journal of Embryology and Experimental Morphology*, Vol. 87, pp. 27—45.

Evans, M.J. & Kauffman, M., (1981). Establishment in culture of pluripotent cells from mouse embryos. *Nature*, Vol. 292, pp. 154-156.

Fairchild, P.J.; Brook, F.A.; Gardner, R.L.; Graça, L.; Strong, V.; Tone, Y.; Tone, M.; Nolan, K.F. & Waldmann, H., (2000). Directed differentiation of dendritic cells from mouse embryonic stem cells. *Curr Biol.*, Vol. 10(23): pp. 1515-1518.

Fry, R.C.; Batt, P.A.; Fairclough, R.J & Parr, R.A., (1992). Human leukemia inhibitory factor improves the viability of cultured ovine embryos. *Biol. Reprod.*, Vol. 46, pp. 470-474.

Gardner, R.L. & Papaioanou, V.E., (1975). Differentiation of the trophectoderm and inner cell mass. *Early Development of Mammals*, p107-132; Ed. Balls, M. and Wild, A.E., (The second symposium of the British Society of Developmental Biologists, Cambridge University Press).

Gossler, A.; Doetschmann, T.; Korn, R.; Serfling, E. & Kembler, R., (1986). Transgenesis by means of blastocyst-derived embryonic stem cell lines. *Proc. Natl. Acad. Sci. USA*, Vol. 83, pp. 9065-9069.

Gutierrez-Aranda, I.; Ramos-Mejía, V.; Bueno, C.; Muñoz-López, M.; Real, P.J.; Macía, A.; Sánchez, L.; Ligero, G.; García-Pérez, J.L. & Menendez, P., (2010). Human Induced Pluripotent Stem Cells Develop Teratoma More Efficiently and Faster than Human Embryonic Stem Cells Regardless the Site of Injection. *Stem Cells*, Vol. 28(9), pp. 1568-70.

Hanna, J. ; Markoulaki, S.; Mitalipova, M.; Jaenisch, R., et al. (2009). Metastable Pluripotent States in NOD-Mouse-Derived ESCs. *Cell Stem Cell*, Vol. 4, pp. 513–524.

Jacob, H.J.; Lazar, J.; Dwinell, M.R.; Moreno, C. & Geurts, A.M., (2010). Gene targeting in the rat: advances and opportunities. *Trends in Genetics*, Vol. 26(11), pp. 459-492.

Jurisicova, A.; Ben-Chetrit, A.; Varmuza, S.L. & Casper, R.F., (1995). Recombinant human leukemia inhibitory factor does not enhance in vitro human blastocyst formation. *Fertil. Steril.*, Vol. 64, pp. 999-1002.

Kauma, S.W. & Matt, D.W., (1995). Coculture cells that express leukemia inhibitory factor (LIF) enhance mouse blastocyst development in vitro. *J Assist Reprod. Genet.*, Vol. 12(2), pp. 153-6.

Kawazoe, S.; Ikeda, N.; Miki, K.; Shibuya, M.; Morikawa, K.; Nakano, S.; Oshimura, M.; Hisatome, I. & Shirayoshi, Y., (2009). Extrinsic factors derived from mouse embryonal carcinoma cell lines maintain pluripotency of mouse embryonic stem cells through a novel signal pathway. *Develop. Growth Differ.*, Vol. 51, pp. 81–93.

Kim, I.; Saunders, T.L. & Morrison, S.J., (2007). Sox17 Dependence Distinguishes the Transcriptional Regulation of Fetal from Adult Hematopoietic. *Stem Cells*, Vol. 130(3), pp. 470-83.

Kleinsmith, L.J. & Pierce, G.B., (1964). Multipotentiality of single embryonal carcinoma cells. *Cancer Res.*, Vol. 24, pp. 1544-1552.

Koller, B.H.; Hagemann, L.J.; Doetschman, T., et al, (1989). Germ-line transmission of a planned alteration made in a hypoxanthine phosphoribosyltransferase gene by homologous recombination in embryonic stem cells. *Proc. Natl. Acad. Sci. USA*, Vol. 86, pp. 8927-8931.

Kurosawa, H., (2007). Methods for inducing embryoid body formation: in vitro differentiation system of embryonic stem cells. *J. Biosci. Bioeng.*, Vol. 103(5), pp. 389-98.

Labosky, P.A.; Barlow, D.P. & Hogan, B.L., (1994). Mouse embryonic germ(EG) cell lines: transmission through the germ line and differences in the methylation imprint of insulin-like growth factor 2 receptor(Igf2r) gene compound with embryonic stem cell lines. *Development*, Vol. 120, pp. 3197-3204.

Li, P.; Tong, C.; Mehrian-Shai, R.; Ying, Q.L., et al, (2008). Germline competent embryonic stem cells derived from rat blastocysts. *Cell*, Vol. 135, pp. 1299-1310.

Lodge, P.; McWhir, J.; Gallagher, E., et al, (2005). Increased gp130 signaling in combination with inhibition of the MEK/ERK pathway facilitates embryonic stem cell isolation from normally refractory murine CBA blastocysts. *Cloning Stem Cells* Vol. 7, pp. 2-7.

Lowry, W.E.; Richter, L.; Plath, K. et al, (2008). Generation of human induced pluripotent stem cells from dermal fibroblasts. *Proc. Natl. Acad. Sci. USA*, Vol. 105, pp. 2883-2888.

Marquant-Le, G.B.; Guienne, B.; Humblot, P.; Gullion, N. & Thibier, M., (1993). Murine LIF improves the development of IVF cultured bovine morulae. *J. Reprod. Fertil. Abstr.*, Vol. 12, pp. 61.

Martin, G.R., (1981). Isolation of a pluripotent cell line from early mouse embryos cultured in medium conditioned by teratocarcinoma stem cells. *Proc Natl Acad Sci U S A*, Vol. 78(12), pp. 7634-8.

Matsuda, T.; Nakamura, T.; Nakao, K.; Arai, T.; Katsuki, M.; Heike, T. & Yokota, T., (1999). STAT3 activation is sufficient to maintain an undifferentiated state of mouse embryonic stem cells. *EMBO J.*, Vol. 18(15), pp. 4261-9.

McWhir, J.; Schnieke, A.E.; Ansell, R.; Wallace, H.; Colman, A.; Scott, A.R. & Kind, A.J., (1996). Selective ablation of differentiated cells permits isolation of embryonic stem cell lines from murine embryos with a non-permissive genetic background. *Nat. Genet.*, Vol. 14(2), pp. 223-6.

Miller-Hance, W. C.; LaCorbiere, M.; Fuller, S. J.; Evans, S. M.; Lyons, G.; Schmidt, C.; Robbins, J. & Chien, K. R. (1993). In vitro chamber specification during embryonic stem cell cardiogenesis. Expression of the ventricular myosin light chain_2 gene is independent of heart tube formation. *Journal of Biological Chemistry*, Vol. 268, pp. 25244–25252.

Nagafuchi, S.; Katsuta, H.; Kogawa, K.; Akashi, T.; Kondo, S.; Sakai, Y.; Tsukiyama, T.; Kitamura, D.; Niho, Y. & Watanabe, T., (1999). Establishment of an embryonic stem (ES) cell line derived from a non-obese diabetic (NOD) mouse: In vivo differentiation into lymphocytes and potential for germ line transmission. *FEBS Lett.*, Vol. 455, pp. 101–104.

Nichols, J.; Chambers, I. & Smith, A., (1994). Derivation of Germline Competent Embryonic Stem Cells with a Combination of Interleukin-6 and Soluble Interleukin-6 Receptor. *Exp. Cell Res.*, Vol. 215(1), pp. 237-239.

Nichols, J.; Jones, K.; Phillips, J.M.; Newland, S.A.; Roode, M.; Mansfield, W.; Smith, A. & Cooke, A., (2009). Validated germline-competent embryonic stem cell lines from nonobese diabetic mice. *Nature Medicine*, Vol. 15(7), pp. 814-818.

Nichols, J. & Smith, A., (2009). Naïve and primed pluripotent states. *Cell Stem Cell*, Vol. 4, pp. 487-492.

Niwa, H.; Burdon, T.; Chambers, I. & Smith, A., (1998). Self-renewal of pluripotent embryonic stem cells is mediated via activation of STAT3. *Genes Dev.*, Vol. 12(13), pp. 2048-60.

Ogawa, K.; Saito, A.; Matsui, H.; Suzuki, H.; Ohtsuka, S.; Shimosato, D.; Morishita, Y.; Watabe, T.; Niwa, H. & Miyazono, K., (2006). Activin-Nodal signaling is involved in propagation of mouse embryonic stem cells. *Journal of Cell Science*, Vol. 120 (1), pp. 55-65.

Okada, M.; Oka, M. & Yoneda, Y., (2010). Effective culture conditions for the induction of pluripotent stem cells. *Biochemica et Biophysica Acta*, Vol. 1800(9), pp. 956-963.

Okita, K.; Ichisaka, T. & Yamanaka, S., (2007). Generation of germline-competent induced pluripotent stem cells. *Nature*, Vol. 448(7151), pp. 313-7.

Qi, X.; Li, T.G.; Hao, J.; Hu, J.; Wang, J.; Simmons, H.; Miura, S.; Mishina, Y. & Zhao, G.Q., (2004). BMP4 supports self-renewal of embryonic stem cells by inhibiting mitogen-activated protein kinase pathways. *Proc. Natl. Acad. Sci., U S A*, Vol. 101, pp. 6027–6032.

Roach, M.L.; Stock, J.L.; Byrum, R.; Koller, B.H. & McNeish, J.D., (1995). A new embryonic stem cell line from DBA/1lacJ mice allows genetic modification in a murine model of human inflammation. *Exp. Cell Res.* Vol. 221(2), pp. 520-5.

Robertson, E.J., (1987). Teratocarcinoma and Embryonic Stem cells: A practical Approach, pp. 71-112.

Robertson, E.; Bradley, A.; Kuehn, M. & Evans, M., (1986). Germ-line transmission of genes introduced into cultured pluripotential cells by retroviral vector. *Nature*, Vol. 323, pp. 445-448.

Rungsiwiwut, R.; Rungarunlert, S.; Numchaisrika, P.; Pruksananonda, K.; Techakumphu, M. & Virutamasen, P., (2008). Effect of leukemia inhibitory factor (LIF) on the quality of in vitro produced mouse blastocysts and subsequent derivation of embryonic stem (ES) cells. *J. Med. Assoc. Thai.*, Vol. 91(5), pp. 608-14.

Schmitt, R. M.; Bruyns, E. & Snodgrass, H. R., (1991). Hematopoietic development of embryonic stem cells in vitro: cytokine and receptor gene expression. *Genes and Development* Vol. 5, pp. 728—740.

Schoonjans, L.; Kreemers, V.; Danloy, S.; Moreadith, R.W.; Laroche, Y. & Collen, D., (2003). Improved generation of germ-line competent embryonic stem cell lines from inbred mouse strains. *Stem Cells,* Vol. 21, pp. 90-97.

Sherman, M.I., (1975). The culture of cells derived from mouse blastocysts. *Cell,* Vol. 5, pp. 343-349.

Skoudy, A.; Rovira, M.; Savatier, P.; Martin, F.; León-Quinto, T.; Soria, B. & Real, F.X., (2004). Transforming growth factor (TGF)beta, fibroblast growth factor (FGF) and retinoid signalling pathways promote pancreatic exocrine gene expression in mouse embryonic stem cells. *Biochem J,* Vol. 379(Pt 3), pp. 749-56.

Smith, A.G.; Heath, J.K.; Donaldson, D.D.; Wong, G.G.; Moreau, J.; Stahl, M. & Rogers, D., (1988). Inhibition of pluripotential embryonic stem cell differentiation by purified polypeptides. *Nature,* Vol. 336, pp. 688–690.

Smithies, O.; Gregg, R.G.; Boggs, S.S.; Koralewski, M.A. & Kucherlaparti, R.S., (1985). Insertion of DNA sequences into human chromosomal beta-globulin locus by homologous recombination. *Nature,* Vol. 317, pp. 230-234.

Seong, E.; Saunders, T.L.; Stewart, C.L., & Burmeister, M., (2004). To knockout in 129 or in C57BL/6: That is the question. *Trends in Genetics,* Vol. 20, pp. 59-62.

Stevens, L.C., (1958). Studies on transplantable testicular teratomas of strain 129 mice. *J. Nat. Cancer Institute,* Vol. 20, pp. 1257-1276.

Stevens, L.C., (1967). Origin of testicular teratomas from primordial germ cells in mice. *J. Natl. Cancer Inst.,* Vol. 38, pp. 549-552.

Stewart, C.L.; Kaspar, P.; Brunet, L.J.; Bhatt, H.; Gadi, I.; Köntgen, F. & Abbondanzo, S.J., (1992). Blastocyst implantation depends on maternal expression of leukaemia inhibitory factor. *Nature,* Vol. 359(6390), pp. 76-9.

Stewart, C.L.; Gadi, I. & Bhatt, H., (1994). Stem cells from primordial germ cells can reenter the germ line. *Dev. Biol.,* Vol. 161, pp. 626-628.

Takahashi, K. & Yamanaka, S., (2006). Induction of pluripotent stem cells from mouse embryonic and adult fibroblast cultures by defined factors. *Cell,* Vol. 126, pp. 663-676.

Takahashi, K.; Tanabe, K.; Yamanaka, S. et al; (2007). Induction of pluripotent stem cells from adult human fibroblasts by defined factors. *Cell,* Vol. 131, pp. 861-872.

Teasr, P.J., (2005). Derivation of germ-line competent embryonic stem cell lines from pre-blastocyst mouse embryos. *Proc. Natl. Acad. Sci. USA,* Vol. 102, pp. 8239-8244.

Tesar, P.J.; Chenoweth, J.G.; Brook, F.A.; Davies, T.J.; Evans, E.P.; Mack, D.L.; Gardner, R.L. & Mckay, R.D., (2007). New cell lines from mouse epiblast share defining features with human embryonic stem cells. *Nature,* Vol. 448, pp. 196-199.

Thomas, K.R. & Capecchi, M.R., (1990). Targeted disruption of the murine Int-1 proto-oncogene resulting in severe abnormalities in midbrain and cerebellar development. *Nature,* Vol. 346, pp. 847-850.

Thomson, J.A.; Kalishman, J.; Golos, T.G.; Durning, M.; Harris, C.P.; Becker, R.A. & Hearn, J.P., (1995). Isolation of a primate embryonic stem cell line. *Proc. Natl. Acad. Sci. USA*, Vol. 92, pp. 7844-7848.

Thomson, J.A.; Itskovitz-Eldor, J.; Shapiro, S.S.; Waknitz, M.A.; Swiergiel, J.J.; Marshall, V.S. & Jones, J.M., (1998). Embryonic Stem Cell Lines Derived from Human Blastocysts. *Science*, Vol. 282(5391), pp. 1145 - 1147. Erratum in: *Science*, Vol. 282(5395), pp. 1827.

Umehara, H.; Kimura, T.; Ohtsuka, S.; Nakamura, T.; Kitajima, K.; Ikawa, M.; Okabe, M.; Niwa, H. & Nakano, T., (2007). Efficient derivation of embryonic stem cells by inhibition of glycogen synthase kinase-3. *Stem Cells*, Vol. 25(11), pp. 2705-2711.

Wernig, M.; Meissner, A.; Hochedlinger, K.; Jaenisch, R. et al, (2007). In vitro reprogramming of fibroblasts into a pluripotent ES-cell-like state. *Nature*, Vol. 448, pp. 318-324.

Williams, R.L.; Hilton, D.J.; Pease, S.; Willson, T.A.; Stewart, C.L.; Gearing, D.P.; Wagner, E.F.; Metcalf, D.; Nicola, N.A. & Gough, N.M., (1988). Myeloid leukaemia inhibitory factor maintains the developmental potential of embryonic stem cells. *Nature*, Vol. 336, pp. 684–687.

Wobus, A.M.; Wallukat, G. & Hescheler, J., (1991). Pluripotent mouse embryonic stem cells are able to differentiate into cardiomyocytes expressing chronotropic responses to adrenergic and cholinergic agents and Ca2+ channel blockers. *Differentiation*, Vol. 48(3), pp.173-82.

Yang, J.; van Oosten, A.L.; Theunissen, T.W.; Guo, G.; Silva, J.C.R. & Smith, A., (2010). Stat3 activation is limiting for reprogramming to ground state pluripotency. *Cell Stem Cell*, Vol. 7, pp. 319-328.

Yang, W.; Wei, W.; Shi, C.; Zhu, J.; Ying, W.; Shen, Y.; Ye, X.; Fang, L.; Duo, S.; Che, J.; Shen, H.; Ding, S. & Deng, H., (2009). Pluripotin combined with Leukemia Inhibitory Factor greatly promotes the derivation of Embryonic Stem Cell lines from refractory strains. *Stem Cells*, Vol. 27, pp. 383–389.

Ying, Q.L.; Nichols, J.; Chambers, I. & Smith, A., (2003). BMP induction of Id proteins suppresses differentiation and sustains embryonic stem cell self-renewal in collaboration with STAT3. *Cell*, Vol. 115, pp. 281–292.

Ying, Q.L.; Wray, J.; Nichols, J.; Batlle-Morera, L.; Dobie, B.; Woodgett, J.; Cohen, P. & Smith, A., (2008). The ground state of embryonic stem cell self-renewal. *Nature*, Vol. 453, pp. 519-523.

Yu, J.; Vodyanik, M.A.; Smuga-Otto, K.; Stewart, R. et al, (2007). Induced pluripotent stem cells derived from human somatic cells. *Science*, Vol. 318, pp. 1917-1920.

Zhao, X.Y.; Li, W.; Lv, Z.; Liu, L.; Tong, M.; Hai, T.; Hao, J.; Guo, C.L.; Ma, Q.W.; Wang, L.; Zeng, F. & Zhou, Q., (2009). iPS cells produce viable mice through tetraploid complementation. *Nature*, Vol. 461(7260), pp. 86-90.

2

Molecular Mechanisms of Pluripotency in Murine Embryonic Stem Cells

Elisa A. Casanova, Kurt Bürki and Paolo Cinelli
Institute of Laboratory Animal Science, University of Zurich
Switzerland

1. Introduction

Embryonic stem cells (ESCs) are pluripotent cells, which can be isolated from the inner cell mass (ICM) of blastocyst stage embryos. They are defined by two properties: they can indefinitely self-renew *in vitro* and contribute to the formation of all cells of an adult organism, including functional gametes for genome transmission. Due to their pluripotent state ESCs can be used for various applications, like the generation of knockout or transgenic animals, and potentially as a cell source for cell therapy in regenerative medicine. Alternatively self-renewing cells with pluripotent potential can also be generated by specifying germ cells with extrinsic factors (Matsui, 1992) or by reprogramming somatic cells using gene transfection to generate the so called induced pluripotent stem cells (iPSCs) (Takahashi and Yamanaka, 2006).

Due to the fundamental characteristics of ESCs, developmental biology, regenerative medicine and cancer biology are more and more interested in understanding the molecular mechanisms controlling stem cells. Even though a lot of efforts have been made in the past years to elucidate the factors that regulate stem cell self-renewal and pluripotency, the precise mechanism of how these processes are regulated remains largely unknown. In this chapter we will give an overview on the known molecular pathways and transcription factors involved in maintaining pluripotency, by especially focusing on the core transcription factors OCT3/4, SOX2, Nanog, and STAT3. Subsequently we will discuss the epigenetics of ESCs and iPSCs and finally conclude this chapter with remarks and discussions.

1.1 Establishing and maintaining ESCs *in vitro*

ESC lines are derived from the preimplantation embryo, precisely from the ICM of the blastocysts. Although the presence of self-renewing, pluripotent, cells is transient *in vivo*, apparently immortal cell lines with these properties can be obtained *in vitro*. Once removed from the blastocyst, the cells of the ICM can be cultured under special conditions, into ESCs, which maintain an undifferentiated status (Evans and Kaufman, 1981; Martin, 1981). Under other conditions, for example by giving the right extrinsic and intrinsic signals, ESCs can differentiate into all the cell types that make up the organism. Pluripotent ESCs are characterized by the expression of specific cell surface glycoproteins such as the stage-specific embryonic antigen 1 (SSEA-1) (Solter and Knowles, 1978) as well as by the presence of transcription factors such as OCT3/4 (Schöler, 1991; Schöler et al., 1989) and Nanog

(Chambers et al., 2003). High expression levels of alkaline phosphatase also characterize ESCs. Furthermore, ESCs exhibit a short G1 phase of the cell cycle (Rohwedel et al., 1996) and a high telomerase activity (Thomson et al., 1998). Other essential properties of ESCs include growth as multicellular colonies, normal and stable karyotypes, and prolonged undifferentiated culture.

Derivation and maintenance of murine ESCs *in vitro* was originally achieved by using mitotically inactivated embryonic fibroblasts (feeders) and/or the cytokine leukaemia inhibitory factor (LIF) in combination with fetal calf serum and/or the growth factor bone morphogenetic protein (BMP) (Smith et al., 1988; Ying et al., 2003). However, the same culture conditions are not sufficient for derivation of ESCs from most of the mouse strains and not at all from the rat. The genetic background strongly affects the efficiency of ESC isolation and almost all lines in use are derived from the strain 129. Even though ESCs were discovered more than 25 years ago only limited number of ESCs of proven ability to colonize the germ-line have been obtained and only a few mouse strains other than 129 (Simpson et al., 1997). Only after adjusting the culture conditions, germ line competent inbred ESCs could be established, e.g. from C57BL6/J (Keskintepe et al., 2007; Ledermann and Bürki, 1991), DBA/1lacJ (Roach et al., 1995), BALB/c (Kawase et al., 1994; Noben-Trauth et al., 1996), and CBA mice (Lodge et al., 2005). Recently it has been shown that extrinsic stimuli are dispensable for derivation and maintenance of the pluripotent state. Culture conditions free from feeders, serum and cytokines were established by using a combination of small-chemical molecules, which inhibit the fibroblast growth factor (FGF)/mitogen-activated protein kinase (MEK)/extracellular signal-related kinase (ERK1/2) and the glycogen synthase kinase 3 (GSK3) (Ying et al., 2008). These culture conditions are known as 3i or 2i and have been applied for derivation of ESCs from non-permissive mouse strains like non-obese diabetic (NOD) mice (Nichols et al., 2009) but also from rat embryos, resulting in the production of the first germ-line competent rat ESCs (Buehr et al., 2008; Li et al., 2008).

The finding that by using the 2i conditions it is nowadays possible to derive ESCs from almost all the mouse strains and importantly also from the rat is surely of high importance in the field of ESC research.

In the following paragraphs we will review the most important molecular and cellular mechanisms that regulate stem cell self-renewal and pluripotency prevalently in mouse ESCs, since they are the most investigated ESCs.

2. Signalling through cytokine receptors: LIF/gp130 pathway

In 1988 Austin Smith and colleagues isolated a soluble glycoprotein that prevents stem cell differentiation and established that ESC self-renewal was dependent on paracrine signals produced from the feeders on which ESCs were cultivated (Smith et al., 1988). The principal factor required for self-renewal was shown to be leukaemia inhibitory factor (LIF) (Gearing et al., 1987; Williams et al., 1988). *Lif* knockout feeders were reported to be unable to support ESCs self-renewal (Stewart et al., 1992), indicating that supply of LIF was a key attribute of feeders. LIF belongs to the family of interleukin (IL)-6-type cytokines and exerts its effects by binding to a two-part receptor complex, which consists of the low-affinity LIF receptor (LIFRβ) and the glycoprotein 130 (gp130). LIF induces heterodimerization of the LIFRβ and gp130 resulting predominantly in the activation of the JAK/STAT signal transduction pathway, which promotes self-renewal in ESCs. Several studies showed that LIFRβ receptor is not sufficient to mediate the signal to maintain ESCs self-renewal, whereas gp130 is (Niwa et al., 1998; Starr et al., 1997). These results indicate that gp130 is the main component of the activated LIFRβ/gp130 receptor. As we will see in the following chapters activation of LIFRβ/gp130 receptor through the binding of LIF leads also to the activation of the mitogen-activated protein kinase (MAPK) and the phosphatidylinositol-3 phosphate kinase (PI3K) pathways, which together with the JAK/STAT3 pathway are essential for regulating biological responses in ESCs.

In the embryo the epiblast is the transient population of cells from which the foetus is derived. Because of their characteristics, ESCs seem to be the *in vitro* counter part of the epiblast cells *in vivo*. However, in contrast to the LIF dependency of ESCs, early epiblast cells do not require LIF stimulation, since *Lif* -/- embryos develop normally into later stages (Stewart et al., 1992) and embryos carrying mutations on the LIFβR and gp130 receptor develop normally, at least until mid-gestation (Li et al., 1995; Nakashima et al., 1999; Ware et al., 1995). Nevertheless, it has been shown that the embryos do express LIF, LIFβR and the gp130 mRNA indicating a possible function of this pathway also *in vivo*.

Mice can temporary arrest embryogenesis at the blastocyst stage; this phenomenon is called diapause and has evolved in certain mammals to overcome sub-optimal conditions for pregnancy. During diapause the embryos develop to the hatched blastocyst stage but then stop their development remaining unimplanted in the uterus. This situation can persist in mice for weeks, a period during which the epiblast cells have to be maintained pluripotent till the development of the embryo is restored. Interestingly ESCs were first established from diapause embryos (Evans and Kaufman, 1981). It has been shown that development arrested embryos carrying mutation on the LIFβR and the gp130 receptors fail to restore normal embryogenesis (Nichols et al., 2001). These findings highlight the absolute requirement for LIF/gp130 signalling in the epiblast during diapause and give also an explanation why ESCs are LIF dependent. However this is a facultative situation, because the pathway is dispensable for early development without diapause.

2.1 JAK/STAT3 signalling and self-renewal

Binding of the cytokine LIF to the receptor results in conformational changes in the intracellular part of the receptor. Cytosolic tyrosine kinases of the JAK family are then recruited to the receptor. The activated receptor phosphorylates the tyrosine residues in the kinase molecule that become docking sites for the STAT3 transcription factors. When bound to the receptor, STAT3 molecules are phosphorylated on the tyrosine 705 (Tyr705) residues and dimerize with another phosphorylated STAT3. The dimers are then translocated to the nucleus in a regulated manner where they bind to promoter and enhancer regions of their target genes (Fig.1).

Active STAT3 is necessary and sufficient for maintaining pluripotency in ESCs (Cinelli et al., 2008; Matsuda et al., 1999; Niwa et al., 1998). Matsuda and colleagues used a chimeric STAT3-estrogen receptor (STAT3-ER) composed of the entire coding region of STAT3 and the ligand-binding domain of the estrogen receptor. Dimerization of the chimeric STAT3 was activated after treatment with the estrogen derived 4-hydroxy-tamoxifen (4-OHT). ESCs cultivated in presence of 4-OHT were able to self-renew in absence of LIF. In a recent study, by using the same STAT3 inducible system we were able to generate with high efficiency ESCs from the non-permissive FVB/N mouse strain (Cinelli et al., 2008). Wild type FVB/N ESCs derived in presence of LIF were not able to generate chimeras whereas cells derived from transgenic ICMs overexpressing STAT3-ER, in absence of LIF and in presence of 4-OHT, were able to generate germline competent ESCs (Cinelli et al., 2008).

The *Socs* genes (Suppressor of cytokine signalling) are well characterized STAT3 target genes, whose encoded proteins generally act in a negative feedback loop to suppress further STAT3 signalling (O'Sullivan et al., 2007) (Fig 1.). Although overexpression of STAT3 promotes stem cell self-renewal and maintenance of pluripotency in the absence of LIF and in presence of serum (Cinelli et al., 2008; Matsuda et al., 1999), inactivation of STAT3 in LIF-maintained ESCs promotes spontaneous differentiation (Niwa et al., 1998). Even though these lines of evidence establish STAT3 as an essential component of the LIF-dependent self-renewal in ESCs, the downstream target genes of activated STAT3 have remained elusive. In order to isolate these genes, several studies based on chromatin immunoprecipitation (ChIP) analysis or on microarray technology have been performed.

Recently Cartwright et al. indicated a role for the transcription factor c-MYC in self-renewal by functioning as a key target of LIF/STAT3 signalling (Cartwright et al., 2005). Like other genes involved in the maintenance of pluripotency, such as *Nanog* (Chambers et al., 2003), *Klf2*, (Hall et al., 2009), *Pem/Rhox5* (Cinelli et al., 2008; Fan et al., 1999) and *Pramel7* (Cinelli et al., 2008), constitutive expression of c-MYC renders self-renewal independent of LIF (Cartwright et al., 2005). On the other hand expression of a dominant negative form of c-MYC promotes differentiation (Cartwright et al., 2005). c-MYC is a transcription factor that controls many different biological processes, such as cell proliferation, growth, differentiation and apoptosis. Several studies have shown the importance of this gene in embryonic development, since homozygous deletion of *c-Myc* in the mouse results in embryonic lethality before E10.5 of gestation (Davis et al., 1993). Interestingly, overexpression or amplification of the *c-Myc* gene has been detected in numerous solid tumours and blood malignancies (Dang et al., 1999).

Cartwright and colleagues demonstrated by ChIP analysis that the *c-Myc* gene is a direct transcriptional target of the LIF/STAT3 pathway in ESCs and that the sole overexpression of c-MYC is sufficient to maintain self-renewal in absence of LIF (Cartwright et al., 2005). Not surprisingly, c-MYC is one of the four transcription factors found to be able, together with OCT3/4, KLF4, and SOX2, to reprogram somatic cells into undifferentiated, induced pluripotent stem cells (iPSCs) (Takahashi and Yamanaka, 2006).

In a recent microarray study we could identify a group of 26 hypothetical STAT3 target genes (Cinelli et al., 2008). In situ hybridization experiments indicated that the expression of four of the up-regulated genes was restricted to the ICM of the blastocysts, pointing to their potential role in maintaining pluripotency in ESCs. Overexpression in ESCs of two of these genes, namely the *Pem/Rhox5* homeobox-containing gene and the *Pramel7* gene, were shown to be sufficient to maintain expression of pluripotency markers in absence of LIF. These results suggest that these two genes could represent possible STAT3 transcriptional target genes, involved in self-renewal and maintenance of pluripotency in ESCs (Cinelli et al., 2008).

The fact that overexpression of the transcription factor Nanog does not increase significantly the level of phosphorylated STAT3, and *vice versa* the overexpression of STAT3 seems not to affect Nanog expression leads to the conclusions that Nanog is not a direct transcriptional target of STAT3, nor does it regulate STAT3 activity (Chambers et al., 2003). Moreover increased STAT3 activity maintains pluripotency even when Nanog expression is reduced (Bourillot et al., 2009), confirming that these two transcription factors are regulated through different signals. Nevertheless, in a recent study it was shown that 55% of the putative STAT3 target genes display binding sites for Nanog, and 41% of the putative Nanog target genes display binding sites for STAT3 (Chen et al., 2008). These results suggest that both transcription factors co-regulate the expression of a large number of target genes, whose expression is involved in the maintenance of the undifferentiated state in ESCs. Bourillot et al. identified in a microarray study twenty-four STAT3 target genes which showed binding sites for STAT3 and/or Nanog (Bourillot et al., 2009). Knockdown experiments showed increased differentiation in LIF-supportive conditions, confirming that these STAT3 target genes contribute to the maintenance of the undifferentiated state (Bourillot et al., 2009). Among these there were some already identified as STAT3 targets, like *Pim-1* and *Pim-3* kinases (Aksoy et al., 2007) and two genes of the Krüppel-factors family, namely *Klf4* and *Klf5*. The *Klf* genes have been proposed as downstream targets of LIF/STAT3 also by other research groups (Hall et al., 2009; Li et al., 2005; Niwa et al., 2009). Recently Hall et al. showed that OCT3/4 in addition to the LIF/STAT3 signalling activates the Krüppel-factors KLF4 and KLF2 and that their overexpression reduces LIF dependence (Hall et al., 2009). However only KLF2 was able to sustain pluripotency in absence of either LIF or *Stat3*, and was shown to be OCT3/4 induced. KLF4 was shown to be selectively induced by LIF/STAT3 but was not sufficient in absence of the LIF/STAT3 signalling to sustain prolonged ESCs self-renewal. Interestingly, like c-MYC, also KLF4 is one of the four transcription factors that are able to reprogram somatic cells into undifferentiated, self-renewing cells (Takahashi and Yamanaka, 2006).

In the molecular mechanisms involved in the maintenance of pluripotency in ESCs extrinsic stimuli converge with intrinsic circuitries in a synergistic manner propagating the undifferentiated and self-renewing state in ESCs. STAT3 is an important regulator of mouse ESC self-renewal and it is known to inhibit differentiation into both mesoderm and endoderm lineages (Ying et al., 2003) by preventing the activation of lineage-specific differentiation programs. However its mechanisms of action remain to be better elucidated.

2.2 gp130-dependent activation of the MAPK/ERK pathway

Self-renewal and differentiation converge downstream from the LIFβR/gp130 receptor. The binding of the cytokine LIF to the heterodimeric receptor not only activates the JAK/STAT3 pathway but also the mitogen-activated protein kinase (MAPK) pathway that culminates in the activation of the extracellular signal-regulated kinases (ERK1/2). Because it has been shown that ERK regulates early differentiation *in vivo* and *in vitro* (Kunath et al., 2007; Nichols et al., 2009b), the balance between self-renewal and differentiation has to be maintained in order to preserve the undifferentiated state of ESCs.

Active gp130 receptor can also associate with the protein tyrosine phosphatase SHP-2 (Fukada et al., 1996), which acts as a positive effector of the MAPK signalling cascade. Interaction between active gp130 receptor and SHP-2 phosphatase induces the recruitment of GAB1. The complex formed by the gp130 receptor, SHP-2 and GAB1 proteins, through the activation of further kinases (RAS/RAF and MEK) results in the activation of the ERK1 and ERK2 kinases (Takahashi-Tezuka et al., 1998) (Fig.1).

Burdon et al. confirmed that after stimulation with LIF, ERK1 and ERK2 were activated through phosphorylation of SHP-2 (Burdon et al., 1999). Surprisingly they also showed that suppression of the SHP-2/ERK signalling was not affecting propagation of stem cells, but on the contrary it was enhancing ESCs self-renewal (Burdon et al., 1999). This study indicates that SHP-2/ERK signalling activation is a necessary component of the normal differentiation processes. Differentiation of ESCs into embryoid bodies *in vitro* was associated with an induction of expression of G1 cyclins, a lengthening of the G1 phase and a decrease in the rate of cell division (Savatier et al., 1996). As we mentioned before, ESCs typically exhibit a short G1 phase of the cell cycle (Rohwedel et al., 1996) and high rate of cell divisions. Entrance into the G1 phase of the cell cycle is a prerogative for cell differentiation. ERK signalling is known to regulate proliferation and survival of somatic cells (Lloyd, 1998), and *in vivo* phosphorylated ERK has been detected from the 2-cell stage till the blastocyst stage (Wang et al., 2004). Incubation of 2-cell stage embryos with an ERK inhibitor results in a developmental arrest at the four-cell stage embryo, however, normal embryo development can be restored once the inhibitor is removed (Maekawa et al., 2007). This confirms that the ERK1/2 pathway is required for progression of early cell division cycles in the preimplantation embryo. In contrast to many mammalian cells, where ERK activity is essential for the cell cycle progression from G0/G1 to S phase (Lewis et al., 1998), during the development from the 2-cell to the 8-cell stage embryo, ERK signalling seems to be essential in the G2/M transition (Maekawa et al., 2007).

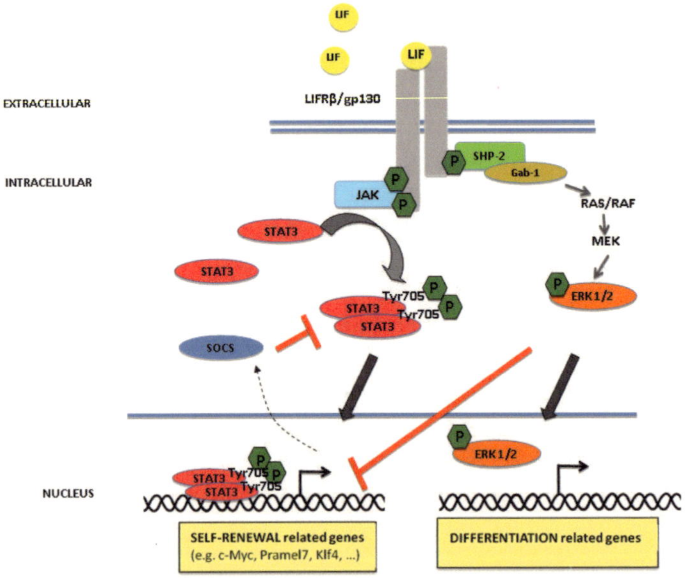

Fig. 1. gp130/LIF dependent STAT3 and MAPK/ERK signalling

Interestingly the ERK pathway may be able to inhibit the JAK/STAT3 pathway at the level of STAT3. It has been shown that ESCs knockout for *Shp2* phosphatase after LIF stimulation showed an increased phosphorylation of STAT3 when compared to the wild type cells (Chan et al., 2003). These data support the evidence that the two pathways seem to converge and thereby determine the choice between self-renewal and differentiation (Fig. 1).

Recently Ying et al. cultivated ESCs in a serum-free medium containing B27- and N2-supplement in presence of selective small-chemical inhibitors of the FGF receptor and the ERK kinase in combination with a GSK3 inhibitor, the so called 3i or 2i conditions (Ying et al., 2008). While the first two inhibitors are involved in selectively blocking differentiation signals induced by the ERK pathway, the third inhibitor is used for blocking the negative regulation on biosynthetic pathways driven by the GSK3 protein (Ying et al., 2008). This work indicates that by inhibiting differentiation-inducing signals from the MAPK pathway by 3i/2i it is possible to maintain self-renewal in absence of LIF/STAT3 stimulation. This was confirmed by the fact that under these conditions neither STAT3 nor SOCS3 activation was detected. Moreover it was possible to establish *Stat3* null cells, which did not show morphologically differences when compared with the wild type cells. All the different cell lines established under the 3i conditions expressed the typical pluripotency markers like OCT3/4 and Nanog and were able to contribute to chimera formation and germ line transmission (Ying et al., 2008).

2.3 gp130-dependent activation of the PI3K signalling

There are three distinct classes of phosphatidylinositol-3 phosphate kinase (PI3K); members of the class I_A family of PI3Ks are activated via the LIF/gp130 receptor (Fig. 2.). The products of PI3K transmit the signals through downstream effectors including the serine/threonine protein kinase B (PKB, also known as AKT). AKT has been implicated in many cellular processes like regulation of the cell cycle progression, cell death, adhesion, migration, metabolism and tumorigenesis (for review see Brazil et al., 2004).

Initially it was shown that this pathway was implicated in the control of proliferation in ESCs (Jirmanova et al., 2002; Takahashi et al., 2003). Paling et al. demonstrated that LIF induced PI3K signal activation in ESCs is involved not only in the regulation of cell proliferation but also in their self-renewal (Paling et al., 2004). ESCs incubated with LIF and a small chemical PI3K inhibitor showed less alkaline phosphatase activity compared to the untreated control cells, indicating a reduced ability of LIF to promote self-renewal. However these cells did not show altered levels of phosphorylated STAT3 when compared to the control cells. This result led to the conclusion that *Stat3* is not a target of PI3K action. The loss of self-renewal and the consequently differentiation of the cells after inhibition of PI3K was explained by an increase in ERK phosphorylation upon LIF stimulation (Paling et al., 2004). These findings are consistent with other studies were it was reported PI3K playing a role in negatively regulating ERK activity in ESCs (Hallmann et al., 2003). Paling et al. demonstrated that self-renewal was restored after incubation with both ERK and PI3K inhibitor, therefore confirming that the regulation of ERK activity by PI3K signalling contributes to the determination of ESCs self-renewal.

The involvement of PI3K/AKT signalling in the regulation of stem cell systems has been also proposed from studies of conditional *Pten*-deficient mice. PTEN is a lipid phosphatase that antagonizes the physiological and pathological processes of PI3K/AKT pathway (Stiles et al., 2004). It was shown for example that self-renewal of neural stem cells was increased in brain-specific mutant mice lacking *Pten* gene (Groszer et al., 2001). It was therefore interesting to investigate if a constitutive active form of AKT was able to support self-renewal in ESCs. Watanabe and colleagues investigated this hypothesis by generating ESCs expressing a myristoylated, active form of AKT and they demonstrated that constitutive expression of AKT liberates ESCs from LIF-dependence. These cells showed no alteration in the level of phosphorylated STAT3 compared to the control cells, but interestingly showed an increased ERK phosphorylation (Watanabe et al., 2006). These findings are in contrast with the results from Paling et al. that observed an increased ERK phosphorylation upon PI3K inhibition thereby attributing to the PI3K signalling an involvement in controlling ERK phosphorylation and therefore blocking differentiation.

GSK3β (glycogen synthase kinase 3β) is a common target from the PI3K/AKT pathway and the WNT pathway, which phorphorylation leads to its inactivation (Fig. 4.). WNT blocks proteasome-mediated degradation of β-catenin through the inhibition of GSK3β (Moon et al., 2004). Watanabe et al. postulated an involvement of the β-catenin signalling in the AKT-mediated maintenance of the undifferentiated state. Although in the transgenic ESCs expressing a constitutive active form of AKT, GSK3β phosphorylation was enhanced, β-catenin signalling was not activated (Watanabe et al., 2006). They concluded that AKT-dependent maintenance of the undifferentiatied state in ESCs was independent from the WNT/β-catenin signalling. Long term treatment of ESCs with the PI3K inhibitor LY294002 was shown to inhibit the proliferation and to induce cell cycle arrest at the G1 phase (Jirmanova et al., 2002). It is known that PI3K/AKT controls cell-cycle regulation: AKT promotes the G1 to S phase transition by facilitating the formation of cyclin/CDK complexes (Brazil et al., 2004). ESCs lack cell-cycle inhibitory mechanisms which are acquired only upon differentiation (Burdon et al., 2002). It is therefore possible that AKT-mediated maintenance of self-renewal in ESCs is due to its ability to block the cell-cycle inhibitory mechanisms and consequently block differentiation (Watanabe et al., 2006).

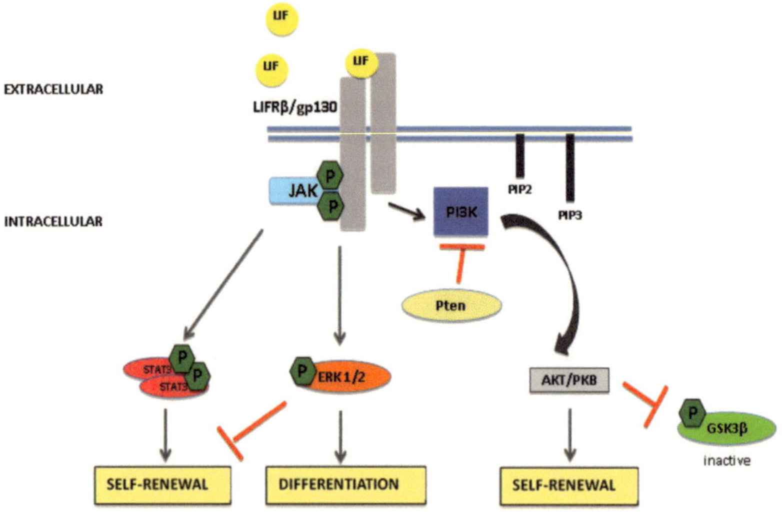

Fig. 2. gp130/LIF dependent PI3K/AKT signalling

As mentioned before GSK3 inhibition together with ERK inhibition promotes self-renewal in absence of LIF and serum. It can be assumed that both the ERK and the PI3K pathways are important for controlling self-renewal and differentiation in ESCs (Fig. 2.).

3. Signalling through the TGF-β pathway

Transforming Growth Factor β (TGF-β) signalling controls diverse sets of cellular processes, including cell proliferation, recognition, differentiation, apoptosis, and specification of developmental fate, during embryogenesis as well as in mature tissues. Moreover it has been implicated in the progression of many cancers, functioning both as an antiproliferative and as a tumor-promoting factor. The TGF-β family members bone morphogenetic proteins (BMPs), Nodal and Activin have been implicated in the development and maintenance of various stem cells, including ESCs.

BMP, Nodal and Activin act through the transmembrane type I and type II serine/threonine kinase receptors, leading to their dimerization. The activated receptor recruits SMAD molecules, which carry then the signal from the cell surface to the nucleus. SMAD molecules can be divided into three classes: the receptor-regulated SMAD (R- SMAD), the Co- SMAD, and the inhibitory SMAD. The R- SMAD are transcription factors and the group is composed of the SMAD1, -2, -3, -5 and -8. They are activated through direct phosphorylation by the type I receptor and one of their functions is to translocate to the cell nucleus, where they form heteromeric complexes with SMAD4, the only member of the Co-SMAD. Formation of the complex in the nucleus induces association with many DNA binding partners, thereby regulating the transcription of target genes both positively and negatively. The members of the inhibitory SMADs are SMAD6 and SMAD7, they antagonize the TGF-β signalling by binding to the receptor-regulated SMAD and therefore limiting their ability to interact with Co-SMAD and form the transcriptionally active complex (Fig. 3.).

In the following paragraphs we will review the most important functions of the BMP and the Nodal/Activin pathways in regulating pluripotency in ESCs.

3.1 The BMP/SMAD pathway

It has been shown that in serum-free culture, LIF is insufficient to maintain the undifferentiated state of murine ESCs (Ying et al., 2003). On the other hand the overexpression of STAT3 alone is enough to sustain pluripotency in ESCs in a LIF-independent manner, however in presence of serum and feeders (Cinelli et al., 2008; Matsuda et al., 1999). These observations suggest that there must be other factors in the serum or produced by the feeders, which suppress differentiation and concomitantly, efficiently sustain self-renewal in ESCs. One of these signals was shown to be the bone morphogenetic proteins (BMPs). BMPs bind to the Activin receptor-like kinases (ALKs) ALK2, ALK3, and ALK6, and activate the Inhibitor of differentiation (*Id*) genes through the activation of the receptor-regulated SMAD2, SMAD5 and SMAD8 (Fig. 3.) (Ying et al., 2003).

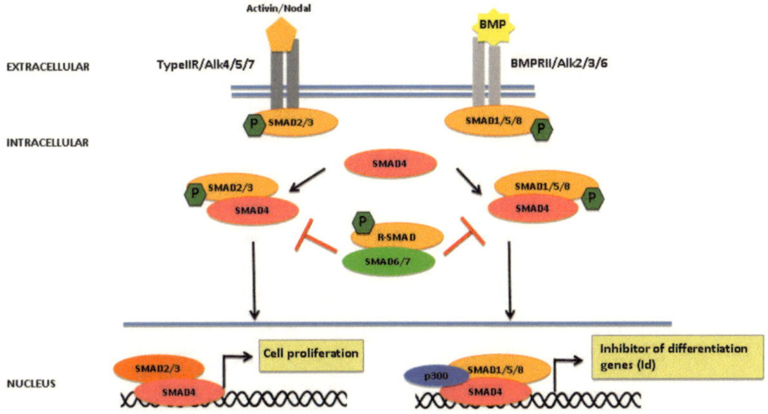

Fig. 3. Signalling through the TGF-β pathway: Nodal, Activin and BMP

In a serum-free culture, ESCs differentiate into neural precursors under the influence of autocrine FGF signal even in presence of LIF (Ying and Smith, 2003) indicating that other signals are required for suppressing neural differentiation (Ying et al., 2003). The combination of LIF and BMP4 or BMP2 was found to be sufficient to maintain the undifferentiated state of ESCs in serum-free medium (Ying et al., 2003). However, withdrawal of LIF and retention of BMP causes differentiation into epithelial-like cells, leading to the conclusion that the self-renewal response to BMP is dependent on continuous LIF signalling and that the BMP main function is therefore to antagonize the neural differentiation induced by LIF in absence of serum (Ying and Smith, 2003). This was corroborated by the fact that ID overexpression in ESCs enables serum-free self-renewal in the sole presence of LIF. Upon LIF withdrawal, ID overexpressing cells differentiated into non-neuronal precursors, therefore demonstrating that these genes contribute to the ESCs self-renewal by complementing the blockade of other lineages induced by STAT3. Interestingly, overexpression of Nanog rendered BMP or serum requirement dispensable. These findings were partially explained by a constitutively high *Id* expression in these cells, which probably allowed bypassing the BMP signalling thus maintaining ESCs undifferentiated (Ying and Smith, 2003). However these cells were induced to differentiate upon BMP stimulation, indicating that Nanog cannot inhibit in an unlimited manner SMAD-induced differentiation pathways.

These findings clearly indicate that the cooperation between LIF/STAT3 and BMP/SMAD pathway is required for maintaining ESCs in serum-free media. Nevertheless the balance has to be critically regulated, since depletion of the LIF/STAT3 signalling induces BMP-dependent promotion of differentiation and overexpression of SMAD1/4 overrides the effect of LIF and causes non-neural differentiation (Ying and Smith, 2003). Recapitulating, high expression of BMP leads to differentiation into mesoderm and endoderm lineages, whereas neural differentiation is suppressed. On the other side, low level of BMP also induces differentiation into mesoderm. Therefore, support of pluripotency by BMP pathway is highly dose-dependent and needs to be counter-regulated by STAT3 signalling.

However, STAT3 is not the only factor that interacts with BMP, Nanog was also proven to bind to SMAD1 thereby suppressing the formation of the active complex with SMAD4 (Suzuki et al., 2006b). Nanog blocks BMP-induced mesoderm differentiation by binding to SMAD1 and therefore interfering with the recruitment of the co-activator and blocking the transcription of downstream targets responsible for BMP-induced mesodermal specification (Suzuki et al., 2006b). The Nanog promoter contains predicted binding-sites for both STAT3 and Brachyury T, an early mesoderm marker. ESCs transfection with a luciferase construct containing the regulatory sequence, which included both STAT3- and Brachyury T-binding sites, resulted in LIF concentration dependent increase in transcription. The enhancer activity of this region was lost when one or the other binding-site on the Nanog promoter was mutated. Immunoprecipitation assays confirmed then the association of Brachyury T and STAT3, interestingly only after stimulation of the cells with LIF. These results demonstrated that through a cooperative action STAT3 and Brachyury T induced Nanog expression in cells that are initiating to differentiate into mesodermal lineage (Suzuki et al., 2006b).

The p300 protein is an important nuclear factor in the BMP pathway. It interacts with the active R-SMAD and SMAD4 and regulates gene transcription (Fig. 3.). Nanog was found to interfere with the recruitment of the co-activator p300 to the SMAD complex, therefore negatively regulating BMP signalling. Suzuki et al. demonstrated that Nanog inhibits Brachyury T expression by inhibiting the formation of active SMAD/p300 complexes, and therefore maintaining the undifferentiated state of ESCs (Suzuki et al., 2006b). This study, in contrast to the results shown by Ying et al., links BMP signalling more to the differentiation processes than to the maintenance of pluripotency. Ying et al. worked with low concentration of BMP and in presence of LIF; these conditions maintain pluripotency and inhibit mesoderm differentiation (Ying et al., 2003) whereas Suzuki et al. demonstrated with their studies that this mesoderm-inhibition is in part due to a negative effect mediated by Nanog and Brachyury T (Suzuki et al., 2006b).

BMP proteins can bind to the type I ALK3 receptor and to the type II BMPRII receptor (Fig. 3.). Different studies showed that *in vivo* embryos lacking one of these two receptors die because they are unable to undergo gastrulation (Beppu et al., 2000; Mishina et al., 1995), highlighting the essential role of BMP during embryogenesis. The impossibility of generating embryos carrying these mutations leads to the lack of the respective ESCs with the only exception of *Smad4* knockout embryos which die before embryonic day 7.5 and therefore allow the isolation of blastocyst stage embryos and the establishment of viable ESCs (Sirard et al., 1998). SMAD4 as mentioned before is the co-activator molecule, which binds to R-SMADs for forming the active transcriptional complex. These results suggest that there is a SMAD-independent mode of BMP action in the embryos but also in the ESCs. *Alk3* receptor null blastocysts are morphologically identical to the wild type blastocysts; however ICMs of the knockout embryos fail to expand once in culture. By incubating *Alk3* knockout ESCs with an inhibitor for ERK and p38 MAPK pathways Qi and colleagues established and expanded these cells *in vitro* in absence of BMP4. They concluded that BMP4 supports self-renewal by inhibiting the mitogen-activated protein kinase MAPK/ERK pathway (Qi et al., 2004). Anyway the mediators of this inhibition have still to be identified. A more recent study suggests that in the BMP4-induced maintenance of the undifferentiated ESCs status, together with the SMAD proteins, also the PI3K/AKT and the WNT pathway are involved (Lee et al., 2009).

All together these data confirm the complexity of the BMP signalling, where depending on its expression level, presence of active interaction partners in different pathways, and timing of expression during embryogenesis; it strongly influences cell fate decisions both *in vivo* and *in vitro*.

3.2 The Nodal/Activin pathway

Nodal and Activin ligands bind to the type I and II Activin receptor-like kinases ALKs (ALK4, -5, and -7), which phosphorylate SMAD proteins to regulate gene expression. Nodal/Activin signals are received by the transcription factors SMAD2 and SMAD3, which form complexes with the Co-SMAD4 and are then translocated to the nucleus where they associate with other cofactors to regulate target gene transcription (Fig. 3.). Nodal/Activin signalling is antagonized by Lefty, an extracellular molecule or by the classical TGF-β inhibitor SMAD7. *In vivo* this pathway has been shown to be essential for the induction of mesoderm and endoderm lineages and for the determination of the left-right axis during embryogenesis. *Nodal* knockout show a reduced epiblast cell population that display very low expression of the pluripotency marker OCT3/4 and arrest the development before gastrulation (Robertson et al., 2003).

Nodal expression in mouse ESCs is high, and was found to build an active signalling together with SMAD2. Thus, stimulation of ESCs with Activin or Nodal leads to an increase in SMAD2 phosphorylation and a higher ESCs proliferation, whereas inhibition of SMAD2 activation reduces cell proliferation (Ogawa et al., 2007).

In serum-containing medium, both BMP and Nodal/Activin pathways are autonomously activated in ESCs. Ogawa and colleagues showed that after overexpression of the inhibitors SMAD6 and SMAD7, ESCs proliferation significantly decreased, the effect in SMAD7 transfected ESCs being much more dramatic. It is known that SMAD6 predominantly inhibits BMP-mediated signalling, whereas SMAD7 blocks both BMP and Nodal/Activin pathways. The SMAD7-dependent inhibition of cell proliferation was reversible after excision of the transgene; moreover this was not affecting ESCs pluripotency since injection

of the SMAD7-reverted cells produced live chimeras (Ogawa et al., 2007). In serum-free medium the SMAD7 induced blockade of proliferation is more reduced, leading to the conclusion that soluble TGF-β-related molecules in the serum are also involved in the SMAD7-dependent growth inhibition (Ogawa et al., 2007). In fact, addition of exogenous Activin or Nodal to the serum-free medium caused SMAD2 activation and restored cell proliferation; this was not observed when soluble BMP4 was added to the medium. These results suggested that Nodal and Activin promoted ESCs proliferation through the canonical pathway. Further analyses demonstrated that ESCs autonomously activate Nodal/Activin signalling by producing these ligands in serum-free conditions; in presence of serum the soluble TGF-β-related molecules might increase the endogenous Nodal/Activin activity leading to an enhanced cell proliferation (Ogawa et al., 2007).
Another recent work showed that Nodal signalling acts also through the BMP pathway for regulating self-renewal in ESCs (Galvin et al., 2010). In this study it was observed that treatment of the cells with a chemical SMAD2 inhibitor resulted in an enhanced BMP signalling. The authors found SMAD7 to be the critical component of the Nodal pathway that influences BMP signalling through the regulation of *Id* genes (Galvin et al., 2010).
Although many efforts have been made in order to elucidate the function of Nodal/Activin in mouse ESCs, future work will be needed to determine the molecular mechanisms by which this pathway maintains pluripotency. If the involvement of the Nodal/Activin pathway in maintaining mouse ESCs has still not been completely elucidated, this is not the case for human ESCs (hESCs). In this article we review the molecular mechanisms of pluripotency in ESCs generated from the mouse, being murine ESCs the most investigated cells. Nevertheless it is to mention that the Nodal/Activin pathway plays a fundamental role in the maintenance of pluripotency in hESCs. Like mouse ESCs, hESCs do express all the components of the LIF/STAT3 pathway but in contrast to the mouse cells, cannot be maintained pluripotent in presence of LIF (Humphrey et al., 2004). This indicates that signalling through this pathway is insufficient to prevent differentiation of hESCs and suggests the existence of other pathways involved in the regulation of pluripotency in hESCs. Interestingly, mouse epiblast stem cells derived from the E5.5-E6.5 post-implantation embryos, can also not be maintained undifferentiated in presence of LIF or BMP4. Like hESCs, mouse epiblast stem cells require FGF4 and Nodal/Activin signalling for self-renewal (Brons et al., 2007; Tesar et al., 2007).
In conclusion, even though all the components of the Nodal/Activin pathway are highly expressed in both hESCs and mouse ESCs, the outcome of these signalling is different in the two species. More studies are needed for a better determination of the target genes activated by the Nodal/Activin pathway; nevertheless it is clear that the TGF-β signalling is a critical regulator of stem cell functions.

4. The canonical WNT pathway

The WNT pathway plays crucial roles in controlling genetic programs of embryonic development and adult homeostasis. WNT signals are transduced depending on their functions through different receptors and members: The canonical WNT pathway is known to be involved in transmitting signals for cell fate determination, whereas the non-canonical WNT pathway is involved in controlling cell movements and tissue polarity. In the context of maintaining pluripotency in ESCs, we will focus in the next paragraph only on the canonical pathway.

Canonical WNT signalling starts when the extracellular WNT ligand binds members of the Frizzled and LDL receptor family. The main player of the cascade is the cytoplasmic protein β-catenin. When the WNT ligand activates the pathway, β-catenin translocates to the nucleus where it interacts with other members of the signalling pathway for activating target genes. The tumour suppressor adenomatous polyposis coli (APC) and AXIN are two components of the cytosolic destruction complex responsible for β-catenin degradation in absence of WNT ligand. Casein kinase Iα (CKIα) and glycogen synthase kinase 3β (GSK3β) kinases are also part of the destruction complex and specifically phosphorylate β-catenin (Fig. 4.). This is the signal for ubiquitin ligases, which target β-catenin for proteasomal degradation: Therefore in absence of ligands β-catenin is constantly degraded. On the contrary, in presence of WNT ligands, the membrane receptor LRP5/6 is phosphorylated and bound by Dishevelled (Dsh) and AXIN, which cannot form the destruction complex therefore leading to an accumulation of cytosolic β-catenin. Free β-catenin can enter the nucleus where it binds to the T-cell factor/lymphoid enhancer factor (TCF/LEF) family of transcription factors (Fig. 4.). The β-catenin-TCF/LEF complex can then activate the transcription of WNT target genes. In absence of β-catenin the TCF/LEF factors are associated with histone deacetylase 1 (HDAC1), forming a complex that promotes gene expression silencing (Kioussi et al., 2002). Once β-catenin is again present it replaces HDAC1 promoting gene expression.

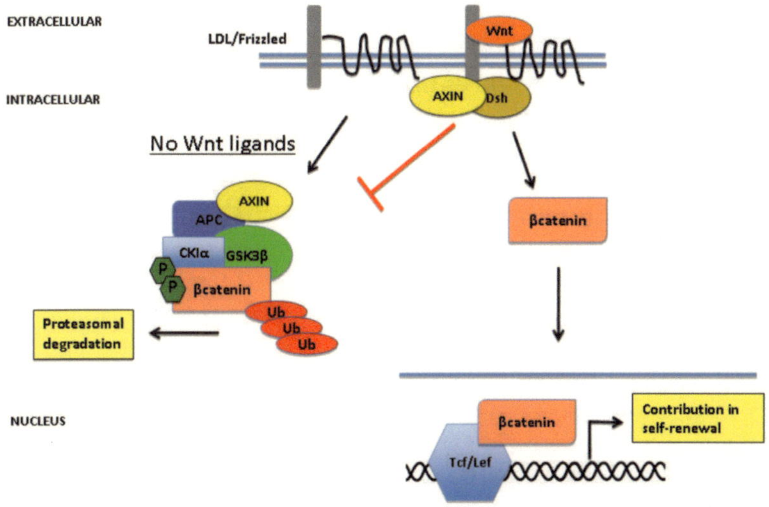

Fig. 4. The canonical WNT pathway

Several publications support the role of the WNT/β-catenin pathway in maintaining pluripotency in ESCs, although the precise mode of action needs still to be clarified. A potential source of WNT ligands is represented by the feeders used for the cultivation of ESCs. A functional screening of different sub-lines of STO feeders, which showed variable ability in maintaining ESCs self-renewal, identified WNT5A and WNT6 ligands to be the determinant factors. Their overexpression in STO feeders is sufficient to maintain ESCs undifferentiated in serum-containing medium and in absence of LIF (Hao et al., 2006). But also WNT3 and WNT3A are able to prevent ESCs differentiation under the same conditions; an effect that can be neutralized by WNT inhibitor. Moreover, ESCs conditionally expressing a constitutive active form of β-catenin were able to maintain the undifferentiated state in a LIF-independent manner (Hao et al., 2006). This study demonstrated that the WNT-mediated maintenance of pluripotency was depending on β-catenin stabilization. Moreover constitutive activation β-catenin induced an upregulation of STAT3 mRNA and protein. In serum-free medium the WNT pathway is not sufficient for sustaining ESCs self-renewal, however if 10 U/ml of LIF are added to the medium, ESCs form compact colonies and show OCT3/4 expression suggesting a synergistic effect of both pathways; the WNT pathway effectively upregulating STAT3 mRNA and the LIF pathway phosphorylating the protein and finally activating STAT3 target gene expression (Hao et al., 2006).

Nevertheless cultivation of ESCs in a feeder- and serum-independent manner with the sole presence of WNT3A conditioned medium is also not sufficient for maintaining ESCs undifferentiated; this is only possible in combination with LIF (10U/ml). The same holds true when ESCs constitutively express an active form of β-catenin (Ogawa et al., 2006).

In *β-catenin* knockout ESCs the expression of the core pluripotency factors Nanog, OCT3/4 and SOX2 is still present indicating that β-catenin regulates the expression of several stemness genes, but is not directly required for maintenance of pluripotency (Anton et al., 2007). Undifferentiated ESCs were shown to accumulate β-catenin in the nucleus (Naito et al., 2006; Takao et al., 2007). It was therefore investigated if the β-catenin-mediated expression of pluripotency factors was controlled through the interaction with TCF/LEF transcription factors. The TCF/β-catenin-mediated gene regulation in ESCs was measured by comparing endogenous TOPFlash activity with a FOPFlash reporter. Interestingly a very minimal TCF/β-catenin activity was detected with this method, indicating that probably β-catenin might interact with factors other than TCF for inducing pluripotency-related gene expression (Anton et al., 2007). Actually other reports demonstrated the interaction of β-catenin with well-known pluripotency factors, such as KLF4, SOX2, and OCT3/4. The expression level of β-catenin was found not to be regulated through the LIF/STAT3 pathway, since LIF depletion for 4 days did not change the mRNA and protein level of β-catenin (Takao et al., 2007). However in the same cells, the amount of β-catenin found exclusively in the nuclei was reduced when compared to the one of ESCs cultivated in presence of LIF. Also by forced expression of an active form of β-catenin, the same authors could cultivate ESCs in absence of LIF. Interestingly, they found Nanog and OCT3/4 to be upregulated in the β-catenin transgenic ESCs compared to the wild type. On the other hand they noticed that after withdrawal of LIF for 3 and 6 days, Nanog and β-catenin expression disappeared at day 3 whereas OCT3/4 expression was decreasing at day 6. These observations suggested the possibility that Nanog could be a potential target of β-catenin (Takao et al., 2007). By luciferase assay *Nanog* promoter activity was examined in ESCs carrying the constitutive active form of β-catenin. In presence of LIF these cells showed higher promoter activity compared to the wild type cells. Furthermore, upon LIF withdrawal the *Nanog* promoter was shown to be still highly active, whereas wild type cells lost its expression and differentiated. Nanog mRNA level correlated with the promoter activity, leading to the hypothesis that *Nanog* promoter contains a β-catenin-responsive element (Takao et al., 2007). Moreover, it was previously shown that the *Nanog* promoter contains OCT3/4-SOX2 binding sites at the promoter region (-322/+50) (Kuroda et al., 2005; Rodda et al., 2005). Mutations in this promoter region in ESCs expressing the active form of β-catenin showed reduced *Nanog* reporter activity in both conditions, with and without LIF, suggesting a reduced β-catenin transcriptional activity. The presence of OCT3/4-SOX2 binding region on the *Nanog* promoter is indispensable for the β-catenin-mediated upregulation of Nanog. These results aroused the question if β-catenin could physically interact with OCT3/4. Pull-down assay confirmed this hypothesis when OCT3/4 was found co-precipitated with β-catenin (Takao et al., 2007). The discrepancy in the results between the Takao and the Ogawa work could be due to the use of different cell lines.

We can conclude that the WNT/β-catenin signalling helps in the maintenance of pluripotency, nevertheless β-catenin alone is not sufficient to maintain the cells in a completely undifferentiated state.

5. Transcription factors that regulate pluripotency in ESCs

A critical role in maintaining ESC identity is played by a set of transcription factors centred on the octamer binding protein 3/4 (OCT3/4) (Nichols et al., 1998; Schöler et al., 1989), the SRY-related HMG-box gene 2 (SOX2) (Yuan et al., 1995), and Nanog (Chambers et al., 2003; Mitsui et al., 2003). Genome-wide studies have highlighted the co-localization of these three transcription factors in ESCs chromatin increasing the complexity of the transcriptional networks that direct ESC identity. The following paragraphs review the current knowledge on the molecular functions and regulation of the transcription factors OCT3/4, SOX2, and Nanog.

5.1 The transcription factor OCT3/4

OCT3/4 is a member of the POU transcription factor family; it recognizes an 8-base pairs DNA sequence found in the promoters and enhancer regions of many ubiquitously expressed and cell-specific genes (Ruvkun and Finney, 1991). The gene encoding OCT3/4 is named *Pou5f1*. OCT3/4 was first identified as an active binding factor in the extract of undifferentiated embryonic stem and embryonal carcinoma cells (Lenardo et al., 1989; Schöler et al., 1989). The presence of OCT3/4 protein in ESCs and embryonal carcinoma cells first suggested an association with the early stage of mouse embryogenesis. During the mouse development, it is first detected in oocytes and its expression declines during the first two embryonic divisions, but it reappears at the 4-8 cell-stage, where it is expressed in all the nuclei. Subsequently it is reduced in the trophectoderm and becomes restricted to the ICM. In the post-implantation embryos, OCT3/4 is localized in the epiblast but disappears as cells undergo differentiation, with expression persisting in the germ cells (Palmieri et al., 1994). The importance of OCT3/4 during early embryogenesis was also highlighted by the fact that embryos lacking the *Pou5f1* gene die after implantation (Nichols et al., 1998) due to the absence of the ICM. Using a conditional expression and repression system in ESCs, the requirement of OCT3/4 in the maintenance of developmental potency was determined (Niwa et al., 2000). A critical amount of OCT3/4 is needed to sustain self-renewal, and up- or downregulation induces differentiation of the ESCs. Overexpression of this factor promotes differentiation into primitive endoderm and mesoderm, whereas repression of it causes loss of pluripotency and differentiation into trophectoderm (Fig. 5.). Effectively, *in vivo* OCT3/4 is abundant in the ICM cells of the blastocyst and down-regulated in the trophectoderm, whereas in the primitive endoderm the level of expression increases (Palmieri et al., 1994). These findings highlight the fundamental function of OCT3/4 in preventing dedifferentiation of epiblast cells or of ESCs into trophectoderm lineage.

Exogenous expression only of OCT3/4 is sufficient to generate pluripotent stem cells from mouse neural stem cells (Kim et al., 2009) and together with three other transcription factors it is essential in the reprogramming of somatic cells into the pluripotent state (Nakagawa et al., 2008; Takahashi and Yamanaka, 2006). While the other factors involved in reprogramming are replaceable by family members or other factors, without OCT3/4 no reprogramming occurs (Nakagawa et al., 2008). Moreover knockdown experiments in ESCs showed a very dramatic change in gene expression compared to the one caused for example by Nanog or SOX2 knockdown (Ivanova et al., 2006) implying OCT3/4 as a chief selector for ESC fate decision.

LIF/STAT3 signalling does not directly support ESCs renewal by maintaining OCT3/4 expression, but it has been shown that STAT3 and OCT3/4 share some target genes involved in maintaining pluripotency. It is the case for the embryonic ectoderm development (*Eed*) gene, which was shown to be a common downstream target of both transcription factors (Ura et al., 2008). EED is a major component of the Polycomb repressive complex 2 (PRC2), which is involved in the methylation of lysine 27 on the histone H3 (H3K27). As we will describe later epigenetic mechanisms, like dynamical modification of the chromatin, play a fundamental role in maintaining pluripotency and controlling differentiation. EED function is to silence the expression of differentiation-associated genes in self-renewing ESCs and its downregulation induces differentiation (Boyer et al., 2006). Expression of a dominant negative form of STAT3 was shown to induce downregulation of EED, whereas overexpression of STAT3 caused an upregulation. Similar results were found when OCT3/4 was downregulated, suggesting *Eed* to be a common downstream target of both transcription factors (Ura et al., 2008). Further analyses confirmed that STAT3 and OCT3/4 directly bind the promoter region of the *Eed* gene, regulating its expression and thereby inducing silencing of differentiation-associated genes.

Ura et al. found that self-renewal correlates with the presence of high concentration of methylated H3K27 that leads to the transcriptional repression of the gene carrying this signal. Elimination of STAT3 and OCT3/4 causes a reduction of this methylation on the promoter of differentiation marker genes. These results suggested that EED is involved in STAT3- and OCT3/4-dependent gene silencing of differentiation markers and highlight another important function of OCT3/4 transcription factor in maintaining and controlling pluripotency in ESCs.

Of the known OCT3/4 target genes, four have been shown to be essential for the maintenance of pluripotency: *Sox2*, the undifferentiated transcription factor 1 (*Utf1*), *Rex1/Zfp42* and *Nanog*. The regulatory regions of these genes contain an octamer element capable of binding OCT3/4. As we will describe in the following paragraphs, *Sox2* is not only an OCT3/4 target gene but also serves as a cofactor for OCT3/4 (Okumura-Nakanishi et al., 2005). Although OCT3/4 and SOX2 have independent roles in determining other cell types, in pluripotent cells they act synergistically to drive transcription of their target genes. Furthermore, this complex was found to promote transcription of the Nanog homeoprotein (Rodda et al., 2005) (Fig. 5.).

5.2 The transcription factor SOX2

SOX2 is a member of the sex-determining region of the Y chromosome-related (SRY-related) high-mobility group (HMG) box (SOX) family of transcription factors. SOX2 has an expression pattern similar to that of OCT3/4 through the mouse preimplantation development. SOX2 expression is also associated with precursor cells of the developing central nervous system and indeed can be used to isolate these cells. Like OCT3/4, downregulation of SOX2 correlates with a commitment to differentiation and is no longer expressed in cell types with restricted developmental potential. The phenotypic lethal consequences of the absence of SOX2 expression in the embryo are in contrast to OCT3/4, visible only after implantation (Avilion et al., 2003). This is due to an accumulation of maternal SOX2 in the cytoplasm of the oocytes, which persist in all cells at least until the blastocyst stage, and not like OCT3/4 maternal transcripts, which last only till the 2-cell stage embryo (Palmieri et al., 1994). Thus, mutant embryos lacking the *Sox2* gene die presumably when the maternal SOX2 becomes diluted causing the differentiation of the epiblast cells into trophectoderm or extraembryonic ectoderm (Avilion et al., 2003) (Fig. 5.).

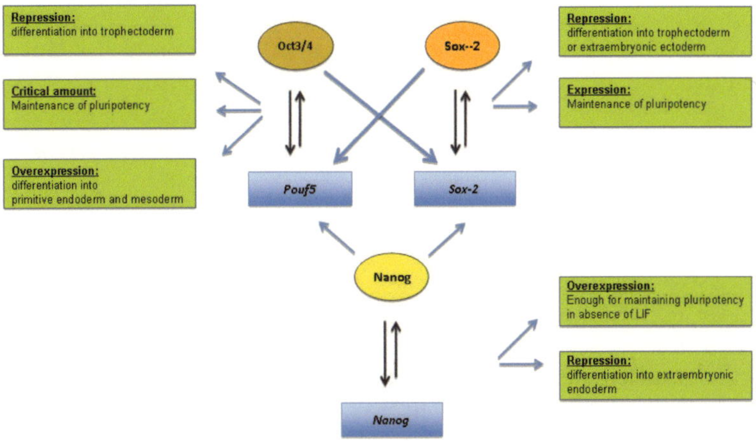

Fig. 5. OCT3/4, SOX2, Nanog core transcription factors: regulatory circuitry

As mentioned before, loss of OCT3/4 causes differentiation of the epiblast cells into trophectoderm derivatives (Niwa et al., 2000). Avilion et al. proposed that the presence of both transcription factors is required for the formation and maintenance of epiblast cells. An upregulation of OCT3/4 and the accompanying downregulation of SOX2 lead to the extraembryonic ectoderm differentiation.

We know that the regulatory regions of the *Sox2* gene contain an octamer element capable of binding OCT3/4 (Tomioka et al., 2002; Yuan et al., 1995). A possible interaction of these two factors was also proposed by the fact that almost all the SOX2-OCT3/4 target genes have both the octamer and sox heptamer elements separated by either 0 or 3 base pairs (Reményi et al., 2003; Williams et al., 2004). The *Pou5f1* gene has different regulatory regions that are important for its expression. The distal enhancer, which contains the conserved region CR4, was shown to be required for ESC-specific OCT3/4 expression (Nordhoff et al., 2001). Two regulatory regions (the SRR1 and SRR2), known to confer ESC-specific expression, were also found in the *Sox2* sequence. Using a ChIP assay Chew et al. first demonstrated that SOX2 and OCT3/4 interact with the enhancers of *Pou5f1* and *Sox2* genes. Moreover OCT3/4 and SOX2 knockdown experiments showed a reduced enhancer activity, confirming that both factors positively control their reciprocal expression (Chew et al., 2005). The same authors hypothesized a transcriptional regulatory network consisting of auto-regulatory and multi-component loops. In an auto-regulatory system the gene product binds to its own regulatory element allowing its continued and stable expression. In a multi-component system, the OCT3/4 factor binds to the *Sox2* regulatory element and *vice versa* generating a bi-stable system with the possibility to switch between the two different states (Chew et al., 2005) (Fig. 5.).

We described before that *Sox2* null embryo die after implantation, because of differentiation of the epiblast cells. The same findings are reproducible also *in vitro*, where ESCs lacking *Sox2* gene differentiate primarily into trophectoderm-like cells. Thus, SOX2 was defined to be indispensable for maintaining ESCs pluripotency. However, the transcription of many OCT3/4-SOX2 target genes was not affected in ESCs null for *Sox2*. These findings suggested that SOX2 regulates the expression of OCT3/4 through the regulation of multiple transcription factors (Masui et al., 2007). So, it seems that the main contribution of SOX2 in ESCs is to maintain OCT3/4 expression. Consistent with this idea is the finding that enforced expression of OCT3/4 can rescue ESCs from differentiation induced by the loss of *Sox2* (Masui et al., 2007). Not to forget is also that SOX2 is one of the four transcription factors, which together are able to induce reprogramming to pluripotency in differentiated cells (Takahashi and Yamanaka, 2006). Furthermore, large-scale ChIP studies have shown the OCT3/4-SOX2 complex closely localized to another important regulator of pluripotency, the Nanog transcription factor.

5.3 The transcription factor Nanog

Nanog is a homeodomain protein, which acts as an intrinsic effector of ESCs self-renewal. Nanog was detected in ESCs, embryonic carcinoma cells, and in the embryonic germ cells. During ESCs differentiation, Nanog mRNA declines markedly, and its expression is retained only in undifferentiated cells. In the mouse embryo its first expression appears in the compacted morula and it is localized to the interior cells, the future ICM. In blastocysts, the expression is confined to the ICM and absent from the trophectoderm. In the later blastocyst, Nanog is further restricted to the epiblast and excluded from the primitive endoderm; it is then down regulated after implantation (Chambers et al., 2003; Mitsui et al., 2003). At day E11.5-E12.5 Nanog expression marks the pluripotent germ cells.

Overexpression of Nanog in ESCs was found to allow proliferation of undifferentiated ESCs in absence of LIF and as well as in the presence of the LIF-antagonist hLIF-05, which blocks the activity of all known LIFR ligands (Chambers et al., 2003). After transfection with Cre-recombinase, in which the *Nanog* expression cassette had been eliminated, these cells reverted to LIF-dependence, demonstrating that this phenotype was directly attributable to Nanog overexpression. Following exposure to a differentiation-promoting agent, like retinoic acid, cells overexpressing Nanog remained, in contrast to the Cre-reverted cells, morphologically undifferentiated and expressed OCT3/4. Cre-reverted cells were injected into mouse blastocysts and contributed to the generation of germline competent chimeras. Even though cells overexpressing Nanog self-renew in a cytokine independent manner, the presence of LIF confers to these cells an enhanced self-renewing capacity (Chambers et al., 2003).

A reduction in the level of Nanog causes ESCs to differentiate to extraembryonic endoderm lineages (Chambers et al., 2007; Ivanova et al., 2006) (Fig. 5.). *In vivo* the absence of *Nanog* during embryo development results in early lethality. At developmental day E5.5 the mutant embryos showed disorganized extraembryonic tissues with no discernible epiblast or extraembryonic ectoderm. ICM of *Nanog* null blastocysts failed to proliferate and differentiated into parietal endoderm-like cells, demonstrating that its expression is essential for maintenance of pluripotency of the ICM cells (Mitsui et al., 2003). Interestingly, ESCs upon targeted deletion of the *Nanog* gene can self-renew indefinitely and can contribute to the generation of chimeric animals (Chambers et al., 2007). This indicates that loss of *Nanog* in ESCs does not affect pluripotency once this was established. Nevertheless, even though

Nanog null cells colonize the germ layers of the chimeric mice and they are recruited to the germ line, primordial germ cells lacking *Nanog* fail to mature on reaching the genital ridge (Chambers et al., 2007). These data confirm that Nanog is specifically required for the formation of germ cells.

The fact that STAT3 activation in presence of LIF was not increased in cells overexpressing Nanog (Mitsui et al., 2003) and the fact that these cells did not differentiate in presence of a JAK/STAT3 inhibitor (Chambers et al., 2003), clearly demonstrate the existence of a parallel and LIF/STAT3-independent mechanism sustaining pluripotency in ESCs. *In vivo* these findings are confirmed by the fact that Nanog is absolutely required for epiblast formation, while STAT3 is dispensable. Further analysis showed also that Nanog expression was not altered after incubation of the cells with a MEK inhibitor, indicating that Nanog-dependent maintenance of pluripotency was not due to a repression of the MEK/ERK pathway (Chambers et al., 2003).

Several studies showed that Nanog and OCT3/4 factors overlap substantially in their target genes, causing sometimes, also by cooperation, the activation or suppression of the expression of target genes. Mapping of the binding sites of OCT3/4 and Nanog in the mouse ESC genome showed that a substantial proportion of the genes were occupied by both factors, in some cases Nanog-OCT3/4 were found to co-localize, in other cases they bound indepentently (Loh et al., 2006). Screening for the presence of OCT3/4 and Nanog binding sites after differentiation revealed an enrichment of OCT3/4 or Nanog-bound genes that were induced and repressed upon differentiation. The majority of these genes were repressed and the ones that exhibited the strongest downregulation were bound by both factors (Loh et al., 2006). In the same study it could be shown that the OCT3/4-regulated target genes predominantly repress trophectoderm markers whereas Nanog activates the transcription of *Pou5f1* and *Sox2* evidencing a possible role of Nanog in controlling the levels of both OCT3/4 and SOX2. These factors in turn control the downstream genes involved in the maintenance of pluripotency or inhibition of differentiation (Loh et al., 2006). Interestingly promoter-sequence analyses showed the presence of a Sox-Oct element on the *Nanog* promoter. Two different groups demonstrated that both OCT3/4 and SOX2 can bind to the *Nanog* promoter therefore driving its transcription (Kuroda et al., 2005; Rodda et al., 2005).

ESCs under conventional culture conditions (namely with serum and LIF) are in a dynamic state, that fluctuates between a stable state, in which Nanog expression is high, and an unstable state, where Nanog expression levels are low (Kalmar et al., 2009; Singh et al., 2007). As a consequence ESCs form a heterogeneous population where pluripotent ESCs exhibit a highly variegated gene expression pattern. When induced to differentiate only ESCs with low levels of Nanog expression are able to commit in a stable manner (Chambers et al., 2003). This hypothesis was confirmed by a study where the two different populations were sorted and exposed to differentiation conditions for 3 days (Kalmar et al., 2009). The low-Nanog cells did not change their Nanog expression level, but lost their ability to form colonies and showed neuronal differentiation. On the contrary only a very small part of the high-Nanog cells differentiated and showed reduced Nanog expression. Nevertheless, it seems that both cell types are able to switch from one to the other condition. Moreover the smaller proportion of low-Nanog cells suggests that this could be a transient, short-lived event in the cells. Cultivation of low-Nanog cells population for 24h in conventional culture conditions proved this hypothesis. A subpopulation of cells showed Nanog distribution like the high-Nanog population, confirming that low-Nanog expression is an instable state of the cells. The same experiment performed only with the high-Nanog cells did not give rise to a subpopulation expressing low level of Nanog, confirming that this is the most stable state. The levels of Nanog expression in an ESC are related to its probability to differentiate or not. In this model the advantage is that in a heterogeneous population, there is always a subpopulation of cells pre-primed for differentiation. Such priming would be an advantage in situations where the cells must be ready within a short period to respond to diverse signalling (Kalmar et al., 2009).

Although there are other transcription factors associated with ESC self-renewal and pluripotency, a large number of studies support the notion that the trio SOX2, OCT3/4 and Nanog are the main regulators that generate and maintain the pluripotent state *in vivo* and *in vitro*. This idea is also corroborated by the fact that all three factors play an essential role in reprogramming somatic cells (Meissner et al., 2007; Okita et al., 2007; Takahashi and Yamanaka, 2006).

6. Regulation of pluripotency by epigenetic mechanisms

In the following paragraph we will discuss some of the emerging evidence of the interaction between the core transcription factors and the epigenetic machinery, which together regulate stem cell pluripotency and differentiation. We will focus on the principal epigenetic modifications and how these influence and regulate the effect of pluripotency factors like Nanog, OCT3/4 and STAT3.

Epigenetic is the study of inherited changes in phenotype or gene expression caused by mechanisms other than changes in the underlying DNA sequence. Non-genetic factors cause cell's genes to behave differently and these changes may remain through cell divisions. The main known epigenetic mechanisms affecting transcription of the genes are: DNA methylation of the CpG islands, modifications of the histones, chromatin remodelling, and small RNA molecules. These modifications induce conformational changes in the chromatin structure: High-condensed chromatin, the so-called heterochromatin, is resistant to transcriptional activation, whereas the more open euchromatin allows transcription factors to bind on the DNA and drive the transcription of genes. Transcription factor regulation of gene expression and chromatin-controlled epigenetic memory systems are closely cooperating in establishing the pluripotent state of ESCs and in maintaining the cell fate decisions throughout development of an organism. ESCs are, in contrast to differentiated cells, highly transcriptional active. This implies that differentiation processes involve downregulation of general transcriptional activity. Therefore differentiation leads to a decrease in the euchromatic nature of the chromosome rendering it more condensed and less accessible for the transcriptional machinery; the resulting heterochromatin leads to a loss of pluripotency.

Histone proteins are subjected to various posttranslational modifications. The N-terminal tails of core histones H2A, H2B; H3 and H4 can be for instance acetylated, methylated, ubiquitinated, and phosphorylated. All these modifications modulate and control the conformation of chromatin, and therefore the transcription of the genes (Margueron et al., 2005). For example two important heterochromatic markers are increased trimethylated lysine 9 on the histone 3 (TriMeK9 H3) and decreased acetylated histone H3 and H4 (AcH3 and AcH4) (Kimura et al., 2004). When these conditions are global, they cause gene repression. These modifications can spread over kilobases of genomic DNA and are transmitted to daughter cells: This is the concept of the epigenetic inheritance (Cavalli and

Paro, 1999). Two other important histone modifications are methylation of lysine 27 on the histone H3 (MeK27 H3), which leads to repression of transcription; and the methylation of lysine 4 on the histone H3 (MeK4 H3) that positively regulates gene transcription by recruiting nucleosome remodelling enzymes and histone acetylases. Methylation of these lysines is catalyzed by thrithorax-and Polycomb-group proteins, which are involved in mitotic inheritance of lineage-specific gene expression programs and play therefore key functions during development (Plath et al., 2003; Ringrose and Paro, 2004).

Bernstein et al. mapped the histone methylation patterns in mouse ESCs and found a novel chromatin modification pattern, the "bivalent domains", which consist of large DNA regions of K27 methylation and smaller regions of K4 methylation that silence developmental-associated genes in ESCs while keeping them in a standby mode for activation. The bivalent histone modification represses gene expression because of the dominant repressive effect of K27 methylation over the K4 methylation. Thus, when ESCs differentiate these genes can rapidly be activated for transcription and the typical bivalent status is lost. Once the cell is committed it shows either K27 or K4 methylation (Bernstein et al., 2006).

The Polycomb protein complexes (PcG) are methyl-transferases responsible for methylation of H3K27. The main function of these complexes is to repress transcription and therefore maintain the cellular identity. They are fundamental during embryonic development and have been associated with many lineage-control gene loci in human and murine ESCs. Recent studies have demonstrated that in ESCs the target genes of PcG highly overlap with those of the core pluripotency markers: Nanog, OCT3/4, and SOX2 (Boyer et al., 2006; Lee et al., 2006). PcG complexes functions as two distinct components, the Polycomb repressive complex 1 (PRC1) and the PRC2 (Francis and Kingston, 2001). Microarray analyses with wild type cells and *Eed* (embryonic ectoderm development) null cells, a component of the PRC2, showed an upregulation of PcG target genes, like for example *Gata3*, *Gata4*, and *Gata6* genes. Moreover most of the PcG target genes were found upregulated upon cell differentiation, highlighting a fundamental role of PcG complex in silencing differentiation-associated genes in ESCs, and therefore contributing to the maintenance of pluripotency (Boyer et al., 2006).

Among the enzymes that modify histones there are also the histone demethylases, which are responsible for removing the methyl groups on the lysine 4 in the histone H3 (Agger et al., 2008). The Jumonji protein family, JARID1 was shown to specifically demethylate the H3K4 lysine, which is usually associated with active transcription (Mikkelsen et al., 2007). The bivalent histone modification theory suggests a mechanism where a cross talk between PcG and demethylases regulates transcriptional and developmental programs. Pasini and colleagues demonstrated that the H3K4 demethylase RBP2 (JARID1A) binds together with the PRC2 complex on PcG target genes effectively regulating their expression. The binding of RBP2 to the target genes is required for maintaining the PRC2-dependent transcriptional silencing (Pasini et al., 2008).

We mentioned before that ESCs show a unique chromatin status that is characterized by hyperdynamic and open chromatin environment. The maintenance of transcriptional permissive chromatin is achieved with general low level of the repressive marks H3K9 methylation and enrichment in active marks like methylation of H3K4 and acetylation of H3 and H4 (Meshorer and Misteli, 2006). These epigenetic processes are required for the maintenance of ESCs pluripotency and plasticity. STAT3 together with OCT3/4 transcription factors can regulate the expression of EED (Ura et al., 2008). However, the enzymes that catalyze demethylation of the histones also play a fundamental role. OCT3/4 binding sites studies revealed that this factor controls also the architecture of the chromatin by controlling the expression of genes that encode for histone-modifying enzymes (Loh et al., 2006). It was shown with ChIP assay that OCT3/4 clusters within the *Jmjd1a* and *Jmjd2c* histone demethylase genes and after OCT3/4 knockdown their expression was decreased (Loh et al., 2007). Interestingly RNAi-depletion of Nanog had no or little effect on the expression of both JMJD1a and JMJD2c, confirming both genes as OCT3/4 targets. Both JMJD1a and JMJD2c enzymes are involved in the demethylation of the repressive mark H3K9. ESCs differentiated upon knockdown of either JMJD1a or JMJD2c and showed fibroblast-like morphology (Loh et al., 2007). Search for downstream targets identified different genes that were down- or up-regulated depending on which Jmjd was silenced; only few genes were found to overlap between JMJD1a and JMJD2c (Loh et al., 2007). JMJD1a was found to regulate the expression of the *Tcl1* gene, which is known to be required in pluripotent ESCs (Ivanova et al., 2006; Ivanova et al., 2002). Depletion of JMJD1a induced an increase of H3K9Me2 at the *Tcl1* promoter, resulting in its repression and differentiation of the ESCs (Loh et al., 2007). Moreover it is known that *Tcl1* promoter is bound by the OCT3/4 transcription factor (Loh et al., 2006) and it was shown that an increase in H3K9Me2 abolished this binding. Thus, the loss of OCT3/4 binding was probably the cause for downregulation of TCL1 after JMJD1a knockdown. Interestingly overexpression of TCL1 in ESCs rescued the effect of JMJD1a depletion, confirming *Tcl1* as a key effector of JMJD1a. Microarray analyses showed that after JMJD2c knockdown Nanog expression was reduced. With ChIP assay after JMJD2c depletion an enrichment of H3K9Me3 on the *Nanog* promoter region was observed and further analyses demonstrated that JMJD2c directly binds to the *Nanog* promoter, positively regulating its expression (Loh et al., 2007). Like for TCL1 and JMJD1a, overexpression of Nanog in JMJD2c knockdown ESCs was enough for maintaining the undifferentiated state of the cells, demonstrating that Nanog could compensate the loss of JMJD2c. In the end, this study provided the evidence that the histone demethylases JMJD1a and JMJD2c help in the maintenance of pluripotency through the regulation of downstream genes that encode for self-renewal factors.

Histone tails can also be acetylated by acetyl-transferases. The p300 acetyl-transferase catalyzes the acetylation of K9 and K14 of histone H3 which are markers of transcriptionally active chromatin and is required for ESCs to undergo early differentiation (Zhong and Jin, 2009). Interestingly, deletion of p300 did not affect ESCs self-renewal and the expression of Nanog, OCT3/4 and SOX2. However, these cells exhibited an increase in the endoderm marker GATA6 and a reduction of the ectoderm marker FGF5. Embryoid bodies generated from the *p300* null cells had a higher expression of GATA6, resulting in an overgrowth of extra-embryonic endoderm-like cells. Moreover Nanog but not OCT3/4 expression, was much more reduced both at the protein and mRNA levels compared to the wild type embryoid bodies. This suggested that transcription of Nanog during differentiation is dependent on the presence of p300. Overexpression of Nanog in *p300* -/- embryoid bodies could in part rescue the aberrant expression of GATA6, but not for instance the one of FGF5 (Zhong and Jin, 2009). Therefore the unusual phenotype of the knockout embryoid bodies could be caused by an improper regulation of Nanog during differentiation. Reporter activity assay demonstrated that p300 was able to control Nanog expression at the transcriptional level in a dose-dependent manner and independently of the presence or not of LIF in the medium and suggested that p300 might regulate Nanog expression through a regulatory element located at -3.8 to -4.8 kb. Interestingly, previous works showed that

STAT3 and Brachyury T bind to this region of the *Nanog* promoter, regulating early mesoderm differentiation (Suzuki et al., 2006a). By ChIP assays it was confirmed that p300 binds to the same regulatory region and differentiated *p300* knockout ESCs show a significant reduction in H3K9 acetylation at the Nanog/Brachyury T/STAT3 binding site (Zhong and Jin, 2009). In conclusion this study demonstrated that p300 controls transcriptional regulation of Nanog during differentiation through epigenetic modifications of histone acetylation at the Brachyury T/STAT3 binding site on the *Nanog* promoter.
Understanding how epigenetic mechanisms interact with pluripotency factors is of high interest also for the better understanding of reprogramming processes.

6.1 Induced pluripotent stem cells: an epigenetic revolution

Due to the exceptional characteristics of ESCs, the possibility to generate pluripotent cells from fully differentiated cells by resetting their epigenetic memory has fascinated the scientific world for many years. Originally it was demonstrated that activated enucleated oocytes were able to reprogram differentiated cell nuclei and give rise to viable offspring (Wilmut et al., 1997) or that it was possible to achieve reprogramming of mature cells through fusion with ESCs (Tada et al., 2001). Recently genetic strategies were developed for manipulating and finally re-activating pluripotency mechanisms in terminally differentiated cells (Takahashi and Yamanaka, 2006; Yamanaka 2008). In the following paragraph we will discuss a few examples of epigenetic manipulations, which were shown to help lineage-committed cells to re-activate pluripotency-related genes, giving rise to induced pluripotent stem cells (iPSCs). Yamanaka and colleagues, by transducing four transcription factors (OCT3/4, SOX2, KLF4, and c-MYC) into mouse fibroblasts, could reprogram these cells into iPSCs, which showed an ESC-like transcriptional circuitry and epigenetic landscape. Moreover these cells were demonstrated to be pluripotent being able *in vitro* to differentiate into cell types of the three germ layers and *in vivo* to contribute to germline competent chimeras (Maherali and Hochedlinger, 2008).
Undifferentiated cells usually show transcriptionally highly-active chromatin, which is gradually condensed and therefore less transcriptionally active upon differentiation. It is then easy to understand that reprogramming induces marked epigenetic changes in the differentiated cells making them again competent to respond to pluripotency factors.
We already discussed the importance of the transcription factors OCT3/4, SOX2, KLF4, and c-MYC in the context of ESCs, we therefore refer to the original paper of the Yamanaka laboratory for more information about how these genes were selected for generating iPSCs (Takahashi and Yamanaka, 2006; Yamanaka, 2008). Interestingly, the capacity of reprogramming differentiated cells seems not to be an exclusive characteristic of the four transcription factors (OCT3/4, SOX2, KLF4, and c-MYC). Other studies, in which the same reprogramming technique was used, demonstrated that some of the Yamanaka factors could be replaced (Feng et al., 2009) or used in combination with other factors (Mikkelsen et al., 2008). Importantly only OCT3/4 was found to be indispensable for a proper reprogramming independently from the origin of the parental cells. Moreover by overexpression or knockdown of additional factors, or by using small chemical compounds it was possible to increase the efficiency of the reprogramming process.
As mentioned before DNA methylation plays a fundamental role in regulating gene expression, since highly methylated genes are normally transcriptionally inactive. Thus, the first important change that has to happen in somatic cells undergoing reprogramming is a remodelling of the methylated sites on the chromatin. Moreover for achieving a fully

reprogrammed state iPSCs have to re-activate the endogenous expression of pluripotency-related genes, establish again an open chromatin structure, and re-activate the Polycomb machinery in order to repress lineage specific gene expression.

It has been shown that with the Yamanaka factors, only a small population of the reprogrammed cells showed reactivation of endogenous Nanog expression, contribution to chimera formation, and formation of teratoma. These results demonstrated that most of the infected cells were only partially de-differentiated and their epigenetic landscape was only to some extent changed but did not completely resemble the one of ESCs (Mikkelsen et al., 2008). For example examination of the methylation state of high-CpG-Promoters (HCP), which are usually bivalently marked in ESCs, revealed that partially reprogrammed fibroblasts showed 70% more of HCP with bivalent chromatin structure when compared to the original parental cells. However a comparison with fully reprogrammed iPSCs showed that partially reprogrammed iPSC had 40% less bivalent chromatin structure (Mikkelsen et al., 2008). Further analyses confirmed that in partially reprogrammed cells, the pluripotency-related genes exhibited DNA hypermethylation instead of harbouring the bivalent markers typical for ESC chromatin. The first major reprogramming barrier was shown to be an incorrect induction of DNA methylation changes. Incomplete demethylation of promoters of fundamental pluripotency-related genes such *Nanog* or *Oct3/4* leads to an unsuccessful reprogramming of the cells. Mikkelsen and colleagues demonstrated that treatment of partially reprogrammed iPSCs with a DNA methyltransferase inhibitor (5-aza-cytidine) rapidly and fully reprogrammed these cells into authentic iPSCs (Mikkelsen et al., 2008). Analyses of the methylation state of CpGs near to pluripotency genes such as *Nanog* and *Utf1* confirmed significant demethylation. Identical results were achieved by transient knockdown of DNMT1 in partially reprogrammed cells. Treatment with DNA methyltransferase inhibitors during the reprogramming processes led to an increased efficiency in the generation of ESC-like colonies, however only if the treatment was started 8 days after infection of the cells with the four reprogramming factors (Mikkelsen et al., 2008). Another reason why incomplete reprogramming might occur is the inappropriate activation or failure in the repression of endogenous cell-fate specific transcription factors, which in turn fail to reactivate hypermethylated pluripotency-related genes. Once more these results highlight the importance of a coordinated regulation of transcription factors, which have to be stabilized by correct epigenetic remodelling mechanisms.

Using a chemical approach, aimed at the optimization of the reprogramming efficiency and at the reduction of genetic manipulations due to the integration in the genome of the reprogramming factors, it was demonstrated that by inhibiting G9a histone methyltransferase (BIX-01294) in combination with a L-calcium channel agonist (BayK), the reprogramming process was successfully achieved only by transfecting mouse fibroblast with *Oct3/4* and *Klf4* (Shi et al., 2008). The generated iPSCs were then expanded without the two chemical compounds, demonstrating that once the reprogramming has taken place, they are dispensable. G9a is a histone methyltransferase known to repress gene activity by transferring methyl groups on the histone H3 lysine 9 (H3K9) at target promoters. Bisulphite sequencing analyses of the *Nanog* promoter region confirmed that these iPSCs had demethylated this region allowing the re-activation of Nanog transcription. These data clearly show that inhibition of methyltransferases renders the chromatin structure of lineage-restricted cells more susceptible to the reprogramming, therefore facilitating the re-activation of the transcription of pluripotency factors. This concept was corroborated by another study where it was shown that treatment with histone deacetylase (HDAC) inhibitors such as valproic acid, after transfection with four or only three reprogramming factors also improved the reprogramming efficiency (Huangfu et al., 2008a). In addition, the combination of HDAC inhibitor and the DNA methyltransferase inhibitor significantly enhanced the kinetics and efficiency of the reprogramming. Furthermore by using HDAC inhibitors it was possible to reprogram human fibroblasts transfecting the cells only with two factors (*Oct3/4* and *Sox2*) instead of four (Huangfu et al., 2008b).

In conclusion, despite having found the essential factors needed for reprogramming differentiated cells, still much has to be clarified before this technique can be used for therapeutical purposes. Reprogramming is a balanced interplay between epigenetic changes and re-activation or suppression of specific gene expression. Controlling these sophisticated processes will surely be the topic for future studies. Nevertheless the combined chemical and genetic approach seems to be nowadays a good combination for driving and facilitating this complicated but powerful process. Moreover a combined approach permits a more precise and time controlled manipulation of the reprogramming processes.

7. Conclusions

We aimed in this chapter at reviewing the current knowledge regarding maintenance of pluripotency in embryonic and induced pluripotent stem cells. Even though in the past years many different key factors could be identified, how they interact between each other and control the pluripotent cell identity is still largely unknown. Understanding these mechanisms is essential because ESCs and iPSCs hold great promise for the therapeutic treatment of human diseases. Successful reprogramming of differentiated cells into iPSCs needs faithful remodelling of epigenetic modifications, because epigenetic aberrations often induce pathological conditions such as cancer. Therefore a thorough understanding of the epigenetic remodelling involved in reprogramming somatic cells is of fundamental interest. The field of stem cell research is one of the great challenges of our times and in the next future extensive studies aimed at the clarification of the correlations between pluripotency and epigenetic will be necessary. The advent of the modern high throughput technologies will open new possibilities in understanding the mechanisms ruling cell fate. All this will lead to the discovery not only of new mechanisms and key players determining pluripotency, but also will allow expanding the knowledge how to drive controlled differentiation of pluripotent cells towards pure populations of precursors and terminally differentiated cells, by identifying the right signals and culture environments.

8. References

Agger, K.; Christensen, J.; Cloos, P. A. C. & Helin, K. (2008). The emerging functions of histone demethylases. *Curr Opin Genet Dev*, 18, 2, 159-168

Aksoy, I.; Sakabedoyan, C.; Bourillot, P.-Y.; Malashicheva, A. B.; Mancip, J.; Knoblauch, K.; Afanassieff, M. & Savatier, P. (2007). Self-renewal of murine embryonic stem cells is supported by the serine/threonine kinases pim-1 and pim-3. *Stem Cells*, 25, 12, 2996-3004

Anton, R.; Kestler, H. A. & Kühl, M. (2007). Beta-catenin signaling contributes to stemness and regulates early differentiation in murine embryonic stem cells. *FEBS Lett*, 581, 27, 5247-5254

Avilion, A. A.; Nicolis, S. K.; Pevny, L. H.; Perez, L.; Vivian, N. & Lovell-Badge, R. (2003). Multipotent cell lineages in early mouse development depend on SOX2 function. *Genes Dev*, 17, 1, 126-140

Beppu, H.; Kawabata, M.; Hamamoto, T.; Chytil, A.; Minowa, O.; Noda, T. & Miyazono, K. (2000). BMP type II receptor is required for gastrulation and early development of mouse embryos. *Dev Biol*, 221, 1, 249-258

Bernstein, B. E.; Mikkelsen, T. S.; Xie, X.; Kamal, M.; Huebert, D. J.; Cuff, J.; Fry, B.; Meissner, A.; Wernig, M.; Plath, K.; Jaenisch, R.; Wagschal, A.; Feil, R.; Schreiber, S. L. & Lander, E. S. (2006). A bivalent chromatin structure marks key developmental genes in embryonic stem cells. *Cell*, 125, 2, 315-326

Bourillot, P.; Aksoy, I.; Schreiber, V.; Wianny, F.; Schulz, H.; Hummel, O.; Hubner, N. & Savatier, P. (2009). Novel STAT3 target genes exert distinct roles in the inhibition of mesoderm and endoderm differentiation in cooperation with Nanog. *Stem Cells*,

Boyer, L. A.; Plath, K.; Zeitlinger, J.; Brambrink, T.; Medeiros, L. A.; Lee, T. I.; Levine, S. S.; Wernig, M.; Tajonar, A.; Ray, M. K.; Bell, G. W.; Otte, A. P.; Vidal, M.; Gifford, D. K.; Young, R. A. & Jaenisch, R. (2006). Polycomb complexes repress developmental regulators in murine embryonic stem cells. *Nature*, 441, 7091, 349-353

Brazil, D. P.; Yang, Z.-Z. & Hemmings, B. A. (2004). Advances in protein kinase B signalling: AKTion on multiple fronts. *Trends Biochem Sci*, 29, 5, 233-242

Brons, I. G. M.; Smithers, L. E.; Trotter, M. W. B.; Rugg-Gunn, P.; Sun, B.; Chuva de Sousa Lopes, S. M.; Howlett, S. K.; Clarkson, A.; Ahrlund-Richter, L.; Pedersen, R. A. & Vallier, L. (2007). Derivation of pluripotent epiblast stem cells from mammalian embryos. *Nature*, 448, 7150, 191-195

Buehr, M.; Meek, S.; Blair, K.; Yang, J.; Ure, J.; Silva, J.; McLay, R.; Hall, J.; Ying, Q.-L. & Smith, A. (2008). Capture of authentic embryonic stem cells from rat blastocysts. *Cell*, 135, 7, 1287-1298

Burdon, T.; Smith, A. & Savatier, P. (2002). Signalling, cell cycle and pluripotency in embryonic stem cells. *Trends Cell Biol*, 12, 9, 432-438

Burdon, T.; Stracey, C.; Chambers, I.; Nichols, J. & Smith, A. (1999). Suppression of SHP-2 and ERK signalling promotes self-renewal of mouse embryonic stem cells. *Dev Biol*, 210, 1, 30-43

Cartwright, P.; McLean, C.; Sheppard, A.; Rivett, D.; Jones, K. & Dalton, S. (2005). LIF/STAT3 controls ES cell self-renewal and pluripotency by a Myc-dependent mechanism. *Development*, 132, 5, 885-896

Cavalli, G. & Paro, R. (1999). Epigenetic inheritance of active chromatin after removal of the main transactivator. *Science*, 286, 5441, 955-958

Chambers, I.; Colby, D.; Robertson, M.; Nichols, J.; Lee, S.; Tweedie, S. & Smith, A. (2003). Functional expression cloning of Nanog, a pluripotency sustaining factor in embryonic stem cells. *Cell*, 113, 5, 643-655

Chambers, I.; Silva, J.; Colby, D.; Nichols, J.; Nijmeijer, B.; Robertson, M.; Vrana, J.; Jones, K.; Grotewold, L. & Smith, A. (2007). Nanog safeguards pluripotency and mediates germline development. *Nature*, 450, 7173, 1230-1234

Chan, R. J.; Johnson, S. A.; Li, Y.; Yoder, M. C. & Feng, G.-S. (2003). A definitive role of Shp-2 tyrosine phosphatase in mediating embryonic stem cell differentiation and hematopoiesis. *Blood*, 102, 6, 2074-2080

Chen, X.; Xu, H.; Yuan, P.; Fang, F.; Huss, M.; Vega, V. B.; Wong, E.; Orlov, Y. L.; Zhang, W.; Jiang, J.; Loh, Y.-H.; Yeo, H. C.; Yeo, Z. X.; Narang, V.; Govindarajan, K. R.; Leong, B.; Shahab, A.; Ruan, Y.; Bourque, G.; Sung, W.-K.; Clarke, N. D.; Wei, C.-L. & Ng, H.-H. (2008). Integration of external signaling pathways with the core transcriptional network in embryonic stem cells. *Cell*, 133, 6, 1106-1117

Chew, J.-L.; Loh, Y.-H.; Zhang, W.; Chen, X.; Tam, W.-L.; Yeap, L.-S.; Li, P.; Ang, Y.-S.; Lim, B.; Robson, P. & Ng, H.-H. (2005). Reciprocal transcriptional regulation of Pou5f1 and Sox2 via the Oct4/Sox2 complex in embryonic stem cells. *Molecular and cellular biology*, 25, 14, 6031-6046

Cinelli, P.; Casanova, E. A.; Uhlig, S.; Lochmatter, P.; Matsuda, T.; Yokota, T.; Rülicke, T.; Ledermann, B. & Bürki, K. (2008). Expression profiling in transgenic FVB/N embryonic stem cells overexpressing STAT3. *BMC Dev Biol*, 8, 57

Dang, C. V.; Resar, L. M.; Emison, E.; Kim, S.; Li, Q.; Prescott, J. E.; Wonsey, D. & Zeller, K. (1999). Function of the c-Myc oncogenic transcription factor. *Exp Cell Res*, 253, 1, 63-77

Davis, A. C.; Wims, M.; Spotts, G. D.; Hann, S. R. & Bradley, A. (1993). A null c-myc mutation causes lethality before 10.5 days of gestation in homozygotes and reduced fertility in heterozygous female mice. *Genes Dev*, 7, 4, 671-682

Evans, M. J. & Kaufman, M. H. (1981). Establishment in culture of pluripotential cells from mouse embryos. *Nature*, 292, 5819, 154-156

Feng, B.; Jiang, J.; Kraus, P.; Ng, J.-H.; Heng, J.-C. D.; Chan, Y.-S.; Yaw, L.-P.; Zhang, W.; Loh, Y.-H.; Han, J.; Vega, V. B.; Cacheux-Rataboul, V.; Lim, B.; Lufkin, T. & Ng, H.-H. (2009). Reprogramming of fibroblasts into induced pluripotent stem cells with orphan nuclear receptor Esrrb. *Nat Cell Biol*, 11, 2, 197-203

Francis, N. J. & Kingston, R. E. (2001). Mechanisms of transcriptional memory. *Nat Rev Mol Cell Biol*, 2, 6, 409-421

Fukada, T.; Hibi, M.; Yamanaka, Y.; Takahashi-Tezuka, M.; Fujitani, Y.; Yamaguchi, T.; Nakajima, K. & Hirano, T. (1996). Two signals are necessary for cell proliferation induced by a cytokine receptor gp130: involvement of STAT3 in anti-apoptosis. *Immunity*, 5, 5, 449-460

Galvin, K. E.; Travis, E. D.; Yee, D.; Magnuson, T. & Vivian, J. L. (2010). Nodal signaling regulates the bone morphogenic protein pluripotency pathway in mouse embryonic stem cells. *J Biol Chem*, 285, 26, 19747-19756

Gearing, D. P.; Gough, N. M.; King, J. A.; Hilton, D. J.; Nicola, N. A.; Simpson, R. J.; Nice, E. C.; Kelso, A. & Metcalf, D. (1987). Molecular cloning and expression of cDNA encoding a murine myeloid leukaemia inhibitory factor (LIF). *EMBO J*, 6, 13, 3995-4002

Groszer, M.; Erickson, R.; Scripture-Adams, D. D.; Lesche, R.; Trumpp, A.; Zack, J. A.; Kornblum, H. I.; Liu, X. & Wu, H. (2001). Negative regulation of neural stem/progenitor cell proliferation by the Pten tumor suppressor gene in vivo. *Science*, 294, 5549, 2186-2189

Hall, J.; Guo, G.; Wray, J.; Eyres, I.; Nichols, J.; Grotewold, L.; Morfopoulou, S.; Humphreys, P.; Mansfield, W.; Walker, R.; Tomlinson, S. & Smith, A. (2009). Oct4 and LIF/Stat3 Additively Induce Krüppel Factors to Sustain Embryonic Stem Cell Self-Renewal. *Cell stem cell*, 5, 6, 597-609

Hallmann, D.; Trümper, K.; Trusheim, H.; Ueki, K.; Kahn, C. R.; Cantley, L. C.; Fruman, D. A. & Hörsch, D. (2003). Altered signaling and cell cycle regulation in embryonal stem cells with a disruption of the gene for phosphoinositide 3-kinase regulatory subunit p85alpha. *J Biol Chem*, 278, 7, 5099-5108

Hao, J.; Li, T.-G.; Qi, X.; Zhao, D.-F. & Zhao, G.-Q. (2006). WNT/beta-catenin pathway up-regulates Stat3 and converges on LIF to prevent differentiation of mouse embryonic stem cells. *Dev Biol*, 290, 1, 81-91

Huangfu, D.; Maehr, R.; Guo, W.; Eijkelenboom, A.; Snitow, M.; Chen, A. & Melton, D. (2008a). Induction of pluripotent stem cells by defined factors is greatly improved by small-molecule compounds. *Nat Biotechnol*,

Huangfu, D.; Osafune, K.; Maehr, R.; Guo, W.; Eijkelenboom, A.; Chen, S.; Muhlestein, W. & Melton, D. (2008b). Induction of pluripotent stem cells from primary human fibroblasts with only Oct4 and Sox2. *Nat Biotechnol*,

Humphrey, R. K.; Beattie, G. M.; Lopez, A. D.; Bucay, N.; King, C. C.; Firpo, M. T.; Rose-John, S. & Hayek, A. (2004). Maintenance of pluripotency in human embryonic stem cells is STAT3 independent. *Stem Cells*, 22, 4, 522-530

Ivanova, N.; Dobrin, R.; Lu, R.; Kotenko, I.; Levorse, J.; DeCoste, C.; Schafer, X.; Lun, Y. & Lemischka, I. R. (2006). Dissecting self-renewal in stem cells with RNA interference. *Nature*, 442, 7102, 533-538

Ivanova, N. B.; Dimos, J. T.; Schaniel, C.; Hackney, J. A.; Moore, K. A. & Lemischka, I. R. (2002). A stem cell molecular signature. *Science*, 298, 5593, 601-604

Kalmar, T.; Lim, C.; Hayward, P.; Muñoz-Descalzo, S.; Nichols, J.; Garcia-Ojalvo, J. & Martinez Arias, A. (2009). Regulated fluctuations in nanog expression mediate cell fate decisions in embryonic stem cells. *PLoS Biol*, 7, 7, e1000149

Kawase, E.; Suemori, H.; Takahashi, N.; Okazaki, K.; Hashimoto, K. & Nakatsuji, N. (1994). Strain difference in establishment of mouse embryonic stem (ES) cell lines. *Int J Dev Biol*, 38, 2, 385-390

Kim, J. B.; Sebastiano, V.; Wu, G.; Araúzo-Bravo, M. J.; Sasse, P.; Gentile, L.; Ko, K.; Ruau, D.; Ehrich, M.; van den Boom, D.; Meyer, J.; Hübner, K.; Bernemann, C.; Ortmeier, C.; Zenke, M.; Fleischmann, B. K.; Zaehres, H. & Schöler, H. R. (2009). Oct4-induced pluripotency in adult neural stem cells. *Cell*, 136, 3, 411-419

Kimura, H.; Tada, M.; Nakatsuji, N. & Tada, T. (2004). Histone code modifications on pluripotential nuclei of reprogrammed somatic cells. *Mol Cell Biol*, 24, 13, 5710-5720

Kioussi, C.; Briata, P.; Baek, S. H.; Rose, D. W.; Hamblet, N. S.; Herman, T.; Ohgi, K. A.; Lin, C.; Gleiberman, A.; Wang, J.; Brault, V.; Ruiz-Lozano, P.; Nguyen, H. D.; Kemler, R.; Glass, C. K.; Wynshaw-Boris, A. & Rosenfeld, M. G. (2002). Identification of a Wnt/Dvl/beta-Catenin --> Pitx2 pathway mediating cell-type-specific proliferation during development. *Cell*, 111, 5, 673-685

Kuroda, T.; Tada, M.; Kubota, H.; Kimura, H.; Hatano, S.-y.; Suemori, H.; Nakatsuji, N. & Tada, T. (2005). Octamer and Sox elements are required for transcriptional cis regulation of Nanog gene expression. *Mol Cell Biol*, 25, 6, 2475-2485

Lee, M. Y.; Lim, H. W.; Lee, S. H. & Han, H. J. (2009). Smad, PI3K/Akt, and Wnt-dependent signaling pathways are involved in BMP-4-induced ESC self-renewal. *Stem Cells*, 27, 8, 1858-1868

Lee, T. I.; Jenner, R. G.; Boyer, L. A.; Guenther, M. G.; Levine, S. S.; Kumar, R. M.; Chevalier, B.; Johnstone, S. E.; Cole, M. F.; Isono, K.-i.; Koseki, H.; Fuchikami, T.; Abe, K.;

Murray, H. L.; Zucker, J. P.; Yuan, B.; Bell, G. W.; Herbolsheimer, E.; Hannett, N. M.; Sun, K.; Odom, D. T.; Otte, A. P.; Volkert, T. L.; Bartel, D. P.; Melton, D. A.; Gifford, D. K.; Jaenisch, R. & Young, R. A. (2006). Control of developmental regulators by Polycomb in human embryonic stem cells. *Cell*, 125, 2, 301-313

Lenardo, M. J.; Staudt, L.; Robbins, P.; Kuang, A.; Mulligan, R. C. & Baltimore, D. (1989). Repression of the IgH enhancer in teratocarcinoma cells associated with a novel octamer factor. *Science*, 243, 4890, 544-546

Lewis, T. S.; Shapiro, P. S. & Ahn, N. G. (1998). Signal transduction through MAP kinase cascades. *Adv Cancer Res*, 74, 49-139

Li, M.; Sendtner, M. & Smith, A. (1995). Essential function of LIF receptor in motor neurons. *Nature*, 378, 6558, 724-727

Li, P.; Tong, C.; Mehrian-Shai, R.; Jia, L.; Wu, N.; Yan, Y.; Maxson, R. E.; Schulze, E. N.; Song, H.; Hsieh, C.-L.; Pera, M. F. & Ying, Q.-L. (2008). Germline competent embryonic stem cells derived from rat blastocysts. *Cell*, 135, 7, 1299-1310

Lloyd, A. C. (1998). Ras versus cyclin-dependent kinase inhibitors. *Curr Opin Genet Dev*, 8, 1, 43-48

Lodge, P.; McWhir, J.; Gallagher, E. & Sang, H. (2005). Increased gp130 signaling in combination with inhibition of the MEK/ERK pathway facilitates embryonic stem cell isolation from normally refractory murine CBA blastocysts. *Cloning Stem Cells*, 7, 1, 2-7

Loh, Y.-H.; Wu, Q.; Chew, J.-L.; Vega, V. B.; Zhang, W.; Chen, X.; Bourque, G.; George, J.; Leong, B.; Liu, J.; Wong, K.-Y.; Sung, K. W.; Lee, C. W. H.; Zhao, X.-D.; Chiu, K.-P.; Lipovich, L.; Kuznetsov, V. A.; Robson, P.; Stanton, L. W.; Wei, C.-L.; Ruan, Y.; Lim, B. & Ng, H.-H. (2006). The Oct4 and Nanog transcription network regulates pluripotency in mouse embryonic stem cells. *Nat Genet*, 38, 4, 431-440

Loh, Y.-H.; Zhang, W.; Chen, X.; George, J. & Ng, H.-H. (2007). Jmjd1a and Jmjd2c histone H3 Lys 9 demethylases regulate self-renewal in embryonic stem cells. *Genes Dev*, 21, 20, 2545-2557

Maekawa, M.; Yamamoto, T.; Kohno, M.; Takeichi, M. & Nishida, E. (2007). Requirement for ERK MAP kinase in mouse preimplantation development. *Development*, 134, 15, 2751-2759

Maherali, N. & Hochedlinger, K. (2008). Guidelines and techniques for the generation of induced pluripotent stem cells. *Cell stem cell*, 3, 6, 595-605

Margueron, R.; Trojer, P. & Reinberg, D. (2005). The key to development: interpreting the histone code? *Curr Opin Genet Dev*, 15, 2, 163-176

Martin, G. R. (1981). Isolation of a pluripotent cell line from early mouse embryos cultured in medium conditioned by teratocarcinoma stem cells. *Proc Natl Acad Sci USA*, 78, 12, 7634-7638

Masui, S.; Nakatake, Y.; Toyooka, Y.; Shimosato, D.; Yagi, R.; Takahashi, K.; Okochi, H.; Okuda, A.; Matoba, R.; Sharov, A. A.; Ko, M. S. H. & Niwa, H. (2007). Pluripotency governed by Sox2 via regulation of Oct3/4 expression in mouse embryonic stem cells. *Nat Cell Biol*, 9, 6, 625-635

Matsuda, T.; Nakamura, T.; Nakao, K.; Arai, T.; Katsuki, M.; Heike, T. & Yokota, T. (1999). STAT3 activation is sufficient to maintain an undifferentiated state of mouse embryonic stem cells. *EMBO J*, 18, 15, 4261-4269

Matsui, Y. (1992). [Regulation of murine primordial germ cell (PGC) growth by peptide growth factors]. *Tanpakushitsu Kakusan Koso*, 37, 15, 2935-2946

Meissner, A.; Wernig, M. & Jaenisch, R. (2007). Direct reprogramming of genetically unmodified fibroblasts into pluripotent stem cells. *Nat Biotechnol*, 25, 10, 1177-1181

Meshorer, E. & Misteli, T. (2006). Chromatin in pluripotent embryonic stem cells and differentiation. *Nat Rev Mol Cell Biol*, 7, 7, 540-546

Mikkelsen, T. S.; Hanna, J.; Zhang, X.; Ku, M.; Wernig, M.; Schorderet, P.; Bernstein, B. E.; Jaenisch, R.; Lander, E. S. & Meissner, A. (2008). Dissecting direct reprogramming through integrative genomic analysis. *Nature*, 454, 7200, 49-55

Mikkelsen, T. S.; Ku, M.; Jaffe, D. B.; Issac, B.; Lieberman, E.; Giannoukos, G.; Alvarez, P.; Brockman, W.; Kim, T.-K.; Koche, R. P.; Lee, W.; Mendenhall, E.; O'Donovan, A.; Presser, A.; Russ, C.; Xie, X.; Meissner, A.; Wernig, M.; Jaenisch, R.; Nusbaum, C.; Lander, E. S. & Bernstein, B. E. (2007). Genome-wide maps of chromatin state in pluripotent and lineage-committed cells. *Nature*, 448, 7153, 553-560

Mishina, Y.; Suzuki, A.; Ueno, N. & Behringer, R. R. (1995). Bmpr encodes a type I bone morphogenetic protein receptor that is essential for gastrulation during mouse embryogenesis. *Genes Dev*, 9, 24, 3027-3037

Mitsui, K.; Tokuzawa, Y.; Itoh, H.; Segawa, K.; Murakami, M.; Takahashi, K.; Maruyama, M.; Maeda, M. & Yamanaka, S. (2003). The homeoprotein Nanog is required for maintenance of pluripotency in mouse epiblast and ES cells. *Cell*, 113, 5, 631-642

Moon, R. T.; Kohn, A. D.; De Ferrari, G. V. & Kaykas, A. (2004). WNT and beta-catenin signalling: diseases and therapies. *Nat Rev Genet*, 5, 9, 691-701

Naito, A. T.; Shiojima, I.; Akazawa, H.; Hidaka, K.; Morisaki, T.; Kikuchi, A. & Komuro, I. (2006). Developmental stage-specific biphasic roles of Wnt/beta-catenin signaling in cardiomyogenesis and hematopoiesis. *Proc Natl Acad Sci USA*, 103, 52, 19812-19817

Nakagawa, M.; Koyanagi, M.; Tanabe, K.; Takahashi, K.; Ichisaka, T.; Aoi, T.; Okita, K.; Mochiduki, Y.; Takizawa, N. & Yamanaka, S. (2008). Generation of induced pluripotent stem cells without Myc from mouse and human fibroblasts. *Nat Biotechnol*, 26, 1, 101-106

Nakashima, K.; Yanagisawa, M.; Arakawa, H. & Taga, T. (1999). Astrocyte differentiation mediated by LIF in cooperation with BMP2. *FEBS Lett*, 457, 1, 43-46

Nichols, J.; Chambers, I.; Taga, T. & Smith, A. (2001). Physiological rationale for responsiveness of mouse embryonic stem cells to gp130 cytokines. *Development (Cambridge, England)*, 128, 12, 2333-2339

Nichols, J.; Jones, K.; Phillips, J.; Newland, S.; Roode, M.; Mansfield, W.; Smith, A. & Cooke, A. (2009). Validated germline-competent embryonic stem cell lines from nonobese diabetic mice. *Nat Med*,

Nichols, J.; Zevnik, B.; Anastassiadis, K.; Niwa, H.; Klewe-Nebenius, D.; Chambers, I.; Schöler, H. & Smith, A. (1998). Formation of pluripotent stem cells in the mammalian embryo depends on the POU transcription factor Oct4. *Cell*, 95, 3, 379-391

Niwa, H.; Burdon, T.; Chambers, I. & Smith, A. (1998). Self-renewal of pluripotent embryonic stem cells is mediated via activation of STAT3. *Genes Dev*, 12, 13, 2048-2060

Niwa, H.; Miyazaki, J. & Smith, A. G. (2000). Quantitative expression of Oct-3/4 defines differentiation, dedifferentiation or self-renewal of ES cells. *Nat Genet*, 24, 4, 372-376

Noben-Trauth, N.; Köhler, G.; Bürki, K. & Ledermann, B. (1996). Efficient targeting of the IL-4 gene in a BALB/c embryonic stem cell line. *Transgenic Res*, 5, 6, 487-491

Nordhoff, V.; Hübner, K.; Bauer, A.; Orlova, I.; Malapetsa, A. & Schöler, H. R. (2001). Comparative analysis of human, bovine, and murine Oct-4 upstream promoter sequences. *Mamm Genome*, 12, 4, 309-317

O'Sullivan, L. A.; Liongue, C.; Lewis, R. S.; Stephenson, S. E. M. & Ward, A. C. (2007). Cytokine receptor signaling through the Jak-Stat-Socs pathway in disease. *Mol Immunol*, 44, 10, 2497-2506

Ogawa, K.; Nishinakamura, R.; Iwamatsu, Y.; Shimosato, D. & Niwa, H. (2006). Synergistic action of Wnt and LIF in maintaining pluripotency of mouse ES cells. *Biochem Biophys Res Commun*, 343, 1, 159-166

Ogawa, K.; Saito, A.; Matsui, H.; Suzuki, H.; Ohtsuka, S.; Shimosato, D.; Morishita, Y.; Watabe, T.; Niwa, H. & Miyazono, K. (2007). Activin-Nodal signaling is involved in propagation of mouse embryonic stem cells. *J Cell Sci*, 120, Pt 1, 55-65

Okita, K.; Ichisaka, T. & Yamanaka, S. (2007). Generation of germline-competent induced pluripotent stem cells. *Nature*, 448, 7151, 313-317

Okumura-Nakanishi, S.; Saito, M.; Niwa, H. & Ishikawa, F. (2005). Oct-3/4 and Sox2 regulate Oct-3/4 gene in embryonic stem cells. *J Biol Chem*, 280, 7, 5307-5317

Paling, N. R. D.; Wheadon, H.; Bone, H. K. & Welham, M. J. (2004). Regulation of embryonic stem cell self-renewal by phosphoinositide 3-kinase-dependent signaling. *J Biol Chem*, 279, 46, 48063-48070

Palmieri, S. L.; Peter, W.; Hess, H. & Schöler, H. R. (1994). Oct-4 transcription factor is differentially expressed in the mouse embryo during establishment of the first two extraembryonic cell lineages involved in implantation. *Dev Biol*, 166, 1, 259-267

Pasini, D.; Hansen, K. H.; Christensen, J.; Agger, K.; Cloos, P. A. C. & Helin, K. (2008). Coordinated regulation of transcriptional repression by the RBP2 H3K4 demethylase and Polycomb-Repressive Complex 2. *Genes Dev*, 22, 10, 1345-1355

Plath, K.; Fang, J.; Mlynarczyk-Evans, S. K.; Cao, R.; Worringer, K. A.; Wang, H.; de la Cruz, C. C.; Otte, A. P.; Panning, B. & Zhang, Y. (2003). Role of histone H3 lysine 27 methylation in X inactivation. *Science*, 300, 5616, 131-135

Qi, X.; Li, T.-G.; Hao, J.; Hu, J.; Wang, J.; Simmons, H.; Miura, S.; Mishina, Y. & Zhao, G.-Q. (2004). BMP4 supports self-renewal of embryonic stem cells by inhibiting mitogen-activated protein kinase pathways. *Proc Natl Acad Sci USA*, 101, 16, 6027-6032

Reményi, A.; Lins, K.; Nissen, L. J.; Reinbold, R.; Schöler, H. R. & Wilmanns, M. (2003). Crystal structure of a POU/HMG/DNA ternary complex suggests differential assembly of Oct4 and Sox2 on two enhancers. *Genes Dev*, 17, 16, 2048-2059

Ringrose, L. & Paro, R. (2004). Epigenetic regulation of cellular memory by the Polycomb and Trithorax group proteins. *Annu Rev Genet*, 38, 413-443

Roach, M. L.; Stock, J. L.; Byrum, R.; Koller, B. H. & McNeish, J. D. (1995). A new embryonic stem cell line from DBA/1lacJ mice allows genetic modification in a murine model of human inflammation. *Exp Cell Res*, 221, 2, 520-525

Robertson, E. J.; Norris, D. P.; Brennan, J. & Bikoff, E. K. (2003). Control of early anterior-posterior patterning in the mouse embryo by TGF-beta signalling. *Philos Trans R Soc Lond, B, Biol Sci*, 358, 1436, 1351-1357; discussion 1357

Rodda, D. J.; Chew, J.-L.; Lim, L.-H.; Loh, Y.-H.; Wang, B.; Ng, H.-H. & Robson, P. (2005). Transcriptional regulation of nanog by OCT4 and SOX2. *J Biol Chem*, 280, 26, 24731-24737

Rohwedel, J.; Sehlmeyer, U.; Shan, J.; Meister, A. & Wobus, A. M. (1996). Primordial germ cell-derived mouse embryonic germ (EG) cells in vitro resemble undifferentiated stem cells with respect to differentiation capacity and cell cycle distribution. *Cell Biol Int*, 20, 8, 579-587

Ruvkun, G. & Finney, M. (1991). Regulation of transcription and cell identity by POU domain proteins. *Cell*, 64, 3, 475-478

Sato, N.; Meijer, L.; Skaltsounis, L.; Greengard, P. & Brivanlou, A. H. (2004). Maintenance of pluripotency in human and mouse embryonic stem cells through activation of Wnt signaling by a pharmacological GSK-3-specific inhibitor. *Nat Med*, 10, 1, 55-63

Savatier, P.; Lapillonne, H.; van Grunsven, L. A.; Rudkin, B. B. & Samarut, J. (1996). Withdrawal of differentiation inhibitory activity/leukemia inhibitory factor up-regulates D-type cyclins and cyclin-dependent kinase inhibitors in mouse embryonic stem cells. *Oncogene*, 12, 2, 309-322

Schöler, H. R. (1991). Octamania: the POU factors in murine development. *Trends Genet*, 7, 10, 323-329

Schöler, H. R.; Balling, R.; Hatzopoulos, A. K.; Suzuki, N. & Gruss, P. (1989). Octamer binding proteins confer transcriptional activity in early mouse embryogenesis. *EMBO J*, 8, 9, 2551-2557

Shi, Y.; Desponts, C.; Do, J. T.; Hahm, H. S.; Schöler, H. R. & Ding, S. (2008). Induction of pluripotent stem cells from mouse embryonic fibroblasts by Oct4 and Klf4 with small-molecule compounds. *Cell stem cell*, 3, 5, 568-574

Simpson, E. M.; Linder, C. C.; Sargent, E. E.; Davisson, M. T.; Mobraaten, L. E. & Sharp, J. J. (1997). Genetic variation among 129 substrains and its importance for targeted mutagenesis in mice. *Nat Genet*, 16, 1, 19-27

Singh, A. M.; Hamazaki, T.; Hankowski, K. E. & Terada, N. (2007). A heterogeneous expression pattern for Nanog in embryonic stem cells. *Stem Cells*, 25, 10, 2534-2542

Sirard, C.; de la Pompa, J. L.; Elia, A.; Itie, A.; Mirtsos, C.; Cheung, A.; Hahn, S.; Wakeham, A.; Schwartz, L.; Kern, S. E.; Rossant, J. & Mak, T. W. (1998). The tumor suppressor gene Smad4/Dpc4 is required for gastrulation and later for anterior development of the mouse embryo. *Genes Dev*, 12, 1, 107-119

Smith, A. G.; Heath, J. K.; Donaldson, D. D.; Wong, G. G.; Moreau, J.; Stahl, M. & Rogers, D. (1988). Inhibition of pluripotential embryonic stem cell differentiation by purified polypeptides. *Nature*, 336, 6200, 688-690

Solter, D. & Knowles, B. B. (1978). Monoclonal antibody defining a stage-specific mouse embryonic antigen (SSEA-1). *Proc Natl Acad Sci USA*, 75, 11, 5565-5569

Starr, R.; Novak, U.; Willson, T. A.; Inglese, M.; Murphy, V.; Alexander, W. S.; Metcalf, D.; Nicola, N. A.; Hilton, D. J. & Ernst, M. (1997). Distinct roles for leukemia inhibitory factor receptor alpha-chain and gp130 in cell type-specific signal transduction. *J Biol Chem*, 272, 32, 19982-19986

Stewart, C. L.; Kaspar, P.; Brunet, L. J.; Bhatt, H.; Gadi, I.; Köntgen, F. & Abbondanzo, S. J. (1992). Blastocyst implantation depends on maternal expression of leukaemia inhibitory factor. *Nature*, 359, 6390, 76-79

Stiles, B.; Groszer, M.; Wang, S.; Jiao, J. & Wu, H. (2004). PTENless means more. *Dev Biol*, 273, 2, 175-184

Suzuki, A.; Raya, A.; Kawakami, Y.; Morita, M.; Matsui, T.; Nakashima, K.; Gage, F. H.; Rodríguez-Esteban, C. & Belmonte, J. C. I. (2006a). Maintenance of embryonic stem cell pluripotency by Nanog-mediated reversal of mesoderm specification. *Nature clinical practice Cardiovascular medicine*, 3 Suppl 1, S114-122

Suzuki, A.; Raya, A.; Kawakami, Y.; Morita, M.; Matsui, T.; Nakashima, K.; Gage, F. H.; Rodríguez-Esteban, C. & Izpisúa Belmonte, J. C. (2006b). Nanog binds to Smad1 and blocks bone morphogenetic protein-induced differentiation of embryonic stem cells. *Proc Natl Acad Sci USA*, 103, 27, 10294-10299

Tada, M.; Takahama, Y.; Abe, K.; Nakatsuji, N. & Tada, T. (2001). Nuclear reprogramming of somatic cells by in vitro hybridization with ES cells. *Curr Biol*, 11, 19, 1553-1558

Takahashi, K. & Yamanaka, S. (2006). Induction of pluripotent stem cells from mouse embryonic and adult fibroblast cultures by defined factors. *Cell*, 126, 4, 663-676

Takahashi-Tezuka, M.; Yoshida, Y.; Fukada, T.; Ohtani, T.; Yamanaka, Y.; Nishida, K.; Nakajima, K.; Hibi, M. & Hirano, T. (1998). Gab1 acts as an adapter molecule linking the cytokine receptor gp130 to ERK mitogen-activated protein kinase. *Molecular and cellular biology*, 18, 7, 4109-4117

Takao, Y.; Yokota, T. & Koide, H. (2007). Beta-catenin up-regulates Nanog expression through interaction with Oct-3/4 in embryonic stem cells. *Biochem Biophys Res Commun*, 353, 3, 699-705

Tesar, P. J.; Chenoweth, J. G.; Brook, F. A.; Davies, T. J.; Evans, E. P.; Mack, D. L.; Gardner, R. L. & McKay, R. D. G. (2007). New cell lines from mouse epiblast share defining features with human embryonic stem cells. *Nature*, 448, 7150, 196-199

Thomson, J. A.; Itskovitz-Eldor, J.; Shapiro, S. S.; Waknitz, M. A.; Swiergiel, J. J.; Marshall, V. S. & Jones, J. M. (1998). Embryonic stem cell lines derived from human blastocysts. *Science*, 282, 5391, 1145-1147

Tomioka, M.; Nishimoto, M.; Miyagi, S.; Katayanagi, T.; Fukui, N.; Niwa, H.; Muramatsu, M. & Okuda, A. (2002). Identification of Sox-2 regulatory region which is under the control of Oct-3/4-Sox-2 complex. *Nucleic Acids Res*, 30, 14, 3202-3213

Ura, H.; Usuda, M.; Kinoshita, K.; Sun, C.; Mori, K.; Akagi, T.; Matsuda, T.; Koide, H. & Yokota, T. (2008). STAT3 and Oct-3/4 control histone modification through induction of Eed in embryonic stem cells. *J Biol Chem*,

Wang, Y.; Wang, F.; Sun, T.; Trostinskaia, A.; Wygle, D.; Puscheck, E. & Rappolee, D. A. (2004). Entire mitogen activated protein kinase (MAPK) pathway is present in preimplantation mouse embryos. *Dev Dyn*, 231, 1, 72-87

Ware, C. B.; Horowitz, M. C.; Renshaw, B. R.; Hunt, J. S.; Liggitt, D.; Koblar, S. A.; Gliniak, B. C.; McKenna, H. J.; Papayannopoulou, T. & Thoma, B. (1995). Targeted disruption of the low-affinity leukemia inhibitory factor receptor gene causes placental, skeletal, neural and metabolic defects and results in perinatal death. *Development (Cambridge, England)*, 121, 5, 1283-1299

Watanabe, S.; Umehara, H.; Murayama, K.; Okabe, M.; Kimura, T. & Nakano, T. (2006). Activation of Akt signaling is sufficient to maintain pluripotency in mouse and primate embryonic stem cells. *Oncogene*, 25, 19, 2697-2707

Williams, D. C.; Cai, M. & Clore, G. M. (2004). Molecular basis for synergistic transcriptional activation by Oct1 and Sox2 revealed from the solution structure of the 42-kDa

Oct1.Sox2.Hoxb1-DNA ternary transcription factor complex. *J Biol Chem*, 279, 2, 1449-1457

Williams, R. L.; Hilton, D. J.; Pease, S.; Willson, T. A.; Stewart, C. L.; Gearing, D. P.; Wagner, E. F.; Metcalf, D.; Nicola, N. A. & Gough, N. M. (1988). Myeloid leukaemia inhibitory factor maintains the developmental potential of embryonic stem cells. *Nature*, 336, 6200, 684-687

Wilmut, I.; Schnieke, A. E.; McWhir, J.; Kind, A. J. & Campbell, K. H. (1997). Viable offspring derived from fetal and adult mammalian cells. *Nature*, 385, 6619, 810-813

Yamanaka, S. (2008). Induction of pluripotent stem cells from mouse fibroblasts by four transcription factors. *Cell Prolif*, 41 Suppl 1, 51-56

Ying, Q.-L. & Smith, A. G. (2003). Defined conditions for neural commitment and differentiation. *Meth Enzymol*, 365, 327-341

Ying, Q.-L.; Wray, J.; Nichols, J.; Batlle-Morera, L.; Doble, B.; Woodgett, J.; Cohen, P. & Smith, A. (2008). The ground state of embryonic stem cell self-renewal. *Nature*, 453, 7194, 519-523

Ying, Q. L.; Nichols, J.; Chambers, I. & Smith, A. (2003). BMP induction of Id proteins suppresses differentiation and sustains embryonic stem cell self-renewal in collaboration with STAT3. *Cell*, 115, 3, 281-292

Yuan, H.; Corbi, N.; Basilico, C. & Dailey, L. (1995). Developmental-specific activity of the FGF-4 enhancer requires the synergistic action of Sox2 and Oct-3. *Genes Dev*, 9, 21, 2635-2645

Zhong, X. & Jin, Y. (2009). Critical roles of coactivator p300 in mouse embryonic stem cell differentiation and Nanog expression. *J Biol Chem*, 284, 14, 9168-9175

3

The LIF/STAT3 Pathway in ES Cell Self-renewal

Hiroshi Koide and Takashi Yokota
Department of Stem Cell Biology,
Graduate School of Medical Science,
Kanazawa University
Japan

1. Introduction

Embryonic stem (ES) cells are pluripotent cells derived from the inner cell mass (ICM) of blastocysts (Evans & Kaufman, 1981; Martin, 1981). Recent progress in establishing human ES cells and human induced pluripotent stem (iPS) cells opens the possibility of utilizing these stem cells in regenerative medicine (Thomson et al., 1998; Takahashi et al., 2007; Yu et al., 2007). There are, however, several issues that remain to be resolved in relation to the application of pluripotent stem cells to cell therapy. One critical issue is how to efficiently amplify these stem cells, since their pluripotent potential means that under commonly used culture conditions they tend to spontaneously differentiate rather than undergo continued self-renewal. Cell therapy will require a large number of stem cells, and it will be necessary to understand the molecular mechanisms of ES cell self-renewal in order to establish efficient *in vitro* expansion systems.

Since the discovery that leukemia inhibitory factor (LIF) can support self-renewal of mouse ES cells (Smith et al., 1988; Williams et al., 1988), researchers have sought to understand the underlying molecular mechanisms. It is now well established that signal transducer and activator of transcription 3 (STAT3), a transcription factor downstream of LIF, plays an indispensable role in the self-renewal of mouse ES cells. Extensive studies have identified many interesting molecules downstream of LIF/STAT3 signaling, including transcription factors, epigenetic regulators, and oncogenes. In this chapter, we will introduce these downstream molecules and discuss their roles in ES cell self-renewal.

2. The LIF signaling pathway and ES cell self-renewal

LIF belongs to the interleukin-6 cytokine family. LIF binds to a heterodimeric receptor consisting of the low-affinity LIF receptor and gp130, with downstream signals being transmitted through gp130. There are a number of signaling pathways downstream of gp130, including the STAT3, phosphatidylinositol 3-kinase (PI3K) and Ras/Erk pathways. In the STAT3 pathway, signaling through gp130 leads to activation of Janus-associated tyrosine kinases (JAKs), which in turn phosphorylate STAT3. Phosphorylated STAT3 dimerizes and translocates to the nucleus, where it functions as a transcription factor. In the PI3K pathway, PI3K phosphorylates phosphoinositides on the 3-OH position of the inositol ring to generate the second messengers phosphatidylinositol-3,4-bisphosphate (PI-3,4-P2) and phosphatidylinositol-3,4,5-trisphosphate (PIP3), leading to activation of the serine/threonine protein kinase Akt (also known as protein kinase B or PKB). Activated Akt then phosphorylates its target molecules, such as glycogen synthase kinase (GSK)-3 and pro-apoptotic BCL2-antagonist of death (BAD) protein. In the Ras/Erk pathway, Ras stimulates sequential activation of the Raf/MEK/Erk kinase cascade, leading to phosphorylation of Erk target molecules, including the transcription factor Elk-1, pro-apoptotic protein caspase-9, and p90 ribosomal S6 protein kinase (RSK).

In mouse ES cells, the STAT3 pathway plays a critical role in the maintenance of self-renewal. When the function of STAT3 is abrogated by expression of a dominant-negative mutant or gene disruption, ES cells undergo differentiation (Boeuf et al., 1997; Niwa et al., 1998; Ying et al., 2008). Artificial activation of the STAT3 pathway can maintain ES cell self-renewal even in the absence of LIF and can improve the efficiency of establishment of ES cells from ICM (Matsuda et al., 1999; Cinelli et al., 2008). These observations underscore the importance of STAT3 in self-renewal of mouse ES cells. The importance of the LIF/STAT3 pathway in self-renewal has been reported also for chicken and rat ES cells (Horiuchi et al. 2004; Buehr et al., 2008; Li et al., 2008). Similarly, the PI3K pathway positively regulates self-renewal (Paling et al., 2004; Watanabe et al., 2006). On the other hand, activation of the Ras/Erk pathway leads to ES cell differentiation into the endoderm lineage (Yoshida-Koide et al, 2004; Hamazaki et al. 2006), while suppression of Ras/Erk signaling promotes self-renewal (Burdon et al., 1999). Thus, the balance among the three pathways allows fine-tuning of the LIF-mediated maintenance of ES cell self-renewal.

In contrast to mouse ES cells, LIF and STAT3 signaling have no effect on self-renewal of monkey and human ES cells (Dahéron et al., 2004; Sumi et al., 2004). When these findings were reported, it was generally thought that the difference between mouse and primate ES cells reflected a genuine species difference. However, studies from stem cells established from mouse epiblasts (EpiSC) have offered an alternative explanation, since these cells show greater similarity to human ES cells than to mouse ES cells (Brons et al., 2007; Tesar et al., 2007). For example, both human ES cells and mouse EpiSC form flatter colonies than mouse ES cells; require the presence of fibroblast growth factor (FGF)-2 and activin A, but not LIF, for their self-renewal; and cannot contribute significantly to chimeras. These results suggest that human ES cells are, in fact, human EpiSC and that the observed difference between mouse and human ES cells actually reflects the difference between ES cells and EpiSC.

In addition to STAT3, the transcription factors Oct3/4 and Sox2 are also indispensable for ES cell self-renewal (Niwa et al., 2000; Masui et al., 2007). Oct3/4 is a POU family transcription factor encoded by the *pou5f1* gene, and deficiency of this transcription factor during development results in loss of ICM (Nichols et al., 1998). Sox2 is an SRY-related HMG-box protein and *Sox2*-deficient mouse embryos die shortly after implantation (Avilion et al., 2003). In ES cells, both transcription factors show self-renewal-specific expression and are downregulated upon LIF removal. When the expression of Oct3/4 or Sox2 is shut off, ES cells differentiate into trophectoderm cells or trophectoderm-like cells (Niwa et al., 2000; Masui et al., 2007). Interestingly, Oct3/4 and Sox2 have been identified as reprogramming factors that can generate iPS cells from fibroblast cells (Takahashi & Yamanaka, 2006), suggesting a strong role for these factors in determining pluripotency. It has been proposed that STAT3, Oct3/4, and Sox2, together with several other important transcription factors such as Nanog, form transcriptional networks to maintain the self-renewal of ES cells (Boyer et al., 2005; Loh et al., 2006; Chen et al., 2008).

3. Target molecules of the LIF/STAT3 pathway in ES cells

When LIF is removed from the culture medium, mouse ES cells rapidly lose the capacity for self-renewal and differentiate into a variety of cell types. During this process, the expression of self-renewal genes declines and the expression of differentiation-associated genes increases. This suggests that LIF signaling in ES cells promotes expression of self-renewal genes and simultaneously suppresses induction of differentiation-associated genes. How is the STAT3 pathway involved in LIF-mediated maintenance of self-renewal? ChIP-on-chip and ChIP-seq analyses have revealed that STAT3 binds to the regulatory regions of several self-renewal genes in ES cells (Chen et al., 2008; Kidder et al., 2008). Which STAT3 target molecules are involved in the maintenance of ES cell self-renewal? Several groups including ourselves have extensively searched for downstream target genes of STAT3 that regulate ES cell self-renewal and identified a range of genes, as discussed below.

3.1 Transcription factors

GA-binding protein (GABP) belongs to the Ets transcription factor family and forms a heterotetramer consisting of two α-subunits and two β-subunits (Rosmarin et al., 2004). The α-subunit (GABPα) mediates DNA binding, while the β-subunit (GABPβ) enhances the transcriptional activity of the α-subunit. GABPα-deficient embryos die prior to implantation and fail to form a blastocyst (Ristevski et al., 2004). Knockdown of GABPα in ES cells results in the downregulation of Oct3/4 expression and the induction of differentiation-associated genes (Kinoshita et al., 2007). Although expression of GABPα alone is not sufficient for the maintenance of ES cells, overexpression of this transcription factor delays the downregulation of Oct3/4 that occurs during differentiation (Kinoshita et al., 2007). These findings suggest that GABPα is involved in the maintenance of ES cells by positively regulating expression of Oct3/4.

JunB is a member of the AP-1 transcription factor family (Shaulian & Karin, 2001). Like other Jun proteins (c-Jun and JunD), JunB forms either homo- or heterodimers with members of the Fos and ATF protein families. Although JunB-null mouse embryos die between E8.5 and E10.0, JunB-null ES cells are viable and have normal growth potential (Schorpp-Kistner et al, 1999), suggesting that JunB is not necessary for the self-renewal of ES cells.

Krüppel-like factor (Klf)-4 belongs to the KLF zinc finger protein family that shares homology with the *Drosophila* Krüppel segmentation protein (Rowland & Peeper, 2006). Klf4 is expressed in a variety of tissues and plays an important role in many physiological processes, including cell proliferation and terminal differentiation. Klf4 can either activate or repress transcription, depending on the specific target gene, and it can function as an oncogene or a tumor suppressor gene in certain cellular contexts. In ES cells, Klf4 acts as a co-factor for Oct3/4 and Sox2 (Nakatake et al., 2006). Klf4 activates the transcription of Sox2 (Niwa et al., 2009). Overexpression of Klf4 delays or prevents ES cell differentiation (Li et al., 2005; Niwa et al., 2009). Moreover, Klf4 has been shown to be a reprogramming factor (Takahashi & Yamanaka, 2006), and overexpression of Klf4 can drive the conversion of EpiSC to ES cells (Guo et al., 2009). These observations strongly suggest that Klf4 is important in ES cell self-renewal. However, knockdown of Klf4 has no effect on ES cell self-renewal (Nakatake et al., 2007; Jiang et al., 2008), possibly due to redundancy within the Klf family, since simultaneous depletion of Klf2, Klf4 and Klf5 leads to ES cell differentiation (Jiang et al., 2008).

The oncoprotein c-Myc binds to the regulatory regions of numerous self-renewal genes in ES cells (Chen et al., 2008; Kidder et al., 2008; Lin et al., 2009a). Expression of a dominant-negative c-Myc induces ES cell differentiation, and artificial activation of this transcription factor maintains self-renewal even in the absence of LIF (Cartwright et al., 2005), indicating the importance of c-Myc in ES self-renewal. In addition, it has been shown that c-Myc regulates expression of several miRNAs to promote ES cell self-renewal (Lin et al., 2009b).

Pem (also known as Rhox5) is an X-linked homeobox-containing gene product. Overexpression of Pem is sufficient for maintenance of ES cells in the absence of LIF (Fan et al., 1999; Cinelli et al., 2008). On the other hand, Pem-null ES cells do not differ from wild-type ES cells in their morphology and Oct3/4 expression (Fan et al., 1999), suggesting that, although Pem can promote ES cell self-renewal, it is dispensable. Interestingly, when Pem-null ES cells are allowed to differentiate, the level of Oct3/4 does not decline as rapidly as in wild-type ES cells (Fan et al., 1999). These results suggest that Pem plays important roles in both self-renewal and differentiation.

Zinc finger protein (Zfp)-57 is a transcription factor containing a zinc finger motif and a Krüppel-associated box (KRAB) domain. Zfp57 was originally identified as an undifferentiated state-specific gene in embryonal carcinoma cells (Okazaki et al., 1994). Expression of Zfp57 in ES cells is also self-renewal-specific and is lost following LIF removal (Akagi et al., 2005). The successful establishment of Zfp57-null ES cells, however, indicates that expression of this transcription factor is dispensable for the maintenance of ES cell self-renewal (Akagi et al., 2005). Interestingly, detailed analysis of Zfp57-null ES cells has revealed that this transcription factor is involved in imprinting during development (Mackay et al., 2008; Li et al., 2008).

β-catenin is a downstream molecule in the Wnt signaling pathway and plays an important role in tumorigenesis (Reya & Clevers, 2005). In the absence of Wnt signaling, the so-called "degradation complex", which contains GSK-3, phosphorylates β-catenin, leading to its rapid degradation via the ubiquitin/proteasome pathway. On the other hand, Wnt signaling leads to the accumulation of a cytoplasmic pool of β-catenin, which translocates into the nucleus and functions as a co-activator of the T-cell factor (TCF) family proteins. Several reports have shown that activation of the Wnt/ β-catenin pathway promotes ES cell self-renewal (Kielman et al. 2002; Sato et al., 2004; Hao *et al.*, 2006; Ogawa *et al.*, 2006; Singla *et al.*, 2006).

In addition to the Wnt signal, LIF can also enhance nuclear accumulation of β-catenin (Takao et al., 2007), suggesting an additional pathway by which LIF regulates ES cell self-renewal. The mechanism by which LIF signaling stabilizes β-catenin protein is not known. A LIF-stimulated pathway leading to inactivation of GSK-3 through the PI3K/Akt pathway has been described in ES cells (Paling et al., 2004), although the importance of this pathway is unclear since treatment with a PI3K inhibitor did not inhibit the activity of β-catenin and expression of constitutively active Akt fails to promote the nuclear localization and activity of β-catenin (Watanabe et al., 2006). Thus, it is not yet clear whether LIF regulates the stability of β-catenin through activation of the PI3K/Akt pathway in ES cells.

It is likely that multiple molecular mechanisms contribute to the promotion of ES cell self-renewal by β-catenin (Fig. 1). Activation of β-catenin results in the upregulation of STAT3 mRNA (Hao et al., 2006), and c-Myc is also a target gene of the β-catenin/TCF complex. Furthermore, β-catenin physically associates with Oct-3/4 and can upregulate Nanog in an Oct3/4-dependent manner (Takao et al., 2007). Although several studies have shown the importance of β-catenin in the self-renewal of ES cells, it is not essential since the establishment of β-catenin-null ES cells has been reported (Huelsken et al., 2000). This may reflect functional compensation by γ-catenin, a protein highly homologous to β-catenin (Takao et al., 2007).

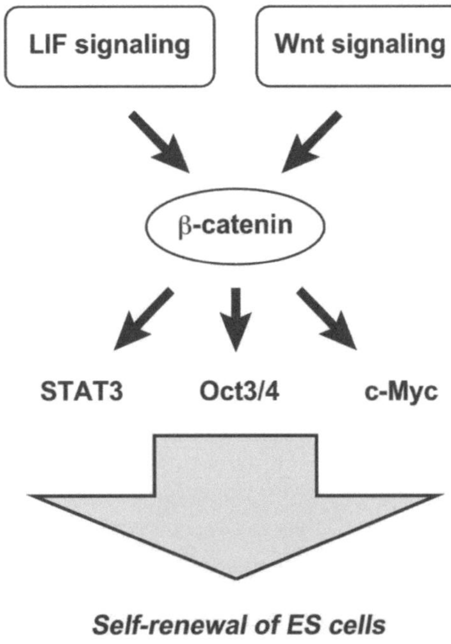

Fig. 1. LIF and Wnt signalings promote ES cell self-renewal by enhancing the stabilization of β-catenin. As a result, β-catenin translocates to the nucleus and acts as a co-activator of TCF, leading to the induction of c-Myc. Activation of β-catenin also results in the induction of STAT3 by unknown mechanism. In addition, β-catenin forms a complex with Oct3/4, which can stimulate Nanog expression

3.2 Transcriptional repressors

The orphan nuclear receptor Dax1 (DSS-AHC critical region on the X-chromosome gene 1; also known as Nr0b1 and Ahch) was originally identified as a gene responsible for the congenital disease dosage-sensitive sex reversal (DSS) and for adrenal hypoplasia congenita (AHC); in humans, gene duplication causes male-to-female sex reversal, while mutations in DAX1 result in AHC (Niakan & McCabe, 2005). Dax1 also plays an important role in the establishment and maintenance of steroid-producing tissues such as testis and adrenal cortex. Dax1 is expressed in self-renewing ES cells under the regulation of STAT3 and Oct3/4 (Clipsham et al., 2004; Sun et al., 2008), and its importance in ES cell self-renewal has been suggested by both knockdown and knockout experiments (Niakan et al., 2006; Wang et al., 2006; Khalfallah et al., 2009). Interestingly, Dax1 binds Oct3/4, leading to suppression of the DNA binding activity and transcriptional activity of Oct3/4 (Sun et al., 2009). The observation that Dax1-overexpressing ES cells have a similar phenotype to Oct3/4-knockdown ES cells further supports the role of Dax1 as a negative regulator of Oct3/4 in ES cells. Since overexpression of Oct3/4 activity results in ES cell differentiation (Niwa et al., 2000), Dax1 may have a major role in neutralizing the excess activity of Oct3/4. The existence of a regulatory loop between Dax1 and Oct3/4 raises the possibility that Dax1 is not just a negative regulator but a fine-tuner of Oct3/4 activity (Fig. 2).

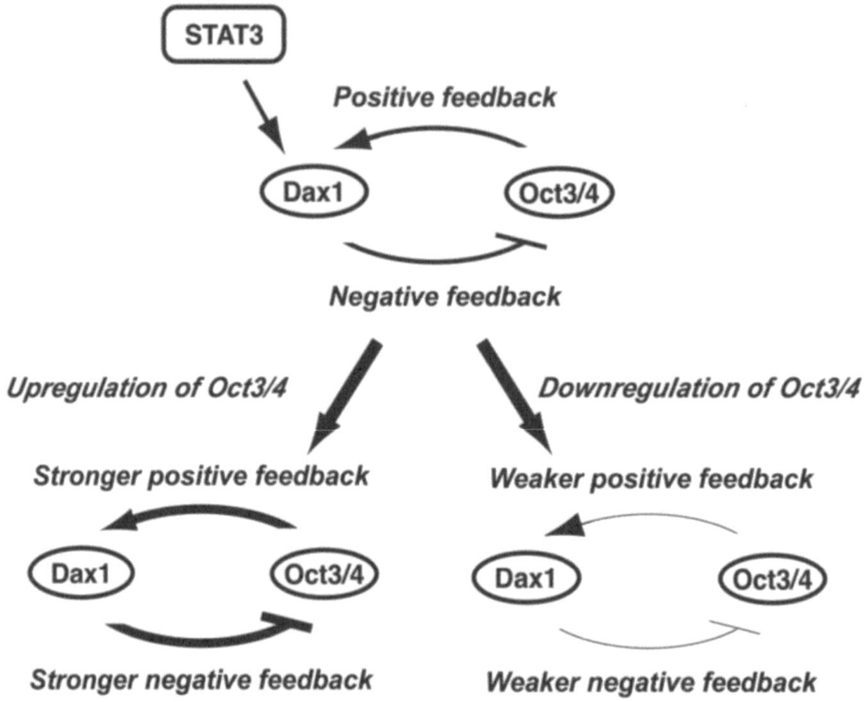

Fig. 2. Dax1 may function as a fine-tuner of Oct3/4 activity in ES cells. Dax1 negatively regulates the transcriptional activity of Oct3/4, whereas Oct3/4 and STAT3 positively regulate the expression of Dax1. When an excess amount of Oct3/4 exists in ES cells, it triggers the induction of Dax1 protein, which in turn suppresses the activity of Oct3/4. On the other hand, when Oct3/4 is downregulated, the expression level of its negative regulator, Dax1, also decreases, which allows Oct3/4 to maintain its activity at the normal level. Thus, the suppression of Oct3/4 by Dax1 depends on the activity of Oct3/4, and this dependency enables Dax1 to act as a "fine-tuner" that maintains the Oct3/4 activity at an appropriate level in ES cells.

Sall4 is a zinc-finger transcription factor that was originally cloned based on sequence homology to *Drosophila spalt* (*sal*), which is a homeotic gene essential for the development of posterior-head and anterior-tail segments (de Celis & Barrio, 2009). In humans, mutations in SALL4 cause an autosomal dominant disorder known as Okihiro syndrome or Duane-radial ray syndrome. Sall4 functions as a strong transcriptional repressor by associating with the Mi-2/Nucleosome remodeling and deacetylase (NuRD) complex (Lu et al., 2009; Yuri et al., 2009). Sall4-null ICM proliferates poorly, and Sall4-deficient mice die shortly after implantation (Sakaki-Yumoto et al., 2006). In agreement with these findings, Sall4-null ES cells exhibit retarded proliferation and express the trophectodermal marker Cdx2. ChIP-on-chip analysis reveals that Sall4 binds to a broad variety of genes that may be important for stem cell functions (Yang et al., 2008). Furthermore, Sall4 forms a complex (or complexes) with Nanog and Oct3/4 (Wang et al., 2006; Wu et al., 2006; Liang et al., 2008). On the other hand, Sall4-null

ES cells can be maintained for long periods, express Oct3/4 at normal level, and generate chimeric mice (Sakaki-Yumoto et al., 2006; Yuri et al., 2009). These observations suggest that Sall4 is essential for stabilization, but not for pluripotency, of ES cells.

Amino-terminal enhancer of split (Aes1, also known as Grg5) belongs to the Gro/TLE family and is a direct target of STAT3 in ES cells (Sekkaï et al., 2005). Gro/TLE family proteins function as transcriptional repressors and negatively regulate several signaling pathways such as Wnt and Notch (Gasperowicz & Otto, 2005). Aes1 lacks the C-terminal region common to most Gro/TLE family proteins and under certain circumstances can block Gro/TLE-mediated transcriptional repression in a dominant-negative fashion. Aes1-null mice show no severe phenotype except postnatal growth retardation (Mallo et al., 1995), suggesting that Aes1 is dispensable for ES cell self-renewal.

3.3 Epigenetic modulators

Embryonic ectoderm development (Eed) was originally identified as a gene that regulates early ectoderm development. Eed belongs to the polycomb family and forms the PRC2 complex together with polycomb proteins Suz12 and Ezh2 (Simon & Kingston, 2009). The PRC2 complex can suppress gene expression by catalyzing the methylation of Lys-27 on histone H3. Eed deficiency in ES cells results in the loss of PRC2 activity and the induction of a variety of differentiation-associated genes (Azura et al., 2006; Boyer et al., 2006; Ura et al., 2008), suggesting that the role of Eed in ES cells is to suppress the expression of differentiation-associated genes via the PRC2 complex. Since expression of Eed is also regulated by Oct3/4 (Ura et al., 2008), it is likely that STAT3 and Oct3/4 both regulate PRC2 activity by controlling Eed expression (Fig. 3). Despite the expression of differentiation-associated genes, Eed-null ES cells can be cultured for a long time and maintain the expression level of Oct3/4, suggesting that Eed is required for the stabilization of ES cell self-renewal.

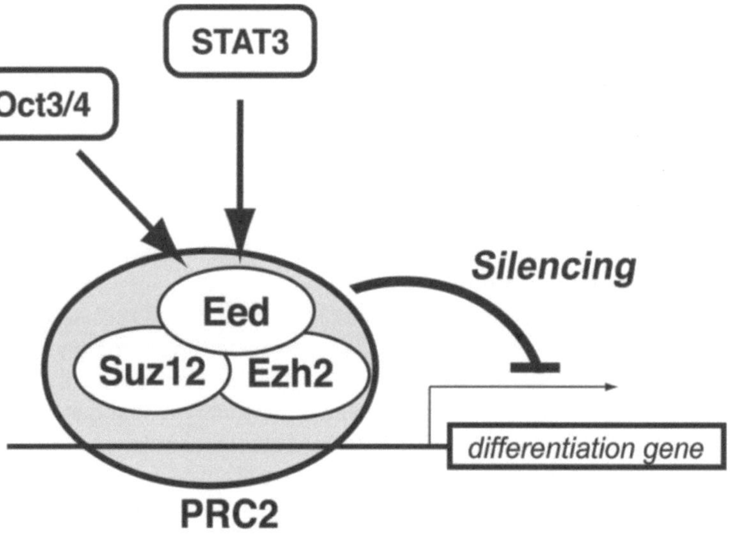

Fig. 3. Eed is involved in the silencing of differentiation-associated genes. PRC2 mainly consists of three subunits, Eed, Ezh2 and Suz12. STAT3 and Oct3/4 regulate Eed expression to maintain the activity level of PRC2, and thereby suppress the induction of differentiation-associated genes

In addition to Eed, which regulates methylation of histone H3, STAT3 controls the expression of jumonji domain containing 1A (Jmjd1a), a demethylase of Lys-9 of histone H3 (Ko et al., 2006). Lys-9 methylation also contributes to the suppression of gene expression, indicating that the demethylase activity of Jmjd1a may be involved in the activation of gene expression in ES cells. Indeed, Jmjd1a is known to regulate expression of self-renewal genes in ES cells, including Tcl1, Zfp57 and Tcfcp2l1 (Loh et al, 2007). Knockdown of Jmjd1a results in ES cell differentiation, which can be partially reversed by overexpression of Tcl1. Since Tcl1 has been shown to regulate self-renewal of ES cells (Ivanova et al. 2006; Matoba et al. 2006), these results suggest that Jmjd1a plays an important role in the maintenance of ES cells by upregulating Tcl1 expression.

3.4 Kinases

In addition to transcription-related genes, the LIF/STAT3 pathway regulates expression of other classes of genes, for example, kinases. The *pim* genes encode serine/threonine kinases, Pim-1, Pim-2 and Pim-3, which regulate cell growth and apoptosis (Bachmann & Möröy, 2005). The expression of both Pim-1 and Pim-3 is regulated by STAT3 in ES cells (Aksoy et al., 2007). Overexpression of Pim-1 and Pim-3 promotes self-renewal and knockdown of these genes increases the rate of spontaneous differentiation and apoptosis (Aksoy et al., 2007), suggesting that these kinases have important roles in ES cell self-renewal.

The Akt-related kinase serum/glucocorticoid-regulated kinase (SGK) can phosphorylate and inactivate GSK-3, a kinase that is known to regulate β-catenin levels in ES cells (Tessier & Woodgett, 2006). SGK is downstream of both the STAT3 and PI3K pathways (Kobayashi & Cohen, 1999; Park et al., 1999; Bourillot et al., 2009). As described above (Section 3.1), the molecular mechanism behind the LIF-mediated activation of β-catenin is unknown. The ability of SGK to inactivate GSK-3 raises the intriguing possibility that the LIF/STAT3 pathway can stabilize β-catenin protein through SGK-mediated inactivation of GSK-3.

3.5 Other target molecules

Stem cell-derived differentiation regulator (Sddr, also known as Moep19, Ooep and Floped) contains the domain similar to KH domain, which binds to RNA. Sddr is highly expressed in self-renewing ES cells, and is downregulated when ES cells undergo differentiation (Miura et al., 2010). Despite the self-renewal-specific expression, studies with Sddr-null ES cells indicate that this protein is dispensable for self-renewal (Miura et al., 2010). However, since Sddr deficiency promotes ES cell differentiation, Sddr may play a role in switching from self-renewal and differentiation.

Embryonic stem cell-specific gene 1 (Esg1, also known as H34, Ecat2 and Dppa5) is another KH domain-containing protein that is specifically expressed in self-renewing ES cells (Tanaka et al., 2002). Similar to the case of Sddr, Esg1 is likely to be dispensable for self-renewal, since Esg1-null ES cells can be established and show no abnormality (Amano et al., 2006).

Suppressor of cytokine signaling (Socs)-3 is an SH2 domain-containing protein that plays an important role in placental development and immunological processes (Kubo et al., 2003). Socs3 inhibits JAK tyrosine kinase activity through a kinase inhibitory region in its N-terminal domain. As a result, Socs3 functions as a feedback regulator of LIF/STAT3 signaling and its overexpression blocks the self-renewal of ES cells (Ying et al, 2003). It was therefore surprising that Socs3-null ES cells also exhibit impaired self-renewal and increased differentiation into primitive endoderm (Forrai et al., 2006). Detailed analysis revealed that Socs3 deficiency enhances Ras/Erk signaling in addition to its effect on STAT3, and that the self-renewal activity of Socs3-null ES cells can be recovered by inhibiting the Ras/Erk pathway. It is likely therefore that Socs3 maintains the balance between the STAT3 and the Ras/Erk pathways in ES cells.

STAT3 also regulates the expression of the plasma membrane-associated molecule CD9 (Oka et al., 2002). ES cell colony formation and cell viability are reduced by the addition of anti-CD9 antibody to the culture medium (Oka et al., 2002). However, CD9-null ES cells exhibit normal morphology and growth properties, and normal expression of self-renewal markers (Akutsu et al., 2009), indicating that expression of CD9 is not essential for ES cell self-renewal.

4. The role of the LIF/STAT3 pathway in ES cell self-renewal

Identification of STAT3 targets helps us to understand the molecular mechanisms by which the LIF/STAT3 pathway regulates ES cell self-renewal. The known STAT3 targets can be divided into several groups based on their functions and it is likely that the LIF/STAT3 pathway plays multiple roles in the maintenance of self-renewal (Fig. 4).

One major role of the LIF/STAT3 pathway is to form transcriptional networks with other key transcription factors such as Oct3/4, Sox2, Nanog, c-Myc, Klf4, Esrrb and Sall4. Some of these transcription factors are themselves regulated by the LIF/STAT3 pathway, including c-Myc, Klf4, and Sall4. Although Oct3/4 is not a downstream molecule of the LIF/STAT3 pathway, its activity and expression are regulated in various ways by STAT3 targets. For example, the expression and transcriptional activity of Oct3/4 are positively regulated by GABPα and β-catenin, respectively, while Oct3/4 activity is negatively regulated by Dax1. In turn, STAT3-regulated transcription factors bind to the regulatory regions of other important factors and induce their expression. In this way, the LIF/STAT3 pathway participates in the formation of transcriptional networks.

The LIF/STAT3 pathway also regulates chromatin structure. The chromatin in self-renewing ES cells exhibits increased accessibility due to fewer and more loosely bound histones and architectural proteins (Meshorer and Misteli, 2006). When ES cells undergo differentiation, their chromatin structure changes dynamically in response to global histone modifications, and this mechanism regulates gene activation and repression during development (Kouzarides, 2007). For example, methylation of Lys-4 of histone H3 is associated with transcriptional activation, whereas methylation of Lys-9 or Lys-27 of histone H3 is linked to transcriptional silencing. The LIF/STAT3 pathway regulates expression of molecules such as Eed and Jmjd1a that are involved in histone modifications, and thereby controls chromatin structure in ES cells.

In addition to transcription factors, the LIF/STAT3 pathway also communicates with other signaling pathways. For example, there may be cross-talk between the LIF/STAT3 and Wnt signaling pathways due to their shared effects on β-catenin (Fig. 1). In fact, LIF and Wnt exhibit a synergistic effect on the maintenance of ES cell self-renewal (Ogawa et al., 2006). The LIF/STAT3 and Wnt pathways may interact with each other through Aes1, which is believed to exert dominant-negative activity on Gro/TLE family proteins that, in turn, can negatively regulate the Wnt pathway. The LIF/STAT3 pathway may also communicate with PI3K through its ability to regulate the expression of Tcl1 through Jmjd1a. Tcl1 can bind to Akt and enhance its kinase activity (Laine et al., 2000; Pekarsky et al., 2000), suggesting a plausible link between the PI3K/Akt and LIF/STAT3 pathways. The identification of SGK as a downstream molecule of STAT3 raises the possibility that this kinase may act as an additional connector molecule between STAT3, PI3K and Wnt, since the activity of SGK is regulated by the PI3K pathway, and SGK can control the Wnt pathway through GSK-3.

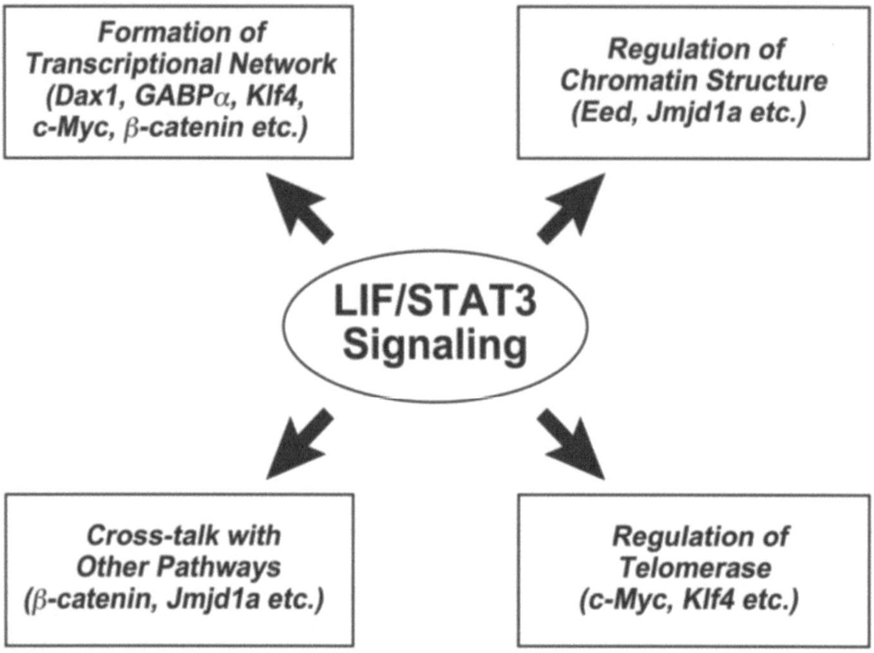

Fig. 4. Role of the LIF/STAT3 pathway in the self-renewal of mouse ES cells. The LIF/STAT3 pathway regulates the expression of a variety of genes, and thereby plays multiple roles in the maintenance of self-renewal in mouse ES cells

Unlike most somatic cells, ES cells are immortal and capable of indefinite self-renewal. This unique ability is supported by high levels of telomerase activity in self-renewing ES cells. The expression of telomerase reverse transcriptase (Tert) is essential for this telomerase activity (Liu et al., 2000), although the mechanisms that regulate Tert expression in ES cells are not known. Among the STAT3 targets described above, c-Myc has been shown to regulate Tert expression in primary human fibroblasts (Greenberg et al., 1999) and Klf4 can stimulate Tert expression in human cancer cells and ES cells (Wong et al., 2010). GABPα and c-Myc have also been identified as candidate regulators of Tert in mouse ES cells, based on RNAi screening (Coussens et al., 2010). Therefore, the LIF/STAT3 pathway may regulate expression of Tert through its target genes such as c-Myc.

5. Conclusion

This chapter summarizes current knowledge on the role of the LIF/STAT3 pathway in the self-renewal of mouse ES cells. The LIF/STAT3 pathway stimulates the expression of several classes of genes, which play important roles in self-renewal through a range of activities such as the formation of transcriptional networks, suppressing expression of differentiation-associated genes, and communicating with other signaling pathways. In addition, we would also add that STAT3 can function as a transcriptional repressor. STAT3 associates with coactivators, such as steroid receptor co-activator 1 (SRC1/NcoA) and p300/cAMP response element binding protein (CBP), but it can also associate with co-repressors, including KRAB-associated protein-1 (Kap1/Tif1β), to inhibit transcription (Tsuruma et al., 2008). STAT3 binds to the regulatory regions of many differentiation-associated genes (Chen et al., 2008; Kidder et al., 2008) and it is therefore likely that STAT3 suppresses expression of differentiation-associated genes in ES cells by both direct and indirect effects on gene expression.

Our understanding of the LIF/STAT3 pathway is not restricted to ES cell self-renewal; for example, STAT3 is activated in a variety of tumors and has been described as an oncogene (Yu & Jove, 2004). It is likely that ES cells and cancer cells may use several STAT3 targets in common for their growth, and this hypothesis is supported by the fact that the list of STAT3 target genes contains well-established tumor-related genes such as c-Myc, β-catenin, and Pim. Since the discovery of cancer stem cells, it has been demonstrated that tumor growth and stem cell self-renewal share several signaling pathways including the Wnt pathway (Reya & Clevers, 2005). Therefore, it would be interesting to perform a detailed comparison of the key molecular mechanisms in STAT3-mediated self-renewal of ES cells and STAT3-mediated tumorigenesis.

Due to the finding that human ES cells do not respond to LIF, the importance of the LIF/STAT3 pathway has been neglected. However, we now know that so-called human ES cells are not true ES cells but EpiSC. Furthermore, it has been shown recently that human "ES" cells, which are LIF-dependent and show greater similarity to ES cells than to EpiSC, can be generated, although we still need some genetic manipulations (Hanna et al., 2010). Understanding the role of the LIF/STAT3 pathway in mouse ES cells may give us valuable clues to support the development of an efficient expansion system for human ES and iPS cells in future.

6. Acknowledgements

This work has been supported in part by the Naito Foundation and Grants-in-Aid for Scientific Research from the Ministry of Education, Culture, Sports, Science, and Technology of Japan.

7. References

Akagi, T.; Usuda, M.; Matsuda, T.; Ko, M.S.H.; Niwa, H.; Asano, M.; Koide, H. & Yokota, T. (2005) Identification of Zfp-57 as a downstream molecule of STAT3 and Oct-3/4 in embryonic stem cells. *Biochem. Biophys. Res. Commun.*, 331, 23-30.

Aksoy, I.; Sakabedoyan, C.; Bourillot, P.Y.; Malashicheva, A.B.; Mancip, J.; Knoblauch, K.; Afanassieff, M. & Savatier, P. (2007) Self-renewal of murine embryonic stem cells is supported by the serine/threonine kinases Pim-1 and Pim-3. *Stem Cells*, 25, 2996-3004.

Akutsu, H.; Miura, T.; Machida, M.; Birumachi, J.; Hamada, A.; Yamada, M.; Sullivan, S.; Miyado, K. & Umezawa, A. (2009) Maintenance of pluripotency and self-renewal ability of mouse embryonic stem cells in the absence of tetraspanin CD9. *Differentiation*, 78, 137-142.

Amano, H.; Itakura, K.; Maruyama, M.; Ichisaka, T.; Nakagawa, M. & Yamanaka, S. (2006) Identification and targeted disruption of the mouse gene encoding ESG1 (PH34/ECAT2/DPPA5). *BMC Dev. Biol.*, 6, 11.

Avilion, A.A.; Nicolis, S.K.; Pevny, L.H.; Perez, L.; Vivian, N. & Lovell-Badge, R. (2003) Multipotent cell lineages in early mouse development depend on SOX2 function. *Genes Dev.*, 17, 126-140.

Azuara, V.; Perry, P.; Sauer, S.; Spivakov, M.; Jorgensen, H.F.; John, R.M.; Gouti, M.; Casanova, M.; Warnes, G.; Merkenschlager, M. & Fisher, A.G. (2006) Chromatin signatures of pluripotent cell lines. *Nat. Cell Biol.*, 8, 532-538.

Bachmann, M. & Möröy, T. (2005) The serine/threonine kinase Pim-1. *Int. J. Biochem. Cell Biol.*, 37, 726-730.

Boeuf, H.; Hauss, C.; Graeve, F.D.; Baran, N. & Kedinger, C. (1997) Leukemia inhibitory factor-dependent transcriptional activation in embryonic stem cells. *J. Cell Biol.*, 138, 1207-1217.

Bourillot, P.-Y.; Aksoy, I .; Schreiber, V.; Wianny, F.; Shulz, H.; Hummel, O.; Hubner, N. & Savatier, P. (2009) Novel STAT3 target genes exert distinct roles in the inhibition of mesoderm and endoderm differentiation in cooperation with Nanog. *Stem Cells*, 27, 1760-1771.

Boyer, L.A.; Lee, T.I.; Cole, M.F.; Johnstone, S.E.; Levine, S.S.; Zucker, J.P.; Guenther, M.G.; Kumar, R.M.; Murray, H.L.; Jenner, R.G.; Gifford, D.K.; Melton, D.A.; Jaenisch, R. & Young, R.A. (2005) Core transcriptional regulatory circuitry in human embryonic stem cells. *Cell*, 122, 947-956.

Boyer, L.A.; Plath, K.; Zeitlinger, J.; Brambrink, T.; Medeiros, L.A.; Lee, T.I.; Levine, S.S.; Wernig, M.; Tajonar, A.; Ray, M.K.; Bell, G.W.; Otte, A.P.; Vidal, M.; Gifford, D.K.; Young, R.A. & Jaenisch, R. (2006) Polycomb complexes repress developmental regulators in murine embryonic stem cells. *Nature*, 441, 349-353.

Brons, I.G.; Smithers, L.E.; Trotter, M.W.; Rugg-Gunn, P.; Sun, B.; Chuva de Sousa Lopes, S.M.; Howlett, S.K.; Clarkson, A.; Ahrlund-Richter, L.; Pedersen. R.A. & Vallier, L. (2007) Derivation of pluripotent epiblast stem cells from mammalian embryos. *Nature*, 448, 191-195.

Buehr, M.; Meek, S.; Blair, K.; Yang, J.; Ure, J.; Silva, J.; McLay, R.; Hall, J.; Ying, Q.L. & Smith, A. (2008) Capture of authentic embryonic stem cells from rat blastocysts. *Cell*, 135, 1287-1298.

Burdon, T.; Stracey, C.; Chambers, I.; Nichols, J. & Smith, A. (1999) Suppression of SHP-2 and ERK signalling promotes self-renewal of mouse embryonic stem cells. *Dev. Biol.*, 210, 30-43.

Cartwright, P.; McLean, C.; Sheppard, A.; Rivett, D.; Jones, K. & Dalton, S. (2005) LIF/STAT3 controls ES cell self-renewal and pluripotency by a Myc-dependent mechanism. *Development*, 132, 885-896.

de Celis, J.F. & Barrio, R. (2009) Regulation and function of Spalt proteins during animal development. *Int. J. Dev. Biol.*, 53, 1385-1398.

Chen, X.; Xu, H.; Yuan, P.; Fang, F.; Huss, M.; Vega, V.B.; Wong, E.; Orlov, Y.L.; Zhang, W.: Jiang, J.; Loh, Y.H.; Yeo, H.C.; Yeo, Z.X.; Narang, V.; Govindarajan, K.R.; Leong, B.; Shahab, A.; Ruan, Y.; Bourque, G.; Sung, W.K.; Clarke, N.D.; Wei, C.L. & Ng, H.H. (2008) Integration of external signaling pathways with the core transcriptional network in embryonic stem cells. *Cell*, 133, 1106-1117.

Cinelli, P.; Casanova, E.A.; Uhlig, S.; Lochmatter, P.; Matsuda, T.; Yokota, T.; Rülicke, T.; Ledermann, B. & Bürki, K. (2008) Expression profiling in transgenic FVB/N embryonic stem cells overexpressing STAT3. *BMC Developmental Biology*, 8, 57.

Clipsham, R.; Niakan, K. & McCabe, E.R. (2004) Nr0b1 and its network partners are expressed early in murine embryos prior to steroidogenic axis organogenesis. *Gene Expr. Patterns*, 4, 3-14.

Coussens, M.; Davy, P.; Brown, L.; Foster, C.; Andrews, W.H.; Nagata, M. & Allsopp, R. (2010) RNAi screen for telomerase reverse transcriptase transcriptional regulators identifies HIF1alpha as critical for telomerase function in murine embryonic stem cells. *Proc. Natl. Acad. Sci. USA*, 107, 13842-13847.

Dahéron, L.; Opitz, S.L.; Zaehres, H.; Lensch, M.W.; Andrews, P.W.; Itskovitz-Eldor, J. & Daley, G.Q. (2004) LIF/STAT3 signaling fails to maintain self-renewal of human embryonic stem cells. *Stem Cells*, 22, 770-778.

Evans, M.J. & Kauman, M.H. (1981) Establishment in culture of pluripotential cells from mouse embryos. *Nature*, 292, 154–156.

Fan, Y.; Melhem, M.F. & Chaillet, J.R. (1999) Forced expression of the homeobox-containing gene Pem blocks differentiation of embryonic stem cells. *Dev Biol.*, 210, 481-496.

Forrai, A.; Boyle, K.; Hart, A.H.; Hartley, L.; Rakar, S.; Willson, T.A.; Simpson, K.M.; Roberts, A.W.; Alexander, W.S.; Voss, A.K. & Robb, L. (2006) Absence of suppressor of cytokine signalling 3 reduces self-renewal and promotes differentiation in murine embryonic stem cells. *Stem Cells*, 24, 604-614.

Gasperowicz, M. & Otto, F. (2005) Mammalian Groucho homologs: redundancy or specificity? *J. Cell Biochem.*, 95, 670-687.

Greenberg, R.A.; O'Hagan, R.C.; Deng, H.; Xiao, Q.; Hann, S.R.; Adams, R.R.; Lichtsteiner, S.: Chin, L.; Morin, G.B. & DePinho, R.A. (1999) Telomerase reverse transcriptase gene is a direct target of c-Myc but is not functionally equivalent in cellular transformation. *Oncogene*, 18, 1219-1226.

Guo, G.; Yang, J.; Nichols, J.; Hall, J.S.; Eyres, I.; Mansfield, W. & Smith, A. (2009) Klf4 reverts developmentally programmed restriction of ground state pluripotency. *Development*, 136, 1063-1069.

Hamazaki, T.; Kehoe, S.M.; Nakano, T. & Terada, N. (2006) The Grb2/Mek pathway represses Nanog in murine embryonic stem cells. *Mol. Cell. Biol.*, 26, 7539-7549.

Hanna, J.; Cheng, A.W.; Saha, K.; Kim, J.; Lengner, C.J.; Soldner, F.; Cassady, J.P.; Muffat, J.; Carey, B.W. & Jaenisch, R. (2010) Human embryonic stem cells with biological and epigenetic characteristics similar to those of mouse ESCs. *Proc. Natl. Acad. Sci. USA*, 107, 9222-9227.

Hao, J.; Li, T.G.; Qi, X.; Zhao, D.F. & Zhao, G.O. (2006) WNT/beta-catenin pathway up-regulates Stat3 and converges on LIF to prevent differentiation of mouse embryonic stem cells. *Dev. Biol.*, 290, 81–91.

Horiuchi, H.; Tategaki, A.; Yamashita, Y.; Hisamatsu, H.; Ogawa, M.; Noguchi, T.; Aosasa, M.; Kawashima, T.; Akita, S.; Nishimichi, N.; Mitsui, N.; Furusawa, S. & Matsuda, H. (2004) Chicken leukemia inhibitory factor maintains chicken embryonic stem cells in the undifferentiated state. *J. Biol. Chem.*, 279, 24514–24520.

Huelsken, J.; Vogel, R.; Brinkmann, V.; Erdmann, B.; Birchmeier, C. & Birchmeier, W. (2000) Requirement for beta-catenin in anterior-posterior axis formation in mice. *J. Cell. Biol.*, 148, 567–578.

Ivanova, N.; Dobrin, R.; Lu, R.; Kotenko, I.; Levorse, J.; DeCoste, C.; Schafer, X.; Lun, Y. & Lemischka, I.R. (2006) Dissecting self-renewal in stem cells with RNA interference. *Nature*, 442, 533-538.

Jiang, J.; Chan, Y.S.; Loh, Y.H.; Cai, J.; Tong, G.Q.; Lim, C.A.; Robson, P.; Zhong, S. & Ng, H.H. (2008) A core Klf circuitry regulates self-renewal of embryonic stem cells. *Nat. Cell Biol.*, 10, 353-360.

Khalfallah, O.; Rouleau, M.; Barbry, P.; Bardoni, B. & Lalli, E. (2009) Dax-1 knockdown in mouse embryonic stem cells induces loss of pluripotency and multilineage differentiation. *Stem Cells*, 27, 1529-1537.

Kidder, B.L.; Yang, J. & Palmer, S. (2008) STAT3 and c-Myc genome-wide promoter occupancy in embryonic stem cells. *PLoS ONE*, 3, e3932.

Kielman, M.F.; Rindapää, M.; Gaspar, C.; van Poppel, N.; Breukel, C.; van Leeuwen, S.; Taketo, M.M.; Roberts, S.; Smits, R. & Fodde, R. (2002) Apc modulates embryonic stem-cell differentiation by controlling the dosage of beta-catenin signaling. *Nat. Genet.*, 32, 594-605.

Kinoshita, K.; Ura,H.; Akagi,T.; Usuda, M.; Koide, H. & Yokota, T. (2007) GABPα regulates Oct-3/4 expression in mouse embryonic stem cells. *Biochem. Biophys. Res. Commun.*, 353, 686-691.

Ko, S.Y.; Kang, H.Y.; Lee, H.S.; Han, S.Y. & Hong, S.H. (2006) Identification of Jmjd1a as a STAT3 downstream gene in mES cells. *Cell Struct. Funct.*, 31, 53-62.

Kobayashi, T. & Cohen, P. (1999) Activation of serum- and glucocorticoid-regulated protein kinase by agonists that activate phosphatidylinositide 3-kinase is mediated by 3-phosphoinositide-dependent protein kinase-1 (PDK1) and PDK2. *Biochem. J.*, 339, 319–328.

Kouzarides, T. (2007) Chromatin modifications and their function. *Cell*, 128, 693-705

Kubo, M.; Hanada, T. & Yoshimura, A. (2003) Suppressors of cytokine signaling and immunity. *Nat. Immunol.*, 4, 1169-1176.

Laine, J.; Kunstle, G.; Obata, T.; Sha, M. & Noguchi, M. (2000) The proto-oncogene TCL1 is an Akt kinase coactivator. *Mol. Cell* 6, 395-407.

Li, P.; Tong, C.; Mehrian-Shai, R.; Jia, L.; Wu, N.; Yan, Y.; Maxson, R.E.; Schulze, E.N.; Song, H.; Hsieh, C.L.; Pera, M.F. & Ying, Q.L. (2008) Germline competent embryonic stem cells derived from rat blastocysts. *Cell*, 135, 1299-1310.

Li, X.; Ito, M.; Zhou, F.; Youngson, N.; Zuo, X.; Leder, P. & Ferguson-Smith, A.C. (2008) A maternal-zygotic effect gene, zfp57, maintains both maternal and paternal imprints. *Dev. Cell*, 15, 547-557.

Li, Y.; McClintick, J.; Zhong, L.; Edenberg, H.J.; Yoder, M.C. & Chan, R.J. (2005) Murine embryonic stem cell differentiation is promoted by SOCS-3 and inhibited by the zinc finger transcription factor Klf4. *Blood*, 105, 635-637.

Liang, J.; Wan, M.; Zhang, Y.; Gu, P.; Xin, H.; Jung, S.Y.; Qin, J.; Wong, J.; Cooney, A.J.; Liu, D. & Songyang, Z. (2008) Nanog and Oct4 associate with unique transcriptional repression complexes in embryonic stem cells. *Nat. Cell Biol.*, 10, 731-739.

Lin, C.H.; Lin, C.; Tanaka, H.; Fero, M.L. & Eisenman, R.N. (2009a) Gene regulation and epigenetic remodeling in murine embryonic stem cells by c-Myc. *PLoS One*, 4, e7839.

Lin, C.H.; Jackson, A.L.; Guo, J.; Linsley, P.S. & Eisenman, R.N. (2009b) Myc-regulated microRNAs attenuate embryonic stem cell differentiation. *EMBO J.*, 28, 3157-3170.

Liu, Y.; Snow, B.E.; Hande, M.P.; Yeung, D.; Erdmann, N.J.; Wakeham, A.; Itie, A.; Siderovski, D.P.; Lansdorp, P.M.; Robinson, M.O. & Harrington, L. (2000) The telomerase reverse transcriptase is limiting and necessary for telomerase function in vivo. *Curr. Biol.*, 10, 1459-1462.

Loh, Y.H.; Wu, Q.; Chew, J.L.; Vega, V.B.; Zhang, W.; Chen, X.; Bourque, G.; George, J.; Leong, B.; Liu, J.; Wong, K.Y.; Sung, K.W.; Lee, C.W.; Zhao, X.D.; Chiu, K.P.; Lipovich, L.; Kuznetsov, V.A.; Robson, P.; Stanton, L.W.; Wei, C.L.; Ruan, Y.; Lim, B. & Ng, H.H. (2006) The Oct4 and Nanog transcription network regulates pluripotency in mouse embryonic stem cells. *Nat. Genet.*, 38, 431-440.

Loh, Y.H.; Zhang, W.; Chen, X.; George, J. & Ng, H.H. (2007) Jmjd1a and Jmjd2c histone H3 Lys9 demethylases regulate self-renewal in embryonic stem cells. *Genes Dev.*, 21, 2545-2557.

Lu, J.; Jeong, H.W.; Kong, N.; Yang, Y.; Carroll, J.; Luo, H.R.; Silberstein, L.E.; YupoMa & Chai, L. (2009) Stem cell factor SALL4 represses the transcriptions of PTEN and SALL1 through an epigenetic repressor complex. *PLoS One*, 4, e5577.

Mackay, D.J.; Callaway, J.L.; Marks, S.M.; White, H.E.; Acerini, C.L.; Boonen, S.E.; Dayanikli, P.; Firth, H.V.; Goodship, J.A.; Haemers, A.P.; Hahnemann, J.M.; Kordonouri, O.; Masoud, A.F.; Oestergaard, E.; Storr, J.; Ellard, S.; Hattersley, A.T.; Robinson, D.O. & Temple, I.K. (2008) Hypomethylation of multiple imprinted loci in individuals with transient neonatal diabetes is associated with mutations in ZFP57. *Nat. Genet.*, 40, 949-951.

Mallo, M.; Gendron-Maguire, M.; Harbison, M.L. & Gridley, T. (1995) Protein characterization and targeted disruption of Grg, a mouse gene related to the groucho transcript of the Drosophila Enhancer of split complex. *Dev. Dyn.*, 204, 338-347.

Martin, G.R. (1981) Isolation of a pluripotent cell line from early mouse embryos cultured in medium conditioned by teratocarcinoma stem cells. *Proc. Natl. Acad. Sci. USA*, 78, 7634-7638.

Masui, S.; Nakatake, Y.; Toyooka, Y.; Shimosato, D.; Yagi, R.; Takahashi, K.; Okochi, H.; Okuda, A.; Matoba, R.; Sharov, A.A.; Ko, M.S. & Niwa, H. (2007) Pluripotency governed by Sox2 via regulation of Oct3/4 expression in mouse embryonic stem cells. *Nat. Cell Biol.*, 9, 625-635.

Matoba, R.; Niwa, H.; Masui, S.; Ohtsuka, S.; Carter, M.G.; Sharov, A.A. & Ko, M.S. (2006) Dissecting Oct3/4-regulated gene networks in embryonic stem cells by expression profiling. *PLoS One*, 1, e26.

Matsuda, T.; Nakamura, T.; Nakao, N.; Arai, T.; Katsuki, M.; Heike, T. & Yokota, T. (1999) STAT3 activation is sufficient to maintain an undifferentiated state of mouse embryonic stem cells. *EMBO J.*, 18, 4261-4269.

Meshorer, E. & Misteli, T. (2006) Chromatin in pluripotent embryonic stem cells and differentiation. *Nat. Rev. Mol. Cell. Biol.*, 7, 540-546

Miura, M.; Ueda, A.; Takao, Y.; Koide, H. & Yokota, T. (2010) A stem cell-derived gene, Sddr, negatively regulates differentiation of embryonic stem cells. *Int. J. Dev. Biol.*, 54, 33-39.

Nakatake, Y.; Fukui, N.; Iwamatsu, Y.; Masui, S.; Takahashi, K.; Yagi, R.; Yagi, K.; Miyazaki, J.; Matoba, R.; Ko, M.S. & Niwa, H. (2006) Klf4 cooperates with Oct3/4 and Sox2 to activate the Lefty1 core promoter in embryonic stem cells. *Mol. Cell. Biol.*, 26, 7772-7782.

Niakan, K.K. & McCabe, E.R. (2005) DAX1 origin, function, and novel role. *Mol. Genet. Metab.*, 86, 70-83.

Niakan, K.K.; Davis, E.C.; Clipsham, R.C.; Jiang, M.; Dehart, D.B.; Sulik, K.K. & McCabe, E.R. (2006) Novel role of the orphan nuclear receptor Dax1 in embryogenesis, different from steroidgenesis. *Mol. Genet. Metab.*, 88, 261-271.

Nichols, J.; Zevnik, B.; Anastassiadis, K.; Niwa, H.; Klewe-Nebenius, D.; Chambers, I.; Schöler, H. & Smith, A. (1998) Formation of pluripotent stem cells in the mammalian embryo depends on the POU transcription factor Oct4. *Cell*, 95, 379-391.

Niwa, H.; Burdon, T.; Chambers, I. & Smith, A. (1998) Self-renewal of pluripotent embryonic stem cells is mediated via activation of STAT3. *Genes Dev.*, 12, 2048-2060.

Niwa, H.; Miyazaki, J. & Smith, A.G. (2000) Quantitative expression of Oct-3/4 defines differentiation, dedifferentiation or self-renewal of ES cells. *Nat. Genet.*, 24, 372-376.

Niwa, H.; Ogawa, K.; Shimosato, D. & Adachi, K. (2009) A parallel circuit of LIF signaling pathways maintains pluripotency of mouse ES cells. *Nature*, 460, 118-122.

Ogawa, K.; Nishinakamura, R.; Iwamatsu, Y.; Shimosato, D. & Niwa, H. (2006) Synergistic action of Wnt and LIF in maintaining pluripotency of mouse ES cells. *Biochem. Biophys. Res. Commun.*, 343, 159–166.

Oka, M.; Tagoku, K.; Russell, T.L.; Nakano, Y.; Hamazaki, T.; Meyer, E.M.; Yokota, T. & Terada, N. (2002) CD9 is associated with leukemia inhibitory factor-mediated maintenance of embryonic stem cells. *Mol. Biol. Cell*, 2002 13, 1274-1281.

Okazaki, S.; Tanase, S.; Choudhury, B.K.; Setoyama, K.; Miura, R.; Ogawa, M. & Setoyama, C. (1994) A novel nuclear protein with zinc fingers down-regulated during early mammalian cell differentiation. *J. Biol. Chem.*, 269, 6900-6907.

Paling, N.R.; Wheadon, H.; Bone, H.K. & Welham, M.J. (2004) Regulation of embryonic stem cell self-renewal by phosphoinositide 3-kinasedependent signaling. *J. Biol. Chem.*, 279, 48063–48070.

Park, J.; Leong, M.L.L.; Buse, P.; Maiyar, A.C.; Firestone, G.L. & Hemmings, B.A. (1999) Serum and glucocorticoid-inducible kinase (SGK) is a target of the PI 3-kinase-stimulated signaling pathway. *EMBO J.*, 18, 3024–3033.

Pekarsky, Y.; Koval, A.; Hallas, C.; Bichi, R.; Tresini, M.; Malstrom, S.; Russo, G.; Tsichlis, P. & Croce, C.M. (2000) Tcl1 enhances Akt kinase activity and mediates its nuclear translocation. *Proc. Natl. Acad. Sci. USA*, 97, 3028-3033.

Reya, T. & Clevers, H. (2005) Wnt signalling in stem cells and cancer. *Nature*, 434, 843–850.

Ristevski, S.; O'Leary, D.A.; Thornell, A.P.; Owen, M.J.; Kola, I. & Hertzog, P.J. (2004) The ETS transcription factor GABPα is essential for early embryogenesis. *Mol. Cell. Biol.*, 24, 5844-5849.

Rowland, B.D. & Peeper, D.S. (2006) KLF4, p21 and context-dependent opposing forces in cancer. *Nat. Rev. Cancer*, 6, 11-23.

Rosmarin, A.G.; Resendes, K.K.; Yang, Z.; McMillan, J.N. & Fleming, S.L. (2004) GA-binding protein transcription factor: a review of GABP as an integrator of intracellular signaling and protein-protein interactions. *Blood Cells Mol. Dis.*, 32, 143-154.

Sakaki-Yumoto, M.: Kobayashi, C.; Sato, A.; Fujimura, S.; Matsumoto, Y.; Takasato, M.; Kodama, T.; Aburatani, H.; Asashima, M.; Yoshida, N. & Nishinakamura, R. (2006) The murine homolog of SALL4, a causative gene in Okihiro syndrome, is essential for embryonic stem cell proliferation, and cooperates with Sall1 in anorectal, heart, brain and kidney development. *Development*, 133, 3005-3013.

Sato, N.; Meijer, L.; Skaltsounis, L.; Greengard, P. & Brivanlou, A.H. (2004) Maintenance of pluripotency in human and mouse embryonic stem cells through activation of Wnt signaling by a pharmacological GSK-3-specific inhibitor. *Nat. Med.*, 10, 55–63.

Shaulian, E. & Karin, M. (2001) AP-1 in cell proliferation and survival. *Oncogene*, 20,2390-2400.

Schorpp-Kistner, M.; Wang, Z.-Q.; Angel, P. & Wagner, E.F. (1999) JunB is essential for mammalian placentation. *EMBO J.*, 18, 934 - 948.

Sekkaï, D.; Gruel, G.; Herry, M.; Moucadel, V.; Constantinescu, S.N.; Albagli, O.; Tronik-Le Roux, D.; Vainchenker, W. & Bennaceur-Griscelli, A. (2005) Microarray analysis of LIF/Stat3 transcriptional targets in embryonic stem cells. *Stem Cells*, 23, 1634-1642.

Simon, J.A. & Kingston, R.E. (2009) Mechanisms of polycomb gene silencing: knowns and unknowns. *Nat. Rev. Mol. Cell Biol.*, 10, 697-708.

Singla, D.K.; Schneider, D.J.; LeWinter, M.M. & Sobel, B.E. (2006) Wnt3a but not wnt11 supports self-renewal of embryonic stem cells. *Biochem. Biophys. Res. Commun.*, 345, 789–795.

Smith, A.G.; Heath, J.K.; Donaldson, D.D.; Wong, G.G.; Moreau, J.; Stahl, M. & Rogers, D. (1998) Inhibition of pluripotential embryonic stem cell differentiation by purified polypeptides. *Nature*, 336, 688-690

Sumi, T.; Fujimoto, Y.; Nakatsuji, N. & Suemori, H. (2004) STAT3 is dispensable for maintenance of self-renewal in nonhuman primate embryonic stem cells. *Stem Cells*, 22, 861-872.

Sun, C.; Nakatake, Y.; Ura, H.; Akagi, T.; Niwa, H.; Koide, H. & Yokota, T. (2008) Stem cell-specific expression of Dax1 is conferred by STAT3 and Oct3/4 in embryonic stem cells. *Biochem. Biophys. Res. Commun.*, 372, 91-96.

Sun, C.; Nakatake, Y.; Akagi, T.; Ura, H.; Matsuda, T.; Nishiyama, A.; Koide, H.; Ko, M.S.; Niwa, H. & Yokota, T. (2009) Dax1 binds to Oct3/4 and inhibits its transcriptional activity in embryonic stem cells. *Mol. Cell. Biol.*, 29, 4574-4583.

Takahashi, K. & Yamanaka, S. (2006) Induction of pluripotent stem cells from mouse embryonic and adult fibroblast cultures by defined factors. *Cell*, 126, 663-676.

Takahashi, K.; Tanabe, K.; Ohnuki, M.; Narita, M.; Ichisaka, T.; Tomoda, K. & Yamanaka, S. (2007) Induction of pluripotent stem cells from adult human fibroblasts by defined factors. *Cell*, 131, 861-872.

Takao, Y.; Yokota, T. & Koide, H. (2007) β-catenin up-regulates Nanog expression through interaction with Oct-3/4 in embryonic stem cells. *Biochem. Biophys. Res. Commun.*, 353, 699-705.

Tanaka, T.S.; Kunath, T.; Kimber, W.L.; Jaradat, S.A.; Stagg, C.A.; Usuda, M.; Yokota, T.; Niwa, H.; Rossant, J. & Ko, M.S. (2002) Gene expression profiling of embryo-derived stem cells reveals candidate genes associated with pluripotency and lineage specificity. *Genome Res.*, 12, 1921-1928.

Tesar, P.J.; Chenoweth, J.G.; Brook, F.A.; Davies, T.J.; Evans, E.P.; Mack, D.L.; Gardner, R.L. & McKay, R.D. (2007) New cell lines from mouse epiblast share defining features with human embryonic stem cells. *Nature*, 448, 196-199.

Tessier, M. & Woodgett, J.R. (2006) Serum and Glucocorticoid-Regulated Protein Kinases: Variations on a Theme. *J. Cell. Biochem.*, 98,1391–1407.

Thomson, J.A.; Itskovitz-Eldor, J.; Shapiro, S.S.; Waknitz, M.A.; Swiergiel, J.J.; Marshall, V.S. & Jones, J.M. (1998) Embryonic stem cell lines derived from human blastocysts. *Science*, 282, 1145-1147.

Tsuruma, R.; Ohbayashi, N.; Kamitani, S.; Ikeda, O.; Sato, N.; Muromoto, R.; Sekine, Y.; Oritani, K. & Matsuda, T. (2008) Physical and functional interactions between STAT3 and KAP1. *Oncogene*, 27, 3054-3059.

Ura, H.; Usuda, M.; Kinoshita, K.; Sun, C.; Mori, K.; Akagi, T.; Matsuda, T.; Koide, H. & Yokota, T. (2008) STAT3 and Oct-3/4 control histone modification through induction of Eed in embryonic stem cells. *J. Biol. Chem.*, 283, 9713-9723.

Wang, J.; Rao, S.; Chu, J.; Shen, X.; Levasseur, D.N.; Theunissen, T.W. & Orkin, S.H. (2006) A protein interaction network for pluripotency of embryonic stem cells. *Nature*, 444, 364-368.

Watanabe, S.; Umehara, H.; Murayama, K.; Okabe, M.; Kimura, T. & Nakano, T. (2006) Activation of Akt signaling is sufficient to maintain pluripotency in mouse and primate embryonic stem cells. *Oncogene*, 25, 2697-2707.

Williams, R.L.; Hilton, D.J.; Pease, S.; Willson, T.A.; Stewart, C.L.; Gearing, D.P.; Wagner, E.F.; Metcalf, D.; Nicola, N.A. & Gough, N.M. (1988) Myeloid leukaemia inhibitory factor maintains the developmental potential of embryonic stem cells. *Nature*, 336, 684-687.

Wong, C.W.; Hou, P.S.; Tseng, S.F.; Chien, C.L.; Wu, K.J.; Chen, H.F.; Ho, H.N.; Kyo, S. & Teng, S.C. (2010) KLF4 Contributes to Maintenance of Telomerase Activity in Stem Cells. *Stem Cells* (in press)

Wu, Q.; Chen, X.; Zhang, J.; Loh, Y.H.; Low, T.Y.; Zhang, W.; Zhang, W.; Sze, S.K.; Lim, B. & Ng, H.H. (2006) Sall4 interacts with Nanog and co-occupies Nanog genomic sites in embryonic stem cells. *J. Biol. Chem.*, 281, 24090-24094.

Yang, J.; Chai, L.; Fowles, T.C.; Alipio, Z.; Xu, D.; Fink, L.M.; Ward, D.C. & Ma, Y. (2008) Genome-wide analysis reveals Sall4 to be a major regulator of pluripotency in murine-embryonic stem cells. *Proc. Natl. Acad. Sci. USA*, 105, 19756-19761.

Ying, Q.L.; Nichols, J.; Chambers, I. & Smith, A. (2003) BMP induction of Id proteins suppresses differentiation and sustains embryonic stem cell self-renewal in collaboration with STAT3. *Cell*, 115, 281-292.

Ying, Q.L.; Wray, J.; Nichols, J.; Batlle-Morera, L.; Doble, B.; Woodgett, J.; Cohen, P. & Smith, A. (2008) The ground state of embryonic stem cell self-renewal. *Nature*, 453, 519-523.

Yoshida-Koide, U.; Matsuda, T.; Saikawa, K.; Nakanuma, Y.; Yokota, T.; Asashima, M. & Koide, H. (2004) Involvement of Ras in extraembryonic endoderm differentiation of embryonic stem cells. *Biochem. Biophys. Res. Commun.*, 313, 475-481.

Yu, H. & Jove, R. (2004) The STATs of cancer--new molecular targets come of age. *Nat. Rev. Cancer*, 4, 97-105.

Yu, J.; Vodyanik, M.A.; Smuga-Otto, K.; Antosiewicz-Bourget, J.; Frane, J.L.; Tian, S.; Nie, J.; Jonsdottir, G.A.; Ruotti, V.; Stewart, R.; Slukvin, I.I. & Thomson, J.A. (2007) Induced pluripotent stem cell lines derived from human somatic cells. *Science*, 318, 1917-1920.

Yuri, S.; Fujimura, S.; Nimura, K.; Takeda, N.; Toyooka, Y.; Fujimura, Y.; Aburatani, H.; Ura, K.; Koseki, H.; Niwa, H. & Nishinakamura, R. Sall4 is essential for stabilization, but not for pluripotency, of embryonic stem cells by repressing aberrant trophectoderm gene expression. *Stem Cells*, 27, 796-805.

4

Effects of Recombinant Leukemia Inhibitory Factor (LIF) on Functional Status of Mouse Embryonic Stem Cells

Ludmila M. Mezhevikina[1], Raushaniya R. Petrova[1], Marina P. Borisova[2], Olga Yu. Seraya[2] and E.E. Fesenko[1]
[1]*Institute of Cell Biophysics Russian Academy of Sciences, Institutskaya 3, Pushchino, Moscow Region, 142290*
[2]*Institute of Theoretical and Experimental Biophysics, Russian Academy of Sciences, Institutskaya 3, Pushchino, Moscow Region, 142290*
Russia

1. Introduction

LIF (Leukemia Inhibitory Factor) protein isolated and characterized at early 1990 as a cytokine of the IL-6 family (Williams et al., 1988; Smith et al., 1988; Gearing et al., 1989) is now one of the most popular factors used for investigations with steam cells and embryos (Nagy et al., 1993; Thomson, Marshal, 1998; Metcalf, 2003; Gonzalez et al., 2004; Paling et al., 2004; Ratajaczak et al., 2009), an a potential therapy for oncology and reproductive medicine (Kurek, 2000; Cheng et al., 2001; Gunawardana et al., 2003; Wobus, Boheler, 2005; Guney et al., 2008; Novatny et al., 2009; Aghajanova, 2010). Its molecular structure (amphiphilic protein composed mainly from hydrophobic positively charged amino acids) and molecular mechanism of its action are known in details today (Haines et al., 1999, 2000; Auernhammer, Melmed, 2000; Heinrich et al., 2003; Giese et al., 2005). As to current knowledge, LIF binds with LIF-receptor (LIF-R) of a plasma membrane and with gp130 glycoprotein, a transmembrane transporter causing activation of JAK/STAT3, MAPK and PI3P paths of intercellular transduction of a trigger signal (Hirano et al., 1997; Raz et al., 1999; Auernhammer, Melmed, 2000; Cheng et al., 2001; Park et al., 2003; Paling et al., 2004; Gonzalez et al., 2004; Boniani, Scholer, 2005). Targets genes activated by these signal systems response for proliferation, apoptosis, cell cycles and differentiation.

Acting appropriate receptor complexes (LIF-R-gp130), LIF may cause various cell reactions. In mouse embryonic stem cells (ESC) it blocks processes of spontaneous differentiation and formation of embryoid bodies (EB, analogs of early mammal embryos) maintaining thus the stem cells in pluripotent state *in vitro* (Gearing et al., 1989; Raz et al., 1999; Nagy et al., 1993; Thomson, Marshal, 1998; Wobus, Boheler, 2005). High dependence of mouse ESC lines from LIF is evidenced by the fact that at a LIF-free culture medium the cells spontaneously differentiate into EB loosing thus their pluripotent properties. Main mechanisms of LIF effects, through cell membrane receptors, are comprehensively investigated (Haines et al., 1999, 2000; Heinrich et al., 2003; Gonzalez et al., 2004; Giese et al., 2005) and widely recognized; nevertheless dynamics of events at early stages of the protein interaction with ESC needs further clarifying. First, LIF is known to have three isoforms one of which (t-LIF) is not secreted but acts within a cell without contacting with outer-cell membrane receptors (Haines et al., 1999); second, mutant mouse ESC lines free of LIF-receptor complex proteins, LIFR and/or GP130, require presence of LIF in a culture medium (Ware et al., 1995; Li et al., 1995; Yoshida et al., 1996; Dani et al., 1998; Boniani, Scholer, 2005). These facts tell some other (not receptor based) mechanisms of LIF mouse ESC regulation.

Possibility of direct effects of recombinant LIF on properties of biological membranes is evidenced by our investigations with bilayer lipid membranes (Borisova et al., 2009), and membranes of intact cells highly dependent on the protein (Lobanok et al., 2008, 2009). We found that LIF acts as a membranotropic agent: it affects viscosity of membrane lipids, surface charge and conductivity of cell membranes. Basing these results, we supposed LIF-mediated changes of cytoplasm membranes structure and functions to be related with maintenance of mouse ESC proliferative activity and pluripotent properties *in vitro.*

2. Molecular mechanisms of LIF's effects on cell membranes

Possibility of non-receptor mechanism of stem cells LIF-regulation has been evidenced by studies with bilayer lipid membranes (BLM). We tried to answer the following questions: 1) what are effects of recombinant LIF proteins on a lipid bilayer and 2) whether mechanism of action varies according to a source of the protein origin (prokaryote or eukaryote). LIF is known to be produced in eukaryotic cells in a secrete form with 32-67 kDa molecular mass (Hilton et al., 1988; Haines et al., 2000; Heinrich et al., 2003); in bacterial cells recombinant LIF molecular masses are not higher than 20 kDa corresponding to theoretical estimations of LIF cDNA nucleotide sequence (Gearing et al., 1989; Hinds et al., 1998). Prokaryotic and eukaryotic LIFs have the same primary structures composed mainly by hydrophobic positively charged amino acids; differences between the recombinant proteins occur during post-translation events (Hinds et al., 1998; Hilton, Gough, 1998; Hirano et al., 2000).

Our studies have demonstrated that eukaryotic LIF(eu) and prokaryotic LIF(pro) effect lipids of cell membranes in similar manner shifting their surface charge towards positive values (fig. 1). Sorption of recombinant LIF on a lipid bilayer depends on its concentration and pH of an external solution. pH 7.0 enhances LIF effects on a surface potential (fig. 1 A). At pH 6.0, effects of the same level need higher concentrations of the protein (fig. 1 Б). From fig.1 one can see that at use of 10 ng/ml and higher LIF, the potential jumps become less pH-dependent evidencing surface-active properties of the protein. From these data it emerges that recombinant LIF in concentrations used routinely for ESC cultivation sorbs at a surface of a lipid bilayer resulting in increase of the membrane positive charge. Calcium ions affect cell membranes in a similar way.

Studies with BLM have shown that all used recombinant LIF proteins, regardless their origin, affect both surface potential (fig. 1) and conductivity of a lipid bilayer (fig. 2-4). A distinctive feature of LIF(pro) at its action on a lipid bilayer is spontaneous conductivity fluctuations without stationary state. At 100 mV, current through a lipid membrane changes chaotically and do not cause stable conductivity levels in a millisecond measurements range (fig.2, *a*); with S-shaped volt-ampere characteristics (fig. 2, *b*).

Chaotic and non-regular character of productivity increase at presence of LIF(pro) is levelled at addition of 10мM $CaCl_2$ into external solution. The observed changes of a lipid bilayer show dependence from presence of Ca^{2+}. At that, bilayer conductivity is decreased and short

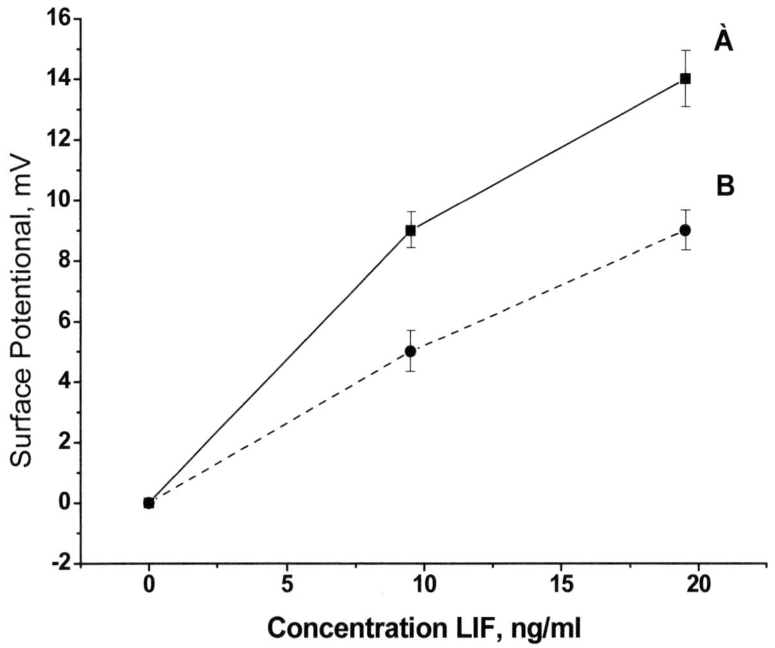

Fig. 1. Effect of mouse recombinant LIF protein from a prokaryotic expression system, LIF(pro), on a surface potential of a bilayer phosphatidylcholine membrane: A - pH 7.0, B - pH 6.0.

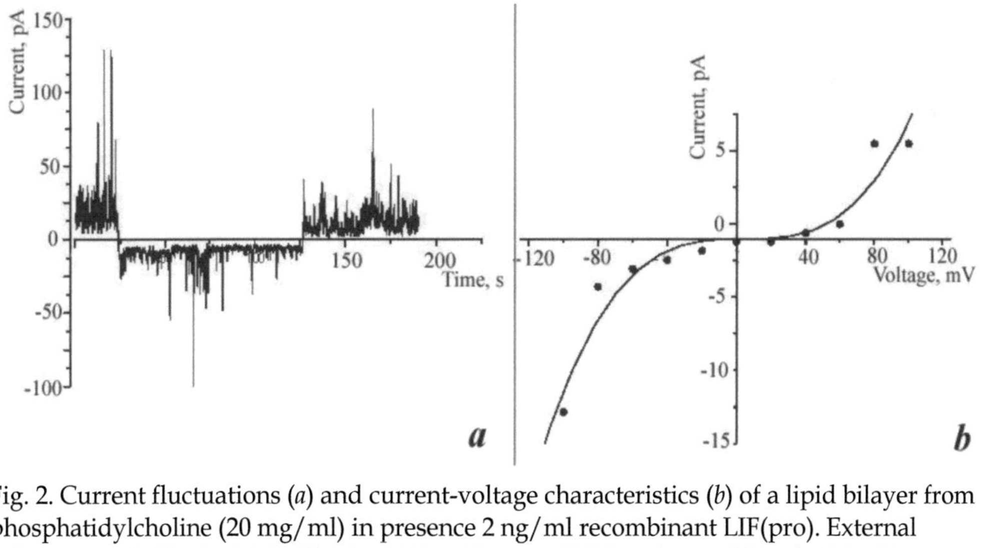

Fig. 2. Current fluctuations (*a*) and current-voltage characteristics (*b*) of a lipid bilayer from phosphatidylcholine (20 mg/ml) in presence 2 ng/ml recombinant LIF(pro). External solution: 1M KCl, 10 мM $CaCl_2$; 40 мM Tris-Hepes, pH 7.2; voltage 100 and -100 mV.

current disconnections vanish. No steady-state conductivity and moreover no relatively low rise of it could be approached at use of both LIF(eu) and LIF(pro), hence, effects of the proteins on a bilayer membrane may be evaluated only by quality characteristics of integral current change.

A similar pattern is registered at BLM-testing of a potentially therapeutic protein LIF(eu) isolated from *lif*-transfected Cos-1 cells (Petrova et al., 2006). But in this case current increases by jumps of varying heights; such changes are usually registered when current passes through an ion channel (fig.3 *a*, current jumps 13.6, 2.4, 3.2 and 3.6 pA are pointed with arrows). Heights of current jumps in presence of LIF(eu) depend on ion strength of a external solution: the less is salt concentration the higher is current passing through the current-conducting unit. For example, in 0.1M KCl at 100 mV impressed voltage its mean is 90±20 pA, in 0.2M KCl it is 22±6 pA and in 1M KCl it is 7±1 pA. A mean conductivity of a unit channel depends, besides, on a potential sign: at a negative value a mean amplitude of jumps is higher, than at a positive one (in 0.1M KCl - 175±19 pA and 1M KCl it is 90±20 pA versus 90 and 7 pA, respectively). Presented at fig. 3 *b* data show that in presence of LIF(eu) in 0.1M KCl solutions current depends linearly on voltage; in 0.2M KCl there is a hyperfunction under positive potential and overall current at 100 mV is almost twice greater than at -100 mV. In 1M KCl overall dependence of current has an opposite character but conductivity of the same channel also remains higher under a negative potential (fig. 4).

The presented data tell that recombinant LIF proteins, regardless of a producer (prokaryote or eukaryote), affect a surface potential and conductivity of a lipid bilayer. LIF(eu) has an additional activity: affecting a BLM, it forms stable current-conducting structures, ion channels. Characteristics of current through the channels depend on value and sign of a membrane potential, ionic strength of a surrounding solution, and on a lipid composition of the bilayer. Secreted sequences (20 amino acid residues) which are activated at posttranslational modifications of mature protein may be responsible for formation of the ion channels (Petrova et al., 2006). Bacterial producers have no these systems, so we suppose LIF(pro) to lack channel-forming possibilities. Participation of glycated LIF(eu) sites in forming of current-conducting structures cannot be exempted either, though their function is considered to be stabilisation of molecules and their protection of photolytic degradation (Hinds et al., 1998; Hilton, Gough, 1998; Heinrich et al., 1998).

Thus, electrophysiological studies show a possibility of direct effect of LIF recombinant molecules on lipid matrix of cell membranes involving not only LIF-receptor complex for pass of signals into cells. We suppose modification of cytoplasm membranes by ion-channels forming as one of the mechanisms of LIF regulatory action. This may play an important role in cell-cell interaction and metabolic cooperation of stem cells at their cultivation *in vitro*.

3. Comparison of biological activity of LIF(eu) and LIF(pro) with mouse ESC cultures

Recombinant LIF proteins from pro- and eukaryotic expression systems were tested on cells of R1 line, characterized by their dependence from feeder and LIF in a culture medium presence (Nagy et al., 1993). To prevent effects of feeder cells on ESC, R1 cells were cultivated on 0.1% gelatine cover. This way of cultivation resulted no reliable differences in performances of LIF(pro) and LIF(eu) effects on growth and renovation of R1 line cells were registered (table 1, fig. 5). During 48 h both proteins in 10-13 ng/ml concentrations maintained pluripotent properties of the cells and their relatively high rate of growth in

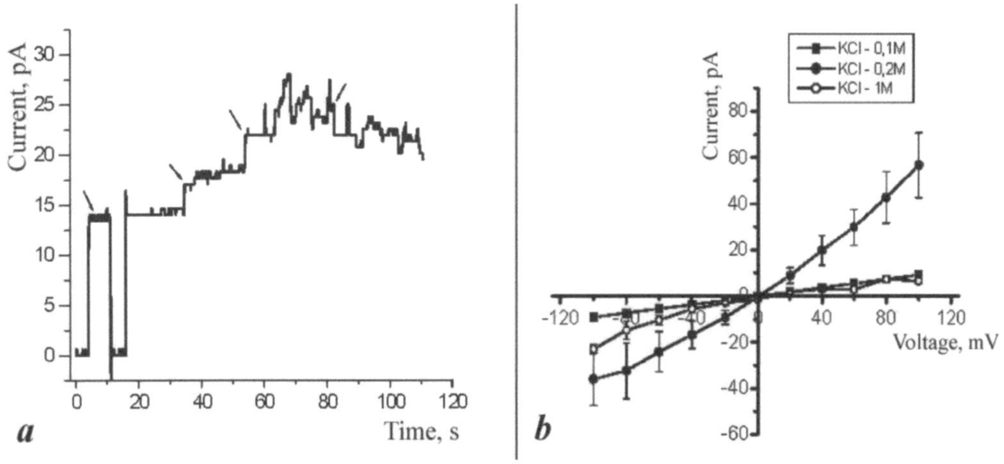

Fig. 3. Current jumps (0.1M KCl; 100 mV) (*a*) and integral current-voltage characteristics at different ionic strengths (*b*); 6.5 ng/ml LIF(eu); phosphatidilcholine membrane.

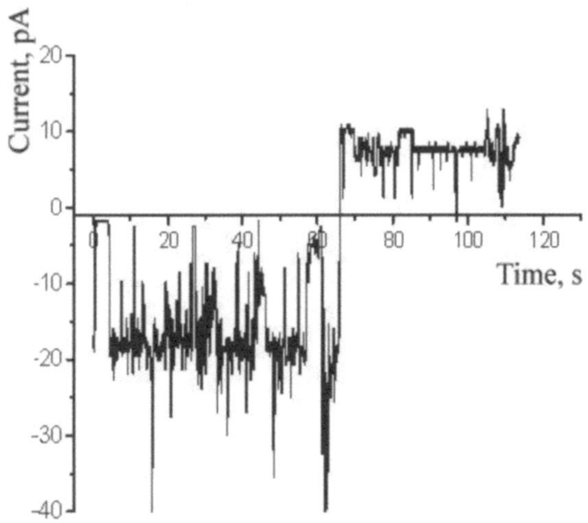

Fig. 4. Current fluctuations at -100 and at 100 mV; phosphatidilcholine membrane (20 mg/ml); 1M KCl; and 19.5 ng/ml LIF(eu).

form of colonies without morphological signs of differentiation and EB formation (table 1, fig. 6 *c*). Correlation between differentiated and non-differentiated (pluripotent) colonies with high endogenous alkaline phosphatase (AP) activity was kept at a high level and did not depend on a source of LIF origin (fig. 5). At the absence of LIF: cell growth was inhibited; time of populations doubling increased (table 1) and EB were found in the medium pointing thus prevalence of embryonic differentiation processes over proliferation ones (fig. 5, fig. 6, *d*).

Culture medium	cell numbers (x10^5)		growth rate	ESC population doubling time (h)
	48 h	72 h		
LIF(eu)	2.0±0.52	3.4±0.85*	4.0±0.31*	13.6±2.4
LIF(pro)	2.1±0.40	2.1±0.43	3.7±0.25*	13.2±2.7
Control (LIF-free)	1.1±0.20	1.2±0.80	2.4±0.17	22.2±6.3**

Table 1. Comparative proliferative activity of R1 line mouse embryonic stem cells cultivated on 0.1 gelatine in presence of recombinant LIF(eu) and LIF(pro). Notice. Final concentration of LIF proteins in medium: 10-13 ng/ml. * - differences confidence (P≤0.05); **- differences confidence by time of cells populations doubling in LIF-supplemented and LIF-free media (P≤0.01).

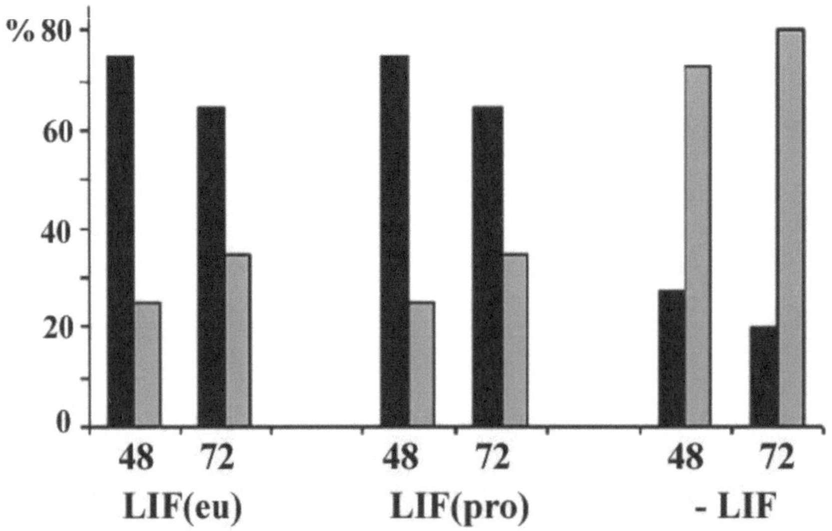

Fig. 5. Percentage of differentiated (▢) and non-differentiated (■) colonies of R1 line embryonic cells cultivated on 0.1% gelatine in media with recombinant LIF proteins (10-13 ng/ml).

Regulatory functions of LIF are apparently dose dependent (fig. 7); this was demonstrated by our experiments with techniques of R1 line cells cultivation on a feeder from mouse primary embryonic fibroblasts (PEF) and media supplemented with recombinant LIF protein in various concentrations. These cultivation conditions enhance the general tendency of pluripotent cells: development by colonies without visible signs of spontaneous differentiation and formation of EB. It is well known, that cells of mouse PEF can produce their own LIF (Williams et al., 1988) and can additionally saturate a nutrient medium with this factor. But in spite of such a possibility, use of recombinant LIF in small concentrations (2.5-5.0 ng/ml) and of feeder cultivation does not provide a pluripotent status of R1 line cells in culture completely (fig. 7). To 48 h number of damaged cells in populations with low

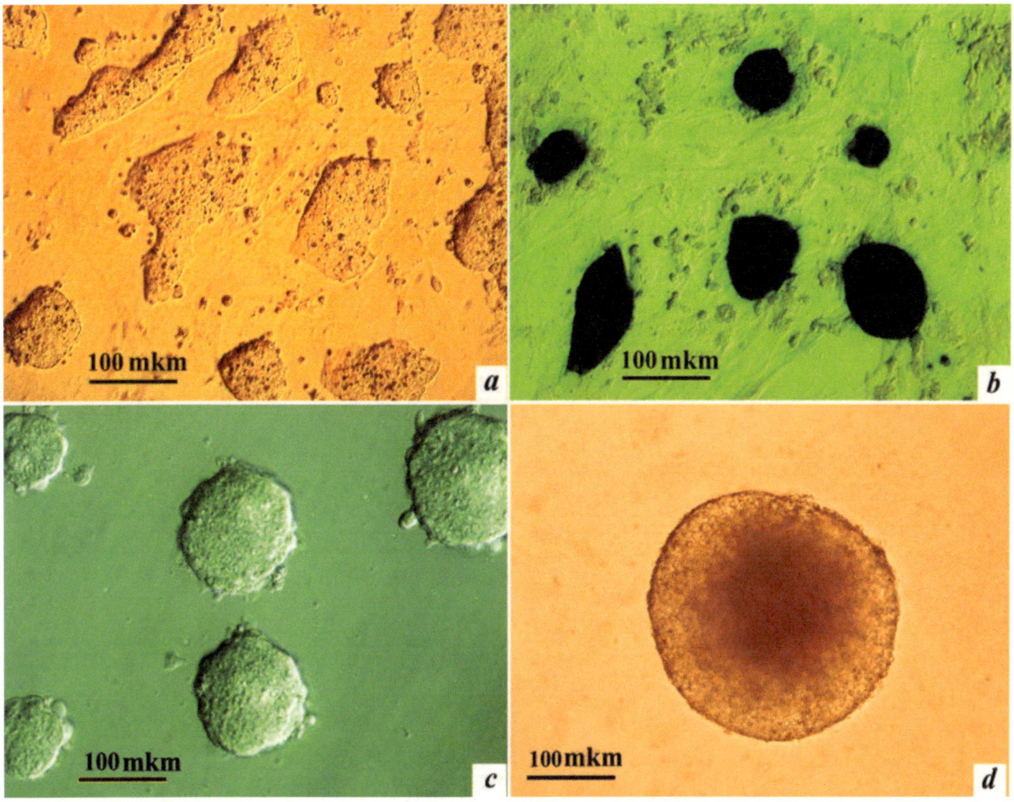

Fig. 6. Morphology of colonies of R1 line embryonic stem cells in a medium with recombinant LIF(pro) after 48 h cultivation on PEF feeder (a) and on 0.1% gelatine (*c*); detecting of alkaline phosphatase activity (b), embryoid body in a LIF-free medium (d).

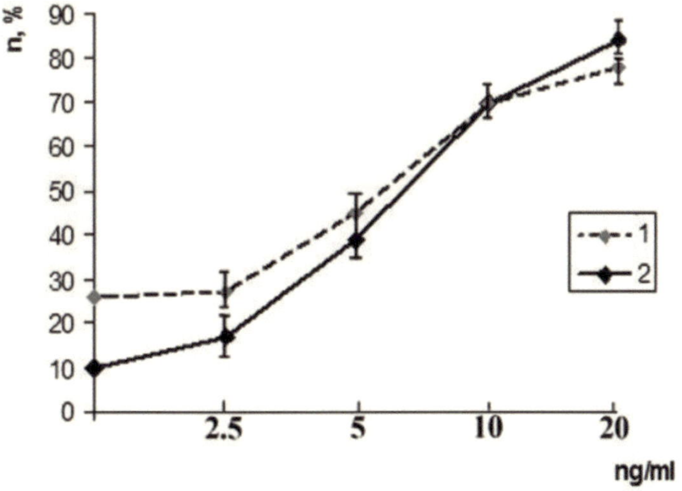

Fig. 7. Dependence of formation of non-differentiated pluripotent colonies of R1 embryonic stem cells from LIF(pro) concentration in 24 (1) and 48 h (2) of PEF feeder cultivation.

content of exogenous LIF protein is increased; rate of cell divisions is inhibited and EB occur (table 2). We registered such a situation at growing of R1 cells on a gelatine support in a LIF-free medium (table 1, fig. 5, 6 д).

LIF concentration (ng/ml)	Cell number (x10^5)		Growth rate	Population doubling time (h)	N/D (%)
	48 h	72 h			
5	19.7±1.2	27.7±1.4	3.9±0.2	12.2±0.9	42.75
10	21.3±0.7*	34.0±1.1*	4.3±0.2	10.0±0,5**	67.45
20	34.3±1.8*	32.3±1.5*	6.9±0.3*	7.0±0.7**	82.90
Control (LIF-free)	15.6±3.1	19.0±8.0	3.1±0.6	15.0±1.8	33.70

Table 2. Proliferative activity and growth of R1 embryonic stem cells colonies in dependence on recombinant LIF(pro) concentration at cultivation on PEF feeder. Notice. N/D - non-differentiated pluripotent colonies of embryonic stem cells with high activity of endogenous alkaline phosphatase; * - reliable increase of proliferative activity and rate of colonies growth under effect of 10-20 ng/ml of LIF ($P \leq 0.05$); ** - reliability of differences of cell populations doubling time in media with 10-20 ng/ml LIF and without LIF ($P \leq 0.01$).

Stem cells of R1 line require relatively high concentrations of recombinant LIF in a culture medium (10-20 ng/ml) that decreases their doubling times up to 10 hours in average and increases rate of their growth 1.5-2.0 folds relatively cells developing in a LIF-free medium (table 1, 2). At different conditions of R1 cultivation (gelatine or PEF feeder), in the presence of LIF in optimal concentrations, approximately the same share (70-80%) of pluripotent colonies can be registered (fig. 5, table 2). Hence, we may postulate that LIF(pro) and LIF(eu) are equally effective in their support of ESC pluripotency and high proliferative activity *in vitro*. Our observations tell that only R1 stem cells with active proliferation keep their pluripotent potential in the presence recombinant LIF.

4. Effects on cell death (apoptosis) and cell cycles

When cell differentiation is difficult, correlation of two forms of stem cells response, proliferation/cell death, may serve as an essential parameter of their reaction on effects of regulatory LIF protein and reflect self-renewal potential of population. For example, our investigations showed that 5-10 ng/ml concentrations of LIF decrease share of apoptotic cells in R1 population to 2.0% in comparison with 7.2 % following 18 hours of incubation in LIF-free medium (Lobanok et al., 2008, 2009). At 5 ng/ml concentration LIF is more effective as anti-apoptotic agent than at 10 ng/ml. It may be caused by that apoptosis is an active form of cell reaction not only on unfavourable conditions, like deficiency of growth factors in cultivation medium but on their physiological concentrations activating cell growth. In case of R1 ESC this concentration is 10-20 ng/ml LIF (table 1 and 2, fig. 7).

Another essential characteristic of mouse ESC response on effect of LIF regulatory molecules reflecting potential of the cells for renovation and reproduction is distribution of cells by phases of cell cycle. Presented histograms: G1 (content of 2C DNA), S (from 2 to 4C) and

G2+M (4C) tell that if presence 5 and 10 ng/ml of recombinant LIF(pro) presents in a medium, share of R1 cells in S-phase of cell-cycle decreases (fig. 8). A specific feature of mouse ESC lines worth mentioning: their cells spend the most part of their life-time in S-phase, during which DNA synthesis occurs. Contrary to somatic cells of tissue culture, ESC do not need external stimuli to initiate DNA replication processes (Smith, 2001). It can explain why ESC are hardly moved out from cell cycle into G1 or G0 stages.

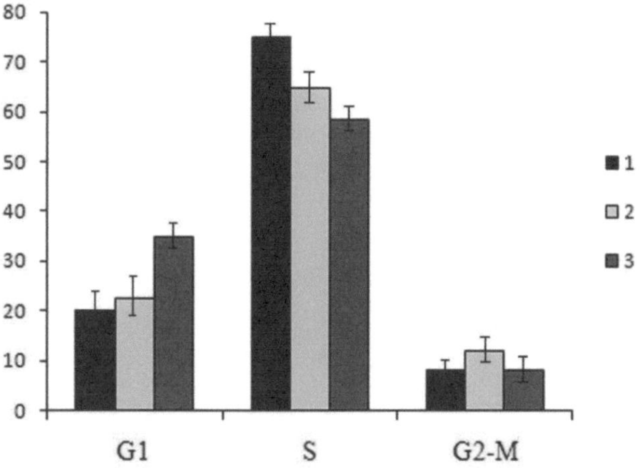

Fig. 8. Effect of recombinant LIF(pro) concentration on distribution of R1 mouse embryonic stem cells by phases of a cell cycle after 18 h cultivation on PEF feeder. 1) Control (LIF-free), 2) 5 and 3) 10 ng/ml LIF in culture medium.

Nevertheless, from fig. 8 one can see that LIF in 10 ng/ml concentration, optimal for ESC, activates these processes. In growing populations of R1 after 18 hours of incubation with 10 ng/ml LIF(pro) share of stem cells in G1 phase increases in comparison with cells that did not get the protein or got it in a smaller concentration (fig. 8). Increase of G1-phase cells number may tell about their spontaneous differentiation into EB and about enhanced production of D-cyclins inhibiting cell division (Savatier et al., 1996). But we must keep in mind that too short G1 period is typical for ESC and therefore mechanisms of their activation may differ somewhat from those of somatic cells (Savatier et al., 1996; Rohwedel et al., 1996; Wianny et al., 1998). For example, some part of mouse ESC keep their ability to differentiate into EB and to grow by colonies in LIF-free medium, but rates of their division become shorter in comparison with growing with LIF (table 1 and 2, fig. 5).

ESC, being cultivated, may stop at one of the cell cycle stages, therefore increase of their number in S synthetic phase does not always reflect proliferative activity of population in whole; this is exemplified by fig. 8. In this connection, we calculated rate of cells in S phase share to G_2+M as S/(G_2+M). The calculations shows that after 18 hours of incubation in a LIF-free medium the most part of R1 cells (more than 70 %) are in S phase and S/(G_2+M) equation reflecting proliferative potential of a culture equals to 10.9. Affected by 5 ng/ml of LIF, the parameter decreases to 4.7, whereas 10 ng/ml decreases it only to 8.3. It may be provided by the following: in presence of 10 ng/ml LIF, mouse ESC pass an S-phase more rapidly and about 10 % of cells stop at a G1 restriction point (fig. 8). Small doses of LIF effects exit of cells from S-phase, increasing thus a proliferative index of a population and share of cells with DNA content ≥2C.

5. Conclusion

Thus comparative analysis of recombinant mouse LIF from pro- and eukaryotic expression systems made in the investigation with bilayer lipid membranes and cultures of ESC (R1 line) showed that both protein types, LIF(pro) and LIF(eu), have membranotropic effects. Independently on an expression system, molecules of recombinant LIF are incorporated into lipid bilayer, causing increase of surface potential and membrane electric conductivity. LIF-induced changes of lipid matrix of cell membranes may play a key role in maintenance of pluripotent properties of mouse ESC *in vitro*. We suppose binary mechanism of LIF actions on both receptors (LIF-R and gp130) and lipid matrix of cell membranes to underlie its functional redundancy. For example, our investigations show that recombinant mouse LIF affects ESC proliferation, apoptosis and cells distribution by cell cycle phases (decreases $S/(G_2+M)$ ratio of a cell cycle and time of ESC population doubling).

6. References

Aghajanova, L. (2010). Update on the role leukemia inhibitory factor in assisted reproduction. *Curr. Opin Obstet Gynecol.*, Vol. 22, No 3, 213-219.

Auernhammer, C.J. & Melmed, S. (2000). Leukemia-inhibitory factor - neuroimmune modulator of endocrine function. *Endocrine Reviews*, Vol. 21, No 3, 313-345.

Boiani, M., & Scholer, H.R. (2005). Regulatory networks in embryo-derived pluripotent stem cells. *Molecular Cell Biology*, Vol. 6, 872-884.

Borisova, M.P., Mezhevikina L.M., Petrova, R.R. & Fesenko, E.E. (2009). Action of Leukemia Inhibitory Factor on a lipid bilayer. *Biophisics*, Vol. 54, 486-489.

Cheng, J.G., Chen, J.R., Hernandez L., Alvord, W.C. & Stewart, C.L. (2001). Dual control LIF expression and LIF receptor function regulate Stat3 activation at the onset of uterine receptivity and embryo implantation. *Proc. Natl. Acad. Sci.* USA, Vol. 98, 8680-8685.

Dani, C., Chambers, I., Johnston, S., Robertson, M., Ebrahimi, B., Saito, M., Taga, T., Li M., Burdon, T., Nichols, J. & Smith, A. (1998). Paracrine induction of stem sell renewal by LIF-deficient cells: a new ES cell regulatory pathway. *Develop. Biol.*, Vol. 203, No 1, 149-162.

Gearing, D., Nicola, N.A., Metcalf, D., Foote, S., Gough, N.M., & Williams, R.L. (1989). Production of leukemia inhibitory factor (LIF) in Escherichia coli and its use in embryonic stem (ES) cell culture. *Biotechnology*, vol. 7, 1157-1161.

Giese, B., Roderburg, C., Sommerauer, M. et al. (2005). Demonization of the cytokine receptors gp130 and LIFR analyzed in single cells. *J. Cell Sci.*, Vol. 118, 5129-5140.

Gonzalez, R.R., Rueda, B.R., Ramos, M.P., Littell, R.D., Glasser, S., & Leavis, P.C. (2004). Leptin-induced increase in Leukemia Inhibitory Factor and its Receptor by human endometrium is partially mediated by interleukin 1 receptor signalling. *Endocrinology*, Vol. 145, № 8, 3850-3857.

Gunawardana, D.H., Basser, R.L., Davis, I.D., Cebon, J., Mitchell P., Underhill, C., Kilpatrick, T.J. et al. (2003). A Phase I Study of Recombinant Human Leukemia Inhibitory Factor in Patients with Advanced Cancer. *Clinical Cancer Research*, Vol. 9, 2056-2065.

Guney, M., Erdemoglu, E, Oral, B., Karahan, N., & Mungan, T. (2008). Leukemia inhibitory factor (LIF) is immunohistochemically localized in tubal ectopic pregnancy. *Acta Histochem.*, Vol. 110, No 4, 319-323.

Haines B.P., Voyle, R.B., Peiton, T.A., Forrest, R., & Rathjen, P.D. (1999). Complex conserved organization of the mammalian LIF gene: a novel mechanism for regulated expression of intracellular and extracellular cytokines. *J. Immunol.*, Vol. 162, 4637-4646.

Haines, B.P., Voyle, R.B. & Rathjen, P.D. (2000). Intracellular and extracellular Leukemia inhibitory factor proteins have different cellular activities that are mediated by distinct protein motifs. *Mol. Biol. Cell*, Vol. 11, No 4, 1369-1383.

Heinrich, P.C., Behrmann, I., Haan, S., Hermanns, H.M., Multer-Newen, G. & Schaper, F. (2003). Principles of interleukin (IL)-6-type cytokine signalling and its regulation. *Biochem. J.*, Vol. 374, 1-20.

Hilton, D.J., Nicola, N.A. & Metcalf, D. (1988). Purification of a murine leukemia inhibitory factor from Krebs ascites conditioned cells. *Analytical Biochemistry*, Vol. 173, 359-367.

Hilton D.J., & Gough N.M. (1998). Leukemia inhibitory factor, In: *Cytokines,* A. Mire-Sluis & R. Thorpe (eds.), Academic Press Lim., London, 277-296.

Hinds, M.G., Maurer, T., Zhang, J.-G., Nicola, N.A., & Norton, R.S. (1998). Solution structure of leukemia inhibitory factor. *J. Biol. Chem.*, Vol. 273, No 22, 1373-1378.

Hirano, T., Nakajiama, K. & Hibi, M. (1997). Signalling mechanisms through gp130: A model of the cytokine system. *Cytokine Growth Factor Rev.*, Vol. 8, 241-252.

Hirano, T., Ishihara, K. & Hibi, M. (2000). Roles STAT3 in mediating the cell growth, differentiation and survival signals relayed through the IL-6 family of cytokine receptor. *Oncogene*, Vol. 19, 2548-2556.

Kurek, J. (2000). AM424: history of a novel drug candidate. *Clin. Exp. Pharmacol. Physiol.*, Vol. 27, 553-557.

Li, M., Sendtner, M., & Smith, A. (1995). Essential function of LIF receptor in motor neurons. *Nature*, vol. 378, 724-727.

Lobanok, E.S., Belyanovich, L.M., Mezhevikina, L.M., Volotovsky, I.D., & Fesenko, E.E. (2008). Effect of the stem cell factor on the morphology and functional state of mouse embryonic stem cells. *Biofizika* (Russia), Vol. 53, No 4, 646-651.

Lobanok, E.S., Mezhevikina, L.M., Belyanovich, L.M., Petrova, R.R., Vasilevich, I.B., Volotovsky, I.D., & Fesenko, E.E. (2009). The influence of LIF (Leukemia Inhibitory Factor) on the functional status of R1 mouse embryonic stem cells. *Biochemistry* (suppl. Series B: Biomedical Chemistry), Vol. 3, No 1, 54-57.

Metcalf, D. (2003). The unsolved enigmas of leukemia inhibitory factor. *Stem Cells*, Vol. 21, No 1, 5-14.

Nagy, A., Rossant, J., Nagy, R., Abramow-Newerly, W., & Roder, J.C. (1993). Derivation of completely cell culture-derived mice from early passage embryonic stem cells. *Proc. Natl. Acad. Sci.* USA, Vol. 90, No 18, 8424-8428.

Novatny, Z., Krizan, J., Sima, R., Sima, P., Uher, P., Zech, N. et al. (2009). Diagnosed with unexplained infertility and endometriosis have a negative impact on the IVF outcome a pilot study. *Folia Biol.* (Praha), vol. 55, 92-97.

Paling, N. R. D., Wheadon, H., Bone, H.K. & Welham, M. J. (2004). Regulation of embryonic stem cell self-renewal byphosphoinositides 3-Kinase-dependent signalling. *J. Biol. Chem.*, Vol. 279, No 46, 48063–48070.

Park, J.I., Strock, C.J., Ball, D.W., & Nelkin, B.D. (2003). The Ras/Raf/Mek/Extracellular signal-regulated kinase pathway induces autocrine-paracrine growth inhibition via the Leukemia Inhibitory Factor/JAK/Stat pathway. *Mol. Cell. Biol* (Jap.), Vol. 23, No 2, 543-554.

Petrova, R.R., Mezhevikina, L.M., & Kaloshin, A.A. (2006). An expression system for production of a recombinant protein with the properties of cytokine LIF in Cos-1 cells. *Biotechnology (Russia)*, Vol., 5, 3-11.

Ratajaczak, M.Z., Shin, D.-M. & Kucia, M. (2009). Very small embryonic/epiblast-like stem cells. A missing link to support the germ line hypothesis of cancer development? *Am. J. Pathology*, Vol. 174, No 6, 1985-1992.

Raz, R., Lee, C.-K., Cannizzaro, L.A., d'Eustachio, P., & Levy, D.E. (1999). Essential role of STAT3 for embryonic stem cell pluripotency. *Proc. Natl. Acad. Sci.* USA, Vol. 96, No 6, 2846-2851.

Rohwedel, J., Sehlmeyer, U., Shan, J. et al. (1996). Primordial germ cell-derived mouse embryonic germ (EG) cells in vitro resemble undifferentiated stem cells with respect to differentiation capacity and cell cycle distribution. *Cell Biol. Int.*, Vol. 20, 579-587.

Savatier, P., Lapillonne, H., van Grunsven, L.A., Rudkin, B.B., & Samarut, J. (1996). Withdrawal of differentiation inhibitory activity leukemia inhibitory factor up-regulates D-type cyclins and cyclin-dependent kinase inhibitors in mouse embryonic stem cells. *Oncogene*, Vol. 12, 309-322.

Smith, A.G., Heath, J.K., Donaldson, D.D., Wong, G.G., Morean, J., Stahl, M. & Rogers, D. (1988). Inhibition of pluripotental embryonic stem cell differentiation by purified polypeptides. *Nature*, Vol. 336, 688-690.

Thomson, J.A. & Marshal, V.S. (1998). Primate embryonic stem cells. Current Topics in Developmental Biology, Vol. 38, 133-165.

Ware, C.B., Horowitz, M.C., Renshaw, B.R. et al. (1995). Targeted disruption of the low-affinity leukemia inhibitory factor receptor gene causes placental, skeletal, neural and metabolic defects and results in perinatal death. *Development*, Vol. 121, 1283-1299.

Wianny, F., Real, F.X., Mummery, C.L. et al. (1998). G-1 phase regulators. Cyclin D1, Cyclin D2, Cyclin D3: up-regulation at gastrulation and dynamic expression during neurolation. Dev. Dyn., Vol. 212, 49-62.

Williams, R.L., Hilton, D.J., Pease, S., Willson, T.A., Stewart, C.L., Gering, D.P., Wagner, E.F., Metcalf, D., Nicola, N.A., &

Gough, N.M. (1988). Myeloid leukaemia inhibitory factor maintains the development potential of embryonic stem cells. *Nature*, Vol. 336, 684-687.

Wobus, A.M., & Boheler, K.R. (2005). Embryonic Stem Cells: Prospects for Developmental Biology and Cell Therapy. *Physiol. Rev.*, Vol. 85, 635-678.

Yoshida, K., Taga, T., Saito, M. et al. (1996). Targeted disruption of gp130, a common signal transducer for the interleukin 6 family of cytokines, leads to myocardial and haematological disorders. *Proc. Natl. Acad. Sci.* USA, Vol. 93, 407-411.

5

Maintaining Embryonic Stem Cells and Induced Pluripotent Stem Cells

Qiang-Sun Zheng[1], Dong-Bo Ou[1], Xiong-Tao Liu and Jing-Jing Guo
Department of Cardiology, Tangdu Hospital, Fourth Military Medical University, Xi' an China

1. Introduction

Early in 1981, the pluripotential cells were first established from mouse embryos by Evans et al (Evans and Kaufman, 1981). In the same year, Martin et al (Martin, 1981) named these pluripotential cells embryonic stem (ES) cells. In 1990s, human ES cells were first established by Thomson et al (Thomson et al., 1998). Such ES cells isolated from inner cell masses of blastocysts present the unique property of self-renewal and the ability to generate differentiated progeny in all embryonic lineages both in vitro and in vivo. The pluripotency of these ES cells was demonstrated conclusively by the observation that subclonal cultures, derived from isolated single cells, can differentiate into a wide variety of cell types including gut epithelium (endoderm); cartilage, bone, smooth muscle, and striated muscle (mesoderm); and neural epithelium, embryonic ganglia, and stratified squamous epithelium (ectoderm). Due to these unique properties, ES cells may become an exceptional source of tissues for transplantation and have great potential for the therapy of incurable diseases.

2. The derivation of embryonic stem (ES) cells and induced pluripotent stem (iPS) cells

Recently, Laurent et al. (Laurent et al., 2010) determined the ethnic origins of the 47 commonly used human ES cell lines by genome-wide SNP genotyping and Bayesian analysis of population structure and found that the large majority of human ES cell lines (43 of 47) were of European and East Asian ethnicity. There was a notable lack of cell lines representing African ethnicity. Mosher and colleagues (Mosher et al., 2010) also described the lack of population diversity in widely distributed human ES cell lines and suggested deriving and disseminating new ES cell lines based on underrepresented populations or diverse donors to increase the ethnic diversity in human ES cell lines. Stem cell lines with a greater ethnic genetic diversity must be developed to optimize the use of such cells as research tools and in future therapies.
To increase the diversity of human ES cell lines, however, more emphasis should be put on the protocols involved in technique system, such as those for derivation, propagation, and long-term potency maintenance of human ES cells, which have still to be improved. Blastocyst-stage embryos donated for research after assisted reproductive techniques were used for new ES cell isolation. However, ethical or technical limitations restrict the research projects in derivation of new human ES cell lines. Establishment of human ES cells from

---

[1] The first 2 authors contributed equally to this work.

discarded poor-quality embryos (Lerou et al., 2008) (Chen et al., 2009) minimizes the ethical problem but requires further technical improvement and financial support. Defining protocols to derive and propagate high quality human ES cells from embryos should be well developed and promoted, especially in underdeveloped African countries.

Because of technical complications and ethical controversies of establishing human ES cell lines from embryos, the somatic cell in vitro reprogramming approach has become the most efficient and practical way to produce large banks of pluripotent cells. Recent breakthrough studies using a combination of four factors to reprogram somatic cells into induced pluripotent stem (iPS) cells without using embryos or eggs have led to an important revolution in stem cell research (Fig.1). Comparative analysis of human iPS cells and human ES cells using assays for morphology, cell surface marker expression, gene expression profiling, epigenetic status, and differentiation potential has revealed a remarkable degree of similarity between these two pluripotent stem cell types. These advances in reprogramming will enable the creation of patient-specific stem cell lines to study various disease mechanisms, offer valuable tools for drug discovery, and provide great potential to design customized patient-specific stem cell therapies with economic feasibility.

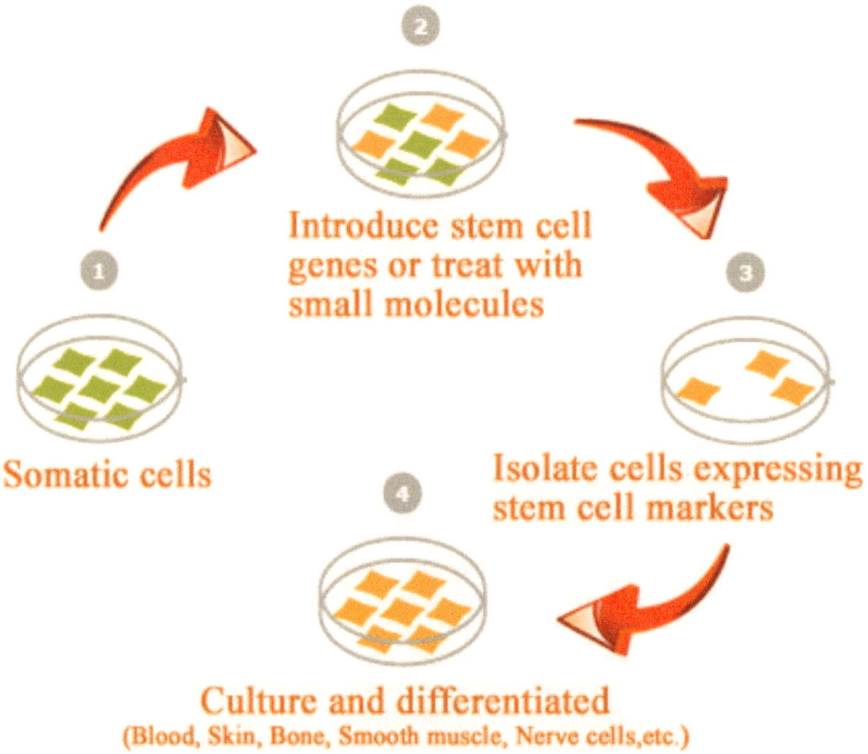

Fig. 1. Reprogramming somatic cells into induced pluripotent stem (iPS) cells.

Patient-specific pluripotent cell lines may provide a limitless source for human cell therapeutic application. However, although human ES cells and human iPS cells have been shown to share a number of similarities, there are still differences electrophysiology properties between human ES cells and human iPS cells (Jiang et al., 2010). It has been showed that foreign genes were silenced or removed after reprogramming, but those approaches have low reprogramming efficiency, and either leave residual vector sequences, or require tedious steps. Whether reprogramming methods can be improved will depend on a better understanding on the molecular cell biology of pluripotent stem cells.

3. The cultivation of ES cells and iPS cells

ES cells were initially established and maintained by coculture with murine embryonic fibroblast (MEF) feeder cells (Evans and Kaufman, 1981). Subsequent studies identified that fibroblasts secrete multiple factors, including Leukemia inhibitory factor (LIF), fibroblast growth factors (FGFs), transforming growth factor b (TGFb), Activin, Wnts, insulin-like growth factor (IGF), and antagonists of BMP signaling. ES cells are normally derived and maintained in media containing these factors in combination. In stem cell cultures LIF is the essential media supplement for the maintenance of pluripotency of ES and iPS cells.

Practically, ES cells were cultured in Dulbecco's modified Eagle's minimal essential medium (DMEM) supplemented with 10~15% refined fetal calf serum, 0.1 mM nonessential amino acids, 0.1 mM β-mercaptoethanol, and 100U/mL leukemia inhibitory factor (LIF). Generally, pluripotent embryonic cells require a co-culture environment for their self-renewal in monolayer expansion, achieved by culturing on a layer of feeder cells. Mouse embryonic fibroblast cells obtained from 13.5-day embryos of mice were treated for 3 hours with mitomycin C (10μg/mL) as feeders for mouse or human ES cells and iPs cells (Fig.2). In our study, we have developed and validated a feeder-free culture model for ES cells propagation maintaining their pluripotency (Fig.3).

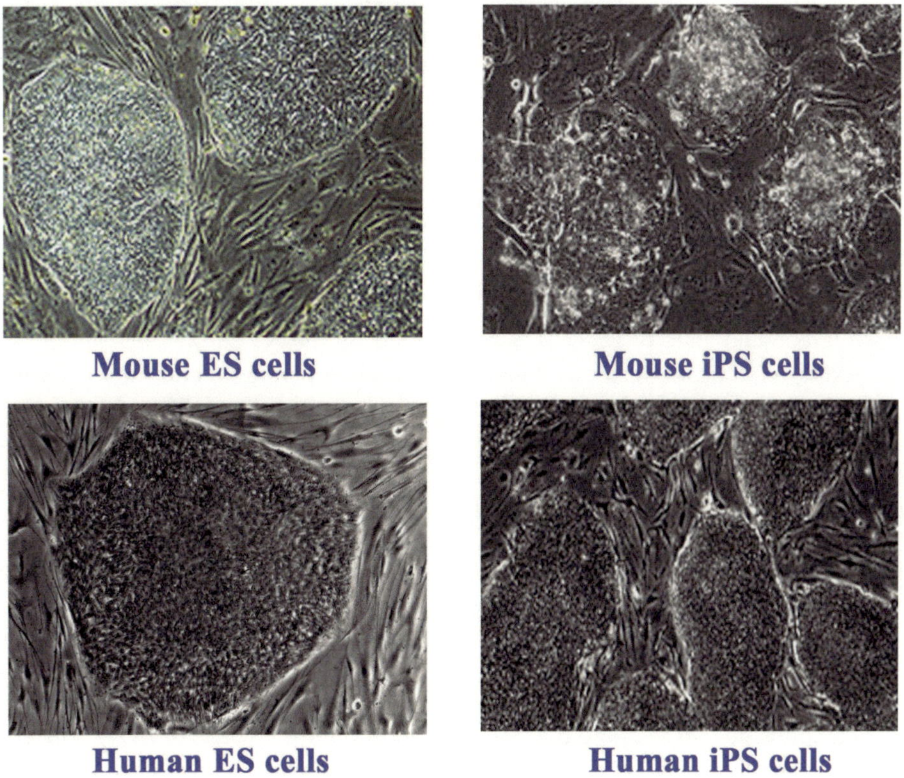

Fig. 2. The culture of ES cells and iPS cell with feeder cells.

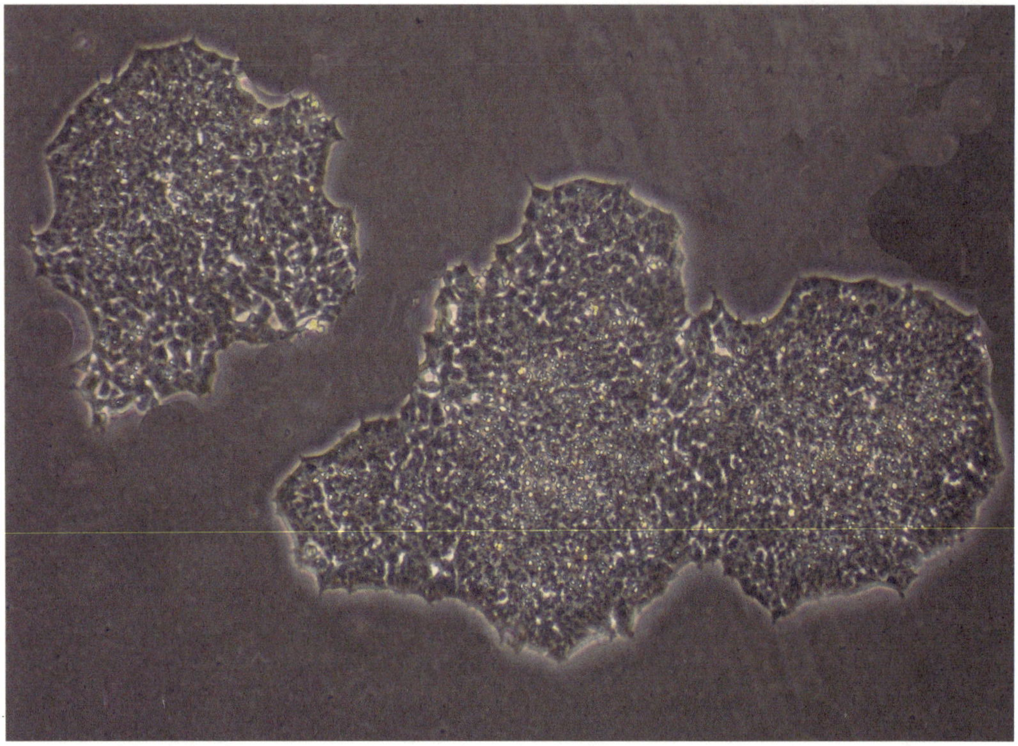

Fig. 3. The culture of mouse ES cells without feeder cells.

The ES cells grew in culture flask precoated with 0.1% gelatin and were maintained at less than 60% confluency to keep an undifferentiated phenotype. Once the cells were to reach 60% confluency, passage of cell was conducted at a 1:8 subculture ratio.

4. The molecular mechanisms and signaling pathways leading to maintain ES cells, the differences between mouse ES cells and human ES cells

The undifferentiated state of ES cells is maintained by the action of transcription factors, some of which are mouse specific and some are common to human and mouse. Oct4, Sox2, and Nanog are master transcription factors for maintenance of the undifferentiated state and self-renewal of ES cells. Regulatory mechanisms pertaining to the self-renewal of stem cells remain incompletely understood.

Signaling in stem cells maintenance includes LIF/Stat3 signaling, Wnt/b-catenin signaling, BMP signaling, and FGF signaling. The maintenance of mouse ES cells is synergistic signaling process. We have investigated the synergistic effect of retinol and leukemia inhibitory factor (LIF) on maintaining pluripotency of mouse ES cells and found that retinol showed a synergistic effect in maintaining pluripotency of mouse ES cells when combined with LIF in moderate concentration and the effect may be attributable to the over-expression of Nanog under retinol stimulation (Fig.4).

The key components that regulate the self-renewal of mouse ES cells have been deciphered and they are largely dependent on two key signaling pathways involving LIF and BMP signaling (Niwa et al., 1998) (Ying et al., 2003). However, human ES cells have significant differences from mouse ES cells such as variations in the stage-specific antigens and in the ability of leukemia inhibitory factor (LIF) to maintain the undifferentiated state. In human ES cells, LIF receptors are expressed, and LIF can stimulate activation of Stat3 under experiment condition, but this pathway is not activated in the undifferentiated state, suggesting that the maintenance of human ES cells is Stat3 independent (Humphrey et al., 2004). The factors involved in human ES cells self-renewal still have not been elucidated, although significant progress has been made in recent years.

In contrast to mouse ES cells, human ES cells can induce trophoblast differentiation by BMP4 (Xu et al., 2005). The effect may at least partly owing to Smad 1/5/8 activation moderately represses Sox2 (Greber et al., 2008). Activin A, another TGFβ family member, is necessary and sufficient for the maintenance of self-renewal and pluripotency of human ES cells. It can induce the expression of Oct4, Nanog, Nodal, Wnt3, bFGF, and FGF8, and suppresses the BMP signal, support long-term growth of human ES cells on Matrigel coated flasks without either feeder cells or conditioned medium (Xiao et al., 2006).

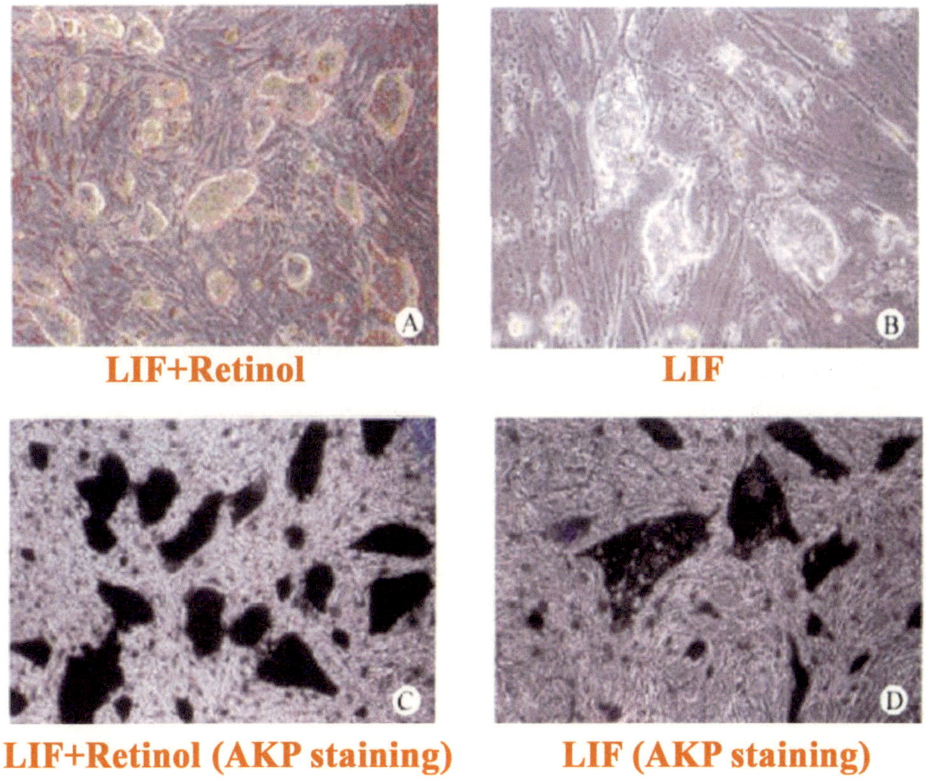

Fig. 4. Morphological analysis and alkaline phosphatase (AKP) assay of ES cell-S19 cultured for 14 days.

5. Support self-renewal and maintain potency of ES cells and iPS cells

Conventionally, pluripotent embryonic cells require a co-culture environment for their self-renewal in monolayer expansion, which is achieved by culturing on a layer of feeder cells. Mouse embryonic fibroblast (MEF) is used as typical feeders for ES and iPS cells. Many studies have explored several alternative cell sources as feeders to support human ES cells culture in monolayer and thereby limiting cross-species contaminations, which includes human embryo derived fibroblasts, foreskin fibroblasts, adult bone marrow cells, and visceral-endoderm (VE)-like cells (Amit et al., 2003) (Hovatta et al., 2003) (Cheng et al., 2003) (Richards et al., 2002) (Mummery et al., 2003).

A microporous poly (ethylene terephthalate) membrane-based indirect co-culture system for human pluripotent stem cells propagation in prolonged culture, which allows real-time conditioning of the culture medium with human fibroblasts while maintaining the complete separation of the two cell types, have developed and validated. This co-culture system is a significant advance in human pluripotent stem cells culture methods, providing a facile stem cell expansion system with continuous medium conditioning while preventing mixing of human pluripotent stem cells and feeder cells (Abraham et al., 2010).

Optimization and development of better defined culture methods for human ES and iPS cells will provide an invaluable contribution to the field of regenerative medicine. Recently, extracellular matrix supporting more undifferentiated growth of feeder-free human ES and iPS cells upon passaging was investigated (Pakzad et al., 2010). Extracellular proteome is found to maintain ES cells. It is reported that the pigment epithelium-derived factor receptor-Erk1/2 signaling pathway activated by the pigment epithelium-derived factor is sufficient to maintain the self-renewal of pluripotent human ES cells (Gonzalez et al., 2010). Synthetic substrate for culturing human ES cells and maintaining pluripotency was also developed (Mahlstedt et al., 2010).

6. Long-term culture and GMP standards

Prolonged culture of human ES cells may lead to adaptation and the acquisition of chromosomal abnormalities (Narva et al., 2010). In our study, the vulnerability of human ES cells and human iPS cells to apoptosis causes a low plating efficiency upon passaging was found. So far, no such small molecular events that promote self-regulation of human ES and iPS cells over a prolonged period of time have been reported in the literature. Maintaining the long-term potency of human ES and iPS cells in well state and produce more homogenous cell clones is still a grand challenge.

Therapeutic application of stem cell derivatives requires large quantities of cells produced in defined media that cannot be produced via conventional adherent culture. The use of feeder cells as well as animal-based products in ES or iPS cells culture may introduce batch-to-batch variations. The ideal culture method is developing feeder-free culture condition (Xu et al., 2001), even chemically defined culture conditions (Ludwig et al., 2006) for human ES cells or iPS cells expansion. But so far few human ES or iPS cell lines were produced without any exposure to animal-derived compounds or in accordance with good manufacturing practices (GMP) standards (Loser et al., 2010).

Therefore, culturing human ES or iPS cells in complete xeno-free conditions to reduce the risk of cross-transfer of pathogens without loss pluripotency would be a crucial prerequisite for clinical-grade applications. Nagaoka et al. (Nagaoka et al., 2010) cultured human pluripotent stem cells using completely defined conditions on a recombinant E-cadherin substratum, which should facilitate growth of stem cells using GMP standards. Olmer et al. (Olmer et al., 2010) have applied a scalable suspension culture to expand single cell of undifferentiated human ES cells, which represents a critical step towards standardized production in stirred bioreactors. Rajala et al. (Rajala et al., 2010) demonstrated that human ES cells, iPS cells can be maintained in the same defined xeno-free medium formulation for a prolonged period of time while maintaining their characteristics, demonstrating the applicability of the simplified xeno-free medium formulation for the production of clinical-grade stem cells.

For cryopreservation, an effective serum- and xeno-free chemically defined freezing procedure for human embryonic and induced pluripotent stem cells is also needed (Holm et al., 2010).

7. Challenges and prospects

Research on the ES and iPS cells to develop stem cell-based regenerative medicine is still in its early stages and there are still many challenges, including standardization of protocols for cell derivation and cultivation, identification of specific molecular markers, development of new approaches for directed differentiation etc. Among them, culture scale-up ensuring maintenance of cell pluripotency is a central issue, because cell therapy is far more complex and resource-consuming process as compared to drug-based medicine; pluripotent stem cell biology and technology is in need of further investigation and development before these cells can be used in clinics safely and successfully.

In addition, to minimize the ethical controversies, unify guidelines for reviewing ES cell research is also important. Mandatory registration is required for stem cell lines.

8. References

Abraham S., Sheridan S.D., Laurent L.C., Albert K., Stubban C., Ulitsky I., Miller B., Loring J.F. & Rao R.R. (2010). Propagation of human embryonic and induced pluripotent stem cells in an indirect co-culture system. Biochem Biophys Res Commun, 393, 211-216

Amit M., Margulets V., Segev H., Shariki K., Laevsky I., Coleman R. & Itskovitz-Eldor J. (2003). Human feeder layers for human embryonic stem cells. Biol Reprod, 68, 2150-2156

Chen A.E., Egli D., Niakan K., Deng J., Akutsu H., Yamaki M., Cowan C., Fitz-Gerald C., Zhang K., Melton D.A. & Eggan K. (2009). Optimal timing of inner cell mass isolation increases the efficiency of human embryonic stem cell derivation and allows generation of sibling cell lines. Cell Stem Cell, 4, 103-106

Cheng L., Hammond H., Ye Z., Zhan X. & Dravid G. (2003). Human adult marrow cells support prolonged expansion of human embryonic stem cells in culture. Stem Cells, 21, 131-142

Evans M.J. & Kaufman M.H. (1981). Establishment in culture of pluripotential cells from mouse embryos. Nature, 292, 154-156

Gonzalez R., Jennings L.L., Knuth M., Orth A.P., Klock H.E., Ou W., Feuerhelm J., Hull M.V., Koesema E., Wang Y., Zhang J., Wu C., Cho C.Y., Su A.I., Batalov S., Chen H., Johnson K., Laffitte B., Nguyen D.G., Snyder E.Y., Schultz P.G., Harris J.L. & Lesley S.A. (2010). Screening the mammalian extracellular proteome for regulators of embryonic human stem cell pluripotency. Proc Natl Acad Sci U S A, 107, 3552-3557

Greber B., Lehrach H. & Adjaye J. (2008). Control of early fate decisions in human ES cells by distinct states of TGFbeta pathway activity. Stem Cells Dev, 17, 1065-1077

Holm F., Strom S., Inzunza J., Baker D., Stromberg A.M., Rozell B., Feki A., Bergstrom R. & Hovatta O. (2010). An effective serum- and xeno-free chemically defined freezing procedure for human embryonic and induced pluripotent stem cells. Hum Reprod, 25, 1271-1279

Hovatta O., Mikkola M., Gertow K., Stromberg A.M., Inzunza J., Hreinsson J., Rozell B., Blennow E., Andang M. & Ahrlund-Richter L. (2003). A culture system using human foreskin fibroblasts as feeder cells allows production of human embryonic stem cells. Hum Reprod, 18, 1404-1409

Humphrey R.K., Beattie G.M., Lopez A.D., Bucay N., King C.C., Firpo M.T., Rose-John S. & Hayek A. (2004). Maintenance of pluripotency in human embryonic stem cells is STAT3 independent. Stem Cells, 22, 522-530

Jiang P., Rushing S.N., Kong C.W., Fu J., Lieu D.K., Chan C.W., Deng W. & Li R.A. (2010). Electrophysiological properties of human induced pluripotent stem cells. Am J Physiol Cell Physiol, 298, C486-495

Laurent L.C., Nievergelt C.M., Lynch C., Fakunle E., Harness J.V., Schmidt U., Galat V., Laslett A.L., Otonkoski T., Keirstead H.S., Schork A., Park H.S. & Loring J.F. (2010). Restricted ethnic diversity in human embryonic stem cell lines. Nat Methods, 7, 6-7

Lerou P.H., Yabuuchi A., Huo H., Takeuchi A., Shea J., Cimini T., Ince T.A., Ginsburg E., Racowsky C. & Daley G.Q. (2008). Human embryonic stem cell derivation from poor-quality embryos. Nat Biotechnol, 26, 212-214

Loser P., Schirm J., Guhr A., Wobus A.M. & Kurtz A. (2010). Human embryonic stem cell lines and their use in international research. Stem Cells, 28, 240-246

Ludwig T.E., Levenstein M.E., Jones J.M., Berggren W.T., Mitchen E.R., Frane J.L., Crandall L.J., Daigh C.A., Conard K.R., Piekarczyk M.S., Llanas R.A. & Thomson J.A. (2006). Derivation of human embryonic stem cells in defined conditions. Nat Biotechnol, 24, 185-187

Mahlstedt M.M., Anderson D., Sharp J.S., McGilvray R., Munoz M.D., Buttery L.D., Alexander M.R., Rose F.R. & Denning C. (2010). Maintenance of pluripotency in human embryonic stem cells cultured on a synthetic substrate in conditioned medium. Biotechnol Bioeng, 105, 130-140

Martin G.R. (1981). Isolation of a pluripotent cell line from early mouse embryos cultured in medium conditioned by teratocarcinoma stem cells. Proc Natl Acad Sci U S A, 78, 7634-7638

Mosher J.T., Pemberton T.J., Harter K., Wang C., Buzbas E.O., Dvorak P., Simon C., Morrison S.J. & Rosenberg N.A. (2010). Lack of population diversity in commonly used human embryonic stem-cell lines. N Engl J Med, 362, 183-185

Mummery C., Ward-van Oostwaard D., Doevendans P., Spijker R., van den Brink S., Hassink R., van der Heyden M., Opthof T., Pera M., de la Riviere A.B., Passier R. & Tertoolen L. (2003). Differentiation of human embryonic stem cells to cardiomyocytes: role of coculture with visceral endoderm-like cells. Circulation, 107, 2733-2740

Nagaoka M., Si-Tayeb K., Akaike T. & Duncan S.A. (2010). Culture of human pluripotent stem cells using completely defined conditions on a recombinant E-cadherin substratum. BMC Dev Biol, 10, 60

Narva E., Autio R., Rahkonen N., Kong L., Harrison N., Kitsberg D., Borghese L., Itskovitz-Eldor J., Rasool O., Dvorak P., Hovatta O., Otonkoski T., Tuuri T., Cui W., Brustle O., Baker D., Maltby E., Moore H.D., Benvenisty N., Andrews P.W., Yli-Harja O. & Lahesmaa R. (2010). High-resolution DNA analysis of human embryonic stem cell lines reveals culture-induced copy number changes and loss of heterozygosity. Nat Biotechnol, 28, 371-377

Niwa H., Burdon T., Chambers I. & Smith A. (1998). Self-renewal of pluripotent embryonic stem cells is mediated via activation of STAT3. Genes Dev, 12, 2048-2060

Olmer R., Haase A., Merkert S., Cui W., Palecek J., Ran C., Kirschning A., Scheper T., Glage S., Miller K., Curnow E.C., Hayes E.S. & Martin U. (2010). Long term expansion of undifferentiated human iPS and ES cells in suspension culture using a defined medium. Stem Cell Res, 5, 51-64

Pakzad M., Totonchi M., Taei A., Seifinejad A., Hassani S.N. & Baharvand H. (2010). Presence of a ROCK inhibitor in extracellular matrix supports more undifferentiated growth of feeder-free human embryonic and induced pluripotent stem cells upon passaging. Stem Cell Rev, 6, 96-107

Rajala K., Lindroos B., Hussein S.M., Lappalainen R.S., Pekkanen-Mattila M., Inzunza J., Rozell B., Miettinen S., Narkilahti S., Kerkela E., Aalto-Setala K., Otonkoski T., Suuronen R., Hovatta O. & Skottman H. (2010). A defined and xeno-free culture method enabling the establishment of clinical-grade human embryonic, induced pluripotent and adipose stem cells. PLoS One, 5, e10246

Richards M., Fong C.Y., Chan W.K., Wong P.C. & Bongso A. (2002). Human feeders support prolonged undifferentiated growth of human inner cell masses and embryonic stem cells. Nat Biotechnol, 20, 933-936

Thomson J.A., Itskovitz-Eldor J., Shapiro S.S., Waknitz M.A., Swiergiel J.J., Marshall V.S. & Jones J.M. (1998). Embryonic stem cell lines derived from human blastocysts. Science, 282, 1145-1147

Xiao L., Yuan X. & Sharkis S.J. (2006). Activin A maintains self-renewal and regulates fibroblast growth factor, Wnt, and bone morphogenic protein pathways in human embryonic stem cells. Stem Cells, 24, 1476-1486

Xu C., Inokuma M.S., Denham J., Golds K., Kundu P., Gold J.D. & Carpenter M.K. (2001). Feeder-free growth of undifferentiated human embryonic stem cells. Nat Biotechnol, 19, 971-974

Xu R.H., Peck R.M., Li D.S., Feng X., Ludwig T. & Thomson J.A. (2005). Basic FGF and suppression of BMP signaling sustain undifferentiated proliferation of human ES cells. Nat Methods, 2, 185-190

Ying Q.L., Nichols J., Chambers I. & Smith A. (2003). BMP induction of Id proteins suppresses differentiation and sustains embryonic stem cell self-renewal in collaboration with STAT3. Cell, 115, 281-292

6

The Function of Glycan Structures for the Maintenance and Differentiation of Embryonic Stem Cells

Shoko Nishihara
Department of Bioinformatics, Faculty of Engineering, Soka University
Japan

1. Introduction

Various types of glycoprotein and glycolipid are present on the cell surface and function to regulate cell-cell interactions, cell-extracellular matrix interactions, and signals from extrinsic factors, for instance, Wnt, fibroblast growth factor (FGF), Hedgehog (Hh) and bone morphogenetic protein (BMP). The glycan structures on proteins and lipids change dramatically during differentiation. Some of these glycoproteins and glycolipids can be used as markers for the identification of embryonic stem cells (ES cells), such as stage-specific embryonic antigen-1 (SSEA-1) (Atwood et al., 2008; Muramatsu & Muramatsu, 2004), SSEA-3, TRA-1-60 antigen and TRA-1-81 antigen (Adewumi et al., 2007).

ES cells were originally isolated from the inner cell mass (ICM) of blastocysts and have the essential characteristics of pluripotency and self-renewal. Pluripotency enables the cells to differentiate into all the cell types that constitute the adult body. In 1981, the first mouse ES cell lines were established, and these have proved invaluable as a tool for gene-targeting strategies in mice (Evans & Kaufman, 1981; Martin, 1981). Since the establishment of human ES cell lines in 1998, they have been used in a large number of research studies looking at potential applications for regenerative medicine (Thomson et al., 1998). Thus, ES cells are promising tools for biotechnology and possess key features that should allow their exploitation in the development of cell replacement therapies. To exploit the potential of ES cells for these various purposes, a better understanding of the molecular mechanisms that control self-renewal and pluripotency and also direct differentiation of ES cells is required.

Several signaling cascades activated by extrinsic factors such as leukemia inhibitory factor (LIF) (Smith et al., 1988; Williams et al., 1988), BMP (Ying et al., 2003) and Wnt (Miyabayashi et al., 2007; Sasaki et al., 2008; Sato et al., 2004) and the expression of intrinsic factors, such as Oct3/4 and Nanog, maintain self-renewal and pluripotency in ES cells (Boiani & Scholer, 2005). Likewise, Wnt, FGF, Hh and BMP signaling also play key roles in the determination of cell fate during the differentiation of ES cells (Kunath et al., 2007; Sasaki et al., 2009).

Proteoglycans are one form of sulfated glycoprotein and consist of several different types of core protein and glycosaminoglycans (sulfated glycans). There are two types of glycosaminoglycan: heparan sulfate (HS) and chondroitin sulfate (CS). They are ubiquitously present on the surface of many different types of cell and are known to play crucial roles in regulating several signaling pathways (Bishop et al., 2007). A large number of physiologically important molecules can bind to specific sulfated regions of HS and CS. Genetic analyses using *Drosophila* have demonstrated that HS is involved in signal regulatory pathways that respond to extrinsic factors, such as Wnt, FGF, Hh and BMP (Yan & Lin, 2009). On the other hand, self-renewal and pluripotency of ES cells are maintained by a balance among several signaling pathways, such as Wnt, FGF and BMP, and the differentiation of ES cells into a specific lineage is induced by disruption of this balance. Through these signals, HS and CS contribute to the maintenance and differentiation of ES cells (Kraushaar et al., 2010; Lanner et al., 2010; Sasaki et al., 2009; Sasaki et al., 2008). The aims of this chapter are (1) to describe the use of carbohydrate antigens as markers of ES cells, (2) to consider the function of HS in the maintenance of self-renewal and pluripotency of ES cells, and (3) to outline the function of HS and CS in the differentiation of ES cells.

2. Carbohydrate antigens can be used as markers of ES cells

Mouse ES cells express SSEA-1, also known as Lewis X carbohydrate antigen (LeX), which has the structure Galβ1,4(Fucα1,3)GlcNAc. It is found on both glycoproteins and glycolipids (Atwood et al., 2008; Muramatsu & Muramatsu, 2004). SSEA-1 is a marker of mouse ES cells and use of an anti-SSEA-1 antibody enables positive staining of ES cells and of the ICM, the origin of mouse ES cells.

However, human ES cells do not express SSEA-1. Instead, they express SSEA-3, SSEA-4, TRA-1-60 antigen and TRA-1-81 antigen, and these can be used as specific markers for human ES cells (Adewumi et al., 2007). The carbohydrate structures of SSEA-3 and SSEA-4 are R-3GalNacβ1,3Galα1,4R′ and NeuAcα2,3Galβ1,3GalNacβ1,3Galα1,4R′, respectively, and they are carried on globo-series glycolipids (Kannagi et al., 1983). Both TRA-1-60 antigen and TRA-1-81 antigen are present on keratan sulfate, a sulfated poly-*N*-acetyllactosamine. The TRA-1-60 epitope, but not that of TRA-1-81, includes sialic acid (Badcock et al., 1999). These carbohydrate structures are also expressed in the human ICM. Recently, a glycome analysis of *N*-linked glycans on human ES cells reported abundant expression of LeX and H type 2 antennae in sialylated complex-type *N*-linked glycans (Satomaa et al., 2009). LeX would not be recognized by an anti-SSEA-1 antibody when it is presented on a biantennary *N*-glycan antenna. However, the biological functions of these cell surface markers have not been fully elucidated.

3. Heparan sulfate (HS) is mainly expressed on mouse ES cells, and both HS and chondroitin sulfate (CS) increase during differentiation of embryoid bodies (EBs)

In comparison to other glycosaminoglycans (sulfated glycans), HS is highly expressed on mouse ES cells (Fig. 1A and B) (Nishihara, 2009; Sasaki et al., 2009; Sasaki et al., 2008). For example, the ratio of HS to CS is almost 5:1. Thus, the main glycosaminoglycan expressed on mouse ES cells is HS. However, the amounts of HS and CS on the cell surface increase more than five-fold during differentiation from ES cells to EBs (Fig. 2B) (Nairn et al., 2007). Therefore, it is expected that HS will be the principal contributor to the maintenance of mouse ES cells and that CS, as well as HS, will contribute to the differentiation of mouse ES cells.

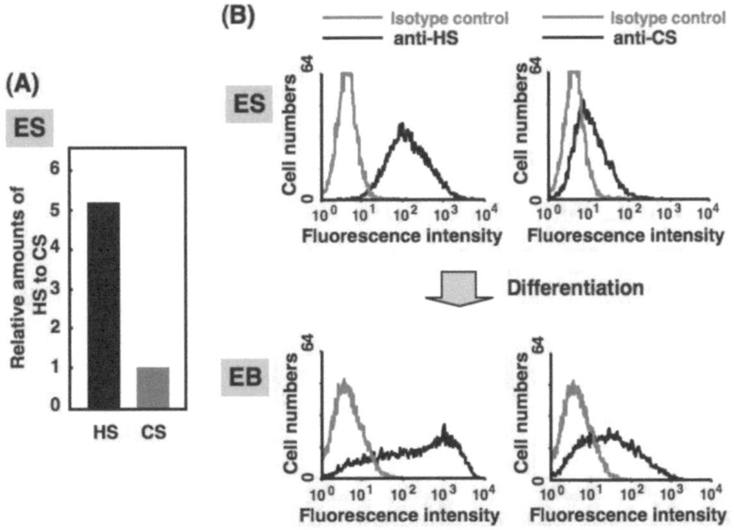

Fig. 1. The expression of heparan sulfate (HS) and chondroitin sulfate (CS) on mouse ES cells and embryoid bodies (EBs).
(A) The relative amounts of HS and CS in mouse ES cells. HS is approximately 5 times more abundant in mouse ES cells than CS. The histograms show the outcome of an HPLC analysis for unsaturated disaccharides.
(B) HS and CS on mouse ES cells and EBs. A FACS analysis of mouse ES cells and EBs using an anti-HS antibody (HepSS-1) and anti-CS antibody (2H6) shows that HS is more highly expressed on mouse ES cells and that both HS and CS increase during differentiation of EBs from ES cells.

4. Heparan sulfate (HS) and chondroitin sulfate (CS) are synthesized in the Golgi apparatus

Both HS and CS are synthesized in the Golgi apparatus by a series of glycosyltransferases and sulfotransferases (Fig. 2A). PAPS is a donor substrate for sulfotransferases; it is synthesized in the cytosol by PAPS synthases, and is translocated into the Golgi apparatus by the PAPS transporters, PAPST1 and PAPST2 (Goda et al., 2006; Kamiyama et al., 2006; Kamiyama et al., 2003). Therefore, if the expression of PAPS transporters is regulated, then sulfation of both HS and CS can be regulated (Sasaki et al., 2009).
HS has repeating disaccharide units of D-glucuronic acid-*N*-acetyl-D-glucosamine (GlcA-GlcNAc) that are modified differentially by epimerization and sulfation (Fig. 2B). The disaccharide repeats are synthesized by members of the EXT protein family, including EXT1, and sulfated by a series of sulfotransferases (Bishop et al., 2007). The first step in this series of sulfation reactions is catalyzed by *N*-deacetylase/*N*-sulfotransferase (NDST). Of the four known *NDSTs*, *NDST1* and *NDST2* are expressed in mouse ES cells (Nairn et al., 2007). Therefore, if we regulate the expression of *EXT1, NDST1* and *NDST2*, it should be possible to regulate elongation of the HS chain and HS-specific sulfation (Sasaki et al., 2009; Sasaki et al., 2008). A large number of physiologically important molecules can bind to specific sulfated regions of HS (Fig. 2A). As mentioned above, genetic analyses using *Drosophila* have demonstrated that HS is involved in signal regulatory pathways responding

to extrinsic factors, such as FGF, Wnt, Hh and BMP (Tabata & Takei, 2004; Ueyama et al., 2008).

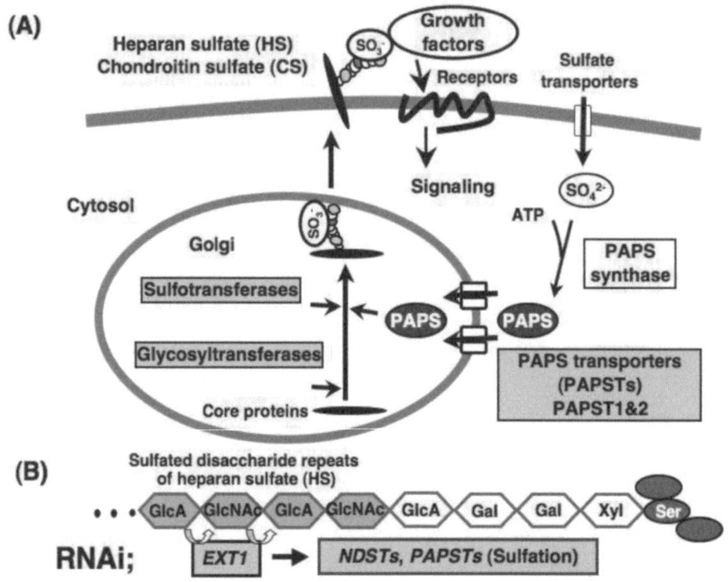

Fig. 2. Outline of the synthetic pathways for heparan sulfate (HS) and chondroitin sulfate (CS) in Golgi apparatus.
(A) HS and CS are synthesized and sulfated in the Golgi apparatus.
(B) A diagrammatic representation of the structure of HS and the role of EXT1, NDST and PAPSTs in its synthesis. The major components of HS are sulfated disaccharide repeats that are covalently bound to Ser residues of specific core proteins through the glycosaminoglycan-protein linkage region GlcAβ1,3Galβ1,3Galβ1,4Xyl-*O*-Ser. (GlcA, D-glucuronic acid; Gal, galactose; Xyl, xylose; GalNAc, *N*-acetyl-D-galactosamine; GlcNAc, *N*-acetyl-D-glucosamine).

We analyzed the function of HS and sulfation of HS and CS by using RNA interference (RNAi) to knockdown (KD) *EXT1, NDST, PAPST1* and *PAPST2* (Fig. 2B) (Sasaki et al., 2009; Sasaki et al., 2008). Although the KD efficiency was less than 100%, we used this method rather than performing gene knockouts (KO) because, in addition to the direct effects of gene knockouts, secondary effects may also be observed that are caused by adaptation of the cells during long-term culture. For example, the expression of a novel gene might be induced that has secondary effects on the mouse ES cells. If the targets are essential for cell survival and proliferation, analysis of the knockout cells may be complicated by cell death. Thus, knockout of some genes that are related to HS sulfation, e.g., *6-O-endosulfatase, C5-epimerase* and *HS2ST*, leads to a number of unexpected changes in the structure of sulfated glycosaminoglycans, presumably due to secondary effects (Lamanna et al., 2006; Li et al., 2003; Merry et al., 2001). By contrast, RNAi knockdown of gene expression gave a specific effect for each gene. For instance, knockdown of *EXT1* expression in mouse ES cells resulted in the shortening of the HS chain (Sasaki et al., 2008); knockdown of *NDST1* and *NDST2* specifically reduced HS sulfation (Sasaki et al., 2009); and knockdown of *PAPST1* or *PAPST2* reduced both HS and CS sulfation (Sasaki et al., 2009).

5. Heparan sulfate (HS) contributes to the maintenance of self-renewal and pluripotency of mouse ES cells

5.1 Wnt, bone morphogenetic protein (BMP) and leukemia inhibitory factor (LIF) signals work on the maintenance of self-renewal and pluripotency in mouse ES cells, while fibroblast growth factor 4 (FGF4) works on the exit from the undifferentiated ground state

As described above, self-renewal and pluripotency of mouse ES cells are maintained by several signaling cascades from both extrinsic factors, such as LIF (Smith et al., 1988; Williams et al., 1988), BMP (Ying et al., 2003) and Wnt (Miyabayashi et al., 2007; Sasaki et al., 2008; Sato et al., 2004), and intrinsic factors, such as Oct3/4 and Nanog (Boiani & Scholer, 2005).

In mouse ES cells, LIF molecules interact with the heteromeric receptor gp130 and the low affinity LIF receptor to induce activation of STAT3 (Boeuf et al., 1997; Matsuda et al., 1999; Niwa et al., 1998; Raz et al., 1999) and then upregulate the expression level of Myc (Cartwright et al., 2005). BMP induces the expression of inhibitor of differentiation (*Id*) genes, which suppress expression of genes for neural differentiation (Ying et al., 2003), through activation of Smad signaling. Thus, BMP suppresses neural differentiation and, in combination with LIF, is sufficient to maintain self-renewal of mouse ES cells without feeder cells and serum factors.

In contrast, Wnt signals play a role in the regulation of self-renewal of both mouse and human ES cells independently of LIF/STAT3 signaling (Miyabayashi et al., 2007; Sasaki et al., 2008; Sato et al., 2004). The binding of Wnt to its cognate receptor, Frizzled, results in the inhibition of glycogen synthase kinase-3β (GSK3β). In turn, inhibition of GSK3β allows the stabilization and accumulation of β-catenin in the nucleus that is required for transcription of downstream genes. The canonical Wnt pathway maintains the expression of downstream Nanog, a transcription factor that is essential for the maintenance of the ICM and of ES cell pluripotency (Cole et al., 2008; Miyabayashi et al., 2007; Sasaki et al., 2008; Sato et al., 2004). The activation of Nanog sustains ES cell self-renewal without feeder cells or treatment with LIF (Sato et al., 2004).

On the other hand, FGF4 is produced in an autocrine fashion in mouse ES cells and functions in the exit from the undifferentiated ground state (Kunath et al., 2007; Ying et al., 2008). FGF4/extracellular signal-regulated kinase (ERK) signaling contributes to differentiation into neural and mesodermal lineages. However, the mechanism that regulates extrinsic signaling in ES cells has not been fully elucidated.

5.2 Heparan sulfate (HS) and its sulfation are important for self-renewal, pluripotency and proliferation of mouse ES cells

In order to analyze the function of HS and its sulfation in mouse ES cells, we constructed siRNA expression plasmids that targeted *EXT1*, *NDST1*, *NDST2*, *PSPST1* or *PAPST2* (Fig. 2A) by inserting the corresponding short hairpin RNA (shRNA) sequence into pSilencer 3.1-H1 vector, which has a PolIII promoter and a puromycin resistance gene. The construct was then transfected into mouse ES cells. After puromycin selection, we confirmed the specific knockdown of the targeted gene and the reduction of its product. Knockdown (KD) of *EXT1* expression inhibited the elongation of HS chains and resulted in a reduction in the size (~35kDa) of the HS chain compared to control cells (50~150 kDa) (Sasaki et al., 2008). Knockdown of *NDST1* or *NDST2* specifically reduced sulfation of HS. Knockdown of *PAPST1* or *PAPST2* reduced sulfation of both HS and CS (Sasaki et al., 2009; Sasaki et al., 2008).

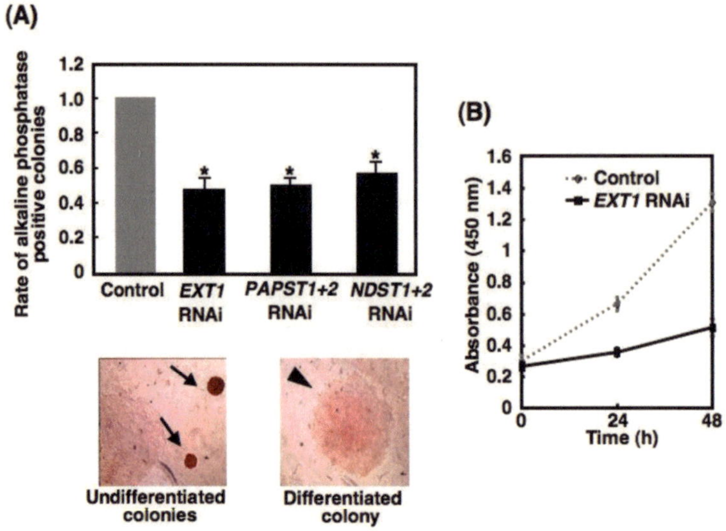

Fig. 3. The effect of HS and its sulfation on self-renewal and proliferation in mouse ES cells. (A) Self-renewal assay. In *EXT1*-KD ES cells, the number of AP positive (undifferentiated) colonies was reduced to 50% of that of control cells, even in the presence of LIF and serum. *PAST1&2*-double KD ES cells and *NDST1&2*-double KD ES cells also showed a reduction in the number of AP positive colonies. The rate of AP positive colonies is shown after normalization against control cells (value=1).
(B) Proliferation assay. The rate of proliferation of *EXT1*-KD cells decreased significantly compared to control cells.
The values shown are the means ± SD from three independent experiments and significant values are indicated; *P < 0.01, in comparison to control cells.

In *EXT1*-KD ES cell cultures, the number of alkaline phosphatase (AP) positive colonies, indicative of the undifferentiated state, fell to 50% of that in control cultures, even in the presence of LIF and serum (Fig. 3A). *PAST1&2*-double KD ES cells and *NDST&2*-double KD ES cell cultures also showed a reduction in the number of AP positive colonies. *EXT1-KD* ES cells showed a reduction in their rate of proliferation (Fig. 3B), as did *PAST1&2*-double KD ES cells and *NDST1&2*-double KD ES cells (data not shown). Our analyses clearly demonstrated that HS, and in particular its sulfation, has a significant role in the self-renewal and proliferation of mouse ES cells (Sasaki et al., 2009; Sasaki et al., 2008).
Four days after *EXT1* knockdown, even in the presence of LIF, *EXT1*-KD ES cells displayed a flattened and differentiated morphology reminiscent of the stellate appearance of parietal endoderm cells (Fig. 4A). *PAPST1* or *PAPST2*-KD ES cells, *PAPST1&2*-double KD ES cells and *NDST1&2*-double KD ES cells showed a similar morphology to the *EXT1*-KD ES cells. Real time PCR analysis for various germ layer markers showed that expression of Oct3/4 and Nanog, undifferentiated state markers, was significantly decreased in the *EXT1*-KD ES cells (Fig. 4B). All of the other types of KD-ES cells also showed reduced expression of Oct3/4 and Nanog. After withdrawal of LIF to allow further differentiation, *EXT1*-KD ES cells and the other KD-ES cells showed increased expression of markers of the extraembryonic endoderm lineage compared to control cells, and lost pluripotency (Fig. 4B).

These results indicate that HS, and in particular sulfation of HS, has a role in the maintenance of pluripotency in mouse ES cells (Sasaki et al., 2009; Sasaki et al., 2008).

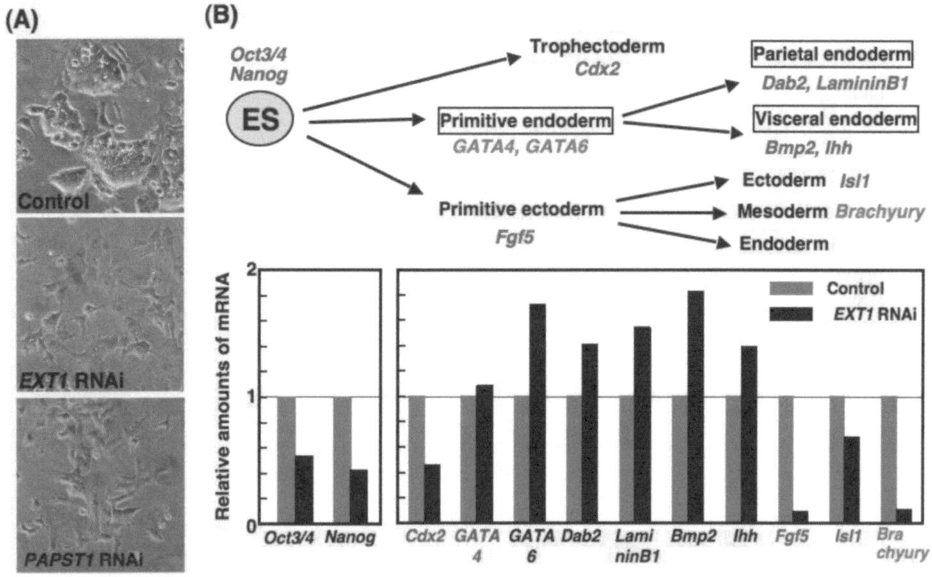

Fig. 4. Reduction in HS and in its sulfation induce mouse ES cells to spontaneously differentiate into extraembryonic endoderm cells.

(A) Photomicrographs of cells at four days after transfection with an siRNA expression plasmid vectors targeting *EXT1* and *PAPST1* in the presence of LIF. Almost all of the *EXT1*-KD ES cells and *PAPST1*-KD ES cells exhibited a flattened, differentiated morphology. Similar results were obtained by single *PAPST1* or *PAPST2* knockdown, *PAPST1&2* double knockdown and *NDST1&2* double knockdown.

(B) Real time PCR analysis of several differentiation markers four days after transfection in the presence of LIF. Oct3/4 and Nanog expression was significantly decreased in *EXT1*-KD cells compared to control cells. Increased expression of markers of the extraembryonic endoderm lineage (Gata4, Gata6, Dab2, Laminin B1, Bmp2 and Ihh) was observed in *EXT1*-KD cells. The results are shown after normalization against control cells (value=1). Similar results were obtained by *PAPST1* or *PAPST2* knockdown, *PAPST1&2* double knockdown and *NDST1&2* knockdown.

5.3 Heparan sulfate (HS) and its sulfation on mouse ES cells regulate BMP/Smad signaling

HS and its sulfation play a role in the self-renewal and pluripotency of mouse ES cells (see section 5.2 above). Various extrinsic factors, such as BMP, Wnt and LIF, affect the maintenance of self-renewal and pluripotency in mouse ES cells (Sasaki et al., 2008; Sato et al., 2004; Smith et al., 1988; Williams et al., 1988; Ying et al., 2003), (and see section 5.1 above). BMP4 acts in synergy with LIF to maintain self-renewal via the Smad-mediated induction of *Id* (inhibitor of differentiation) gene expression (Ying et al., 2003) and inhibition of p38 mitogen-activated protein kinase (Qi et al., 2004). We therefore examined the level of phosphorylation of Smad1 in *EXT1*-KD ES cells to determine whether BMP/Smad1 signaling was altered in these KD ES cells.

No reduction in the level of phosphorylation of Smad1 could be detected in *EXT1*-KD ES cells compared to control cells (Fig. 5A), suggesting that even a short HS chain can bind to BMP4 and contribute to BMP4 signaling (Sasaki et al., 2008). However, *PAPST1* or *PAPST2*-KD ES cells, *PAPT1&2*-double KD ES cells and *NDSTT1&2*-double KD ES cells show

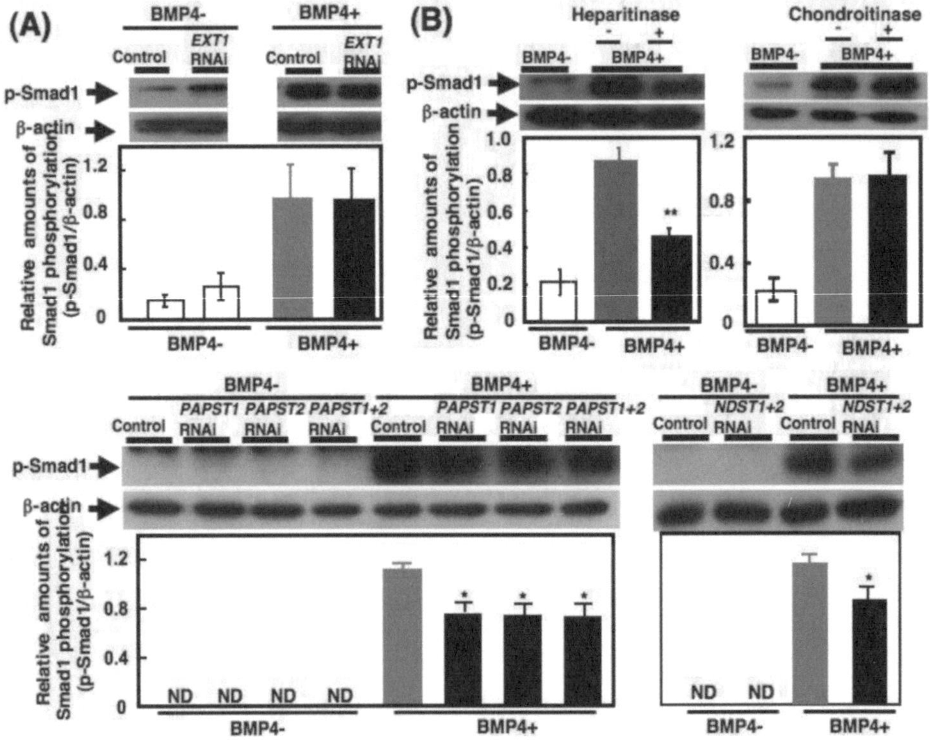

Fig. 5. HS and its sulfation regulate BMP/Smad signaling.

(A) Western blot analysis of *EXT1*-KD ES cells, *PAPST1* or *PAPST2*-KD ES cells, *PAPST1&2*-double KD ES cells and *NDST1&2*-double KD ES cells stimulated with BMP4. The stimulation of *EXT1*-KD and control cells with BMP4 gave rise to similar increases in Smad1 phosphorylation. But *PAPST1* or *PAPST2*-KD ES cells, *PAPST1&2*-double KD ES cells and *NDST1&2*-double KD ES cells showed reduction of phosphorylated Smad1. The histograms show mean densitometric readings ± SD of the phosphorylated Smad1/loading controls. Values were obtained from duplicate measurements of two independent experiments and significant values are indicated; $^{*}P < 0.05$ in comparison to control cells.

(B) Western blot analysis of mouse ES cells stimulated with BMP4 after treatment with heparitinase and chondroitinase. HS-depletion by heparitinase treatment led to a reduction in BMP/Smad signaling while CS-depletion by chondroitinase treatment did not. The histograms show mean densitometric readings ± SD of the phosphorylated Smad1/loading controls. Values were obtained from duplicate measurements of two independent experiments and significant values are indicated; $^{**}P < 0.01$, in comparison to untreated control cells.

significantly reduced levels of Smad1 phophorylation demonstrating that HS, and in particular sulfation of HS, is important for BMP/Smad signaling (Sasaki et al., 2009). A significantly reduced signal was also observed after heparitinase treatment while no reduction was observed after chondroitinase treatment (Fig. 5B). Thus, HS and sulfation of HS (but not of CS) on the surface of mouse ES cells regulate BMP/Smad signaling to maintain self-renewal and pluripotency.

In a similar fashion, we analyzed LIF signaling in *EXT1*-KD ES cells and in other KD ES cells, but we did not observe any reduction in signal intensity, indicating that HS and its sulfation on the cell surface does not contribute to LIF/STAT signaling (Sasaki et al., 2009; Sasaki et al., 2008).

5.4 Heparan sulfate (HS) and its sulfation on mouse ES cells regulate autocrine/paracrine Wnt/β-catenin signaling

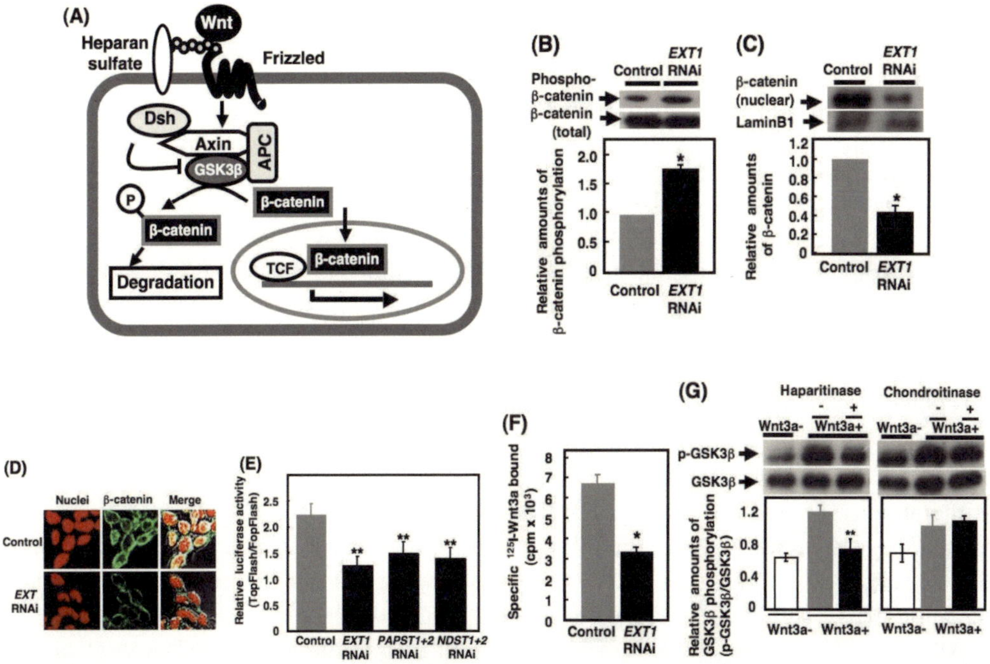

Fig. 6. HS, and its sulfation, regulates autocrine/paracrine Wnt/β-catenin signaling.

(A) Schematic diagram of the Wnt/β-catenin signaling pathway.

(B) Western blot analysis of *EXT1*-KD ES cells starved of LIF and serum for 4hr. In *EXT1*-KD cells, a significant increase in β-catenin phosphorylation was observed. The histograms show mean densitometric readings ± SD of the phospho-β-catenin/total β-catenin after normalization against control cells (value=1). Values were obtained from duplicate measurements of two independent experiments and significant values are indicated; *$P < 0.01$, in comparison to control cells.

(C) Western blot analysis of nuclear extracts of *EXT1*-KD ES cells. The histograms show mean densitometric readings ± SD of the β-catenin/Lamin B_1 after normalization against control cells (value=1). Values are obtained from duplicate measurements of two

independent experiments and significant values are indicated; *P < 0.01, in comparison to control cells.
(D) Confocal images of *EXT1*-KD ES cells. Accumulation of β-catenin was significantly decreased in nuclei of *EXT1*-KD cells. (β-catenin, green; nucleus, red; merged image of β-catenin and nucleus, yellow).
(E) Luciferase reporter assay. The autocrine/paracrine luciferase activity was significantly decreased in *EXT1*-KD ES cells, *PAPST1&2*-double KD ES cells and *NDST1&2*-double KD ES cells. Relative luciferase activities (TopFlash/FopFlash) are shown as means ± SD from three independent experiments, and significant values are indicated; **P < 0.05, in comparison to control cells.
(F) Binding assay of ^{125}I-labeled Wnt3a to *EXT1*-KD ES cells. *EXT1*-KD ES cells exhibited significantly lower specific ^{125}I-Wnt3a binding. The value of specific ^{125}I-Wnt3a binding (total cpm minus cpm bound in the presence of 100 μg/ml free heparin) was the mean ± SD of three independent experiments, and significant values are indicated; *P < 0.01, in comparison to control.
(G) Western blot analysis of mouse ES cells stimulated with Wnt3a after treatment with heparitinase and chondroitinase. HS-depletion by heparitinase treatment reduced the GSK3β in response to Wnt3a while CS-depletion by chondroitinase treatment did not. The histograms show mean densitometric readings ± SD of the phosphorylated GSK3β/GSK3β. Values were obtained from duplicate measurements of two independent experiments and significant values are indicated; **P < 0.01, in comparison to untreated control cells.

We next examined the potential influence of HS on the Wnt signaling pathway (Fig. 6A) using various gene-KD ES cells, including *EXT1*-KD ES cells. The culture medium included BMP, a serum component, and LIF, but did not include Wnt. As a first step, we performed an RT-PCR of mouse ES cells for Wnt expression and found that several Wnts, such as Wnt2, Wnt3a, Wnt4, Wnt5a, Wnt7a and Wnt10b, were expressed (Sasaki et al., 2008).
A Western blot analysis for β-catenin, which transmits Wnt signals, showed that there was a significantly higher level of phosphorylation of β-catenin in *EXT1*-KD cells than control cells in the absence of feeder cells (Fig. 6B) (Sasaki et al., 2008). We also observed significant reductions in the level of β-catenin in the nucleus (Fig. 6C and D) and of luciferase reporter activity (Fig. 6E). These results demonstrate that autocrine/paracrine Wnt/β-catenin signaling is reduced in *EXT1*-KD ES cells, indicating that HS contributes to Wnt/β-catenin signaling. In addition to *EXT1*-KD ES cells, *PAPST1&2*-double KD ES cells and *NDST1&2*-double KD ES cells showed a reduction in luciferase reporter activity, also indicating that sulfation of HS plays a key role in Wnt/β-catenin signaling (Sasaki et al., 2009).
Next, we analyzed the binding of ^{125}I-Wnt3a and found a lower level of binding on *EXT*1-KD ES cells compared to control cells (Fig. 6F) (Sasaki et al., 2008). Thus, HS contributes to the binding of Wnt3a to the mouse ES cell surface. We also analyzed phosphorylation of GSK3β after heparitinase or chondroitinase treatment and observed a reduction in level of GSK3β phosphorylation only after HS-depletion by heparitinase treatment and not after CS-depletion by chondroitinase treatment (Fig. 6G) (Sasaki et al., 2009). This finding confirms that HS, but not CS, contributes to Wnt/β-catenin signaling. Our results demonstrate that HS and its sulfation regulates autocrine/paracrine Wnt/β-catenin signaling in mouse ES cells by enhancing the binding of Wnt to its cognate receptor, Frizzled.

5.5 Heparan sulfate (HS) and its sulfation on mouse ES cells regulate FGF4/ERK signaling

HS and its sulfation play a role in the self-renewal and pluripotency of mouse ES cells (see section 5.2). Extrinsic factors that affect the maintenance of self-renewal and pluripotency in mouse ES cells are affected by HS; this has been shown for BMP (see section 5.3) and Wnt (see section 5.4), although does not apply to LIF (see section 5.3) (Sasaki et al., 2008; Sato et al., 2004; Smith et al., 1988; Williams et al., 1988; Ying et al., 2003). In contrast to these extrinsic factors, FGF4 works on the exit from the undifferentiated ground state (Kunath et al., 2007). Mouse ES cells express FGF1, bFGF(FGF2), FGF4, FGF8, FGF9, FGF10, GHG15, FGF18 and FGFR1~4 (Sasaki et al., 2009). Thus FGF4 is produced in an autocrine fashion in mouse ES cells and is known to bind to HS (Sugaya et al., 2008). Therefore, we analyzed FGF4 signaling in gene-KD mouse ES cells showing reduction in HS and in its sulfation.

The level of phosphorylation of extracellular signal-regulated kinase (ERK) following treatment with FGF4 was significantly reduced in *PAPST1*-KD or *PAPST2*-KD ES cells, *PAPT1&2*-double KD ES cells and *NDSTT1&2*-double KD ES cells, compared to control cells (Fig.7A) (Sasaki et al., 2009). Moreover, an even larger reduction in ERK phosphorylation was observed after treating mouse ES cells with heparitinase, which digests HS on the surface of the cells, while no reduction was observed after chondroitinase treatment (Fig. 7B). These observations indicate that HS and its sulfation, but not CS, on the surface of ES cells regulate FGF4/ERK signaling.

5.6 Heparan sulfate (HS) and its sulfation on mouse ES cells regulate bFGF/ERK signaling

In addition to FGF4 (see section 5.5), bFGF (FGF2) is produced in an autocrine fashion in mouse ES cells and is known to bind to HS (Sugaya et al., 2008). Therefore, we analyzed bFGF signaling in various HS-related gene-KD mouse ES cells. The level of phosphorylation of extracellular signal-regulated kinase (ERK) following treatment with bFGF was significantly reduced in all of the HS related gene-KD mouse ES cells which were tested, compared to control cells (Fig. 8A) (Sasaki et al., 2009; Sasaki et al., 2008). Moreover, a larger reduction in ERK phosphorylation was observed after treating mouse ES cells with heparitinase, while no reduction was observed after chondroitinase treatment (Fig. 8B). These observations indicate that HS and its sulfation, but not CS, on the surface of ES cells regulate bFGF/ERK signaling.

To date, LIF, Activin/Nodal and bFGF have been reported to contribute to mouse ES cell proliferation (Dvorak et al., 1998; Ogawa et al., 2007; Smith et al., 1988; Williams et al., 1988). It is known that FGF signaling mediated by HS, contributes to the proliferation of various types of cell (Lin, 2004). Furthermore, the proliferation of mouse ES cells treated with SU5402, an inhibitor of FGFR1 tyrosine phosphorylation, was reduced compared to that of control cells (Fig. 8C), demonstrating that autocrine/paracrine FGF signaling mediated by FGFR1 contributes to mouse ES cell proliferation (Sasaki et al., 2009). Therefore, the above results suggest that the reduced proliferation of *EXT1*-KD cells, as well as other HS related gene-KD mouse ES cells (Fig. 3B), is due to a reduction in autocrine/paracrine bFGF signaling, which in turn is caused by reduced HS chain sulfation (Fig. 8A and B).

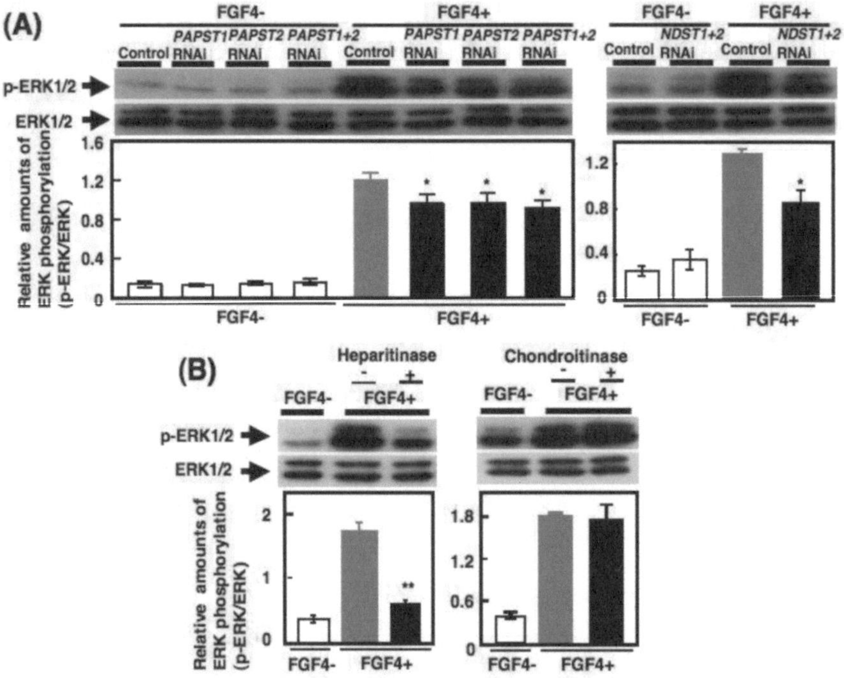

Fig. 7. HS and its sulfation regulate FGF4/ERK signaling.

(A) Western blot analysis of *PAPST1*-KD or *PAPST2*-KD ES cells, *PAPST1&2*-double KD ES cells and *NDST1&2*-double KD ES cells stimulated with FGF4. All of the gene-KD cells showed significant reductions in ERK phosphorylation. The histograms show mean densitometric readings ± SD of phosphorylated ERK/ERK. Values were obtained from duplicate measurements of two independent experiments and significant values are indicated; $*P < 0.05$ in comparison to control cells.

(B) Western blot analysis of mouse ES cells stimulated with FGF4 after treatment with heparitinase and chondroitinase. HS-depletion by heparitinase treatment led to a reduction in FGF4/ERK signaling while CS-depletion by chondroitinase treatment did not. The histograms show mean densitometric readings ± SD of phosphorylated ERK/ERK. Values were obtained from duplicate measurements of two independent experiments and significant values are indicated; $**P < 0.01$, in comparison to untreated control cells.

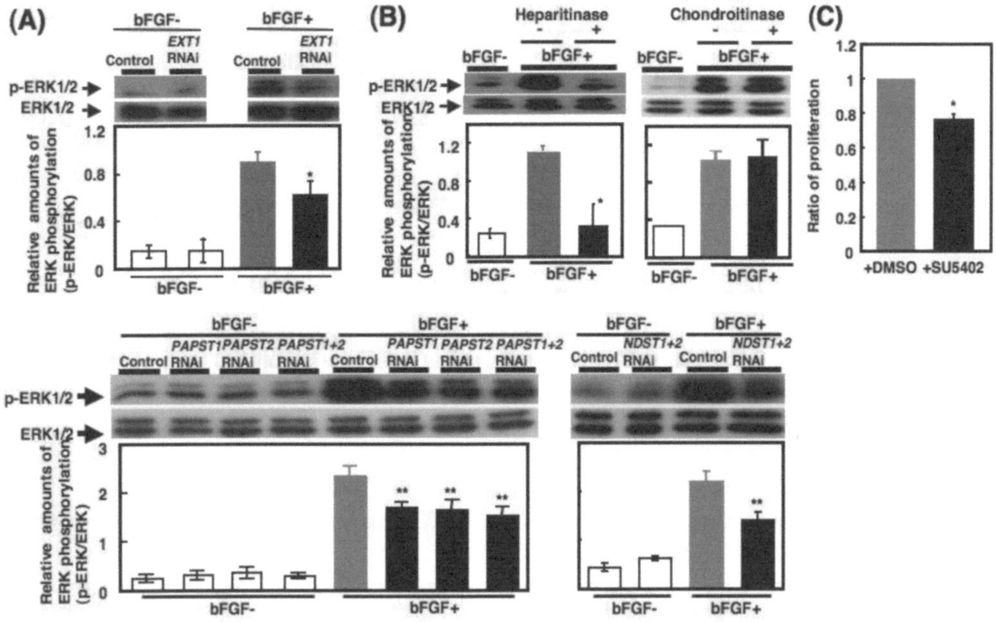

Fig. 8. HS and its sulfation regulate bFGF/ERK signaling.
(A) Western blot analysis of *EXT1*-KD ES cells, *PAPST1*-KD or *PAPST2*-KD ES cells, *PAPST1&2*-double KD ES cells and *NDST1&2*-double KD ES cells stimulated with bFGF. All of the gene-KD ES cells showed significant reduction in ERK phosphorylation. The histograms show mean densitometric readings ± SD of phosphorylated ERK/ERK. Values were obtained from duplicate measurements of two independent experiments and significant values are indicated; *$P < 0.01$ and **$P < 0.05$ in comparison to control cells.
(B) Western blot analysis of mouse ES cells stimulated with bFGF after treatment with heparitinase and chondroitinase. HS-depletion by heparitinase treatment led to a reduction in bFGF/ERK signaling, while CS-depletion by chondroitinase treatment did not. The histograms show mean densitometric readings ± SD of the phosphorylated ERK/ERK. Values were obtained from duplicate measurements of two independent experiments and significant values are indicated; *$P < 0.01$, in comparison to untreated control cells.
(C) Proliferation assay of mouse ES cells treated with SU5402. The rate of proliferation after 48h of culture is shown; the values were normalized against those of DMSO-treated cells (value=1). The values shown are the means ± SD from three independent experiments and significant values are indicated; *$P < 0.01$, in comparison to DMSO-treated cells.

5.7 HS and its sulfation contribute to the maintenance of mouse ES cells by regulating the balance between Wnt, FGF4 and BMP4 signaling

In undifferentiated mouse ES cells, FGF4 signal, which is the key signal for the exit from the undifferentiated ground state and the initiation of differentiation (Kunath et al., 2007; Ying et al., 2008), is inhibited by BMP4 (Qi et al., 2004). Thus, differentiation is inhibited. Wnt also upregulates Nanog expression and maintains the capacity for self-renewal(Cole et al., 2008; Miyabayashi et al., 2007; Sasaki et al., 2008; Sato et al., 2004).

In HS-related gene-KD ES cells, and in cells with reduction of HS or its sulfation, the FGF4 signal is reduced but is still present. As a consequence, inhibition of the FGF4 signal by

BMP4 is reduced, and also the upregulation of Nanog is reduced (Sasaki et al., 2009; Sasaki et al., 2008). Therefore, differentiation is induced in HS-related gene-KD ES cells. In a similar manner, HS and its sulfation, but not CS, contribute to the maintenance of mouse ES cells by regulating the balance between Wnt, FGF4 and BMP4 signaling.

Recently, HS-null mouse ES cells (*EXT1-knock out* (KO) ES cells) were reported to show no defects in pluripotency (Johnson et al., 2007), but failed to transit from the self-renewal state to the initiation of differentiation following the removal of LIF (Kraushaar et al., 2010). In addition, *NDST1&2* double-KO ES cells fail to initiate differentiation but adopt a more naive pluripotent Nanog/KLF4/Tbx3 positive state due to the reduction of FGF4/ERK signaling (Lanner et al., 2010). In these cases, complete depletion of HS completely shuts down the induction of differentiation due to the total depletion of FGF4 signaling; these KO ES cells stay in the undifferentiated ground state.

6. Sulfation of heparan sulfate (HS) and chondroitin sulfate (CS) contribute to the differentiation of mouse ES cells

6.1 Reduced sulfation induces abnormal differentiation into three germ layers during embryoid body (EB) formation in mouse ES cells

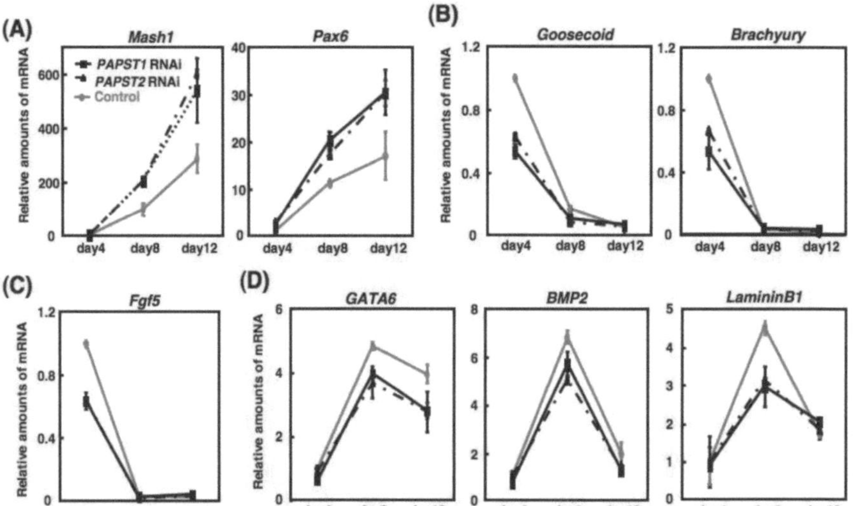

Fig. 9. Abnormal differentiation was observed in *PAPST1*-KD or *PAPST2*-KD cells during EB formation. EB formation was induced by transferring the cells to low cell binding dishes and culturing in ESC medium without LIF.

(A)-(D) Real time PCR analysis of germ layer markers at 4, 8 and 12 days after EB formation (A, neurectoderm marker; B, mesoderm marker; C, primitive ectoderm marker; D, extraembryonic endoderm (ExE) marker). The results are shown after normalization against the values obtained with control EBs on day 4 (value=1). The values shown are the means ± SD of two independent experiments.

Several extrinsic factors, such as BMP, FGF and Wnt, play important roles in the differentiation of mouse ES cells, in addition to their involvement in self-renewal (Loebel et al., 2003). BMP/Smad signaling is essential for the decision between ectodermal and mesodermal fates. Antagonizing the BMP/Smad signal, for example, by exposure of mouse ES cells to Noggin or transfection with a Noggin-encoding plasmid, promotes neurectodermal differentiation via EB formation (Finley et al., 1999; Loebel et al., 2003). FGF4 is produced in an autocrine fashion in mouse ES cells and FGF4/ERK signaling contributes to differentiation into neural and mesodermal lineages (Kunath et al., 2007) (see section 5.1 above). Wnt/β-catenin signaling inhibits neural differentiation via EB formation. Neural differentiation in mouse ES cells can be inhibited by either inactivation of the adenomatous polyposis coli (APC) protein, which regulates the phosphorylation of β-catenin through GSK3β (Fig. 6A), or the introduction of a dominant active form of β-catenin (Haegele et al., 2003). The Wnt antagonist Sfrp2 is expressed during the neural differentiation of EBs and expression of Sfrp2 enhances neuronal differentiation (Aubert et al., 2002). During EB formation, both CS and HS are upregulated (Fig. 1). Therefore, we analyzed the influence of their simultaneous sulfation in the differentiation of EBs (which are comprised of the three germ layers, endoderm, mesoderm and ectoderm) using *PAPST1*-KD or *PAPST2*-KD ES cells.

To maintain the knockdown effects during the long culture period required for EB formation, we used stable *PAPST1*-KD or *PAPST2*-KD ES cells. Before EB formation, both *PAPST1*-KD and *PAPST2*-KD cells showed an approximately 50% reduction in the targeted mRNA compared to control cells. We examined the expression of several germ layer markers by real time PCR after EB formation (Fig. 9) (Sasaki et al., 2009). The expression of neurectoderm markers (Mash1, Pax6) increased in a time-dependent manner and the expression in *PAPST1*-KD or *PAPST2*-KD cells was higher than in control cells, indicating that neurectodermal differentiation was promoted in these KD cells. The expression of early mesoderm markers (Brachyury, Goosecoid) and a primitive ectoderm marker (Fgf5) decreased in a time-dependent manner and the expression in *PAPST1*-KD or *PAPST2*-KD cells was lower than in control cells, indicating that differentiation of primitive ectodermal and mesodermal cells was inhibited in *PAPST1*-KD or *PAPST2*-KD cells. Expression of ExE lineage markers (Gata6, Laminin B1 and Bmp2) initially increased and reached a maximum level 8 days after EB formation, after which it decreased. The expression of these genes was lower in *PAPST1*-KD or *PAPST2*-KD cells than in control cells, indicating that endodermal differentiation decreased in *PAPST1*-KD or *PAPST2*-KD cell cultures. These results indicate that *in vitro* differentiation in *PAPST1*-KD or *PAPST2*-KD cells is abnormal and that sulfation contributes to differentiation of mouse ES cells.

6.2 Reduced sulfation promotes neurogenesis

The observation that *PAPST1* or *PAPST2* knockdown promoted the differentiation of mouse ES cells into neurectoderm (see section 6.1 above) prompted us to investigate neural differentiation in *PAPST1*-KD or *PAPST2*-KD cells (Sasaki et al., 2009). The expression of several neural markers, such as neural stem/progenitor cell markers (Nestin, Musashi-1) and proneural markers (Mash1, Math1, NeuroD1 and NeuroD2) was higher in *PAPST1*-KD or *PAPST2*-KD cells than control cells in the presence of all-trans retinoic acid (RA) (Fig.10A). Similar results were obtained in the absence of RA. These findings indicate there is a higher frequency of neural stem/neural progenitor cells and neural precursor cells amongst the *PAPST1*- or *PAPST2*-KD -KD cells.

The ability of *PAPST1*-KD or *PAPST2*-KD cells to differentiate into neurons was then examined. At 6 days after replating EBs treated with RA, *PAPST1*-KD or *PAPST2*-KD cells generated dense networks of neurite outgrowths as compared to control cells (Fig. 10B). Western blotting analysis showed that the level of βIII-tubulin in *PAPST1*-KD or *PAPST2*-KD cells was quantitatively higher than that in control cells (Sasaki et al., 2009). FACS analysis also showed that βIII-tubulin positive cells were more abundant in *PAPST1*-KD or *PAPST2*-KD cell cultures than in control cells (Fig. 10C). Similar results were obtained for EBs not treated with RA. These findings demonstrate that sulfation contributes to neurogenesis in mouse ES cells.

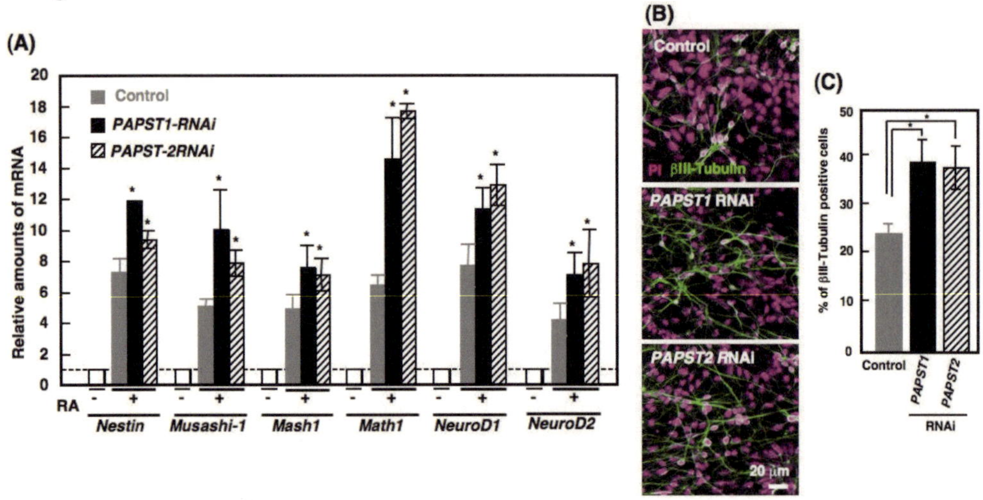

Fig. 10. Neurogenesis is promoted in *PAPST1*-KD or *PAPST2*-KD cells.
Neuronal differentiation was induced by addition of RA on days 4 and 6 after EB formation. On day 8, EBs were replated onto PDL/laminin-coated dishes in DMEM-F12 containing N2 supplement. The medium was replaced every other day and the cells were incubated for 6 days (Bain et al., 1995).
(A) Real time PCR analysis of neural differentiation markers 8 days after EB formation in the presence of RA. The results are shown after normalization against the values obtained with control cells not treated with RA (value=1). The values shown are the means ± SD of duplicate measurements from two independent experiments and significant values are indicated; *$P < 0.05$, in comparison to the control.
(B) Immunocytochemical staining 6 days after replating of EBs treated with RA. Representative confocal images from two independent experiments are shown. (βIII-tubulin, green; PI, purple). Scale bar, 20 μm.
(C) FACS analysis using an anti-βIII-tubulin at 6 days after replating of EBs treated with RA. Three independent experiments were performed and representative results are shown. The histograms show the ratio of the mean fluorescence intensity within the area representing βIII-tubulin positive cells to the mean fluorescence intensity over the total area ± SD of three independent experiments. Significant values are indicated: *$P < 0.01$, in comparison to the control

Recently it was reported that ES cells from HS-null mice show abnormal neural differentiation due to defects in FGF4 signaling (Johnson et al., 2007). The protocol used to induce neural differentiation in this study differs from the one used in our laboratory. Johnson used a Sox1-EGFP reporter cell line in an adherent cell culture; under these conditions, neural differentiation is induced by autocrine FGF4 signaling (Johnson et al., 2007). In contrast, we used EB formation plus RA treatment; under these conditions autocrine signaling by BMP and Wnt inhibits neural differentiation (Aubert et al., 2002; Haegele et al., 2003). Thus, the variant outcomes with respect to neural differentiation might be caused by the use of dissimilar culture conditions that induce neural differentiation by different signaling pathways.

6.3 Sulfation of both HS and CS regulates several signaling pathways required for the correct differentiation of mouse ES cells during EB formation

Several signaling pathways, such as the BMP, FGF and Wnt pathways, play important roles in the mouse embryo during early embryogenesis and mouse ES cell differentiation (Loebel et al., 2003) (see section 6.1 above). We therefore examined whether defects in these signaling pathways contribute to the abnormal differentiation of *PAPST1*-KD or *PAPST2*-KD EBs (Fig. 9), especially the promotion of neurogenesis (Fig.10) (Sasaki et al., 2009).

The nuclear accumulation of β-catenin and the levels of phosphorylated ERK1/2 and Smad1 were reduced in *PAPST1*-KD or *PAPST2*-KD cells as compared to control cells (Fig.11A), indicating that Wnt/β-catenin, FGF/ERK and BMP/Smad signaling were reduced in these EBs. Furthermore, all of these signals were reduced in EBs depleted for HS and CS chains in the absence of RA (Fig. 11B). In the presence of RA, HS depletion reduced signaling via all of these pathways as compared with non-depleted EBs. By contrast, CS depletion reduced FGF/ERK and BMP/Smad signaling to a similar extent as HS depletion but promoted Wnt/β-catenin signaling (Fig. 11B).

Next, we performed a surface plasmon resonance (SPR) analysis for Wnt3a and BMP4 against heparin, a structural analogue of HS chains, and CS-E (GlcAβ1,3GalNAc(4,6SO_3)), a particular form of CS chain. Wnt3a and BMP4 bind to both heparin and CS-E (Table 1) (Sasaki et al., 2009; Sasaki et al., 2008). The KD values for the binding of bFGF and FGF4 to HS and to CS-E have been also determined (Deepa et al., 2002; Sugaya et al., 2008). The analysis clearly shows that the sulfate groups of HS and CS contribute to the binding of Wnt3a, BMP4, bFGF and FGF4 to both HS and CS.

These results demonstrate that sulfation on both HS and CS regulates BMP/Smad, FGF/ERK and Wnt/β-catenin signaling during EB formation. In addition, the reduction in such signaling contributes to the abnormal differentiation of *PAPST1*-KD or *PAPST2*-KD cells, such as the promotion of neurogenesis.

Ligand	GAG	k_a ($M^{-1}Sec^{-1}$)	k_d (Sec^{-1})	K_D (nM)
Wnt3a	Heparin	2.22×10^5	5.77×10^{-3}	26.0
Wnt3a	CS-E	8.26×10^5	2.26×10^{-2}	27.3
BMP4	Heparin	2.76×10^5	1.92×10^{-2}	69.4
BMP4	CS-E	1.44×10^5	4.33×10^{-3}	30.0

The ka, kd and KD values were determined by SPR analysis.

Table 1. The apparent association (ka), dissociation (kd) rate constants and equilibrium dissociation constants (KD) for the interaction of Wnt3a and BMP4 with immobilized heparin or CS-E.

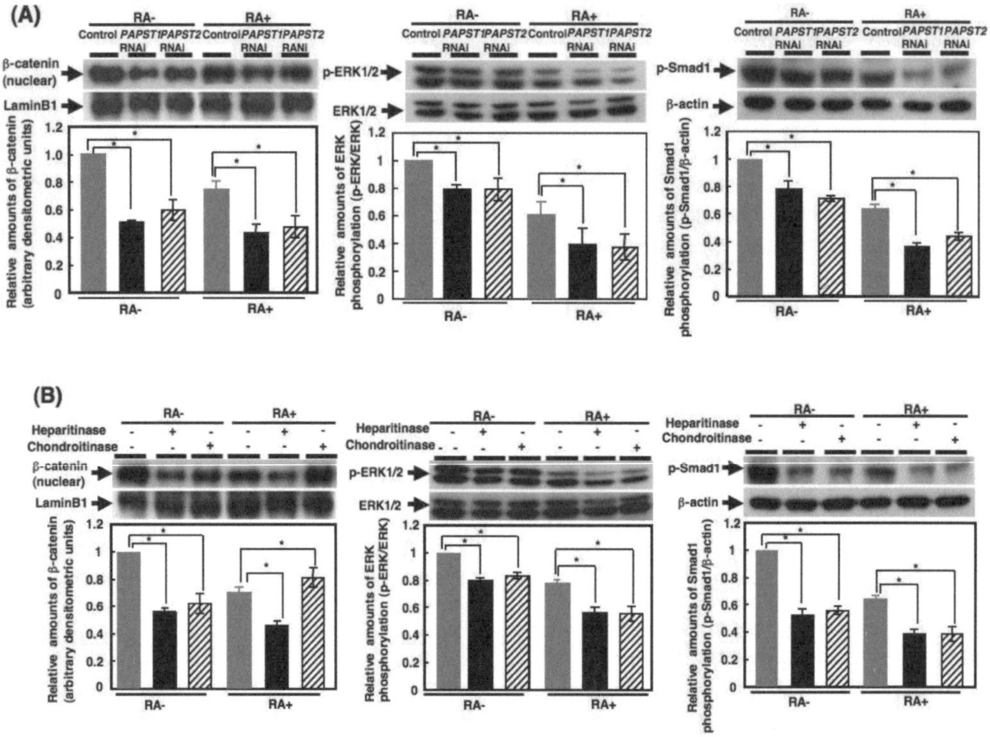

Fig. 11. *PAPST1*- or *PAPST2*-KD cells show decreased signaling in a number of pathways during EB formation.

(A) Western blot analysis of several signaling molecules in EBs on day 8. Two independent experiments were performed and representative results are shown. The histograms show mean densitometric readings ± SD of β-catenin or the phosphorylated proteins/loading controls after normalization against the values obtained with control cells in the absence of RA (value=1). Values were obtained from duplicate measurements of two independent experiments and significant values are indicated; *P < 0.01, in comparison to the control.

(B) Western blot analysis of several signaling molecules in EBs on day 8 after heparitinase or chondroitinase treatment. Two independent experiments were performed and representative results are shown. The histograms show mean densitometric readings ± SD of β-catenin or the phosphorylated proteins/loading controls after normalization against the values obtained with cells not treated with RA and enzyme (value=1). Values were obtained from duplicate measurements of two independent experiments and significant values are indicated; *P < 0.01, in comparison to cells not treated with enzyme.

7. Conclusion

Glycan structures on mouse ES cells have roles in the maintenance and differentiation of mouse ES cells. In particular, the sulfated glycans, HS and CS, function in the biologically important signaling pathways involving Wnt, BMP and FGF. Through these signaling pathways, glycan structures function in the maintenance of self-renewal and pluripotency and also in differentiation (Fig. 12).

The pluripotency of mouse ES cells in adherent cell culture is maintained by a balance among extrinsic signaling pathways, such as LIF, BMP, Wnt and FGF signaling, and also by a combination of extrinsic and intrinsic factors, such as Oct3/4 and Nanog (Boiani & Scholer, 2005; Chambers & Smith, 2004; Ivanova et al., 2006; Sasaki et al., 2008; Zhang & Li, 2005). In undifferentiated mouse ES cells, the sulfation of HS, but not of CS, regulates extrinsic signaling by BMP4, Wnt, bFGF and FGF4 (Fig. 12B). BMP4 and autocrine/paracrine Wnt maintain self-renewal and pluripotency by inhibiting neural differentiation (Ying et al., 2003; Zhang & Li, 2005) and maintaining the Nanog expression level, respectively (Sasaki et al., 2008; Sato et al., 2004). Autocrine/paracrine bFGF signaling contributes to the growth of mouse ES cells (Dvorak et al., 1998), while FGF4 signaling contributes to initiation of differentiation by mouse ES cells (Kunath et al., 2007).

In EBs, both HS and CS regulate signaling by BMP, FGF or Wnt and, through their interaction with these signaling pathways, they regulate the differentiation of EBs (Fig. 12C). Wnt and BMP signaling inhibit ectodermal differentiation and contribute to mesodermal and definitive endodermal differentiation (Aubert et al., 2002; Finley et al., 1999; Gratsch & O'Shea, 2002; Haegele et al., 2003; Loebel et al., 2003; Yoshikawa et al., 1997). Sulfation of both HS and CS contributes to the decision between ectodermal and mesodermal fates by regulating these signals. FGF/ERK and FGF/Akt signaling contribute to mesodermal and definitive endodermal differentiation and primitive ectodermal and visceral endodermal differentiation, respectively (Loebel et al., 2003) (Chen et al., 2000; Kimelman, 2006).

During neural differentiation of EBs after RA treatment, HS and CS regulate the extrinsic signaling by BMP, Wnt and FGF that inhibits or is required for neuronal differentiation (Fig. 12D). The transduction of extrinsic signals is dependent on the sulfation of both HS and CS and results in neuronal differentiation. Wnt and BMP signaling inhibit neurogenesis in mouse ES cells via EB formation (Aubert et al., 2002; Haegele et al., 2003). In contrast, FGF (e.g., bFGF) signaling may promote neurogenesis. On the other hand, CS has a negative effect on Wnt signaling, presumably by sequestering Wnt proteins and preventing them interacting with Wnt receptors. Thus, we propose that CS promotes the differentiation of neural stem/progenitor cells into the neuronal lineage.

In this chapter, we have outlined the functions of the sulfated glycans that are present on mouse ES cells. Some of these functions should also hold true for human ES and iPS cells. This conclusion is supported by our observation that sulfation has a similar function during neuronal differentiation in human iPS cells as in mouse ES cells (Sasaki et al., 2010). However, bFGF/ERK signaling is also regulated by HS and contributes to proliferation in mouse ES cells, whereas in human ES cells, FGF signals have a role in the maintenance of self-renewal and pluripotency (Xu et al., 2005). Indeed, human ES cells are stimulated to proliferate by addition of heparin or HS proteoglycan to the culture medium (Furue et al., 2008; Levenstein et al., 2008). In light of the currently available information, it is clear that further studies on the roles of the sulfated glycans, HS and CS, should provide greater insight into and understanding of the maintenance of self-renewal and pluripotency in human ES and iPS cells as well as in mouse ES cells and should also help to elucidate the regulation of differentiation in stem cells.

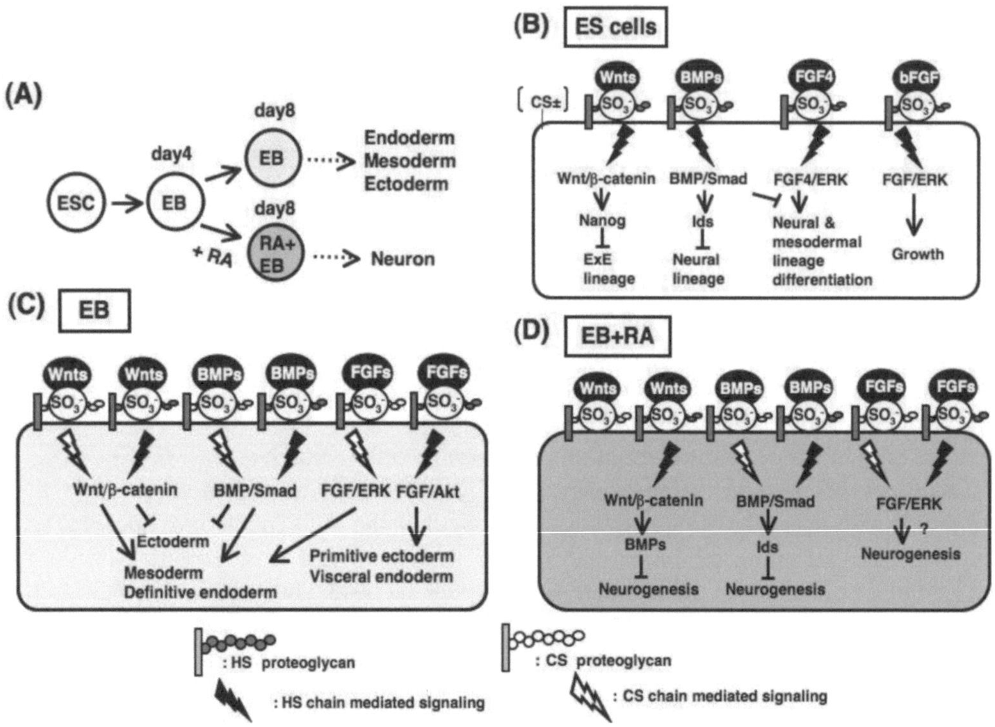

Fig. 12. Sulfated glycans contribute to the maintenance and differentiation of mouse ES cells. (A) *In vitro* differentiation flowchart of mouse ES cells. EBs that are not treated with RA produce cells from all three germ layers (endoderm, mesoderm and ectoderm), whereas RA-treated EBs produce neurons after further adherent culture.

(B) HS regulates the extrinsic signaling (by BMP and Wnt) that is required for the pluripotency of mouse ES cells. In undifferentiated mouse ES cells, the transduction of extrinsic signals is dependent on the sulfation of HS, but not CS, and this maintains pluripotency, the undifferentiated state and growth. Autocrine/paracrine bFGF signaling contributes to the growth of mouse ES cells, while FGF4 signaling contributes to the initiation of differentiation in the mouse ES cells.

(C) Both HS and CS regulate the extrinsic signaling (by BMP, FGF and Wnt) that is required for normal differentiation of EBs. During EB differentiation into the three germ layers, the transduction of the extrinsic signals is dependent on the sulfation of both HS and CS. Wnt and BMP signaling inhibit ectodermal differentiation and contribute to mesodermal and definitive endodermal differentiation. FGF/ERK and FGF/Akt signaling contribute to mesodermal and definitive endodermal differentiation and primitive ectodermal and visceral endodermal differentiation, respectively.

(D) HS and CS regulate the extrinsic signaling (by BMP, Wnt and FGF) that inhibits or is required for neuronal differentiation of RA-treated EBs. During RA-treated EB differentiation, the transduction of extrinsic signals is dependent on the sulfation of both HS and CS and results in neuronal differentiation. Wnt and BMP signaling inhibit neurogenesis and FGF (e.g., bFGF) signaling may promote neurogenesis. CS regulates Wnt signaling negatively, presumably by sequestering Wnt proteins and preventing them interacting with Wnt receptors.

8. References

Adewumi, O.; Aflatoonian, B.; Ahrlund-Richter, L.; Amit, M.; Andrews, P. W.; Beighton, G.; Bello, P. A.; Benvenisty, N.; Berry, L. S.; Bevan, S., *et al.* (2007). Characterization of human embryonic stem cell lines by the International Stem Cell Initiative. *Nat Biotechnol*, 25, 7, 803-816, 1087-0156

Atwood, J. A., 3rd; Cheng, L.; Alvarez-Manilla, G.; Warren, N. L.; York, W. S. & Orlando, R. (2008). Quantitation by isobaric labeling: applications to glycomics. *J Proteome Res*, 7, 1, 367-374, 1535-3893

Aubert, J.; Dunstan, H.; Chambers, I. & Smith, A. (2002). Functional gene screening in embryonic stem cells implicates Wnt antagonism in neural differentiation. *Nat Biotechnol*, 20, 12, 1240-1245, 1087-0156

Badcock, G.; Pigott, C.; Goepel, J. & Andrews, P. W. (1999). The human embryonal carcinoma marker antigen TRA-1-60 is a sialylated keratan sulfate proteoglycan. *Cancer Res*, 59, 18, 4715-4719, 0008-5472

Bain, G.; Kitchens, D.; Yao, M.; Huettner, J. E. & Gottlieb, D. I. (1995). Embryonic stem cells express neuronal properties *in vitro*. *Dev Biol*, 168, 2, 342-357, 0012-1606

Bishop, J. R.; Schuksz, M. & Esko, J. D. (2007). Heparan sulphate proteoglycans fine-tune mammalian physiology. *Nature*, 446, 7139, 1030-1037, 0028-0836

Boeuf, H.; Hauss, C.; Graeve, F. D.; Baran, N. & Kedinger, C. (1997). Leukemia inhibitory factor-dependent transcriptional activation in embryonic stem cells. *J Cell Biol*, 138, 6, 1207-1217, 0021-9525

Boiani, M. & Scholer, H. R. (2005). Regulatory networks in embryo-derived pluripotent stem cells. *Nat Rev Mol Cell Biol*, 6, 11, 872-884, 1471-0072

Cartwright, P.; McLean, C.; Sheppard, A.; Rivett, D.; Jones, K. & Dalton, S. (2005). LIF/STAT3 controls ES cell self-renewal and pluripotency by a Myc-dependent mechanism. *Development*, 132, 5, 885-896, 0950-1991

Chambers, I. & Smith, A. (2004). Self-renewal of teratocarcinoma and embryonic stem cells. *Oncogene*, 23, 43, 7150-7160, 0950-9232

Chen, Y.; Li, X.; Eswarakumar, V. P.; Seger, R. & Lonai, P. (2000). Fibroblast growth factor (FGF) signaling through PI 3-kinase and Akt/PKB is required for embryoid body differentiation. *Oncogene*, 19, 331, 3750-3756, 0950-9232

Cole, M. F.; Johnstone, S. E.; Newman, J. J.; Kagey, M. H. & Young, R. A. (2008). Tcf3 is an integral component of the core regulatory circuitry of embryonic stem cells. *Genes Dev*, 22, 6, 746-755, 0890-9369

Deepa, S. S.; Umehara, Y.; Higashiyama, S.; Itoh, N. & Sugahara, K. (2002). Specific molecular interactions of oversulfated chondroitin sulfate E with various heparin-binding growth factors. Implications as a physiological binding partner in the brain and other tissues. *J Biol Chem*, 277, 46, 43707-43716, 0021-9258

Dvorak, P.; Hampl, A.; Jirmanova, L.; Pacholikova, J. & Kusakabe, M. (1998). Embryoglycan ectodomains regulate biological activity of FGF-2 to embryonic stem cells. *J Cell Sci*, 111 (Pt 19), 2945-2952, 0021-9533

Evans, M. J. & Kaufman, M. H. (1981). Establishment in culture of pluripotential cells from mouse embryos. *Nature*, 292, 5819, 154-156, 0028-0836

Finley, M. F.; Devata, S. & Huettner, J. E. (1999). BMP-4 inhibits neural differentiation of murine embryonic stem cells. *J Neurobiol*, 40, 3, 271-287, 0022-3034

Furue, M. K.; Na, J.; Jackson, J. P.; Okamoto, T.; Jones, M.; Baker, D.; Hata, R.; Moore, H. D.; Sato, J. D. & Andrews, P. W. (2008). Heparin promotes the growth of human embryonic stem cells in a defined serum-free medium. *Proc Natl Acad Sci U S A*, 105, 36l, 13409-13414, 0027-8424

Goda, E.; Kamiyama, S.; Uno, T.; Yoshida, H.; Ueyama, M.; Kinoshita-Toyoda, A.; Toyoda, H.; Ueda, R. & Nishihara, S. (2006). Identification and characterization of a novel *Drosophila* 3'-phosphoadenosine 5'-phosphosulfate transporter. *J Biol Chem*, 281, 39, 28508-28517, 0021-9258

Gratsch, T. E. & O'Shea, K. S. (2002). Noggin and chordin have distinct activities in promoting lineage commitment of mouse embryonic stem (ES) cells. *Dev Biol*, 245, 1, 83-94, 0012-1606

Haegele, L.; Ingold, B.; Naumann, H.; Tabatabai, G.; Ledermann, B. & Brandner, S. (2003). Wnt signalling inhibits neural differentiation of embryonic stem cells by controlling bone morphogenetic protein expression. *Mol Cell Neurosci*, 24, 3, 696-708, 1044-7431

Ivanova, N.; Dobrin, R.; Lu, R.; Kotenko, I.; Levorse, J.; DeCoste, C.; Schafer, X.; Lun, Y. & Lemischka, I. R. (2006). Dissecting self-renewal in stem cells with RNA interference. *Nature*, 442, 7102, 533-538, 0028-0836

Johnson, C. E.; Crawford, B. E.; Stavridis, M.; Ten Dam, G.; Wat, A. L.; Rushton, G.; Ward, C. M.; Wilson, V.; van Kuppevelt, T. H.; Esko, J. D.; Smith, A.; Gallagher, J. T. & Merry, C. L. (2007). Essential alterations of heparan sulfate during the differentiation of embryonic stem cells to Sox1-enhanced green fluorescent protein-expressing neural progenitor cells. *Stem Cells*, 25, 8, 1913-1923, 1066-5099

Kamiyama, S.; Sasaki, N.; Goda, E.; Ui-Tei, K.; Saigo, K.; Narimatsu, H.; Jigami, Y.; Kannagi, R.; Irimura, T. & Nishihara, S. (2006). Molecular cloning and characterization of a novel 3'-phosphoadenosine 5'-phosphosulfate transporter, PAPST2. *J Biol Chem*, 281, 16, 10945-10953, 0021-9258

Kamiyama, S.; Suda, T.; Ueda, R.; Suzuki, M.; Okubo, R.; Kikuchi, N.; Chiba, Y.; Goto, S.; Toyoda, H.; Saigo, K.; Watanabe, M.; Narimatsu, H.; Jigami, Y. & Nishihara, S. (2003). Molecular cloning and identification of 3'-phosphoadenosine 5'-phosphosulfate transporter. *J Biol Chem*, 278, 28, 25958-25963, 0021-9258

Kannagi, R.; Cochran, N. A.; Ishigami, F.; Hakomori, S.; Andrews, P. W.; Knowles, B. B. & Solter, D. (1983). Stage-specific embryonic antigens (SSEA-3 and -4) are epitopes of a unique globo-series ganglioside isolated from human teratocarcinoma cells. *EMBO J*, 2, 12, 2355-2361, 0261-4189

Kimelman, D. (2006). Mesoderm induction: from caps to chips. *Nat Rev Genet*, 7, 5, 360-372, 1471-0056

Kraushaar, D. C.; Yamaguchi, Y. & Wang, L. (2010). Heparan sulfate is required for embryonic stem cells to exit from self-renewal. *J Biol Chem*, 285, 8l, 5907-5916, 0021-9258

Kunath, T.; Saba-El-Leil, M. K.; Almousailleakh, M.; Wray, J.; Meloche, S. & Smith, A. (2007). FGF stimulation of the Erk1/2 signalling cascade triggers transition of pluripotent embryonic stem cells from self-renewal to lineage commitment. *Development*, 134, 16, 2895-2902, 0950-1991

Lamanna, W. C.; Baldwin, R. J.; Padva, M.; Kalus, I.; Ten Dam, G.; van Kuppevelt, T. H.; Gallagher, J. T.; von Figura, K.; Dierks, T. & Merry, C. L. (2006). Heparan sulfate 6-*O*-endosulfatases: discrete in vivo activities and functional co-operativity. *Biochem J*, 400, 1, 63-73, 0264-6021

Lanner, F.; Lee, K. L.; Sohl, M.; Holmborn, K.; Yang, H.; Wilbertz, J.; Poellinger, L.; Rossant, J. & Farnebo, F. (2010). Heparan sulfation-dependent fibroblast growth factor signaling maintains embryonic stem cells primed for differentiation in a heterogeneous state. *Stem Cells*, 28, 2, 191-200, 1066-5099

Levenstein, M. E.; Berggren, W. T.; Lee, J. E.; Conard, K. R.; Llanas, R. A.; Wagner, R. J.; Smith, L. M. & Thomson, J. A. (2008). Secreted proteoglycans directly mediate human embryonic stem cell-basic fibroblast growth factor 2 interactions critical for proliferation. *Stem Cells*, 26, 12, 3099-3107, 1066-5099

Li, J. P.; Gong, F.; Hagner-McWhirter, A.; Forsberg, E.; Abrink, M.; Kisilevsky, R.; Zhang, X. & Lindahl, U. (2003). Targeted disruption of a murine glucuronyl C5-epimerase gene results in heparan sulfate lacking L-iduronic acid and in neonatal lethality. *J Biol Chem*, 278, 31, 28363-28366, 0021-9258

Lin, X. (2004). Functions of heparan sulfate proteoglycans in cell signaling during development. *Development*, 131, 24, 6009-6021, 0950-1991

Loebel, D. A.; Watson, C. M.; De Young, R. A. & Tam, P. P. (2003). Lineage choice and differentiation in mouse embryos and embryonic stem cells. *Dev Biol*, 264, 1, 1-14, 0012-1606

Martin, G. R. (1981). Isolation of a pluripotent cell line from early mouse embryos cultured in medium conditioned by teratocarcinoma stem cells. *Proc Natl Acad Sci U S A*, 78, 12, 7634-7638, 0027-8424

Matsuda, T.; Nakamura, T.; Nakao, K.; Arai, T.; Katsuki, M.; Heike, T. & Yokota, T. (1999). STAT3 activation is sufficient to maintain an undifferentiated state of mouse embryonic stem cells. *EMBO J*, 18, 15, 4261-4269, 0261-4189

Merry, C. L.; Bullock, S. L.; Swan, D. C.; Backen, A. C.; Lyon, M.; Beddington, R. S.; Wilson, V. A. & Gallagher, J. T. (2001). The molecular phenotype of heparan sulfate in the Hs2st-/- mutant mouse. *J Biol Chem*, 276, 38, 35429-35434, 0021-9258

Miyabayashi, T.; Teo, J. L.; Yamamoto, M.; McMillan, M.; Nguyen, C. & Kahn, M. (2007). Wnt/beta-catenin/CBP signaling maintains long-term murine embryonic stem cell pluripotency. *Proc Natl Acad Sci U S A*, 104, 13, 5668-5673, 0027-8424

Muramatsu, T. & Muramatsu, H. (2004). Carbohydrate antigens expressed on stem cells and early embryonic cells. *Glycoconj J*, 21, 1-2, 41-45, 0282-0080

Nairn, A. V.; Kinoshita-Toyoda, A.; Toyoda, H.; Xie, J.; Harris, K.; Dalton, S.; Kulik, M.; Pierce, J. M.; Toida, T.; Moremen, K. W. & Linhardt, R. J. (2007). Glycomics of proteoglycan biosynthesis in murine embryonic stem cell differentiation. *J Proteome Res*, 6, 11, 4374-4387, 1535-3893

Nishihara, S. (2009). The function of glycan structures expressed on embryonic stem cells. *Trends in Glycoscience and Glycotechnology*, 21, 120, 207-218, 1883-2113

Niwa, H.; Burdon, T.; Chambers, I. & Smith, A. (1998). Self-renewal of pluripotent embryonic stem cells is mediated via activation of STAT3. *Genes Dev*, 12, 13, 2048-2060, 0890-9369

Ogawa, K.; Saito, A.; Matsui, H.; Suzuki, H.; Ohtsuka, S.; Shimosato, D.; Morishita, Y.; Watabe, T.; Niwa, H. & Miyazono, K. (2007). Activin-Nodal signaling is involved in propagation of mouse embryonic stem cells. *J Cell Sci*, 120, Pt 1, 55-65, 0021-9533

Qi, X.; Li, T. G.; Hao, J.; Hu, J.; Wang, J.; Simmons, H.; Miura, S.; Mishina, Y. & Zhao, G. Q. (2004). BMP4 supports self-renewal of embryonic stem cells by inhibiting mitogen-activated protein kinase pathways. *Proc Natl Acad Sci U S A*, 101, 16, 6027-6032, 0027-8424

Raz, R.; Lee, C. K.; Cannizzaro, L. A.; d'Eustachio, P. & Levy, D. E. (1999). Essential role of STAT3 for embryonic stem cell pluripotency. *Proc Natl Acad Sci U S A*, 96, 6, 2846-2851, 0027-8424

Sasaki, N.; Hirano, T.; Ichimiya, T.; Wakao, M.; Hirano, K.; Kinoshita-Toyoda, A.; Toyoda, H.; Suda, Y. & Nishihara, S. (2009). The 3'-phosphoadenosine 5'-phosphosulfate transporters, PAPST1 and 2, contribute to the maintenance and differentiation of mouse embryonic stem cells. *PLoS One*, 4, 12, e8262, 1932-6203

Sasaki, N.; Okishio, K.; Ui-Tei, K.; Saigo, K.; Kinoshita-Toyoda, A.; Toyoda, H.; Nishimura, T.; Suda, Y.; Hayasaka, M.; Hanaoka, K.; Hitoshi, S.; Ikenaka, K. & Nishihara, S. (2008). Heparan sulfate regulates self-renewal and pluripotency of embryonic stem cells. *J Biol Chem*, 283, 6, 3594-3606, 0021-9258

Sasaki, N.; Hirano, T.; Kobayashi, K.; Toyoda, M.; Miyakawa, Y.; Okita, H.; Kiyokawa, N.; Akutsu, H.; Umezawa, A.; Nishihara S. (2010). Chemical inhibition of sulfation accelerates neural differentiation of mouse embryonic stem cells and human induced pluripotent stem cells. Biochem Biophys Res Commun, 401, 3, 480-486.

Sato, N.; Meijer, L.; Skaltsounis, L.; Greengard, P. & Brivanlou, A. H. (2004). Maintenance of pluripotency in human and mouse embryonic stem cells through activation of Wnt signaling by a pharmacological GSK-3-specific inhibitor. *Nat Med*, 10, 1, 55-63, 1078-8956

Satomaa, T.; Heiskanen, A.; Mikkola, M.; Olsson, C.; Blomqvist, M.; Tiittanen, M.; Jaatinen, T.; Aitio, O.; Olonen, A.; Helin, J., *et al.* (2009). The *N*-glycome of human embryonic stem cells. *BMC Cell Biol*, 10, 42, 1471-2121

Smith, A. G.; Heath, J. K.; Donaldson, D. D.; Wong, G. G.; Moreau, J.; Stahl, M. & Rogers, D. (1988). Inhibition of pluripotential embryonic stem cell differentiation by purified polypeptides. *Nature*, 336, 6200, 688-690, 0028-0836

Sugaya, N.; Habuchi, H.; Nagai, N.; Ashikari-Hada, S. & Kimata, K. (2008). 6-*O*-sulfation of heparan sulfate differentially regulates various fibroblast growth factor-dependent signalings in culture. *J Biol Chem*, 283, 16, 10366-10376, 0021-9258

Tabata, T. & Takei, Y. (2004). Morphogens, their identification and regulation. *Development*, 131, 4, 703-712, 0950-1991

Thomson, J. A.; Itskovitz-Eldor, J.; Shapiro, S. S.; Waknitz, M. A.; Swiergiel, J. J.; Marshall, V. S. & Jones, J. M. (1998). Embryonic stem cell lines derived from human blastocysts. *Science*, 282, 5391, 1145-1147, 0036-8075

Ueyama, M.; Takemae, H.; Ohmae, Y.; Yoshida, H.; Toyoda, H.; Ueda, R. & Nishihara, S. (2008). Functional analysis of proteoglycan galactosyltransferase II RNA interference mutant flies. *J Biol Chem*, 283, 10, 6076-6084, 0021-9258

Williams, R. L.; Hilton, D. J.; Pease, S.; Willson, T. A.; Stewart, C. L.; Gearing, D. P.; Wagner, E. F.; Metcalf, D.; Nicola, N. A. & Gough, N. M. (1988). Myeloid leukaemia inhibitory factor maintains the developmental potential of embryonic stem cells. *Nature*, 336, 6200, 684-687, 0028-0836

Xu, R. H.; Peck, R. M.; Li, D. S.; Feng, X.; Ludwig, T. & Thomson, J. A. (2005). Basic FGF and suppression of BMP signaling sustain undifferentiated proliferation of human ES cells. *Nat Methods*, 2, 3, 185-190, 1548-7091

Yan, D. & Lin, X. (2009). Shaping morphogen gradients by proteoglycans. *Cold Spring Harb Perspect Biol*, 1, 3, a002493, 1943-0264

Ying, Q. L.; Nichols, J.; Chambers, I. & Smith, A. (2003). BMP induction of Id proteins suppresses differentiation and sustains embryonic stem cell self-renewal in collaboration with STAT3. *Cell*, 115, 3, 281-292, 0092-8674

Ying, Q. L.; Wray, J.; Nichols, J.; Batlle-Morera, L.; Doble, B.; J., W.; Cohen, P. & Smith, A. (2008). The ground state of embryonic stem cell self-renewal. *Nature*, 453, 7194, 519-523, 0028-0836

Yoshikawa, Y.; Fujimori, T.; McMahon, A. P. & Takada, S. (1997). Evidence that absence of Wnt-3a signaling promotes neuralization instead of paraxial mesoderm development in the mouse. *Dev Biol*, 183, 2, 234-242, 0012-1606

Zhang, J. & Li, L. (2005). BMP signaling and stem cell regulation. *Dev Biol*, 284, 1, 1-11, 0012-1606

7

The Function of E-cadherin in ES Cell Pluripotency

Kate Hawkins and Christopher M Ward
The University of Manchester
United Kingdom

1. Introduction

Stem cells have the potential to revolutionise medicine, providing treatment options for a wide range of diseases that currently lack a long term control strategy. In order for us to harness this potential, however, we need a thorough understanding of their self-renewal and differentiation capacities. Differentiation capacity, or 'potency', can be defined as the ability of a cell to give rise to every cell type within the developing embryo and its supporting tissues (totipotency), every tissue of the three germ layers: endoderm, ectoderm and mesoderm (pluripotency, Fig. 1.) or a restricted range of cell types (multipotency). The potency of a cell is dictated to some extent by the stage at which it is isolated from the embryo, which in turn affects its gene expression profile. For example, whereas embryonic stem (ES) cells are isolated from the inner cell mass (ICM) of the pre-implantation blastocyst, EpiS cells are derived from the later stage epiblast and exhibit significantly altered gene expression compared to mouse ES (mES) cells (Tesar *et al.*, 2007).

mES cells were first isolated from the ICM of the mouse blastocyst by two independent groups in 1981 (Evans and Kaufman, 1981; Martin, 1981) and have since been used as a model system in which to study mechanisms of development and disease. ES cells have also been isolated from other species including pig (Notarianni *et al.*, 1990), rabbit (Graves and Moreadith, 1993) and chicken (Pain *et al.*, 1996). In addition, pluripotent stem cells have been derived from cleavage-stage embryos, individual blastomeres (Chung *et al.*, 2006; Klimanskaya *et al.*, 2006; Wakayama *et al.*, 2007), parthenogenic embryos (Lin *et al.*, 2007; Mai *et al.*, 2007; Revazova *et al.*, 2007), trophectoderm (Tanaka *et al.*, 1998) and extraembryonic endoderm (Kunath *et al.*, 2005). Pioneering work in 1998 by Thomson and colleagues (Thomson *et al.*) resulted in the derivation of human embryonic stem (hES) cells from human blastocysts. Whilst mES and hES cells differ greatly in their gene expression profile (Tesar *et al.*, 2007) and respond to distinct pluripotency-inducing signals (Daheron *et al.*, 2004; Vallier *et al.*, 2005), the core pluripotency regulatory network of Oct4, Sox2 and Nanog is conserved between the two species. Numerous studies have demonstrated the formation of an Oct4/Sox2 heterodimeric complex which is then responsible for activating the expression of a multitude of pluripotency-associated genes (Yuan *et al.*, 1995; Botquin *et al.*, 1998; Nishimoto *et al.*, 1999). Many of the Oct4/Sox2 gene targets have been shown to be shared with Nanog in both mES (Loh *et al.*, 2006) and hES cells (Boyer *et al.*, 2005). Insight into this core

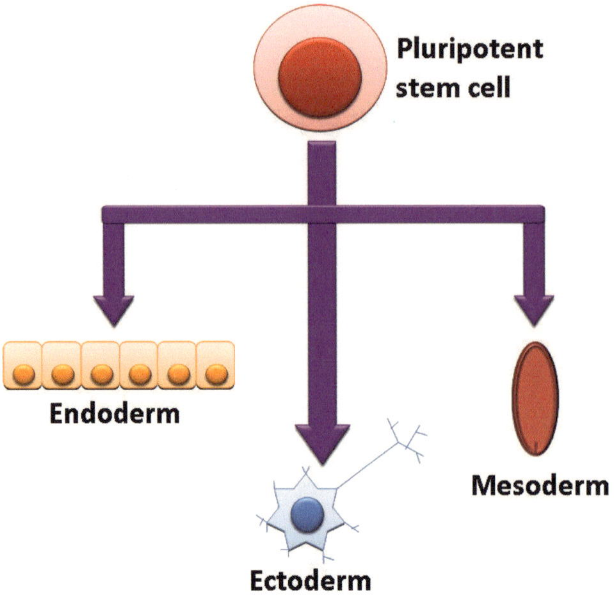

Fig. 1. Pluripotent stem cells are defined as possessing the ability to form all cells of the three primary germ layers (endoderm, ectoderm and mesoderm). Multipotent stem cells can form some of the cells of the germ layers and, therefore, exhibit more restricted potency compared to pluripotent cells.

regulatory network that governs the pluripotent state has recently led to the generation of induced pluripotent stem (iPS) cells from mouse (Takahashi and Yamanaka, 2006) and human somatic cells (Takahashi and Yamanaka, 2006; Yu *et al.*, 2007; Hanna *et al.*, 2008; Stadtfeld *et al.*, 2008). These cells, obtained by overexpression of a cocktail of pluripotency-associated genes (most commonly Oct4, Sox2, c-Myc and Klf4), have the potential to provide patient-specific cells for regenerative medicine strategies. The primary challenge of this research at present is the tumorigenicity of the transplanted cells, in one study 20% of mice injected with these cells developed tumors, likely to be due to the presence of the c-myc oncogene (Takahashi and Yamanaka, 2006). More recently, however, iPS cell generation from murine embryonic fibroblasts (MEFs) has been achieved without genetic alteration using recombinant proteins (Zhou *et al.*, 2009). These cells, termed piPS (protein-induced pluripotent stem) cells, demonstrated both self-renewal and pluripotent capacities both *in vitro* (i.e. embryoid body formation) and *in vivo* (chimera generation). iPS cells have also been generated from patients suffering from a range of disorders including diabetes mellitus and Parkinson's disease (Park *et al.*, 2008), with such cells providing a useful system with which to study the progression of specific disorders. Whereas iPS cells are more likely to be suitable for therapeutic applications (largely due to the ease of matching HLA types of donors to patients), ES cell research continues to provide valuable data about the pluripotent state to inform iPS cell research. This knowledge will be key to the translation of iPS cell technology from the laboratory to a clinical setting.

In recent years, evidence for the function of E-cadherin in regulating self-renewal signalling pathways, cell surface localisation of receptors and survival of ES and iPS cells has emerged.

In this chapter, we discuss the function of E-cadherin in regulating pluripotent signalling pathways in ES cells, the importance of E-cadherin expression in hES cells and a mesenchymal-epithelial transition event that appears to dictate efficient derivation of iPS cells. In addition, we discuss epithelial-mesenchymal transition during human and mouse ES cell differentiation and highlight similarities between this process and tumour cell metastasis.

2. Embryonic stem cell self-renewal

2.1 mES cell self-renewal

Mouse ES cells were originally isolated and maintained in an undifferentiated state by co-culture with mitotically inactivated MEFs. More recently, however, advances have been made to more clearly define the specific combinations of signals that activate key pluripotency pathways. The interleukin-6 family cytokine leukaemia inhibitory factor (LIF) has been shown to be crucial for mES cell self-renewal *in vitro*. Upon LIF engagement, gp130 forms a heterodimeric complex with the LIF receptor β subunit allowing its activation and the subsequent stimulation of three parallel signalling cascades; phosphatidylinositol-3-OH kinase (PI3K)/Akt, Janus kinase (Jak)/signal transducer and activator of transcription 3 (STAT3) and Grb2/mitogen activated protein kinase (MAPK). PI3K/Akt and Jak/STAT3 pathways converge to activate the core pluripotency mediators Sox2 and Nanog whilst the Grb/MAPK pathway inhibits Tbx3-mediated stimulation of Nanog and Sox2 (Niwa *et al.*, 2009) (Fig. 2.).

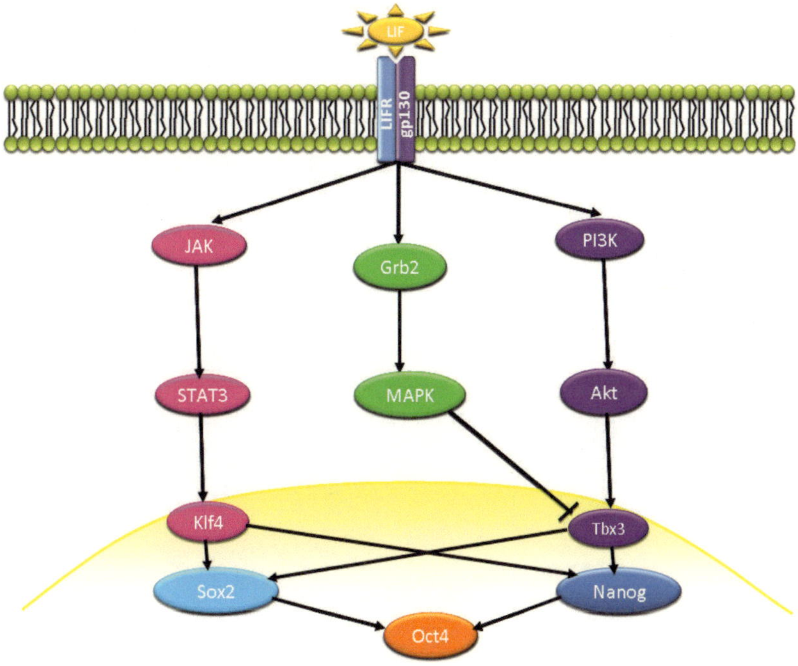

Fig. 2. The LIFR/gp130 signalling pathway stimulates expression of Sox2 and Nanog in mouse ES cells via various intermediate components. Adapted from Niwa et al, 2009 (Niwa *et al.*, 2009).

Whilst LIF has been shown to prevent mesodermal and endodermal differentiation, the transforming growth factor β (Tgf β) family member bone morphogenic protein 4 (Bmp4) is also required to block neuroectoderm lineage specification within mES cell cultures. Bmp4 has been shown to perform this function by activating transcription of the inhibitor of differentiation (Id) genes via Smads1/5/8 (Ying *et al.*, 2003). Together, LIF and BMP4 are the major components of a media formulation known as Clonal Grade Medium (Millipore, Watford, UK) which can maintain mES cell pluripotency in the absence of animal serum. mES cells can also be maintained in an undifferentiated state in the absence of LIF and Bmp4 in medim supplemented with antagonists of MAPK and glycogen synthase 3 (Ying *et al.*, 2008). Additional pathways that have been linked to self-renewal include the PI3K/Akt signalling cascade, likely due to its key role in LIF signal propagation, and the canonical Wnt pathway. Canonical Wnt pathway activation has been shown to support self-renewal in both mES and hES cells (Sato *et al.*, 2004) and, more recently, to derive FAB stem cells, so-called due to the factors used in their isolation (Fibroblast growth factor 2 (Fgf2), Activin and BIO, the latter a Wnt pathway activator) (Chou *et al.*, 2008).

2.2 hES cell self-renewal

Unlike mES cells, hES cell pluripotency is not reliant upon the LIF signalling network (Daheron *et al.*, 2004). Instead, hES cell self-renewal has been shown to be supported by the Tgfβ family ligands Activin and Nodal, in combination with Fgf2 (Vallier *et al.*, 2005) (Fig. 3.). When bound to their Activin-like kinase (Alk) receptors, Activin, Nodal and Tgfβ initiate a signalling cascade involving the phosphorylation of Smads2/3. These transcription factors

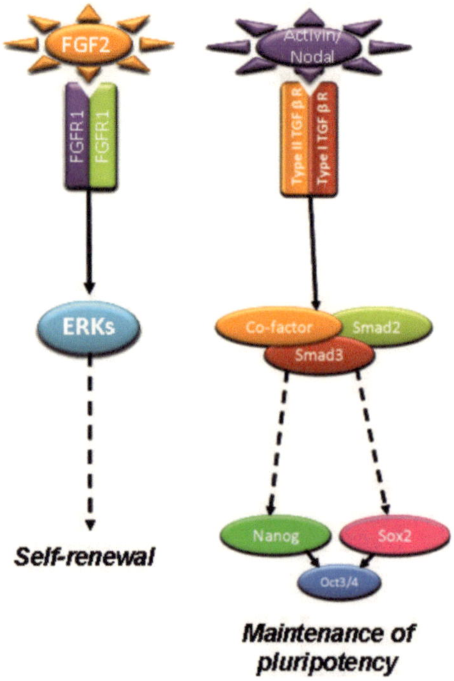

Fig. 3. Pathways associated with pluripotency and self-renewal in hES, mouse E-cadherin-/-ES cells and EpiSCs.

then form a complex with Smad4 allowing them to translocate to the nucleus and activate target genes (Massague and Chen, 2000), including Nanog (Vallier *et al.*, 2009). Interestingly, mouse EpiS cells have been shown to require the same molecular milieu as hES cells and human iPS cells to maintain pluripotency (Brons *et al.*, 2007) and their global gene expression pattern also more closely resembles the hES cell transcriptome than that of mES cells (Tesar *et al.*, 2007).

3. E-cadherin

Both mES and hES cells grow as tight colonies and their integrity is maintained by cell-cell contacts, such as adherens junctions (AJ), of which a major component is E-cadherin (Fig. 4.). E-cadherin is a well-characterised member of the calcium-dependent cadherin superfamily. E-cadherin is a glycoprotein and is categorised as a type 1 classical cadherin due to its possession of a histidine-alanine-valine (HAV) sequence in the extracellular domain. The role of cell-cell adhesion remains fundamental throughout embryogenesis and E-cadherin plays a critical role in the sorting of mixed cell populations to allow tissue segregation (Cavallaro and Christofori, 2004). This key developmental role is evidenced by studies in E-cadherin null (-/-) mutant embryos, which fail to form a trophectodermal epithelium or compact and form a blastocyst cavity due to the loss of cell-cell contact (Larue *et al.*, 1994). The role of E-cadherin in cell-cell adhesion is facilitated by its five extracellular (EC) domains and a cytoplasmic region, the latter of which allows stabilisation of the molecule at the cell membrane via its interaction with $p120^{ctn}$, β-catenin and α-E-catenin

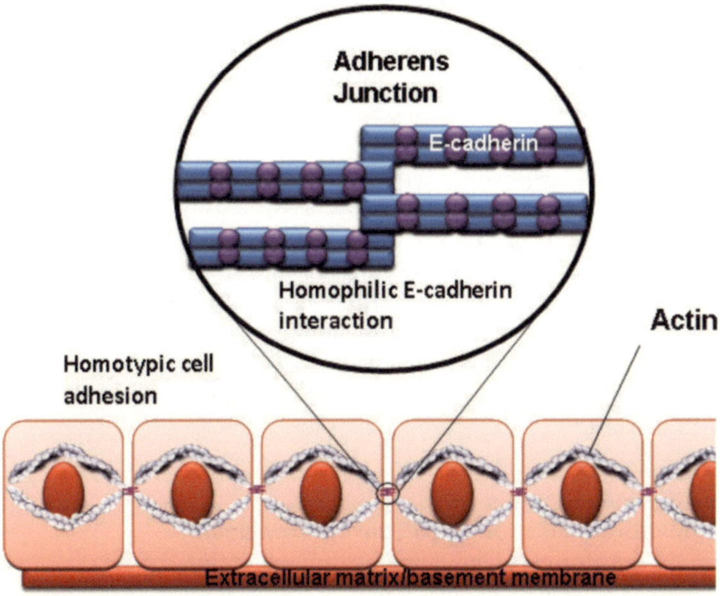

Fig. 4. Diagrammatic representation of adherens junctions formed by E-cadherin interactions. E-cadherin cis-homodimers interact with identical homodimers on neighbouring cells (homophilic interaction) with cell-cell adhesion most often associated with identical cells (homotypic adhesion). For simplicity, only E-cadherin is represented in the AJ.

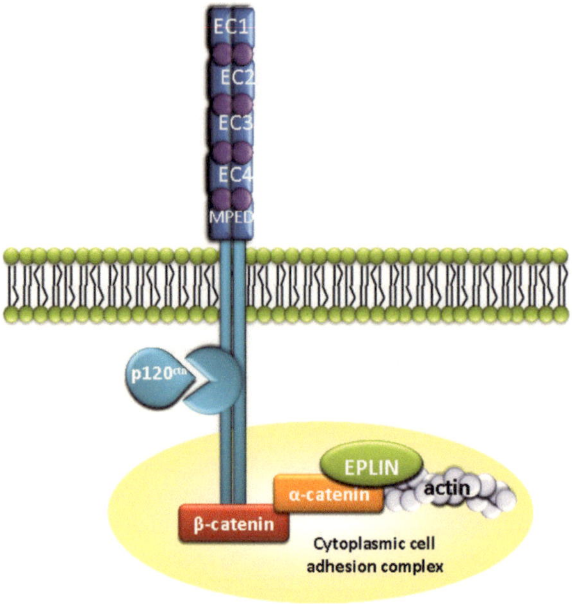

Fig. 5. Diagrammatic representation of E-cadherin cis-homodimers and the cytoplasmic cell adhesion complex. EC - extracellular domain; MPED - membrane proximal extracellular domain.

(possibly via Epithelial Protein Lost In Neoplasm). Together these molecules make up the Cytoplasmic Cell adhesion Complex (CCC) which anchors E-cadherin to the actin cytoskeleton (Fig. 5.). There is some controversy surrounding the roles of specific regions of E-cadherin in cell-cell adhesion, however, there is evidence for a critical role of the HAV domain, located on residues 79-81 of EC1, in cell-cell contact. The HAV domain is thought to form a hydrophobic pocket into which a Tryptophan residue (Trp2) from an adjacent E-cadherin molecule docks (Cavallaro and Christofori, 2004). Mutations of Trp2 and the Alanine residue of the HAV domain, W2A and A80I respectively, were shown to abolish trans- but not cis-homodimerisation of E-cadherin molecules, thus demonstrating the key roles of these amino acids in the formation of E-cadherin-mediated cell-cell contact (Pertz *et al.*, 1999). Additional roles for EC4 and the membrane proximal extracellular domain (MPED) of E-cadherin in homophilic adhesion have been evidenced by the E-cadherin targeting antibody DECMA-1, which abolishes cell-cell contact by interacting with these regions (Ozawa *et al.*, 1990).

3.1 Function of E-cadherin in ES cells

3.1.1 E-cadherin regulates localisation of cell surface molecules

Loss of E-cadherin in ES cells imparts significant alterations within the cellular architecture and can result in mis-localisation of cell surface proteins. For example, we have demonstrated that loss of E-cadherin in mouse and hES cells results in translocation of the promigratory molecule 5T4 from the cytoplasm to the plasma membrane and that this is associated with altered actin cytoskeleton arrangement and induction of cell polarisation (Eastham *et al.*, 2007; Spencer *et al.*, 2007). In addition, it has been reported that E-cadherin mediated cell-cell contact regulates expression of Eph receptors and Ephrins (Orsulic and

Kemler, 2000). Therefore, E-cadherin appears to play a critical role in regulating cellular architecture and, thus, localisation of plasma membrane proteins. This is supported by our unpublished observations which suggest that E-cadherin also functions to maintain plasma membrane localisation of a range of proteoglycans in ES cells (Soncin et al, unpublished data). As a result, E-cadherin should not be seen as simply a cell adhesion protein but an important molecule for maintaining the integrity of epithelium and ES cell colonies. Surprisingly, little has been reported on the function of altered E-cadherin expression in cell surface protein localisation. Since loss of E-cadherin is a defining event in tumour cell metastasis it is possible that cell surface proteins regulated by E-cadherin expression could provide novel targets for tumour therapies.

3.1.2 E-cadherin expression regulates signalling pathways in pluripotent cells

We have recently shown that E-cadherin-/- mouse ES cells do not respond to LIF when cultured in medium supplemented with foetal bovine serum (FBS). Although E-cadherin-/- ES cells exhibit an undifferentiated phenotype in medium supplemented with FBS/LIF, the cells maintain pluripotency and self-renewal by utilising Activin, Nodal and Fgf2 present within ES-screened FBS (Soncin *et al.*, 2009) (Fig. 3.). E-cadherin-/- ES cells can be cultured in an undifferentiated state in serum-free medium supplemented with Activin A, Nodal and Fgf2 and exposure of the cells to the Activin-like kinase receptors (Alks)-4, -5 and -7 inhibitor (SB431542) under these conditions induces their differentiation (Soncin *et al.*, 2009). Mutant E-cadherin protein analysis in our lab demonstrated that the β-catenin binding domain of E-cadherin is critical for LIF/Bmp-mediated pluripotency in ES cells. Although E-cadherin-/- ES cells utilise Activin/Nodal signalling as a default pluripotent pathway in serum-supplemented medium they are also able to maintain pluripotency via LIF/BMP signals in the absence of Activin/Nodal (Soncin *et al.*, 2009). Therefore, E-cadherin-/- ES cells possess a functional "ground state" pluripotent signalling network (Ying *et al.*, 2008) as well as the ability to circumvent this pathway by utilising Activin and Nodal. Our results suggested that mES cells exhibit a hierarchy of pluripotency signalling pathways and this was confirmed by our observation that reversible Activin/Nodal-dependent pluripotency could be induced in wild type (wt)ES cells by their treatment with an E-cadherin homodimerisation-inhibiting peptide (CHAVC). Interestingly, β-catenin null (-/-) ES cells are also able to maintain an undifferentiated state in the absence of LIF and, similar to that observed in E-cadherin-/- ES cells, this is achieved via Activin, Nodal and Fgf2 signalling. Therefore, the CCC appears to play a major function in regulating LIF-dependent pluripotency and self-renewal of mES cells.

We have performed global gene array analysis of E-cadherin-/- mES cells and detected >2000 transcript alterations compared to the wild-type parental cell line (Soncin et al, unpublished). Surprisingly, the altered transcripts detected in the array of E-cadherin-/- ES cells were not only confined to cell adhesion and motility but also genes associated with a range of biological functions, such as primary metabolic processes, catabolism, apoptosis and differentiation (Soncin et al, unpublished data). E-cadherin-/- ES cells also exhibited a transcriptional phenotype more similar to epiblast-derived cells, suggesting that loss of E-cadherin expression may determine the ICM-to-epiblast transition. Therefore, E-cadherin does not function solely as an adhesion molecule in mES cells but also to regulate transcription associated with a diverse range of cell functions, maintain appropriate growth factor responsiveness of the cells and retain plasma membrane localisation of a range of molecules.

3.1.3 Culture of mES cells in shake flask suspension culture using an E-cadherin neutralising antibody

Adherent methods for the culture of ES cells are cumbersome, result in significant batch-to-batch variation and are costly and labour-intensive. Given the requirement in regenerative medicine to reproducibly derive sufficient numbers of cells of a consistent quality in a cost-effective manner we have been investigating the use of bioreactors for the suspension culture of ES cells. The advantage of such an approach over existing adherent methods is the ability to design a scaleable, non-intensive and relatively homogeneous high cell volume density microenvironment that can be monitored *in silico*. E-cadherin has been demonstrated to be the cause of aggregation of ES cells in suspension culture (Fok and Zandstra, 2005) and several groups have suggested that abrogation of E-cadherin in ES cells results in low cell viability (Fok and Zandstra, 2005) and could adversely affect the pluripotent status of the cells when cultured in bioreactors (Dang *et al.*, 2004). However, our data in E-cadherin-/- ES cells demonstrated that these cells could be cultured in static suspension culture and maintain pluripotent marker expression over 30d (Mohamet *et al.*, 2010). Furthermore, we have recently demonstrated that wild type mES cells can be cultured as a near single cell suspension over prolonged periods in scalable shake flasks in the absence of additional media supplements (Mohamet *et al.*, 2010). Using the E-cadherin neutralising antibody DECMA-1 we have shown that wtES cells exhibit doubling times of 15.6h±4.7 and mean-fold increase in viable cell numbers over 48h of 16±0.9. Under these conditions, wtES cells could be cultured for 15d whilst maintaining expression of pluripotency markers and high cell viability. In addition, the cells exhibited a normal karyotype and, subsequently, were able to differentiate to cells representative of the three primary germ layers (Mohamet *et al.*, 2010). Culture of ES cells in shake flasks provides a useful cost-effective method which significantly decreases the requirement for technical input and plastic consumables associated with current adherent methods.

3.1.4 E-cadherin expression enhances hES cell colony formation and self-renewal

Several studies have identified the Rho-associated kinase (ROCK) inhibitor Y-27632 as a potent factor for increasing the survival of dissociated hES cells (Watanabe *et al.*, 2007). Recently, it has emerged that Y-27632 is likely to function by stabilising E-cadherin protein at the plasma membrane, allowing re-aggregation of the dissociated cells (Li *et al.*, 2009). Stabilisation of E-cadherin at the cell surface of disaggregated hES cells was associated with expression of the apoptotic inhibitory gene Bcl-XL and inhibition of the pro-apoptotic gene Caspase-3 (Li *et al.*, 2009). Forced expression of E-cadherin in dissociated hES cells increased clonogenicity up to 20-fold and cells lacking E-cadherin were shown to undergo cell death or differentiation within 48h. Similarly, Xu and colleagues (Xu *et al, 2009*) identified a small molecule, Thiazovivin (Tzv), which was observed to promote cell aggregation in suspension by inhibiting endocytosis of E-cadherin. The authors identified ROCK as a direct target of Tzv and concluded that inhibition of the ROCK pathway reflected the increased survival of hES cells treated with Tzv. It has also been shown that functional interactions between small GTPase Rap1 and E-cadherin is responsible for regulating self-renewal of hES cells (Li *et al, 2009.*). Colony formation and self-renewal of hES cells was found to be suppressed by inhibition of Rap1 as a consequence of altered endocytic recycling of E-cadherin. We have observed that culture of hES cells with the E-cadherin neutralising antibody SHE78.7 results in decreased proliferation of the cells (Eastham *et al.*, 2007), perhaps reflecting degradation of Rap1 in these cells (Li *et al, 2009*). Therefore, stabilisation of E-cadherin at the plasma membrane appears to be critical for the survival of disaggregated hES cells.

Several studies have utilised plates coated with E-cadherin-Fc protein to demonstrate increased survival of hES cells (Nagaoka *et al, 2010* ; Xu *et al, 2009*). Culture of hES or iPS cells on E-cadherin-Fc coated plates exhibited a normal karyotype and maintained pluripotent marker expression over 60 days (Nagaoka *et al, 2010*). These cells were subsequently induced to form cells representative of the three primary germ layers by their culture as embryoid bodies or following formation of teratomas. The plating efficiency of disaggregated hES cells was found to be decreased by proteolytic degradation of cell surface E-cadherin which could be prevented by the use of non-proteolytic dissociation buffer. As a result, E-cadherin-Fc coated plates may represent a useful method to provide a defined substratum for the growth of human ES and iPS cells.

3.1.5 E-cadherin expression enhances iPS cell derivation

Derivation of iPS cells represents a useful alterative to the use of human embryos for isolation of pluripotent cells as well as allowing the investigation of genetic mutations during cell lineage formation. Recently it has been suggested that nuclear reprogramming of mouse fibroblasts to iPS cells requires a mesenchymal-epithelial event (MET) (Li *et al, 2010.*). When MET was blocked during iPS cell-induction the derivation of such cells was found to be significantly reduced. In addition, low chimaera forming ability of iPS cells was associated with low levels of E-cadherin protein expression. However, recent evidence has suggested that low chimaera-forming ability of cells is associated with inefficient incorporation of the cells into the ICM, rather than a true lack of pluripotency of the cells (Li *et al., 2010*; Chou *et al.*, 2008).

Improved iPS cell derivation has been demonstrated by Chen and colleagues (Chen *et al.*, 2010) who isolated two small molecules that enhanced E-cadherin expression. To confirm the function of E-cadherin in the iPS-derivation procedure they forced expression of E-cadherin in MEFs and showed that this increased iPS-derivation rates 4-fold. Inhibition of E-cadherin expression using RNAi or an inhibitory peptide during iPS cell derivation lead to decreased iPS cell isolation. Interestingly, in contrast to signalling pathway alterations described by Soncin et al (Soncin *et al.*, 2009), the β-catenin binding domain of E-cadherin was not required for efficient iPS cell derivation and instead was dependent on the extracellular domain of the protein. We have previously demonstrated that the β-catenin domain of E-cadherin is required to restore cell-cell contact in E-cadherin-/- ES cells (Soncin *et al.*, 2009), suggesting that the function of E-cadherin in iPS cell derivation reflects a requirement for cell-extracellular matrix (ECM) interaction rather than cell-cell adhesion. Therefore, E-cadherin appears to function via two discreet mechanisms; firstly, as a regulator of pluripotent signalling pathways via the CCC and, secondly, as an enhancer of ES cell-ECM interactions to aid cell survival.

4. Loss of E-cadherin during ES cell differentiation

4.1 Epithelial-mesenchymal transition

Epithelial-mesenchymal transition (EMT) events involve the conversion of a cell from an epithelial to a more motile mesenchymal phenotype. This process is associated with a switch from E-cadherin expression to a less adhesive cadherin, such as N-cadherin. Coordinated EMT events are important during embryogenesis, for example, to allow the ingression of epiblast cells within the primitive streak during gastrulation. However, aberrant EMT-like events have also been implicated in tumorigenesis whereby a cell is transformed to a more metastatic phenotype, leading to tumor cell invasion of the surrounding tissues. Therefore, E-cadherin is considered to be a metastasis suppressor gene (Vleminckx *et al.*, 1991). The role of E-cadherin down-regulation during tumorigenesis is likely to be two-fold; loss of E-cadherin-mediated cell-cell contact initiates morphological changes as well as affecting the signal transduction status of the cell (Mohamet *et al.*, 2011). EMT has been shown to be stimulated by various signalling cascades and loss of E-cadherin has been shown to affect signalling via a variety of receptor tyrosine kinases, such as, the epidermal growth factor receptor and the hepatocyte growth factor receptor c-Met (reviewed in Cavallaro and Christofori, 2004). The upregulation of N-cadherin that occurs during EMT events can also affect signalling pathways such as those mediated by Fgf receptor 1 and may contribute to the increased cell survival and invasiveness that is characteristic of cells that have undergone an EMT event (Suyama *et al.*, 2002; Cavallaro and Christofori, 2004).

4.1.1 EMT during ES cell differentiation

As well as their potential use in regenerative therapies ES cells are an excellent model system for elucidating mechanisms involved in development and disease. We have demonstrated that an EMT event occurs during mouse (Spencer *et al.*, 2007) and human (Eastham *et al.*, 2007) ES cell differentiation and that this exhibits striking similarity to processes associated with ingression of epiblast cells within the primitive streak and tumour cell metastasis. The ES cell EMT event is associated with an E- to N-cadherin switch (Fig. 6.), upregulation of Snail, Slug and Sip1 (E-cadherin transcript repressors) and increased gelatinase activity and cellular motility (Fig. 7.). Whilst cell density can affect the extent to which the E- to N-cadherin switch occurs in adherent ES cell culture, EMT is often observed during embryoid body culture (Ward, unpublished).

The exact reason for the EMT event during ES cell differentiation remains unclear, although it is likely to reflect early events associated with embryogenesis. Furthermore, EMT during ES cell differentiation appears to be a regulated event, unlike the process associated with tumour cell metastasis. We have shown that loss of E-cadherin during ES cell EMT is associated with lack of pluripotent transcripts in E-cadherin-negative cells (Spencer *et al.*, 2007). As a result, E-cadherin expression can be used as a useful non-invasive tool to assess the pluripotent status of ES cells (Spencer *et al.*, 2011). Since EMT/metastasis is difficult to study *in* vivo, ES cells may provide a useful model system for the study of this process (Eastham *et al.*, 2007; Spencer *et al.*, 2007).

Our analysis of the EMT event during mES cell differentiation has enabled an understanding of the hierarchy of this process. For example, whilst induced loss of E-cadherin expression in ES cells resulted in increased cellular motility it did not result in an EMT-like event, with transcripts for Snail, Slug, Sip1 and MMP2/9 remaining unchanged (Spencer *et al.*, 2007). This suggests that loss of E-cadherin is not responsible for induction of EMT in ES cells. We also assessed EMT during differentiation of N-cadherin-/- mES cells and showed that these cells exhibited a characteristic EMT event similar to that of the parental cell line. Therefore, localisation of N-cadherin at the plasma membrane, and possible subsequent activation of signalling cascades during this process, are unlikely to be critical for the onset of EMT in mES cells, although N-cadherin-/- ES cells did exhibit significantly decreased motility compared to wtES cells. Therefore, whilst loss of E-cadherin and gain of N-cadherin are associated with an EMT event during ES cell differentiation these proteins do not appear to initiate nor control this process.

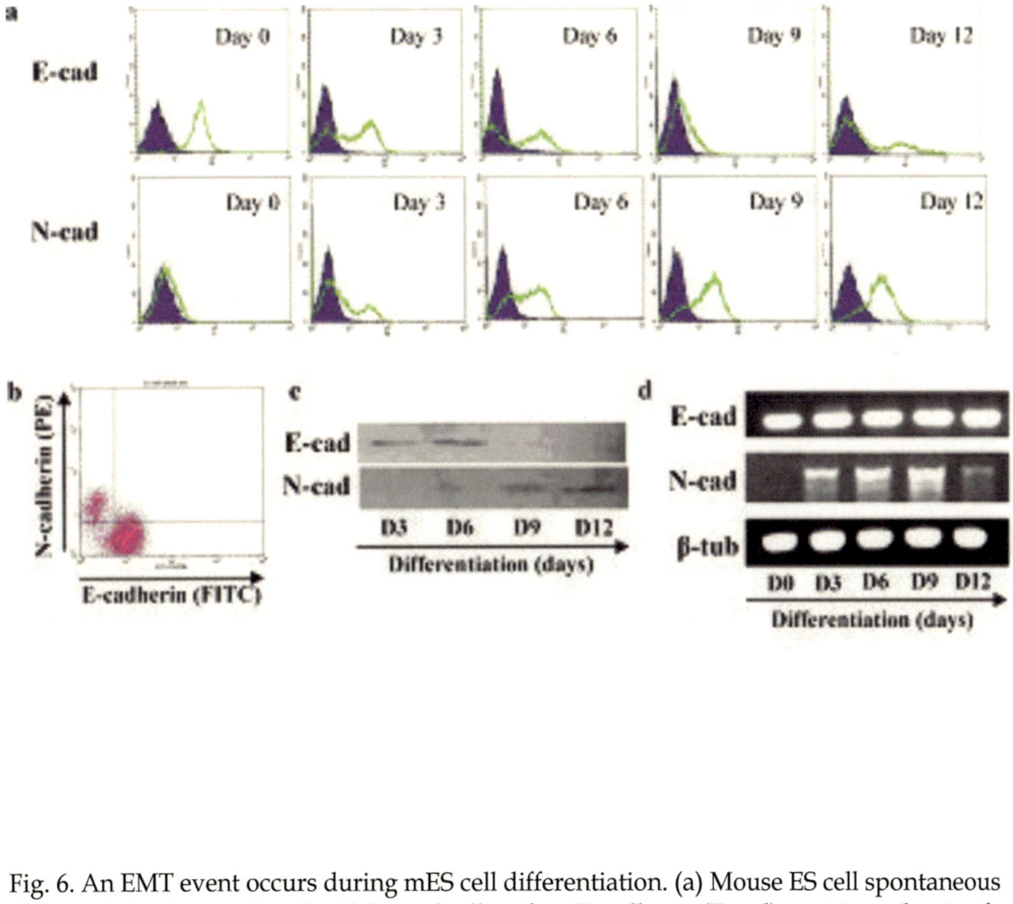

Fig. 6. An EMT event occurs during mES cell differentiation. (a) Mouse ES cell spontaneous differentiation is associated with loss of cell surface E-cadherin (E-cad) protein and gain of cell surface N-cadherin (N-cad) protein. (b) Fluorescent flow cytometry dual staining for E- and N-cadherin on ES cells differentiated for 3d. (c) Western blot analysis of total cellular E- (E-cad) or N-cadherin (N-cad) proteins in ES cells differentiated for 3, 6, 9 and 12 d. (d) RT-PCR analysis of E- (E-cad) and N-cadherin (N-cad) and β-tubulin (β-tub; control) transcript expression during ES cell differentiation. Images in these figures have been previously published in full or in part by ASCB (http://www.molbiolcell.org/cgi/content/full/18/8/2838#F1)

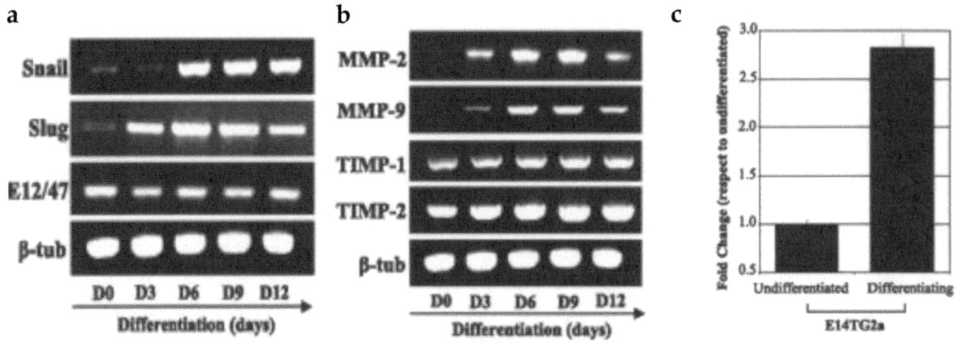

Fig. 7. Mouse ES cell differentiation is associated with expression of the E-cadherin repressors Snail and Slug, matrix metalloproteinase activity, and increased motility. (a) Transcript expression of Snail, Slug, E12/E47, and β-tubulin (β-tub; control) in undifferentiated ES cells (day 0) and in cells differentiated for 3, 6, 9, and 12 d. (b) Transcript expression of MMP-2 and -9, tissue inhibitor of metalloproteinase (TIMP)-1 and -2 and β-tubulin (β-tub; control) in undifferentiated and differentiating ES cells. (c) Cellular motility of undifferentiated and differentiating (3 d in the absence of LIF) wild-type ES cells was assessed using Costar Transwell 5-μm pore size plates. Data represents the fold change in motility compared with undifferentiated cells. Images in these figures have been previously published in full or in part by ASCB (http://www.molbiolcell.org/cgi/content/full/18/8/2838#F1)

5. Conclusion

E-cadherin is emerging as a key regulator of human and mouse ES cell pluripotency and self-renewal and is challenging long-held views of signalling pathways in these cells. For example, we have demonstrated that mES cells possess at least two functional pluripotent pathways that are dependent upon E-cadherin protein expression levels (Soncin *et al.*, 2009). Whilst the exact mechanisms controlling the switch between LIF/BMP and Activin/Nodal pathways in mES cells are not fully understood it is clear that E-cadherin functions to maintain the hierarchy of these independent pathways. We have also observed that abrogation of E-cadherin in hES cells can alter dependence of the cells to FGF2 signalling (Ward, unpublished), suggesting that E-cadherin also functions to regulate pluripotency pathways in human cells.

Current evidence suggests that E-cadherin plays several roles in ES cells. Firstly, E-cadherin functions to maintain cell-cell contact and this is likely to effect localisation of plasma membrane proteins. Our demonstration that E-cadherin inhibits cell surface localisation of the pro-migratory factor 5T4 (Eastham *et al.*, 2007; Spencer *et al.*, 2007) suggests that E-cadherin expression exerts a physical effect on the localisation of plasma membrane proteins. Unpublished data in our lab has revealed that loss of E-cadherin in mES cells does not affect cell surface expression of gp130 or the LIFR (Hawkins, unpublished), suggesting that alterations in signalling pathways in E-cadherin-/- mES cells are not due to changes in the localisation of pluripotency-associated receptors. The exact reason for the switch from LIF/BMP to Activin/Nodal dependent pluripotency in mES cells lacking E-cadherin is not clear. Indeed, we have found that inhibition of E-cadherin-mediated cell-cell contact using the neutralising antibody DECMA-1 does not stimulate Activin/Nodal-dependent pluripotency in mES cells. Therefore, the exact region(s) of E-cadherin which regulate LIF/BMP-dependent pluripotency in mES cells remain unknown. The third function of E-cadherin appears to be in regulating the expression of several thousand transcripts in mES cells. Whilst many of these mRNA changes are likely to reflect indirect transcriptional regulation by E-cadherin, it serves to highlight the importance of this protein in maintaining cellular homeostasis. The fourth function of E-cadherin in ES cells appears to be specific to human pluripotent cells: Enhancement of cell survival in dissociated cell populations. mES cells do not exhibit significant cell death when dissociated into single cells, indeed, this is desirable for maintenance of an undifferentiated population, and E-cadherin-/- ES cells exhibit almost 2-fold increased proliferation compared to wt mES cells (Soncin *et al.*, 2009). The effect of loss of E-cadherin in disaggregated hES cells appears to reflect endocytic cycling of the protein leading to decreased stability at the cell surface. However, hES cells are able to proliferate, albeit more slowly, upon inhibition of E-cadherin and maintain expression of pluripotent markers (Eastham *et al.*, 2007). The key difference between loss of E-cadherin expression in hES and mES cells, therefore, appears to be cellular disaggregation in the absence of a substratum. For example, whilst hES cells maintain viability in adherent culture following treatment with the E-cadherin neutralising antibody SHE78.7, the majority of the population die within 24h when cultured in suspension (Mohamet *et al.*, 2010). In summary, E-cadherin exhibits a range of functions in ES cells which result in stabilisation of epithelial integrity and associated signalling pathways. As a result, E-cadherin should no longer be classed as a mere cell adhesion protein but as a fundamental regulator of ES cell homeostasis and identity.

6. References

Botquin, V., Hess, H., Fuhrmann, G., Anastassiadis, C., Gross, M.K., Vriend, G., and Scholer, H.R. (1998). New POU dimer configuration mediates antagonistic control of an osteopontin preimplantation enhancer by Oct-4 and Sox-2. Genes & development *12*, 2073-2090.

Boyer, L.A., Lee, T.I., Cole, M.F., Johnstone, S.E., Levine, S.S., Zucker, J.P., Guenther, M.G., Kumar, R.M., Murray, H.L., Jenner, R.G., Gifford, D.K., Melton, D.A., Jaenisch, R., and Young, R.A. (2005). Core transcriptional regulatory circuitry in human embryonic stem cells. Cell *122*, 947-956.

Brons, I.G., Smithers, L.E., Trotter, M.W., Rugg-Gunn, P., Sun, B., Chuva de Sousa Lopes, S.M., Howlett, S.K., Clarkson, A., Ahrlund-Richter, L., Pedersen, R.A., and Vallier, L. (2007). Derivation of pluripotent epiblast stem cells from mammalian embryos. Nature *448*, 191-195.

Cavallaro, U., and Christofori, G. (2004). Cell adhesion and signalling by cadherins and Ig-CAMs in cancer. Nat Rev Cancer *4*, 118-132.

Chen, T., Yuan, D., Wei, B., Jiang, J., Kang, J., Ling, K., Gu, Y., Li, J., Xiao, L., and Pei, G. (2010). E-cadherin-Mediated Cell-Cell Contact is Critical for Induced Pluripotent Stem Cell Generation. Stem Cells *doi: stem.456*.

Chou, Y.F., Chen, H.H., Eijpe, M., Yabuuchi, A., Chenoweth, J.G., Tesar, P., Lu, J., McKay, R.D., and Geijsen, N. (2008a). The growth factor environment defines distinct pluripotent ground states in novel blastocyst-derived stem cells. Cell *135*, 449-461.

Chou, Y.F., Chen, H.H., Eijpe, M., Yabuuchi, A., Chenoweth, J.G., Tesar, P., Lu, J., McKay, R.D.G., and Geijsen, N. (2008b). The Growth Factor Environment Defines Distinct Pluripotent Ground States in Novel Blastocyst-Derived Stem Cells. Cell *135*, 449-461.

Chung, Y., Klimanskaya, I., Becker, S., Marh, J., Lu, S.J., Johnson, J., Meisner, L., and Lanza, R. (2006). Embryonic and extraembryonic stem cell lines derived from single mouse blastomeres. Nature *439*, 216-219.

Daheron, L., Opitz, S.L., Zaehres, H., Lensch, M.W., Andrews, P.W., Itskovitz-Eldor, J., and Daley, G.Q. (2004). LIF/STAT3 signaling fails to maintain self-renewal of human embryonic stem cells. Stem Cells *22*, 770-778.

Dang, S.M., Gerecht-Nir, S., Chen, J., Itskovitz-Eldor, J., and Zandstra, P.W. (2004). Controlled, scalable embryonic stem cell differentiation culture. Stem Cells *22*, 275-282.

Eastham, A.M., Spencer, H., Soncin, F., Ritson, S., Merry, C.L.R., Stern, P.L., and Ward, C.M. (2007). Epithelial-mesenchymal transition events during human embryonic stem cell differentiation. Cancer Research *67*, 11254-11262.

Evans, M.J., and Kaufman, M.H. (1981). Establishment in culture of pluripotential cells from mouse embryos. Nature *292*, 154-156.

Fok, E.Y.L., and Zandstra, P.W. (2005). Shear-controlled single-step mouse embryonic stem cell expansion and embryoid body-based differentiation. Stem Cells *23*, 1333-1342.

Graves, K.H., and Moreadith, R.W. (1993). Derivation and characterization of putative pluripotential embryonic stem cells from preimplantation rabbit embryos. Molecular reproduction and development *36*, 424-433.

Hanna, J., Markoulaki, S., Schorderet, P., Carey, B.W., Beard, C., Wernig, M., Creyghton, M.P., Steine, E.J., Cassady, J.P., Foreman, R., Lengner, C.J., Dausman, J.A., and Jaenisch, R. (2008). Direct reprogramming of terminally differentiated mature B lymphocytes to pluripotency. Cell *133*, 250-264.

Klimanskaya, I., Chung, Y., Becker, S., Lu, S.J., and Lanza, R. (2006). Human embryonic stem cell lines derived from single blastomeres. Nature *444*, 481-485.

Kunath, T., Arnaud, D., Uy, G.D., Okamoto, I., Chureau, C., Yamanaka, Y., Heard, E., Gardner, R.L., Avner, P., and Rossant, J. (2005). Imprinted X-inactivation in extra-embryonic endoderm cell lines from mouse blastocysts. Development *132*, 1649-1661.

Larue, L., Ohsugi, M., Hirchenhain, J., and Kemler, R. (1994). E-cadherin null mutant embryos fail to form a trophectoderm epithelium. Proceedings of the National Academy of Sciences of the United States of America *91*, 8263-8267.

Li, L., Wang, B.H., Wang, S.A., Moalim-Nour, L., Mohib, K., Lohnes, D., and Wang, L.S. Individual Cell Movement, Asymmetric Colony Expansion, Rho-Associated Kinase, and E-Cadherin Impact the Clonogenicity of Human Embryonic Stem Cells. Biophysical Journal *98*, 2442-2451.

Li, L., Wang, S., Jezierski, A., Moalim-Nour, L., Mohob, K., Parks, R., Francesco Retta, S., and Wang, L. (2009). A Unique Interplay between Rap1 and E-Cadherin in the Endocytic Pathway Regulates Self-Renewal of Human Embryonic Stem Cells. Stem Cells *Epub print Dec 28.*

Li, L., Wang, S.A., Jezierski, A., Moalim-Nour, L., Mohib, K., Parks, R.J., Retta, S.F., and Wang, L.S (2009). A Unique Interplay Between Rap1 and E-Cadherin in the Endocytic Pathway Regulates Self-Renewal of Human Embryonic Stem Cells. Stem Cells *28*, 247-257.

Li, R., Liang, J., Ni, S., Zhou, T., Qing, X., Li, H., He, W., Chen, J., Li, F., Zhuang, Q., Qin, B., Xu, J., Li, W., Yang, J., Gan, Y., Qin, D., Feng, S., Song, H., Yang, D., Zhang, B., Zeng, L., Lai, L., Esteban, M.A., and Pei, D (2010). A mesenchymal-to-epithelial transition initiates and is required for the nuclear reprogramming of mouse fibroblasts. Cell Stem Cell *7*, 51-63.

Lin, G., OuYang, Q., Zhou, X., Gu, Y., Yuan, D., Li, W., Liu, G., Liu, T., and Lu, G. (2007). A highly homozygous and parthenogenetic human embryonic stem cell line derived from a one-pronuclear oocyte following in vitro fertilization procedure. Cell research *17*, 999-1007.

Loh, Y.H., Wu, Q., Chew, J.L., Vega, V.B., Zhang, W., Chen, X., Bourque, G., George, J., Leong, B., Liu, J., Wong, K.Y., Sung, K.W., Lee, C.W., Zhao, X.D., Chiu, K.P., Lipovich, L., Kuznetsov, V.A., Robson, P., Stanton, L.W., Wei, C.L., Ruan, Y., Lim, B., and Ng, H.H. (2006). The Oct4 and Nanog transcription network regulates pluripotency in mouse embryonic stem cells. Nature genetics *38*, 431-440.

Mai, Q., Yu, Y., Li, T., Wang, L., Chen, M.J., Huang, S.Z., Zhou, C., and Zhou, Q. (2007). Derivation of human embryonic stem cell lines from parthenogenetic blastocysts. Cell research *17*, 1008-1019.

Martin, G.R. (1981). Isolation of a pluripotent cell line from early mouse embryos cultured in medium conditioned by teratocarcinoma stem cells. Proceedings of the National Academy of Sciences of the United States of America *78*, 7634-7638.

Massague, J., and Chen, Y.G. (2000). Controlling TGF-beta signaling. Genes & development *14*, 627-644.

Mohamet, L., Hawkins, K., and Ward, C.M. (2011). Loss of function of E-cadherin in embryonic stem cells and the relevance to models of tumorigenesis. J of Oncology *In press.*

Mohamet, L., Lea, M., and Ward, C.M. (2010). Abrogation of E-cadherin-mediated cellular aggregation allows proliferation of pluripotent mouse embryonic stem cells in scaleable shake flask suspension bioreactors. Plos One *In Press.*

Nagaoka, M., Si-Tayeb, K., Akaike, T., and Duncan, S.A. Culture of human pluripotent stem cells using completely defined conditions on a recombinant E-cadherin substratum. Bmc Developmental Biology *10.*

Nishimoto, M., Fukushima, A., Okuda, A., and Muramatsu, M. (1999). The gene for the embryonic stem cell coactivator UTF1 carries a regulatory element which selectively interacts with a complex composed of Oct-3/4 and Sox-2. Molecular and cellular biology *19*, 5453-5465.

Niwa, H., Ogawa, K., Shimosato, D., and Adachi, K. (2009a). A parallel circuit of LIF signalling pathways maintains pluripotency of mouse ES cells. Nature *460*, 118-122.

Notarianni, E., Laurie, S., Moor, R.M., and Evans, M.J. (1990). Maintenance and differentiation in culture of pluripotential embryonic cell lines from pig blastocysts. Journal of reproduction and fertility *41*, 51-56.

Orsulic, S., and Kemler, R. (2000). Expression of Eph receptors and ephrins is differentially regulated by E-cadherin. Journal of Cell Science *113*, 1793-1802.

Ozawa, M., Hoschutzky, H., Herrenknecht, K., and Kemler, R. (1990). A possible new adhesive site in the cell-adhesion molecule uvomorulin. Mechanisms of development *33*, 49-56.

Pain, B., Clark, M.E., Shen, M., Nakazawa, H., Sakurai, M., Samarut, J., and Etches, R.J. (1996). Long-term in vitro culture and characterisation of avian embryonic stem cells with multiple morphogenetic potentialities. Development *122*, 2339-2348.

Park, I.H., Arora, N., Huo, H., Maherali, N., Ahfeldt, T., Shimamura, A., Lensch, M.W., Cowan, C., Hochedlinger, K., and Daley, G.Q. (2008). Disease-specific induced pluripotent stem cells. Cell *134*, 877-886.

Pertz, O., Bozic, D., Koch, A.W., Fauser, C., Brancaccio, A., and Engel, J. (1999). A new crystal structure, Ca2+ dependence and mutational analysis reveal molecular details of E-cadherin homoassociation. The EMBO journal *18*, 1738-1747.

Revazova, E.S., Turovets, N.A., Kochetkova, O.D., Kindarova, L.B., Kuzmichev, L.N., Janus, J.D., and Pryzhkova, M.V. (2007). Patient-specific stem cell lines derived from human parthenogenetic blastocysts. Cloning and stem cells *9*, 432-449.

Sato, N., Meijer, L., Skaltsounis, L., Greengard, P., and Brivanlou, A.H. (2004). Maintenance of pluripotency in human and mouse embryonic stem cells through activation of Wnt signaling by a pharmacological GSK-3-specific inhibitor. Nat Med *10*, 55-63.

Soncin, F., Mohamet, L., Eckardt, D., Ritson, S., Eastham, A.M., Bobola, N., Russell, A., Davies, S., Kemler, R., Merry, C.L.R., and Ward, C.M. (2009). Abrogation of E-Cadherin-Mediated Cell-Cell Contact in Mouse Embryonic Stem Cells Results in Reversible LIF-Independent Self-Renewal. Stem Cells *27*, 2069-2080.

Spencer, H., Keramari, M., and Ward, C.M. (2011). Using Cadherin Expression to Assess Spontaneous Differentiation of Embryonic Stem Cells. Methods in Molecular Biology *690*, In press.

Spencer, H.L., Eastham, A.M., Merry, C.L., Southgate, T.D., Perez-Campo, F., Soncin, F., Ritson, S., Kemler, R., Stern, P.L., and Ward, C.M. (2007). E-Cadherin Inhibits Cell Surface Localization of the Pro-Migratory 5T4 Oncofetal Antigen in Mouse Embryonic Stem Cells. Molecular biology of the cell *18*, 2838-2851.

Stadtfeld, M., Brennand, K., and Hochedlinger, K. (2008). Reprogramming of pancreatic beta cells into induced pluripotent stem cells. Curr Biol *18*, 890-894.

Suyama, K., Shapiro, I., Guttman, M., and Hazan, R.B. (2002). A signaling pathway leading to metastasis is controlled by N-cadherin and the FGF receptor. Cancer Cell *2*, 301-314.

Takahashi, K., and Yamanaka, S. (2006). Induction of pluripotent stem cells from mouse embryonic and adult fibroblast cultures by defined factors. Cell *126*, 663-676.

Tanaka, S., Kunath, T., Hadjantonakis, A.K., Nagy, A., and Rossant, J. (1998). Promotion of trophoblast stem cell proliferation by FGF4. Science *282*, 2072-2075.

Tesar, P.J., Chenoweth, J.G., Brook, F.A., Davies, T.J., Evans, E.P., Mack, D.L., Gardner, R.L., and McKay, R.D. (2007). New cell lines from mouse epiblast share defining features with human embryonic stem cells. Nature *448*, 196-199.

Thomson, J.A., Itskovitz-Eldor, J., Shapiro, S.S., Waknitz, M.A., Swiergiel, J.J., Marshall, V.S., and Jones, J.M. (1998). Embryonic stem cell lines derived from human blastocysts. Science *282*, 1145-1147.

Vallier, L., Alexander, M., and Pedersen, R.A. (2005). Activin/Nodal and FGF pathways cooperate to maintain pluripotency of human embryonic stem cells. J Cell Sci *118*, 4495-4509.

Vallier, L., Mendjan, S., Brown, S., Chng, Z., Teo, A., Smithers, L.E., Trotter, M.W., Cho, C.H., Martinez, A., Rugg-Gunn, P., Brons, G., and Pedersen, R.A. (2009). Activin/Nodal signalling maintains pluripotency by controlling Nanog expression. Development *136*, 1339-1349.

Vleminckx, K., Vakaet, L., Jr., Mareel, M., Fiers, W., and van Roy, F. (1991). Genetic manipulation of E-cadherin expression by epithelial tumor cells reveals an invasion suppressor role. Cell *66*, 107-119.

Wakayama, S., Hikichi, T., Suetsugu, R., Sakaide, Y., Bui, H.T., Mizutani, E., and Wakayama, T. (2007). Efficient establishment of mouse embryonic stem cell lines from single blastomeres and polar bodies. Stem Cells *25*, 986-993.

Watanabe, K., Ueno, M., Kamiya, D., Nishiyama, A., Matsumura, M., Wataya, T., Takahashi, J.B., Nishikawa, S., Nishikawa, S., Muguruma, K., and Sasai, Y. (2007). A ROCK inhibitor permits survival of dissociated human embryonic stem cells. Nature Biotechnology *25*, 681-686.

Xu, Y., Zhu, X.W., Hahm, H.S., Wei, W.G., Hao, E.G., Hayek, A., and Ding, S. Revealing a core signaling regulatory mechanism for pluripotent stem cell survival and self-renewal by small molecules. Proceedings of the National Academy of Sciences of the United States of America *107*, 8129-8134.

Yang, J., Mani, S.A., Donaher, J.L., Ramaswamy, S., Itzykson, R.A., Come, C., Savagner, P., Gitelman, I., Richardson, A., and Weinberg, R.A. (2004). Twist, a master regulator of morphogenesis, plays an essential role in tumor metastasis. Cell *117*, 927-939.

Ying, Q.L., Nichols, J., Chambers, I., and Smith, A. (2003). BMP induction of Id proteins suppresses differentiation and sustains embryonic stem cell self-renewal in collaboration with STAT3. Cell *115*, 281-292.

Ying, Q.L., Wray, J., Nichols, J., Batlle-Morera, L., Doble, B., Woodgett, J., Cohen, P., and Smith, A. (2008). The ground state of embryonic stem cell self-renewal. Nature *453*, 519-U515.

Yu, J., Vodyanik, M.A., Smuga-Otto, K., Antosiewicz-Bourget, J., Frane, J.L., Tian, S., Nie, J., Jonsdottir, G.A., Ruotti, V., Stewart, R., Slukvin, II, and Thomson, J.A. (2007). Induced pluripotent stem cell lines derived from human somatic cells. Science *318*, 1917-1920.

Yuan, H., Corbi, N., Basilico, C., and Dailey, L. (1995). Developmental-specific activity of the FGF-4 enhancer requires the synergistic action of Sox2 and Oct-3. Genes & development *9*, 2635-2645.

Zhou, H., Wu, S., Joo, J.Y., Zhu, S., Han, D.W., Lin, T., Trauger, S., Bien, G., Yao, S., Zhu, Y., Siuzdak, G., Scholer, H.R., Duan, L., and Ding, S. (2009). Generation of induced pluripotent stem cells using recombinant proteins. Cell Stem Cell *4*, 381-384.

8

Proteomic Analysis of Mouse ES Cells

Akira Kurisaki[1,2,3], Yasuhiro Seki[1] and Atsushi Intoh[1]
[1]Research Center for Stem Cell Engineering, National Institute of Advanced Industrial Science and Technology (AIST),
[2]PRESTO, Japan Science and Technology Agency,
[3]Graduate School of Life and Environmental Sciences, Tsukuba University, Japan

1. Introduction

Embryonic stem (ES) cells have generated enormous interest because of their capacity to self-renew and differentiate into various cell types *in vitro*. Although numerous problems are encountered in the use of ES cells for regenerative medicine, such as ethical issues associated with the use of stem cells established from terminated human embryos and immunorejection due to transplantation of allogenic ES cell-derived cells into patients, recent technologies to generate induced pluripotent stem (iPS) cells from adult somatic cells have provided alternative ways to access pluripotent stem cells (Takahashi et al., 2007). However, the practical application of these pluripotent stem cells has yet to emerge, and regulatory mechanisms are not well known. Moreover, precise differentiation methodologies of ES and iPS cells have not been developed. These problems cause difficulties in the manipulation of pluripotent stem cells and derivation of functionally differentiated cells. Detailed analysis of the transcriptome has allowed elucidation of transcription networks that regulate the pluripotency of these stem cells. However, the specific nuclear infrastructures that maintain the pluripotent stem cell-specific transcription network have not yet been elucidated. We used proteomics to analyze the nuclear protein machinery in stem cells and identified some crucial components for the maintenance of pluripotent stem cells. In addition, various growth factors and extracellular matrix components regulate the pluripotency and differentiation of stem cells. Therefore, the cell surface receptors that bind these regulatory factors are important for the precise regulation of stem cells. We have also explored stem cell-specific cell-surface markers by proteomic analysis of mouse ES cells. These cell-surface membrane proteins can be useful to manipulate pluripotent stem cells. In this chapter, we describe some examples of new findings elucidated by proteomic analysis of ES cells.

2. Quantitative analysis of proteins by proteomics

To identify proteins from complex samples, there are three major approaches (Fig. 1). One is a 2-dimensional polyacrylamide gel electrophoresis (2D-PAGE)-based method, and the other two are shotgun-based methods using 2D-liquid chromatography (2D-LC). The first approach, which involves 2D-gel electrophoresis followed by identification of isolated proteins by mass spectrometry (MS), has been used for a long time. To obtain comparative data of two different samples, 2-dimensional difference gel electrophoresis (2D-DIGE) has been developed (Unlu, et al., 1997). In this method, protein samples that are covalently conjugated with different fluorescent dyes are combined, resolved by 2D-PAGE, and differentially expressed protein spots are quantified according to the intensity of each fluorescent color by a laser scanner. Differentially expressed protein spots are then excised from the gel, trypsinized, and subjected to MS analysis. To identify the proteins, molecular weight information of the digested peptides are compared with databases such as the National Center for Biotechnology Information (NCBI) and Swiss-Prot.

In contrast, in the case of the shotgun-based methods, protein samples are first digested with a specific protease, and the resulting huge number of peptides are fractionated by 2D-capillary chromatography followed by automated analysis by MS (Nagele, et al., 2004, Wu, et al., 2006). For quantitative analysis by the shotgun method, isobaric tag for relative and absolute quantitation (iTRAQ) have recently been developed as a labeling reagent for the digested peptide fragments (Ross, et al., 2004). During tandem mass spectrometric (MS/MS) analysis, the isobaric tags are readily cleaved from the peptides, and the generated reporter fragments from different samples give different molecular weight peaks. By comparing the intensity of these reporter fragments, the relative protein quantity of each protein sample can be calculated. In contrast, peptide fragments without reporters are used to identify peptide sequence by MS/MS analysis. Alternatively, SILAC (stable isotope labeling with amino acids in culture) can be used when comparing proteins in cells cultured under different conditions, such as with or without growth factors, chemicals, or at different time points. In this case, cells are cultured in normal medium or medium replaced with selected amino acids synthesized with ^{13}C and ^{15}N (Chen, et al., 2000, Ong, et al., 2002). Under these conditions, the labeled "heavy" amino acid will be incorporated into most of the proteins in the cell. After harvesting, these labeled and non-labeled cells are combined and processed as the normal shotgun method without labeling. Quantification of proteins can be done by comparing the intensity of light and heavy MS peaks of each peptide. In the present study, we applied quantitative proteomics to identify critical proteins in regulating the pluripotency-specific transcription network.

3. Proteomic analysis of pluripotency-specific nuclear proteins expressed in mouse embryonic stem cells

To identify specific proteins involved in the regulation of pluripotent stem cells, we used mouse ES cells. Although mouse ES cells are normally cultured on mouse embryonic fibroblast feeder cells, some mouse ES cell lines, such as D3 cells, can be maintained without feeder cells. We used D3 cell line for our analysis to avoid contamination of feeder cells. Leukemia inhibitory factor (LIF) is a crucial factor for maintenance of undifferentiated mouse ES cells. Culturing ES cells without LIF on a gelatin-coated dish for 7 days induces spontaneous differentiation. These pluripotent ES cells and the differentiated cells cultured without LIF were disrupted in a hypotonic buffer, centrifuged, and nuclear and cytoplasmic fractions were prepared. Proteins in these fractions were analyzed by 2D-DIGE. Proteins prepared from pluripotent cells or differentiated cells were labeled with different fluorescent dyes, separated by 2D-PAGE, and expression of the resolved protein spots were quantitatively analyzed by scanning the fluorescent intensity of the labeled protein spots. Differentially expressed protein spots were extracted and identified by MS (Fig. 2). More

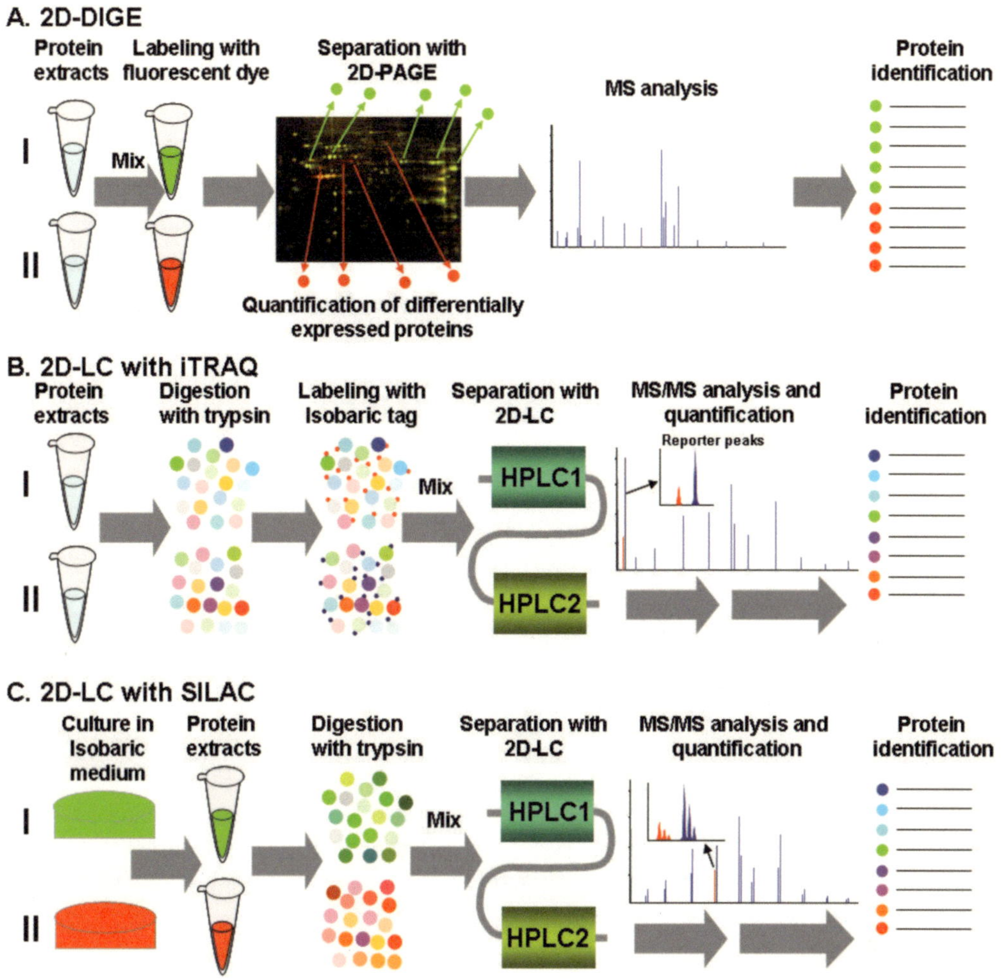

Fig. 1. Three major approaches for quantitative proteomic analysis.
(A) 2D-DIGE is a 2D-PAGE-based method using fluorescently labeled protein samples. Quantification of protein expression is performed on a 2D-PAGE gel and differentially expressed proteins are further analyzed by MS. (B) 2D-LC method with iTRAQ reagents. Protein samples are first digested with a protease and then labeled with isobaric tag reagents. Labeled peptide samples are mixed, fractionated by 2D-LC, and analyzed by MS/MS. Quantification is based on the relative intensity of the reporter fragments of iTRAQ reagents. (C) 2D-LC method with metabolically labeled samples. Cells are cultured with isobaric amino acids, and harvested cells are combined and processed as in (B) without the labeling step. Quantification of proteins is based on the intensity of each MS peak.

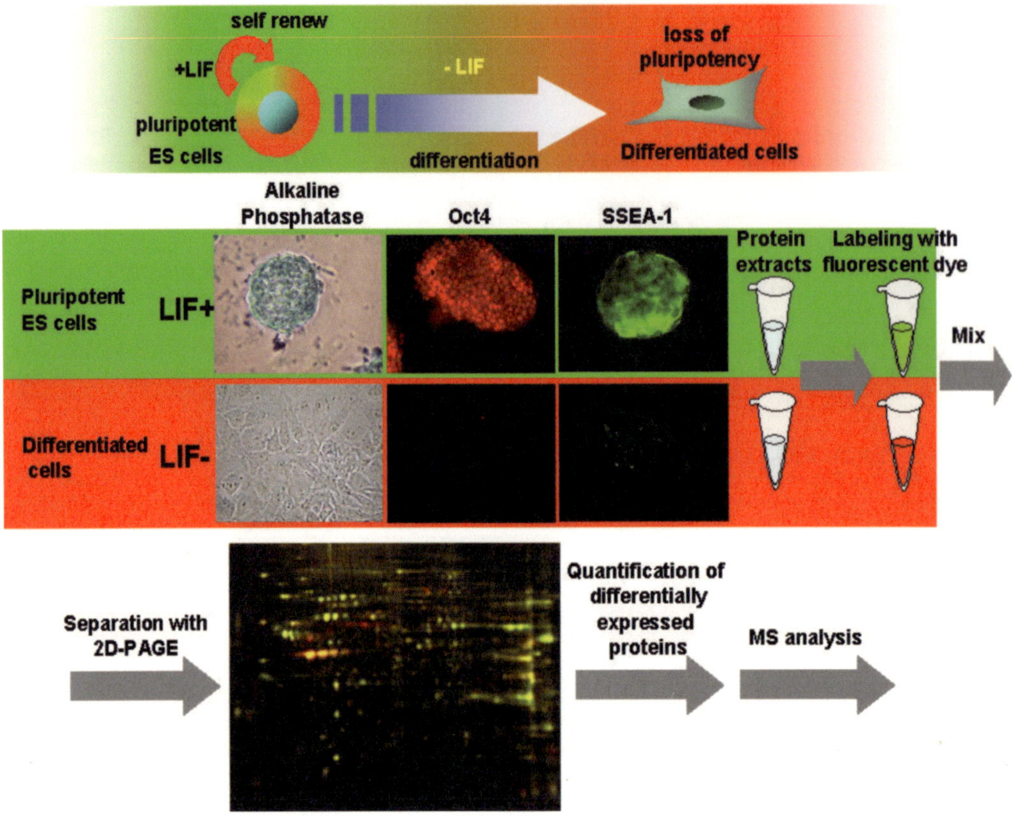

Fig. 2. Proteomic analysis of pluripotency-specific proteins expressed in mouse ES cells. Mouse ES cells cultured with or without LIF for one week were used as the source of pluripotent ES cells and differentiated cells, respectively. Pluripotent ES cells showed alkaline phosphatase activity and were positive for Oct4 and SSEA-1. In contrast, the differentiated cells were negative for these markers. The extracted proteins were labeled with green or red fluorescent dyes, mixed, and analyzed by 2D-DIGE.

than 100 proteins specifically expressed in pluripotent ES cells were identified from nuclear and cytoplasmic extracts. This study was the first detailed proteomic analysis of pluripotency-specific nuclear proteins in mouse ES cells (Kurisaki, et al., 2005).

Interestingly, many of the pluripotent stem cell-specific nuclear proteins were related to chromatin functions. For example, we identified the 60-kDa subunit of the SWI/SNF complex (BAF60a/Smarcd1), a component of a chromatin-remodeling complex, which slides nucleosomes along the DNA helix in an ATP-dependent manner and functions to expose genomic DNA to transcription factors or chromatin modifiers (Roberts, et al., 2004). We also identified one of the high-mobility group proteins (HMG-B2), which loosens the DNA helix thereby enhancing the accessibility to chromatin-remodeling complexes and possibly to transcription factors (Travers, 2003). Amine oxidase flavin-containing domain protein 2 (AOF2), also known as lysine-specific demethylase 1 (LSD1), is a demethylase for histone H3K4. Methylation of histone H3K4 is linked to active transcription. A recent report indicated that LSD1 regulates the expression and appropriate timing of key developmental regulators during early embryonic development (Foster, et al., 2010). Transcriptional intermediary factor

1β (TIF1β/KAP1/Trim28) has been reported to be a universal corepressor that forms a complex with histone methyltransferase SETDB1, which methylates histone H3K9 within euchromatin (Ivanov, et al., 2007, Sripathy, et al., 2006). RbAp48 is a histone-binding protein, which is often found in various histone modifying enzyme complexes (Wolffe, et al., 2000). MSH2 is a DNA mismatch repair protein, and recent reports have suggested that this protein could function as a coactivator of transcription (Wada-Hiraike, et al., 2005).

In the same year, a large proteomic dataset of proteins expressed in the E14 mouse ES cell line was analyzed by 2D-LC-based proteomics (Nagano, et al., 2005). Although the expression of the identified proteins was not systematically compared with other non-ES cells or differentiated cells, 1790 proteins in total were identified, including 365 potential nuclear proteins, such as pluripotency-specific transcription factors Oct4 and Sox2, as well as chromatin-related proteins TIF1β and Smarcd1. Very recently, another group has performed extensive proteomic analysis using ES-like embryonic carcinoma cells (F9) and differentiated cells (NIH3T3), and identified a number of chromatin-remodeling factors highly expressed in F9 cells (Singhal, et al., 2010).

4. TIF1β regulates the pluripotency of embryonic stem cells

To isolate crucial regulatory components for the maintenance of ES cell pluripotency, the proteins identified by MS analysis were stably expressed in ES cells and further functional screening was performed according to their prolonged alkaline phosphatase activity in the absence of LIF. Among these chromatin-related proteins, we found TIF1β as a functional regulator of pluripotency, which prolonged the pluripotency of ES cells after withdrawal of LIF. Recently, other groups have also identified TIF1β as an essential gene for mouse ES cells by RNAi-based screening (Fazzio, et al., 2008, Hu, et al., 2009). However, the mechanism by which TIF1β regulates ES cell pluripotency has not been well elucidated.

When TIF1β was knocked down in mouse ES cells, the cells lost their tight, compact morphology and adopted a stretched-out shape even in the presence of LIF. The growth of TIF1β-knockdown ES cells was significantly decreased, and the expression of pluripotency markers SSEA1 and Nanog was diminished. In contrast, expression of the primitive ectoderm marker gene, Fgf5, and relatively weak expression of the extraembryonic ectoderm marker gene, Eomes, was induced. These results indicate that TIF1β is an indispensable factor for the maintenance of pluripotency in ES cells, which preferentially inhibits the differentiation of ES cells into primitive ectoderm cells.

When we carefully analyzed the expression of TIF1β in ES cells, we found that TIF1β is specifically phosphorylated at serine 824 (S824), which induces active relaxation of chromatin (Ziv, et al., 2006). Phosphorylation at S824 was dramatically decreased when the cells were differentiated. As shown in Fig. 3, TIF1β was highly phosphorylated in the inner cell mass of embryos during early development, from which ES cells can be established (Seki et al., 2010). Ataxia telangiectasia mutated (ATM), a serine/threonine protein kinase, has been reported to be a specific kinase for S824 of TIF1β upon DNA double-strand breakage (Ziv, et al., 2006). We confirmed that ATM could phosphorylate the C-terminus of TIF1β as transfection of ATM shRNA significantly decreased the phosphorylation of TIF1β. Concomitantly, protein levels of Oct4 and Nanog were also decreased by knockdown of ATM in mouse ES cells, further supporting the importance of TIF1 β phosphorylation.

Unexpectedly, TIF1β had distinct effects on ES cells in a phosphorylation-dependent manner. First, phosphorylated TIF1β promotes the pluripotency of mouse ES cells. Stable

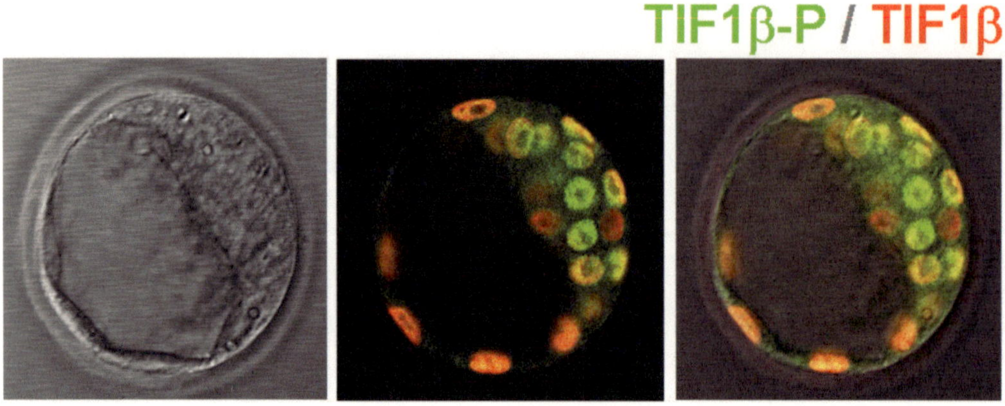

Fig. 3. Specific localization of phosphorylated TIF1β in mouse embryos. Mouse embryos (E3.5) were immunostained with TIF1β (red) and phosphorylated TIF1β (green) antibodies. Phosphorylated TIF1β is highly concentrated in the inner cell mass of blastocysts (quoted from Seki et al., 2010).

expression of the TIF1β-S824D mutant, which mimics phosphorylated TIF1β and induces constitutive chromatin relaxation, maintained Oct4 and Nanog protein expression after 8 days in culture without LIF. In contrast, expression of TIF1β-S824A, which cannot be phosphoryated, in ES cells did not show these effects. Second, TIF1β inhibits the differentiation of mouse ES cells in a C-terminal phosphorylation-dependent manner. Embryoid bodies prepared from dissociated ES cells were cultured in suspension in serum-free medium and then differentiated by adhesion culture on a poly-L-lysine, laminin, and fibronectin coated dish. Although the cells transfected with the control vector and TIF1β-S824A-expressing cells showed significant outgrowth of neurofilament-200- and TuJ1-positive neurites projected from embryoid bodies, TIF1β-S824D-expressing cells did not show such outgrowth. Third, phosphorylation of TIF1β is important for induction of iPS cells from somatic cells. Retroviral gene transfer of TIF1β-S824A dramatically decreased the generation of ES-like colonies from mouse embryonic fibroblasts infected with four transcription factors, Oct4, Sox2, Klf4, and c-Myc. Moreover, the colonies were very difficult to establish as iPS cell lines, and ES cell-specific markers were not expressed in these cells. In contrast, the induction of iPS cells was increased when TIF1β-S824D was introduced in addition to the above four transcription factors. Interstingly, the induced iPS cells generated with TIF1β-S824D showed more complete expression of ES cell-specific marker genes. The phosphorylation of TIF1β, which induces relaxation of chromatin, seems to affect the efficiency of iPS induction and the quality of established clones of iPS cells.

TIF1β has been reported to be involved in transcriptional repression with heterochromatin protein 1 (HP1) and localized to heterochromatin foci in differentiated cells (Sripathy, et al., 2006). However, in pluripotent ES cells, TIF1β is diffusely localized in the nucleoplasm and is not localized with heterochromatin foci. On the other hand, phosphorylated TIF1β (S824) showed punctate staining in the nucleus. This characteristic localization of phosphorylated TIF1β was partially colocalized with transcriptionally activated euchromatin markers such as histone $H3K4me^3$ and H3K9Ac, but not with the heterochromatin markers histone $H3K9me^3$ and HP1α. These data suggest that TIF1β could play a distinct role in pluripotent ES cells as compared to other differentiated cells.

What is the function of phosphorylated TIF1β in the specific nuclear spots? We hypothesized that phosphorylated TIF1β might be involved in transcriptional activation of pluripotency-specific genes in the activated chromatin foci of pluripotent stem cells. In fact, ectopic expression of TIF1β-S824D, but not TIF1β-S824A, in ES cells markedly induced various pluripotency-specific genes such as Nanog, Sox2, and Dax1. Moreover, knockdown of TIF1β resulted in an increased number of $H3K9me^3$ and HP1α foci, suggesting that endogenous TIF1β inhibits $H3K9me^3$ and HP1α foci formation in pluripotent mouse ES cells. These results suggest that TIF1β can selectively activate the expression of various pluripotency-specific genes in a phosphorylation-dependent manner.

Biochemical analysis suggests that TIF1β functions as a transcriptional activator of pluripotency specific genes by forming a complex with pluripotency-specific transcriptional factors at promoter regions. Indeed, co-immunoprecipitation assays revealed that TIF1β specifically forms a complex with Oct4 in a C-terminal phosphorylation-dependent manner. In contrast, another pluripotency-specific transcription factor, Sox2, was not detected in this complex. In addition to these transcription factors, Smarcd1/BAF60a was found to interact with TIF1β, although this interaction was not dependent on the phosphorylation of TIF1β. As mentioned above, our quantitative proteomic analysis has previously identified Smarcd1 as a highly expressed protein in undifferentiated ES cells (Kurisaki, et al., 2005). Smarcd1 functions as an ATP-dependent SWI/SNF chromatin remodeling factor that modulates chromatin structure. Recently, Smarcd1 has been shown to be a component of esBAF, an ES-specific BAF (Brg/Brahma-associated factors) ATP-dependent chromatin remodeling complex, which is essential for ES cell self-renewal and pluripotency (Ho, et al., 2009). Endogenous TIF1β also formed a complex with other esBAF components, such as Brg-1 and BAF155. TIF1β-S824D specifically induced transcriptional activation of a proximal Nanog promoter reporter construct with Oct4. A ChIP assay confirmed that TIF1β forms a complex on the endogenous Nanog promoter in a phosphorylation-dependent manner. TIF1β-S824D was shown to recruit endogenous Oct4 to form an active complex on the Nanog promoter. Microarray analysis revealed that one third of Oct4 target genes are specifically regulated by phosphorylated TIF1β. Several chromatin remodeling factors such as Suz12, Chd9, Pcaf, and Smarcad1 were also induced by TIF1β-S824D. Our data suggest that phosphorylated TIF1β forms a unique complex with Oct4 on pluripotency-specific genes and promotes expression of pluripotency-specific transcriptional factors such as Nanog, Sox2, and Oct4. Recent studies by ChIP-ChIP analysis revealed that TIF1β interacts with half of the promoters occupied by Oct4 and Sox2 (Jin, et al., 2007), suggesting that TIF1β might be an important regulator of Oct4-dependent transcription in ES cells.

The SWI/SNF2-like chromatin complex (BAF complex) is a huge complex with ATP-dependent chromatin remodeling activity, which is about 2 M Da in size. Components in this complex, such as Brg-1, BAF155, and BAF47, are essential in early development. Knock-out mice of these genes lead to peri-implantation lethality and failure to generate both the inner cell mass and trophoblast. Brg has been purified from Xenopus egg extracts and promoted the reprogramming of somatic cells into ES-like stem cells (Hansis, et al., 2004). Ho et al. reported that the ES-specific BAF complex (esBAF), which is required for self-renewal and pluripotency of mouse ES cells, was different from that in mouse embryonic fibroblasts (MEFs) or newborn mouse brain. The esBAF contains Brg-1, BAF155, and BAF60a, but not Brm, BAF170, or BAF60c. Pluripotency-specific transcription factors Oct4 and Sox2 also associated with the esBAF complex (Ho, et al., 2009). Recently, other groups have reported that a couple of proteins, which are components of the chromatin remodeling complex, promote the establishment of iPS cells from MEFs (Singhal, et al., 2010). They performed differential proteomics of the nuclear proteins extracted from NIH3T3 cells and F9 embryonic carcinoma cells by SILAC method. After computer-assisted Gene Ontology analysis of more than 5,000 identified proteins, they found specific expression of chromatin remodelling factors in pluripotent ES cell. Among them, Brg-1 and BAF155, which have been suggested to be crucial factors for maintenance of pluripotency of ES cells, promoted demethylation of pluripotency-specific transcription factor promoters such as the Oct4 and Nanog locus during the induction of iPS cells. Their results also supported the importance of the chromatin remodeling protein complex for the establishment of iPS cells from somatic cells.

5. Interactome of transcription factors identified by proteomics

In addition to describing a total set of proteins expressed in a population, proteomics are quite powerful for the analysis of protein components in a certain complex. Identification of functional protein complexes of transcription factors is important to elucidate transcription networks. Wang et al. constructed a protein interaction network surrounding the pluripotency factor Nanog (Wang, et al., 2006) by taking advantage of unique ES cells that express *Escherichia coli* biotin ligase, BirA. In this ES cell line, Nanog protein was N-terminally tagged with Flag and a biotin acceptor sequence was biotinylated *in vivo*. Thus, the Nanog protein complex was readily isolated by tandem purification with Flag antibody beads and streptavidin beads. In total, 266 proteins were identified. The constructed network was enriched for nuclear factors important for maintenance of the ES cell state and co-regulation of differentiation. Recently, an extended protein network that interacts with Oct4 was also reported using a different epitope-tagging affinity purification strategy (Pardo, et al., van den Berg, et al., 2010). A combination of the 3× FLAG epitope and a calmodulin binding peptide (CBP) separated by a TEV cleavage site was inserted into the C-terminus of the Oct4 coding region of a BAC clone containing full-length Oct4. This BAC construct was then integrated into the Hprt locus of mouse ES cells. In this system, expression levels of the Oct4-fusion protein were less than that of endogenous Oct4. In another study, van den Berg et al. used a single N-terminally 3× FLAG epitope-tagged Oct4 as the probe. Both studies succeeded in describing a detailed Oct4-centered interactome network in mouse ES cells. Although Oct4, Sox2, and Nanog form a positive feedback loop to maintain the ES cell-specific transcription network, such protein interactomes constructed by proteomics helps to locate Oct4 networks in known signaling pathways. Interestingly, both studies identified chromatin-related proteins, such as SWI/SNF chromatin remodelling factors, the NuRD complex, and the LSD1 complex, rather than pluripotency-enriched transcription factors. These results suggest the importance of epigenetic modifying complex associated with transcription factors in ES cells.

6. Application of pluripotent stem cells for regenerative medicine

Recently developed iPS cell technology enabled to obtain patient-derived pluripotent stem cells by reprogramming adult somatic cells with four transcription factors that can reorganize the ES-like transcription network (Takahashi, et al., 2007). Although iPS technology can overcome two critical problems, such as immunorejection due to mismatch of HLA types and ethical issues associated with using ES cells established from destroyed human embryos, there are still many problems that need to be addressed. First, these stem cells are not standardized; not all pluripotent stem cells can efficiently differentiate into specific cells. Moreover, insufficient differentiation methodologies for stem cells pose an increasing problem. Differentiation of pluripotent stem cells has been mainly studied with mouse and human ES cells. However, the specificity and efficiency of differentiation of these stem cells by current methods is still insufficient. Moreover, most of the differentiated cells are embryonic-type cells, not adult-type, and may not effectively function in patients. Therefore, in addition to establishing quantitative and specific differentiation methods, maturation protocols for differentiated cells may be required to obtain the targeting cells *in vitro*.

One of the most serious concerns for their application to regenerative medicine is the tumorigenic property of iPS cells. After transplantation of differentiated cells derived from iPS cells, occasionally teratomas can form from transplanted cells. Two major reasons have been suggested for this phenomenon. One is due to reactivation of the integrated tumorigenic transgene c-Myc in iPS cells. Human iPS cells have been established by the stable introduction of crucial transcription factors, Oct4, Sox2, Klf4, and c-Myc into the genome using retroviral or lentiviral gene transfer. Induction of iPS cells from fibroblasts would be much less efficient if c-Myc was excluded. However, genomic incorporation of transgenes by these viruses itself destroys multiple loci of the endogenous genome, which increases the potential risk of tumor formation by disruption of tumor suppression genes. Recently developed alternative integration-free induction methods, such as protein (Kim, et al., 2009, Zhou, et al., 2009), RNA (Fusaki, et al., 2009, Nishimura, et al., 2011, Warren, et al., 2010), and chemical compound-based induction methods (Li, et al., 2009) are expected to overcome this problem. In addition to reactivation of the Myc transgene during differentiation, residual contamination of undifferentiated iPS cells in the differentiated cell population may also lead to tumor formation (Nakagawa, et al., 2008). Transplantation of pluripotent ES cells into immunodeficient mice by definition generates teratomas *in vivo* that contain tridermically differentiated tissues. For the differentiation of both human and mouse ES/iPS cells, an embryoid body-based method has been widely adopted. Differentiation by embryoid body formation is an effective and convenient method for ES/iPS cell differentiation. However, there could be some residual differentiation-resistant pluripotent stem cells inside the embryoid body, and these stem cells may continue to grow after transplantation into the patient. Therefore, contamination of a small number of pluripotent stem cells causes high risk of tumor formation after transplantation (Miura, et al., 2009).

7. Tumorigenicity of pluripotent stem cells

To avoid tumor formation caused by contamination of undifferentiated stem cells, some approaches have been proposed. For example, prolonged *in vitro* differentiation of human ES cells into dopaminergic neurons is effective for preventing the formation of teratomas (Brederlau, et al., 2006). Another example is the control of stem cell fate by "stem cell suicide genes". Stable introduction of a suicide gene such as the herpes simplex virus thymidine kinase gene has been reported (Schuldiner, et al., 2003), which makes ES cells highly sensitive to ganciclovir at low concentrations. The expression of thymidine kinase under the control of the Oct4 promotor is effective for the ablation of undifferentiated ES cells *in vivo* that may produce teratomas (Hara, et al., 2008). However, the long-term stability of exogenous suicide genes *in vivo* has yet to be evaluated. An alternative approach is separation of residual pluripotent stem cells by cell-surface markers. Several cell-surface markers specific to pluripotent ES/iPS cells, such as SSEA4, TRA-1-60, and TRA-1-81, have been widely used. SSEA-4 is a glycolipid carbohydrate epitope specifically detected in human ES/iPS/EC cells. SSEA-4 is also expressed in some populations of mesenchymal stem cells (Gang, et al., 2007) and spermatogonial cells (Conrad, et al., 2008). Both TRA-1-60 and TRA-1-81 are antigens that are carbohydrate chains of a pericellular matrix proteoglycan, podocalyxin. They were first identified as cell-surface markers expressed in embryonic carcinomas. TRA-1 family proteins are specifically expressed both in human and rodent ES/iPS cells. However, these antigens can be lost by digestion of proteoglycan peptide with trypsin or collagenase treatment, which is required for dissociation of cells after differentiation. Podocalyxin without glycosylation is no longer a pluripotency-specific marker. On the other hand, insufficient dissociation of differentiated cell mixtures in embryoid bodies could increase the risk of contamination of undifferentiated cells that cause uncontrolled growth and teratoma formation after transplantation. Thus, efficient and complete removal of residual ES/iPS cells seems to be important for transplantation of differentiated cells. Mouse and human ES cells are considerably different from each other with respect to growth factor requirements and cell surface markers. Recent reports have suggested that human ES cells are somewhat similar to mouse Epi-stem cells (Epi-SC). Human ES/iPS cells have difficulty surviving after complete dissociation by proteases without a ROCK inhibitor (Watanabe, et al., 2007), and their growth rate is relatively slow. This property of human pluripotent stem cells might be helpful to lower the risk of teratoma formation after transplantation of differentiated cells. However, recently developed culture conditions for human ES cells (Hanna, et al., 2010) may allow easier handling of human ES/iPS cells as mouse ES cells and concomitantly increase the danger of tumor formation.

8. Proteomics approaches to identify cell-surface markers in pluripotent stem cells

Protein analyses by comprehensive proteomics are a potent strategy to elucidate not only identification of marker proteins, but also the molecular dynamics of proteins and their biological implications in cells and tissues. In particular, approximately 20-30% of vertebrate genes encode integral membrane proteins, and especially cell-surface membrane proteins are involved in critical cellular processes and are considered major pharmaceutical drug targets. Despite their biological significance and benefits as cell-surface markers, proteomic analysis of membrane proteins has technical challenges because of their relatively higher molecular weight and hydrophobicity. 2D-PAGE in combination with multicolor fluorescent labeling of proteins for quantitative analysis has limitations in terms of the dynamic range of protein quantity and the molecular mass range of detectable proteins. In addition, difficulties in fully automating 2D-PAGE coupled with MS have hampered its widespread use for large-scale proteomic analysis. Shotgun-based techniques that digest protein samples into small peptides prior to MS analysis have recently been developed for high-throughput proteomic analysis. Protease treatment of protein extracts generates a huge number of peptides. Therefore, a digested peptide mixture should be well fractionated and concentrated for efficient protein identification with low background noise. We have previously reported an improved method for MALDI-TOF based proteomic analysis of membrane proteins with a combination of zwitterionic hydrophilic interaction liquid chromatography and reverse-phase chromatography, which is a more effective separation method for digested peptide mixtures (Intoh, et al., 2009a). Using this method, we have performed proteomic analysis of membrane proteins specifically expressed in pluripotent mouse ES cells. For quantitative comparison, we used isobaric tags reagents, iTRAQ,

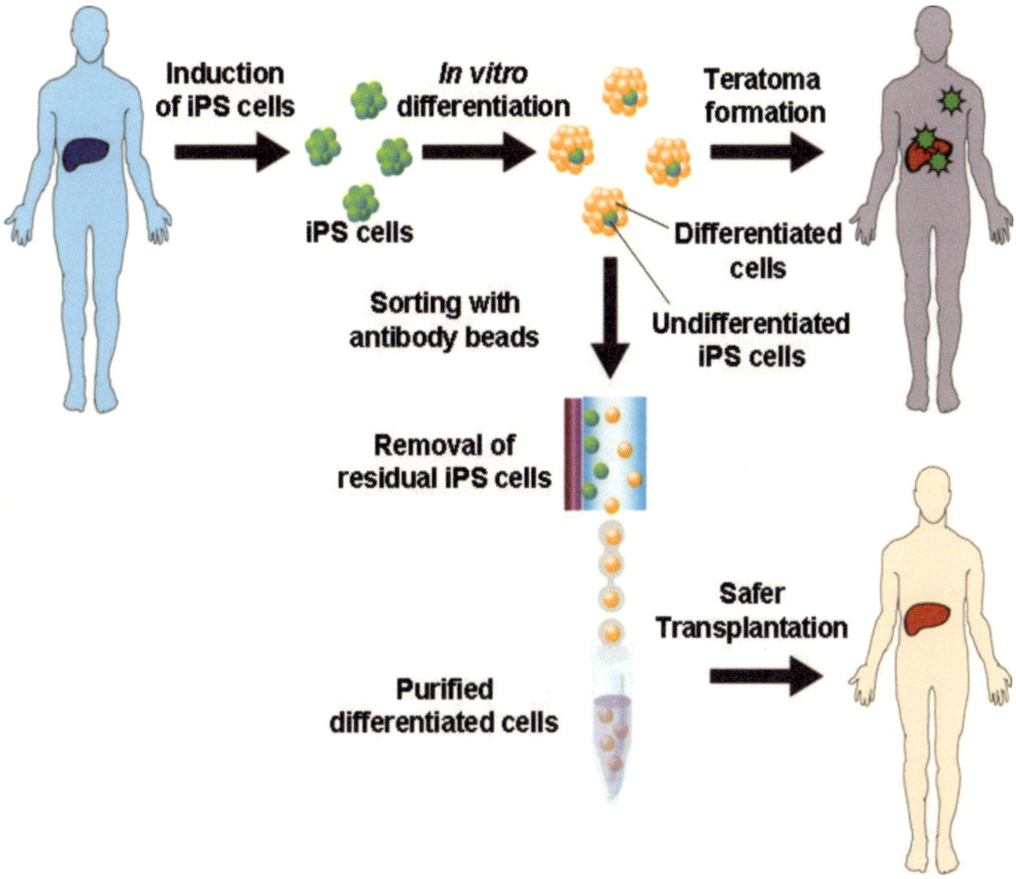

Fig. 4. Safe regenerative medicine using differentiated cells without contamination of tumor-forming pluripotent stem cells. The iPS cells established from patients are processed to differentiate into target cells. However, contamination of a small number of differentiation-resistant pluripotent stem cells in the differentiated cell pool can result in teratoma formation. Pluripotency-specific cell-surface markers may be useful to remove residual pluripotent iPS cells from the differentiated cell pool before transplantation.

to label trypsinized peptides prepared from different cell sources. Simultaneously, we also tested the conventional 2D-PAGE method for identification of membrane proteins for comparison, where extracted proteins were labeled with different fluorescent dyes, combined, and separated by 2D-PAGE. As such, we identified various candidate proteins that are highly expressed on the plasma membrane of pluripotent ES cells and significantly down-regulated during differentiation (Intoh, et al., 2009b). Some of the identified cell-surface proteins were also highly expressed in human ES/iPS cells and down-regulated during differentiation via embryoid body formation. These highly expressed membrane proteins could be useful for separation of residual ES/iPS cells from *in vitro* differentiated cell mixtures. The ultimate goals of safe regenerative medicine using human pluripotent stem cells require purification of targeting progenitors or differentiated cells. They also request complete removal of residual pluripotent stem cells, which could continue to grow and form teratomas *in vivo*. Development of more efficient differentiation methods as well as identification of specific cell-surface markers for both pluripotent and differentiated cells will contribute to safe and effective regenerative medicine in the near future (Fig. 4).

9. Conclusion

In this chapter, we have introduced some examples of proteomic analysis of ES cells. Transcriptome analysis using microarrays or direct sequencing of transcripts reveals the expression of mRNAs. However, the transcriptome does not necessarily correlate with expression levels of the corresponding protein. Moreover, analysis of the protein interactome by proteomics provides significant insight into understanding the mechanisms of how transcription factors function to establish pluripotency-specific functions in ES cells. Thus, proteomics approaches are important for further understanding the regulatory mechanisms of pluripotency and differentiation of ES/iPS cells.

10. References

Brederlau, A., Correia, A. S., Anisimov, S. V., Elmi, M., Paul, G., Roybon, L., Morizane, A., Bergquist, F., Riebe, I., Nannmark, U., Carta, M., Hanse, E., Takahashi, J., Sasai, Y., Funa, K., Brundin, P., Eriksson, P. S. and Li, J. Y., (2006). Transplantation of human embryonic stem cell-derived cells to a rat model of parkinson's disease: Effect of in vitro differentiation on graft survival and teratoma formation, *Stem Cells,* 24, 6, 1433-1440, ISSN 1066-5099.

Chen, X., Smith, L. M. and Bradbury, E. M., (2000). Site-specific mass tagging with stable isotopes in proteins for accurate and efficient protein identification, *Anal Chem,* 72, 6, 1134-1143, ISSN 0003-2700

Conrad, S., Renninger, M., Hennenlotter, J., Wiesner, T., Just, L., Bonin, M., Aicher, W., Buhring, H. J., Mattheus, U., Mack, A., Wagner, H. J., Minger, S., Matzkies, M., Reppel, M., Hescheler, J., Sievert, K. D., Stenzl, A. and Skutella, T., (2008). Generation of pluripotent stem cells from adult human testis, *Nature,* 456, 7220, 344-349, ISSN 1476-4687

Fazzio, T. G., Huff, J. T. and Panning, B., (2008). An RNAi screen of chromatin proteins identifies Tip60-p400 as a regulator of embryonic stem cell identity, *Cell,* 134, 1, 162-174, ISSN 1097-4172

Foster, C. T., Dovey, O. M., Lezina, L., Luo, J. L., Gant, T. W., Barlev, N., Bradley, A. and Cowley, S. M., (2010). Lysine-specific demethylase 1 regulates the embryonic transcriptome and corest stability, *Mol Cell Biol,* 30, 20, 4851-4863, ISSN 1098-5549

Fusaki, N., Ban, H., Nishiyama, A., Saeki, K. and Hasegawa, M., (2009). Efficient induction of transgene-free human pluripotent stem cells using a vector based on Sendai virus, an RNA virus that does not integrate into the host genome, *Proc Jpn Acad Ser B Phys Biol Sci,* 85, 8, 348-362, ISSN 1349-2896

Gang, E. J., Bosnakovski, D., Figueiredo, C. A., Visser, J. W. and Perlingeiro, R. C., (2007). SSEA-4 identifies mesenchymal stem cells from bone marrow, *Blood,* 109, 4, 1743-1751, ISSN 0006-4971

Hanna, J., Cheng, A. W., Saha, K., Kim, J., Lengner, C. J., Soldner, F., Cassady, J. P., Muffat, J., Carey, B. W. and Jaenisch, R., (2010). Human embryonic stem cells with biological and epigenetic characteristics similar to those of mouse ESCs, *Proc Natl Acad Sci U S A,* 107, 20, 9222-9227, ISSN 1091-6490

Hansis, C., Barreto, G., Maltry, N. and Niehrs, C., (2004). Nuclear reprogramming of human somatic cells by Xenopus egg extract requires Brg1, *Curr Biol,* 14, 16, 1475-1480, ISSN 0960-9822

Hara, A., Aoki, H., Taguchi, A., Niwa, M., Yamada, Y., Kunisada, T. and Mori, H., (2008). Neuron-like differentiation and selective ablation of undifferentiated embryonic stem cells containing suicide gene with Oct-4 promoter, *Stem Cells Dev,* 17, 4, 619-627, ISSN 1557-8534

Ho, L., Ronan, J. L., Wu, J., Staahl, B. T., Chen, L., Kuo, A., Lessard, J., Nesvizhskii, A. I., Ranish, J. and Crabtree, G. R., (2009). An embryonic stem cell chromatin remodeling complex, esBAF, is essential for embryonic stem cell self-renewal and pluripotency, *Proc Natl Acad Sci U S A,* 106, 13, 5181-5186, ISSN 1091-6490

Hu, G., Kim, J., Xu, Q., Leng, Y., Orkin, S. H. and Elledge, S. J., (2009). A genome-wide RNAi screen identifies a new transcriptional module required for self-renewal, *Genes Dev,* 23, 7, 837-848, ISSN 1549-5477

Intoh, A., Kurisaki, A., Fukuda, H. and Asashima, M., (2009a). Separation with zwitterionic hydrophilic interaction liquid chromatography improves protein identification by matrix-assisted laser desorption/ionization-based proteomic analysis, *Biomed Chromatogr,* 23, 6, 607-14, 1099-0801

Intoh, A., Kurisaki, A., Yamanaka, Y., Hirano, H., Fukuda, H., Sugino, H. and Asashima, M., (2009b). Proteomic analysis of membrane proteins expressed specifically in pluripotent murine embryonic stem cells, *Proteomics,* 9, 1, 126-37, ISSN 1615-9861

Ivanov, A. V., Peng, H., Yurchenko, V., Yap, K. L., Negorev, D. G., Schultz, D. C., Psulkowski, E., Fredericks, W. J., White, D. E., Maul, G. G., Sadofsky, M. J., Zhou, M. M. and Rauscher, F. J., 3rd, (2007). PHD domain-mediated E3 ligase activity directs intramolecular sumoylation of an adjacent bromodomain required for gene silencing, *Mol Cell,* 28, 5, 823-837, ISSN 1097-2765

Jin, V. X., O'geen, H., Iyengar, S., Green, R. and Farnham, P. J., (2007). Identification of an Oct4 and Sry regulatory module using integrated computational and experimental genomics approaches, *Genome Res,* 17, 6, 807-817, ISSN 1549-5469

Kim, D., Kim, C. H., Moon, J. I., Chung, Y. G., Chang, M. Y., Han, B. S., Ko, S., Yang, E., Cha, K. Y., Lanza, R. and Kim, K. S., (2009). Generation of human induced pluripotent stem cells by direct delivery of reprogramming proteins, *Cell Stem Cell,* 4, 6, 472-476, ISSN 1875-9777

Kurisaki, A., Hamazaki, T. S., Okabayashi, K., Iida, T., Nishine, T., Chonan, R., Kido, H., Tsunasawa, S., Nishimura, O., Asashima, M. and Sugino, H., (2005). Chromatin-related proteins in pluripotent mouse embryonic stem cells are downregulated after removal of leukemia inhibitory factor, *Biochem Biophys Res Commun,* 335, 3, 667-675, ISSN 0006-291X

Li, W., Wei, W., Zhu, S., Zhu, J., Shi, Y., Lin, T., Hao, E., Hayek, A., Deng, H. and Ding, S., (2009). Generation of rat and human induced pluripotent stem cells by combining genetic reprogramming and chemical inhibitors, *Cell Stem Cell,* 4, 1, 16-19, ISSN 1875-9777

Miura, K., Okada, Y., Aoi, T., Okada, A., Takahashi, K., Okita, K., Nakagawa, M., Koyanagi, M., Tanabe, K., Ohnuki, M., Ogawa, D., Ikeda, E., Okano, H. and Yamanaka, S., (2009). Variation in the safety of induced pluripotent stem cell lines, *Nat Biotechnol,* 27, 8, 743-745, ISSN 1546-1696

Nagano, K., Taoka, M., Yamauchi, Y., Itagaki, C., Shinkawa, T., Nunomura, K., Okamura, N., Takahashi, N., Izumi, T. and Isobe, T., (2005). Large-scale identification of proteins expressed in mouse embryonic stem cells, *Proteomics,* 5, 5, 1346-1361, ISSN 1615-985

Nagele, E., Vollmer, M., Horth, P. and Vad, C., (2004). 2D-LC/MS techniques for the identification of proteins in highly complex mixtures, *Expert Rev Proteomics,* 1, 1, 37-46, ISSN 1478-9450

Nakagawa, M., Koyanagi, M., Tanabe, K., Takahashi, K., Ichisaka, T., Aoi, T., Okita, K., Mochiduki, Y., Takizawa, N. and Yamanaka, S., (2008). Generation of induced pluripotent stem cells without Myc from mouse and human fibroblasts, *Nat Biotechnol,* 26, 1, 101-106, ISSN 1546-1696

Nishimura, K., Sano, M., Ohtaka, M., Furuta, B., Umemura, Y., Nakajima, Y., Ikehara, Y., Kobayashi, T., Segawa, H., Takayasu, S., Sato, H., Motomura, K., Uchida, E., Kanayasu-Toyoda, T., Asashima, M., Nakauchi, H., Yamaguchi, T., Nakanishi, M., (2011). Development of defective and persistent sendai virus vector: a unique gene delivery/expression system ideal for cell reprogramming. *J Biol Chem, in press,* ISSN 0021-9258

Ong, S. E., Blagoev, B., Kratchmarova, I., Kristensen, D. B., Steen, H., Pandey, A. and Mann, M., (2002). Stable isotope labeling by amino acids in cell culture, SILAC, as a simple and accurate approach to expression proteomics, *Mol Cell Proteomics,* 1, 5, 376-386, ISSN 1535-9476

Pardo, M., Lang, B., Yu, L., Prosser, H., Bradley, A., Babu, M. M. and Choudhary, J., An expanded Oct4 interaction network: Implications for stem cell biology, development, and disease, *Cell Stem Cell,* 6, 4, 382-395, ISSN 1875-9777

Roberts, C. W. and Orkin, S. H., (2004). The SWI/SNF complex--chromatin and cancer, *Nat Rev Cancer,* 4, 2, 133-142, ISSN 1474-175X

Ross, P. L., Huang, Y. N., Marchese, J. N., Williamson, B., Parker, K., Hattan, S., Khainovski, N., Pillai, S., Dey, S., Daniels, S., Purkayastha, S., Juhasz, P., Martin, S., Bartlet-Jones, M., He, F., Jacobson, A. and Pappin, D. J., (2004). Multiplexed protein quantitation in Saccharomyces cerevisiae using amine-reactive isobaric tagging reagents, *Mol Cell Proteomics,* 3, 12, 1154-1169, ISSN 1535-9476

Schuldiner, M., Itskovitz-Eldor, J. and Benvenisty, N., (2003). Selective ablation of human embryonic stem cells expressing a "Suicide" Gene, *Stem Cells,* 21, 3, 257-265, ISSN 1066-5099

Singhal, N., Graumann, J., Wu, G., Arauzo-Bravo, M. J., Han, D. W., Greber, B., Gentile, L., Mann, M. and Scholer, H. R., (2010). Chromatin-remodeling components of the BAF complex facilitate reprogramming, *Cell,* 141, 6, 943-955, ISSN 1097-4172

Sripathy, S. P., Stevens, J. and Schultz, D. C., (2006). The kap1 corepressor functions to coordinate the assembly of de novo HP1-demarcated microenvironments of heterochromatin required for KRAB zinc finger protein-mediated transcriptional repression, *Mol Cell Biol,* 26, 22, 8623-8638, ISSN 0270-7306

Takahashi, K., Tanabe, K., Ohnuki, M., Narita, M., Ichisaka, T., Tomoda, K. and Yamanaka, S., (2007). Induction of pluripotent stem cells from adult human fibroblasts by defined factors, *Cell,* 131, 5, 861-872, ISSN 0092-8674

Travers, A. A., (2003). Priming the nucleosome: A role for HMGB proteins?, *EMBO Rep,* 4, 2, 131-136, ISSN 1469-221X

Unlu, M., Morgan, M. E. and Minden, J. S., (1997). Difference gel electrophoresis: A single gel method for detecting changes in protein extracts, *Electrophoresis,* 18, 11, 2071-2077, ISSN 0173-0835

Van Den Berg, D. L., Snoek, T., Mullin, N. P., Yates, A., Bezstarosti, K., Demmers, J., Chambers, I. and Poot, R. A., (2010). An Oct4-centered protein interaction network in embryonic stem cells, *Cell Stem Cell,* 6, 4, 369-381, ISSN 1875-9777

Wada-Hiraike, O., Yano, T., Nei, T., Matsumoto, Y., Nagasaka, K., Takizawa, S., Oishi, H., Arimoto, T., Nakagawa, S., Yasugi, T., Kato, S. and Taketani, Y., (2005). The DNA mismatch repair gene hMSH2 is a potent coactivator of oestrogen receptor alpha, *Br J Cancer,* 92, 12, 2286-2291, ISSN 0007-0920

Wang, J., Rao, S., Chu, J., Shen, X., Levasseur, D. N., Theunissen, T. W. and Orkin, S. H., (2006). A protein interaction network for pluripotency of embryonic stem cells, *Nature,* 444, 7117, 364-368, ISSN 1476-4687

Warren, L., Manos, P. D., Ahfeldt, T., Loh, Y. H., Li, H., Lau, F., Ebina, W., Mandal, P. K., Smith, Z. D., Meissner, A., Daley, G. Q., Brack, A. S., Collins, J. J., Cowan, C., Schlaeger, T. M. and Rossi, D. J., (2010). Highly efficient reprogramming to pluripotency and directed differentiation of human cells with synthetic modified mRNA, *Cell Stem Cell,* (in press) ISSN 1875-9777

Watanabe, K., Ueno, M., Kamiya, D., Nishiyama, A., Matsumura, M., Wataya, T., Takahashi, J. B., Nishikawa, S., Muguruma, K. and Sasai, Y., (2007). A ROCK inhibitor permits survival of dissociated human embryonic stem cells, *Nat Biotechnol,* 25, 6, 681-686, ISSN 1087-0156

Wolffe, A. P., Urnov, F. D. and Guschin, D., (2000). Co-repressor complexes and remodelling chromatin for repression, *Biochem Soc Trans,* 28, 4, 379-386, ISSN 0300-5127

Wu, W. W., Wang, G., Baek, S. J. and Shen, R. F., (2006). Comparative study of three proteomic quantitative methods, DIGE, iCAT, and iTRAQ, using 2D gel- or LC-MALDI TOF/TOF, *J Proteome Res,* 5, 3, 651-658, ISSN 1535-3893

Zhou, H., Wu, S., Joo, J. Y., Zhu, S., Han, D. W., Lin, T., Trauger, S., Bien, G., Yao, S., Zhu, Y., Siuzdak, G., Scholer, H. R., Duan, L. and Ding, S., (2009). Generation of induced pluripotent stem cells using recombinant proteins, *Cell Stem Cell*, 4, 5, 381-384, ISSN 1875-9777

Ziv, Y., Bielopolski, D., Galanty, Y., Lukas, C., Taya, Y., Schultz, D. C., Lukas, J., Bekker-Jensen, S., Bartek, J. and Shiloh, Y., (2006). Chromatin relaxation in response to DNA double-strand breaks is modulated by a novel ATM- and Kap-1 dependent pathway, *Nat Cell Biol*, 8, 8, 870-876, ISSN 1097-6256

9

Molecular Biomarkers of Embryonic Stem Cells

Takumi Miura, Akihiro Umezawa and Hidenori Akutsu
Department of Reproduction, National Research Institute for Child Health and Development, Tokyo, Japan

1. Introduction

Embryonic stem cells (ESCs) are pluripotent cells capable of both limitless selfrenewal and differentiation into all embryonic lineages, and thus ESCs can give rise to any adult cell type. When ESCs are stably maintained in culture and their pluripotency is strictly enforced, they can serve as an unlimited source for tissue replacement in regenerative medicine for degenerative diseases such as neural disorders, heart disease, and type I diabetes. They also offer enormous potential for drug discovery and toxicology, human developmental biology, and cancer research. Studies of human ESCs (hESCs) biology have developed rapidly since the first reports of their derivation in 1998 (Thomson *et al.*, 1998). Many studies have tried to manipulate the growth and differentiation conditions of hESCs with variable success (Biswas and Hutchins, 2007; Hoffman and Carpenter, 2005). hESCs have been differentiated into the derivatives of all three germ layers: ectoderm, mesoderm, and endoderm. Specifically, these derivatives include cardiomyocytes, neural cells, hepatocyte-like cells, endothelial cells, pancreatic hormone expressing endocrine cells, and hematopoietic progenitor cells (Barberi *et al.*, 2007; Carpenter *et al.*, 2003; D'Amour *et al.*, 2006; Levenberg *et al.*, 2007; Lu *et al.*, 2007; Roy *et al.*, 2006; Wang *et al.*, 2007), and thus hESCs have great potential for use in regenerative medicine to restore heart disease, neuronal functions, hepatic disease, blood vessels, and type I diabetes. In addition, mouse ESCs (mESCs) can generate hepatocytes (Gouon-Evans *et al.*, 2006; Soto-Gutierrez *et al.*, 2007), insulin-producing cells (Schroeder *et al.*, 2006), cerebellar neurons (Salero and Hatten, 2007), and even germ cells (West *et al.*, 2006) *in vitro*, suggesting that hESC can be applied much more widely to regenerative medicine in the future. On October 2010, Geron corporation in United States announced plans to initiate the phase I clinical trial of hESC-derived oligodendrocyte progenitor cells. However, the clinical application of hESCs is restricted thus far for alleged ethical and scientific reasons. First, hESC research often faces opposition from those who object to the destruction of human embryos. Second, ESC therapy potentially poses the risk of tumorigenesis. ESCs frequently form teratocarcinomas when transplanted into mice. Moreover, the ability of ESCs to provide differentiated cells for regenerative medicine will require continual maintenance of the undifferentiated stem cells for long periods in culture. However, chromosomal stability during extended cell passage cannot be guaranteed, and recent cytogenetic studies of ESCs have revealed karyotypic aberrations (Baker *et al.*, 2007). Third, cell replacement therapies have been limited by the availability of sufficient quantities of cells for transplantation. Although there are many reports describing a method to maintain ESC properties in culture, the large-scale culture of

ESC lines is still problematic and susceptible to substantial challenges at present (Thomson, 2007). Fourth, the potential for immunorejection should be a concern in its therapeutic use, and thus histocompatible ESCs will be required. Genetically, matched pluripotent ESCs generated *via* somatic cell nuclear transfer or parthenogenesis are a potential source of patient-derived histocompatible cells and tissues for transplantation (Kim *et al.*, 2007; Menendez *et al.*, 2005; Yang *et al.*, 2007). Selected hESCs can serve as a source of histocompatible tissues for transplantation (Kim *et al.*, 2007). The largest impact on recent ESC biology is the generation of ESC-like cells termed "induced pluripotent stem cells (iPSCs)" from fibroblasts that are created by introducing four genes, Oct4, Sox2, c-Myc, and Klf4 (Takahashi and Yamanaka, 2006). Although the first report described the generation of mouse iPSCs, human iPSCs have also been generated by introducing the same four genes as the mouse iPSCs derived from adult human dermal fibroblasts (Takahashi *et al.*, 2007) or the other distinct genes, Oct4, Sox2, Nanog, and LIN28, from human fetal fibroblasts (Yu *et al.*, 2007). These cells could differentiate into cell types of the three germ layers *in vitro* and produce teratomas, suggesting that iPSCs have the potential to generate patient- and disease-specific stem cells. Despite the importance of our knowledge of ESCs both in cell biology and clinical medicine, the molecular mechanism underlying the cell biological characteristics of ESCs such as the mechanism that maintains pluripotency and that regulates ESC differentiation, remains largely unknown.

Recent we showed the function of CD9, which is highly expressed in undifferentiated state in ESCs, as well as in embryos using the conventional gene targeting strategy to reveal whether CD9 can serve as a molecular marker to detect, classify, and isolate a particular subpopulation of ESCs and to monitor their state of differentiation (Akutsu *et al.*, 2009). This chapter also reviews other ESC molecular markers (Oct4, Sox2, Nanog, Klf4 and Rex1) in addition to CD9. The accumulation of these ES molecular marker studies will be provided a more detailed view of ESCs and facilitated our understanding of early embryonic development and cell-based therapies.

2. The membrane protein CD9

CD9 is a member of the transmembrane 4 superfamily, also known as the tetraspanin family. Most of these members are cell-surface proteins. CD9 is expressed on the cell surface of mouse and rat male germline stem cells and of neural stem cells. Therefore, CD9 may be involved in the common machinery in stem cells of many self-renewing tissues. In addition, CD9 is also involved in cell development, growth, motility, cell differentiation, and egg-sperm fusion (Hadjiargyrou and Patterson, 1995; Kanatsu-Shinohara *et al.*, 2004b; Kaprielian *et al.*, 1995; Miyado *et al.*, 2000; Miyado *et al.*, 2008). The expression of CD9 in embryonic as well as adult stem cells populations may indicate a role of CD9 in stem cell self-renewal. Oka et al. have been reported that *CD9* is highly expressed in undifferentiated ESCs but rapidly down-regulated after cells differentiation (Oka *et al.*, 2002). Upon application of an antibody against CD9, mouse ESCs can not form compact ES-like colonies. Moreover, ESCs are dead in the presence of the anti-CD9 antibody. Therefore, CD9 may play a role in maintenance of undifferentiated mouse ESCs (Oka *et al.*, 2002). Despite high potential role of CD9 in mouse ESCs, however, *CD9* null mice are born healthy and grew normally. Therefore, the question whether CD9 has a role in pluripotent cells of the inner cell mass has not been addressed. Based on these findings, we recently reported that CD9 is dispensable for maintenance of an undifferentiated state and pluripotency (Figure 1)

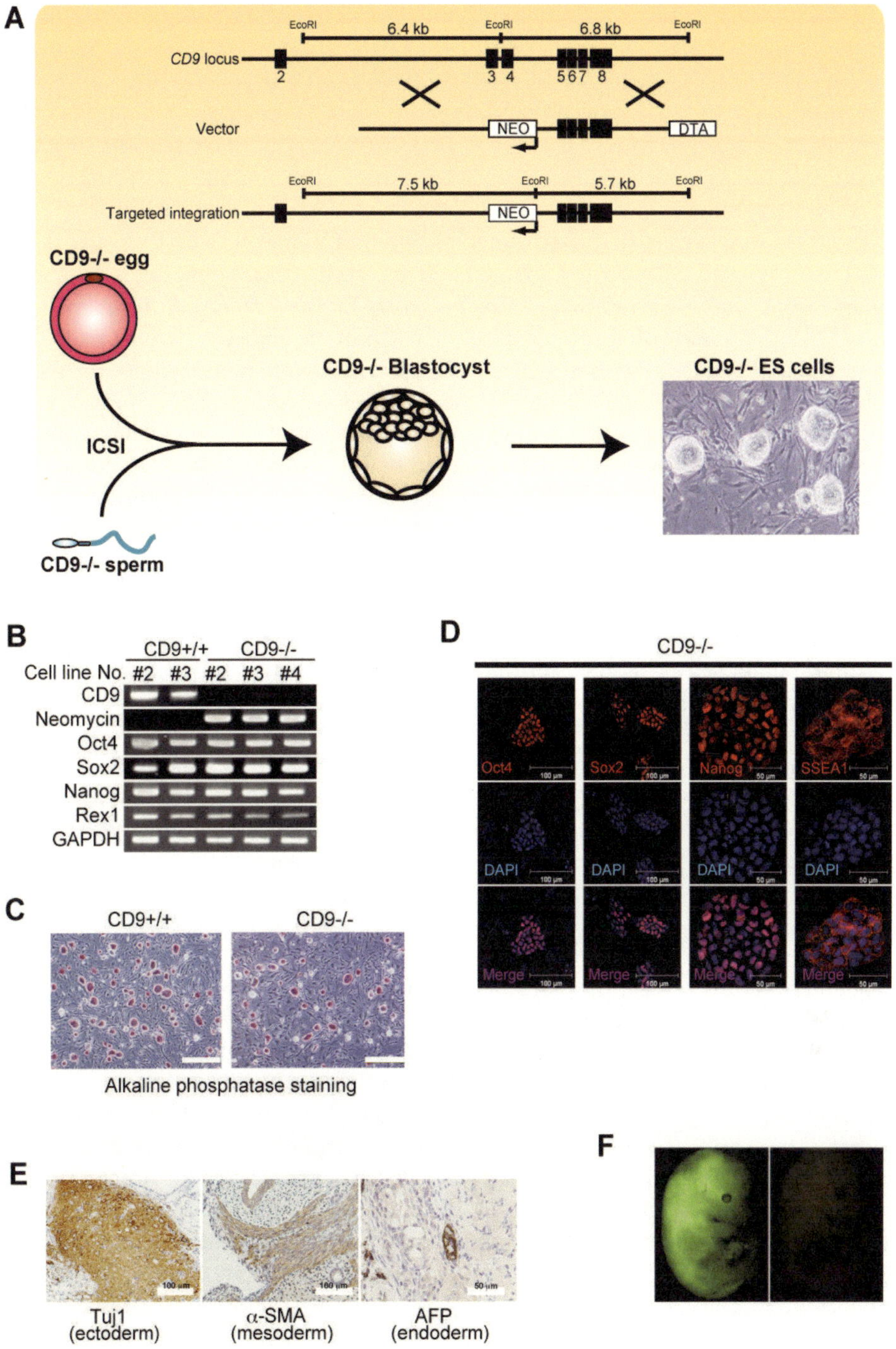

Fig. 1. **Generation of CD9 knockout ESCs.**

A. Strategy for generation of CD9-KO ES cells. The schematic maps of the CD9 allele (top), the KO vector carrying the neomycin cassette (middle), and the KO allele generated by homologous recombination (bottom) were shown in scale. *CD9-/-* fertilized eggs can not be obtained by a cross of *CD9-/-* female and male mice. To address this issue, we used ICSI to insert *CD9-/-* sperm directly into the cytoplasm of *CD9-/-* egg and bypass the fusion step. As a result, *CD9-/-* ES cells were successfully isolated from blastocyst of the *CD9-/-* fertilized egg. **B**. RT-PCR analysis of ES cell-marker genes. Transcripts of *Oct4*, *Sox2*, *Nanog*, and *Rex1* were detected in both *CD9-/-* and *CD9+/+* ESCs without feeder cells. A neomycin-resistance gene was targeted to delete a part of the third exon and all of the fourth exon of *CD9*. Therefore, neomycin gene was detected in *CD9-/-* ES cell lines. **C**. Alkaline phosphatase staining shows undifferentiated *CD9-/-* ESCs as well as *CD9+/+* ESCs. Bar = 500 μm. **D**. *CD9-/-* and *CD9+/+* ESCs were fixed and stained with antibodies against Oct4, Nanog and Sox2. Nuclei were counterstained with DAPI. Bar = 50 μm. **E**. Teratomas of *CD9* knockout ESCs containing all three germ layers. **F**. Chimeric embryos derived from *CD9* knockout ESCs. When EGFP-positive *CD9-/-* ESCs, which were homozygotes for the partially deleted *CD9* allele and marked by the constitutively-active *EGFP* transgene, were injected into blastocysts, the embryos developed to chimeras at E13.5 in which widespread contributions of GFP-positive cells were observed in fluorescent stereomicroscopic observation (left panel). Right panel is the control embryo, showing an absence of fluorescence.

(Akutsu *et al.*, 2009). In this report, we established mouse ESCs lacking *CD9* by gene targeting. These *CD9-/-* ESCs exhibited the morphology and growth properties of ESCs, which express the ES marker factors Oct4, Sox2 and Nanog and have the ability to give rise to teratomas composed of tissues from all three germ layers. *CD9-/-* ESCs also generated mouse chimeras, contributing to various tissues. However, it has been reported that CD9 strongly expresses in mouse and human ES cells, suggesting that CD9 may be one marker of pluripotent stem cells (Nash *et al.*, 2007; Oka *et al.*, 2002). Therefore, our *CD9* knockout ESCs may explain the role of CD9 as a hallmark trait of stem cells-self-renewal and differentiation capacity. Thus, we should consider that CD9 might be one of markers for identification of pluripotent stem cells without functional significance like Oct4.

3. The transcription factor OCT4

Oct4 (octamer-binding transcription factor 4) also known as POU5F1 (POU domain, class 5, transcription factor 1) is a protein that is expressed by all pluripotent cells during mouse embryogenesis, and is also abundantly expressed by undifferentiated mouse ESCs and ECC cell lines (Okamoto *et al.*, 1990; Rosner *et al.*, 1990; Scholer *et al.*, 1989a; Scholer *et al.*, 1989b), as well as in EGC cell lines (Donovan and de Miguel, 2003). So far, however, experiments show that *Oct4* expression is generally weaker in germline stem cells (GSCs) (Kanatsu-Shinohara *et al.*, 2004a). Oct4 has also been established as a marker for human pluripotent ESCs. Therefore, downregulation of *Oct4* is required for the differentiation of somatic lineages. *Oct4*-deficient mouse embryos only develop to a stage that looks like a blastocyst, and although cells are allocated to the interior, these blastocysts are actually only composed of trophectoderm cells. As these structures lack a genuine ICM, they cannot be used to produce ESC cell lines (Nichols *et al.*, 1998). Oct4 has therefore been viewed as being involved in preventing trophectoderm and perhaps somatic-cell differentiation from the ICM, as well as being crucial for maintaining the pluripotent state during embryonic development. Recently, it has been reported that Oct4 only is sufficient to reprogram human neural stem cells to pluripotency (Kim *et al.*, 2009). Therefore, Oct4 is a master gene for pluripotency. In mouse ESCs, the manipulation of *Oct4* expression through inducible or repressible *Oct4* transgenes indicates that the relative amount of Oct4 protein ultimately determines cell fate (Niwa *et al.*, 2000). However, the target genes that are actually responsible for implementing Oct4 decisions are only partly known (Du *et al.*, 2001; Saijoh *et al.*, 1996). Similarly, the potential interactions of Oct4 with other (co)factors, except for Sox2 (SRYrelated high-mobility group (HMG)-box protein-2; (Pevny and Lovell-Badge, 1997)), remain unclear.

4. The transcription factor SOX2

SRY (sex determining region Y)-box 2, also known as Sox2, is a transcription factor. Sox2 is a member of the HMG-domain DNA-binding-protein family that is implicated in the regulation of transcription and chromatin architecture (Pevny and Lovell-Badge, 1997). Sox2 forms a ternary complex with either Oct4 or the ubiquitous Oct1 protein on the enhancer DNA sequences of *Fgf4* (Yuan *et al.*, 1995). This allows Sox2 to participate in the regulation of the ICM and its progeny or derivative cells. Consistent with this role, *Sox2* is expressed in ESCs, but it is also expressed in neural stem cells. Therefore, Sox2 is essential to maintain self-renewal of undifferentiated embryonic stem cells. When gene targeting was used to inactivate *Sox2*, the primitive ectoderm was defective, but it could be rescued (albeit only to survive longer) by the injection of wild-type ESCs into the *Sox2-/-* blastocysts (Avilion *et al.*, 2003). Reduction of *Sox2* expression induces mouse ESCs to differentiate cells into the trophoectoderm lineage, indicating that *Sox2* function is essential for maintaining pluripotency. These results also are suggested by *Sox2* ablation *in vivo*. Interestingly, the forced expression of *Oct4* rescues the pluripotency of *Sox2*-null ESCs (Masui *et al.*, 2007). These findings indicate that Sox2 has a unique function in maintaining the pluripotency of ESCs that is related to the transcriptional activation of *Oct4*.

5. The transcription factor NANOG

Nanog is a highly divergent homeodomain-containing protein commonly accorded a central position in the transcriptional network of pluripotency (Boyer *et al.*, 2005; Cole *et al.*, 2008; Loh *et al.*, 2006; Wang *et al.*, 2006). It is essential for early embryonic development (Mitsui *et al.*, 2003). Undifferentiated, wild-type ESCs normally express Nanog. However, the physiological levels of Nanog in ESCs do not prevent their differentiation after LIF withdrawal. So, under physiological conditions, Nanog seems to be one of several factors that are expressed in pluripotent cells and are downregulated at the onset of differentiation. *Nanog-/-* mouse ESCs differentiate slowly into extra-embryonic endoderm lineages, which is consistent with the absence of a primitive ectoderm in *Nanog-/-* embryos that were analysed at E5.5 *in vivo* (Mitsui *et al.*, 2003). So Nanog expression is responsible for the maintenance of a primitive ectoderm in the embryo. Unlike wild-type ESCs and those forced to express *Oct4*, mouse ESCs that are overexpressing *Nanog* are resistant, but not completely refractory, to the spontaneous differentiation that occurs after LIF withdrawal or by chemical induction (for example, after treatment with 3-methoxybenzamide or all-*trans* retinoic acid). The persistence of Nanog therefore seems to delay, rather than block, the differentiation of ESCs; that is, the threshold of differentiation is increased rather than abolished. In contrast to Nanog overexpression, the reduced expression seen in *Nanog+/-* ESCs results in labile pluripotency whereby spontaneous differentiation is more likely to occur after longer times spent in culture ('passages') (Hatano *et al.*, 2005). So, the amount of Nanog per cell is crucial for stably maintaining an undifferentiated state even in the presence of LIF. In addition, Nanog is not one of the Yamanaka 4 factors employed to reprogram mouse fibroblasts (Maherali *et al.*, 2007; Okita *et al.*, 2007; Takahashi *et al.*, 2007; Takahashi and Yamanaka, 2006; Wernig *et al.*, 2007). Moreover, addition of Nanog to these 4 factors has not been reported to increase efficiencies. However, Nanog is expressed weakly or not at all in incompletely reprogrammed cells that fail to activate properly the endogenous pluripotent transcriptional circuitry (Silva and Smith, 2008; Sridharan *et al.*, 2009; Takahashi and Yamanaka, 2006). Selection or screening for activation of endogenous Nanog expression facilitates isolation of fully reprogrammed iPSCs that can contribute to adult chimeras and give germline transmission (Okita *et al.*, 2007). Furthermore, in human cells Nanog does facilitate molecular reprogramming (Yu *et al.*, 2007). It has also been shown that Nanog promotes the transfer of pluripotency after ES cell fusion (Silva *et al.*, 2006). However, conditional gene deletion in ESCs revealed that Nanog is not essential for propagation of pluripotency *ex vivo* (Chambers *et al.*, 2007). *Nanog* null ESCs are more prone to differentiate but can be maintained indefinitely. Moreover, they contribute extensively to somatic chimeras, similar to *CD9* null ESCs.

6. Other transcriptional factors KLF4 and REX1

The mechanism by which Klf4 regulates ES cell self-renewal was first revealed by its identification as a highly up-regulated target gene of LIF signaling in ES cells (Li *et al.*, 2005). ES cells overexpressing Klf4 had a great propensity for self-renewal based on secondary embryoid body (EB) formation. *Klf4*-transduced EBs expressed higher levels of Oct4, consistent with the notion that Klf4 regulates ES cell self-renewal (Li *et al.*, 2005). The role of Klf4 in regulating pluripotency of ES cells is further revealed by global analysis of promoter occupancy by Yamanaka 4 factors (Kim *et al.*, 2008a). The results identified a transcriptional hierarchy within the four reprogramming factors with both auto-regulatory and feed-forward regulation. In addition, the study indicated that Klf4 is an upstream regulator of a large feed-forward loop that contains Oct4, Sox2, and c-Myc, as well as other common downstream factors including Nanog (Kim *et al.*, 2008a). Combining the results of these studies, it appears that Klf4 exerts a crucial role in somatic cell reprogramming and maintenance of ES cell self-renewal. On the other hand, Klf4 also exhibits both cytostatic and anti-apoptotic effect that is context-dependent. The ability of Klf4 in maintaining immortality of iPSCs maybe explained in part by the requirement of c-Myc as a member of reprogramming factor. Thus, in a manner similar to the cooperation between Klf4 and Ras to affect transformation (Rowland and Peeper, 2006), Klf4 and c-Myc cooperate to affect iPS cell self-renewal. Thus, Klf4 may suppress apoptosis induced by c-Myc and c-Myc neutralizes Klf4's cytostatic effect by suppressing p21WAF1/CIP1 (Yamanaka, 2007). In this manner, the balance between Klf4 and c-Myc might play a critical role in the establishment of an immortalized state of iPSCs. In addition, re-expression of Klf4 in an appropriate environment can regenerate the naïve ground state from mouse epiblast stem cells (EpiSCs), which are derived from columnar epithelial epiblast of the early post-implantation embryo (Brook and Gardner, 1997; Hanna *et al.*, 2010; Hanna *et al.*, 2009; Tesar *et al.*, 2007). Therefore, the essential requirement of Klf4 for reprogramming of somatic cells has subsequently been substantiated (Di Stefano *et al.*, 2009; Shi *et al.*, 2008). On the other hand, *Klf4+/-* mice were phenotypically and histologically normal (Katz *et al.*, 2005; Segre *et al.*, 1999). *Klf4-/-* mice were born at the expected Mendelian ration. Therefore, Klf4 is also dispensable for maintenance of self-renewal and pluripotency of ESCs.

In addition to Oct4, Sox2, Nanog and Klf4, other putative transcription factors expressing pluripotent stem cells in stem-cell-specific manner have been also identified by several investigators. For example, Rex1 (for reduced expression-1, also known as *Zfp42*) was first identified a gene that expresses in F9 embryonal carcinoma (EC) cells and is down-regulated after retinoic acid (RA) treatment to induce differentiation (Hosler *et al.*, 1989). This gene encodes a C2H2 zinc-finger protein that is closely similar to Yy1, an evolutionally-conserved component of polycomb- related complex 2 (Gordon *et al.*, 2006). Its highly-specific expression in pluripotent stem cells has been confirmed in mouse and human ESCs (Eiges *et al.*, 2001; Rogers *et al.*, 1991), making it one of the most famous markers of pluripotency tested in various stem cells such as multipotent adult progenitor cells (Jiang *et al.*, 2002) and amniotic fluid cells (Karlmark *et al.*, 2005). Moreover, Rex1 is also known as a marker of the naïve ground state (Nichols and Smith, 2009). This has been argued that the blastocyst origin of human ESCs is evidenced by their expression of Rex1. However, its function in ESCs has not yet been characterized well although it has been reported that a targeted deletion of *Rex1* results in loss of the ability to differentiate into visceral endoderm induced by RA in F9 EC cells (Thompson and Gudas, 2002), and that a gene silencing by RNA interference for Rex1 results in loss of capacity to self-renew in ESCs (Zhang *et al.*, 2006). In addition, it has been recently reported that over-expression of Rex1 in ESCs neither induces differentiation in the presence of LIF nor maintains self-renewal in the absence of LIF. *Rex1-/-* ESCs can be established and contribute whole embryos after blastocyst injection, indicating that they possess proper pluripotency. Moreover, *Rex1-/-* mice were produced by the intercross of heterozygotes, and both male and female homozygotes were normal and fertile (Masui *et al.*, 2008). These findings support that Rex1 is also dispensable for maintenance of pluripotency in ESCs,

7. Concluding remarks

ESCs can bring unique application to medical and pharmaceutical research. Of note, recent advances in ESC biology have led to the successful generation of iPSCs, which could solve many scientific and ethical problems associated with regenerative medicine and cell-based therapies for degenerative human diseases. Thus, the understanding molecular biomarkers for ESCs is becoming increasingly important for the detection, classification, and isolation of a particular population of ES/iPS cells, and for monitoring the state of differentiation. This chapter discusses that Oct4 only is functionally essential for maintenance of pluripotency in ESCs (Table 1). This is consistent with evidence that Oct4 alone is able to reprogram mouse and human neural stem cells (Kim *et al.*, 2009; Kim *et al.*, 2008b). Therefore, other molecular biomarker highly expressed in ESCs might be markers for identification of pluripotent stem cells without functional significance like Oct4.

Although we have mainly focused here on studies using mouse ESCs, it will be important to understand how these findings relate to human ES cell studies. Studies on human ESCs may be best compared with studies on pluripotent mouse EpiSC lines, which have been established from post-implantation embryos (Brons *et al.*, 2007; Tesar *et al.*, 2007). ESCs and EpiSCs differ from one another in their factor requirements in vitro and in their capacity to incorporate into developing chimaeras. The recent demonstration of revertibility of primed EpiSC state to naïve ESC state is reported in mouse and human (Hanna *et al.*, 2010; Hanna *et al.*, 2009). In the near future, naïve human ESCs will need to be generated from blastocyst

Disrupted genes	Knockout ESCs	Passed pluripotency test	References
Oct4	not established[a]	N/D	Okamoto et al., 1990
Sox2	established[b]	chimera	Masui et al., 2007
Nanog	established	chimera	Chambers et al., 2007
Klf4	established	chimera	Nakatake et al., 2006
Rex1	established	chimera	Masui et al., 2008
CD9	established	chimera	Akutsu et al., 2009

N/D: not done
[a]Oct4 deficient-blastocysts lack ICM.
[b]Sox2 null ESCs are maintenanced by the forced expression of Oct4 .

Table 1. Function of the best-characterized ES cell markers

embryos. Because the naïve human pluripotent stem cells will provide a critical tool to model the earliest steps in human embryonic development. Understanding how pluripotent molecular biomarker assemblies change as cells move from one pluripotent compartment to another will allow us to view how the dynamic alterations in cell phenotype that underlie developmental transitions are dictated, which will be surely enhanced our knowledge of ESCs and of early embryonic development and cell-based therapies.

8. References

Akutsu, H.; Miura, T.; Machida, M.; Birumachi, J.; Hamada, A.; Yamada, M.; Sullivan, S.; Miyado, K. & Umezawa, A. (2009). Maintenance of pluripotency and self-renewal ability of mouse embryonic stem cells in the absence of tetraspanin CD9. *Differentiation*, Vol. 78, pp. 137-142.

Avilion, A. A.; Nicolis, S. K.; Pevny, L. H.; Perez, L.; Vivian, N. & Lovell-Badge, R. (2003). Multipotent cell lineages in early mouse development depend on SOX2 function. *Genes Dev*, Vol. 17, pp. 126-140.

Baker, D. E.; Harrison, N. J.; Maltby, E.; Smith, K.; Moore, H. D.; Shaw, P. J.; Heath, P. R.; Holden, H. & Andrews, P. W. (2007). Adaptation to culture of human embryonic stem cells and oncogenesis in vivo. *Nat Biotechnol*, Vol. 25, pp. 207-215.

Barberi, T.; Bradbury, M.; Dincer, Z.; Panagiotakos, G.; Socci, N. D. & Studer, L. (2007). Derivation of engraftable skeletal myoblasts from human embryonic stem cells. *Nat Med*, Vol. 13, pp. 642-648.

Biswas, A. & Hutchins, R. (2007). Embryonic stem cells. *Stem Cells Dev*, Vol. 16, pp. 213-222.

Boyer, L. A.; Lee, T. I.; Cole, M. F.; Johnstone, S. E.; Levine, S. S.; Zucker, J. P.; Guenther, M. G.; Kumar, R. M.; Murray, H. L.; Jenner, R. G.; Gifford, D. K.; Melton, D. A.; Jaenisch, R. & Young, R. A. (2005). Core transcriptional regulatory circuitry in human embryonic stem cells. *Cell*, Vol. 122, pp. 947-956.

Brons, I. G.; Smithers, L. E.; Trotter, M. W.; Rugg-Gunn, P.; Sun, B.; Chuva de Sousa Lopes, S. M.; Howlett, S. K.; Clarkson, A.; Ahrlund-Richter, L.; Pedersen, R. A. & Vallier, L.

(2007). Derivation of pluripotent epiblast stem cells from mammalian embryos. *Nature*, Vol. 448, pp. 191-195.

Brook, F. A. & Gardner, R. L. (1997). The origin and efficient derivation of embryonic stem cells in the mouse. *Proc Natl Acad Sci U S A*, Vol. 94, pp. 5709-5712.

Carpenter, M. K.; Rosler, E. & Rao, M. S. (2003). Characterization and differentiation of human embryonic stem cells. *Cloning Stem Cells*, Vol. 5, pp. 79-88.

Chambers, I.; Silva, J.; Colby, D.; Nichols, J.; Nijmeijer, B.; Robertson, M.; Vrana, J.; Jones, K.; Grotewold, L. & Smith, A. (2007). Nanog safeguards pluripotency and mediates germline development. *Nature*, Vol. 450, pp. 1230-1234.

Cole, M. F.; Johnstone, S. E.; Newman, J. J.; Kagey, M. H. & Young, R. A. (2008). Tcf3 is an integral component of the core regulatory circuitry of embryonic stem cells. *Genes Dev*, Vol. 22, pp. 746-755.

D'Amour, K. A.; Bang, A. G.; Eliazer, S.; Kelly, O. G.; Agulnick, A. D.; Smart, N. G.; Moorman, M. A.; Kroon, E.; Carpenter, M. K. & Baetge, E. E. (2006). Production of pancreatic hormone-expressing endocrine cells from human embryonic stem cells. *Nat Biotechnol*, Vol. 24, pp. 1392-1401.

Di Stefano, B.; Prigione, A. & Broccoli, V. (2009). Efficient genetic reprogramming of unmodified somatic neural progenitors uncovers the essential requirement of Oct4 and Klf4. *Stem Cells Dev*, Vol. 18, pp. 707-716.

Donovan, P. J. & de Miguel, M. P. (2003). Turning germ cells into stem cells. *Curr Opin Genet Dev*, Vol. 13, pp. 463-471.

Du, Z.; Cong, H.&Yao, Z. (2001). Identification of putative downstream genes of Oct-4 by suppression-subtractive hybridization. *Biochem Biophys Res Commun*, Vol. 282, pp. 701-706.

Eiges, R.; Schuldiner, M.; Drukker, M.; Yanuka, O.; Itskovitz-Eldor, J. & Benvenisty, N. (2001). Establishment of human embryonic stem cell-transfected clones carrying a marker for undifferentiated cells. *Curr Biol*, Vol. 11, pp. 514-518.

Gordon, S.; Akopyan, G.; Garban, H. & Bonavida, B. (2006). Transcription factor YY1: structure, function, and therapeutic implications in cancer biology. *Oncogene*, Vol. 25, pp. 1125-1142.

Gouon-Evans, V.; Boussemart, L.; Gadue, P.; Nierhoff, D.; Koehler, C. I.; Kubo, A.; Shafritz, D. A. & Keller, G. (2006). BMP-4 is required for hepatic specification of mouse embryonic stem cell-derived definitive endoderm. *Nat Biotechnol*, Vol. 24, pp. 1402-1411.

Hadjiargyrou, M. & Patterson, P. H. (1995). An anti-CD9 monoclonal antibody promotes adhesion and induces proliferation of Schwann cells in vitro. *J Neurosci*, Vol. 15, pp. 574-583.

Hanna, J.; Cheng, A. W.; Saha, K.; Kim, J.; Lengner, C. J.; Soldner, F.; Cassady, J. P.; Muffat, J.; Carey, B. W. & Jaenisch, R. (2010). Human embryonic stem cells with biological and epigenetic characteristics similar to those of mouse ESCs. *Proc Natl Acad Sci U S A*, Vol. 107, pp. 9222-9227.

Hanna, J.; Markoulaki, S.; Mitalipova, M.; Cheng, A. W.; Cassady, J. P.; Staerk, J.; Carey, B. W.; Lengner, C. J.; Foreman, R.; Love, J.; Gao, Q.; Kim, J. & Jaenisch, R. (2009). Metastable pluripotent states in NOD-mouse-derived ESCs. *Cell Stem Cell*, Vol. 4, pp. 513-524.

Hatano, S. Y.; Tada, M.; Kimura, H.; Yamaguchi, S.; Kono, T.; Nakano, T.; Suemori, H.; Nakatsuji, N. & Tada, T. (2005). Pluripotential competence of cells associated with Nanog activity. *Mech Dev*, Vol. 122, pp. 67-79.

Hoffman, L. M. & Carpenter, M. K. (2005). Characterization and culture of human embryonic stem cells. *Nat Biotechnol*, Vol. 23, pp. 699-708.

Hosler, B. A.; LaRosa, G. J.; Grippo, J. F. & Gudas, L. J. (1989). Expression of REX-1, a gene containing zinc finger motifs, is rapidly reduced by retinoic acid in F9 teratocarcinoma cells. *Mol Cell Biol*, Vol. 9, pp. 5623-5629.

Jiang, Y.; Jahagirdar, B. N.; Reinhardt, R. L.; Schwartz, R. E.; Keene, C. D.; Ortiz-Gonzalez, X. R.; Reyes, M.; Lenvik, T.; Lund, T.; Blackstad, M.; Du, J.; Aldrich, S.; Lisberg, A.; Low, W. C.; Largaespada, D. A. & Verfaillie, C. M. (2002). Pluripotency of mesenchymal stem cells derived from adult marrow. *Nature*, Vol. 418, pp. 41-49.

Kanatsu-Shinohara, M.; Inoue, K.; Lee, J.; Yoshimoto, M.; Ogonuki, N.; Miki, H.; Baba, S.; Kato, T.; Kazuki, Y.; Toyokuni, S.; Toyoshima, M.; Niwa, O.; Oshimura, M.; Heike, T.; Nakahata, T.; Ishino, F.; Ogura, A. & Shinohara, T. (2004a). Generation of pluripotent stem cells from neonatal mouse testis. *Cell*, Vol. 119, pp. 1001-1012.

Kanatsu-Shinohara, M.; Toyokuni, S. & Shinohara, T. (2004b). CD9 is a surface marker on mouse and rat male germline stem cells. *Biol Reprod*, Vol. 70, pp. 70-75.

Kaprielian, Z.; Cho, K. O.; Hadjiargyrou, M. & Patterson, P. H. (1995). CD9, a major platelet cell surface glycoprotein, is a ROCA antigen and is expressed in the nervous system. *J Neurosci*, Vol. 15, pp. 562-573.

Karlmark, K. R.; Freilinger, A.; Marton, E.; Rosner, M.; Lubec, G.&Hengstschlager, M. (2005). Activation of ectopic Oct-4 and Rex-1 promoters in human amniotic fluid cells. *Int J Mol Med*, Vol. 16, pp. 987-992.

Katz, J. P.; Perreault, N.; Goldstein, B. G.; Actman, L.; McNally, S. R.; Silberg, D. G.; Furth, E. E. & Kaestner, K. H. (2005). Loss of Klf4 in mice causes altered proliferation and differentiation and precancerous changes in the adult stomach. *Gastroenterology*, Vol. 128, pp. 935-945.

Kim, J.; Chu, J.; Shen, X.; Wang, J. & Orkin, S. H. (2008a). An extended transcriptional network for pluripotency of embryonic stem cells. *Cell*, Vol. 132, pp. 1049-1061.

Kim, J. B.; Greber, B.; Arauzo-Bravo, M. J.; Meyer, J.; Park, K. I.; Zaehres, H. & Scholer, H. R. (2009). Direct reprogramming of human neural stem cells by OCT4. *Nature*, Vol. 461, pp. 649-643.

Kim, J. B.; Zaehres, H.; Wu, G.; Gentile, L.; Ko, K.; Sebastiano, V.; Arauzo-Bravo, M. J.; Ruau, D.; Han, D. W.; Zenke, M. & Scholer, H. R. (2008b). Pluripotent stem cells induced from adult neural stem cells by reprogramming with two factors. *Nature*, Vol. 454, pp. 646-650.

Kim, K.; Lerou, P.; Yabuuchi, A.; Lengerke, C.; Ng, K.; West, J.; Kirby, A.; Daly, M. J. & Daley, G. Q. (2007). Histocompatible embryonic stem cells by parthenogenesis. *Science*, Vol. 315, pp. 482-486.

Levenberg, S.; Zoldan, J.; Basevitch, Y. & Langer, R. (2007). Endothelial potential of human embryonic stem cells. *Blood*, Vol. 110, pp. 806-814.

Li, Y.; McClintick, J.; Zhong, L.; Edenberg, H. J.; Yoder, M. C. & Chan, R. J. (2005). Murine embryonic stem cell differentiation is promoted by SOCS-3 and inhibited by the zinc finger transcription factor Klf4. *Blood*, Vol. 105, pp. 635-637.

Loh, Y. H.; Wu, Q.; Chew, J. L.; Vega, V. B.; Zhang, W.; Chen, X.; Bourque, G.; George, J.; Leong, B.; Liu, J.; Wong, K. Y.; Sung, K. W.; Lee, C. W.; Zhao, X. D.; Chiu, K. P.; Lipovich, L.; Kuznetsov, V. A.; Robson, P.; Stanton, L. W.; Wei, C. L.; Ruan, Y.; Lim, B. & Ng, H. H. (2006). The Oct4 and Nanog transcription network regulates pluripotency in mouse embryonic stem cells. *Nat Genet*, Vol. 38, pp. 431-440.

Lu, S. J.; Feng, Q.; Caballero, S.; Chen, Y.; Moore, M. A.; Grant, M. B. & Lanza, R. (2007). Generation of functional hemangioblasts from human embryonic stem cells. *Nat Methods*, Vol. 4, pp. 501-509.

Maherali, N.; Sridharan, R.; Xie, W.; Utikal, J.; Eminli, S.; Arnold, K.; Stadtfeld, M.; Yachechko, R.; Tchieu, J.; Jaenisch, R.; Plath, K. & Hochedlinger, K. (2007). Directly reprogrammed fibroblasts show global epigenetic remodeling and widespread tissue contribution. *Cell Stem Cell*, Vol. 1, pp. 55-70.

Masui, S.; Nakatake, Y.; Toyooka, Y.; Shimosato, D.; Yagi, R.; Takahashi, K.; Okochi, H.; Okuda, A.; Matoba, R.; Sharov, A. A.; Ko, M. S. & Niwa, H. (2007). Pluripotency governed by Sox2 via regulation of Oct3/4 expression in mouse embryonic stem cells. *Nat Cell Biol*, Vol. 9, pp. 625-635.

Masui, S.; Ohtsuka, S.; Yagi, R.; Takahashi, K.; Ko, M. S. & Niwa, H. (2008). Rex1/Zfp42 is dispensable for pluripotency in mouse ES cells. *BMC Dev Biol*, Vol. 8, pp. 45.

Menendez, P.; Bueno, C.; Wang, L. & Bhatia, M. (2005). Human embryonic stem cells: potential tool for achieving immunotolerance? *Stem Cell Rev*, Vol. 1, pp. 151-158.

Mitsui, K.; Tokuzawa, Y.; Itoh, H.; Segawa, K.; Murakami, M.; Takahashi, K.; Maruyama, M.; Maeda, M. & Yamanaka, S. (2003). The homeoprotein Nanog is required for maintenance of pluripotency in mouse epiblast and ES cells. *Cell*, Vol. 113, pp. 631-642.

Miyado, K.; Yamada, G.; Yamada, S.; Hasuwa, H.; Nakamura, Y.; Ryu, F.; Suzuki, K.; Kosai, K.; Inoue, K.; Ogura, A.; Okabe, M. & Mekada, E. (2000). Requirement of CD9 on the egg plasma membrane for fertilization. *Science*, Vol. 287, pp. 321-324.

Miyado, K.; Yoshida, K.; Yamagata, K.; Sakakibara, K.; Okabe, M.; Wang, X.; Miyamoto, K.; Akutsu, H.; Kondo, T.; Takahashi, Y.; Ban, T.; Ito, C.; Toshimori, K.; Nakamura, A.; Ito, M.; Miyado, M.; Mekada, E. & Umezawa, A. (2008). The fusing ability of sperm is bestowed by CD9-containing vesicles released from eggs in mice. *Proc Natl Acad Sci U S A*, Vol. 105, pp. 12921-12926.

Nakatake, Y.; Fukui, N.; Iwamatsu, Y.; Masui, S.; Takahashi, K.; Yagi, R.; Yagi, K.; Miyazaki, J.; Matoba, R.; Ko, M. S. & Niwa, H. (2006). Klf4 cooperates with Oct3/4 and Sox2 to activate the Lefty1 core promoter in embryonic stem cells. *Mol Cell Biol*, Vol. 26, pp. 7772-7782.

Nash, R.; Neves, L.; Faast, R.; Pierce, M. & Dalton, S. (2007). The lectin Dolichos biflorus agglutinin recognizes glycan epitopes on the surface of murine embryonic stem cells: a new tool for characterizing pluripotent cells and early differentiation. *Stem Cells*, Vol. 25, pp. 974-982.

Nichols, J. & Smith, A. (2009). Naive and primed pluripotent states. *Cell Stem Cell*, Vol. 4, pp. 487-492.

Nichols, J.; Zevnik, B.; Anastassiadis, K.; Niwa, H.; Klewe-Nebenius, D.; Chambers, I.; Scholer, H. & Smith, A. (1998). Formation of pluripotent stem cells in the mammalian embryo depends on the POU transcription factor Oct4. *Cell*, Vol. 95, pp. 379-391.

Niwa, H.; Miyazaki, J. & Smith, A. G. (2000). Quantitative expression of Oct-3/4 defines differentiation, dedifferentiation or self-renewal of ES cells. *Nat Genet*, Vol. 24, pp. 372-376.

Oka, M.; Tagoku, K.; Russell, T. L.; Nakano, Y.; Hamazaki, T.; Meyer, E. M.; Yokota, T. & Terada, N. (2002). CD9 is associated with leukemia inhibitory factor-mediated maintenance of embryonic stem cells. *Mol Biol Cell*, Vol. 13, pp. 1274-1281.

Okamoto, K.; Okazawa, H.; Okuda, A.; Sakai, M.; Muramatsu, M. & Hamada, H. (1990). A novel octamer binding transcription factor is differentially expressed in mouse embryonic cells. *Cell*, Vol. 60, pp. 461-472.

Okita, K.; Ichisaka, T. & Yamanaka, S. (2007). Generation of germline-competent induced pluripotent stem cells. *Nature*, Vol. 448, pp. 313-317.

Pevny, L. H. & Lovell-Badge, R. (1997). Sox genes find their feet. *Curr Opin Genet Dev*, Vol. 7, pp. 338-344.

Rogers, M. B.; Hosler, B. A. & Gudas, L. J. (1991). Specific expression of a retinoic acid-regulated, zinc-finger gene, Rex-1, in preimplantation embryos, trophoblast and spermatocytes. *Development*, Vol. 113, pp. 815-824.

Rosner, M. H.; Vigano, M. A.; Ozato, K.; Timmons, P. M.; Poirier, F.; Rigby, P. W. & Staudt, L. M. (1990). A POU-domain transcription factor in early stem cells and germ cells of the mammalian embryo. *Nature*, Vol. 345, pp. 686-692.

Rowland, B. D. & Peeper, D. S. (2006). KLF4, p21 and context-dependent opposing forces in cancer. *Nat Rev Cancer*, Vol. 6, pp. 11-23.

Roy, N. S.; Cleren, C.; Singh, S. K.; Yang, L.; Beal, M. F. & Goldman, S. A. (2006). Functional engraftment of human ES cell-derived dopaminergic neurons enriched by coculture with telomerase-immortalized midbrain astrocytes. *Nat Med*, Vol. 12, pp. 1259-1268.

Saijoh, Y.; Fujii, H.; Meno, C.; Sato, M.; Hirota, Y.; Nagamatsu, S.; Ikeda, M. & Hamada, H. (1996). Identification of putative downstream genes of Oct-3, a pluripotent cell-specific transcription factor. *Genes Cells*, Vol. 1, pp. 239-252.

Salero, E. & Hatten, M. E. (2007). Differentiation of ES cells into cerebellar neurons. *Proc Natl Acad Sci U S A*, Vol. 104, pp. 2997-3002.

Scholer, H. R.; Balling, R.; Hatzopoulos, A. K.; Suzuki, N. & Gruss, P. (1989a). Octamer binding proteins confer transcriptional activity in early mouse embryogenesis. *Embo J*, Vol. 8, pp. 2551-2557.

Scholer, H. R.; Hatzopoulos, A. K.; Balling, R.; Suzuki, N. & Gruss, P. (1989b). A family of octamer-specific proteins present during mouse embryogenesis: evidence for germline-specific expression of an Oct factor. *Embo J*, Vol. 8, pp. 2543-2550.

Schroeder, I. S.; Rolletschek, A.; Blyszczuk, P.; Kania, G. & Wobus, A. M. (2006). Differentiation of mouse embryonic stem cells to insulin-producing cells. *Nat Protoc*, Vol. 1, pp. 495-507.

Segre, J. A.; Bauer, C. & Fuchs, E. (1999). Klf4 is a transcription factor required for establishing the barrier function of the skin. *Nat Genet*, Vol. 22, pp. 356-360.

Shi, Y.; Desponts, C.; Do, J. T.; Hahm, H. S.; Scholer, H. R. & Ding, S. (2008). Induction of pluripotent stem cells from mouse embryonic fibroblasts by Oct4 and Klf4 with small-molecule compounds. *Cell Stem Cell*, Vol. 3, pp. 568-574.

Silva, J.; Chambers, I.; Pollard, S. & Smith, A. (2006). Nanog promotes transfer of pluripotency after cell fusion. *Nature*, Vol. 441, pp. 997-1001.

Silva, J. & Smith, A. (2008). Capturing pluripotency. *Cell*, Vol. 132, pp. 532-536.

Soto-Gutierrez, A.; Navarro-Alvarez, N.; Zhao, D.; Rivas-Carrillo, J. D.; Lebkowski, J.; Tanaka, N.; Fox, I. J. & Kobayashi, N. (2007). Differentiation of mouse embryonic stem cells to hepatocyte-like cells by co-culture with human liver nonparenchymal cell lines. *Nat Protoc*, Vol. 2, pp. 347-356.

Sridharan, R.; Tchieu, J.; Mason, M. J.; Yachechko, R.; Kuoy, E.; Horvath, S.; Zhou, Q. & Plath, K. (2009). Role of the murine reprogramming factors in the induction of pluripotency. *Cell*, Vol. 136, pp. 364-377.

Takahashi, K.; Tanabe, K.; Ohnuki, M.; Narita, M.; Ichisaka, T.; Tomoda, K. & Yamanaka, S. (2007). Induction of pluripotent stem cells from adult human fibroblasts by defined factors. *Cell*, Vol. 131, pp. 861-872.

Takahashi, K. & Yamanaka, S. (2006). Induction of pluripotent stem cells from mouse embryonic and adult fibroblast cultures by defined factors. *Cell*, Vol. 126, pp. 663-676.

Tesar, P. J.; Chenoweth, J. G.; Brook, F. A.; Davies, T. J.; Evans, E. P.; Mack, D. L.; Gardner, R. L. & McKay, R. D. (2007). New cell lines from mouse epiblast share defining features with human embryonic stem cells. *Nature*, Vol. 448, pp. 196-199.

Thompson, J. R. & Gudas, L. J. (2002). Retinoic acid induces parietal endoderm but not primitive endoderm and visceral endoderm differentiation in F9 teratocarcinoma stem cells with a targeted deletion of the Rex-1 (Zfp-42) gene. *Mol Cell Endocrinol*, Vol. 195, pp. 119-133.

Thomson, H. (2007). Bioprocessing of embryonic stem cells for drug discovery. *Trends Biotechnol*, Vol. 25, pp. 224-230.

Thomson, J. A.; Itskovitz-Eldor, J.; Shapiro, S. S.; Waknitz, M. A.; Swiergiel, J. J.; Marshall, V. S. & Jones, J. M. (1998). Embryonic stem cell lines derived from human blastocysts. *Science*, Vol. 282, pp. 1145-1147.

Wang, J.; Rao, S.; Chu, J.; Shen, X.; Levasseur, D. N.; Theunissen, T. W. & Orkin, S. H. (2006). A protein interaction network for pluripotency of embryonic stem cells. *Nature*, Vol. 444, pp. 364-368.

Wang, Z. Z.; Au, P.; Chen, T.; Shao, Y.; Daheron, L. M.; Bai, H.; Arzigian, M.; Fukumura, D.; Jain, R. K. & Scadden, D. T. (2007). Endothelial cells derived from human embryonic stem cells form durable blood vessels in vivo. *Nat Biotechnol*, Vol. 25, pp. 317-318.

Wernig, M.; Meissner, A.; Foreman, R.; Brambrink, T.; Ku, M.; Hochedlinger, K.; Bernstein, B. E. & Jaenisch, R. (2007). In vitro reprogramming of fibroblasts into a pluripotent ES-cell-like state. *Nature*, Vol. 448, pp. 318-324.

West, J. A.; Park, I. H.; Daley, G. Q. & Geijsen, N. (2006). In vitro generation of germ cells from murine embryonic stem cells. *Nat Protoc*, Vol. 1, pp. 2026-2036.

Yamanaka, S. (2007). Strategies and new developments in the generation of patient-specific pluripotent stem cells. *Cell Stem Cell*, Vol. 1, pp. 39-49.

Yang, X.; Smith, S. L.; Tian, X. C.; Lewin, H. A.; Renard, J. P. & Wakayama, T. (2007). Nuclear reprogramming of cloned embryos and its implications for therapeutic cloning. *Nat Genet*, Vol. 39, pp. 295-302.

Yu, J.; Vodyanik, M. A.; Smuga-Otto, K.; Antosiewicz-Bourget, J.; Frane, J. L.; Tian, S.; Nie, J.; Jonsdottir, G. A.; Ruotti, V.; Stewart, R.; Slukvin, II & Thomson, J. A. (2007). Induced pluripotent stem cell lines derived from human somatic cells. *Science*, Vol. 318, pp. 1917-1920.

Yuan, H.; Corbi, N.; Basilico, C. & Dailey, L. (1995). Developmental-specific activity of the FGF-4 enhancer requires the synergistic action of Sox2 and Oct-3. *Genes Dev*, Vol. 9, pp. 2635-2645.

Zhang, J. Z.; Gao, W.; Yang, H. B.; Zhang, B.; Zhu, Z. Y. & Xue, Y. F. (2006). Screening for genes essential for mouse embryonic stem cell self-renewal using a subtractive RNA interference library. *Stem Cells*, Vol. 24, pp. 2661-2668.

10

Target Identification of MicroRNAs Expressed Highly and Regulated by Activin A in Human Embryonic Stem Cells

Steven Shoei-Lung Li
Kaohsiung Medical University
Taiwan

1. Introduction

Since 1998, when the first human embryonic stem (hES) cell lines were reported (Thomson et al., 1998), a large number of hES cell lines have been derived from the inner cell mass of preimplantation embryos donated at *in vitro* fertilization (IVF) clinics (Guhr et al., 2006). These hES cell lines possess remarkable ability of both unlimited self-renewal and pluripotency to generate any cell type differentiated from three germ-layers ectoderm, mesoderm and endoderm. Thus, these hES cell lines have great potentials for cell therapies in regenerative medicine and experimental models for drug discovery and toxicity testing in addition to basic studies on stem cell biology and molecular embryogenesis (Wobus & Boheler, 2005).

The proliferation of undifferentiated hES cells can be maintained on either mitotically inactivated mouse embryonic fibroblasts (MEF) as feeder or Matrigel-coated plastic surfaces in MEF-conditioned medium (Xu et al., 2001). The continuous culture of undifferentiated hES cells either on MEF feeder or in the MEF-conditioned medium bears the risk of transmitting animal pathogens, and limits future medical applications of hES cells. A few human cell systems, including hES-derived fibroblast-like cells as feeder (Stojkovic et al., 2005), with capacity to support the growth of undifferentiated hES cells have been developed to replace the use of MEF. Activin A was previously reported to be necessary and sufficient for the maintenance of self-renewal and pluripotency of hES cells in long-term feeder- and serum-free culture (Xiao et al., 2006). It would be of interest to compare the gene expression profiles of these undifferentiated hES cells grown under these different conditions in order to better understand their common molecular mechanisms of unlimited self-renewal and pluripotency.

The genome-wide gene expression analyses using high-throughput microarray have been used to identify key "stemness" genes responsible for the unlimited self-renewal and pluripotency of hES cells (Ivanova et al., 2002; Ramalho-Santos et al., 2002; Sperger et al., 2003). A meta-analysis of 20 previously reported transcriptomes had identified 48 genes overexpressed in hES cells compared to differentiated cell types (Assou et al., 2007), and these 48 genes, including transcription factors such as OCT4 (also known as POU5F1), SOX2 and NANOG, may be responsible for the unlimited self-renewal and pluripotency of hES cells. However, molecular mechanisms involved in unlimited self-renewal and pluripotency of hES cells remain to be fully understood.

Recently, microRNAs (miRNAs) have been shown to play important roles in mammalian embryo development and cell differentiation. Mammalian genomes encode many hundreds of miRNAs, which are predicted to regulate expression of as many as 30% of protein-coding genes (Bartel, 2004; Griffiths-Jones et al., 2006; John et al., 2004; Landgraf et al., 2007). Although the biological functions of most miRNAs are unknown, some miRNAs appear to participate in determination of cell fate, in pattern formation in embryonic development, and in control of cell proliferation, cell differentiation and cell apoptosis in animals (Alvarez-Garcia & Miska, 2005; Kloosterman & Plasterk, 2006). The miRNAs negatively regulate the posttranscriptional expression by translational repression and/or destabilization of protein-coding mRNAs. The impact of miRNAs on protein output was recently shown that although some targets were repressed without detectable changes in mRNA levels, those translationally repressed by more than a third also displayed detectable mRNA destabilization, and, for the more highly repressed targets, mRNA destabilization usually comprised the major component of repression (Baek et al., 2008; Selbach et al., 2008). In human and mouse, several ES cell-specific miRNAs were identified (Houbaviy et al., 2003; Laurent et al., 2008; Suh et al., 2004), and these miRNAs were shown to play crucial roles in ES cell differentiation, lineage specification, and organogenesis, especially neurogenesis and cardiogenesis (Chen et al., 2007). It would be important to identify the target mRNAs of hES cell-specific miRNAs that will lead us to understand the complex and interesting networks of regulation in hES cells.

In my laboratory at Kaohsiung Medical University in Taiwan, five hES cell lines have been derived with institutional review board approval from preimplantation embryos donated at *in vitro* fertilization clinics in Taiwan (Li et al., 2006), and these lines have since been continuously cultured on mitotically inactivated MEF feeder in the hES medium for more than 44 passages and underwent freezing/thawing processes. All of five hES cell lines expressed characteristic undifferentiated hES markers such as SSEA-4, TRA-1-81, alkaline phosphatase, TERT, transcription factors POU5F1 (OCT4) and NANOG. The hES cell lines T1 and T3 possess normal female karyotypes, whereas lines T4 and T5 are normal male, but line T2 is male trisomy 12 (47XY,+12). The hES cell lines T1, T2, T3 and T5 were able to produce teratomas in SCID mice, and line T4 could only form embryoid bodies *in vitro*. Global gene expression profiling of these five hES cell lines indicated that the TGFβ/activin branch components inhibin BC, ACVR2A, ACVR1 (ALK2), TGFBR1 (ALK5) and SMAD2 were found to be highly expressed in undifferentiated states of these five hES cell lines and decreased upon differentiation. These hES cell lines are useful for drug development and toxicity testing in addition to basic studies on human stem cell biology.

2. Target identification of hES cell-specific miRNAs

The undifferentiated state of hES-T3 cells grown on MEF feeder layer was indicated by positive staining of OCT4 and NANOG, and these undifferentiated hES-T3 cells were designated as T3MF. An autogeneic feeder cells with capacity to support the growth of undifferentiated hES cells was established according to the previously published procedure (Stojkovic et al., 2005). hES-T3 (passage 19) cells were transferred into feeder-free and noncoated plate (10 cm) in DMEM supplemented with 10% FBS under 5% CO_2 at 37°C. After 10 days, cells appeared as fibroblast-like morphology, that is, flat cells with elongated nucleus and branching pseudopodia, and these differentiated fibroblast-like cells are designated as T3DF. These T3DF cells were passaged using trypsin (0.05%) every 4 days or cryopreserved. After inactivation using mitomycin C (10 ug/ml), T3DF cells (passage 8) as feeder have been shown to support the undifferentiated growth of hES-T3 cells for more than 14 passages, and the hES-T3 colonies grown on T3DF feeder were stained positively for OCT4 and NANOG [Li et al. 2009].

The expression profiles of 250 human miRNAs in T3MF and T3DF cells were quantitated using TagMan MicroRNA Assays as described previously (Chen et al., 2005; Liang et al., 2007; Li et al., 2009), and the expression level of each miRNA was indicated as folds over U6 snRNA. The four hES cell-specific miRNAs miR-302d, miR-367, miR-372 and miR-200c were found to express abundantly (more than 20-fold U6 snRNA) in T3MF cells, but little (0.03-fold U6 snRNA) in T3DF cells (Table 1). These four miRNAs were also reported to be upregulated in hES cells (Laurent et al., 2008).

The genome-wide mRNA expression of T3MF and T3DF cells was determined using Affymetrix human genome U133 plus 2.0 GeneChip (Li et al., 2009). The original data had been deposited to NCBI database, and the GEO series number is GSE9440. The most expressed genes of T3MF cells include the undifferentiated hES markers such as OCT4, SOX2 and NANOG, and many of the most expressed mRNAs of T3MF cells were found to be in common with the 48 over-expressed genes in undifferentiated hES cells reported previously (Assou et al., 2007). The most expressed genes of differentiated T3DF cells include genes TRPS1, KLF13, MBNL2, MTMR3, NF1B, RAB6A, MARCKS, DDEF1, MBNL1 and QKI, which were later identified to be targets of four ES cell-specific miRNAs miR-302d, miR-372, miR-200c and/or miR-367 (Table 2).

miRNAs	T3MF	T3DF	T3MF/ T3DF	Specificity	Chromosome
hsa-miR-302d	204.60	0.01	20460	hES	4q25
hsa-miR-367	136.40	0.01	13640	hES	4q25
hsa-miR-372	27.80	0.01	2780	hES	19q13.42
hsa-miR-200c	27.73	0.03	937	hES	12p13.31
hsa-miR-20b	44.68	0.06	791		Xq26.2
hsa-miR-26b	24.02	0.07	349		2q35
hsa-miR-302c*	13.82	0.01	1382	hES	4q25
hsa-miR-302a*	1.29	0.01	129	hES	4q25
hsa-miR-302b*	0.11	0.01	11	hES	4q25
hsa-miR-371	3.25	0.01	325	hES	19q13.42
hsa-miR-373*	0.21	0.01	21	hES	19q13.42
hsa-miR-368	0.01	0.01	1	hES	14q32.31

Table 1. Expression levels of hES cell-specific miRNAs in T3ES and T3DF cells. This table is modified from Li et al. (2009).

The targets of the four hES cell-specific miRNAs miR-302d, miR-367, miR-372 and miR-200c miRNAs in hES cells are not known previously (Griffiths-Jones et al., 2006; John et al., 2004; Landgraf et al., 2007), and their potential targets were predicted using the methods of PicTar and TargetScanS (Sethupathy et al., 2006). The expression levels of these predicted target mRNAs in T3MF and T3DF cells were analyzed by the Volcano plot, and the differentially expressed genes were defined by more than 3-folds of changes (T3DF/T3MF) and a p-value cutoff of 0.05. The 58 differentially expressed genes were found to express more than 3-folds

miR-	miR-	miR-	miR-	Gene Symbol	T3MF	T3DF	T3DF/ T3MF	Description	UniGene
302d	372	200c		TRPS1	4.21	81.92	19.5	trichorhinophalangeal syndrome I	Hs.657018
302d	372	200c		KLF13	6.13	113.7	18.5	Kruppel-like factor 13	Hs.525752
302d	372	200c		MBNL2	1.47	19.69	13.4	muscleblind-like 2 (Drosophila)	Hs.657347
302d	372	200c		MTMR3	0.99	11.28	11.4	myotubularin related protein 3	Hs.474536
302d	372		367	NFIB	0.67	12.88	19.3	nuclear factor I/B	Hs.644095
302d	372			RAB6A	2.37	16.23	6.9	RAB6A, member RAS oncogene family	Hs.503222
302d	372			ZNF238	1.76	9.84	5.6	zinc finger protein 238	Hs.69997
302d	372			MEF2C	0.95	9.62	10.1	myocyte enhancer factor 2C	Hs.654474
302d	372			PURB	1.03	8.49	8.3	purine-rich element binding protein B	Hs.349150
302d	372			NR4A2	1.52	7.61	5	nuclear receptor subfamily 4, group A, member 2	Hs.563344
302d	372			TAL1	1.67	7.46	4.5	T-cell acute lymphocytic leukemia 1	Hs.658150
302d	372			PPP3R1	1.21	5.62	4.7	isoform	Hs.280604
302d	372			FNDC3A	1.34	5.02	3.8	fibronectin type III domain containing 3A	Hs.508010
302d	372			ZFP91	0.65	4.79	7.3	zinc finger protein 91 homolog (mouse)	Hs.524920
302d	372			C11orf9	0.81	4.54	5.6	chromosome 11 open reading frame 9	Hs.473109
302d	372			IQSEC2	0.99	4.38	4.4	IQ motif and Sec7 domain 2	Hs.496138
302d	372			RBL1	1.24	4.31	3.5	retinoblastoma-like 1 (p107)	Hs.207745
302d	372			ZMYND11	0.82	3.76	4.6	zinc finger, MYND domain containing 11	Hs.292265
302d	372			ESR1	1.14	3.47	3	estrogen receptor 1	Hs.208124
302d	372			LMO7	0.99	3.13	3.2	LIM domain 7	Hs.207631
302d		200c		ZFHX4	0.27	8.46	31.8	zinc finger homeodomain 4	Hs.458973
302d			367	RUNX1	1.16	4.02	3.5	leukemia 1)	Hs.149261
302d				UBE3A	1.24	4.16	3.3	small nuclear ribonucleoprotein polypeptide N	Hs.654383
		200c	367	KLF4	0.52	5.15	10	Kruppel-like factor 4 (gut)	Hs.376206
		200c	367	C6orf62	0.87	3.97	4.6	chromosome 6 open reading frame 62	Hs.519930
		200c	367	SYNJ1	1.14	3.83	3.3	synaptojanin 1	Hs.473632
		200c		MARCKS	1.58	26.97	17.1	myristoylated alanine-rich protein kinase C substrate	Hs.519909
		200c		DDEF1	1.51	12.34	8.2	development and differentiation enhancing factor 1	Hs.655552
		200c		MBNL1	0.67	10.55	15.7	muscleblind-like (Drosophila)	Hs.478000
		200c		ARHGDIA	0.89	10.17	11.4	Rho GDP dissociation inhibitor (GDI) alpha	Hs.159161
		200c		QKI	0.94	9.55	10.2	quaking homolog, KH domain RNA binding (mouse)	Hs.510324
		200c		BAT3	1.07	7.32	6.9	HLA-B associated transcript 3	Hs.440900
		200c		TSC22D2	0.99	6.51	6.6	TSC22 domain family, member 2	Hs.644065
		200c		MAP2	0.57	5.58	9.7	microtubule-associated protein 2	Hs.368281
		200c		CNOT4	0.94	5.54	5.9	CCR4-NOT transcription complex, subunit 4	Hs.490224
		200c		SOX1	1.44	4.33	3	SRY (sex determining region Y)-box 1	Hs.202526
		200c		ETS1	0.81	4.09	5	(avian)	Hs.369438
		200c		TMEFF2	0.58	3.8	6.5	domains 2	Hs.144513
		200c		ANKRD15	0.92	3.55	3.9	DKFZp451J1819)	Hs.306764
		200c		RAP2C	0.84	3.39	4.1	RAP2C, member of RAS oncogene family	Hs.119889
		200c		SNAP25	0.89	3.36	3.8	synaptosomal-associated protein, 25kDa	Hs.167317
		200c		DMD	1.05	3.35	3.2	types)	Hs.495912
		200c		FBXO33	0.68	3.33	4.9	F-box protein 33	Hs.324342
		200c		VASH1	0.6	3.32	5.6	vasohibin 1	Hs.525479
		200c		SEMA6D	0.61	3.27	5.3	(semaphorin) 6D	Hs.511265
		200c		GIT2	0.95	3.15	3.3	CDNA clone IMAGE:5272062	Hs.434996
		200c		GATA2	0.76	3.1	4.1	GATA binding protein 2	Hs.367725

Table 2. Expression levels of 58 target mRNAs predicted by four miRNAs. This table is modified from Li et al. (2009).

miR-	miR-	miR-	miR-	Gene Symbol	T3MF	T3DF	T3DF/ T3MF	Description	UniGene
			367	SMAD6	1.5	6.23	4.1	SMAD family member 6	Hs.153863
			367	CPEB4	0.56	4.76	8.6	cytoplasmic polyadenylation element binding protein 4	Hs.127126
			367	WWP2	1.08	4.28	4	WW domain containing E3 ubiquitin protein ligase 2	Hs.408458
			367	CACNA1C	0.97	4.2	4.3	subunit	Hs.118262
			367	TEF	0.73	4.16	5.7	thyrotrophic embryonic factor	Hs.181159
			367	FMR1	0.98	4.05	4.1	fragile X mental retardation 1	Hs.103183
			367	DNAJB12	0.87	4.02	4.6	DnaJ (Hsp40) homolog, subfamily B, member 12	Hs.696014
			367	ZNF287	1.05	3.81	3.6	zinc finger protein 287	Hs.99724
			367	HAS3	1.15	3.79	3.3	hyaluronan synthase 3 isoform b	Hs.592069
			367	ATXN3	1.12	3.48	3.1	ataxin 3	Hs.532632

Table 2. continued.

of overall mean in T3DF cells (Table 2). Therefore, these 58 abundantly differentially expressed mRNAs are very likely to be the targets of four abundantly expressed miRNAs miR-302d, miR-372, miR-200c and/or miR-367 in hES cells (Li et al. 2009). Recently, 253 target mRNAs of mouse miR-290 cluster (corresponding to human miR-302 cluster) were found by the transcriptome analysis of ES cells lacking Dicer of miRNA processing enzyme (Sinkkonen et al., 2008). A comparison of these mouse 253 targets with the 58 targets of human miR-302d, miR-367, miR-372 and/or miR-200c revealed four common targets TRPS1, MBNL1, MBNL2 and NF1B.

The three genes TRPS1, KLF13 and MBNL2 were found to express extremely abundantly (more than 20-fold overall mean) in T3DF cells in which miR-302d, miR-372 and miR-200c had almost no expression (Tables 1 and 2), whereas these three genes TRPS1, KLF13 and MBNL2 exhibited relatively low mRNA expression in T3MF cells in which miR-302d, miR-372 and miR-200c were expressed abundantly (more than 20-fold U6 snRNA). Therefore, the TRPS1, KLF13 and MBNL2 genes are very likely to be the common targets of three abundantly expressed miRNAs miR-302d, miR-372 and miR-200c in hES cells. To demonstrate directly whether TRPS1, KLF13 and MBNL2 were indeed the targets of miR-302d and miR-372, the luciferase reporter vectors harboring the amplified 3'UTR fragments of target genes were cotransfected with and without miRNA precursor of either miR-302d or miR-372 in HEK293T cells (Li et al., 2009). The potential binding structures of the 3'UTR of TRPS1, KLF13 and MBNL2 genes by miRNAs miR-302d and miR-372 were predicted using PicTar program. The base-pairings between miRNAs and their targets, as well as the constructions of luciferase reporter vectors, were shown in Fig. 1A and B. The miR-302d specifically suppressed the luciferase activity to 57, 62 and 80% of reporter vector harboring TRPS1, KLF13 and MBNL2, respectively. The miR-372 inhibited the luciferase activity to 86, 37 and 47% for TRPS1, KLF13 and MBNL2, respectively (Fig. 1C). These results implied that miR-302d and miR-372 were able to regulate negatively the same targets due to their same seed sequence. Furthermore, in order to confirm the predicted miRNA binding sites, the 4-bp mutations of miR-302d and miR-372 binding sites on KLF13 and MBNL2 genes were generated by site-directed mutagenesis. The results of these reporter assays showed that miR-302d and miR-372 were indeed able to suppress specifically the luciferase activities of the reporter vectors harboring the predicted binding sites of wild-type 3'UTR of KLF13 and MBNL2, but not mutant-types (Fig. 1D).

The GeneOntology of the 58 abundantly differentially expressed targets by hES cell-specific miRNAs miR-302d, miR-372, miR-200c and/or miR-367 was analyzed (Li et al., 2009). 48

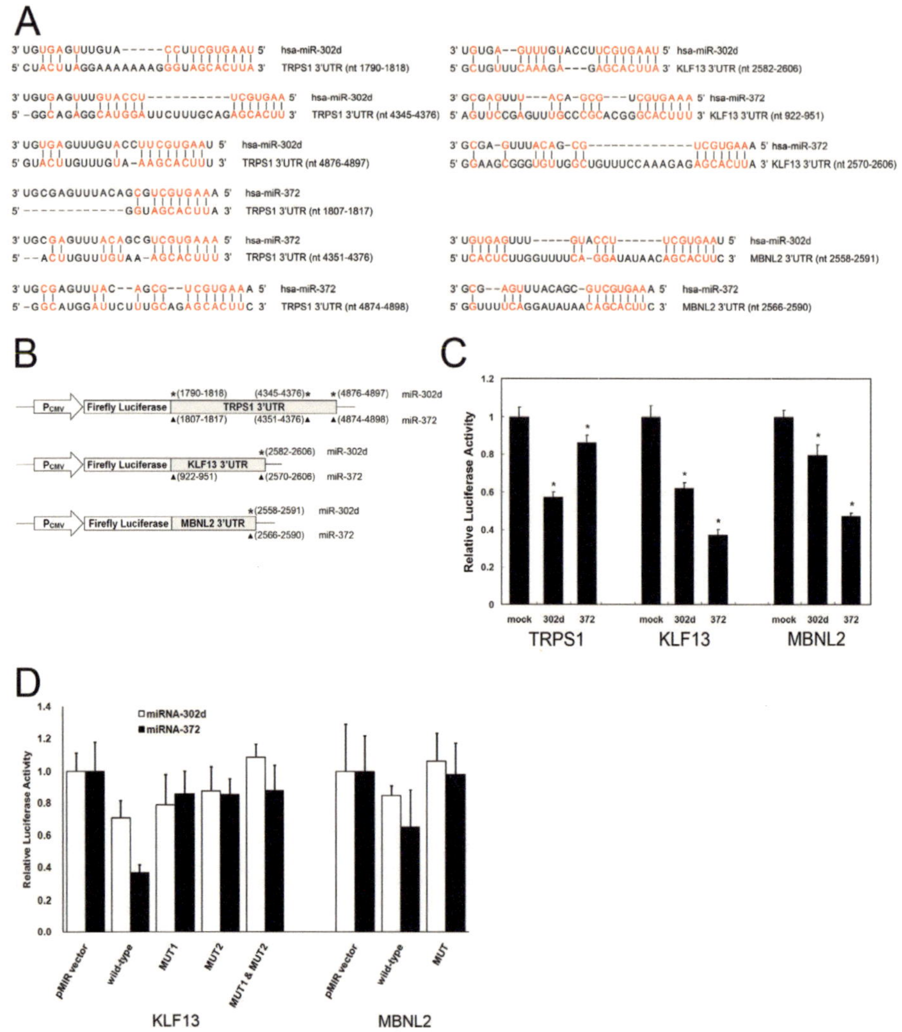

Fig. 1. TRPS1, KLF13 and MBNL2 targets of miR-302d and miR-372.
(A) Predicted binding sites of miR-302d and miR-372 within the 3'UTRs of TRPS1, KLF13 and MBNL2. (B) The construction of luciferase reporter vectors. (C) The effects of miR-302d and miR-372 on the luciferase activity of TRPS1, KLF13 and MBNL2 reporter vectors. (D) The effects of miR-302d and miR-372 on the luciferase activity of reporter vectors containing wild-type and mutant 3'UTRs of KLF13 and MBNL2 .
This figure is adapted from Li et al. (2009).

(p-value of 6.09E-10) of these 58 targets are localized in nucleus, and 39 (p-value of 4.38E-10) of them are involved in regulation of gene expression processes. The molecular functions of the 58 targets of four miRNAs expressed abundantly in hES cells were analyzed using GeneSpring GeneOntology Browser and MetaCore Analytical Suite. The top four molecular functions of eight categories obtained by GeneSpring were found to be the same to the top four molecular functions of eight categories obtained by MetaCore (Table 3). All of the top four molecular functions are involved in gene transcription. Among the other different categories between GeneSpring and MetaCore, zinc ion binding and sequence- specific DNA binding are also involved in gene transcription. In other words, the molecular functions of approximately half of these 58 targets are involved in gene transcription in hES cells.

	GeneSpring		MetaCore	
Molecular functions	**Genes**	**p-Value**	**Genes**	**p-Value**
Transcription factor activity	18	1.03E-08	28	8.87E-16
Transcription regulator activity	20	2.81E-07	30	1.40E-13
Nucleic acid binding	29	2.78E-06	38	5.03E-10
DNA binding	22	1.80E-05	31	2.01E-09
Zinc ion binding	19	0.000595		
Calmodulin binding	4	0.0039		
Ciliary neurotrophic factor receptor binding	1	0.00495		
Calmodulin inhibitor activity	1	0.00495		
Protein N-terminus binding			6	2.65E-07
Estrogen receptor activity			3	2.67E-07
Sequence-specific DNA binding			12	1.91E-06
Nitric-oxide synthase regulator activity			3	5.53E-06

Table 3. Molecular functions of 58 abundantly differentially expressed targets.
This table is adapted from Li et al. (2009).

3. Target identification of miRNAs up- and down-regulated by activin A

The hES-T3 cells (passage 36) were cultured on the inactivated MEF feeder (designated as T3MF) in hES medium (containing 4 ng/ml bFGF) and feeder-free Matrigel in MEF-conditioned medium with additional 4 ng/ml bFGF (designated as T3CM) for 14 and 12 more passages, respectively. The T3MF and T3CM cells were stained positively for OCT4 and NANOG, indicating that both T3MF and T3CM cell populations contained very high proportions of undifferentiated hES cells. When hES-T3 cells were grown on feeder-free Matrigel in hES medium (without bFGF) supplemented with 5, 10 and 25 ng/ml activin A, many cells around the edges, as well as the center, of colonies differentiated into fibroblast-like cells with much less staining of OCT4 and NANOG, and these cells were able to proliferate only two more passages, indicating that bFGF is indispensable for maintenance of self-renewal of hES cells. However, the hES cells cultured on feeder-free Matrigel in hES medium containing both 4 ng/ml bFGF and 5 ng/ml activin A (designated as T3BA) were able to proliferate for seven more passages, and most of these T3BA cells were stained positively for both OCT4 and NANOG (Tsai et al., 2010).

The genome-wide mRNA expression of T3MF, T3CM and T3BA cells was determined using Affymetrix human genome U133 plus 2.0 GeneChip (Tsai et al., 2010). The original data have been deposited to NCBI database, and the GEO series number is GSE16910. The average values of expressed mRNAs from T3MF, T3CM and T3BA cells were compared by scatter plots (Fig. 2A, B). The Pearson correlation coefficient of R^2 = 0.9934 between T3MF and T3CM cells indicates their very similar expression profiles of mRNAs (Fig. 2A), and only 49 and 17 genes were found to be abundantly (more than 3-folds of overall mean)

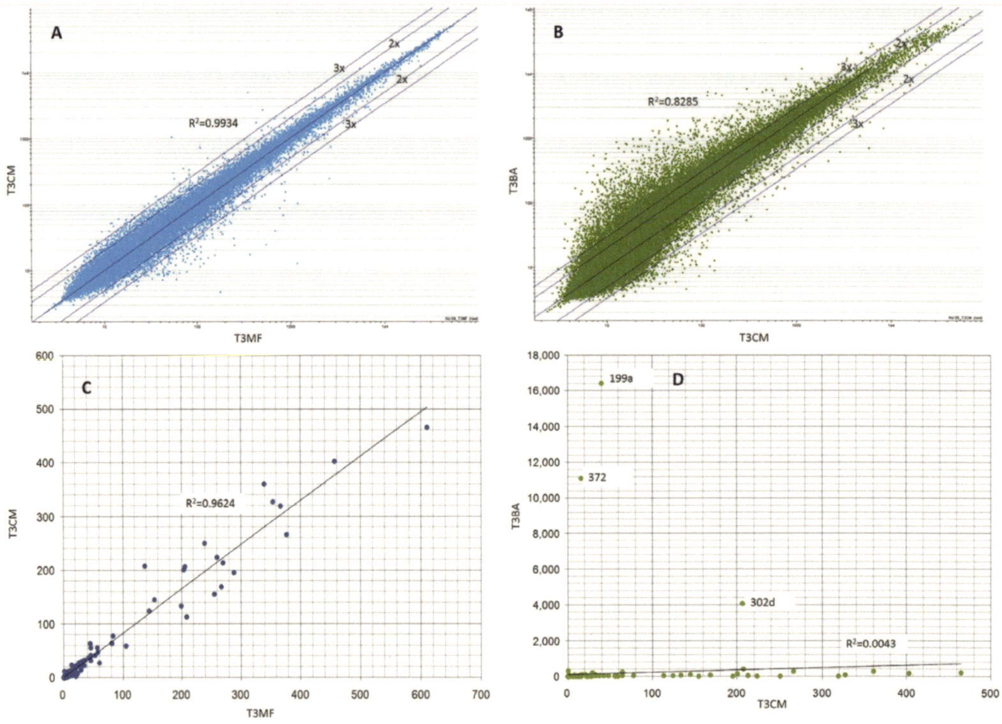

Fig. 2. Scatter plot and correlation analyses of mRNAs and miRNAs among T3MF, T3CM and T3BA cells.
(A) mRNAs of T3MF and T3CM; (B) mRNAs of T3BA and T3CM;
(C) miRNAs of T3MF and T3CM; (D) miRNAs of T3BA and T3CM.
This figure is adapted from Tsai et al. (2010).

differentially (more than 3-folds of changes) expressed in T3MF and T3CM cells, respectively. These results indicated that the unlimited self-renewal and pluripotency of hES-T3 cells can be maintained by continuous culture on either inactivated MEF feeder in hES medium (containing 4 ng/ml bFGF) or feeder-free Matrigel in MEF-conditioned medium (containing additional 4 ng/ml bFGF). The correlation value of $R^2 = 0.8285$ between T3BA and T3CM cells suggests less similarity between their mRNA profiles (Fig. 2B), and 589 and 58 genes were abundantly differentially expressed in T3BA and T3CM cells, respectively.

The expression profiles of 250 human miRNAs in T3MF, T3CM and T3BA cells were quantitated using TaqMan microRNA Assays as described previously (Chen et al., 2005; Liang et al., 2007; Tsai et al., 2010), and the expression level of each miRNA was indicated as folds over U6 snRNA. The Pearson correlation coefficient of $R^2 = 0.9624$ between T3MF and T3CM cells indicates their very similar miRNA expression profiles (Fig. 2C), while no correlation ($R^2 = 0.0043$) was found between T3BA and T3CM cells (Fig. 2D). When the three highly expressed miRNAs miR-199a, miR-372 and miR-302d were excluded from analysis, low correlation ($R^2 = 0.3541$) was still observed for the remaining 247 miRNAs of T3BA and T3CM cells. These results indicate that four hES cell-specific miRNAs miR-372, miR-302d, miR-367 and miR-200c, as well as three other miRNAs miR-199a, miR-217 and miR-19a, were over-expressed in T3BA cells, whereas five miRNAs miR-19b, miR-221, miR-222, let-7b and let-7c were under-expressed in T3BA cells compared with T3CM cells (Table 4).

miRNAs	T3MF	T3CM	T3BA	T3BA/ T3CM	T3CM/ T3BA	Specificity
A group (up-regulated)						
hsa-miR-372	27.97	15.65	11079.38	707.9	0	hES
hsa-miR-302d	205.43	206.53	4071.21	19.7	0.1	hES
hsa-miR-367	136.63	207.63	397.69	1.9	0.5	hES
hsa-miR-200c	27.76	28.94	201.26	7	0.1	hES
hsa-miR-199a	44.84	39.81	16380.51	411.5	0	
hsa-miR-217	0.01	0.01	296.13	29613	0	
hsa-miR-19a	43.74	63.94	233.47	3.7	0.3	
B group (down-regulated)						
hsa-miR-19b	366.63	319.07	2.56	0	124.6	
hsa-miR-221	254.57	154.9	2.93	0	52.9	
hsa-miR-222	287.47	195.53	6.73	0	29.1	
hsa-let-7b	259.6	223.2	12.12	0.1	18.4	
hsa-let-7c	237.9	250.27	15.17	0.1	16.5	

Table 4. Levels of miRNAs expressed highly in T3MF, T3CM and T3BA cells
This table is modified from Tsai et al. (2010).

Gene Symbol	miR- 372	302d	367	200c	19a	199a	217	Gene Description
NR4A2	372	302d			19a		217	nuclear receptor subfamily 4, group A, member 2
ERBB4	372	302d			19a			v-erb-a erythroblastic leukemia viral oncogene homolog 4
CXCR4	372	302d						chemokine (C-X-C motif) receptor 4
PCDH9			367					protocadherin 9
TMEFF2				200c				transmembrane protein with EGF- & 2 follistatin-like dom 2
CD24					19a			CD24 molecule
COX6A1						199a		cytochrome c oxidase subunit VIa polypeptide 1
TAL1	372	302d						T-cell acute lymphocytic leukemia 1
GRIA3			367					glutamate receptor, ionotrophic, AMPA 3
PRDM1				200c	19a			PR domain containing 1, with ZNF domain
MYT1				200c	19a			myelin transcription factor 1
EIF4G2					19a			eukaryotic translation initiation factor 4 gamma, 2
CHMP4B					19a			chromatin modifying protein 4B

Table 5. The 13 genes targeted by the activin A up-regulated miRNAs in T3BA cells.
This table is modified from Tsai et al. (2010).

Gene Symbol	miR- 19b	221	222	let- 7b	7c	Gene Description
ACTN1	19b					actinin, alpha 1
ADRB2				7b	7c	adrenergic, beta-2-, receptor, surface
AMMECR1					7c	Alport syndrome, mental retardation, gene 1
AMT				7b		aminomethyltransferase
ARFGEF1	19b					ADP-ribosylation factor guanine nucleotide-exchange factor 1
ARIH2	19b					ariadne homolog 2 (Drosophila)
ATXN7L1	19b					ataxin 7-like 1
CALB1				7b	7c	calbindin 1, 28kDa
CCND2	19b			7b	7c	cyclin D2
CDC25A				7b	7c	cell division cycle 25 homolog A (S. pombe)
CHD7		221	222	7b	7c	chromodomain helicase DNA binding protein 7
CLASP2	19b			7b	7c	cytoplasmic linker associated protein 2
CLASP2	19b			7b	7c	cytoplasmic linker associated protein 2
COIL				7b	7c	coilin
CPNE8		221	222			copine VIII
DACH1		221	222			dachshund homolog 1 (Drosophila)
DTNA	19b					dystrobrevin, alpha
EFNB2	19b					ephrin-B2
EIF2C1	19b			7b	7c	eukaryotic translation initiation factor 2C, 1
FAM46B	19b					family with sequence similarity 46, member B
FIGN	19b				7c	fidgetin
FRS2	19b	221				fibroblast growth factor receptor substrate 2
GRSF1	19b					G-rich RNA sequence binding factor 1
GULP1	19b					GULP, engulfment adaptor PTB domain containing 1
HIC2				7b	7c	hypermethylated in cancer 2
ILF3	19b					interleukin enhancer binding factor 3, 90kDa
KCMF1	19b	221	222			potassium channel modulatory factor 1
KLF13	19b					Kruppel-like factor 13
KRAS	19b					v-Ki-ras2 Kirsten rat sarcoma viral oncogene homolog
MPPED2	19b					metallophosphoesterase domain containing 2
MYCL1	19b					v-myc myelocytomatosis viral oncogene homolog 1,
NAV3	19b					neuron navigator 3
NRG1	19b		222			neuregulin 1
NRK	19b	221	222	7b	7c	Nik related kinase
PCDHA9	19b	221	222			protocadherin alpha 9
PGM2L1				7b	7c	phosphoglucomutase 2-like 1
POGZ		221	222	7b	7c	pogo transposable element with ZNF domain
PRPF38B				7b	7c	PRP38 pre-mRNA processing factor 38 domain containing B
PTBP2		221				polypyrimidine tract binding protein 2
PTEN		221				phosphatase and tensin homolog
RGS16				7b	7c	regulator of G-protein signalling 16
SLC1A2	19b		222			solute carrier family 1 , member 2
SMARCD1				7b	7c	SWI/SNF related, matrix associated, member 1
SOCS1	19b					suppressor of cytokine signaling 1
SOX4	19b		222			SRY (sex determining region Y)-box 4
STRBP				7b	7c	spermatid perinuclear RNA binding protein
VANGL2			222	7b		vang-like 2 (van gogh, Drosophila)
VAV3	19b			7b	7c	vav 3 oncogene
ZBTB10	19b			7b	7c	zinc finger and BTB domain containing 10
ZNF518						zinc finger protein 518

Table 6. The 50 genes targeted by the activin A down-regulated miRNAs
This table is modified from Tsai et al. (2010).

The potential targets of four abundantly expressed hES cell-specific miRNAs miR-372, miR-302d, miR-367 and miR-200c, as well as three other over-expressed miRNAs miR-19a, miR-199a and miR-217, were predicted by the PicTar (4-way) and TargetScanS with a cutoff p value less than 0.05. The expression levels of 13 target mRNAs were found to be inversed to their miRNAs (Tsai et al., 2010). Seven abundantly (more than 3-folds of overall mean) differentially (more than 3-folds of changes) expressed genes NR4A2, ERBB4, CXCR4, PCDH9, TMEFF2, CD24 and COX6A1, as well as six other genes TAL1, GRIA3, PRDM1, MYT1, EIF4G2 and CHMP4B, were found to be targets of miR-372, miR-302d, miR-367, miR-200c, miR-19a, miR-199a and/or miR-217 (Table 5). It may be noted that the NR4A2, TMEFF2 and TAL1 were also included in the 58 target genes of hES cell-specific miR-372, miR-302d, miR-367 and/or miR-200c (Li et al., 2009). The five miRNAs miR-19b, miR-221, miR-222, let-7b and let-7c were under-expressed in T3BA cells compared with T3CM cells, and their 50 target mRNAs (Table 6) were also found to exhibit inverse expression levels (Tsai et al. 2010).

The two target genes NR4A2 and ERBB4 were chosen to be validated by luciferase assay, since they expressed abundantly (more than four folds of overall mean in T3CM cells) and highly differentially (more than 7-fold changes of T3CM/T3BA) (Table 5). The potential miRNA binding sites of the 3′ UTRs of NR4A2 and ERBB4 genes were predicted using PicTar and TargetScanS programs. The 3′ UTR of NR4A2 was found by both methods to contain two potential miR-372 binding sites, four miR-302d binding sites and one miR-19a binding site, but miR-217 binding site was predicted by TargetScanS only. The 3′ UTR of ERBB4 was found by both methods to have one site for miR-302d and two sites for miR19a, but miR-372 binding site was predicted by TargetScanS only. The base-pairing between miRNAs and their target mRNAs, as well as the construction of luciferase reporter vectors, are shown in Fig. 3A, B. To demonstrate directly whether NR4A2 and ERBB4 genes were indeed the targets of miR-372, miR-302d, miR-19a and/or miR-217, the luciferase reporter vectors harboring the 3′UTRs of target genes were cotransfected with pSilencer vector containing miRNAs precursor or not in HEK293T cells (Tsai et al., 2010). The luciferase activity of reporter vector harboring NR4A2 3′UTR was inhibited to 37%, 38%, 53% or 33% by miR-372, miR-302d, miR-19a or miR-217, respectively. In the case of ERBB4, the luciferase activity was suppressed to 58%, 74% or 81% by miR-372, miR-302d or miR-19a, respectively (Fig. 3C). These results implied that NR4A2 gene was inhibited much stronger than ERBB4 gene by miR-372, miR-302d and miR-19a.

The mRNAs expressed more than three-folds of overall mean from T3BA and T3CM cells were also analyzed for the network and signaling pathways by using MetaCore Analytical Suite (Tsai et al., 2010). Besides 969 common genes, 1,396 and 153 genes were found to be unique for T3BA and T3CM cells, respectively (Fig. 4A), and the top 3 scored pathways (with lowest p-value) involved in regulation of cell cycle were highly regulated by activin A (Fig. 4B). Activin A has been reported to inhibit cell proliferation and activate cell differentiation, although it was also shown to participate in maintenance of pluripotency (Beattie et al., 2005; James et al., 2005). Activin A through its receptor ActR2B increased the expression of OCT4 and NANOG, as well as LEFTY-1 and LEFTY-2, to regulate stem cell maintenance. The abundantly differentially expressed target genes NR4A2, ERBB4 and CXCR4 of miRNAs miR-372, miR-302d, miR-19a and/or miR-217 highly induced by activin A in T3BA cells were further analyzed using IPA for their involvements in network and signaling pathways. The NR4A2 in nucleus and CXCR4 associated with plasma membrane

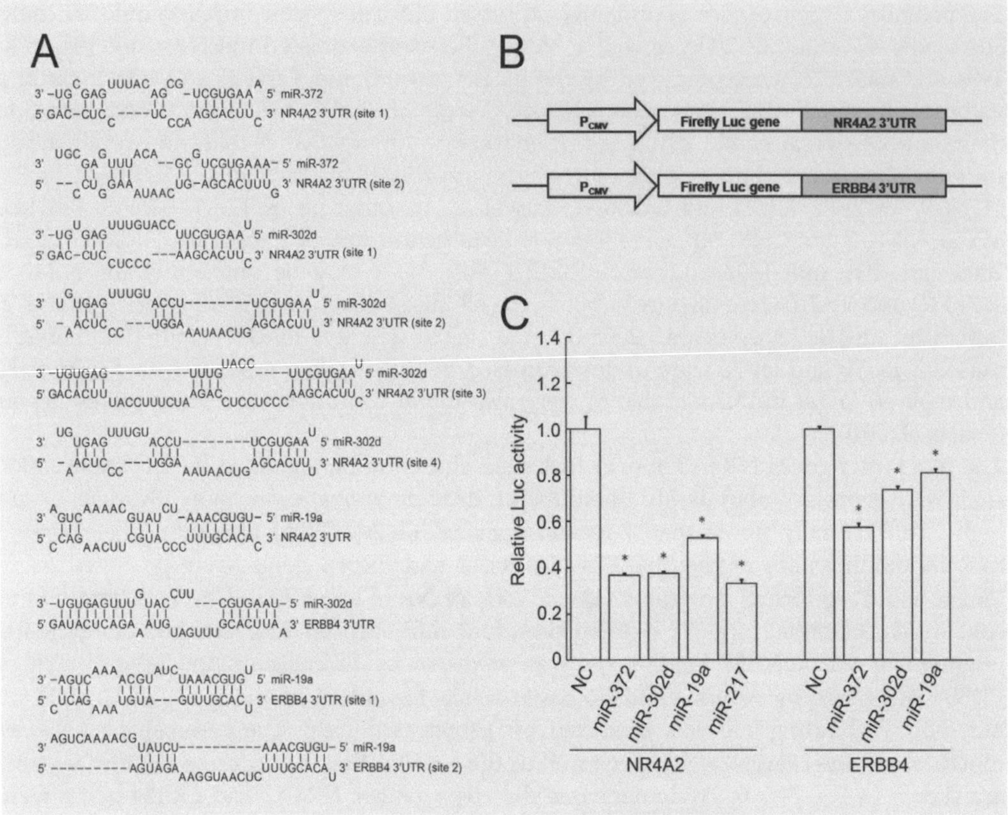

Fig. 3. Leuciferase experiments to validate NR4A2 and ERBB4 targets
(A) Predicted binding sites of miR-372, miR-302d and miR-19a within the 3'UTRs of NR4A2 and ERBB4. (B) The construction of luciferase reporter vectors. (C) The effects of miR-372, miR-302d, miR-19a and/or miR-217 on the luciferase activity of NR4A2 and ERBB4 reporter vectors.
This figure is adapted from Tsai et al. (2010).

were found to be commonly regulated by both EGF and TNF, while the membrane associated ERBB4 is regulated by EGF only. The NR4A2, also known as NURR1, is essential for the differentiation of the midbrain dopaminergic neurons, and it was reported to cooperate with PITX3 in promoting the terminal maturation of human and murine embryonic stem cell cultures to a dopamine neuron phenotype, and neither factor alone induced differentiation (Martinat et al., 2006). The ERBB4 is a transmembrane receptor tyrosine kinase that regulates cell proliferation and differentiation. The ERBB4 and its ligand heregulin are essential for neuronal development. The ERBB4 was reported to express at high levels in rat subventricular zone and rostral migatory system and to play a role in neuroblast tangential migration and olfactory interneuronal placement (Anton et al., 2004). The CXCR4, a chemokine receptor, is a crucial effector of the transcriptional pathway specifying mouse ventral motor neurons, and it controls the precision of initial motor axon trajectories (Lieberam et al., 2005). The CXCR4 has also been used as a biomarker of definite endoderm which is induced by activin A.

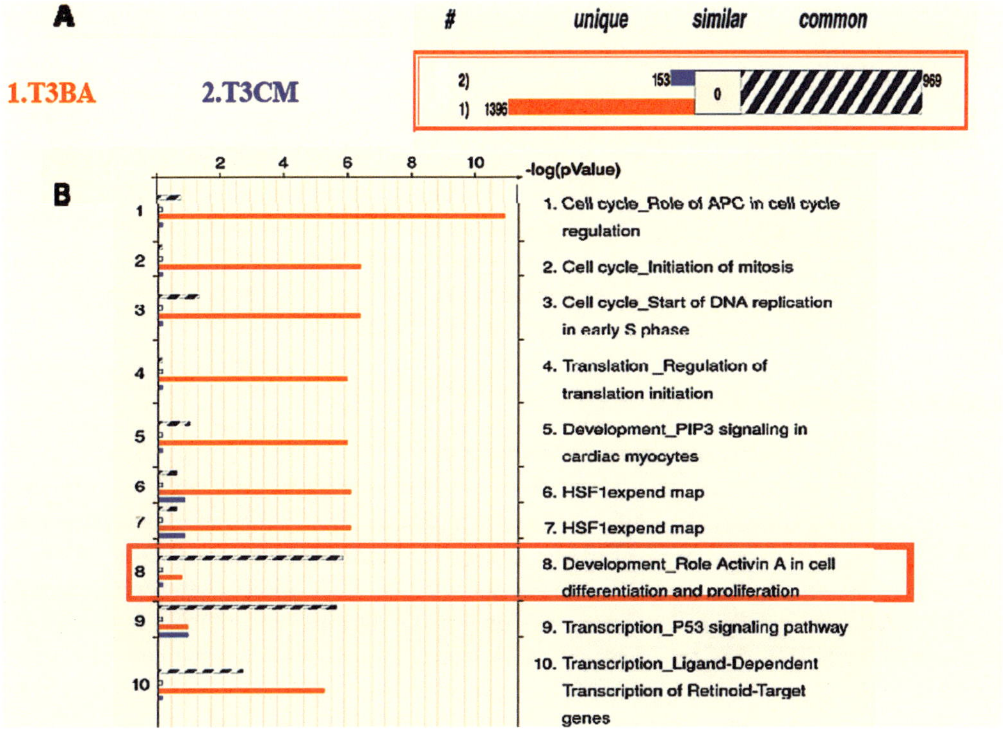

Fig. 4. Comparison of gene expression and GeneGo canonical pathway maps among T3CM and T3BA cells.
(A) The parameters for comparison are set at threshold of 3 with p-value of 0.05. The common genes are indicated by blue/white strips. The unique genes are marked as color band: (1) T3BA, orange; (2) T3CM, blue. (B) The top 10 GeneGo canonical pathway maps among T3BA and T3CM cells. The degree of "relevance" to different GeneGo ontology categories is defined by p-value, so that the lower random p-value gets higher priority. This figure is adapted from Tsai et al. (2010).

The seven miRNAs miR-372, miR-302d, miR-367, miR-200c, miR-19a, miR-199a and miR-217 were found to be highly up-regulated by activin A in T3BA cells (Table 4 and Fig. 2D). The expression of hES cell-specific miR-302 cluster was previously shown to be regulated by OCT4 and SOX2 (Card et al., 2008). Thus, activin A through its receptor ActR2B increased indirectly the expression of OCT4 to induce the expression of hES cell-specific miR-372, miR-302d, miR-367 and miR-200c. The miR-199a was found to be most abundant in T3BA cells, and its target COX6A1 expressed extremely abundantly in T3MF and T3CM cells was down-regulated inT3BA cells. The miR-199a is encoded by duplicated genes located within the intron of dynamin genes on chromosomes 1 and 19. Since the miR-199a genes are positioned in opposite direction to the dynamin genes, they may be transcribed from their own promoters. The expression of miR-199a was also reported to be controlled by transcription factor TWIST-1 via an E-box promoter element (Lee et al., 2009), and how activin A indirectly regulate TWIST-1 remains to be elucidated. It may be noted that miR-199a and miR-199a* (processed from the same miRNA precursor) were recently reported to down-regulate the MET proto-oncogene and its downstream effector extracellular signal-regulated kinase 2 (ERK2) gene resulting in inhibiting cell proliferation of tumor cells (Kim et al., 2008). The miR-199a and miR-199a* were also shown to inhibit the mRNA translation of IκB kinase β required for NF-κB activation in ovarian cancer cells (Chen et al., 2008). The miR-217 was also found to be highly expressed in T3BA, but not at all in T3MF and T3CM cells (Table 4). It is of interest that the miR-217 was reported to be linked to tumorigenesis in pancreatic ductal adenocarcinoma (Szafranska et al., 2007) and used as one of biomarkers to discriminate benign and malignant pancreatic tissues (Szafranska et al., 2008).

4. Conclusions

The hES-T3 cells with normal female karyotype, one of five hES cell lines derived in my laboratory in Taiwan, were maintained their undifferentiated growth on the inactivated MEF feeder (T3MF), in MEF-conditioned medium (T3CM) and hES medium supplemented with both bFGF and activin A (T3BA). Autogeneic feeder fibroblast (T3DF) cells with capacity to support the undifferentiated growth of hES cells were spontaneously differentiated from hES-T3 cells. The expression profiles of miRNAs and mRNAs from these four cell types (T3MF, T3CM, T3BA & T3DF) were quantitatively determined, and many target mRNAs of miRNAs expressed highly and regulated by activin A in hES cells were identified by the inverse expression levels of miRNAs to their target mRNAs. Approximately half of the 58 targets by hES cell-specific miRNAs miR-372, miR-302d, miR-367 and/or miR-200c were involved in gene transcription. Four hES cell-specific microRNAs miR-372, miR-302d, miR-367 and miR-200c, as well as three other microRNAs miR-199a, miR-19a and miR-217, were found to be up-regulated, whereas five miRNAs miR-19b, miR-221, miR-222, let-7b and let-7c were down-regulated by activin A. Thirteen abundantly differentially expressed mRNAs, including NR4A2, ERBB4, CXCR4, PCDH9, TMEFF2, CD24 and COX6A1 genes, targeted by seven over-expressed miRNAs were identified. The NR4A2, ERBB4 and CXCR4 target genes were further found to be regulated by EGF and/or TNF. The 50 abundantly differentially expressed genes targeted by five under-expressed miRNAs were also identified. The abundantly expressed mRNAs were also analyzed for the network and signaling pathways, and roles of activin A in cell proliferation and differentiation were found. These findings will help elucidate the complex signaling networks which maintain the unlimited self-renewal and pluripotency of hES cells. The autogeneic fibroblasts derived from hES cells will be useful to reduce the risk of animal pathogens in the future medical applications of hES cells.

5. Acknowledgements

I thank Drs. P.-C. Yang, S.-L. Yu, S. Singh and Y.-H. Liu for their productive collaborations; Z.-Y. Tsai, L.-P. Kao , B.-Z. Chen and B.-C. Ho for their hard work; and M.P. Li for copy editing. The tables and figures are modified/adapted with permission from Li et al. (2009) and Tsai et al. (2010). My research has been supported by grants from National Science Council of Taiwan.

6. References

Alvarez-Garcia, I. & Miska, E.A. (2005). MicroRNA functions in animal development and human disease. Development 132:4653-62.

Anton, E.S.; Ghashghaei, H.T.; Weber, J.L,; McCann, C.; Fischer, T.M.; Cheung, I.D.; Gassmann, M.; Messing, A.; Klein, R.; Schwab, M.H.; Lloyd, K.C. & Lai, C. (2004) Receptor tyrosine kinase ErbB4 modulates neuroblast migration and placement in the adult forebrain. Nat Neurosci 7:1319-28.

Assou, S.; Le Carrour, T.; Tondeur, S.; Strom, S.; Gabelle, A.; Marty, S.; Nadal, L.; Pantesco, V.; Reme, T.; Hugnot, J.P.; Gasca, S.; Hovatta, O.; Hamamah, S.; Klein, B. & De Vos, J. (2007) A meta-analysis of human embryonic stem cells transcriptome integrated into a web-based expression atlas. Stem Cells 25:961-73.

Baek, D.; Villen, J.; Shin, C.; Camargo, F.D.; Gygi, S.P & Bartel, D.P. (2008) The impact of microRNAs on protein output. Nature 455:64-71.

Bartel, D.P. (2004) MicroRNAs: genomics, biogenesis, mechanism, and function. Cell 116:281-97.

Beattie, G.M.; Lopez, A.D.; Bucay, N.; Hinton, A.; Firpo, M.T.; King, C.C. & Hayek, A. (2005) Activin A maintains pluripotency of human embryonic stem cells in the absence of feeder layers. Stem Cells 23:489-95.

Card, D.A.; Hebbar, P.B.; Li, L.; Trotter, K.W.; Komatsu, Y.; Mishina, Y. & Archer. T.K. (2008) Oct4/Sox2-regulated miR-302 targets cyclin D1 in human embryonic stem cells. Mol Cell Biol 28:6426-38.

Chen, C.; Ridzon, D.A.; Broomer, A.J.; Zhou, Z.; Lee, D.H.; Nguyen, J.T.; Barbisin, M.; Xu, N.L.; Mahuvakar, V.R.; Andersen, M.R.; Lao, K.Q.; Livak, K.J. & Guegler, K.J. (2005) Real-time quantification of microRNAs by stem-loop RT-PCR. Nucleic Acids Res 33:e179.

Chen, C.; Ridzon, D.; Lee, C.T.; Blake, J.; Sun, Y. & Strauss, W.M. (2007) Defining embryonic stem cell identity using differentiation-related microRNAs and their potential targets. Mamm Genome 18:316-27.

Chen, R.; Alvero, A.B.; Silasi, D.A.; Kelly, M.G.; Fest, S.; Visintin, I.; Leiser, A.; Schwartz, P.E.; Rutherford, T. & Mor, G. (2008) Regulation of IKKbeta by miR-199a affects NF-kappaB activity in ovarian cancer cells. Oncogene 27:4712-23.

Griffiths-Jones, S.; Grocock, R.J.; van Dongen, S.; Bateman, A. & Enright, A.J. (2006) miRBase: microRNA sequences, targets and gene nomenclature. Nucleic Acids Res 34:D140-4.

Guhr, A.; Kurtz, A.; Friedgen, K. & Loser, P. (2006) Current state of human embryonic stem cell research: an overview of cell lines and their use in experimental work. Stem Cells 24:2187-91.

Houbaviy, H.B.; Murray, M.F. & Sharp, P.A. (2003) Embryonic stem cell-specific MicroRNAs. Dev Cell 5:351-8.

Ivanova, N.B.; Dimos, J.T.; Schaniel, C.; Hackney, J.A.; Moore, K.A. & Lemischka, I.R. (2002) A stem cell molecular signature. Science 298:601-4.

James, D.; Levine, A.J.; Besser, D. & Hemmati-Brivanlou, A. (2005) TGFbeta/activin/nodal signaling is necessary for the maintenance of pluripotency in human embryonic stem cells. Development 132:1273-82.

John, B.; Enright, A.J.; Aravin, A.; Tuschl, T.; Sander, C. & Marks, D.S. (2004) Human microRNA targets. PLoS Biol 2:e363.

Kim, S.; Lee, U.J.; Kim, M.N.; Lee, E.J.; Kim, J.Y.; Lee, M.Y.; Choung, S.; Kim, Y.J. & Choi, Y.C. (2008) MicroRNA miR-199a* regulates the MET proto-oncogene and the downstream extracellular signal-regulated kinase 2 (ERK2). J Biol Chem 283:18158-66.

Kloosterman, W.P. & Plasterk, R.H. (2006) The diverse functions of microRNAs in animal development and disease. Dev Cell 11:441-50.

Landgraf, P.; Rusu, M.; Sheridan, R.; Sewer, A.; Iovino, N.; Aravin, A.; Pfeffer, S.; Rice, A.; Kamphorst, A.O.; Landthaler, M.; Lin, C.; Socci, N.D.; Hermida, L.; Fulci, V.; Chiaretti, S.; Foa, R.; Schliwka, J.; Fuchs, U.; Novosel, A.; Muller, R.U.; Schermer, B.; Bissels, U.; Inman, J.; Phan, Q.; Chien, M.; Weir, D.B.; Choksi, R.; De Vita, G.; Frezzetti, D.; Trompeter, H.I.; Hornung, V.; Teng, G.; Hartmann, G.; Palkovits, M.; Di Lauro, R.; Wernet, P.; Macino, G.; Rogler, C.E.; Nagle, J.W.; Ju, J.; Papavasiliou, F.N.; Benzing, T.; Lichter, P.; Tam, W.; Brownstein, M.J.; Bosio, A.; Borkhardt, A.; Russo, J.J.; Sander, C.; Zavolan, M. & Tuschl, T. (2007) A mammalian microRNA expression atlas based on small RNA library sequencing. Cell 129:1401-14.

Laurent, L.C.; Chen, J.; Ulitsky, I.; Mueller, F.J.; Lu, C.; Shamir, R.; Fan, J.B. & Loring, J.F. (2008) Comprehensive microRNA profiling reveals a unique human embryonic stem cell signature dominated by a single seed sequence. Stem Cells.

Lee, Y.B.; Bantounas, I.; Lee, D.Y.; Phylactou, L.; Caldwell, M.A. & Uney, J.B. (2009) Twist-1 regulates the miR-199a/214 cluster during development. Nucleic Acids Res 37:123-8.

Li, S.S.; Liu, Y.-H.; Tseng, C.-N.; Chung, T.-L.; Lee, T.-Y. & Singh, S. (2006) Characterization and gene expression profiling of five new human embryonic stem cell lines derived in Taiwan. Stem Cells Dev. 15:532-55.

Li, S.S.; Yu, S.L.; Kao, L.P.; Tsai, Z.Y.; Singh, S.; Chen, B.Z.; Ho, B.C.; Liu, Y.H. & Yang, P.C. (2009.) Target identification of microRNAs expressed highly in human embryonic stem cells. J Cell Biochem 106:1020-30.

Liang, Y.; Ridzon, D.; Wong, L. & Chen, C. (2007) Characterization of microRNA expression profiles in normal human tissues. BMC Genomics 8:166.

Lieberam, I.; Agalliu, D.; Nagasawa, T.; Ericson, J. & Jessell, T.M. (2005) A Cxcl12-CXCR4 chemokine signaling pathway defines the initial trajectory of mammalian motor axons. Neuron 47:667-79.

Martinat, C.; Bacci, J.J.; Leete, T.; Kim, J.; Vanti, W.B.; Newman, A.H.; Cha, J.H.; Gether, U.; Wang, H. & Abeliovich, A. (2006) Cooperative transcription activation by Nurr1 and Pitx3 induces embryonic stem cell maturation to the midbrain dopamine neuron phenotype. Proc Natl Acad Sci U S A 103:2874-9.

Ramalho-Santos, M.; Yoon, S.; Matsuzaki, Y.; Mulligan, R.C. & Melton, D.A. (2002) "Stemness": transcriptional profiling of embryonic and adult stem cells. Science 298:597-600.

Selbach, M.; Schwanhausser, B.; Thierfelder, N.; Fang, Z.; Khanin, R. & Rajewsky, N. (2008) Widespread changes in protein synthesis induced by microRNAs. Nature 455:58-63.

Sethupathy, P.; Megraw, M. & Hatzigeorgiou, A.G. (2006) A guide through present computational approaches for the identification of mammalian microRNA targets. Nat Methods 3:881-6.

Sinkkonen, L.; Hugenschmidt, T.; Berninger, P.; Gaidatzis, D.; Mohn, F.; Artus-Revel, C.G.; Zavolan, M.; Svoboda, P. & Filipowicz, W. (2008) MicroRNAs control de novo DNA methylation through regulation of transcriptional repressors in mouse embryonic stem cells. Nat Struct Mol Biol 15:259-67.

Sperger, J.M.; Chen, X.; Draper, J.S.; Antosiewicz, J.E.; Chon, C.H.; Jones, S.B.; Brooks, J.D.; Andrews, P.W.; Brown, P.O. & Thomson, J.A. (2003) Gene expression patterns in human embryonic stem cells and human pluripotent germ cell tumors. Proc Natl Acad Sci U S A 100:13350-5.

Stojkovic, P.; Lako, M.; Stewart, R.; Przyborski, S.; Armstrong, L.; Evans, J.; Murdoch, A.; Strachan, T. & Stojkovic, M. (2005) An autogeneic feeder cell system that efficiently supports growth of undifferentiated human embryonic stem cells. Stem Cells 23:306-14.

Suh, M.R.; Lee, Y.; Kim, J.Y.; Kim, S.K.; Moon, S.H.; Lee, J.Y.; Cha, K.Y.; Chung, H.M.; Yoon, H.S.; Moon, S.Y.; Kim, V.N. & Kim, K.S. (2004) Human embryonic stem cells express a unique set of microRNAs. Dev Biol 270:488-98.

Szafranska, A.E.; Davison, T.S.; John, J.; Cannon, T.; Sipos, B.; Maghnouj, A.; Labourier, E. & Hahn, S.A. (2007) MicroRNA expression alterations are linked to tumorigenesis and non-neoplastic processes in pancreatic ductal adenocarcinoma. Oncogene 26:4442-52.

Szafranska, A.E.; Doleshal, M.; Edmunds, H.S.; Gordon, S.; Luttges, J.; Munding, J.B.; Barth, R.J. Jr.; Gutmann, E.J.; Suriawinata, A.A.; Marc Pipas, J.; Tannapfel, A.; Korc, M.; Hahn, S.A.; Labourier, E. & Tsongalis, G.J. (2008) Analysis of microRNAs in pancreatic fine-needle aspirates can classify benign and malignant tissues. Clin Chem 54:1716-24.

Thomson, J.A.; Itskovitz-Eldor, J.; Shapiro, S.S.; Waknitz, M.A.; Swiergiel, J.J.; Marshall, V.S. & Jones, J.M. (1998) Embryonic stem cell lines derived from human blastocysts. Science 282:1145-7.

Tsai, Z.Y., Yu, S.-L.; Singh, S.; Kao, L.-P.; Chen, B.Z.; Yang, P.-C. & Li. S.S. (2010) Identification of microRNAs regulated by activin A in human embryonic stem cells. J. Cell. Biochem. 109:93-102.

Wobus, A.M. & Boheler, K.R. (2005) Embryonic stem cells: prospects for developmental biology and cell therapy. Physiol Rev 85:635-78.

Xiao, L.; Yuan, X. & Sharkis, S.J. (2006) Activin A maintains self-renewal and regulates fibroblast growth factor, Wnt, and bone morphogenic protein pathways in human embryonic stem cells. Stem Cells 24:1476-86.

Xu, C.; Inokuma, M.S.; Denham, J.; Golds, K.; Kundu, P.; Gold, J.D. & Carpenter, M.K. (2001) Feeder-free growth of undifferentiated human embryonic stem cells. Nat Biotechnol 19:971-4.

11

Smads – the Intracellular Hubs of Signalling in Regulation of Pluripotency and Differentiation of Stem Cells

Dmitry A. Ovchinnikov and Ernst J. Wolvetang
Australian Institute for Bioengineering and Nanotechnology, University of Queensland
Australia

1. Introduction

Signalling by the members of one of the largest groups of peptide signalling molecules, the transforming growth factor beta (TGFβ) superfamily, has been implicated in the regulation of aspects of essentially all events in the life and death of an animal cell. Smad proteins are versatile intracellular mediators of those signals, responsible for their direct transmission from the TGFβ superfamily receptor complexes on the membrane into the nucleus, resulting in specific changes in gene expression. Explicitly, two different Smad groups, transmitting BMP (bone morphogenetic protein)/GDF (growth and differentiation factor) and TGFβ/activin signals, respectively, have been shown to be involved in the maintenance of pluripotency in the mouse and human and all other characterised vertebrate embryonic stem cells.

Smad proteins are subject to extensive post-translational modifications, which are often a result of activation of other important cellular signalling pathways, rendering Smads important hubs of the major signalling pathways. Signalling by the members of the TGFβ superfamily starts with their binding to the complexes of type I and II (and in case of TGFβs type III) receptors, resulting in phosphorylation of the type I receptor by type II, in its turn leading to the phosphorylation of pathway-specific regulatory Smads (R-Smads). R-Smads then enter the nucleus in complexes with the co-Smad, Smad4, and activate or repress transcription of target genes, often after binding other transcription factors. Much of the intracellular regulation is achieved via regulation of the levels of Smad proteins available for signal transduction, mostly by competition for receptor binding, degradation or changes in phosphorylation status. The latter regulation is often achieved by phosphorylation by the kinases activated by other important signalling pathways, notably by cytokines FGF/EGF (fibroblast and epidermal growth factors, respectively) and Wnt proteins. It was recently discovered that some types of Smad linker phosphorylation accompany Smad activation, and act to ensure the transient nature of activated Smad action, thus maintaining constant sensitivity of the cell to changes in the levels of the TGFβ/BMP signal.

2. The Smad family

The name of the family is a combination of designations of the first identified members of this family of intracellular effectors of signalling, *sma* ("small", in nematode worm

Caenorhabditis elegans) and *mad* ("mothers against decapentaplegic") from the fruitfly *Drosophila melanogaster*. As apparent from the fruit fly gene's name, it was identified in the mutant screen for modifiers of the mutation in the *Drosophila* BMP2/4 homologue, *decapenthaplegic*, thus revealing its involvement in this signalling pathway (Derynck et al 1998). There are 3 types of *mad*-related Smads: receptor-regulated Smads which, in response to the ligand, are directly phosphorylated by type I receptors and shuttle into the nucleus after associating with the second type of Smad, the co-Smad, and finally the anti-Smads, acting as negative regulators of the signalling pathway (Shi & Massague 2003). All members of the Smad family share certain structural similarities: they are proteins of around 500 amino acids in length, consist of two globular domains connected by a linker. Smads of all three classes possess a C-terminal "Mad homology 2" domain 2 (MH2) that mediates protein-protein interactions (see Figure 1), and an N-terminal DNA-binding Mad homology domain 1 which is present in R-Smads and the co-Smad. In mammals, Smad genes represent a highly paralogous group of 8 genes. Not unexpectedly, considering its role in binding to the activated receptors and oligomerisation of phosphorylated Smads, the MH2 domain bears certain structural similarity to the phosphopeptide-binding forkhead-associated (FHA) domain (Durocher et al 2000). Conservation within classes of Smads is so high that human R-Smad-expressing transgenes were found to elicit similar phenotypes to their endogenous fruit fly's counterparts in the *Drosophila* embryo (Marquez et al 2001). Human Smad1 protein shares a remarkable 82% of its amino acid sequence with its *Drosophila* Mad orthologue (Attisano & Lee-Hoeflich 2001). To highlight the incredible evolutionary conservation of the Smad signalling pathway, it has to be noted that organisms as basic and evolutionarily removed from mammals as pseudocoelomates, *e.g.* nematode *C. elegans,* possess a complement of Smads of all three classes: R-Smad, co-Smad and anti-Smad (Newfeld et al 1999). Regulation of Smad function is achieved mostly at the posttranscriptional level, at the level of Smad proteins, their post-translational modifications, most significantly R-Smad activation by receptor-dependent phosphorylation. Thus, at the transcriptional level, R-Smads and the co-Smad appear to be expressed relatively uniformly, at moderate levels in most tissues (*e.g.* see GNF BioGPS data, biogps.gnf.org). Less is known about post-transcriptional regulation of the inhibitory Smad activity, levels of transcription of these Smad genes appears more variable, for instance with significantly higher levels of *SMAD7* transcripts in placenta and natural killer (NK) immune cells compared to majority of other tissues.

2.1 Domain structure of Smad proteins

In addition to the above mentioned Mad homology (MH) domains 1 and 2, involved in DNA and protein binding, respectively, increased attention is being drawn to the central proline-rich linker region, thought to be a major site of post-translational modifications modulating the function of the Smad proteins (Burch et al 2010, Wrighton et al 2009). While the linker is only moderately conserved across Smad classes, a high degree of conservation is observed within a given class and especially a subclass across the animal kingdom, all the way down to insects and even nematodes (Newfeld et al 1999). Another feature, until recently often considered to be a part of the MH2 domain, is the receptor phosphorylation domain containing the $SSxS_{COO^-}$ consensus, present at the C-termini of R-Smads. This domain is specifically targeted for phosphorylation by the type I TGFβ/BMP receptor serine/threonine kinases (Shi & Massague 2003). A number of other Smad protein features and motifs, mostly involved in regulation of Smad localisation and degradation, will be discussed in detail in corresponding sections of this review (Figure 1).

2.2 Receptor-regulated, or R-Smads

In mammals, this group of Smads consists of 2 subclasses, or subfamilies: one primarily involved in the mediation of the BMP/GDF signalling (BMP Smads), and the others transmitting TGFβ/activin/nodal signals (TGFβ Smads). The first group is represented by 3 members - the closely related *Smad1* and *Smad5* genes and slightly more divergent *Smad8*. On the phylogenetic tree, BMP Smads cluster with the *D. melanogaster* Mad protein, the archetypal Smad involved in signalling by the product of the fly's orthologue of mammalian BMPs 2 and 4, *decapentaplegic* (Derynck et al 1998). Members of the other group, encompassing the TGFβ/activin/nodal Smads, *Smad2* and *Smad3*, are more closely related to their Drosophila homologue, dSmad2, than to proteins from different classes of Smads (Newfeld et al 1999). It is interesting to note that even in insects, as it is known to be the case in the fruit fly and the bee, there exists a paralogue for the Smads of both BMP and TGFβ subfamilies in higher organisms. Consistent with the concept of those two groups of Smads transmitting different signals, they are thought to serve as substrates for different sets of type I receptor kinases: BMP Smads are primarily phosphorylated by the activin receptor-like kinase (ALK) 3 and ALK6 (also known as BMP receptor type I A and B) and in some cases ALK2, while TGFβ Smads 2 and 3 are activated by ALKs 4, 5 and 7 (Derynck & Zhang 2003). Interestingly, in some cell types ALK1 was shown as being able to transmit TGFβ signals via BMP Smads 1 and 5, highlighting the complexity of the signalling by the members of the TGFβ superfamily (Goumans et al 2002). Also at the N-terminal end of the MH2 domain is situated a "basic pocket", which functions to enable binding of the R-Smads to GS-domain of the type I receptors which have been activated by phosphorylation (Wu et al 2000). In R-Smads this pocket is utilised to accommodate R-Smads' phosphorylated SSxP motif during the Smad oligomerisation, and it is believed that a similar motif exists in the co-Smad's MH2 domain (Massague et al 2005).

Upon phosphorylation by the receptor, the R-Smads are thought to normally form first homodimeric, and then heterotrimeric, complexes with incorporation of the co-Smad, Smad4, thus consisting of 2 R-Smad and 1 co-Smad proteins (ten Dijke & Hill 2004). It is believed that while the MH1 domain of Smads (specifically shown for 3 and 4) confers specific binding to the specific sequence called the Smad-binding element (SBE), recently defined as 5'-GTCT-3' (or its complement), the binding is relatively weak, and relies upon complexing with other transcription factors, thus achieving activation or repression of downstream gene promoters (Derynck et al 1998, Massague et al 2005, Shi et al 1998). It is believed that Smads 1 and 5 complexed with Smad4 bind an assymetrical site composed of one SBE and one GC-rich consensus, 5'-GRCGNC-3' (Pyrowolakis et al 2004).

2.3 Common mediator, or co-Smad

Structurally, co-Smad protein is very similar to the R-Smads, as it consists of the N-terminal MH1 domain, central proline-rich linker and the C-terminal MH2 domain. Co-Smad's function is two-fold: to serve as an oligomerisation partner facilitating nuclear translocation of activated R-Smads, and to augment binding of R-Smad-containing complexes to target genes in the nucleus (Massague et al 2005). The important function of the recruitment of the transcription activating or repressing factors is performed by the so-called "Smad4 activation domain" (SAD), residing in the N-terminal portion of the MH2 domain.

This class of Smads is very well conserved evolutionarily, with clear orthologues of vertebrate Smad4 genes present in *Drosophila (medea)* and *C. elegans (sma-4)* (Newfeld et al 1999). Human Smad4 is a well-established tumour suppressor gene, often found mutated in human pancreatic and intestinal cancers, and was originally called "deleted in pancreatic cancer 4", DPC4 (Hahn et al 1996, Howe et al 1998). As one would expect, based on a proposed function as a co-factor for R-Smads, the *Smad4* gene is expressed ubiquitously and uniformly across all tissues in both mouse and human (*e.g.* GFN SymAtlas public access and authors' own data).

2.4 Inhibitory Smads

Inhibitory Smads (I-Smads) are the more "distantly related" group of Smads, lacking the DNA-binding MH1 domain and displaying a higher degree of internal divergence. In contrast to the case with other Smads, *Drosophila's* inhibitory Smad, *Dad,* appears to be more phylogenetically distant from vertebrate genes than one of the genes from *C. elegans,* 1L81 (Newfeld et al 1999). While the MH2 domain bears substantial homology to analogous regions of R-Smads and co-Smad, the N-terminal half of I-Smads is divergent enough not to be considered as a functional MH1. At the same time, curiously, a reliable alignment of N-terminal halves of all Smads could be performed, demonstrating few homologous motifs, including a stretch of basic amino acids, mostly lysine, interrupted by an insertion of a glutamate residue present in I-Smads and co-Smad and possibly compromising the ability of this region to function as a potential nuclear localisation signal. As the name of the family indicates, Smads of this group act as antagonists of the signalling by the TGFβ superfamily members, and in mammals it consists of two family members: Smads 6 and 7. Currently, there are up to four different mechanisms by which inhibitory, or anti-Smads, exert their effects. Initially they were found to both compete with R-Smads for binding to activated type I receptors (Hayashi et al 1997) and promote proteasomal degradation of those receptors via recruitment of ubiquitin ligases, Smurf (Shi & Massague 2003). I-Smads were also found to be capable of mediating dephosphorylation of the type I TGFβ superfamily receptors by recruitment of complexes of GADD34 with the catalytic subunit of the protein phosphatase 1 (Shi et al 2004). In addition, I-Smads were found to potentially have a role to play in the nucleus, where Smad6 has been shown to promote repression of the BMP target genes via interaction with co-repressor CtBP (Lin et al 2003).

3. Modulation of Smad signalling

Last decade saw the discovery of a significant number of mechanisms involved in the regulation of TGFβ signalling at the Smad level, from the regulation of Smad binding to the activated receptors on the outer cell membrane to modification of the repertoire of transcriptional co-factors binding to Smad complexes in the nucleus. This does not come as a surprise, since the importance of the pathway in the regulation of the cell's behaviour is difficult to overestimate. In this section, we will discuss some of the better understood mechanisms that cells employ to modulate Smad activity.

3.1 Regulation of receptor-Smad interactions

Transmission of the TGFβ/BMP signal is dependent on direct association of the unphosphorylated, inactive R-Smad with the receptor complex. This process is essential for phosphorylation of the R-Smads, and is facilitated by the SARA (Smad anchor for receptor activation) protein (Shi 2001, Wu et al 2000). The Introduction of a specific point mutation, into the MH2 domain of Smad 2 which is known to be critical for its efficient interaction with SARA, appeared to abolish its ability to transmit TGFβ signalling (Wu et al 2000).

Some mutations in the SARA protein were shown to lead to a mislocalisation of Smad2 and consequently to compromised Smad2-mediated signalling, highlighting an essential role for this SARA-mediated R-Smad receptor targeting. The SARA protein also contains a FYVE phospholipid-binding domain, with a particularly high affinity towards phosphatidylinositol-3'-phosphate, a phospholipid highly enriched on endosomal membranes. This is consistent with the model in which the majority of Smad activation by the receptor complex might be taking place at the early endosome, where most of the SARA protein appears to be localised (Di Guglielmo et al 2003, Itoh & ten Dijke 2007). Another FYVE domain-containing protein, Hgs, has been implicated in promoting Smad phosphorylation in cooperation with SARA (Miura et al 2000). A whole array of factors, proposed to amplify the interactions of Smads with the receptor complexes, has been identified, including Disabled-2, Dok-1, Axin, ELF β-spectrin and cytoplasmic PML (Lin et al 2004, Massague et al 2005). Other proteins, such as TRAP1 (TGFβ receptor-associated protein 1) and TRAP1-like protein (TLP), have been proposed as adaptors interacting with the inactive receptor complexes whilst also promoting the formation of heteromeric R-Smad-coSmad complexes by acting as Smad4 chaperones (Wurthner et al 2001). It has to be noted that there is still a lack of reliable, especially genetic, evidence of the requirement of these proteins for normal Smad signalling, in particular for the BMP pathway, as the role for SARA and other proteins has been demonstrated for the receptors and R-Smads normally associated with transmission of the TGFβ signal.

A different mechanism of regulation of R-Smad activity by its sequestration from binding to the activated type I TGFβ/BMP receptor was described in the interaction of Akt with Smad3 (Conery et al 2004, Remy et al 2004). Recently, a different model was put forward for a mechanism explaining this Akt-Smad interaction, proposing that an Akt kinase substrate, mTOR (mammalian target of rapamycin), acts to inhibit Smad3 phosphorylation by ALK5, the main type I receptor for TGFβ molecules 1-3 (Song et al 2006).

3.2 Modulation of Smad function by phosphorylation

Phosphorylation of the C-terminus of R-Smads by the type I receptor is the central event in intracellular transmission of the TGFβ and BMP signals, as it serves as a major trigger for the oligomerisation of Smads and ensuing regulation of gene expression by Smad complexes upon their translocation into the nucleus (Massague et al 2005). The specificity of the signal transmitted from the membrane is determined by the interaction of specific domains in both the type I receptor and the R-Smad. From its inactive form, with the unphosphorylated GS domain serving as a binding site for FKBP12 molecules, the type I receptor becomes an active serine/threonine kinase upon phosphorylation of the GS domain and ensuing release of FKBP12 (Shi & Massague 2003). The phosphorylated GS domain then acts as one of the R-Smad binding sites, while the specificity of receptor-R-Smad interaction is determined by an L45 loop on the receptor's intracellular kinase domain and the L3 near the C-terminus of the MH2 domain of the R-Smad.

In addition to the C-terminal phosphorylation by the receptors, a significant portion of the activated R-Smads in the cell is subject to secondary phosphorylation events, often thought to ensure a limited lifespan or attenuate the function of the active form (Itoh & ten Dijke 2007, Wrighton et al 2009). Most of those phosphorylation events take place in the linker region, while a few occur at the N-terminus of the R-Smad protein, in the MH1 domain. In the majority of cases, kinases of two families are implicated in these phosphorylation events: the mitogen-activated protein kinases (MAPK) and cyclin-dependent kinases (CDKs).

MAPKs that were specifically shown to phosphoryllate the R-Smad linker include p38 MAPK, c-Jun N-terminal kinase (JNK), extracellular signal-regulated kinase (ERK), Rho-associated protein kinase (ROCK) and MEKK-1 (currently known as MAP3K1). Among other known R-Smad linker kinases are protein kinase C (PKC), G protein-coupled receptor kinase GRK2, Ca2+/calmodulin-dependent protein kinase (CaMKII) and casein kinases CK1γ2 and CK1ε (e.g. Wrighton et al 2009).

Interestingly, nearly all of the phosphorylated Smad residues are serines with few known threonine targets and not a single tyrosine, continuing the common theme for the whole pathway of utilisation of serine/threonine kinases for propagation and attenuation of the signal. It might be possible to surmise that this serves as means of separation of this pathway from interference from many other cellular signalling pathways utilising tyrosine kinases. At the same time, a wide variety of serine-threonine kinases have been shown to be capable of Smad phosphorylation, providing potentially physiologically-important inputs from other pathways. There also appear to be "layers" of phosphorylation, with some events dependent on the preceding modifications (Wrighton et al 2009). A good example of such an arrangement is phosphorylation by the enzymes of the MAPK family, which primes the Smad1 linker for further phosphorylation by the glycogen synthase kinase 3β (GSK-3β), which targets the Smad for ubiquitination and proteasomal degradation (Fuentealba et al 2007). C-terminally phosphorylated Smads are translocated to the nucleus, where the secondary phosphorylation by the MAPKs (Erk, p38 and JNK) occurs. This event is thought to promote the nuclear exclusion, and tertiary phosphorylation by the GSK-3β kinase maximises the affinity of the "triple-phosphorylated" Smad for a Smurf or NEDD ubiquitin ligases, eventuating in the efficient proteasome-dependent degradation. Strong support for the notion that MAPK phosphorylation has a negative effect on Smad-mediated signalling comes from the phenotypic analysis of mice bearing mutations disabling phosphorylation at 6 of the most frequently utilised sites in the linker of Smad1 (Aubin et al 2004). Analysis of the localisation of MAPK phosphorylation-resistant Smad1 revealed its abnormal preferential concentration on the membrane, suggesting that MAPK phosphorylation might be also important for proper subcellular localisation.

Traditionally, it was assumed that most Smad phosphorylation events taking place outside of the C-terminal SSxS domain, predominantly at the linker region, act to down-modulate Smad action (Itoh & ten Dijke 2007, Wrighton et al 2009). However, very recently it was demonstrated that some of those phosphorylation events could act to enhance nuclear functions of Smads, while ensuring their transient nature by priming Smads for rapid degradation (Alarcon et al 2009). In this case, the Smad linker is phosphorylated by one of the cyclin-dependent kinases (CDK8 or 9), and this phosphorylation event correlates with C-terminal Smad phosphorylation, and just like the C-terminal activation is induced by an agonist of the pathway, *i.e.* a member of the TGFβ superfamily of signalling factors. This event, called the agonist-induced linker phosphorylation (ALP), occurs precisely at the serine and threonine residues, the phosphorylation of which is known to facilitate proteosomal degradation of Smads transmitting both TGFβ and BMP signals (Gao et al 2009, Sapkota et al 2007). Experimental evidence supports the model in which double (C-terminally and ALP)-phosphorylated Smads are efficiently targeted by the ubiquitin ligases Smurf1 and NEDD4L for proteasomal degradation. Interestingly, while in the nucleus, activity of the double-phosphorylated Smads is augmented by ALP-dependent binding of the transcriptional co-activator Yap, and this interaction appears to play an important role in regulation of downstream targets, specifically demonstrated for the BMP-induced Smad1-

regulated genes of the *Id* family of repressors of gene activity (Alarcon et al 2009). Smad4 appears to exist in most cells in a constitutively-active state, and while most phosphorylation sites are unknown, phosphorylation by ERK at Thr277 have been shown to be required for efficient nuclear translocation of Smad4-containing complexes (Roelen et al 2003).

3.3 Regulation of Smad activity by phosphatases

The presence of phosphatases capable of specific dephosphorylation of the Smad linker region adds yet another level to the dynamic regulation of the levels of Smad signalling. Originally, the major focus in the interpretation of the phospho-regulation of R-Smad function was on their nuclear export and subsequent degradation. The presence of a large pool of Smads that are at least partially dephosphorylated prior to their export from the nucleus was acknowledged only relatively recently (Inman et al 2002), indicating the existence of phosphatases acting on phospho-Smads. It was also demonstrated that nuclear export-mediating proteins display a preference for the C-terminally dephosphorylated R-Smads (Xu et al 2002). One of the first Smad C-terminal phosphatases was identified in an RNAi screen in *Drosophila* for modifiers of the BMP-Smad mutant, *Mad,* and turned out to be the pyruvate dehydrogenase phosphatase (PDP), previously known for its function in mitochondria (Chen et al 2006). Interestingly, this phosphatase appears to be specific for BMP signal-transmitting Smads in mammals, and has no activity on TGFβ Smads 2 and 3.
The phosphatase that acts on Smads relaying TGFβ/Activin/nodal signals is protein phosphatase 1A (PPM1A, aka PP2Cα), which was found to also function as a facilitator of the nuclear export of C-terminally dephosphorylated Smads 2 and 3 (Lin et al 2006). Depletion of PPM1A lead to an enhanced TGFβ response, confirming the role for this phosphatase in the normal modulation of TGFβ signalling which, in its turn, is known to regulate PPM1A stability (Bu et al 2008). Interestingly, stabilisation of PPM1A is performed by an important regulator of normal, and especially abnormal, cell growth PTEN (phosphatase and tensin homologue), known for its strong positive correlation with carcinogenesis and malignancy.
Another important group of Smad phosphatases is the so-called small C-terminal domain phosphatases (SCPs 1-3). While they have been originally identified for their ability to dephosphorylate the C-termini of BMP signal-transmitting R-Smads (primarily Smad1), SCPs were also able to dephosphorylate the linker regions of all R-Smads (Knockaert et al 2006, Sapkota et al 2006), thus being able to completely erase the phosphorylation marks from the Smad1 protein. Curiously, the evidence that SCPs can dephosphorylate the linker is much more consistent than that on its ability to erase activating C-terminal signatures, as in some systems there was no evidence of such activity of the SCPs (Wrighton et al 2006).

3.4 Targeting Smads for degradation

Control of Smad levels via proteosomal degradation is a well-established important mechanism for regulation of the Smad availability and consequently their signalling. It is likely that, together with controlled dephosphorylation, it could serve as one of the key mechanisms by which the cell maintains and/or limits its sensitivity to the changing level(s) of extracellular ligands. It is important to note that both C-terminally unphosphorylated and phosphorylated (activated) forms of Smads are targeted for degradation by the 26S proteasome (Itoh & ten Dijke 2007, Wrighton et al 2006). A number of E3-class ubiquitin ligases were found to be capable of mediating Smad degradation: the HECT-domain ligases Smurfs (Smad-ubiquitin regulatory factors) 1 and 2 and related proteins, including NEDDs, Itch, WPP/Tuil1, as well as a few other ubiquitin-ligases such as CHIP, Skp1-Cul-F-box (SCF)/Roc1 complex and Arkadia, a U-box E3 ligase (reviewed in Itoh & ten Dijke 2007, Izzi & Attisano 2004).

One of the first observations linked the activated phosphorylation status of Smad2 with degradation, ensuring the turnover of the active form and serving as an example of the negative feedback ensuring transient nature of the activation of the pathway (Lin et al 2000, Lo & Massague 1999). Smurf2, which was identified as the protein performing this function, binds to both Smad2 and Smad3, but was found not to degrade the closely-related Smad3 itself (Bonni et al 2001). To add even more complexity to the subject, many of the Smad2-binding ubiquitin-ligases were found to stimulate degradation of this Smad and antagonise its activity (e.g. WWP1/Tiul1, NEDD4-2), some, e.g. Itch, were not shown to cause degradation and actually enhanced Smad2 signalling (Itch) (Bai et al 2004, Izzi & Attisano 2004). Other ubiquitin ligases were found to mediate the degradation of Smad3 but, curiously, most of them seem to either preferentially degrade the unphosphorylated form (Axin/GSK-3β), or display no preference (CHIP, SCF/Roc1) (Guo et al 2008, Izzi & Attisano 2004).

Arkadia, a nuclear RING-domain E3 ubiquitin ligase, was found to be involved in the degradation of activated Smads 2 and 3 in mouse embryonic tissues (Mavrakis et al 2007). Surprisingly, inactivation of Arkadia activity led to a nuclear accumulation of phospho-Smad2/3 accompanied by a decrease in the pathway's activity. Conversely, overexpression of Arkadia did not repress, and in some settings actually activated, the pathway (embryonic stem muse cells), while leading to the decline in total abundance of the Smads (Mavrakis et al 2007).

Unlike Smurf2, which is thought to interact with both TGFβ and BMP Smads, Smurf1 appears to specifically target Smads 1 and 5, while showing a strong preference for linker-phosphorylated forms, resulting in both exclusion from the nucleus and degradation (Sapkota et al 2007).

Much less is known about the regulation of the proteasome-dependent degradation of co-Smad Smad4, which is thought to exist in at least a partially-phosphorylated form. It was recently shown that phosphorylation by the JNK/p38 kinases, often enhanced for some oncogenic Smad4 mutants, has been shown to promote Smad4's degradation via the SCF complex-mediated polyubiquitination mechanism (Yang et al 2006). Other ubiquitin E3 ligases performing this function include Jab1 and CHIP (Itoh & ten Dijke 2007). Many other ligases can degrade Smad4 via Smad7 interaction, including Smurfs 1 and 2, NEDD4-2 and Tiul1 (Moren et al 2005). Another important RING-type ubiquitin ligase, Ectodermin (also known as TIF1γ), was found to efficiently antagonise TGFβ/BMP signalling by degrading Smad4 in settings ranging from developing embryos to transformed neoplastic cells (Dupont et al 2005).

In *Drosophila*, an unexpected player-translation initiation factor eIF4A-was found to directly interact with the fly's Smad1 and 4 homologues and enhance their degradation, acting synergistically but independently of the *Drosophila*'s Smurf homologue (Li & Li 2006).

3.5 Controlling the nucleo-cytoplasmic shuttling of Smads

One of the important properties of the intracellular distribution of Smads is the very dynamic nature of their shuttling between the nucleus and the cytoplasm. Even in a cell with active Smad signalling, when most Smads and their complexes are localised to the nucleus, a fraction of them appear to be dephosphorylated and exported from the nucleus (Xu et al 2002).

Despite the presence of what appears to be a nuclear localisation signal (NLS) in the MH1 domain, it has been shown that the nuclear import of co- and R-Smads does not involve the participation of the conventional nuclear transport factors, importins (Xu et al 2000, Xu et al 2002). It becomes possible due to the ability of Smads to directly interact with nucleoporins, the proteins forming the actual nuclear pore complex, as has been specifically shown for nucleoporins Nup153 and Nup214 (Xu et al 2000, Xu et al 2002). The so-called hydrophobic corridor, a contiguous set of hydrophobic areas of the MH2 domain, of R-Smads is responsible for direct interaction with the FG repeat region on nucleoporin (normally interacting with importins) (Xu et al 2002).

Some evidence suggests the existence of the Smad shuttling involving conventional nuclear transport receptors. For instance, Smad3 was found to undergo importin-β dependent translocation, but direct comparison of the contribution of importin-dependent and -independent processes revealed significant dominance of the latter (Xu et al 2003). Another important example of Smad "assisted" redistribution is CRM1-dependent export of Smad4 (Pierreux et al 2000). This process is known to depend on a leucine-rich nuclear export signal (NES), located in the N-terminal portion of the linker region (Watanabe et al 2000). Smad4 NES mutant variants are indeed retained in the nucleus permanently (Watanabe et al 2000). The nuclear import of Smad4 also differs from that of R-Smads, with its basic bipartite NLS binding to importin-α leading to nuclear localisation following the conventional pathway (Reguly & Wrana 2003).

3.6 Sequestration of Smads from signalling

By definition, essentially any high-affinity interaction may be construed as a sequestration from signalling if it renders Smads unable to transmit signal by regulating the activity of the downstream target gene. Many of these interactions act to prevent shuttling of Smads into the nucleus, while others target them to particular locales in the cell, such as membranes and the nuclear envelope, or compete for binding, for instance between R-Smads and the co-Smad. Some of the better-characterised cases include the transcriptional repressor SnoN (Ski-related novel protein N), known to cause retention of Smads in the cytoplasm (Krakowski et al 2005). One of the integral proteins of the internal nuclear membrane, Man1 (also known as LEMD3), can sequester R-Smads to the inner nuclear membrane and down-modulate the levels of TGFβ signalling (Lin et al 2005, Pan et al 2005).

Another important negative regulator of Smad activity, Ectodermin (TIF1γ), was identified recently in the haematopoietic system, where it plays a pivotal role in controlling haematopoietic stem cell proliferation and differentiation. Using proteomics analysis, it was found to act as a competing binding partner with Smad4 for the phosphorylated form of Smad2 (He et al 2006). As we mentioned earlier, Ectodermin also appears to function as a ubiquitin ligase for Smad4 itself, thus antagonising the Smad4 function via two distinct mechanisms (Itoh & ten Dijke 2007).

3.7 Alternative splicing of *Smads*

In recent years, an understanding of the important role of alternative transcripts, their functions and expression patterns has received increased attention. As a result, a significant amount of data has been accumulated on alternative splice isoforms of the Smad genes (reviewed in Tao and Sampath, 2010). We will draw on a few examples that illustrate how even minimal changes in splicing patterns can lead to the generation of proteins with

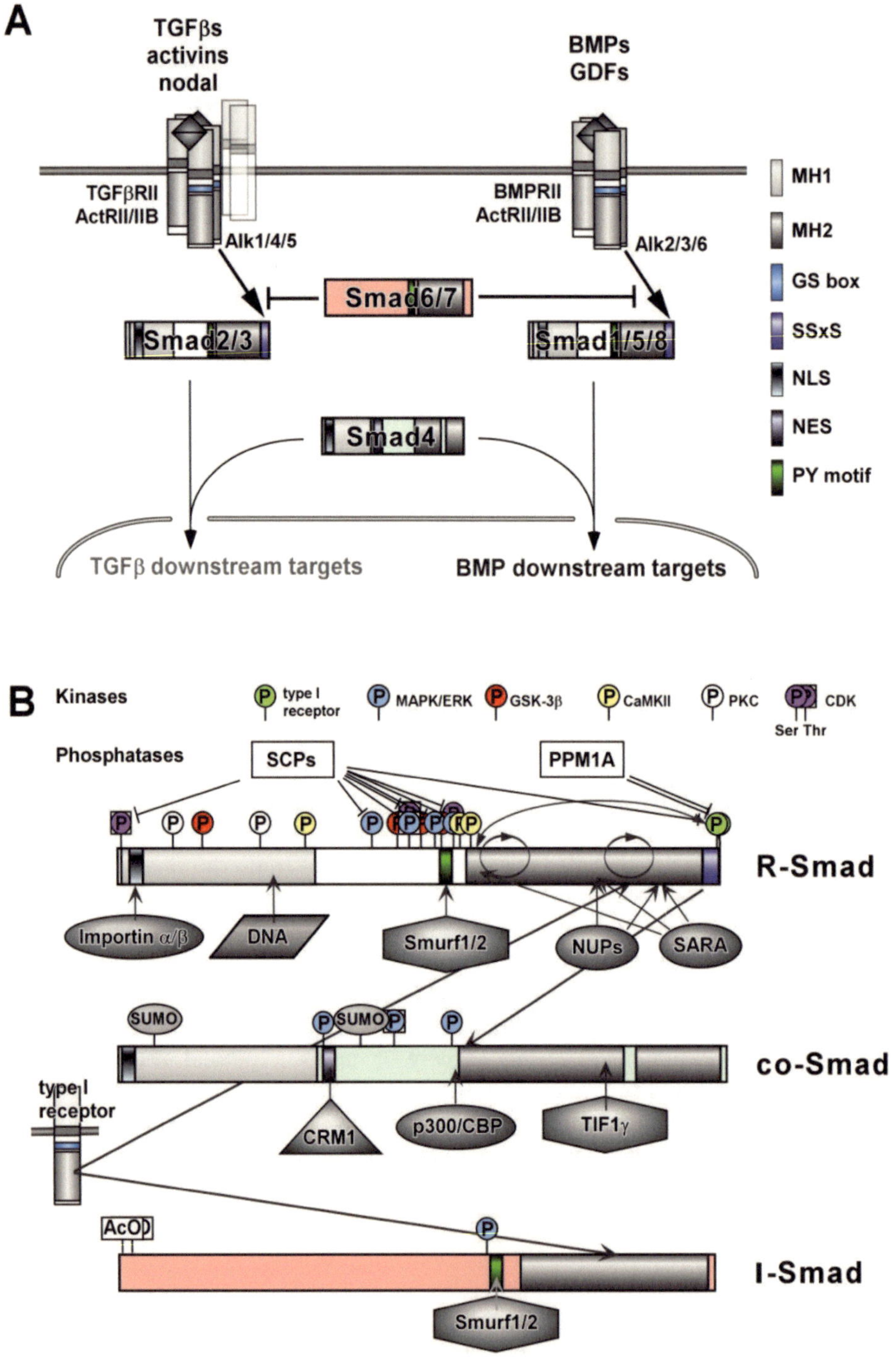

Fig. 1. **TGFβ superfamily signalling and Smad protein interactions. A.** Schematic of the canonical, Smad-mediated TGFβ superfamily signalling. Annotation of the labelled domains in the type I receptor and Smads shown on the right. **B.** Tree types of Smad protein, their

modifications and interactions. Kinases targeting known colour-coded phosphorylation sites in Smads are shown on the top of the panel. Smad domain coding as in panel A. Hammerheaded lines show specificity of phosphatases towards particular phosphoserine/threonine residues. Selected major interacting proteins shown under the Smad of each type. Oval-shaped proteins positively, and triangle shaped-negatively regulate the Smad signalling. Two shown ubiquitin-ligases (Smurf and TIF1γ/ectodermin) shown as hexagons. Arrows indicate interacting region of proteins, arrowed circle - regions involved in homotypic interactions. Abbreviations: NUPs-nucleoporins, SUMO-sumoylation and AcO-acetylation sites on Smad4 and inhibitory Smads, respectively. Modifications and interactions often differ for different Smads of the same type, and only selected examples shown. Drawings are not to exact scale. See text for more details.

significantly different properties. The first example is the retention of exon 3 in the majority of *Smad2* transcripts. As this is indeed the prevalent transcript form, it was designated *Smad2*, even though the corresponding protein appears to lack any DNA-binding activity due to an insertion of the polypeptide encoded by exon 3 near the β-hairpin in the MH1 domain, normally responsible for this interaction (Shi et al 1998). Interestingly, despite the fact that the *Smad2* transcript is normally 3 to 10 times more abundant than the one lacking exon 3 (*Smad2Δ3*), mice homozygous for the allele where exon 3 is deleted appear to be viable, unlike *Smad2*-deficient animals, suggesting that *Smad2* transcript-encoded protein does not play a unique vital role in development (Dunn 2002). This raises questions about the actual function of the longer Smad2 protein, as its *Drosophila* orthologue, *DSmad2*, also contains a similar "insertion" (Brummel et al 1999). It was reported that the long Smad2 can form an active transcription activation complex (with Smad4 and a co-factor FoxH1/FAST1) at the promoter of a downstream gene, *Mix2*, however it is difficult to imagine it making a significant contribution to DNA binding. It could serve as an "adaptor" enhancing the interaction of other transcription factors, or it might even have a role as a competitive inhibitor of Smad2Δ3/Smad3 signalling.

Another interesting example is the splicing isoform of Smad4 lacking the nuclear export signal, NES, encoded by exon 3, or the adjacent sequences in exon 4, shown to have a similar effect on nuclear retention (Pierreux et al 2000). A number of splicing isoforms, including those lacking the NES, can be detected at a comparable, albeit somewhat lower, to the normal form levels in various tissues. The protein forms derived from transcripts lacking either exons 3 or 4 appeared to enhance the response to TGFβ signals (Pierreux et al 2000). In the absence of the TGFβ/BMP signal Smad4 is present in and shuttling between the nucleus and the cytoplasm, and nuclear Smad4 is thought to act to ensure low basal level of expression of the Smad target genes by forming complexes at their promoters with the SnoN transcriptional repressor (Stroschein et al 1999). Thus, expression of a population of the NES-deficient, nuclear-bound Smad4 might serve a dual purpose-to sensitize cells to the TGFβ/BMP signal by minimizing the basal signal level and amplifying the response upon activation of the R-Smads.

Recent advent of the alternative transcript-detecting microarrays will greatly aid in shedding some light on regulation of the alternative promoter and splicing pattern usage in Smad signalling, with a particular interest in the signalling in pluripotent and undergoing directed or undirected differentiation stem cells.

4. Smads in the differentiation and maintenance of pluripotency

Both the maintenance of pluripotency by embryonic stem cells and the differentiation decisions they make are known to be greatly dependent on signalling by the members of the TGFβ superfamily. This appears to be true for both *in vitro* and *in vivo* settings. To date, all embryonic stem cells are known to be dependent on signalling by either molecules of the BMP/GDF family, (as is the case with mouse ES cells) or the TGFβ/activin/Nodal family (in case of human ES cells) to maintain their pluripotent state. Many of these signals are converted into cellular responses (such as changes in gene expression or cytoskeletal reorganisation) via the canonical, or Smad-mediated, signalling pathways. In this section, we will focus on the specific evidence linking roles of the Smad proteins to the maintenance of pluripotency and the differentiation of particular germ layers and cell types, drawn from both *in vivo* (to a large extent on the mouse model) and *in vitro* studies.

4.1 Role of Smads in early embryo patterning and development of germ layers

The advent of mouse molecular genetics, and in particular homologous recombination-based gene knockouts (KOs), enabled researchers to unequivocally address questions about

	Smad knockout							
Phenotype	1	2	3	4	5	6	7	8
Lethality at (dpc)	10.5	8.5	-	7	10	-	-	-
Epiblast proliferation				X				
Gastrulation				X				
Left-right axis defects	X				X			
Mesoderm formation				X				
Cardiac development					X	X		
Angiogenesis					X			
Vasculogenesis					X	X		
Hematopoietic system			X				X	
CNS development				X				
Definitive endoderm		X						
Primordial germ cells	X				X			
Extraembryonic tissues	X	X						
Tumours in hets (adult)			X	X				

Table 1. A brief summary of single knockout phenotypes for mouse *Smad* genes. See text for more details.

the essential roles of genes in mammalian development. The limitations of conventional KOs, associated with the early lethality of complete gene KO, or the functional redundancy of the target gene, can be overcome by the use of recombinase-dependent conditional knockouts (cKOs) and/or genetic intercrossing approaches. However, to the date, our insight into the requirements of Smads in mammalian development is limited to the phenotypes of single mutants, some of their combinations and selected tissue-specific cKOs. Simultaneous cKO in various tissues of all R-Smads involved in the transmission of the TGFβ/activin/nodal (Smads 2/3) or BMP/GDF (1/5/8) signals will provide us with important insight into the interplay of signalling by these branches of the TGFβ superfamily, addressing one of the most interesting questions in the field.

Manipulation of Smad4 function in mice has been extremely fruitful in allowing us to assess the function of the canonical, Smad-mediated TGFβ superfamily signalling as a whole, as this Smad is thought to function as a co-Smad in all R-Smad-mediated events. Full knockout of Smad4 is early embryonic lethal (Sirard et al 1998, Yang et al 1998), with defects in epiblast proliferation and gastrulation. Interestingly, a conditional knockout of Smad4 in epiblast resulted in a much milder phenotype, and demonstrated that BMP signal-mediating Smad signalling is dispensable for some aspects of gastrulation (Chu et al 2004). Consistent with the role in carcinogenesis, mice heterozygous for the Smad4-null allele exhibited a high incidence of intestinal tumours (Yang et al 1998).

Smad1-deficient mice die at ~10.5 dpc due to implantation defects, and chimera experiments show that this Smad's function is essential in extraembryonic tissues and allantois (Tremblay et al 2001). Apart from a prominent defect in germ cell formation, the embryo proper appears to develop normally, suggesting a functional rescue by another Smad, e.g. Smad5, as BMP signalling is known to play a prominent role in early embryo patterning (Tremblay et al 2001).

Smad2 was found to be indispensable for the development of the endodermal lineages, with a clear inability of Smad2-deficient cells to contribute to definitive and visceral endoderm (Tremblay et al 2000). Interestingly, those experiments also confirmed a functional redundancy of Smad2 with other (almost certainly Smad3, expressed in the embryo proper and known as another TGFβ/nodal Smad) transmitters of Nodal signalling in the embryo. At the same time, in the visceral endoderm the phenotype became apparent due to a known lack of potentially compensatory Smad3 expression in that tissue (Tremblay et al 2000). Another study suggested a role for Smad2 in the extraembryonic tissues, important for gastrulation, as well as in the regulation of embryo rotation and the later development of anterior structures (Heyer et al 1999).

Analysis of Smad3-deficient mice revealed that its function is dispensable during embryonic development, but is important for various aspects of normal immune responses, such as T-cell and splenocyte functions. These mice also show an improved wound healing response, defects in articular cartilage (Ashcroft et al 1999, Yang et al 1999b) and had a higher propensity to develop metastatic colorectal cancers (Zhu et al 1998).

Interestingly, ablation of the Smad5 protein in mice lead to a defect in germ cell development similar to that observed in Smad1-null mice, suggesting that either the maximal level of BMP signalling specifying those cells requires both Smad 1 and 5 functions or, despite the high similarity of the protein structure these Smads perform not fully overlapping (and thus redundant) functions (Chang et al 1999, Chang & Matzuk 2001). This hypothesis was tested by removing one copy of each of *Smad1* and *Smad5* genes, with the resulting double heterozygous animals displaying primordial germ cell, cardiovascular and allantoic development defects (Arnold et al 2006). Other abnormalities found in Smad5-deficient mice include angiogenesis and vasculogenesis defects, left-right patterning abnormalities and multiple other extraembryonic and embryonic defects (Chang et al 1999, Chang & Matzuk 2001, Yang et al 1999a).

Unlike other Smad knockouts, Smad8 deficiency in mice does not lead to any obvious phenotype, nor does it enhance the phenotype of either Smad 1or 5 heterozygous mice (Arnold et al 2006). Unlike Smads1/5, Smad8's expression appears to be very restricted, with only some expression in the visceral yolk sac during early development and very specific expression domains observed later on (Arnold et al 2006). It has to be noted that it is quite possible that in different vertebrates this highly conserved gene has an important and unique role to play, since the expression domains and levels are quite likely to vary between different taxa. Smad8 is somewhat divergent from the highly homologous Smads 1 and 5, and may perform other functions. For instance, epigenetic silencing of Smad8 was uncovered in 30% of breast and colon cancer samples (Cheng et al 2004), pointing to a potential role for Smad8 in modulating cell growth rate.

Importantly, inhibitory Smads 6 and 7 differ from other Smads in that their regulation often takes place at the transcriptional level. Consistent with this notion, observed mutant phenotypes are normally expected to be confined to those tissues with high levels of Smad 6 and 7 expression. In the case of Smad6, defects in mutant mice appear to be restricted to the cardiac mesenchyme and vasculature, and include cardiac valve and septation defects, hypertension and aortic ossifications (Galvin et al 2000). Similarly, mice with a hypomorphic mutation of Smad7 appear to display only B-cell specific abnormal up-regulation of TGFβ signalling (Li et al 2006). Some interesting evidence about the importance of Smad7 comes from human sclerodermal tissues and fibroblasts, which have been found to be deficient in Smad7 function (Dong et al 2002). The restoration of Smad7 levels by adenoviral expression appears to be sufficient in this system to down-regulate the TGFβ signalling to normal levels (Dong et al 2002).

4.2 Smad signalling in pluripotency

Until recently, it was presumed that much, if not all, of the TGFβ superfamily signalling known to be required for the maintenance of pluripotency in all known embryonic stem cells (ESCs) is transmitted via the canonical Smad pathway. However, it is only recently that specific evidence supporting this notion has started to appear.

Significant progress has been made in understanding the molecular basis of the differential requirement for the TGFβ superfamily members for the maintenance of pluripotency in the *in vitro* cultures of the two main ESC models, mouse and human ESCs. The two differ greatly in that while BMP/GDF activity is necessary in addition to leukemia inhibitory factor (LIF) for the maintenance of the pluripotent state of mouse ESCs (Ying et al 2003), it induces trophoectodermal differentiation of human ESCs (Ying et al 2003, Ying et al 2008). Conversely, signalling by members of another branch of the TGFβ superfamily, TGFβ/activin, is essential for the maintenance of the pluripotent state of human ESCs in combination with bFGF (FGF2) (Rao & Zandstra 2005). Our own studies show that other members of the TGFβ superfamily, capable of inducing sufficient activation of Smads2/3 (namely GDF11 and GDF8/myostatin), are capable of substituting for the TGFβ/activin activity originating from the feeder fibroblast cells (Hannan et al 2009). This is not totally unexpected, considering the notion that while mouse ESCs, as well as ESCs from most other organisms, represent an equivalent of cells of the inner cell mass of the blastocyst, the human ESCs more closely resemble epiblast cells of a more advanced stage embryo (Nichols & Smith 2009).

The homeobox-containing transcription factor Nanog is a well-established key player, or even a master-regulator factor, in the maintenance, and as a marker, of pluripotency in human, mouse and most other ESCs (Chambers et al 2003). Loss of Nanog function leads to the unavoidable differentiation of ESCs into extraembryonic tissues, in both human and mouse models (Hyslop et al 2005, Mitsui et al 2003). Recently, a specific mechanism for the maintenance of pluripotency was put forward (and tested in mouse ESCs) implicating one of the BMP signal-transmitting Smads, Smad1, and Nanog (Suzuki et al 2006). In this model, Nanog, induced by LIF/Stat3 signalling, prevents persistence of the BMP mesodermal differentiation-inducing signal by directly binding to Smad1 and modulating its activity in transcriptional complexes. Interestingly, more recent transcriptomics studies showed a high coincidence of the Nanog and Smad1/5-binding sites in promoters of many genes involved in the regulation of pluripotency and early differentiation events in the mouse highlighting the possibility of a more complex interplay between the two (Chen et al 2008).

In human ESCs, NANOG plays an equally crucial a role in the maintenance of pluripotency and, just like in the mouse, its enforced expression is sufficient to maintain the pluripotent state of cells even in the absence of extrinsic factors, such as TGFβ/activin and FGF2 (Chambers et al 2003). The *NANOG* locus was found to be a direct target of Smad-mediated TGFβ/activin signalling in human ESCs, with SMADs 2 and 3 shown to directly interact with the promoter (Xu et al 2008). At the same time, BMP signalling induced repression of *NANOG* expression coincidental with the binding of Smad1/5 to its promoter.

Another important recent finding related to Smad2/3 signalling is the elucidation of the mechanism negating the well-known mesendoderm-inducing effect of TGFβ/activin while simultaneously preserving its pluripotency-maintaining action. It was discovered that one of the Smad-interacting transcriptional repressor proteins, SIP1, is capable of repressing Smad2/3-driven mesendodermal differentiation via direct interaction with the said Smads at the target genes' promoters (Verschueren et al 1999).

Interestingly, BMP signalling (albeit possibly at different intensity levels) is thought to play at least one similar role in both mouse and human ESC systems. That role is repression of the neuroectodermal differentiation via Smad-mediated up-regulation of factors of the *Id (inhibitors of differentiation)* family, antagonising the activity of bHLH (basic helix-loop-helix) factors promoting neural differentiation (Ying et al 2003).

Some recent evidence, however, highlights an omnipresent complexity of the effects of Smad signalling on ESCs. Knockdown of Smad4, the co-Smad commonly required by both Smad2/3 and Smad1/5/8 signalling branches, in human ESCs appears to indicate dispensability of the Smad-mediated TGFβ/*activin* signalling for the maintenance of pluripotency (Avery et al 2010). The most logical explanation of such an effect is that the sole purpose of enhanced Smad2/3 signalling is the suppression of the signalling by Smads1/5/8, as SMAD4 knockdown will negate both. Consistent with the earlier mentioned role for BMP signalling in ESCs, cells with the diminished Smad4 activity displayed an up-regulation of genes associated with neural differentiation (*Pax6, NeuroD1, HASH1*). Significantly, in the absence of Smad4 activity, inactivation of the TGFβ/activin signalling using small molecule receptor kinase inhibitor SB431542 did not result in differentiation, strengthening the possibility that human ESC differentiation is driven by the canonical Smad-mediated BMP signalling. Simultaneous dramatic up-regulation of the BMP target genes (*e.g. Msx1*) and down-regulation of Smad2/3 targets (*LeftyA*/B) were observed in human ESCs during SB431542-induced differentiation. However, when SMAD2/3 and SMAD4 signalling was inactivated simultaneously, the expression of pluripotency genes *OCT4* (*POU5F1*) and *NANOG* was not significantly decreased, challenging the postulate that expression of those genes requires active canonical SMAD2/3, or for that matter any SMAD, signalling (Xu et al 2008). Further experimentation involving specific inactivation of the BMP SMAD1/5/8 signalling, either by receptor- or gene-specific inactivation approaches, will be required for further clarification of the interplay of the signalling by the two branches in the maintenance of pluripotency and early differentiation choices of human and, probably, other ESCs. Another important question to be addressed concerns the possible role of SMAD-independent signalling, as the differentiation is induced by inhibition of the type I receptor kinase activity, known to signal through the pathways not involving the Smads.

Consistent with the idea that BMP signalling is not directly involved in the maintenance of the pluripotency-controlling network, genome-wide mapping of Smad1 and Smad4 promoter occupancy indicates direct involvement in the regulation of genes responsible for lineage commitments rather than the maintenance of pluripotency (Fei et al 2010). Interestingly, the Smad1/4 binding mapped predominantly to the genes enriched in bivalent histone methylation marks ($H3K27^{me3}$ and $H3K4^{me3}$) typically repressed in the pluripotent state but primed for quick induction upon differentiation. Described in the study by Fei at al. lack of the change in the pluripotency gene (*NANOG* and *OCT4*/*POU5F1*) expression as a result of the knockdown of *Smad4* and (to lesser degree) *Smad1* points once again to the dispensable nature of the function of Smads for regulation of key genes in the pluripotency network.

5. Smads as regulators of cellular behaviour

5.1 Context-dependent signalling by the Smad transcriptional complexes

Smads bind DNA via their N-terminal MH1 with a relatively low affinity and, unlike most other transcription factors, are thought to require association with DNA-binding co-factors, either repressors or activators for recruitment of the transcription initiation machinery (Massague et al 2005, Ross & Hill 2008). Most of the interactions with those co-factors, as well as other proteins, are thought to be mediated by the MH2 domain, and are tightly regulated by posttranslational modifications, including phosphorylation, sumoylation and, in case of inhibitory Smads, acetylation (see Figure 1B). It also has to be remembered that the majority of *Smad2* gene product contains a DNA binding-disrupting insert in the MH1 domain, and thus is likely to act as an auxiliary adaptor protein (Massague et al 2005).

Smads responsive to different branches of signalling, TGFβ/activin/nodal vs. BMP/GDF, tend to depend on different co-factor sets for recruitment to the promoters of regulated genes. Smads 2 and 3, transmitting the TGFβ/activin signals, tend to rely on members of the forkhead transcription factors (FoxH/FAST) and of the Mix transcription factor (Bix, Mixer) families. The partners of BMP Smads1/5/8 are less characterised, with notable exceptions of the large zinc-finger-containing Schnurri protein in *Drosophila* and Runx1-3 Runt-domain factors in mammals (Ross & Hill 2008). A common theme with all Smad-dependent transcription complexes is the requirement for the recruitment of chromatin-remodelling factors such as p300/CBP and Swi/SNF complex (Ross et al 2006). Significantly, core Smad transcriptional complexes can exist as either heterodimers or heterotrimers, depending on the promoter and cellular state (Massague et al 2005). Some of the better studied examples include activation of the ARE (activin response element) promoters by the Smad2-Smad4 heterotrimeric activating complex (Ross et al 2006). In these cases, p300 and Swi/SNF recruitment was accompanied by the nucleosome remodelling via acetylation of histone H3, allowing for initiation of transcription by the Pol II machinery. In the case of repression of gene expression by the Smad3-Smad4 heterodimer, recruitment of HDAC4/5 by the bound Smad-containing complex lead to deacetylation of histone H4 and formation of the restrictive chromatin on the promoter region, rendering it transcriptionally inactive (Kang et al 2005).

Specificity of the promoter binding by the Smad complexes depends on many factors, including availability and identity of the transcriptional co-factors recruited to the Smad complex (Massague et al 2005). For instance, Smad complexes recruiting coactivators often contain the before-mentioned FoxH and Mix and related factors, as well as Runx, OAZ and few others. Corepressor-recruiting complexes start with E2F factors 4 or 5, Runx2 and Nkx3.2 homeobox factors (Massague et al 2005). As eluded to earlier, the chromatin-remodelling coactivators normally are p300/CBP, P/CAF, and corepressors include HDACs (1,4,5) or p107.

5.2 Smad signalling and cytoskeletal dynamics

A few lines of evidence point to the possibility that Smad signalling might be regulated by and itself regulates the changes in cytoskeletal dynamics. For instance, Smad protein levels were found to be induced in osteocytes under mechanical loading, and integrin signalling was proposed to interact with that signalling from the BMP receptors (Jadlowiec et al 2006, Rath et al 2008). BMPs are also involved in chemosensing and axon guidance by a number of cell types, and dynamics of the distribution of BMP type I and II receptors is a tightly controlled process, including by the BMPs themselves (Liu et al 2003, Nishita et al 2006, Sammar et al 2009).

5.3 Smad signalling and the cell cycle

Some of the original roles attributed to the TGFβ1-3 molecules were cytostatic and pro-apoptotic functions (Ten Dijke et al 2002). Most of these effects are mediated by regulation of the TGFβ target genes with anti-proliferative or pro-apoptotic functions. One of the best-known examples is activation of the promoter of p15INK4b gene, a potent cell cycle inhibitor (Gomis et al 2006). Regulation of this promoter also represents an interesting example of attenuation of the Smad activity, with normally present C/EBPβ transcriptional co-activator's function inhibited by its regulated alternatively spliced form (LIP), acting as a dominant-negative form (Gomis et al 2006). Some of the recently uncovered mechanisms for regulation of the cell cycle progression via regulation of microRNA function are discussed in the next section.

5.4 Smads and microRNAs

MicroRNAs (miRNAs) have recently been identified as omnipresent and versatile post-transcriptional regulators of gene expression, with established roles in induction and maintenance of pluripotency (Viswanathan & Daley 2010). They are thought to exert their effects by targeting specific mRNAs, typically in their 3' untranslated regions, in complexes with proteins of the Argonaute (Ago) family (Bartel 2009). This normally results in translational repression or degradation of the targeted mRNA. One of the well-established mechanisms is the pluripotency-promoting factor Lin28, often used in the generation of induced pluripotent stem cells (iPSCs), a known RNA-binding protein found to specifically degrade the miRNA let-7 (Viswanathan & Daley 2010).

Remarkably, Smads have recently been implicated in the regulation, in a ligand-dependent manner, of miRNA activation via processing (Davis 2008, 2010). Smads were originally found to control miRNA maturation via an effect on Drosha, an RNAse III enzyme, miRNA-processing complex (Davis et al 2008). It has recently been established that this processing is highly miRNA sequence-dependent, with Smads binding specifically to the SBE-like sequence in a double-stranded stem region in most cases, thus mimicking the normal DNA sequence recognised by the Smads. Some target miRNAs contained a more GC-rich consensus, resembling the recognition site for the BMP Smads1/5/8 (Davis et al 2010).

Of particular interest amongst the microRNA targets of regulation by Smads is miR-21, known to be specifically involved in the maintenance of pluripotency. In mouse ESCs, it was found that a transcription factor REST/NRST (RE-1 silencing transcription factor, a neuronal repressor) acts to promote maintenance of pluripotency via suppression of the miR-21 (Singh et al 2008). miR-21 was shown to specifically suppress the self-renewal of the mouse ESCs, and decrease the levels of expression of core pluripotency markers *Oct4/Pou5f1, Nanog* and *Sox2.*

Amongst other Smad-regulated microRNAs are miR-105, shown to regulate the cell cycle and apoptosis by affecting expression of PCNA and cyclin B1, and p53, respectively (Sirotkin et al 2010). miR-214 was found to regulate the polycomb group genes, in particular Ezh2, known to mediate epigenetic gene silencing, including in ESCs (Juan et al 2009). miR-215, another BMP/Smad-regulated microRNA, has a well-established role in regulation of cell cycle progression and checkpoint (Georges et al 2008).

Of the microRNAs containing the GC-rich Smad-binding consensus, of particular relevance is miR-23b, which is known to feed back and regulate Smads themselves in an adult stem cell setting (Rogler et al 2009). It is particularly interesting as this microRNA was found to be induced by BMP/TGFβ signal themselves, suggesting the possibility of a novel negative feedback mechanism.

5.5 Role of the inhibitory Smads

Inhibitory Smads exist in all organisms, and as the name indicates act to antagonise canonical Smad signalling. In mammals, Smad6 is believed to largely target the BMP signalling, while Smad7 seems to inhibit both BMP and TGFβ/activin branches (Massague et al 2005). So far, four major mechanisms have been identified: competition with R-Smads for type I receptor binding, targeting type I receptors for degradation, competing against R-Smad for the interaction with co-Smad, and recently proposed role in regulation of the nuclear Smad complex formation (Itoh & ten Dijke 2007, Massague et al 2005). Being themselves targets of TGFβ/BMP signalling-induced up-regulation, inhibitory Smads are believed thus to constitute an important part of the feedback mechanism, acting to limit strength and amplitude of the TGFβ/BMP signal (Massague et al 2005).

6. Conclusion and future directions

In this review, we have tried to exemplify the wide range of processes regulated by Smads and regulating them in return, as well as the major mechanisms by which that regulation is achieved.

One of the important considerations for the field that we want to bring forward is the need for a careful interpretation of the phenotypes elicited by ligand stimulation, receptor inhibition and Smad manipulation. The ambiguity in interpretations stems from the promiscuous nature of the receptors (towards both ligands, receptor partners and even R-Smads transmitting the signal), limited specificity of the small molecule inhibitors, and lastly, but very significantly, substantial amount of non-canonical, Smad-independent signalling from the receptors.

It does appear that Smads indeed act as the intracellular "hubs" of the signalling, as they integrate inputs from all major cellular pathways, including the TGFβ/BMP signalling itself, the mitogen/receptor tyrosine kinase/MAPK pathway, JAK/Stat mediated pathways and the Wnt signalling pathway. It is possible then that Smads themselves can act as the key transcriptional switches. Indeed, the Smad promoter occupancy studies have identified a limited number of gene targets, many of which are the "master"regulators, positioned high in the hierarchy of transcriptional networks responsible for activation of distinct developmental programs. In particularly in a naïve pluripotent stem cell it appears to be possible to manipulate lineage and cell-fate decisions by altering states of the two branches of the Smad signalling, TGFβ/activin and BMP (e.g. Chng et al 2010). It is particularly interesting that maintenance of the balance of the signalling through the two branches appears not to significantly disturb the maintenance of pluripotency, supporting the idea of embryonic stem cell's "ground state", while the disbalance causes differentiation with a strong and consistent slant in lineage choices (Figure 2, Avery et al 2010, Ying et al 2008).

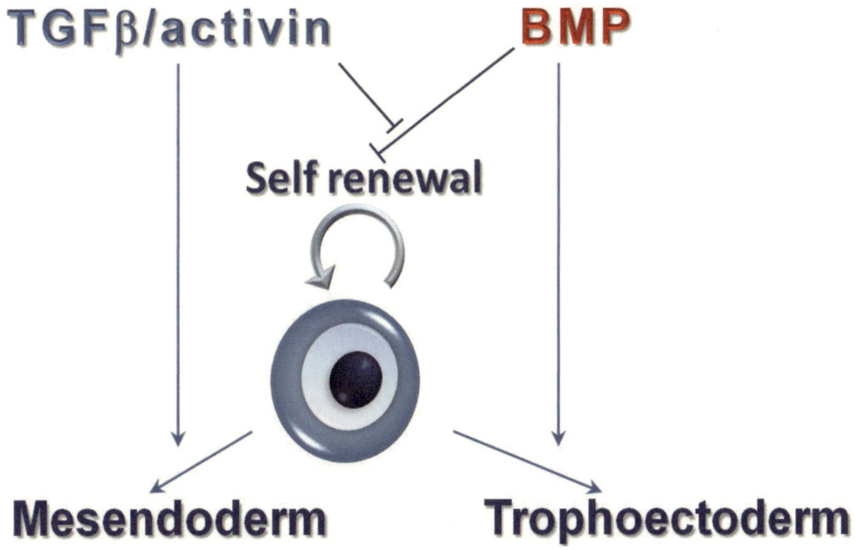

Fig. 2. Recent evidence suggests that Smad-mediated signalling (from either TGFβ/activin or BMPs) is dispensable for maintenance of the pluripotent "ground state" of the embryonic stem cell. See text for discussion.

Some of the most intriguing questions include dissection of the "Smad code", i.e. different levels of the TGFβ/activin and BMP Smad activities, for the various lineage-specification programs for early lineage commitment and downstream cell-type differentiation decisions. It is possible, however, that a significant input from the non-canonical, Smad-independent signalling is required in establishment and execution of those differentiation programs. It is also important to perform a genome-wide occupancy assessment for the TGFβ/activin and BMP Smads at various levels of inputs from other interfering pathways, such as mitogen/ERK, Wnt and JAK/Stat-mediated signals.

To properly understand the nature of Smad activity in particular tissues, a comprehensive analysis of alternative transcript abundance might prove fruitful. For instance, the question about the biological role of the major, exon 3-retaining form of *Smad2* remains open, as it is believed this form does not bind DNA. The recent avaialability of microarrays containing probes for various significantly represented transcript forms will prove extremely useful in that task.

7. Acknowledgement

Dr Deanne Whitworth is greatly acknowledged for proofreading the manuscript and help with editing.

8. References

Alarcon C, Zaromytidou AI, Xi Q, Gao S, Yu J, et al. 2009. Nuclear CDKs drive Smad transcriptional activation and turnover in BMP and TGF-beta pathways. *Cell* 139: 757-69

Arnold SJ, Maretto S, Islam A, Bikoff EK, Robertson EJ. 2006. Dose-dependent Smad1, Smad5 and Smad8 signaling in the early mouse embryo. *Dev Biol* 296: 104-18

Ashcroft GS, Yang X, Glick AB, Weinstein M, Letterio JL, et al. 1999. Mice lacking Smad3 show accelerated wound healing and an impaired local inflammatory response. *Nat Cell Biol* 1: 260-6

Attisano L, Lee-Hoeflich ST. 2001. The Smads. *Genome Biol* 2: REVIEWS3010

Aubin J, Davy A, Soriano P. 2004. In vivo convergence of BMP and MAPK signaling pathways: impact of differential Smad1 phosphorylation on development and homeostasis. *Genes Dev* 18: 1482-94

Avery S, Zafarana G, Gokhale PJ, Andrews PW. 2010. The role of SMAD4 in human embryonic stem cell self-renewal and stem cell fate. *Stem Cells* 28: 863-73

Bai Y, Yang C, Hu K, Elly C, Liu YC. 2004. Itch E3 ligase-mediated regulation of TGF-beta signaling by modulating smad2 phosphorylation. *Mol Cell* 15: 825-31

Bartel DP. 2009. MicroRNAs: target recognition and regulatory functions. *Cell* 136: 215-33

Bonni S, Wang HR, Causing CG, Kavsak P, Stroschein SL, et al. 2001. TGF-beta induces assembly of a Smad2-Smurf2 ubiquitin ligase complex that targets SnoN for degradation. *Nat Cell Biol* 3: 587-95

Brummel T, Abdollah S, Haerry TE, Shimell MJ, Merriam J, et al. 1999. The Drosophila activin receptor baboon signals through dSmad2 and controls cell proliferation but not patterning during larval development. *Genes Dev* 13: 98-111

Bu S, Kapanadze B, Hsu T, Trojanowska M. 2008. Opposite effects of dihydrosphingosine 1-phosphate and sphingosine 1-phosphate on transforming growth factor-beta/Smad signaling are mediated through the PTEN/PPM1A-dependent pathway. *J Biol Chem* 283: 19593-602

Burch ML, Zheng W, Little PJ. 2010. Smad linker region phosphorylation in the regulation of extracellular matrix synthesis. *Cell Mol Life Sci*

Chambers I, Colby D, Robertson M, Nichols J, Lee S, et al. 2003. Functional expression cloning of Nanog, a pluripotency sustaining factor in embryonic stem cells. *Cell* 113: 643-55

Chang H, Huylebroeck D, Verschueren K, Guo Q, Matzuk MM, Zwijsen A. 1999. Smad5 knockout mice die at mid-gestation due to multiple embryonic and extraembryonic defects. *Development* 126: 1631-42

Chang H, Matzuk MM. 2001. Smad5 is required for mouse primordial germ cell development. *Mech Dev* 104: 61-7

Chen HB, Shen J, Ip YT, Xu L. 2006. Identification of phosphatases for Smad in the BMP/DPP pathway. *Genes Dev* 20: 648-53

Chen X, Xu H, Yuan P, Fang F, Huss M, et al. 2008. Integration of external signaling pathways with the core transcriptional network in embryonic stem cells. *Cell* 133: 1106-17

Cheng KH, Ponte JF, Thiagalingam S. 2004. Elucidation of epigenetic inactivation of SMAD8 in cancer using targeted expressed gene display. *Cancer Res* 64: 1639-46

Chng Z, Teo A, Pedersen RA, Vallier L. 2010. SIP1 mediates cell-fate decisions between neuroectoderm and mesendoderm in human pluripotent stem cells. *Cell Stem Cell* 6: 59-70

Chu GC, Dunn NR, Anderson DC, Oxburgh L, Robertson EJ. 2004. Differential requirements for Smad4 in TGFbeta-dependent patterning of the early mouse embryo. *Development* 131: 3501-12

Conery AR, Cao Y, Thompson EA, Townsend CM, Jr., Ko TC, Luo K. 2004. Akt interacts directly with Smad3 to regulate the sensitivity to TGF-beta induced apoptosis. *Nat Cell Biol* 6: 366-72

Davis BN, Hilyard AC, Lagna G, Hata A. 2008. SMAD proteins control DROSHA-mediated microRNA maturation. *Nature* 454: 56-61

Davis BN, Hilyard AC, Nguyen PH, Lagna G, Hata A. 2010. Smad proteins bind a conserved RNA sequence to promote microRNA maturation by Drosha. *Mol Cell* 39: 373-84

Derynck R, Zhang Y, Feng XH. 1998. Smads: transcriptional activators of TGF-beta responses. *Cell* 95: 737-40

Derynck R, Zhang YE. 2003. Smad-dependent and Smad-independent pathways in TGF-beta family signalling. *Nature* 425: 577-84

Di Guglielmo GM, Le Roy C, Goodfellow AF, Wrana JL. 2003. Distinct endocytic pathways regulate TGF-beta receptor signalling and turnover. *Nat Cell Biol* 5: 410-21

Dong C, Zhu S, Wang T, Yoon W, Li Z, et al. 2002. Deficient Smad7 expression: a putative molecular defect in scleroderma. *Proc Natl Acad Sci U S A* 99: 3908-13

Dupont S, Zacchigna L, Cordenonsi M, Soligo S, Adorno M, et al. 2005. Germ-layer specification and control of cell growth by Ectodermin, a Smad4 ubiquitin ligase. *Cell* 121: 87-99

Durocher D, Taylor IA, Sarbassova D, Haire LF, Westcott SL, et al. 2000. The molecular basis of FHA domain:phosphopeptide binding specificity and implications for phospho-dependent signaling mechanisms. *Mol Cell* 6: 1169-82

Fei T, Xia K, Li Z, Zhou B, Zhu S, et al. 2010. Genome-wide mapping of SMAD target genes reveals the role of BMP signaling in embryonic stem cell fate determination. *Genome Res* 20: 36-44

Fuentealba LC, Eivers E, Ikeda A, Hurtado C, Kuroda H, et al. 2007. Integrating patterning signals: Wnt/GSK3 regulates the duration of the BMP/Smad1 signal. *Cell* 131: 980-93

Galvin KM, Donovan MJ, Lynch CA, Meyer RI, Paul RJ, et al. 2000. A role for smad6 in development and homeostasis of the cardiovascular system. *Nat Genet* 24: 171-4

Gao S, Alarcon C, Sapkota G, Rahman S, Chen PY, et al. 2009. Ubiquitin ligase Nedd4L targets activated Smad2/3 to limit TGF-beta signaling. *Mol Cell* 36: 457-68

Georges SA, Biery MC, Kim SY, Schelter JM, Guo J, et al. 2008. Coordinated regulation of cell cycle transcripts by p53-Inducible microRNAs, miR-192 and miR-215. *Cancer Res* 68: 10105-12

Gomis RR, Alarcon C, Nadal C, Van Poznak C, Massague J. 2006. C/EBPbeta at the core of the TGFbeta cytostatic response and its evasion in metastatic breast cancer cells. *Cancer Cell* 10: 203-14

Goumans MJ, Valdimarsdottir G, Itoh S, Rosendahl A, Sideras P, ten Dijke P. 2002. Balancing the activation state of the endothelium via two distinct TGF-beta type I receptors. *EMBO J* 21: 1743-53

Guo X, Ramirez A, Waddell DS, Li Z, Liu X, Wang XF. 2008. Axin and GSK3- control Smad3 protein stability and modulate TGF- signaling. *Genes Dev* 22: 106-20

Hahn SA, Schutte M, Hoque AT, Moskaluk CA, da Costa LT, et al. 1996. DPC4, a candidate tumor suppressor gene at human chromosome 18q21.1. *Science* 271: 350-3

Hannan NR, Jamshidi P, Pera MF, Wolvetang EJ. 2009. BMP-11 and myostatin support undifferentiated growth of human embryonic stem cells in feeder-free cultures. *Cloning Stem Cells* 11: 427-35

Hayashi H, Abdollah S, Qiu Y, Cai J, Xu YY, et al. 1997. The MAD-related protein Smad7 associates with the TGFbeta receptor and functions as an antagonist of TGFbeta signaling. *Cell* 89: 1165-73

He W, Dorn DC, Erdjument-Bromage H, Tempst P, Moore MA, Massague J. 2006. Hematopoiesis controlled by distinct TIF1gamma and Smad4 branches of the TGFbeta pathway. *Cell* 125: 929-41

Heyer J, Escalante-Alcalde D, Lia M, Boettinger E, Edelmann W, et al. 1999. Postgastrulation Smad2-deficient embryos show defects in embryo turning and anterior morphogenesis. *Proc Natl Acad Sci U S A* 96: 12595-600

Howe JR, Roth S, Ringold JC, Summers RW, Jarvinen HJ, et al. 1998. Mutations in the SMAD4/DPC4 gene in juvenile polyposis. *Science* 280: 1086-8

Hyslop L, Stojkovic M, Armstrong L, Walter T, Stojkovic P, et al. 2005. Downregulation of NANOG induces differentiation of human embryonic stem cells to extraembryonic lineages. *Stem Cells* 23: 1035-43

Inman GJ, Nicolas FJ, Hill CS. 2002. Nucleocytoplasmic shuttling of Smads 2, 3, and 4 permits sensing of TGF-beta receptor activity. *Mol Cell* 10: 283-94

Itoh S, ten Dijke P. 2007. Negative regulation of TGF-beta receptor/Smad signal transduction. *Curr Opin Cell Biol* 19: 176-84

Izzi L, Attisano L. 2004. Regulation of the TGFbeta signalling pathway by ubiquitin-mediated degradation. *Oncogene* 23: 2071-8

Jadlowiec JA, Zhang X, Li J, Campbell PG, Sfeir C. 2006. Extracellular matrix-mediated signaling by dentin phosphophoryn involves activation of the Smad pathway independent of bone morphogenetic protein. *J Biol Chem* 281: 5341-7

Juan AH, Kumar RM, Marx JG, Young RA, Sartorelli V. 2009. Mir-214-dependent regulation of the polycomb protein Ezh2 in skeletal muscle and embryonic stem cells. *Mol Cell* 36: 61-74

Kang JS, Alliston T, Delston R, Derynck R. 2005. Repression of Runx2 function by TGF-beta through recruitment of class II histone deacetylases by Smad3. *EMBO J* 24: 2543-55

Knockaert M, Sapkota G, Alarcon C, Massague J, Brivanlou AH. 2006. Unique players in the BMP pathway: small C-terminal domain phosphatases dephosphorylate Smad1 to attenuate BMP signaling. *Proc Natl Acad Sci U S A* 103: 11940-5

Krakowski AR, Laboureau J, Mauviel A, Bissell MJ, Luo K. 2005. Cytoplasmic SnoN in normal tissues and nonmalignant cells antagonizes TGF-beta signaling by sequestration of the Smad proteins. *Proc Natl Acad Sci U S A* 102: 12437-42

Li J, Li WX. 2006. A novel function of Drosophila eIF4A as a negative regulator of Dpp/BMP signalling that mediates SMAD degradation. *Nat Cell Biol* 8: 1407-14

Li R, Rosendahl A, Brodin G, Cheng AM, Ahgren A, et al. 2006. Deletion of exon I of SMAD7 in mice results in altered B cell responses. *J Immunol* 176: 6777-84

Lin F, Morrison JM, Wu W, Worman HJ. 2005. MAN1, an integral protein of the inner nuclear membrane, binds Smad2 and Smad3 and antagonizes transforming growth factor-beta signaling. *Hum Mol Genet* 14: 437-45

Lin HK, Bergmann S, Pandolfi PP. 2004. Cytoplasmic PML function in TGF-beta signalling. *Nature* 431: 205-11

Lin X, Duan X, Liang YY, Su Y, Wrighton KH, et al. 2006. PPM1A functions as a Smad phosphatase to terminate TGFbeta signaling. *Cell* 125: 915-28

Lin X, Liang M, Feng XH. 2000. Smurf2 is a ubiquitin E3 ligase mediating proteasome-dependent degradation of Smad2 in transforming growth factor-beta signaling. *J Biol Chem* 275: 36818-22

Lin X, Liang YY, Sun B, Liang M, Shi Y, et al. 2003. Smad6 recruits transcription corepressor CtBP to repress bone morphogenetic protein-induced transcription. *Mol Cell Biol* 23: 9081-93

Liu J, Wilson S, Reh T. 2003. BMP receptor 1b is required for axon guidance and cell survival in the developing retina. *Dev Biol* 256: 34-48

Lo RS, Massague J. 1999. Ubiquitin-dependent degradation of TGF-beta-activated smad2. *Nat Cell Biol* 1: 472-8

Marquez RM, Singer MA, Takaesu NT, Waldrip WR, Kraytsberg Y, Newfeld SJ. 2001. Transgenic analysis of the Smad family of TGF-beta signal transducers in Drosophila melanogaster suggests new roles and new interactions between family members. *Genetics* 157: 1639-48

Massague J, Seoane J, Wotton D. 2005. Smad transcription factors. *Genes Dev* 19: 2783-810

Mavrakis KJ, Andrew RL, Lee KL, Petropoulou C, Dixon JE, et al. 2007. Arkadia enhances Nodal/TGF-beta signaling by coupling phospho-Smad2/3 activity and turnover. *PLoS Biol* 5: e67

Mitsui K, Tokuzawa Y, Itoh H, Segawa K, Murakami M, et al. 2003. The homeoprotein Nanog is required for maintenance of pluripotency in mouse epiblast and ES cells. *Cell* 113: 631-42

Miura S, Takeshita T, Asao H, Kimura Y, Murata K, et al. 2000. Hgs (Hrs), a FYVE domain protein, is involved in Smad signaling through cooperation with SARA. *Mol Cell Biol* 20: 9346-55

Moren A, Imamura T, Miyazono K, Heldin CH, Moustakas A. 2005. Degradation of the tumor suppressor Smad4 by WW and HECT domain ubiquitin ligases. *J Biol Chem* 280: 22115-23

Newfeld SJ, Wisotzkey RG, Kumar S. 1999. Molecular evolution of a developmental pathway: phylogenetic analyses of transforming growth factor-beta family ligands, receptors and Smad signal transducers. *Genetics* 152: 783-95

Nichols J, Smith A. 2009. Naive and primed pluripotent states. *Cell Stem Cell* 4: 487-92

Nishita M, Yoo SK, Nomachi A, Kani S, Sougawa N, et al. 2006. Filopodia formation mediated by receptor tyrosine kinase Ror2 is required for Wnt5a-induced cell migration. *J Cell Biol* 175: 555-62

Pan D, Estevez-Salmeron LD, Stroschein SL, Zhu X, He J, et al. 2005. The integral inner nuclear membrane protein MAN1 physically interacts with the R-Smad proteins to repress signaling by the transforming growth factor-{beta} superfamily of cytokines. *J Biol Chem* 280: 15992-6001

Pierreux CE, Nicolas FJ, Hill CS. 2000. Transforming growth factor beta-independent shuttling of Smad4 between the cytoplasm and nucleus. *Mol Cell Biol* 20: 9041-54

Pyrowolakis G, Hartmann B, Muller B, Basler K, Affolter M. 2004. A simple molecular complex mediates widespread BMP-induced repression during Drosophila development. *Dev Cell* 7: 229-40

Rao BM, Zandstra PW. 2005. Culture development for human embryonic stem cell propagation: molecular aspects and challenges. *Curr Opin Biotechnol* 16: 568-76

Rath B, Nam J, Knobloch TJ, Lannutti JJ, Agarwal S. 2008. Compressive forces induce osteogenic gene expression in calvarial osteoblasts. *J Biomech* 41: 1095-103

Reguly T, Wrana JL. 2003. In or out? The dynamics of Smad nucleocytoplasmic shuttling. *Trends Cell Biol* 13: 216-20

Remy I, Montmarquette A, Michnick SW. 2004. PKB/Akt modulates TGF-beta signalling through a direct interaction with Smad3. *Nat Cell Biol* 6: 358-65

Roelen BA, Cohen OS, Raychowdhury MK, Chadee DN, Zhang Y, et al. 2003. Phosphorylation of threonine 276 in Smad4 is involved in transforming growth factor-beta-induced nuclear accumulation. *Am J Physiol Cell Physiol* 285: C823-30

Rogler CE, Levoci L, Ader T, Massimi A, Tchaikovskaya T, et al. 2009. MicroRNA-23b cluster microRNAs regulate transforming growth factor-beta/bone morphogenetic protein signaling and liver stem cell differentiation by targeting Smads. *Hepatology* 50: 575-84

Ross S, Cheung E, Petrakis TG, Howell M, Kraus WL, Hill CS. 2006. Smads orchestrate specific histone modifications and chromatin remodeling to activate transcription. *EMBO J* 25: 4490-502

Ross S, Hill CS. 2008. How the Smads regulate transcription. *Int J Biochem Cell Biol* 40: 383-408

Sammar M, Sieber C, Knaus P. 2009. Biochemical and functional characterization of the Ror2/BRIb receptor complex. *Biochem Biophys Res Commun* 381: 1-6

Sapkota G, Alarcon C, Spagnoli FM, Brivanlou AH, Massague J. 2007. Balancing BMP signaling through integrated inputs into the Smad1 linker. *Mol Cell* 25: 441-54

Sapkota G, Knockaert M, Alarcon C, Montalvo E, Brivanlou AH, Massague J. 2006. Dephosphorylation of the linker regions of Smad1 and Smad2/3 by small C-terminal domain phosphatases has distinct outcomes for bone morphogenetic protein and transforming growth factor-beta pathways. *J Biol Chem* 281: 40412-9

Shi W, Sun C, He B, Xiong W, Shi X, et al. 2004. GADD34-PP1c recruited by Smad7 dephosphorylates TGFbeta type I receptor. *J Cell Biol* 164: 291-300

Shi Y. 2001. Structural insights on Smad function in TGFbeta signaling. *Bioessays* 23: 223-32

Shi Y, Massague J. 2003. Mechanisms of TGF-beta signaling from cell membrane to the nucleus. *Cell* 113: 685-700

Shi Y, Wang YF, Jayaraman L, Yang H, Massague J, Pavletich NP. 1998. Crystal structure of a Smad MH1 domain bound to DNA: insights on DNA binding in TGF-beta signaling. *Cell* 94: 585-94

Singh SK, Kagalwala MN, Parker-Thornburg J, Adams H, Majumder S. 2008. REST maintains self-renewal and pluripotency of embryonic stem cells. *Nature* 453: 223-7

Sirard C, de la Pompa JL, Elia A, Itie A, Mirtsos C, et al. 1998. The tumor suppressor gene Smad4/Dpc4 is required for gastrulation and later for anterior development of the mouse embryo. *Genes Dev* 12: 107-19

Sirotkin AV, Laukova M, Ovcharenko D, Brenaut P, Mlyncek M. 2010. Identification of microRNAs controlling human ovarian cell proliferation and apoptosis. *J Cell Physiol* 223: 49-56

Song K, Wang H, Krebs TL, Danielpour D. 2006. Novel roles of Akt and mTOR in suppressing TGF-beta/ALK5-mediated Smad3 activation. *EMBO J* 25: 58-69

Stroschein SL, Wang W, Zhou S, Zhou Q, Luo K. 1999. Negative feedback regulation of TGF-beta signaling by the SnoN oncoprotein. *Science* 286: 771-4

Suzuki A, Raya A, Kawakami Y, Morita M, Matsui T, et al. 2006. Nanog binds to Smad1 and blocks bone morphogenetic protein-induced differentiation of embryonic stem cells. *Proc Natl Acad Sci U S A* 103: 10294-9

Ten Dijke P, Goumans MJ, Itoh F, Itoh S. 2002. Regulation of cell proliferation by Smad proteins. *J Cell Physiol* 191: 1-16

ten Dijke P, Hill CS. 2004. New insights into TGF-beta-Smad signalling. *Trends Biochem Sci* 29: 265-73

Tremblay KD, Dunn NR, Robertson EJ. 2001. Mouse embryos lacking Smad1 signals display defects in extra-embryonic tissues and germ cell formation. *Development* 128: 3609-21

Tremblay KD, Hoodless PA, Bikoff EK, Robertson EJ. 2000. Formation of the definitive endoderm in mouse is a Smad2-dependent process. *Development* 127: 3079-90

Verschueren K, Remacle JE, Collart C, Kraft H, Baker BS, et al. 1999. SIP1, a novel zinc finger/homeodomain repressor, interacts with Smad proteins and binds to 5'-CACCT sequences in candidate target genes. *J Biol Chem* 274: 20489-98

Viswanathan SR, Daley GQ. 2010. Lin28: A microRNA regulator with a macro role. *Cell* 140: 445-9

Watanabe M, Masuyama N, Fukuda M, Nishida E. 2000. Regulation of intracellular dynamics of Smad4 by its leucine-rich nuclear export signal. *EMBO Rep* 1: 176-82

Wrighton KH, Lin X, Feng XH. 2009. Phospho-control of TGF-beta superfamily signaling. *Cell Res* 19: 8-20

Wrighton KH, Willis D, Long J, Liu F, Lin X, Feng XH. 2006. Small C-terminal domain phosphatases dephosphorylate the regulatory linker regions of Smad2 and Smad3 to enhance transforming growth factor-beta signaling. *J Biol Chem* 281: 38365-75

Wu G, Chen YG, Ozdamar B, Gyuricza CA, Chong PA, et al. 2000. Structural basis of Smad2 recognition by the Smad anchor for receptor activation. *Science* 287: 92-7

Wurthner JU, Frank DB, Felici A, Green HM, Cao Z, et al. 2001. Transforming growth factor-beta receptor-associated protein 1 is a Smad4 chaperone. *J Biol Chem* 276: 19495-502

Xu L, Alarcon C, Col S, Massague J. 2003. Distinct domain utilization by Smad3 and Smad4 for nucleoporin interaction and nuclear import. *J Biol Chem* 278: 42569-77

Xu L, Chen YG, Massague J. 2000. The nuclear import function of Smad2 is masked by SARA and unmasked by TGFbeta-dependent phosphorylation. *Nat Cell Biol* 2: 559-62

Xu L, Kang Y, Col S, Massague J. 2002. Smad2 nucleocytoplasmic shuttling by nucleoporins CAN/Nup214 and Nup153 feeds TGFbeta signaling complexes in the cytoplasm and nucleus. *Mol Cell* 10: 271-82

Xu RH, Sampsell-Barron TL, Gu F, Root S, Peck RM, et al. 2008. NANOG is a direct target of TGFbeta/activin-mediated SMAD signaling in human ESCs. *Cell Stem Cell* 3: 196-206

Yang L, Wang N, Tang Y, Cao X, Wan M. 2006. Acute myelogenous leukemia-derived SMAD4 mutations target the protein to ubiquitin-proteasome degradation. *Hum Mutat* 27: 897-905

Yang X, Castilla LH, Xu X, Li C, Gotay J, et al. 1999a. Angiogenesis defects and mesenchymal apoptosis in mice lacking SMAD5. *Development* 126: 1571-80

Yang X, Letterio JJ, Lechleider RJ, Chen L, Hayman R, et al. 1999b. Targeted disruption of SMAD3 results in impaired mucosal immunity and diminished T cell responsiveness to TGF-beta. *EMBO J* 18: 1280-91

Yang X, Li C, Xu X, Deng C. 1998. The tumor suppressor SMAD4/DPC4 is essential for epiblast proliferation and mesoderm induction in mice. *Proc Natl Acad Sci U S A* 95: 3667-72

Ying QL, Nichols J, Chambers I, Smith A. 2003. BMP induction of Id proteins suppresses differentiation and sustains embryonic stem cell self-renewal in collaboration with STAT3. *Cell* 115: 281-92

Ying QL, Wray J, Nichols J, Batlle-Morera L, Doble B, et al. 2008. The ground state of embryonic stem cell self-renewal. *Nature* 453: 519-23

Zhu Y, Richardson JA, Parada LF, Graff JM. 1998. Smad3 mutant mice develop metastatic colorectal cancer. *Cell* 94: 703-14

12

Determination and Regulation of 'Stemness' by MicroRNAs

Kye-Seong Kim[1] and Jaesang Kim[2]
[1]*Department of Anatomy and Cell Biology, Hanyang University, College of Medicine, Seoul 133-791,*
[2]*Division of Life and Pharmaceutical Sciences, Ewha Womans University, Seoul 120-750, Republic of Korea*

1. Introduction

Embryonic stem cells (ESCs) are pluripotent cells derived from the inner cell mass of pre-implantation blastocysts, and by definition they are the cells that can self-renew indefinitely and differentiate into diverse cell types found in three embryonic germ layers (Thomson et al., 1998). In addition to allowing investigation of early events of embryonic development, the remarkable attributes of ESCs raise the possibility of using them as potential source of raw material for cell-based therapies (Mantel et al., 2007; Yu and Thomson, 2008). Molecular mechanisms of self-renewal and differentiation have been intensively studied during last two decades, and significant progresses have been achieved with regards to several core issues. Transcriptional regulation is one area in which key findings have been made culminating in generation of induced pluripotent stem cells (iPS) (Lee et al., 2006). Still, it goes without saying that much remains unknown about the proliferation and differentiation essential nature of 'stemness' at the molecular level.

Recent years have witnessed the establishment of microRNAs (miRNAs) as important regulators of development, homeostasis and metabolism of plants and animals (Bartel, 2004). These noncoding small RNAs have been discovered in virtually all eukaryotic organisms, from the brown algae (Cock et al., 2010) to the human, and their list is growing rapidly, in excess of 5000 having been identified so far. Over 1,000 miRNAs have been experimentally validated in mammals, and evolutionary conservation seen amongst miRNAs and across species strongly indicates functional and mechanistic significance of these small molecules (Bentwich et al., 2005). The best known post-transcriptional role of miRNAs is inhibition of translation which occurs via annealing to their target sequence in 3'UTR often accompanied by degradation of target mRNAs. It has been proposed that up to 60% of human genes are regulated by miRNAs which implies that each miRNA can target several different mRNAs as confirmed by both computational and experimental studies (Friedman et al., 2009). The expression pattern of the miRNAs frequently shows a consistent and significant correlation with the phenotypic transitions of the cells, such as proliferation, differentiation and apoptosis. Accordingly, several recent studies have demonstrated critical involvement of miRNAs in early developmental processes such as neurogenesis (Makeyev et al., 2007), myogenesis (Rao et al., 2006), cardiogenesis (Zaho et al., 2005) and hematopoiesis (Chen et al., 2004). Mammalian cells thus appear to produce miRNAs in a cell-type-specific manner to regulate unique subsets of genes specifically expressed in the host cells.

Of importance is that the cell-type specificity rule of miRNA expression and function applies to ESCs as well. Several recent studies reported expression of unique clusters of miRNAs in ESCs. These include miRNAs of human and mouse miR-302 clusters, mouse miR-290 cluster and human miR-371 cluster (Houbaviy et al., 2003; Suh et al., 2004; Landgraf et al., 2008). Based on the ESC-specific expression of these miRNAs, it was hypothesized that some of the miRNAs from these clusters may function as determinants or regulators of 'stemness', a word embodying the salient characteristics of ESCs: self-renewal and pluripotency (Blakaj and Lin, 2008; Kim et al., 2009). As the latest reports on regulation and function indicate that the hypothesis is likely to be more than mere speculation (Marson et al., 2008; Lin et al., 2008; Lee et al., 2008; Card et al., 2008; Barroso-Deljesus et al., 2008; Lee et al., 2010), it has been inevitably envisioned that miRNAs may be useful tools for controlling the proliferation and the differentiation of ESCs. In addition, other lines of investigation showed that ES-like stem cells including iPS and multipotent spermatogonial stem cells (mSSCs) as well as embryonal carcinoma cells (ECCs) share to a varying extent the expression profile of miRNAs. This in turn seems to suggest that expression profiling of miRNAs can be a way of 'measuring' the 'stemness' of a given type of cells. In this chapter, we will introduce the recent discovery of miRNAs in ESC and ES-like stem cells, describe molecular and functional characterization of identified miRNAs and provide perspectives on issues and the research agenda of this flourishing field.

2. Identification of the ESC specific miRNAs

Houbaviy and coworkers reported in 2003 the isolation of 15 novel miRNAs from a mouse ESC cDNA library which included members of miR-290 cluster and predicted the existence of human homologues (Houbaviy et al., 2003). Using a similar approach, Suh et al. (2004) reported the cloning of 17 novel miRNAs belonging to miR-302 and miR-371 clusters from human ESCs. Remarkably, the miRNAs identified from the three clusters showed varying degrees of conservation between the two species which immediately led to a speculation about significant and specific functions in ESCs. In addition, murine miR-367 was isolated through a computational approach by Sewer et al. (2005), and Mineno et al. (2006) identified mouse miR-302b, miR-302d and miR-367 using a massive parallel signature sequencing technique. More recently, Landgraf et al. (2008) cloned and validated additional novel miRNAs from human and mouse ESCs (human miR-302d* and miR-367*; mouse miR-302b*, miR-302b, miR-302c*, miR-302c, miR-302a* and miR-302d; * denotes the minor miRNA species in case two excised miRNAs are generated, one from each arm) by a large-scale cloning. Altogether, the miRNA registry (miRBase 15; http://www.mirbase.org/) currently lists 10 miRNAs in the human miR-302 cluster and 8 miRNAs in the mouse miR-302 cluster (Table 1). ESC-specific expression of some of these miRNAs (miR-371, miR-372, miR-373 and miR-373*) has been confirmed by Northern blotting (Suh et al. 2004). Subsequently, Landgraf et al. (2008) reported identification of another member of the miR-371 cluster, miR-371-5p and corrected the sequence of miR-371 yielding miR-371-3p (Table 2). Currently, there are five miRNAs in the miR-371 cluster.

Of the 15 novel miRNAs reported by Houbaviy et al. (2003), 8 miRNAs belong to the miR-290 cluster (miR-290, miR-291-s, miR-291-as, miR-292-s, miR-292-as, miR-293, miR-294 and miR-295). Later, Sewer et al. (2005) isolated two additional miRNAs (miR-291b-5p and miR-291b-3p) in the same cluster, through a bioinformatics analysis. More recently, 4 more miRNAs (miR-290-3p, miR-293*, miR-294*, and miR-295*) have been added to the miR-290 cluster by direct cloning from mouse ESCs and ECCs (F19) (Landgraf et al., 2008). Currently, miRNA registry (miRBase 15) lists 14 miRNAs in the mouse miR-290 cluster (Table 3).

ID	Accession[a]		Sequences[a]	Genome Context	Expression[b]
	Stem-loop	Mature			
hsa-miR-302b	MI0000772	hsa-miR-302b* (MIMAT0000714)	acuuuaacauggaagugcuuuc	4: 113789090-113789162 [-] LARP7; intron 8	hESC[1,2], NT2/D1[1,2], PA1[2], NCCIT[2]
		hsa-miR-302b (MIMAT0000715)	uaagugcuuccauguuuuaguag		
hsa-miR-302c	MI0000773	hsa-miR-302c* (MIMAT0000716)	uuuaacaugggggguaccugcug	4: 113788968-113789035 [-] LARP7; intron 8	hESC[1,2], NT2/D1[1,2], PA1[2], NCCIT[2]
		hsa-miR-302c (MIMAT0000717)	uaagugcuuccauguuucagugg		
hsa-miR-302a	MI0000738	hsa-miR-302a* (MIMAT0000683)	acuuaaacguggauguacuugcu	4: 113788788-113788856 [-] LARP7; intron 8	hESC[1,2], NT2/D1[1,2], PA1[2], NCCIT[2]
		hsa-miR-302a (MIMAT0000684)	uaagugcuuccauguuuuggguga		
hsa-miR-302d	MI0000774	hsa-miR-302d* (MIMAT0004685)	acuuuaacauggaggcacuugc	4: 113788609-113788676 [-] LARP7; intron 8	hESC[1,2], NT2/D1[1,2], PA1[2], NCCIT[2]
		hsa-miR-302d (MIMAT0000718)	uaagugcuuccauguuugagugu		
hsa-miR-367	MI0000775	hsa-miR-367* (MIMAT0004686)	acuguugcuaauaugcaacucu	4: 113788479-113788546 [-] LARP7; intron 8	hESC[1,2], NT2/D1[1,2], PA1[2], NCCIT[2]
		hsa-miR-367 (MIMAT0000719)	aauugcacuuuagcaaugguga		
mmu-miR-302b	MI0003716	mmu-miR-302b* (MIMAT0003373)	acuuuaacaugggaaugcuuucu	3: 127248146-127248219 [+] Larp7; intron8	mESC[2], teratocarcinoma (P19)[2]
		mmu-miR-302b (MIMAT0003374)	uaagugcuuccauguuuuaguag		
mmu-miR-302c	MI0003717	mmu-miR-302c* (MIMAT0003375)	gcuuuaacaugggguuaccugc	3: 127248281-127248348 [+] Larp7; intron8	mESC[2], teratocarcinoma (P19)[2]
		mmu-miR-302c (MIMAT0003376)	aagugcuuccauguuucagugg		
mmu-miR-302a	MI0000402	mmu-miR-302a* (MIMAT0004579)	acuuaaacgugguuguacuugc	3: 127248414-127248482 [+] Larp7; intron8	mESC[2], teratocarcinoma (P19)[2]
		mmu-miR-302a (MIMAT0000380)	uaagugcuuccauguuuuggguga		
mmu-miR-302d	MI0003718	mmu-miR-302d (MIMAT0003377)	uaagugcuuccauguuugagugu	3: 127248542-127248607 [+] Larp7; intron8	mESC[2,3], teratocarcinoma (P19)[2]
mmu-miR-367	MI0003531	mmu-miR-367 (MIMAT0003181)	aauugcacuuuagcaaugguga	3: 127248651-127248725 [+] Larp7; intron8	mESC[2], teratocarcinoma (P19)[2]

[a] Accession number and sequences from the miRBase 15 (http://www.mirbase.org/)
[b] References, 1; Suh MR et al., (2004), 2; Landgraf P et al., (2007), 3; Houbaviy HB et al., (2003), 4; Voorhoeve P. M et al., (2006)

Table 1. Current miRBase 15 registered mature miRNA sequences of miR-302 cluster. Accession numbers for precursors andmature miRNAs, current mature miRNA sequences, genomic locations, expressed cell lines and references are listed.

ID	Accession[a]		Sequences[a]	Genome Context	Expression[b]
	Stem-loop	Mature			
hsa-miR-371	MI0000779	hsa-miR-371-5p (MIMAT0004687)	acucaaacugugggggcacu	19: 58982741-58982807 [+] intergenic	hESC[1], 833KE1[4], Tera1[4]
		hsa-miR-371-3p (MIMAT0000723)	aagugccgccaucuuuugagugu		
hsa-miR-372	MI0000780	hsa-miR-372 (MIMAT0000724)	aaagugcugcgacauuugagcgu	19: 58982956-58983022 [+] intergenic	hESC[1], 2102EP[4], 833KE[4], Tera1[4]
hsa-mir-373	MI0000781	hsa-miR-373* (MIMAT0000725)	acucaaaaugggggcgcuuucc	19: 58983771-58983839 [+] intergenic	hESC[1], 2102EP[4], 833KE[4], Tera1[4]
		hsa-miR-373 (MIMAT0000726)	gaagugcuucgauuuuggggugu		

[a] Accession number and sequences from the miRBase 15 (http://www.mirbase.org/)
[b] References, 1; Suh MR et al., (2004), 2; Landgraf P et al., (2007), 3; Houbaviy HB et al., (2003), 4; Voorhoeve P. M et al., (2006)

Table 2. Current miRBase 15 registered mature miRNA sequences of miR-371 cluster. Accession numbers for precursors andmature miRNAs, current mature miRNA sequences, genomic locations, expressed cell lines and references are listed.

ID	Accession[a]		Sequences[a]	Genome Context	Expression[b]
	Stem-loop	Mature			
mmu-miR-290	MI0000388	mmu-miR-290-5p (MIMAT0000366)	acucaaacuaugggggcacuuu	7: 3218627-3218709 [+] intergenic	mESC[4], teratocarcinoma (F9, P19)[3]
		mmu-miR-290-3p (MIMAT0004572)	aaagugccgccuaguuuuaagccc		
mmu-miR-291a	MI0000389	mmu-miR-291a-5p (MIMAT0000367)	caucaaaguggaggcccucucu	7: 3218920-3219001 [+] intergenic	mESC[4], teratocarcinoma (F9, P19)[3]
		mmu-miR-291a-3p (MIMAT0000368)	aaagugcuuccacuuugugugc		
mmu-miR-292	MI0000390	mmu-miR-292-5p (MIMAT0000369)	acucaaacugggggcucuuuug	7: 3219190-3219271 [+] intergenic	mESC[4], teratocarcinoma (F9, P19)[3]
		mmu-miR-292-3p (MIMAT0000370)	aaagugccgccagguuuugagugu		
mmu-miR-291b	MI0003539	mmu-miR-291b-5p (MIMAT0003189)	gaucaaaguggaggcccucucc	7: 3219483-3219561 [+] intergenic	mESC[4], teratocarcinoma (F9, P19)[3]
		mmu-miR-291b-3p (MIMAT0003190)	aaagugcauccauuuuguuugu		
mmu-miR-293	MI0000391	mmu-miR-293* (MIMAT0004573)	acucaaacugugugacauuuug	7: 3220344-3220423 [+] intergenic	mESC[4], teratocarcinoma (F9, P19)[3]
		mmu-miR-293 (MIMAT0000371)	agugccgcagaguuuguagugu		
mmu-miR-294	MI0000392	mmu-miR-294* (MIMAT0004574)	acucaaaauggaggcccuaucu	7: 3220642-3220725 [+] intergenic	mESC[4], teratocarcinoma (F9, P19)[3]
		mmu-miR-294 (MIMAT0000372)	aaagugcuucccuuuugugugu		
mmu-miR-295	MI0000393	mmu-miR-295* (MIMAT0004575)	acucaaauguggggcacacuuc	7: 3220774-3220842 [+] intergenic	mESC[4], teratocarcinoma (F9, P19)[3]
		mmu-miR-295 (MIMAT0000373)	aaagugcuacuacuuuugagucu		

[a] Accession number and sequences from the miRBase 15 (http://www.mirbase.org/)
[b] References, 1; Suh MR et al., (2004), 2; Landgraf P et al., (2007), 3; Houbaviy HB et al., (2003), 4; Voorhoeve P. M et al., (2006)

Table 3. Current miRBase 15 registered mature miRNA sequences of miR-290 cluster. Accession numbers for precursors andmature miRNAs, current mature miRNA sequences, genomic locations, expressed cell lines and references are listed.

3. Molecular characterization of the ESC specific miRNAs

3.1 miR-302 cluster members

Since the discovery in mouse and human, the existence of a conserved miR-302 cluster has been proposed in multiple species. For *Gallus gallus* (International Chicken Genome Sequencing Consortium, 2004), *Monodelphis domestica* (Devor & Samollow, 2008) , *Macaca mulatta* (Yue et al., 2008), *Pan troglodytes* (Baev et al., 2009), *Bos Taurus* (Artzi et al., 2008), *Canis familiaris* (Artzi et al., 2008), *Pongo pygmaeus* (Brameier, 2010), *Equus caballus* (Zhou et al., 2009), and *Ornithorhynchus anatinus* (Murchison et al., 2008), the miR-302 cluster has been identified by genomic sequence analysis, whereas in *Xenopus tropicalis,* the expression in early blastula has been experimentally validated (Watanabe et al., 2005).

Most of the detailed studies have been carried out with the members of human and mouse miR-302 clusters which show remarkably high degree of conservation. The precursor forms of orthologous miRNAs show the homology rates from 85% to 93%, while except for miR-302c and miR-367 all orthologous pairs show a perfect match in the mature form. Individual miRNAs also show strong homology to one another, consistent with the common origin, with the membership of miR-367 being somewhat controversial (Kim et al., 2009). According to miRBase 15, miR-367 belongs to a separate family, even though it is generated from the same polycistronic primary transcript as the rest of miRNAs of miR-302 cluster (Suh et al., 2004). Based on the weak homology, it is proposed that miR-302 family members and miR-367 target different sets of mRNAs and consequently perform different functions.

Human miR-302 cluster locus is located within a 700-bp region on chromosome 4, and the corresponding mouse cluster is found within a region of about 600 bp on chromosome 3. Interestingly, both human and mouse miR-302 loci are found within intron 8 of the La ribonucleoprotein domain family, member 7 (LARP7) gene, which is known to act through the 7SK ribonucleoprotein system to negatively regulate polymerase II-mediated transcription (Markert et al., 2008). It is not currently determined whether miRNAs of miR-302 cluster and LARP7 have regulatory interaction or functional coordination.

3.2 miR-290 and miR-371 cluster members

miRNAs of mouse miR-290 cluster appear to be expressed in ESC-specific manner, conserved across mammalian species and clustered within 2.2 kb of one another. Similarity of sequence to members of the mouse miR-290 cluster led to the prediction of the existence of miRNAs of human miR-371 cluster which was demonstrated experimentally. Human miR-371 cluster is located in an intergenic region, spanning approximately 1050 bp, on chromosome 19. The miR-371 cluster has also been identified in *Macaca mulatta* (Yue et al., 2008), *Pan troglodytes* (Baev et al., 2009), *Bos Taurus* (Artzi et al., 2008), *Canis familiaris* (Artzi et al., 2008), *Pongo pygmaeus* (Brameier, 2010), and *Equus caballus* (Zhou et al., 2009) by genomic sequence analysis while the homologous miR-290 cluster has been identified in *Rattus norvegicus* (Linsen et al., 2010).

Although miRNAs belonging to miR-371 cluster and miR-290 cluster are all classified as miR-290 family members at miRBase, homology rates among precursors and mature forms are relatively low compared to intra- and inter-species homology seen in miR-302 cluster members (Kim et al., 2009). Nevertheless, based on that seed region are much better conserved, it was hypothesized that at least some regulatory targets are shared. This awaits further examination as functional analyses miR-371 and miR-290 clusters reported thus far do not fully support this notion

3.3 Expression of miRNAs in ES-like pluripotent stem cells

An important insight into the significance of ESC-specific expression of the miRNAs in fact came from their expression in other pluripotent cell types. Specifically, expression of the miR-302 cluster has been found in ECCs derived from testicular and ovarian teratocarcinoma tissues (PA1) (Zeuthen et al., 1980), leading to the generalization that the miR-302 cluster is specific for cells capable of forming embryoid bodies (EBs) and differentiating into diverse cell types.

Somatic cells have been successfully reprogrammed by fusion with ES cells, and it has thus been suggested that some factors in ES cells are capable of reprogramming somatic cells into pluripotent stem cells (Cowan et al., 2005; Yu et al., 2006). This was dramatically demonstrated by Takahashi and Yamanaka in 2006 when they identified Oct-3/4, Sox2, c-Myc, and Klf4 as the pluripotency-inducing factors by introducing these genes into mouse embryonic or adult fibroblasts and generating iPS. Studies on miRNA expression pattern in various iPS and multipotent spermatogonial stem cells (mSSCs) are beginning to be reported. For example, Chin et al. (2009) examined if miRNAs are expressed in human iPS at the level and in a pattern as in hESCs. Expression profiling of all known miRNAs was performed on hESCs, iPS, and the fibroblasts from which iPS had been derived. Hierarchical clustering with the 105 miRNAs expressed showed that there is little difference in miRNA expression pattern between the pluripotent cells, i.e. iPS and hESCs. Conversely, all of the pluripotent cell lines had vastly different miRNA expression profiles from those of fibroblasts. These observations are overall highly suggestive of that a strong correlation exists between essential features of stem cells and the miRNA expression profiles. Still, the situation is likely to be much more complex. Specifically, it was noticed that a few miRNAs were consistently expressed differently between iPS and hESCs. This finding was also confirmed by another group using a different set of hESCs and iPS (Wilson et al., 2009), indicating that a distinct miRNA pattern is essentially reproducible among different reprogrammed cells and that iPS have a miRNA signature that distinguishes them from hESCs. Study by Wilson et al. (2009) in particular highlights the differences between the two pluripotent cell types with respect to the expression of the miR-371/372/373 cluster.

3.4 miRNA expression profile as an indicator of 'stemness'

Stadler et al. (2010) analyzed changes in the expression of miRNAs and mRNAs in nine different hESC lines during early commitment and also examined the expression of key ESC-enriched miRNAs in early developmental stages in several species. They demonstrated that several previously defined hESC-specific miRNA group members (belonging to the miR-302, -17, and - 515 families and the miR-371-373 cluster) and several other hESC-enriched miRNAs are down-regulated rapidly in response to differentiation. They further found that mRNAs upregulated upon differentiation are enriched in terms of harboring potential target sites for the down-regulated miRNAs. Interestingly, they also observed that the expression of ESC-enriched miRNAs bearing common seed sequences was modulated in a discordant manner while the cells transitioned through early embryonic states. In human and monkey ESCs, as well as human iPS, the miR-371-373 cluster was consistently upregulated while the miR-302 family was mildly downregulated when the cells were chemically treated to regress to an earlier developmental state. Similarly, miR-302b, but not miR-295, was expressed at higher levels in murine epiblast stem cells (mEpiSC) as compared to mouse ESCs which represents an earlier developmental state. These data for the first time raise the possibility that miRNAs bearing identical seed sequences could have distinct functions from one another and during sub-stages of early embryonic development. Perhaps more importantly, these results suggest that the relative expression levels of related miRNAs may serve as indicators for defining the developmental state of ESCs and other stem cell lines, such as iPS.

Another series of reprogramming experiments are also worthy of close attention. The conversion of spermatogonial stem cells (SSCs) from neonatal hybrid (Kanatsu-Shinohara et al., 2004), neonatal inbred (Kim et al., 2010), adult inbred (Oh et al., 2009), and adult transgenic mice (Guan et al., 2006; Izadyar et al., 2008; Seandel et al., 2007) into ES-like cells (multipotent spermatogonial stem cells, mSSCs) has been reported. Interestingly, SSCs and mSSCs exhibit significant differences in gene expression and epigenetic properties even in the cases of identical genetic origin (Kanatsu-Shinohara et al., 2008). In particular, mSSCs have lost the expression of germ stem cell-specific genes and showed enhanced expression of pluripotent stem cell-specific genes. Importantly, Zovoislis et al. (2010) reported that miRNAs of miR-290 and 302 clusters are expressed in mSSCs and that the 290-miRNA family is functionally connected with Oct-4 in maintenance of the pluripotent state. It should be noted however that detection of miR-302 family members likely results from the presence of a proportion of spontaneously differentiating cells as members of the 302-miRNA family have been shown to be induced during first stages of in vitro differentiation in all pluripotent cell types tested. In sum, in addition to histone modification, promoter methylation, and transcription factor expression, cell type specific expression and dynamic modulation of miRNAs reflect a layer of regulatory control in ESCs and at the same time represent signature of 'stemness'.

3.5 Transcriptional regulation of expression by ESC-specific transcription factors

ESC-specific expression of the miRNAs of miR-302, miR-290, and miR-371 clusters has gained an added significance from a series of recent studies (Marson et al., 2008; Card et al., 2008; Barroso-Deljesus et al., 2008). Strikingly, transcription factors that have been shown to be critical determinants of 'stemness' of ESC were also demonstrated to be involved in the regulation of these miRNA cluster. Specifically, in the study by Marson et al. (2008), potential binding sites for Nanog, Oct3/4, Sox2 and Tcf3 have been shown to be present in the 5′ region of the miR-302 clusters. Furthermore, down-regulation of Oct3/4 led to reduced expression of the miRNAs of this cluster. Card et al. (2008) also showed that Oct3/4, Sox2 and Nanog bind to the chromosomal region 5′ to the miR-302 cluster and that Oct3/4 is required for the expression of the miR-302 transcript. Rex1, another ESC-specific transcription factor, was also predicted to bind to the same region (Barroso-Deljesus et al., 2008). Similar transcriptional regulation seems to be in operation for miR-290 and miR-371 clusters. Marson et al. (2008) showed that the two clusters have potential binding sites for Oct4, Sox2, Nanog and Tcf3 and that inhibition of Oct3/4 leads to down-regulation of miR-290 expression. Together, these studies strongly suggest that miRNAs expressed specifically in ESCs are in fact integral elements of the regulatory network, on one hand controlled by the ESC-specific transcriptional program and on the other likely controlling characteristics of ESC at the molecular level.

4. Role of the ESC-specific miRNAs

4.1 Studying the function of miRNAs in ESC

The current understanding of the molecular function of ESC-specific miRNAs is, simply but, limited. This in part stems from the technical demand in handling ESCs, especially those of human origin. Ectopically expressing cDNAs, siRNAs, or miRNAs in human ESCs is difficult as these cells typically grow as aggregates. Delivery of foreign DNA or RNA can be much more easily achieved with mouse ESCs which are initially seeded as individual cells that can be grown into clones. How miRNAs determine or affect the self-renewal and pluripotency, the essential characteristics of stem cells, can be tested using mouse ESCs, and much can be inferred about the function of orthologous human miRNAs. It goes without saying that species difference must be kept in mind even in the case of unambiguous orthologues. There are additional difficulties even with mouse ESCs. For example miRNAs originating from the same cluster likely have overlapping target mRNAs which would make determining the function of individual miRNAs difficult. At any rate, an efficient delivery system and inducible vector constructs for small RNA molecules would be clearly useful. Recently developed advanced tools such as miRNA- or small interfering RNA (siRNA)-expressing lentiviral or retroviral systems with pseudotyping feature for a broad host cell range should be valuable tools (Chang et al., 2006). Additional features such as inducibility and cell type specificity can be incorporated in principle assisting precise functional analysis of miRNAs in murine and human ESCs (Chang et al., 2006). Identification of potential targets initially with bioinformatics approach and subsequently with 'wet-lab' validation also represents an important part of functional analyses. There are challenges here as well. Target prediction algorithms must be continually refined based on results of validation experiments, but exhaustive and systematic efforts are rather limited. Also, these prediction programs should ideally be linked with various gene-expression pattern databases, but again at this point a user-friendly in silico analysis program with such features does not exist.

4.2 Function of miR-302 cluster members

Functional conservation of the miR-302 cluster in ESCs of various species can be readily hypothesized based on the high degree of homology, conservation of genomic loci and cell type specific expression, as well as the regulation by ESC-specific transcription factors. In spite of the aforementioned difficulties, direct functional tests in ESC of miR-302 cluster are beginning to be reported. One notable study by Card et al., (2008) showed that ectopic expression of individual miR-302 members in both primary and transformed cell lines leads to increase in the proportion of cells in the S phase and decrease in the proportion of cells in the G1 phase. Consistently, bare expression led to the opposite phenotype. Cyclin D1, a G1 regulator, was predicted as a target of multiple members of miR-302 cluster, and miR-302 miRNA expression indeed led to inhibition of Cyclin D1 translation. These data indicate that one of the primary functions of miR-302 in ESC is cell cycle regulation. Subsequently, Lee et al., (2008) found that miR-302b indirectly regulates Oct4 and directly targets Cyclin D2, developmental regulators during gastrulation. Such observation suggests that at least miR-302b participates in maintaining the pluripotency of ECCs and likely of ESCs. A noteworthy study on the function and potential application of miR-302 has been reported by Lin et al. in 2008. They showed that expression of miR-302 induced reprogramming of human skin cancer cells into ESC-like cells (Lin et al., 2008). These miRNA-induced pluripotent stem cells shared 86% of their expressed genes with ESCs including the hallmark ESC marker genes such as Oct3/4, Sox2, SSEA-2 and SSEA-4. Furthermore, the miRNA-induced pluripotent cells were able to differentiate into multiple types of cells including neurons, chondrocytes, fibroblasts and spermatogonial cells in vitro. More recently, Scheel et al. ran a functional screen using a large miRNA expression library and reported identification of the miR-302 cluster as a potent suppressor of p63 accumulation in various cell types. It was shown that miR-302 miRNAs reduce mRNA and proteins levels of p63 through two target sites within the 3' UTR. The role of miR-302 members was also confirmed in testicular cancer cells in which the endogenous miR-302 contributes to the suppression of p63. It was also proposed that miRNAs of the miR-302 cluster also contribute to the elimination of p63 mRNA in mature oocytes. Thus, miR-302 appears to be a part of the stringent regulatory mechanism for p63 in germ cells, reminiscent of the tight control for p53 levels in somatic cells.

4.3 Function of miR-290 and miR-371 clusters

The function of miRNAs of the miR-290 cluster has been studied in diverse contexts. First, a definitive loss-of-function analysis was carried out with gene-targeted mouse. Homozygous loss of miR-290–295 locus, which resulted in frequent embryonic lethality, led to infertility among female survivors. Notably they lacked germ cells, indicating that at least some of the members may be responsible for the maintenance of pluripotency (Giraldez et al., 2006; Marson et al., 2008). Benetti et al. (2008) showed that miRNAs of the miR-290 cluster silence Rbl2 which functions as a transcriptional repressor of the Dnmt3a and Dnmt3b enzymes and that these enzymes as well as miRNAs are down-regulated in Dicer1 -/- cells. In an independent study, Sinkkonen et al., (2008) demonstrated that de novo DNA methylation is defective in Dicer1-deficient ESCs consistent with indirect control of the expression of DNA methyltransferases by the miRNAs of miR-290 cluster. The mechanistic link between members of miR-290 cluster and de novo DNA methylation in ESCs indicates that these miRNAs are involved in the epigenetic control of gene expression. In addition, Hayashi et al., (2008) reported the requirement of miRNA biogenesis in primordial germ cell development and early spermatogenesis. miR-290 cluster members were among the most highly expressed in male germline cells suggesting a potential role in this particular developmental program. Wang et al. (2008) reported that members of the miR-290 family rescue cellular proliferation defect seen in *Dgcr8-/-* ESCs. Apparently, these miRNAs function by suppressing several key regulators of the G1/S transition which leads to release of cells from the arrest in G1 phase. Finally, Judson et al. (2009) demonstrated that introduction of miR-291-3p, miR-294 and miR-295 into fibroblasts enhances the production of mouse iPS cells by Oct4, Sox2 and Klf4.

Several recent reports describe the function of the miRNAs of miR-371 cluster. Most notably, Voorhoeve et al., (2006) demonstrated that miR-372/373 miRNAs promote cellular transformation in cooperation with oncogenes. It was also shown that the expression of the miR-372/373 cluster was seen in subsets of ECC lines such as Tera1, 2102Ep and 833KE but not in NT2 and NCCIT lines. Voorhoeve et al. noted a correlation between the expression of the miR-372/373 cluster and p53 status in these ECC lines. Specifically, whereas all three expresser cell lines contain high levels of wt-p53, NT2 expresses wt-p53 at a low-level and NCCIT has only single mutated allele and no wild type allele. Based on these findings, they proposed that miRNAs of the miR-372/373 cluster contribute to tumorigenesis from cells that retain wt-p53. Interestingly, Duale et al. (2007) showed that cisplatin represses the oncogenic properties of the miR-372/373 cluster, and Huang et al. (2008) demonstrated that miR-373 and miR-520 stimulated migration and invasion of cancer cells in vitro and in vivo. Related reports on the function of miR-372/373 members are also noteworthy. It was reported that E-cadherin and CSDC2 contain potential target sites highly complementary to miR-373 within their promoters. Transfection of miR-373 readily induced the expression of the two genes, and induction was specifically dependent on the presence of both miR-373 and the proposed target promoter sites (Place et al., 2008).

5. Conclusion

Recently reported lines of evidence, some of which have been described thus far, all point to that miRNAs belonging to the three clusters, miR-302, miR-209, and miR-371, represent major regulators of pluripotent stem cells. They most likely participate in controlling all key aspects of 'stemness' including pluripotency, proliferation, and differentiation (Fig. 1). The latest series of reports indicate that ESC-specific miRNAs are regulated by ESC-specific transcription factors and that these miRNAs in turn control ESC-specific transcription factors. This fact alone should be sufficient to conclude that ESC-specific miRNAs are an integral part of the regulatory network of ESCs. More importantly, functional analyses based on the expression and inhibition of miRNAs in ESCs have begun to be reported, and preliminary data are consistent with that miRNAs is one of the determinants of the 'stemness'. As miRNAs are established as key regulators of numerous cellular events, ways of utilizing them as molecular tools are also being envisioned. This certainly also applies to ESCs: if miRNAs are regulators of 'stemness', we should be able to use them to control proliferation and differentiation of ESCs. Another use for miRNAs may be in reprogramming primary somatic cells into ESC-like cells with self-renewal and pluripotency. It is possible that miRNAs, organized in a single polycistronic transcript, may be a more convenient or efficient tool than a series of transcription factors. Furthermore, direct delivery of RNA into somatic cells would represent a safer alternative than transduction of retroviral vectors. It goes without saying that through the target analysis of miRNAs, we should also be able to gain insight into the gene regulatory networks and signaling pathways involved in particular characteristics or behaviors of ESCs, which is the fundamental prerequisite for applying ESCs for the regenerative therapy.

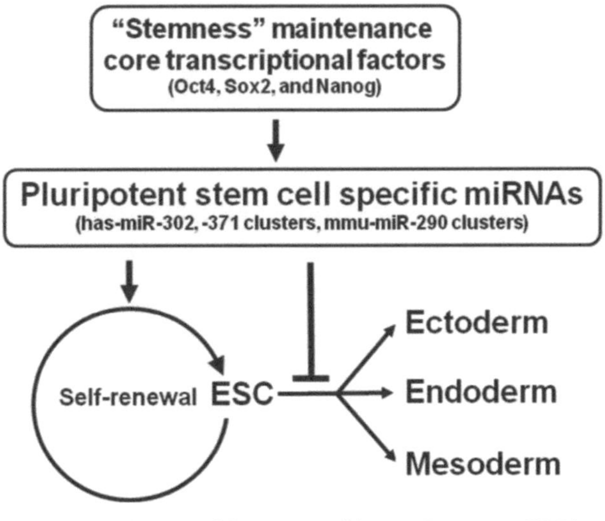

Fig. 1. Determination and regulation of "stemness" by miRNAs. miRNAs are regulated by core transcription factors of ESCs and in turn promote self-renewal and inhibit differentiation of ESCs .

6. Acknowledgements

We are grateful to Hye In Moon for editing the manuscript. This work was supported by a grant (SC-2220) from Stem Cell Research Center funded by Ministry of Education, Science and Technology, Republic of Korea to KSK.

7. References

Ambros V & Chen X. (2007). The regulation of genes and genomes by small RNAs. *Development,* 134, 1635-1641.

Artzi S, Kiezun A & Shomron N. (2008). miRNAminer: a tool for homologous microRNA gene search. *BMC Bioinformatics,* 23,9-39.

Baev V, Daskalova E & Minkov I. (2009). Computational identification of novel microRNA homologs in the chimpanzee genome. *Comput Biol Chem,* 33(1), 62-70.

Barroso-Deljesus A, Romero-Lopez C, Lucena-Aguilar G, Melen GJ, Sanchez L, Ligero G, Berzal-Herranz A & Menendez P. (2008). Embryonic stem cell specific miR302-367 cluster: Human Gene Structure and Functional Characterization of its Core Promoter. *Mol Cell Biol,* 28, 6609-6619.

Bartel DP. (2004). MicroRNAs: genomics, biogenesis, mechanism, and function. *Cell,* 116, 281-297.

Benetti R, Gonzalo S, Jaco I, Munoz P, Gonzalez S, Schoeftner S, Murchison E, Andl T, Chen T, Klatt P, Li E, Serrano M, Millar S, Hannon G & Blasco MA. (2008). A mammalian microRNA cluster controls DNA methylation and telomere recombination via Rbl2-dependent regulation of DNA methyltransferases. *Nat Struct Mol Biol,* 15, 268-279.

Bentwich, I., Avniel, A., Karov, Y., Aharonov, R., Gilad, S., Barad, O., Barzilai, A., Einat, P., Einav, U. & Meiri, E. (2005). Identification of hundreds of conserved and nonconserved human microRNAs. *Nat Genet,* 37, 766-770.

Blakaj A & Lin H. (2008). Piecing together the mosaic of early mammalian development through microRNAs. *J Biol Chem,* 283, 9505-9508.

Brameier M. (2010). Genome-wide comparative analysis of microRNAs in three non-human primates. *BMC Res Notes,* 9, 3-64.

Card DA, Hebbar PB, Li L, Trotter KW, Komatsu Y, Mishina Y & Archer TK. (2008). Oct4/Sox2-regulated miR-302 targets cyclin D1 in human embryonic stem cells. *Mol Cell Biol,* 28, 6426-6438.

Chang K, Elledge SJ & Hannon GJ. (2006). Lessons from Nature: microRNA-based shRNA libraries. *Nat Methods,* 3, 707-714.

Chen CZ, Li L, Lodish HF & Bartel DP. (2004). MicroRNAs modulate hematopoietic lineage differentiation. *Science,* 303, 83-86.

Chin MH, Mason MJ, Xie W, Volinia S, Singer M, Peterson C, Ambartsumyan G, Aimiuwu O, Richter L, Zhang J, Khvorostov I, Ott V, Grunstein M, Lavon N, Benvenisty N, Croce CM, Clark AT, Baxter T, Pyle AD, Teitell MA, Pelegrini M, Plath K & Lowry WE. (2009). Induced Pluripotent Stem Cells and Embryonic Stem Cells Are Distinguishedby Gene Expression Signatures. *Cell Stem Cell,* 5, 111–123.

Cock JM, Sterck L, Rouzé P, Scornet D, Allen AE, Amoutzias G, Anthouard V, Artiguenave F, Aury JM, Badger JH, Beszteri B, Billiau K, Bonnet E, Bothwell JH, Bowler C, Boyen C, Brownlee C, Carrano CJ, Charrier B, Cho GY, Coelho SM, Collén J, Corre E, Da Silva C, Delage L, Delaroque N, Dittami SM, Doulbeau S, Elias M, Farnham G, Gachon CM, Gschloessl B, Heesch S, Jabbari K, Jubin C, Kawai H, Kimura K, Kloareg B, Küpper FC, Lang D, Le Bail A, Leblanc C, Lerouge P, Lohr M, Lopez PJ,

Martens C, Maumus F, Michel G, Miranda-Saavedra D, Morales J, Moreau H, Motomura T, Nagasato C, Napoli CA, Nelson DR, Nyvall-Collén P, Peters AF, Pommier C, Potin P, Poulain J, Quesneville H, Read B, Rensing SA, Ritter A, Rousvoal S, Samanta M, Samson G, Schroeder DC, Ségurens B, Strittmatter M, Tonon T, Tregear JW, Valentin K, von Dassow P, Yamagishi T, Van de Peer Y, & Wincker P. (2010). The Ectocarpus genome and the independent evolution of multicellularity in brown algae. *Nature*, 465 (7298), 617–621.

Cowan CA, Atienza J, Melton DA & Eggan K. (2005). Nuclear reprogramming of somatic cells after fusion with human embryonic stem cells. *Science*, 309, 1369-1373.

Devor EJ & Samollow PB. (2008). In vitro and in silico annotation of conserved and nonconserved microRNAs in the genome of the marsupial Monodelphis domestica. *J Hered*, 99, 66-72.

Duale N, Lindeman B, Komada M, Olsen AK, Andreassen A, Soderlund EJ & Brunborg G. (2007). Molecular portrait of cisplatin induced response in human testis cancer cell lines based on gene expression profiles. *Mol Cancer*, 6, 53.

Friedländer MR, Chen W, Adamidi C, Maaskola J, Einspanier R, Knespel S & Rajewsky N. (2008). Discovering microRNAs from deep sequencing data using miRDeep. *Nat Biotechnol*, 26(4), 407-15.

Friedman, R. C., Farh, K. K., Burge, C. B & Bartel, D. P. (2009). Most mammalian mRNAs are conserved targets of microRNAs. *Genome Res*, 19, 92-105.

Giraldez AJ, Cinalli RM, Glasner ME, Enright AJ, Thomson JM, Baskerville S, Hammond SM, Bartel DP & Schier AF. (2005). MicroRNAs regulate brain morphogenesis in zebrafish. *Science*, 308, 833-838.

Giraldez AJ, Mishima Y, Rihel J, Grocock RJ, Van Dongen S, Inoue K, Enright AJ & Schier AF. (2006). Zebrafish MiR-430 promotes deadenylation and clearance of maternal mRNAs. *Science*, 312, 75-79.

Guan K, Nayernia K, Maier LS, Wagner S, Dressel R, Lee JH, Nolte J, Wolf F, Li M, Engel W & Hasenfuss G. (2006). Pluripotency of spermatogonial stem cells from adult mouse testis. *Nature*, 440, 1199-1203.

Hayashi K, Chuva de Sousa Lopes SM, Kaneda M, Tang F, Hajkova P, Lao K, O'Carroll D, Das PP, Tarakhovsky A, Miska EA & Surani MA. (2008). MicroRNA biogenesis is required for mouse primordial germ cell development and spermatogenesis. *PLoS ONE*, 3, e1738.

Houbaviy HB, Murray MF & Sharp PA. (2003). Embryonic stem cell-specific MicroRNAs. *Dev Cell*, 5, 351-358.

Huang Q, Gumireddy K, Schrier M, le Sage C, Nagel R, Nair S, Egan DA, Li A, Huang G, Klein-Szanto AJ, Gimotty PA, Katsaros D, Coukos G, Zhang L, Pure E & Agami R. (2008). The microRNAs miR-373 and miR-520c promote tumour invasion and metastasis. *Nat Cell Biol*, 10, 202-210.

International Chicken Genome Sequencing Consortium, Sequence and comparative analysis of the chicken genome provide unique perspectives on vertebrate evolution. (2004). *Nature*, 432, 695-716.

Izadyar F, Pau F, Marh J, Slepko N, Wang T, Gonzalez R, Ramos T, Howerton K, Sayre C & Silva F. (2008). Generation of multipotent cell lines from a distinct population of male germ line stem cells. *Reproduction*, 135, 771-784.

Judson RL, Babiarz JE, Venere M & Blelloch R. (2009). Embryonic stem cell–specific microRNAs promote induced pluripotency. *Nat Biotechnol*, 27, 459-461.

Kanatsu-Shinohara M, Inoue K, Lee J, Yoshimoto M, Ogonuki N, Miki H, Baba S, Kato T, Kazuki Y, Toyokuni S, Toyoshima M, Niwa O, Oshimura M, Heike T, Nakahata T,

Ishino F, Ogura A & Shinohara T. (2004). Generation of pluripotent stem cells from neonatal mouse testis. *Cell*, 119, 1001-1012.

Kanatsu-Shinohara M, Lee J, Inoue K, Ogonuki N, Miki H, Toyokuni S, Ikawa M, Nakamura T, Ogura A & Shinohara T. (2008). Pluripotency of a single spermatogonial stem cell in mice. *Biol Reprod*, 78, 681-687.

Kanellopoulou C, Muljo SA, Kung AL, Ganesan S, Drapkin R, Jenuwein T, Livingston DM & Rajewsky K. (2005). Dicer-deficient mouse embryonic stem cells are defective in differentiation and centromeric silencing. *Genes Dev*, 19, 489-501.

Kim HJ, Lee HJ, Lim JJ, Kwak KH, Kim JS, Kim JH, Han YM, Kim KS & Lee DR. (2010). Identification of intermediate state as spermatogonial stem cells reprogram to multipotent cells. *Mol Cells*, 29, 519-526.

Kim KS, Kim JS, Lee MR, Jeong HS & Kim J. (2009) A Study of microRNAs *in silico* and *in vivo*: Emerging Regulators of Embryonic Stem Cells. *FEBS Journal*, 276, 2140-2149.

Landgraf P, Rusu M, Sheridan R, Sewer A, Iovino N, Aravin A, Pfeffer S, Rice A, Kamphorst AO, Landthaler M, Lin C, Socci ND, Hermida L, Fulci V, Chiaretti S, Foa R, Schliwka J, Fuchs U, Novosel A, Muller RU, Schermer B, Bissels U, Inman J, Phan Q, Chien M, Weir DB, Choksi R, De Vita G, Frezzetti D, Trompeter HI, Hornung V, Teng G, Hartmann G, Palkovits M, Di Lauro R, Wernet P, Macino G, Rogler CE, Nagle JW, Ju J, Papavasiliou FN, Benzing T, Lichter P, Tam W, Brownstein MJ, Bosio A, Borkhardt A, Russo JJ, Sander C,

Lee MR, Kim JS & Kim KS. (2010). miR-124a is Important for Migratory Cell Fate Transition During Gastrulation of Human Embryonic Stem Cells. Stem Cells, 28, 1550-1559.

Lee NS, Kim JS, Cho WJ, Lee MR, Steiner R, Gompers A, Ling D, Zhang J, Strom P, Behlke M, Moon SH, Salvaterra PM, Jove R & Kim KS. (2008). miR-302b maintains "stemness" of human embryonal carcinoma cells by post-transcriptional regulation of Cyclin D2 expression. *Biochem Biophys Res Commun*, 377, 434-440.

Lee TI, Jenner RG, Boyer LA, Guenther MG, Levine SS, Kumar RM, Chevalier B, Johnstone SE, Cole MF, Isono K, Koseki H, Fuchikami T, Abe K, Murray HL, Zucker JP, Yuan B, Bell GW, Herbolsheimer E, Hannett NM, Sun K, Odom DT, Otte AP, Volkert TL, Bartel DP, Melton DA, Gifford DK, Jaenisch R & Young RA. (2006). Control of developmental regulators by Polycomb in human embryonic stem cells. *Cell*, 125, 301-313.

Lewis BP, Burge CB & Bartel DP. (2005). Conserved seed pairing, often flanked by adenosines, indicates that thousands of human genes are microRNA targets. *Cell*, 120, 15-20.

Linsen SE, de Wit E, de Bruijn E & Cuppen E. (2010). Small RNA expression and strain specificity in the rat, *BMC Genomics*, 11, 249.

Lin SL, Chang DC, Chang-Lin S, Lin CH, Wu DT, Chen DT & Ying SY. (2008). Mir-302 reprograms human skin cancer cells into a pluripotent ES-cell-like state. *RNA*, 14, 115-2124.

Makeyev EV, Zhang J, Carrasco MA & Maniatis T. (2007). The MicroRNA miR-124 promotes neuronal differentiation by triggering brain-specific alternative pre-mRNA splicing. *Mol Cell*, 27, 435-448.

Markert A, Grimm M, Martinez J, Wiesner J, Meyerhans A, Meyuhas O, Sickmann A & Fischer U. (2008). The La-related protein LARP7 is a component of the 7SK ribonucleoprotein and affects transcription of cellular and viral polymerase II genes. *EMBO Rep*, 9, 569-575.

Mantel C, Guo Y, Lee MR, Kim MK, Han MK, Shibayama H, Fukuda S, Yoder MC, Pelus LM, Kim KS & Broxmeyer HE. (2007). Checkpoint-apoptosis uncoupling in human and mouse embryonic stem cells: a source of karyotpic instability. *Blood*, 109, 4518-4527.

Marson A, Levine SS, Cole MF, Frampton GM, Brambrink T, Johnstone S, Guenther MG, Johnston WK, Wernig M, Newman J, Calabrese JM, Dennis LM, Volkert TL, Gupta S, Love J, Hannett N, Sharp PA, Bartel DP, Jaenisch R & Young RA. (2008). Connecting microRNA genes to the core transcriptional regulatory circuitry of embryonic stem cells. *Cell*, 134, 521-533.

Mineno J, Okamoto S, Ando T, Sato M, Chono H, Izu H, Takayama M, Asada K, Mirochnitchenko O, Inouye M & Kato I. (2006). The expression profile of microRNAs in mouse embryos. *Nucleic Acids Res*, 34, 1765-1771.

Murchison EP, Kheradpour P, Sachidanandam R, Smith C, Hodges E, Xuan Z, Kellis M, Grützner F, Stark A & Hannon GJ. (2008). Conservation of small RNA pathways in platypus. *Genome Res*, 18(6), 995-1004.

Obernosterer G, Leuschner PJ, Alenius M & Martinez J. (2006). Post-transcriptional regulation of microRNA expression. *RNA*, 12, 1161-1167.

Oh SH, Jung YH, Gupta MK, Uhm SJ & Lee HT. (2009). H19 gene is epigenetically stable in mouse multipotent germline stem cells. *Mol Cells*, 27, 635-640.

Palmieri A, Pezzetti F, Brunelli G, Zollino I, Scapoli L, Martinelli M, Arlotti M & Carinci F. (2007). Differences in osteoblast miRNA induced by cell binding domain of collagen and silicate-based synthetic bone. *J Biomed Sci*, 14, 777-782.

Place RF, Li LC, Pookot D, Noonan EJ & Dahiya R. (2008). MicroRNA-373 induces expression of genes with complementary promoter sequences. *Proc Natl Acad Sci U S A*, 105, 1608-1613.

Rao PK, Kumar RM, Farkhondeh M, Baskerville S & Lodish HF. (2006). Myogenic factors that regulate expression of muscle-specific microRNAs. *Proc Natl Acad Sci U S A*, 103, 8721-8726.

Seandel M, James D, Shmelkov SV, Falciatori I, Kim J, Chavala S, Scherr DS, Zhang F, Torres R, Gale NW, Yancopoulos GD, Murphy A, Valenzuela DM, Hobbs RM, Pandolfi PP & Rafii S. (2007). Generation of functional multipotent adult stem cells from GPR125+ germline progenitors. *Nature*, 449, 346-350.

Sewer A, Paul N, Landgraf P, Aravin A, Pfeffer S, Brownstein MJ, Tuschl T, van Nimwegen E & Zavolan M. (2005). Identification of clustered microRNAs using an ab initio prediction method. *BMC Bioinformatics*, 6, 267.

Sinkkonen L, Hugenschmidt T, Berninger P, Gaidatzis D, Mohn F, Artus-Revel CG, Zavolan M, Svoboda P & Filipowicz W. (2008). MicroRNAs control de novo DNA methylation through regulation of transcriptional repressors in mouse embryonic stem cells. *Nat Struct Mol Biol*, 15, 259-267.

Suh MR, Lee Y, Kim JY, Kim SK, Moon SH, Lee JY, Cha KY, Chung HM, Yoon HS, Moon SY, Kim VN & Kim KS. (2004). Human embryonic stem cells express a unique set of microRNAs. *Dev Biol*, 270, 488-498.

Stadler B, Ivanovska I, Mehta K, Song S, Nelson A, Tan Y, Mathieu J, Darby C, Blau CA, Ware C, Peters G, Miller DG, Shen L, Cleary MA & Ruohola-Baker H. (2010). Characterization of microRNAs Involved in Embryonic Stem Cell States. *Stem Cells Dev*,19,935-950.

Takahashi K, & Yamanaka S. (2006). Induction of pluripotent stem cells from mouse embryonic and adult fibroblast cultures by defined factors. Cell, 126, 663-676.

Thomson JA, Itskovitz-Eldor J, Shapiro SS, Waknitz MA, Swiergiel JJ, Marshall VS & Jones JM. (1998). Embryonic stem cell lines derived from human blastocysts. *Science*, 282, 1145-1147.

Ventura A, Young AG, Winslow MM, Lintault L, Meissner A, Erkeland SJ, Newman J, Bronson RT, Crowley D, Stone JR, Jaenisch R, Sharp PA & Jacks T. (2008). Targeted deletion reveals essential and overlapping functions of the miR-17 through 92 family of miRNA clusters. *Cell,* 132, 875-886.

Voorhoeve PM, le Sage C, Schrier M, Gillis AJ, Stoop H, Nagel R, Liu YP, van Duijse J, Drost J, Griekspoor A, Zlotorynski E, Yabuta N, De Vita G, Nojima H, Looijenga LH & Agami R. (2006). A genetic screen implicates miRNA-372 and miRNA-373 as oncogenes in testicular germ cell tumors. *Cell,* 124, 1169-1181.

Wagner W, Horn P, Castoldi M, Diehlmann A, Bork S, Saffrich R, Benes V, Blake J, Pfister S, Eckstein V & Ho AD. (2008). Replicative senescence of mesenchymal stem cells: a continuous and organized process. *PLoS ONE,* 3, e2213.

Wang Y, Baskerville S, Shenoy A, Babiarz JE, Baehner L & Blelloch R. (2008). Embryonic Stem Cell Specific MicroRNAs Regulate the G1/S Transition and Promote Rapid Proliferation. *Nat Genet,* 40,1478–1483.

Watanabe T, Takeda A, Mise K, Okuno T, Suzuki T, Minami N & Imai H. (2005). Stage-specific expression of microRNAs during Xenopus development. *FEBS Lett,* 579, 318-324.

Wilson KD, Venkatasubrahmanyam S, Jia F, Sun N, Butte AJ & Wu JC. (2009). MicroRNA Profiling of Human-Induced Pluripotent Stem Cells. *Stem cell Dev,* 18, 749-757.

Yeom KH, Lee Y, Han J, Suh MR & Kim VN. (2006). Characterization of DGCR8/Pasha, the essential cofactor for Drosha in primary miRNA processing. *Nucleic Acids Res,* 34, 4622-4629.

Yu J, Vodyanik MA, He P, Slukvin II & Thomson JA. (2006). Human embryonic stem cells reprogram myeloid precursors following cell-cell fusion. *Stem Cells,* 24, 168-176.

Yu J & Thomson JA. (2008). Pluripotent stem cell lines. *Genes Dev,* 22, 1987-1997.

Yue J, Sheng Y & Orwig KE. (2008). Identification of novel homologous microRNA genes in the rhesus macaque genome. *BMC Genomics,* 9, 8.

Zeuthen J, Norgaard JO, Avner P, Fellous M, Wartiovaara J, Vaheri A, Rosen A & Giovanella BC. (1980). Characterization of a human ovarian teratocarcinoma-derived cell line. *Int J Cancer,* 25, 19-32.

Zhao Y, Samal E & Srivastava D. (2005). Serum response factor regulates a muscle-specific microRNA that targets Hand2 during cardiogenesis. *Nature,* 436, 214-220.

Zhou M, Wang Q, Sun J, Li X, Xu L, Yang H, Shi H, Ning S, Chen L, Li Y, He T & Zheng Y. (2009). In silico detection and characteristics of novel microRNA genes in the Equus caballus genome using an integrated ab initio and comparative genomic approach. *Genomics,* 94(2), 125-31.

Zovoilis A, Pantazi A, Smorag L, Opitz L, Salinas Riester G, Wolf M, Zechner U, Holubowska A, Stewart CL & Engel W. (2010). Embryonic stem cell related miRNAs are involved in differentiation of pluripotent cells originating from the germ line. *Mol Hum Reprod,* doi: 10.1093/molehr/gaq053.

Part 2

Hormonal Signals that Regulate Early Embryogenesis

13

Dynamic Changes in Gene Expression during Early Trophoblast Differentiation from Human Embryonic Stem Cells Treated with BMP4

Toshihiko Ezashi[1], Bhanu Prakash V.L. Telugu[1] and R. Michael Roberts[1,2]
[1]*Division of Animal Sciences,*
[2]*Department of Biochemistry; University of Missouri, Columbia, MO, USA*

1. Introduction

Embryonic stem cells (ESC) have routinely been established from the blastocyst stage embryo in rodents and primates, including the human. Whereas mouse (m) ESC are dependent on LIF and BMP4 for maintenance of pluripotency and represent the *"ground"* or *"naïve"* pluripotent state, human (h) ESC require FGF2 and TGFB/ACTIVIN/NODAL signaling and are analogous to stem cells derived from advanced epiblast stage of mouse embryos. Such, epiblast-derived stem cells (EpiSC) are often categorized as *"primed"*. Moreover, while the former are not generally prone to spontaneous trophoblast differentiation *in vitro* except in the face of perturbations of *Pou5f1* expression and Ras signaling, hESC are highly susceptible to both spontaneous and BMP4-mediated conversion to trophoblast (TR), particularly under high oxygen (O_2) conditions. Spontaneous differentiation in hESC has generally been counteracted by a variety of strategies to overcome or quench BMP4 activity, a growth factor that is present in most medium formulations. Here, we shall review these phenomena, as well as the opposing roles of FGF2 and BMP4 in mediating differentiation of hESC into TR. We shall describe how omission of FGF2 not only accelerates TR differentiation but drives the process unidirectionally, avoiding up-regulation of lineages representing endoderm, ectoderm, and mesoderm. In addition we shall summarize data from microarray analyses that have allowed gene expression changes associated with temporal changes in TR differentiation to be examined from early time points after BMP4 exposure for several days when extensive differentiation into specialized sub-lineages has occurred. Such analyses have also permitted the examination of the contrasting effects of low (4 %) and high (20 %) O_2 on gene regulation, colony morphology, and the production of progesterone and chorionic gonadotropin over time. These experiments illustrate the value of the BMP4/hESC model for studying emergence of TR from pluripotent precursor cells to fully differentiated syncytiotrophoblast and extravillous trophoblast and possibly for investigating the etiology of pregnancy disorders such as preeclampsia.

2. Embryonic stem cells (ESC) and trophoblast stem cells (TSC)

Mouse embryonic stem cells (mESC) were established from d 3.5 blastocyst stage embryos of certain strains almost 30 years ago (Evans and Kaufman 1981; Martin 1981). They require LIF/STAT3 signaling for maintenance of pluripotency (Hall, Guo et al. 2009), while BMP4 provides additional support for self renewal and resistance to differentiation (Ying, Nichols et al. 2003). However, mESC are generally recognized as being pluripotent rather than totipotent and only capable of contributing to the embryonic lineage and not to elements of extraembryonic tissues derived from trophectoderm (TE) (Beddington and Robertson 1989; Nagy, Gocza et al. 1990), presumably because they are already stably committed offshoots of the ICM and/or its derivative, the epiblast (Rossant and Spence 1998; Hemberger, Nozaki et al. 2003).

ESC established from monkey and human blastocysts (Thomson, Kalishman et al. 1995; Thomson, Itskovitz-Eldor et al. 1998), have markedly different characteristics to mESC. They exhibit a more flattened colony morphology and are dependent on FGF2 and TGFB/ ACTIVIN/NODAL signaling rather than LIF/STAT3 for maintenance of their pluripotency (Dvorak and Hampl 2005). These and other differences suggest a departure of hESC from the ground state pluripotency evident in LIF-dependent mESC, and, as we shall see, readily give rise to cells with the features of trophoblast. Intriguingly, however, ESC derived from a later gastrulation (egg-cylinder) stage mouse embryos, so-called epiblast stem cells (EpiSC) share striking similarities to hESC in terms of their flattened morphology, lack of dependence on LIF, but a requirement for FGF2 and ACTIVIN signals to maintain pluripotency (Brons, Smithers et al. 2007; Tesar, Chenoweth et al. 2007). Additionally, the EpiSC from mice lack demonstrated competence to contribute to fetal tissues after being introduced into pre-implantation embryos, although they can donate to the three main germ layers, ectoderm, mesoderm, and endoderm, in embryoid bodies and teratomas, and are, therefore regarded as pluripotent. EpiSC therefore, may represent a more "advanced" state and are frequently termed "*primed*" ESC. Based on their similarities, hESC and mEpiSC may embody physiologically equivalent states.

In addition to their abilities to give rise to pluripotent ESC, blastocysts have provided a source of multipotent TSC, most likely from actively proliferating polar TE. Such cells are, therefore, of extra-embryonic ectoderm origin. Such TSC can be cultured in a continuously self-renewing state *in vitro* in the presence of FGF4 plus heparin, but also require NODAL signaling, which is usually derived from a supporting feeder layer of primary embryonic fibroblasts (Tanaka, Kunath et al. 1998; Rossant 2001; Rossant 2007). However, it should be noted that dependency on FGF4 may not be a universal feature of TSC from all species (Vandevoort, Thirkill et al. 2007; Grigor'eva, Shevchenko et al. 2009).

There have been many excellent reviews on embryonic and trophoblast stem cells (TSC), and the differences that distinguish them. Our goal here is rather different from those of such earlier publications and is intended to illustrate the various ways whereby ESC can be diverted towards the trophoblast (TR) lineage. In the second half of the manuscript, we concentrate on work from our own laboratory on a unique hESC-based model to derive TR *de novo*. As well as reviewing past work, we include some previously unpublished data to emphasize the value of this model system for studying both the origins of TR and the differentiation of its sub-lineages.

3. Differentiation potential of murine and human ESC into trophoblast

3.1 Genetic manipulation

TE, the precursor of placental TR becomes apparent as the blastocyst emerges and the blastocoel cavity expands. The formation of TE and its segregation from the pluripotent inner cell mass marks the first visual evidence of separation of cell lineages during mammalian conceptus development. Although, *naïve* mESC are incapable of differentiating into TR when introduced into early mouse embryos (they instead populate the inner cell mass) and do not form TR when permitted to form embryoid bodies or teratomas, genetic manipulations of some of the key transcription factors that maintain pluripotency can result in TR differentiation *in vitro*. For example, down regulation of master regulators of pluripotency such as *Pou5f1* (Niwa, Miyazaki et al. 2000; Velkey and O'Shea 2003; Hay, Sutherland et al. 2004; Hough, Clements et al. 2006; Ivanova, Dobrin et al. 2006) and *Sox2* (Masui, Nakatake et al. 2007) in mESC by RNAi or by other method leads to TR differentiation (Table 1). Alternatively, upregulation of key TSC markers such as *Cdx2* (Niwa, Toyooka et al. 2005; Tolkunova, Cavaleri et al. 2006) and *Eomes* (Niwa, Toyooka et al. 2005), as well as an upstream transcription factor of *Cdx2* known as *Tead4* (Nishioka, Inoue et al. 2009), *Lef1* a component of WNT signaling pathway (He, Pant et al. 2008), and finally, Ras (Lu, Yabuuchi et al. 2008) a signaling molecule can lead to TR differentiation in mESC (Table 1).

Primed ESC, on the other hand, represented by primate ESC and mEpiSC are susceptible for spontaneous differentiation into TR *in vitro* (Thomson, Kalishman et al. 1995; Thomson, Itskovitz-Eldor et al. 1998; Brons, Smithers et al. 2007; Tesar, Chenoweth et al. 2007). For instance, chorionic gonadotropin (CG), a secreted protein produced by TR is readily detectable in the culture media of primate ESC (Thomson, Kalishman et al. 1995; Thomson, Itskovitz-Eldor et al. 1998). Although hESC are clearly distinct from mESC, down regulation of *POU5F1* still leads to TR differentiation (Hay, Sutherland et al. 2004; Matin, Walsh et al. 2004; Zaehres, Lensch et al. 2005; Babaie, Herwig et al. 2007) (Table 1). Additionally, silencing of the expression of NANOG, a component of the triad of transcription factors that also includes POU5F1 and SOX2 and is essential for maintenance of the pluripotent state (Boyer, Lee et al. 2005; Loh, Wu et al. 2006), induces TR and extraembryonic endoderm-associated genes (Hyslop, Stojkovic et al. 2005; Zaehres, Lensch et al. 2005). This similarity in outcomes in both mESC and hESC suggests that TR is the default state when the core transcriptional circuitry associated with pluripotency is disturbed. The conversion of hESC and especially mEpiSC to TR is intriguing, given that both cell types are assumed to represent a more "advanced" ESC state representing the later forming epiblast rather than the inner cell mass, a stage in embryogenesis that occurs well after TE and its accompanying TSC have been specified.

3.2 Manipulation of culture conditions

In addition to the genetic manipulations, TR differentiation from ESC has also been accomplished by modifying the culture conditions in which the cells are grown. As mentioned earlier, hESC, in particular, are relatively more susceptible for manipulation compared to mESC. There are two principal means of achieving TR differentiation from hESC (Table 1). They are:

	Grouud, Naïve type	Primed, Epiblast type
Genetic manipulation, down-regulation	Oct4 (Niwa 2000 Velkey 2003 Hay 2004 Hough 2006 Ivanova2006)	OCT4 (Hay 2004 Matin 2004 Zaehres 2005 Babaie 2007)
	NANOG (Hough 2006*)	NANOG (Zaehres 2005, Hyslop 2005)
	Sox2 (Masui 2007)	
Genetic manipulation, over-regulation	Cdx2 (Niwa 2005, Tolkunova2006)	
	Eomes (Niwa 2005)	
	Tead4 (Nishioka 2009)	
	Ras (Lu 2008)	
	Wnt3a (He 2008)	
Culture condition modifcation	BMP4-Laminin (Hayashi 2010)	BMP4-Matrigel (Xu 2002, Das 2007, Wu 2008, Chen 2008)
	Collagen IV (Schenke-Layland 2007)	EB in semi-solid medium (Golos2006, Gerami-Naini 2004, Peiffer 2007)
		Enrichment of hCGb producing cells (Harun 2006)
		inhibition of Activin/Nodal signaling (Wu 2008)

Table 1. Differentiation potential of murine and human ESC into trophoblast by manipulations of genetic (upper) and culture conditions (lower). *Reduction in *Nanog* expression correlated with induction of extraembryonic endoderm genes *GATA4*, *GATA6*, and *Lamb1*, with subsequent generation of groups of cells with parietal endoderm phenotype.

Embryoid body (EB) based differentiation into TR: This approach has been used with hESC, which are first dissociated and converted into EB in a semi-solid medium (Gerami-Naini, Dovzhenko et al. 2004; Golos, Pollastrini et al. 2006) when they spontaneously differentiate and form TR, particularly in their outer layers. It has proved possible to select EBs secreting the highest levels of CG to provide multicellular structures for TR (Harun, Ruban et al. 2006). Alternatively, EB that attach readily to the substratum are also enriched in TR (Peiffer, Belhomme et al. 2007).

Perturbation of BMP4 and/ or ACTIVIN/NODAL signaling axis: Treatment of hESC with BMP4 on a medium conditioned by mouse embryonic fibroblasts and on a Matrigel substratum under otherwise standard feeder-free culture conditions for ESC maintenance induces differentiation of hESC towards the TR lineage (Xu, Chen et al. 2002; Das, Ezashi et al. 2007; Chen, Ye et al. 2008; Wu, Zhang et al. 2008), as does inhibition of ACTIVIN/NODAL signaling through use of SB-431542, which binds to the activin receptor-like kinase ALK5/4/7 (Wu, Zhang et al. 2008). In the inhibitor approach, it is unclear whether the outcome is either the direct result of interfering with the signaling pathway or due to the observed up-regulation of BMP4 that occurs concurrently (Wu, Zhang et al. 2008). Not just hESC are susceptible to BMP4 driven differentiation towards TR, mouse EpiSC (Brons, Smithers et al. 2007), as well as the recently reported porcine EpiSC (Alberio, Croxall et al. 2010) also respond in a somewhat similar manner. These results indicate that BMP4 mediated differentiation into TR may be a conserved phenomenon across all FGF2-dependent, i.e. "*primed*" stem cell lines and not just restricted to hESC. Confusing this straightforward interpretation, however, a recent report has demonstrated that *naïve* mESC are also prone to TR differentiation in response to BMP4 provided that they are grown on a laminin or fibronectin rather than a collagen I-, or poly-D-lysine (PDL)-coated culture surface (Hayashi, Furue et al.). At first glance, such an outcome is counterintuitive since it contradicts the established dogma that exposure of mESC to BMP4 favors pluripotency. On the other hand, it is becoming increasingly clear that the extracellular matrix provides directional cues and plays an important role in controlling the fate of stem cells and their differentiation (Gerecht-Nir, Ziskind et al. 2003; Schenke-Layland, Angelis et al. 2007; Xiao, Zeng et al. 2007). Conceivably, the laminin substratum shifted the phenotype of the mESC from the *naïve* to the *primed* state, and hence made the cells susceptible to BMP4-driven differentiation to TR.

The general features of TR differentiation in colonies of hESC responding to BMP4 were first described by Xu et al. (Xu, Chen et al. 2002). The process was strongly dependent on the concentration of BMP4 supplied, and occurred much more rapidly at 100 ng/ml than at 10 ng/ml. As differentiation proceeded, cells within the colonies gradually changed their morphologies to a more flattened epithelioid form. Differentiation proceeded from the periphery inwards, so that a central core of smaller cells in the larger colonies remained for as long as 5 days. These overt changes in gross morphology correlated with changes in gene expression reflecting the emergence and later differentiation of TR. Several TR expressed genes, such as *TCFAP2C, HERVW, GATA3, MSX2, CGA, CGB, GCM1, HASH2, HLA-G, MMP9* and *KRT7*, were highly up-regulated by the time that multinuclear cells presumed to correspond to syncytial TR, became evident, approximately a week after addition of BMP4. Accompanying these changes in gene expression, the colonies markedly increased their production of the placental hormones hCG, progesterone, and estradiol-17β. Based on the weight of evidence, this model system appears to provide a more powerful means of assessing the emergence and differentiation of human TR than the one based on EB. In the sections that follow we summarize some of our results employing this model system.

3.3 Role of FGF2 in the BMP-driven differentiation of trophoblast from hESC

As BMP signaling activity is detectable in the "serum replacement", a component of standard hESC culture medium (Xu, Peck et al. 2005), it is perhaps no surprise that hESC maintained under such conditions tend to undergo some differentiation toward TR over time.

Consequently, media with elevated FGF2 concentrations and containing the BMP antagonist, NOGGIN, provide better sustained proliferation of undifferentiated hESC than standard conditions (Wang, Zhang et al. 2005; Xu, Peck et al. 2005). Such high concentrations of FGF2 appear to conteract BMP signaling activity and also support feeder-independent growth of hESC (Ludwig, Bergendahl et al. 2006). Keeping in mind the ability of FGF2 to antagonize differentiation of hESC, we hypothesized that elimination of this growth factor from the differentiation medium would expedite TR formation, as it proved to do (Das, Ezashi et al. 2007). Complete elimination of FGF2 from the differentiation medium accelerated the response of hESC to BMP4 as evident from earlier changes in cell morphology and elevated hCG and progesterone release over time (Das, Ezashi et al. 2007). In absence of of FGF2, 10 ng/ml BMP4 was about as effective as the 50 ng/ml in inducing morphological differentiation of the cells over time (Das, Ezashi et al. 2007). Furthermore, these culture conditions led to unidirectional TR differentiation as evidenced by microarray analysis of the transcripts present (GEO GSE10469), with complete lack of up-regulation of gene markers for primitive (yolk sac) and definitive endoderm, e.g. *AFP* (alpha-fetoprotein), *RBP4* (retinol binding protein 4), *FGG* (fibrinogen-γ), which had been noted to increase in expression when FGF2 was present (Xu, Chen et al. 2002; Zhang, Li et al. 2007). Expression of lineage markers for mesoderm, e.g. *WNT3A, T* (Brachyury), *MEOX2, MIXL1;* for endoderm, e.g. *FOXA2, SOX7, ONECUT1, ATBF1, DPF3;* and for ectoderm, e.g. *PAX6, MEIS1, HOXB1, OTX1, SOX1, ZIC1* (Cai, Chen et al. 2006; Zhang, Li et al. 2007) was either absent or very low in both control hESC and in the cells after BMP4 addition in absence of FGF2 (GEO GSE10469). There was a modest 2-fold up-regulation of WNT3, a potential mesoderm marker, after 24 h, but since WNT signaling is probably involved in directing TR emergence, such a change was not unanticipated. Taking these results into consideration, our laboratory now routinely omits FGF2 and employs 10 ng/ml BMP4 in the differentiation medium.

3.4 Effects of oxygen and time on the BMP-driven differentiation of TR from hESC

Human pre-implantation embryos only experience a low O_2 environment *in utero* (as discussed by Ezashi et al. 2005), while hESC are routinely cultured by most groups in 20% (atmospheric) O_2 conditions. We were concerned that hESC cultured under such non-physiological conditions might be either poised for differentiation, or, worse, already cryptically differentiated. Such a concern was justified by the fact that cells maintained under 20 % O_2 produced considerable amounts of hCG and progesterone after about five days in a standard culture condition, suggesting that lineage commitment had indeed progressed even before signs of overt differentiation became evident. By contrast, hESC which grow equally as efficiently under 4% O_2 conditions as under ~20 % O_2 (Ezashi, Das et al. 2005), produced very little of the two placental hormones. In addition, growth under low O_2 provided significant protection against subsequent differentiation when the same cells were passaged into high O_2 conditions (Ezashi, Das et al. 2005). Since O_2 tension slows the proliferation and enhances differentiation of villous and extra-villous TR (Genbacev, Joslin et al. 1996; Genbacev, Zhou et al. 1997; James, Stone et al. 2006), we included O_2 tension as an additional variable in our experiments. As described below, the expression of TR markers was markedly accelerated under 20 % relative to low O_2 conditions (Das, Ezashi et al. 2007; Schulz, Ezashi et al. 2008).

Over the initial days of exposure to BMP4, smaller ESC colonies rapidly differentiate to TR. In the case of larger colonies, however, cells in the center begin to pile up, possibly because they continue to divide in space that is progressively restricted in area (Fig 1, also see Fig 2 and 5 of Das et al 2007). These cells remain POU5F1-positive and are negative for KRT7, a common marker of TR. Whether such cells have retained their original pluripotent potential remains unclear. However, the cells that ring this core demonstrate overt signs of

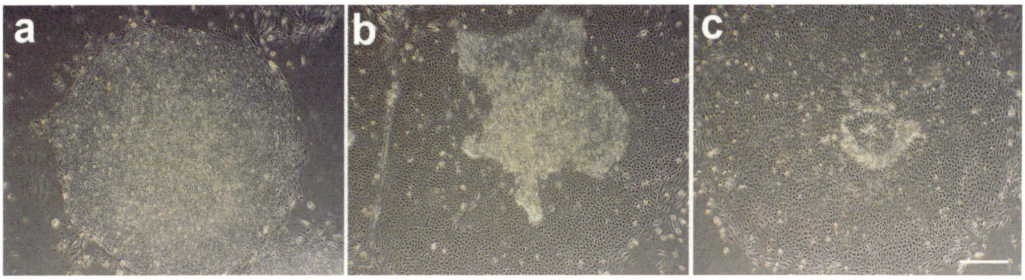

Fig. 1. Effects of treatment with 10 ng/ml BMP4 on hESC colony morphology under a low (4%, **a** & **b**) and high (20%, **c**) O_2 atmosphere at day 3. (**a**) Control cells on standard medium under 4 % O_2. (**b**) BMP4 treated cells under 4 % O_2. (**c**) BMP4 treated cells under 20 % O_2. Bar, 0.5 mm. Note that the colonies treated with BMP4 are larger than ones without treatment (**a**), due to spreading of differentiated cells at periphery.

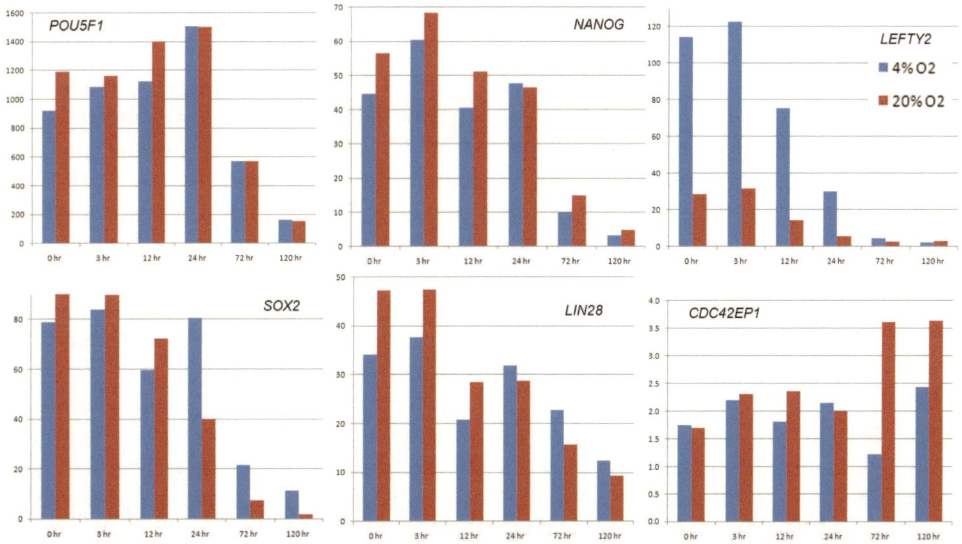

Fig. 2. Changes in gene expression as assessed by microarray (with values normalized to the median intensity of the array) for five genes associated with a pluripotent phenotype (*POU5F1, NANOG, SOX2, LIN28* and *LEFTY2*) and one implicated with cell morphogenesis (*CDC42EP1*). H1 hESC were exposed to BMP4 (10 ng/ml) under either 4% (blue) or 20% (red) oxygen. RNA isolated at 0, 3, 12, 24, 72 and 120 h. These data were obtained on Agilent Whole Human Genome Oligo microarrays. Genes that were changing significantly over time within each oxygen condition were identified by ANOVA. The value represents a normalized expression (1.0 being the signal at the 50th percentile). The genes shown here were selected from 33,814 genes of which 5,194 changed expression more than 2-fold at $P<0.01$ in at least one time point. Changes in gene expression have been verified by real-time PCR with cDNA from H1 and H9 cells for *POU5F1, NANOG, SOX2, LEFTY2, CGA* and *CGB*.

morphological differentiation, taking on a more flattened appearance and becoming negative for POU5F1 and positive for KRT7. Why differentiation proceeds in this directional manner is uncertain, as all the cells in the colony are exposed to BMP4. However, one explanation, albeit unexpected, for the apparent directional differentiation from the periphery to the center of the colony is that it is due to migration of differentiating TR outwards from the interior. In this regard, the product of the *Cdc42ep1* (*Borg5*) gene, CDC42EP1, sometimes known as either CDC42 effector protein or binder of Rho guanosine 5′-triphosphatase 5, has been proposed as directing such movement (Vong, Liu et al. 2010). Reducing *Cdc42ep1* expression by using an RNAi approach lowered the number of colonies with peripheral TR and led to an increase in patches of TR scattered within the interior of the colonies. Our microarray data (Fig. 2) are somewhat consistent with this hypothesis in the sense that under high O_2 there was a increase in expression of *CDC42EP1* at 72 h when many of the main TR markers were being up-regulated (Schulz, Ezashi et al. 2008). On the other hand, visible differentiation had also occurred in the outer parts of the colony by this stage and even a modest up-regulation of gene expression was not observed under physiological O_2 conditions. Obviously, clarifying a role for CDC42EP1 in hESC differentiation will require a much more sophisticated approach than transcript profiling.

In order to explore the temporal changes in gene expression following BMP4 supplementation of hESC under both low and high O_2 environments, microarray analysis was performed on BMP4-treated H1 hESC with RNA collected at 3, 12, 24, 72, and 120 h of BMP4 treatment under both sets of O_2 conditions. Biotinylated cRNA was prepared and hybridized to Agilent-014850 Whole Human Genome 4x44K Microarrays. Slides were washed and scanned on an Agilent G2565 Microarray scanner, and data analyzed with Agilent Feature Extraction and GeneSpring GX v7.3.1 softwares. Approximately 32,000 probe sets (out of 40,391) provided a signal above background at one of the time points. Of these, 14,478 showed a ≥2-fold (plus or minus) change ($p<0.05$) under 4 % O_2, and 18,580 under 20 % O_2. When more stringent selection was made (2-fold; $p<0.005$) 4547 genes were found to be regulated over time of exposure to BMP4 and 739 by O_2. Adjustment of the data to $p<0.05$ with the extremely stringent Beniamini & Hochberg False Discovery Rate in a two-way ANOVA indicated that 5,173 genes were differentially regulated on at least one time-point. Full access to the microarray data set is available at GEO GSE10469.

As noted by Xu et al. (Xu, Chen et al. 2002), we observed a decline in the expression of transcription factors associated with pluripotency (Boyer, Lee et al. 2005; Cai, Chen et al. 2006; Loh, Wu et al. 2006; Wang, Rao et al. 2006) over time after exposure to BMP4 (Fig. 2). In general, the decrease did not become obvious in the first 24 h but became pronounced between 24 and 72 h. The O_2 atmosphere had no affect the changes observed. Interestingly, one gene *LEFTY2*, which is highly expressed in hESC and is considered to be under the transcriptional control of the POU5F1/NANOG/SOX2 triad (Boyer, Lee et al. 2005; Loh, Wu et al. 2006; Babaie, Herwig et al. 2007), showed a much more rapid drop in expression than POU5F1 and the other pluripotency genes. It also had a much lower expression in cells under 20% O_2 than in those under 4% O_2 (Fig. 2). We have previously hypothesized that the *LEFTY* genes may play a central role guarding against differentiation and that their low concentrations in ESC maintained under 20% O_2 could be a reflection of the cryptically differentiated state of such cells (Westfall, Sachdev et al. 2008).

During the first 3 h of BMP4 exposure more than 500 genes showed greater than a 2-fold up-regulation. Since transcription factors likely underpin emergence of new cell lineages from pluripotent precursors, we examined expression of transcription factor genes that were upregulated during the first 3 h of exposure to BMP4 under both O_2 conditions. An additional criterion was that the genes should have a normalized expression of > 0.1 at 3 h (1.0 being the signal at the 50th percentile) under at least one of the two conditions. The following 15 transcription factor genes were identified as significantly up-regulated at 3 h: *ID2*, 10.2-fold; *MSX2*, 11.9-fold; *GATA5*, 8.4-fold; *GATA3*, 8.4-fold; *LEF1*, 6.4-fold (under high O_2 only); *DLX2*, 3.0-fold; *TBX3*, 3.0-fold; *FOXF2*, 5.8-fold; *HEY1*, 4.9-fold; *GATA2*, 4.9-fold; *TCFAP2A*, 3.3-fold; *TCFAP2C*, 2.2-fold; *HHEX*, 9.4-fold; *RUNX3*, 2.7-fold; *CDX2*, 3.6-fold. A DAVID analysis of the full list of up-regulated genes suggested that WNT as well as NOTCH signaling was an immediate early response following BMP4 treatment as early as 3 h of exposure, as evidenced by up-regulation of *WIF1, FRCB, MEGF10, MYL4, LEF1, LUM, WISP1* and *FRAT1*.

We also examined the temporal expression of a number of genes encoding transcription factors that have been implicated in the emergence of TR in genetic studies in mice and mouse TSC. The caudal-related transcription factor CDX2 is one such gene product. It is the best marker for distinguishing TE from ICM in a variety of species (Roberts, Yong et al. 2006; Kuijk, Du Puy et al. 2008; Harvey, Armant et al. 2009; Katayama, Ellersieck et al. 2010). *Cdx2* -/- mouse conceptuses fail to implant (Chawengsaksophak, James et al. 1997), and ectopic expression of CDX2 in mouse ESC down-regulates POU5F1 (Niwa, Toyooka et al. 2005) and suppresses the expression of genes that are components of pluripotency network, thereby causing the cells to adopt a TR phenotype. The gene is up-regulated immediately in ESC in response to BMP4 (Fig. 3), reaching maximal expression between 24 and 72 h, the stage at which *POU5F1* transcripts are in rapid decline, an observation consistent with the reciprocal relationship observed for these transcription factors in embryos. Unlike *LEFTY2*, its expression is higher under 20% O_2 than under 4 % O_2 (Fig. 3).

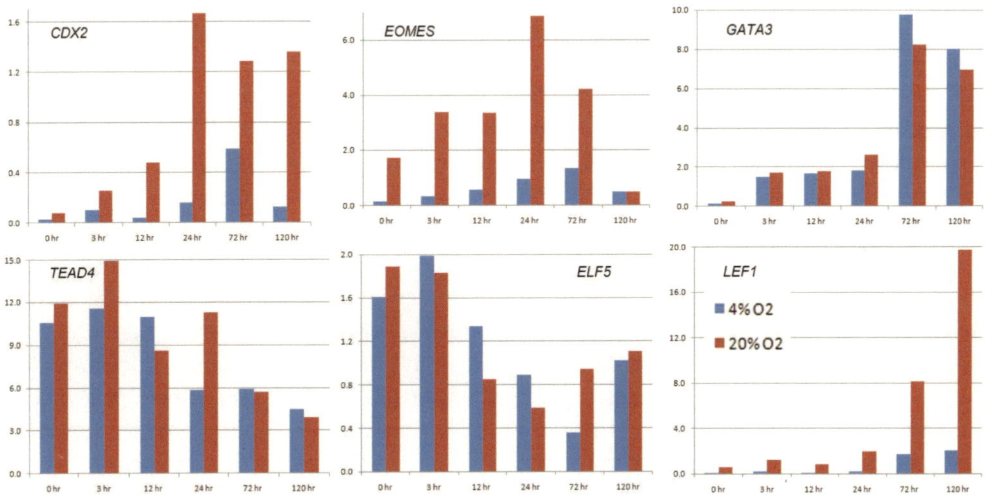

Fig. 3. Changes in gene expression as assessed by microarray (with values normalized to the median intensity of the array) for six genes implicated in the lineage decisions leading to acquisition of a trophoblast phenotype (*CDX2, EOMES, GATA3, TEAD4, ELF5* and *LEF1*). H1 hESC were exposed to BMP4 (10 ng/ml) in either 4% (blue) or 20% (red) oxygen and RNA collected after 0, 3, 12, 24, 72 and 120 h as described in Fig 2.

The product of the T-box gene, *Eomes*, is the earliest acting transcription factor known to be required for immediate post-implantation lineage commitment steps, as mice lacking the *Eomes* gene implant but arrest at a blastocyst-like stage of development (Russ, Wattler et al. 2000). Also such conceptuses are unable to form TSC. *EOMES* is not, however, markedly responsive to treatment of ESC with BMP4 and its expression remains low under 4 % O_2 (Fig. 3).

Like *Cdx2*, the *Gata3* gene is transcribed in TE, but not in the ICM of mouse blastocysts (Home, Ray et al. 2009), and can force emergence of TR when expressed ectopically in mouse ESC (Ralston, Cox et al. 2010). Although both genes are regulated by TEAD4 (see below), they appear to operate semi-independently, specifying TR fate via different pathways and targets (Ralston, Cox et al. 2010). *GATA3* expression is barely detectable in control hESC, but is highly responsive to BMP4 (Fig. 3) and exhibits a further up-regulation between 24 and 72 h to a level that is maintained until 120 h. It seems that GATA3 has a dual role, one in TR emergence, the other in driving the differentiation of on or more specialized sublineages.

Mouse embryos mutant for the *Tead4* cannot form blastocysts (Yagi, Kohn et al. 2007; Nishioka, Yamamoto et al. 2008) and are unable to maintain CDX2 expression. Instead, the entire conceptus consists of ICM-like cells expressing POU5F1 and NANOG. Importantly, TSC cannot be isolated from either *Cdx2*$^{-/-}$ or *Tead4*$^{-/-}$ blastocysts, while ES cells can. Moreover, forcing *Tead4* transcription in mouse ESC induces *CDX2* expression and a TR phenotype (Nishioka, Inoue et al. 2009). Thus, *Tead4* is probably the earliest known gene that is required for trophoblast specification. On the other hand, *TEAD4* is expressed in hESC before BMP4 treatment, shows only minor responsiveness to BMP4 at early time points, and then declines in activity over time (Fig. 3), Although these results do not rule out a role for TEAD4 in directing the TR lineage, e.g. by controlling expression of the *CDX2* and *GATA3* genes, it is clear that the microarray results are not easily interpreted and that TEAD4, like many other transcription factors, is probably pleiotropic in action and involved in multiple developmental events.

ELF5, an ETS-domain transcription factor, like all the other transcription factors discussed in this section, is implicated in early placental development of the mouse. *Elf5*$^{-/-}$ conceptuses implant, form an ectoplacental cone, but extraembryonic ectoderm, the site of the TSC niche, is not detectable. TSC cannot be derived from these embryos suggesting a fundamental defect in their generation or self-renewal (Donnison, Beaton et al. 2005). One proposed role for ELF5 is in transcriptional control of *Cdx2* and *Eomes*, with the three transcription factors forming a positive feedback loop that reinforces the early stages of TSC commitment (Ng, Dean et al. 2008). Interestingly, ELF5/CDX2 double-positive cells have been identified amongst human villous cytotrophoblasts and may define a TSC-like subpopulation (Hemberger, Udayashankar et al. 2010). *ELF5* is expressed in hESC and transcript levels are only barely affected by exposure to BMP4 (Fig. 3). As with *TEAD4*, the data do not allow inferences to be made about any role for ELF5 in TR emergence, but suggests that this transcription factor is also pleiotropic in its actions.

Lymphoid enhance factor-1 (LEF1), a downstream component of the WNT signaling pathway, has been implicated in modulating *Cdx2* expression (He, Pant et al. 2008). However, *Lef1*$^{-/-}$ mouse conceptuses are born alive, suggesting that the placenta is functional despite the fact that several other organs are defective in the pups (van Genderen, Okamura et al. 1994). *LEF1* transcripts are not detectable in hESC grown under physiological O_2 conditions and only barely so under 4 % O_2 (Fig. 3). Expression rose markedly over time (over 200-fold by 5 days) and was again much more pronounced at high O_2. Such data suggest that LEF1 likely plays an important role in human TR development, particularly in later lineage differentiation.

Clearly microarray data on these transcription factors implicated in TR lineage determination have only limited value, but they form a starting point for genetic manipulation studies in which their expression can either be silenced or over-expressed and effects on cellular phenotype evaluated. Several other transcription factors, such as GCM1, HAND1, and DLX3, although not up-regulated immediately and therefore not implicated in "early" responses, show major increases in expression over time of exposure to BMP4 (data available at GEO GSE10469).

Finally, this model system for studying TR emergence, permits a study of the events accompanying differentiation of more advanced lineages, such as syncytiotrophoblast and extravillous TR. As BMP4 treatment progressed beyond 72 h, many genes considered general TR markers such as *KRT7* and *KRT8*, as well as ones whose expression is associated with syncytiotrophoblast, e.g. *CGA, CGB, HERVW*, became highly up-regulated, especially under 20 % O_2 (Fig. 4). These changes correlated well with the emergence of extensive patches of syncytial TR appearing on the periphery of the colonies. These hCG-positive syncytial cells have multiple nuclei in a continuous cytoplasm and materialize much more rapidly under 20 % O_2 than under 4 % O_2 (Das, Ezashi et al. 2007). Consistent with the immunohistochemistry and gene expression studies, abundant hCG is released into the medium under these condition (Das, Ezashi et al. 2007). Also evident by 72 h is the expression of HLA-G, a marker of extravillous TR.

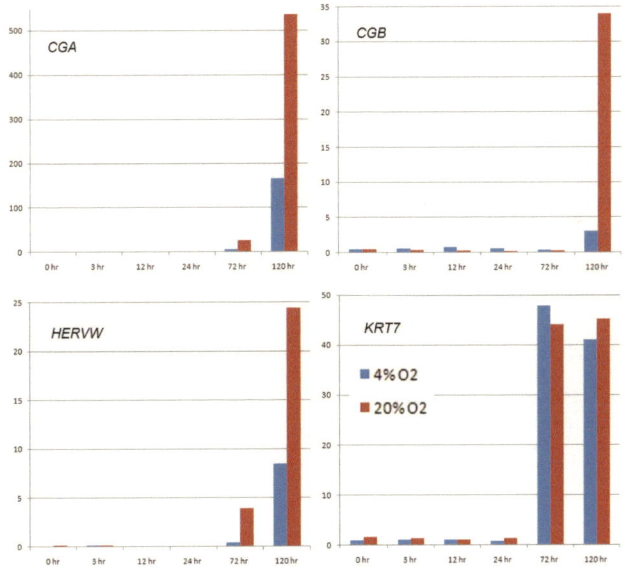

Fig. 4. Changes in in gene expression as assessed by microarray of markers associated with trophoblast differentiation (*CGA, CGB, HERVW,* and *KRT7)* after 0, 3, 12, 24, 72 and 120 h of BMP4 treatment as described in Fig. 2.

4. Concluding remarks

These results strongly suggest that only TR emerges when hESC are driven to differentiate by BMP4 in absence of FGF2, that high O_2 promotes and low oxygen impedes TR differentiation, and that markers representing all specialized TR cells are represented in BMP4 supplemented hESC culture by 3-5 days of treatment. This model system also affords an opportunity to follow the origin or "birth" of TR *in vitro*, obviating the need for obtaining first trimester primary human TR from selective or spontaneous abortions, which until recently have represented the only credible means of investigating early differentiation events (Golos, Pollastrini et al. 2006). The model also affords a means to investigate both extrinsic factors, such as O_2 and various hormones and growth factors, and intrinsic genetic and epigenetic mechanisms that control the emergence and differentiation of TR. Therefore, the BMP4/hESC model system may prove useful for investigating the etiology of pregnancy disorders, such as preeclampsia (Golos, Giakoumopoulos et al. 2010; Das, Ezashi et al. 2007; Schulz, Ezashi et al. 2008).

5. Acknowledgements

This work was supported by HD055231 and NIH HD21896.

6. References

Alberio, R., N. Croxall, et al. (2010). "Pig Epiblast Stem Cells Depend on Activin/Nodal Signaling for Pluripotency and Self-Renewal." Stem Cells Dev.

Babaie, Y., R. Herwig, et al. (2007). "Analysis of Oct4-dependent transcriptional networks regulating self-renewal and pluripotency in human embryonic stem cells." Stem Cells 25(2): 500-10.

Beddington, R. S. and E. J. Robertson (1989). "An assessment of the developmental potential of embryonic stem cells in the midgestation mouse embryo." Development 105(4): 733-7.

Boyer, L. A., T. I. Lee, et al. (2005). "Core transcriptional regulatory circuitry in human embryonic stem cells." Cell 122(6): 947-56.

Brons, I. G., L. E. Smithers, et al. (2007). "Derivation of pluripotent epiblast stem cells from mammalian embryos." Nature 448(7150): 191-5.

Cai, J., J. Chen, et al. (2006). "Assessing self-renewal and differentiation in human embryonic stem cell lines." Stem Cells 24(3): 516-30.

Chawengsaksophak, K., R. James, et al. (1997). "Homeosis and intestinal tumours in Cdx2 mutant mice." Nature 386(6620): 84-7.

Chen, G., Z. Ye, et al. (2008). "Trophoblast differentiation defect in human embryonic stem cells lacking PIG-A and GPI-anchored cell-surface proteins." Cell Stem Cell 2(4): 345-55.

Das, P., T. Ezashi, et al. (2007). "Effects of FGF2 and oxygen in the BMP4-driven differentiation of trophoblast from human embryonic stem cells." Stem Cell Research 1(1): 61-74.

Donnison, M., A. Beaton, et al. (2005). "Loss of the extraembryonic ectoderm in Elf5 mutants leads to defects in embryonic patterning." Development 132(10): 2299-308.

Dvorak, P. and A. Hampl (2005). "Basic fibroblast growth factor and its receptors in human embryonic stem cells." Folia Histochem Cytobiol 43(4): 203-8.

Evans, M. J. and M. H. Kaufman (1981). "Establishment in culture of pluripotential cells from mouse embryos." Nature 292(5819): 154-6.

Ezashi, T., P. Das, et al. (2005). "Low O2 tensions and the prevention of differentiation of hES cells." Proc Natl Acad Sci U S A 102(13): 4783-4788.

Genbacev, O., R. Joslin, et al. (1996). "Hypoxia alters early gestation human cytotrophoblast differentiation/invasion in vitro and models the placental defects that occur in preeclampsia." J Clin Invest 97(2): 540-50.

Genbacev, O., Y. Zhou, et al. (1997). "Regulation of human placental development by oxygen tension." Science 277(5332): 1669-72.

Gerami-Naini, B., O. V. Dovzhenko, et al. (2004). "Trophoblast differentiation in embryoid bodies derived from human embryonic stem cells." Endocrinology 145(4): 1517-24.

Gerecht-Nir, S., A. Ziskind, et al. (2003). "Human embryonic stem cells as an in vitro model for human vascular development and the induction of vascular differentiation." Lab Invest 83(12): 1811-20.

Golos, T. G., M. Giakoumopoulos, et al. (2010). "Embryonic stem cells as models of trophoblast differentiation: progress, opportunities, and limitations." Reproduction 140(1): 3-9.

Golos, T. G., L. M. Pollastrini, et al. (2006). "Human embryonic stem cells as a model for trophoblast differentiation." Semin Reprod Med 24(5): 314-21.

Grigor'eva, E. V., A. I. Shevchenko, et al. (2009). "FGF4 independent derivation of trophoblast stem cells from the common vole." PLoS One 4(9): e7161.

Hall, J., G. Guo, et al. (2009). "Oct4 and LIF/Stat3 additively induce Kruppel factors to sustain embryonic stem cell self-renewal." Cell Stem Cell 5(6): 597-609.

Harun, R., L. Ruban, et al. (2006). "Cytotrophoblast stem cell lines derived from human embryonic stem cells and their capacity to mimic invasive implantation events." Hum Reprod 21(6): 1349-58.

Harvey, A. J., D. R. Armant, et al. (2009). "Inner cell mass localization of NANOG precedes OCT3/4 in rhesus monkey blastocysts." Stem Cells Dev 18(10): 1451-8.

Hay, D. C., L. Sutherland, et al. (2004). "Oct-4 knockdown induces similar patterns of endoderm and trophoblast differentiation markers in human and mouse embryonic stem cells." Stem Cells 22(2): 225-35.

Hayashi, Y., M. K. Furue, et al. (2010). "BMP4 induction of trophoblast from mouse embryonic stem cells in defined culture conditions on laminin." In Vitro Cell Dev Biol Anim 46(5): 416-30.

He, S., D. Pant, et al. (2008). "Lef1 Mediated Wnt Signaling Promotes the Initiation of Trophoblast Lineage Differentiation in Mouse Embryonic Stem Cells." Stem Cells.

He, S., D. Pant, et al. (2008). "Lymphoid enhancer factor 1-mediated Wnt signaling promotes the initiation of trophoblast lineage differentiation in mouse embryonic stem cells." Stem Cells 26(4): 842-9.

Hemberger, M., T. Nozaki, et al. (2003). "Parp1-deficiency induces differentiation of ES cells into trophoblast derivatives." Dev Biol 257(2): 371-81.

Hemberger, M., R. Udayashankar, et al. (2010). "ELF5-enforced transcriptional networks define an epigenetically regulated trophoblast stem cell compartment in the human placenta." Hum Mol Genet 19(12): 2456-67.

Home, P., S. Ray, et al. (2009). "GATA3 is selectively expressed in the trophectoderm of peri-implantation embryo and directly regulates Cdx2 gene expression." J Biol Chem 284(42): 28729-37.

Hough, S. R., I. Clements, et al. (2006). "Differentiation of mouse embryonic stem cells after RNA interference-mediated silencing of OCT4 and Nanog." Stem Cells 24(6): 1467-75.

Hyslop, L., M. Stojkovic, et al. (2005). "Downregulation of NANOG induces differentiation of human embryonic stem cells to extraembryonic lineages." Stem Cells 23(8): 1035-43.

Ivanova, N., R. Dobrin, et al. (2006). "Dissecting self-renewal in stem cells with RNA interference." Nature 442(7102): 533-8.

James, J. L., P. R. Stone, et al. (2006). "The regulation of trophoblast differentiation by oxygen in the first trimester of pregnancy." Hum Reprod Update 12(2): 137-44.

Katayama, M., M. R. Ellersieck, et al. (2010). "Development of Monozygotic Twin Mouse Embryos from the Time of Blastomere Separation at the Two-Cell stage to Blastocyst." Biol Reprod 82: 1237-47.

Kuijk, E. W., L. Du Puy, et al. (2008). "Differences in early lineage segregation between mammals." Dev Dyn 237(4): 918-27.

Loh, Y. H., Q. Wu, et al. (2006). "The Oct4 and Nanog transcription network regulates pluripotency in mouse embryonic stem cells." Nat Genet 38(4): 431-40.

Lu, C. W., A. Yabuuchi, et al. (2008). "Ras-MAPK signaling promotes trophectoderm formation from embryonic stem cells and mouse embryos." Nat Genet 40(7): 921-6.

Ludwig, T. E., V. Bergendahl, et al. (2006). "Feeder-independent culture of human embryonic stem cells." Nat Methods 3(8): 637-46.

Martin, G. R. (1981). "Isolation of a pluripotent cell line from early mouse embryos cultured in medium conditioned by teratocarcinoma stem cells." Proc Natl Acad Sci U S A 78(12): 7634-8.

Masui, S., Y. Nakatake, et al. (2007). "Pluripotency governed by Sox2 via regulation of Oct3/4 expression in mouse embryonic stem cells." Nat Cell Biol 9(6): 625-35.

Matin, M. M., J. R. Walsh, et al. (2004). "Specific knockdown of Oct4 and beta2-microglobulin expression by RNA interference in human embryonic stem cells and embryonic carcinoma cells." Stem Cells 22(5): 659-68.

Nagy, A., E. Gocza, et al. (1990). "Embryonic stem cells alone are able to support fetal development in the mouse." Development 110(3): 815-21.

Ng, R. K., W. Dean, et al. (2008). "Epigenetic restriction of embryonic cell lineage fate by methylation of Elf5." Nat Cell Biol 10(11): 1280-90.

Nishioka, N., K. Inoue, et al. (2009). "The Hippo signaling pathway components Lats and Yap pattern Tead4 activity to distinguish mouse trophectoderm from inner cell mass." Dev Cell 16(3): 398-410.

Nishioka, N., S. Yamamoto, et al. (2008). "Tead4 is required for specification of trophectoderm in pre-implantation mouse embryos." Mech Dev 125(3-4): 270-83.

Niwa, H., J. Miyazaki, et al. (2000). "Quantitative expression of Oct-3/4 defines differentiation, dedifferentiation or self-renewal of ES cells." Nat Genet 24(4): 372-6.

Niwa, H., Y. Toyooka, et al. (2005). "Interaction between Oct3/4 and Cdx2 determines trophectoderm differentiation." Cell 123(5): 917-929.

Peiffer, I., D. Belhomme, et al. (2007). "Simultaneous differentiation of endothelial and trophoblastic cells derived from human embryonic stem cells." Stem Cells Dev 16(3): 393-402.

Ralston, A., B. J. Cox, et al. (2010). "Gata3 regulates trophoblast development downstream of Tead4 and in parallel to Cdx2." Development 137(3): 395-403.

Roberts, R., H. Yong, et al. (2006). "What drives the formation of trophectoderm during early embryonic development?" J Reprod & Dev 52: S87-97.

Rossant, J. (2001). "Stem cells from the Mammalian blastocyst." Stem Cells 19(6): 477-82.

Rossant, J. (2007). "Stem cells and lineage development in the mammalian blastocyst." Reprod Fertil Dev 19(1): 111-8.

Rossant, J. and A. Spence (1998). "Chimeras and mosaics in mouse mutant analysis." Trends Genet 14(9): 358-63.

Russ, A. P., S. Wattler, et al. (2000). "Eomesodermin is required for mouse trophoblast development and mesoderm formation." Nature 404(6773): 95-9.

Schenke-Layland, K., E. Angelis, et al. (2007). "Collagen IV induces trophoectoderm differentiation of mouse embryonic stem cells." Stem Cells 25(6): 1529-38.

Schulz, L. C., T. Ezashi, et al. (2008). "Human Embryonic Stem Cells as Models for Trophoblast Differentiation." Placenta 29S: 10-16.

Tanaka, S., T. Kunath, et al. (1998). "Promotion of trophoblast stem cell proliferation by FGF4." Science 282(5396): 2072-5.

Tesar, P. J., J. G. Chenoweth, et al. (2007). "New cell lines from mouse epiblast share defining features with human embryonic stem cells." Nature 448(7150): 196-9.

Thomson, J. A., J. Itskovitz-Eldor, et al. (1998). "Embryonic stem cell lines derived from human blastocysts." Science 282(5391): 1145-7.

Thomson, J. A., J. Kalishman, et al. (1995). "Isolation of a primate embryonic stem cell line." Proc Natl Acad Sci U S A 92(17): 7844-8.

Tolkunova, E., F. Cavaleri, et al. (2006). "The caudal-related protein cdx2 promotes trophoblast differentiation of mouse embryonic stem cells." Stem Cells 24(1): 139-44.

van Genderen, C., R. M. Okamura, et al. (1994). "Development of several organs that require inductive epithelial-mesenchymal interactions is impaired in LEF-1-deficient mice." Genes Dev 8(22): 2691-703.

Vandevoort, C. A., T. L. Thirkill, et al. (2007). "Blastocyst-derived trophoblast stem cells from the rhesus monkey." Stem Cells Dev 16(5): 779-88.

Velkey, J. M. and K. S. O'Shea (2003). "Oct4 RNA interference induces trophectoderm differentiation in mouse embryonic stem cells." Genesis 37(1): 18-24.

Vong, Q. P., Z. Liu, et al. (2010). "A role for borg5 during trophectoderm differentiation." Stem Cells 28(6): 1030-8.

Wang, G., H. Zhang, et al. (2005). "Noggin and bFGF cooperate to maintain the pluripotency of human embryonic stem cells in the absence of feeder layers." Biochem Biophys Res Commun 330(3): 934-42.

Wang, J., S. Rao, et al. (2006). "A protein interaction network for pluripotency of embryonic stem cells." Nature 444(7117): 364-8.

Westfall, S. D., S. Sachdev, et al. (2008). "Identification of Oxygen-Sensitive Transcriptional Programs in Human Embryonic Stem Cells." Stem Cells Dev.

Wu, Z., W. Zhang, et al. (2008). "Combinatorial signals of activin/nodal and bone morphogenic protein regulate the early lineage segregation of human embryonic stem cells." J Biol Chem 283(36): 24991-5002.

Xiao, Q., L. Zeng, et al. (2007). "Stem cell-derived Sca-1+ progenitors differentiate into smooth muscle cells, which is mediated by collagen IV-integrin alpha1/beta1/alphav and PDGF receptor pathways." Am J Physiol Cell Physiol 292(1): C342-52.

Xu, R. H., X. Chen, et al. (2002). "BMP4 initiates human embryonic stem cell differentiation to trophoblast." Nat Biotechnol 20(12): 1261-4.

Xu, R. H., R. M. Peck, et al. (2005). "Basic FGF and suppression of BMP signaling sustain undifferentiated proliferation of human ES cells." Nat Methods 2(3): 185-90.

Xu, R. H., R. M. Peck, et al. (2005). "Basic FGF and suppression of BMP signaling sustain undifferentiated proliferation of human ES cells." Nat Methods 2(3): 185-190.

Yagi, R., M. J. Kohn, et al. (2007). "Transcription factor TEAD4 specifies the trophectoderm lineage at the beginning of mammalian development." Development 134(21): 3827-36.

Ying, Q. L., J. Nichols, et al. (2003). "BMP induction of Id proteins suppresses differentiation and sustains embryonic stem cell self-renewal in collaboration with STAT3." Cell 115(3): 281-92.

Zaehres, H., M. W. Lensch, et al. (2005). "High-efficiency RNA interference in human embryonic stem cells." Stem Cells 23(3): 299-305.

Zhang, P., J. Li, et al. (2007). "Short-term BMP-4 treatment initiates mesoderm induction in human embryonic stem cells." Blood.

14

Human Embryonic Stem Cells: A Model System for Delineating the Molecular Basis of Human Embryogenesis and Aging-related Diseases

Craig S. Atwood and Sivan Vadakkadath Meethal
University of Wisconsin, Departments of Medicine and Neurological Surgery and Geriatric Research, Education and Clinical Center, Veterans Administration Hospital, Madison, WI 53705
U.S.A.

1. Introduction

The recent development of techniques to culture human embryonic stem cells (hESC) has allowed the study of reproductive (pregnancy) hormones involved in the growth and development of the early embryo. Until the advent of hESC culture techniques, no model system existed that could readily assess the requirement for pregnancy hormones in the growth and development of the human embryo. Hormonal manipulation of developing embryoes *in utero* was technically cumbersome and complicated by competing *in vivo* maternal hormonal signals. This chapter describes recent experimental studies, utilizing hESC, aimed at identifying physiologically relevant signals that promote cell division, differentiation and apoptosis during early embryogenesis and summarizes our current knowledge of how reproductive hormones direct growth and development during embryogenesis. It also describes the potential for using hESC, embryoid bodies (EBs) and neuroectodermal rosettes to gain insights into how reproductive endocrine dyscrasia associated with menopause/andropause drives aberrant cell cycle signalling mechanisms leading to age-related diseases including neurodegeneration and associated cognitive decline.

2. Human embryogenesis

2.1 Hormonal regulation of pregnancy

Embryogenesis is a complex coordinated series of molecular and cellular changes that takes place within a well-defined internal environment. Human embryogenesis is orchestrated by a complex array of endocrine signals that commences with conception, is followed by the growth and development of the zygote into a blastocyst, its implantation into the endometrium, and the subsequent growth and development of the blastocyst into the neonate. Following conception, the developing embryo (zygote/morula/blastocyst) has 7-14 days in which to produce sufficient human chorionic gonadotropin (hCG), and subsequently progesterone (P_4; from both embryonic and corpus luteal sources), to allow implantation and halt degradation and discharging of the endometrium (menstruation) (Gupta, et al. 2007). The upregulation of hCG and P_4 not only allows for the maintenance of the endometrium, blastocyst attachment and synctiotrophoblast proliferation into the endometrium (Larson et al. 2003; Licht et al. 2001; Pepe and Albrecht 1995), but also prevents ovulation and prepares the immune, metabolic and psychological systems of the mother for pregnancy.

P_4 production is an *absolute* requirement for the maintenance of pregnancy (Larson et al. 2003). Indeed, administration of RU-486 (mifepristone), an anti-P_4 and anti-glucocorticosteroid agent to humans is used for the medical termination of pregnancies of up to 49 days gestation (up to 63 days gestation in Britain and Sweden), and in combination with prostaglandin E1, for termination of pregnancies between 13 and 24 weeks gestation (Fiala and Gemzel-Danielsson 2006). Inhibition of P_4 signaling using RU-486, a P_4 receptor (PR) competitive antagonist, results in endometrial decidual degeneration, trophoblast detachment and decreased syncytiotrophoblast production of hCG which in turn decreases P_4 production by the corpus luteum. In addition, RU-486 induces cervical softening and dilatation, release of endogenous prostaglandins and an increase in the sensitivity of the myometrium to the contractile effects of prostaglandins leading to the expulsion of the embryo/fetus (Gemzell-Danielsson et al. 2006).

Despite our understanding of the endocrinology of pregnancy, the lack of an appropriate model system limited experimentally our ability to answer fundamental questions such as what endocrine/paracrine/juxtacrine/autocrine factors 1) regulate embryonic cell division, 2) regulate cell migration, 3) specify differentiation into particular lineages, and 4) regulate apoptosis, during early embryogenesis.

2.2 Human embryonic stem cells: A model system for understanding the cellular and molecular mechanisms regulating early human embryogenesis

Thompson and colleagues first isolated pluripotent hESC lines from surplus embryos donated by individuals undergoing infertility treatment (Thomson et al. 1998). Inner mass cells isolated from these embryos were allowed to develop to the blastocyst stage and then passaged in defined media (to maintain pluripotency) to increase cell numbers. Five diploid cell lines (H1, H7, H9, H13 and H14) were obtained from 14 blastocysts. These cells are Oct-4, SSEA1, SSEA-3, SSEA-4, TRA 1-60, TRA 1-81 and alkaline phosphatase positive.

hESC derived from the inner cell mass of the blastocyst can be differentiated into EBs which resemble the early post-implantation embryo (blastocyst containing all 3 germ layers) (O'Shea 1999). hESC also can be differentiated into columnar neuroectodermal cells and mimics *in vivo* neuroectodermal development in terms of timing and morphology (Li and Zhang 2006). *In vitro*, hESC differentiate into primitive neuroectodermal (or neural precursor) cells at around day 10 and then neuroectodermal cells that exhibit neural tube-like rosettes in 14–17 days of differentiation in a chemically defined neural induction media (**Fig. 1**; Gallego et al., 2010; Zhang et al. 2001). These structures are predominantly composed of neuroectodermal cells akin to those that form the neural tube and are neural precursor cells/neural stem cells that can be further differentiated into different neural lineages. Considering hESC are equivalent to a 5-6 days embryo, development of the neuroectoderm *in vitro* takes about 18-20 days, the time window when the neural tube forms in a human embryo (Muller and O'Rahilly 1985; Zhang 2003). The ability to manipulate the hormonal milieu of cultured hESC, embryroid bodies or neuroectodermal rosettes during this time period, either positively or negatively, allows for the identification of signaling pathways involved in cell division, differentiation and apoptosis during embryogenesis.

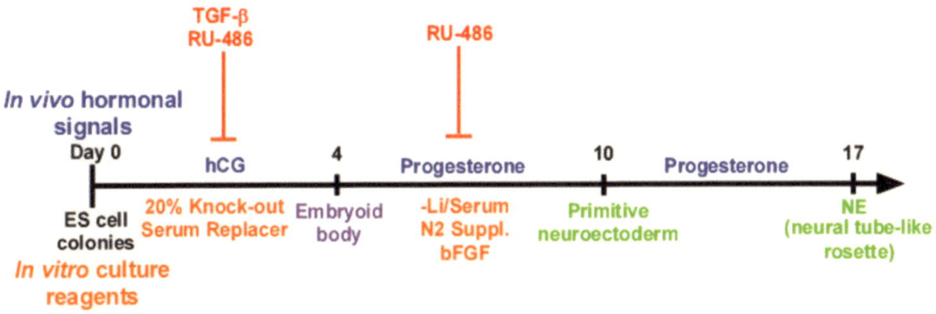

Fig. 1. Time course for the induction of embryoid bodies and neuroectodermal rosettes by hCG and P_4. hESC can be differentiated into primitive blastocyst-like structures (embryoid bodies (EBs) that contain the 3 germ layers) and then into primitive neuroectodermal cells (NE; or neural precursor cells) at ~day 10 and then NE that exhibit neural tube-like rosettes in 14–17 days of differentiation (Okabe et al. 1996; Zhang et al. 2001). Above Line: Physiological hormones secreted by trophoblasts demonstrated to induce cell division (hCG) during EB formation and cell differentiation (P_4) during neuroectodermal rosette formation. Below Line: Current unspecified medias used to generate EBs and neuroectodermal rosettes. These processes are negatively regulated by TGF-β signaling and by blocking signaling via PR

2.3 Trophoblastic hormone secretion during early embryogenesis

Zygotic division into a blastocyst establishes the extra-embryonic tissues (trophoblast layer or outer cell mass) and hypoblast (extraembryonic endoderm) that support the embryonic epiblast (inner cell mass) early in embryogenesis (Gilbert 2003). Trophoblasts secrete an array of hormones (Cemerikic et al. 1994; DiPirro and Kristal 2004; Gallego et al. 2009, 2010; Pidoux et al. 2007; Zhuang and Li 1991) including P_4, endorphins, hCG, 17β-estradiol (E_2) and gonadotropin-releasing hormone (GnRH) (**Fig. 2**). The dramatic elevation in the production of hCG by the trophoblastic layer of the blastocyst during the early embryonic stage (from 5 to ≥1000 mIU/mL in the maternal serum; Braunstein 1976; Pidoux et al. 2007) signals both the corpus lutea and trophoblast (Golos, et al. 2006) to synthesize and secrete P_4 (Bukovsky et al. 1995; Carr et al. 1982; Casper and Yen 1979; Duncan et al. 1996; Richardson and Masson 1985; Strauss et al. 2000). Trophoblastic secretion of these hormones appears to occur during the migration of the blastocyst through the fallopian tube and its implantation into the endometrium and subsequently from the placental tissues during later stages of embryogenesis. This is most clearly demonstrated by the elevation in maternal hCG during the early embryonic period.

The secretion of P_4, endorphins, hCG, E_2 and GnRH by trophoblasts that lie adjacent to the embryoblast in the blastocyst suggests that these hormones may *directly* signal the growth and development of the embryoblast. Evidence supporting this notion includes the presence of placental opioid-enhancing factor in amniotic fluid and placenta, and that the ingestion of placenta potentiates delta- and kappa-opioid antinociception (DiPirro and Kristal 2004). Likewise, trophoblastic and corpa luteal production of hCG/P_4 is markedly elevated post-conception and is obligatory for the maintenance of pregnancy (Larson et al. 2003). An autocrine/paracrine role for hCG secreted from invasive extravillous cytotrophoblasts (Handschuh et al. 2007) in the induction of neoangiogenesis during endometrial vascularization has previously been proposed (Licht et al. 2001). hCG signaling

via full-length LH/hCG receptors (LHCGR) on trophoblasts has been shown to modulate differentiation of the trophoblasts for subsequent villus projection and placentation.
Given the close spatial localization of trophoblasts to the embryoblast, and the availability of hESC that can be cultured continuously, i.e. embryoblast-derived stem cells, it has become possible to explore trophoblastic hormone function in the development of the embryo.

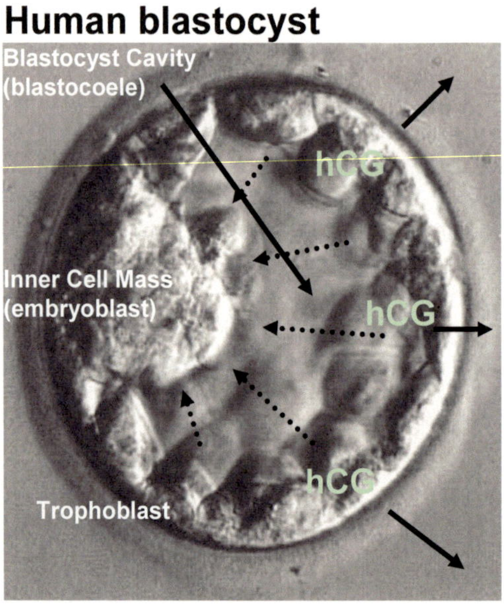

Fig. 2. Trophoblastic hCG secretion from the human blastocyst. Illustration of hCG autocrine (dotted line) and paracrine/endocrine (full line) secretion from the trophopblastic layer of a human blastocyst. Modified from: Muckle, C, Feinberg, E, Glob. libr. women's med., (ISSN: 1756-2228) 2008; DOI 10.3843/GLOWM.10002

3. Trophoblastic hormones and early human embryogenesis

3.1 Regulation of blastulation by human chorionic gonadotropin, progesterone and opioids

The first evidence for a function of any trophoblastic hormone in the regulation of human embryogenesis was ironically demonstrated by the finding that hCG induces the expression of the adhesion and neuritogenic protein amyloid-β precursor protein (AβPP; Porayette et al. 2007), a protein normally associated with the neurodegenerative pathology of Alzheimer's disease (AD; Hardy and Selkoe 2002). AβPP expression was detected at the transcriptional and translational levels (Porayette et al. 2007, 2009). These results indicated a critical molecular signaling link between the hormonal environment of pregnancy and the expression of AβPP in hESC that was suggestive of an important function for this protein during early human embryogenesis prior to the formation of neural precursor cells (Porayette et al. 2007).
That hCG could induce changes in epiblast protein expression was subsequently supported by the finding that hESCs express mRNA and protein for the full-length mature LHCGR (Gallego et al. 2008, 2010). That LHCGR is expressed on hESC implicates hCG as an

important signaling molecule in the growth and development of the embryo. LHCGR expression did not alter upon differentiation into EBs (structures that resemble early post-implantation embryos containing all three germ layers) (O'Shea 1999), or into neuroectodermal rosettes, which consist of >90% columnar neural precursor cells (NPC) and are the *in vitro* equivalent of a rudimentary neural tube (Gallego et al. 2008, 2010; Li and Zhang 2006). The comparable level of LHCGR expression between the different cell lineages was suggestive of 1) the existence of a tight regulatory system for the maintenance of hCG signaling during embryonic stem cell division and differentiation, and 2) a basal requirement for LH/hCG signaling during this early stage of embryogenesis. Indeed, hCG signaling via its hESC receptor was found to be essential for the proliferation of hESC (**Fig. 1**); inhibition of LHCGR signaling with P-antisense oligonucleotides suppressed hESC proliferation, as did a specific blocking antibody against the extracellular activation site of LHCGR, an effect that was reversed by treatment with hCG (Gallego et al. 2008, 2010). These data are supported by the known proliferative properties of (hyperglycosylated) hCG, which has been demonstrated to act as an autocrine factor on extravillous invasive cytotrophoblast cells to initiate and control invasion as occurs at implantation of pregnancy and the establishment of hemochorial placentation, and malignancy as occurs in invasive hydatidiform mole and choriocarcinoma (Cole 2009).

In addition to its cell cycle signaling activity, signaling of hCG via the hESC receptor rapidly upregulated steroidogenic acute regulatory protein (StAR)-mediated cholesterol transport and the synthesis of P_4, a neurogenic steroid (**Fig. 3**; Brewer et al. 1993; Wang et al. 2005). StAR, a key rate-limiting step in the production of sex steroids in reproductive tissues, was detected in hESC at both mRNA and protein (37-kDa, 30-kDa and 20-kDa variants) levels. hCG treatment dose-dependently suppressed the expression of these StAR variants, while P_4 treatment decreased the truncation of the 37-kDa to the 30/32-kDa variants of StAR, indicative of decreased cholesterol transport across the mitochondrial membrane for steroidogenesis (Epstein and Orme-Johnson 1991; Krueger and Orme-Johnson 1983; Pon et al. 1986; Stocco 2001; Stocco and Chen 1991; Yamazaki et al. 2006). Importantly, hCG treatment markedly increased P_4 secretion 15-fold, indicating that embryoblast-derived hESC already possess the machinery to transport cholesterol and synthesize sex steroids. Together, these findings indicate negative feedback pathways exist for the regulation of hCG/LH signaling and mitochondrial cholesterol uptake for the synthesis of sex steroids in hESC and differentiating lineages.

hESC and EBs express P_4 receptor A (PR-A; Gallego et al. 2010; Hong et al. 2004; Sauter et al. 2005) implying P_4 signals hESC differentiation. The requirement for P_4 in the differentiation of hESC into EBs was confirmed by the finding that RU-486, a PR competitive antagonist (Fiala 2006), potently inhibited the differentiation of hESC into EBs (**Fig. 1**). RU-486 treated colonies failed to form normal cystic structures after 10 days in culture, and instead formed solid irregular spheres that were ~20% the size of normal spheroidal EBs (Gallego et al. 2008, 2009, 2010).

Trophoblastic production of endorphins (Zhuang and Li 1991) also is crucial for embryogenesis (Gallego et al. 2009). Treatment of hESC colonies with the delta opioid receptor selective antagonist ICI 174,864 (Corbett et al. 1984; Paterson et al. 1984) inhibits the formation of the EB cystic structure, and instead forms non-spherical structures ~40% the size of normal spheroidal EBs. The mechanism by which opioid signaling promotes blastulation is unclear, however delta opioid antagonists may function to inhibit embryogenesis by regulating hCG release (Cemerikic et al. 1992) required for P_4 production. Thus, the tight regulation of hCG signaling, and sex steroid and opioid synthesis and signaling, is required to coordinate hESC proliferation and differentiation during gastrulation.

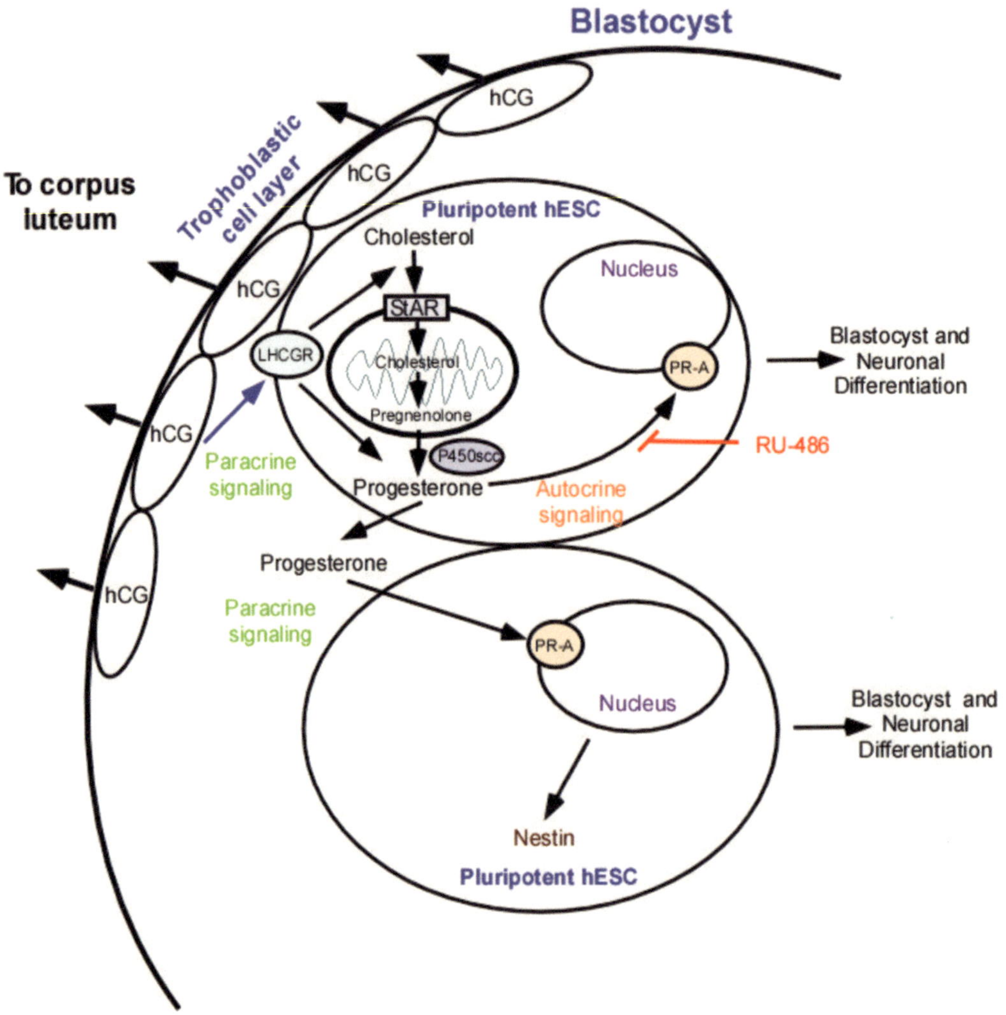

Fig. 3. Model of the autocrine and paracrine pathways regulating blastulation and neurulation. Putative autocrine and endocrine signalling pathways involved in cell proliferation, steroidogenesis and differentiation of the blastocyst and primitive neural tube. For details see Gallego et al., (2010)

3.2 Regulation of neurulation by human chorionic gonadotropin and progesterone

In addition to the obligatory signaling of P_4 for gastrulation, P_4 signaling also is required for the specification of NPCs from hESC (**Fig. 1**; Gallego et al. 2008, 2009, 2010). hCG treatment suppresses expression of the pluripotent marker Oct-3/4, suggesting hCG, or steroid production initiated by hCG signaling, could direct lineage commitment (Gallego et al. 2008,

2010). In the presence of P_4, hESC colonies differentiate into spherical structures containing a minimum of three neuroectodermal rosettes inside of the cavity, while hESC colonies treated without P_4 or with RU-486 failed to form rosettes with columnar neuroectodermal cells after 17 days in culture (Gallego et al. 2008, 2009, 2010). Morphological changes were more severe in the absence of P_4 than with RU-486. P_4, and to a lesser extend E_2, were found to increase the expression of nestin, an early marker of NPC formation, in hESC. RU-486 completely suppressed nestin expression. Interestingly, 'E_2 priming' is required for induction of PR expression in other tissues (Atwood et al. 2000; Mylonas et al. 2007). Thus, the increase in nestin expression with E_2 treatment may reflect increased PR expression together with endogenous P_4 signaling, and explain the current requirement for serum priming of hESC colonies in the preparation of neuroectodermal rosettes.

Interestingly, hESCs default towards a primitive neural stem cell fate if maintained for any length of time in culture (Munoz-Sanjuan and Brivanlou 2002). Since hCG treatment induces nestin expression in hESC, endogenous gonadotropin production by hESC or trophoblastic cells (Golos et al. 2006) may be sufficient for NPC formation, thereby explaining the intrinsic hormonal signals regulating the 'default pathway' of hESC differentiation into neuronal lineages (Munoz-Sanjuan and Brivanlou 2002).

These results suggest that trophoblastic hCG production adjacent to the embryoblast is required not only for trophoblast steroidogenesis and attachment of the blastocyst to the uterine wall (Wahabi et al. 2007), but also for signaling normal proliferation and differentiation of the epiblast (**Fig. 3**). hCG-induced P_4 synthesis therefore has, in addition to its role in uterine decidualization for the implantation and maintenance of pregnancy, an obligatory role prior to the formation of neural precursor cells, as well as an inductive role in the directed differentiation and specification of the first neuronal cell types (organogenesis) and the formation of the neural tube. While the structural importance of P_4 and alloprogesterone has previously been recognized by its early synthesis (by at least day 13) within the developing rat central nervous system (Pomata et al. 2000), these results demonstrate an early (within the first 7 days) and absolute requirement for P_4 during human blastulation and neurulation. In this respect, it has been shown that P_4 is necessary and sufficient (in Neurobasal media) for the maintenance and differentiation of primary hippocampal/cortical/striatal neurons *in vitro* (Brewer et al. 1993). That P_4 is the hormone regulating these key events is perhaps not surprising given its location high in the steroidogenic synthetic pathway; P_4 is the first steroid synthesized from pregnenolone, the precursor to all other steroids.

Opioid signaling also is required for neuroectodermal rosette formation since ICI 174,864 (Corbett et al. 1984; Paterson et al. 1984) inhibits normal neuroectodermal rosette formation and nestin expression (Gallego et al. 2009, 2010). Previous data has implicated P_4 as acting in the arcuate nucleus and anteroventral periventricular nucleus through beta-endorphin and dynorphin B neurons to affect preoptic area GnRH neurons and gonadotropin secretion (Dufourny et al. 2005; Gu and Simerly 1994). Thus, delta opioid receptor signaling is required both for normal human blastulation and neurulation, but it remains to be determined if there is crosstalk between opioid signaling and the regulation of GnRH/gonadotropin secretion.

Previous studies have demonstrated the importance of P_4 and related steroids as neurotrophic agents that promote adult neurogenesis, neuronal survival and neuroprotection (Brewer et al. 1993; Ciriza et al. 2004; Cutler et al. 2006; Guo et al. 2006;

Mauch et al. 2001; Schumacher et al. 2003, 2004; VanLandingham et al. 2006; Wang et al. 2005). Clinical studies supporting the neurotrophic actions of P_4 administration are demonstrated by the decrease in mortality rate and improved outcome following acute traumatic brain injury in humans (Wright et al. 2007).

Dependent upon the timing of administration during pregnancy, suppression of P_4 signaling with RU-486 (Fiala and Gemzel-Danielsson 2006) aside from intrauterine disruptive functions (decidual breakdown and trophoblast detachment) also will disrupt time-sensitive developmental processes. The requirement of P_4 during cavitation processes indicates the structural influence of these molecular pathways on the developing embryo within the first 7 days, but also on the formation of the neural tube at around day 17-19, which will influence future neural connectivity. The relative binding affinity of RU-486 for the PR is twice that of P_4 (Heikinheimo et al. 2003), and is used at a dose of 200-600 mg for the termination of pregancies (Fiala 2006) (this equates to ~6-19 μM, equivalent to that used in Gallego et al. 2008, 2009, 2010). Thus, the abortifacient effects of RU-486 in blocking PR signaling also extend to blocking blastulation and neurulation and the normal growth and development of the embryo.

3.3 Regulation of organogenesis by human chorionic gonadotropin and progesterone

Aside from the induction of blastulation and neurulation early in embryogenesis, hCG/LH and P_4 signaling may play a role in the development of other tissues (LH is the adult hCG homolog with 83 % homology and binds the same receptor - LHCGR). LHCGR and PR have been identified on numerous reproductive and non-reproductive tissues (Ascoli et al. 2002; Bouchard 1999; Bukovsky et al. 2003; Mulac-Jericevic and Conneely 2004). With regard hCG/LH, the free glycoprotein α-subunit of gonadotropins has been shown to stimulate differentiation of prolactin cells in the pituitary (Avsian-Kretchmer and Hsueh 2004) and endometrial stromal cell decidualization in the placenta (Blithe et al. 1991). Although it has not been demonstrated if hCG/LH has a developmental function during organogenesis, hyperglycosylated hCGβ has potent cell growth and invasion properties as observed in early pregnancy, gestational choriocarcinoma and testicular cancers (Cole 2009). Interestingly, in the adult brain, subcutaneous administration of LH has been shown to induce neurogenesis in the hippocampus of the adult mouse (Mak et al. 2007), while in sheep there is evidence that GnRH directly, or indirectly via LH, induces neurogenesis in the hippocampus (Hawken et al. 2009). hCG also is known to promote angiogenesis by inducing the up-regulation of vascular endothelial growth factor (Berndt et al. 2006; Licht et al. 2002; Zygmunt et al. 2002) and P_4 (Rogers et al. 2009).

The potential for P_4 to regulate organogenesis has been reported during puberty and adulthood, where P_4 is obligatory for the development of the tertiary ducts on the mammary gland, and the functional differentiation of the lobuloalveolar system from the lobular buds (Atwood et al. 2000). In the adult, P_4 and related metabolites have been demonstrated to regulate bone formation (Prior 1990), promote angiogenesis and arteriogenesis (Rogers et al. 2009), promote formation of the placenta and promote neurogenesis by increasing rat neuroprogenitor cell proliferation and human neural stem cell proliferation (Wang et al. 2005). While our knowledge of hCG/LH and P_4 during organogenesis is rudimentary at this point, the above evidence indicates these hormones likely play important functions in many tissues during organogenesis.

4. Utilizing human embryonic stem cells as a model system to understand the molecular basis of aging-related diseases

It is becoming clear that the hormonal mechanisms that regulate the coordinated division and differentiation of cells early in life become dysregulated later in life following menopause and during andropause (which starts at ~ 30 years of age; see Vadakkadath Meethal et al. 2005; Atwood and Bowen 2011), and that these dysregulated hormonal mechanisms drive aberrant re-entry of cells into the cell cycle and their abortive death leading to tissue compromise and ultimately disease (Atwood et al. 2005).

The reproductive endocrine dyscrasia associated with menopause and andropause are intimately associated with disease (Bowen and Atwood 2004; Atwood and Bowen, 2011). The decline in sex steroid production by the gonads following menopause and during andropause leads to a loss of hypothalamic feedback inhibition that stimulates GnRH and gonadotropin production (Carr 1998). In addition, the decrease in gonadal inhibin production at this time (Reichlin 1998) results in decreased activin receptor inhibition, and together with the increase in bioavailable activin (Gray et al. 2002) leads to a further increase in the secretion of GnRH and gonadotropins (MacConell et al. 1999; Schwall et al. 1988; Weiss et al. 1993). Thus, the lack of negative feedback from the ovary (P_4, E_2 and inhibin) is responsible for the unopposed and marked elevations in the secretion of GnRH and gonadotropins with ovarian and testicular senescence (Atwood et al. 2005; Chakravarti et al. 1976; Neaves et al. 1984; Reame et al. 1996; Schmidt et al. 1996).

The concentration of brain sex steroids, including P_4, is a mixture of peripherally derived sex steroids, converted peripheral steroids, and neuro-sex steroids. The contribution of peripheral sex steroids to total brain sex steroids is unknown, but post-reproductive declines in peripheral P_4 would be expected to impact brain P_4 concentrations. While elevations in GnRH and gonadotropin concentrations might promote brain neurosteroid production (see Vadakkadath Meethal et al. 2009), including P_4, it is not known if this is sufficient to counter the loss of peripherally-derived sex steroids. The consequences of these hormonal alterations are discussed below in the context of AD.

4.1 Alzheimer's disease

Dementia accounts for 3% of deaths in the USA (CDC National Vital Statistics Report, 2009) although by the age of 85 ~45% of the population has some form of dementia. AD accounts for ~70% of all dementia cases and is characterized neurologically by age-related progressive memory loss, impairments in behavior, language, and visuo-spatial skills (Atwood et al. 2005; Vadakkadath Meethal et al. 2005).

Unlike development, where mitogenic and differentiative signaling are both elevated, senescence is associated with elevated mitogenic (i.e. gonadotropins, GnRH) signaling but decreased differentiative signaling (i.e. sex steroids).

The age-related loss of P_4 is of particular importance given the differentiative properties of this steroid described above. Also of importance are the aging-related elevations in serum GnRH, FSH and LH, especially given the known proliferative properties of hCG/LH (Cole 2009; Gallego et al. 2008, 2010). In this regard, LH/hCG is known to have powerful mitogenic properties in certain reproductive tissues (Davies et al. 1999; Harris et al. 2002; Horiuchi et al. 2000; Sriraman et al. 2001; Webber and Sokoloff 1981), the brain (Mak et al., 2007; Hawken et al., 2009) and are frequently expressed by tumor cells (Krichevsky et al.

1995; Whitfield and Kourides 1985; Yokotani et al. 1997 and reviewed in Cole 2009). Thus, these multiple changes in hormonal signaling with menopause/andropause, i.e. increased mitogenic signaling but decreased differentiation signaling (*dyotic signaling*) might be expected to impact normal cell cycle dynamics. Indeed, accumulating evidence suggests there is a reactivation of the cell cycle with aging (see Bowen and Atwood 2004) as has been demonstrated for post-mitotic pyramidal neurons of the brains of aged individuals with AD (Herrup and Yang 2007; Raina et al. 2000). This data includes 1) the ectopic expression of cell cycle proteins in those regions of the brain affected by AD, but not in areas unaffected by AD pathology or in control brains, 2) chromosomal replication (endoreduplication) in differentiated AD neurons, demonstrating entry into S-phase of the cell cycle, 3) elevated cytoplasmic mitochondrial DNA and Cox-1 expression, suggestive of *de novo* mitochondrion synthesis, and 4) upregulated growth factor signal transduction pathways. Importantly, the spatio-temporal expression of sex hormone receptors throughout the brain is in those areas of AD neurodegeneration. Other parallels between embryonic neurogenesis and adult neurodegeneration include the expression of AβPP, secretases and tau, together with the processing of AβPP either towards the amyloidogenic or non-amyloidogenic pathways, and the phosphorylation of tau (Porayette et al. 2009) (Atwood and Porayette, unpublished data). Similarly, the fetal brain has been reported to display a number of biochemical similarities to the AD brain, namely the presence of Aβ and AβPP (Arai et al. 1997; Takashima et al. 1990), presenilin-1 expression (Berezovska et al. 1997) and hyper-phosphorylated tau (Goedert et al. 1993). The phosphorylation of tau is a mitogenic-associated event that normally occurs during metaphase of neuronal division, and is observed during differentiation of neurons in the fetal brain (Goedert et al. 1993; Liu, et al. 2004).

This increased developmental protein expression in the AD brain suggests reactivation of the cell cycle in differentiated neurons of the AD brain (Herrup and Yang 2007) and explains the majority of the biochemical and pathological features associated with the disease (Atwood et al. 2005; Meethal et al. 2005). Two recent studies support this claim. Forced cell cycle activation in terminally differentiated neurons via conditional expression of the simian virus 40 large T antigen (oncogene) forms Aβ deposits and tau pathology in the mouse cortex (Park et al. 2007). Similarly, forced cell cycle activation in primary neurons is accompanied by tau phosphorylation (McShea et al. 2007). These data suggest that AD neuropathology is a result of an imbalance in cell cycle regulation in the adult brain.

4.2 Amyloid-β precursor protein and neurogenesis

That amyloidogenic pathways are involved in neurogenesis has recently been reported by a number of workers (Calafiore et al. 2006; Heo et al. 2007; Liu et al. 2004; Lopez-Toledano and Shelanski 2004). In this context, an increase in neurogenesis has been reported in young transgenic mice overexpressing human mutant AβPP (Jin et al. 2004; Lopez-Toledano and Shelanski 2007). Moreover, the overexpression of wild-type or FAD mutant AβPP, which promotes Aβ generation (Citron et al. 1997), also has been shown to promote the re-entry of primary neurons into the cell cycle, as demonstrated by the induction of DNA synthesis and cell cycle markers (McPhie et al. 2003). Not surprisingly, AβPP has structural similarity to growth factors (Trapp and Hauer 1994) and modulates several important neurotrophic functions, including neuritogenesis, synaptogenesis, and synaptic plasticity (Gralle and Ferreira 2007).

4.3 Hormonal regulation of neurogenesis via modulation of amyloid-β precursor protein metabolism

In hESC, the differential processing of AβPP via secretase enzymes regulates the proliferation and differentiation of hESC; processing towards the amyloidogenic pathway is associated with cell proliferation, processing towards the non-amyloidogenic pathway is associated with cell specification and differentiation. Specifically, P_4 induces processing of AβPP towards the production of the soluble AβPPα in hESC (Porayette et al. 2009), which has known differentiative properties (Milward et al. 1992). Similarly, E_2 has been shown to stimulate the processing of AβPP by the nonamyloidogenic α-secretory pathway and reduces cellular Aβ production in both nonneuronal (Jaffe et al. 1994) and neuronal cultures (Greenfield et al. 2002; Manthey et al. 2001; Xu et al. 1998). Conversely, the loss of sex steroids and elevation in gonadotropins following ovariectomy has been shown to increase Aβ generation in non-transgenic animals (Petanceska et al. 2000). Importantly, we have demonstrated that LH promotes the processing of AβPP towards the amyloidogenic pathway *in vitro*, while suppression of serum LH in mice using GnRH agonists decreases the concentration of brain Aβ1-40 and Aβ1-42, the 2 major variants that deposit in the AD brain (Bowen et al. 2004). It is therefore plausible that the interaction of these hormonal pathways on the modulation of the processing of AβPP may regulate cell cycle events throughout life, with dyotic signaling by these hormones leading to the reactivation of the cell cycle in differentiated neurons of the AD brain (Herrup and Yang 2007). This aberrant, albeit unsuccessful, re-entry of neurons into the cell cycle leads to synapse contraction and neuron death (see Atwood et al. 2005; Herrup et al. 2004; Raina et al. 2000 for reviews). In addition to the loss of neurons following the reactivation of the cell cycle in differentiated neurons, it is possible that dyotic signaling prevents normal neurogenesis from resident neural stem cells, thereby preventing replacement of neurons. Further studies are required to determine whether post-reproductive levels of GnRH/gonadotropins are sufficient to induce neurosteroidogenesis in neural stem cells and neuronal cell types, and whether the level of post-reproductive neurosteroid synthesis dictates normal or dyotic signaling in these cell types.

5. Conclusion

The advent of the human embryonic stem cell era has allowed experimental determination of the physiological hormone requirements for early embryogenesis. In this respect, although progestagens are often considered primarily reproductive hormones with maternal influences, it is now clear that progestagens and hCG are essential for the growth and development of the embryo as well as the normal health of the brain throughout life. Paracrine/juxtacrine signaling of hCG (and opioids) for mobilization of cholesterol for P_4 production by the epiblast/synctiotrophoblast following conception is obligatory for human blastulation and neurulation (**Fig. 3**). This paracrine/juxtacrine signaling by extraembryonic tissues is the commencement of trophic support by placental tissues in the growth and development of the human embryo. The identification of these hormones that regulate cell proliferation and differentiation of hESC *in vitro* will help direct the development of medias that most closely reflect their *in utero* environment for *in vitro* culture of these cells and their differentiation towards various cell lineages. Such medias could be used to further delineate the molecular basis of embryogenesis and organogenesis, as well as establish *ex utero* embryonic and fetal cultures.

The above discussion also indicates how hESC can provide insight into neurodegenerative disease (and likely many other aging-related diseases). While appropriate gonadotropin/ GnRH and P_4 signaling is necessary for normal growth and development during embryogenesis, fetal life and childhood, and for the maintenance of brain health during adult reproductive life, the unopposed elevations in GnRH/gonadotropins with the loss of sex steroids following menopause/andropause appears to lead to dysregulation of cell cycle events (Bowen and Atwood 2004).

It will be possible in the future to use hESC cells as a model for understanding how endocrine, paracrine and autocrine signals regulate cell cycle progression, from entry to exit, and how dysregulated signaling leads to entry but no exit from the cell cycle, i.e. leading to endoreduplication such as is apparent in the pyramidal neurons of the AD brain. Thus, hESC are a useful cell model system for examining questions related to early embryonic neurogenesis, adult neuroregeneration and neurodegeneration.

6. References

Arai Y, Suzuki A, Mizuguchi M & Takashima S 1997 Developmental and aging changes in the expression of amyloid precursor protein in Down syndrome brains. *Brain Dev* 19 290-294.

Ascoli M, Fanelli F & Segaloff DL 2002 The lutropin/choriogonadotropin receptor, a 2002 perspective. *Endocr Rev* 23 141-174.

Atwood C & Bowen R 2011 The Reproductive-Cell Cycle Theory of Aging: An Update. *Experimental Gerontology* In press.

Atwood CS, Hovey RC, Glover JP, Chepko G, Ginsburg E, Robison WG & Vonderhaar BK 2000 Progesterone induces side-branching of the ductal epithelium in the mammary glands of peripubertal mice. *J Endocrinol* 167 39-52.

Atwood CS, Meethal SV, Liu T, Wilson AC, Gallego M, Smith MA & Bowen RL 2005 Dysregulation of the hypothalamic-pituitary-gonadal axis with menopause and andropause promotes neurodegenerative senescence. *J Neuropathol Exp Neurol* 64 93-103.

Avsian-Kretchmer O & Hsueh AJ 2004 Comparative genomic analysis of the eight-membered ring cystine knot-containing bone morphogenetic protein antagonists. *Mol Endocrinol* 18 1-12.

Berezovska O, Xia MQ, Page K, Wasco W, Tanzi RE & Hyman BT 1997 Developmental regulation of presenilin mRNA expression parallels notch expression. *J Neuropathol Exp Neurol* 56 40-44.

Berndt S, Perrier d'Hauterive S, Blacher S, Pequeux C, Lorquet S, Munaut C, Applanat M, Herve MA, Lamande N, Corvol P, et al. 2006 Angiogenic activity of human chorionic gonadotropin through LH receptor activation on endothelial and epithelial cells of the endometrium. *Faseb J* 20 2630-2632.

Blithe DL, Richards RG & Skarulis MC 1991 Free alpha molecules from pregnancy stimulate secretion of prolactin from human decidual cells: a novel function for free alpha in pregnancy. *Endocrinology* 129 2257-2259.

Bouchard P 1999 Progesterone and the progesterone receptor. *J Reprod Med* 44 153-157.

Bowen RL & Atwood CS 2004 Living and dying for sex. A theory of aging based on the modulation of cell cycle signaling by reproductive hormones. *Gerontology* 50 265-290.

Bowen RL, Verdile G, Liu T, Parlow AF, Perry G, Smith MA, Martins RN & Atwood CS 2004 Luteinizing hormone, a reproductive regulator that modulates the processing of amyloid-beta precursor protein and amyloid-beta deposition. *J Biol Chem* 279 20539-20545.

Braunstein GD, Rasor J., Danzer H. 1976 Serum human chorionic gonadotropin levels throughout normal pregnancy. *Am J Obstet Gynecol* 126 678.

Brewer GJ, Torricelli JR, Evege EK & Price PJ 1993 Optimized survival of hippocampal neurons in B27-supplemented Neurobasal, a new serum-free medium combination. *J Neurosci Res* 35 567-576.

Bukovsky A, Caudle MR, Keenan JA, Wimalasena J, Upadhyaya NB & Van Meter SE 1995 Is corpus luteum regression an immune-mediated event? Localization of immune system components and luteinizing hormone receptor in human corpora lutea. *Biol Reprod* 53 1373-1384.

Bukovsky A, Indrapichate K, Fujiwara H, Cekanova M, Ayala ME, Dominguez R, Caudle MR, Wimalsena J, Elder RF, Copas P, et al. 2003 Multiple luteinizing hormone receptor (LHR) protein variants, interspecies reactivity of anti-LHR mAb clone 3B5, subcellular localization of LHR in human placenta, pelvic floor and brain, and possible role for LHR in the development of abnormal pregnancy, pelvic floor disorders and Alzheimer's disease. *Reprod Biol Endocrinol* 1 46.

Calafiore M, Battaglia G, Zappala A, Trovato-Salinaro E, Caraci F, Caruso M, Vancheri C, Sortino MA, Nicoletti F & Copani A 2006 Progenitor cells from the adult mouse brain acquire a neuronal phenotype in response to beta-amyloid. *Neurobiol Aging* 27 606-613.

Carr BR 1998 Williams Textbook of Endocrinology. pp 751-817. Ed FD Wilson JD, Kronenberg HM, Larsen PR. Philadelphia PA: WB Saunders Co.

Carr BR, MacDonald PC & Simpson ER 1982 The role of lipoproteins in the regulation of progesterone secretion by the human corpus luteum. *Fertil Steril* 38 303-311.

Casper RF & Yen SS 1979 Induction of luteolysis in the human with a long-acting analog of luteinizing hormone-releasing factor. *Science* 205 408-410.

Cemerikic B, Schabbing R & Ahmed MS 1992 Selectivity and potency of opioid peptides in regulating human chorionic gonadotropin release from term trophoblast tissue. *Peptides* 13 897-903.

Cemerikic B, Zamah R & Ahmed MS 1994 Opioids regulation of human chorionic gonadotropin release from trophoblast tissue is mediated by gonadotropin releasing hormone. *J Pharmacol Exp Ther* 268 971-977.

Chakravarti S, Collins WP, Forecast JD, Newton JR, Oram DH & Studd JW 1976 Hormonal profiles after the menopause. *Br. Med. J.* 2 784-787.

Ciriza I, Carrero P, Azcoitia I, Lundeen SG & Garcia-Segura LM 2004 Selective estrogen receptor modulators protect hippocampal neurons from kainic acid excitotoxicity: differences with the effect of estradiol. *J Neurobiol* 61 209-221.

Citron M, Westaway D, Xia W, Carlson G, Diehl T, Levesque G, Johnson-Wood K, Lee M, Seubert P, Davis A, et al. 1997 Mutant presenilins of Alzheimer's disease increase production of 42-residue amyloid beta-protein in both transfected cells and transgenic mice. *Nat Med* 3 67-72.

Cole LA 2009 New discoveries on the biology and detection of human chorionic gonadotropin. *Reprod Biol Endocrinol* 7 8.

Corbett AD, Gillan MG, Kosterlitz HW, McKnight AT, Paterson SJ & Robson LE 1984 Selectivities of opioid peptide analogues as agonists and antagonists at the delta-receptor. *Br J Pharmacol* 83 271-279.

Cutler SM, VanLandingham JW, Murphy AZ & Stein DG 2006 Slow-release and injected progesterone treatments enhance acute recovery after traumatic brain injury. *Pharmacol Biochem Behav* 84 420-428.

Davies BR, Finnigan DS, Smith SK & Ponder BAAogsponmoseGE 1999 Administration of gonadotropins stimulates proliferation of normal mouse ovarian surface epithelium. *Gynecol. Endocrinol.* 13 75-81.

DiPirro JM & Kristal MB 2004 Placenta ingestion by rats enhances delta- and kappa-opioid antinociception, but suppresses mu-opioid antinociception. *Brain Res* 1014 22-33.

Dufourny L, Caraty A, Clarke IJ, Robinson JE & Skinner DC 2005 Progesterone-receptive beta-endorphin and dynorphin B neurons in the arcuate nucleus project to regions of high gonadotropin-releasing hormone neuron density in the ovine preoptic area. *Neuroendocrinology* 81 139-149.

Duncan WC, McNeilly AS, Fraser HM & Illingworth PJ 1996 Luteinizing hormone receptor in the human corpus luteum: lack of down-regulation during maternal recognition of pregnancy. *Hum Reprod* 11 2291-2297.

Epstein LF & Orme-Johnson NR 1991 Regulation of steroid hormone biosynthesis. Identification of precursors of a phosphoprotein targeted to the mitochondrion in stimulated rat adrenal cortex cells. *J Biol Chem* 266 19739-19745.

Fiala C 2006 Exelgyn, (February 2006). "Mifegyne UK Summary of Product Characteristics (SPC)" Retrieved on 9-9-08; http://emc.medicines.org.uk/emc/assets/c/html/displaydoc.asp?documentid=617.

Fiala C & Gemzel-Danielsson K 2006 Review of medical abortion using mifepristone in combination with a prostaglandin analogue. *Contraception* 74 66-86.

Gallego MJ, Porayette P, Kaltcheva M, Bowen RL, Vadakkadath Meethal S & Atwood CS 2008 Trophoblastic hormones direct early human embryogenesis. *Available from Nature Precedings* http://hdl.handle.net/10101/npre.2008.2671.1.

Gallego MJ, Porayette P, Kaltcheva M, Bowen RL, Vadakkadath Meethal S & Atwood CS 2010 The pregnancy hormones hCG and progesterone induce human embryonic stem cell proliferation and differentiation into neuroectodermal rosettes. *Stem Cell Research and Therapy* 1:28.

Gallego MJ, Porayette P, Kaltcheva MM, Meethal SV & Atwood CS 2009 Opioid and progesterone signaling is obligatory for early human embryogenesis. *Stem Cells Dev* 18 737-740.

Gemzell-Danielsson K, Bygdeman M & Aronsson A 2006 Studies on uterine contractility following mifepristone and various routes of misoprostol. *Contraception* 74 31-35.

Gilbert S 2003 *Developmental Biology*. Sunderland: Sinauer Associates.

Goedert M, Jakes R, Crowther RA, Six J, Lubke U, Vandermeeren M, Cras P, Trojanowski JQ & Lee VM 1993 The abnormal phosphorylation of tau protein at Ser-202 in Alzheimer disease recapitulates phosphorylation during development. *Proc Natl Acad Sci U S A* 90 5066-5070.

Golos TG, Pollastrini LM & Gerami-Naini B 2006 Human embryonic stem cells as a model for trophoblast differentiation. *Semin Reprod Med* 24 314-321.

Gralle M & Ferreira ST 2007 Structure and functions of the human amyloid precursor protein: the whole is more than the sum of its parts. *Prog Neurobiol* 82 11-32.

Gray PC, Bilizikjian LM & Vale W 2002 Antagonism of activin by inhibin and inhibin receptors: a functional role for betaglycan. *Mol. Cell. Endocrinol.* 188 254-260.

Greenfield JP, Leung LW, Cai D, Kaasik K, Gross RS, Rodriguez-Boulan E, Greengard P & Xu H 2002 Estrogen lowers Alzheimer beta-amyloid generation by stimulating trans-Golgi network vesicle biogenesis. *J Biol Chem* 277 12128-12136.

Gu G & Simerly RB 1994 Hormonal regulation of opioid peptide neurons in the anteroventral periventricular nucleus. *Horm Behav* 28 503-511.

Guo Q, Sayeed I, Baronne LM, Hoffman SW, Guennoun R & Stein DG 2006 Progesterone administration modulates AQP4 expression and edema after traumatic brain injury in male rats. *Exp Neurol* 198 469-478.

Gupta AK, Holzgreve W & Hahn S 2007 Decrease in lipid levels of syncytiotrophoblast micro-particles reduced their potential to inhibit endothelial cell proliferation. *Arch Gynecol Obstet.*

Handschuh K, Guibourdenche J, Tsatsaris V, Guesnon M, Laurendeau I, Evain-Brion D & Fournier T 2007 Human chorionic gonadotropin expression in human trophoblasts from early placenta: comparative study between villous and extravillous trophoblastic cells. *Placenta* 28 175-184.

Hardy J & Selkoe DJ 2002 The amyloid hypothesis of Alzheimer's disease: progress and problems on the road to therapeutics. *Science* 297 353-356.

Harris D, Bonfil D, Chuderland D, Kraus S, Seger R & Naor Z 2002 Activation of MAPK cascades by GnRH: ERK and Jun N-terminal kinase are involved in basal and GnRH-stimulated activity of the glycoprotein hormone LHbeta-subunit promoter. *Endocrinology* 143 1018-1025.

Hawken PA, Jorre TJ, Rodger J, Esmaili T, Blache D & Martin GB 2009 Rapid induction of cell proliferation in the adult female ungulate brain (Ovis aries) associated with activation of the reproductive axis by exposure to unfamiliar males. *Biol Reprod* 80 1146-1151.

Heikinheimo O, Kekkonen R & Lahteenmaki P 2003 The pharmacokinetics of mifepristone in humans reveal insights into differential mechanisms of antiprogestin action. *Contraception* 68 421-426.

Heo C, Chang KA, Choi HS, Kim HS, Kim S, Liew H, Kim JA, Yu E, Ma J & Suh YH 2007 Effects of the monomeric, oligomeric, and fibrillar Abeta42 peptides on the proliferation and differentiation of adult neural stem cells from subventricular zone. *J Neurochem* 102 493-500.

Herrup K, Neve R, Ackerman SL & Copani A 2004 Divide and die: cell cycle events as triggers of nerve cell death. *J Neurosci* 24 9232-9239.

Herrup K & Yang Y 2007 Cell cycle regulation in the postmitotic neuron: oxymoron or new biology? *Nat Rev Neurosci* 8 368-378.

Hong SH, Nah HY, Lee YJ, Lee JW, Park JH, Kim SJ, Lee JB, Yoon HS & Kim CH 2004 Expression of estrogen receptor-alpha and -beta, glucocorticoid receptor, and progesterone receptor genes in human embryonic stem cells and embryoid bodies. *Mol Cells* 18 320-325.

Horiuchi A, Nikaido T, Yoshizawa T, Itoh K, Kobayashi Y, Toki T, Konishi I & Fujii S 2000 HCG promotes proliferation of uterine leiomyomal cells more strongly than that of myometrial smooth muscle cells in vitro. *Mol. Hum. Reprod.* 6 523-528.

Jaffe AB, Toran-Allerand CD, Greengard P & Gandy SE 1994 Estrogen regulates metabolism of Alzheimer amyloid beta precursor protein. *J Biol Chem* 269 13065-13068.

Jin K, Galvan V, Xie L, Mao XO, Gorostiza OF, Bredesen DE & Greenberg DA 2004 Enhanced neurogenesis in Alzheimer's disease transgenic (PDGF-APPSw,Ind) mice. *Proc Natl Acad Sci U S A* 101 13363-13367.

Krichevsky A, Campbell-Acevedo EA, Tong JY & Acevedo HF 1995 Immunological detection of membrane-associated human luteinizing hormone correlates with gene expression in cultured human cancer and fetal cells. *Endocrinology* 136 1034-1039.

Krueger RJ & Orme-Johnson NR 1983 Acute adrenocorticotropic hormone stimulation of adrenal corticosteroidogenesis. Discovery of a rapidly induced protein. *J Biol Chem* 258 10159-10167.

Larson P, Kronenberg H, Melmed S & Polonsky K 2003 *Williams Textbook of Endocrinology.* Philadelphia, PA: Saunders.

Li X & Zhang S 2006 In vitro differentiation of neural precursors from human embryonic stem cells. *Methods in Molecular Biology* 331 169-177.

Licht P, Russu V, Lehmeyer S, Moll J, Siebzehnrubl E & Wildt L 2002 Intrauterine microdialysis reveals cycle-dependent regulation of endometrial insulin-like growth factor binding protein-1 secretion by human chorionic gonadotropin. *Fertil Steril* 78 252-258.

Licht P, Russu V & Wildt L 2001 On the role of human chorionic gonadotropin (hCG) in the embryo-endometrial microenvironment: implications for differentiation and implantation. *Semin Reprod Med* 19 37-47.

Liu T, Perry G, Chan HW, Verdile G, Martins RN, Smith MA & Atwood CS 2004 Amyloid-beta-induced toxicity of primary neurons is dependent upon differentiation-associated increases in tau and cyclin-dependent kinase 5 expression. *J Neurochem* 88 554-563.

Lopez-Toledano MA & Shelanski ML 2004 Neurogenic effect of beta-amyloid peptide in the development of neural stem cells. *J Neurosci* 24 5439-5444.

Lopez-Toledano MA & Shelanski ML 2007 Increased Neurogenesis in Young Transgenic Mice Overexpressing Human APP_{Sw, Ind}. *J Alzheimers Dis* 12 229-240.

MacConell LA, Lawson MA, Mellon PL & Roberts VJ 1999 Activin A regulation of gonadotropin-releasing hormone synthesis and release in vitro. *Neuroendocrinology* 70 246-254.

Mak GK, Enwere EK, Gregg C, Pakarainen T, Poutanen M, Huhtaniemi I & Weiss S 2007 Male pheromone-stimulated neurogenesis in the adult female brain: possible role in mating behavior. *Nat Neurosci* 10 1003-1011.

Manthey D, Heck S, Engert S & Behl C 2001 Estrogen induces a rapid secretion of amyloid beta precursor protein via the mitogen-activated protein kinase pathway. *Eur J Biochem* 268 4285-4291.

Mauch DH, Nagler K, Schumacher S, Goritz C, Muller EC, Otto A & Pfrieger FW 2001 CNS synaptogenesis promoted by glia-derived cholesterol. *Science* 294 1354-1357.

McPhie DL, Coopersmith R, Hines-Peralta A, Chen Y, Ivins KJ, Manly SP, Kozlowski MR, Neve KA & Neve RL 2003 DNA synthesis and neuronal apoptosis caused by familial Alzheimer disease mutants of the amyloid precursor protein are mediated by the p21 activated kinase PAK3. *J Neurosci* 23 6914-6927.

McShea A, Lee HG, Petersen RB, Casadesus G, Vincent I, Linford NJ, Funk JO, Shapiro RA & Smith MA 2007 Neuronal cell cycle re-entry mediates Alzheimer disease-type changes. *Biochim Biophys Acta* 1772 467-472.

Meethal SV, Liu T, Chan HW, Ginsburg E, Wilson AC, Gray DN, Bowen RL, Vonderhaar BK & Atwood CS 2009 Identification of a regulatory loop for the synthesis of neurosteroids: a steroidogenic acute regulatory protein-dependent mechanism involving hypothalamic-pituitary-gonadal axis receptors. *J Neurochem* 110 1014-1027.

Meethal SV, Smith MA, Bowen RL & Atwood CS 2005 The gonadotropin connection in Alzheimer's disease. *Endocrine* 26 317-326.

Milward EA, Papadopoulos R, Fuller SJ, Moir RD, Small D, Beyreuther K & Masters CL 1992 The amyloid protein precursor of Alzheimer's disease is a mediator of the effects of nerve growth factor on neurite outgrowth. *Neuron* 9 129-137.

Mulac-Jericevic B & Conneely OM 2004 Reproductive tissue selective actions of progesterone receptors. *Reproduction* 128 139-146.

Muller F & O'Rahilly R 1985 The first appearance of the neural tube and optic primordium in the human embryo at stage 10. *Anat Embryol (Berl)* 172 157-169.

Munoz-Sanjuan I & Brivanlou AH 2002 Neural induction, the default model and embryonic stem cells. *Nat Rev Neurosci* 3 271-280.

Mylonas I, Jeschke U, Shabani N, Kuhn C, Kunze S, Dian D, Friedl C, Kupka MS & Friese K 2007 Steroid receptors ERalpha, ERbeta, PR-A and PR-B are differentially expressed in normal and atrophic human endometrium. *Histol Histopathol* 22 169-176.

Neaves WB, Johnson L, Porter JC, Parker CR, Jr. & Petty CS 1984 Leydig cell numbers, daily sperm production, and serum gonadotropin levels in aging men. *J Clin Endocrinol Metab* 59 756-763.

O'Shea KS 1999 Embryonic stem cell models of development. *Anat Rec* 257 32-41.

Okabe S, Forsberg-Nilsson K, Spiro AC, Segal M & McKay RD 1996 Development of neuronal precursor cells and functional postmitotic neurons from embryonic stem cells in vitro. *Mech Dev* 59 89-102.

Park KH, Hallows JL, Chakrabarty P, Davies P & Vincent I 2007 Conditional neuronal simian virus 40 T antigen expression induces Alzheimer-like tau and amyloid pathology in mice. *J Neurosci* 27 2969-2978.

Paterson SJ, Corbett AD, Gillan MG, Kosterlitz HW, McKnight AT & Robson LE 1984 Radioligands for probing opioid receptors. *J Recept Res* 4 143-154.

Pepe GJ & Albrecht ED 1995 Actions of placental and fetal adrenal steroid hormones in primate pregnancy. *Endocr Rev* 16 608-648.

Petanceska SS, Nagy V, Frail D & Gandy S 2000 Ovariectomy and 17beta-estradiol modulate the levels of Alzheimer's amyloid beta peptides in brain. *Neurology* 54 2212-2217.

Pidoux G, Gerbaud P, Tsatsaris V, Marpeau O, Ferreira F, Meduri G, Guibourdenche J, Badet J, Evain-Brion D & Frendo JL 2007 Biochemical characterization and modulation of LH/CG-receptor during human trophoblast differentiation. *J Cell Physiol* 212 26-35.

Pomata PE, Colman-Lerner AA, Baranao JL & Fiszman ML 2000 In vivo evidences of early neurosteroid synthesis in the developing rat central nervous system and placenta. *Brain Res Dev Brain Res* 120 83-86.

Pon LA, Epstein LF & Orme-Johnson NR 1986 Acute cAMP stimulation in Leydig cells: rapid accumulation of a protein similar to that detected in adrenal cortex and corpus luteum. *Endocr Res* 12 429-446.

Porayette P, Gallego MJ, Kaltcheva MM, Bowen RL, Vadakkadath Meethal S & Atwood CS 2009 Differential processing of amyloid-beta precursor protein directs human embryonic stem cell proliferation and differentiation into neuronal precursor cells. *J Biol Chem* 284 23806-23817.

Porayette P, Gallego MJ, Kaltcheva MM, Meethal SV & Atwood CS 2007 Amyloid-beta precursor protein expression and modulation in human embryonic stem cells: a novel role for human chorionic gonadotropin. *Biochem Biophys Res Commun* 364 522-527.

Prior JC 1990 Progesterone as a bone-trophic hormone. *Endocr Rev* 11 386-398.

Raina AK, Zhu X, Rottkamp CA, Monteiro M, Takeda A & Smith MA 2000 Cyclin' toward dementia: cell cycle abnormalities and abortive oncogenesis in Alzheimer disease. *J Neurosci Res* 61 128-133.

Reame NE, Kelche RP, Beitins IZ, Yu MY, Zawacki CM & Padmanabhan V 1996 Age effects of follicle-stimulating hormone and pulsatile luteinizing hormone secretion across the menstrual cycle of premenopausal women. *J Clin Endocrinol Metab* 81 1512-1518.

Reichlin S 1998 Williams Textbook of Endocrinology. In *Williams Textbook of Endocrinology. 10 ed*, pp 165-248. Eds JD Wilson, DW Foster, HM Kronenberg & PR Larsen. Philadelphia PA: WB Saunders Co.

Richardson MC & Masson GM 1985 Lack of direct inhibitory action of oxytocin on progesterone production by dispersed cells from human corpus luteum. *J Endocrinol* 104 149-151.

Rogers PA, Donoghue JF, Walter LM & Girling JE 2009 Endometrial angiogenesis, vascular maturation, and lymphangiogenesis. *Reprod Sci* 16 147-151.

Sauter CN, McDermid RL, Weinberg AL, Greco TL, Xu X, Murdoch FE & Fritsch MK 2005 Differentiation of murine embryonic stem cells induces progesterone receptor gene expression. *Exp Cell Res* 311 251-264.

Schmidt PJ, Gindoff PR, Baron DA & Rubinow DR 1996 Basal and stimulated gonadotropin levels in the perimenopause. *Am J Obstet Gynecol* 175 643-650.

Schumacher A, Arnhold S, Addicks K & Doerfler W 2003 Staurosporine is a potent activator of neuronal, glial, and "CNS stem cell-like" neurosphere differentiation in murine embryonic stem cells. *Mol Cell Neurosci* 23 669-680.

Schumacher M, Guennoun R, Robert F, Carelli C, Gago N, Ghoumari A, Gonzalez Deniselle MC, Gonzalez SL, Ibanez C, Labombarda F, et al. 2004 Local synthesis and dual actions of progesterone in the nervous system: neuroprotection and myelination. *Growth Horm IGF Res* 14 Suppl A 18-33.

Schwall RH, Szonyi E, Mason AJ & Nikolics K 1988 Activin stimulates secretion of follicle-stimulating hormone from pituitary cells desensitized to gonadotropin-releasing hormone. *Biochem Biophys Res Commun* 151 1099-1104.

Sriraman V, Rao VS, Rajesh N, Vasan SS & Rao AJ 2001 A preliminary study on the possible role for luteinizing hormone in androgen independent growth of prostate. *Reprod Biomed Online* 3 6-13.

Stocco DM 2001 StAR protein and the regulation of steroid hormone biosynthesis. *Annu Rev Physiol* 63 193-213.

Stocco DM & Chen W 1991 Presence of identical mitochondrial proteins in unstimulated constitutive steroid-producing R2C rat Leydig tumor and stimulated nonconstitutive steroid-producing MA-10 mouse Leydig tumor cells. *Endocrinology* 128 1918-1926.

Strauss JF, 3rd, Christenson LK, Devoto L & Martinez F 2000 Providing progesterone for pregnancy: control of cholesterol flux to the side-chain cleavage system. *J Reprod Fertil Suppl* 55 3-12.

Takashima S, Kuruta H, Mito T, Nishizawa M, Kunishita T & Tabira T 1990 Developmental and aging changes in the expression patterns of beta-amyloid in the brains of normal and Down syndrome cases. *Brain Dev* 12 367-371.

Thomson JA, Itskovitz-Eldor J, Shapiro SS, Waknitz MA, Swiergiel JJ, Marshall VS & Jones JM 1998 Embryonic stem cell lines derived from human blastocysts. *Science* 282 1145-1147.

Trapp BD & Hauer PE 1994 Amyloid precursor protein is enriched in radial glia: implications for neuronal development. *J Neurosci Res* 37 538-550.

Vadakkadath Meethal S, Atwood CS. (2005) The role of hypothalamic-pituitary-gonadal hormones in the normal structure and functioning of the brain. *Cell Mol Life Sci.* 62 257-70.

VanLandingham JW, Cutler SM, Virmani S, Hoffman SW, Covey DF, Krishnan K, Hammes SR, Jamnongjit M & Stein DG 2006 The enantiomer of progesterone acts as a molecular neuroprotectant after traumatic brain injury. *Neuropharmacology* 51 1078-1085.

Wahabi H, Abed Althagafi N & Elawad M 2007 Progestogen for treating threatened miscarriage. *Cochrane Database Syst Rev* CD005943.

Wang JM, Johnston PB, Ball BG & Brinton RD 2005 The neurosteroid allopregnanolone promotes proliferation of rodent and human neural progenitor cells and regulates cell-cycle gene and protein expression. *J Neurosci* 25 4706-4718.

Webber RJ & Sokoloff L 1981 In vitro culture of rabbit growth plate chondrocytes. 1. Age-dependence of response to fibroblast growth factor and "chondrocyte growth factor". *Growth* 45 252-268.

Weiss J, Crowley WF, Jr., Halvorson LM & Jameson JL 1993 Perifusion of rat pituitary cells with gonadotropin-releasing hormone, activin, and inhibin reveals distinct effects on gonadotropin gene expression and secretion. *Endocrinology* 132 2307-2311.

Whitfield GK & Kourides IA 1985 Expression of chorionic gonadotropin alpha- and beta-genes in normal and neoplastic human tissues: relationship to deoxyribonucleic acid structure. *Endocrinology* 117 231-236.

Wright DW, Kellermann AL, Hertzberg VS, Clark PL, Frankel M, Goldstein FC, Salomone JP, Dent LL, Harris OA, Ander DS, et al. 2007 ProTECT: a randomized clinical trial of progesterone for acute traumatic brain injury. *Ann Emerg Med* 49 391-402, 402 e391-392.

Xia X & Zhang SC 2009 Differentiation of neuroepithelia from human embryonic stem cells. *Methods Mol Biol* 549 51-58.

Xu H, Gouras GK, Greenfield JP, Vincent B, Naslund J, Mazzarelli L, Fried G, Jovanovic JN, Seeger M, Relkin NR, et al. 1998 Estrogen reduces neuronal generation of Alzheimer beta-amyloid peptides. *Nat Med* 4 447-451.

Yamazaki T, Matsuoka C, Gendou M, Izumi S, Zhao D, Artemenko I, Jefcoate CR & Kominami S 2006 Mitochondrial processing of bovine adrenal steroidogenic acute regulatory protein. *Biochim Biophys Acta* 1764 1561-1567.

Yokotani T, Koizumi T, Taniguchi R, Nakagawa T, Isobe T, Yoshimura M, Tsubota N, Hasegawa K, Ohsawa N, Baba S, et al. 1997 Expression of alpha and beta genes of human chorionic gonadotropin in lung cancer. *Int J Cancer* 71 539-544.

Zhang SC 2003 Embryonic stem cells for neural replacement therapy: prospects and challenges. *J Hematother Stem Cell Res* 12 625-634.

Zhang SC, Wernig M, Duncan ID, Brustle O & Thomson JA 2001 In vitro differentiation of transplantable neural precursors from human embryonic stem cells. *Nat Biotechnol* 19 1129-1133.

Zhuang LZ & Li RH 1991 Study on reproductive endocrinology of human placenta (II)--Hormone secreting activity of cytotrophoblast cells. *Sci China B* 34 1092-1097.

Zygmunt M, Herr F, Keller-Schoenwetter S, Kunzi-Rapp K, Munstedt K, Rao CV, Lang U & Preissner KT 2002 Characterization of human chorionic gonadotropin as a novel angiogenic factor. *J Clin Endocrinol Metab* 87 5290-5296.

15

Directed Differentiation of Mesendoderm Derivatives from Embryonic Stem Cells

Miriam Gordillo, Ritu Kumar, Harma K. Turbendian and Todd Evans
Weill Cornell Medical College of Cornell University
USA

1. Introduction

In amniotes, mesoderm and endoderm arise during gastrulation, the process that derives the three primary germ layers and establishes the basic body plan of the embryo. However, in recent years there has been a new appreciation for a very early stage of development, when some blastomeres are bipotential and may still contribute to either mesoderm or endoderm (but not ectoderm). This tissue has been termed "mesendoderm" (or sometimes "endomesoderm", but we will use the more common term). Specifically, experiments in nematodes, sea urchins, frogs, or zebrafish showed that when certain single cells were marked at the mid-blastula stage, the labeled cell can contribute to both mesoderm (e.g. blood, heart, muscle) and endoderm (e.g. gut, liver, pancreas) derivatives. Remarkably, the signaling molecules and genetic programs appear to be well conserved across these species (reviewed in Rodaway and Patient, 2001; Wardle and Smith, 2006). Most prominently this involves the nodal signaling pathway (Schier, 2003) and several families of regulatory proteins, including those encoding T-box and GATA transcription factors (Fig. 1). Since zebrafish and frogs are vertebrates, it seems likely that the same developmental programs should function in other vertebrates, including mouse and man.

In the mouse, the three germ layers are derived from the epiblast through gastrulation beginning at approximately day 6.5 of gestation. After implantation, the blastocyst, comprising the inner cell mass inside the trophectoderm, develops into an elongated structure composed of the ectoplacental cone, the extraembryonic ectoderm, the visceral endoderm and the epiblast. Gastrulation begins with the formation of a transient structure known as the primitive streak (PS) in the presumptive posterior end of the embryo through which uncommitted epiblast cells mobilize and egress to form the mesoderm and the endoderm (Tam et al., 2007). On the basis of developmental potential and gene expression patterns, the PS can be divided into anterior, mid and posterior regions, with mesoderm developing from the posterior region and the endoderm developing from the most anterior domain. While the close developmental association between endoderm and mesoderm supports the notion that mesendoderm also generates these two germ layers in mammals, the concept is most strongly supported by studies in the embryonic stem cell system (Tada et al., 2005).

Mouse embryonic stem (ES) cells generated from the blastocyst inner cell mass can be maintained and expanded as a pure undifferentiated population of cells when grown on mouse feeder cells in media containing leukemia inhibitory factor (LIF) and serum (Evans and Kaufman, 1981). More importantly, ES cells are pluripotent and can be differentiated to a broad spectrum of lineages in vitro, providing a putative source of replacement cells for regenerative therapies. The isolation of human ES cells (hES) increased the interest in the possibility of cell therapies using embryonic stem cells (Thomson et al., 1998). In vitro differentiation of ES cells is induced by removing the ES cells from the feeder layer, or by removing LIF from the culture medium. The cells can be differentiated in aggregates of cells called embryoid bodies (EBs) that are grown in suspension (Fig. 2), in monolayer on extracellular matrix proteins, or in layers cultured on supportive stromal cells (Murry and Keller, 2008). Alternatively, methods have been developed for ES cell differentiation using directed differentiation, involving the addition of factors or small molecules that promote the development, differentiation, and maturation of the ES cells toward specific lineages.

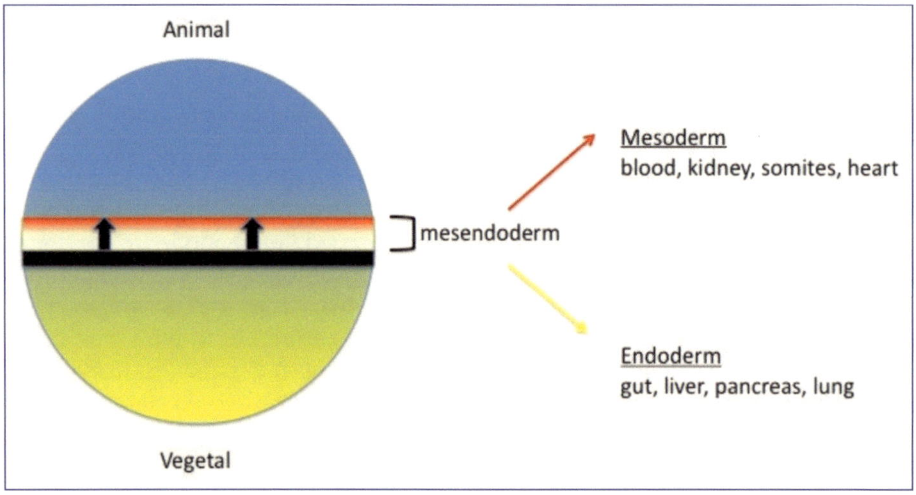

Fig. 1. Shown is a schematized mid-blastula stage zebrafish embryo with the cellularized epiblast toward the animal pole at the top (blue), and the non-cellular yolk cell (yellow) toward the vegetal pole. Unknown signals from the yolk cell induce nodal signaling (black arrows) from the margin (black) to induce bipotential mesendoderm blastomeres. These will generate mesoderm (red) and endoderm (yellow) and associated tissues as indicated by the examples listed (lung, of course, by analogy for mammals)

Directed differentiation provides great promise for generating clinically relevant cell types, including neurons, cardiomyocytes, hepatocytes, insulin-secreting pancreatic beta cells, etc. However, it is clearly not efficient to derive these specific cell types directly from ES cells. Rather, the best strategy is to exploit our knowledge of how the embryo normally generates the cell types during embryonic development, and then applying this knowledge to educate and transition the ES cells toward the desired fate. The fact of a mesendodermal transition stage presents special challenges for harvesting abundant and pure populations of specific progenitors or differentiated cell types from mesoderm or endoderm. During this stage, the same signals are essential for progenitor specification and commitment for multiple different cell types. In this review, we consider progress in directed differentiation for three mesendodermal derivatives: cardiomyocytes (from mesoderm), lung, and liver (both from closely related endoderm). For each tissue, we review what is known about the normal developmental program (signaling pathways and transcriptional regulators), and then discuss progress in using mouse and human ES cells (and induced pluripotent, or iPS cells) to recapitulate these programs in vitro. Finally, we briefly describe the strategy we are taking to better understand the role of GATA factors as key components of specificity for derivation of mesendodermal cell types.

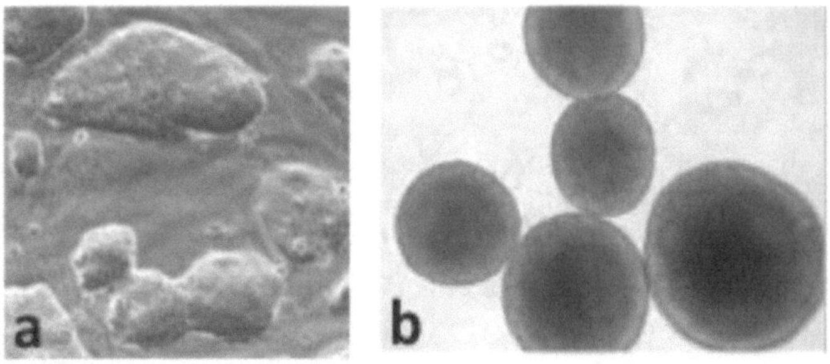

Fig. 2. a) Representative colonies of pluripotent murine ES cells. b) When ES cells are harvested by dissociation and plated back into culture in the absence of LIF, they form round aggregates called embryoid bodies, as shown here. Cells commit to fates from all three germ layers, and differentiate in EBs with a time frame roughly equivalent to normal mouse development

2. Directed differentiation of cardiac cells

Directed differentiation of stem cells toward cardiac tissue forms the basis for a diverse field of research focused on cellular-based therapeutics for myocardial infarction, heart failure, cardiomyopathies, and congenital cardiac defects. Relatively efficient protocols are documented that derive cardiovascular progenitor cells (CPCs) from murine and human ES and iPS cells. Under the appropriate conditions, the pluripotent cells generate CPCs and differentiated cardiomyocytes (CMs) (Boheler et al., 2002; Irion et al., 2008; Mauritz et al., 2008; Wei et al., 2005). Differentiation within the CPC model systems mirrors that of embryonic development, thus providing an accurate and accessible *in vitro* system that facilitates translational research. Major challenges remain to generate in a rationale manner the diversity of cardiac cells, to generate mature and functioning cardiac cells, and to integrate these cells in a productive manner into cardiac tissue.

2.1 Normal cardiogenesis

Cardiac development in mammalian embryos is defined by the expression of a set of conserved temporal and spatial markers. CPCs are derived from mesoderm, marked by the transcription factor brachyury and arising from the PS during gastrulation under the influence of the Wnt, BMP, and Nodal pathways (Tam and Behringer, 1997). Two waves of epiblast cells that are double-positive for both brachyury and fetal liver kinase-1 (Flk1) (also called kinase insert domain receptor, KDR) exit the PS. The first group is c-Kit positive and represents the hematopoietic mesoderm, and the second is the c-Kit negative cardiac mesoderm (Irion et al., 2008). In both hES and mES, early activation of the WNT, BMP, and Nodal pathways induces formation of cardiac mesoderm (Gadue et al., 2006; Laflamme et al., 2007; Lindsley et al., 2006). However, additional studies demonstrated that subsequent WNT pathway inhibition is required to specify cardiac mesoderm, likely at a time-point after exit from the PS (Naito et al., 2006; Ueno et al., 2007). The CPCs within cardiac mesoderm are identified by their expression of Flk1/KDR and the transcription factor Nkx2.5 (Moretti et al., 2006; Wu et al., 2006). They proceed to the anterior region of the embryo and form a structure known as the cardiac crescent, which fuses to form the primitive heart tube. In addition to Nkx2.5, the GATA and T-box families of transcription factors are key regulators of cardiac development. Their precise roles are yet to be defined; however, the GATA4/5/6 and the TBX1-5, 18, and 20 genes are required for cardiogenesis in all vertebrate models (Peterkin et al., 2005; Plageman and Yutzey, 2005). Combinations of GATA, TBX and NKX2 genes are likely essential to specify cardiac fate from precardiac mesoderm. This has been demonstrated through loss-of-function experiments in both zebrafish (Holtzinger and Evans, 2007) and mouse (Zhao et al., 2008). Loss of single genes leads to morphological defects, because sister genes can functionally compensate for an earlier function in cell specification. Recently, it was shown that expression of Gata4, Tbx5, and Mef2c is sufficient to reprogram fibroblasts to a cardiac fate (Ieda et al., 2010), supporting key functions for these genes in cardiac cell specification. There are two subpopulations of CPCs within the developing heart - the first heart field, which gives rise to the left ventricle and atria, and the second heart field, which expresses Isl1 and Fgf10 and contributes to the right ventricle, outflow tract, and atria (Buckingham et al., 2005).

2.2 Directed differentiation techniques and cardiac subtype generation

Directed differentiation techniques based on recapitulating normal developmental pathways have been employed to enhance CPC generation and CM differentiation in both ES & iPS systems. Spontaneously beating CMs form within murine and human embryoid bodies in suspension cultures at a relatively low efficiency (Doetschman et al., 1985). Co-culture of hES with END-2 cells, which is thought to provide paracrine signals specifying cardiac fate, results in a relatively heterogeneous population of differentiated cells (Mummery et al., 2003). Fetal calf serum formed the basis of cell culture and CM differentiation media in many early studies; however, this is wrought with the inherent inter-batch variability of cytokines. To avoid this potential pitfall, serum-free cultures were developed and enhanced with ActivinA (a Nodal pathway activator), BMP, and Wnt inhibitors, to preferentially form CMs in both mES and hES. This strategy has been further adapted to a high-density monolayer, feeder cell-free system for hES differentiation (Laflamme et al., 2007).

Mouse ES cells display a temporal response to BMP, Wnt, and Activin, wherein there are defined time points of responsiveness and non-responsiveness to these inducing factors (Jackson et al., 2010). The presence of BMP and Wnt appears to be critical for CM specification between days 1.5 through 3 of differentiation. Nodal signaling is required from differentiation day 2 through 5. Thus, the addition of inducing factors can be optimized for CM development. Furthermore, directed differentiation would ideally permit the efficient production of individual subtypes of cardiac cells and their progenitors for therapeutic application. Nkx2.5-expressing mES-derived CMs have been shown to form ventricular, atrial, and pacemaker type cells (Hidaka et al., 2003). The precise details of individual lineage development remain to be elucidated. Lineage specification seems to occur at a relatively early time point, as several groups have shown differential induction consistent with particular CM subtypes (He et al., 2003; Kolossov et al., 2005; Mummery et al., 2003; Satin et al., 2004). Cell structure (as observed by electron microscopy) and gene expression profiles in CPC model systems progress in parallel to known developmental, electrophysiological, and contractile maturation profiles, and several groups have been able to generate and select for subtype-specific CMs using fluorescent protein tagging techniques. For example, ventricular-like CMs were isolated by selection using eGFP expression under the control of the myosin light chain (MLC) - 2v promoter (Muller et al., 2000). Likewise, Nkx2.5 and Isl1 positive progenitors differentiate into ventricular CMs (Domian et al., 2009), while connexin 40 and 45 positive pacemaker-like CMs are generated under the influence of endothelin (Gassanov et al., 2004).

2.3 Comparison of model systems

Cardiac development is highly conserved between mouse and human models. Findings in mES models have been generally reproducible in hES models and vice versa. Since their initial description in 2006, iPS cells (Takahashi and Yamanaka, 2006) appear to be a comparable model to ES systems for generating CMs, as shown by electrophysiological and CM-specific protein expression profiles (Mauritz et al., 2008). Nonetheless, there are differences among these *in vitro* model systems that represent potential confounders for generalizing results to *in vivo* systems and considering clinical applications. CMs derived from hES proliferate to a much higher degree than those derived from mES, suggesting that there are additional poorly understood growth signaling pathways involved (McDevitt et al., 2005; Snir et al., 2003; Xu et al., 2002). However, hES differentiate at a slower pace and with a lower efficiency. The miPS are similarly slow to differentiate, form smaller CMs, and may have a predilection for a ventricular phenotype (Kuzmenkin et al., 2009; Mauritz et al., 2008).

2.4 Clinical applications

ES cell-based therapeutics for cardiac pathology rely on the ability to develop non-immunogenic, non-neoplastic, functioning CMs or CM progenitors with high fidelity and high efficiency that can be localized to a specific target region. The mES and hES cell derived CMs form stable myocardial grafts in a variety of immuno-compromised animal hosts (Dai et al., 2007; Kehat et al., 2004; Klug et al., 1996; Laflamme et al., 2005) Moreover, hES derived CMs have been shown to restore function in cardiac damaged models. They improve contractility for the infarcted murine heart and provide pacing activity in pig hearts that have undergone atrioventricular node ablation (Cai et al., 2007a; Kehat et al., 2004). Regardless of their direct therapeutic potential, both hES and hiPS already have great value as *in vitro* models for pharmaceutical testing, providing a method of noninvasive assessment of potential cardiotoxicity and arrhythmogenicity (Vidarsson et al., 2010).

3. Directed differentiation of hepatocytes

The liver is the largest internal organ with a mass that accounts for 2% to 5% of body weight. The liver is also the main detoxifying organ of the body and performs numerous essential metabolic, exocrine and endocrine functions. Metabolic functions include synthesis, storage, metabolism and redistribution of nutrients, carbohydrates, fats and vitamins and secretion of plasma proteins including albumin, clotting factors, and apolipoproteins. Endocrine functions include the secretion of insulin-like growth factors, angiotensinogen, and thrombopoietin. Exocrine secretion is mainly in the form of bile. In humans, the liver is composed of two lobes each subdivided into lobules, the basic architectural unit of the liver. Lobules are roughly hexagonal or pentagonal cylinders about 2mm high and 1mm in diameter formed by single cell sheets of hepatocytes lined by sinusoidal capillaries that radiate toward a small branch of the hepatic vein that extends to the center of each lobule (Si-Tayeb et al., 2010a). At the periphery of each lobule, the interlobular bile ducts, the portal vein, and the hepatic artery run in parallel, forming the portal triad. The basal surface of hepatocytes absorbs metabolites and toxins from blood flowing from the portal vein and the hepatic artery through the sinusoidal capillaries, while bile is secreted from the apical surface of adjoining hepatocytes into the bile canaliculi, and then flows through the interlobular bile ducts to the extrahepatic bile ducts and into the gall bladder.

The primary functional cell types of the liver are the hepatocytes and the cholangiocytes (biliary epithelial cells). Hepatocytes are polarized epithelial cells that account for near 80% of the liver volume (Blouin et al., 1977), and carry out most of the liver functions. Cholangiocytes account for almost 3% of the liver cell population and line the bile ducts contributing to bile transportation. Other liver cell types include endothelial cells from the sinusoids and other liver vasculature, Pit cells (liver natural killer cells), Kupffer cells (resident liver macrophages), and stellate cells.

3.1 Embryonic endoderm development

Hepatocytes and cholangiocytes are derived from the embryonic definitive endoderm (DE), while the remaining liver cell types are derived from mesoderm. In vertebrates, the different populations of the PS are dependent on different levels of Nodal signaling, with the anterior region that contains the DE requiring a higher and more sustained period of Nodal signaling for its specification (Lowe et al., 2001). A conserved network of transcription factors acting downstream of Nodal signaling drives DE specification. With variations in some species, this network includes FoxA2, Gata4 and Gata6, the T-box protein Eomesodermin (Eomes), Mix-like proteins, and Sox17. FoxA2, a member of the Forkhead family of transcription factors, is essential for endoderm differentiation. FoxA2 null embryos can form hindgut but not foregut or midgut endoderm (Dufort et al., 1998), and conditional inactivation has shown that this factor is required for the development of various endoderm-derived organs (Gao et al., 2008; Lee et al., 2005; Wan et al., 2004). Gata4, Gata5 and Gata6 are involved in the specification and differentiation of the endoderm throughout evolution (Woodland and Zorn, 2008). In zebrafish and Xenopus, GATA factors are involved in endoderm patterning downstream of nodal, while in mouse Gata4 and Gata6 have an additional role regulating extraembryonic endoderm development (Zorn and Wells, 2009). However, a triple knockout mouse of all three GATA genes in epiblast-derived tissues has not been reported, leaving open the possibility that these genes are redundantly required for early endoderm development. There is precedence, since they are functionally redundant in a similar way at later stages (Holtzinger and Evans, 2007; Zhao et al., 2008). Eomes is also required for proper endoderm formation in the mouse (Arnold et al., 2008), whereas mouse Mixl1 is essential for definitive endoderm possibly through the control of mesendoderm development (Hart et al., 2002). Sox17 is also a key gene in mouse endoderm development that appears to be essential only for posterior endoderm. In the knockout mutant, anterior endoderm is generated, but posterior and lateral endoderm from midgut and hindgut are reduced and fail to expand (Kanai-Azuma et al., 2002).

Initial endoderm patterning is coincident with the formation of DE during gastrulation. The cells emerging earlier from the PS will form the foregut and those emerging later will give rise to the embryonic midgut and hindgut during morphogenesis. Morphogenetic movements of the primitive gut begin when the sheet of epithelial DE surrounding the ventral side of the mouse embryo folds over to form two gut pockets (Lewis and Tam, 2006). Anterior axial and lateral endoderm folds ventrally to form the foregut pocket, whose opening moves posterior as the foregut develops, while the hindgut pocket opening moves anterior. Concurrently, the lateral endoderm folds ventrally to meet with the anterior and posterior folds at the yolk stalk, completing the formation of the gut tube (Zorn and Wells, 2009). Thus, morphogenetic movements during gut tube formation allow the convergence of the lateral and medial endoderm progenitors from three spatially separated embryonic domains. The primitive gut tube is patterned along the anterior-posterior axis in a manner that presages the subsequent budding of gut-derived organs along the dorsal and ventral aspect of the foregut (Fig. 3). In addition, morphogenesis results in the close apposition of the foregut progenitors with liver potential and mesoderm from the lateral plate. This relationship is essential for liver specification by inductive signals from this nearby mesoderm.

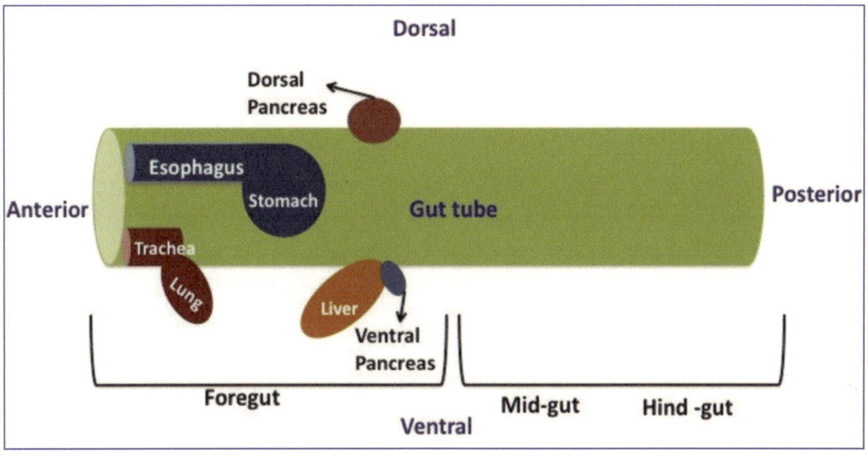

Fig. 3. Schematic illustrating the positional relationship for the various organ domains that emerge by budding from the foregut of the endoderm-derived gut tube at E 9.5 of mouse development

3.2. Liver development

Interaction between foregut endoderm and adjacent mesoderm appears to control two different steps at the onset of liver organogenesis. Initially, signals from the mesoderm regulate regional identity of the endoderm and establish foregut progenitors that are competent to develop into liver (Zorn and Wells, 2009). Genetic and chromatin occupancy studies indicate that FoxA2 and Gata4 factors either mark or help maintain the competence of the endoderm to activate liver genes such as albumin in response to inductive signals (Zaret et al., 2008). Competent cells then respond to several extracellular signals including members of the FGF and BMP families that induce specification to a hepatic fate. FGFs secreted by the cardiogenic mesoderm specifically induce mitogen-activated protein kinase (MAPK) signaling, but not phosphatidylinositol 3-kinase (PI3K) signaling during hepatic gene induction (Calmont et al., 2006). The role of FGF at the onset of liver development is evolutionarily conserved, as FGF also shows hepatogenic properties in frogs and fish. FGF signaling is not sufficient at this stage and BMP signaling from the septum transversum mesenchyme is also required for hepatic induction. BMP signaling may be mediated at least in part by maintaining the expression of Gata4 that is essential for expression of albumin and other hepatic genes (Rossi et al., 2001). WNT signaling has also been implicated during hepatic induction but its role is complex and may not be evolutionarily conserved (Si-Tayeb et al., 2010a).

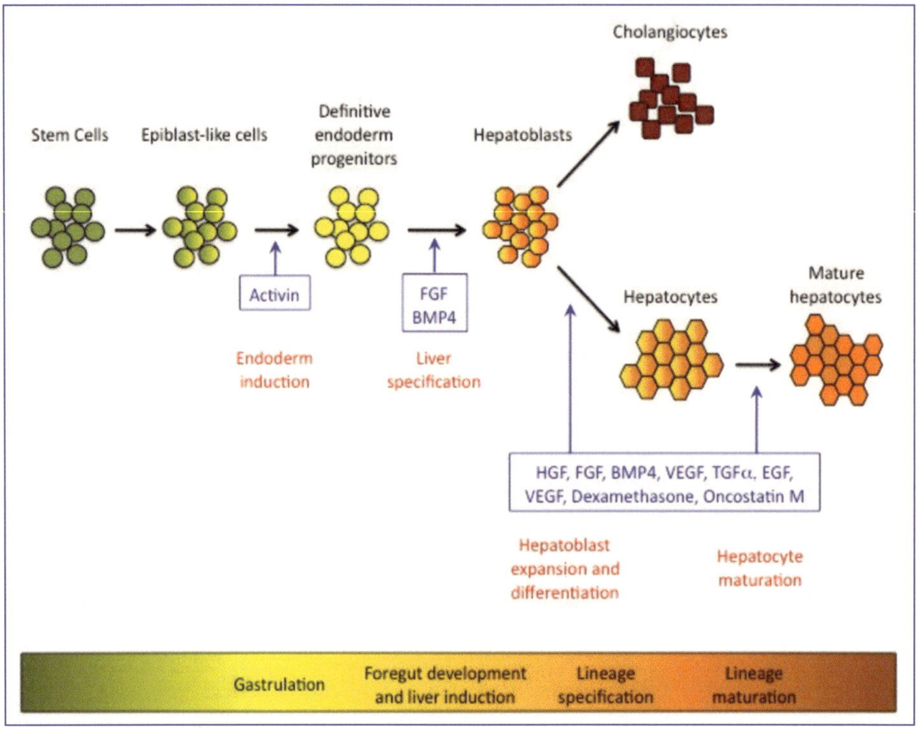

Fig. 4. The schematic illustrates the transition stages that occur in ES cell-based hepatic differentiation cultures (top) relative to normal mouse development (bottom). Also shown are key signaling factors that have been found to strongly promote the transition *in vitro* (blue), based on the known developmental pathways that are required *in vivo*. This strategy therefore demonstrates an effective directed differentiation protocol [as developed by Gouon-Evans, et al. (2006)]

Inductive signals from the cardiac mesoderm and septum transversum mesenchyme regulate the expression of a network of transcription factors, including Hex, Hnf1B, FoxA1, FoxA2 and Gata4, that is critical for the onset of hepatogenesis. At this stage the endoderm in the ventral floor of the foregut becomes thicker and expands to form an outgrowing bud of polarized proliferating cells that is separated from the surrounding mesenchyme by a basement membrane (Lemaigre, 2009). When the hepatic endoderm is specified and the liver bud begins to grow, the cells are referred to as hepatoblasts (Fig. 4). Soon after the liver bud is apparent, the basement membrane is lost and cells delaminate from the foregut and migrate as cords of hepatoblasts into the septum transversum mesenchyme. This process requires the transcription factor Prox1 (Sosa-Pineda et al., 2000). Generation of the liver bud is also accompanied by development of the hepatic vasculature. Interactions with endothelial cells derived from mesoderm are crucial for this budding phase (Matsumoto et al., 2001). Further proliferation enlarges the liver bud. FGF and BMP are also required at this stage, together with HGF. Genetic studies have revealed that mutations suppressing proliferation and promoting apoptosis result in liver hypoplasia. Signal transduction molecules that are needed for continued fetal liver growth and to prevent apoptosis include members of the anti-apoptotic NFkB complex, integrins and the SAPK/JNK, Ras/Raf-1 and WNT/beta-catenin pathways (Tanimizu and Miyajima, 2007). Transcription factors expressed in the septum transversum mesenchyme (Hlx, N-myc and Jumonji) and hepatoblasts (Foxm1b and Xbp1) are also necessary for growth and proliferation of hepatoblasts at this stage.

Although hepatoblasts already express some genes specific for fully differentiated hepatocytes such as serum albumin, hepatoblasts will give rise to both definitive hepatocytes and cholangiocytes. Differentiation toward cholangiocytes is promoted by Notch and antagonized by HGF, that instead promotes hepatocyte differentiation (Lemaigre and Zaret, 2004). The role of Notch signaling in cholangiocyte development is conserved in fish and in humans. Alagille syndrome, characterized by reduction in intrahepatic bile ducts, is caused by mutations in the Notch ligand Jagged 1 (Li et al., 1997; Oda et al., 1997). In addition, TGF-beta and WNT signaling promotes cholangiocyte development. In response to these and other unknown mesenchymal signals, the expression of cholangiocyte transcription factors OC1, OC2, and HNF1b is increased, while expression levels of the pro-hepatic factors Hnf4a, Tbx3 and C/EBPa are down-regulated. Continued Notch, EGF and HGF signaling is essential for ductal plate remodeling. During mid-gestation, the hematopoietic cells in the liver secrete the cytokine oncostatin M, which in combination with glucocorticoid hormones, HGF and WNT promotes hepatocyte maturation. These factors act in part by regulating liver enriched transcription factors such as C/EBPa, Hnf1a, FoxA factors and Hnf4a (Zorn and Wells, 2009). The polarization of the hepatic epithelium allows the liver to develop its final architecture. After the hepatic epithelium polarizes, canaliculi are created and the basal layer becomes juxtaposed to the fenestrated endothelium that lines the sinusoids.

3.3 Hepatocyte differentiation from embryonic stem cells

The first report of spontaneous differentiation of ES cells into endodermal cells showed expression in EBs of alpha-fetoprotein (Afp), transthyretin and albumin after more than 8–10 days in culture (Abe et al., 1996). After this, reports of spontaneous differentiation were mainly focused on confirming that these markers were expressed by hepatocyte-like cells derived from definitive endoderm and not by cells in yolk sack-like structures derived from visceral endoderm that is also found in EBs (Kumashiro et al., 2005). There is a substantial overlap in markers and regulatory pathways controlling both primitive and definitive endoderm. In addition to identification and analysis of liver-specific markers, these studies characterized hepatocyte-like cells using electron microscopy, and by functional assays such as indocyanine green uptake, urea synthesis, and Periodic acid-Schiff staining for glycogen.

Because of the poorly defined nature of serum, variability between lots, and the low efficiency of protocols containing only serum, subsequent studies aimed to mimic the *in vivo* hepatic differentiation process using combinations of various factors and culture conditions. Hamazaki et al. (2001) presented the first report using growth factors, aFGF and HGF, in addition to serum, to direct the differentiation of embryonic stem cells toward a hepatic fate. Subsequently, other combinations of serum, growth factors and adherent matrices have been tested with varying success (Heo et al., 2006; Hu et al., 2004; Imamura et al., 2004; Kania et al., 2004; Shirahashi et al., 2004; Teratani et al., 2005). In general, all these protocols direct growth of EBs in serum for several days before plating on an adherent matrix, usually collagen since this protein is found in the connective tissue of the septum transversum that harbors and promotes hepatoblast proliferation. Most protocols include FGF to mimic the secretion of this factor by the cardiac mesoderm, HGF to support mid-stage hepatogenesis, oncostatin M that is produced by hematopoietic stem cells in the embryo liver and induces maturation of fetal hepatocytes, and dexamethasone, which is a synthetic glucocorticoid hormone that induces enzymes involved in gluconeogenesis. A matrix of gelatin and HGF, insulin, transferrin and selenious acid have also been used. In addition to these growth factors and supplements, other protocols add a co-culture step with mesodermal derived cells, nonparenchymal hepatic cells or adult hepatocytes (Cho et al., 2008; Ishii et al., 2005; Shiraki et al., 2008; Soto-Gutierrez et al., 2006a).

The creation of reporter ES cells to identify and monitor early developmental stages allowed the reduction or elimination of serum and the implementation of more efficient protocols in serum-free fully defined media. To monitor DE development, genes expressed in the anterior primitive streak during embryo gastrulation, i.e., Brachyury, Goosecoid, FoxA2 and Sox17, were targeted with reporter molecules. Using these cells it was found that ActivinA, a TGF-beta family member that signals through the same receptor as nodal, efficiently induces DE progenitors *in vitro* (Kubo et al., 2004). These reporter cells also allowed the discovery of the surface receptors Cxcr4, c-Kit, E-cadherin and PDGFRa as useful markers to enrich DE progenitors differentiated from ES cells that have not been genetically modified. By translating findings from embryo development to ES cell culture (Fig. 4), FGF and BMP4 were shown to efficiently induce hepatic fate in the ActivinA-induced endoderm (Gouon-Evans et al., 2006). VEGF was also used during the specification period because the endothelial lineage promotes liver bud growth during embryo development. The hepatoblasts obtained were later expanded and matured in hepatic plating media containing EGF, bFGF, HGF, TGFa, VEGF and dexamethasone. Reporter cells have also been created to monitor and enrich differentiated hepatocyte-like cells (see below).

Efficient hepatic differentiation protocols have also been reported for human embryonic stem (hES) cells. Early reports showed that after EB formation hES cells also differentiate spontaneously into derivatives of the three germ layers, among them endodermal cells and hepatocyte-like cells expressing Afp and albumin (Itskovitz-Eldor et al., 2000; Lavon et al., 2004). An early study that aimed to direct hepatocyte differentiation from hES cells used sodium butyrate as the inducing agent (Rambhatla et al., 2003). Around 15% of the differentiated cells presented morphological features similar to that of primary hepatocytes and expressed albumin, a-1-antitrypsin, cytokeratin 8 and 18, accumulated glycogen and had inducible cytochrome P450 activity. Subsequently, differentiation toward hepatic-like cells was pursued using a variety of protocols that involve the use of growth factors, EB formation, co-culturing with hepatic and non-hepatic cell types, culture in collagen scaffold 3D culture systems, genetic manipulation, and epigenetic modifications (Agarwal et al., 2008; Baharvand et al., 2006; Basma et al., 2009; Cai et al., 2007b; Shirahashi et al., 2004; Touboul et al., 2010). Growth factors and other supplements included different combinations of aFGF, bFGF, EGF, HGF, WNT, all-trans-retinoic acid, oncostatin M, dexamethasone, insulin, transferrin and selenium. Protocols including ActivinA and using monolayer growth instead of EB formation have increased the yield of hES cell-derived definitive endoderm (D'Amour et al., 2005). These protocols have led to higher efficiency in the production of cells displaying hepatic functions including ureagenesis, glycogen storage, albumin, fibrinogen and fibronectin secretion, and CYP activity. Recently, a protocol that initially enriched definitive endoderm precursors to more than 85% purity generated a population with 90% of cells expressing albumin after hepatocyte differentiation (Duan et al., 2010). Importantly, this population appeared to have complete metabolic function and had the capacity to metabolize drugs at levels that were comparable to primary hepatocytes. Ethical and safety debates about the use of embryonic stem cells for research and therapy have stimulated the search of alternative approaches to generate pluripotent stem cells. Thus, hES cells may soon be outpaced by iPS cells, generated from postnatal cells by viral-mediated transfer of reprogramming genes (Yamanaka and Blau, 2010). Human iPS cells also open the possibility of creating customized pluripotent stem cell lines which may not only be critical in cell therapy but also in assessing human drug toxicity. More recently iPS cells have been used successfully to model inherited metabolic disorders of the liver (Rashid et al., 2010). Hepatocyte-like cells have been generated from human-iPS cells, including cells obtained by reprogramming of human primary hepatocytes (Liu et al., 2010; Si-Tayeb et al., 2010b; Song et al., 2009; Sullivan et al., 2010; Touboul et al., 2010). Most of these protocols used fully defined culture conditions with growth factors and cytokines that recapitulate essential stages of liver development. In fact, in some cases iPS cells were able to generate hepatocytes more robustly than hES cells, suggesting efficacy of these protocols with pluripotent stem cells of diverse origins (Sullivan et al., 2010). Importantly, iPS cells derived from human foreskin fibroblasts were efficiently induced to form hepatocyte-like cells in culture that were able to integrate into the hepatic parenchyma *in vivo* (Si-Tayeb et al., 2010b).

3.4 Clinical applications and challenges for treating liver disease

Due to the critical roles of the liver in many metabolic processes, liver disease leads to high rates of morbidity and mortality. Liver disease is the fourth leading cause of death among middle-aged adults in the United States. Treatment for patients with acute hepatic failure or end-stage liver disease relies on liver transplantation with approximately 6,000 liver transplant operations performed in the United States every year. The strong regenerative capacity of the liver makes it possible for a healthy donor to provide a portion of liver for transplantation, although scarcity of organ donors limits the efficacy. Therefore, transplantation of hepatocytes is an option to replace whole liver transplantation. Unfortunately, the source of hepatocytes is also limited and cell therapy using hepatocytes or hepatocyte-like cells differentiated from embryonic stem cells or the recently developed iPS cells, is therefore an attractive alternative. While the use of hepatocyte-like cells from ES or iPS cells in the clinic may be a long-term goal, their utility in the short term resides in the assessment of drug toxicity and safety during drug development. Due to the central role that the liver plays in drug metabolism, the measurement of liver toxicity in cell-based models is a critical step in studying the safety of new compounds (Gomez-Lechon et al., 2010).

All available *in vitro* differentiation protocols using ES or iPS cells result in a heterogeneous cell population that includes not only cells with the desired phenotype but also cells with undesired phenotypes, including potentially undifferentiated stem cells. Therefore, ES cell-derived hepatocyte-like cells need to be selected from contaminating undifferentiated and possible tumor forming cells. To this end, selection based on fluorescence-activated cell sorting (FACS) of cells expressing enhanced green fluorescent protein (eGFP) under control of the Afp or albumin promoters has been used (Drobinskaya et al., 2008; Heo et al., 2006; Ishii et al., 2007; Soto-Gutierrez et al., 2006b; Teratani et al., 2005; Yin et al., 2002). In some of these studies, hepatocyte-like cells were purified using albumin promoter–based cell sorting and transplanted into mouse models for liver injury, or implanted subcutaneously as a bioartificial liver (BAL) device, into mice subjected to 90% hepatectomy, reversing liver failure to variable degrees. Other approaches include the generation of a transgenic ES cell line that expresses a fusion of the hygromycin resistance gene and eGFP under the control of an Afp promoter, or ES cells with bicistronic expression of the eGFP and the puromycin resistance, also under the control of an Afp promoter. Antibiotic selection resulted in an enriched population of cells with hepatoblast and more mature hepatocyte phenotypes that were used in cell replacement experiments. In human cells, reporter genes expressed under the control of the albumin or Afp genes have been used to improve differentiation protocols and to purify differentiated populations for determining global transcriptional profiles (Chiao et al., 2008; Lavon et al., 2004). Differentiated human hepatocyte-like cells have also been enriched by transduction with a lentivirus vector containing the eGFP gene driven by the alpha-1-antitrypsin promoter (Duan et al., 2007).

ES cell-derived hepatocyte-like cells have been tested by transplantation in mouse models of liver disease or damage. Initially, *in vivo* engraftment and proliferation after transplantation was inefficient and teratoma formation accompanied engraftment. However, recent reports show improved results (Gouon-Evans et al., 2006; Heo et al., 2006; Ishii et al., 2007). After transplantation, hepatocyte-like cells not only engrafted and proliferated but they also participated significantly in liver repair and survival. The transplanted cells also showed differentiation into mature hepatocytes and were responsive to normal growth regulation, and proliferated at the same rate as the host hepatocytes after additional growth stimulus from recurrent liver injury. Although in one of these studies teratomas appeared after sixty days (Ishii et al., 2007), no teratomas were observed in the other study for up to 82 days after transplantation (Heo et al., 2006). Recently promising results have been obtained using hepatocyte-like cells in BAL devices to remove toxins from the blood and supply physiologically active molecules important for recovery of hepatic function. In these instances, embryonic derived hepatocytes implanted subcutaneously as a BAL reversed liver failure in mice subjected to 90% hepatectomy (Soto-Gutierrez et al., 2006a) or were able to improve survival in rats with liver failure induced by galactosamine (Cho et al., 2008). The use of BAL devices decreases the risk of teratoma formation. Human hepatocyte-like cells obtained by different protocols have also been transplanted in mouse models. The transplanted cells survive, engraft, and present some functional characteristics of hepatocytes including secretion of liver specific proteins; teratoma formation was reported in some cases. Clearly, novel approaches will need to be developed for the selection of pure hepatic cell populations from *in vitro* culture systems (for example without using transgenic reporter genes) before transplant protocols will be feasible.

4. Directed differentiation of lung epithelium

The lung develops from DE that is closely related to the liver primordium. Again, this raises a special challenge for directed differentiation, since all of the protocols needed for generating lung progenitors will also generate prehepatic cells. Biasing the protocol too much away from liver may promote the generation of more anterior DE-derived tissue, such as thyroid. A recommended strategy is to use our knowledge of normal mesendoderm development to promote DE, and then, while providing lung-dependent signals, inhibit the progression of non-lung DE-derived progenitors. The normal developmental programs that derive anterior DE were discussed above in relation to liver. These are equally relevant to lung, but will not be reiterated in this section.

4.1 Phases of lung development

The lung is a respiratory organ with the principal function of transporting oxygen from the atmosphere to blood cells, and to release carbon dioxide from the blood into the atmosphere. Mammals have two lungs with great cellular diversity, with multiple resident epithelial and mesenchymal cell lineages. The lung is a complex system of branched epithelial tubules (airways), which connect to a network of gas exchanging units called alveoli. Pulmonary alveoli are spherical outcroppings that consist of an epithelial layer and extracellular matrix surrounded by capillaries (Ten Have-Opbroek, 1991). The respiratory epithelium in the mammalian lung is composed of a variety of cell types with distinct morphologic and biochemical characteristics. Epithelial cell lineages are arranged in a distinct proximo-distal spatial pattern in the airways. The proximal conducting airways are lined with a ciliated pseudostratified epithelium, but in the distal region they are replaced by a very thin squamous epithelial cell lining and simple cuboidal epithelial cells known as type I and type II airway epithelial cells, respectively (Breeze and Wheeldon, 1977). Type I squamous alveolar cells are responsible for gas exchange in the alveoli and cover a majority (>95%) of the alveolar surface area (Williams, 2003). Type II alveolar cells produce pulmonary surfactant proteins (SP) A, B, C and D, a complex mixture of phospholipids and proteins that is critical for reducing surface tension at the air-liquid interface of the distal lung, to prevent alveolar collapse upon expiration (Fehrenbach, 2001). Non-ciliated Clara cells are present in the small airways and trachea. Clara cells are secretory and the source of Clara cell secretory protein (CCSP) (Bishop and Polak, 2006). Clara cells and type I alveolar cells are unable to replicate and are susceptible to toxic insults. In the event of damage, type II cells can proliferate and/or differentiate into type I cells to compensate for the loss (Evans et al., 1975).

The development and differentiation of lung epithelial cells is accomplished through the proper coordination of various transcription factors including Forkhead box A1 (FoxA1), FoxA2, Foxj1, homeodomain Nkx2.1 (Ttf1), homeodomain box (Hox) A5, the zinc finger Gli transcription factors, basic helix loop helix (bHLH) factors, and GATA transcription factors (Cardoso and Lu, 2006; Costa et al., 2001; Warburton et al., 2000). Also essential are various autocrine and paracrine signaling events, initiated by several families of secreted factors including FGF and TGF-beta family members, for both lung development and morphogenesis. Lung development in the mouse and human (Fig. 5) can be divided into five overlapping stages (Cardoso and Lu, 2006; Costa et al., 2001; Warburton et al., 2000).

i. **Embryonic phase -** Following gastrulation, definitive endoderm undergoes complex morphogenetic changes that ultimately leads to the formation of the primitive gut tube as described above. The most anterior part of the primitive gut tube is the foregut, while the midgut and hindgut are located at progressively more posterior regions. The respiratory system arises from the ventral foregut endoderm. The process initiates with the establishment of respiratory cell fate in the ventral foregut. The foregut endoderm differentiates into various epithelial cell types, which line the inner surface of the developing lung and trachea. This is followed by the development of a tree-like system of epithelial tubules and vascular structures (Metzger et al., 2008) that ultimately gives rise to the mature airways and alveoli.

ii. **Pseudoglandular stage -** This stage is characterized by formation of the bronchial and respiratory bronchiole tree, lined with undifferentiated epithelial cells juxtaposed to the splanchnic mesoderm. By day 12 of mouse lung development branching of the bronchial buds gives rise to the left lung lobes and the four lobes of the right lung.

iii. **Canalicular stage -** During this stage there is extensive branching of the distal epithelium and mesenchyme resulting in formation of terminal sacs lined with epithelial cells integrating with the mesoderm-derived vasculature.

iv. **Saccular stage -** There is a coordinated increase in terminal sac formation and vasculogenesis in conjugation with the differentiation of alveolar epithelial type I and type II cells.

v. **Alveolar stage -** Finally, postnatal lung development features maturation of the terminal respiratory sacs into alveolar ducts and sacs.

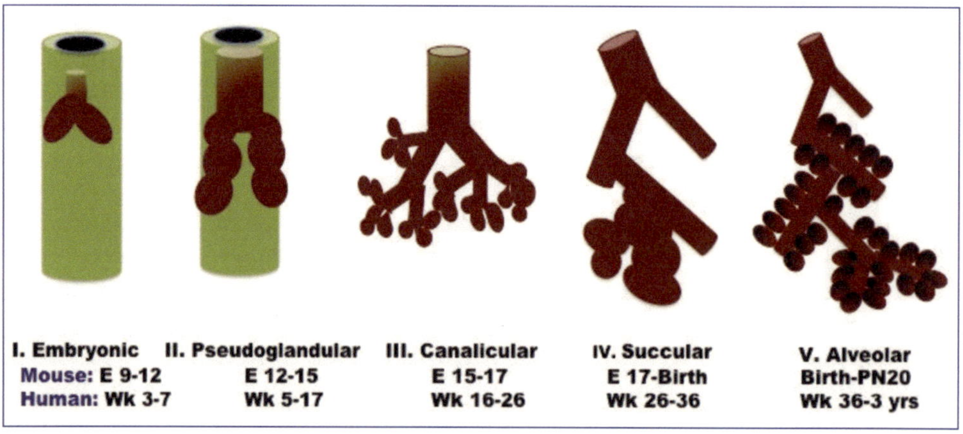

Fig. 5. Shown are the developmental stages of lung morphogenesis for mammals. The schematic primarily depicts airway morphogenesis

4.2 Establishment of endodermal cell fate

As alluded to above, as gastrulation proceeds, the DE cells migrate anteriorly, displacing the extraembryonic visceral endoderm (Lawson and Pedersen, 1987) and this tissue subsequently begins to fold at the anterior and posterior ends, forming the anterior and caudal intestinal portals (Tam et al., 2007). In mouse, by embryonic day 9.5 budding of various domains of endodermally derived organ from the distinct sections of the gut tube begins, and progressively these domains obtain their specific architecture and eventually start differentiating into corresponding organs. The foregut gives rise to lung, thyroid, thymus, esophagus, trachea, stomach, liver and pancreas. Midgut forms the small intestine, and hindgut gives rise to large intestine. Initially, the generation of pre-lung tissue is generally under the control of the same programs responsible for pre-hepatic tissue (see Figs. 3 and 4).

4.3 Establishment of lung epithelial cell fate

Primordial lung bud development originates at E9.5 from the anterior foregut endoderm. As compared to liver and pancreas, the molecular mechanisms responsible for primary lung bud induction are less clear. Precisely when cells fated to lung epithelium originate in the foregut endoderm is not well defined. Respiratory progenitors are first identified by *in situ* hybridization as a group of Thyroid transcription factor 1 (Titf1, also known as Nkx2.1)-expressing endodermal cells in the prospective lung/tracheal region of the foregut, at E9 (Minoo et al., 1999). Titf1 expression has been reported in respiratory progenitors of the foregut endoderm even before the lung or tracheal primordium can be identified (Serls et al., 2005). However Titf1 is not a specific marker for lung, as endoderm-derived thyroid progenitors also express Titf1 (Lazzaro et al., 1991). The surfactant protein C (SPC) gene is the most specific marker of lung epithelial cells lineage, but its expression is consistently detected only by E10-10.5, after the primary buds form (Wert et al., 1993). FGFs secreted by the cardiac mesoderm influence cell fate decision in the adjacent ventral foregut endoderm in a concentration-dependent manner. Explant cultures of 2-5 somite stage ventral foregut in the presence of high concentrations of FGF2 result in lung lineage specification marked by the expression of Titf1 and Sptfc, whereas a low concentration of FGF2 results in expression of liver markers. The high concentration of FGF2 appears to be instructive for lung specification since dorsal midgut endoderm, which ordinarily does not give rise to lung, also responds to FGF2 induction by expression of Titf1.

4.4 Formation of primary lung bud formation

Once respiratory fate is established in the foregut endoderm, the initial pool of lung progenitors expand giving rise to the primary lung bud. In mammals, signaling by Fgf10 from the surrounding mesoderm and the expression of its receptor FGfr2b on the subset of Titf1-expressing endodermal cells is crucial for lung bud formation. Deletion of either Fgf10 or Fgfr2b results in lung agenesis and multiple abnormalities (Min et al., 1998; Sekine et al., 1999). It has been reported that expression of Fgf10 and formation of lung bud is controlled by retinoic acid signaling (Desai et al., 2006). Disruption of retinoic acid signaling in the foregut affects lung most dramatically resulting is several lung abnormalities including lung agenesis (Mollard et al., 2000). Retinoic acid seems to act as a major regulatory signal integrating the WNT and TGF-beta pathways in the control of Fgf10 induction (Chen et al., 2010).

4.5 Mesenchymal-epithelial interaction

Once the primary lung buds have formed, they extend into the surrounding mesenchyme and begin the process of branching morphogenesis. Branching morphogenesis requires specific interactions between the endodermal epithelium and its mesenchyme (Shannon and Hyatt, 2004). In chick it has been shown that if the lung mesenchyme is removed from the lung rudiments prior to branching, all development is blocked, but if mesenchyme is removed after primary and secondary branching, the subsequent branching morphogenesis is normal. With the optimization of culture techniques, this phenomenon was shown to hold true in mammals as well. Non-lung mesenchyme is capable of inducing formation of lung bud in the gut tube, but this bud does not branch further, again indicating the requirement of lung mesenchyme in branching morphogenesis. Lung mesenchyme is capable of inducing branching morphogenesis in non-lung epithelium such as tracheal epithelium (Wessells, 1970). Since tracheal epithelial cells, just like lung epithelial cells, arise from endoderm, the ability of lung mesenchyme to reprogram tracheal cells may not be surprising. However, lung mesenchyme is also able to induce lung-like patterning in ureteric bud (Lin et al., 2001) and salivary gland (Lawson, 1974).

4.6 Growth factors that regulate lung development

Fibroblast growth factors (FGFs). Although FGF1, FGF2, FGF7, FGF9, FGF10 and FGF18 are all expressed in the developing lung, only FGF10 is essential for lung development. FGF10 is expressed in the mesenchymal cells surrounding distal lung epithelial cells (Bellusci et al., 1997) and its receptor FGFR2IIIb is found throughout the embryonic lung epithelium (Orr-Urtreger et al., 1993). Because FGF10 is produced by the lung mesenchyme and its receptor is on epithelial cells, it is considered a key mediator of the epithelial mesenchymal interaction. FGF10 null mice have no lung development below the trachea (Sekine et al., 1999). *In vitro*, FGF10 causes budding of embryonic lung epithelium in mesenchyme- free cultures, but the ability of FGF10 to sustain expression of SPC was not determined (Bellusci et al., 1997). A high concentration of FGF10 in mesenchymal-free culture conditions caused budding in tracheal epithelium, but again in this study expression of lung specific markers was not studied (Ohtsuka et al., 2001). Rat tracheal epithelium was shown to express lung specific markers (SPC) in the presence of FGF10 and in the absence of lung mesenchyme (Shannon et al., 1999). Other than FGF10, two other members of the FGF family have been shown to influence lung development: FGF9 (Colvin et al., 2001) and FGF7 (Simonet et al., 1995).

Bone morphogenic protein (BMP) 4. In early mouse development, BMP4 is expressed in the splanchnic mesenchyme surrounding the gut tube where the future lung buds will form (Weaver et al., 1999). After lung buds have formed, BMP4 continues to be expressed in the more proximal mesenchyme and also in the epithelium of the distal tips. Over-expression of BMP4 in the distal lung epithelium results in lung hypoplasia with a decrease in the number of SPC-positive cells (Bellusci et al., 1996). Inhibiting BMP4 either by driving noggin expression under the control of the SPC promoter, or by expression of a dominant negative form of the BMP receptor, also resulted in a decrease in the number of SPC positive cells, but with an expansion of epithelial cells expressing proximal markers CC10 and Foxj1 (Weaver et al., 1999). Addition of exogenous BMP4 on either lung (Weaver et al., 2000) or tracheal (Hyatt et al., 2002) epithelium in mesenchyme-free cultures inhibited proliferation. These data suggest that BMP4 is a signal that derives from both epithelial and mesenchymal tissue compartments.

Epidermal growth factor (EGF). EGF receptor (EGFR) signaling has been shown to stimulate murine embryonic branching morphogenesis, epithelial and mesenchymal cell proliferation, and Nkx2.1 and SPC expression (Schuger et al., 1996; Warburton et al., 1993). Similarly, inhibition of EGFR signaling results in decreased branching morphogenesis in culture and a neonatal pulmonary lethal phenotype in the null mutant, associated with decreased branching morphogenesis, and decreased Nkx2.1 and SPC expression levels (Miettinen et al., 1997).

Hepatocyte growth factor (HGF). HGF is expressed in primitive lung mesenchyme, while its receptor, the c-met tyrosine kinase, is expressed in primitive lung epithelium, suggesting the possibility of inductive mesenchymal-epithelial interactions (Sonnenberg et al., 1993).

Platelet derived growth factor (PDGF). PDGF peptides are dimeric ligands formed from two peptide chains, A and B. PDGF-AA and PDGF–BB, and the receptor PDGFR, are expressed in embryonic mouse lung. Abrogation of PDGF-A decreases the size of early embryonic mouse lungs in culture, as well as affecting early branch point formation (Souza et al., 1995). On the other hand, abrogation of PDGF-B reduces the size of the epithelial component of early embryonic mouse lung explants, but does not affect branching.

4.7 Key transcription factors in lung development

Thyroid transcription factor 1 (Ttf1/ Nkx2.1). Ttf1 is a homeodomain protein (also known as thyroid transcription factor-1 (Ttf-1). Although Ttf1 is the earliest known marker for the respiratory system, Ttf1-null mutant mice do have lungs. However, these lungs are highly abnormal, consisting of two main bronchi. In these mice epithelial cells fail to express any of the surfactant proteins or Clara cell CC-10 protein (Minoo et al., 1999), although proximal differentiation is normal. Lack of Ttf1 also affects branching morphogenesis for unknown reasons. Studies have suggested that Ttf1 is essential for the differentiation of distal lung epithelial cells and is required for the expression of several lung markers such as SPC (Kelly et al., 1996). Ttf1 consensus recognition sites are found in the 5′ promoters of several important peripheral lung genes including SPA, B, C, D, CC-10 and Ttf1 itself. It has been demonstrated that Ttf1 promoter activity is stimulated by HNF-3β (Ikeda et al., 1996) and GATA6 (Shaw-White et al., 1999).

Gata6. Unlike in liver, Gata6 is the only GATA factor expressed in the developing lung and this is restricted to distal lung epithelium. Gata6 expression is observed as early as E10.5 (Morrisey et al., 1996). Binding sites for Gata6 are present in several lung-specific promoters including those for the SPA, C and Ttf1 genes. Gata6 is able to trans-activate these promoters in non-lung cells, suggesting a role in the transcriptional regulation of these genes (Bruno et al., 2000; Shaw-White et al., 1999). The key role for Gata6 in the development of lung epithelial cells became clear when Gata6$^{-/-}$ cells were shown to be unable to contribute to the airway epithelium of Gata6$^{-/-}$/C57BL6 chimeric mice (Morrisey et al., 1998). The lungs of these chimeric mice showed defects in lung morphogenesis and down-regulation of lung-specific genes including SPC. The role of Gata6 in distal lung epithelium was studied by expressing a Gata6-Engrailed dominant-negative fusion protein exclusively in distal lung epithelium (Yang et al., 2002). In these mice, airway epithelial cell differentiation was defective, as type I cells were completely absent, while certain aspects of type II cells were also affected. Proximal airway development was also disrupted as shown by decreased levels of CC-10 expression. Lineage tracing of knockout Gata6-/- cells showed that the gene is not required for specification of endoderm, but that it is essential intrinsic to the endoderm for branching morphogenesis and differentiation of epithelium (Keijzer et al., 2001).

Forkhead homologue hepatocyte nuclear family (HNF) proteins. Hnf-3α and Hnf3β are known to have role in regulating transcription of Surfactant protein (SP) and Clara cell secretory protein (CCSP) (Bohinski et al., 1994; Clevidence et al., 1994). The Hnf3β null mutation results in an early embryonic lethal phenotype with complete failure of the primitive foregut to close into a tube (Ang and Rossant, 1994). During lung development HNF3β protein is expressed at higher levels in epithelial cells lining the proximal airways and at lower levels in the distal type II epithelium (Zhou et al., 1996). Consensus HNF binding sites are found in the 5′ promoter regions of several lung specific genes including SPA, B, C, D, and CC-10 in close proximity to Nkx2.1 sites, and it has been reported that HNF-3 activates transcription of Nkx2.1 in respiratory epithelial cells (Ikeda et al., 1996).

Gli transcription factors. Targeted disruption of the Gli2 gene results in diminished lung proliferation and branching. Gli2$^{-/-}$ mice have hypoplastic trachea (Mo et al., 1997). A more severe lung defect is observed with the Gli gene deficiency analyzed in a Gli3 heterozygous background. Gli2$^{-/-}$, Gli3$^{+/-}$ embryonic mouse lungs are more hypoplastic, and the right and left lobes fail to separate (Motoyama et al., 1998). These mice resemble the phenotype observed with Ttf1$^{-/-}$ mice. The most severe phenotype was observed in Gli2$^{-/-}$, Gli3$^{-/-}$ mice, which display a complete absence of lung, trachea, and esophagus and smaller stomach, liver and pancreas.

4.8 Differentiation of ES cells into lung epithelial cells

Compared to the other derivatives we have discussed, relatively few protocols have been reported that promote the differentiation of ES cells into lung epithelial cells, and even with these, the generation of lung epithelial cells is quite inefficient. To date there are no published protocols capable of generating a homogenous population of functional lung epithelial cells. An early report showed that SPC-positive lung epithelial cells could be derived from mouse ES cells (Ali et al., 2002) by culturing ES cells in a commercially available small airway growth media (SAGM) that supports the growth of lung epithelial cells *in vitro*. However, the ES cells were not responsive to the growth media until they had achieved a relatively advanced state of differentiation, suggesting that the system selected out spontaneously differentiated type II alveolar epithelial cells, rather than actively promoting their differentiation.

Studies from Rippon et al. (2006) further optimized the derivation of distal lung epithelial progenitors from murine embryonic stem cells using a novel three-step protocol. In this strategy, definitive endoderm was induced using a high concentration of ActivinA, followed by culturing the floating embryoid bodies in knockout serum replacement (KOSR) media for 10 days. Subsequently, EBs were attached to gelatinized plates and cultured in the presence of KOSR for an additional 10 days. KOSR was replaced with SAGM and cultured for 14 more days. However, the lung epithelial cells derived by this method had a gene expression profile resembling that of the lung committed precursor ($SPC^{+}Ttf1^{+}$) found in foregut endoderm, rather than mature alveolar epithelium. While trying to optimize serum free conditions for lung epithelial cell differentiation, Winkler et al. (2008) observed in the presence of serum relatively high expression levels of surfactant proteins A, B, C and D, CCSP and aquaporin 5, suggesting that epithelial cells expressing the markers for mature type II alveolar epithelium cells could be generated in serum-containing conditions. Surprisingly, a late treatment with ActivinA following serum activation resulted in significantly higher expression levels of SPC compared to an early induction with ActivinA.

Subsequently, other strategies have been used to promote the formation of pulmonary epithelium from ES cells. Co-culture of murine EBs with mesenchyme dissected from the distal tips of mid-gestation murine fetal lungs directed the differentiation of epithelial cells expressing Ttf1 and SPC (Van Vranken et al., 2005). It has also been reported that mouse ES cells can differentiate into type II alveolar epithelial cells when cultured with extracts from murine pneumocytes (Qin et al., 2005). The idea of transitioning lung epithelia through known normal developmental stages, which has been very successful for hepatocytes, may also be applied to lung. For example, initial generation of anterior endoderm using ActivinA, coupled with Wnt3a to suppress primitive endoderm, was successful using a subsequent step of FGF2 induction, to generate SPC-positive cells characteristic of type II alveolar epithelium (Roszell et al., 2009). For the generation of human lung epithelial cells from human ES cells there is to date only a single published protocol (Samadikuchaksaraei et al., 2006). In this case the investigators used commercially available SAGM to support the differentiation. When differentiation of hES was performed in the presence of SAGM, there was up-regulation of SPC levels. Clearly, much progress has been made in understanding the directive differentiation of endoderm derived organs like liver and pancreas, while the generation of effective protocols for the differentiation of ES cells into lung epithelial cells is still at an early stage, and represents an area of great opportunity. Considerable information is now available concerning the normal programs that regulate lung epithelial development and morphogenesis, and this should continue to be exploited in the near future for additional progress.

5. A strategy to dissect the function of GATA factors for specificity of mesendodermal derivatives

In our own laboratory we are studying organ development and regeneration, and have focused considerable attention toward the role of GATA transcription factors, using zebrafish embryo (Heicklen-Klein et al., 2005) and murine ES cell (Evans, 2008) experimental models. It is particularly intriguing that GATA factors seem to play key roles throughout the various stages of organ development. For example, members of the Gata4/5/6 subfamily may be involved in mesendoderm development (our unpublished results), but also have requirements for subsequent developmental transitions related to both mesoderm and endoderm. This includes, for example, cardiomyocyte specification from precardiac mesoderm (Holtzinger and Evans, 2007), heart tube looping and gut-derived organ budding (Holtzinger and Evans, 2005), and myocardial regeneration in adult zebrafish (Kikuchi et al., 2010). Understanding specificity of function is complicated since the genes are sometimes (but not always) functionally redundant, and they may function cell autonomous, or non-cell-autonomous through activation of secreted signaling factors.

One strategy we are taking is to develop conditional expression systems for specific GATA factors in the context of ES cell directed differentiation protocols, to determine the temporal effect and the ability for the genes to promote specific developmental transition stages. For this purpose we have been using a conditional system established by Kyba, Daley and colleagues (Ting et al., 2005), in which the gene encoding the reverse tetracycline transactivator protein is pre-targeted in the Ainv ES cell line to the constitutive Rosa26 locus, and a tet-operator sequence is pre-targeted just upstream of a loxP site (Fig. 6). This allows the insertion, by homologous recombination, of the coding sequences for any gene, which can then be activated conditionally by the addition of the tetracycline analogue doxycycline into the culture media.

This system provides significant advantages for studying genes like GATA factors that function at multiple stages of embryonic development. For example, forced expression of Gata4 in ES cells efficiently directs the development of primitive endoderm. This is useful for studying primitive endoderm, but abrogates the ability to test functions for subsequent stages, such as cardiomyocyte specification. Using this system we could show that expression of Gata4 during EB development is sufficient to generate DE (although primitive endoderm also forms) and that either type of endoderm is capable of inducing, by a non-cell-autonomous mechanism, the generation of cardiomyocyte progenitors (Holtzinger et al., 2010). Using conditional expression of different GATA factors in this system, we are currently testing their ability to influence mesendoderm development, mesoderm and endoderm patterning, the specification of different anterior DE derivatives including liver and lung, and the maturation of differentiated cell types. A major advantage of the strategy is that the induction is reversible, simply by washing doxycycline out of the media. This turns out to be important, since the transcription factor may have both positive and negative influences, depending on the time it is expressed (Zafonte et al., 2007). Finally, we have also adapted the system for loss-of-function, by conditional expression of gene-specific shRNAs.

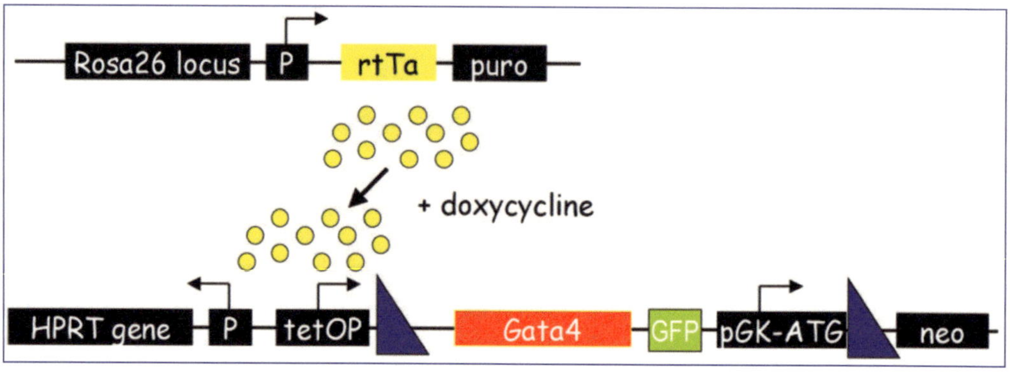

Fig. 6. In the Ainv ES cell line, the reverse tet-transactivator protein (rtTa, yellow balls) is expressed from the Rosa26 locus, and a loxP site (blue-triangle) is located downstream of a tet operator site. We used Cre-mediated recombination to target each flag-tagged Gata4/5/6 cDNA and IRES:GFP cassette into the loxP site, placing e.g. Gata4 (and GFP) under conditional (and reversible) control

6. Conclusion

Recent studies, particularly in zebrafish and frogs, revealed an ancient bipotential germ layer called mesendoderm, regulated by pathways conserved all the way to nematodes. It seems clear that this developmental stage is conserved also in mammals, providing both opportunities and challenges for directing the generation of defined cell types from ES cell cultures. A very successful strategy has been to define the normal developmental pathways that direct progenitors to mesodermal or endodermal fates, and then to recapitulate these pathways in the context of the ES cell *in vitro* system. A major challenge is that the same pathways regulate many diverse programs. However, this challenge can be overcome using conditional strategies (transgenic, or ultimately with small molecules) to fine-tune the timing and dose of the induction for optimization of each desired transitional stage (specification, commitment, proliferation, differentiation, etc.). Perhaps the major challenge moving forward will be to optimize and scale the generation of functional mature cell types, so that these can be applied successfully for therapeutic purposes in regenerative medicine protocols.

7. References

Abe, K., Niwa, H., Iwase, K., Takiguchi, M., Mori, M., Abe, S.I., and Yamamura, K.I. (1996). Endoderm-specific gene expression in embryonic stem cells differentiated to embryoid bodies. *Exp Cell Res* 229, 27-34.

Agarwal, S., Holton, K.L., and Lanza, R. (2008). Efficient differentiation of functional hepatocytes from human embryonic stem cells. *Stem Cells* 26, 1117-1127.

Ali, N.N., Edgar, A.J., Samadikuchaksaraei, A., Timson, C.M., Romanska, H.M., Polak, J.M., and Bishop, A.E. (2002). Derivation of type II alveolar epithelial cells from murine embryonic stem cells. *Tissue Eng* 8, 541-550.

Ang, S.L., and Rossant, J. (1994). HNF-3 beta is essential for node and notochord formation in mouse development. *Cell* 78, 561-574.

Arnold, S.J., Hofmann, U.K., Bikoff, E.K., and Robertson, E.J. (2008). Pivotal roles for eomesodermin during axis formation, epithelium-to-mesenchyme transition and endoderm specification in the mouse. *Development* 135, 501-511.

Baharvand, H., Hashemi, S.M., Kazemi Ashtiani, S., and Farrokhi, A. (2006). Differentiation of human embryonic stem cells into hepatocytes in 2D and 3D culture systems in vitro. *Int J Dev Biol* 50, 645-652.

Basma, H., Soto-Gutierrez, A., Yannam, G.R., Liu, L., Ito, R., Yamamoto, T., Ellis, E., Carson, S.D., Sato, S., Chen, Y., *et al.* (2009). Differentiation and transplantation of human embryonic stem cell-derived hepatocytes. *Gastroenterology* 136, 990-999.

Bellusci, S., Grindley, J., Emoto, H., Itoh, N., and Hogan, B.L. (1997). Fibroblast growth factor 10 (FGF10) and branching morphogenesis in the embryonic mouse lung. *Development* 124, 4867-4878.

Bellusci, S., Henderson, R., Winnier, G., Oikawa, T., and Hogan, B.L. (1996). Evidence from normal expression and targeted misexpression that bone morphogenetic protein (Bmp-4) plays a role in mouse embryonic lung morphogenesis. *Development* 122, 1693-1702.

Bishop, A.E., and Polak, J.M. (2006). Pulmonary epithelium. *Methods Enzymol* 418, 333-349.

Blouin, A., Bolender, R.P., and Weibel, E.R. (1977). Distribution of organelles and membranes between hepatocytes and nonhepatocytes in the rat liver parenchyma. A stereological study. *J Cell Biol* 72, 441-455.

Boheler, K.R., Czyz, J., Tweedie, D., Yang, H.T., Anisimov, S.V., and Wobus, A.M. (2002). Differentiation of pluripotent embryonic stem cells into cardiomyocytes. *Circ Res* 91, 189-201.

Bohinski, R.J., Di Lauro, R., and Whitsett, J.A. (1994). The lung-specific surfactant protein B gene promoter is a target for thyroid transcription factor 1 and hepatocyte nuclear factor 3, indicating common factors for organ-specific gene expression along the foregut axis. *Mol Cell Biol* 14, 5671-5681.

Breeze, R.G., and Wheeldon, E.B. (1977). The cells of the pulmonary airways. *Am Rev Respir Dis* 116, 705-777.

Bruno, M.D., Korfhagen, T.R., Liu, C., Morrisey, E.E., and Whitsett, J.A. (2000). GATA-6 activates transcription of surfactant protein A. *J Biol Chem* 275, 1043-1049.

Buckingham, M., Meilhac, S., and Zaffran, S. (2005). Building the mammalian heart from two sources of myocardial cells. *Nat Rev Genet* 6, 826-835.

Cai, J., Yi, F.F., Yang, X.C., Lin, G.S., Jiang, H., Wang, T., and Xia, Z. (2007a). Transplantation of embryonic stem cell-derived cardiomyocytes improves cardiac function in infarcted rat hearts. *Cytotherapy* 9, 283-291.

Cai, J., Zhao, Y., Liu, Y., Ye, F., Song, Z., Qin, H., Meng, S., Chen, Y., Zhou, R., Song, X., *et al.* (2007b). Directed differentiation of human embryonic stem cells into functional hepatic cells. *Hepatology* 45, 1229-1239.

Calmont, A., Wandzioch, E., Tremblay, K.D., Minowada, G., Kaestner, K.H., Martin, G.R., and Zaret, K.S. (2006). An FGF response pathway that mediates hepatic gene induction in embryonic endoderm cells. *Dev Cell* 11, 339-348.

Cardoso, W.V., and Lu, J. (2006). Regulation of early lung morphogenesis: questions, facts and controversies. *Development* 133, 1611-1624.

Chen, F., Cao, Y., Qian, J., Shao, F., Niederreither, K., and Cardoso, W.V. (2010). A retinoic acid-dependent network in the foregut controls formation of the mouse lung primordium. *J Clin Invest* 120, 2040-2048.

Chiao, E., Elazar, M., Xing, Y., Xiong, A., Kmet, M., Millan, M.T., Glenn, J.S., Wong, W.H., and Baker, J. (2008). Isolation and transcriptional profiling of purified hepatic cells derived from human embryonic stem cells. *Stem Cells* 26, 2032-2041.

Cho, C.H., Parashurama, N., Park, E.Y., Suganuma, K., Nahmias, Y., Park, J., Tilles, A.W., Berthiaume, F., and Yarmush, M.L. (2008). Homogeneous differentiation of hepatocyte-like cells from embryonic stem cells: applications for the treatment of liver failure. *FASEB J* 22, 898-909.

Clevidence, D.E., Overdier, D.G., Peterson, R.S., Porcella, A., Ye, H., Paulson, K.E., and Costa, R.H. (1994). Members of the HNF-3/forkhead family of transcription factors exhibit distinct cellular expression patterns in lung and regulate the surfactant protein B promoter. *Dev Biol* 166, 195-209.

Colvin, J.S., White, A.C., Pratt, S.J., and Ornitz, D.M. (2001). Lung hypoplasia and neonatal death in Fgf9-null mice identify this gene as an essential regulator of lung mesenchyme. *Development* 128, 2095-2106.

Costa, R.H., Kalinichenko, V.V., and Lim, L. (2001). Transcription factors in mouse lung development and function. *Am J Physiol Lung Cell Mol Physiol* 280, L823-838.

D'Amour, K.A., Agulnick, A.D., Eliazer, S., Kelly, O.G., Kroon, E., and Baetge, E.E. (2005). Efficient differentiation of human embryonic stem cells to definitive endoderm. *Nat Biotechnol* 23, 1534-1541.

Dai, W., Field, L.J., Rubart, M., Reuter, S., Hale, S.L., Zweigerdt, R., Graichen, R.E., Kay, G.L., Jyrala, A.J., Colman, A., *et al.* (2007). Survival and maturation of human embryonic stem cell-derived cardiomyocytes in rat hearts. *J Mol Cell Cardiol* 43, 504-516.

Desai, T.J., Chen, F., Lu, J., Qian, J., Niederreither, K., Dolle, P., Chambon, P., and Cardoso, W.V. (2006). Distinct roles for retinoic acid receptors alpha and beta in early lung morphogenesis. *Dev Biol* 291, 12-24.

Doetschman, T.C., Eistetter, H., Katz, M., Schmidt, W., and Kemler, R. (1985). The in vitro development of blastocyst-derived embryonic stem cell lines: formation of visceral yolk sac, blood islands and myocardium. *J Embryol Exp Morphol* 87, 27-45.

Domian, I.J., Chiravuri, M., van der Meer, P., Feinberg, A.W., Shi, X., Shao, Y., Wu, S.M., Parker, K.K., and Chien, K.R. (2009). Generation of functional ventricular heart muscle from mouse ventricular progenitor cells. *Science* 326, 426-429.

Drobinskaya, I., Linn, T., Saric, T., Bretzel, R.G., Bohlen, H., Hescheler, J., and Kolossov, E. (2008). Scalable selection of hepatocyte- and hepatocyte precursor-like cells from culture of differentiating transgenically modified murine embryonic stem cells. *Stem Cells* 26, 2245-2256.

Duan, Y., Catana, A., Meng, Y., Yamamoto, N., He, S., Gupta, S., Gambhir, S.S., and Zern, M.A. (2007). Differentiation and enrichment of hepatocyte-like cells from human embryonic stem cells in vitro and in vivo. *Stem Cells* 25, 3058-3068.

Duan, Y., Ma, X., Zou, W., Wang, C., Bahbahan, I.S., Ahuja, T.P., Tolstikov, V., and Zern, M.A. (2010). Differentiation and characterization of metabolically functioning hepatocytes from human embryonic stem cells. *Stem Cells* 28, 674-686.

Dufort, D., Schwartz, L., Harpal, K., and Rossant, J. (1998). The transcription factor HNF3beta is required in visceral endoderm for normal primitive streak morphogenesis. *Development* 125, 3015-3025.

Evans, M.J., Cabral, L.J., Stephens, R.J., and Freeman, G. (1975). Transformation of alveolar type 2 cells to type 1 cells following exposure to NO2. *Exp Mol Pathol* 22, 142-150.

Evans, M.J., and Kaufman, M.H. (1981). Establishment in culture of pluripotential cells from mouse embryos. *Nature* 292, 154-156.

Evans, T. (2008). Embryonic Stem Cells as a Model for Cardiac Development and Disease. *Drug Discov Today Dis Models* 5, 147-155.

Fehrenbach, H. (2001). Alveolar epithelial type II cell: defender of the alveolus revisited. *Respir Res* 2, 33-46.

Gadue, P., Huber, T.L., Paddison, P.J., and Keller, G.M. (2006). Wnt and TGF-beta signaling are required for the induction of an in vitro model of primitive streak formation using embryonic stem cells. *Proc Natl Acad Sci U S A* 103, 16806-16811.

Gao, N., LeLay, J., Vatamaniuk, M.Z., Rieck, S., Friedman, J.R., and Kaestner, K.H. (2008). Dynamic regulation of Pdx1 enhancers by Foxa1 and Foxa2 is essential for pancreas development. *Genes Dev* 22, 3435-3448.

Gassanov, N., Er, F., Zagidullin, N., and Hoppe, U.C. (2004). Endothelin induces differentiation of ANP-EGFP expressing embryonic stem cells towards a pacemaker phenotype. *FASEB J* 18, 1710-1712.

Gomez-Lechon, M.J., Tolosa, L., Castell, J.V., and Donato, M.T. (2010). Mechanism-based selection of compounds for the development of innovative in vitro approaches to hepatotoxicity studies in the LIINTOP project. *Toxicol In Vitro.*

Gouon-Evans, V., Boussemart, L., Gadue, P., Nierhoff, D., Koehler, C.I., Kubo, A., Shafritz, D.A., and Keller, G. (2006). BMP-4 is required for hepatic specification of mouse embryonic stem cell-derived definitive endoderm. *Nat Biotechnol* 24, 1402-1411.

Hamazaki, T., Iiboshi, Y., Oka, M., Papst, P.J., Meacham, A.M., Zon, L.I., and Terada, N. (2001). Hepatic maturation in differentiating embryonic stem cells in vitro. *FEBS Lett* 497, 15-19.

Hart, A.H., Hartley, L., Sourris, K., Stadler, E.S., Li, R., Stanley, E.G., Tam, P.P., Elefanty, A.G., and Robb, L. (2002). Mixl1 is required for axial mesendoderm morphogenesis and patterning in the murine embryo. *Development* 129, 3597-3608.

He, J.Q., Ma, Y., Lee, Y., Thomson, J.A., and Kamp, T.J. (2003). Human embryonic stem cells develop into multiple types of cardiac myocytes: action potential characterization. *Circ Res* 93, 32-39.

Heicklen-Klein, A., McReynolds, L.J., and Evans, T. (2005). Using the zebrafish model to study GATA transcription factors. *Semin Cell Dev Biol* 16, 95-106.

Heo, J., Factor, V.M., Uren, T., Takahama, Y., Lee, J.S., Major, M., Feinstone, S.M., and Thorgeirsson, S.S. (2006). Hepatic precursors derived from murine embryonic stem cells contribute to regeneration of injured liver. *Hepatology* 44, 1478-1486.

Hidaka, K., Lee, J.K., Kim, H.S., Ihm, C.H., Iio, A., Ogawa, M., Nishikawa, S., Kodama, I., and Morisaki, T. (2003). Chamber-specific differentiation of Nkx2.5-positive cardiac precursor cells from murine embryonic stem cells. *FASEB J* 17, 740-742.

Holtzinger, A., and Evans, T. (2005). Gata4 regulates the formation of multiple organs. *Development* 132, 4005-4014.

Holtzinger, A., and Evans, T. (2007). Gata5 and Gata6 are functionally redundant in zebrafish for specification of cardiomyocytes. *Dev Biol* 312, 613-622.

Holtzinger, A., Rosenfeld, G.E., and Evans, T. (2010). Gata4 directs development of cardiac-inducing endoderm from ES cells. *Dev Biol* 337, 63-73.

Hu, A.B., Cai, J.Y., Zheng, Q.C., He, X.Q., Shan, Y., Pan, Y.L., Zeng, G.C., Hong, A., Dai, Y., and Li, L.S. (2004). High-ratio differentiation of embryonic stem cells into hepatocytes in vitro. *Liver Int* 24, 237-245.

Hyatt, B.A., Shangguan, X., and Shannon, J.M. (2002). BMP4 modulates fibroblast growth factor-mediated induction of proximal and distal lung differentiation in mouse embryonic tracheal epithelium in mesenchyme-free culture. *Dev Dyn* 225, 153-165.

Ieda, M., Fu, J.D., Delgado-Olguin, P., Vedantham, V., Hayashi, Y., Bruneau, B.G., and Srivastava, D. (2010). Direct reprogramming of fibroblasts into functional cardiomyocytes by defined factors. *Cell* 142, 375-386.

Ikeda, K., Shaw-White, J.R., Wert, S.E., and Whitsett, J.A. (1996). Hepatocyte nuclear factor 3 activates transcription of thyroid transcription factor 1 in respiratory epithelial cells. *Mol Cell Biol* 16, 3626-3636.

Imamura, T., Cui, L., Teng, R., Johkura, K., Okouchi, Y., Asanuma, K., Ogiwara, N., and Sasaki, K. (2004). Embryonic stem cell-derived embryoid bodies in three-dimensional culture system form hepatocyte-like cells in vitro and in vivo. *Tissue Eng* 10, 1716-1724.

Irion, S., Nostro, M.C., Kattman, S.J., and Keller, G.M. (2008). Directed differentiation of pluripotent stem cells: from developmental biology to therapeutic applications. *Cold Spring Harb Symp Quant Biol* 73, 101-110.

Ishii, T., Yasuchika, K., Fujii, H., Hoppo, T., Baba, S., Naito, M., Machimoto, T., Kamo, N., Suemori, H., Nakatsuji, N., *et al.* (2005). In vitro differentiation and maturation of mouse embryonic stem cells into hepatocytes. *Exp Cell Res* 309, 68-77.

Ishii, T., Yasuchika, K., Machimoto, T., Kamo, N., Komori, J., Konishi, S., Suemori, H., Nakatsuji, N., Saito, M., Kohno, K., *et al.* (2007). Transplantation of embryonic stem cell-derived endodermal cells into mice with induced lethal liver damage. *Stem Cells* 25, 3252-3260.

Itskovitz-Eldor, J., Schuldiner, M., Karsenti, D., Eden, A., Yanuka, O., Amit, M., Soreq, H., and Benvenisty, N. (2000). Differentiation of human embryonic stem cells into embryoid bodies compromising the three embryonic germ layers. *Mol Med* 6, 88-95.

Jackson, S.A., Schiesser, J., Stanley, E.G., and Elefanty, A.G. (2010). Differentiating embryonic stem cells pass through 'temporal windows' that mark responsiveness to exogenous and paracrine mesendoderm inducing signals. *PLoS One* 5, e10706.

Kanai-Azuma, M., Kanai, Y., Gad, J.M., Tajima, Y., Taya, C., Kurohmaru, M., Sanai, Y., Yonekawa, H., Yazaki, K., Tam, P.P., *et al.* (2002). Depletion of definitive gut endoderm in Sox17-null mutant mice. *Development* 129, 2367-2379.

Kania, G., Blyszczuk, P., Jochheim, A., Ott, M., and Wobus, A.M. (2004). Generation of glycogen- and albumin-producing hepatocyte-like cells from embryonic stem cells. *Biol Chem* 385, 943-953.

Kehat, I., Khimovich, L., Caspi, O., Gepstein, A., Shofti, R., Arbel, G., Huber, I., Satin, J., Itskovitz-Eldor, J., and Gepstein, L. (2004). Electromechanical integration of cardiomyocytes derived from human embryonic stem cells. *Nat Biotechnol* 22, 1282-1289.

Keijzer, R., van Tuyl, M., Meijers, C., Post, M., Tibboel, D., Grosveld, F., and Koutsourakis, M. (2001). The transcription factor GATA6 is essential for branching morphogenesis and epithelial cell differentiation during fetal pulmonary development. *Development* 128, 503-511.

Kelly, S.E., Bachurski, C.J., Burhans, M.S., and Glasser, S.W. (1996). Transcription of the lung-specific surfactant protein C gene is mediated by thyroid transcription factor 1. *J Biol Chem* 271, 6881-6888.

Kikuchi, K., Holdway, J.E., Werdich, A.A., Anderson, R.M., Fang, Y., Egnaczyk, G.F., Evans, T., Macrae, C.A., Stainier, D.Y., and Poss, K.D. (2010). Primary contribution to zebrafish heart regeneration by gata4(+) cardiomyocytes. *Nature* 464, 601-605.

Klug, M.G., Soonpaa, M.H., Koh, G.Y., and Field, L.J. (1996). Genetically selected cardiomyocytes from differentiating embronic stem cells form stable intracardiac grafts. *J Clin Invest* 98, 216-224.

Kolossov, E., Lu, Z., Drobinskaya, I., Gassanov, N., Duan, Y., Sauer, H., Manzke, O., Bloch, W., Bohlen, H., Hescheler, J., *et al.* (2005). Identification and characterization of embryonic stem cell-derived pacemaker and atrial cardiomyocytes. *FASEB J* 19, 577-579.

Kubo, A., Shinozaki, K., Shannon, J.M., Kouskoff, V., Kennedy, M., Woo, S., Fehling, H.J., and Keller, G. (2004). Development of definitive endoderm from embryonic stem cells in culture. *Development* 131, 1651-1662.

Kumashiro, Y., Asahina, K., Ozeki, R., Shimizu-Saito, K., Tanaka, Y., Kida, Y., Inoue, K., Kaneko, M., Sato, T., Teramoto, K., *et al.* (2005). Enrichment of hepatocytes differentiated from mouse embryonic stem cells as a transplantable source. *Transplantation* 79, 550-557.

Kuzmenkin, A., Liang, H., Xu, G., Pfannkuche, K., Eichhorn, H., Fatima, A., Luo, H., Saric, T., Wernig, M., Jaenisch, R., *et al.* (2009). Functional characterization of cardiomyocytes derived from murine induced pluripotent stem cells in vitro. *FASEB J* 23, 4168-4180.

Laflamme, M.A., Chen, K.Y., Naumova, A.V., Muskheli, V., Fugate, J.A., Dupras, S.K., Reinecke, H., Xu, C., Hassanipour, M., Police, S., *et al.* (2007). Cardiomyocytes derived from human embryonic stem cells in pro-survival factors enhance function of infarcted rat hearts. *Nat Biotechnol* 25, 1015-1024.

Laflamme, M.A., Gold, J., Xu, C., Hassanipour, M., Rosler, E., Police, S., Muskheli, V., and Murry, C.E. (2005). Formation of human myocardium in the rat heart from human embryonic stem cells. *Am J Pathol* 167, 663-671.

Lavon, N., Yanuka, O., and Benvenisty, N. (2004). Differentiation and isolation of hepatic-like cells from human embryonic stem cells. *Differentiation* 72, 230-238.

Lawson, K.A. (1974). Mesenchyme specificity in rodent salivary gland development: the response of salivary epithelium to lung mesenchyme in vitro. *J Embryol Exp Morphol* 32, 469-493.

Lawson, K.A., and Pedersen, R.A. (1987). Cell fate, morphogenetic movement and population kinetics of embryonic endoderm at the time of germ layer formation in the mouse. *Development* 101, 627-652.

Lazzaro, D., Price, M., de Felice, M., and Di Lauro, R. (1991). The transcription factor TTF-1 is expressed at the onset of thyroid and lung morphogenesis and in restricted regions of the foetal brain. *Development* 113, 1093-1104.

Lee, C.S., Friedman, J.R., Fulmer, J.T., and Kaestner, K.H. (2005). The initiation of liver development is dependent on Foxa transcription factors. *Nature* 435, 944-947.

Lemaigre, F., and Zaret, K.S. (2004). Liver development update: new embryo models, cell lineage control, and morphogenesis. *Curr Opin Genet Dev* 14, 582-590.

Lemaigre, F.P. (2009). Mechanisms of liver development: concepts for understanding liver disorders and design of novel therapies. *Gastroenterology* 137, 62-79.

Lewis, S.L., and Tam, P.P. (2006). Definitive endoderm of the mouse embryo: formation, cell fates, and morphogenetic function. *Dev Dyn* 235, 2315-2329.

Li, L., Krantz, I.D., Deng, Y., Genin, A., Banta, A.B., Collins, C.C., Qi, M., Trask, B.J., Kuo, W.L., Cochran, J., *et al.* (1997). Alagille syndrome is caused by mutations in human Jagged1, which encodes a ligand for Notch1. *Nat Genet* 16, 243-251.

Lin, Y., Zhang, S., Rehn, M., Itaranta, P., Tuukkanen, J., Heljasvaara, R., Peltoketo, H., Pihlajaniemi, T., and Vainio, S. (2001). Induced repatterning of type XVIII collagen expression in ureter bud from kidney to lung type: association with sonic hedgehog and ectopic surfactant protein C. *Development* 128, 1573-1585.

Lindsley, R.C., Gill, J.G., Kyba, M., Murphy, T.L., and Murphy, K.M. (2006). Canonical Wnt signaling is required for development of embryonic stem cell-derived mesoderm. *Development* 133, 3787-3796.

Liu, H., Ye, Z., Kim, Y., Sharkis, S., and Jang, Y.Y. (2010). Generation of endoderm-derived human induced pluripotent stem cells from primary hepatocytes. *Hepatology* 51, 1810-1819.

Lowe, L.A., Yamada, S., and Kuehn, M.R. (2001). Genetic dissection of nodal function in patterning the mouse embryo. *Development* 128, 1831-1843.

Matsumoto, K., Yoshitomi, H., Rossant, J., and Zaret, K.S. (2001). Liver organogenesis promoted by endothelial cells prior to vascular function. *Science* 294, 559-563.

Mauritz, C., Schwanke, K., Reppel, M., Neef, S., Katsirntaki, K., Maier, L.S., Nguemo, F., Menke, S., Haustein, M., Hescheler, J., *et al.* (2008). Generation of functional murine cardiac myocytes from induced pluripotent stem cells. *Circulation* 118, 507-517.

McDevitt, T.C., Laflamme, M.A., and Murry, C.E. (2005). Proliferation of cardiomyocytes derived from human embryonic stem cells is mediated via the IGF/PI 3-kinase/Akt signaling pathway. *J Mol Cell Cardiol* 39, 865-873.

Metzger, R.J., Klein, O.D., Martin, G.R., and Krasnow, M.A. (2008). The branching programme of mouse lung development. *Nature* 453, 745-750.

Miettinen, P.J., Warburton, D., Bu, D., Zhao, J.S., Berger, J.E., Minoo, P., Koivisto, T., Allen, L., Dobbs, L., Werb, Z., *et al.* (1997). Impaired lung branching morphogenesis in the absence of functional EGF receptor. *Dev Biol* 186, 224-236.

Min, H., Danilenko, D.M., Scully, S.A., Bolon, B., Ring, B.D., Tarpley, J.E., DeRose, M., and Simonet, W.S. (1998). Fgf-10 is required for both limb and lung development and exhibits striking functional similarity to Drosophila branchless. *Genes Dev* 12, 3156-3161.

Minoo, P., Su, G., Drum, H., Bringas, P., and Kimura, S. (1999). Defects in tracheoesophageal and lung morphogenesis in Nkx2.1(-/-) mouse embryos. *Dev Biol* 209, 60-71.

Mo, R., Freer, A.M., Zinyk, D.L., Crackower, M.A., Michaud, J., Heng, H.H., Chik, K.W., Shi, X.M., Tsui, L.C., Cheng, S.H., *et al.* (1997). Specific and redundant functions of Gli2 and Gli3 zinc finger genes in skeletal patterning and development. *Development* 124, 113-123.

Mollard, R., Ghyselinck, N.B., Wendling, O., Chambon, P., and Mark, M. (2000). Stage-dependent responses of the developing lung to retinoic acid signaling. *Int J Dev Biol* 44, 457-462.

Moretti, A., Caron, L., Nakano, A., Lam, J.T., Bernshausen, A., Chen, Y., Qyang, Y., Bu, L., Sasaki, M., Martin-Puig, S., *et al.* (2006). Multipotent embryonic isl1+ progenitor cells lead to cardiac, smooth muscle, and endothelial cell diversification. *Cell* 127, 1151-1165.

Morrisey, E.E., Ip, H.S., Lu, M.M., and Parmacek, M.S. (1996). GATA-6: a zinc finger transcription factor that is expressed in multiple cell lineages derived from lateral mesoderm. *Dev Biol* 177, 309-322.

Morrisey, E.E., Tang, Z., Sigrist, K., Lu, M.M., Jiang, F., Ip, H.S., and Parmacek, M.S. (1998). GATA6 regulates HNF4 and is required for differentiation of visceral endoderm in the mouse embryo. *Genes Dev* 12, 3579-3590.

Motoyama, J., Liu, J., Mo, R., Ding, Q., Post, M., and Hui, C.C. (1998). Essential function of Gli2 and Gli3 in the formation of lung, trachea and oesophagus. *Nat Genet* 20, 54-57.

Muller, M., Fleischmann, B.K., Selbert, S., Ji, G.J., Endl, E., Middeler, G., Muller, O.J., Schlenke, P., Frese, S., Wobus, A.M., *et al.* (2000). Selection of ventricular-like cardiomyocytes from ES cells in vitro. *FASEB J* 14, 2540-2548.

Mummery, C., Ward-van Oostwaard, D., Doevendans, P., Spijker, R., van den Brink, S., Hassink, R., van der Heyden, M., Opthof, T., Pera, M., de la Riviere, A.B., *et al.* (2003). Differentiation of human embryonic stem cells to cardiomyocytes: role of coculture with visceral endoderm-like cells. *Circulation* 107, 2733-2740.

Murry, C.E., and Keller, G. (2008). Differentiation of embryonic stem cells to clinically relevant populations: lessons from embryonic development. *Cell* 132, 661-680.

Naito, A.T., Shiojima, I., Akazawa, H., Hidaka, K., Morisaki, T., Kikuchi, A., and Komuro, I. (2006). Developmental stage-specific biphasic roles of Wnt/beta-catenin signaling in cardiomyogenesis and hematopoiesis. *Proc Natl Acad Sci U S A* 103, 19812-19817.

Oda, T., Elkahloun, A.G., Pike, B.L., Okajima, K., Krantz, I.D., Genin, A., Piccoli, D.A., Meltzer, P.S., Spinner, N.B., Collins, F.S., *et al.* (1997). Mutations in the human Jagged1 gene are responsible for Alagille syndrome. *Nat Genet* 16, 235-242.

Ohtsuka, N., Urase, K., Momoi, T., and Nogawa, H. (2001). Induction of bud formation of embryonic mouse tracheal epithelium by fibroblast growth factor plus transferrin in mesenchyme-free culture. *Dev Dyn* 222, 263-272.

Orr-Urtreger, A., Bedford, M.T., Burakova, T., Arman, E., Zimmer, Y., Yayon, A., Givol, D., and Lonai, P. (1993). Developmental localization of the splicing alternatives of fibroblast growth factor receptor-2 (FGFR2). *Dev Biol* 158, 475-486.

Peterkin, T., Gibson, A., Loose, M., and Patient, R. (2005). The roles of GATA-4, -5 and -6 in vertebrate heart development. *Semin Cell Dev Biol* 16, 83-94.

Plageman, T.F., Jr., and Yutzey, K.E. (2005). T-box genes and heart development: putting the "T" in heart. *Dev Dyn* 232, 11-20.

Qin, M., Tai, G., Collas, P., Polak, J.M., and Bishop, A.E. (2005). Cell extract-derived differentiation of embryonic stem cells. *Stem Cells* 23, 712-718.

Rambhatla, L., Chiu, C.P., Kundu, P., Peng, Y., and Carpenter, M.K. (2003). Generation of hepatocyte-like cells from human embryonic stem cells. *Cell Transplant* 12, 1-11.

Rashid, S.T., Corbineau, S., Hannan, N., Marciniak, S.J., Miranda, E., Alexander, G., Huang-Doran, I., Griffin, J., Ahrlund-Richter, L., Skepper, J., *et al.* (2010). Modeling inherited metabolic disorders of the liver using human induced pluripotent stem cells. *J Clin Invest* 120, 3127-3136.

Rippon, H.J., Polak, J.M., Qin, M., and Bishop, A.E. (2006). Derivation of distal lung epithelial progenitors from murine embryonic stem cells using a novel three-step differentiation protocol. *Stem Cells* 24, 1389-1398.

Rodaway, A., and Patient, R. (2001). Mesendoderm. an ancient germ layer? *Cell* 105, 169-172.

Rossi, J.M., Dunn, N.R., Hogan, B.L., and Zaret, K.S. (2001). Distinct mesodermal signals, including BMPs from the septum transversum mesenchyme, are required in combination for hepatogenesis from the endoderm. *Genes Dev* 15, 1998-2009.

Roszell, B., Mondrinos, M.J., Seaton, A., Simons, D.M., Koutzaki, S.H., Fong, G.H., Lelkes, P.I., and Finck, C.M. (2009). Efficient derivation of alveolar type II cells from embryonic stem cells for in vivo application. *Tissue Eng Part A* 15, 3351-3365.

Samadikuchaksaraei, A., Cohen, S., Isaac, K., Rippon, H.J., Polak, J.M., Bielby, R.C., and Bishop, A.E. (2006). Derivation of distal airway epithelium from human embryonic stem cells. *Tissue Eng* 12, 867-875.

Satin, J., Kehat, I., Caspi, O., Huber, I., Arbel, G., Itzhaki, I., Magyar, J., Schroder, E.A., Perlman, I., and Gepstein, L. (2004). Mechanism of spontaneous excitability in human embryonic stem cell derived cardiomyocytes. *J Physiol* 559, 479-496.

Schier, A.F. (2003). Nodal signaling in vertebrate development. *Annu Rev Cell Dev Biol* 19, 589-621.

Schuger, L., Skubitz, A.P., Gilbride, K., Mandel, R., and He, L. (1996). Laminin and heparan sulfate proteoglycan mediate epithelial cell polarization in organotypic cultures of embryonic lung cells: evidence implicating involvement of the inner globular region of laminin beta 1 chain and the heparan sulfate groups of heparan sulfate proteoglycan. *Dev Biol* 179, 264-273.

Sekine, K., Ohuchi, H., Fujiwara, M., Yamasaki, M., Yoshizawa, T., Sato, T., Yagishita, N., Matsui, D., Koga, Y., Itoh, N., *et al.* (1999). Fgf10 is essential for limb and lung formation. *Nat Genet* 21, 138-141.

Serls, A.E., Doherty, S., Parvatiyar, P., Wells, J.M., and Deutsch, G.H. (2005). Different thresholds of fibroblast growth factors pattern the ventral foregut into liver and lung. *Development* 132, 35-47.

Shannon, J.M., Gebb, S.A., and Nielsen, L.D. (1999). Induction of alveolar type II cell differentiation in embryonic tracheal epithelium in mesenchyme-free culture. *Development* 126, 1675-1688.

Shannon, J.M., and Hyatt, B.A. (2004). Epithelial-mesenchymal interactions in the developing lung. *Annu Rev Physiol* 66, 625-645.

Shaw-White, J.R., Bruno, M.D., and Whitsett, J.A. (1999). GATA-6 activates transcription of thyroid transcription factor-1. *J Biol Chem* 274, 2658-2664.

Shirahashi, H., Wu, J., Yamamoto, N., Catana, A., Wege, H., Wager, B., Okita, K., and Zern, M.A. (2004). Differentiation of human and mouse embryonic stem cells along a hepatocyte lineage. *Cell Transplant* 13, 197-211.

Shiraki, N., Umeda, K., Sakashita, N., Takeya, M., Kume, K., and Kume, S. (2008). Differentiation of mouse and human embryonic stem cells into hepatic lineages. *Genes Cells* 13, 731-746.

Si-Tayeb, K., Lemaigre, F.P., and Duncan, S.A. (2010a). Organogenesis and development of the liver. *Dev Cell* 18, 175-189.

Si-Tayeb, K., Noto, F.K., Nagaoka, M., Li, J., Battle, M.A., Duris, C., North, P.E., Dalton, S., and Duncan, S.A. (2010b). Highly efficient generation of human hepatocyte-like cells from induced pluripotent stem cells. *Hepatology* 51, 297-305.

Simonet, W.S., DeRose, M.L., Bucay, N., Nguyen, H.Q., Wert, S.E., Zhou, L., Ulich, T.R., Thomason, A., Danilenko, D.M., and Whitsett, J.A. (1995). Pulmonary malformation in transgenic mice expressing human keratinocyte growth factor in the lung. *Proc Natl Acad Sci U S A* 92, 12461-12465.

Snir, M., Kehat, I., Gepstein, A., Coleman, R., Itskovitz-Eldor, J., Livne, E., and Gepstein, L. (2003). Assessment of the ultrastructural and proliferative properties of human embryonic stem cell-derived cardiomyocytes. *Am J Physiol Heart Circ Physiol* 285, H2355-2363.

Song, Z., Cai, J., Liu, Y., Zhao, D., Yong, J., Duo, S., Song, X., Guo, Y., Zhao, Y., Qin, H., *et al.* (2009). Efficient generation of hepatocyte-like cells from human induced pluripotent stem cells. *Cell Res* 19, 1233-1242.

Sonnenberg, E., Meyer, D., Weidner, K.M., and Birchmeier, C. (1993). Scatter factor/hepatocyte growth factor and its receptor, the c-met tyrosine kinase, can mediate a signal exchange between mesenchyme and epithelia during mouse development. *J Cell Biol* 123, 223-235.

Sosa-Pineda, B., Wigle, J.T., and Oliver, G. (2000). Hepatocyte migration during liver development requires Prox1. *Nat Genet* 25, 254-255.

Soto-Gutierrez, A., Kobayashi, N., Rivas-Carrillo, J.D., Navarro-Alvarez, N., Zhao, D., Okitsu, T., Noguchi, H., Basma, H., Tabata, Y., Chen, Y., *et al.* (2006a). Reversal of mouse hepatic failure using an implanted liver-assist device containing ES cell-derived hepatocytes. *Nat Biotechnol* 24, 1412-1419.

Soto-Gutierrez, A., Navarro-Alvarez, N., Rivas-Carrillo, J.D., Chen, Y., Yamatsuji, T., Tanaka, N., and Kobayashi, N. (2006b). Differentiation of human embryonic stem cells to hepatocytes using deleted variant of HGF and poly-amino-urethane-coated nonwoven polytetrafluoroethylene fabric. *Cell Transplant* 15, 335-341.

Souza, P., Kuliszewski, M., Wang, J., Tseu, I., Tanswell, A.K., and Post, M. (1995). PDGF-AA and its receptor influence early lung branching via an epithelial-mesenchymal interaction. *Development* 121, 2559-2567.

Sullivan, G.J., Hay, D.C., Park, I.H., Fletcher, J., Hannoun, Z., Payne, C.M., Dalgetty, D., Black, J.R., Ross, J.A., Samuel, K., *et al.* (2010). Generation of functional human hepatic endoderm from human induced pluripotent stem cells. *Hepatology* 51, 329-335.

Tada, S., Era, T., Furusawa, C., Sakurai, H., Nishikawa, S., Kinoshita, M., Nakao, K., and Chiba, T. (2005). Characterization of mesendoderm: a diverging point of the definitive endoderm and mesoderm in embryonic stem cell differentiation culture. *Development* 132, 4363-4374.

Takahashi, K., and Yamanaka, S. (2006). Induction of pluripotent stem cells from mouse embryonic and adult fibroblast cultures by defined factors. *Cell* 126, 663-676.

Tam, P.P., and Behringer, R.R. (1997). Mouse gastrulation: the formation of a mammalian body plan. *Mech Dev* 68, 3-25.

Tam, P.P., Khoo, P.L., Lewis, S.L., Bildsoe, H., Wong, N., Tsang, T.E., Gad, J.M., and Robb, L. (2007). Sequential allocation and global pattern of movement of the definitive endoderm in the mouse embryo during gastrulation. *Development* 134, 251-260.

Tanimizu, N., and Miyajima, A. (2007). Molecular mechanism of liver development and regeneration. *Int Rev Cytol* 259, 1-48.

Ten Have-Opbroek, A.A. (1991). Lung development in the mouse embryo. *Exp Lung Res* 17, 111-130.

Teratani, T., Yamamoto, H., Aoyagi, K., Sasaki, H., Asari, A., Quinn, G., Terada, M., and Ochiya, T. (2005). Direct hepatic fate specification from mouse embryonic stem cells. *Hepatology* 41, 836-846.

Thomson, J.A., Itskovitz-Eldor, J., Shapiro, S.S., Waknitz, M.A., Swiergiel, J.J., Marshall, V.S., and Jones, J.M. (1998). Embryonic stem cell lines derived from human blastocysts. *Science* 282, 1145-1147.

Ting, D.T., Kyba, M., and Daley, G.Q. (2005). Inducible transgene expression in mouse stem cells. *Methods Mol Med* 105, 23-46.

Touboul, T., Hannan, N.R., Corbineau, S., Martinez, A., Martinet, C., Branchereau, S., Mainot, S., Strick-Marchand, H., Pedersen, R., Di Santo, J., *et al.* (2010). Generation of functional hepatocytes from human embryonic stem cells under chemically defined conditions that recapitulate liver development. *Hepatology* 51, 1754-1765.

Ueno, S., Weidinger, G., Osugi, T., Kohn, A.D., Golob, J.L., Pabon, L., Reinecke, H., Moon, R.T., and Murry, C.E. (2007). Biphasic role for Wnt/beta-catenin signaling in cardiac specification in zebrafish and embryonic stem cells. *Proc Natl Acad Sci U S A* 104, 9685-9690.

Van Vranken, B.E., Romanska, H.M., Polak, J.M., Rippon, H.J., Shannon, J.M., and Bishop, A.E. (2005). Coculture of embryonic stem cells with pulmonary mesenchyme: a microenvironment that promotes differentiation of pulmonary epithelium. *Tissue Eng* 11, 1177-1187.

Vidarsson, H., Hyllner, J., and Sartipy, P. (2010). Differentiation of human embryonic stem cells to cardiomyocytes for in vitro and in vivo applications. *Stem Cell Rev* 6, 108-120.

Wan, H., Xu, Y., Ikegami, M., Stahlman, M.T., Kaestner, K.H., Ang, S.L., and Whitsett, J.A. (2004). Foxa2 is required for transition to air breathing at birth. *Proc Natl Acad Sci U S A* 101, 14449-14454.

Warburton, D., Lee, M., Berberich, M.A., and Bernfield, M. (1993). Molecular embryology and the study of lung development. *Am J Respir Cell Mol Biol* 9, 5-9.

Warburton, D., Schwarz, M., Tefft, D., Flores-Delgado, G., Anderson, K.D., and Cardoso, W.V. (2000). The molecular basis of lung morphogenesis. *Mech Dev* 92, 55-81.

Wardle, F.C., and Smith, J.C. (2006). Transcriptional regulation of mesendoderm formation in Xenopus. *Semin Cell Dev Biol* 17, 99-109.

Weaver, M., Dunn, N.R., and Hogan, B.L. (2000). Bmp4 and Fgf10 play opposing roles during lung bud morphogenesis. *Development* 127, 2695-2704.

Weaver, M., Yingling, J.M., Dunn, N.R., Bellusci, S., and Hogan, B.L. (1999). Bmp signaling regulates proximal-distal differentiation of endoderm in mouse lung development. *Development* 126, 4005-4015.

Wei, H., Juhasz, O., Li, J., Tarasova, Y.S., and Boheler, K.R. (2005). Embryonic stem cells and cardiomyocyte differentiation: phenotypic and molecular analyses. *J Cell Mol Med* 9, 804-817.

Wert, S.E., Glasser, S.W., Korfhagen, T.R., and Whitsett, J.A. (1993). Transcriptional elements from the human SP-C gene direct expression in the primordial respiratory epithelium of transgenic mice. *Dev Biol* 156, 426-443.

Wessells, N.K. (1970). Mammalian lung development: interactions in formation and morphogenesis of tracheal buds. *J Exp Zool* 175, 455-466.

Williams, M.C. (2003). Alveolar type I cells: molecular phenotype and development. *Annu Rev Physiol* 65, 669-695.

Winkler, M.E., Mauritz, C., Groos, S., Kispert, A., Menke, S., Hoffmann, A., Gruh, I., Schwanke, K., Haverich, A., and Martin, U. (2008). Serum-free differentiation of murine embryonic stem cells into alveolar type II epithelial cells. *Cloning Stem Cells* 10, 49-64.

Woodland, H.R., and Zorn, A.M. (2008). The core endodermal gene network of vertebrates: combining developmental precision with evolutionary flexibility. *Bioessays* 30, 757-765.

Wu, S.M., Fujiwara, Y., Cibulsky, S.M., Clapham, D.E., Lien, C.L., Schultheiss, T.M., and Orkin, S.H. (2006). Developmental origin of a bipotential myocardial and smooth muscle cell precursor in the mammalian heart. *Cell* 127, 1137-1150.

Xu, C., Police, S., Rao, N., and Carpenter, M.K. (2002). Characterization and enrichment of cardiomyocytes derived from human embryonic stem cells. *Circ Res* 91, 501-508.

Yamanaka, S., and Blau, H.M. (2010). Nuclear reprogramming to a pluripotent state by three approaches. *Nature* 465, 704-712.

Yang, H., Lu, M.M., Zhang, L., Whitsett, J.A., and Morrisey, E.E. (2002). GATA6 regulates differentiation of distal lung epithelium. *Development* 129, 2233-2246.

Yin, Y., Lim, Y.K., Salto-Tellez, M., Ng, S.C., Lin, C.S., and Lim, S.K. (2002). AFP(+), ESC-derived cells engraft and differentiate into hepatocytes in vivo. *Stem Cells* 20, 338-346.

Zafonte, B.T., Liu, S., Lynch-Kattman, M., Torregroza, I., Benvenuto, L., Kennedy, M., Keller, G., and Evans, T. (2007). Smad1 expands the hemangioblast population within a limited developmental window. *Blood* 109, 516-523.

Zaret, K.S., Watts, J., Xu, J., Wandzioch, E., Smale, S.T., and Sekiya, T. (2008). Pioneer factors, genetic competence, and inductive signaling: programming liver and pancreas progenitors from the endoderm. *Cold Spring Harb Symp Quant Biol* 73, 119-126.

Zhao, R., Watt, A.J., Battle, M.A., Li, J., Bondow, B.J., and Duncan, S.A. (2008). Loss of both GATA4 and GATA6 blocks cardiac myocyte differentiation and results in acardia in mice. *Dev Biol* 317, 614-619.

Zhou, L., Lim, L., Costa, R.H., and Whitsett, J.A. (1996). Thyroid transcription factor-1, hepatocyte nuclear factor-3beta, surfactant protein B, C, and Clara cell secretory protein in developing mouse lung. *J Histochem Cytochem* 44, 1183-1193.

Zorn, A.M., and Wells, J.M. (2009). Vertebrate endoderm development and organ formation. *Annu Rev Cell Dev Biol* 25, 221-251.

16

BNP is a Novel Regulator of Embryonic Stem Cell Proliferation

Abdelalim[1,2], EM and Tooyama[1], I
[1]*Molecular Neuroscience Research Center, Shiga University of Medical Science,*
[2]*Dep. of Cytology & Histology, Fac. of Vet. Medicine, Suez Canal University*
[1]*Japan,*
[2]*Egypt*

1. Introduction

Embryonic stem (ES) cells are derived from pre-implantation blastocysts. The blastocyst consists of an outer layer of trophoblast cells and an inner cell population called the inner cell mass (ICM). The ICM gives rise to all tissues of the body and some extraembryonic tissues, and it is from these cells that ES cells are derived (Suda et al., 1987). ES cells have two defining properties: self-renewal and pluripotency, and these make them a promising source for cell transplantation therapies (Suda et al., 1987). The precise mechanism that regulates stem cell self-renewal and pluripotency remains largely unknown. Thus investigation into the molecular and cellular mechanisms of stem cell self-renewal and pluripotency provide the necessary tools to harness the regenerative potential of ES cells for therapeutic purposes.

The two most striking features related to murine ES cell proliferation rate; (1) their unusual cell cycle structure and (2) their rapid rate of cell division (Savatier et al., 1994; Burdon et al., 2002). Murine ES cells divide with an unusually short generation time of approximately 8-10 hours (Savatier et al., 1994; Burdon et al., 2002), and have unusual cell cycle structure, consist of high proportion of cells in the S phase and, a short G1 phase. Since most cell types spend the majority of their time in G1, the short G1 phase of murine ES cells can account for their rapid rate of cell division since the length of S phase is similar within cell types.

The natriuretic peptides (NPs) are a family of three peptides: atrial NP (ANP), brain NP (BNP), and C-type NP (CNP) (Sudoh et al., 1988; Brenner et al., 1990; Sudoh et al., 1990). BNP is produced predominately in the heart (Minamino et al., 1988; Abdelalim et al., 2006a; Abdelalim et al., 2006b). The biological actions of NPs are mediated by binding to cell-surface receptors. These include NP receptor type A (NPR-A or GC-A), which is sensitive to ANP and BNP (Garbers, 1992), NP receptor type B (NPR-B), which is highly specific for CNP (Koller et al., 1991), and NP receptor type C (NPR-C), which comprises up to 95% of the total NPR population (Maack, 1992) and is known to bind all NPs with similar affinity (Levin et al. 1998). Hormone binding to NPR-A and NPR-B activates guanylyl cyclase (GC) and produces cyclic guanosine monophosphate (cGMP), the secondary messenger for a number of biological responses associated with NPs (Garber, 1992; Potter et al., 2006).

NPs are released into the circulation from cardiac cells to act as hormones to control fluid volume homeostasis and blood pressure by causing natriuresis, diuresis, vasorelaxation and inhibition of the renin–angiotensin–aldosterone system (Minamino et al., 1988; Abdelalim et al., 2006a,b; Levin et al., 1998). In addition, cell-based studies have shown that ANP and BNP exhibit important autocrine and paracrine functions such as modulating myocyte growth, apoptosis and proliferation of smooth-muscle cells (Abell et al., 1989; Silberbach and Roberts, 2001) and cardiac myocytes (Horio et al., 2000), and suppressing cardiac-fibroblast proliferation (Redondo et al., 1998) and extracellular-matrix secretion (Redondo et al., 1998; Tsuruda et al., 2002).

BNP-transgenic mice exhibit overgrowth of the growth-plate cartilage through a cGMP-dependent mechanism (Suda et al., 1998). Furthermore, signaling through NPR-A plays a pivotal role in tumor growth (Kong et al., 2008). Although little is known about the role of NPs in pre-implantation embryo development, one study reported that NPR-B-deficient mice were sterile due to a lack of development of the reproductive system, and the majority (75%) of the NPR-B-deficient mice studied died before 100 days of age (Tamura et al., 2004). In addition, exogenous BNP can enhance clonal propagation in murine ES cells (Ogawa et al., 2004) suggesting the presence of functional NP receptors.

In this chapter, we focus on the evidence that BNP is expressed by undifferentiated murine ES cells and has a role in the regulation of proliferation and survival of murine ES cells. We also discuss the mechanisms by which BNP affects ES cells.

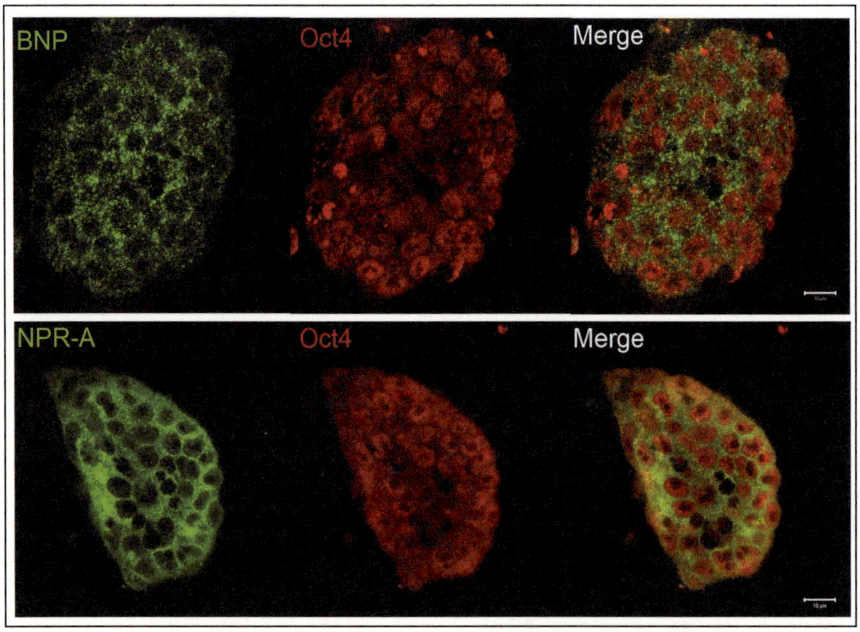

Fig. 1. Expression of BNP and NPR-A in undifferentiated ES cells. Double-immunofluorescence images of ES cells cultured in the presence of LIF, stained with antibodies against the ES-cell marker Oct4 and BNP *(upper panels)*, or Oct4 and NPR-A *(lower panels)*. Scale bar = 10 μm

2. Expression of BNP in murine ES cells and pre-implantation embryos

2.1 BNP and NPR-A are expressed in undifferentiated ES cells

The pluripotent identity of ES cells is controlled by a group of transcription factors (Niwa et al., 2000; Mitsui et al., 2003; Avilion et al., 2003; Chambers et al., 2003; Loh et al., 2006). The transcription factors Oct4, Nanog and Sox2 contribute to the hallmark characteristics of ES cells by activating target genes that encode pluripotency and self-renewal mechanisms and by repressing signaling pathways that promote differentiation (Koller et al., 1991). Also, ES cell self-renewal and pluripotency require inputs from extrinsic factors and their downstream effectors (Chambers and Smith, 2004). Self-renewal of murine ES cells under conventional culture conditions depends on the leukemia inhibitory factor (LIF). Withdrawal of LIF induces differentiation (Burdon et al., 2002).

To determine whether BNP exists in undifferentiated ES cells, the BNP expression in murine ES cells was examined under self-renewal conditions (+LIF) and differentiation conditions (-LIF). Immunofluorescence analysis revealed that BNP was expressed in undifferentiated ES cells (Oct4-positive cells) cultured in the presence of LIF, and expression was down-regulated upon differentiation induced by culturing ES cells without LIF for five days. The differentiated cells, which were negative for Oct4 expression, were also negative for BNP expression (Fig. 1). Similar results were obtained by Western blotting, flow cytometry, and reverse transcription-polymerase chain reaction (RT-PCR). These findings indicate that BNP is expressed specifically in self-renewing ES cells.

2.2 BNP and its receptors are expressed in murine blastocyst

Since ES cells are derived from the ICM of pre-implantation embryos, the expression of BNP and its receptor were examined in pre-implantation embryos (Abdelalim and Tooyama, 2009). The blastocyst stage of murine pre-implantation development occurs approximately 3.5 days post-fertilization. At this stage of embryonic development the first cell differentiation step has occurred. The blastocyst is composed of the epithelial

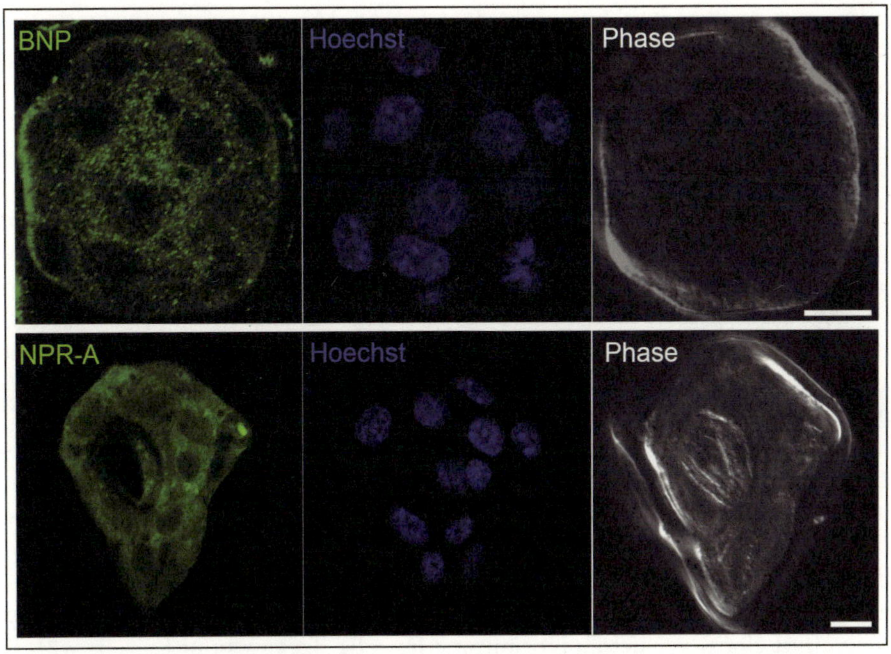

Fig. 2. Expression of BNP and NPR-A in pre-implantation embryos. Immunofluorescence images of 3.5-day-old blastocysts stained with antibodies against BNP *(upper panels)*, or NPR-A *(lower panels)*. Scale bar = 10 μm

trophectoderm, which is the layer of cells that develops into the placenta, and the ICM, which consists of the pluripotent cells that give rise to the embryo. Prior to implantation, the developing embryo is dependent on signals produced by embryonic and maternal growth factors. These growth factors are known to regulate cellular proliferation and differentiation during pre-implantation development (Weil et al., 1996).

Immunofluorescence analysis showed that BNP and NPR-A were expressed in 3.5-day-old murine blastocysts (Fig. 2). The expression of BNP and NPR-A were co-localized to those of Oct4 (Abdelalim and Tooyama, 2009). Oct4 expression is required to maintain the pluripotent-cell population of the ICM and epiblast (Nichols et al., 1998). Also, Oct4 is a crucial regulator of ES-cell pluripotency, and acts as a gatekeeper to prevent ES-cell differentiation (Nichols et al., 1998). Reduction of Oct4 expression below 50% of normal levels induces differentiation of ES cells into trophectoderm (Niwa et al., 2000). The localization of BNP and NPR-A in Oct4-positive cells in the pre-implantation embryos suggests that BNP may be involved in early embryo development.

3. Effect of BNP on ES cell proliferation

The use of the ribonucleic acid (RNA) interference (siRNA)-based technique to specifically knockdown BNP expression in ES cells is one approach for investigating the importance of BNP signaling in ES cells. We (Abdelalim and Tooyamna, 2009) specifically knocked down BNP expression in murine ES cells using two siRNAs targeting different regions. The efficiency of BNP knockdown was examined by RT-PCR and Western blotting at 48 hours after transfection. Silencing of BNP caused a significant reduction in ES-cell number and colony size as examined at 48 hours post-transfection. Morphological examination showed no signs of differentiation (Fig. 3).

The ability of ES cells to self-renew is maintained through promotion of proliferation, and prevention of differentiation and cell death. Because it appeared that the knockdown of BNP did not promote differentiation, we were interested to know whether its knockdown would affect deoxy-ribonucleic acid (DNA) synthesis. Therefore, the proliferation, and bromo-deoxyuridine (BrdU) incorporation, assays were performed to analysis the rate of proliferation. Immunofluorescence and flow-cytometry analyses showed a significant reduction in the number of BrdU-positive cells in the BNP siRNA-treated cells compared with control siRNA-treated cells. These findings suggest that knockdown of BNP signaling has a significant effect in decreasing the rate of DNA synthesis (Abdelalim and Tooyama, 2009).

The role of BNP in proliferation has been observed in other studies (Suda et al., 1998; Ogawa et al., 2004). BNP-transgenic mice exhibit marked skeletal overgrowth, and studies using in vitro organ culture of mouse tibia demonstrated that BNP increases cGMP production and activates the proliferation of growth-plate chondrocytes via GC-coupled NP receptors (Suda et al., 1998).

Undifferentiated murine ES cells express high levels of receptors specific for BNP, suggesting that BNP signaling can be transduced in ES cells via its receptors. Exogenous BNP may affect clonal proliferation of murine ES cells. Treatment of ES cells with BNP in low-density culture and in serum-free (ES cell KSR) medium for six days resulted in an increase in ES-cell propagation. Similar findings were reported by Ogawa et al. (2004). Thus, clonal propagation of murine ES cells is increased in the presence of BNP.

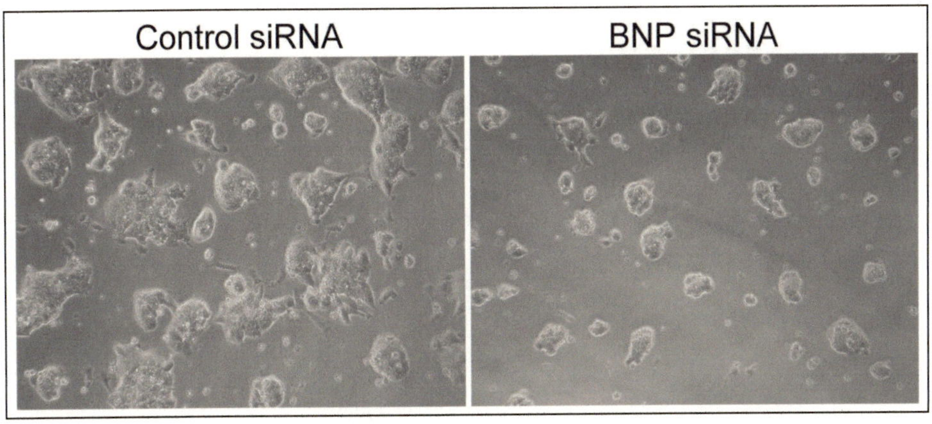

Fig. 3. BNP knockdown suppresses ES-cell proliferation. Morphologies of murine ES cells 48 hours after transfection with control siRNA or BNP siRNA

Flow-cytometric analysis of cell-cycle distribution revealed a significant reduction in the proportion of cells in S phase, and an increase in the proportion of cells in G1 and G2/M phases in ES cells treated with BNP siRNA (Fig. 5a, b). However, cell-cycle analysis showed no significant difference between the phases of the cell cycle of differentiated ES cells treated with BNP siRNA and control siRNA. These results indicate BNP function in undifferentiated ES cells but not in differentiated cells and indicate that BNP is a critical factor in sustaining viability and proliferation of murine ES cells.

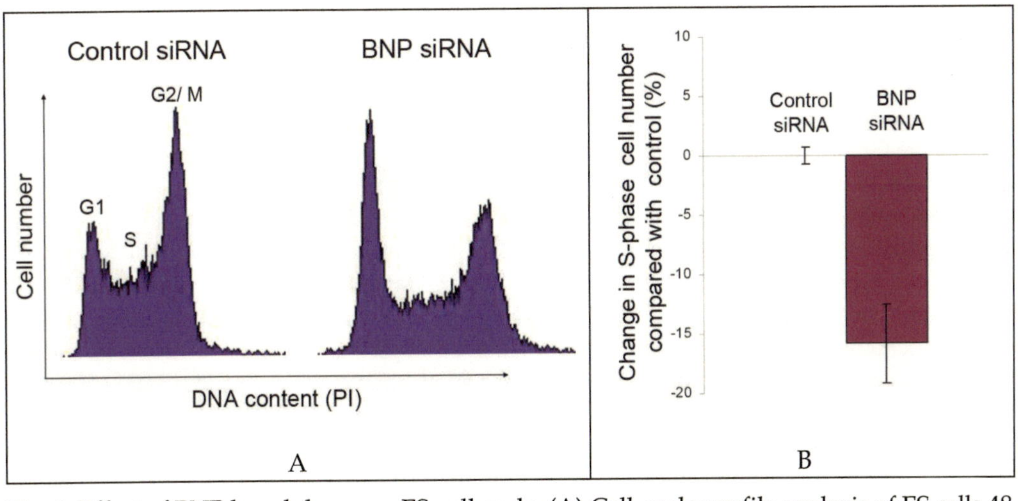

Fig. 4. Effect of BNP knockdown on ES-cell cycle. (A) Cell-cycle profile analysis of ES cells 48 hours after transfection with control siRNA or BNP siRNA. (B) Quantitative analysis of **A** showing the percent change of cells in S phase (n = 3). This figure is reproduced from (Abdelalim and Tooyama, 2009)

The reduction in ES-cell proliferation caused by abrogation of BNP signaling had no effect on the undifferentiated status of the ES cells as determined by morphologic examination (Fig. 3). This was confirmed by measurement of alkaline-phosphatase activity (Fig. 5), which revealed that levels of self-renewal in the control siRNA- and BNP siRNA-treated cells were identical. Also, the expression levels of Oct4 mRNA and Nanog mRNA did not differ between control siRNA-treated cells and BNP siRNA-treated cells, suggesting that BNP signaling is not involved in the pluripotency of ES cells (Abdelalim and Tooyama, 2009).

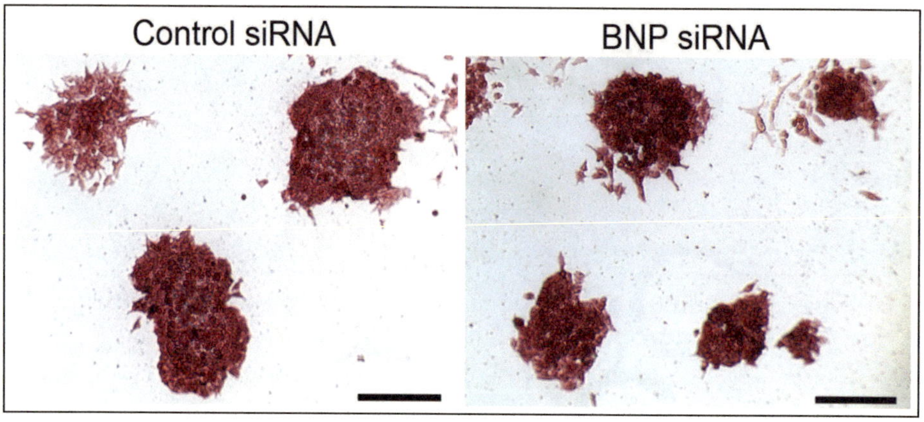

Fig. 5. Effect of BNP knockdown on alkaline-phosphatase (AP) activity. Self-renewal assay AP-staining of the ES cells treated with control siRNA or BNP siRNA four days after transfection. Scale bar = 10 μm

4. BNP signaling is mediated by GC-coupled receptors

BNP exerts its biological effects by intracellular accumulation of cGMP through the activation of particulate GCs (NPR-A and NPR-B) (Yasoda et al., 1998; Potter et al., 2006). We found that NPR-A and NPR-B are expressed in undifferentiated ES cells and pre-implantation embryos (Abdelalim and Tooyama, 2009), suggesting the involvement of particulate GCs in the signaling pathway of BNP in murine ES cells.

The intracellular levels of cGMP were measured to determine whether the cGMP pathway is involved in the effects of BNP on ES cells. Levels of cGMP were reduced significantly in ES cells treated with BNP siRNA compared with ES cells treated with the control siRNA. These findings suggest that BNP promotes ES-cell proliferation via its binding to GC-coupled NP receptors. Furthermore, the expression of NPR-B mRNA was dramatically decreased after BNP knockdown in ES cells, and no change was observed in levels of NPR-A mRNA (Abdelalim and Tooyama, 2009). These findings suggest that the reduction in cGMP levels after BNP knockdown in ES cells is a reflection of the reduced NPR-B mRNA level. Tamura et al. (2004) reported that mice lacking in NPR-B were sterile due to a lack of development of the reproductive tract, and that the majority of the NPR-B-deficient mice studied died before 100 days of age. These results indicate that the signaling pathway through NPR-B is a possible regulator for ES-cell self-renewal.

Although ANP and BNP have similar affinity for known GC-coupled NP receptors (Koller et al., 1991; Suga et al., 1992), transgenic mice overexpressing BNP exhibit skeletal overgrowth (Suda et al., 1998), while no skeletal defects have been reported in transgenic mice overexpressing ANP (Steinhelper et al., 1990). Furthermore, mice with targeted deletion of BNP (Tamura et al., 2000) exhibit a different phenotype than ANP-deficient mice (Hohn et al., 1995). Mice without BNP do not have hypertension; instead they show focal ventricular fibrosis (Tamura et al., 2000). NPR-A-deficient mice display salt-resistant hypertension and cardiac hypertrophy without skeletal defects (Oliver et al., 1997). BNP-Tg/GC-A-/- mice continue to exhibit marked longitudinal growth of vertebrae and long bones comparable with that in BNP-Tg mice (Chusho et al., 2000). Given that ANP-transgenic mice do not display the skeletal phenotype, these reports suggest that BNP can signal through pathways independent of NPR-A. However, in cultures of embryonic-mouse tibia, BNP and CNP increased bone growth and stimulated cGMP production by signaling through NPR-B (Yasoda et al., 1998).These findings support the idea that BNP modulates ES-cell proliferation through NPR-B/cGMP-dependent mechanisms.

5. Effect of BNP on ES cell viability

The G1 checkpoint provides time for somatic cells to repair damaged DNA and to prevent cells with damaged DNA from entering S phase. One consequence of a checkpoint arrest is that cells with repaired DNA are less subject to apoptosis. Thus, restoration of a G1 checkpoint in ES cells predicts that these cells would be protected from cell death.

ES cells do not undergo cell-cycle arrest at the G1 and G2 checkpoints in response to DNA damage or nucleotide depletion, although they synthesize abundant quantities of transcription active p53 (Aladjem et al., 1998; Prost et al., 1998). Therefore, ES cells appear to lack the p53-dependent G1 and G2 checkpoints that characterize normal somatic cells. Several factors may account for the inability of ES cells to arrest growth at the G1-S or G2-M transitions, and how these cells maintain genome integrity in the absence of cell-cycle checkpoints remains to be clarified.

To determine the effect of BNP on ES-cell survival, apoptosis was examined by staining control siRNA-treated ES cells and BNP siRNA-treated ES cells for annexin V, a marker of apoptosis, 48 hours after transfection (Abdelalim and Tooyama, 2009). The percentage of apoptotic cells significantly increased in BNP siRNA-treated ES cells compared with control siRNA-treated cells (Fig. 6). This result suggests that endogenous BNP protects ES cells from apoptosis. However, up-regulation of gamma-aminobutyric acid type A receptors ($GABA_AR$), a downstream target of BNP, does not cause apoptosis in ES cells (Andang et al., 2008), suggesting the involvement of other factors in apoptosis induced by BNP knockdown. The phosphoinositide 3-kinase (PI3K) signaling pathway stimulates the G1-S phase transition and is critical for the maintenance of murine ES cell self-renewal, and its inhibition induces apoptosis in ES cells (Paling et al., 2004; Storm et al., 2007). Protein analysis showed no change in the level of phospho-Akt (ser 473), a marker for PI3K, suggesting that BNP signaling does not interfere with PI3K pathway.

The transcription factor Ets-1 is required for normal survival of T cells, and Ets-1-/- T cells display a severe proliferative defect and demonstrate increased rates of spontaneous apoptosis indicating an anti-apoptotic role of Ets-1 (Bories et al., 1995; Muthusamy et al., 1995). Furthermore, Ets-1 participates in the activation of the BNP gene (Pikkarainen et al., 2003). In addition, NPR-A gene transcription and GC activity of the receptor are critically regulated by Ets-1 in target cells (Kumar et al., 2009). Knockdown of BNP in ES cells leads to a marked down-regulation in the expression of Ets-1 (Abdelalim and Tooyama, 2009). Although, no data are available on the role of Ets-1 in ES cell proliferation, the effect of BNP knockdown on Ets-1 suggest that Ets-1 may be involved in the anti-apoptotic and/or proliferative role of endogenous BNP in ES cells.

The role of Ets-1 in proliferation and apoptosis is somewhat controversial. Ets-1 expression regulates endothelial-cell proliferation during angiogenesis and is essential for normal coronary and myocardial development (Seth et al., 2005; Lie-Venema et al., 2003; Lelievre et al., 2000). These data demonstrate that BNP promotes ES-cell survival in part through regulation of Ets-1. Also, Ets-1 has an anti-apoptotic effect in smooth-muscle cells (Zhang et al., 2003). In contrast Ets-1 overexpression in human umbilical vein endothelial cells can stimulate apoptosis by modulating the expression of apoptotic genes (Teruyama et al., 2001).

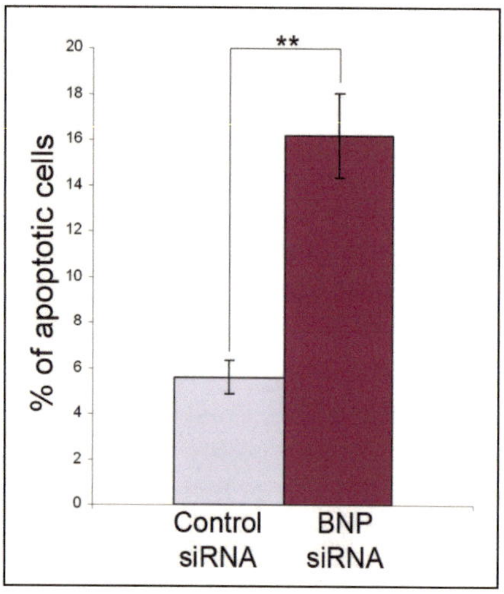

Fig. 6. Effect of BNP knockdown on survival of murine ES cells. The percentage of Annexin-V-positive cells (early apoptotic marker), as measured by flow cytometry 48 hours after transfection with control siRNA or BNP siRNA. Data represent mean ± s.d. (n = 2); ***P* < 0.01 (two-tailed *t*-test). This figure is reproduced from (Abdelalim and Tooyama, 2009)

6. BNP regulates GABA$_A$R genes in ES cells

The GABA$_A$R genes were examined to identify the genes regulated by BNP. Murine ES cells synthesize GABA and express functional GABA$_A$Rs. Activation of GABA$_A$R by muscimol (a GABA$_A$R agonist) inhibits murine ES cell proliferation (Andang et al., 2008; Schwirtlich et al., 2010). The relationship between BNP and GABA$_A$R has been previously determined; BNP suppresses GABA$_A$R currents in retinal bipolar cells (Yu et al., 2006). Therefore, the effects of BNP knockdown on expression of GABA$_A$Rα1 and GABA$_A$Rβ3, the major subunits of GABA$_A$Rs in ES cells (Andang et al., 2008), were examined. As expected, BNP knockdown led to up-regulation of GABA$_A$R genes. Also, treatment of ES cells with muscimol significantly reduced the BNP expression that is associated with reductions in ES-cell proliferation, indicating a link between endogenous BNP and GABA$_A$R signaling in the control of ES-cell proliferation (Abdelalim and Tooyama, 2009). These data suggest that endogenous BNP signaling is essential for maintaining the appropriate level of GABA$_A$R in ES cells to promote ES-cell proliferation.

The effect of GABA$_A$R on ES-cell proliferation is mediated by phosphorylation of H2AX in cell-cycle dependent and DNA-damage independent manners (Ichijima et al., 2005; Andang et al., 2008). In addition, the reduction of BNP levels using siRNA increases the accumulation of γ-H2AX nuclear foci in ES cells. γ-H2AX is a critical factor in the S/G2 DNA-damage checkpoint complex (Fernandez-Capetillo et al., 2004) and for the surveillance of genome integrity (Celeste et al., 2003). Down-regulation of H2AX in ES cells using siRNA increases ES-cell proliferation (Andang et al., 2008). Of note, $GABA_AR$ may not be the only factor that mediates the BNP-induced proliferative activity. Since BNP down-regulation caused an increase in the rate of apoptosis, the accumulation of γ-H2AX after BNP knockdown in ES cells may be the result of activation of $GABA_ARs$ and/or apoptosis.

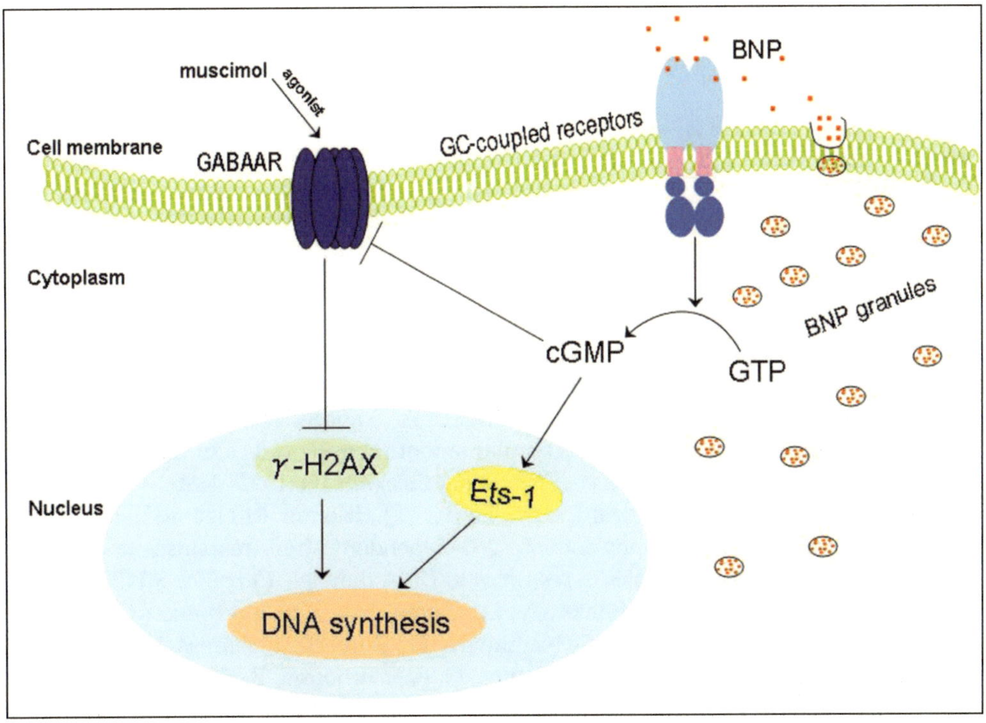

Fig. 7. Simplified schematic presentation of the role of BNP in murine ES cells. BNP and its receptors are expressed by undifferentiated ES cells. BNP activates cGMP through GC-coupled receptors, which inhibit the expression of $GABA_AR$. Inhibition of $GABA_AR$ subsequently reduces the accumulation of γ-H2AX in the nucleus, increasing DNA synthesis. BNP also activates Ets-1 expression, which may contribute to survival and proliferation of ES cells

7. Conclusion

BNP signaling is an essential signaling pathway for the proliferation and survival of murine ES cells. BNP and its receptors are expressed by undifferentiated ES cells. Knockdown of BNP in ES cells leads to suppression of ES-cell proliferation as well as increases apoptosis. The suppression of ES-cell proliferation is due to the activation of $GABA_AR$, γ-H2AX accumulation and Ets-1 inhibition. We propose two pathways that may explain the involvement of BNP in the control of proliferation of ES cells (Figure 7). In pathway 1, BNP controls the level of $GABA_AR$, which is a negative regulator of ES-cell proliferation. In the other pathway, BNP stimulates the transcription factor Ets-1, which may be involved in the survival and proliferation of ES cells. These findings will greatly enhance our understanding of the pathways involved in the regulation of self-renewal of ES cells. Future investigation will help delineate the mechanisms by which BNP contributes to the regulation of ES cell self-renewal.

8. Acknowledgments

This work was supported by a Grant-in-Aid for Scientific Research (No. 21-09133) from the Japan Society for the Promotion of Science.

9. References

Abdelalim EM, Takada T, Toyoda F, Omatsu-kanbe M., Matsuura H, Tooyama, I. & Torii, R. (2006a). In vitro expression of natriuretic peptides in cardiomyocytes differentiated from monkey embryonic stem cells. *Biochem Biophys Res Commun* 340: 689-95.

Abdelalim, E. M. & Tooyama, I. (2009). BNP is crucial for embryonic stem cell proliferation. *PloS ONE* 4: e5341.

Abdelalim, E. M., Takada, T., Torii, R. & Tooyama, I. (2006b). Molecular cloning of BNP from heart and its immunohistochemical localization in the hypothalamus of monkey. *Peptides* 27: 1886-1893.

Abell, T. J., Richards, A. M., Ikram, H., Espiner, E. A., Yandle, T. (1989). Atrial natriuretic factor inhibits proliferation of vascular smooth muscle cells stimulated by platelet-derived growth factor. *Biochem Biophys Res Commun* 160:1392– 1396.

Aladjem, M. I., Spike, B. T., Rodewald, L.W., Hope, T. J., Klemm, M., Jaenisch, R., Wahl, G. M. (1998). ES cells do not activate p53-dependent stress responses and undergo p53-independent apoptosis in response to DNA damage. *Curr Biol* 8:145-155.

Andang, M., Hjerling-Leffler, J., Moliner, A., Lundgren, T. K., Castelo-Branco, G.,Nanou, E., Pozas, E., Bryia, V., Halliez, S., Nishimaru, H., Wilbertz, J., Arenas, E., Koltzenburg, M., Charnay, P., El Manira, A., Ibanez, C. F. & Ernofors, P. (2008). Histone H2AX-dependent $GABA_A$ receptor regulation of stem cell proliferation. *Nature* 451: 460-464.

Avilion, A. A., Nicolis, S. K., Pevny, L. H., Perez, L., Vivian, N. & Lovell-Badge, R. (2003). Multipotent cell lineages in early mouse development depend on SOX2 function. *Genes Dev* 17:126- 140.

Bories, J. C., Willerford, D. M., Grevin, D., Davidson, L., Camus, A., Martin, P., StAhelin, D. & Alt, F. W. (1995). Increased T-cell apoptosis and terminal B-cell differentiation induced by inactivation of the Ets-1 pro-oncogene. *Nature* 377: 635-638.

Brenner, B. M., Ballermann, B. J., Gunning, M. E.& Zeidel, M. L. (1990). Diverse biological actions of atrial natriuretic peptide. *Physiol Rev* 70:665-699.

Burdon, T., Smith, A. & Savatier, P. (2002). Signalling, cell cycle and pluripotency in embryonic stem cells. *Trends Cell Biol* 12: 432 - 438.

Celeste, A., Difilippantonio, S., Difilippantonio, M. J., Fernandez-Capetillo, O., Pilch, D. R., Sedelnikova, O. A., Eckhaus, M., Ried, T., Bonner, W. M. & Nussenzweig, A. (2003). H2AX haploinsufficiency modifies genomic stability and tumor susceptibility. *Cell* 114: 371-383.

Chambers, I. & Smith, A. (2004). Self-renewal of teratocarcinoma and embryonic stem cells. Oncogene 23: 7150- 7160.

Chambers, I., Colby, D., Robertson, M., Nichols, J., Lee, S., Tweedie, S. & Smith, A. (2003). Functional expression cloning of Nanog, a pluipotency sustaining factor in embryonic stem cells. *Cell* 113: 643-655.

Chusho H, Ogawa, Y., Tamura, N., Suda, M., Yasoda, A., Miyazawa, T., Kishimoto, I., Komatsu, Y., Itoh, H., Tanaka, K., Saito, Y., Garbers, D., V. & Nakao, K. (2000). Genetic models reveal that brain natriuretic peptide can signal through different tissue-specific receptor-mediated pathways. *Endocrinology* 141: 3807-3813.

Fernandez-Capetillo, O., Lee, A., Nussenzweig, M. & Nussenzweig, A. (2004). H2AX: the histone guardian of the geneome. *DNA Repair (Amst.)* 3, 959-967.

Garbers, D. L. (1992). Guanylyl cyclase receptors and their endocrine, paracrine, and autocrine ligands. *Cell* 71: 1-4.

Hohn, S. W. M., Krege, J. H., Oliver, P. M., Hagaman, J. R., Hodgin, J. B., Pang, S. C., Flynn, T. G. & Smithies, O. (1995). Genetic decreases in atrial natriuretic peptide and salt-sensitive hypertension. *Science* 267: 679-681.

Horio, T., Nishikimi, T., Yoshihara, F., Matsuo, H., Takishita, S. & Kangawa, K. (2000) Inhibitory regulation of hypertrophy by endogenous atrial natriuretic peptide in cultured cardiac myocytes. *Hypertension* 35:19-24.

Ichijima, Y., Sakasai, R., Okita, N., Asahina, K., Mizutani, S. & Teraoka, H. (2005). Phosphorylation of histone H2AX at M phase in human cells without DNA damage response. *Biochem Biophys Res Commun* 336: 807-812.

Koller, K. J., Lowe, D. G. & Bennett, G. L. (1991). Selective activation of the B natriuretic peptide receptor by c-type natriuretic peptide (CNP). *Science* 252: 120–123.

Kong, X., Wang, X., Xu, W., Behera, S., Hellermann, G., Kumar, A., Lockey, R. F., Mohapatra, S. & Mohapatra, S. S. (2008). Natriuretic peptide receptor a as a novel anticancer target. *Cancer Res* 68: 249-56.

Kumar, P., Bolden, G., Arise, K. K., Krazit, S. T. & Pandey, K. N. (2009). Regulation of natriuretic peptide receptor-A gene expression and stimulation of its guanylyl cyclase activity by transcription factor Ets-1. *Biosci Rep* 29: 57-70.

Lelievre, E., Mattot, V., Huber, P., Vandenbunder, B. & Soncin, F. (2000). Ets1 lowers capillary endothelial cell density at confluence and induces the expression of ve-cadherin. *Oncogene* 19:2438 -2446.

Levin, E. R., Gardner, D. G. & Samson WK (1998). Natriuretic peptides. *N Engl J Med* 339: 321-328.

Lianguzova, M. S., Chuykin, I. A., Nordheim, A. & Pospelov, V. A. (2007). Phosphoinositide 3-kinase inhibitor LY294002 but not serum withdrawal suppresses proliferation of murine embryonic stem cells. *Cell Biol Int* 31: 330-337.

Lie-Venema, H., Gittenberger-de Groot, A. C., van Empel, L. J., Boot, M. J., Kerkdijk, H., de Kant, E. & DeRuiter, M. C. (2003). Ets-1 and Ets-2 transcription factors are essential for normal coronary and myocardial development in chicken embryos. *Circ Res* 92:749 -756.

Loh, Y. H., Wu, Q., Chew, J. L., Vega, V. B., Zhang, W., Chen, X., Bourque, G., George, J., Leong, B., Liu, J., Wong, K. Y., Sung, K. W., Lee, C. W., Zhao, X. D., Chiu, K. P., Lipovich, L., Kuznetsov, V. A., Robson, P., Stanton, L. W., Wei, C. L., Ruan, Y., Lim,

B. & Ng, H. H. (2006). The Oct4 and Nanog transcription network regulates pluripotency in mouse embryonic stem cells. *Nat Genet* 38: 431-440.

Maack, T. (1992). Receptors of atrial natriuretic factor. *Annu Rev Physiol* 54:11-27.

Minamino, N., Aburaya, M., Ueda, S., Kangawa, K. & Matsuo, H. (1988). The presence of brain natriuretic peptide of 12,000 daltons in porcine heart. *Biochem Biophys Res Commun* 155: 470-746.

Mitsui, K., Tokuzawa, Y., Itoh, H., Segawa, K., Murakami, M., Takahashi, K., Maruyama, M., Maeda, M. & Yamanaka, S. (2003). The homeoprotein Nanog is required for maintenance of pluripotency in mouse epiblast and ES cells. *Cell* 113: 631-642.

Muthusamy, N., Barton, K. & Leiden, J. M. (1995). Defective activation and survival of T cells lacking the Ets-1 transcription factor. *Nature* 377: 639-642.

Nichols, J., Zevnik, B., Anastassiadis, K., Niwa, H., Klewe-Nebenius, D., Chambers, I, Scoler, H. & Smith, A. (1998). Formation of pluripotent stem cells in the mammalian embryo depends on the POU transcription factor Oct4. *Cell* 95: 379-391.

Niwa, H., Miyazaki, J. & Smith, A. G. (2000). Quantitative expression of *Oct-3/4* defines differentiation, dedifferentiation or self-renewal of ES cells. *Nature Genet* 24: 372-376.

Ogawa, K., Matsui, H., Ohtsuka, S., Niwa, H. (2004). A novel mechanism for regulating clonal propagation of mouse ES cells. *Genes Cells* 9: 471-477.

Oliver, P. M., Fox, J. E., Kim, R., Rockman, H. A., Kim, H., Reddick, R. L., Pandey, K. N., Milgram, S. L. & Maeda, N. (1997). Hypertension, cardiac hypertrophy, and sudden death in mice lacking natriuretic peptide receptor A. *Proc Nalt Acad Sci U S A* 94: 14730 - 14735.

Paling, N. R., Wheadon, H., Bone, H. K., Welham, M. J. (2004). Regulation of embryonic stem cell self-renewal by phosphoinositide 3-kinase-dependent signalling. *J Biol Chem* 279: 48063- 48070.

Pikkarainen, S., Tokola, H., Kerkelä, R., Majalahti-Palviainen, T., Vuolteenaho, O. & Ruskoaho, H. (2003). Endothelin-1-specific activation of B-type natriuretic peptide gene via p38 mitogen-activated protein kinase and nuclear ETS factors. *J Biol Chem* 278: 3969–3975.

Potter, L. R., Abbey-Hosch, S. & Dickey, D. M. (2006). Natriuretic peptides, their receptors, and cyclic guanosine monophosphate-dependent signaling functions. *Endocr Rev* 27: 47-72.

Prost, S., Bellamy, C. O., Clarke, A. R., Wyllie, A. H., Harrison, D. J. (1998). P53-independent DNA repair and cell cycle arrest in embryonic stem cells. *FEBS Lett* 425: 499-504.

Redondo, J., Bishop, J. E. & Wilkins, M. R. (1998). Effect of atrial natriuretic peptide and cyclic GMP phosphodiesterase inhibition on collagen synthesis by adult cardiac fibroblasts. *Br J Pharmacol.* 124:1455–1462.

Savatier, P, Huang, S, Szekely, L, Wiman, K. G. & Samarut, J. (1994). Contrasting patterns of retinoblastoma protein expression in mouse embryonic stem cells and embryonic fibroblasts. *Oncogene* 9: 809 - 818.

Schwirtlich, M., Emri, Z., Antal, K., Máté, Z., Katarova, Z. & Szabó, G. (2010). $GABA_A$ and $GABA_B$ receptors of distinct properties affect oppositely the proliferation of mouse embryonic stem cells through synergistic elevation of intracellular Ca^{2+}. *FASEB J.* 24; 1218-1228.

Seth, A. & Watson, D. K. (2005). Ets transcription factors and their emerging roles in human cancer. *Eur J Cancer* 41:2462–2478.

Silberbach, M. & Roberts, C. T. Jr (2001). Natriuretic peptide signaling: molecular and cellular pathways to growth regulation. *Cell Signal* 13: 221-231.

Steinhelper, M. E., Cochrane, K. L. & Field, L. J. (1990). Hypotension in transgenic mice expressing atrial natriuretic factor fusion genes. *Hypertension* 16: 301–307.

Storm, M. P., Bone, H. K., Beck, C. G., Bourillot, P. Y., Schreiber, V., Damiano, T. , Nelson, A., Savatier, P. & Welham, M. J. (2007). Regulation of Nanog expression by phosphoinsitide 3-kinase-dependent signaling in murine embryonic stem cells. *J Biol Chem* 282: 6265-6273.

Suda, M., Ogawa, Y., Tanaka, K., Tamura, N., Yasoda, A., Takigawa, T., Uehira, M., Nishimoto, H., Itoh, H., Saito, Y., Shiota, K. & Nakao, K. (1998). Skeletal overgrowth in transgenic mice that overexpress brain natriuretic peptide. *Proc Natl Acad Sci USA* 95: 2337-2342.

Suda, Y., Suzuki, M., Ikawa, Y. & Aizawa, S. (1987). Mouse embryonic stem cells exhibit indefinite proliferative potential. *J Cell Physiol* 133: 197–201.

Sudoh, T., Kangawa, K., Minamino, N. & Matsuo, H. (1988). A new natriuretic peptide in porcine brain. *Nature* 332:78-81.

Sudoh, T., Minamino, N., Kangawa, K. & Matsuo, H. (1990). C-type natriuretic peptide (CNP): a new member of natriuretic peptide family identified in porcine brain. *Biochem Biophys Res Commun* 168:863-870.

Suga, S., Nakao, K., Hosoda, K., Mukoyama, M., Ogawa, Y., Shirakami, G., Arai, H., Saito, Y., Kambayashi, Y. & Inouye, K. (1992). Receptor selectivity of natriuretic peptide family, atrial natriuretic peptide, brain natriuretic peptide, and C-type natriuretic peptide. *Endocrinology* 130: 229–239.

Tamura, N., Doolittle, L. K., Hammer, R. E., Shelton, J. M., Richardson, J. A. & Garbers, D. L. (2004). Critical roles of the guanylyl cyclase B receptor in endochondral ossification and development of female reproductive organs. *Proc Natl Acad Sci USA* 101: 17300–17305.

Tamura, N., Ogawa, Y., Chusho, H., Nakamura, K., Nakao, K., Suda, M., Kasahara, M., Hashimoto, R, Katsuura, G., Mukoyama, M., Itoh, H., Tanaka, I., Otani, H. & Katsuki, M. (2000). Cardiac fibrosis in mice lacking brain natriuretic peptide. *Proc Nalt Acad Sci U S A* 97:4239-4244.

Teruyama, K., Abe, M., Nakano, T., Iwasaka-Yagi, C., Takahashi, S., Yamada, S., Sato, Y. (2001). Role of transcription factor Ets-1 in the apoptosis of human vascular endothelial cells. *J Cell Physiol* 188: 243-252.

Tsuruda, T., Boerrigter, G., Huntley, B. K., Noser, J. A., Cataliotti, A. & Costello-Boerrigter, L. C., Chen, H. H. & Burnett, J. C. (2002). Brain natriuretic peptide is produced in cardiac fibroblasts and induces matrix metalloproteinases. *Circ Res* 91: 1127- 1134.

Weil, M., Jacobson, M.D., Coles, H.S.R., Davies, T. J., Gardner, R.L., Raff, K. D., Raff, M. C. (1996). Constitutive expression of the machinery for programmed cell death. *J Cell Biol* 133, 1053–1059.

Yasoda, A., Ogawa, Y., Suda, M., Tamura, N., Mori, K., Sakuma, Y., Chusho, H., Shiota, K., Tanaka, K. & Nakao, K. (1998). Natriuretic peptide regulation of endochondral ossification. Evidence for possible roles of the C-type natriuretic peptide/guanylyl cyclase-B pathway. *J Biol Chem* 273:11695–700.

Yu, C. Y., Cao, L. H. & Yang, X. L. (2006). Modulation by brain natriuretic peptide of GABA receptors on retinal ON-type bipolar cells. *J Neurosci* 26: 696-707.

Zhang, C., Kavurma, M. M., Lai, A. & Khachigian, L. M. (2003). Ets-1 protects vascular smooth muscle cells from undergoing apoptosis by activating p21WAF1/Cip1: ETS-1 regulates basal and inducible p21WAF1/Cip1 transcription via distinct cis-acting elements in the p21WAF/Cip1 promotor. *J Biol Chem* 278: 27903-9

17

Non-classical Signalling Mechanisms in Stem Cells

Alice Pébay[1, 2], Hitesh Peshavariya[1],
Raymond C.B. Wong[3] and Gregory J Dusting[1,4]
[1]O'Brien Institute, Fitzroy, VIC 3065,
[2]Centre for Neuroscience, Department of Pharmacology, University of Melbourne, Parkville VIC 3010,
[3]Department of Biological Chemistry, University of California Irvine, Irvine, CA 92617,
[4]Department of Surgery, University of Melbourne, Parkville VIC 3010,
[1,2,4]Australia
[3]USA

1. Introduction

There are many gaps in our understanding of stem cell biology that need to be explored in order to realize the full potential of stem cells for therapies. Understanding how stem cells communicate within their microenvironment is important to be able to manipulate their functions. Indeed, the stem cell microenvironment is critical for their maintenance, and communication between neighbouring cells plays an important part in determining cell fate. Most studies of the stem cell niche focus on paracrine or juxtacrine cell interactions, particularly the influence of cytokines and various G-protein coupled receptor ligands, influences which have been reviewed elsewhere (Kobayashi et al 2010). Other signalling mechanisms that have received less attention include oxidant signalling through various protein kinase pathways, and intercellular communication through gap junctions. Not surprisingly gap junctional intercellular communication (GJIC) has been implicated in regulating crucial biological events in many stem cells, including proliferation, differentiation and apoptosis. Understanding and modulating GJIC in stem cells could potentially lead to the development of novel methods for expanding stem cells *in vitro* and directing their differentiation into functional mature cells. In this review we have summarized current knowledge on the identified roles of gap junctions and of reactive oxygen species (ROS) in stem cells, and speculate on how these may be exploited to develop the therapeutic potential of stem cells.

2. Gap junctions and gap junction intercellular communication

The classical background on gap junctions and GJIC has been covered extensively in a number of excellent recent reviews and therefore will only be briefly discussed here (Martin and Evans, 2004). Gap junctions are intercellular junctions found to be either in an opened or a closed conformation. They are the only intercellular junctions that allow direct transfer of signalling molecules and metabolites to adjacent cells (Alexander and Goldberg, 2003; Kumar and Gilula, 1996). Gap junctions are hydrophilic channels consisting of two connexons, hemichannels localized in the membrane of adjacent cells, each of them consisting of six connexins (Cx) (Kumar and Gilula, 1996; Sosinsky and Nicholson, 2005) which can be assembled from either a single type or multiple types of connexins. Importantly, the connexin constitution of the gap junctions will define the pore size of the junction, hence allowing different permeability for the transfer of molecules (Saez et al., 2003). It is however generally accepted that only molecules less than 1-1.5kDa diffuse through gap junctions (De Maio et al., 2002; Evans et al., 2006). It is now also suggested that unpaired connexon hemichannels can mediate intercellular communication without forming gap junctions (Ebihara 2003; Goodenough and Paul 2003; Evans, De Vuyst et al. 2006).

3. Gap junctions in undifferentiated stem cells

3.1 Mouse embryonic stem cells

Mouse embryonic stem cells (mESC) display functional GJIC, express mRNA transcripts of various connexins: Cx26, Cx30.3, Cx31, Cx32 and Cx37, Cx43 and Cx45 but only Cx31, Cx43 and Cx45 proteins, suggesting a translational regulation of connexins in mESC (Nishi et al., 1991; Oyamada et al., 1996; Worsdorfer et al., 2008). Studies so far seem to indicate that Cx45 does not play a fundamental role in mESC regulation of pluripotency and differentiation and suggest a role of Cx43, although still unclear, in these processes. Indeed, Cx45-null mESC are able to differentiate into cells of the three germ layers following embryoid body formation (Egashira et al., 2004). In contrast, some studies showed that Cx43-knock down mESC display decreased cell proliferation, down-regulation of several stem cell markers and up-regulation of differentiation markers and inability to form embryoid bodies (Todorova et al., 2008; Worsdorfer et al., 2008). However, other data suggest the opposite where the down-regulation of Cx43 resulted in Src phosphorylation and increase in cell proliferation (Kim et al., 2010). In the same study, the adenosine analogue 5'-N-ethylcarboxamide (NECA) was shown to stimulate Cx43 phosphorylation and to inhibit GJIC in mESC, through the activation of PI3K/Akt, PKC, MAPK and NFκB signalling pathways. The closure of GJIC would subsequently lead to Src-induced cell proliferation and migration (Kim et al., 2010). Lastly, Cx43-null mESC show not only a defective differentiation to oligodendrocytes but also an increase in differentiation to astrocytes (Parekkadan et al., 2008). Hence there are conflicting findings on the role of GJIC in mESC, with some data suggesting that Cx43 and open gap junctions are necessary for mESC proliferation, for survival and for maintaining them in the pluripotent state, while other data suggest that the closure of gap junctions is in fact a signal for proliferation and migration. Clearly further work is required to properly define the roles GJIC in mESC under different conditions.

3.2 Human embryonic stem cells

Human embryonic stem cells (hESC) are also coupled through functional gap junctions (Bhattacharya et al., 2004; Carpenter et al., 2004; Wolvetang et al., 2007; Wong et al., 2006; Wong et al., 2004) (Figure 1) and express almost all transcripts of human connexin isoforms, except Cx40.1 and Cx50 (Huettner et al., 2006). Functional GJIC seems to be a common characteristic of hESC maintained in different culture conditions (Wong et al., 2006; Wong et al., 2004). So far, only a few factors have been shown to inhibit GJIC in hESC such as bone morphogenetic protein (BMP)-4 (Wong et al., 2006), endothelin (ET) 1 and ET-2 (Wong et al.,

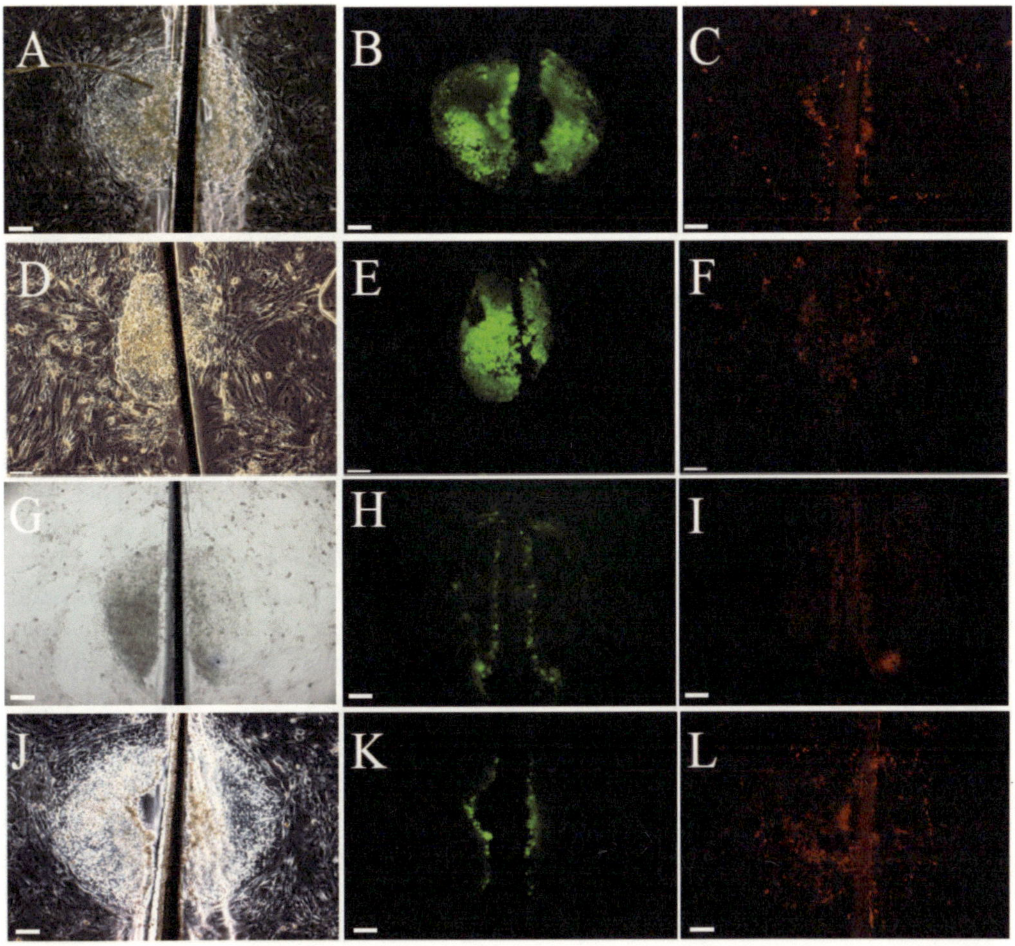

Fig. 1. GJIC in hESC. **(A, D, G, J):** Light and fluorescence micrographs with Lucifer yellow **(B, E, H, K)** and rhodamine-dextran **(C, F, I, L)** in HES-3 cells. Rhodamine-dextran was used as a negative control, showing no dye transfer across to the neighboring cell. Cells were incubated in the presence **(A–C)** or absence **(D–F)** of Ca2+Mg2+ or in the presence of phorbol 12-myristate 13-acetate **(G–I)** or U0126 **(J–L)**. Scale bars = 100 μm. Reproduced, with permission, from (Wong et al., 2004).

2009). Similar to mESC, studies suggest that gap junctions play an important role in the regulation of hESC pluripotency and differentiation. Indeed, long-term chemical inhibition of GJIC in hESC can increase apoptosis (Wong et al., 2006) and in pro-differentiation conditions (such as the absence of a feeder cell layer, and non-conditioned medium), hESC do not communicate through gap junctions (Wong et al., 2006). Furthermore, the pro-differentiation factor BMP-4 (Pera et al., 2004; Xu et al., 2002) inhibits GJIC in hESC, an effect that can be prevented by the addition of the BMP antagonist noggin (Wong et al., 2006). The inhibitory effect of ET-1 is short-lasting and is not associated with changes in colony size, morphology or stem cell marker expression (Wong et al., 2009). These studies suggest that open gap junctions are required for the maintenance of hESC pluripotency and that closure of gap junctions is associated with differentiation or cell death. Further work is however needed to understand the precise implications of GJIC for maintaining hESC pluripotency.

3.3 Somatic stem cells

The inhibition of GJIC in somatic stem cells prevents regeneration of the planarian flatworm, suggesting a conserved role of gap junctions in regulating stem cell fate (Oviedo and Levin, 2007). However, not all somatic stem cells communicate through functional gap junctions. Some somatic stem and putative progenitor/stem somatic cells lack connexin expression and/or functional GJIC : keratinocyte stem cells (Matic et al., 2002), corneal epithelial stem cells (Matic et al., 1997), pancreatic ductal epithelial stem cells (Tai et al., 2003), neural-glial stem cells (Dowling-Warriner and Trosko, 2000), bovine mammary gland progenitor cells (Holland et al., 2003), human breast epithelial stem cells (Kao et al., 1995) and human kidney epithelial stem cells (Chang et al., 1987). Other somatic stem cells, in particular hematopoietic stem cells (HSC), mesenchymal stem cells (MSC) and neural stem/progenitor cells (NS/PC) express connexins and/or possess functional gap junctions that seem to play a role in regulating their homeostasis, pluripotency and/or differentiation. Current knowledge is however sparse. It has been proposed that Cx32, Cx43 and GJIC play a role in HSC maintenance and differentiation (Hirabayashi et al., 2007b; Ploemacher et al., 2000; Rosendaal et al., 1997). Indeed, Cx32 knockout mice exhibit more undifferentiated HSC and fewer progenitor cells, suggesting a role of Cx32 in maturation of HSC to progenitor cells (Hirabayashi et al., 2007a; Hirabayashi et al., 2007b). Moreover, Cx43-deficient mice demonstrate defects in blood cell formation (Montecino-Rodriguez et al., 2000). Cx43 mRNA is not expressed in undifferentiated and quiescent HSC (Montecino-Rodriguez et al., 2000) but is up-regulated in adult mouse bone marrow upon stem cell division (Rosendaal et al., 1994). These data thus suggest an important role of Cx43 in hematopoiesis, but the precise molecular mechanisms remain to be elucidated. Human MSC can communicate through GJIC and express Cx40, Cx43 and Cx45 (Lin et al., 2007; Valiunas et al., 2004). Moreover, human MSC have been demonstrated to form Cx43-mediated GJIC with umbilical vein endothelial cells, which is of importance for their osteogenic differentiation (Villars et al., 2002). Lastly, rat brain derived NS/PC express Cx43 and Cx45 and communicate through GJIC, which is essential for their survival and proliferation (Cai et al., 2004). In these cells, Cx32 and Cx43 are upregulated during differentiation (Yang et al., 2005). Similar data were observed in mouse fetal NS/PC, where the closure of gap junctions decreases cell proliferation and reduces cell survival (Cheng et al., 2004; Duval et al., 2002). Interestingly the overexpression of Cx43 stimulates proliferation of these cells (Cheng et al., 2004), findings somewhat different to the observations in rat-derived NS/PC. Furthermore, in mouse embryonic NS/PC, Cx43 and GJIC allow interkinetic nuclear migration, an apical basal movement of the nucleus observed during cell cyle and necessary for corticogenesis (Liu et al., 2010). In addition, NS/PC from other species have also been demonstrated to express Cx43 and communicate via GJIC (Wen et al., 2008) (Russo et al., 2008), suggesting a conserved and critical role of GJIC in NS/PC self-renewal and differentiation. Finally, it is notable that gap junctions appear to be important for establishing functional interactions between grafted NS/PC and host (Jaderstad et al., 2010).

4. What goes through gap junctions and how can this modify stem cell fate?

GJIC refers to the passive diffusion of intracellular molecules through gap junctions to a neighboring cell (Kumar and Gilula, 1996). Numerous cytoplasmic molecules can diffuse through gap junction channels, including small ions (Na^+, K^+, Ca^{2+}, H^+, Cl^-), second messengers (cyclic nucleotides, inositol triphosphate), amino acids (glycine, glutamate), cellular metabolites (glucose, glutathione, adenosine, AMP, ADP, ATP), short interfering RNA (siRNA) and peptides involved in cross-presentation of major histocompatibility complex class I molecules (Alexander and Goldberg, 2003; Krysko et al., 2005; Neijssen et al., 2005; Valiunas et al., 1997). In adult cells and some tissue systems, GJIC has long been known as crucial for certain cellular functions, such as electrical synchronization, intercellular buffering of cytoplasmic ions, cell metabolism, control of cell migration and cell fate including carcinogenesis (De Maio et al., 2002; Krysko et al., 2005; Mesnil et al., 2005; Parekkadan et al., 2008; Todorova et al., 2008; Vine and Bertram, 2002). However, few studies have addressed the importance of gap junctions for promulgating intercellular Ca^{2+} waves in stem or progenitor cells. Early studies suggested that IP_3 can induce intracellular Ca^{2+} release from endoplasmic-reticulum, and both Ca^{2+} and IP_3 can permeate gap junction channels to the neighbouring cells (Boitano et al., 1992; Saez et al., 1989). Such diffusion of IP_3 and Ca^{2+} through gap junctions effectively forms a positive feedback loop allowing intercellular communication between distant cells. Alternatively, other studies have demonstrated that such intercellular Ca^{2+} waves can also be maintained by the paracrine messenger ATP (Guthrie et al., 1999). It was demonstrated that IP_3 can trigger ATP release to the extracellular space through connexon hemichannels, and act on G-protein coupled receptors on neighbouring cells, leading to phospholipase C activation, IP_3 production and subsequent Ca^{2+} release in the neighbouring cells (Braet et al., 2003; Ebihara et al., 2003; Guthrie et al., 1999). Gap junction-mediated transmission of Ca^{2+} waves has been inferred in some progenitor cells. In NS/PC cells, transmission via Ca^{2+} waves appears to control their proliferation in the ventricular zone (Weissman et al., 2004) as well as in retinal neural progenitor cells (Pearson et al., 2005). Moreover, transient increases in intracellular Ca^{2+} can also stimulate differentiation and neurite outgrowth of different NS/PC cells, but whether such Ca^{2+} signalling occurs through gap junction-mediated waves remains to be determined (Carey and Matsumoto, 1999; Gomez and Spitzer, 1999; Gu and Spitzer, 1995). Recent transplantantion studies by Jaderstad et al. (2010) showed that establishment of GJIC that allows synchronized Ca^{2+} waves is important for grafted NS/PC to integrate functionally to the host neural circuitry. Such GJIC between grafted NS/PC and host cells thus provided a neuroprotective effect in mouse models of neurodegeneration (Jaderstad et al., 2010). Interestingly, diffusion of Ca^{2+} via gap junctions has also been suggested to modulate differentiation of other somatic stem cells. Previous studies by Muller-Borer et al. (2004) suggested that synchronized Ca^{2+} signalling via GJIC co-cultured with neonatal cardiomyocytes induces rat liver stem cells to express cardiac transcription factors and acquire a phenotype resembling cardiomyocytes (Anderson et al., 2007; Muller-Borer et al., 2004). Similarly, we have recently shown gap junctions established between human adipose-derived MSC and neonatal rat cardiomyocytes in co-culture, which induces the expression of cardiac genes and spontaneous cardiomyocyte-like contractions in the human cells (Choi et al., 2010a) (Figure 2). All this points to a growing appreciation of the role of gap-junction-mediated Ca^{2+} waves in modulating gene expression and promoting differentiation of stem cells. Other metabolites that can diffuse through gap junction include cyclicAMP and glutamate, which were previously shown to act as signals for cell death (Amsterdam et al., 1996; Ozog et al., 2002). Although it remains to be proven, it is possible that cyclicAMP and glutamate might play a role in propagation of death signals in stem cells that possess GJIC, such as hESC and NS/PC. Finally, interesting candidates amongst many that might yet be exposed to diffuse through gap junctions, are small RNAs. Previous studies provided evidence that exogenous siRNA can diffuse through gap junction to the neighbouring cells

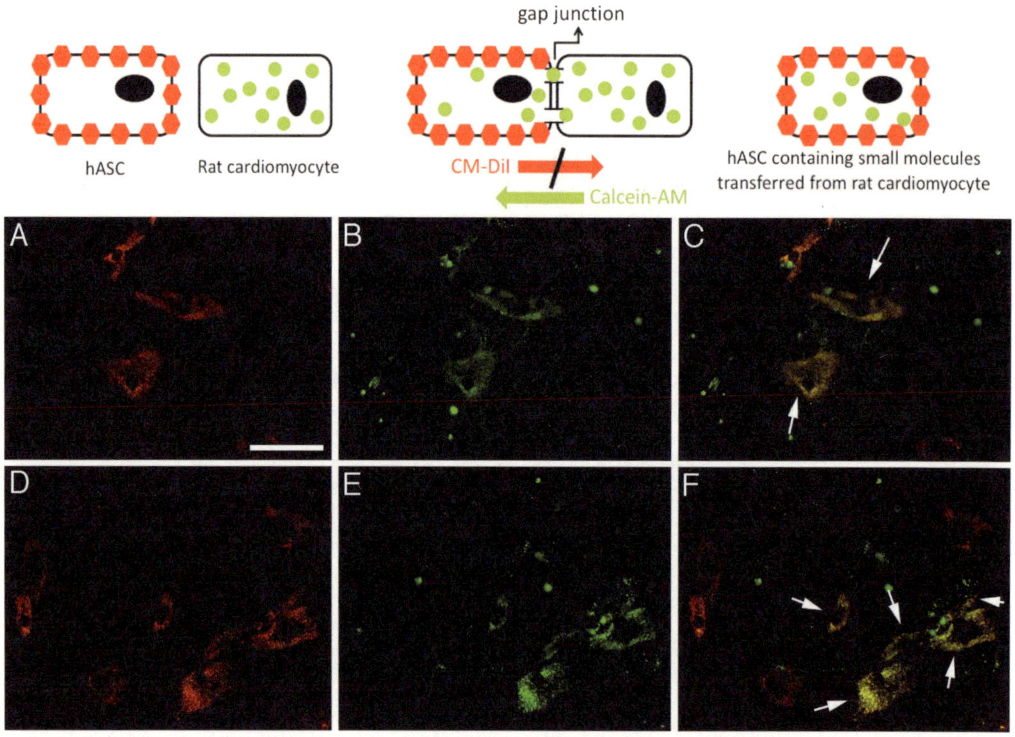

Fig. 2. Functional gap junctions formed between human adipose-derived MSC and rat neonatal cardiomyocytes co-cultured for 24 hours. The human stem cells are labeled red with DiI, which is unable to transfer between cells, and the rat cardiomyocytes were stained by calcein-AM which can transfer from the cytosol of one cell to the adjacent ones through gap junctions. After co-culture, double labeled cells (human cells - cardiomyocyte-like-differentiated from MSC) can be seen as indicated by the arrows (Choi, Dusting, Dilley et al - unpublished, similar to Choi et al 2010a)

(Valiunas et al., 2005; Wolvetang et al., 2007) to silence genes therein. Thus, it is possible that this mechanism may also apply to endogenous microRNAs that provide post-transcriptional regulation of a diverse array of genes. Given the emerging role of microRNA in regulating various physiological processes in hESC and other somatic stem cells (Liu and Zhao, 2009; Mallanna and Rizzino, 2010; Navarro and Lieberman, 2010), future work is likely to explore gap junction-mediated transfer of microRNA in modulating the fate of adjacent stem cells.

5. Oxidant signalling and NADPH oxidase in proliferation and survival of vascular cells

Over-production of reactive oxygen species (ROS) and diminished antioxidant systems (e.g superoxide dismutase, catalase, glutathione peroxidase and glutathione) may lead to oxidative stress, and this is known to contribute to the pathogenesis of several diseases. These include ischemic-reperfusion injury (e.g. heart attack and stroke), atherosclerosis, hypertension, ischemic heart disease, cancers and neurodegeneration. However, given that all types of cells generate low but detectable amounts of ROS under different circumstances, it is likely that ROS serve as important mediators under physiological conditions. In fact, the production of ROS is tightly regulated by antioxidant systems, which maintain redox homeostasis within the cellular environment. As a consequence, ROS have distinct functional effects, which are dependent on a number of factors such as the type of cell within which ROS are generated, and the type and ultimate concentration of ROS at sub-cellular sites where they may modulate enzyme activity and influence gene expression. One of the most important ROS in the vasculature is superoxide anion, formed enzymatically and non-enzymatically, by the univalent reduction of oxygen. The best characterized source of superoxide is the mitochondrial electron transport chain, but many other intracellular enzymes such as xanthine oxidase (XO), cyclooxygenase (COX), nitric oxide synthase (NOS), cytochrome P_{450} oxidase and NADPH oxidase are capable of producing this radical. All these enzymes, save NADPH oxidase, have important cellular functions apart from superoxide generation, whereas the NADPH oxidase enzyme complex is the only known enzyme dedicated to production of ROS, using intracellular NADPH as the "substrate" and electron donor. Since the mid 1990's it has become evident that many cells produce superoxide constitutively by an enzyme with all the characteristics of the NADPH oxidase previously shown to be present in dedicated phagocytic or inflammatory cells. Although constitutively active, NADPH oxidases in blood vessels can be further activated by stimuli such as angiotensin-II (Ang-II), tumour necrosis fector-alpha (TNFα), TGFβ, thrombin, platelet-derived growth factor (PDGF) and by specific ROS themselves (Barry-Lane et al., 2001; Lassegue et al., 2001; Li and Shah, 2003; Moe et al., 2006; Patterson et al., 1999; Suh et al., 1999). The NADPH oxidase is comprised of a membrane-bound heterodimeric unit called flavocytochrome *b558,* composed of small subunit p22phox and gp91phox (aka Nox2). The catalytic moiety of gp91phox (Nox2) contains flavin-adenine dinucleotide (FAD) binding site, two heme components and one NADPH binding site. In the presence of stimuli such as phorbol ester, bacterial lipopolysaccharides or formyl-methionyl-leucyl-phenylalanine (fMLP), protein kinase C (PKC) causes phosphorylation of p47phox and initiates the translocation of p47phox, and its associated proteins p67phox, p40phox and small G-protein Rac1 to the membrane to bind to the cytochrome *b558* complex. The fully assembled complex allows NADPH to bind to gp91phox on the cytoplasmic side of the membrane to initiate a series of electron transfers starting from NADPH to FAD then to heme and finally to oxygen to produce two molecules of superoxide anion radical. To date five isoforms of the catalytic subunit Nox have been identified (Nox1 to Nox5). In addition two homologs of the associated intracellular proteins p47phox and p67phox known as NoxA1 (NADPH oxidase activator1) and NoxO1 (NADPH oxidase organizer), respectively, have been identified (Babior et al., 1973; Lassegue and Clempus, 2003; Li and Shah, 2004). We and others have shown that NADPH oxidase-derived ROS have important functions in survival and proliferation of vascular cells (Ago et al., 2004; Chen et al., 2008; Peshavariya et al., 2009; Petry et al., 2006). Petry et al. showed that suppression of either Nox2 or Nox4 reduce endothelial cell proliferation in vitro, whereas over-expression of these isoforms increased cell proliferation (Petry et al., 2006). Similarly, we have shown that suppression of Nox4 only reduces proliferation, whereas suppression of Nox2 increases apoptosis and therefore also effectively promotes proliferation of endothelial cells (Peshavariya et al., 2009). Vascular smooth muscle cells (VSMC) from different sources express highly both Nox4 and Nox1, but do not express Nox2 (Chan et al., 2009). It is well documented that several growth factors increase Nox1 expression or activity or both, and they also enhance proliferation of VSMC (Lassegue et al., 2001; Suh et al., 1999). However the role of Nox4 in VSMC proliferation is complicated. For instance, it has been shown that transforming growth factor-beta (TGFβ) - induced Nox4 is important in pulmonary smooth muscle cell proliferation (Sturrock et al., 2006). In contrast, the expression of Nox4 increased under quiescent (serum-deprived) conditions and this leads to aortic VSMC differentiation rather than proliferation under these conditions (Clempus et al., 2007). Despite such inconsistencies in the literature regarding the role of Nox4-derived ROS in proliferation versus differentiation, almost all studies indicate that Nox4 is involved in migration of VSMC and endothelial cells (Datla et al., 2007; Lyle et al., 2009; Sturrock et al., 2006). Taken together these findings suggest that NADPH oxidase and its isoforms have distinct roles in vascular cell survival and proliferation, and the disparity of functions may be due to the different ligand receptor couplings and sub-cellular localization of the NADPH oxidase complexes involved.

6. Intracellular kinase pathways induced by oxidant signalling promote cell survival or proliferation.

There is compelling evidence that low levels of oxidants activate several cell signalling pathways and regulate cell survival and proliferation, whereas high levels of oxidants stimulate stress-activated signalling pathways leading to cell death. For example, nanomolar to sub-micromolar concentrations of hydrogen peroxide (H_2O_2) stimulate the proliferation of several cell types, but higher concentrations of H_2O_2 (>100μM) leads to cell death (Giorgio et al., 2007; Kim et al., 2009; Stone and Yang, 2006). However, the effect of H_2O_2 is cell type dependent: H_2O_2 (100μM) increased VSMC proliferation whereas the same concentration inhibits the proliferation of endothelial and fibroblast cells (Rao and Berk, 1992). It is important to note, however, that provision of exogenous H_2O_2 may produce very different effects from stimulation of endogenous H_2O_2 release at particular subcellular sites (Forman, 2007). Ligand-receptor interactions also result in the generation of H_2O_2 by mammalian cells, but these may have different effects on their downstream signalling pathways and thus exert distinct effects on cell survival, proliferation and differentiation (Chen et al., 2008; Datla et al., 2007; Suh et al., 1999; Wang et al., 2000). ROS have several targets such as transcription factors, phosphatases, and enzymes of the receptor kinase family. It has become evident that ROS regulate activity of MAP kinases (MAPKs), a family of serine/threonine kinases, including extracellular signal-regulated kinases 1 and 2 (ERK1/2), c-Jun *N*-terminal kinases (JNKs, also termed stress activated protein kinase; SAPKs) and p38MAPK (Figure 3) Several growth factors (Vascular endothelial growth factor, VEGF; epidermal growth factor, EGF; PDGF and thrombin) and cytokines (Ang-II and TNF-α) induce proliferation of endothelial and VSMC via ROS-mediated activation of MAPK family members (Chen et al., 2008; Datla et al., 2007; Lassegue et al., 2001; Li et al., 2005; Li and Shah, 2003; Park et al., 2009; Patterson et al., 1999; Suh et al., 1999; Ushio-Fukai et al., 1998; Ushio-Fukai et al., 1999; Ushio-Fukai et al., 2002). For example, VEGF, EGF, and TNF-α stimulate proliferation or angiogenesis of endothelial cells which is dependent upon ROS-induced phosphorylation of ERK (Chen et al., 2004; Chen et al., 2008; Datla et al., 2007; Li et al., 2005; Ushio-Fukai et al., 2002). The effect of TNF-α on endothelial cell proliferation is concentration dependent: lower concentrations of TNF-α induce angiogenesis whereas higher concentrations induce apoptosis of endothelial cells (Chen et al., 2004; Deshpande et al., 2000). Similarly, Ang-II and PDGF induced the phosphorylation of p38MAPK, JNK and ERK, but only the phosphorlyation of p38MAPk and JNK are ROS sensitive and lead to VSMC growth (Lassegue et al., 2001). Furthermore, in human aortic VSMC 7-ketocholesterol induces ROS via Nox4 at the endoplasmic reticulum and activates pJNK, but this leads to cell death (Pedruzzi et al., 2004). In contrast to the MAPK, Akt/protein kinase B has been identified as an important component of a pro-survival signalling pathway. Addition of either exogenous H_2O_2 or receptor-mediated intracellular H_2O_2 leads to the phosphoryaltion of Akt (Esposito et al., 2003; Ushio-Fukai et al., 1999). Recently, it has been demonstrated that ROS-mediated activation of the PI3kinase/Akt pathway increased the production of

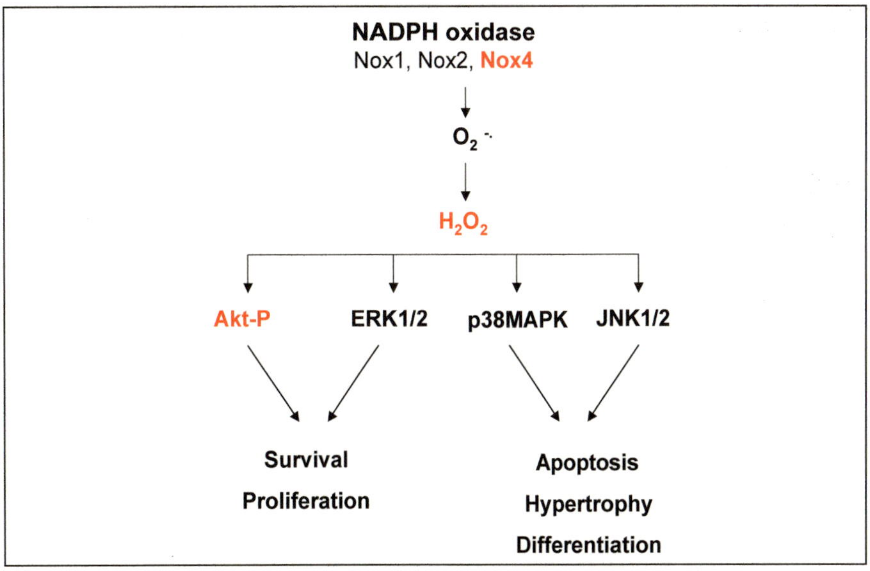

Fig. 3. NAPDH oxidase and ROS signalling. Extracellular ligand-activated membrane receptors (including GPCRs) linked to NADPH oxidase produce ROS intracellularly (superoxide anion or H2O2 in the case of Nox4), which in turn, lead to the phosphorylation and activation of the MAP kinases indicated. The different kinases have different effects in

adult vascular smooth muscle and endothelial cells, as indicated, some of which are shared in stem cells (see text).

nitric oxide and again promoted survival of endothelial cells (Bodiga et al.; Dhanasekaran et al., 2009). Previously, we showed that the proliferative state of endothelial cells exhibits higher ROS production and phosphorylation of Akt compared to quiescent cells, and inhibition of either ROS production or the PI3kinase/Akt pathway reduces endothelial cell proliferation (Peshavariya et al., 2009). ROS-mediated activation of Phospho-Akt (pAkt) promoting cell survival has also been reported in other cell types, such as HeLa, NIH3T3 cells (Wang et al., 2000) and hepatocytes (Kim et al., 2008) . Thus it emerges that activation of ROS-dependent signalling pathways are influenced by several factors such as ligand-receptor interaction, the type of cell, sub-cellular localisation of ROS producing enzymes and the antioxidant status of cells. Therefore, ROS mediated signalling pathways are fine-tuned to differential functions such as proliferation, survival or apoptosis in different cell types under different conditions.

7. Intracellular kinase pathways promoting survival of stem cells and oxidant signalling inducing differentiation

Given that ROS signalling clearly has important roles in the proliferation, survival and differentiation of several cell types, it should be considered whether or not this applies to stem cells. In hESC, the PI3k/Akt and ERK1/2 pathways are constitutively active in most culture media used for the maintenance of hESC (Lin et al., 2007). Moreover, the pro-maintenance factors bFGF, neutrophins, S1P and PDGF activate these kinase signalling pathways, suggesting an essential role of these pathways in hESC maintenance and pluripotency. Furthermore, inhibition of either pathway results in differentiation of hESC or cell death (Armstrong et al., 2006; McLean et al., 2007; Wong et al., 2007). Recently the link to ROS activation of these pathways has been demonstrated in hESC, where ROS stimulation leads to differentiation to mesodermal cells (Ji et al., 2010). Recently we have explored intracellular pathways promoting cell survival after hypoxic pre-conditioning of adipose-derived mesenchymal stem cells (MSC). In these cells it was clear that VEGFA was the major cytoprotective cytokine released during hypoxia, and again VEGF acted via Akt1 phosphorylation to protect these MSC from apoptosis during subsequent severe ischaemia, for the protective effect of the preconditioning was blocked by a VEGF antibody and the PI3 kinase inhibitor LY294002 (Stubbs et al., 2010). Interestingly this paracrine protective effect could be imparted to endothelial cells subjected to hypoxia, revealing an interesting way that adipose-derived MSC that we have utilised in tissue engineering of cardiac tissue, could promote the growth of any complex tissue that requires a vasculature (Chan et al., 2009; Choi et al., 2010b). Several other studies have suggested that ROS signalling can determine the fate of stem cells (Lee et al., 2009; Li et al., 2006; Li and Marban, 2010). For example an early study showed that mESC generate intracellular ROS, and addition of exogenous H_2O_2 promotes their differentiation to cardiomyocytes. This study suggested that PI3 kinase/Akt pathway is upstream of ROS, for inhibition of PI3 kinase reduced ROS formation and cardiac cell differentiation (Sauer et al., 2000; Sauer and Wartenberg, 2005). The enzymatic source of ROS and their downstream signalling pathways have been further explored in differentiation of ESC down the cardiac lineage. Interestingly, Nox4 emerged as the main source of ROS involved in cardiac differentiation, and it appears to regulate phosphorylation of p38MAPK and the cardiac differentiation markers Nkx2.5 and myocyte enhancer factor 2C (MEF2C; (Li et al., 2006)). Mechanical strain induces cardiac differentiation of mESC and both ROS and their downstream signalling pathways were shown to be involved. Antioxidants N-(2-mercapto-propionyl-glycine (NMPG) and vitamin E suppress mechanical strain-induced ROS and reduces the critical downstream signalling pathways p38MAPK, ERK and JNK and also compromises cardiac differentiation and vasculogenesis. Thus it seems that mechanical strain activates these three members of the MAPK family via ROS and there is no specificity at this level (Schmelter et al., 2006). Intracellular ROS derived from mESC not only enhanced the differentiation to cardiomyocytes but also increased their proliferation, underlining the importance of ROS in cardiomyogenesis (Buggisch et al., 2007). Finally, ESC-derived ROS signalling is not limited to cardiac differentiation but is also involved in differentiation to VSMC and vascular stabilization. Xiao et al (2009) demonstrated that Nox4 over-expressing ESC showed enhanced differentiation to VSMC, involving transcription factors including serum response factor (SRF) and myocardin.

8. Conclusion

We are just beginning to understand the signalling that occurs between stem cells, both adult mesenchymal and embryonic, and soon studies will be focused on these mechanisms in induced pluripotent stem cells. There is emerging evidence on the importance of intercellular and intracellular signalling mechanisms that use gap junctions and oxidant signalling to regulate maintenance of stemness and proliferation, or alternatively trigger differentiation or apoptosis to grow new tissues. Defining these mechanisms will lead to greater efficiency in developing stem cell therapies for the clinic, and we amongst others are using this to build larger, more robust constructs from stem cells through tissue engineering (eg Chan et al 2009). The fruits of this signalling research will enhance approaches to regenerative medicine in many fields, and hopefully allow the great promise of stem cells to realise its full potential.

9. Acknowledgements

The authors' work and Fellowship (GJD) was supported by grants from the National Health and Medical Research Council of Australia.
RCB Wong is currently supported by the California Institute of Regenerative Medicine (RC1-00110-1).

10. References

Ago, T., Kitazono, T., Ooboshi, H., Iyama, T., Han, Y. H., Takada, J., Wakisaka, M., Ibayashi, S., Utsumi, H., Iida, M. (2004). Nox4 as the major catalytic component of an endothelial NAD(P)H oxidase. *Circulation*. 109, 227-33.

Alexander, D. B., Goldberg, G. S. (2003). Transfer of biologically important molecules between cells through gap junction channels. *Curr Med Chem*. 10, 2045-58.

Amsterdam, A., Keren-Tal, I., Aharoni, D. (1996). Cross-talk between cAMP and p53-generated signals in induction of differentiation and apoptosis in steroidogenic granulosa cells. *Steroids*. 61, 252-6.

Anderson, P. A., Muller-Borer, B. J., Esch, G. L., Coleman, W. B., Grisham, J. W., Malouf, N. N. (2007). Calcium signals induce liver stem cells to acquire a cardiac phenotype. *Cell Cycle*. 6, 1565-9.

Armstrong, L., Hughes, O., Yung, S., Hyslop, L., Stewart, R., Wappler, I., Peters, H., Walter, T., Stojkovic, P., Evans, J., Stojkovic, M. & Lako, M. (2006). The role of PI3K/AKT, MAPK/ERK and NFkappabeta signalling in the maintenance of human embryonic stem cell pluripotency and viability highlighted by transcriptional profiling and functional analysis, *Hum Mol Genet* Vol.(11): 1894-913.

Babior, B. M., Kipnes, R. S., Curnutte, J. T. (1973). Biological defense mechanisms. The production by leukocytes of superoxide, a potential bactericidal agent. *J Clin Invest*. 52, 741-4.

Barry-Lane, P. A., Patterson, C., van der Merwe, M., Hu, Z., Holland, S. M., Yeh, E. T., Runge, M. S. (2001). p47phox is required for atherosclerotic lesion progression in ApoE(-/-) mice. *J Clin Invest*. 108, 1513-22.

Bhattacharya, B., Miura, T., Brandenberger, R., Mejido, J., Luo, Y., Yang, A. X., Joshi, B. H., Ginis, I., Thies, R. S., Amit, M., Lyons, I., Condie, B. G., Itskovitz-Eldor, J., Rao, M. S., Puri, R. K. (2004). Gene expression in human embryonic stem cell lines: unique molecular signature. *Blood*. 103, 2956-64.

Bodiga, S., Gruenloh, S. K., Gao, Y., Manthati, V. L., Dubasi, N., Falck, J. R., Medhora, M., Jacobs, E. R. 20-HETE-induced nitric oxide production in pulmonary artery endothelial cells is mediated by NADPH oxidase, H2O2, and PI3-kinase/Akt. *Am J Physiol Lung Cell Mol Physiol*. 298, L564-74.

Boitano, S., Dirksen, E. R., Sanderson, M. J. (1992). Intercellular propagation of calcium waves mediated by inositol trisphosphate. *Science*. 258, 292-5.

Braet, K., Vandamme, W., Martin, P. E., Evans, W. H., Leybaert, L. (2003). Photoliberating inositol-1,4,5-trisphosphate triggers ATP release that is blocked by the connexin mimetic peptide gap 26. *Cell Calcium*. 33, 37-48.

Buggisch, M., Ateghang, B., Ruhe, C., Strobel, C., Lange, S., Wartenberg, M., Sauer, H. (2007). Stimulation of ES-cell-derived cardiomyogenesis and neonatal cardiac cell proliferation by reactive oxygen species and NADPH oxidase. *J Cell Sci*. 120, 885-94.

Cai, J., Cheng, A., Luo, Y., Lu, C., Mattson, M. P., Rao, M. S., Furukawa, K. (2004). Membrane properties of rat embryonic multipotent neural stem cells. *J Neurochem*. 88, 212-26.

Carey, M. B., Matsumoto, S. G. (1999). Spontaneous calcium transients are required for neuronal differentiation of murine neural crest. *Dev Biol*. 215, 298-313.

Carpenter, M. K., Rosler, E. S., Fisk, G. J., Brandenberger, R., Ares, X., Miura, T., Lucero, M., Rao, M. S. (2004). Properties of four human embryonic stem cell lines maintained in a feeder-free culture system. *Dev Dyn*. 229, 243-58.

Chan, E. C., Jiang, F., Peshavariya, H. M., Dusting, G. J. (2009). Regulation of cell proliferation by NADPH oxidase-mediated signaling: potential roles in tissue repair, regenerative medicine and tissue engineering. (Review). *Pharmacology Therapeutics*. 122, 97-108.

Chang, C. C., Trosko, J. E., el-Fouly, M. H., Gibson-D'Ambrosio, R. E., D'Ambrosio, S. M. (1987). Contact insensitivity of a subpopulation of normal human fetal kidney epithelial cells and of human carcinoma cell lines. *Cancer Res.* 47, 1634-45.

Chen, J. X., Chen, Y., DeBusk, L., Lin, W., Lin, P. C. (2004). Dual functional roles of Tie-2/angiopoietin in TNF-alpha-mediated angiogenesis. *Am J Physiol Heart Circ Physiol.* 287, H187-95.

Chen, K., Kirber, M. T., Xiao, H., Yang, Y., Keaney, J. F., Jr. (2008). Regulation of ROS signal transduction by NADPH oxidase 4 localization. *J Cell Biol.* 181, 1129-39.

Cheng, A., Tang, H., Cai, J., Zhu, M., Zhang, X., Rao, M., Mattson, M. P. (2004). Gap junctional communication is required to maintain mouse cortical neural progenitor cells in a proliferative state. *Dev Biol.* 272, 203-16.

Choi, Y. S., Dusting, G. J., Stubbs, S., Arunothayaraj, S., Han, X. L., Collas, P., Morrison, W. A., Dilley, R. J. (2010a). Differentiation of human adipose-derived stem cells into beating cardiomyocytes. *J Cell Mol Med.* 14, 878-889.

Choi, Y. S., Matsuda, K., Dusting, G. J., Morrison, W. A., Dilley, R. J. (2010b). Engineering cardiac tissue in vivo from human adipose-derived stem cells. *Biomaterials.* 31, 2236-42.

Clempus, R. E., Sorescu, D., Dikalova, A. E., Pounkova, L., Jo, P., Sorescu, G. P., Schmidt, H. H., Lassegue, B., Griendling, K. K. (2007). Nox4 is required for maintenance of the differentiated vascular smooth muscle cell phenotype. *Arterioscler Thromb Vasc Biol.* 27, 42-8.

Datla, S. R., Peshavariya, H., Dusting, G. J., Mahadev, K., Goldstein, B. J., Jiang, F. (2007). Important role of Nox4 type NADPH oxidase in angiogenic responses in human microvascular endothelial cells in vitro. *Arterioscler Thromb Vasc Biol.* 27, 2319 -24.

De Maio, A., Vega, V., Contreras, J. (2002). Gap junctions, homeostasis, and injury. *J Cell Physiol.* 191, 269-282.

Deshpande, S. S., Angkeow, P., Huang, J., Ozaki, M., Irani, K. (2000). Rac1 inhibits TNF-alpha-induced endothelial cell apoptosis: dual regulation by reactive oxygen species. *Faseb J.* 14, 1705-14.

Dhanasekaran, A., Bodiga, S., Gruenloh, S., Gao, Y., Dunn, L., Falck, J. R., Buonaccorsi, J. N., Medhora, M., Jacobs, E. R. (2009). 20-HETE increases survival and decreases apoptosis in pulmonary arteries and pulmonary artery endothelial cells. *Am J Physiol Heart Circ Physiol.* 296, H777-86.

Dowling-Warriner, C. V., Trosko, J. E. (2000). Induction of gap junctional intercellular communication, connexin43 expression, and subsequent differentiation in human fetal neuronal cells by stimulation of the cyclic AMP pathway. *Neuroscience.* 95, 859-68.

Duval, N., Gomes, D., Calaora, V., Calabrese, A., Meda, P., Bruzzone, R. (2002). Cell coupling and Cx43 expression in embryonic mouse neural progenitor cells. *J Cell Sci.* 115, 3241-51.

Ebihara, L., Liu, X., Pal, J. D. (2003). Effect of external magnesium and calcium on human connexin46 hemichannels. *Biophys J.* 84, 277-86.

Egashira, K., Nishii, K., Nakamura, K., Kumai, M., Morimoto, S., Shibata, Y. (2004). Conduction abnormality in gap junction protein connexin45-deficient embryonic stem cell-derived cardiac myocytes. *Anat Rec A Discov Mol Cell Evol Biol.* 280, 973-9.

Esposito, F., Chirico, G., Montesano Gesualdi, N., Posadas, I., Ammendola, R., Russo, T., Cirino, G., Cimino, F. (2003). Protein kinase B activation by reactive oxygen species is independent of tyrosine kinase receptor phosphorylation and requires SRC activity. *J Biol Chem.* 278, 20828-34.

Evans, W. H., De Vuyst, E., Leybaert, L. (2006). The gap junction cellular internet: connexin hemichannels enter the signalling limelight. *Biochem J.* 397, 1-14.

Forman, H. J. (2007). Use and abuse of exogenous H2O2 in studies of signal transduction. *Free Radic Biol Med.* 42, 926-32.

Giorgio, M., Trinei, M., Migliaccio, E., Pelicci, P. G. (2007). Hydrogen peroxide: a metabolic by-product or a common mediator of ageing signals? *Nat Rev Mol Cell Biol.* 8, 722-8.

Gomez, T. M., Spitzer, N. C. (1999). In vivo regulation of axon extension and pathfinding by growth-cone calcium transients. *Nature.* 397, 350-5.

Gu, X., Spitzer, N. C. (1995). Distinct aspects of neuronal differentiation encoded by frequency of spontaneous Ca2+ transients. *Nature.* 375, 784-7.

Guthrie, P. B., Knappenberger, J., Segal, M., Bennett, M. V., Charles, A. C., Kater, S. B. (1999). ATP released from astrocytes mediates glial calcium waves. *J Neurosci.* 19, 520-8.

Hirabayashi, Y., Yoon, B. I., Tsuboi, I., Huo, Y., Kodama, Y., Kanno, J., Ott, T., Trosko, J. E., Inoue, T. (2007a). Membrane channel connexin 32 maintains Lin(-)/c-kit(+) hematopoietic progenitor cell compartment: analysis of the cell cycle. *J Membr Biol.* 217, 105-13.

Hirabayashi, Y., Yoon, B. I., Tsuboi, I., Huo, Y., Kodama, Y., Kanno, J., Ott, T., Trosko, J. E., Inoue, T. (2007b). Protective role of connexin 32 in steady-state hematopoiesis, regeneration state, and leukemogenesis. *Exp Biol Med (Maywood).* 232, 700-12.

Holland, M. S., Tai, M. H., Trosko, J. E., Griffin, L. D., Stasko, J. A., Cheville, N. C., Holland, R. E. (2003). Isolation and differentiation of bovine mammary gland progenitor cell populations. *Am J Vet Res.* 64, 396-403.

Huettner, J. E., Lu, A., Qu, Y., Wu, Y., Kim, M., McDonald, J. W. (2006). Gap junctions and connexon hemichannels in human embryonic stem cells. *Stem Cells.* 24, 1654-67.

Jaderstad, J., Jaderstad, L. M., Li, J., Chintawar, S., Salto, C., Pandolfo, M., Ourednik, V., Teng, Y. D., Sidman, R. L., Arenas, E., Snyder, E. Y., Herlenius, E. (2010). Communication via gap junctions underlies early functional and beneficial interactions between grafted neural stem cells and the host. *Proc Natl Acad Sci U S A.* 107, 5184-9.

Ji, A. R., Ku, S. Y., Cho, M. S., Kim, Y. Y., Kim, Y. J., Oh, S. K., Kim, S. H., Moon, S. Y., Choi, Y. M. (2010). Reactive oxygen species enhance differentiation of human embryonic stem cells into mesendodermal lineage. *Exp Mol Med.* 42, 175-86.

Kao, C. Y., Nomata, K., Oakley, C. S., Welsch, C. W., Chang, C. C. (1995). Two types of normal human breast epithelial cells derived from reduction mammoplasty:

phenotypic characterization and response to SV40 transfection. *Carcinogenesis.* 16, 531-8.

Kim, H. S., Loughran, P. A., Rao, J., Billiar, T. R., Zuckerbraun, B. S. (2008). Carbon monoxide activates NF-kappaB via ROS generation and Akt pathways to protect against cell death of hepatocytes. *Am J Physiol Gastrointest Liver Physiol.* 295, G146-G152.

Kim, J. H., Choi, W., Lee, J. H., Jeon, S. J., Choi, Y. H., Kim, B. W., Chang, H. I., Nam, S. W. (2009). Astaxanthin inhibits H2O2-mediated apoptotic cell death in mouse neural progenitor cells via modulation of P38 and MEK signaling pathways. *J Microbiol Biotechnol.* 19, 1355-63.

Kim, M. O., Lee, Y. J., Han, H. J. (2010). Involvement of Cx43 phosphorylation in 5'-N-ethylcarboxamide-induced migration and proliferation of mouse embryonic stem cells. *J Cell Physiol.* 224, 187-94.

Krysko, D. V., Leybaert, L., Vandenabeele, P., D'Herde, K. (2005). Gap junctions and the propagation of cell survival and cell death signals. *Apoptosis.* 10, 459-69.

Kumar, N. M., Gilula, N. B. (1996). The gap junction communication channel. *Cell.* 84, 381-8.

Lassegue, B., Clempus, R. E. (2003). Vascular NAD(P)H oxidases: specific features, expression, and regulation. *Am J Physiol Regul Integr Comp Physiol.* 285, R277-97.

Lassegue, B., Sorescu, D., Szocs, K., Yin, Q., Akers, M., Zhang, Y., Grant, S. L., Lambeth, J. D., Griendling, K. K. (2001). Novel gp91(phox) homologues in vascular smooth muscle cells : nox1 mediates angiotensin II-induced superoxide formation and redox-sensitive signaling pathways. *Circ Res.* 88, 888-94.

Lee, S. H., Na, S. I., Heo, J. S., Kim, M. H., Kim, Y. H., Lee, M. Y., Kim, S. H., Lee, Y. J., Han, H. J. (2009). Arachidonic acid release by H2O2 mediated proliferation of mouse embryonic stem cells: involvement of Ca2+/PKC and MAPKs-induced EGFR transactivation. *J Cell Biochem.* 106, 787-97.

Li, J., Stouffs, M., Serrander, L., Banfi, B., Bettiol, E., Charnay, Y., Steger, K., Krause, K. H., Jaconi, M. E. (2006). The NADPH oxidase NOX4 drives cardiac differentiation: Role in regulating cardiac transcription factors and MAP kinase activation. *Mol Biol Cell.* 17, 3978-88.

Li, J. M., Fan, L. M., Christie, M. R., Shah, A. M. (2005). Acute tumor necrosis factor alpha signaling via NADPH oxidase in microvascular endothelial cells: role of p47phox phosphorylation and binding to TRAF4. *Mol Cell Biol.* 25, 2320-30.

Li, J. M., Shah, A. M. (2003). Mechanism of endothelial cell NADPH oxidase activation by angiotensin II. Role of the p47phox subunit. *J Biol Chem.* 278, 12094-100.

Li, J. M., Shah, A. M. (2004). Endothelial cell superoxide generation: regulation and relevance for cardiovascular pathophysiology. *Am J Physiol Regul Integr Comp Physiol.* 287, R1014-30.

Li, T. S., Marban, E. (2010). Physiological levels of reactive oxygen species are required to maintain genomic stability in stem cells. *Stem Cells.* 28, 1178-85.

Lin, T. M., Chang, H. W., Wang, K. H., Kao, A. P., Chang, C. C., Wen, C. H., Lai, C. S., Lin, S. D. (2007). Isolation and identification of mesenchymal stem cells from human lipoma tissue. *Biochem Biophys Res Commun.* 361, 883-9.

Liu, C., Zhao, X. (2009). MicroRNAs in adult and embryonic neurogenesis. *Neuromolecular Med.* 11, 141-52.

Liu, X., Hashimoto-Torii, K., Torii, M., Ding, C., Rakic, P. (2010). Gap junctions/hemichannels modulate interkinetic nuclear migration in the forebrain precursors. *J Neurosci.* 30, 4197-209.

Lyle, A. N., Deshpande, N. N., Taniyama, Y., Seidel-Rogol, B., Pounkova, L., Du, P., Papaharalambus, C., Lassegue, B., Griendling, K. K. (2009). Poldip2, a novel regulator of Nox4 and cytoskeletal integrity in vascular smooth muscle cells. *Circ Res.* 105, 249-59.

Mallanna, S. K., Rizzino, A. (2010). Emerging roles of microRNAs in the control of embryonic stem cells and the generation of induced pluripotent stem cells. *Dev Biol.* 344, 16-25.

Martin, P. E., Evans, W. H. (2004). Incorporation of connexins into plasma membranes and gap junctions. *Cardiovasc Res.* 62, 378-87.

Matic, M., Evans, W. H., Brink, P. R., Simon, M. (2002). Epidermal stem cells do not communicate through gap junctions. *J Invest Dermatol.* 118, 110-6.

Matic, M., Petrov, I. N., Chen, S., Wang, C., Dimitrijevich, S. D., Wolosin, J. M. (1997). Stem cells of the corneal epithelium lack connexins and metabolite transfer capacity. *Differentiation.* 61, 251-60.

McLean, A. B., D'Amour, K. A., Jones, K. L., Krishnamoorthy, M., Kulik, M. J., Reynolds, D. M., Sheppard, A. M., Liu, H., Xu, Y., Baetge, E. E. & Dalton, S. (2007). Activin a efficiently specifies definitive endoderm from human embryonic stem cells only when phosphatidylinositol 3-kinase signaling is suppressed, *Stem Cells* Vol.(1): 29-38.

Mesnil, M., Crespin, S., Avanzo, J. L., Zaidan-Dagli, M. L. (2005). Defective gap junctional intercellular communication in the carcinogenic process. *Biochim Biophys Acta.* 1719, 125-45.

Moe, K. T., Aulia, S., Jiang, F., Chua, Y. L., Koh, T. H., Wong, M. C., Dusting, G. J. (2006). Differential upregulation of Nox homologues of NADPH oxidase by tumor necrosis factor-alpha in human aortic smooth muscle and embryonic kidney cells. *J Cell Mol Med.* 10, 231-9.

Montecino-Rodriguez, E., Leathers, H., Dorshkind, K. (2000). Expression of connexin 43 (Cx43) is critical for normal hematopoiesis. *Blood.* 96, 917-24.

Muller-Borer, B. J., Cascio, W. E., Anderson, P. A., Snowwaert, J. N., Frye, J. R., Desai, N., Esch, G. L., Brackham, J. A., Bagnell, C. R., Coleman, W. B., Grisham, J. W., Malouf, N. N. (2004). Adult-derived liver stem cells acquire a cardiomyocyte structural and functional phenotype ex vivo. *Am J Pathol.* 165, 135-45.

Navarro, F., Lieberman, J. (2010). Small RNAs guide hematopoietic cell differentiation and function. *J Immunol.* 184, 5939-47.

Neijssen, J., Herberts, C., Drijfhout, J. W., Reits, E., Janssen, L., Neefjes, J. (2005). Cross-presentation by intercellular peptide transfer through gap junctions. *Nature.* 434, 83-8.

Nishi, M., Kumar, N. M., Gilula, N. B. (1991). Developmental regulation of gap junction gene expression during mouse embryonic development. *Dev Biol.* 146, 117-30.

Oviedo, N. J., Levin, M. (2007). smedinx-11 is a planarian stem cell gap junction gene required for regeneration and homeostasis. *Development*. 134, 3121-31.

Oyamada, Y., Komatsu, K., Kimura, H., Mori, M., Oyamada, M. (1996). Differential regulation of gap junction protein (connexin) genes during cardiomyocytic differentiation of mouse embryonic stem cells in vitro. *Exp Cell Res*. 229, 318-26.

Ozog, M. A., Siushansian, R., Naus, C. C. (2002). Blocked gap junctional coupling increases glutamate-induced neurotoxicity in neuron-astrocyte co-cultures. *J Neuropathol Exp Neurol*. 61, 132-41.

Parekkadan, B., Berdichevsky, Y., Irimia, D., Leeder, A., Yarmush, G., Toner, M., Levine, J. B., Yarmush, M. L. (2008). Cell-cell interaction modulates neuroectodermal specification of embryonic stem cells. *Neurosci Lett*. 438, 190-195.

Park, B. C., Thapa, D., Lee, J. S., Park, S. Y., Kim, J. A. (2009). Troglitazone inhibits vascular endothelial growth factor-induced angiogenic signaling via suppression of reactive oxygen species production and extracellular signal-regulated kinase phosphorylation in endothelial cells. *J Pharmacol Sci*. 111, 1-12.

Patterson, C., Ruef, J., Madamanchi, N. R., Barry-Lane, P., Hu, Z., Horaist, C., Ballinger, C. A., Brasier, A. R., Bode, C., Runge, M. S. (1999). Stimulation of a vascular smooth muscle cell NAD(P)H oxidase by thrombin. Evidence that p47(phox) may participate in forming this oxidase in vitro and in vivo. *J Biol Chem*. 274, 19814-22.

Pearson, R. A., Dale, N., Llaudet, E., Mobbs, P. (2005). ATP released via gap junction hemichannels from the pigment epithelium regulates neural retinal progenitor proliferation. *Neuron*. 46, 731-44.

Pedruzzi, E., Guichard, C., Ollivier, V., Driss, F., Fay, M., Prunet, C., Marie, J. C., Pouzet, C., Samadi, M., Elbim, C., O'Dowd, Y., Bens, M., Vandewalle, A., Gougerot-Pocidalo, M. A., Lizard, G., Ogier-Denis, E. (2004). NAD(P)H oxidase Nox-4 mediates 7-ketocholesterol-induced endoplasmic reticulum stress and apoptosis in human aortic smooth muscle cells. *Mol Cell Biol*. 24, 10703-17.

Pera, M. F., Andrade, J., Houssami, S., Reubinoff, B., Trounson, A., Stanley, E. G., Ward-van Oostwaard, D., Mummery, C. (2004). Regulation of human embryonic stem cell differentiation by BMP-2 and its antagonist noggin. *J Cell Sci*. 117, 1269-80.

Peshavariya, H., Dusting, G. J., Jiang, F., Halmos, L. R., Sobey, C. G., Drummond, G. R., Selemidis, S. (2009). NADPH oxidase isoform selective regulation of endothelial cell proliferation and survival. *Naunyn Schmiedebergs Arch Pharmacol*. 380, 193-204.

Petry, A., Djordjevic, T., Weitnauer, M., Kietzmann, T., Hess, J., Gorlach, A. (2006). NOX2 and NOX4 mediate proliferative response in endothelial cells. *Antioxid Redox Signal*. 8, 1473-84.

Ploemacher, R. E., Mayen, A. E., De Koning, A. E., Krenacs, T., Rosendaal, M. (2000). Hematopoiesis: Gap Junction Intercellular Communication is Likely to be Involved in Regulation of Stroma-dependent Proliferation of Hemopoietic Stem Cells. *Hematology*. 5, 133-147.

Rao, G. N., Berk, B. C. (1992). Active oxygen species stimulate vascular smooth muscle cell growth and proto-oncogene expression. *Circ Res*. 70, 593-9.

Rosendaal, M., Green, C. R., Rahman, A., Morgan, D. (1994). Up-regulation of the connexin43+ gap junction network in haemopoietic tissue before the growth of stem cells. *J Cell Sci.* 107 (Pt 1), 29-37.

Rosendaal, M., Mayen, A., de Koning, A., Dunina-Barkovskaya, T., Krenacs, T., Ploemacher, R. (1997). Does transmembrane communication through gap junctions enable stem cells to overcome stromal inhibition? *Leukemia.* 11, 1281-9.

Russo, R. E., Reali, C., Radmilovich, M., Fernandez, A., Trujillo-Cenoz, O. (2008). Connexin 43 delimits functional domains of neurogenic precursors in the spinal cord. *J Neurosci.* 28, 3298-309.

Saez, J. C., Berthoud, V. M., Branes, M. C., Martinez, A. D., Beyer, E. C. (2003). Plasma Membrane Channels Formed by Connexins: Their Regulation and Functions. *Physiol Rev.* 83, 1359-1400.

Saez, J. C., Connor, J. A., Spray, D. C., Bennett, M. V. (1989). Hepatocyte gap junctions are permeable to the second messenger, inositol 1,4,5-trisphosphate, and to calcium ions. *Proc Natl Acad Sci U S A.* 86, 2708-12.

Sauer, H., Rahimi, G., Hescheler, J., Wartenberg, M. (2000). Role of reactive oxygen species and phosphatidylinositol 3-kinase in cardiomyocyte differentiation of embryonic stem cells. *FEBS Lett.* 476, 218-23.

Sauer, H., Wartenberg, M. (2005). Reactive oxygen species as signaling molecules in cardiovascular differentiation of embryonic stem cells and tumor-induced angiogenesis. *Antioxid Redox Signal.* 7, 1423-34.

Schmelter, M., Ateghang, B., Helmig, S., Wartenberg, M., Sauer, H. (2006). Embryonic stem cells utilize reactive oxygen species as transducers of mechanical strain-induced cardiovascular differentiation. *Faseb J.* 20, 1182-4.

Sosinsky, G. E., Nicholson, B. J. (2005). Structural organization of gap junction channels. *Biochim Biophys Acta.* 1711, 99-125.

Stone, J. R., Yang, S. (2006). Hydrogen peroxide: a signaling messenger. *Antioxid Redox Signal.* 8, 243-70.

Stubbs, S., Dilley, R., Peshavariya, H., Dusting, G. (2010). Hypoxic Preconditioning Improves Survival of Human Adipose-derived Stem Cells for Cardiac Tissue Engineering. *Basic & Clinical Pharmacology & Toxicology.* 107 (s), 128.

Sturrock, A., Cahill, B., Norman, K., Huecksteadt, T. P., Hill, K., Sanders, K., Karwande, S. V., Stringham, J. C., Bull, D. A., Gleich, M., Kennedy, T. P., Hoidal, J. R. (2006). Transforming growth factor-beta1 induces Nox4 NAD(P)H oxidase and reactive oxygen species-dependent proliferation in human pulmonary artery smooth muscle cells. *Am J Physiol Lung Cell Mol Physiol.* 290, L661-L673.

Suh, Y. A., Arnold, R. S., Lassegue, B., Shi, J., Xu, X., Sorescu, D., Chung, A. B., Griendling, K. K., Lambeth, J. D. (1999). Cell transformation by the superoxide-generating oxidase Mox1. *Nature.* 401, 79-82.

Tai, M. H., Olson, L. K., Madhukar, B. V., Linning, K. D., Van Camp, L., Tsao, M. S., Trosko, J. E. (2003). Characterization of gap junctional intercellular communication in immortalized human pancreatic ductal epithelial cells with stem cell characteristics. *Pancreas.* 26, e18-26.

Todorova, M. G., Soria, B., Quesada, I. (2008). Gap junctional intercellular communication is required to maintain embryonic stem cells in a non-differentiated and proliferative state. *J Cell Physiol.* 214, 354-62.

Ushio-Fukai, M., Alexander, R. W., Akers, M., Griendling, K. K. (1998). p38 Mitogen-activated protein kinase is a critical component of the redox-sensitive signaling pathways activated by angiotensin II. Role in vascular smooth muscle cell hypertrophy. *J Biol Chem.* 273, 15022-9.

Ushio-Fukai, M., Alexander, R. W., Akers, M., Yin, Q., Fujio, Y., Walsh, K., Griendling, K. K. (1999). Reactive oxygen species mediate the activation of Akt/protein kinase B by angiotensin II in vascular smooth muscle cells. *J Biol Chem.* 274, 22699-704.

Ushio-Fukai, M., Tang, Y., Fukai, T., Dikalov, S. I., Ma, Y., Fujimoto, M., Quinn, M. T., Pagano, P. J., Johnson, C., Alexander, R. W. (2002). Novel role of gp91(phox)-containing NAD(P)H oxidase in vascular endothelial growth factor-induced signaling and angiogenesis. *Circ Res.* 91, 1160-7.

Valiunas, V., Bukauskas, F. F., Weingart, R. (1997). Conductances and selective permeability of connexin43 gap junction channels examined in neonatal rat heart cells. *Circ Res.* 80, 708-19.

Valiunas, V., Doronin, S., Valiuniene, L., Potapova, I., Zuckerman, J., Walcott, B., Robinson, R. B., Rosen, M. R., Brink, P. R., Cohen, I. S. (2004). Human mesenchymal stem cells make cardiac connexins and form functional gap junctions. *J Physiol.* 555, 617-26.

Valiunas, V., Polosina, Y. Y., Miller, H., Potapova, I. A., Valiuniene, L., Doronin, S., Mathias, R. T., Robinson, R. B., Rosen, M. R., Cohen, I. S., Brink, P. R. (2005). Connexin-specific cell-to-cell transfer of short interfering RNA by gap junctions. *J Physiol.* 568, 459-68.

Villars, F., Guillotin, B., Amedee, T., Dutoya, S., Bordenave, L., Bareille, R., Amedee, J. (2002). Effect of HUVEC on human osteoprogenitor cell differentiation needs heterotypic gap junction communication. *Am J Physiol Cell Physiol.* 282, C775-85.

Vine, A. L., Bertram, J. S. (2002). Cancer chemoprevention by connexins. *Cancer Metastasis Rev.* 21, 199-216.

Wang, X., McCullough, K. D., Franke, T. F., Holbrook, N. J. (2000). Epidermal growth factor receptor-dependent Akt activation by oxidative stress enhances cell survival. *J Biol Chem.* 275, 14624-31.

Weissman, T. A., Riquelme, P. A., Ivic, L., Flint, A. C., Kriegstein, A. R. (2004). Calcium waves propagate through radial glial cells and modulate proliferation in the developing neocortex. *Neuron.* 43, 647-61.

Wen, C. M., Cheng, Y. H., Huang, Y. F., Wang, C. S. (2008). Isolation and characterization of a neural progenitor cell line from tilapia brain. *Comp Biochem Physiol A Mol Integr Physiol.* 149, 167-80.

Wolvetang, E. J., Pera, M. F., Zuckerman, K. S. (2007). Gap junction mediated transport of shRNA between human embryonic stem cells. *Biochem Biophys Res Commun.* 363, 610-5.

Wong, R. C., Davidson, K. C., Leung, J., Pera, M. F., Pebay, A. (2009). Acute effect of endothelins on intracellular communication of human embryonic stem cells. *Journal of Stem Cells*. 4, 47-55.

Wong, R. C., Dottori, M., Koh, K. L., Nguyen, L. T., Pera, M. F., Pebay, A. (2006). Gap junctions modulate apoptosis and colony growth of human embryonic stem cells maintained in a serum-free system. *Biochem Biophys Res Commun*. 344, 181-8.

Wong, R. C., Pebay, A., Nguyen, L. T., Koh, K. L., Pera, M. F. (2004). Presence of functional gap junctions in human embryonic stem cells. *Stem Cells*. 22, 883-9.

Worsdorfer, P., Maxeiner, S., Markopoulos, C., Kirfel, G., Wulf, V., Auth, T., Urschel, S., von Maltzahn, J., Willecke, K. (2008). Connexin expression and functional analysis of gap junctional communication in mouse embryonic stem cells. *Stem Cells*. 26, 431-9.

Xu, R. H., Chen, X., Li, D. S., Li, R., Addicks, G. C., Glennon, C., Zwaka, T. P., Thomson, J. A. (2002). BMP4 initiates human embryonic stem cell differentiation to trophoblast. *Nat Biotechnol*. 20, 1261-4.

Yang, S. R., Cho, S. D., Ahn, N. S., Jung, J. W., Park, J. S., Jo, E. H., Hwang, J. W., Jung, J. Y., Kim, T. Y., Yoon, B. S., Lee, B. H., Kang, K. S., Lee, Y. S. (2005). Role of gap junctional intercellular communication (GJIC) through p38 and ERK1/2 pathway in the differentiation of rat neuronal stem cells. *J Vet Med Sci*. 67, 291-4.

18

Telomerase and Oxidative Stress in Embryonic Stem Cells

Gabriele Saretzki
Newcastle University,
United Kingdom

1. Introduction

Embryonic stem (ES) cells are derivatives of the inner cell mass from pre-implantation blastocysts of early mammalian embryos. They are uncommitted and pluripotent cells with a characteristic high self renewal potential. If propagated under appropriate tissue culture conditions they can proliferate indefinitely while maintaining a high genomic stability and are in fact immortal. An important contributor to these properties is a high amount of telomerase activity. This occurs mainly via the telomere maintenance function of telomerase. However, recently it has also been shown that telomerase can localise to mitochondria and decrease intracellular oxidative stress.

The final goal for the use of human embryonic stem cells (ESC) is to generate differentiated cells and products for cell replacement therapies. For this purpose, embryonic stem cells can be induced to differentiate into cells from all three germ layers: mesoderm, entoderm and ectoderm and eventually give rise to all somatic cell type of the human body. *In vitro*, this requires specialised growth conditions and factors which are often defined empirically rather than due to scientific evidence. For example, it emerges recently that growing ES cells under low, physiological oxygen conditions rather than atmospheric oxygen (21%) are greatly beneficial for the properties and application of these cells.

This chapter will summarise our current knowledge on telomerase expression and oxygen conditions and their involvement in the maintenance of self-renewal, pluripotency and stem cell differentiation.

2. Telomerase expression in ES cells

There exist several factors which are known to contribute to ESCs pluripotency. Among them there are the transcription factors Oct4, Sox2, NANOG as well as a constitutively high level of telomerase expression. All these factors are down regulated during differentiation of ESCs into various cells and tissues. The only exception to that are germ line cells which continue to express high amounts of telomerase. The mechanisms of maintaining pluripotency differ in mouse and human ES cells. While the murine system requires both LIF and BMP4 for self-renewal human ESC culture needs to be supplemented with basic fibroblast factors and the TGFβ/Activin/Nodal pathway has to be activated.

The ribonucleoprotein telomerase is a unique reverse transcriptase that consists of two main parts: TERT, the catalytic subunit and TR or TERC, the RNA component that contains the template region for the synthesis of telomeric repeats. The main and best studied function of telomerase is the extention of telomeres. Telomeres are the end of chromosomes and protect them from genomic instabilities. Telomeres in cells without or with rather low levels of telomerase shorten continuously (Harley et al., 1990) due to loss of sequences caused by inability of conventional polymerases to fill the gap of the last RNA primer at the lagging strand synthesis as well as to oxidative stress (von Zglinicki et al., 1995). There is an important difference between mouse and humans regarding the expression of telomerase in somatic cells and tissues. While telomerase expression persists in many adult tissues in mice in humans telomerase levels are high during early embryonic development, but then down regulated during development. Therefore, only some cell and tissue types in the adult human body retain telomerase activity such as endothelial cells and lymphocytes while adult stem cells are able to activate telomerase upon activation. However, a stable maintenance of telomeres is necessary for ongoing cellular proliferation and immortality. Embryonic stem cells share a high level of telomerase expression with cancer cells but they are characterised by a high genomic stability. The physiological regulation and fine-tuning of telomerase expression occurs at multiple levels including transcriptional, post-transcriptional, post-translational and different sub-cellular localisations.

In addition to telomere maintenance that is of particular significance in vigorously proliferating cells, non-telomeric functions of telomerase have been described recently (for review see Saretzki, 2009). These include interference of the catalytic subunit TERT with important signalling pathways of stem cells and early embryonic development such as WNT and myc (Choi et al., 2008). These pathways are also important for embryonic stem cells. Recently it has been demonstrated that TERT can function as a transcriptional co-factor that modulates chromatin structure and activates genes downstream of the WNT signalling pathway (Park et al., 2009). No RNA component or catalytic function are necessary for this non-canonical function of telomerase. Although these studies have been performed in adult stem cells one can speculate that telomerase/TERT might have additional, non-telomere roles in embryonic stem cells as well. However, so far no studies have been performed to address this issue.

It is also known that the presence of telomerase correlates to cellular survival under conditions of DNA damage and stress. While over- expression of telomerase protects various cell types including embryonic stem cells (Fu et al., 2000, Zhu et al., 2000, Zhang et al., 2003, Armstrong et al., 2005, Ahmed et al., 2008, Yang et al., 2008) telomerase inhibition can lead to increased apoptosis and increased sensitivity against cellular stress including treatment with chemotherapeutic drugs (Kondo et al., 1998, Saretzki et al., 2001, Ludwig et al., 2001, Yang et al.2008).

In addition, it has been demonstrated that telomerase can be excluded from the nucleus under oxidative stress and localise to the mitochondria (Santos et al., 2004, Ahmed et al., 2008, Haendeler et al., 2009). This can improve mitochondrial function, decrease cellular oxidative stress and protect cells from apoptosis and DNA damage (Ahmed et al., 2008, Haendeler et al., 2009). However, no detailed studies on the relationship between telomerase expression and mitochondrial properties, including generation of reactive oxygen species (ROS) have been performed on embryonic stem cells although we have preliminary data that show a mitochondrial localisation of TERT protein within mitochondria in human embryonic stem cells (unpublished data). It is likely that other factors than oxidative stress can promote mitochondrial localisation of telomerase in order to provide a pro-survival function. Since telomerase is able to shuttle from the nucleus to the mitochondria it can carry out nuclear functions such as telomere maintenance or interference with gene expression and transcription in parallel with any potential function at other cellular locations including mitochondria. However, this fact also makes it difficult to separate the various functions of telomerase within a cell. Over-expression of telomerase in mouse and human embryonic stem cells improved many stem cell parameters including proliferation and colony forming abilities (Armstrong et al., 2005, Yang et al., 2008). However, the differentiated progeny from TERT over-expressing stem cells maintained high telomerase levels and showed less intracellular oxidative stress as well as higher apoptosis resistance (Armstrong et al., 2005, Lee et al., 2005, Yang et al., 2008).

In contrast, down regulation of TERT expression in ES cells correlated either with immediate spontaneous differentiation or decreased proliferation rates demonstrating the importance of a high telomerase for pluripotency in human embryonic stem cells (Yang et al., 2008).

3. Telomerase activity and differentiation

A constant high telomerase expression is a hallmark of pluripotent human stem cells (Amit et al., 2000). In contrast, it is lower, although strongly inducible in adult human stem cells (Hiyama and Hiyama, 2007). Down regulation of telomerase expression strongly correlates with the *in vitro* differentiation process in both embryonic and adult stem cells resembling the situation *in vivo* very closely (Saretzki et al., 2008, O'Connor et al., 2009).

We have shown previously that down regulation of telomerase as well as other pluripotency factors such as Sox2, Oct4 and NANOG and increase of oxidative stress, DNA damage and mitochondrial biogenesis in embryonic stem cells coincided in time when cells were differentiating (Armstrong et al., 2000, Saretzki et al., 2008). However, it is not known whether there is a causal interplay between these factors during the differentiation process. The differentiated progeny of ESCs down regulated telomerase within a week, shorten telomeres and accumulate DNA damage (Saretzki et al., 2008). The mechanism for this down regulation seem to be epigenetic processes such as methylation and acetylation of the TERT and, perhaps to a lesser extend, the TERC promoters (Lopatina et al., 2003, Atkinson et al., 2005, Saretzki et al., 2008). However, additional mechanisms for down regulation of telomerase such as polyunsaturated fatty acids have been described as well (Eitsuka et al., 2005).

In general, it seems important that endogenous telomerase is down regulated during differentiation of ESCs as well as cancer cells (Bagheri et al., 2006, Li et al., 2005). However, it is not entirely clear whether down regulation of telomerase activity is merely associated or even necessary for differentiation of ESC.

Therefore studies which analyse over-expression or inhibition of telomerase in ESCs are very informative to address this question. Over-expression was achieved by constitutively over-expressing TERT, the catalytic subunit of telomerase in ES cells. Over-expression of TERT further potentiated the pluripotency and proliferation capacities of mouse and human ESCs (Armstrong et al., 2005, Yang et al., 2008). TERT over-expression resulted in a higher proliferation rate due to an increased fraction of cells in S-phase in mouse as well as human ES cells (Armstrong et al., 2005, Yang et al., 2008). However, differentiation capabilities after TERT over-expression varied between mouse and human ESCs and between different studies (Armstrong et al., 2005, Lee et al., 2005, Yang et al., 2008). While endogenous telomerase is normally down regulated during differentiation of ES cells ectopic telomerase remained stably expressed in the differentiated progeny (Armstrong et al., 2005, Yang et al.,

2008). TERT over-expressing mouse ESCs showed an almost 5 fold increased efficiency to generate hematopoietic progenitors (CFU-GEMM) as well as other myeloid lineages (Armstrong et al., 2005). Lee et al., (2005) did not find any changes in proliferation and differentiation in mouse ESC when they over-expressed mTERT. However, they found a protection of ES cells against cell death during differentiation and increased resistance to apoptosis induced by oxidative stress and other genotoxic insults that depends on the catalytic activity of mTERT.

In addition, TERT over-expression resulted in a decrease in spontaneous as well as induced apoptosis rates in ESCs and in embryoid bodies (EBs) at day 6 of differentiation. At the same time also the increase in oxidative stress levels during differentiation into EBs was significantly attenuated (Armstrong et al., 2005). However, TERT over-expression in human ESCs only promoted proliferation and accelerated cell cycle progression, but suppressed differentiation properties (number of hematopoietic colonies formed) of TERT over-expressing cells although they could form all three germ layers *in vivo* (Yang et al., 2008). The reasons for these differences in the differentiation capacity of mouse and human ESC after ectopic TERT expression are not clear but could be explained with general differences between mouse and human telomere and telomerase biology. Conversely, when TERT and telomerase activity levels are severely reduced due to over-expression of a transcriptional inhibitor of TERT (Zap) cell growth as well as differentiation potential are significantly compromised (Armstrong et al., 2004).

Comparison of gene expression between normal and TERT over-expressing mESCs revealed an up regulation of cell cycle genes (CyclinD1, which was also up regulated in the human experiment, Yang et al, 2008), detoxification and differentiation as well as a down regulation of two DNA damage and stress-related downstream genes of p53: p21 and Gadd45a (Armstrong et al., 2005). These results correspond well with a decreased oxidative stress and apoptosis sensitivity as well as with a higher proliferation rate and differentiation capacity towards hematopoietic lineages in TERT over-expressing murine ES cells. Interestingly, Tsai et al (2010) published recently that over expression of TERT in human mesenchymal stem cells increases proliferation and differentiation potential and brings those cells close to human ES cells (hESC) in their gene expression and methylation pattern (Tsai et al., 2010).

Together, these results show that ectopically over-expressed telomerase can mitigate increases in ROS and oxidative stress but not diminish them entirely. The mechanism could be the delayed down regulation of genes involved in antioxidant defences as well as various heat shock factors (Armstrong et al., 2005). This data suggests that stably high levels of telomerase might be responsible for the maintenance of high levels of stress defence capacities in accordance with findings from our and other groups on hTERT over-expressing cells (Fu et al., 200, Zhu et al., 2000, Zhang et al., 2003, Ahmed et al., 2008).

The data on a protective function of telomerase in differentiated progeny of ES cells are in good accordance with telomeric protection of telomerase as well as newly described non-telomeric functions. In addition to telomere maintenance TERT can also directly change gene expression in a telomere-independent fashion by functioning as a chromosomal modulator and transcriptional c-factor of the Wnt pathway (Choi et al., 2008, Park et al., 2009). TERT therefore can actively influence gene expression levels and interact with important signalling pathways in stem cells although this observation has not been extended to embryonic stem cells yet. A connection between Wnt signalling and the activation of mitochondrial biogenesis and OXPHOS gene expression involved in oxidative phosphorylation has been described in immortalised mouse myoblast cells recently (Yoon et al., 2010).

In summary, these results also suggest that down regulation of telomerase is usually associated with differentiation of ES cells, however, it does not seem to be essential for the differentiation process since ES cells stably over- expressing telomerase retain their differentiation capacity.

4. Long term cultivation and karyotypic stability

In order to obtain enough stem cells for any potential application in regenerative medicine and cell replacement therapies a large amount of cells and consequently, an extensive expansion of stem cell cultures would be required. Contradictory data exist about genetic instabilities in embryonic stem cells during long term cultivation. ES cells maintain a constant telomere length due to the presence of high telomerase levels and also show lower rates of genomic mutation than somatic cells even after prolonged *in vitro* culture. Some stem cell lineages that had been grown for a long time in culture retained a normal karyotype (Rosler et al., 2004, Buzzard et al., 2004) as well as their epigenetic stability (Rugg-Gunn et al., 2005) while others described an accumulation of chromosomal abnormalities when ES cells were cultured for a prolonged time (Longo et al., 1997, Maitra et al., 2005, Armstrong et al, 2005). This might depend on the species (mouse or human) as well as the (often sub-optimal) culture conditions.

Various groups had cultivated human embryonic stem cell lines for around 100 passages (Maitra et al., 2005, Park et al., 2008, Xie et al., 2010). Park et al. (2008) and Xie et al. (2010) did not find any gross morphological changes or different expression levels for stem cell markers between early and late passages. Both groups consistently reported an increased proliferation of ESCs at higher passages, no changes in telomerase activity as well as a reduced differentiation ability.

However, while Park et al. (2008) found no karyotypic changes Xie et al (2010) described some karyotypic abnormalities (translocation) in one of the two hES cell lines analysed while Maitra and co-workers (2005) and found karyotypic anomalies in 8 out of 9 analysed human ESC lines at high passage numbers. These included copy number changes (amplifications and deletions of chromosomes, 45%) and gene promoter methylation (90%). Interestingly, many of those changes are commonly known to occur in cancer. Both Maitra and co-authors (2005) as well as Xie et al., (2010) observed in addition to genomic instabilities also changes in mitochondrial sequences and function. Maitra et al. (2005) found changes in mitochondrial DNA sequence six heteroplasmic sequence alterations in two of nine (22%) late-passage hESC lines, most of them in protein coding regions such as in the ATPase6 gene which most likely probably affected the protein function. Xie et al., (2010) described a compromised mitochondrial function as well as an increased level of ROS production and a higher membrane potential and mitochondrial mass. These two studies emphasise that in addition to nuclear changes there seem to occur changes in mitochondrial sequence and function as well. Decline in mitochondrial function is a hallmark of cellular senescence as well as ageing (Passos et al., 2006, Passos et al., 2010). In general, little research has been conducted on changes in mitochondrial properties of ES cells during prolonged culture.

Since it has been shown recently that telomerase can influence mitochondrial function (Santos et al., 2004, Ahmed et al., 2008, Haendeler et al, 2009) it is possible that in addition to telomere-maintenance by telomerase its intracellular localisation could also be important for proper mitochondrial function in ES cells.

In summary, although contradictory results exist regarding the frequency of various genomic and epigenetic changes that might occur during long-term cultivation of human ES cells a careful karyotypic analysis has to be performed for cells that are intended to be used for regenerative purposes. More data on genome instability will be discussed in connection with various oxygen conditions.

5. Mitochondria, oxidative stress and pluripotency in ES cells: lessons for optimal culture conditions

Mitochondria are important organelles that produce the energy of a cell. In addition, they play an essential role for many additional cellular processes including cellular survival and differentiation. Therefore, a tight regulation of mitochondrial function and integrity is important for all cells, including embryonic stem cells. In contrast, compromising of mitochondrial function results in cellular senescence, ageing and age-related degenerative diseases (Passos et al., 2006, Passos et al., 2010, Wallace, 2005). New connections and mechanisms between mitochondrial properties and stem cells emerge recently and will be summarised here.

Mammalian blastocysts and embryonic stem cells of pre-implantation embryos exist under highly hypoxic conditions. While early stage embryos develop low oxygen conditions (1.5% in rhesus monkeys, slightly higher in rabbits: 3.5% and hamster: 5.3%) oxygen content increases upon implantation of the embryo into the uterus (Fisher and Bavister, 1993). Thus, they have a limited oxidative capacity and immature mitochondria. They generate most of their ATP via aerobic respiration and glycolysis. Only when these cells differentiate do their mitochondria mature, increase in number and switch to oxidative phosphorylation for energy production (Cho et al., 2006). Mitochondrial content increases dramatically when oocytes mature (Shoubridge and Wai, 2007). Decreased mitochondrial content in oocytes correlates to poor fertilisation efficencies (Santos et al., 2006). During early embryogenesis the initial amount of mitochondria (around 150 000-300 000 per human oocyte) gets distributed to the dividing cells and is therefore decreased in number per cell. Consequently, the mitochondrial number is also low in undifferentiated embryonic stem cells that derived from the blastocyst stage embryos. At later stages of pre-implantation development mitochondrial DNA replication and transcription are activated. These processes coincide with mitochondrial maturation into morphologically typical and metabolically active organelles that use oxidative phoshorylation and aerobic respiration for ATP generation (Trimarchi et al., 2000). However, at the blastocyst stage *in vivo* cells of the trophoectoderm start mitochondrial maturation and differentiation while the pluripotent cells of the inner cell mass have still immature mitochondria and mainly rely on glycolysis for energy production (Houghton 2006). Mammalian mitochondrial DNA is around 16kb in size and encodes 13 mitochondrial proteins as well as various RNAs (tRNAs and rRNAs). All other proteins and factors are encoded in the nucleus and then imported into mitochondria. The mitochondrial genome is double stranded and contains no introns. Mitochondrial genomes form complexes-nucleoids that reside in the matrix close to the inner mitochondrial membrane.

Undifferentiated embryonic stem cells from both mouse and humans have very small cytoplasm (Thomson et al, 1998, Varum et al., 2009) and seem to conserve characteristics of small, undeveloped mitochondria which is also reflected in their low mitochondrial mass and low mitochondrial copy number (Saretzki et al., 2004, Saretzki et al., 2008, Facucho-

Oliveira et al., 2007). In addition, these cells have low levels of ATP (Cho et al., 2006) as well as low levels of oxygen consumption (Kondoh et al., 2007).

Facucho-Oliveira et al., (2007) suggested that the maturation of the mitochondrial network and pluripotency are mutually exclusive. The authors showed that activation of aerobic respiration during differentiation of ESCs coincided with down regulation of pluripotency markers. The authors claim that in mouse ES cells the expression of pluripotency markers decreased prior to the expansion and maturation of mitochondria during spontaneous differentiation. However, they used different pluripotency markers from the classical (Oct4, Sox2, NANOG). Surprisingly, they even detected an increased expression of their markers (Dppa5, Pramel 7 and NDP52L1) during early differentiation days (up to day3), only after that the expression of those markers decreased while changes in mitochondrial copy number as well as expression of mitochondrial replication and transcription related factors started to increase at or after day 6 respectively. In addition, no direct experimental prove was given that the initial decrease of pluripotency markers was indeed causal for mitochondrial maturation or just a co-incidence. The same authors claim that a steady-state expression of PolG, the mitochondrial DNA polymerase is essential for the maintenance of a pluripotent state in ES cells (Facucho-Oliveira et al., 2007). It has been demonstrated by various groups that once ES cells differentiated their mitochondrial number increased (Saretzki et al., 2008, St. John et al., 2005). This process coincided with the loss of pluripotency during differentiation of ESC. St. John et al. (2005) suggest that this is in accordance with a concept that each stage of cellular development adjusts its metabolic and energetic demands to it oxidative capacity. However, it is not clear whether the correlation between pluripotency of the inner cell mass and immature mitochondria is purely associative or mutually dependent on each other.

In accordance with the hypoxic nature in the pre-implantation embryo there is accumulating evidence that favours low oxygen for embryonic development *in vitro* and *in vivo*.

Rinaudo et al. (2006) found that mouse embryos cultivated under 5% oxygen showed a gene expression pattern which was much more similar to that of *in vivo* embryos when compared with those cultivated under 20% oxygen. This data corresponds to observations from Harvey et al. (2004) who found a significant increase in the proportion of ICM cells compared with trophectoderm cells when embryos were cultured under 2% oxygen instead of 21% oxygen. These results are in agreement with findings that pluripotent cells display an increased proliferation under reduced oxygen concentrations (see below). There is evidence accumulating that suggests that reducing the oxygen concentration towards physiological levels *in vivo* is beneficial for the *in vitro* maintenance and pluripotency of hES cells. This includes a decrease of spontaneous differentiation while promoting self-renewal and maintaining a stable karyotype (Ezashi et al. 2005, Westfall et al. 2008, Forsyth et al. 2006). Westfall and co-workers compared gene expression patterns in stem cells under low and high oxygen conditions of cultivating stem cells (Westfall et al. 2008). The authors identified various genes whose expression was sensitive to oxygen concentrations, many of them under the control of hypoxia-inducing factors and genes encoding enzymes involved in carbohydrate catabolism and cellular redox state. Although genes associated with pluripotency, including OCT4, SOX2, and NANOG were not directly affected, some downstream genes such as LEFTY2 were decreased under 20% oxygen (Westfall et al. 2008). Zachar and co-workers (2010) analysed human ESC up to 42 month under 20% and 5% oxygen and found that while under 20% many cells spontaneously differentiated the cultures under hypoxia (5%) better maintained their pluripotency: they continued to express high levels of Oct4 and NANOG as well as telomerase activity. The cultures also retained their normal karyotype and were able to differentiate into all 3 germ layers.

However, other studies found no apparent advantages of culturing hES cells under reduced oxygen tension when cells are passaged sufficiently frequent (Chen et al. 2009).

At least mouse ES cells seem to be not very sensitive against oxidative stress that is even higher than 21% oxygen levels (Saretzki et al., 2004) while hESCs might be more delicate in their requirements and many studies show an advantage of hypoxic culture conditions. Ezashi et al. (2005) found that cultivating human ESC under low oxygen (5%) compared to 21% of ambient oxygen reduced the amount of spontaneous differentiation dramatically.

Several lines of evidence seem to explain the beneficial effect of hypoxia with the induction of. hypoxia-inducible factors (HIF). For example, Ji et al. (2009) demonstrated the positive effect of transcription factor HIF-1alpha on pluripotency and the maintenance of an undifferentiated phenotype in human ESCs. The authors co-cultured human embryonic cells with human feeder cells that stably expressed HIF-1alpha. Their results demonstrated that HIF-1alpha was critical for preventing differentiation of hES cells in culture (Ji et al., 2009). At the same time hypoxia also upregulated the expression of NANOG and Oct-4 and of some soluble factors including βFGF and SDF-1alpha which are released into the microenvironment to maintain the undifferentiated status of hES cells (Ji et al., 2009).

These results have been confirmed and extended recently by Forrestal et al. (2010). The authors also found a faster proliferation and larger colony size in human ES cells cultivated up to 12 month under hypoxia (5%) compared to normoxia (21%). Consequently, hES cells cultured under 20% oxygen had to be passaged more frequently prior to confluency in order to remove differentiated areas, while hES cells under low oxygen were only passaged when they reached confluency. Importantly, upon exposure to hypoxic conditions, cells displayed a physiological response that ensures availability of sufficient oxygen levels for oxygen-dependent processes. This response is regulated by hypoxia inducible factors (HIFs) which regulate the expression of many downstream genes including various involved in erythropoiesis, apoptosis and proliferation (Semenza, 2000). HIFs are transcription factors consisting of three oxygen-dependent subunits: HIF1α, HIF2α and HIF3α. Although HIF1α in mouse is the main regulator of hypoxic responses in human ES cells, HIF1α is only transiently expressed after hypoxic exposure and predominantly regulates different (mainly glycolytic) target genes than HIF2α. The authors identified HiF2α as the directly responsible one for the maintenance of pluripotency under hypoxic (5%) oxygen (Forrestal et al., 2010) corresponding to earlier findings in mouse ES cells that HIF2α is a direct upstream regulator of OCT4 (Covello et al.2006). This data correlated well with a 60-80% reduction in expression of pluripotency markers under normoxia. Inhibition of HIF2α and HIF3α, but not HIF1α decreased the expression of the pluripotency markers Oct4, Sox2 and NANOG. However, only HIF2A silencing correlated to morphological changes of more differentiation, a proliferation stop and loss of pluripotency markers (Forrestal et al., 2010). Taken together, HIF2α regulated the long-term hypoxic response by controlling hES cell pluripotency and proliferation.

In contrast to other cell types, HIF1α is only responsible for the initial adaptation of cells to hypoxia in embryonic stem cells and it was observed in mouse ES cells that expression of HIF1α under hypoxia suppressed the LIF-STAT signalling pathway resulting in the inhibition of self renewal and the promotion of cell differentiation (Jeong et al. 2007). This data provide a greater understanding of the mechanisms which regulate ES cell function and the pluripotent state and confirm that environmental oxygen tension has a critical role for maintaining of pluripotency in hES cells.

A direct influence of hypoxia and HIF1α on telomerase expression in mouse ES cells was recently established by Coussens et al. (2010). A role for HIF1α for the activation of telomerase had already been established in cancer cells previously (Nishi et al., 2004, Yatabe et al., 2004). Coussens et al. (2010) used a RNAi screen in order to identify factors that are responsible for maintaining a high TERT expression in murine ES cells. Knocking down HIF1α did not result in any differences in pluripotency, morphology or growth rate in mES cells but it down regulated TERT expression as well as telomerase activity in those cells and consequently, also telomeres shortened over 120 population doublings (PDs). The authors identified a functional HIF1α binding site at position +1 of the TERT promoter. The mechanism for up-regulation of Hif1α levels in cells in response to hypoxia involves a posttranslational stabilization of Hif1α via posttranslational modification of specific prolyl residues in the oxygen-dependent degradation domain (Ulrich, 2007), however the factors that are needed for HIF1α stabilisation are not well understood yet. Coussen et al. (2010) observed that under hypoxic growth conditions both, the targeted and non targeted mES cells induced Hif1α levels as well as TERT levels and telomerase activity. The ongoing telomere shortening in HIF1α targeted mES cells under long term cultivation resulted eventually in a decreased growth rate, increased γH2A.X levels and greatly enhanced p21 levels-indicative for the occurrence of senescence (although it was not directly shown). However, growing Hif1α-targeted cells under hypoxic conditions restored telomerase levels and telomere length. This study is the first to demonstrate a direct connection between hypoxic culture conditions and telomerase expression that is an important factor for pluripotency and unlimited proliferation capacity of undifferentiated stem cells. The authors also speculate that HIF1α might be required to sustain high telomerase levels at the inner cell mass of the blastocyst since early embryos develop under hypoxic conditions.

Another paper that seems at odds with the conclusion that a low oxygen atmosphere is beneficial for the maintenance of pluripotency was published recently by Varum and colleagues (2009). Normal cells decrease the amount of oxidative phosphorylation under hypoxic conditions, therefore the authors hypothesised that a further decrease of mitochondrial function in ES cells might directly promote pluripotency. They inhibited complex III of the mitochondrial respiratory chain pharmacologically and detected a high preservation of stem cell features such as high nuclear/cytoplasmic ratio and a specific increase in NANOG expression while other pluripotency markers (Oct4 and Sox2) or general stemness markers were not changed. In accordance with earlier observations that NANOG over-expression prevented neuroectoderm differentiation (Vallier et al., 2009) the authors found that expression of genes involved in ectoderm differentiation was significantly reduced in their system due to antimycin A. However, antimycin A treatment had a specific interaction with FGFβ signalling in human ES cells. Complex III is directly involved in the electron flow and ATP production of mitochondria, but has also been shown to directly generate ROS (Hamanaka and Chandel, 2009). Treatment of ESCs with antimycin A resulted in impairing of mitochondrial respiration and to a shift from oxidative phosphorylation to glycolysis while ATP levels were maintained at a constant level. In addition, this treatment increased ROS generation while apoptosis and cell proliferation were unchanged (Varum et al., 2010). There seemed to be a direct correlation between ROS generation after antimycin A treatment and NANOG expression since quenching ROS by using a MnSOD mimetic diminished any increase in NANOG expression. These results suggest that ROS generated at complex III could be at least partially responsive for the effects of antimycin A on NANOG expression levels.

While many other studies had demonstrated an increase of pluripotency under low oxygen levels Varum and co-authors (2010) showed that a certain level of ROS seemed necessary specifically for a high expression of NANOG. From these data it could be speculated that modulation of mitochondrial function has a rather specific effect on skewing the differentiation probability of specific, in this case ectodermic, lineages. It could also mean that specific oxidative conditions favour the differentiation of certain tissues, however, more evidence for that has to be gathered before any meaningful conclusion can be drawn.

A possible explanation for the apparent difference between the papers that showed a benefical effect of hypoxia on pluripotency of ES cells and the one published by Varum et al. (2009) is that it had been shown that ROS generation at mitochondria is responsible for the propagation of a hypoxic signal (Chandel et al., 1998). Complex III produces increased amounts of ROS under hypoxia which is similar to inhibition of complex III with antimycin A (Hamanaka and Chandel, 2009). Hypoxia-induced mitochondrial ROS inhibit HIF-α subunit turnover. Therefore it seems that mitochondrial ROS are required for hypoxic activation of HIFs and that ROS generation at mitochondrial complex III in particular is critical for hypoxia signalling. While complex I inhibitors such as rotenone inhibit the induction of HIF-1α, antimycin A does not (Chandel et al., 1998, Chandel et al., 2000). Guzy et al. (2005) found that a functional complex III of the mitochondrial electron transport chain (ETC) is required for the hypoxic stabilization of HIF-1α and HIF-2α and that an increase in ROS from complex III results in a stabilisation of HIF-α subunits. These results suggest that mitochondria function as cellular oxygen sensors and signal hypoxic HIF-α stabilisation by releasing ROS to the cytosol. Thus, mitochondrial ROS seem to act upstream of prolyl hydroxylases in regulating the stability of both HIF-1α and HIF- 2α in mammalian cells.

However, whether oxidative stress and ROS production is either decreased (Richter and von Zglinicki, 2007) or increased (Hanaka and Chandel,2009) under hypoxia is still controverse and might to a large degree depend on the actual oxygen percentage (1-5%) as well as changes in oxygen concentration during cell handling which might favour re-oxygenation that leads to increased ROS.

Evidence for very specific influences of mitochondria on stem cells has been recently demonstrated by Todd et al. (2010). The authors used loss and gain of function experiments in order to show that the growth factor Gfer plays a very specific role in stem cells, but not in somatic cells. The factor is present in mitochondria as well as in the nucleus and cytoplasm and seems to be highly enriched in various types of stem cells including embryonic. By influencing the mitochondrial fission factor Drp1 Gfer promotes survival and pluripotency in mouse ESCs (Todd et al., 2010). Down regulating Gfer in mouse ESCs decreased pluripotency markers, mitochondrial membrane potential and resulted in mitochondrial fission and fragmentation, a sign of cellular stress. At the same time the damaged mitochondria were eliminated by autophagy and apoptosis. This data suggests that dysfunctional mitochondria are effectively cleared from embryonic stem cells thereby ensuring a high quality of mitochondria remaining within the stem cells. While the level of Drp1 was increased when Gfer levels were decreased a decrease of Drp1 rescued mitochondrial dysfunction, pluripotency and apoptosis in Gfer depleted cells (Todd et al., 2010). Although the authors have not directly characterised glycolysis and energy generation in those ES cells their data would suggest that embryonic stem cells rely on functional and healthy mitochondria rather than undeveloped and non functioning ones. However, nothing is known about the role of Gfer in human ES cells and more research is needed to clarify and understand this apparent contradiction as well as the direct role of mitochondria for pluripotency.

6. Redox regulation in ES cells

Oxidative stress can be generated endogenously by dysfunctional mitochondria and NADH oxidases or applied externally via cell culture conditions or drugs and agents. It recently has been shown to control self renewal and differentiation of embryonic stem cells. ROS are generated in cells by enzymatic as well as non-enzymatic pathways. NADPH oxidases and xanthine oxidase generate superoxide anions by reducing molecular oxygen. In mitochondria superoxide leaks during electron transport from complexes I and III. However, at the same time there are many different ways in a cell to counteract oxidative stress. A first defence are antioxidant enzymes such as superoxide dismutases that dismutate the superoxide anions to hydrogen peroxide, which then is further scavenged by peroxidases, catalase or other molecular scavengers such as vitamins, carotinoids etc. Therefore, there generally exists a very balanced homeostasis of oxidative stress and redox state within a cell. Disturbing that balance can lead to increased oxidative stress and result in damage of DNA, lipids and proteins. On the other hand, ROS are also an important intracellular signalling factors (Finkel, 2001, Schieke and Finkel, 2006 Starkov 2008).

A better understanding of the influence of culture and environmental conditions should help to optimise growth conditions for undifferentiated and differentiated ES cells. Thus, it will also impinge on the successful application of those cells for the purpose of regenerative medicine.

Evidence is emerging currently that distinct functions of stem cells are under the influence of a redox potential. For example, low oxygen keeps hematopoietic stem cells in a quiescent state while higher ROS promote their differentiation (Hosokawa et al., 2007).

Recently, Yanes and colleagues (2010) uncovered a mechanism how increased oxidative stress not only accompanies differentiation of ES cells as described earlier (Saretzki et al., 2004, Saretzki et al., 2008) but seems to drive it actively. The authors performed an untargeted metabolomics approach in order to compare pluripotent mouse embryonic stem cells and their differentiated progeny. They also used metabolites involved in and produced by oxidative pathways in order to characterise their role in pluripotency and differentiation. They found that embryonic stem cells are characterised by an abundance of metabolites with highly unsaturated structures that decrease during differentiation. Some of the most up-regulated metabolites were secondary lipid messengers as well as inflammatory mediators. These unsaturated metabolites are preferentially susceptible to pro-oxidative events that occur due to intracellular changes in redox status. As evidence the authors monitored the ratio of reduced and oxidised glutathione (GSH/GSSG) during the course of differentiation. Oxidative stress is known to change that ratio due to conversion of GSH into GSSG. While the GSH/GSSG levels decreased by day 4 of differentiation ascorbic acid (an important antioxidant) levels increased during the same time pattern. The authors conclude that cell differentiation is associated with an increase of oxidative stress, an observation corresponding to findings from our laboratory (Saretzki et al., 2004, Saretzki et al., 2008). To prove that activation of an oxidative pathway is indeed necessary for cellular differentiation they inhibited the eicosanoid signalling pathway. That resulted in a promotion of pluripotency (increase of Oct4 and NANOG expression) and a reduction of differentiation in a dose dependent manner of inhibitors. Accordingly, when treated with eicosanoids, differentiation was promoted. These pathways acted similarly in human ESCs (Yanes et al., 2010). These results correspond well to an increased pluripotency in ESCs grown under physiological (5%) oxygen instead of normoxia (see above). They also correspond to findings of Eitsuka et al. (2005) who found that unsaturated, but not saturated fatty acids decreased telomerase activity and hTERT expression. In particular, polyunsaturated fatty acids (PUFAs) like EPA and DHA, were identified as powerful telomerase inhibitors mediated via protein kinase C (Eitsuka et al. 2005). These results identify an additional mechanism (in addition to epigenetic changes) how telomerase might be down regulated during ES cell differentiation and adds another component to the study (Yanes et al. 2010).

Powers et al (2008) analysed mouse ES cells under various oxygen levels and found that these cells adapted well to all oxygen conditions while maintaining pluripotency markers and an undifferentiated state. The authors found that the cells adapted their metabolism accordingly and generated constant ATP levels, although to a different degree by aerobic respiration in dependence on the oxygen conditions (Powers et al., 2008). However, once differentiated oxygen conditions had a substantial effect on the function and properties of those cells. Only extremely low oxygen concentrations lower or equal 1% induced differentiation of mouse ES cells. That could mean that a certain amount of oxygen is needed to maintain a pluripotent and undifferentiated state in mouse ES cells confirming the results obtained by Varum et al., (2009). The results from Powers et al (2008) are in accordance with our findings that mouse ES cells proliferate undistinguishable under normoxia (21%, ambient oxygen) and under hyperoxia (40% oxygen) (Saretzki et al., 2004). Powers et al (2008) did not find any difference in DNA damage due to different oxygen concentration, but could correlate the damage levels with increasing amounts of hydrogen peroxide. However, these results seem to be at odds with the findings of Li and Marban, (2010) who found a four fold higher DNA damage levels under 20% oxygen than under 5%. However, while Powers et al (2008) used mouse ES cells, Li and Marban (2010) used human stem cells which might explain the differences in DNA damage sensitivities.

In addition to growing human ES cells under physiological oxygen (5%) Li and Marban (2010) as well supplemented the culture media with various antioxidants. The latter were much more powerful in reducing intracellular ROS levels than hypoxia. Surprisingly, there was a biphasic dosage-dependent response to the treatment with antioxidants: while at low concentrations they reduced DNA damage, high concentrations promoted DNA damage. Importantly, the authors found that high doses of antioxidants decreased the level of proteins involved in DNA damage response such as DNA damage mediating kinases ATM and ATR and other downstream DNA repair factors such as Rad50, Rad 51, Chk1 and Chk2 (Li and Marban, 2010). The authors discussed this observation as a concept of "oxidative optimum" and concluded that a certain level of ROS are necessary in order to activate the DNA repair pathway and maintain chromosomal stability and genomic integrity, at least in stem cells. This data suggests that ROS not also induce DNA damage but at the same time are also able to promote the DNA damage response that in turn induces DNA repair.

Thus, instead of a low oxygen content (hypoxia) also antioxidants could be supplemented to the culture medium, however their concentrations have to be carefully optimised in order to avoid any detrimental effect as demonstrated by Li and Marban (2010).

To read more about the current knowledge of cultivating ES cells under low oxygen see Millman et al, (2009) for review.

Napier et al. (2010) showed recently that 20% oxygen concentration down regulated the levels of the RNA component hTR of the telomerase complex in endothelial cells thus hampering telomere maintenance by telomerase. These findings correspond well with the up regulation of TERT expression due to HIF1α under hypoxia (Coussens et al., 2010).

7. P53 and DNA damage response in ES cells

Cells are constantly under environmental and metabolic conditions that may cause genetic damage. Various environmental factors can lead to DNA damage that, if not repaired lead to either apoptosis and senescence or to genomic instabilities.

It is imperative for ES cells which give rise to all cells of the body and, most importantly, any germ cells, to guarantee a low mutation load and minimise any influence of genomic instability and DNA damage. ES cells have extremely stringent mechanisms in place that ensure a high degree of genomic integrity. Spontaneous mutation rates are around 100 fold lower in ES cells than in differentiated somatic cells (Cervantes et al, 2002). One of the most remarkable properties of ES cells is their high stress resistance (Saretzki et al., 2004, Saretzki et al., 2008, Guo et al, 2010). Mouse ES cells are very resistant against an increase in external oxidative stress and proliferated with the same rate in 20% and 40% oxygen (Saretzki et al., 2004). In addition, when compared to other immortal mouse cells such as 3T3 which both express high and comparable telomerase levels, mES cells accumulated less DNA damage when gamma-irradiated and had a better or faster DNA repair capacity (Saretzki et al., 2004). This seems to reflect a better DNA repair capacity as well as a high amount of antioxidant enzymes as shown at the gene expression level for mouse and human ES cells alike (Saretzki et al., 2004, Saretzki et al., 2008). In general, there seem to be two main pathways how ES cells cope with DNA damage: apoptosis and differentiation.

The tumour suppressor molecule p53 is one of the most important response mechanisms that operate in most somatic cells and play an important role in maintaining genomic integrity. Without any endogenous or exogenous stress p53 is expressed at a low level and remains inactive due to interaction with MDM2 and rapid ubiquitinylation and degradation of the protein. Only after stress occurs it is quickly stabilised and activated. The pathway ensures that cells that underwent DNA damage can halt their cell cycle in order to repair the damage or alternatively, if the damage is too high and cannot be repaired, drive cells into apoptosis or senescence. However, the p53 signalling pathway seems very differently regulated in ES cells compared to somatic cells. p21 is considered to be an important component in the senescence pathway and thought to correlate with ROS generation (Macip et al., 2002, Passos et al., 2010). In somatic cells the DNA damage response pathway activates p53 which in turn induces the expression of its downstream target genes such as p21. Consequently, high levels of p21 following extensive DNA damage had been implicated as a hallmark for DNA damage-induced premature senescence (Waldman et al., 1995).

Aladjem et al. (1998) have challenged mouse ES cells with an antimetabolite that depleted ribonucleotides and DNA damaging agents. They found that mouse ES cells have a high level of p53 protein but it was mainly localised in the cytoplasm and did not localise to the nucleus as in differentiated cells. P53 was therefore not able to act as a transcription factor for downstream genes such as p21 in the nucleus. Consequently, it did not activate a cell cycle arrest after rNTP depletion or DNA damage. Instead, cells went into apoptosis that was independent of p53 (Aladjem et al., 1998). The authors showed that the p53 checkpoint became operational again once the ES cells were differentiated (Aladjem et al., 1998). On the basis of this and other papers (Hong and Stambrook, 2004) it was thought that mouse ES cells lack a G1 dependent cell cycle arrest and have a deficient p53 DNA damage response. Miura et al. (2004) suggested that p53 is not required for ES cell self renewal and early embryogenesis and that also the RB/E2F pathway is inactive in ES cells thereby uncoupling mitogenic signalling from cell cycle progression.

It is well accepted that mouse and human ESC have a unique cell cycle structure different from somatic cells. They show only a very short and rather transient G1 and G2 cell cycle phases while most cells are in S-phase (Becker et al., 2006). Becker et al. (2006) described the shortened G1 cell cycle phase of human embryonic stem cells as necessary for their self renewal properties. Due to their very short G1 phase it was believed for some time that ES cells lack a G1 dependent growth arrest upon DNA damage.

However, recently a transient cell cycle arrest has been described in mouse and human ES cells. Barta and colleagues (2010) exposed human ES cells to UVC irradiation in G1 phase and found a cell cycle arrest just before DNA synthesis in S-phase. They found a down regulation of the cyclin-dependent kinase 2 activity depending on the checkpoint kinases Chk1 and Chk2 (Barta et al., 2010). The authors conclude that human ES cells are able to prevent entry into S-phase upon DNA damage by activating a G1/S arrest that is independent of p53 and p21 activation.

Lin et al. (2005) have described a role for p53 in the maintenance of a high genomic stability of ES cells by driving damaged ES cells into differentiation and thereby eliminating them from the ES cell pool. The mechanism for that seemed to be a direct down regulation of NANOG expression by p53 in mouse ES cells after DNA damage. NANOG is an important pluripotency associated transcription factor and essential for embryonic survival. The authors found a marked increase in p53 expression 4 hours after the treatment when apoptosis was not detectable yet. The authors also showed the activation of downstream genes such as Noxa and Puma proving that p53 was transcriptionally active. The decrease was specific for NANOG and not due to differentiation since other pluripotency markers such as Oct3/4 were stably highly expressed. P53 directly bound the NANOG promoter with higher efficiency than the p21 promoter. They also demonstrated that p21 was increased during differentiation of mouse ES cells due to active p53 (Ser 315 phosphorylation) although the overall p53 protein level decreased (Lin et al., 2005). Most importantly, down regulation of NANOG as well as successful differentiation induced by retinoid acid was dependent on Ser 315 phosphorylation of p53. Apoptosis induction was dependent on the presence and activity of a functional p53 (Lin et al., 2005) contradicting findings from others that a p53 dependent cell cycle arrest and apoptosis induction might not be operating in ES cells after various DNA damage inducers (Aladjem et al.,1998, Guo et al., 2010).

Mora-Castilla et al (2010) recently confirmed the role of p53 Ser315 phosphorylation for NANOG down regulation during differentiation of mouse and human ES cells following the exposure to NO-inducing agents. They showed that both p53 Ser315 and trimethylated histone H3 bound and repressed the NANOG promoter. Epigenetic modification of Oct4 and NANOG had been previously demonstrated during ESC differentiation (Wang et al., 2009).

Qin and colleagues (2006) demonstrated a direct link between p53 and apoptosis induction in human ES cells. They confirmed that treatment with DNA damaging agents accumulated p53 but did not activate its transcriptional targets such as p21. The authors demonstrated that this could be due to differential modification/phosphorylation of p53 in ES cells although p53 still binds to the promoters of p21 and Noxa after DNA damage. p53 was still able to bind to the promoters and down regulate the pluripotency genes Oct4 and NANOG. However, instead of transcriptional regulation p53 induced apoptosis in ES cells via a transcription-independent mitochondrial pathway. Like others the authors also observed that p53 is involved in spontaneous differentiation. This study proves the important role of p53 for the survival and differentiation of ES cells. P53 seemed also directly responsible for down regulating Oct4 and NANOG after DNA damage treatment via p53 binding to the promoters of both genes, while decreasing p53 levels correlated with an amelioration of this down regulation. Over-expressing Mdm2 could inhibit p53 accumulation and reduced spontaneous apoptosis levels. The authors also suggested that a decrease of p53 levels due to lowering stress during cultivation such as free oxygen radicals could be beneficial for growth and expansion of ES cells in culture. In summary, the study of Qin et al (2006) suggests that there is a very specific mechanism of p53 regulation that is active in ES cells but not in differentiated and somatic cells.

In general, it seems that p53 is a suppressor of pluripotency in ESC because it suppresses self-renewal of ESC after DNA damage. Thus, p53 seems to play an important role for the maintenance of genomic stability of ES cells by influencing their DNA damage response and self-renewal properties (Zhao and Xu, 2010).

Both mouse and human embryonic stem cells (ESCs) are unique among all stem cells by replicating continuously and being immortal in the absence of replicative senescence (Zheng and Rao, 2007, Guo et al., 2010).

8. Telomerase and mitochondria during the induction of pluripotency from somatic cells

Induced pluripotency cells (iPSCs) are a stem cell population that is experimentally derived from somatic cells by forced expression of various transcription factors that are normally expressed in embryonic stem cells. These cells acquire many properties of normal ES cells, including a high self-renewal potential, gene expression, high telomerase activity and differentiation potential as well as low mitochondrial number (Armstrong et al., 2010).

The generation of iPS cells is very promising in order to analyse and model detailed disease mechanisms *in vitro*. In addition, this procedure opens even greater regenerative potential than normal ES cells since in theory a somatic cell can be taken from the same donor who then receives the transplant from differentiated iPS cells. This type of custom tailored cell generation has enormous potential, but several processes have not yet been understood sufficiently and have to be solved in order for any potential cell replacement to be safe.

Induced pluripotency cells (iPSCs) were for the first time generated from mouse somatic cells by Takahashi and Yamanaka in 2006 (Takahashi Yamanaka, 2006). The authors used a cocktail of 4 essential transcription factors: Oct4, Sox2, KLF4 and c-myc. The latter two are thought to induce telomerase (Wong et al., 2010) however myc seems not really necessary for iPS induction. Meanwhile many groups repeated and refined this method in the mouse and the human system (Marion et al., 2009, Armstrong et al., 2010). In principle the 4 transcription factors re-programme the epigenetic signature of a somatic cell into that of an undifferentiated embryonic stem cell, including telomeres (Marion et al., 2009).

These properties are also reflected during the generation of induced pluripotency cells (iPSCs). One striking example is the activation of telomerase during the induction of pluripotency. It occurs in human, but is much less robust in mouse ES cells (Mathew et al., 2010). iPS cells, as normal stem cells are characterised by a high level of telomerase expression. Apparently, a high telomerase expression level in necessary in human iPS cells in order to re-elongate telomeres to the length of embryonic stem cells. A stable long telomere length is an essential prerequisite for self-renewal capacities of embryonic as well as iPS cells. Maria Blascos' group has provided a direct evidence for the importance of telomere length reset in iPS cells. Efficiency of iPS generation was dramatically reduced when cells from late generation telomerase (TERC) knockout mice were used that do not have telomerase activity and have very short telomeres (Marion et al., 2009). However, they also noticed, that some iPS clones from G2 and G3 generation TERC knock-out mice could have activated telomerase-independent telomere elongation mechanisms. The importance of telomere length and telomerase activity is evidenced by the finding that re-introduction of telomerase into telomerase knock-out mice is able to re-adjust telomere length and promote efficient iPS cell derivation. However, more important than the average telomere length for iPS cell derivation is probably the absence of very short individual telomeres. The authors also found that during the normal iPS cell derivation process telomeres acquire embryonic stem cell properties including telomere length and epigenetic changes (Marion et al., 2009).

Mathew et al. (2010) demonstrated in detail that telomerase activation and TERT promoter activation occurred at discrete steps during reprogramming of somatic cells. Comparing mouse and human cells they found a very modest threefold increase of the mTERT promoter activity during reprogramming while the hTERT promoter was induced between 10 fold (partial reprogramming) to hundred fold (complete reprogramming) in different colonies. While many clones had telomerase activity up regulated only those with the highest levels were able to maintain their telomeres (Mathew et al., 2010). Since mice have much longer telomeres in their somatic cells than humans it seems highly likely that human cells have to activate hTERT and telomerase to a much larger extent than it is necessary in mouse cells.

Many promoters such as the one from NANOG were hypomethylated in ES cells and demethylated during iPS cell generation (Mathew et al., 2010). However, the hTERT promoter was not highly methylated in somatic cells and demethylation events did not play an essential role in its re-activation while the histon modification of the promoter changed substantially during re-programming (Mathew et al., 2010). A general similarity but not identity had been shown for gene expression patterns of ES cells and iPS cells (Prigione et al., 2010). Similarities were described for most pluripotency and self-renewal genes while differences could be due to a conservation of some donor cell characteristics since both cell types did not derive from the same individual.

Progione et al. (2010) recently demonstrated that iPS cells derived from human skin also shared similarities with ES cells regarding their reduced p53 activity and mitochondrial activity including low ROS levels. A decrease in mitochondrial copy number and mass were also confirmed by others (Armstrong et al., 2010).

Since all mitochondria within iPS cells derive from a somatic cell the authors were interested in the analysis of how mitochondria from somatic cells are able to acquire stem cell like properties. They showed that somatic mitochondria re-acquire an immature phenotype which resembled that of ES cells. This included an ESC like state of morphology, biogenesis factors, mitochondrial copy number and a decreased level of ATP as well as oxidative stress generation (Prigione et al., 2010). iPS cells like ES cells relied on glycolysis rather than oxidative phosphorylation for their energy generation.

Progione and co-authors also showed a lower amount of oxidatively modified DNA lesions (oxodG) in undifferentiated ES cells and iPS cells and confirmed also a higher resistance to oxidative stress and DNA damage shown by our group previously (Saretzki et al., 2004, Saretzki et al., 2008, Armstrong et al., 2010). However, while Prigione et al. (2010) found a decrease in the expression of antioxidant enzymes such as catalase, GPX1 and all three SODs in undifferentiated iPS and ES cells that increased during differentiation our group detected the opposite: high antioxidant gene expression in undifferentiated ES cells, but a decrease during differentiation (Saretzki et al., 2004, Saretzki et al., 2008). Further analysis, such as comparing the expression at the protein or activity levels seem necessary to resolve this discrepancy.

Prigione et al. (2010) concluded that reprogramming restored a cellular state characterised by immature mitochondria and low oxidative stress. It is an intriguing finding that it seems to be possible to rejuvenate mitochondria so that they escape cellular senescence. Mitochondria and mitochondrially generated oxidative stress are thought to be important components in the induction of cellular senescence and human ageing alike (Wallace, 2005, Passos et al., 2010). In contrast, long telomeres, low p53 and p21 activity as well as low levels of oxidative stress and immature mitochondria are hallmarks of embryonic stem cells (Saretzki et al., 2004, Saretzki et al., 2008, Facuchi-Olivera et al., 2007). Furthermore, there is evidence to suggest that telomerase that resides inside the mitochondria actively decreases oxidative stress although the mechanism remains elusive yet (Ahmed et al., 2008, Haendeler et al., 2009).

Recapitulation of immature mitochondria might also suggest that this state is important for the generation of iPS cells and possibly causally involved in the establishment of pluripotency. However, further research is necessary to establish a direct connection between mitochondrial state and pluripotency.

However, some mechanisms that enhance the efficiency of iPS cell derivation at the same time also raise concern about the possible risks of application of cells derived from iPS cells generated after inhibition of the p53 pathway. A decrease of p53 has been shown to increase the efficiency of iPS cell derivation (Hong et al., 2009). However, it remains elusive whether the p53 induced senescence or apoptosis are involved in this block of reprogramming. Since many cancer types inactivate this important tumour suppressor pathway it remains controversial whether this could increase the cancer risk for patients treated with iPS cells that have been derived by compromising the p53 checkpoint (Zhao and Xu, 2010).

9. Conclusion

High telomerase expression as well as a special mitochondrial state are both characteristic properties of embryonic stem cells in addition to pluripotency. Our group was the first to analyse oxidative stress and stress resistance in normal and TERT over-expressing ES cells (Saretzki et al., 2004, Armstrong et al., 2005, Saretzki et al., 2008, Yang et al., 2008).

Recently, evidence emerges that favours hypoxic condtitions for the improvement of proliferation, pluripotency and differentiation capacities in ES cells. A direct link between HIF1α and telomerase function provides some rational for that (Coussens et al., 2010).

More data and possible mechanisms emerged on the role of reactive oxygen species (ROS) and the differentiation and pluripotency of ES cells. In particular, the role of mitochondria for ES cell properties gains more and more interest. Specific mitochondrial regulatory pathways that operate only in embryonic stem cells have been demonstrated recently (Todd et al., 2010) and ROS have been identified as possible differentiation inducer (Yanes et al., 2010).

Other new results emerge recently to explain why the tumour suppressor checkpoint p53 operates via a mitochondrial pathway that promotes apoptosis rather than cell cycle arrest or premature senescence as in somatic cells rather than via transcriptional control (Qin et al., 2007). Others have shown that p53 promotes differentiation of ES cells via downregulation of NANOG (Lin et al., 2005).

Induced pluripotency derived cells recapitulate mitochondrial numbers and properties known from early embryonic development under low oxygen conditions. However, no direct causal relationship between mitochondria and pluripotency has been investigated to date.

My group was the first to show a direct correlation between telomerase that shuttles to mitochondria and improved mitochondrial properties and a decreased oxidative stress (Ahmed et al., 2008). Other non-canonical functions of telomerase include its involvement in gene expression and Wnt signalling (Choi et al., 2008, Park et al, 2009). Interestingly, Yoon et al (2010) uncovered a new link between Wnt signalling and mitochondrial biogenesis and gene expression involved in oxidative phosphorylation. Bringing the different components involved in regulation of stem cell properties including mitochondria and telomerase together will certainly result in a better understanding of ES cell properties.

10. References

Ahmed, S.; Passos, J F; Birket, M J; Beckmann, T.; Brings, S.; Peters, H.; Birch-Machin, M A.; von Zglinicki, T. & Saretzki, G. (2008). Telomerase does not counteract telomere shortening but protects mitochondrial function under oxidative stress. *J Cell Science*, 17, 1046-1053.

Aladjem, M. I.; Spike, B.T.; Rodewald, L.W.; Hope, T.J.; Klemm, M.; Jaenisch, R. & Wahl, G.M. (1998). ES cells do not activate p53-dependent stress responses and undergo p53-independent apoptosis in response to DNA damage. *Curr. Biol.*, 8, 145–155.

Amit, M.;Carpenter, M.K.; Inokuma, M.S.; Chiu, C.P.; Harris, C.P.; Waknitz, M.A.; Itskovitz-Eldor, J.& Thomson, J.A. (2000). Clonally derived human embryonic stem cell lines maintain pluripotency and proliferative potential for prolonged periods of culture. *Developmental Biology*, 227, 271-278.

Armstrong, L.; Lako, M.; Lincoln, J.; Cairns, P.M.& Hole, N. (2000). mTert expression correlates with telomerase activity during the differentiation of murine embryonic stem cells. *Mechanisms of Development*, 97, (1-2), 109-16.

Armstrong, L.; Lako, M.; Van Herpe, I.; Saretzki, G.& Hole, N. (2004). Reduction of telomerase activity during differentiation of embryonal stem cells is mediated by transcriptional repression of murine telomerase reverse transcriptase (mTert). *Mechanisms of Development*, 121 (12), 1509-22.

Armstrong, L.; Saretzki, G.; Evans, J.; Hole, N.; von Zglinicki, T.& Lako, M. (2005).Overexpression of telomerase confers growth advantage, stress resistance and enhanced differentiation of ES cells towards the haematopoietic lineage. *Stem Cells*, 23 (4), 516-29.

Armstrong, L.; Tilgner, K.; Saretzki, G.; Atkinson, S.P.; Stojkovic, M.; Moreno, R.;

Pryzborski, S.& Lako, M. (2010). Human induced pluripotent stem cell line show similar stress defense mechanisms and mitochondrial regulation to human embryonic stem cells, *Stem Cells*, 28 (4), 661-673.

Atkinson, S.P.; Hoare, S.F.; Glasspool, R.M.& Keith, W.N. (2005). Lack of telomerase gene expression in alternative lengthening of telomere cells is associated with chromatin remodeling of the hTR and hTERT gene promoters. *Cancer Res.*, 65 (17), 7585-7590.

Bagheri, S.; Nosrati, M.; Li, S.; Fong, S.; Torabian, S.; Rangel, J.; Moore, D.H.; Federman, S.; Laposa, R.R.; Baehner, F.L.; Sagebiel, R.W.; Cleaver, J.E.; Haqq, C.; Debs, R.J.; Blackburn, E.H.& Kashani-Sabet, M. (2006). Genes and pathways downstream of telomerase in melanoma metastasis. *Proc Natl Acad Sci U S A*,. 103 (30), 11306-11.

Bárta, T.; Vinarský, V.; Holubcová, Z.; Dolezalová, D.; Verner, J.; Pospísilová, S.; Dvorák, P.& Hampl, A. (2010). Human embryonic stem cells are capable of executing G1/S checkpoint activation. *Stem Cells*, 28 (7), 1143-52.

Becker, K.A.; Ghule, P.N.; Therrien, J.A.; Lian, J.B.; Stein, J.L.; van Wijnen, A.J.& Stein, G.S. (2006). Self-renewal of human embryonic stem cells is supported by a shortened G1 cell cycle phase. *J Cell Physiol.*, 209 (3), 883-93.

Buzzard, J.J.; Gough, N.M.; Crook, J.M. & Colman, A. (2004). Karyotype of human ES cells during extended culture. *Nat Biotec.*, 22(4), 381-382.

Chandel, N.S.; Maltepe, E.; Goldwasser, E.; Mathieu, C.E.; Simon, M.C. &Schumacker, P.T. (1998). Mitochondrial reactive oxygen species trigger hypoxia-induced transcription. *Proc Natl Acad Sci U S A*, 95, 11715-11720

Chandel, N.S.; McClintock, D.S.; Feliciano, C.E.; Wood, T.M.; Melendez, J.A.; Rodriguez, A.M. & Schumacker, P.T. (2000). Reactive oxygen species generated at mitochondrial complex III stabilize hypoxia-inducible factor-1alpha during hypoxia: a mechanism of O2 sensing. *J Biol Chem*, 275, 25130-25138.

Cervantes, R. B.; Stringer, J. R.; Shao, C.; Tischfield, J. A. & Stambrook, P. J. (2002) Embryonic stem cells and somatic cells differ in mutation frequency and type *Proc. Natl. Acad. Sci. USA*, 99, 3586–3590.

Chen, H.F.; Kuo, H.C.; Chen, W.; Wu, F.C.; Yang, Y.S. & Ho, H.N. (2009). A reduced oxygen tension (5%) is not beneficial for maintaining human embryonic stem cells in the undifferentiated state with short splitting intervals. *Hum Reprod.*, 24 (1), 71-80.

Cho, Y. M.; Kwon, S.; Pak, Y. K.; Seol, H.W.; Choi, Y.M.; Park, do J.; Park, K.S. & Lee HK. (2006). Dynamic changes in mitochondrial biogenesis and antioxidant enzymes during the spontaneous differentiation of human embryonic stem cells. *Biochemical and Biophysical Research Communications*, 348, 1472–1478.

Choi. J.; Southworth, L.K.; Sarin, K.Y.; Venteicher, A.S.; Ma, W.; Chang, W.; Cheung, P.;Jun, S.; Artandi, M.K.; Shah, N.; Kim, S.K.& Artandi, S.E. (2008). TERT promotes epithelial proliferation through transcriptional control of a Myc- and Wnt-related developmental program. *PLoS Genet.*, 1, e10.

Coussens, M.; Davy, P.; Brown, L.; Foster, C.; Andrews, W.H.; Nagata, M. & Allsopp R. (2010) RNAi screen for telomerase reverse transcriptase transcriptional regulators identifies HIF1alpha as critical for telomerase function in murine embryonic stem cells. *Proc Natl Acad Sci U S A.*, 107 (31), 13842-7.

Covello, K.L.; Kehler, J.; Yu, H.; Gordan, J.D.; Arsham, A.M.; Hu, C.J.; Labosky, P.A. & Simon, M.C. (2006). Keith B. HIF-2alpha regulates Oct-4: effects of hypoxia on stem cell function, embryonic development, and tumor growth. *Genes Dev.*, 2006 20(5), 557-70.

Eitsuka, T.; Nakagawa, K.; Suzuki, T. & Miyazawa, T. (2005). Polyunsaturated fatty acids inhibit telomerase activity in DLD-1 human colorectal adenocarcinoma cells: a dual mechanism approach. *Biochim Biophys Acta*, 1737 (1), 1-10.

Ezashi, T.; Das, P.& Roberts, R.M. Low O2 tensions and the prevention of differentiation of hES cells. Proc Natl Acad Sci U S A, 102 (13), 4783-4788.

Facucho-Oliveira, J.M.; Alderson, J.; Spikings, E. C.; Egginton, S. & St John, J. C. (2007). Mitochondrial DNA replication during differentiation of murine embryonic stem cells. *Journal of Cell Science*, 120, 4025–4034.

Finkel, T. (2001). Reactive oxygen species and signal transduction. *IUBMB Life*, 52(1-2), 3-6.

Fischer, B.& Bavister, B.D. (1993). Oxygen tension in the oviduct and uterus of rhesus monkeys, hamsters and rabbits. *J. Reprod. Ferti.,l* 99 (2), 673–679.

Forristal, C.E.; Wright, K.L.; Hanley, N.A.; Oreffo, R.O. & Houghton FD. (2010). Hypoxia inducible factors regulate pluripotency and proliferation in human embryonic stem cells cultured at reduced oxygen tensions. *Reproduction*, 139(1), 85-97.

Forsyth, N.R.; Musio, A.; Vezzoni, P.; Simpson, A.H.; Noble, B.S.& McWhir, J. (2006). Physiologic oxygen enhances human embryonic stem cell clonal recovery and reduces chromosomal abnormalities. *Cloning Stem Cells*, 8 (1), 16-23.

Fu, W.; Killen, M.; Culmsee, C.; Dhar, S.; Pandita, T.K. & Mattson, M.P. (2000). The catalytic subunit of telomerase is expressed in developing brain neurons and serves a cell survival-promoting function. *J. Mol. Neurosci.*, 14, 3-15.

Guo, Y.L.; Chakraborty, S.; Rajan, S.S.; Wang, R.& Huang, F. (2010). Effects of oxidative stress on mouse embryonic stem cell proliferation, apoptosis, senescence, and self-renewal. *Stem Cells Dev.*, 19 (9), 1321-31.

Guzy, R.D.; Hoyos, B.; Robin, E.; Chen, H.; Liu, L.; Mansfield, K.D.; Simon, M.C.; Hammerling, U. & Schumacker, P.T. (2005). Mitochondrial complex III is required for hypoxia-induced ROS production and cellular oxygen sensing. *Cell Metab.*, 1 (6), 401-8.

Haendeler. J.; Dröse, S.; Büchner, N.; Jakob, S.; Altschmied, J.; Goy, C.; Spyridopoulos, I.; Zeiher, A.M.; Brandt, U.& Dimmler, S. (2009). Mitochondrial telomerase reverse transcriptase binds to and protects mitochondrial DNA and function from damage. *Arterioscler Thromb Vasc Biol.*, 29 (6), 929-35

Hamanaka, R.B. & Chandel, N.S. (2009). Mitochondrial reactive oxygen species regulate hypoxic signaling. *Curr Opin Cell Biol.*, 21 (6), 894-9.

Harley, C.B.; Futcher, A.B. & Greider, C.W. (1990). Telomeres shorten during ageing of human fibroblasts. , 345 (6274), 458-60.

Harvey, A.J.; Kind, K.L.& Thompson, J.G. (2002). Effect of the oxidative phosphorylation uncoupler 2,4- dinitrophenol on hypoxia-inducible factor-regulated gene expression in bovine blastocysts. , 16(7):665-73.

Hiyama, E. &Hiyama, K. (2007). Telomere and telomerase in stem cells. *Br. J. Cancer*, 96 (7), 1020-4.

Hong, H.; Takahashi, K.; Ichisaka, T.; Aoi, T.; Kanagawa, O.; Nakagawa, M.; Okita, K. & Yamanaka, S. (2009). Suppression of induced pluripotent stem cell generation by the p53-p21 pathway. *Nature*, 460 (7259),1132-5.

Hong, Y. & Stambrook, P.J. (2004). Restoration of an absent G1 arrest and protection from apoptosis in embryonic stem cells after ionizing radiation. *Proc Natl Acad Sci U S A*, 101 (40), 14443-8.

Hosokawa, K.; Arai, F.; Yoshihara, H.; Nakamura, Y.; Gomei, Y.; Iwasaki, H.; Miyamoto, K.; Shima, H.; Ito, K. & Suda, T. (2007). Function of oxidative stress in the regulation of hematopoietic stem cell-niche interaction. *Biochem Biophys Res Commun.*, 363 (3), 578-83.

Houghton, F.D. (2006). Energy metabolism of the inner cell mass and trophectoderm of the mouse blastocyst. *Differentiation*, 74 (1), 11-8.

Jeong, C.H.; Lee, H.J.; Cha, J.H.; Kim, J.H.; Kim KR, Kim, J.H.; Yoon, D.K. & Kim, K.W. (2007). Hypoxia-inducible factor-1 alpha inhibits self-renewal of mouse embryonic stem cells in Vitro via negative regulation of the leukemia inhibitory factor-STAT3 pathway. *J Biol Chem.*, 282 (18), 13672-9.

Ji, L.; Liu, Y.X.; Yang, C.; Yue, W.; Shi, S.S.; Bai, C.X.; Xi, J.F.; Nan, X. & Pei, X.T. (2009). Self-renewal and pluripotency is maintained in human embryonic stem cells by co-culture with human fetal liver stromal cells expressing hypoxia inducible factor 1alpha. *J Cell Physiol.*, 221 (1), 54-66.

Kondo, Y.; Kondo, S.; Tanaka, Y.; Haqqi, T.; Barna, B.P.& Cowell, J.K. (1998). Inhibition of telomerase increases the susceptibility of human malignant glioblastoma cells to cisplatin-induced apoptosis. *Oncogene*, 16, 2243-2248.

Kondoh, H.; Lleonart, M. E.; Nakashima, Y.; Yokode, M.; Tanaka, M.; Bernard, D.; Gil, J. & Beach, D. (2007). A high glycolytic flux supports the proliferative potential of murine embryonic stem cells. *Antioxidants & Redox Signalling*, 9, 293–299.

Lee, M.K.; Hande, M.P.& Sabapathy, K. (2005). Ectopic mTERT expression in mouse embryonic stem cells does not affect differentiation but confers resistance to differentiation- and stress-induced p53-dependent apoptosis. *J Cell Sci.*, 118 (4), 819-29.

Li, T.S. & Marbán, E. (2010). Physiological levels of reactive oxygen species are required to maintain genomic stability in stem cells. *Stem Cells*, 28 (7), 1178-85.

Li, S.; Crothers, J.; Haqq, C.M. & Blackburn, E.H. (2005). Cellular and gene expression responses involved in the rapid growth inhibition of human cancer cells by RNA interference-mediated depletion of telomerase RNA. *J. Biol. Chem.*, 280, 23709-23717.

Lin, T.; Chao, C.; Saito, S.; Mazur, S.J.; Murphy, M.E.; Appella, E. & Xu, Y. (2005). p53 induces differentiation of mouse embryonic stem cells by suppressing Nanog expression. *Nat Cell Biol.*, 7 (2), 165-71

Longo, L.; Bygrave, A.; Grosveld, F.G.; Pandolfi, P.P. (1997). The chromosome make-up of mouse embryonic stem cells is predictive of somatic and germ cell chimaerism. *Transgenic Res.*, 6, 321–328

Lopatina, N.G.; Poole, J.C.; Saldanha, S.N.; Hansen, N.J.; Key, J.S.; Pita, M.A.; Andrews, L.G. & Tollefsbol, T.O. (2003). Control mechanisms in the regulation of telomerase reverse transcriptase expression in differentiating human teratocarcinoma cells. *Biochem Biophys Res Commun.*, 306 (3), 650-9.

Ludwig, A.; Saretzki, G.; Holm, P.S.; Tiemann, F.; Lorenz, M.; Emrich, T.; Harley, C.B. &von Zglinicki, T.(2001). Ribozyme cleavage of telomerase mRNA sensitizes breast epithelial cells to inhibitors of topoisomerase. *Cancer Res.*, 61 (7), 3053-61.

Macip, S.; Igarashi, M.; Fang, L.; Chen, A.; Pan, Z. Q.; Lee, S. W. & Aaronson, S. A. (2002). Inhibition of p21-mediated ROS accumulation can rescuep21-induced senescence. *EMBO J*, 21:2180–2188.

Maitra, A.; Arking, D.E.; Shivapurkar, N.; Ikeda, M.; Stastny, V.; Kassauei, K.; Sui, G.; Cutler, D.J.; Liu, Y.; Brimble, S.N.; Noaksson, K.; Hyllner, J.; Schulz, T.C.; Zeng, X.; Freed, W.J.; Crook, J.; Abraham, S.; Colman, A.; Sartipy, P.; Matsui, S.; Carpenter, M.; Gazdar, A.F.; Rao, M. & Chakravarti, A. (2005). Genomic alterations in cultured human embryonic stem cells. *Nat Genet.*, 37 (10), 1099-103.

Marion, R.M.; Strati, K.; Li, H.; Tejera, A.; Schoeftner, S.; Ortega, S.; Serrano, M. & Blasco, M.A. (2009). Telomeres acquire embryonic stem cell characteristics in induced pluripotent stem cells. *Cell Stem Cell,* 4 (2), 141-54

Millman, J.R.; Tan, J.H.& Colton, C.K. (2009). The effects of low oxygen on self-renewal and differentiation of embryonic stem cells. *Curr Opin Organ Transplant.*, 14 (6), 694-700.

Mathew, R.; Jia, W.; Sharma, A.; Zhao, Y.; Clarke, L.E.; Cheng, X.; Wang, H.; Salli, U. & Vrana, K.E.(2004). Robertson GP, Zhu J, Wang S. Robust activation of the human but not mouse telomerase gene during the induction of pluripotency. *FASEB J.*, 24 (8), 2702-15.

Miura, T.; Mattson, M.P. & Rao, M.S. (2004). Cellular lifespan and senescence signaling in embryonic stem cells. *Aging Cell,* 3 (6), 333-43.

Mora-Castilla, S.; Tejedo, J.R.; Hmadcha, A.; Cahuana, G.M.; Martín, F.; Soria, B. & Bedoya, F.J. (2010). Nitric oxide repression of Nanog promotes mouse embryonic stem cell differentiation. *Cell Death Diffe.*, 17 (6), 1025-33.

Napier, C.E.; Veas, L.A.; Kan, C.-Y.; Taylor, L.M.; Yuan, J.;, Wen, V.W.;, James, A.; O'Brien, T.A.; Lock, R.B. & MacKenzie, K.L. (2010). Mild hyperoxia limits hTR levels, telomerase activity, and telomere length maintenance in hTERT-transduced bone marrow endothelial cells. *Biochimica et Biophysica Acta - Molecular Cell* Research, 1803 (10), 1142 - 1153

Nishi, H.; Nakada, T.; Kyo, S.; Inoue, M.; Shay, J.W. & Isaka, K. (2004). Hypoxia-inducible factor 1 mediates upregulation of telomerase (hTERT). *Mol. Cell. Biol.*, 24, 6076–6083.

O'Connor, M.S.; Carlson, M.E. & Conboy, I.M. (2009). Differentiation rather than aging of muscle stem cells abolishes their telomerase activity. *Biotechnol Prog.* 25 (4), 1130-7.

Park, Y.B.; Kim, Y.Y.; Oh, S.K.; Chung, S.G.; Ku, S.Y.; Kim, S.H.; Choi, Y.M. & Moon, S.Y. (2008). Alterations of proliferative and differentiation potentials of human embryonic stem cells during long-term culture. *Exp. Mo.l Med.*, 40 (1), 98-108.

Park, J.I.; Venteicher, A.S.; Hong, J.Y.; Choi, J.; Jun, S.; Shkreli, M.; Chang, W.; Meng, Z.; Cheung, P.; Ji, H.; McLaughlin, M,; Veenstra, T.D.; Nusse, R.; McCrea, P.D.& Artandi S.E. (2009). Telomerase modulates Wnt signalling by association with target gene chromatin. *Nature*, 460 (7251):66-72.

Passos, J.F.; von Zglinicki, T. & Saretzki, G. (2006). Mitochondrial dysfunction and cell senescence:cause or consequence? *Rejuvenation Research*, 9 (1), 64-8

Passos, J.F.; Nelson, G.; Wang, C.; Richter, T.; Simillion, C.; Proctor, C.J.; Miwa, S.; Olijslagers, S.;Hallinan, J.; Wipat, A.; Saretzki, G.; Rudolph, K.L.; Kirkwood, T.B.L. & von Zglinicki, T. (2010). Feedback between p21 and reactive oxygen production is necessary for cell senescence. *Mol. Syst. Biol.*, 6, 347

Powers, D.E.; Millman, J.R.; Huang, R.B.& Colton, C.K. (2008). Effects of oxygen onmouse embryonic stem cell growth, phenotype retention, and cellular energetics. *Biotechnol. Bioeng.*, 101 (2), 241-54.

Prigione, A.; Fauler, B. Lurz, R.; Lehrach, H. & Adjaye, J. (2010). The senescence-related mitochondrial/oxidative stress pathway is repressed in human induced pluripotent stem cells. *Stem Cells*, 28 (4), 721-33.

Qin H, Yu T, Qing T, Liu Y, Zhao Y, Cai J, Li J, Song Z, Qu X, Zhou P, Wu J, Ding M, Deng H. Regulation of apoptosis and differentiation by p53 in human embryonic stem cells. J Biol Chem. 2007 Feb 23;282(8):5842-52

Richter, T. & von Zglinicki, T. (2007). A continuous correlation between oxidative stress and telomere shortening in fibroblasts. *Exp. Gerontol.*, 42 (11), 1039-42.

Rinaudo, P.F.; Giritharan, G.; Talbi, S.; Dobson, A.T.& Schultz, R.M. (2006). Effects of oxygen tension on gene expression in preimplantation mouse embryos. *Fertil. Steril.*, 86 (4 Suppl), 1252-1265.

Rosler, E.S.; Fisk, G.J.; Ares, X.; Irving, J.; Miura, T.; Rao, M.S. & Carpenter, M.K. (2004). Long-term culture of human embryonic stem cells in feeder-free conditions. *Dev Dyn.*, 229 (2), 259-274.

Rugg-Gunn, P.J.; Ferguson-Smith ,A.C.& Pedersen, R.A. (2005). Epigenetic status of human embryonic stem cells. *Nature Genetic*, 37 (6), 585-587.

Santos, J.H.; Meyer, J.N.; Skorvaga, M.; Annab, L.A. & Van Houten, B. (2004). Mitochondrial hTERT exacerbates free-radical-mediated mtDNA damage. *Aging Cell*, 3 (6), 399-411

Santos, T. A.; El Shourbagy, S. & St John, J. C. (2006). Mitochondrial content reflects oocyte variability and fertilization outcome. *Fertility and Sterility*, 85, 584–591.

Saretzki, G.; Ludwig, A.; von Zglinicki, T. & Runnebaum, I.B. (2001). Ribozyme-mediated telomerase inhibition induces immediate cell loss but not telomere shortening in ovarian cancer cells. *Cancer Gene Ther.* 8(10):827-34.

Saretzki, G.; Armstrong, L.; Leake, A.; Lako, M. & von Zglinicki, T. (2004). Stress defence in murine embryonic stem cells is superior to that of various differentiated murine cells. *Stem Cells,* 2004, 22(6):962-71

Saretzki, G.; Walter, T.; Atkinson, S.; Passos, J.F.; Bareth, B.; Keith, W.N.; Stewart, R.; Hoare, S.; Stojkovic, M.; Armstrong, L.; von Zglinicki, T. &. Lako, M. (2008). Down-regulation of multiple stress defence mechanisms during differentiation of human embryonic stem cells. *Stem Cells,* 26 (2), 455-465

Saretzki, G. (2009). Telomerase, mitochondria and oxidative stress. *Experimental Gerontology,* 44, 485-492

Schieke, S.M.& Finkel, T. (2006). Mitochondrial signaling, TOR, and life span. *Biol. Chem.,* 387 (10-11), 1357-61.

Semenza, G.L. (2000). HIF-1: mediator of physiological and pathophysiological responses to hypoxia. *J. App.l Physiol.,* 88 (4), 1474-80.

Shoubridge, E. A. & Wai, T. (2007). Mitochondrial DNA and the mammalian oocyte *Curr. Top. Dev. Biol.* 77, 87-111.

Starkov, A. A. (2008). The role of mitochondria in reactive oxygen species metabolism and signaling. *Ann. N. Y. Acad. Sci.* 1147:37–52.

St John, J. C.; Ramalho-Santos, J.; Gray, H. L.; Petrosko, P.; Rawe, V. Y.; Navara, C. S. Simerly, C. R. & Schatten, G. P. (2005). The expression of mitochondrial DNA transcription factors during early cardiomyocyte in vitro differentiation from human embryonic stem cells. *Cloning Stem Cells,* 7, 141-153.

Takahashi, K.;& Yamanaka, S. (2006). Induction of pluripotent stem cells from mouse embryonic and adult fibroblast cultures by defined factors. *Cell,* 126(4):663-76.

Thomson, J. A.; Itskovitz-Eldor, J., Shapiro, S. S.; Waknitz, M. A.; Swiergiel, J. J.; Marshall, V. S. &Jones, J. M. (1998). Embryonic stem cell lines derived from human blastocysts. *Science,* 282, 1145-1147.

Todd, L.R.; Damin, M.N.; Gomathinayagam, R.; Horn, S.R.; Means, A.R. & Sankar, U. (2010). Growth factor erv1-like modulates Drp1 to preserve mitochondrial dynamics and function in mouse embryonic stem cells. *Mol. Biol. Cell,* 21 (7), 1225-36.

Trimarchi, J. R.; Liu, L.; Porterfield, D. M.; Smith, P. J. &Keefe, D. L. (2000). Oxidativephosphorylation-dependent and -independent oxygen consumption by individual preimplantationmouse embryos. *Biology of Reproduction,* 62, 1866–1874.

Tsai, C.C.; Chen, C.L.; Liu, H.C.; Lee, Y.T.; Wang, H.W.; Hou, L.T. & Hung, S.C. (2010) Overexpression of hTERT increases stem-like properties and decreases spontaneous differentiation in human mesenchymal stem cell lines. *Biomed. Sci.,* 17, 64.

Ulrich, H.D. (2007). SUMO teams up with ubiquitin to manage hypoxia. *Cell* 131:446–447.

Vallier, L.; Touboul, T.; Chng, Z.; Brimpari, M.; Hannan, N.; Millan, E.; Smithers, L.E.; Trotter, M.; Rugg-Gunn, P.; Weber, A.& Pedersen, R.A. (2009). Early cell fate decisions of human embryonic stem cells and mouse epiblast stem cells are controlled by the same signalling pathways. *PLoS One,* 4(6), e6082.

Varum, S.; Momčilović, O.; Castro, C.; Ben-Yehudah, A.; Ramalho-Santos, J. & Navara, C.S. (2009). Enhancement of human embryonic stem cellpluripotency through inhibition of the mitochondrial respiratory chain. *Stem Cell Research,* 3, 142–156 von Zglinicki, T.; Saretzki, G.; Docke, W.; Lotze, C. (1995). Mild hyperoxia shortens telomeres and inhibits proliferation of fibroblasts: a model for senescence? *Exp. Cell. Res.*, 220 (1), 186-93.

Waldman, T.; Kinzler, K. W. & Vogelstein, B. (1995). p21 is necessary for the p53-mediated G1 arrest in human cancer cells. *Cancer Res.*, 55:5187–5190.

Wallace, D.C. (2005). A mitochondrial paradigm of metabolic and degenerative diseases, aging, and cancer: a dawn for evolutionary medicine. *Annu Rev Genet.* 39, 359-407.

Wang, K.; Chen, Y.; Chang, E.A.; Knott, J.G. & Cibelli, J.B. (2009). Dynamic epigenetic regulation of the Oct4 and Nanog regulatory regions during neural differentiation in rhesus nuclear transfer embryonic stem cells. *Cloning Stem Cells*, 11 (4), 483-96.

Westfall, S.D.; Sachdev, S.; Das, P.; Hearne, L.B.; Hannink, M.; Roberts, R.M. & Ezashi, T. (2008). Identification of oxygen-sensitive transcriptional programs in human embryonic stem cells. *Stem Cells Dev.*, 17 (5), 869-81.

Wong, C.W.; Hou, P.S.; Tseng, S.F.; Chien, C.L.; Wu, K.J.; Chen, H.F.; Ho, H.N.; Kyo, S. & Teng, S.C. (2010). KLF4 Contributes to Maintenance of Telomerase Activity in Stem Cells. Stem Cells. 2010 Jul 13. [Epub ahead of print]

Xie, X.; Hiona, A.; Lee, A.S.; Cao, F.; Huang, M.; Li, Z.; Cherry, A.; Pei, X.T.& Wu, J.C. (2010). Effects of Long Term Culture on Human Embryonic Stem Cell Aging. *Stem Cells Dev.*, 2010 Jul 14. [Epub ahead of print]

Yanes, O.; Clark, J.; Wong, D.M.; Patti, G.J.; Sánchez-Ruiz A.; Benton, H.P.; Trauger, S.A.,; Desponds, C.; Ding, S. & Siuzdak, G. (2010). Metabolic oxidation regulates embryonic stem cell differentiation. *Nat. Chem. Biol.*, 6 (6), 411-7.

Yang, C.; Pryzborski, S.; Cooke, M.J.; Zhang, X.; Stewart, R.; Atkinson, S.; Saretzki, G.; Armstrong, L. & Lako, M. (2008). A key role for telomerase reverse transcriptase unit (TERT) in modulating human ESC proliferation, cell cycle dynamics and in vitro differentiation. *Stem Cells,* 26 (4), 850-63.

Yatabe, N.; Kyo, S.; Maida, Y.; Nishi, H.; Nakamura, M.; Kanaya, T.; Tanaka, M.; Isaka,K.; Ogawa, S. & Inoue, M. (2004). HIF-1-mediated activation of telomerase in cervical cancer cells. *Oncogene,* 23,3708–3715.

Yoon, J.C.; Ng, A.; Kim, B.H.; Bianco, A.; Xavier, RJ.& Elledge, S.J. (2010). Wnt signaling regulates mitochondrial physiology and insulin sensitivity. *Genes Dev.*, 24(14):1507-18.

Zachar, V.; Prasad, S.M.; Weli, S.C.; Gabrielsen, A.; Petersen, K.; Petersen, M.B. & Fink T (2010). The effect of human embryonic stem cells (hESCs) long-term normoxic and hypoxic cultures on the maintenance of pluripotency. *In Vitro Cell. Dev. Biol. Anim.*, 46 (3-4), 276-83..

Zeng, X. & Rao, M.S. (2007). Human embryonic stem cells: long term stability, absence of senescence and a potential cell source for neural replacement. *Neuroscience*, 145 (4), 1348-58.

Zhao, T. & Xu, Y. (2010). p53 and stem cells: new developments and new concerns. *Trends in Cell Biology*, 20 (3), 170-175

Zhang, P.; Chan, S.L.; Fu, W.; Mendoza, M. &Mattson, M.P. (2003). TERT suppresses apoptosis at a premitochondrial step by a mechanism requiring reverse transcriptase activity and 14-3-3 protein-binding ability. *FASEB J* 17, 767-769.

Zhu, H.; Fu, W. & Mattson, M.P. (2000). The catalytic subunit of telomerase protects neurons against amyloid beta-peptide-induced apoptosis. *J. Neurochem.* 75, 117-124.

Part 3

Hormonal Signals that Regulate the Differentiation of Ectodermal Cells - Neurogenesis

19

Embryonic Neural Stem Cell Differentiation to Aldynoglia Induced by Olfactory Bulb Ensheathing Cell-Conditioned Medium

Daniel Ortuño-Sahagún,[1] Argelia E. Rojas-Mayorquín,[2] Antoni Camins[3] and Mercè Pallàs,[3]

[1]*Laboratorio de Desarrollo y Regeneración Neural, Instituto de Neurobiología, Departamento de Biología Celular y Molecular, CUCBA, Universidad de Guadalajara, Guadalajara, Jalisco,*

[2]*Instituto de Geriatría, Institutos Nacionales de Salud, México City,*

[3]*Unitat de Farmacologia i Farmacognòsia Facultat de Farmàcia, Institut de Biomedicina (IBUB), Centros de Investigación Biomédica en Red de Enfermedades Neurodegenerativas (CIBERNED), Universitat de Barcelona, Nucli Universitari de Pedralbes, Barcelona,*

[1,2]*México*

[3]*Spain*

1. Introduction

Although the relevance of glial cells in regulating brain activity was predicted by Ramon y Cajal more than a century ago (García-Marín et al., 2007), it was not until almost fifty years ago that initial descriptions of a close functional relationship between neuroglia and neuronal perikarya (Hyden, 1962) or axonal processes (Blunt et al, 1965) began to reveal that neurons and glial cells operate as functional units in the central nervous system (CNS). However, this functional interaction has only been more carefully studied and analysed in the last few decades, generating a substantial increase in research on the roles of neuron-glia interactions in the control of brain function. Glial cells have subsequently been implicated in many functions, including: guiding the migration of neurons in early development, axonal guidance and being responsible for their integrity, forming the necessary scaffold for neuronal architecture and neural protection and proliferation by trophic effects, modulating neurodegenerative processes, and also being critical participants in synaptic transmission, and key regulators of neurotransmitter release.

Interactions between axons and glia are dynamic and reciprocal. Experiments carried out in several laboratories have shown that glial cells directly participate in synaptic signalling and potentially regulate synaptic plasticity and network excitability (e.g. Vijayaraghavan, 2007). There is also functional coupling between neurons and glia (Álvarez-Maubecín et al., 2000), and a variety of close anatomical relationship; on the one hand, white matter glia preserve axonal integrity (Edgar & Nave, 2009), and on the other hand neurons regulate migration, survival and proliferation of the glial cells that they need for guidance (Learte & Hidalgo, 2007). Hence, there are a plenty of examples regarding morphological and physiological interactions between glial cells and neurons (Wigley, 2007). Such, communication between glial cells and neurons has led to the concept of the neuron–glial functional unit as the substrate of integration in the CNS.

Various glial cells types are involved in neuronal functioning. During the development of the CNS, the reciprocal communication between neurons and oligodendrocytes is essential for the generation of myelin (Simons & Trajkovic, 2006). These interconnections at multiple levels show how neurons and glia cooperate to build a complex network during development. In addition, although neurons are indispensable for brain function, an emerging alternative view holds that astrocytes, the dominant glial cell type, coordinate synaptic networks. Through the release of glutamate, astrocytes locally excite neurons, and via adenosine, which accumulates due to the hydrolysis of released ATP, astrocytes suppress distant synapses (Fellin et al, 2006).

Astrocytes respond to neuronal activity and neurotransmitters through the activation of metabotropic receptors, and can release gliotransmitters (ATP, d-serine, and glutamate), which act on neurons participating in a tripartite synapse (consisting of classic pre- and postsynaptic elements and an associated astrocytic process) (Halassa & Haydon, 2010). Moreover, astrocytes release several transmitters that affect neuronal activity and involve glial signalling in the modulation of mammalian behaviour (Halassa et al, 2009). By controlling neuronal A1-receptor signalling, astrocytes modulate mammalian sleep homeostasis and are essential for mediating the cognitive consequences of sleep deprivation. Slow-signalling glia modulate fast synaptic transmission and neuronal firing to impact behavioural output (Halassa et al, 2009). Cell type-specific molecular genetics has allowed a new level of examination of the role of astrocytes in brain function and has revealed an important role of these glial cells in gliotransmission, which is mainly mediated by adenosine accumulation, in the control of sleep and in cognitive impairments that follow sleep deprivation (Halassa & Haydon, 2010). Thus, a new challenge is to try to understand the functioning and rules of the neuron–glial networks (e.g. Araque & Navarrete, 2010).

These close functional interactions between glia and neurons are not completely unexpected given their intimate ontogenic relationship, despite the ancient and frequent postulation of classical neuroscience from the last millennium that neurons and glial cells were derived from different groups of precursor cells and that no new neurons were produced once development was completed. Henceforward, we used the term "precursor cell" to generally refer to a cell that precedes others in time and space and that gives rise to a variant, specialized or more mature form of cell in a cell lineage, instead of the term "progenitor cell" which vaguely indicates a biologically related ancestor.

Radial glia are now known to be the main source of neurons in several regions of the CNS, notably in the cerebral cortex (Malatesta et al, 2008), and they share several features with neuroepithelial cells, but also with astrocytes in the mature brain (Mori et al., 2005). However, they can be distinguished from neuroepithelial precursors by the expression of astroglial markers (Malatesta et al, 2008). During the last decade, the role of radial glia has been radically revaluated, and they are no longer considered to be a mere structural component serving to guide newborn neurons towards their final destinations.

Radial glial cells can be defined by morphological, cell biological and molecular criteria as true glial cells, like astroglia, they appear during neural development as a precursor cell intermediate between immature neuroepithelial cells and differentiating progeny, and they subsequently generate different cell lineages. In addition they persist into adulthood in various vertebrate classes ranging from fish to birds, while neurogenic glial cells become restricted to a few small regions of the adult forebrain in mice and humans (Pinto & Gôtz, 2007). This leads us to consider the molecular mechanisms involved in the regulation of the heterogeneity of the radial glial cells lineage, including their diversity in distinct regions of the CNS, their identity, heterogeneity and their functions.

Surprisingly, when the primary precursor cells (stem cells) of new neurons in neurogenic regions were identified, they exhibited the structural and biological markers of astrocytes (Ihrie & Alvarez-Buylla, 2008), and subsequently became for the focus of research related to the mechanisms of control of the proliferation and differentiation of these precursor cells.

One of the key advances in the field of neurobiology was the discovery that astroglial cells can generate neurons not only during development, but also throughout adult life and potentially even after brain lesion. Interestingly, only some astrocytes maintain their neurogenic potential and continue to generate neurons throughout life (Mori et al., 2005). At late-embryonic and postnatal stages, radial glial cells give rise to the neural stem cells responsible for adult neurogenesis. Embryonic pluripotent radial glia and adult neural stem cells may be clonally linked, thus representing a lineage displaying stem cell features in both the developing and mature CNS (Malatesta et al, 2008).

Adult neural precursor cells appear to be strongly influenced by their local microenvironment, while also contributing significantly to the architecture of germinal zones. However, environment alone does not seem to be sufficient to induce non-germinal astrocytes to behave as neural precursor cells. Although emerging evidence suggests that there is significant heterogeneity within populations of germinal zone astrocytes, the way that these differences are encoded remains unclear (Ihrie & Alvarez-Buylla, 2008).

Glial function is fundamental to the CNS, however the *in vitro* differentiation of embryonic stem cells into glia has received relatively limited attention when compared with the interest in the generation of neurons. This was the case until the last decade, when the study of astrocytes, oligodendrocytes and other macroglial cells lineages became a very interesting and productive area of research. The further characterization of glial cells should eventually provide a body of knowledge central to the understanding of cell differentiation during brain development and also of brain response during disease.

2. Aldynoglia: not a particular cell lineage but a glial cell phenotype

The former view that glial cells can be classified based on the expression of various biochemical markers has been overturned by the recognition that different cell types in the brain have common precursors (Alvarez-Buylla et al., 2001) that can be phenotypically interconvertible.

All the neurons and glial cells of the CNS are generated from the neuroepithelial cells in the walls of the embryonic neural tube, the "embryonic neural stem cells". The stem cells seem to be equivalent to the radial glial cells, which for many years were regarded as a specialized type of glial cell (Kessaris et al, 2008).

Glial cells are derived from precursor cells that mature through specific stages of development to generate fully differentiated astrocytes and oligodendrocytes as well as other macroglial cell types. Several types of intermediate precursors have been described and in some cases lineage relationships have been identified, although these remain controversial.

Glial precursor cells (GPCs) comprise the most abundant population of precursor cells in the adult human brain. They are responsible for CNS remyelination, and may contribute to the astrogliotic response to brain injury and degeneration. Adult human GPCs are biased to differentiate as oligodendrocytes and elaborate new myelin, and yet they retain multilineage plasticity, and can give rise to neurons as well as astrocytes and oligodendrocytes once removed from the adult parenchymal environment. GPCs retain the capacity for cell-autonomous self-renewal, and yet both their phenotype and fate may be dictated by their microenvironment (Sim et al, 2009).

Various other glial precursor cells have been partially described, probably all deriving from earlier appearing precursor cells but segregating at different stages in development. These include; motoneuron-oligodendrocyte precursors (MNOPs), white matter precursor cells (WMPCs), polydendrocytes, glial restricted precursors (GRPs), astrocyte precursor cells (APCs), and oligodendroblasts (e.g. Liu & Rao, 2004). Some of these precursors persist in the adult, and there are also intermediate glial precursors, rather than stem cells, that respond after injury and participate in the repairing process. Nevertheless, it remains unclear which specific glial precursor responds under different situations, and therefore new consensus sets of markers need to be identified, in order to improve our ability to define more clearly the different stages of glial maturation as well as the different glial phenotypes.

Besides the range of glia in the CNS, aldynoglia have recently emerged as a novel concept (Gudiño Cabrera & Nieto-Sampedro, 2000; Rojas-Mayorquín et al., 2008; 2010) and are beginning to be recognized as a specific glial phenotype (Panicker and Rao, 2001; Boyd et al., 2003; Liu and Rao, 2004; Rojas-Mayorquín et al., 2008; 2010) that includes and describes a group of macroglial cells that could constitute a separate functional group of CNS macroglia. These possess the ability to myelinate dorsal root ganglia neurites *in vitro* and provide the scaffold to guide neurons to their ultimate destinations (Gudiño Cabrera & Nieto-Sampedro, 1996; 2000). Functionally, aldynoglia differ from astrocytes and oligodendrocytes, but maintain the expression of some molecular markers that are common to both, and also to Schwann cells, which can be considerered as peripheral aldynoglia (Table 1). The prototype of this central glial cell is the olfactory bulb ensheathing cell (OBEC) (Gudiño-Cabrera & Nieto-Sampedro, 1999).

Gene name	OBECs	OB	SCs	ACs	OLs
GFAP	+	+	+	+	-
S100a6	+	+	+	+	-
N-myc	+	+	+	+	-
Adcy5	+	+	+	-	-
p75 (NGFR)	+	+	+	-	-
Cdkn2b	+	+	-	+	+
Lyn	+	+	-	-	+
Mtmr2	+	-	+	-	-
F2	+	-	-	+	+
Abcc1	+	-	-	+	-
Apbb1	+	-	-	+	-
Eotaxin	+	-	-	-	+

Table 1. Genes expressed in common among glial cells. OBECs - olfactory bulb ensheathing cells in culture, OB – olfactory bulb, SCs – Schwann cells, ACs – astrocytes, OLs – oligodendrocytes, (+) reported as expressed, (-) not yet reported or expression not yet verified. All (+) expression from OBECs was reported in Rojas-Mayorquín et al, 2008, and all other expression was from bona fide literature.

On the basis of an analysis of the expression profile by microarrays (Rojas-Mayorquín et al, 2008), we propose that OBECs are more closely related to Schwann cells, and that astrocytes are also more closely related to OBECs than to oligodendrocytes, although there are some similarities between OBECs and oligodendrocytes that are consistent with their respective phenotype as well as behavioural characteristics.

It should be emphasized that the concept of the aldynoglia refers to a glial cell phenotype and not a particular cell lineage (Rojas-Mayorquín et al., 2008; 2010).

The lineages of both astrocytes and oligodendrocytes have received more attention in the last decade, because the source of these cells in the mature CNS is relevant to the study of the cellular response to CNS injury. Some authors have argued that there is a common glial precursor cell from which both differentiate and that the microenvironment surrounding the injury determines the fate of the stimulated precursor cell (Rao & Mayer-Proschel, 1997; Rao et al., 1998). However, the precise origin of these glial cells is still not completely understood, though it appears that they derive from multiple regions of the CNS rather than from a single location (Kessaris et al., 2006; Richardson et al., 2006). A significant amount of evidence suggests that resident precursor cells proliferate and differentiate into mature glial cells that facilitate tissue repair and recovery. Additionally, the re-entry of mature astrocytes into the cell cycle can also contribute to the pool of new astrocytes that are observed following CNS injury. Better understanding of the origin of new glial cells in the injured CNS will facilitate the development of therapeutics targeted at altering the glial response in a beneficial way (e.g. Carmen et al, 2007).

Among the intermediate glial precursor cells, NG2 cells, first described more than two decades ago, are glial cells that specifically express the NG2 chondroitin sulphate proteoglycan (CSPG) and platelet-derived growth factor receptor alpha. Also known as synantocytes or polydendrocytes, to reflect their multi-processed morphology and lineal relationship to oligodendrocytes, they constitute a population of CNS cells distinct from neurons, mature oligodendrocytes, astrocytes, other macroglia and microglia.

Because they differentiate into oligodendrocytes *in vitro*, NG2-expressing glia were considered to be oligodendrocyte precursor cells (OPC; e.g. Nishiyama et al, 2009). Moreover, these cells persist in the adult CNS to generate oligodendrocytes throughout life (Hermann et al, 2010). However, they are widespread in the CNS, and in the adult human cerebral cortex and white matter they represent 10–15% of non-neuronal cells (Butt et al, 2005). Labelling techniques have shown them to have more than just precursor functions. A large proportion of NG2 cells do not appear to divide or generate oligodendrocytes, instead they form interactive domains with astrocytes and neurons (Wigley & Butt, 2009).

The morphology and distribution of NG2 glia are similar to, but distinct from, both microglia and astrocytes. The antigenic profile and morphology of NG2 glia in human tissues are consistent with an OPC function and this been well established in rodent models (Staugaitis & Trapp, 2009). NG2 proteoglycan expression could label newly generated cells or be inherited by resident cell populations that produce oligodendrocytes for remyelination, astrocytes that provide trophic support and other cells that contribute to CNS function (Sellers & Horner, 2005).

During embryonic development, multipotential neural precursor cells (MNP) possess the ability to self-renew and to generate the major CNS cell types: neurons, astrocytes and oligodendrocytes. However, the molecular mechanisms that control MNP fate specification are not yet fully understood. Recent studies have provided evidence that soluble protein mediators such as cytokines and transcriptional factors play critical roles in cell fate determination.

Furthermore, it has become apparent that epigenetic gene regulation plays an important intracellular role as cell-intrinsic programs in the specification of cell lineages (Abematsu et al, 2006). This process appears to gradually reduce the neurogenic potential of the astrocytic progeny, through mechanisms involving the epigenetic silencing of neurogenic fate determinants (Hirabayashi et al., 2009). The epigenetic "programming" of precursor cells into oligodendrocytes, for example, moves through three sequential stages of lineage progression (Liu & Casaccia, 2010), first from pluripotent precursor cells to MNP, from this to the oligodendrocyte precursor, and finally to differentiation into myelin-forming cells.

Consequently, it is possible to depict a model that combines genetic and epigenetic clues to define the differentiation pathway in such a way that the intermediate stages transiently reveal certain phenotypes, allowing parallels to be drawn with a river crossing, where the process of differentiation itself is the river and the cells that undergo differentiation need to cross from one riverbank to the other. Hence, we can consider that during embryonic development differentiation of neural cells initially starts from pluripotent precursor cells (ectodermal stem cells) to MNP (both located at one side of the river).

As shown in Figure 1, the first step in neuroglial differentiation in the neuroepithelium is characterized by the repression of pluripotency genes and restriction of the lineage potential to the neural fate, which is equivalent to starting to cross the river. The next step involves the generation, sequentially or in parallel, of a large amount of intermediate precursor cells from MNP with a neural fate. This step is associated with the progressive loss of plasticity and the expression/repression of neuronal- or glial-specific genes, for astrocytes, oligodendrocytes, polydendrocytes, aldynoglial and other phenotypes accordingly.

These intermediate precursors are crossing the river of differentiation; they are at different stages, and most of them (if not all) are interconvertible depending on the environmental signals that they receive, which indicates location and timing. And finally, differentiation into a particular phenotype comes when the cells cross the river and arrive at the other side, defined by the expression/repression of a particular set of genes that confers the diverse phenotypes (Figure 1).

Therefore, it seem feasible that CNS macroglial cells such as astrocytes and olygodendrocytes, or even polydendrocytes (NG2+), aldynoglia (p75+,GFAP+) and others, could be identified at one time as functionally segregated cell populations, but that during development or under certain circumstances, such as response to injury or neurodegenerative diseases, their precursors could constitute interconvertible cell populations.

3. Microarrays and proteomics as tools for the study of the mechanisms of differentiation

The mechanisms of differentiation are very complex, and thus microarrays and subtractive cDNA libraries, as well as proteomics as experimental approaches for their initial analysis, provide powerful techniques for studying differential gene expression.

It is therefore obvious at this point that new fate-mapping tools must be developed to allow us to unambiguously distinguish between different multipotential, pluripotential or even more restricted intermediate neural precursor cell populations in the adult cerebral cortex and neurogenic zones. With this in mind, one needs first to identify genes that are selectively expressed in one population, but not in others. To accomplish this it is quite important to use an experimental model in which the participation of different proteins can be reproducibly tested.

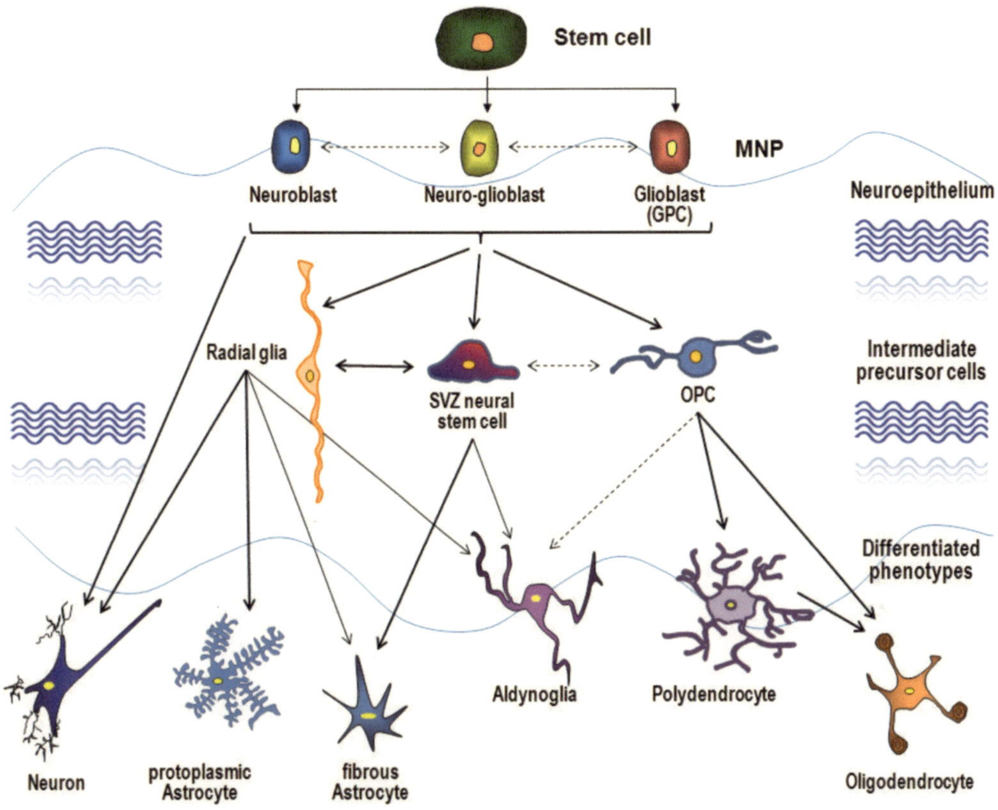

Fig. 1. Crossing the river of differentiation. The strength of a lineage relationship is indicated by bold (generally accepted), solid (recognized by some) or dashed arrows (hypothetical). MNP - multipotential neural precursor; GPC - glial precursor cell; OPC - oligodendroglial precursor cell; SVZ - subventricular zone.

As a first step, microarrays provide a powerful technique with which to study differential gene expression as they permit the expression of thousands of genes to be assessed in parallel in a single experiment (Figure 2). This approach has identified genes with potential relevance to cell differentiation in several cases (Roupioz et al., 2005), and in particular for MNP (Sousa et al., 2007).

Since cell differentiation is a complex process that can be regulated by intrinsic signals as well as by the extrinsic environment, a detailed analysis of microarray results (Rojas-Mayorquín et al., 2008) has enabled us to generate an expression profile for cells before and after they are subjected to different conditions or treatments. Along with other integral approaches, such as subtractive cDNA libraries (Rojas-Mayorquín et al., 2010) and proteomics, microarrays will help to establish a more comprehensive model of the molecular mechanisms involved in MNP differentiation that are directed by OBEC-conditioned media.

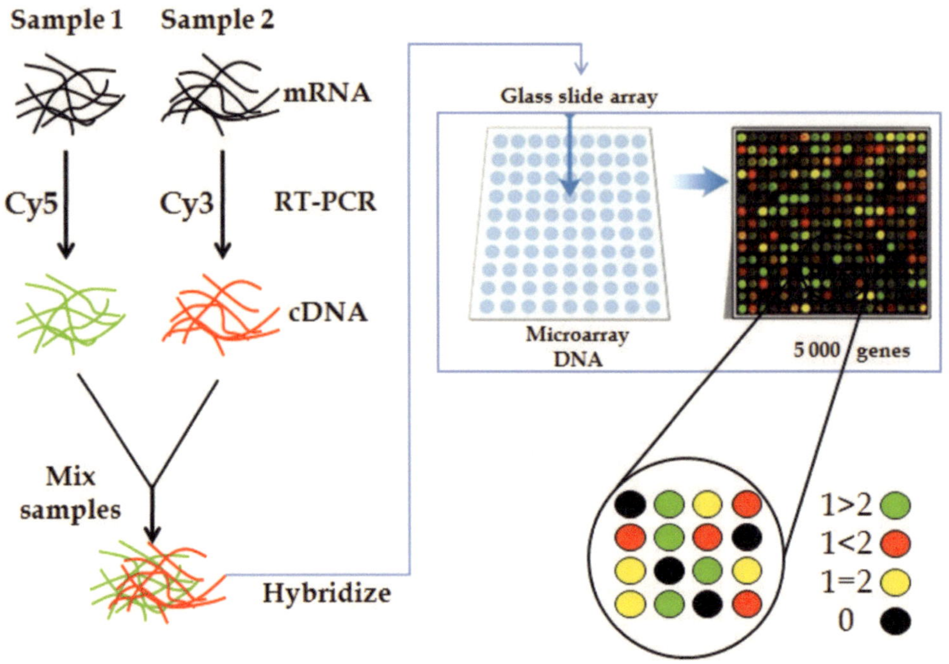

Fig. 2. Microarray procedure. This is used to compare the gene expression profile of thousand of genes from two different cell populations simultaneously.

4. OBEC-conditioned medium is a suitable model for inducing and directing stem cell differentiation *in vitro* towards an aldynoglia phenotype

Initially, OBEC-conditioned medium was used to elicit neuritogenesis *in vitro* via the co-cultures of neurons and EC (Kafitz & Greer, 1999; Sonigra et al., 1999; Wang et al., 2003), and more recently, we and other groups have been studying the capacity of the OBEC-conditioned medium to induce and direct *in vitro* differentiation from neural stem cells (Rojas-Mayorquín et al., 2008; 2010; Zhang et al., 2008; Doncel-Pérez et al., 2009) or even from other stem cells (Wang et al., 2007).

Given that OBECs are able to produce, secrete and also respond to a variety of growth factors (Table 2) and because the OB displays constant neuroglial replacement (Altman, 1969; Graziadei and Monti Graziadei, 1978), it is therefore reasonable to propose that the OBEC could induce its own differentiation from stem cell precursors located in the OB *in vivo*, or even that of other stem cells *in vitro* (Rojas-Mayorquín et al., 2008; 2010).

To analyse the differentiation process we employed a cell culture model that can be applied to study the possible molecular mechanisms that direct embryonic stem cell differentiation through aldynoglial cells. It also provides new insight into the molecular events and biological features that occur during the transition from MNP toward an aldynoglia phenotype, particularly when induced by EC-conditioned medium. In this model, we first obtained primary cultures of EC from OB, and then performed an immunopurification of ECs that express the p75 receptor, and finally obtained from these ECs the conditioned medium to be used to induce the differentiation of a heterogenic embryonic stem cell population from the striatum (figure 3).

Growth factor	Receptor	*In vivo* expression	*In vitro* expression	*In vitro* response
BDNF			(4) (19) (21) (28) (30)	
BMP				
BMP-4		(20)		
CNTF			(19) (29)	
	CNTFR-a		(29)	
Estrogen	**ER-alpha**		(14)	
FGF1		(7) (17)		
FGF2		(7)	(22)	(2) (32)
	FGFr1	(16)		
GDNF			(19) (22) (30)	
	GFRalpha-1		(30)	
	GFRalpha-2		(30)	
HGF				(33)
IGF-1		(27)	(15)	
	IGF-1r			(32)
Neuregulin (EGF-like)		(5) (25)	(4) (26) (30)	(9) (24) (26)
	ErbB-2	(23)	(9)	
	ErbB-3	(5)	(9)	
	ErbB-4		(9)	
GGF2			(8)	(8)
Neurturin (NTN)			(30)	
NGF			(4)(6)(19)(22)(28)(30)	
	NGFR		(30)	
NT-3			(4) (28)	(3)
NT-4/5	**Trk B**		(30)	
	p75 (Ngfr)	(11) (12)	(10) (13)	
PDGF		(18)		(32)
VEGF			(1)	

Table 2. Growth factors and their receptors expressed by OBEC both *in vivo* and *in vitro* and their response *in vitro*.

(1) Au & Roskams, 2003; (2) Barraud et al., 2007; (3) Bianco et al., 2004; (4) Boruch et al., 2001; (5) Bovetti et al., 2006; (6) Cao 2007; (7) Chuah & Teague 1999; (8) Chuah et al., 2000; (9) De Mello et al., 2007; (10) Franceschini & Barnett, 1996; (11) Gómez-Pinilla et al., 1987; (12) Gong et al., 1994; (13) Gudiño-Cabrera & Nieto-Sampedro, 1996; (14) Gudiño-Cabrera & Nieto-Sampedro, 1999; (15) Gudiño-Cabrera & Nieto-Sampedro, 2000; (16) Hsu et al., 2001; (17) Key et al., 1996; (18) Kott et al., 1994; (19) Lipson et al., 2003; (20) Liu et al., 2010; (21) Pastrana et al., 2007; (22) Pellitteri et al., 2007; (23) Perroteau et al., 1998; (24) Pollock et al.,1999; (25) Salehi-Ashtiani & Farbman, 1996; (26) Thompson et al., 2000; (27) Vicario-Abejón et al., 2003; (28) Vincent et al., 2003; (29) Wewetzer et al., 2001; (30) Woodhall et al., 2001; (31) Woodhall et al., 2003; (32) Yan et al., 2001a; (33) Yan et al., 2001b.

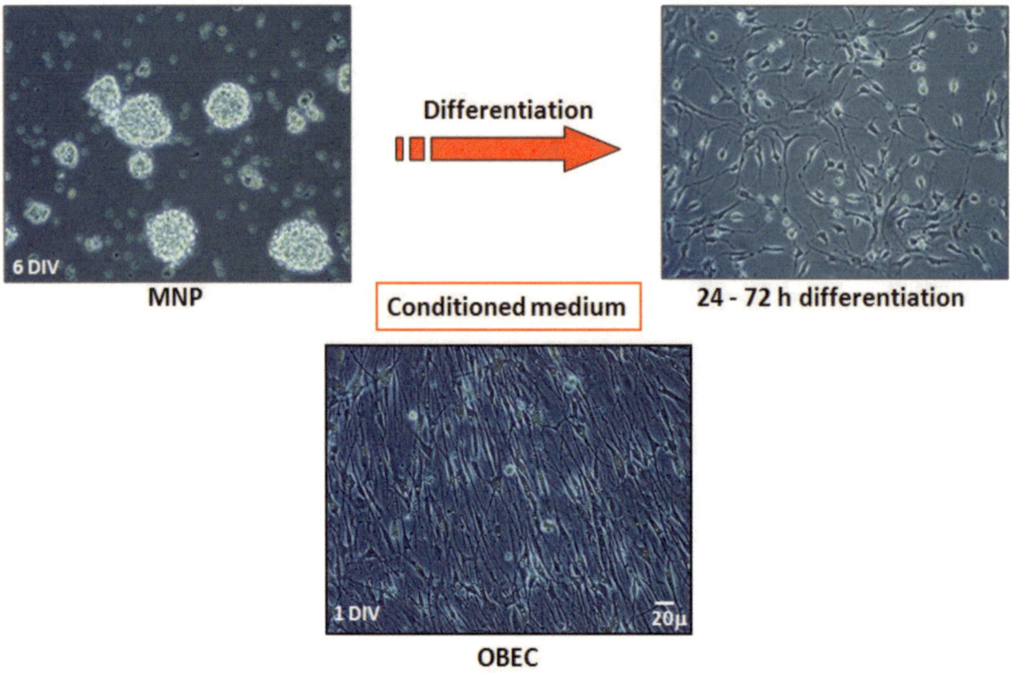

Fig. 3. Cell culture model of differentiation. First, a proliferating and undifferentiated embryonic neural stem cell culture of MNP was maintained. Then, neurospheres were cultured in conditioned medium from an immunopurified EC culture for 24 to 72 hours to achieve almost complete differentiation to an aldynoglia phenotype. MNP–multipotential neural precursors, OBECs–olfactory bulb ensheathing cells, DIV–days *in vitro*. Bar 20 μm.

5. Differentiation of the aldynoglia phenotype can be achieved by blocking Wnt signalling and activating the BMP pathway, with the involvement of IGF-1

Glial cells were long considered to be supportive cells with a different origin from neurons. New studies have shown that some glial cells function as primary precursors or MNP, deviating from the classical view that glia and neurons ontogenically develop from early separate lineages and demonstrating that during development and in the adult brain, many neurons and glial cells are not the direct progeny of neural stem cells (NSC), but instead originate from transit-amplifying, or intermediate precursor cells (IPC) (e.g. Kriegstein & Alvarez-Buylla, 2009). Thus, a clearer identification of NSC and IPC is critical to understand brain development and adult neurogenesis and to develop new strategies for brain repair.
Recent findings have revealed part of the underlying mechanistic basis for this preferential differentiation into astroglia. The more oxidized state of pathological brain tissue leads to upregulation of the protein deacetylase sirtuin 1 (Sirt1). Sirt1 appears to stabilize a co-repressor complex of Hairy/enhancer of split (Hes)1, thereby suppressing expression of the proneural transcription factor Mash1, and directs precursor cell differentiation towards the glial lineage. Sirt1 upregulated by CNS inflammation may also inhibit neuronal differentiation. Myelin-associated inhibitors such as Nogo, acting through the Nogo-66 receptor (NgR), also appear to promote neural stem/precursor cell differentiation into

astrocytes. Understanding the molecular basis of the glial lineage restriction of neural precursors in the injured or diseased CNS would provide clues to improving the success of stem cell-based transplantation therapy (Teng et al, 2009).

The specification of cells into the oligodendrocyte lineage is largely the result of interplay between the bone morphogenetic protein, sonic hedgehog and Notch signalling pathways, which regulate the expression of transcription factors dictating the spatial and temporal aspects of oligodendrogenesis. Many of these transcription factors and others then direct oligodendrocyte development through to a mature myelinating oligodendrocyte both in the spinal cord and brain (Nicolay et al, 2007).

Based on our results and those reported in the literature, we propose a gene expression profile that accompanies the induced differentiation of MNP into the aldynoglia phenotype. Under the conditions tested, OBEC-conditioned medium could induce MNP differentiation by inactivating Wnt signalling and promoting BMP signalling through the BMP1rA receptor with the possible involvement of IGF-1 (Rojas-Mayorquín et al., 2008; 2010).

IGF-1 is produced by the OBECs (Gudiño-Cabrera & Nieto-Sampedro, 2000) and it regulates the continuous generation of OBECs from local precursor cells within the OB (Vicario-Abejón et al., 2003). IGF-1 also induces Igfbp-5 expression, which is associated with the ECM (Jones et al., 1993), thereby promoting IGF activity (Cheng & Feldman, 1997; Duan et al., 1999) and resulting in positive regulation. In fact, Igfbp-5 was strongly expressed by immunopurified OBEC *in vitro*, and its expression was induced by conditioned medium following MNP differentiation (Rojas-Mayorquín et al., 2010). Hence, it seems probable that IGF-1 participates in the initial induction of MNP differentiation by OBEC-conditioned medium.

Moreover, it has been reported that IGF-1 also induces the expression of Tn-C (Kenney et al., 2003), which is also strongly expressed in the OBEC *in vitro*, and is induced in differentiated MNP in OBEC-conditioned medium (Rojas-Mayorquín et al., 2010). Tn-C could in turn be involved in the inactivation of the Wnt signalling pathway (Kakinuma et al., 2004) that we suggested could be inactivated during induced MNP differentiation (Rojas-Mayorquín et al., 2008). Additionally, the expression of BMPr1A is also increased (Rojas-Mayorquín et al., 2008), this being a BMP receptor that preferentially binds BMP-2 and BMP-4 (Chen et al., 2004). It has also been reported that BMP-2 increases IGF-1 binding to growth plate chondrocytes, suggesting that the BMPs may modulate the action of IGF-I via the type 1 IGF receptor and/or IGF binding proteins (Takahashi et al., 2007).

Given that cell differentiation is a complex process that can be regulated by intrinsic signals as well as by the extrinsic environment, the data obtained from integral approaches such as microarray analysis (Rojas-Mayorquín et al., 2008), subtractive cDNA libraries (Rojas-Mayorquín et al., 2010) and proteomics, will help to establish a more comprehensive model of the molecular mechanisms involved in MNP differentiation directed by OBEC-conditioned medium.

6. Perspectives and future use of aldynoglia cell transplants as a complementary therapy for neurodegenerative disorders

In the CNS of mammals, axonal regeneration is limited by two main factors: first, the low intrinsic regenerative potential of adult CNS neurons, and second the inhibitory influences of the glial and extracellular environment. Myelin-associated inhibitors of neurite growth as well as some properties of the so called "reactive astrocytes" contribute to the non-

permissiveness of CNS tissue for axonal growth. In contrast, the peripheral nervous system environment is supportive of regeneration because Schwann cells provide suitable substrates for regrowing axons (Hirsch & Bähr, 1999; Lavdas et al, 2008).
Knowledge of the mechanisms that direct stem cell differentiation to diverse cell phenotypes raises the possibility of generating *in vitro* a large population of cells of specific phenotypic characteristics that can be used to promote regeneration after a CNS injury by means of autologous transplantation from previously differentiated precursor cells (Franklin, 2003), in our case this particular phenotype is represented by the aldynoglia, which can be obtained by inducing MNP cells to differentiate using OBEC-conditioned medium. It would also be feasible to attempt to generate new cells endogenously by activating specific signalling pathways that elicit the production of regeneration-promoting glial cells (Kulbatski et al., 2008) of the aldynoglia phenotype.
There are two experimental strategies for developing cell-based therapies for neurodegenerative diseases: replacing degenerated cells (neurons and glia) via transplantation of cells that have been expanded *in vitro* and subsequently specified into the desired cell type, or recruiting endogenous cells for brain repair, either from endogenous adult neural stem cells that reside in neurogenic zones such as the adult subependymal zone lining the lateral ventricle (Saghatelyan et al., 2004), or from stem cell-like potential cells that are present to a varying degree throughout the brain (Nakatomi et al, 2002; Nunes et al, 2003).
The capacity to instruct endogenous precursor cells to generate neurons takes advantage of the important pool of endogenous and consequently autologous cells, thereby circumventing many of the problems associated with therapeutic strategies based on transplantation. Moreover, it has been shown that astroglia can be reprogrammed *in vitro* by forced expression of neurogenic transcription factors to transgress their lineage restriction and stably acquire a neuronal identity (e.g. Berninger, 2010).
In conclusion, it is clear that to treat complex pathologies in neurodegenerative diseases such as Alzheimer's, Parkinson's disease or multiple sclerosis, among others, it is not sufficient to use just one approach, and that it would be more helpful and convenient to combine therapies, making it possible to take care of patients with neurodegenerative diseases with a combination of neuronal replacement, either from endogenous or exogenous precursor cells, and the use of regenerative glia such as aldynoglia, in transplants that promote functional restoration.

7. References

Abematsu, M.; Smith, I. & Nakashima, K. (2006). Mechanisms of neural stem cell fate determination: extracellular cues and intracellular programs. *Curr Stem Cell Res Ther*,1.,2.,(may 2006)(267-277),

Altman, J. (1969). Autoradiographic and histological studies of postnatal neurogenesis. IV. Cell proliferation and migration in the anterior forebrain, with special reference to persisting neurogenesis in the olfactory bulb. *J. Comp. Neurol.*,137.,4.,(1969)(433-457),

Alvarez-Buylla, A.; García-Verdugo, J.M. & Tramontin, A.D. (2001). A unified hypothesis on the lineage of neural stem cells. *Nat Rev Neurosci.*,2.,4.,(apr 2001)(287-293),

Alvarez-Maubecin, V.; Garcia-Hernandez, F.; Williams, J.T. & Van Bockstaele, E.J. (2000). Functional coupling between neurons and glia. *J Neurosci.*,20.,11.,(jun 2000)(4091-4098.

Araque, A. & Navarrete, M. (2010). Glial cells in neuronal network function. *Philos Trans R Soc Lond B Biol Sci.*, 365.,1551.,(aug 2010)(2375-2381),

Au, E. & Roskams A.J. (2003). Olfactory ensheathing cells of the lamina propria in vivo and in vitro. *Glia*,41.,3.,(feb 2003)(224-236),

Barraud, P.; He, X.; Zhao, C.; Ibanez, C.; Raha-Chowdhury, R.; Caldwell, M.A. & Franklin, R.J. (2007). Contrasting effects of basic fibroblast growth factor and epidermal growth factor on mouse neonatal olfactory mucosa cells. *Eur J Neurosci*,26.,12.,(dec 2007)(3345-3357),

Berninger, B. (2010). Making neurons from mature glia: a far-fetched dream? *Neuropharmacology*,58.,6.,(may 2010)(894-902),

Bianco, J.I.; Perry, C.; Harkin, D.G.; Mackay-Sim, A. & Féron, F. (2004). Neurotrophin 3 promotes purification and proliferation of olfactory ensheathing cells from human nose. *Gli*,45.,2.,(jan 2004)(111-123),

Blunt, M. J.; Wendell-Smith, C. P. & Baldwin, F. (1965). Glia-nerve fibre relationships in mammalian optic nerve. *J. Anat.*,99.,(1965)(1-11),

Boruch, A.V.; Conners, J.J.; Pipitone, M.; Deadwyler, G.; Storer, P.D.; Devries, G.H. & Jones, K.J. (2001). Neurotrophic and migratory properties of an olfactory ensheathing cell line. *Glia*,33.,3.,(mar 2001)(225-229),

Bovetti, S.; De Marchis, S.; Gambarotta, G.; Fasolo, A.; Perroteau, I.; Puche, A.C. & Bovolin, P. (2006). Differential expression of neuregulins and their receptors in the olfactory bulb layers of the developing mouse. *Brain Res*,1077.,1.,(mar 2006)(37-47),

Boyd, J.G.; Skihar, V.; Kawaja, M. & Doucette, R. (2003). Olfactory ensheathing cells: historical perspective and therapeutic potential. *Anat. Rec. B New Anat.*,271.,1.,(2003)(49–60),

Butt, A.M.; Hamilton, N.; Hubbard, P.; Pugh, M. & Ibrahim, M. (2005). Synantocytes: the fifth element. *J. Anat.*,207.,6.,(dec 2005)(695-706),

Cao, L.; Zhu, Y.L.; Su, Z.; Lv, B.; Huang, Z.; Mu, L. & He, C. (2007). Olfactory ensheathing cells promote migration of Schwann cells by secreted nerve growth factor. *Glia*,55.,9.,(jul 2007)(897-904),

Carmen, J.; Magnus, T.; Cassiani-Ingoni, R.; Sherman, L.; Rao, M.S. & Mattson, M.P. (2007). Revisiting the astrocyte-oligodendrocyte relationship in the adult CNS. *Prog Neurobiol.*,82.,3.,(jun 2007)(151-162),

Chen, D.; Zhao, M. & Mundy, G.R. (2004). Bone morphogenetic proteins. *Growth Factors*, 22.,4.,(2004)(233–241),

Cheng, H.L. & Feldman, E.L. (1997). Insulin-like growth factor-I (IGF-I) and IGF binding protein-5 in Schwann cell differentiation. *J. Cell Physiol.*,171.,(1997)(161–167),

Chuah, M.I. & Teague, R. (1999). Basic fibroblast growth factor in the primary olfactory pathway: mitogenic effect on ensheathing cells. *Neuroscience*,88.,4.,(1999)(1043-1050),

Chuah, M.I.; Cossins, J.; Woodhall, E.; Tennent, R.; Nash, G. & West, A.K. (2000). Glial growth factor 2 induces proliferation and structural changes in ensheathing cells. *Brain Res*,857.,1-2.,(feb 2000)(265-274),

De Mello, T.R.; Busfield, S.; Dunlop, S.A. & Plant, G.W. (2007). Culture conditions affect proliferative responsiveness of olfactory ensheathing glia to neuregulins. *Glia*,55.,7.,(may 2007)(734-745),

Doncel-Pérez, E.; Caballero-Chacón, S. & Nieto-Sampedro, M. (2009). Neurosphere cell differentiation to aldynoglia promoted by olfactory ensheathing cell conditioned medium. *Glia*,57.,13.,(oct 2009)(1393-1409),

Duan, C.; Liimatta, M.B. & Bottum, O.L. (1999). Insulin-like growth factor (IGF)-I regulates IGF-binding protein-5 gene expression through the phosphatidylinositol 3-kinase, protein kinase B/Akt, and p70 S6 kinase signaling pathway. *J Biol Chem*,274.,52.,(1999)(37147–37153),

Edgar, J.M. & Nave, K.A. (2009). The role of CNS glia in preserving axon function. *Curr Opin Neurobiol.*,19.,5.,(2009)(498-504),

Fellin, T.; Pascual, O. & Haydon, P.G. (2006). Astrocytes coordinate synaptic networks: balanced excitation and inhibition. *Physiology (Bethesda)*,21.,(jun 2006)(208-215),

Franceschini, I.A. & Barnett, S.C. (1996) Low-affinity NGF-receptor and E-N-CAM expression define two types of olfactory nerve ensheathing cells that share a common lineage. *Dev Biol*,173.,1.,(jan 1996)(327-343),

Franklin, R.J. (2003). Remyelination by transplanted olfactory ensheathing cells. *Anat Rec B New Anat.*,271.,1.,(mar 2003)(71-76),

García-Marín, V.; García-López, P. & Freire, M. (2007). Cajal's contributions to glia research. *Trends Neurosci.*,30.,9.,(sep 2007)(479-487),

Gomez-Pinilla, F.; Cotman, C.W. & Nieto-Sampedro, M. (1987). NGF receptor immunoreactivity in rat brain: topographic distribution and response to entorhinal ablation. *Neurosci Lett*,82.,3.,(dec 1987)(260-266),

Gong, Q.; Bailey, M.S.; Pixley, S.K.; Ennis, M.; Liu, W. & Shipley, M.T. (1994). Localization and regulation of low affinity nerve growth factor receptor expression in the rat olfactory system during development and regeneration. *J Comp Neurol*,15.,(1994)(336-348),

Graziadei, P.P. & Monti Graziadei, G.A. (1978). The olfactory system: a model for the study of neurogenesis and axon regeneration in mammals. In: *Neuronal Plasticity*, Cotman, C.W. (Ed.), (131-153), Raven Press, ISBN 0890042101, New York (USA).

Gudiño-Cabrera, G. & Nieto-Sampedro, M. (1996). Ensheathing cells: Large scale purification from adult olfactory bulb, freeze-preservation and migration of trasplanted cells in adult brain. *Restor. Neurol. Neurosci.*,10.,(1996)(25–34),

Gudiño-Cabrera, G.& Nieto-Sampedro, M. (1999). Estrogen receptor immunoreactivity in schwann-like brain macroglia. *J. Neurobiol.*,40.,4.,(1999)(458–470),

Gudiño-Cabrera, G. & Nieto-Sampedro, M. (2000). Schwann-like macroglia in adult rat brain. *Glia*,30.,1.,(2000)(49–63),

Halassa, M.M.; Fellin, T. & Haydon, P.G. (2009). Tripartite synapses: roles for astrocytic purines in the control of synaptic physiology and behavior. *Neuropharmacology*,57.,4.,(sep 2009)(343-346),

Halassa, M.M. & Haydon, P.G. (2010). Integrated brain circuits: astrocytic networks modulate neuronal activity and behavior. *Annu Rev Physiol.*,17.,72.,(mar 2010)(335-355),

Hermann, A.; Brandt, M.D.; Loewenbrück, K.F. & Storch, A. (2010). "Silenced" polydendrocytes: a new cell type within the oligodendrocyte progenitor cell population? *Cell Tissue Res.*,340.,1.,(apr 2010)(45-50),

Hirsch, S. & Bähr, M. (1999). Growth promoting and inhibitory effects of glial cells in the mammalian nervous system. *Adv Exp Med Biol.*,468.,(1999)(199-205),

Hirabayashi, Y.; Suzki, N.; Tsuboi, M.; Endo, T.A.; Toyoda, T.; Shinga, J.; Koseki, H.; Vidal, M. & Gotoh, Y. (2009). Polycomb limits the neurogenic competence of neural precursor cells to promote astrogenic fate transition. *Neuron*, 63.,(600-613),

Hsu, P.; Yu, F.; Féron, F.; Pickles, J.O.; Sneesby, K. & Mackay-Sim, A. (2001). Basic fibroblast growth factor and fibroblast growth factor receptors in adult olfactory epithelium. *Brain Res*,896.,1-2.,(mar 2001)(188-197),

Hyden, H. (1962). The neuron and its glia-a biochemical and functional unit. *Endeavour*,21.,(1962)(144-155),

Ihrie, R.A. & Alvarez-Buylla, A. (2008). Cells in the astroglial lineage are neural stem cells. *Cell Tissue Res.*331.,1.,(jan 2008)(179-191),

Jones, J.I.; Gockerman, A.; Busby Jr., W.H.; Camacho-Hubner, C. & Clemmons, D.R. (1993). Extracellular matrix contains insulin-like growth factor binding protein-5: potentiation of the effects of IGF-I. *J. Cell Biol.*,121.,(1993)(679–687),

Kakinuma, Y.; Saito, F.; Osawa, S. & Miura, M. (2004). A mechanism of impaired mobility of oligodendrocyte progenitor cells by tenascin C through modification of Wnt signaling. *FEBS Lett*,568.,1–3.,(2004)(60–64),

Kafitz, K.W. & Greer, C.A. (1999). Olfactory ensheathing cells promote neurite extension from embryonic olfactory receptor cells in vitro. *Glia*,25.,2.,(jan 1999)(99-110),

Kenney, M.C.; Zorapapel, N.; Atilano, S.; Chwa, M.; Ljubimov, A. & Brown, D. (2003). Insulin-like growth factor-I (IGF-I) and transforming growth factor-beta (TGFbeta) modulate tenascin-C and fibrillin-1 in bullous keratopathy stromal cells in vitro. *Exp Eye Res*,77.,5.,(2003)(537–546),

Kessaris, N.; Fogarty, M.; Iannarelli, P.; Grist, M.; Wegner, M. & Richardson, W.D. (2006). Competing waves of oligodendrocytes in the forebrain and postnatal elimination of an embryonic lineage. *Nat. Neurosci.*,9.,(2006)(173–179),

Kessaris, N.; Pringle, N. & Richardson, W.D. (2008). Specification of CNS glia from neural stem cells in the embryonic neuroepithelium. *Philos Trans R Soc Lond B Biol Sci.*,363.,1489.,(jan 2008)(71-85),

Key, B.; Treloar, H.B.; Wangerek, L.; Ford, M.D. & Nurcombe, V. (1996). Expression and localization of FGF-1 in the developing rat olfactory system. *J Comp Neurol*,366.,2.,(mar 1996)(197-206),

Kott, J.N.; Westrum, L.E.; Raines, E.W.; Sasahara, M. & Ross, R. (1994). Olfactory ensheathing glia and platelet-derived growth factor B-chain reactivity in the transplanted rat olfactory bulb. *Int J Dev Neurosci*,12.,4.,(jun 1994)(315-323),

Kriegstein, A. & Alvarez-Buylla, A. (2009). The glial nature of embryonic and adult neural stem cells. *Annu Rev Neurosci.*,32.,(2009)149-184,

Kulbatski, I.; Mothe, A.J.; Parr, A.M.; Kim, H.; Kang, C.E.; Bozkurt, G. & Tator, C.H. (2008). Glial precursor cell transplantation therapy for neurotrauma and multiple sclerosis. *Prog Histochem Cytochem.*,43.,3.,(2008)(123-176),

Lavdas, A.A.; Papastefanaki, F.; Thomaidou, D. & Matsas, R. (2008). Schwann cell transplantation for CNS repair. *Curr Med Chem.*,15.,2.,(2008)(151-160),

Learte, A.R. & Hidalgo, A. (2007). The role of glial cells in axon guidance, fasciculation and targeting. *Adv Exp Med Biol.*,621.,(2007)(156-166),

Lipson, A.C.; Widenfalk, J.; Lindqvist, E.; Ebendal, T. & Olson, L. (2003). Neurotrophic properties of olfactory ensheathing glia. *Exp Neurol*,180.,2.,(apr 2003)(167-171),

Liu, J. & Casaccia, P. (2010). Epigenetic regulation of oligodendrocyte identity. *Trends Neurosci.*,33.,4.,(apr 2010)(193-201),

Liu, Q.; Ye, J.; Yu, H.; Li, H.; Dai, C.; Gu, Y.; Zhu, Y. & Zhang, Z. (2010). Survival-enhancing of spiral ganglion cells under influence of olfactory ensheathing cells by direct cellular contact. *Neurosci Lett*,478.,1.,(jun 2010)(37-41),

Liu, Y. & Rao, M.S. (2004). Glial progenitors in the CNS and possible lineage relationships among them. *Biol Cell.*,96.,4.,(may 2004)(279-290),

Malatesta, P.; Appolloni, I. & Calzolari, F. (2008). Radial glia and neural stem cells. *Cell Tissue Res.*,331.,1.,(jan 2008)(165-178),

Mori, T.; Buffo, A. & Götz, M. (2005). The novel roles of glial cells revisited: the contribution of radial glia and astrocytes to neurogenesis. *Curr Top Dev Biol.*,69.,(2005)(67-99),

Nakatomi, H.; Kuriu, T.; Okabe, S.; Yamamoto, S.; Hatano, O.; Kawahara, N.; Tamura, A.; Kirino, T. & Nakafuku, M. (2002). Regeneration of hippocampal pyramidal neurons after ischemic brain injury by recruitment of endogenous neural progenitors. *Cell*,110.,(429-441),

Nicolay, D.J.; Doucette, J.R. & Nazarali, A.J. (2007). Transcriptional control of oligodendrogenesis. *Glia.*,55.,13.,(oct 2007)(1287-1299),

Nishiyama, A.; Komitova, M.; Suzuki, R. & Zhu, X. (2009). Polydendrocytes (NG2 cells): multifunctional cells with lineage plasticity. *Nat Rev Neurosci.*,10.,1.,(jan 2009)(9-22),

Nunes, M.C.; Roy, N.S.; Keyoung, H.M.; Goodman, R.R.; McKhann 2nd, G.; Jiang, L.; Kang, J.; Nedergaard, M. & Goldman, S.A. (2003). Identification and isolation of multipotential neural progenitor cells from the subcortical white matter of the adult human brain. *Nat Med*,9.,(439-447),

Panicker, M. & Rao, M. (2001). Stem cells and neurogenesis. In: *Stem Cell Biology*, Marshak, D.R., Gardner, D.K., y Gottlieb, D. (Ed.), (399–438), Cold Spring Harbor Laboratory Press, ISBN 978-0879695750, Cold Spring Harbor, New York (USA).

Pastrana, E.; Moreno-Flores, M.T.; Avila, J.; Wandosell, F.; Minichiello, L. & Diaz-Nido, J. (2007). BDNF production by olfactory ensheathing cells contributes to axonal regeneration of cultured adult CNS neurons. *Neurochem Int*,50.,3.,(feb 2007)(491-498),

Pellitteri, R.; Spatuzza, M.; Russo, A. & Stanzani, S. (2007). Olfactory ensheathing cells exert a trophic effect on the hypothalamic neurons in vitro. *Neurosci Lett*,417.,1.,(apr 2007)(24-29),

Perroteau, I.; Oberto, M.; Ieraci, A.; Bovolin, P. & Fasolo, A. (1998). ErbB-3 and ErbB-4 expression in the mouse olfactory system. *Ann N Y Acad Sci*,855.,(1998)(255-259),

Pinto, L. & Götz, M. (2007). Radial glial cell heterogeneity--the source of diverse progeny in the CNS. *Prog Neurobiol.*,83.,1.,(sep 2007)(2-23),

Pollock, G.S.; Franceschini, I.A.; Graham, G.; Marchionni, M.A. & Barnett, S.C. (1999). Neuregulin is a mitogen and survival factor for olfactory bulb ensheathing cells and an isoform is produced by astrocyte. *Eur J Neurosci*,11.,(1999)(769-780),

Rao, M.S. & Mayer-Proschel, M. (1997). Glial-restricted precursors are derived from multipotent neuroepithelial stem cells. *Dev. Biol.*,188.,(1997)(48–63),

Rao, M.S.; Noble, M. & Mayer-Proschel, M. (1998). A tripotential glial precursor cell is present in the developing spinal cord. *Proc. Natl. Acad. Sci. U.S.A.*,95.,(1998)(3996-4001),

Richardson, W.D.; Kessaris, N. & Pringle, N. (2006). Oligodendrocyte wars. *Nat. Rev. Neurosci.*,7.,(2006)(11-18),

Rojas-Mayorquín, A.E.; Torres-Ruíz, N.M.; Ortuño-Sahagún, D. & Gudiño-Cabrera, G. (2008). Microarray analysis of striatal embryonic stem cells induced to differentiate by ensheathing cell conditioned media. *Dev Dyn.*,237.,4.,(apr 2008)(979-994),

Rojas-Mayorquín, A.E.; Torres-Ruíz, N.M.; Ortuño-Sahagún, D. & Gudiño-Cabrera, G. (2010). Subtractive hybridization identifies genes differentially expressed by olfactory ensheathing cells and neural stem cells. *Int J Dev Neurosci.*,28.,(feb 2010)(75-82),

Roupioz, Y.; Castel, D.; Pitaval, A.; Baghdoyan, S. & Gidrol, X. (2005). Cell microarrays and functional genomics. *Med Sci (Paris)*,21.,(2005)(535-538),

Salehi-Ashtiani, K. & Farbman, A.I. (1996). Expression of neu and Neu differentiation factor in the olfactory mucosa of rat. *Int J Dev Neurosci*,14.,7-8.,(nov 1996)(801-811),

Sellers, D.L. & Horner, P.J. (2005). Instructive niches: environmental instructions that confound NG2 proteoglycan expression and the fate-restriction of CNS progenitors. *J. Anat.*,207.,6.,(dec 2005)(727-734),

Saghatelyan, A.; de Chevigny, A.; Schachner, M. & Lledo, P.M. (2004). Tenascin-R mediates activity-dependent recruitment of neuroblasts in the adult mouse forebrain. *Nat. Neurosci.*,7.,(347-356),

Sim, F.J.; Windrem, M.S. & Goldman, S.A. (2009). Fate determination of adult human glial progenitor cells. *Neuron Glia Biol.*,5.,3-4.,(nov 2009)(45-55),

Simons, M. & Trajkovic, K. (2006). Neuron-glia communication in the control of oligodendrocyte function and myelin biogenesis. *J Cell Sci.*,1.,119 Pt 21.,(nov 2006)(4381-4389),

Sonigra, R.J.; Brighton, P.C.; Jacoby, J.; Hall, S. & Wigley, C.B. (1999). Adult rat olfactory nerve ensheathing cells are effective promoters of adult central nervous system neurite outgrowth in coculture. *Glia*,25.,3.,(feb 1999)(256-269),

Sousa, K.M.; Mira, H.; Hall, A.C.; Jansson-Sjostrand, L.; Kusakabe, M. & Arenas, E. (2007). Microarray analyses support a role for Nurr1 in resistance to oxidative stress and neuronal differentiation in neural stem cells. *Stem Cells*,25.,(2007)(511-519),

Staugaitis, S.M. & Trapp, B.D. (2009). NG2-positive glia in the human central nervous system. *Neuron Glia Biol.*,5.,3-4.,(nov 2009)(35-44),

Takahashi, T.; Morris, E.A. & Trippel, S.B. (2007). Bone morphogenetic protein-2 and -9 regulate the interaction of insulin-like growth factor-I with growth plate chondrocytes. *Int J Mol Med*,20.,1.,(2007)(53-57),

Teng, F.Y.; Hor, C.H. & Tang, B.L. (2009). Emerging cues mediating astroglia lineage restriction of progenitor cells in the injured/diseased adult CNS. *Differentiation*,77.,2.,(feb 2009)(121-127),

Thompson, R.J.; Roberts, B.; Alexander, C.L.; Williams, S.K. & Barnett, S.C. (2000). Comparison of neuregulin-1 expression in olfactory ensheathing cells, Schwann cells and astrocytes. *J Neurosci Res*,61.,2.,(jul 2000)(172-185),

Vicario-Abejón, C.; Yusta-Boyo, M.J.; Fernández-Moreno, C. & de Pablo, F. (2003). Locally born olfactory bulb stem cells proliferate in response to insulin-related factors and require endogenous insulin-like growth factor-I for differentiation into neurons and glia. *J Neurosci*,23.,3.,(feb 2003)(895-906),

Vijayaraghavan, S. (2009). Glial-neuronal interactions--implications for plasticity and drug addiction. *AAPS J*.11.,1.,(mar 2009)(123-132),

Vincent, A.J.; West, A.K. & Chuah, M.I. (2003). Morphological plasticity of olfactory ensheathing cells is regulated by cAMP and endothelin-1. *Glia*,41.,4.,(mar 2003)(393-403),

Wang, B.; Han, J.; Gao, Y.; Xiao, Z.; Chen, B.; Wang, X.; Zhao, W. & Dai, J. (2007). The differentiation of rat adipose-derived stem cells into OEC-like cells on collagen scaffolds by co-culturing with OECs. *Neurosci Lett.*,421.,3.,(jun 2007)(191-196),

Wang, Y.Z.; Meng, J.H.; Yang, H.; Luo, N.; Jiao, X.Y. & Ju, G. (2003). Differentiation-inducing and protective effects of adult rat olfactory ensheathing cell conditioned medium on PC12 cells. *Neurosci Lett.*,346.,1-2.,(jul 2003)(9-12),

Wewetzer, K.; Grothe, C. & Claus, P. (2001). In vitro expression and regulation of ciliary neurotrophic factor and its alpha receptor subunit in neonatal rat olfactory ensheathing cells. *Neurosci Lett*,306.,3.,(jun 2001)(165-168),

Wigley, R.; Hamilton, N.; Nishiyama, A.; Kirchhoff, F. & Butt, A.M. (2007). Morphological and physiological interactions of NG2-glia with astrocytes and neurons. *J. Anat.*,210.,6.,(jun 2007)(661-670),

Wigley, R. & Butt, A.M. (2009). Integration of NG2-glia (synantocytes) into the neuroglial network. *Neuron Glia Biol.*,5.,1-2.,(may 2009)(21-28),

Woodhall, E.; West, A.K. & Chuah, M.I. (2001). Cultured olfactory ensheathing cells express nerve growth factor, brain-derived neurotrophic factor, glia cell linederived neurotrophic factor and their receptors. *Brain Res. Mol. Brain Res.*, 88.,1-2.,(2001)(203-213),

Woodhall, E.; West, A.K.; Vickers, J.C. & Chuah, M.I. (2003). Olfactory ensheathing cell phenotype following implantation in the lesioned spinal cord. *Cell Mol Life Sci*,60.,10.,(oct 2003)(2241-2253),

Yan, H.; Bunge, M.B.; Wood, P.M. & Plant, G.W. (2001a). Mitogenic response of adult rat olfactory ensheathing glia to four growth factors. *Glia*,33.,4.,(mar 2001)(334-342),

Yan, H.; Nie, X. & Kocsis, J.D. (2001b). Hepatocyte growth factor is a mitogen for olfactory ensheathing cells. *J Neurosci Res*,66.,4.,(nov 2001)(698-704),

Zhang, J.; Wang, B.; Xiao, Z.; Zhao, Y.; Chen, B.; Han, J.; Gao, Y.; Ding, W.; Zhang, H. & Dai, J. (2008). Olfactory ensheathing cells promote proliferation and inhibit neuronal differentiation of neural progenitor cells through activation of Notch signaling. *Neuroscience*,153.,2.,(may 2008)(406-413),

20

Dopaminergic Neurons Derived from Human Embryonic Stem Cell Derived Neural Progenitors: Biological Relevance and Application

Young, A. and Stice, S.L.
University of Georgia
United States of America

1. Introduction

Dopaminergic neurons are studied at length for their role in Parkinson's disease (PD), schizophrenia and addiction (Iversen and Iversen, 2007). While these commonly known roles for dopamine involve a similar neural subtype, the brain areas and identifying genetic markers involved in each pathway differ. These differences lead to selective involvement of each pathway (nigrostriatal, mesolimbic and mesocortical), allowing for derivation of dopaminergic neurons from human embryonic stem cells (hESCs) as well as from human neural progenitors (hNPs) that can be used for drug development or for cell therapy in PD.
Ever since their isolation in 1998, hESCs have been touted as having potential for cell therapies, drug development assays and as a source for studying human development (Thomson et al., 1998). Due to PD effecting 1% of the American population over 65 as well as the specificity of the cell type affected, PD presents as a neurodegenerative disease with potential to be helped with hESCs (Weintraub et al., 2008a). In 2004, the first report of tyrosine hydroxylase (TH) positive neurons derived from hESCs demonstrated that obtaining dopaminergic neurons would be possible in humans (Perrier et al., 2004). The stromal cell-derived inducing activity (SDIA) method enhanced dopaminergic differentiation through co-culture with mouse derived stromal cells, which secreted factors that directed differentiation towards a dopaminergic phenotype. Following the SDIA method, a 5-stage method for deriving dopaminergic neurons from hESCs that did not require co-culture with contaminating feeder layers obtained fewer dopaminergic neurons from hESCs (Schulz et al., 2004). Since this work, there has been limited success in obtaining high levels of TH+ neurons without the addition of feeder layers.
A selective dopaminergic neuron neuroprotectant was discovered in 1993, glial cell-line derived neurotrophic factor (GDNF) (Lin et al., 1993). The potential for this neurotrophic factor to protect substantia nigra dopaminergic neurons was explored, and in rat models of PD, GDNF administration was effective in protecting those cells lost in PD as well as in protecting neural cells transplanted into lesioned rat midbrains (Kearns and Gash, 1995, Hou et al., 1996). Methods for administering GDNF into human patients have been developed and clinical trials utilizing GDNF as a protectant for dopaminergic neurons have

proceeded with unfavorable results due to localization of GDNF administration as well as invasiveness of the surgery required (Kordower et al., 2000, Maswood et al., 2002, Kordower et al., 2008, Su et al., 2009). Methods for GDNF administration other than lesions have not lead to successful results. However, GDNF has potential as a dopaminergic neuroprotection agent in the differentiation of dopaminergic neurons from hESCs or hNPs.

In this chapter, we intend to cover dopaminergic development in the mouse and human brain in order to understand more fully dopaminergic derivation from hESCs and hNPs. We also intend to examine the processes of dopaminergic derivation that have been used as well as the role the GDNF plays in this process. Finally, we intend to cover the potential applications for hNP derived dopaminergic neurons.

2. Parkinson's disease

2.1 Epidemiology

PD is a progressive neurodegenerative disease that is second in prevalence only to Alzheimer's disease (Weintraub et al., 2008a). While typically thought of as a disorder only affecting the elderly population, early onset PD (appearance of symptoms between 45 and 65) currently accounts for 10% of the diagnosed cases of Parkinson's (Rao et al., 2006).

There are two main subtypes of PD, idiopathic and secondary. Idiopathic forms of PD can be either sporadic (90% of cases) or genetic. Most often seen in young onset PD, the most common genetic mutation is in the PARK8 (LRRK2) gene and the second most common is in the PARK1 gene which encodes for the alpha synuclein protein (Obeso et al., 2010). Sporadic PD has no clear etiology but may be caused by environmental factors, toxins or aging. Secondary PD is caused by medications, infection or metabolic disorders (Poewe, 2006, Elbaz and Moisan, 2008).

Diagnosis usually begins with the presentation of motor symptoms which fall into four categories: 1) resting tremor, 2) bradykinesia, 3) rigidity and 4) postural instability (Weintraub et al., 2008a). Younger patients present with tremor as their primary symptom and older patients present with bradykinesia as their primary symptom (Poewe, 2006). The resting tremor appears unilaterally and moves bilateral as the disease progresses. Most often, the tremor is seen in the distal portion of the limbs in the hands or a shaking leg. Bradykinesia, the inability to initiate movement, leads to the shuffling gait associated with PD. Most often this is noticed in the slowness and difficulty a PD patient has when walking, but it can also lead to difficulty in turning in bed or rising from a chair (Poewe, 2006). Rigidity, stiffness of the muscles in the limbs and trunk, often leads to postural instability (Poewe, 2006). Postural instability, the inability to maintain balance and coordination, occurs in the most advanced stages of PD (Poewe, 2006). This affliction often leads to the falls that can lead to rapid decline in a person's quality of life. In addition to decreasing quality of life, postural instability has very little response to the current treatments for PD (Weintraub et al., 2008a).

In addition to the motor symptoms, PD patients are affected by non-motor symptoms. This is due to the large involvement of other neurons in the limbic area of the brain, the compensation for the loss of dopaminergic neurons by other neurons and the connections between the basal ganglia and the frontal cortex. The most prevalent non-motor symptom is depression beyond which would be expected for the average population affected by a debilitating disorder with between 20 and 45% of people with PD being diagnosed with depression post diagnosis of PD (Weintraub et al., 2008b). The second most common non-motor symptom is psychosis, most often manifesting in hallucinations (Weintraub et al., 2008b). Cognitive decline is seen as the disease progresses with memory loss, attention impairment and executive function deficits reported most often (Lim and Lang, 2010). The co-morbidity of these non-motor symptoms with the motor symptoms paints an image of PD as a whole body and mind disorder not just as a motor disorder (Weintraub et al., 2008b, Gaig and Tolosa, 2009).

The first effective treatment for PD and still the leading treatment is a dopamine precursor that crosses the blood brain barrier (BBB) and is converted into dopamine in the brain known as levo-dopa (L-dopa) (Poewe, 2006). However, L-dopa often produces side effects that are worse than the disease itself. In addition, over time, patients require higher and higher dosages to be effective, a concern with younger patients (Rao et al., 2006). The final hallmark of PD, postural instability, is resistant to L-dopa treatment. Another common treatment is dopamine agonists, which can be used in monotherapy or in combination with L-dopa. Argument for their use alone as a first treatment is to delay L-dopa treatment slowing the wearing off of L-dopa (Rao et al., 2006). However, due to the lack of robustness of dopamine agonists, almost all patients will require L-dopa at some point. In the many years since the beginning of a search for treatment, the lack of progress demonstrates the complexity of the disease (Obeso et al., 2010).

2.2 Pathophysiology

The earliest and most studied cause of PD is the degeneration of the dopaminergic neurons in the substantia nigra (SN). In the normal brain, dopaminergic neurons are found in three main areas, the olfactory bulb, the hypothalamus and the midbrain, which consists of the SN and the ventral tegmental area (VTA). From the midbrain, there are three main projections of the dopaminergic neurons. The mesolimbic pathway projects dopaminergic axons from the VTA to the nucleus accumbens, plays a role in addiction and reward and is the pathway most often affected in schizophrenia (Sillitoe and Vogel, 2008). The mesocortical pathway projects axons from the VTA to the frontal cortex and is most often associated with motivation and memory (Sillitoe and Vogel, 2008). The nigrostriatal pathway projects from the SN to the basal ganglia (BG) and is associated with motor control (Smith and Bolam, 1990). This pathway is involved in PD, and will be the focus of this review.

In the normal brain, dopaminergic projections from the substantia nigra pars compacta (SNc) synapse on the striatum, which consists of the caudate nucleus and the putaman (Figure 2.1) (Smith et al., 1998). From the striatum, a direct or an indirect pathway leads to the substantia nigra pars reticula (SNr) (Smith et al., 1998). The direct pathway sends inhibitory GABA and substance P axons to the globus pallidus internal (GPi)/SNr (Mora et al., 2008, Weintraub et al., 2008b). The indirect pathway projects inhibitory GABA and enkephalin axons to the globus pallidus external (GPe) which then sends GABAergic projections to the subthalamic nucleus (STN) which then sends glutamatergic (excitatory) outputs to the GPi/SNr (Mora et al., 2008) The projections to the SNr proceed to the thalamus. From the thalamus, glutamatergic projections head toward the cortex or GABAergic projections proceed to the brain stem and from the brain stem axons project back to the SNc completing the loop (Mora et al., 2008). Both pathways lead to activation of muscle movement and control. Through the activation the motor cortex or brain stem as well as through feedback loops within the basal ganglia, fine motor movements can be controlled (Mora et al., 2008, Gaig and Tolosa, 2009), the decision to move can be separated from the movement itself and other outside inputs can be factored into muscle movement decision.

In PD patients, the dopaminergic projections to the striatum deteriorate. The decline in dopaminergic modulation of the basal ganglia leads to problems in controlling muscle movements and to the symptoms seen in PD (Figure 2.1). Often the symptoms do not present until approximately 60% of the dopaminergic cells in the SNc have died suggesting a compensating mechanism for controlling movement (Gaig and Tolosa, 2009). Proposed mechanisms for this redundancy include the feedback loops located within the basal ganglia as well as movement of serotonergic neurons located nearby into the basal ganglia (Smith et al., 1998). These mechanisms from the serotonergic neurons may be responsible for some of the early non-motor symptoms (Weintraub et al., 2008b). Additionally, the relationship of the basal ganglia with the frontal cortex may be responsible for cognitive decline (Weintraub et al., 2008b). The dopaminergic neurons in the SNc deteriorate selectively in PD, leaving the dopaminergic neurons in the rest of the brain intact and not leading to symptoms typically seen in other dopaminergic disorders.

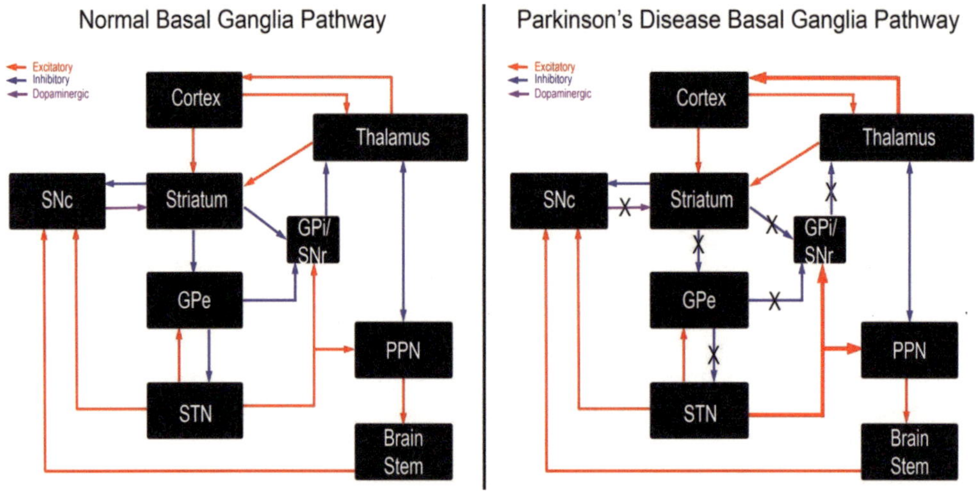

Fig. 2.1. Dopamine Signaling to the Basal Ganglia in Normal and Parkinson's Disease State. In normal state, dopaminergic neurons from the substantia nigra pars compacta (SNc) project onto the striatum. Activation of the striatum leads to motor movement modulation through the direct pathway (globus pallidus internal, GPi) or the indirect pathway (globus pallidus external, GPi). Both pathways lead to the thalamus and then to the cortex and brainstem (Smith et al., 1998). In the Parkinson's disease (PD) state, the dopaminergic neurons from the SNc are absent. This lack of input prevents the inhibitory signaling to the indirect and direct pathways, which causes a disruption in motor control (Obeso et al., 2008).

3. Human embryonic stem cells and derivatives

3.1 Human embryonic stem cells

In 1998, James Thomson and colleagues derived hESCs from the inner cell mast of discarded blastocysts (Thomson et al., 1998). From 14 inner cell masts collected, 5 embryonic stem cell lines could be created. Thomson and colleagues established early characteristics of hESCs which included high nuclear to cytoplasmic ratio, prominent nucleoli, the formation of a distinct colony, high telomerase activity, the ability to form cells from all three germ layers, and teratoma formation in addition to embryonic markers SSEA-3, SSEA-4, Tra-1-60, Tra-1-

81 and alkaline phosphytase (AP) (Thomson et al., 1998). Mouse embryonic fibroblast (MEF) feeder layers were found to support continued proliferation, and hESCs demonstrated the ability to form embryoid bodies, which contain all three germ layers (Thomson et al., 1998).
Mouse embryonic stem cells (mESCs) can be maintained in an undifferentiated state using leukemia inhibitory factor (LIF) alone without feeder layers. LIF activates the signal transducer gp130 there by activating of STAT3 and maintaining the state self-renewal in mESCs (Figure 2.3) (Humphrey et al., 2004). BMPs can be used in the place of serum in addition to LIF to maintain pluripotency through the activation of Id genes (Ying et al., 2003, Humphrey et al., 2004, Rao, 2004). This has not been the case for hESCs as LIF does not maintain the pluripotency of hESCs and is not necessary for maintenance of self-renewal (Xu et al., 2005). Initial attempts at feeder free culture of hESCs expanded upon the knowledge that hESC populations express α6 and β1 integrins leading to successful culture on laminin and Matrigel as extracellular matrices for hESCs in MEF conditioned media with differentiation results similar to what was found in previous studies (Xu et al., 2001). Basic fibroblast growth factor (bFGF) has been used to maintain clonally derived hESCs suggesting potential in a feeder free, serum free culture (Amit et al., 2004). Differentiation studies in which BMPs were blocked in hESC culture initiated neural differentiation. Taking these two together, Xu and colleagues used bFGF and BMP to maintain hESC self-renewal in the absence of MEFs or MEF conditioned media (Xu et al., 2005).
Our lab derived three lines from discarded embryos in 2001. These lines were isolated from the inner cell mass of 19 embryos and resulted in 4 cell lines. These cell lines were maintained in a pluripotent state on MEF feeder layers (Mitalipova et al., 2003). Two of these cell lines (BG01 and BG02) have the ability to form EBs and to differentiate into neural cells and cardiac cells (Mitalipova et al., 2003).

3.2 Neural progenitor cells

Directing the differentiation of hESCs towards neural cells allows for controlled culture system to develop specific neural subtypes including motor neurons, dopaminergic neurons and forebrain neurons. Several groups have attempted to establish a proliferative population of multipotent hNPs, which can be differentiated to neurons, astrocytes or oligodentrocytes.
Differentiation of hESCs toward hNPs occurs through either an embyroid body (EB) or a monolayer culture system. In EB differentiation, hESCs are grown in suspension and allowed to form masses of cells, which form a mixed population that includes hNPs (Schuldiner et al., 2001, Zhang et al., 2001). From these masses, the neural cells were selected and used in further proliferation or differentiation experiments. In monolayer differentiation, hESCs are induced with various morphogens in the tissue culture dish and neural rosette structures are allowed to form. From these structures, neural cells are selected, re-plated and allowed to proliferate or differentiate (Shin et al., 2006).
Each type of differentiation (EB or monolayer) requires various morphogens to direct the differentiation toward a neural multipotent cell. Retinoic acid (RA) plays a role in neural patterning and neural differentiation in the developing embryo (Reubinoff et al., 2001, Carpenter et al., 2003, Maden, 2007). Bone morphogenetic proteins are often inhibited by the antagonist Noggin, which leads to development of the neural phenotype in the mouse (Pera et al., 2004, Itsykson et al., 2005). bFGF signaling maintains the proliferative capacity of neural cells as well as involvement in induction and patterning (Jordan et al., 2009). bFGF was used in a neural differentiation protocol for its known caualzaling factors. Originally,

bFGF was shown to be important in the brain as a neural growth factor that maintained the pluripotency of immortalized NSCs (Figure 2.2) (Li et al., 2000). Later, bFGF has been shown to be a caudalizing factor within the neural plate and the neural floor (Jordan et al., 2009). Epidermal growth factor (EGF) is a mitogen that was used in many hESC neural differentiation protocols to maintain self-renewal potentially through crosstalk with Notch or through EGF's suppression of apoptosis (Carpenter et al., 2003, Elkabetz et al., 2008). LIF is another factor known to maintain proliferation of neural cells (Shin et al., 2006). Several reports of hNP differentiation have used varying combinations of these factors to achieve hNP differentiation from hESCs.

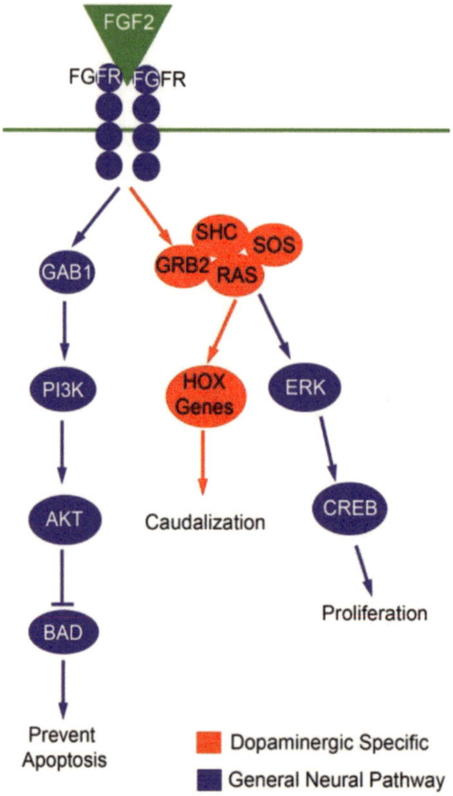

Fig. 2.2. Fibroblast Growth Factor 2 Induces Caudalization and Prevents Apoptosis Fibroblast growth factor 2 (FGF2) binds to the fibroblast growth factor receptor in developing neurons to activate the PI3K pathway or the MAPK pathway. In neural progenitors (hNPs) and neural stem cells (NSCs), FGF2 activates AKT, which blocks BAD signaling to prevent apoptosis (Sato et al., 2010). MAPK activation of the transcription factor CREB increases proliferation in hNPs and NSCs (Sato et al., 2010). These pathways are general neural pathways seen in all neurons prior to specification and represented by the general neural pathway. Additionally, FGF2 activates HOX genes in a gradient within the developing brain to caudalize neurons towards an anterior cell fate, which is the action of FGF2 in the dopaminergic neurons differentiated in this chapter and represented by the dopaminergic specific pathway (Chiba et al., 2005).

Differentiation of these hNPs into specific neural subtypes creates neural cells that are better models for transplanation specific developmental patterns, disease progression or transplantation. Differentiation of motor neurons has required the addition of RA, sonic hedgehog (SHH) and bFGF (Li et al., 2005, Shin et al., 2005). Forebrain differentiation has required *Otx1*, *Otx2* and *Bf1* expression and is thought to involved Wnt signaling (Elkabetz et al., 2008). Serotonergic neuron differentiation requires SHH and fibroblast growth factor 4 (FGF4) (Barberi et al., 2003). Differentiation toward a dopaminergic fate begins with SHH and fibroblast growth factor 8 (FGF8) and will be further explored in this review (Perrier et al., 2004).

In 2006, our lab derived hNPs from hESCs using a monolayer culture system. Neural derivation media was used to induce neural rosette structures from which neural cells were selected and transferred to a monolayer culture (Shin et al., 2006). The combination of bFGF and LIF added to the culture media allowed for the maintenance of neural progenitor cells in a monolayer that could be continually cultured for several (>40 passages) while maintaining a stable karyotype (Shin et al., 2006). We have demonstrated the ability to differentiate these hNPs to motor neurons with the addition of RA (Shin et al., 2005) and to dopaminergic neurons with the addition of GDNF (Young et al., 2010).

4. Factors involved in dopaminergic differentiation

4.1 Sonic hedgehog

In the developing embryo, signaling factors in the developing nervous system control the movement of the different types of neurons in the brain and spinal cord to their correct position. In dopaminergic neuron development, sonic hedgehog (SHH) modulates the dorsal/ventral placement of the midbrain dopaminergic neurons (Hynes et al., 1995). SHH is secreted from the notochord to induce floor plate cells through a decreasing gradient and to signal for the ventral forebrain and midbrain development of serotonergic and dopaminergic neurons (Hynes et al., 1995, Smidt and Burbach, 2007). SHH signaling is closely regulated to ensure proper enlargement of the midbrain area and is turned off to allow for post-mitotic differentiation. Dopaminergic neurons will arise from the pool of neuroepithelial progenitors found in the ventricular floor plate (Smidt and Burbach, 2007). Wnt causes a down regulation of SHH signaling allowing for the end of neural proliferation and the beginning of neurogenesis (Joksimovic et al., 2009). In 1995, SHH was discovered to be important for dopaminergic neural development through its activation of cAMP and PKA (Hynes et al., 1995). Transplantation of floor plate tissue to other areas or induced expression of SHH in other brain areas will cause ventralization of those areas. Over expression of the SHH target Gli1 causes the same effects as SHH itself, further confirming SHH's role in dopaminergic neuron development (Gulino et al., 2007). The ability for floor plate tissue combined with FGF8 beads to induce the formation of midbrain dopaminergic neurons further added increased evidence for SHH in the midbrain/hindbrain organization. SHH activates Patched (Ptc), releasing its negative control on Smoothened (Smo) and activating downstream transcription factors Gli1, Gli2 and Gli3 (Gulino et al., 2007). Each Gli has distinct actions; Gli1 acts to increase SHH activation. Gli2 acts to modulate Wnt, Brachyury, Xhox3 and Bcl-2 genes. Gli3 activates Ptc as a negative control of SHH signaling (Gulino et al., 2007). The decreasing gradient outward from the ventral midbrain signals the induction of neural precursor cells, which is suppressed by Wnt signaling the beginning neurogenesis of the floor plate derived dopaminergic and serotonergic neurons. SHH interacts with FGF8 to induce the correct size pool of dopaminergic neurons (Joksimovic et al., 2009).

4.2 Fibroblast Growth Factor 8

In combination with SHH, FGF8 controls the boundaries of the midbrain-hindbrain organizer (MHO) which direct the area in which dopaminergic neurons will be expressed. FGF8 expression originates at the isthmus and radiates anterior/posterior (Ye et al., 1998). The size of the MHO is determined by outside induction factors including the Hox genes at the anterior edge and FGF4 which signals with SHH for serotonergic neuron development. FGF8 interacts with other early regulatory genes involved in dopaminergic neuron development (Otx2, Gbx2, EN1, EN2, Pax2 and Pax5) to maintain and regulate the dopaminergic field of development (Smits et al., 2006). If ectopically applied, FGF8 and SHH induce a two-dimensional system of midbrain neural precursor cells (Smidt and Burbach, 2007).

4.3 Leukemia Inhibitory Factor

Leukemia inhibitory factor (LIF) is a member of the interleukin 6 family of cytokines and supports cell growth and development. LIF has known function in maintaining the pluripotency of mESCs (Pease et al., 1990). This function does not carry over into hESCs, as the addition of LIF to the culture media for hESCs does not maintain the pluripotency (Xu et al., 2005). In the non-dividing cells of the neural crest, LIF induces sensory neuron development (Murphy et al., 1991). Later it was discovered that LIF also promotes proliferation of the progenitor pool found in the olfactory bulb and in fetal neural stem cells (Satoh and Yoshida, 1997, Galli et al., 2000). This is thought to occur through the gp130 receptor regulation Notch signaling which controls neural stem cell proliferation. LIF has been used to maintain of pluripotency in hNPs derived from hESCs as well as in NSC cultures (Chojnacki et al., 2003).

While it was known that LIF supported glial cell differentiation, in 2003 it was discovered that LIF acts through the ERK pathway to decrease the expression of dopamine beta hydroxylase (DβH) (Figure 2.3) (Dziennis and Habecker, 2003). Mouse and rat mesencephalic derived progenitors were differentiated into dopaminergic neurons using both LIF and GDNF (Storch et al., 2001). These differentiated dopaminergic neurons were maintained in culture for extended periods as well as used for deriving a clonal line (Storch et al., 2001). LIF has also been used in a rat model of PD to increase the number of mesencephalic dopaminergic neurons. Support for the use of LIF as a factor to enhance dopaminergic differentiation from hNPs in this chapter comes from the suppression of DβH by LIF (Figure 2.3) in addition to the known success in a rat and mouse model of dopaminergic differentiation with GDNF and LIF (Ling et al., 1998, Liu and Zang, 2009).

4.4 Glial cell-line Derived Neurotrophic Factor

GDNF was discovered as a neurotrophic factor for dopaminergic neurons in 1993 in rat glial cell cultures (Lin et al., 1993). Since this time, its use as a potential treatment for PD has been explored in several animal models (rat, mouse, non-human primate), cell culture models (rat, mouse, non-human primate, mESCs, hESCs, human fetal tissue) and in human drug trials. GDNF was first tested as a recovery factor in animal models of PD (Bowenkamp et al., 1995, Shults et al., 1996). Rats lesioned with 6-OHDA and then injected with GDNF showed increase in TH expression and a reduction in apomorphine induced turning (Bowenkamp et al., 1995, Shults et al., 1996). Retrograde tracing studies show that GDNF injected into the midbrain was transported back to the SN (Tomac et al., 1995). In C57/B1 mice, injection of

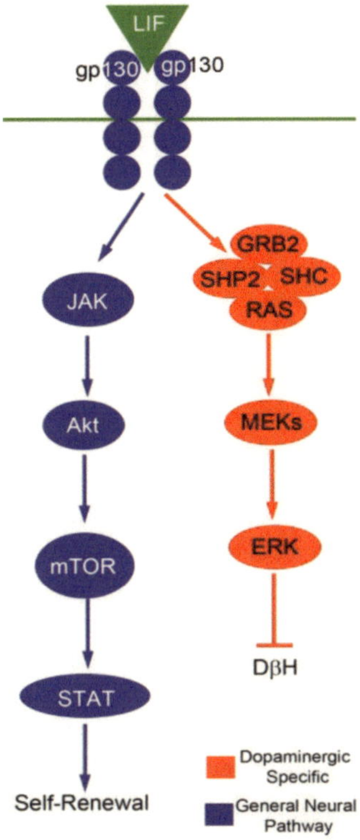

Fig. 2.3. Leukemia Inhibitor Factor action in dopaminergic and non-dopaminergic neurons Leukemia inhibitory factor (LIF) binds to the gp130 receptor on the cell surface causing activation of the JAK/STAT pathway and the MAPK pathway (Matsuda et al., 1999). Activation of the JAK/STAT pathway modulates self-renewal in neurons other than dopaminergic (Matsuda et al., 1999). The MAPK pathway is activated by LIF in dopaminergic neurons to suppress dopamine beta hydroxylase production, which would lead to production of norepinephrine instead of dopamine (Dziennis and Habecker, 2003). General neural pathway refers to any neural subtype other than dopaminergic while dopaminergic specific refers to dopaminergic neurons similar to those derived in this chapter.

GDNF in the SN protects from degeneration and aids in the recovery of dopaminergic neurons in an MPTP model (Hou et al., 1996). The first study of GDNF in a non-human primate, a rhesus monkey, showed recovery in bradykinesia, rigidity and postural instability in an MPTP lesioned striatum that was maintained with injections of GDNF every 4 weeks (Gash et al., 1996). As injections into the brain are not desired for a potential treatment option for PD, fetal mesencephalic neurons from rat brains that excrete GDNF were injected into 6-OHDA lesioned rats and an increase in TH expression and expansion of neurite tracts were seen postmortem (Rosenblad et al., 1996, Wang et al., 1996, Winkler et al., 1996). Adenovirus' created to promote GDNF expression were tested by several groups for their ability to protect the SN from 6-OHDA neurotoxicity with limited results and lack of long-term effectiveness (Choi-Lundberg et al., 1997, Lapchak et al., 1997). Most of the data has shown limited results with long-term data unavailable; however, due to the known protective role for GDNF in DA neurons, enthusiasm is still high.

A 12-week study of GDNF injection into the ventricle of a 6-OHDA lesioned rat showed an increase in response to amphetamine to those seen in normal animals (Sullivan et al., 1998). Injections into the SN are found to be protective where as injections into the putamen are not (Gerhardt et al., 1999). Long-term adenovirus vector administration in the SN led to re-innervation of the striatum after 6 months of administration (Kirik et al., 2000). Lentiviral administration of GDNF into the striatum of both mice and non-human primates that were lesioned 2 weeks later protected TH neurons in both young and old mice (Kozlowski et al., 2000, Georgievska et al., 2002). GDNF administration in a mouse α-synuclein model did not protect the SN from neurodegeneration presenting a complication that remains for GDNF as a therapy for PD (Lo Bianco et al., 2004).

GDNF's role in protecting and recovering neurons in the SN that degenerate in PD led to research into its use as a growth molecule in hESCs. hESCs differentiated towards dopaminergic neurons have been touted as a potential cell therapy in PD. The neurotrophic factor has been used in several differentiation protocols (Buytaert-Hoefen et al., 2004, Perrier et al., 2004, Schulz et al., 2004). The use of GDNF along with co-culture with PA6 cells increased the number of TH positive cells produced over PA6 co-culture alone (Buytaert-Hoefen et al., 2004). Using hESC derived dopaminergic neurons as a model of PD, GDNF provided protection against MPTP toxicity (Zeng et al., 2006).

Another method of improving delivery systems involves genetically altering neural stem cells or astrocytes to release GDNF and injecting these into the brain which protected from parkinsonian motor responses in mouse models of PD (Engele and Franke, 1996, Elsworth et al., 2008). Injection of hNPs modified to secrete GDNF into MPTP monkeys increased axon fibers that express both TH and VMAT2; however, these cells remained at the area of injection and did not travel to the area of need (Emborg et al., 2008). The neuroprotection of GDNF in rat, non-human primate and hESC models of PD demonstrates its robustness as a useful tool for developing future therapies.

Further expanding on the rodent research, intracerebral injections of GDNF into MPTP treated rhesus monkeys induced a 20% increase in dopamine levels and functional recovery with GDNF injections every four weeks (Gash et al., 1996). When GDNF was administered along with the most commonly prescribed L-dopa drug in parkinsonian rhesus monkeys, a significant functional improvement in PD symptoms was seen as well a decrease in the side effects typically accompanying L-dopa drugs (Elsworth et al., 2008). Further study between the relationship of GDNF administration and functional recovery indicates a role for GDNF in modulating dopamine plasticity in the striatum. Safety and efficacy studies in non-human primates demonstrated that the injections do not cause any negative histological effects in the injected brain and that the most notable side effect of GDNF delivery was weight loss (Su et al., 2009). These studies advanced the field towards using GDNF in human clinical trials.

Following determination that human PD patient brains maintained expression of the RET receptor for GDNF, a male patient received intracerebroventricular injections of GDNF which resulted in severe side effects and no functional recovery. Several years later, a randomized double-blind study of intracerebroventricular monthly injections of various dosages of GDNF lead to no parkinsonian symptom improvement with GDNF but an increase in adverse effects including significant weight loss potentially because the GDNF did not reach the target tissues (Kordower et al., 2000). Targeted injections of GDNF into the putamen resulted in significant improvements in PD quality of life scores, dopamine uptake and dyskinesia (Kordower, 2003). In a two-year follow up study, the patients continued to improve with no added side effects. Withdrawal of GDNF injections caused a complete reversal to pre-injection levels of quality of life and symptomatic scores (Kordower et al., 2008). At this stage, GDNF is still being evaluated as a potential treatment for PD but the route of administration and side effect profile are holding back major advances in the field.

5. Genes involved in dopamine development

Protein	Expression (mouse)	Role in Dopaminergic Neurons	References
NURR1	E10.5	• Drive expression of TH, AADC, RET, VMAT2, DAT • Support development of DA neurons • Maintain post-mitotic DA neurons	(Zetterstrom et al., 1997, Saucedo-Cardenas et al., 1998, Wallen et al., 2001)
EN1	E7.5	• Expressed in neuroepithelium • Secreted to maintain mid-/hindbrain boundary • Induced by FGF8 • Maintain post-mitotic DA neurons	(Liu and Joyner, 2001, Sgado et al., 2006)
TH	E11.5	• Driven by NURR1 • Rate limiting enzyme in DA synthesis	(Lehnert and Wurtman, 1993, Maxwell et al., 2005)
PITX3	E11.5	• Drive expression of VMAT2, DAT and RA • Maintain SN neurons	(Lebel et al., 2001, Hwang et al., 2003, Smidt et al., 2004, Jacobs et al., 2007)
DAT	E13.5	• Gives MPTP access to DA neurons • Denser in SN neurons • Removes DA from synapse	(Storch et al., 2004, Schiff et al., 2009)
VMAT2	E18	• Package MPTP to prevent it from damaging cell • Less VMAT2 expression in PD brains	(Harrington et al., 1996, Hansson et al., 1998, Speciale et al., 1998)

Table 2.1 Proteins Expressed in Dopamine Neurons_AADC - Aromatic L-Amino Acid Decarboxylase, DA - Dopamine, DAT – Dopamine Transporter, EN1 – Engrailed 1, FGF8 – Fibroblast Growth Factor 8, MPTP - 1-methyl-4-phenyl-1,2,3,6-tetrahydropyridine, NURR1 - Nuclear Receptor Related 1, PD – Parkinson's Disease, PITX3 - Paired-like Homeodomain Transcription Factor 3 , RA – Retinoic Acid, RET – Rearranged in Transfection, SN – Substantia Nigra, TH – Tyrosine Hydroxylase, VMAT2 - Vesicular Monoamine Transporter 2

5.1 Nuclear receptor related 1

Nuclear receptor related 1 (Nurrl) is a member of the Nur family of proteins which are involved in cell growth and apoptosis. Beginning expression at E10.5 in the mouse, Nurrl is required for normal dopaminergic development (Table 2.1) (Zetterstrom et al., 1997). Nurrl was discovered to be similar to Nur77 already found in the olfactory bulb, cortex, hippocampus, SN and VTA of the mouse (Law et al., 1992, Zetterstrom et al., 1996). Mice administered 6-OHDA not only show a loss in dopaminergic neurons but also in Nurrl expression. Nurrl knockout (Nurrl-/-) mice do not express TH in brain areas A8 (retrorubral nucleus), A9 (SN) and A10 (VTA) (Zetterstrom et al., 1997). Discovery of the need for Nurrl to regulate TH expression led to further examination of the role that Nurrl plays in maintenance of other proteins important for a properly functional dopaminergic neuron (Table 2.1). In Nurrl-/- mice, aromatic L-amino acid decarboxylase (AADC), the enzyme responsible for converting L-dopa to dopamine or 5-hydroxytryptophan to serotonin, was found to be absent in dopaminergic neurons only; however, paired-like homeodomain transcription factor 3 (PITX3), a gene found only in SN dopaminergic neurons, expression was unaffected in Nurr-/- mice (Table 2.1) (Saucedo-Cardenas et al., 1998). As discussed earlier, GDNF signals through co-receptor GFRα1 binding to RET (Table 2.1). Nurrl knockouts are deficient in RET but not in GFRα1 suggesting the importance of Nurrl not only in pathways involved in dopamine production but in neuron maintenance and support (Wallen et al., 2001).

Nurrl is not only important for embryonic development of dopaminergic neurons but also for the maintenance of these neurons in the postnatal and adult brain. Conditional knockouts induced by *Cre* ablation of Nurrl at E13.5 to E15.5 show a loss of TH and the dopamine transporter (DAT) expression in postnatal rats while adult ablation leads to reduction in TH expression in the SN preferentially over the VTA (Table 2.1) (Kadkhodaei et al., 2009). Overall, Nurrl plays a role in activating and maintaining the expression of AADC, TH, RET and DAT. With such an importance in dopaminergic neurons, finding a decrease in Nurrl in PD patients as well as a base pair insertion mutation is not surprising. Further expansion on the role of Nurrl in PD patients may lead to future treatment options.

5.2 Engrailed 1

Engrailed 1 (EN1) is part of a family of homeobox genes consisting of EN1 which is expressed in the VTA and SN and EN2 which is only expressed in a subset of dopaminergic neurons and begins to be expressed later in development than EN1 (Danielian and McMahon, 1996). EN1 is a developmental regulation protein that is expressed in mouse around day E7.5 and plays a role in the development of dopaminergic neurons and the maintenance of those neurons (Table 2.1) (Danielian and McMahon, 1996). EN1 is expressed in the neuroepithelium of the ventral midbrain around the isthmus, which is responsible for controlling the midbrain/hindbrain boundary (Liu and Joyner, 2001). Induction of EN1 by FGF8 maintains the area of the brain that will consist of the dopaminergic neurons. EN1 knockout mice lose the expression of all dopaminergic neurons by birth (Ye et al., 2001). A gain of function study in mice demonstrated that EN1 would induce the midbrain/hindbrain expression in any area in which it was expressed (Table 2.1) (Alberi et al., 2004). Both EN1 and EN2 are necessary for proper induction of midbrain dopaminergic neurons and they can partially compensate for each other (Alberi et al., 2004). The other role of EN1 is in maintaining dopaminergic neurons post-mitotically in the midbrain (Table 2.1) (Sgado et al., 2006). EN1 conditional knockout mice lose their dopaminergic neurons in the midbrain due to caspase 3 induction and apoptosis (Sgado et al., 2006, Sonnier et al., 2007). Heterogeneous EN+/- mice will progressively lose their dopaminergic neurons in a pattern that is similar to that seen in PD patients (Sgado et al., 2006, Sonnier et al., 2007).

5.3 Tyrosine hydroxylase

Tyrosine hydroxylase (TH) is the rate-limiting enzyme in dopamine synthesis, making it the main marker for dopaminergic neurons. In the production of catecholamines, L-tyrosine is converted to L-dopamine by TH. Aromatic L-amino acid decarboxylase (AADC) then converts the L-dopamine into dopamine. In dopaminergic neurons, the process stops there. In noradrenergic neurons, dopamine is converted into norepinephrine by dopamine β hydroxylase (DBH). If the neuron releases epinephrine, then norepinephrine is converted to epinephrine by phenylethanolamine N-methyltransferase (PMNT) (Table 2.1) (Lehnert and Wurtman, 1993). As TH has an important role in several neural subtypes, it cannot be the sole marker for a dopaminergic neuron even if it is an important one. In the midbrain, TH expression occurs at E11.5 in the mouse immediately prior to PITX3 expression (Table 2.1). TH expression is driven by Nurr1, which began its expression at E10.5 (Maxwell et al., 2005).

5.4 Paired-like homeodomain transcription factor 3

Paired-like homeodomain transcription factor 3 (PITX3) is a homeodomain protein which is found only in dopaminergic neurons in the midbrain of the central nervous system. Expression of PITX3 is found outside of the CNS transiently in the eye lens. Expression of PITX3 begins at E11.5 in the mouse, immediately following expression of TH (Table 2.1) (Lebel et al., 2001). PITX3 expression completely overlaps the areas of TH expression in the SN and only a subset of the TH+ neurons in the VTA; additionally, the TH promoter has a PITX3 binding site. Aphakia mutant mice (PITX3-/-) lack SN neurons exclusively starting around E12.5, but not VTA neurons suggesting the molecular mechanism of PITX3 in the VTA differs from that in the SN and this mechanism may provide insight into the selective degeneration of dopaminergic neurons in the SN in PD (Hwang et al., 2003, Nunes et al., 2003, van den Munckhof et al., 2003). Retrograde tracing studies confirm that the absence of dopaminergic neurons in the SN in apkakia mice leads to a lack of the normal connections to the caudate putamen as is seen in PD (Hwang et al., 2003, Nunes et al., 2003, van den Munckhof et al., 2003). The absence of PITX3 in aphakia mice does not lead to an absence in many of the other genes involved in dopaminergic neuron development and maintenance (NURR1, LMX1b, EN1, EN2 and RET) (Smidt et al., 2004). AdH2 expression is affected causing a decrease in retinoic acid (RA; Table 2.1) (Jacobs et al., 2007). Restoring the levels of RA can counteract the developmental deficits seen in these mice suggesting that PITX3's role in dopaminergic neural maintenance is through regulation of RA expression. However, this is only required during early development and does not account for the continued deficits seen in aphakia mice (Jacobs et al., 2007). A possible role for PITX3 in continued deficits is in its control of VMAT2 and DAT expression (Hwang et al., 2009). Aphakia mice lack both VMAT2 and DAT as seen by both in situ hybridization and PCR (Table 2.1) (Smits et al., 2005).

5.5 Dopamine transporter

The dopamine transporter (DAT) is the protein responsible for removing dopamine from the synapse post release and taking it back into the neuron. This allows for recycling of the neurotransmitter as well as halting the activation of the post-synaptic neuron. DAT activity depends on sodium moving down its concentration gradient, dopamine and chloride ions being recognized outside the transporter, dopamine and chloride ion translocation into the cell and unloading, and the transporter returning to its original state (Volz and Schenk, 2005). DAT mRNA is denser in the SN relative to the VTA suggesting a possible role for DAT in the pathology of PD. Over expression of DAT led to excitotoxicity and loss of dopaminergic neurons (Storch et al., 2004). Due to MPP+ entering the dopaminergic neuron through the DAT transporter, DAT knockout mice are insensitive to MPTP toxicity (Table 2.1) (Storch et al., 2004). Variable number tandem repeats (VNTRs) found in DAT occur in patients with various neurological disorders including PD (Haddley et al., 2008). These VNTRs seem to occur prior to symptomology of the disease suggesting that these VNTRs pre-dispose the dopaminergic neurons to susceptibility of the disease (Haddley et al., 2008).

5.6 Vesicular monoamine transporter 2

Vesicular monoamine transporter 2 (VMAT2) is a protein which is responsible for packaging monoamines (dopamine, serotonin, norepinephrine) into vesicles in the cytosol for transmission out of the cell (Harrington et al., 1996). VMAT2 also packages several neurotoxins such as MPTP to prevent them from causing harm to the neuron (Table 2.1) (Harrington et al., 1996). VMAT2 expression starts at E11 in the telencephalon and is seen in the caudate putamen and nucleus accumbens at P1 (Hansson et al., 1998). At E18 expression is found in the SN and VTA (Table 2.1). VMAT2 -/- mice die shortly after birth; however, VMAT2 +/- mice or blockage of the transporter yield results on the transporter function (Hansson et al., 1998). VMAT2+/- mice have a drastic decrease in dopamine despite compensation inside the neuron by more than doubling synthesis (Stankovski et al., 2007). MPTP destruction is more than twice that in normal mice through greater accumulation of the toxin to remain in the cytosol where it can cause damage to the neuron (Harrington et al., 1996). Animals that lack VMAT2 do not lack the neurons themselves, just the monoamines; cells eventually die through lack of use via the caspase 3 and caspase 9 pathways (Stankovski et al., 2007). Brains of PD patients examined postmortem express 88% less VMAT2 in the putamen, 83% less in the caudate and 70% less in the nucleus accumbens compared to brains of people who were not diagnosed with PD (Table 2.1) (Speciale et al., 1998).

6. Dopaminergic differentiation

Due to the lack of success in developing a new therapeutic for PD over the last 30 years combined with the specificity of the cells that deteriorate in PD, differentiating dopaminergic neurons from hESCs for use in cell therapy or drug discovery for PD has been a research focus for many years with the first successful attempt by Perrier and colleagues in 2004. Discovered in 2000 for its ability to induce midbrain dopaminergic neurons from mESCs, stromal cell-derived inducing activity (SDIA) refers to the factors secreted from or imbedded in the cell membrane of PA6 cells or other bone marrow cells which have been shown to promote dopaminergic differentiation (Kawasaki et al., 2000). Studies on fixed PA6 cells and on mitomycin c treated and irradiated cells show a reduction in ability to differentiate to dopaminergic neurons (Vazin et al., 2008). Microarray studies examining the factors secreted from these cells have suggested 8 possible categories (IGF, FGF, Notch,

PDGF, SHH, TGFβ, VEGF, Wnt) for potential secreted factors (Swistowska et al., 2010). Utilizing SDIA, hESCs were co-cultured with stromal cells, SHH and FGF8 to to differentiate them towards a neural fate (Perrier et al., 2004). Removal of SHH and FGF8 and replacement with brain derived neurotrophic factor (BDNF) and ascorbic acid (AA) induced 60-70% TH positive/Tuj positive cells (Perrier et al., 2004). The dopaminergic phenotype of these cells was further confirmed by VMAT2 and EN1 staining (Perrier et al., 2004). GDNF, used in co-culture with SDIA, doubled the number of TH positive cells seen with SDIA activity alone (Zeng et al., 2004). Another hESC line, SA002.5 was differentiated on PA6 cells resulting in up to 37% TH positive/Tuj1 positive neurons. These neurons were transplanted into the nigral-stratial pathway with negative consequences including proliferation following transplantation and terotoma formation (Brederlau et al., 2006). Differentiation of H9 hESCs on a SHH secreting M5S stromal feeder layer with bFGF lead to no teratoma formation when transplanted into the SN, but few TH+ cells survived (Ko et al., 2007). Attempts at differentiation with a bone marrow stromal cell feeder layer and FGF8/SHH lead to 40% TH+ cells but no cells survived the graft. In an effort to differentiate a line that would be post mitotic after injection, H9 and H1 cells were co-cultured with rat astrocytes; however, transplanted cells that survived were still undifferentiated mitotic cells (Roy et al., 2006).

Following induction using the SDIA method, focus on a method using only growth factors and no co-culture methods began and was reported in 2005 by Yan and colleagues. hESCs were differentiated to neural progenitors through an embyroid body (EB) stage. Dopaminergic induction began with 7 days of FGF8 culture followed by 7 days of FGF8 and SHH culture (Yan et al., 2005). Progression to biologically functional dopaminergic neurons required 14 days of culture with dopamine survival factors (GDNF, BDNF), dopamine inducing factor ascorbic acid (AA), neural specification factor cyclic AMP (cAMP) in addition to the FGF8 and SHH. The dopaminergic neurons expressed 31% TH positive neurons after 5 weeks of differentiation (Yan et al., 2005). Another report using all of the above factors plus a dopamine induction factor, TGFβ lead to 43% TH positive cells; however, transplantation lead to few surviving TH+ post mitotic cells and primarily neural precursors that continued to proliferate (Yang et al., 2008). The first report of of hESCs differentiated towards a dopaminergic phenotype being transplanted that resulted in significant improvements in rotational and forepaw stepping also resulted in the formation of tumors (Yang et al., 2008).

The field progressed to promoter systems that express genes known to be involved in dopaminergic development. Lmx1a is induced at E7.5 in mouse by Otx2 (Friling et al., 2009). Lmx1a helps to induce a midbrain dopaminergic neuron identity through controlling NURR1 and PITX3 expression (Chung et al., 2009). An Lmx1a promoter was used in hESCs to promote differentiation of 10 to 20% TH+ neurons (Friling et al., 2009). Efforts to improve the derivation of dopaminergic neurons have included formation of spherical neural masses (SNMs) instead of EBs prior to differentiation (Cho et al., 2008, Vazin et al., 2009). Elucidating the factors expressed in and secreted by stromal cells used to differentiate dopaminergic neurons included microarray studies. One study found that the cell membrane of stromal cells expressed FGF7, hepatocyte growth factor and vascular endothelial growth factor, which were sufficient to induce dopaminergic differentiation (Vazin et al., 2008). A microarray examining the mRNA expression of PA6 cells found IGF2 and several IGF binding proteins, FGF10, DLK1 NGF, SHH, TGF3β, VEGF and Wnt RNAs to be highly expressed in PA6 cells (Swistowska et al., 2010). Additionally, receptors for these genes were more highly expressed in NSCs compared to hESCs. A study using various combinations of factors which activated these receptors or replicated the factors expressed by PA6 cells determined that the combination of factors termed SPIE (SDF-1, PTN, IGF2, and EFNB1) was most effective at differentiating hESCs towards dopaminergic neurons (Cho et al., 2008, Vazin et al., 2009). However, a highly efficient and effective method of dopaminergic differentiation has not been obtained.

In an attempt to improve on the 5 stage method and to remove the feeders from the co-culture system, En-Stem A cells were differentiated with PA6 conditioned media for 4 weeks resulting in 18% TH+ cells compared with 26%TH+ cells derived from H9 derived hNPs cultured in PA6 conditioned media (Swistowska et al., 2010). The time of exposure to PA6 conditioned media was important. Cells exposed to PA6 conditioned media at the neural stem cell stage produced more TH+ neurons than did cells exposed as hESCs or cells exposed later in neural differentiation. Differentiation with FGF-20, a novel neurotrophic factor found to be expressed in the SN of rat brains, on PA6 feeder cells lead to a 5-fold increase (3% to 15%) in TH+ cells and reduced overall cell death via the caspase 8 and BAX pathways (Correia et al., 2007). Foxa2 ventralizes neural progenitors in the developing brain and leads to cell cycle arrest of ventral midbrain cells to promote differentiation over proliferation. Additionally, Foxa2 acts in an auto regulatory loop with SHH to promote dopaminergic neurons and to inhibit GABAergic differentiation (Lin et al., 2009). In order to promote Foxa2+ progenitor cells that mark ventral mesencephalic dopaminergic neurons, a high activity form of SHH and the FGF8a isoform induced dopaminergic neurons (Cooper et al., 2010) Currently research remains ongoing working to improve upon the differentiation protocol used to derive dopaminergic neurons.

7. GDNF and its mechanism of action within the neuron

GDNF belongs to the transforming growth factor-β (TGF-β) superfamily. Within the superfamily is the GDNF family of ligands, which include neurturin (NRTN), artemin (ARTN), persephin (PSPN) and GDNF (Airaksinen and Saarma, 2002). Each of these ligands bind preferentially to GDNF-family receptor-α (GFR-α) co-receptors (GDNF to GFR-α1; NRTN to GFR-α2; ARTN to GFR-α3; PSPN to GFR-α4) prior to binding to receptor tyrosine kinase (RET) protein which is attached to the plasma membrane with a glycosyl phophatidylinositol (GPI) anchor (Airaksinen and Saarma, 2002). In order to activate downstream pathways, the RET-GFRα complex must become associated with a lipid raft, recruited by Src and FRS2 (Airaksinen and Saarma, 2002). This binding activates the PI3K and MAPK pathway involved in neuron survival and neurite outgrowth (Figure 2.4) (Airaksinen and Saarma, 2002). The GDNF interaction with GFR1α1-RET promotes dopamine neuron survival, axon growth and hypertrophy (Figure 2.4) (Airaksinen and Saarma, 2002).

7.1 Mitogen activated protein kinase pathway

The mitogen activated protein kinase (MAPK) pathway consists of a network of kinases that are involved in cell survival, differentiation, proliferation, apoptosis, growth and involved in GDNF signaling (Figure 2.4) (Pimienta and Pascual, 2007), . There are currently three well known MAPK pathways: the c-JUN N-terminal kinase (JNK)/stress activated protein kinase (SAPK), the extracellular signal-regulated kinase (ERK1/2 and ERK5), and the p38 MAPK pathway (Figure2.4) (Roux and Blenis, 2004). MAPKKK1-4 will activate MAPKK 4 and 7, which in turn activates JNK 1, 2 and 3. The JNK pathway is involved in retinoic acid

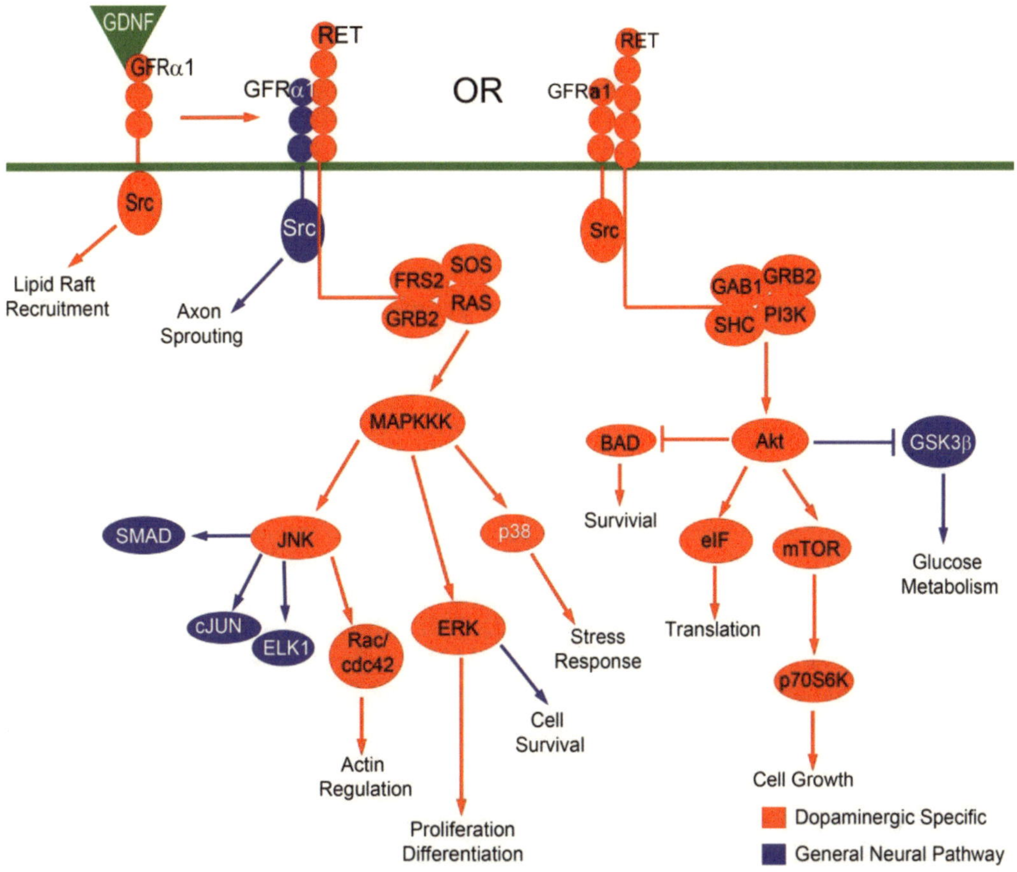

Fig. 2.4. Glial Cell-line Derived Neurotrophic Factor Supports Dopaminergic Differentiation through Activation of Several Pathways Glial cell-line derived neurotrophic factor (GDNF) binds to its co-receptor GFRα1 which then binds to the RET receptor to activate the MAPK, PI3K, JNK and Src pathways. Src recruits RET to the lipid raft for binding (Airaksinen and Saarma, 2002, Sariola and Saarma, 2003). MAPK activation of the JNK pathway regulates actin through RAC/CDC42 activation in its enhancement of the dopaminergic differentiation in this chapter. Other parts of the JNK pathway activated by GDNF not involved in dopaminergic differentiation include SMAD, cJUN and ELK1 (Airaksinen and Saarma, 2002, Sariola and Saarma, 2003). MAPK activation of the ERK pathway leads to enhancement of dopaminergic differentiation and cell survival in the cells in this chapter. The AKT pathway activated by GDNF in the dopaminergic neurons in this pathway leads to cell growth through the mTOR pathway, translation factor activation through the eIF pathway and cell survival through inhibition of BAD. Not involved in the dopaminergic differentiation of the cells in this chapter is activation of GSK3β (Airaksinen and Saarma, 2002, Sariola and Saarma, 2003).

neurogenesis in jnk knockout mESCs (Hayashi et al., 2000). The JNK pathway regulates cellular survival and neuronal migration (Figure2.4) (Garcia-Martinez et al., 2006). The ERK pathway divides into the ERK1/2 pathway and the ERK5 pathway (Nishimoto and Nishida, 2006). The ERK1/2 pathway is activated by MAPKK1/2, which is turned on by aRaf, bRaf or cRaf (Hayashi et al., 2000). The ERK pathway regulates cellular survival (Garcia-Martinez et al., 2006). ERK stimulates transcription factors such as Elk and c-myc and protein kinases such as ribosomal S6 kinase (RSK; Figure 2.4) (Roux and Blenis, 2004). ERK5 is involved in cell survival and proliferation through activation of MAPKKK 1-4 which triggers MAPKK5 (Nishimoto and Nishida, 2006). In vivo mouse models have demonstrated that ERK5 signaling is involved in both cardiovascular and neural development (Roux and Blenis, 2004). The final MAPK pathway, p38 MAPK pathway, is stimulated by MAPKKK1-4 activation of MAPKK3/6 (Figure 2.4) (Roux and Blenis, 2004). In embryonic development, there are two peaks of p38 activity (Roux and Blenis, 2004). The first acts as a switch between cardiovascular and neural development. The later peak modulates neurite formation and neural survival.

The mechanism through which GDNF acts to promote dopaminergic neural survival and differentiation is not entirely known, but it is thought that the MAPK pathway may play a role in promoting neural survival, differentiation or neurite outgrowth (Ohiwa et al., 1997, Nicole et al., 2001). Cultured embryonic rat cortical cells exposed to GDNF increased arborization and neurite outgrowth through activation of the p42/p44 MAPK pathway (Figure 2.4) (Ohiwa et al., 1997, Nicole et al., 2001). RET coupling with the Shc/Grb2 domains leads to downstream activation of the MAPK pathway (Figure 2.4) (Ohiwa et al., 1997, Nicole et al., 2001). Further research needs to be done to determine the involvement of the MAPK pathway in dopaminergic differentiation after RET activation.

7.2 Phosphoinositide 3-kinase pathway

The phosphoinositide 3-kinase (PI3K) pathway is activated by cytokines, growth factors and hormones and is involved in downstream regulation of cell survival, proliferation, apoptosis and regulation of transcription factors. PI3K exerts action on Akt, which acts in cellular functions such as survival, protein synthesis, proliferation, glucose metabolism, and neural signaling through its triggering of several other factors (Duronio, 2008). Akt inhibits pro-apoptotic signals Bad and the Forkhead family thus increasing cell survival (Figure2.4) (Manning and Cantley, 2007). Regulation of glucose metabolism occurs through glycogen synthase kinase 3 (GSK3) activation (Figure 2.4) (Manning and Cantley, 2007). Finally, Akt neural involvement occurs through regulation of the GABA receptor, ataxin-1 and huntingtin in addition to interaction with TGF-β signaling (Figure 2.4) (Manning and Cantly, 2007). Rubinsky and Meyuhas, 2006). mTOR is found in two complexes, mTOR complex 1 (mTORC1) and mTOR complex 2 (mTORC2). mTOR1 integrates signals to encourage cell growth or catabolic processes depending on which condition is more favoured.

Acting along with Akt signaling are pathways involved in translation control (eIF4E and p70 S6K), cell growth and survival (mTOR) and cell cycle regulation [phosphatase and tensin homolog (PTEN) (Figure 2.4)] (Ruvinsky and Meyuhas, 2006). mTORC2 promotes cellular survival and cytoskeletal maintenance (Ruvinsky and Meyuhas, 2006). Mutations in this pathway or deregulation caused by stress leads to complications in protein translation and many wide ranging problems in cellular function (Carnero et al., 2008). The p70 S6K pathway controls phosphorylation of ribosomal protein S6 that is important for cell size and glucose homeostasis (Figure 2.4) (Ruvinsky and Meyuhas, 2006). mTORC2 promotes cellular survival and cytoskeletal maintenance (Wullschleger et al., 2006). Mutations in mTOR signaling are involved in cancer, cardiovascular disease and metabolic disorders (Yap et al., 2008). PTEN is a tumor suppressor through its regulation of cell cycle, cell division and negative regulation of the PI3K/Akt pathway (Carnero et al., 2008).

In the mouse dopaminergic cell line, MN9D, the PI3K inhibitor LY294002 was administered prior to GDNF addition. In these studies, GDNF failed to protect the viability of the neurons exposed to 6-OHDA. In rat primary cultures, GDNF administration phosphorylates Akt (Ugarte et al., 2003). This phosphorylation was completely blocked by pre-incubating the cells with Wortmannin, a PI3K inhibitor (Figure 2.4) (Ugarte et al., 2003). When RET complexes with GAB1 and stimulates CREB, GDNF activates the PI3K pathway preferentially to the MAPK pathway (Figure 2.4) (Maeda et al., 2004).

7.3 Src

Src was the first discovered tyrosine kinase located in the cytoplasm. The family of Src tyrosine kinases (SFK) consists of Fyn, Lyn, Hck, c-Yes, Blk, Fgr, and Lck. SFKs play roles in cell growth, differentiation and survival, as well as cellular adhesion and synaptic transmission (Figure 2.4) (Encinas et al., 2001). When GDNF binds to its co-receptor GFRα1, the glycosyl phosphatidylinositol (GPI) that anchors the GFRα to the membrane recruits RET to the lipid raft and allows for activation of cellular signaling pathways that increase neural survival and differentiation (Figure 2.4) (Tansey et al., 2000). RET activation can occur in *cis* or *trans*. *Cis* activation occurs when a GPI anchored GFRα1 co-localizes on the same cell as the RET and allows for recruitment of a lipid raft in that cell (Tansey et al., 2000). When the GPI anchored GFRα1 is on an adjacent cells (such as a glial cell), the lipid raft is recruited in *trans* (Figure 2.4) (Encinas et al., 2001). *Trans* activation of RET is not sufficient to activated downstream pathways such as MAPK and PI3K. It is not known the reason for the availability of *trans* activation as it leads to decreased differentiation and decreases neural survival (Encinas et al., 2001). RET activation of Src has been shown to increase axon sprouting of dopamine (Akerud et al., 1999).

7.4 c-Jun N-terminal Kinase (JNK) pathway

The c-Jun N-terminal kinase (JNK) pathway is a subfamily of the MAPK pathway. This pathway plays a role in stress response in the cell and is activated by cytokines and environmental stresses (Figure 2.4) (Weston and Davis, 2007). MAPK phosphatases (MKP) negatively regulates the JNK pathway and these MKPs can be inhibited by reactive oxygen species, which causes increased activation of the JNK pathway and can lead to cellular death (Weston and Davis, 2007). There are 3 JNK genes (JNK1-3), but only JNK3 activates neuronal cell death (Sun et al., 2007, Weston and Davis, 2007). JNKs also include a group of scaffold proteins (JIP1-4) which interact with the mechanisms for vesicular transport, axon growth and axon repair after damage (Weston and Davis, 2007). Both Rac1 and Cdc42 activate the JNK pathway (Figure 2.4). Through this activation, they modulate cytoskeletal organization within the neuron as well as aid in neural migration (Sun et al., 2007).

GDNF also activates the JNK pathway. Through GDNF and its co-receptor GFRα1 activating RET, JNK has been shown to modulate neurite outgrowth and extension in dopaminergic neurons (Figure 2.4) (Chiariello et al., 1998). Additionally, this JNK activation causes a cell cycle delay at G2/M to allow actin reorganization to improve cell viability (Figure 2.4) (Fukuda et al., 2005).

8. Conclusion

The high prevalence of PD in the American population combined with the increasing percentage of aging population presents a need for improving upon the treatments currently available for the disease. Currently, the treatments available have not changed from the first largely available compound and the side effects obtained from this drug combined with the lack of long-term response suggest a need for a better treatment option. The models that have been used thus far have been animal models that do not offer a direct comparison to human physiology. hESC derived hNPs that are differentiated to dopaminergic neurons provide an optimal tool for studying the basic biology of dopaminergic neurons as well as for researching new drug options. The methods for deriving these neurons needs to be improved upon in order to provide better treatment options for PD.

9. References

Airaksinen MS, Saarma M (The GDNF family: signalling, biological functions and therapeutic value. Nat Rev Neurosci 3:383-394.2002).

Akerud P, Alberch J, Eketjall S, Wagner J, Arenas E (Differential effects of glial cell line-derived neurotrophic factor and neurturin on developing and adult substantia nigra dopaminergic neurons. J Neurochem 73:70-78.1999).

Alberi L, Sgado P, Simon HH (Engrailed genes are cell-autonomously required to prevent apoptosis in mesencephalic dopaminergic neurons. Development 131:3229-3236.2004).

Amit M, Shariki C, Margulets V, Itskovitz-Eldor J (Feeder layer- and serum-free culture of human embryonic stem cells. Biol Reprod 70:837-845.2004).

Barberi T, Klivenyi P, Calingasan NY, Lee H, Kawamata H, Loonam K, Perrier AL, Bruses J, Rubio ME, Topf N, Tabar V, Harrison NL, Beal MF, Moore MA, Studer L (Neural subtype specification of fertilization and nuclear transfer embryonic stem cells and application in parkinsonian mice. Nat Biotechnol 21:1200-1207.2003).

Bowenkamp KE, Hoffman AF, Gerhardt GA, Henry MA, Biddle PT, Hoffer BJ, Granholm AC (Glial cell line-derived neurotrophic factor supports survival of injured midbrain dopaminergic neurons. J Comp Neurol 355:479-489.1995).

Brederlau A, Correia AS, Anisimov SV, Elmi M, Paul G, Roybon L, Morizane A, Bergquist F, Riebe I, Nannmark U, Carta M, Hanse E, Takahashi J, Sasai Y, Funa K, Brundin P, Eriksson PS, Li JY (Transplantation of human embryonic stem cell-derived cells to a rat model of Parkinson's disease: effect of in vitro differentiation on graft survival and teratoma formation. Stem Cells 24:1433-1440.2006).

Buytaert-Hoefen KA, Alvarez E, Freed CR (Generation of tyrosine hydroxylase positive neurons from human embryonic stem cells after coculture with cellular substrates and exposure to GDNF. Stem Cells 22:669-674.2004).

Carnero A, Blanco-Aparicio C, Renner O, Link W, Leal JF (The PTEN/PI3K/AKT signalling pathway in cancer, therapeutic implications. Curr Cancer Drug Targets 8:187-198.2008).

Carpenter MK, Rosler E, Rao MS (Characterization and differentiation of human embryonic stem cells. Cloning Stem Cells 5:79-88.2003).

Chiariello M, Visconti R, Carlomagno F, Melillo RM, Bucci C, de Franciscis V, Fox GM, Jing S, Coso OA, Gutkind JS, Fusco A, Santoro M (Signalling of the Ret receptor tyrosine kinase through the c-Jun NH2-terminal protein kinases (JNKS): evidence for a divergence of the ERKs and JNKs pathways induced by Ret. Oncogene 16:2435-2445.1998).

Chiba S, Kurokawa MS, Yoshikawa H, Ikeda R, Takeno M, Tadokoro M, Sekino H, Hashimoto T, Suzuki N (Noggin and basic FGF were implicated in forebrain fate and caudal fate, respectively, of the neural tube-like structures emerging in mouse ES cell culture. Exp Brain Res 163:86-99.2005).

Cho MS, Lee YE, Kim JY, Chung S, Cho YH, Kim DS, Kang SM, Lee H, Kim MH, Kim JH, Leem JW, Oh SK, Choi YM, Hwang DY, Chang JW, Kim DW (Highly efficient and large-scale generation of functional dopamine neurons from human embryonic stem cells. Proc Natl Acad Sci U S A 105:3392-3397.2008).

Choi-Lundberg DL, Lin Q, Chang YN, Chiang YL, Hay CM, Mohajeri H, Davidson BL, Bohn MC (Dopaminergic neurons protected from degeneration by GDNF gene therapy. Science 275:838-841.1997).

Chojnacki A, Shimazaki T, Gregg C, Weinmaster G, Weiss S (Glycoprotein 130 signaling regulates Notch1 expression and activation in the self-renewal of mammalian forebrain neural stem cells. J Neurosci 23:1730-1741.2003).

Chung S, Leung A, Han BS, Chang MY, Moon JI, Kim CH, Hong S, Pruszak J, Isacson O, Kim KS (Wnt1-lmx1a forms a novel autoregulatory loop and controls midbrain dopaminergic differentiation synergistically with the SHH-FoxA2 pathway. Cell Stem Cell 5:646-658.2009).

Cooper O, Hargus G, Deleidi M, Blak A, Osborn T, Marlow E, Lee K, Levy A, Perez-Torres E, Yow A, Isacson O (Differentiation of human ES and Parkinson's disease iPS cells into ventral midbrain dopaminergic neurons requires a high activity form of SHH, FGF8a and specific regionalization by retinoic acid. Mol Cell Neurosci.2010).

Correia AS, Anisimov SV, Roybon L, Li JY, Brundin P (Fibroblast growth factor-20 increases the yield of midbrain dopaminergic neurons derived from human embryonic stem cells. Front Neuroanat 1:4.2007).

Danielian PS, McMahon AP (Engrailed-1 as a target of the Wnt-1 signalling pathway in vertebrate midbrain development. Nature 383:332-334.1996).

Duronio V (The life of a cell: apoptosis regulation by the PI3K/PKB pathway. Biochem J 415:333-344.2008).

Dziennis S, Habecker BA (Cytokine suppression of dopamine-beta-hydroxylase by extracellular signal-regulated kinase-dependent and -independent pathways. J Biol Chem 278:15897-15904.2003).

Elbaz A, Moisan F (Update in the epidemiology of Parkinson's disease. Curr Opin Neurol 21:454-460.2008).

Elkabetz Y, Panagiotakos G, Al Shamy G, Socci ND, Tabar V, Studer L (Human ES cell-derived neural rosettes reveal a functionally distinct early neural stem cell stage. Genes Dev 22:152-165.2008).

Elsworth JD, Redmond DE, Jr., Leranth C, Bjugstad KB, Sladek JR, Jr., Collier TJ, Foti SB, Samulski RJ, Vives KP, Roth RH (AAV2-mediated gene transfer of GDNF to the striatum of MPTP monkeys enhances the survival and outgrowth of co-implanted fetal dopamine neurons. Exp Neurol 211:252-258.2008).

Emborg ME, Ebert AD, Moirano J, Peng S, Suzuki M, Capowski E, Joers V, Roitberg BZ, Aebischer P, Svendsen CN (GDNF-secreting human neural progenitor cells increase tyrosine hydroxylase and VMAT2 expression in MPTP-treated cynomolgus monkeys. Cell Transplant 17:383-395.2008).

Encinas M, Tansey MG, Tsui-Pierchala BA, Comella JX, Milbrandt J, Johnson EM, Jr. (c-Src is required for glial cell line-derived neurotrophic factor (GDNF) family ligand-mediated neuronal survival via a phosphatidylinositol-3 kinase (PI-3K)-dependent pathway. J Neurosci 21:1464-1472.2001).

Engele J, Franke B (Effects of glial cell line-derived neurotrophic factor (GDNF) on dopaminergic neurons require concurrent activation of cAMP-dependent signaling pathways. Cell Tissue Res 286:235-240.1996).

Friling S, Andersson E, Thompson LH, Jonsson ME, Hebsgaard JB, Nanou E, Alekseenko Z, Marklund U, Kjellander S, Volakakis N, Hovatta O, El Manira A, Bjorklund A, Perlmann T, Ericson J (Efficient production of mesencephalic dopamine neurons by Lmx1a expression in embryonic stem cells. Proc Natl Acad Sci U S A 106:7613-7618.2009).

Fukuda T, Asai N, Enomoto A, Takahashi M (Activation of c-Jun amino-terminal kinase by GDNF induces G2/M cell cycle delay linked with actin reorganization. Genes Cells 10:655-663.2005).

Gaig C, Tolosa E (When does Parkinson's disease begin? Mov Disord 24 Suppl 2:S656-664.2009).

Galli R, Pagano SF, Gritti A, Vescovi AL (Regulation of neuronal differentiation in human CNS stem cell progeny by leukemia inhibitory factor. Dev Neurosci 22:86-95.2000).

Garcia-Martinez JM, Perez-Navarro E, Gavalda N, Alberch J (Glial cell line-derived neurotrophic factor promotes the arborization of cultured striatal neurons through the p42/p44 mitogen-activated protein kinase pathway. J Neurosci Res 83:68-79.2006).

Gash DM, Zhang Z, Ovadia A, Cass WA, Yi A, Simmerman L, Russell D, Martin D, Lapchak PA, Collins F, Hoffer BJ, Gerhardt GA (Functional recovery in parkinsonian monkeys treated with GDNF. Nature 380:252-255.1996).

Georgievska B, Kirik D, Rosenblad C, Lundberg C, Bjorklund A (Neuroprotection in the rat Parkinson model by intrastriatal GDNF gene transfer using a lentiviral vector. Neuroreport 13:75-82.2002).

Gerhardt GA, Cass WA, Huettl P, Brock S, Zhang Z, Gash DM (GDNF improves dopamine function in the substantia nigra but not the putamen of unilateral MPTP-lesioned rhesus monkeys. Brain Res 817:163-171.1999).

Gulino A, Di Marcotullio L, Ferretti E, De Smaele E, Screpanti I (Hedgehog signaling pathway in neural development and disease. Psychoneuroendocrinology 32 Suppl 1:S52-56.2007).

Haddley K, Vasiliou AS, Ali FR, Paredes UM, Bubb VJ, Quinn JP (Molecular genetics of monoamine transporters: relevance to brain disorders. Neurochem Res 33:652-667.2008).

Hansson SR, Mezey E, Hoffman BJ (Ontogeny of vesicular monoamine transporter mRNAs VMAT1 and VMAT2. II. Expression in neural crest derivatives and their target sites in the rat. Brain Res Dev Brain Res 110:159-174.1998).

Harrington KA, Augood SJ, Kingsbury AE, Foster OJ, Emson PC (Dopamine transporter (Dat) and synaptic vesicle amine transporter (VMAT2) gene expression in the substantia nigra of control and Parkinson's disease. Brain Res Mol Brain Res 36:157-162.1996).

Hayashi H, Ichihara M, Iwashita T, Murakami H, Shimono Y, Kawai K, Kurokawa K, Murakumo Y, Imai T, Funahashi H, Nakao A, Takahashi M (Characterization of intracellular signals via tyrosine 1062 in RET activated by glial cell line-derived neurotrophic factor. Oncogene 19:4469-4475.2000).

Hou JG, Lin LF, Mytilineou C (Glial cell line-derived neurotrophic factor exerts neurotrophic effects on dopaminergic neurons in vitro and promotes their survival and regrowth after damage by 1-methyl-4-phenylpyridinium. J Neurochem 66:74-82.1996).

Humphrey RK, Beattie GM, Lopez AD, Bucay N, King CC, Firpo MT, Rose-John S, Hayek A (Maintenance of pluripotency in human embryonic stem cells is STAT3 independent. Stem Cells 22:522-530.2004).

Hwang DY, Ardayfio P, Kang UJ, Semina EV, Kim KS (Selective loss of dopaminergic neurons in the substantia nigra of Pitx3-deficient aphakia mice. Brain Res Mol Brain Res 114:123-131.2003).

Hwang DY, Hong S, Jeong JW, Choi S, Kim H, Kim J, Kim KS (Vesicular monoamine transporter 2 and dopamine transporter are molecular targets of Pitx3 in the ventral midbrain dopamine neurons. J Neurochem 111:1202-1212.2009).

Hynes M, Porter JA, Chiang C, Chang D, Tessier-Lavigne M, Beachy PA, Rosenthal A (Induction of midbrain dopaminergic neurons by Sonic hedgehog. Neuron 15:35-44.1995).

Itsykson P, Ilouz N, Turetsky T, Goldstein RS, Pera MF, Fishbein I, Segal M, Reubinoff BE (Derivation of neural precursors from human embryonic stem cells in the presence of noggin. Mol Cell Neurosci 30:24-36.2005).

Iversen SD, Iversen LL (Dopamine: 50 years in perspective. Trends Neurosci 30:188-193.2007).

Jacobs FM, Smits SM, Noorlander CW, von Oerthel L, van der Linden AJ, Burbach JP, Smidt MP (Retinoic acid counteracts developmental defects in the substantia nigra caused by Pitx3 deficiency. Development 134:2673-2684.2007).

Joksimovic M, Anderegg A, Roy A, Campochiaro L, Yun B, Kittappa R, McKay R, Awatramani R (Spatiotemporally separable Shh domains in the midbrain define distinct dopaminergic progenitor pools. Proc Natl Acad Sci U S A 106:19185-19190.2009).

Jordan PM, Ojeda LD, Thonhoff JR, Gao J, Boehning D, Yu Y, Wu P (Generation of spinal motor neurons from human fetal brain-derived neural stem cells: role of basic fibroblast growth factor. J Neurosci Res 87:318-332.2009).

Kadkhodaei B, Ito T, Joodmardi E, Mattsson B, Rouillard C, Carta M, Muramatsu S, Sumi-Ichinose C, Nomura T, Metzger D, Chambon P, Lindqvist E, Larsson NG, Olson L, Bjorklund A, Ichinose H, Perlmann T (Nurr1 is required for maintenance of maturing and adult midbrain dopamine neurons. J Neurosci 29:15923-15932.2009).

Kawasaki H, Mizuseki K, Nishikawa S, Kaneko S, Kuwana Y, Nakanishi S, Nishikawa SI, Sasai Y (Induction of midbrain dopaminergic neurons from ES cells by stromal cell-derived inducing activity. Neuron 28:31-40.2000).

Kearns CM, Gash DM (GDNF protects nigral dopamine neurons against 6-hydroxydopamine in vivo. Brain Res 672:104-111.1995).

Kirik D, Rosenblad C, Bjorklund A, Mandel RJ (Long-term rAAV-mediated gene transfer of GDNF in the rat Parkinson's model: intrastriatal but not intranigral transduction promotes functional regeneration in the lesioned nigrostriatal system. J Neurosci 20:4686-4700.2000).

Ko JY, Park CH, Koh HC, Cho YH, Kyhm JH, Kim YS, Lee I, Lee YS, Lee SH (Human embryonic stem cell-derived neural precursors as a continuous, stable, and on-demand source for human dopamine neurons. J Neurochem 103:1417-1429.2007).

Kordower JH (In vivo gene delivery of glial cell line--derived neurotrophic factor for Parkinson's disease. Ann Neurol 53 Suppl 3:S120-132; discussion S132-124.2003).

Kordower JH, Chu Y, Hauser RA, Olanow CW, Freeman TB (Transplanted dopaminergic neurons develop PD pathologic changes: a second case report. Mov Disord 23:2303-2306.2008).

Kordower JH, Emborg ME, Bloch J, Ma SY, Chu Y, Leventhal L, McBride J, Chen EY, Palfi S, Roitberg BZ, Brown WD, Holden JE, Pyzalski R, Taylor MD, Carvey P, Ling Z, Trono D, Hantraye P, Deglon N, Aebischer P (Neurodegeneration prevented by lentiviral vector delivery of GDNF in primate models of Parkinson's disease. Science 290:767-773.2000).

Kozlowski DA, Connor B, Tillerson JL, Schallert T, Bohn MC (Delivery of a GDNF gene into the substantia nigra after a progressive 6-OHDA lesion maintains functional nigrostriatal connections. Exp Neurol 166:1-15.2000).

Lapchak PA, Araujo DM, Hilt DC, Sheng J, Jiao S (Adenoviral vector-mediated GDNF gene therapy in a rodent lesion model of late stage Parkinson's disease. Brain Res 777:153-160.1997).

Law SW, Conneely OM, DeMayo FJ, O'Malley BW (Identification of a new brain-specific transcription factor, NURR1. Mol Endocrinol 6:2129-2135.1992).

Lebel M, Gauthier Y, Moreau A, Drouin J (Pitx3 activates mouse tyrosine hydroxylase promoter via a high-affinity binding site. J Neurochem 77:558-567.2001).

Lehnert H, Wurtman RJ (Amino acid control of neurotransmitter synthesis and release: physiological and clinical implications. Psychother Psychosom 60:18-32.1993).

Li R, Thode S, Zhou J, Richard N, Pardinas J, Rao MS, Sah DW (Motoneuron differentiation of immortalized human spinal cord cell lines. J Neurosci Res 59:342-352.2000).

Li XJ, Du ZW, Zarnowska ED, Pankratz M, Hansen LO, Pearce RA, Zhang SC (Specification of motoneurons from human embryonic stem cells. Nat Biotechnol 23:215-221.2005).

Lim SY, Lang AE (The nonmotor symptoms of Parkinson's disease--an overview. Mov Disord 25 Suppl 1:S123-130.2010).

Lin LF, Doherty DH, Lile JD, Bektesh S, Collins F (GDNF: a glial cell line-derived neurotrophic factor for midbrain dopaminergic neurons. Science 260:1130-1132.1993).

Lin W, Metzakopian E, Mavromatakis YE, Gao N, Balaskas N, Sasaki H, Briscoe J, Whitsett JA, Goulding M, Kaestner KH, Ang SL (Foxa1 and Foxa2 function both upstream of and cooperatively with Lmx1a and Lmx1b in a feedforward loop promoting mesodiencephalic dopaminergic neuron development. Dev Biol 333:386-396.2009).

Ling ZD, Potter ED, Lipton JW, Carvey PM (Differentiation of mesencephalic progenitor cells into dopaminergic neurons by cytokines. Exp Neurol 149:411-423.1998).

Liu A, Joyner AL (EN and GBX2 play essential roles downstream of FGF8 in patterning the mouse mid/hindbrain region. Development 128:181-191.2001).

Liu J, Zang D (Response of neural precursor cells in the brain of Parkinson's disease mouse model after LIF administration. Neurol Res 31:681-686.2009).

Lo Bianco C, Schneider BL, Bauer M, Sajadi A, Brice A, Iwatsubo T, Aebischer P (Lentiviral vector delivery of parkin prevents dopaminergic degeneration in an alpha-synuclein rat model of Parkinson's disease. Proc Natl Acad Sci U S A 101:17510-17515.2004).

Maden M (Retinoic acid in the development, regeneration and maintenance of the nervous system. Nat Rev Neurosci 8:755-765.2007).

Maeda K, Murakami H, Yoshida R, Ichihara M, Abe A, Hirai M, Murohara T, Takahashi M (Biochemical and biological responses induced by coupling of Gab1 to phosphatidylinositol 3-kinase in RET-expressing cells. Biochem Biophys Res Commun 323:345-354.2004).

Manning BD, Cantley LC (AKT/PKB signaling: navigating downstream. Cell 129:1261-1274.2007).

Maswood N, Grondin R, Zhang Z, Stanford JA, Surgener SP, Gash DM, Gerhardt GA (Effects of chronic intraputamenal infusion of glial cell line-derived neurotrophic factor (GDNF) in aged Rhesus monkeys. Neurobiol Aging 23:881-889.2002).

Matsuda T, Nakamura T, Nakao K, Arai T, Katsuki M, Heike T, Yokota T (STAT3 activation is sufficient to maintain an undifferentiated state of mouse embryonic stem cells. EMBO J 18:4261-4269.1999).

Maxwell SL, Ho HY, Kuehner E, Zhao S, Li M (Pitx3 regulates tyrosine hydroxylase expression in the substantia nigra and identifies a subgroup of mesencephalic dopaminergic progenitor neurons during mouse development. Dev Biol 282:467-479.2005).

Mitalipova M, Calhoun J, Shin S, Wininger D, Schulz T, Noggle S, Venable A, Lyons I, Robins A, Stice S (Human embryonic stem cell lines derived from discarded embryos. Stem Cells 21:521-526.2003).

Mora F, Segovia G, Del Arco A (Glutamate-dopamine-GABA interactions in the aging basal ganglia. Brain Res Rev 58:340-353.2008).

Murphy M, Reid K, Hilton DJ, Bartlett PF (Generation of sensory neurons is stimulated by leukemia inhibitory factor. Proc Natl Acad Sci U S A 88:3498-3501.1991).

Nicole O, Ali C, Docagne F, Plawinski L, MacKenzie ET, Vivien D, Buisson A (Neuroprotection mediated by glial cell line-derived neurotrophic factor: involvement of a reduction of NMDA-induced calcium influx by the mitogen-activated protein kinase pathway. J Neurosci 21:3024-3033.2001).

Nishimoto S, Nishida E (MAPK signalling: ERK5 versus ERK1/2. EMBO Rep 7:782-786.2006).

Nunes I, Tovmasian LT, Silva RM, Burke RE, Goff SP (Pitx3 is required for development of substantia nigra dopaminergic neurons. Proc Natl Acad Sci U S A 100:4245-4250.2003).

Obeso JA, Rodriguez-Oroz MC, Benitez-Temino B, Blesa FJ, Guridi J, Marin C, Rodriguez M (Functional organization of the basal ganglia: therapeutic implications for Parkinson's disease. Mov Disord 23 Suppl 3:S548-559.2008).

Obeso JA, Rodriguez-Oroz MC, Goetz CG, Marin C, Kordower JH, Rodriguez M, Hirsch EC, Farrer M, Schapira AH, Halliday G (Missing pieces in the Parkinson's disease puzzle. Nat Med 16:653-661.2010).

Ohiwa M, Murakami H, Iwashita T, Asai N, Iwata Y, Imai T, Funahashi H, Takagi H, Takahashi M (Characterization of Ret-Shc-Grb2 complex induced by GDNF, MEN 2A, and MEN 2B mutations. Biochem Biophys Res Commun 237:747-751.1997).

Pease S, Braghetta P, Gearing D, Grail D, Williams RL (Isolation of embryonic stem (ES) cells in media supplemented with recombinant leukemia inhibitory factor (LIF). Dev Biol 141:344-352.1990).

Pera MF, Andrade J, Houssami S, Reubinoff B, Trounson A, Stanley EG, Ward-van Oostwaard D, Mummery C (Regulation of human embryonic stem cell differentiation by BMP-2 and its antagonist noggin. J Cell Sci 117:1269-1280.2004).

Perrier AL, Tabar V, Barberi T, Rubio ME, Bruses J, Topf N, Harrison NL, Studer L (Derivation of midbrain dopamine neurons from human embryonic stem cells. Proc Natl Acad Sci U S A 101:12543-12548.2004).

Pimienta G, Pascual J (Canonical and alternative MAPK signaling. Cell Cycle 6:2628-2632.2007).

Poewe W (The natural history of Parkinson's disease. J Neurol 253 Suppl 7:VII2-6.2006).

Rao M (Conserved and divergent paths that regulate self-renewal in mouse and human embryonic stem cells. Dev Biol 275:269-286.2004).

Rao SS, Hofmann LA, Shakil A (Parkinson's disease: diagnosis and treatment. Am Fam Physician 74:2046-2054.2006).

Reubinoff BE, Itsykson P, Turetsky T, Pera MF, Reinhartz E, Itzik A, Ben-Hur T (Neural progenitors from human embryonic stem cells. Nat Biotechnol 19:1134-1140.2001).

Rosenblad C, Martinez-Serrano A, Bjorklund A (Glial cell line-derived neurotrophic factor increases survival, growth and function of intrastriatal fetal nigral dopaminergic grafts. Neuroscience 75:979-985.1996).

Roux PP, Blenis J (ERK and p38 MAPK-activated protein kinases: a family of protein kinases with diverse biological functions. Microbiol Mol Biol Rev 68:320-344.2004).

Roy NS, Cleren C, Singh SK, Yang L, Beal MF, Goldman SA (Functional engraftment of human ES cell-derived dopaminergic neurons enriched by coculture with telomerase-immortalized midbrain astrocytes. Nat Med 12:1259-1268.2006).

Ruvinsky I, Meyuhas O (Ribosomal protein S6 phosphorylation: from protein synthesis to cell size. Trends Biochem Sci 31:342-348.2006).

Sariola H, Saarma M (Novel functions and signalling pathways for GDNF. J Cell Sci 116:3855-3862.2003).

Sato T, Shimazaki T, Naka H, Fukami S, Satoh Y, Okano H, Lax I, Schlessinger J, Gotoh N (FRS2alpha Regulates Erk Levels to Control a Self-Renewal Target Hes1 and Proliferation of FGF-Responsive Neural Stem/Progenitor Cells. Stem Cells.2010).

Satoh M, Yoshida T (Promotion of neurogenesis in mouse olfactory neuronal progenitor cells by leukemia inhibitory factor in vitro. Neurosci Lett 225:165-168.1997).

Saucedo-Cardenas O, Quintana-Hau JD, Le WD, Smidt MP, Cox JJ, De Mayo F, Burbach JP, Conneely OM (Nurr1 is essential for the induction of the dopaminergic phenotype and the survival of ventral mesencephalic late dopaminergic precursor neurons. Proc Natl Acad Sci U S A 95:4013-4018.1998).

Schiff M, Weinhold B, Grothe C, Hildebrandt H (NCAM and polysialyltransferase profiles match dopaminergic marker gene expression but polysialic acid is dispensable for development of the midbrain dopamine system. J Neurochem 110:1661-1673.2009).

Schuldiner M, Eiges R, Eden A, Yanuka O, Itskovitz-Eldor J, Goldstein RS, Benvenisty N (Induced neuronal differentiation of human embryonic stem cells. Brain Res 913:201-205.2001).

Schulz TC, Noggle SA, Palmarini GM, Weiler DA, Lyons IG, Pensa KA, Meedeniya AC, Davidson BP, Lambert NA, Condie BG (Differentiation of human embryonic stem cells to dopaminergic neurons in serum-free suspension culture. Stem Cells 22:1218-1238.2004).

Sgado P, Alberi L, Gherbassi D, Galasso SL, Ramakers GM, Alavian KN, Smidt MP, Dyck RH, Simon HH (Slow progressive degeneration of nigral dopaminergic neurons in postnatal Engrailed mutant mice. Proc Natl Acad Sci U S A 103:15242-15247.2006).

Shin S, Dalton S, Stice SL (Human motor neuron differentiation from human embryonic stem cells. Stem Cells Dev 14:266-269.2005).

Shin S, Mitalipova M, Noggle S, Tibbitts D, Venable A, Rao R, Stice SL (Long-term proliferation of human embryonic stem cell-derived neuroepithelial cells using defined adherent culture conditions. Stem Cells 24:125-138.2006).

Shults CW, Kimber T, Martin D (Intrastriatal injection of GDNF attenuates the effects of 6-hydroxydopamine. Neuroreport 7:627-631.1996).

Sillitoe RV, Vogel MW (Desire, disease, and the origins of the dopaminergic system. Schizophr Bull 34:212-219.2008).

Smidt MP, Burbach JP (How to make a mesodiencephalic dopaminergic neuron. Nat Rev Neurosci 8:21-32.2007).

Smidt MP, Smits SM, Bouwmeester H, Hamers FP, van der Linden AJ, Hellemons AJ, Graw J, Burbach JP (Early developmental failure of substantia nigra dopamine neurons in mice lacking the homeodomain gene Pitx3. Development 131:1145-1155.2004).

Smith AD, Bolam JP (The neural network of the basal ganglia as revealed by the study of synaptic connections of identified neurones. Trends Neurosci 13:259-265.1990).

Smith Y, Shink E, Sidibe M (Neuronal circuitry and synaptic connectivity of the basal ganglia. Neurosurg Clin N Am 9:203-222.1998).

Smits SM, Burbach JP, Smidt MP (Developmental origin and fate of meso-diencephalic dopamine neurons. Prog Neurobiol 78:1-16.2006).

Smits SM, Mathon DS, Burbach JP, Ramakers GM, Smidt MP (Molecular and cellular alterations in the Pitx3-deficient midbrain dopaminergic system. Mol Cell Neurosci 30:352-363.2005).

Sonnier L, Le Pen G, Hartmann A, Bizot JC, Trovero F, Krebs MO, Prochiantz A (Progressive loss of dopaminergic neurons in the ventral midbrain of adult mice heterozygote for Engrailed1. J Neurosci 27:1063-1071.2007).

Speciale SG, Liang CL, Sonsalla PK, Edwards RH, German DC (The neurotoxin 1-methyl-4-phenylpyridinium is sequestered within neurons that contain the vesicular monoamine transporter. Neuroscience 84:1177-1185.1998).

Stankovski L, Alvarez C, Ouimet T, Vitalis T, El-Hachimi KH, Price D, Deneris E, Gaspar P, Cases O (Developmental cell death is enhanced in the cerebral cortex of mice lacking the brain vesicular monoamine transporter. J Neurosci 27:1315-1324.2007).

Storch A, Ludolph AC, Schwarz J (Dopamine transporter: involvement in selective dopaminergic neurotoxicity and degeneration. J Neural Transm 111:1267-1286.2004).

Storch A, Paul G, Csete M, Boehm BO, Carvey PM, Kupsch A, Schwarz J (Long-term proliferation and dopaminergic differentiation of human mesencephalic neural precursor cells. Exp Neurol 170:317-325.2001).

Su X, Kells AP, Huang EJ, Lee HS, Hadaczek P, Beyer J, Bringas J, Pivirotto P, Penticuff J, Eberling J, Federoff HJ, Forsayeth J, Bankiewicz KS (Safety evaluation of AAV2-GDNF gene transfer into the dopaminergic nigrostriatal pathway in aged and parkinsonian rhesus monkeys. Hum Gene Ther 20:1627-1640.2009).

Sullivan AM, Opacka-Juffry J, Blunt SB (Long-term protection of the rat nigrostriatal dopaminergic system by glial cell line-derived neurotrophic factor against 6-hydroxydopamine in vivo. Eur J Neurosci 10:57-63.1998).

Sun Y, Yang T, Xu Z (The JNK pathway and neuronal migration. J Genet Genomics 34:957-965.2007).

Swistowska AM, da Cruz AB, Han Y, Swistowski A, Liu Y, Shin S, Zhan M, Rao MS, Zeng X (Stage-specific role for shh in dopaminergic differentiation of human embryonic stem cells induced by stromal cells. Stem Cells Dev 19:71-82.2010).

Tansey MG, Baloh RH, Milbrandt J, Johnson EM, Jr. (GFRalpha-mediated localization of RET to lipid rafts is required for effective downstream signaling, differentiation, and neuronal survival. Neuron 25:611-623.2000).

Tansey MG, McCoy MK, Frank-Cannon TC (Neuroinflammatory mechanisms in Parkinson's disease: potential environmental triggers, pathways, and targets for early therapeutic intervention. Exp Neurol 208:1-25.2007).

Thomson JA, Itskovitz-Eldor J, Shapiro SS, Waknitz MA, Swiergiel JJ, Marshall VS, Jones JM (Embryonic stem cell lines derived from human blastocysts. Science 282:1145-1147.1998).

Tomac A, Widenfalk J, Lin LF, Kohno T, Ebendal T, Hoffer BJ, Olson L (Retrograde axonal transport of glial cell line-derived neurotrophic factor in the adult nigrostriatal system suggests a trophic role in the adult. Proc Natl Acad Sci U S A 92:8274-8278.1995).

Ugarte SD, Lin E, Klann E, Zigmond MJ, Perez RG (Effects of GDNF on 6-OHDA-induced death in a dopaminergic cell line: modulation by inhibitors of PI3 kinase and MEK. J Neurosci Res 73:105-112.2003).

van den Munckhof P, Luk KC, Ste-Marie L, Montgomery J, Blanchet PJ, Sadikot AF, Drouin J (Pitx3 is required for motor activity and for survival of a subset of midbrain dopaminergic neurons. Development 130:2535-2542.2003).

Vazin T, Becker KG, Chen J, Spivak CE, Lupica CR, Zhang Y, Worden L, Freed WJ (A novel combination of factors, termed SPIE, which promotes dopaminergic neuron differentiation from human embryonic stem cells. PLoS One 4:e6606.2009).

Vazin T, Chen J, Lee CT, Amable R, Freed WJ (Assessment of stromal-derived inducing activity in the generation of dopaminergic neurons from human embryonic stem cells. Stem Cells 26:1517-1525.2008).

Volz TJ, Schenk JO (A comprehensive atlas of the topography of functional groups of the dopamine transporter. Synapse 58:72-94.2005).

Wallen AA, Castro DS, Zetterstrom RH, Karlen M, Olson L, Ericson J, Perlmann T (Orphan nuclear receptor Nurr1 is essential for Ret expression in midbrain dopamine neurons and in the brain stem. Mol Cell Neurosci 18:649-663.2001).

Wang Y, Tien LT, Lapchak PA, Hoffer BJ (GDNF triggers fiber outgrowth of fetal ventral mesencephalic grafts from nigra to striatum in 6-OHDA-lesioned rats. Cell Tissue Res 286:225-233.1996).

Weintraub D, Comella CL, Horn S (Parkinson's disease--Part 1: Pathophysiology, symptoms, burden, diagnosis, and assessment. Am J Manag Care 14:S40-48.2008a).

Weintraub D, Comella CL, Horn S (Parkinson's disease--Part 3: Neuropsychiatric symptoms. Am J Manag Care 14:S59-69.2008b).

Weston CR, Davis RJ (The JNK signal transduction pathway. Curr Opin Cell Biol 19:142-149.2007).

Winkler C, Sauer H, Lee CS, Bjorklund A (Short-term GDNF treatment provides long-term rescue of lesioned nigral dopaminergic neurons in a rat model of Parkinson's disease. J Neurosci 16:7206-7215.1996).

Wullschleger S, Loewith R, Hall MN (TOR signaling in growth and metabolism. Cell 124:471-484.2006).

Xu C, Inokuma MS, Denham J, Golds K, Kundu P, Gold JD, Carpenter MK (Feeder-free growth of undifferentiated human embryonic stem cells. Nat Biotechnol 19:971-974.2001).

Xu RH, Peck RM, Li DS, Feng X, Ludwig T, Thomson JA (Basic FGF and suppression of BMP signaling sustain undifferentiated proliferation of human ES cells. Nat Methods 2:185-190.2005).

Yan Y, Yang D, Zarnowska ED, Du Z, Werbel B, Valliere C, Pearce RA, Thomson JA, Zhang SC (Directed differentiation of dopaminergic neuronal subtypes from human embryonic stem cells. Stem Cells 23:781-790.2005).

Yang D, Zhang ZJ, Oldenburg M, Ayala M, Zhang SC (Human embryonic stem cell-derived dopaminergic neurons reverse functional deficit in parkinsonian rats. Stem Cells 26:55-63.2008).

Yap TA, Garrett MD, Walton MI, Raynaud F, de Bono JS, Workman P (Targeting the PI3K-AKT-mTOR pathway: progress, pitfalls, and promises. Curr Opin Pharmacol 8:393-412.2008).

Ye W, Bouchard M, Stone D, Liu X, Vella F, Lee J, Nakamura H, Ang SL, Busslinger M, Rosenthal A (Distinct regulators control the expression of the mid-hindbrain organizer signal FGF8. Nat Neurosci 4:1175-1181.2001).

Ye W, Shimamura K, Rubenstein JL, Hynes MA, Rosenthal A (FGF and Shh signals control dopaminergic and serotonergic cell fate in the anterior neural plate. Cell 93:755-766.1998).

Ying QL, Nichols J, Chambers I, Smith A (BMP induction of Id proteins suppresses differentiation and sustains embryonic stem cell self-renewal in collaboration with STAT3. Cell 115:281-292.2003).

Young A, Assey K, West FD, Sturkie CD, Machacek DW, Stice SL (Glial Cell-line Derived Neurotrophic Factor Enhances in vitro Differentiation of Mid/hindbrain Neural Progenitor Cells to Dopamingeric-like Neurons. Journal of Neuroscience Research.2010).

Zeng X, Cai J, Chen J, Luo Y, You ZB, Fotter E, Wang Y, Harvey B, Miura T, Backman C, Chen GJ, Rao MS, Freed WJ (Dopaminergic differentiation of human embryonic stem cells. Stem Cells 22:925-940.2004).

Zeng X, Chen J, Deng X, Liu Y, Rao MS, Cadet JL, Freed WJ (An in vitro model of human dopaminergic neurons derived from embryonic stem cells: MPP+ toxicity and GDNF neuroprotection. Neuropsychopharmacology 31:2708-2715.2006).

Zetterstrom RH, Solomin L, Jansson L, Hoffer BJ, Olson L, Perlmann T (Dopamine neuron agenesis in Nurr1-deficient mice. Science 276:248-250.1997).

Zetterstrom RH, Williams R, Perlmann T, Olson L (Cellular expression of the immediate early transcription factors Nurr1 and NGFI-B suggests a gene regulatory role in several brain regions including the nigrostriatal dopamine system. Brain Res Mol Brain Res 41:111-120.1996).

Zhang SC, Wernig M, Duncan ID, Brustle O, Thomson JA (In vitro differentiation of transplantable neural precursors from human embryonic stem cells. Nat Biotechnol 19:1129-1133.2001).

21

Dopaminergic Differentiation Potential of Neural Precursor Cells Derived from Embryonic Stem Cells

Gilda Guerrero-Flores and Luis Covarrubias
Department of Developmental Genetics and Molecular Physiology, Instituto de Biotecnología. Universidad Nacional Autónoma de México. Cuernavaca, Mor., México

1. Introduction

One major question in mammalian development is how the diverse cell types of the adult organism are generated from the initial population of undifferentiated cells in the preimplantation embryo. As part of this diversification process, the variety of neurons of the adult brain is generated from neural precursors specified during gastrulation and the posterior neurulation phase. The specification of neural precursors has a definitive influence on the fate of these cells and, consequently, in the neuron type derived after differentiation. Embryonic stem cells (ESCs), pluripotent cells derived from the inner cell mass of the blastocyt, can be used as a source of different neuron types *in vitro*, and the process of neuronal differentiation can thus be studied from the earliest stages starting from naïve non-neural undifferentiated cells. In particular, differentiation of ESCs into dopaminergic neurons has attracted a lot of attention for the relevance it may have in the design of cellular treatments for Parkinson's illness, a neurodegenerative disease characterized by the specific death of these neurons.

The midbrain dopaminergic (mDA) neurons constitute about 75% of all dopaminergic neurons in the adult brain(Wallen & Perlmann, 2003). They are located in the ventral region of the mesencephalon where they are organized laterally in the retrorubral field (RRF) and the substantia nigra pars compacta (SNc), and medially in the ventral tegmental area (VTA). The SNc neurons project to the dorsal striatum, forming the nigrostriatal pathway involved in the control of voluntary movements, and degeneration of this group of mDA neurons provokes the characteristic symptoms of Parkinson's disease (von Bohlen und Halbach, 2004). The neurons of the VTA project to the ventromedial striatum and the subcortical and cortical areas, forming the mesocortical limbic system, which is involved in emotional behaviors and mechanisms of motivation and reward. Misregulation of mesocortical limbic system has been involved in the development of drug addiction and depression (Kelley & Berridge, 2002), and contributes to certain symptoms of schizophrenia (Egan & Weinberger, 1997). Unlike SNc and VTA groups, RRF neurons have not been widely studied but it is known they project to the dorsal striatum and also connect SNc and VTA neurons (Arts et al., 1996).

There have been extensive efforts focused on the generation of mDA neurons from ESC, however, the limited success in this regard is largely due to our still only basic

understanding of the molecular mechanisms underlying mDA neuron development. Presently, it is known that the morphogens Sonic Hedgehog (Shh), Fibroblast growth factor 8 (Fgf8) and Wnt1 (Ye et al., 1998, Prakash et al., 2006) must act on early neural precursors at a particular dose and specific time window, as a failure to do so reduces or inhibits the correct differentiation process. It is known that the sequential and combinatory action of the morphogens activate at least two transcriptional factors, Lmx1a and Foxa2 that interact directly or indirectly to regulate the expression of other transcription factors promoting dopaminergic differentiation or inhibiting alternative neuronal fates (Nakatani et al., 2010, Lin et al., 2009, Ferri et al., 2007, Andersson et al., 2006). The complete potential of ESCs to differentiate into dopaminergic neurons is revealed once embryoid body (EB) cells derived from ESCs, are transplanted to explants of embryonic mesencephalon, where they efficiently and specifically respond to morphogens by differentiating and expressing the genes encoding the relevant transcription factors (Baizabal & Covarrubias, 2009).
Despite recent advances in understanding mDA development, the molecular interaction networks between the extrinsic and intrinsic factors involved are far from being fully understood. In the present chapter we will review how knowledge of mDA development has aided the refinement of differentiation protocols to generate mDA neurons from ESC and, how the ESC-based studies have contributed to understanding the mechanisms of differentiation into mDA neurons. We will discuss these data in the context of the differentiation potential of neural stem cells from different sources, the mechanism of stem cell specification, and the relevance that this has in defining the stem cell population and the differentiation conditions useful for the improvement of protocols for producing mDA neurons for the treatment of Parkinson Disease.

2. Development of midbrain dopaminergic neurons

Midbrain-DA neuronal differentiation initiates in the area rostral to the isthmus organizer and depends on the integration of anteroposterior and dorsoventral signals. These signals promote two related but distinct processes: specification of the general midbrain region, and specification of mDA precursor cells in particular (Fig.1). The distinction between these two processes is essential for the definition of the signaling cascade that leads to the generation of dopaminergic neurons.
The isthmus is formed at the mid-hindbrain border due to the complementary action of the *Otx2* and *Gbx2* genes in the anterior and posterior epiblast, respectively, at the end of gastrulation (E7.5) (Wurst & Bally-Cuif, 2001). Subsequently, *Pax2* is expressed at the interface of the *Otx2/Gbx2* domains, and Wnt1 in the Otx2-positive territories. The expression of *Fgf8* starts at E8 at the caudal *Gbx2* expression domain and later restricts to the isthmus at around E9, forming a mirror image of the *Wnt1* expression domain. The activation of *Fgf8* expression is controlled by Lmx1b which, directly or indirectly, is also important for maintenaning expression of other isthmus genes such as *Wnt1, En1/2, Pax2* and *Gbx2* (Guo et al., 2007). The genes encoding the transcription factors En1/2 are expressed throughout the *Otx2/Gbx2* expression domain interface; there, they are required for the maintenance of *Fgf8* expression, though its early expression is independent of these two genes (Simon et al., 2004). Most of the morphogen activity of the isthmus is due to the productos of *Fgf8* and *Wnt1* genes.
Crosstalk between the signals generated by Fgf8 and Wnt1 contributes to establish the identity and number of precursor cells committed to differentiate into mDA neurons (Lee et

al., 1997, Liu & Joyner, 2001, Matsunaga et al., 2002)(Fig. 1). In addition to their role in the formation of the midbrain, Fgf8 and Wnt1 seem to have a direct function in mDA neuron development. Fgf8 regulates *Wnt1* expression in the isthmus (Chi et al., 2003) but an Fgf activity also controls *Wnt1* expression in the ventral midbrain as revealed in double and triple mutant knockout mice for the genes encoding two or three of the Fgf receptors (Saarimaki-Vire et al., 2007). Importantly, Wnt1 is required for the ectopic induction of mDA neurons by Fgf8 in the forebrain and hindbrain (Prakash et al., 2006).

At the same time as the isthmus is forming, Shh from the notochord induces the estabishment of the floor plate (FP) along the neural tube (Yamada et al., 1991) (Fig.1). *Foxa2* expression in FP cells, via Shh activity-dependent Gli2 transcription factor binding sites in the *Foxa2* promoter (Sasaki et al., 1997), is required for the formation of the FP (Matise et al., 1998). Foxa2, in turn, directly induces *Shh* expression in ventral domains of mesenephalon where is essential for the specification of ventral neural precursors (Jeong & Epstein, 2003). Fate-mapping experiments clearly establish that cells expressing *Shh* in the FP become mDA neurons (Ono et al., 2007). This is a a unique characteristic of mDA neurons, as at other regions along the ventral neural tube, FP cells are non-neurogenic (Joksimovic et al., 2009; Ono, et al., 2007). As expected, lack of Foxa2 and the related transcription factor Foxa1 reduces the number of mDA neurons produced in the midbrain (Lin et al., 2009).

Lmx1a and Lmx1b are among the first markers that identify the mDA precursors and are key regulators of their differentiation. As with Fgf8 and Wnt1, Lmx1b has two separate roles; one is related to the formation and maintenance of the isthmus, as described above, and, consequently is essential for midbrain development; whereas the other seems to be directly involved in mDA neuron generation as *Lmx1b* expression is detected before emergence of mDA neurons (E7.5) and is maintained in the SNc and VTA until adulthood (Smidt et al., 2000). On the other hand, Lmx1a, whose gene is expressed in ventral midline of mesencephalon around E9, appears essential for mDA differentiation in the chick and its absence causes a reduction in dopaminergic neurons in the mouse (Ono et al., 2007, Andersson et al., 2006). Otx2 is necessary for the expression of *Lmx1a* in FP cells but, apparently, not for *Lmx1b* (Omodei et al., 2008, Ono et al., 2007). In addition, recent data suggest that Foxa2 is also required to induce and/or to maintain *Lmx1a* and *Lmx1b* expression (Lin et al., 2009). Therefore, Foxa2 and Otx2 may cooperatively function to define *Lmx1a,* and possibly also *Lmx1b,* expression in ventral midbrain precursors. Until now, it is not clear if Lmx1a and Lmx1b work in independent pathways or if they have redundant activities (see below).

Downstream of Shh signaling, Foxa1/2 inhibit the expression of the gene encoding Nkx2.2 (Lin et al., 2009), a transcription factor present adjacent to the Lmx1a domain at early stages, and later between the dorsal and ventral Nkx6.1 domains in a narrow band that eventually give rise to GABAergic neurons (Nakatani et al., 2007)(Fig.1). On the other hand, Nkx6.1 is present in the mDA domain at the beginning of *Lmx1a* expression, but later its gene is repressed by Msx1; the Shh secreted by the FP activates *Msx1* expression in the ventral midbrain (Andersson et al., 2006). Thus, *Nkx6.1* expression only remains in the regions adjacent to the Lmx1a domain, where motoneurons are generated (Fedtsova & Turner, 2001).

The generation of mDA neurons must be coordinated with the conversion of FP cells into neuronal progenitors. This transition might be due to the action of the canonical Wnt signaling, activated by Otx2 in the midbrain (Ono et al., 2007, Omodei et al., 2008)(Fig. 1). Around E10.5, Wnt1 suppresses *Shh* expression levels, possibly via an Msx1-mediated mechanism (Joksimovic et al., 2009), and consequently, induces the expression of *Ngn2*, encoding Neurogenin 2, a panneural basic helix-loop-helix protein (Ono et al., 2007). Foxa1/2 apparently also contribute to the activation of *Ngn2* expression (Ferri et al., 2007),which reinforces this mDA differentiation regulatory pathway.

The induction of *Ngn2* expression together with that of *Mash1* mark the initiation of neurogenesis itself (Kele et al., 2006)(Fig. 1). Neurogenesis is immediately follow by the expression of *Nurr1* and *Pitx3*, genes encoding transcription factors that are involved in defining the dopaminergic phenotypic characteristics such as the presence of tyrosine hydroxylase (the rate-limiting enzyme in dopamine synthesis; TH), vesicular monoamine transporter 2 (Vmat2) and dopamine transporter (DAT) (Zetterstrom et al., 1997, Smits et al., 2003, Saucedo-Cardenas et al., 1998, Kim et al., 2003b, Ferri et al., 2007)(Fig. 1). Also, Nurr1 and Pitx3 appear to play a role in the maintenance of the mDA neurons (Smidt et al., 2004, Nunes et al., 2003). Foxa1/2 function cooperatively with Lmx1a to regulate *Nurr1* expression in immature neurons, and also directly activate *Th* expression in mature neurons (Fig. 1). Therefore, it is possible that Nurr1 and Foxa2 cooperate to regulate *Th* expression during mDA differentiation. En1 and En2, in addition to their participation in the establishment of the mid-hindbrain border at early stages, also play an important role in the survival of mDA neurons (Alberi et al., 2004, Simon et al., 2001, Simon et al., 2004).

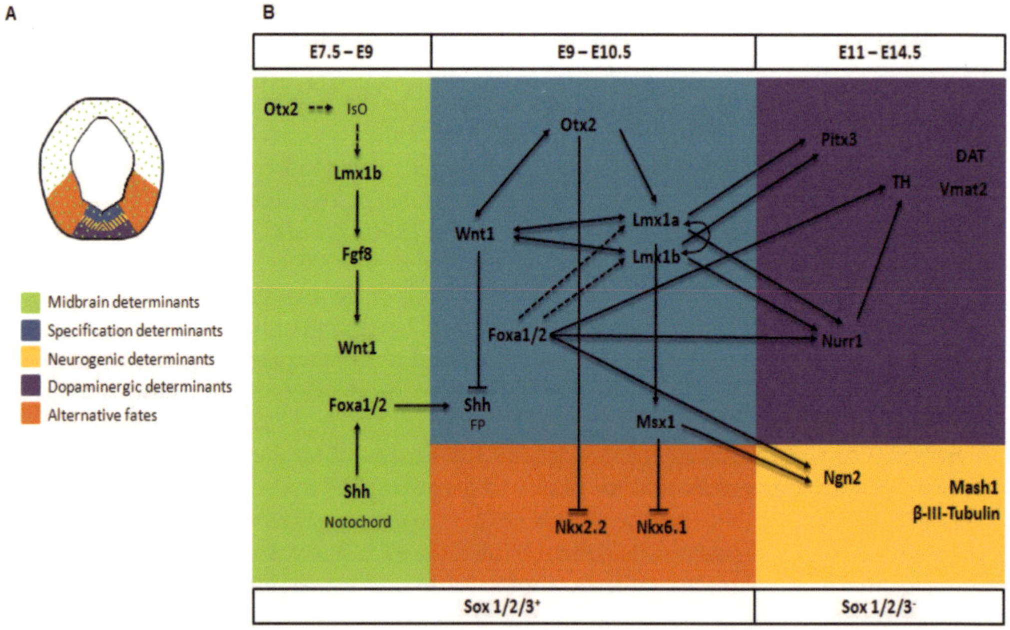

Fig. 1. Molecular network controlling mDA differentiation. (A) Frontal view representation of the embryonic mesencephalon in which the distinct regulatory events during mDA neuron development are shown. (B) Interactions between the distinct determinants of the key events that lead specific mDA differentiation during development (represented by the same colors as in A). The dashed arrows represent indirect or unconfirmed regulations. IsO, Isthmus organizer; FP, floor plate.

3. Neuralization and specification of ESC

The capacity of ESC to generate all cell types present in the embryo has made them a potential source of different neuron types *in vitro,* and so the process of specific neural differentiation can be studied since its very first stage and manipulated in an easier and faster manner than in embryos.

Ectodermal differentiation of ESC through EB formation is commonly used to produce neural precursor cells (NPCs) *in vitro.* The formation of EBs reflects the early developmental stages that lead to the formation of the epiblast and the primitive endoderm, immediately previous to gastrulation. EB cells can be neuralized with retinoic acid and produce about 40% of neurons of several types (Fraichard et al., 1995, Bain et al., 1995). Alternatively, EB cells can be cultured in a neural-defined medium, and the NPC population be expanded with Fgf2; under these conditions, more than 70% of the total cells are NPC which can be differentiated into neurons by removing Fgf2 (Okabe et al., 1996). Retinoic acid not only induces neuralization, but also instructs the cells to posterior fates (Okada et al., 2004), which limits the neuron types that can be derived from EBs; this potential problem can be overcome by treating the cells with low doses of RA. In fact, endogenous RA appears to be required for the neuralization in the absence of a direct RA addition to the medium (Engberg et al., 2010). Bmp signaling represses neural fates *in vivo* (Hemmati-Brivanlou & Melton, 1997, Finley et al., 1999); in concordance, Noggin and Chordin, two Bmp inhibitors, induce neural differentiation of EB cells (Gratsch & O'Shea, 2002). Another condition that promotes neural differentiation of EB cells is to block the Wnt/β-catenin pathway by adding its inhibitor Dkk-1 (Verani et al., 2007) or in the absence of Wnt and Nodal in cultures of EB cells in serum-free conditions (Watanabe et al., 2005).

Neuronal differentiation of ESC can also be induced without forming EBs. One protocol involves culturing the cells at low densities in a chemically defined, serum-free medium with LIF and Fgf2 (Tropepe et al., 2001). The cells of the spherical colonies formed in these conditions express the neural marker Nestin, and can differentiate further into neurons and glia; of note is that even though all cells in the colonies are NPC, very few of the cultured ESC generate colonies (0.2%). It has been proposed that the colonies represent primitive neuroectodermal cells, as similar colonies can be obtained from the embryo (Hitoshi et al., 2004). A similar protocol to generate neural precursors without the formation of EBs is to grow the ESCs as a monolayer at low density in serum-free conditions with N2B27 medium (Ying et al., 2003). After 4 days, 60% of the cells express *Sox1*, one of the first markers of neuroectodermal cells and these Sox1-positive cells, can later differentiate into mature neurons. This latter protocol could represent a simpler and direct system to assess neural differentiation and specification as exogenous growth factors and the formation of EB are not required.

In order to generate a particular type of neurons, it is necessary not only to neuralize the cells but also to expose them to other factors that promote specific neuronal differentiation. For instance, treating EBs with RA (with neuralizing and posteriorizing activities) and Shh (with ventralizing activity) oblige the cells to differentiate into motor neurons of the spinal cord (Wichterle et al., 2002). On the other hand, neural cells generated after inhibition of Wnt and Nodal signaling acquire a telencephalic identity (expression of *Foxg1*), and these neural cells can later be differentiated into subpopulations of telencephalic neurons depending on whether Wnt1 (forming pallial telencephalic neurons $Foxg1^{+}Pax6^{+}$) or Shh (basal telencephalic neurons $Foxg1^{+}Nkx2.1^{+}$) is added to the culture medium (Watanabe et al., 2005).

The fact that, in different culture conditions that promote ESC neuronal differentiation, NPCs emerge prior to neuron formation (as determined by the expression of *Sox1*, *Sox2* and *Nestin*), and that at least some of these NPCs respond to morphogenetic cues, suggest that ESC neural differentiation *in vitro* follows a comparable program to that occurring *in vivo*. Therefore, ESCs are useful as a model to study early neural differentiation.

4. Specification and differentiation of ESCs into mDA neurons

4.1 Response of ESC-derived cells to extrinsic determinants

Shh and Fgf8 were the first morphogens identified that are sufficient to induce the generation of ectopic mDA neurons in explant cultures (Ye et al., 1998). Expectedly, these morphogens were among the first used to induce mDA differentiation of ESC *in vitro* (Fig. 2). EB cells cultured in the absence of growth factors, and the enriched NPC population subsequently expanded in the presence of Fgf8 and Shh, produce between 10-20% of mDA neurons after allowing neuronal differentiation, in comparison with the 1-2% of mDA neurons produced in their absence (expressed as a percentage of the total cells alive at the end of the culture) (Rodriguez-Gomez et al., 2007, Lee et al., 2000, Hedlund et al., 2008, Kim et al., 2002). Although the addition of Shh and Fgf8 after EB formation increases the number of mDA neurons produced, the proportion of mDA neurons generated is still generally low. A similar proportion of mDA neurons are obtained when ESC are grown on a monolayer of PA6 cells (stromal cells derived from skull marrow) (Kawasaki et al., 2000), or when ESC neuralized directly in monolayer culture are treated with Fgf8 and Shh (Ying et al., 2003) (Fig. 2). It is apparent from these data that the population of cells capable of responding to specific environmental cues in culture is limited and/or the culture conditions lack the relevant signals present in the embryonic mesencephalic niche necessary for the efficient differentiation. Alternatively, the time window of action of each growth factor needs to be reproduced as *in vivo* in order to get efficient specific differentiation.

Although ESC are pluripotent and have the full potential to derive into any neuron type, common neuralization protocols might limit the cells ability to differentiate into dopaminergic neurons. This is supported by the fact that most ESC-derived NPCs are unable to respond to cues present in the niche of endogenous differentiation dopaminergic neuron differentiation (Baizabal & Covarrubias, 2009). Fgf8 and Shh do not alter the differentiation potential of the neuralized ESC and only induce the specific dopaminergic differentiation of a small proportion of cells. However, when non-neuralized EB cells are exposed to the embryonic mesencephalic environment, efficient and specific differentiation results (Baizabal & Covarrubias, 2009) (Fig. 2). Therefore, the time window that allows proper dopaminergic specification appears to be around the time at which the cells neuralize.

Since there is no certainty that the adult brain has the proper environment for specific differentiation, transplantation of EB cells at this stage is not the proper way to evaluate their differentiation potential. Nonetheless, differentiation into dopaminergic neurons has been achieved when EB cells are transplanted to the striatum of the adult brain, the target region of dopaminergic neurons of substancia nigra. As expected only few dopaminergic neurons are found several weeks after implantation, and those neurons do not apprently result from induction processes occurring in the adult brain. In fact, the proportion of dopaminergic and serotoninergic neurons is similar to that obtained in vitro or when the cells are transplanted to the kidney capsule (Bjorklund et al., 2002, Deacon et al., 1998).

Transplantation to the embryonic brain could provide a more adequate milieu for specific differentiation of ESC or their derivatives. ESC-derived neural precursors transplanted to the embryonic and neonatal brain efficiently differentiate into neurons and glial cells (Brustle et al., 1997, Wernig et al., 2004, Zhang et al., 2001). However, some ESC-derived neurons that expressed *Th* and *En1* can be found outside the midbrain (Wernig et al., 2004) suggesting that the cells are already committed before transplantation. In the previous type of experiments, cells are transplanted to embryos at stages long after specification of endogenous dopaminergic precursors has completed (E13), and therefore, the environment may not be the best suited to the induction of the dopaminergic differentiation. In contrast, robust specification and differentiation results when EB cells are transplanted to explants of E10.5 mesencephalon, just 1-2 days after specification of the endogenous mDA NPCs. In this condition, nearly 70% of the cells acquire neuronal markers and, of the total transplanted cells, 65% express *Lmx1a* and about 40% *Th* (Baizabal & Covarrubias, 2009). Therefore, at least up to E10.5, all the signals required to instruct EB cells into different mesencephalic phenotypes are present.

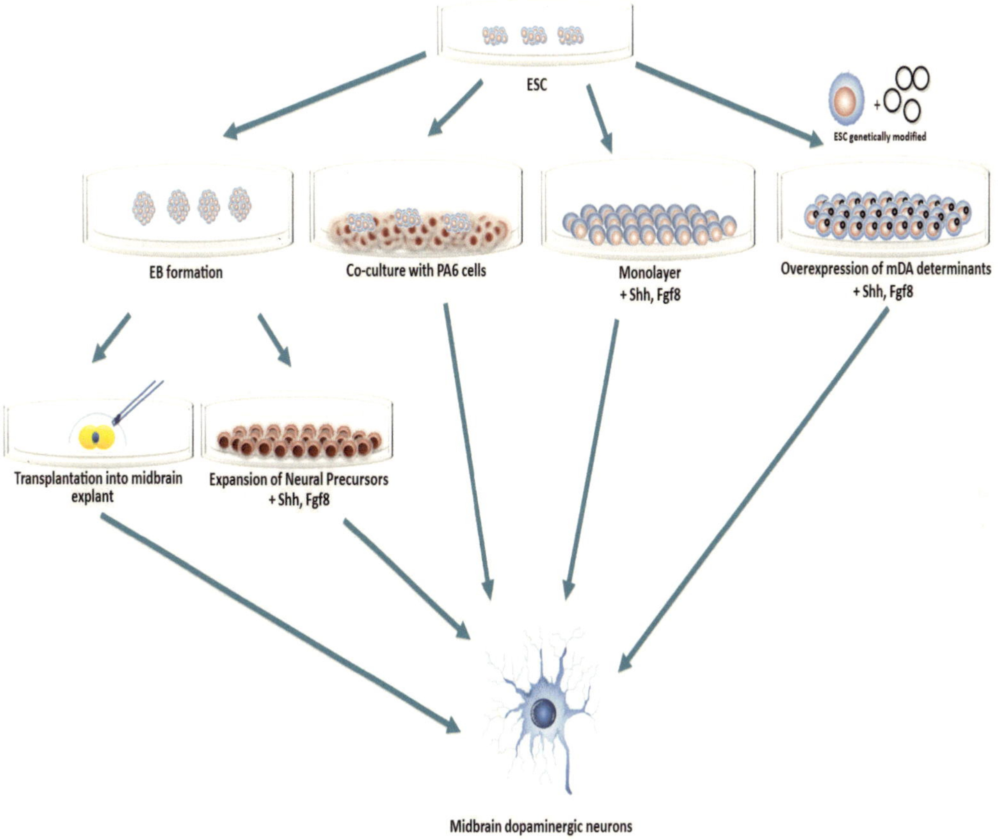

Fig. 2. Schematic representation of diverse strategies used to generate mDA neurons from ESCs.

4.2 Intrinsic determinants

As more is discovered about the transcription factors that control dopaminergic differentiation, the possibilities to induce specific differentiation through the overexpression of one or more transcription factors have increased. It might be predicted that the ectopic expression of a master gene(s) will be sufficient to induce specific neuronal differentiation programmes. However, the success in the searching for such "master genes" in general has been rather low, perhaps due to the requirement for a specific combination of transcription factors to be expressed in a competent recipient stem cell. The 'strength' of a master gene can be defined as a function of its ability to promote specific differentiation of precursor cells from the earliest stages possible. In this sense, the 'strongest' master gene will be that which individually promotes the specific differentiation of undifferentiated ESCs. It is apparent that none of the transcription factors of the dopaminergic network have the ability to induce specific differentiation following expression in undifferentiated ESCs.

Overproduction of transcription factors acting at late stages of dopaminergic differentiation (dopaminergic determinants) have been tested in different cell culture systems. Overexpression of Nurr1 in embryonic NPC from distinct regions of the brain cultured as a monolayer or after forming neurospheres, induces TH expression (Kim et al., 2003a, Kim et al., 2007); however, other characteristics of dopaminergic neurons have not been reported. When overexpression of Nurr1 is induced in NPC generated from ESCs, TH and other dopaminergic neuronal markers, such as DAT, are detected, however, Nurr1 has no influence over transcription factors important for mDA specification (Martinat et al., 2006, Sonntag et al., 2004). Overexpression of Pitx3 in undifferentiated ESC or in ESC-derived neural precursors is not sufficient to induce the complete mDA phenotype, but promotes expression of a subset of markers, particularly Aldh2 (Martinat et al., 2006, Chung et al., 2005). Even Nurr1 and Pitx3 together are not sufficient to induce ESC mDA differentiation and again, they only appear to influence terminal dopaminergic maturation (Martinat et al., 2006). The previous observations are in agreement with a direct role of these transcription factors in the expression of the *Th* gene and other genes that contribute to the dopaminergic phenotype. Nonetheless, an effect of Nurr1 on dopaminergic differentiation appears to occur when is constitutively overexpressed during the differentiation of ESC in the presence of Fgf8 and Shh as it enhances approximately four-fold the quantity of dopaminergic neurons generated (Kim et al., 2002).

Several factors relevant for the neuralization and specification of mDA NPCs have been overexpressed. Overexpression of Msx1 alone in ESC does not induce the generation of mDA neurons, though it is sufficient to induce *Ngn2* expression as it appears to occur *in vivo* (Andersson et al., 2006). In contrast, the overexpression of Lmx1a under the control of the Nestin promoter in ESC treated with Fgf8 and Shh causes an extensive increase in cells expressing *Msx1* and a reduction in those expressing *Nkx6.1* (Fig. 2). Also, there is a robust generation of immature neurons that express *Th*, many of which also express additional genes encoding mDA neuronal markers, including Lmx1a, Lmx1b, En1/2, Nurr1, Pitx3, Foxa2, DAT and Vmat2 (Friling et al., 2009, Andersson et al., 2006). This highly specific differentiation is observed under two distinct *in vitro* protocols suggesting that the effects are due to the action of Lmx1a and not to particular culture conditions. In agreement with a key role for Lmx1a also in human mDA neuron development, overexpression of Lmx1a under the control of the nestin promoter in human ESC generated abundant TH+ cells (Friling et al., 2009). It is worth mentioning that Lmx1a and other transcription factors within the dopaminergic specification network are not able to promote specific differentiation when overexpressed in neurospheres cells, suggesting that these NPCs are not competent to respond to these specification factors (Roybon et al., 2008).

In the previous protocols, the use of the nestin promoter to drive *Lmx1a* expression appears key for the efficient mDA differentiation of ESC-derived neural precursors. The exogenous *Lmx1a* initiates its expression at the time Nestin$^+$ neural precursors first emerge in culture (Andersson et al., 2006, Friling et al., 2009), which may provide the right time window for the action of Lmx1a in which ESC-derived neural precursors are competent to specifically respond and generate mDA neurons. These data reinforces the idea mentioned above proposing that specification and neuralization are interdependent process that must occur simultaneously or within a narrow window during the differentiation process.

Recently, studies with ESC have revealed novel interactions among the different factors participating in mDA differentiation. Based on the transcription factor network determined during mDA development, different combinations of Foxa2, Lmx1a and Otx2 have been introduced into ESC-derived neural precursors and the effect on mDA differentiation determined. When genes encoding the three factors are ectopically expressed in the absence of Shh in the culture medium, robust synergistic induction of mDA markers is observed with the majority of TH$^+$ neurons also coexpressing *Lmx1b* and *Nurr1* (Chung et al., 2009). This highly efficient generation of mDA neurons is not observed when Foxa2 or Otx2 are introduced alone. Therefore, it is apparent that Lmx1a is the key specification determinant of mDA differentiation in ESCs that, in concert with Otx2, could give a midbrain identity to neural precursor cells. Foxa2 might be required to activate the Shh pathwaty relevant for efficient mDA differentiation.

Studies with ESCs have identified a Wnt1-Lmx1a/b loop relevant for the specification of mDA neurons. In ESC Wnt1 induces the activation of Lmx1a independent of both Shh and Otx2, whereas, Lmx1a and Lmx1b upregulate Wnt1 by binding directly to its promoter (Chung et al., 2009). This autoregulatory loop, in turn, directly regulates *Otx2* expression, through the canonical Wnt signaling pathway, and also *Nurr1* and *Pitx3* expression through Lmx1a and Lmx1b. The Wnt1-Lmx1a/b loop also functions in the embryonic midbrain (Chung et al., 2009), demonstrating that this is not a culture artifact and that ESCs is a suitable system to study mDA differentiation. In this system, Lmx1a and Lmx1b appear to regulate each other, and the absence of one compensates the function of the other in the regulation of *Wnt1*, *Nurr1*, *Pitx3* and *Th* expression (Chung et al., 2009).

5. Conclusions

Taking together the data from *in vivo* and *in vitro* studies, at least two major pathways appear to regulate mDA specification, which may be referred to as "Shh-Foxa2" and "Wnt1-Lmx1a/b" (Fig. 1). These pathways, to some extent, are independent, as Wnt1 or Lmx1a do not affect *Shh* or *Foxa2* expression; although maintainance of *Lmx1a* expression is dependent on Foxa2 activity. At a later stage, these two pathways converge on Nurr1 and dopaminergic differentiation itself initiates.

Most stem cell-based differentiation protocols are focused on the addition of growth factors to the culture medium, in part because genetic manipulations are generally not desirable in the generation of cells for therapeutic purposes (Fig. 2). However, finding the right combination of growth factors and the time of action, as well as the competent stem cell population responsive to those factors is a difficult task. On the other hand, as we learn more about the transcriptional network that control specific fates, the use of "master genes" becomes more attractive. Therefore, engineering ESC lines that overproduce transcription factor(s) required for mDA development, in addition to the relevant basic information provided for the understanding of the molecular mechanisms involved in dopaminergic differentiation, it presents a promising alternative to obtain authentic mDA neurons for treating Parkinson's disease (Fig. 3).

It does not appear to be far the time at which there are one or several protocols to produce efficiently mDA neurons *in vitro*. However, the development of a cell therapy adequate for the treatment of patients suffering Parkinson's disease will require more than a protocol to produce mDA neurons, considering that it is unlikely that the adult brain possesses the optimal environment for mDA NPC differentiation and neuron maturation. For instance, it will be relevant to determine the differentiation stage at which implantation and survival of the transplanted cells is more efficient, as well as to identify additional factors needed to reconstitute the nigrostratial pathway.

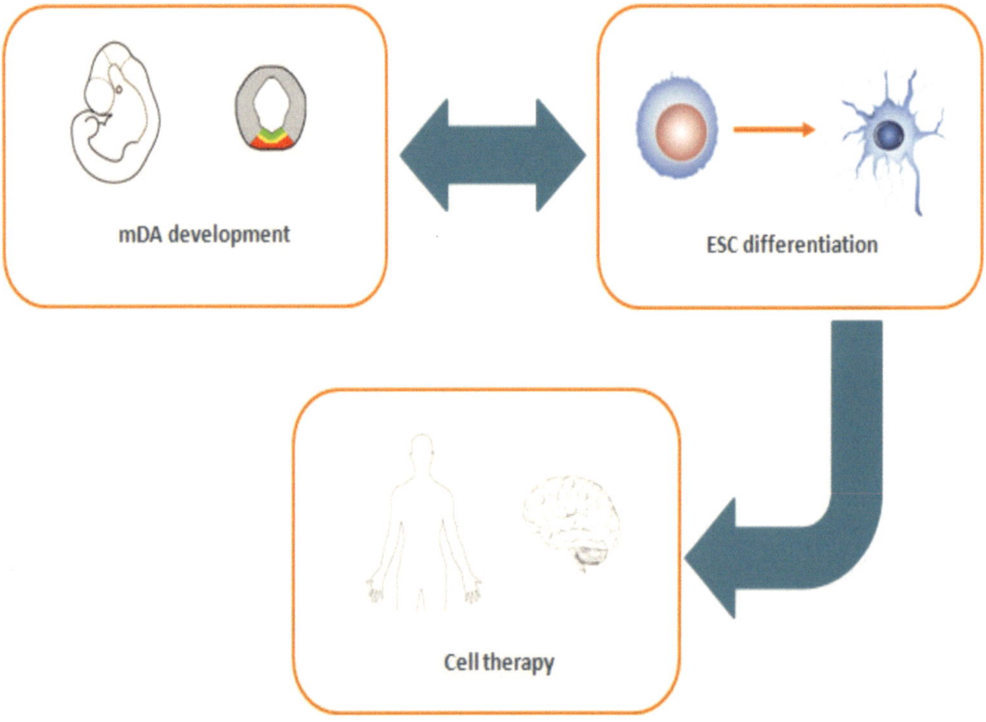

Fig. 3. ESC-based therapies for Parkinson's disease. ESCs are a potential source of mDA neurons useful in cell therapies for Parkinson's disease. The efficiency to generate mDA neurons from ESCs is increasing as we learn more about the factors controlling mDA neuronal differentiation from studies with both embryos and ESCs.

6. Acknowledgements

We thank to present and past members of the Laboratory on Tissue Degeneration and Regeneration for the fruitful discussions that contributed to develop some ideas presented in this review. We are particularly grateful with Christopher Wood for critical reading of the manuscript. This work was supported by Project Program Grant IMPULSA-UNAM 03 and Consejo Nacional de Ciencia y Tecnología (50956-Q).

7. References

Alberi, L., Sgado, P. & Simon, H. H. (2004). Engrailed genes are cell-autonomously required to prevent apoptosis in mesencephalic dopaminergic neurons. *Development.* 131, (Jul 2004) 3229-36, 0950-1991

Andersson, E., Tryggvason, U., Deng, Q., Friling, S., Alekseenko, Z., Robert, B., Perlmann, T. & Ericson, J. (2006). Identification of intrinsic determinants of midbrain dopamine neurons. *Cell.* 124, (Jan 2006) 393-405, 0092-8674

Arts, M. P., Groenewegen, H. J., Veening, J. G. & Cools, A. R. (1996). Efferent projections of the retrorubral nucleus to the substantia nigra and ventral tegmental area in cats as shown by anterograde tracing. *Brain Res Bull.* 40, 1996) 219-28, 0361-9230

Bain, G., Kitchens, D., Yao, M., Huettner, J. E. & Gottlieb, D. I. (1995). Embryonic stem cells express neuronal properties in vitro. *Dev Biol.* 168, (Apr 1995) 342-57, 0012-1606

Baizabal, J. M. & Covarrubias, L. (2009). The embryonic midbrain directs neuronal specification of embryonic stem cells at early stages of differentiation. *Dev Biol.* 325, (Jan 1 2009) 49-59, 0012-1606

Bjorklund, L. M., Sanchez-Pernaute, R., Chung, S., Andersson, T., Chen, I. Y., McNaught, K. S., Brownell, A. L., Jenkins, B. G., Wahlestedt, C., Kim, K. S. & Isacson, O. (2002). Embryonic stem cells develop into functional dopaminergic neurons after transplantation in a Parkinson rat model. *Proc Natl Acad Sci U S A.* 99, (Feb 2002) 2344-9, 0027-8424

Brustle, O., Spiro, A. C., Karram, K., Choudhary, K., Okabe, S. & McKay, R. D. (1997). In vitro-generated neural precursors participate in mammalian brain development. *Proc Natl Acad Sci U S A.* 94, (Dec 23 1997) 14809-14, 0027-8424

Chi, C. L., Martinez, S., Wurst, W. & Martin, G. R. (2003). The isthmic organizer signal FGF8 is required for cell survival in the prospective midbrain and cerebellum. *Development.* 130, (Jun 2003) 2633-44, 0950-1991

Chung, S., Hedlund, E., Hwang, M., Kim, D. W., Shin, B. S., Hwang, D. Y., Jung Kang, U., Isacson, O. & Kim, K. S. (2005). The homeodomain transcription factor Pitx3 facilitates differentiation of mouse embryonic stem cells into AHD2-expressing dopaminergic neurons. *Mol Cell Neurosci.* 28, (Feb 2005) 241-52, 1044-7431

Chung, S., Leung, A., Han, B. S., Chang, M. Y., Moon, J. I., Kim, C. H., Hong, S., Pruszak, J., Isacson, O. & Kim, K. S. (2009). Wnt1-lmx1a forms a novel autoregulatory loop and controls midbrain dopaminergic differentiation synergistically with the SHH-FoxA2 pathway. *Cell Stem Cell.* 5, (Dec 2009) 646-58, 1875-9777

Deacon, T., Dinsmore, J., Costantini, L. C., Ratliff, J. & Isacson, O. (1998). Blastula-stage stem cells can differentiate into dopaminergic and serotonergic neurons after transplantation. *Exp Neurol.* 149, (Jan 1998) 28-41, 0014-4886

Egan, M. F. & Weinberger, D. R. (1997). Neurobiology of schizophrenia. *Curr Opin Neurobiol.* 7, (Oct 1997) 701-7, 0959-4388

Engberg, N., Kahn, M., Petersen, D. R., Hansson, M. & Serup, P. (2010). Retinoic acid synthesis promotes development of neural progenitors from mouse embryonic stem cells by suppressing endogenous, Wnt-dependent nodal signaling. *Stem Cells.* 28, (Sep 2010) 1498-509, 1066-5099

Fedtsova, N. & Turner, E. E. (2001). Signals from the ventral midline and isthmus regulate the development of Brn3.0-expressing neurons in the midbrain. *Mech Dev.* 105, (Jul 2001) 129-44, 0925-4773

Ferri, A. L., Lin, W., Mavromatakis, Y. E., Wang, J. C., Sasaki, H., Whitsett, J. A. & Ang, S. L. (2007). Foxa1 and Foxa2 regulate multiple phases of midbrain dopaminergic neuron development in a dosage-dependent manner. *Development.* 134, (Aug 2007) 2761-9, 0950-1991

Finley, M. F., Devata, S. & Huettner, J. E. (1999). BMP-4 inhibits neural differentiation of murine embryonic stem cells. *J Neurobiol.* 40, (Sep 1999) 271-87, 0022-3034

Fraichard, A., Chassande, O., Bilbaut, G., Dehay, C., Savatier, P. & Samarut, J. (1995). In vitro differentiation of embryonic stem cells into glial cells and functional neurons. *J Cell Sci.* 108 (Pt 10), (Oct 1995) 3181-8, 0021-9533

Friling, S., Andersson, E., Thompson, L. H., Jonsson, M. E., Hebsgaard, J. B., Nanou, E., Alekseenko, Z., Marklund, U., Kjellander, S., Volakakis, N., Hovatta, O., El Manira, A., Bjorklund, A., Perlmann, T. & Ericson, J. (2009). Efficient production of mesencephalic dopamine neurons by Lmx1a expression in embryonic stem cells. *Proc Natl Acad Sci U S A.* 106, (May 2009) 7613-8, 0027-8424

Gratsch, T. E. & O'Shea, K. S. (2002). Noggin and chordin have distinct activities in promoting lineage commitment of mouse embryonic stem (ES) cells. *Dev Biol.* 245, (May 2002) 83-94, 0012-1606

Guo, C., Qiu, H. Y., Huang, Y., Chen, H., Yang, R. Q., Chen, S. D., Johnson, R. L., Chen, Z. F. & Ding, Y. Q. (2007). Lmx1b is essential for Fgf8 and Wnt1 expression in the isthmic organizer during tectum and cerebellum development in mice. *Development.* 134, (Jan 2007) 317-25, 0950-1991

Hedlund, E., Pruszak, J., Lardaro, T., Ludwig, W., Vinuela, A., Kim, K. S. & Isacson, O. (2008). Embryonic stem cell-derived Pitx3-enhanced green fluorescent protein midbrain dopamine neurons survive enrichment by fluorescence-activated cell sorting and function in an animal model of Parkinson's disease. *Stem Cells.* 26, (Jun 2008) 1526-36, 1066-5099

Hemmati-Brivanlou, A. & Melton, D. (1997). Vertebrate embryonic cells will become nerve cells unless told otherwise. *Cell.* 88, (Jan 1997) 13-7, 0092-8674

Hitoshi, S., Seaberg, R. M., Koscik, C., Alexson, T., Kusunoki, S., Kanazawa, I., Tsuji, S. & van der Kooy, D. (2004). Primitive neural stem cells from the mammalian epiblast differentiate to definitive neural stem cells under the control of Notch signaling. *Genes Dev.* 18, (Aug 2004) 1806-11, 0890-9369

Jeong, Y. & Epstein, D. J. (2003). Distinct regulators of Shh transcription in the floor plate and notochord indicate separate origins for these tissues in the mouse node. *Development.* 130, (Aug 2003) 3891-902, 0950-1991

Joksimovic, M., Yun, B. A., Kittappa, R., Anderegg, A. M., Chang, W. W., Taketo, M. M., McKay, R. D. & Awatramani, R. B. (2009). Wnt antagonism of Shh facilitates midbrain floor plate neurogenesis. *Nat Neurosci.* 12, (Feb 2009) 125-31, 1097-6256

Kawasaki, H., Mizuseki, K., Nishikawa, S., Kaneko, S., Kuwana, Y., Nakanishi, S., Nishikawa, S. I. & Sasai, Y. (2000). Induction of midbrain dopaminergic neurons from ES cells by stromal cell-derived inducing activity. *Neuron.* 28, (Oct 2000) 31-40, 0896-6273

Kele, J., Simplicio, N., Ferri, A. L., Mira, H., Guillemot, F., Arenas, E. & Ang, S. L. (2006). Neurogenin 2 is required for the development of ventral midbrain dopaminergic neurons. *Development*. 133, (Feb 2006) 495-505, 0950-1991

Kelley, A. E. & Berridge, K. C. (2002). The neuroscience of natural rewards: relevance to addictive drugs. *J Neurosci*. 22, (May 1 2002) 3306-11, 0270-6474

Kim, H. J., Sugimori, M., Nakafuku, M. & Svendsen, C. N. (2007). Control of neurogenesis and tyrosine hydroxylase expression in neural progenitor cells through bHLH proteins and Nurr1. *Exp Neurol*. 203, (Feb 2007) 394-405, 0014-4886

Kim, J. H., Auerbach, J. M., Rodriguez-Gomez, J. A., Velasco, I., Gavin, D., Lumelsky, N., Lee, S. H., Nguyen, J., Sanchez-Pernaute, R., Bankiewicz, K. & McKay, R. (2002). Dopamine neurons derived from embryonic stem cells function in an animal model of Parkinson's disease. *Nature*. 418, (Jul 2002) 50-6, 0028-0836

Kim, J. Y., Koh, H. C., Lee, J. Y., Chang, M. Y., Kim, Y. C., Chung, H. Y., Son, H., Lee, Y. S., Studer, L., McKay, R. & Lee, S. H. (2003a). Dopaminergic neuronal differentiation from rat embryonic neural precursors by Nurr1 overexpression. *J Neurochem*. 85, (Jun 2003a) 1443-54, 0022-3042

Kim, K. S., Kim, C. H., Hwang, D. Y., Seo, H., Chung, S., Hong, S. J., Lim, J. K., Anderson, T. & Isacson, O. (2003b). Orphan nuclear receptor Nurr1 directly transactivates the promoter activity of the tyrosine hydroxylase gene in a cell-specific manner. *J Neurochem*. 85, (May 2003b) 622-34, 0022-3042

Lee, J., Platt, K. A., Censullo, P. & Ruiz i Altaba, A. (1997). Gli1 is a target of Sonic hedgehog that induces ventral neural tube development. *Development*. 124, (Jul 1997) 2537-52, 0950-1991

Lee, S. H., Lumelsky, N., Studer, L., Auerbach, J. M. & McKay, R. D. (2000). Efficient generation of midbrain and hindbrain neurons from mouse embryonic stem cells. *Nat Biotechnol*. 18, (Jun 2000) 675-9, 1087-0156

Lin, W., Metzakopian, E., Mavromatakis, Y. E., Gao, N., Balaskas, N., Sasaki, H., Briscoe, J., Whitsett, J. A., Goulding, M., Kaestner, K. H. & Ang, S. L. (2009). Foxa1 and Foxa2 function both upstream of and cooperatively with Lmx1a and Lmx1b in a feedforward loop promoting mesodiencephalic dopaminergic neuron development. *Dev Biol*. 333, (Sep 2009) 386-96, 0012-1606

Liu, A. & Joyner, A. L. (2001). EN and GBX2 play essential roles downstream of FGF8 in patterning the mouse mid/hindbrain region. *Development*. 128, (Jan 2001) 181-91, 0950-1991

Martinat, C., Bacci, J. J., Leete, T., Kim, J., Vanti, W. B., Newman, A. H., Cha, J. H., Gether, U., Wang, H. & Abeliovich, A. (2006). Cooperative transcription activation by Nurr1 and Pitx3 induces embryonic stem cell maturation to the midbrain dopamine neuron phenotype. *Proc Natl Acad Sci U S A*. 103, (Feb 2006) 2874-9, 0027-8424

Matise, M. P., Epstein, D. J., Park, H. L., Platt, K. A. & Joyner, A. L. (1998). Gli2 is required for induction of floor plate and adjacent cells, but not most ventral neurons in the mouse central nervous system. *Development*. 125, (Aug 1998) 2759-70, 0950-1991

Matsunaga, E., Katahira, T. & Nakamura, H. (2002). Role of Lmx1b and Wnt1 in mesencephalon and metencephalon development. *Development*. 129, (Nov 2002) 5269-77, 0950-1991

Nakatani, T., Kumai, M., Mizuhara, E., Minaki, Y. & Ono, Y. (2010). Lmx1a and Lmx1b cooperate with Foxa2 to coordinate the specification of dopaminergic neurons and control of floor plate cell differentiation in the developing mesencephalon. *Dev Biol.* 339, (Mar 1 2010) 101-13, 0012-1606

Nakatani, T., Minaki, Y., Kumai, M. & Ono, Y. (2007). Helt determines GABAergic over glutamatergic neuronal fate by repressing Ngn genes in the developing mesencephalon. *Development.* 134, (Aug 2007) 2783-93, 0950-1991

Nunes, I., Tovmasian, L. T., Silva, R. M., Burke, R. E. & Goff, S. P. (2003). Pitx3 is required for development of substantia nigra dopaminergic neurons. *Proc Natl Acad Sci U S A.* 100, (Apr 2003) 4245-50, 0027-8424

Okabe, S., Forsberg-Nilsson, K., Spiro, A. C., Segal, M. & McKay, R. D. (1996). Development of neuronal precursor cells and functional postmitotic neurons from embryonic stem cells in vitro. *Mech Dev.* 59, (Sep 1996) 89-102, 0925-4773

Okada, Y., Shimazaki, T., Sobue, G. & Okano, H. (2004). Retinoic-acid-concentration-dependent acquisition of neural cell identity during in vitro differentiation of mouse embryonic stem cells. *Dev Biol.* 275, (Nov 2004) 124-42, 0012-1606

Omodei, D., Acampora, D., Mancuso, P., Prakash, N., Di Giovannantonio, L. G., Wurst, W. & Simeone, A. (2008). Anterior-posterior graded response to Otx2 controls proliferation and differentiation of dopaminergic progenitors in the ventral mesencephalon. *Development.* 135, (Oct 2008) 3459-70, 0950-1991

Ono, Y., Nakatani, T., Sakamoto, Y., Mizuhara, E., Minaki, Y., Kumai, M., Hamaguchi, A., Nishimura, M., Inoue, Y., Hayashi, H., Takahashi, J. & Imai, T. (2007). Differences in neurogenic potential in floor plate cells along an anteroposterior location: midbrain dopaminergic neurons originate from mesencephalic floor plate cells. *Development.* 134, (Sep 2007) 3213-25, 0950-1991

Prakash, N., Brodski, C., Naserke, T., Puelles, E., Gogoi, R., Hall, A., Panhuysen, M., Echevarria, D., Sussel, L., Weisenhorn, D. M., Martinez, S., Arenas, E., Simeone, A. & Wurst, W. (2006). A Wnt1-regulated genetic network controls the identity and fate of midbrain-dopaminergic progenitors in vivo. *Development.* 133, (Jan 2006) 89-98, 0950-1991

Rodriguez-Gomez, J. A., Lu, J. Q., Velasco, I., Rivera, S., Zoghbi, S. S., Liow, J. S., Musachio, J. L., Chin, F. T., Toyama, H., Seidel, J., Green, M. V., Thanos, P. K., Ichise, M., Pike, V. W., Innis, R. B. & McKay, R. D. (2007). Persistent dopamine functions of neurons derived from embryonic stem cells in a rodent model of Parkinson disease. *Stem Cells.* 25, (Apr 2007) 918-28, 1066-5099

Roybon, L., Hjalt, T., Christophersen, N. S., Li, J. Y. & Brundin, P. (2008). Effects on differentiation of embryonic ventral midbrain progenitors by Lmx1a, Msx1, Ngn2, and Pitx3. *J Neurosci.* 28, (Apr 2008) 3644-56, 0270-6474

Saarimaki-Vire, J., Peltopuro, P., Lahti, L., Naserke, T., Blak, A. A., Vogt Weisenhorn, D. M., Yu, K., Ornitz, D. M., Wurst, W. & Partanen, J. (2007). Fibroblast growth factor receptors cooperate to regulate neural progenitor properties in the developing midbrain and hindbrain. *J Neurosci.* 27, (Aug 2007) 8581-92, 0270-6474

Sasaki, H., Hui, C., Nakafuku, M. & Kondoh, H. (1997). A binding site for Gli proteins is essential for HNF-3beta floor plate enhancer activity in transgenics and can respond to Shh in vitro. *Development.* 124, (Apr 1997) 1313-22, 0950-1991

Saucedo-Cardenas, O., Quintana-Hau, J. D., Le, W. D., Smidt, M. P., Cox, J. J., De Mayo, F., Burbach, J. P. & Conneely, O. M. (1998). Nurr1 is essential for the induction of the dopaminergic phenotype and the survival of ventral mesencephalic late dopaminergic precursor neurons. *Proc Natl Acad Sci U S A*. 95, (Mar 1998) 4013-8, 0027-8424

Simon, H. H., Saueressig, H., Wurst, W., Goulding, M. D. & O'Leary, D. D. (2001). Fate of midbrain dopaminergic neurons controlled by the engrailed genes. *J Neurosci*. 21, (May 2001) 3126-34, 0270-6474

Simon, H. H., Thuret, S. & Alberi, L. (2004). Midbrain dopaminergic neurons: control of their cell fate by the engrailed transcription factors. *Cell Tissue Res*. 318, (Oct 2004) 53-61, 0302-766X

Smidt, M. P., Asbreuk, C. H., Cox, J. J., Chen, H., Johnson, R. L. & Burbach, J. P. (2000). A second independent pathway for development of mesencephalic dopaminergic neurons requires Lmx1b. *Nat Neurosci*. 3, (Apr 2000) 337-41, 1097-6256

Smidt, M. P., Smits, S. M., Bouwmeester, H., Hamers, F. P., van der Linden, A. J., Hellemons, A. J., Graw, J. & Burbach, J. P. (2004). Early developmental failure of substantia nigra dopamine neurons in mice lacking the homeodomain gene Pitx3. *Development*. 131, (Mar 2004) 1145-55, 0950-1991

Smits, S. M., Ponnio, T., Conneely, O. M., Burbach, J. P. & Smidt, M. P. (2003). Involvement of Nurr1 in specifying the neurotransmitter identity of ventral midbrain dopaminergic neurons. *Eur J Neurosci*. 18, (Oct 2003) 1731-8, 0953-816X

Sonntag, K. C., Simantov, R., Kim, K. S. & Isacson, O. (2004). Temporally induced Nurr1 can induce a non-neuronal dopaminergic cell type in embryonic stem cell differentiation. *Eur J Neurosci*. 19, (Mar 2004) 1141-52, 0953-816X

Tropepe, V., Hitoshi, S., Sirard, C., Mak, T. W., Rossant, J. & van der Kooy, D. (2001). Direct neural fate specification from embryonic stem cells: a primitive mammalian neural stem cell stage acquired through a default mechanism. *Neuron*. 30, (Apr 2001) 65-78, 0896-6273

Verani, R., Cappuccio, I., Spinsanti, P., Gradini, R., Caruso, A., Magnotti, M. C., Motolese, M., Nicoletti, F. & Melchiorri, D. (2007). Expression of the Wnt inhibitor Dickkopf-1 is required for the induction of neural markers in mouse embryonic stem cells differentiating in response to retinoic acid. *J Neurochem*. 100, (Jan 2007) 242-50, 0022-3042

von Bohlen und Halbach, O., Schober, A., and Krieglstein, K. (2004). Genes, proteins, and neurotoxins involved in Parkinson's disease. *Prog Neurobiol*. 73, (Jun 2004) 151-77, 0301-0082

Wallen, A. & Perlmann, T. (2003). Transcriptional control of dopamine neuron development. *Ann N Y Acad Sci*. 991, (Jun 2003) 48-60, 0077-8923

Watanabe, K., Kamiya, D., Nishiyama, A., Katayama, T., Nozaki, S., Kawasaki, H., Watanabe, Y., Mizuseki, K. & Sasai, Y. (2005). Directed differentiation of telencephalic precursors from embryonic stem cells. *Nat Neurosci*. 8, (Mar 2005) 288-96, 1097-6256

Wernig, M., Benninger, F., Schmandt, T., Rade, M., Tucker, K. L., Bussow, H., Beck, H. & Brustle, O. (2004). Functional integration of embryonic stem cell-derived neurons in vivo. *J Neurosci*. 24, (Jun 2004) 5258-68, 0270-6474

Wichterle, H., Lieberam, I., Porter, J. A. & Jessell, T. M. (2002). Directed differentiation of embryonic stem cells into motor neurons. *Cell*. 110, (Aug 2002) 385-97, 0092-8674

Wurst, W. & Bally-Cuif, L. (2001). Neural plate patterning: upstream and downstream of the isthmic organizer. *Nat Rev Neurosci*. 2, (Feb 2001) 99-108, 1471-003X

Yamada, T., Placzek, M., Tanaka, H., Dodd, J. & Jessell, T. M. (1991). Control of cell pattern in the developing nervous system: polarizing activity of the floor plate and notochord. *Cell*. 64, (Feb 1991) 635-47, 0092-8674

Ye, W., Shimamura, K., Rubenstein, J. L., Hynes, M. A. & Rosenthal, A. (1998). FGF and Shh signals control dopaminergic and serotonergic cell fate in the anterior neural plate. *Cell*. 93, (May 1998) 755-66, 0092-8674

Ying, Q. L., Stavridis, M., Griffiths, D., Li, M. & Smith, A. (2003). Conversion of embryonic stem cells into neuroectodermal precursors in adherent monoculture. *Nat Biotechnol*. 21, (Feb 2003) 183-6, 1087-0156

Zetterstrom, R. H., Solomin, L., Jansson, L., Hoffer, B. J., Olson, L. & Perlmann, T. (1997). Dopamine neuron agenesis in Nurr1-deficient mice. *Science*. 276, (Apr 1997) 248-50, 0036-8075

Zhang, S. C., Wernig, M., Duncan, I. D., Brustle, O. & Thomson, J. A. (2001). In vitro differentiation of transplantable neural precursors from human embryonic stem cells. *Nat Biotechnol*. 19, (Dec 2001) 1129-33, 1087-0156

22

A Rational Approach to Inducing Neuronal Differentiation in Embryonic Stem Cells

Tadahide Furuno and Mamoru Nakanishi
School of Pharmacy, Aichi Gakuin University
Japan

1. Introduction

ES cells can produce any cell type of a living organism while self-renewing, and have been a favorable tool for cell engineering, transplantation and regenerative medicine (Wobus and Boheler, 2005). There were many issues about the signalling pathways and molecular mechanisms of self-renewal and differentiation in ES cells, however, the discussion has not realized an appropriate goal yet. Thus, it is important to study the signalling pathways, especially in neuronal differentiation, to create a better and novel approach for controlling the differentiation process using mouse model ES cells (Sugimoto et al., 2009; Miyazu et al., 2010).

In mouse ES cells, self-renewal and maintenance of pluripotency requires LIF and bone morphogenic proteins (BMPs) (Smith et al., 1988; Williams et al., 1988; Ying et al., 2003). LIF belongs to the cytokine interleukin (IL)-6 family which utilizes the common gp130 receptor for cellular signaling, and LIF signaling is mediated through a low affinity LIF receptor (LIFRβ) which forms a heterodimer with the gp130 upon LIF binding (Hibi et al., 1990; Stahl et al., 1994). The formation of a high affinity trimeric complex composed of LIFRβ, gp130, and LIF leads to the activation of the janus kinase (JAK)-signal transducers and activators of transcription (STAT) pathway and the mitogen-activated protein (MAP) kinase pathway (Burdon et al., 2002; Heinrich et al., 2003; Ernst and Jenkins, 2004; Krinstensen et al., 2005). The binding of LIF to LIFRβ-gp130 results in the rapid activation of JAK which in turn phosphorylates tyrosine residues of LIFRβ and gp130 (Ernst et al., 1996). The phosphorylated tyrosine residues act as binding sites for signaling molecules such as a transcriptional factor STAT3 and a protein tyrosine phosphatase SHP-2 (Heinrich et al., 1998). Recruited STAT3 and SHP-2 are also phosphorylated at tyrosine residues. Phosphorylation of STAT3 promotes the homodimerization through the interaction of the Src homology (SH) 2 domain and the phosphotyrosine domain. The homodimer of STAT3 translocates to the nucleus and functions as a transcriptional factor (Schindler and Darnell, 1995; Darnell, 1997). On the other hand, phosphorylation of SHP-2 generates binding sites for Grb2 and associates with a scaffold protein Grb2-associated binder protein (Gab) 1. The complex subsequently induces the activation of Ras-MAP kinase cascade (Fukada et al., 1996; Van Vactor et al., 1998). Especially, ERKs, which are members of MAP kinase family proteins, are activated by a large number of ligands. The binding of BMP4 to its receptors triggers the phosphorylation of Smad1 and activates the expression of inhibitor of differentiation (Id) gene family. It is considered that the induction of Id expression through Smad1 activation by BMP4 is the critical contribution to the maintenance of pluripotency in

ES cells (Ying et al., 2003). Hence, mouse ES cells are maintained in an undifferentiated state *in vitro* with LIF, which activates STAT3 and BMP to induce Id proteins (Niwa et al., 1998; Matsuda et al., 1999; Raz et al., 1999; Ying et al., 2003). As described above, the LIF leads to the phosphorylation of STAT3 and ERKs. Actually the phosphorylation form of STAT3 is detected in undifferentiated ES cells cultured in the medium containing LIF. However, the phosphorylation of ERKs does not occur in undifferentiated ES cells. The activation of ERKs is considered to be involved in the differentiation process rather than the maintenance of undifferentiated state, because the removal of LIF leads to the phosphorylation of ERKs and the attenuation of ERKs activation by inhibitor of a MAP/ERK kinase (MEK) activation or expression of ERK phosphatases reduced the level of differentiation of ES cells (Burdon et al., 1999a). Therefore, in ES cells, it has been considered that LIF predominantly activates STAT3 to remain undifferentiated and that the balance of conflicting activation of STAT3 and ERKs might determine the efficiency of the self-renewal and differentiation (Niwa et al., 1998; Burdon et al., 1999b). However, the molecular mechanism for balanced activation of STAT3 and ERKs in ES cells is not yet clear. We recently found that the dephosphorylation of phospho-STAT3 and phosphorylation of ERKs occurred consistently after the removal of LIF and that SOCS3, whose expression was regulated by STAT3, might regulate the activation of ERK pathway (Miyazu et al., 2010).

On the other hand, the removal of LIF is known to induce the formation of EB for ES cells. The most commonly used approach for neural differentiation from ES cells is the formation of EB and the treatment of EB with retinoic acid (Strubing et al., 1995). It has been reported that ES cells overexpressing NeuroD2, a basic helix-loop-helix (bHLH) transcriptional factor, lead to differentiation to neurons efficiently in the absence of retinoic acid after EB formation (Kanda et al., 2004). Because continued culture of EB results in the appearance of ectodermal, mesodermal, and endodermal cell types, the differentiation of EB yields a small fraction of neural cells. The establishment of differentiation system from ES cells to neurons without EB formation is considered to be favorable for the production of neurons in high efficiency. The bHLH transcriptional factors are implicated in the regulation of differentiation in various cell types (Massari and Murre, 2000). NeuroD2 has been reported to play a critical role in the induction of neuronal differentiation, the promotion of neuronal survival, and development of thalamocortical communication (Kume et al., 1996; McCormick et al., 1996; Yasunami et al., 1996; Farah et al., 2000; Olson et al., 2001; Ince-Dunn, et al., 2006; Lin et al., 2006; Noda et al., 2006). NeuroD2 has two basic-rich domains; one is basic region of bHLH and the other is nuclear localization signal which plays a role of protein transduction. In *Xenopus laevis* embryos the expression of NeuroD2 results in ectopic neurogenesis, and in mouse brain NeuroD2 expression starts around embryonic day 11. NeuroD2-deficient mice exhibit small brains, ataxia, reduced seizure threshold, defective thalamacortical synapses, growth failure, and early death. NeuroD2 also induces the differentiation to neurons in mouse neuroblastoma cells, embryonic carcinoma cells, and ES cells (Farah et al., 2000; Kanda et al., 2004; Noda et al., 2006).

Then, we recently established the regulation system of NeuroD2 expression using doxycycline in ES cells (Gossen and Bujard, 1992; Gossen and Bujard, 1995). This system was useful for studying the role of NeuroD2 in the neuronal differentiation because specific transcriptional factors are known to be expressed in a manner consistent with their having a regulatory role in the various stages of ES cell development (Sugimoto et al., 2009). So, in this review we will describe focusing mainly on the SOCS3 signalling and expression of Neuro D2 in mouse ES cells.

2. Materials and methods

2.1 Cell cultures

Mouse 129/sv ES cell lines (passage 15, Dainippon Pharmaceutical Corporation, Osaka, Japan) were routinely cultured on tissue culture plates (Falcon) coated with 0.1% (v/v) gelatin (Dainippon Pharmaceutical Corporation) in Knockout™ Dulbecco's modified Eagle's medium (DMEM) (Invitrogen, Scottland) in the presence of 15% (v/v) Knockout™ serum replacement (Invitrogen), 0.1 mM β-mercaptoethanol, 2 mM L-glutamine, 0.1 mM non-essential amino acids, 100 units/ml penicillin, 0.1 mg/ml streptomycin, and 1000 units/ml mouse LIF (Chemicon) in 5% CO_2 incubator at 37 °C (Amano et al., 2006; Sugimoto et al., 2009; Miyazu et al., 2010). Cells were trypsinized and replated every 3 days.

2.2 Western blotting

To prepare whole cell lysate, collected ES cells were suspended in ice-cold lysis buffer (10 mM Tris-HCl, pH 7.4, 0.5% Nonidet P-40, 1 mM EDTA, 150 mM NaCl, 1 mM Na_3VO_4, 0.5 mM phenylmethylsulfonyl fluoride) and allowed to stand on ice for 30 min. The suspension was clarified by centrifugation (14,000 rpm, 20 min). After centrifugation, the resulting supernatants were solubilized by treatment with Laemmli buffer at 100°C for 5 min and separated by electrophoresis in 10% SDS-polyacrylamide gel. The electrophoresed proteins were transferred to polyvinylidene difluoride membranes with an electroblotter. After blocking with 5% skim milk, the membranes were probed with rabbit anti-phospho STAT3 (Tyr705) antibody (Cell Signaling) (1:1000 dilution), rabbit anti-STAT3 antibody (Cell Signaling) (1:1000 dilution), rabbit anti-phospho c-Raf (Ser338) antibody (Cell Signaling) (1:200), rabbit anti-phospho MEK1/2 (Ser217/221) antibody (Cell Signaling) (1:500), rabbit anti-phospho p44/p42 MAP kinase (Thr202/Tyr204) antibody (Cell Signaling) (1:500), rabbit anti-p44/p42 MAP kinase antibody (Cell Signaling) (1:1000), goat anti-SOCS3 antibody (M-20) (Santa Cruz Biotechnology) (1:100) or rabbit anti-β-actin antibody (clone AC-15) (Sigma) (1:5000 dilution) and treated with 1:1000 dilution HRP-conjugated anti-rabbit IgG (Santa Cruz Biotechnology) or HRP-conjugated donkey anti-goat IgG (Santa Cruz Biotechnology). The amount of HRP-conjugated IgG bound to each protein band was determined by LAS-1000 (Fuji Film, Tokyo, Japan) and analyzed by Image Gauge (Fuji Film). The pixel density in each band was corrected by that in the control protein (STAT3, ERKs or β-actin).

2.3 Immunostaining

ES cells were washed with ice-cold TBS buffer (25 mM Tris, pH 7.4, 137 mM NaCl, and 2.68 mM KCl), attached to microscope slides by Cytospin 4 centrifugation (Thermo Shandon, Pittsburgh, USA) (1,000 rpm, 5 min), and fixed with 4% paraformaldehyde for 15 min. They were permeabilized with 0.2% Triton X-100 for 5 min, and blocked with 1% bovine serum albumin for 30 min. For single staining, fixed cells were incubated with the rabbit anti-phospho STAT3 (Tyr705) antibody (1:100) or rabbit anti-phospho p44/p42 MAP kinase (Thr202/Tyr204) antibody (1:100), rabbit anti-MAP2 antibody (1:400) (Santa Cruz Biotechnology), or rabbit anti-NeuroD2 antibody (1:100) (Sigma), and stained with FITC-conjugated goat anti-rabbit IgG antibody (10 μg/ml) (ICN Pharmaceticals, USA). For double staining, fixed cells were incubated with goat anti-SOCS3 antibody (M-20) (1:50) and stained with rhodamine-conjugated donkey anti-goat IgG (10 μg/ml) (Santa Cruz Biotehnology). Then, the samples were incubated with rabbit anti-phospho MEK1/2 (Ser217/221) antibody (1:200) and stained with FITC-conjugated goat anti-rabbit IgG antibody (10 μg/ml). For staining of nucleus, fixed cells were incubated with propidium iodide (1 μg/ml) (Sigma). Cells were visualized using a confocal laser scanning microscope (CLSM) (LSM510; Carl Zeiss, OberKochen, Germany) (Furuno et al., 1993; Nakanishi, 2003; Nakanishi and Furuno, 2008). FITC was excited at 488 nm by an argon ion laser and its fluorescence was detected through a long pass filter (> 505 nm) for the single staining. In the case of double staining, FITC and rhodamine were excited at 488 nm by an argon ion laser and at 543 nm by a He-Ne laser, respectively. Their fluorescence was detected through a band pass filter (505 - 530 nm) and a long pass filter (> 560 nm), respectively (Suzuki et al., 2004; Furuno et al., 2005).

2.4 Establishment of cell lines

The ES cells (1×10^7 cells / 500 μl) were electroporated in cold K^+-PBS with 30 μg of pTet-On plasmid DNA (Clontech Laboratories, Palo Alto, CA, USA) at 220 V and 475 μF using Gene Pulser (Bio-Rad, Richmond, CA). ES cells transfected with pTet-On were maintained in culture dishes for 48 h and were cloned by the selection with antibiotic G418 (Sigma). To investigate the expression of reverse tetracycline-controlled transactivator (rtTA) in monoclonal ES cells (rtTA-ES cells), luciferase was assayed after the transfection of the plasmid DNA containing tetracycline response element (TRE) to express luciferase (pTRE-Luc) (Clontech).

To obtain NeuroD2 cDNA, poly(A)$^+$ RNA was prepared with a Quickprep Micro mRNA Purification Kit (Amersham Biosciences, Little Chalfont, Buckinghamshire, UK) from mouse cerebellum and served as templates for cDNA synthesis with SuperScript II RT (Invitrogen Corp., Carlsbad, CA, USA). The DNA sequences coding NeuroD2 (GEN Bank accession number: U58471) was obtained by RT-PCR with appropriate primer pairs, 5′-*GTCGAC*ATGCTGACCCGCCTGTTC-3′ (sense; *Acc* I site is italicized) / 5′-*CCGCGG*TCAGTTATGGAAAAATGCGTTGA-3′ (antisense; *Sac* II site is italicized) using a Taq polymerase EX-Taq (Takara, Tokyo, Japan). The PCR products were subcloned into a TA cloning vector pCRII (Invitrogen) and sequenced with an Applied Biosystems 3130xl genetic analyzer to verify the DNA sequence of NeuroD2. The full-length cDNA coding NeuroD2 was ligated with pTRE-Hyg2 expression vector (Clontech) at *Acc* I and *Sac* II sites (pTRE-ND2). pTRE-ND2 plasmid DNA (30 μg) was electroporated to rtTA-ES cells (1×10^7 cells/500 μl), and trandfected ES cells were cultured for 48 hr and then were selected by antibiotic Hygromycin B (Sugimoto et al., 2009).

For genomic DNA PCR, genome DNA was isolated from ES cells (5×10^6 cells) with a Blood and Cell Culture Kit (Qiagen). For RT-PCR, mRNAs were purified from ES cells with a Quickprep Micro mRNA Purification Kit (Amersham Biosciences) and served as templates for cDNA synthesis with SuperScript II RT (Invitrogen). Genomic DNA and cDNA were amplified by PCR using appropriate primers and a Taq polymerase EX-Taq (Takara).

The rtTA-ES cells were transfected with pTRE-Luc and were cultured for 48 hr in the presence or absence of doxycycline (1 μg/ml). Then, the cells were washed three times with PBS and lysed in a cell lysis buffer. The lysate was centrifuged at 12,000×g at 4 °C for 3 min and supernatant was subjected to a luciferase assay. The luciferase assay was carried out using a picagene luciferase assay kit (Toyo Ink, Tokyo, Japan). The relative light unit of chemiluminescence was measured with a luminometer (TD-20/20; Turner Designs; Sunnyvale, CA, USA) and the values were normalized to the amount of protein determined by BCA assay (Inoh et al., 2001).

2.5 Alkaline phosphatase assay

After washing, ES cells were fixed with 4% paraformaldehyde for 15 min at 4 °C and stained with NBT-BCIP solution (Sigma). After incubation for 30 min at 37 °C, the reaction was stopped by EDTA (10 mM). The cells were washed with PBS and examined microscopically.

2.6 Data analysis

Data were presented as mean ± SEM. In statistical analysis, data were analyzed by unpaired Student's *t*-test, 2-tailed. $P<0.05$ were accepted as a level of statistically significant difference.

3. Results

3.1 The relationship between dephosphorylation of phospho-STAT3 and phosphorylation of ERKs

To examine the relationship between activation of STAT3 and ERKs in the differentiation processes in mouse ES cells, we studied time-courses of phosphorylation of STAT3, ERKs, and Raf-1, which is a kinase that acts upstream of the ERK signaling cascade after the removal of LIF. In the presence of LIF, STAT3 was strongly phosphorylated in mouse ES cells, but Raf-1 and ERKs in the same cells were not phosphorylated, as described in previous papers (Qu et al., 1997; Burdon et al., 1999a; Matsuda et al., 1999; Jirmanova et al., 2002; Chan et al., 2003; Yao et al., 2003). When LIF was removed from the culture medium, the level of phospho-STAT3 was decreased, but those of phospho-Raf-1, -MEK, and -ERKs were increased, as shown in Fig. 1a. The level of phospho-STAT3 reached the minimum at ~ 4 h after the removal of LIF, and then, it increased again. In contrast, the phosphorylation of Raf-1 reached the maximum at ~ 4 h, and then decreased to the basal level. The phosphorylation of MEK occurred slowly, and ERKs were subsequently phosphorylated. These results suggested that the time-course of dephosphorylation of phospho-STAT3 was consistent with that of phosphorylation of Raf-1 and that the ERKs were phosphorylated with a lag period after the Raf-1 activation. In addition, immunostaining analysis showed that the fluorescence intensity of phospho-STAT3 in the nucleus in ES cells, which was high in the presence of LIF, decreased at 4 h and then became high at 72 h again after the removal of LIF. On the other hand, the fluorescence intensity of phospho-ERKs increased at 12 h, and fell to the basal level at 72 h. These immunostaining images were corresponding to the data of phosphorylation level observed by Western blotting. These suggested that the removal of LIF led to the differentiation of ES cells by the spatiotemporally balanced activation of STAT3 and ERKs.

3.2 Link of STAT3 and ERKs signaling pathways

Though the phosphorylation of ERKs was detected in ES cells after the removal of LIF, LIF was essentially known to activate the ERK signaling cascade. To study whether STAT3 and ERKs were phosphorylated by LIF stimulation in ES cells, we analyzed the phosphorylations of STAT3 and ERKs after the addition of LIF in ES cells which were pre-cultured in the absence of LIF for 4 hr. LIF induced to the phosphorylation of ERKs as well as STAT3 within 10 min in LIF-starved ES cells. The LIF-mediated phosphorylation of STAT3 lasted 30 min, whereas that of ERKs was transient and almost disappeared at 30 min. Next, to examine the effects of signaling proteins in the ERK signaling cascade on the STAT3 activation, we studied the LIF-mediated phosphorylations of STAT3 and ERKs in LIF-starved ES cells pretreated with a MEK inhibitor PD98059 (20 μM) for 30 min (Alessi et al.,

1995; Furuno et al., 2001). The pretreatment of PD98059 inhibited ERKs phosphorylation after LIF stimulation but not STAT3 phosphorylation. This suggested that the activation level of ERK did not affect the phosphorylation of STAT3. Next, to examine the effects of STAT3 transcriptional activity on the activation of ERKs in ES cells, we investigated the phosphorylation level of ERKs in ES cells cultured in the medium containing LIF and STAT3 inhibitory peptide (STAT3IP) (10 μM), a membrane-permeable inhibitor of STAT3 dimerization (Turkson et al., 2004). When ES cells were cultured in the presence of LIF in the medium containing STAT3IP (10 μM) for 4 days, the phosphorylation of ERKs was found to be accelerated. In contrast, the phosphorylation level of STAT3 was decreased. Immunostaining analysis also showed the increase in fluorescence intensity in the nucleus of phospho-ERKs and the decrease in that of phospho-STAT3 by the treatment with STAT3IP (10 μM) for 4 days. These indicated that STAT3IP led to the phosphorylation of

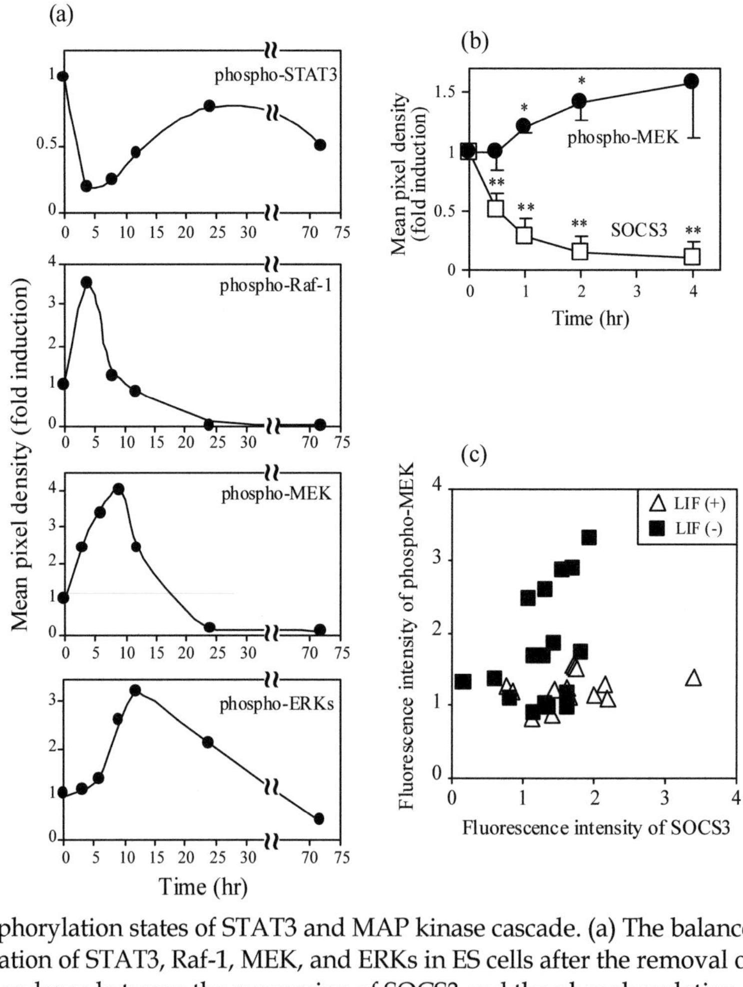

Fig. 1. Phosphorylation states of STAT3 and MAP kinase cascade. (a) The balanced phosphorylation of STAT3, Raf-1, MEK, and ERKs in ES cells after the removal of LIF. (b) The correspondence between the expression of SOCS3 and the phosphorylation of MEK after the removal of LIF. *$P<0.05$ and **$P<0.01$ compared to the value at 0 h. (c) The relation between the expression of SOCS3 and the phosphorylation of MEK at 4 h after the removal of LIF in individual ES cells.

ERKs and the dephosphorylation of STAT3, and that STAT3IP disrupted the balance of the activation of STAT3 and ERKs in ES cells cultured in LIF-containing medium. From the result, the reason why the phosphorylation level of ERKs was not detected in undifferentiated ES cells in the presence of LIF was guessed to be due to the high transcriptional activity of STAT3 mediated by LIF.

3.3 Involvement of SOCS3 in STAT3IP-mediated ERKs phosphorylation

We focused SOCS3 as a molecule involved in STAT3IP-mediated ERKs phosphorylation because the expression of SOCS3 was induced by cytokines including LIF probably through STAT3, and SOCS3 could also act to suppress ERK signaling by competing with SHP-2 for binding to the phosphotyrosine of gp130 (Bousquet et al., 1999; Duval et al., 2000; Nicholson et al., 2000; Schmitz et al., 2000). To confirm that LIF could induce the expression of SOCS3 in ES cells, we examined the expression of SOCS3 after the addition of LIF to LIF-starved ES cells. The expression of SOCS3 was not detected in LIF-starved ES cells, but it increased gradually after the addition of LIF. The pretreatment of LIF-starved ES cells for 60 min with STAT3IP (10 μM) suppressed the LIF-mediated expression of SOCS3.

Then, we studied the expression of SOCS3 in ES cells cultured in the presence of LIF. We found that SOCS3 was expressed in undifferentiated ES cells and that the removal of LIF reduced the expression of SOCS3 time-dependently, as shown in Fig. 1b. SOCS3 was almost undetectable in ES cells at 4 h after the removal of LIF. In contrast, the phosphorylation of MEK proceeded according to the decrease in expression of SOCS3 (Fig. 1b). In addition, double-immunostaining analysis of SOCS3 and phospho-MEK in individual ES cells also showed that SOCS3, which distributed on the plasma membrane, was diminished, and the phosphorylation of MEK increased in the cytoplasm at 4 h after the removal of LIF. In scatter plotting to analyze the correspondence between the expression of SOCS3 and the phophorylation of MEK in individual ES cells, it was shown that the phosphorylation of MEK proceeded in cells where the expression of SOCS3 was down-regulated 4 h after removal of LIF (Fig. 1c). These results strongly suggested that SOCS3, whose expression was regulated by STAT3 transcriptional activity, suppressed the activation of the ERK pathway.

3.4 Establishment of ES cells stably transfected with pTet-On plasmid DNA

As described above, the removal of LIF led to decrease the expression of SOCS3 and activate the ERK cascade in mouse ES cells. This phenomenon is probably the first step for ES cells to differentiate into many kinds of cell types, but it is also important to induce the differentiation into a specific target cell type in high efficiency. Here we tried to differentiate ES cells into neuronal cells using control system of NeuroD2 expression. First, to establish ES cells where the expression of NeuroD2 was regulated by doxycycline, mouse ES cell lines were transfected by electroporation with a pTet-On plasmid DNA. We obtained 21 monoclonal ES cell lines resistant to G418 and investigated the transfection of pTet-On in genomic DNA. PCR for genomic DNA showed that the transfection of pTet-On was detected in the clone #9 ES cell (rtTA-ES cell), as shown in Fig. 2a. When pTRE-Luc plasmid DNA was temporally introduced to rtTA-ES cells, the clone #9 rtTA-ES cell had the highest luciferase activity after the addition of doxycycline among other clones. Thereafter, we transfected pTRE-ND2 plasmid DNA to the clone #9 rtTA-ES cell line and cultured them in the presence of Hygromycin B. After selection by Hygromycin B, polyclonal TRE-ND2-ES cells where pTRE-ND2 was stably transfected were established. Transfection of pTRE-ND2 in genomic DNA was detected in TRE-ND2-ES cells but not in wild type ES cells, as shown in Fig. 2b. TRE-ND2-ES cells also showed alkaline phosphatase activity and Oct3/4 expression, which are features of an undifferentiated state of ES cells, similar to wild type ES cells. These findings indicate that TRE-ND2-ES cells were maintained in an undifferentiated state.

3.5 Expression of NeuroD2 in ES cells after the addition of doxycycline

Next, we confirmed the induction of NeuroD2 expression in TRE-ND2-ES cells by the addition of doxycycline by following mRNA and protein levels. A marked increase in mRNA for NeuroD2 was detected in TRE-ND2-ES cells at 24 h after doxycycline addition by RT-PCR. CLSM images showed that the treatment with doxycycline for 24 hr also led to express NeuroD2 protein in most TRE-ND2-ES cells, as shown in Fig. 2c. The NeuroD2 expression was observed in TRE-ND2-ES cells treated for 48 hr to a similar extent. The results show that doxycycline induced the expression of NeuroD2 in TRE-ND2-ES cells.

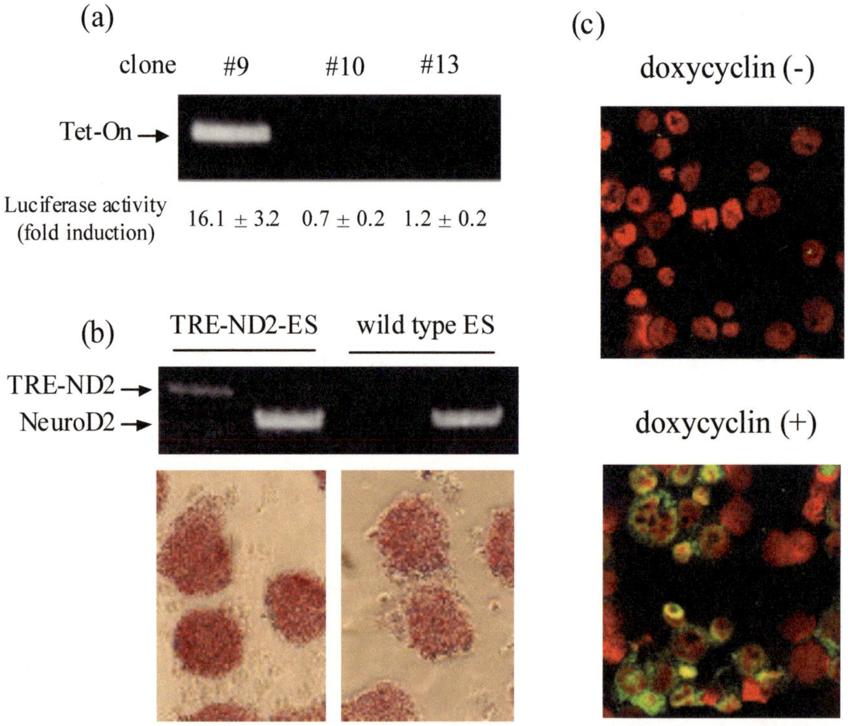

Fig. 2. Establishment of ES cell lines where the expression of NeuroD2 is regulated by doxycycline. (a) A genomic DNA PCR in ES cell clones transfected with pTet-On (rtTA-ES cells). Fold induction of luciferase activity was shown as the value at 24 h after doxycycline addition divided by it at 0 h. (b) A genomic DNA PCR (upper) and an alkaline phosphatase staining (lower) in rtTA-ES cells transfected with pTRE-ND2 (TRE-ND2-ES cells) and wild type ES cells. (c) Immunostaining of NeuroD2 in TRE-ND2-ES cells. Fluorescence derived from NeuroD2 (green) and nucleus (red) was observed in cells treated without and with doxycycline for 24 h.

3.6 Differentiation into neurons after the addition of doxycycline

To check that the induction of NeuroD2 expression elicits the neuronal differentiation, TRE-ND2-ES cells were immunostained with MAP2, a marker protein of neurons. MAP2-positive neuronal cells were significantly developed without formation of EB after 7 days of doxycycline addition, although some spontaneous differentiation was observed in the absence of doxycycline. The fluorescence intensity derived from MAP2 was significantly increased by doxycycline, as shown in Fig. 3. Thus doxycycline-induced NeuroD2 expression leads to neuronal development in TRE-ND2-ES cells.

Finally, we investigated the period of treatment with doxycycline needed to induce clear evidence of neuronal differentiation. MAP2-positive cells were followed over 11 days of culturing; by day 3 of treatment there was sufficient evidence that differentiation into neurons had taken place. Treatment with doxycycline for 1 day did not induce the neuronal differentiation efficiently. In addition, CLSM images showed that NeuroD2 was localized in the cytoplasm but not in the nucleus on day 1, and thereafter proceeded gradually into the nucleus. The expression of NeuroD2 for an appropriate period (~3 days) was therefore necessary for the neuronal differentiation and its intracellular distribution also affected the efficiency of differentiation into neurons.

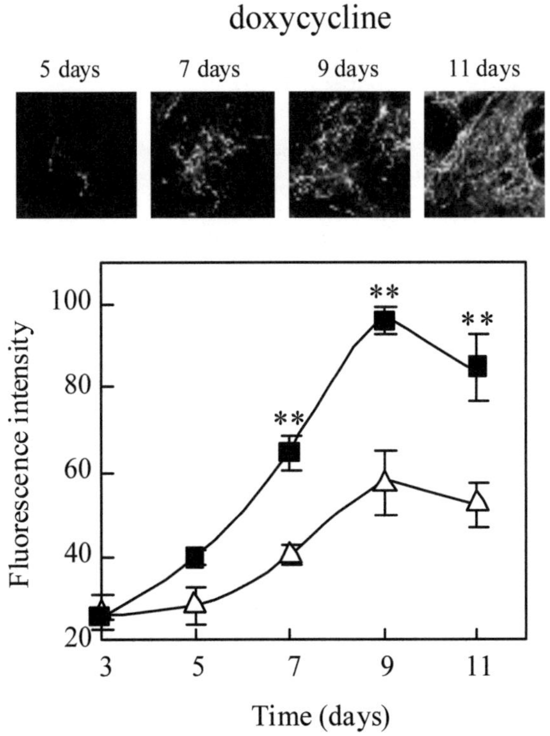

Fig. 3. Differentiation into neurons by doxycycline in TRE-ND2-ES cells. (Upper) Immunostaining of MAP2 in TRE-ND2-ES cells cultured in the presence of doxycycline (100 μM). (Lower) Fluorescence intensity derived from MAP2 in region of interest (ROI) of CLSM images. Averaged fluorescence intensity in ROI was measured from 10 images of TRE-ND2-ES cells cultured in the presence (closed squares) and absence (open triangles) of doxycycline. **P < 0.01 compared to cells without treatment of doxycycline.

4. Discussion

LIF is a member of the family of IL-6 type cytokines, which includes IL-6, IL-11, oncostatin M, ciliary neurotrophic factor, and cardiotrophin-1, and leads to the activation of JAK-STAT and ERK pathways through LIFRβ-gp130 receptors. Because LIF also plays a crucial role in self-renewal and maintenance of pulripotency in mouse ES cells, it is widely used for their propagation. However, STAT3 is constitutively phosphorylated in undifferentiated ES cells in the presence of LIF and functions to maintain their undifferentiated state and pluripotency, but the phosphorylation of ERKs is not detected. The removal of LIF induces not only the dephosphorylation of phospho-STAT3 but also the phosphorylation of ERKs and directs ES cells to differentiate. Therefore, it has been considered that some important connections between the STAT3 and ERK pathways exist in ES cells, but the relationship between those pathways in the process of ES cell differentiation has not been analyzed in detail.

At first, we found that the amount of phospho-STAT3 decreased at ~ 4 hr after the removal of LIF and then gradually increased at ~ 24 hr (Fig. 1). The phosphorylation of Raf-1 occurred at ~ 4 hr, and thereafter MEK and ERKs were subsequently phosphorylated at 8 - 12 h and immediately dephosphorylated. The time-courses of the phosphorylation levels of STAT3 and Raf-1 suggested the interaction between STAT3 and ERK pathways in ES cells. It was reported that the inhibition of ERKs activity resulted in the enhancement of STAT3-mediated transcription in previous papers (Bonni et al., 1997; Ihara et al., 1997; Sengupta et al., 1998). However, the pretreatment with PD98059 affected neither the phosphorylation of STAT3 by LIF addition nor the dephosphorylation of phospho-STAT3 by LIF removal in our present condition. On the other hand, a membrane-permeable inhibitor of STAT3 dimerization, STAT3IP, accelerated ERKs phosphorylation in ES cells cultured in the presence of LIF. These suggested that the phosphorylation of ERK did not affect STAT3 activation, but that the transcriptional activity of STAT3 regulated the ERK activation. From these results, it was considered that the inactivation of STAT3 by the removal of LIF would trigger the activation of the ERKs pathway in ES cells (Miyazu et al., 2010).

We showed here that the expression of SOCS3 depended on the presence of LIF in ES cells. The expression of SOCS3 was increased by the addition of LIF and decreased by the removal of LIF (Fig.1). SOCS3 is reported to exhibit inhibitory function of ERK signaling through the SHP-2 binding site of gp130 (Schmitz et al., 2000). The phosphorylation of MEK was actually promoted in ES cells expressing less SOCS3 in our present results (Fig. 1). In addition, it was reported that the absence of SOCS3 perturbed the balance of activation between STAT3 and ERK pathways and induced the reduction of self-renewal and the promotion of differentiation in ES cells (Forrai et al., 2006). Our data indicated the possibility that SOCS3, which is expressed by constitutively activated STAT3 in the presence of LIF in undifferentiated ES cells, suppressed the activation of ERK pathway, and that the reduction of SOCS3 expression as a result of dephosphorylation of phospho-STAT3 after LIF removal activated ERKs activation and became a trigger for differentiation in mouse ES cells.

Because several signaling pathways are involved in propagation, maintenance of pluripotency, and differentiation in ES cells, its signaling and transcriptional network is very complicated (Chen et al., 2008). Other signaling molecules such as glycogen synthase kinase (GSK) 3 and transcriptional factors including Nanog, Oct-4, Smad1, and NF-κB in addition to ERKs, SOCS3, and STAT3 described in this paper, play important roles in mouse ES cells (Torres and Watt, 2008; Ying et al., 2008). On the other hand, it was reported that ERK signals contributed to the maintenance of self-renewal, and LIF/STAT3 signaling was independent of maintenance of pluripotency in human ES cells (Okita and Yamanaka, 2006; Li et al., 2007). Accordingly, because the signaling pathway to maintain pluripotency and to induce differentiation in mouse ES cells may be different in human ES cells, the experiments to study the signaling pathway in human ES cells are necessary. In any case, the role of SOCS3 in ES cell differentiation needs to be defined in more detail, and our data strongly suggested that SOCS3 is intimately involved in the regulation of balanced activation between the STAT3 and ERK pathways in LIF signaling in mouse ES cells.

In addition, we have tried to get ES cells to differentiate into neurons after the addition of doxycycline using NeuroD2, a bHLH transcriptional factor. NeuroD2 induces ectopic neurogenesis in the ectodermal cells in *Xenopus* embryos and neuronal differentiation in mammalian embryonal carcinoma cells (McCormick et al., 1996; Farah et al., 2000). We have shown that doxycycline induced the expression of NeuroD2 in TRE-ND2-ES cells and that the inducible NeuroD2 expression leads after several days to differentiation into neurons without formation of EBs. Addition of doxycycline significantly promotes neuronal differentiation in TRE-ND2-ES cells (see Fig. 2). About half the cells expressed NeuroD2 after the addition of doxycycline, but the yield of MAP2-positive cells was 25 - 30%. It is known that extracellular matrices affect neural differentiation and growth, and that ES cells cultured on the dishes coated with laminin or fibronectin can be effectively differentiated into neurons (Czyz and Wobus, 2001; Andressen et al., 2005). In addition, several factors such as retinoic acid, basic fibroblast growth factor, sonic hedgehog, bone morphogenetic proteins, and Wnt can also play important roles in accelerating neuronal differentiation (Du and Zhang, 2004; Murashov et al., 2005). To study the essential role of NeuroD2 in neuronal differentiation, we cultured ES cells in gelatin-coated dishes in the absence of supplements such as growth factors and mitogen, in which the efficiency and specificity of differentiation was not completely satisfactory in view of transplantation and regenerative medicine. However, the combination of NeuroD2 expression and growth factor/mitogen addition does increase the specificity and efficiency of neural differentiation in ES cells.

Doxycycline-induced neurons extended long neurites and formed dense neurite network. The number of MAP2-positive neurons reached a plateau at 3, while the MAP2-derived signal increased from 3 to 9 days. This means that ES cells are differentiated into neurons within 3 days after doxycycline addition, and that the differentiated neurons thereafter show neurite extension and network formation. However, although some ES cells spontaneously differentiated into neurons without doxycycline in this experimental condition, their neurites were thin and extend of their outgrowth was small. NeuroD2 was distributed in the cytoplasm of ES cells 1 day after the treatment with doxycycline, but if NeuroD2 could be translocated to the nucleus by some treatment (e.g. cytokines or growth factors), ES cells would be differentiated into neurons without the formation of EB, and raise the possibility of the induction of neuronal differentiation of ES cells in high efficiency by NeuroD2.

5. Conclusion

It is indicated that SOCS3 is involved in maintaining an undifferentiated state mediated by LIF and that the extrinsic expression of NeuroD2 efficiently induced differentiation into neurons in mouse ES cells. These findings provide the possibility to lead to the specific differentiation into target cell types for ES cells.

6. References

Alessi, DR.; Cuenda, A., Cohen, P., Dudley, DT. & Saltiel, AR. (1995). PD 98059 is a specific inhibitor of the activation of mitogen-activated protein kinase kinase in vitro and in vivo. *J. Biol. Chem.*, 270, 27489-27494.

Amano, K.; Furuno, T. & Nakanishi, M. (2006). Conditioned medium from feeder STO cells increases the attachment of mouse embryonic stem cells. *Biol. Pharm. Bull.*, 29, 1747-1750.

Andressen, C., Adrian, S., Fassler, R., Arnhold, S. & Addicks, K. (2005). The contribution of beta1 integrins to neuronal migration and differentiation depends on extracellular matrix molecules. *Eur. J. Cell Biol.*, 84, 973-982.

Bonni, A.; Sun, Y., Nadal-Vicens, M., Bhatt, A., Frank, DA., Rozovsky, I., Stahl, N., Yancopoulos, GD. & Greenberg, ME. (1997). Regulation of gliogenesis in the central nervous system by the JAK-STAT signaling pathway. *Science*, 278, 477-483.

Bousquet, C.; Susini, C. & Melmed, S. (1999). Inhibitory roles for SHP-1 and SOCS-3 following pituitary proopiomelanocortin induction by leukemia inhibitory factor. *J. Clin. Invest.*, 104, 1277-1285.

Burdon, T.; Chambers, I., Stracey, C., Niwa, H. & Smith, A. (1999b). Signaling mechanisms regulating self-renewal and differentiation of pluripotent embryonic stem cells. *Cells Tissues Organs*, 165, 131-143.

Burdon, T.; Smith, A. & Savatier, P. (2002). Signalling, cell cycle and pluripotency in embryonic stem cells. *Trends Cell Biol.*, 12, 432-438.

Burdon, T.; Stracey, C., Chambers, I., Nichols, J. & Smith, A. (1999a). Suppression of SHP-2 and ERK signalling promotes self-renewal of mouse embryonic stem cells. *Dev. Biol.*, 210, 30-43.

Chan. RJ.; Johnson, SA., Li, Y., Yoder, MC. & Feng, GS. (2003). A definitive role of Shp-2 tyrosine phosphatase in mediating embryonic stem cell differentiation and hematopoiesis. *Blood*, 102, 2074-80.

Chen, X,; Xu, H., Yuan, P., Fang, F., Huss, M., Vega, VB., Wong, E., Orlov, YL., Zhang, W., Jiang, J., Loh, YH., Yeo, HC., Yeo, ZX., Narang, V., Govindarajan, KR., Leong, B., Shahab, A., Ruan, Y., Bourque, G., Sung, WK., Clarke, ND., Wei, CL. & Ng, HH. (2008). Integration of external signaling pathways with the core transcriptional network in embryonic stem cells. *Cell*, 133, 1106-1117.

Czyz, J. & Wobus, A. (2001). Embryonic stem cell differentiation: the role of extracellular factors. *Differentiation*, 68, 167-174.

Darnell, JE., Jr. (1997). STATs and gene regulation. *Science*, 277, 1630-1635.

Du, ZW. & Zhang, SC. (2004). Neural differentiation from embryonic stem cells: which way? Stem Cells Dev., 13, 372-381.

Duval, D.; Reinhardt, B., Kedinger, C. & Boeuf, H. (2000). Role of suppressors of cytokine signaling (Socs) in leukemia inhibitory factor (LIF)-dependent embryonic stem cell survival. *FASEB J.*, 14, 1577-1584.

Ernst, M. & Jenkins, BJ. (2004). Acquiring signaling specificity from the cytokine receptor gp130. *Trends Genet.*, 20, 23-32.

Ernst, M.; Oates, A. & Dunn, AR. (1996). gp130-mediated signal transduction in embryonic stem cells involves activation of Jak and Ras/mitogen-activated protein kinase pathways. *J. Biol. Chem.*, 271, 30136-30143.

Farah, MH.; Olson, JM., Sucic, HB., Hume, RI., Tapscott, SJ. & Turner, DL. (2000). Generation of neurons by transient expression of neural bHLH proteins in mammalian cells. *Development*, 127, 693-702.

Forrai, A.; Boyle, K., Hart, AH., Hartley, L., Rakar, S., Willson, TA., Simpson, KM., Roberts, AW., Alexander, WS., Voss, AK. & Robb, L. (2006). Absence of suppressor of cytokine signalling 3 reduces self-renewal and promotes differentiation in murine embryonic stem cells. *Stem Cells*, 24, 604-614.

Fukada, T.; Hibi, M., Yamanaka, Y., Takahashi-Tezuka, M., Fujitani, Y., Yamaguchi, T., Nakajima, K. & Hirano, T. (1996). Two signals are necessary for cell proliferation induced by a cytokine receptor gp130: involvement of STAT3 in anti-apoptosis. *Immunity*, 5, 449-460.

Furuno, T.; Hamano. T. & Nakanishi, M. (1993). Receptor-mediated calcium signal playing a nuclear third messenger in the activation of antigen-specific B cells. *Biophys. J.*, 64, 665-669.

Furuno, T.; Hirashima, N., Onizawa, S., Sagiya, N. & Nakanishi, M. (2001). Nuclear shuttling of mitogen-activated protein (MAP) kinase (extracellular signal-regulated kinase (ERK) 2) was dynamically controlled by MAP/ERK kinase after antigen stimulation in RBL-2H3 cells. *J. Immunol.*, 166, 4416-4421.

Furuno, T.; Ito, A., Koma, Y., Watabe, K., Yokozaki, H., Bienenstock, J., Nakanishi, M. & Kitamura, Y. (2005). The spermatogenic Ig superfamily/synaptic cell adhesion molecule mast-cell adhesion molecule promotes interaction with nerves. *J. Immunol.*, 174, 6934-6942.

Gossen, M. & Bujard, H. (1992). Tight control of gene expression in mammalian cells by tetracycline-responsive promoters. *Proc. Natl. Acad. Sci. USA*, 89, 5547-5551.

Gossen, M. & Bujard, H. (1995). Efficacy of tetracycline-controlled gene expression is influenced by cell type: commentary. *Biotechniques*, 19, 213-216.

Heinrich, PC.; Behrmann, I., Haan, S., Hermanns, HM., Muller-Newen, G. & Schaper, F. (2003). Principles of interleukin (IL)-6-type cytokine signalling and its regulation. *Biochem. J.*, 374, 1-20.

Heinrich, PC.; Behrmann, I., Muller-Newen, G., Schaper, F. & Graeve, L. (1998). Interleukin-6-type cytokine signalling through the gp130/Jak/STAT pathway. *Biochem. J.*, 334, 297-314.

Hibi, M.; Murakami, M., Saito, M., Hirano, T., Taga, T. & Kishimoto, T. (1990). Molecular cloning and expression of an IL-6 signal transducer, gp130. *Cell*, 63, 1149-1157.

Ihara, S.; Nakajima, K., Fukada, T., Hibi, M., Nagata, S., Hirano, T. & Fukui, Y. (1997). Dual control of neurite outgrowth by STAT3 and MAP kinase in PC12 cells stimulated with interleukin-6. EMBO J., 16, 5345-5352.

Ince-Dunn, G.; Hall, BJ., Hu, SC., Ripley, B., Huganir, RL., Olson, JM., Tapscott, SJ. & Ghosh, A. (2006). Regulation of thalamocortical patterning and synaptic maturation by NeuroD2. *Neuron*, 49, 683-695.

Inoh, Y.; Kitamoto, D., Hirashima, N. & Nakanishi, M. (2001). Biosurfactants of MEL-A increase gene transfection mediated by cationic liposomes. *Biochem. Biophys. Res. Commun.*, 289, 57-61.

Jirmanova, L.; Afanassieff, M., Gobert-Gosse, S., Markossian, S. & Savatier, P. (2002). Differential contributions of ERK and PI3-kinase to the regulation of cyclin D1 expression and to the control of the G1/S transition in mouse embryonic stem cells. *Oncogene*, 21, 5515-5528.

Kanda, S.; Tamada, Y., Yoshidome, A., Hayashi, I. & Nishiyama, T. (2004). Over-expression of bHLH genes facilitates neural formation of mouse embryonic stem (ES) cells in vitro. *Int. J. Dev. Neurosci.*, 22, 149-156.

Kristensen, DM.; Kalisz, M. & Nielsen, JH. (2005). Cytokine signalling in embryonic stem cells. *APMIS*, 113, 756-772.

Kume, H.; Maruyama, K., Tomita, T., Iwatsubo, T., Saido, TC. & Obata, K. (1996). Molecular cloning of a novel basic helix-loop-helix protein from the rat brain. *Biochem. Biophys. Res. Commun.*, 219, 526-530.

Li, J., Wang, G., Wang, C., Zhao, Y., Zhang, H., Tan, Z., Song, Z., Ding, M. & Deng, H. (2007). MEK/ERK signaling contributes to the maintenance of human embryonic stem cell self-renewal. *Differentiation*, 75, 299-307.

Lin, CH.; Tapscott, SJ. & Olson, JM. (2006). Congenital hypothyroidism (cretinism) in neuroD2-deficient mice. *Mol. Cell. Biol.*, 26, 4311-4315.

Massari, ME. & Murre, C. (2000). Helix-loop-helix proteins: regulators of transcription in eucaryotic organisms. Mol. Cell. Biol., 20, 429-440.

Matsuda, T.; Nakamura, T., Nakao, K., Arai, T., Katsuki, M., Heike, T. & Yokota, T. (1999). STAT3 activation is sufficient to maintain an undifferentiated state of mouse embryonic stem cells. *EMBO J.*, 18, 4261-4269.

McCormick, MB.; Tamimi, RM., Snider, L., Asakura, A., Bergstrom, D. & Tapscott, SJ. (1996). NeuroD2 and neuroD3: distinct expression patterns and transcriptional activation potentials within the neuroD gene family. *Mol. Cell. Biol.*, 16, 5792-5800.

Miyazu, S.; Furuno, T. & Nakanishi, M. (2010). Phosphorylation states of STAT3 and ERKs in mouse embryonic stem cells. *Cell Biol. Int.*, 34, 485-492.

Murashov, AK.; Pak, ES., Hendricks, WA., Owensby, JP., Sierpinski, PL., Tatko, LM. & Fletcher, PL. (2005). Directed differentiation of embryonic stem cells into dorsal interneurons. *FASEB J.*, 19, 252-254.

Nakanishi, M. (2003). New strategy in gene transfection by cationic transfection lipids with a cationic cholesterol. *Curr. Med. Chem.*, 10, 1289-1296.

Nakanishi, M. & Furuno, T. (2008). Molecular basis of neuroimmune interaction in an in vitro coculture approach. *Cell. Mol. Immunol.*, 5, 249-259.

Nicholson, SE.; De Souza, D., Fabri, LJ., Corbin, J., Willson, TA., Zhang, JG., Silva, A., Asimakis, M., Farley, A., Nash, AD., Metcalf, D., Hilton, DJ., Nicola, NA. & Baca, M. (2000). Suppressor of cytokine signaling-3 preferentially binds to the SHP-2-binding site on the shared cytokine receptor subunit gp130. *Proc. Natl. Acad. Sci. USA*, 97, 6493-6498.

Niwa, H.; Burdon, T., Chambers, I. & Smith, A. (1998). Self-renewal of pluripotent embryonic stem cells is mediated via activation of STAT3. *Genes Dev.*, 12, 2048-2060.

Noda, T.; Kawamura, R., Funabashi, H., Mie, M. & Kobatake, E. (2006). Transduction of NeuroD2 protein induced neural cell differentiation. *J. Biotechnol.*, 126, 230-236.

Okita, K. & Yamanaka, S. (2006). Intracellular signaling pathways regulating pluripotency of embryonic stem cells. *Curr. Stem Cell Res. Ther.*, 1, 103-111.

Olson, JM.; Asakura, A., Snider, L., Hawkes, R., Strand, A., Stoeck, J., Hallahan, A., Pritchard, J. & Tapscott, SJ. (2001). NeuroD2 is necessary for development and survival of central nervous system neurons. *Dev. Biol.*, 234, 174-187.

Qu, CK.; Shi, ZQ., Shen, R., Tsai, FY., Orkin, SH. & Feng, GS. (1997). A deletion mutation in the SH2-N domain of Shp-2 severely suppresses hematopoietic cell development. *Mol. Cell. Biol.*, 17, 5499-5507.

Raz, R.; Lee, CK., Cannizzaro, LA., d'Eustachio, P. & Levy, DE. (1999). Essential role of STAT3 for embryonic stem cell pluripotency. *Proc. Natl. Acad. Sci. USA*, 96, 2846-2851.

Schindler, C. & Darnell, JE., Jr. (1995). Transcriptional responses to polypeptide ligands: The JAK-STAT pathway. *Annu. Rev. Biochem.*, 64, 621-651.

Schmitz, J.; Weissenbach, M., Haan, S., Heinrich, PC. & Schaper, F. (2000). SOCS3 exerts its inhibitory function on interleukin-6 signal transduction through the SHP2 recruitment site of gp130. *J. Biol. Chem.*, 275, 12848-12856.

Sengupta, TK.; Talbot, ES., Scherle, PA. & Ivashkiv, LB. (1998). Rapid inhibition of interleukin-6 signaling and Stat3 activation mediated by mitogen-activated protein kinases. *Proc. Natl. Acad. Sci. USA*, 95, 11107-11112.

Smith, AG.; Heath, JK., Donaldson, DD., Wong, GG., Moreau, J., Stahl, M. & Rogers D. (1988). Inhibition of pluripotential embryonic stem cell differentiation by purified polypeptides. *Nature* 336, 688-690.

Stahl, N.; Boulton, TG., Farruggella, T., Ip, NY., Davis, S., Witthuhn, BA., Quelle, FW., Silvennoinen, O., Barbieri, G., Pellegrini, S., Ihle, JN. & Yancopoulos, GD. (1994). Association and activation of Jak-Tyk kinases by CNTF-LIF-OSM-IL-6β receptor components. *Science*, 263, 92-95.

Strubing, C.; Ahnert-Hilger, G., Shan, J., Wiedenmann, B., Hescheler, J. & Wobus, AM. (1995). Differentiation of pluripotent embryonic stem cells into the neuronal lineage in vitro gives rise to mature inhibitory and excitatory neurons. *Mech. Dev.*, 53, 275-287.

Sugimoto, Y.; Furuno, T. & Nakanishi, M. (2009). Effect of NeuroD2 expression on neuronal differentiation in mouse embryonic stem cells. *Cell Biol. Int.*, 33, 174-179.

Suzuki, A.; Suzuki, R., Furuno, T., Teshima, R. & Nakanishi, M. (2004). N-cadherin plays a role in the synapse-like structures between mast cells and neurites. *Biol. Pharm. Bull.*, 27, 1891-1894.

Torres, J. & Watt, FM. (2008). Nanog maintains pluripotency of mouse embryonic stem cells by inhibiting NFκB and cooperating with Stat3. Nat. Cell Biol., 10, 194-201.

Turkson, J.; Kim, JS., Zhang, S., Yuan, J., Huang, M., Glenn, M., Haura, E., Sebti, S., Hamilton, AD. & Jove, R. (2004). Novel peptidomimetic inhibitors of signal transducer and activator of transcription 3 dimerization and biological activity. *Mol. Cancer Ther.*, 3, 261-269.

Van Vactor, D.; O'Reilly, AM. & Neel, BG. (1998). Genetic analysis of protein tyrosine phosphatases. *Curr. Opin. Genet. Dev.*, 8, 112-126.

Williams, RL.; Hilton, DJ., Pease, S., Willson, TA., Stewart, CL., Gearing, DP., Wagner, EF., Metcalf, D., Nicola, NA. & Gough NM. (1988). Myeloid leukaemia inhibitory factor maintains the developmental potential of embryonic stem cells. *Nature*, 336, 684-687.

Wobus, AM., & Boheler, KR. (2005). Embryonic stem cells: prospects for developmental biology and cell therapy. *Physiol. Rev.*, 85, 635-678.

Yao, Y.; Li, W., Wu, J., Germann, UA., Su, MS., Kuida, K. & Boucher, DM. (2003). Extracellular signal-regulated kinase 2 is necessary for mesoderm differentiation. *Proc. Natl. Acad. Sci. USA*, 100, 12759-12764.

Yasunami, M.;, Suzuki, K., Maruyama, H., Kawakami, H., Nagai, Y., Hagiwara, M. & Ohkubo, H. (1996). Molecular cloning and characterization of a cDNA encoding a novel basic helix-loop-helix protein structurally related to Neuro-D/BHF1. *Biochem. Biophys. Res. Commun.*, 220, 754-758.

Ying, QL.; Nichols, J., Chambers, I. & Smith, A. (2003). BMP induction of Id proteins suppresses differentiation and sustains embryonic stem cell self-renewal in collaboration with STAT3. *Cell*, 115, 281-292.

Ying, QL.; Wray, J., Nichols, J., Batlle-Morera, L., Doble, B., Woodgett, J., Cohen, P. & Smith, A. (2008). The ground state of embryonic stem cell self-renewal. *Nature*, 453, 519-523.

Part 4

Hormonal Signals that Regulate the Differentiation of Mesodermal Cells - Cardiogenesis, Angiogenesis and Osteogenesis

23

ESC Cardiac Differentiation and Applications

Nasser Aghami[1] and Hananeh Fonoudi[2,3]
[1]*Department of regenerative medicine, Royan institute for stem cell biology and technology*
[2]*Department of stem cell and developmental biology, Royan institute for stem cell biology and technology*
[3]*Department of biotechnology, Tehran University*
Iran

1. Introduction

According to the World Health Organization (WHO), cardiovascular diseases are the leading cause of mortality worldwide. In the United States, there are more than three millions patients with a predicted increase to over six million by the year 2030. Unfortunately, current therapeutic methods are not effective and two-thirds of the patients die within five years of diagnosis (Kessler & Byrne, 1999). Thus it is important to discover new drugs which are effective. Application of high-throughput screening in the pharmaceutical industry can enhance procedures for new drug production, from target identification to preclinical compound evaluation. Reducing the lag phase between identification of a new component and production of new drugs requires cell-based methods for testing the efficacy and safety of new compounds. Application of stem cells in drug discovery has afforded new opportunities to better understand the action mechanisms of new targets, increase the safety to control their activity, reduce the amount of *in vivo* testing and evaluate components for different human genetic backgrounds. On the other hand, recently a biomedical approach called regenerative medicine have been developed. In this method a new pool of functional cells will be used in order to enhance the performance of damaged tissue. The ideal cell source for regenerative medicine should have the capacity to restore the organ's function, induce native repair and eliminate the risk of rejection by the host. In recent years, human embryonic stem cells (hES cells) with the ability of self-renewal and differentiation have been considered as ideal therapeutic cell candidates. Although tremendous developments in this field have been achieved, there are still large gaps between recent investigations and successful, safe application of these cells in humans.

In this chapter, a brief definition of ESCs followed by methods for directing their differentiation into a cardiac lineage will be presented followed by a discussion on the best strategies to purify cardiomyocytes. We will also provide a future glimpse into applications of hES cells in drug discovery and regenerative medicine.

2. Human embryonic stem cells

hES cells (Thomson et al., 1998; Reubinoff et al., 2000; Baharvand et al., 2004) are pluripotent stem cell lines which have the ability to differentiate into all three primary germ layers that arise during development (ectoderm, mesoderm and endoderm) even after prolonged culture (Thomson & Marshall, 1998). The origin of these cells is the inner cell mass (ICM) of blastocysts. Blastocysts consist of an outer layer, the trophoectoderm. The ICM of blastocysts are cells within the trophoectoderm which give rise to all cells in the embryo. The trophoectoderm layer has a supporting role for embryonal tissues. In order to derive hES cells the trophoectoderm should be removed, and isolated ICM cells cultured on a feeder layer of mouse embryonic fibroblasts. After cells grow and form colonies they can be selected, passaged and expanded. For humans, ICM generated from IVF-produced embryos not used in the clinic and donated by individuals can be used to form hES cell lines.

hES cells have the ability for self-renewal, a continual high level of telomerase activity, and express Oct4 and Nanog which are two transcription factors that define ESC identity. They maintain a normal diploid karyotype and express undifferentiated markers such as stage specific embryonic antigen 3 (SSEA), SSEA4, Tra-1-60 and Tra-1-81 (Thomson et al., 1998; Amit et al., 2000; Reubinoff et al., 2000).

The pluripotency of these cells has been shown by injection into immunosuppressed mice. In mice, undifferentiated hES cells grew and formed teratoma which was comprised of all three embryonic layers. Pluripotency was further confirmed by the formation of three-dimensional embryoid bodies *in vitro* (Itskovitz-Eldor et al., 2000).

At first, cell populations needed to grow as compact colonies of undifferentiated cells on mouse embryonic fibroblast (MEF) (Thomson et al., 1998), human fetal fibroblast and adult epithelial cells (Richards et al., 2002) or foreskin cells as feeder layers to stay pluripotent, but later experiments have shown that hES cells can stay in an undifferentiated state if cultured on extracellular matrix such as matrigel or laminin, and in medium conditioned by MEF (Xu et al., 2001). Further investigations revealed molecules which are necessary for maintenance of hES cells in undifferentiated state *in vitro*. Basic fibroblast growth factor (bFGF) is one such protein that has been described to inhibit hES cell differentiation in the absence of feeder layers (Xu et al., 2001). Currently, they can be isolated and maintained in chemically defined medium (Ludwig et al., 2006; Yao et al., 2006). Presently, there are more than 400 hES cell lines but only 179 of these have been characterized in detail (Guhr et al., 2006). In an international study (ISCI), 59 hES cell lines were examined for the expression of several markers. Data from this experiment indicated that different cell lines show similar expression patterns despite different genetic backgrounds and derivation techniques (Adewumi et al., 2007). hES cells are attractive sources for studying cardiac differentiation because they can be isolated and survive in culture, expand greatly on feeder layer or feeder-free culture and differentiate into functional cardiomyocytes (Baharvand et al., 2005; Farokhpour et al., 2009; Kehat et al., 2001; Xu et al., 2002; Mummery et al., 2003) and form contractile clusters.

The development of hES cell technology and the ability to culture them *in vitro* has opened a new era in cell transplantation therapy, molecular genetics, drug discovery and developmental studies (Odorico et al., 2001). They also can be considered as a suitable source to determine the physiological and pharmacological properties of human cardiomyocytes and decrease the need for animals as test materials that are time-consuming and cost-prohibitive. However in order to use hES cells in cell therapy and drug screening, development of appropriate culture conditions and differentiation methods which are suitable for scale-up are needed.

3. Differentiation of hES cells to cardiac cells

3.1 Spontaneous differentiation

The most common way to initiate the differentiation of hES cells is embryoid body (EB) formation. These types of cell aggregates provide the essential signaling environment critical for cardiac differentiation. Development of the heart in vertebrates can be divided into four steps. The first step is the formation of an organizing center which secretes signals for endoderm and mesoderm induction. This step leads to activation of canonical wnt/β-catenin members such as wnt3a which then causes accumulation of β-catenin in the nucleus. β-catenin plays a key role in expression of the gene Nodal which induces and controles expression of mesoendodermal markers. In the second step Nodal, a member of the TGF-β family, causes mesoderm induction. Activation of the nodal signaling pathway will then lead to the expression of mesodermal genes, such as brachury (*bra*), in a concentration and duration-dependent manner. In the third step, after formation of the mesoderm, the wnt/β-catenin signal must be inhibited in order to continue with heart formation. Thus, several canonical wnt/β-catenin antagonists will be expressed to induce the precardiac mesoderm, such as DKK1. On the other hand, it has been shown that activation of the noncanonical Wnt signaling pathway stimulates expression of early cardiac markers such as Nkx2.5. The last step is differentiation of cardiac progenitors into beating cardiomyocytes and BMP signaling is critical for end stage gene expression.

The first report of differentiating hES cells into cardiomyocytes was in 2001. In brief, as hES cells cannot survive in single cells, they were dissociated into small clumps of 3-20 cells by collagenase and cultured in low attachment dishes for 7-10 days in suspension in the absence of self-renewal signals provided by MEF feeder layers or basic fibroblast growth factor (bFGF), then plated on gelatin coated plates for further differentiation. As the EBs mature, cells give rise to early embryonic lineages. In this method, four days after plating, 8.1% of the EBs contained beating clusters and showed a beat rate of 30-130 beats per minute. Approximately 30% of the cells in these clusters were actual cardiomyocytes and shown to express cardiac-specific structural genes, such as cardiac troponin I and T, atrial and ventricular myosin light chains (MLCs), atrial natriuretic peptide (ANP) in addition to cardiac transcription factors such as GATA4, Nkx2.5 and MEF2C (Kehat et al., 2001). Investigations indicate that these beating clusters contain cardiac cells which have been shown by several functional assays (extracellular and intracellular electrophysiological recordings, calcium imaging and pharmacological studies). They are structurally and functionally identical with nascent embryonic myocardium (Kehat et al., 2001; Boheler et al., 2002; Xu et al., 2002; He et al., 2003; Mummery et al., 2003; Passier & Mummery, 2005). The patterns of cardiac gene expression in differentiating hES cell derived cardiomyocytes is similar to embryonic cardiogenesis (Xu et al., 2006). During the early stages of differentiation, the expression of pluripotency markers such as Oct3/4, Cripto and telomerase reverse transcriptase (TERT) decrease while the mesoendoderm markers, such as Brachury, gradually increase. Subsequently, early cardiac transcription factors such as Nkx2.5, MEF2C, GATA4 and Tbx5 will be expressed. Later, the expression of cardiac structural proteins such as cardiac α-myosin heavy chain, cardiac β-myosin chain and atrial natriuretic factor appear (Beqqali et al., 2006; Synnergren et al., 2008; Synnergren et al., 2008). The similarity in the timeline of gene expression during cardiac differentiation and cardiac development indicate that data from developmental studies can be used to increase the efficiency of differentiation.

In the year 2002, other embryonic cell lines were differentiated into cardiomyocytes. On the 8th day of differentiation 25% of the EBs could beat while this number increased to 70% by the 20th day (Xu et al., 2002). A third group also reported the derivation of 10-25% beating cardiac cells from hES cells after 30 days (He et al., 2003). The reasons for the differences observed in numbers of contractile clusters are not clear, but it is assumed that the efficiency of different protocols which are applied to direct hES cells into cardiomyocytes depends on the properties of individual cell lines as well as propagation methods prior to differentiation. On the other hand, as individual EBs may contain different cardiac cells, higher beating EBs can not accurately reflect higher efficiency.

To study the events those occur during differentiation toward cardiac cells, whole genome approaches were applied. Initial investigations reported the genetic control of human embryonic development by studying the transcriptional profile of spontaneous differentiation (Brandenberger et al., 2004; Calhoun et al., 2004; Miura et al., 2004).These studies, by comparing undifferentiated hES cells and EBs, have provided important information about signals required for the self-renewal of hES cells. For example, they have indicated that FGF, WNT, NODAL and LIF pathways play important roles in maintaining pluripotency (Brandenberger et al., 2004).The first large-scale microarray has been performed 2, 10 and 30 days after EB formation (Dvash et al., 2004) and identified genes that are expressed during the early and late stages. Although these findings are valuable to understand transcriptional pathway which are critical in embryonic cardiac development, protocols to direct hES cells into cardiomyocytes with higher efficiency than spontaneous differentiation is required.

However because of heterogeneity, requirements for serum and other such limitations, spontaneous differentiation is not a proper method and much effort has been devoted to enhance the efficiency of cardiac differentiation. Thus, several techniques have been introduced to improve the efficiency of cardiomyogenic differentiation *in vitro*. One of the most common reasons for heterogeneity is that individual EBs differ in size and morphology. The first attempt to solve this issue was the application of "hanging drop" method. This method was routinely used for EB formation of mouse ES cells (Yoon et al., 2006). However this method was not successful when translated for hES cells. An alternative method was the forced-aggregation or the "spin-EB". In this method, to control the size of EBs, dissociated cells were centrifuged into V-formed ultra-low adherence 96-well plates (Ng et al., 2005; Burridge et al., 2007). Although this method provided same-size EBs, singly its application could not improve the efficiency of differentiation. Therefore several protocols have been developed to direct the differentiation toward cardiomyocytes. Strategies to induce cardiac specific differentiation rely on co-culture with different cell types, addition of different supplements to the medium and genetic manipulation of hES cells.

3.2 Directed differentiation

3.2.1 Co culture

Early efforts to find a method to enhance the differentiation of hEs cells into cardiomyocytes has led to the introduction of a new protocol by Mummery et al. (Mummery et al., 2002; Mummery et al., 2003; Passier et al., 2005). In this method, hES cells are co-cultured with mitomycin-treated mouse visceral endoderm-like cell line (END2) as a feeder layer and 12 days after co-culture, cells undergo cardiac differentiation (Mummery et al., 2003). The main advantage of this method is that hES cells are in direct contact with visceral endoderm and their surface receptors are exposed to the autocrine and paracrine factors secreted by visceral endoderm, such as activin A and BMPs. As a result, this method mimics natural development because during development, the anterior endoderm provides signals for cells in the adjacent cardiac mesoderm and promotes differentiation into beating cardiomyocytes (Fullilove, 1970; Schultheiss et al., 1995). Furthermore, the anterior endoderm has the ability to induce cardiac differentiation of non-cardiac mesoderm (Sugi & Lough, 1994).

Whole genome microarray analysis has been performed on hES cells which were co-cultured with END2. By cluster analysis of its data, different clusters of genes corresponding to different levels of differentiation have been identified. These clusters consisted of genes which were down regulated after differentiation, such as OCT4 and NANOG, as well as genes attributed to early mesoderm formation, cardiac progenitors and fetal cardiomyocytes.

The presence of serum in differentiation medium is another factor which controls the efficiency of differentiation. Serum has negative effects on cardiac differentiation and by eliminating serum, the efficiency of co-culture with END2 increases more than 20-fold. On the other hand, the presence of serum in culture media converts the protocol into a less reproducible one because the exact component of serum is unknown and varies from one batch to another (Passier et al., 2005). The efficiency of this protocol has been further enhanced by addition of ascorbic acid. These findings have led to a near 25-fold increase in the number of beating areas per cell preparation. The phenotype of nearly 90% of cells derived with this protocol is similar to fetal ventricular cells, although atrial and pacemaker-like cells are also observed (Mummery et al., 2003). In another experiment, the differentiation efficiency has been optimized by removal of serum and insulin, and the addition of prostaglandin I2 to the culture medium (Zaffran & Frasch, 2002; Passier et al., 2005; Graichen et al., 2008; Xu et al., 2008). Co-culture has some advantages that limit the clinical application of this method. One obstacle is the risk of transmission of pathogens, especially viruses. In this situation, the differentiated cadiomyocytes would already be diseased and not suitable for transplantation, thus these cells could not be expanded for drug testing and toxicology.

Another limitation with the use of co-culture is the difficulty of separating co-cultured cell populations. Although cells can be separated by magnetic affinity cell sorting (MACS) (Siegel, 2002), this method does not have the proper purity. Fluorescence-activated cell sorting (FACS) can provide the highest degree of purity but the instrument is expensive, skill-intensive and not available to everyone (Herzenberg et al., 2002). Separation of co-cultured cell populations leads to detachment of the cells from each other and disrupts gap junctions which are necessary for electrical coupling between differentiating cardiomyocytes. These problems can be solved by using condition medium or commercially available Transwell inserts (Giovino et al., 2002).

Since all the previously mentioned protocols produce cardiomyocytes with low efficiency and past experiments with mouse embryonic stem (mES) cells have efficiently resulted to cardiomyocytes, therefore different cardiogenic reagents which have been shown to enhance the cardiac differentiation of mES cells were also tested on hES cells. Unfortunately, no significant improvement was achieved by addition of DMSO, retinoic acid (Kehat et al., 2001; Xu et al., 2002) or BMP-2 (Mummery et al., 2003; Pera et al., 2004). It is unclear whether these protocols need to be manipulated or whether these factors are not important in human cardiac differentiation. Investigators, by evaluation of various growth factors and agents, have provided a diverse array of agents capable of inducing cardiomyogenesis from ES cells.

3.2.2 Growth factors

As spontaneous differentiation is low and both undifferentiated and differentiated hES cells express receptors for different growth factors, the addition of appropriate growth factors in defined time and concentration may enhance differentiation toward cardiomyocytes (Xu et al., 2008).

Data from developmental studies indicate that members of the transforming growth factor (TGF) superfamily (BMPs and activin), the Wnt family, as well as the fibroblast growth factor family which are expressed in the ectoderm and endoderm adjacent to the heart-forming region have key roles in cardiac lineage induction (Filipczyk et al., 2007; Behfar et al., 2002). With the addition of BMP2 and SU5402, a FGF receptor inhibitor, into EB medium Tomescot et al. were successful in increasing expression of mesodermal and cardiac genes by more than three-fold (Tomescot et al., 2007). Burridge et al. indicated that by application of Activin A and FGF2 in the early stages of EB differentiation, the number of beating areas would increase five-fold (Burridge et al., 2007). Both of these experiments have proven that TGF-β superfamily members play an essential role in the cardiac differentiation of hES cells.

It has been shown that BMP signaling promotes commitment of cells into the cardiac lineage by posteriorization of the primitive streak, but it does not play a role in primitive streak induction. This induction is attributed to both the activin/Nodal and the canonical Wnt pathways (Nostro et al., 2008). Thus activin A, which is secreted from the visceral endoderm, has the ability to enhance cardiac differentiation (Yao et al., 2006; Laflamme et al., 2007; Yang et al., 2008). Based on this information, Laflamme et al. reported a differentiation protocol using RPMI that contained 2% B27 supplemented with activin A and BMP4, which obtained over 30% efficiency (Laflamme et al., 2007).

As well as BMP, members of the Wnt signaling pathway have the ability to increase cardiac differentiation efficiency by stabilizing β-catenin which activates cardiac gene expression (Lev et al., 2005). Activation of canonical Wnt signaling in the early stages of differentiation leads to mesoderm formation and enhances cardiac differentiation (Marvin et al., 2001; Schneider & Mercola, 2001; Ueno et al., 2007; Klaus & Birchmeier, 2009). With these results at hand, an improved differentiation protocol has been developed by Tran and co-workers. In this protocol, hES cells at the early stages of EB formation were treated by Wnt-signaling activator Wnt3a or BMP4, followed by reduction in the amount of serum and insulin in the medium (Tran et al., 2009). However, after mesoderm induction this signal inhibited further cardiac differentiation (Cohen et al., 2007; Qyang et al., 2007; Yang et al., 2008). In this stage, activation of non-canonical Wnt signaling such as Wnt11 activates protein kinase C and Jun N-terminal kinase signaling pathway and enhances cardiac differentiation (Terami et al., 2004).

In another stage, the specific differentiation protocol by Yang et al. succeeded in forming cardiac progenitor cells from hES cells by using a combination of activin A, BMP4,FGF2, VEGF and DKK1 in serum-free medium. They were able to identify a population of cardiovascular progenitors which had low expression of KDR and no c-kit (KDR low/c-kit neg) and observed that these cells had the ability to differentiate into cardiomyocytes with greater than 50% efficiency (Yang et al., 2008).

3.2.3 Small molecules

In addition to growth factors, synthetically small molecules have been described to have procardiogenic differentiation effects upon cultured ES cells by affecting a specific signaling pathway. These molecules usually have more half life in the medium and as they are synthetic, they are more structurally and functionally defined. They are stable and inexpensive when compared with polypeptide differentiating agents. To find the best differentiating agent from a library, hES cells which express a reporter gene under the control of a cardiac specific promoter, are used. In this situation, by the addition of compounds from the library, active agents will be identified and the most appropriate one can be chosen. By using this method, Takahashi et al. have screened 880 small molecules and reported that ascorbic acid enhances the efficiency of cardiac differentiation. They noted that this property is independent from its antioxidant property, as other antioxidant agents did not have a positive effect on the efficiency of differentiation (Takahashi et al., 2003).

One factor shown to enhance differentiation into cardiomyocytes is the demethylating agent 5_deoxyazacytidine. Treatment of human EBs with 5_deoxyazacytidine increases expression of cardiac alpha-myosin heavy chain up to two-fold. (Xu et al., 2002).

5-Azacytidine and 5-aza-2′-deoxycytidine are cytosine analogues which are incorporated into DNA and form stable covalent complexes between DNA methyltransferases and DNA. They act as demethylation agents and cause expression of cardiac genes. Therefore treatment of hES cells with 5-aza can improve the efficiency of cardiac differentiation (Xu et al., 2002; Yoon et al., 2006).

Lithium chloride, by inhibition of glycogen synthase kinase 3β and Wnt3a-conditioned medium, increases differentiation efficiency (Nakamura et al., 2003). In addition, other small molecules which are inhibitors of p38 MAP kinase, such as SB203580, have the potential for mesoderm induction and experiments show that addition of such components into differentiation medium almost double the cardiogenic efficiency from 12% to 25% (Graichen et al., 2008). These data indicate that synthetically small molecules are useful agents for directing the differentiation, however experiments in this field still continue.

3.2.4 Suspension

hES cells are routinely cultured as adherent colonies, however culturing hES cells in an adherent position is labor intensive and requires a number of large incubators in order to provide enough cells for cell therapy or drug discovery. Although this kind of cell expansion supports hES cell lines in terms of growth and survival, it limits their application in large scale systems. As a result, several groups have tried to overcome this problem by propagation of hES cells in suspension and expansion of unattached hES cells in spinner flasks for short periods as reported (Lock & Tzanakakis, 2009; Nie et al., 2009; Oh et al., 2009; Krawetz et al., 2010). Recently, Steiner et al. have published a new protocol for prolonged culture of hES cells in suspension in the absence of microcarriers (Steiner et al., 2010). They cultured hES cells in neurobasal medium containing KO-SR, FGF2, activin A, Nutridoma-CS, ECM components and neurotrophic factors. Analysis indicates that cells after 10 passages in suspension culture express pluripotency markers such as SSEA-4,3, TRA-1-60, TRA-1-81 and OCT4; they also express alkaline phosphatase while showing a normal karyotype. These data indicate that cells after 10 weeks in suspension cultures are still pluripotent normal hES cells. These findings indicate that hES cells have the ability for large scale culture in bioreactors and have helped hES cell technology to move toward industrial applications.

hES cells have the ability to survive as single cells in the presence of Rho-associated kinase (ROCK) inhibitor (Watanabe et al., 2007). Thus, by treating hES cells with ROCK inhibitor at least 2 hours before dissociating them from Matrigel, our group succeeded to culture undifferentiated hES cells as spheroids in suspension. Six days after propagation in suspension, the culture medium was changed to a differentiation medium. Differentiation was performed according to the Laflamme protocol (Laflamme et al., 2007). Briefly, cells were treated one day with 100 ng/ml activinA followed by 4 days with 10 ng/ml BMP4 in RPMI/B27 medium, then plated on gelatin-coated plates. In this protocol, beating clusters were observed 10-20 days post-plating (Fig.1). Our data have indicated that 20 days after plating more than 80% of the cells expressed Nkx2.5 when examined by flow cytometery, which is a marker of cardiovascular progenitors. Below, the gene expression profile of hES cells in different differentiation stages is presented (Fig.2). The advantage of this protocol is that it can be used in large scale culturing of cells in bioreactors with the intent to produce with high efficiency large numbers of contractile cardiac cells [unpublished data].

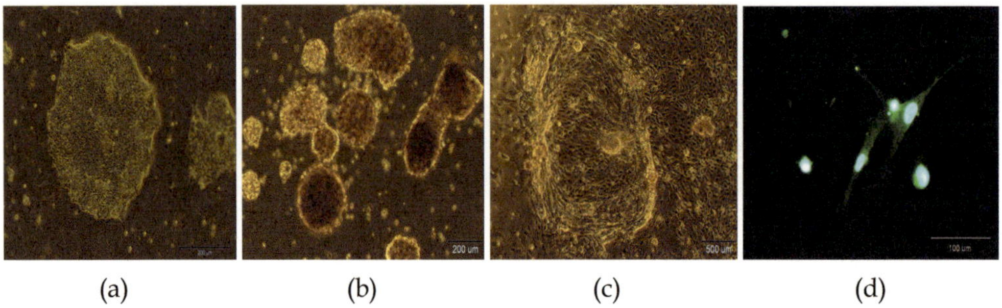

Fig. 1. a) Undifferentiated adherent hES cell colony, b) spheroids c) a beating colony d) immunoflurecence staining of differentiated cells with anti-desmin antibody.

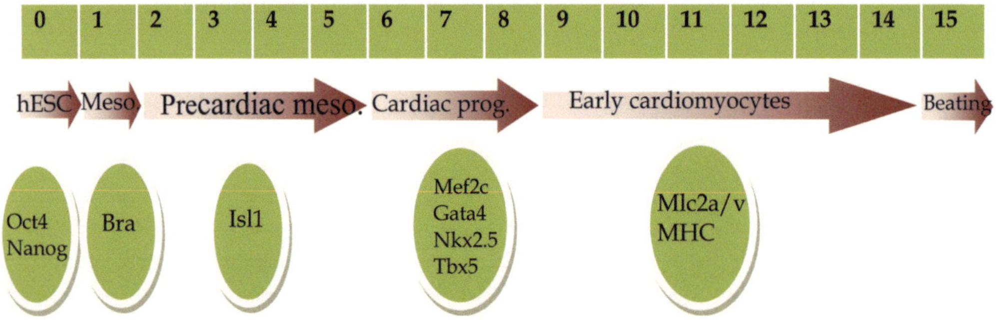

Fig. 2. Gene expression profile of hES cells during differentiation.

3.2.5 Genetic manipulations

Genetic manipulation of hES cells, whether transiently or by stable transformation, can provide a proper tool for understanding cardiac lineage specification and differentiation mechanisms, and lead to the establishment of a protocol for directing hES cells into cardiomyocytes. However, because of the poor transfection efficiency and difficulty of genetic manipulation of hES cells, reports for experiments in this field are limited (Moore et al., 2005). Experiments have demonstrated that transfection efficiency highly depends on the hES cell lines; overall, by the application of lentiviral infection a high efficiency (He et al., 2005), variable efficiency by plasmid transfection (Liew et al., 2007) or low efficiency with the use of adenoviral infection (Smith-Arica et al., 2003) can be achieved. Homologous recombination and electroporation for genetic modification of hES cells has been applied and are the methods of choice for gene targeting in special hES cell lines (Zwaka & Thomson, 2003; Costa et al., 2007). Recently, gene transfers to 12 independent hES cell lines without the loss of stem cell markers has been reported (Braam et al., 2008).

By improving transfection efficiencies, the introduction of a reporter gene into hES cells provides a powerful tool to study gene expression profile during development (Eiges et al., 2001; Matin et al., 2004). Homologous recombination or RNA interference enables us to knockout a specific gene in order to explore its function during differentiation (Zwaka & Thomson, 2003). Overexpression of transcription factors which play important roles in differentiation might direct hES cells to the cardiac lineage (Grepin et al., 1997). As a result, Fijnvandraat et al. overexpressed TBX5 in P19C16 embryonic carcinoma cells and observed that transfected clones began to beat earlier and the number of beating areas tended to increase. Thus, this data indicated that TBX5 plays a role in increasing cardiogenesis (Fijnvandraat et al., 2003). Overexpression of other cardiac transcription factors such as Isl1 (Bu et al., 2009), Gata4 and Nkx2.5 have also been reported, and one of the best results was achieved by using a combination of two transcription factors Nkx2.5 and Tbx5 (Hiroi et al., 2001).

Transdifferentiation to cardiomyocytes is a new topic in cardiac differentiation which by the use of defined factors, different cells can be pushed to express cardiac genes. The transdifferentiation of mouse mesoderm into contractile cardiomyocytes has been reported with the introduction of Gata4, Tbx5 and Baf60c (which is specifically expressed in precardiac mesoderm) into cells (Takeuchi & Bruneau, 2009). Recently, Ieda et al. succeeded in transdifferentiating fibroblasts to cardiac cells. They have shown that with the application of just three cardiac transcription factors (Gata4, Tbx5 and Mef2c), dermal or cardiac fibroblasts can directly be converted to functional cardiomyocytes (Ieda et al., 2010). Although further investigation in this field is required, such experiments are valuable in better understanding both the developmental process and signaling pathways.

3.3 Selection

When a population of ESCs is differentiated into cardiomyocytes, yet another challenge exists with isolation and purification in order to avoid contamination by ESCs. Within EBs, a heterogeneous cell mixture exists; thus, obtaining a homogenous population of cardiomyocytes depends on a selection process. Currently, investigators have developed selection methods to further enrich *in vitro* differentiation populations. These methods fall into two main strategies: mechanical and genetic. In the mechanical strategy, contractile clusters can be isolated manually from a culture plate (Kehat et al., 2004). However, this is not a suitable technique for large-scale, high throughput isolation of cells which are required for pharmaceutical applications and, additionally, the final purity will range from 50-80%. Another mechanical separation method is the Percoll density gradient-based isolation method. Although this strategy improves the throughput of isolation in comparison with manual dissection, the degree of enrichment is insufficient for clinical or research purposes, as the highest purity of cardiomyocytes reported by this method is only 70% (Xu et al., 2002). A third strategy for enrichment of hES cell- derived cardiomyocytes is the formation of cardiac body. After differentiation, cells will be enzymatically dissociated into single cells. In this situation, cardiac cells will re-associate in suspension cultures and form cardiac bodies which contain more than 60% cardiomyocytes (Xu et al., 2006).

Another strategy for the isolation of differentiated cardiomyocytes involves sorting them on the basis of a surface marker expression. Recently, in an experiment cardiac cells were sorted by the endogenously expressed surface marker, ALCAM. However the application of this method is limited due to the lack of cardiomyocyte specific cell surface markers (Rust et al., 2009).

In genetic strategies, selective markers such as an antibiotic-resistance gene, colorimetric markers or cell-surface markers are expressed under the control of a cardiac specific gene. Therefore, when ESCs differentiate into cardiomyocytes and the promoter is activated, selective markers will be expressed and cardiomyocytes can be isolated by exposing cells to an antibiotic or FACS (fluorescence-activated cell sorting). Based on this, Klug et al. have used an antibiotic selection method to purify ES-cell-derived cardiomyocytes. In their research, a fusion gene which consisted of an alpha-cardiac myosin heavy chain promoter followed by an aminoglycoside phosphotransferase gene, an enzyme which causes resistance to the cytotoxic agent geneticin (G418). Once the transfected cells began to differentiate and the cardiac specific promoter was activated, aminoglycoside phosphotransferase was expressed and allowed them to survive under treatment of the culture system with G418. Surviving cells provided a population of cardiomyocytes with approximately 99% purity (Klug et al., 1996). In another experiment, a ventricular-specific promoter ,myosin light chain-2v was linked to enhanced green fluorescent protein (Muller et al., 2000). After differentiation of murine ES cells into cardiomyocytes, GFP protein was expressed and cardiac cells could be collected with 97% purity by FACS (Muller et al., 2000). These data indicate that hES cells can be manipulated in order to obtain enhanced efficacy or purity during cardiac differentiation. Recent studies have proven the benefit of using transgenic approaches, which include cardiac resistance genes alone or in combination with reporter genes (Anderson et al., 2007; Xu et al., 2008; Kita-Matsuo et al., 2009). By the use of lentiviral infection, hES cells were transfected with a construct in which eGFP has been expressed under the control of a cardiac specific promoter (human myosin light chain-2V promoter) After EB formation, cardiomyocytes could be detected by GFP expression and sorted by FACS with more than 90% purity (Huber et al., 2007). By using the same method, cardiomyocytes derived from hES cells could be sorted on the basis of cardiac troponin I expression (Gallo et al., 2008). Another recent study used the same method to isolate cardiac cells based on the expression of dsRed fluorescence reporter gene under the control of MLC2v promoter (Fu et al., 2009).

All the above mentioned studies have selected matured cardiac cells, however due to poor differentiation efficiency and limited proliferative potential of cardiomyocytes, the application of these selected cells have not been fully established (Lyon & Harding, 2007). Therefore, several groups have focused on cardiac progenitors (Yamashita et al., 2005). Yang et al. reported the isolation of a cardiovascular progenitor from differentiated hES cells which were KDR low/c-kit neg in which it was observed that this population had the ability to produce more than 50% contracting cardiomyocytes (Yang et al., 2008). KDR is a marker which can be used for enrichment of early cardiac progenitors (Motoike et al., 2003; Ema et al., 2006; Kattman et al., 2006; Moretti et al., 2006; Yang et al., 2008). However KDR is not a suitable single marker for isolation of cardiac progenitors as it is expressed in different stages of differentiation. Other progenitors which have been selected were ISL1+ cells that give rise to cardiomyocytes, smooth muscle and endothelial cells. In this experiment, ISL1+ cells have been isolated and cultured on mouse embryonic fibroblasts (Bu et al., 2009). The advantage of ISL1 is that it is expressed in a special stage of differentiation and as cells start to complete the process of differentiation, it will be rapidly down regulated. Thus, ISL1 can be considered as a proper pool of cardiac progenitors for different applications.

4. Applications

4.1 Drug discovery

Drug discovery is a long process of about 5-10 years which requires tremendous effort and money. However, all these efforts are necessary to ensure the safety and efficacy of new drugs. Statistics show that more than 40% of new drugs fail to pass phase III clinical trials, which then contribute to waste of time and money (Kessler & Byrne, 1999). Thus, it is critical to discover unacceptable targets in the early stages of development. Pharmaceutical industries need novel technologies to assist with improving the efficiency of drug discovery. High throughput technology in the fields of screening, evaluation and toxicity assays can be beneficial (Kola & Landis, 2004). High throughput screening is the automated screening of a library (more than 100 molecules) in order to find the best molecule for a specific purpose. In this case, ES cells are the proper choice for cell-based screening. In many cases the toxicity of new components will not be discovered until the final stages; in clinical trials or when tested on animals. However, using animals as experimental models are costly and surrounded with ethical and legal considerations. On the other hand, the answers from animal models cannot always be translated into humans. ES cell technology has overcome some of these problems and made the discovery process more efficient.

hES cells have the potential to differentiate into cardiac cells *in vitro*. Phenotypic analysis indicates that they are morphologically and ultrastructurally similar to adult cardiomyocytes, although their myofibrillar and sarcomeric organization are immature (Snir et al., 2003; Olson, 2004; Norstrom et al., 2006; Yoon et al., 2006). The answers from these cells may be different from the *in vivo* situation. Fortunately there are indications which prove the possibility of *in vitro* maturation of hES cell-derived cardiomyocytes (Snir et al., 2003). In this case they can approximate adult cardiomyocytes and can be used as pharmacological models. The technology to direct differentiation toward cardiac cells can reduce the heterogeneity and improve their application in screening. One of the most important advantages of cardiomyocytes derived from hES cells is that they keep their contractile ability in culture for an extended time which is beneficial for testing different components on the same cells. In this situation the answers can be compared easily.

Pharmaceutical industries have began to use hES cell-derived cardiomyocytes as models for evaluation of the biological activities of different components. Functional assays are necessary in different stages of drug discovery; from target identification to end stage detailed pharmacological analysis.

The application of hES cell-derived cardiomyocytes in drug development can be divided into two categories. The first category concerns the necessity of caridac drug discovery in cases where the heart is a diseased organ. One of the applications of hES cells in drug discovery is examining responses of cells with different genetic variations. On the other hand, cardiomyocytes can provide specific disease models (Friedrich Ben-Nun & Benvenisty, 2006). Secondly, when the safety of a new component for the heart needs to be tested, hES cells can be used in both fields of drug development.

Every new drug needs to be tested for safety. From 1991 to 2000 about 30%of the new components failed because of toxicology and clinical safety (Kola & Landis, 2004). A significant portion of the toxicity is due to cardio- and hepatotoxicity, thus it is necessary to

be sure that new components are safe for the heart before their approval for medical applications. Some of the mechanisms for cardiotoxicity are due to the formation of reactive oxygen species, apoptosis, altering proper molecular signaling or cardiac gene expression.
The unavailability of hES cells is one of the most important barriers for the application of hES cells in drug discovery due to the necessity of large scale expansion. On the other hand, after differentiation it is necessary to improve both homogeneity and cardiac cell yields. By overcoming some of the obstacles of hES cells, they can make the process of finding new drugs more efficient and easier as they reduce the need for *in vivo* experiments.

4.2 Cell therapy

Human adult cardiac muscles have low ability to regenerate (Bergmann et al., 2009) and only a limited number of species such as newts and zebrafish are able to renew their myocard, therefore loss of massive cell population as a result of infarction or other heart diseases is irreversible and often leads to heart failure (Kubo et al., 2008). Today, the treatment of choice to prevent death caused by heart failure is heart transplantation. However, because of a shortage of donor organs, the complication of rejection and difficulty of transplantation surgery, this strategy cannot be used easily. On the other hand, to prevent unwanted immune response lifelong immunosuppressive therapy is required and even after successful transplantation failures of donor organs are frequent.
Recent developments in molecular biology, stem cell culturing and tissue engineering have provided a biomedical approach called regenerative medicine. The main objective of cell therapy is to repopulate damaged tissue with a new pool of functional cells in order to perform its normal actions. Several different cells that have been considered as suitable sources are: skeletal myoblasts (Murry et al., 1996; Taylor et al., 1998; Menasche et al., 2003), bone marrow-derived hematopoietic stem cells (Orlic et al., 2001; Zafarghandi et al., 2010) and mesenchymal stem cells (Min et al., 2002; Shake et al., 2002; Toma et al., 2002; Mangi et al., 2003). All these cell types have serious problems and for this reason, have gained limited application in cardiac cell therapy. For example, skeletal myoblasts following transplantation can not express cardiac markers and beat in synchrony with the host myocardium (Reinecke et al., 2000; Leobon et al., 2003; Rubart et al., 2004). Similarly, hematopoietic stem cells and mesenchymal stem cells when injected into the heart rarely differentiate into cardiac cells (Balsam et al., 2004). Other cell sources for regeneration, therefore are necessary. hES cells with their ability to self-renew and differentiation to cardiac cells can be considered as a suitable source for cell therapy. Thus far, several experiments have shown their ability and potential for regenerating infarcted heart as they express molecular components required for electromechanical integration with host cardiomyocytes (Rust et al., 1997; Xu et al., 2002; Mummery et al., 2003; Caspi et al., 2007; Laflamme et al., 2007; Leor et al., 2007; Tomescot et al., 2007; van Laake et al., 2007).
In order to use hES cells in drug development or regeneration, some points need to be considered. First, a scalable system for culture and maintenance of these cells in an undifferentiated state is necessary. Additionally, a well-defined protocol to direct hES cells toward cardiac cells with a high efficiency, a method to isolate and further purify cariomyocytes, a system to check their functionality and finally a high throughput system to test a library of candidate components are all necessary.
The main obstacle of cell therapy is the isolation of purified cardiomyocyte populations suitable for transplant.

As a result, several groups pushed ahead with experiments to evaluate ES cell-derived cardiomyocytes transplanted into the hearts of animal models. Both Gepstein (Kehat et al., 2004) and Li's groups (Xue et al., 2005), in separate experiments, spontaneously transplanted contracting EBs from hES cells into uninjured hearts to show electrochemical integration of transplanted cells with the heart and provide strong evidence of this ability, at least in an uninfarcted heart. In the year 2005, transplantation of hES cell- derived cardiomyocytes into arrythmic rat hearts demonstrate that they can successfully engraft, proliferate and express several cardiac markers, without evidence of teratoma formation up to 4 weeks post-transplantation (Laflamme et al., 2005). Engraftment of these cells into guinea pig hearts have been tested as well and their ability for function and integration with host cardiac tissue has been proven (Xue et al., 2005). Tomescot et al. showed that treatment of hES cells with BMP2 and SU5402 (a fibroblast growth factor receptor inhibitor) followed by transplantation into infarcted rat hearts lead to improvement in the heart performance without teratoma formation (Tomescot et al., 2007). Moreover, electrophysiological analysis indicates that transplanted cells functionally integrated into host myocardium. These data confirmed past evidence that showed electromechanical and structural coupling of transplanted hES cell-derived cardiomyoblasts with the host myocardium in pigs (Kehat et al., 2004). Although these findings suggest that hES cell-derived cardiomyoblast transplantation might be feasible and safe, but before such therapy is applied in humans several concerns must be considered. Some major areas for concern are: immunogenicity, risk of post-transplant cellular misbehavior such as teratoma formation, controlling cell survival and delivery, potential for post-transplant arrhythmogenecity as well as ethical concerns of isolating hES cells from human embryos.

Immunorejection by recipients may decrease the therapeutic potential of transplanted allogeneic hES cells. Because of conflicting evidence by several investigators, it is still unclear whether hES cells are immunoprivileged or not. Drukker et al. have shown that hES cells have low natural killer cell receptor expression (Drukker et al., 2002). In another experiment Li et al. have shown that hES cells did not elicit an immune response when injected into the muscle of immunocompetent mice (Li et al., 2004). In contrast to these findings, some groups have indicated that hES cells can cause an immune response in the host. So, immunogenicity of these cells should be examined before transplantation. One approach to prevent immunorejection of transplanted cells is to minimize allo-antigenic differences between donor and recipient. This can be achieved by establishing a bank of different MHC antigens such as the banks for tissue and organ transplantation. Another approach is to make universal hES cell lines by silencing or modulation of genes associated with the immune response.

Another potential problem of ES cell therapy is the risk of teratoma formation. It is obvious that undifferentiated hES cells have the potential for tumorogenicity. Although further experiments have shown that the pure populations of cardiomyocytes do not create teratomas, there is still the risk of contamination with undifferentiated or dedifferentiated cells. Therefore, selection strategies to separate fully differentiated cardiomocytes from mixed populations are necessary. Another strategy is to genetically manipulate hES cells in order to activate apoptotic signals in undifferentiated cells at special time points after directed differentiation.

One of the most serious problems of cell transplantation is that only a few number of engrafted cells will be survive (Robey et al., 2008). To solve this problem Laflamme et al.

used a "pro-survival cocktail" to improve cell survival after transplantation. This cocktail consists of a mixture of anti-oncotic and anti-apoptotic factors which cause cell survival in the harsh infarct environment (Laflamme et al., 2007). The issue of cell delivery and survival still needs further investigations and optimizations.

Ethical concerns are another serious barrier to the application of hES cells in a clinic setting. Currently, the production of hES cell lines leads to the destruction of embryos which have the potential to live. Thus, new strategies for generating hES cell lines are needed. In the year 2006, the production of ES cells from single cell biopsies of a developing mouse embryo has been reported. If such a strategy can be translated into humans, several ethical concerns will be solved.

Recently, growing evidence suggest that there are multipotent cardiovascular progenitors (Beltrami et al., 2003; Oh et al., 2003; Laugwitz et al., 2005; Smith et al., 2007; Kattman et al., 2006; Yang et al., 2008; Bu et al., 2009).They can be isolated from the adult heart or derived from hES cells. These cells can be separated from adult myocardium on the basis of spheroid or expression of special markers such as c-kit, Isl1 (Laugwitz et al., 2005) , Sca-1 (van Vliet et al., 2008) and Abcg2 (Martin et al., 2004). By injecting c-kit+ population into an infarcted rat model, ventricular function could be improved. In this experiment, transplanted cells were positive for cardiac myosin, but morphologically smaller than cardiomyocytes, yet similar to fibroblast cells (Dawn et al., 2005). The Sca-1+ population can differentiate into cardiac cells by treatment with 5-azacytidine. These populations demonstrated proper survival and integration after transplantation. However, the use of 5-azacytidine in clinical application is not safe as it can cause widespread DNA demethylation. Although these progenitors seem to be suitable tool for cardiac regeneration, still more investigation is required in order to fully characterize them and understand their nature.

Experiments indicate that cardiac stem cells have the ability to functionally integrate with myocardium when injected to infarcted rat and mouse hearts (Bearzi et al., 2007). These populations have less risk of teratoma formation but less differentiated cells have more potential to integrate and survive.

An alternative approach is the use of induced pluripotent stem cells (iPS) which are reprogrammed adult cells. It has been shown that iPS cells derived from dermal fibroblast are able to differentiate into cardiomyocytes (Takahashi & Yamanaka, 2006; Yu et al., 2007). The other application of iPS cells is that they can form models of human disease to study molecular mechanisms involved in disease, drug screening, safety and toxicology on cells of different genetic backgrounds and investigate new treatment opportunities (Dimos et al., 2008; Park et al., 2008).

5. Conclusion

Taken together, the data that has been reviewed above indicate that hES cells are pluripotent cells which have the ability to differentiate into cardiomyocytes *in vitro*. The process of differentiatin can be both spontaneous and directed. Although experiments show that directed differentiation has more efficiency than spontaneous differentiation, however the efficiency is not satisfiable for scientists in order to apply differentiated cells in drug discovery or regenerative medicine. The main obstacles in the way of developing a proper protocol for directing hES cells toward the cardiac fate is our limited knowledge of the exact signaling pathways which control cardiac differentiation. As the best protocol is the one which mimics processes in normal development, by improving our knowledge of developmental processes and finding growth factors which have the ability to induce cardiac differentiation and/or genes involved in natural development, therefore new, nore effective protocols can be achieved which will cause huge improvements in the fields of new drug discoveries for cardiovascular patients and cell therapy.

6. References

Adewumi, O.; Aflatoonian, B.; Ahrlund-Richter, L., et al. (2007). Characterization of human embryonic stem cell lines by the International Stem Cell Initiative. *Nat Biotechnol,* 25, 7, 803-816, 1087-0156 (Print)

Amit, M.; Carpenter, M. K.; Inokuma, M. S., et al. (2000). Clonally derived human embryonic stem cell lines maintain pluripotency and proliferative potential for prolonged periods of culture. *Dev Biol,* 227, 2, 271-278, 0012-1606 (Print)

Anderson, D.; Self, T.; Mellor, I. R., et al. (2007). Transgenic enrichment of cardiomyocytes from human embryonic stem cells. *Mol Ther,* 15, 11, 2027-2036, 1525-0016 (Print)

Baharvand, H.; Ashtiani, S. K.; Valojerdi, M. R., et al. (2004). Establishment and in vitro differentiation of a new embryonic stem cell line from human blastocyst. *Differentiation,* 72, 5, 224-229, 0301-4681 (Print)

Baharvand, H.; Azarnia, M.; Parivar, K., et al. (2005). The effect of extracellular matrix on embryonic stem cell-derived cardiomyocytes. *J Mol Cell Cardiol,* 38, 3, 495-503, 0022-2828 (Print)

Balsam, L. B.; Wagers, A. J.; Christensen, J. L., et al. (2004). Haematopoietic stem cells adopt mature haematopoietic fates in ischaemic myocardium. *Nature,* 428, 6983, 668-673, 1476-4687 (Electronic)

Bearzi, C.; Rota, M.; Hosoda, T., et al. (2007). Human cardiac stem cells. *Proc Natl Acad Sci U S A,* 104, 35, 14068-14073, 0027-8424 (Print)

Behfar, A.; Zingman, L. V.; Hodgson, D. M., et al. (2002). Stem cell differentiation requires a paracrine pathway in the heart. *FASEB J,* 16, 12, 1558-1566, 1530-6860 (Electronic)

Beltrami, A. P.; Barlucchi, L.; Torella, D., et al. (2003). Adult cardiac stem cells are multipotent and support myocardial regeneration. *Cell,* 114, 6, 763-776, 0092-8674 (Print)

Beqqali, A.; Kloots, J.; Ward-van Oostwaard, D., et al. (2006). Genome-wide transcriptional profiling of human embryonic stem cells differentiating to cardiomyocytes. *Stem Cells,* 24, 8, 1956-1967, 1066-5099 (Print)

Bergmann, O.; Bhardwaj, R. D.; Bernard, S., et al. (2009). Evidence for cardiomyocyte renewal in humans. *Science,* 324, 5923, 98-102, 1095-9203 (Electronic)

Boheler, K. R.; Czyz, J.; Tweedie, D., et al. (2002). Differentiation of pluripotent embryonic stem cells into cardiomyocytes. *Circ Res,* 91, 3, 189-201, 1524-4571 (Electronic)

Braam, S. R.; Denning, C.; van den Brink, S., et al. (2008). Improved genetic manipulation of human embryonic stem cells. *Nat Methods,* 5, 5, 389-392, 1548-7105 (Electronic)

Brandenberger, R.; Wei, H.; Zhang, S., et al. (2004). Transcriptome characterization elucidates signaling networks that control human ES cell growth and differentiation. *Nat Biotechnol,* 22, 6, 707-716, 1087-0156 (Print)

Bu, L.; Jiang, X.; Martin-Puig, S., et al. (2009). Human ISL1 heart progenitors generate diverse multipotent cardiovascular cell lineages. *Nature,* 460, 7251, 113-117, 1476-4687 (Electronic)

Burridge, P. W.; Anderson, D.; Priddle, H., et al. (2007). Improved human embryonic stem cell embryoid body homogeneity and cardiomyocyte differentiation from a novel V-96 plate aggregation system highlights interline variability. *Stem Cells,* 25, 4, 929-938, 1066-5099 (Print)

Calhoun, J. D.; Rao, R. R.; Warrenfeltz, S., et al. (2004). Transcriptional profiling of initial differentiation events in human embryonic stem cells. *Biochem Biophys Res Commun,* 323, 2, 453-464, 0006-291X (Print)

Caspi, O.; Huber, I.; Kehat, I., et al. (2007). Transplantation of human embryonic stem cell-derived cardiomyocytes improves myocardial performance in infarcted rat hearts. *J Am Coll Cardiol,* 50, 19, 1884-1893, 1558-3597 (Electronic)

Cohen, E. D.; Wang, Z.; Lepore, J. J., et al. (2007). Wnt/beta-catenin signaling promotes expansion of Isl-1-positive cardiac progenitor cells through regulation of FGF signaling. *J Clin Invest,* 117, 7, 1794-1804, 0021-9738 (Print)

Costa, M.; Dottori, M.; Sourris, K., et al. (2007). A method for genetic modification of human embryonic stem cells using electroporation. *Nat Protoc,* 2, 4, 792-796, 1750-2799 (Electronic)

Dawn, B.; Stein, A. B.; Urbanek, K., et al. (2005). Cardiac stem cells delivered intravascularly traverse the vessel barrier, regenerate infarcted myocardium, and improve cardiac function. *Proc Natl Acad Sci U S A,* 102, 10, 3766-3771, 0027-8424 (Print)

Dimos, J. T.; Rodolfa, K. T.; Niakan, K. K., et al. (2008). Induced pluripotent stem cells generated from patients with ALS can be differentiated into motor neurons. *Science,* 321, 5893, 1218-1221, 1095-9203 (Electronic)

Drukker, M.; Katz, G.; Urbach, A., et al. (2002). Characterization of the expression of MHC proteins in human embryonic stem cells. *Proc Natl Acad Sci U S A,* 99, 15, 9864-9869, 0027-8424 (Print)

Dvash, T.; Mayshar, Y.; Darr, H., et al. (2004). Temporal gene expression during differentiation of human embryonic stem cells and embryoid bodies. *Hum Reprod,* 19, 12, 2875-2883, 0268-1161 (Print)

Eiges, R.; Schuldiner, M.; Drukker, M., et al. (2001). Establishment of human embryonic stem cell-transfected clones carrying a marker for undifferentiated cells. *Curr Biol,* 11, 7, 514-518, 0960-9822 (Print)

Ema, M.; Yokomizo, T.; Wakamatsu, A., et al. (2006). Primitive erythropoiesis from mesodermal precursors expressing VE-cadherin, PECAM-1, Tie2, endoglin, and CD34 in the mouse embryo. *Blood,* 108, 13, 4018-4024, 0006-4971 (Print)

Farokhpour, M.; Karbalaie, K.; Tanhaei, S., et al. (2009). Embryonic stem cell-derived cardiomyocytes as a model system to study cardioprotective effects of dexamethasone in doxorubicin cardiotoxicity. *Toxicol In Vitro,* 23, 7, 1422-1428, 1879-3177 (Electronic)

Fijnvandraat, A. C.; Lekanne Deprez, R. H.; Christoffels, V. M., et al. (2003). TBX5 overexpression stimulates differentiation of chamber myocardium in P19C16 embryonic carcinoma cells. *J Muscle Res Cell Motil,* 24, 2-3, 211-218, 0142-4319 (Print)

Filipczyk, A. A.; Passier, R.; Rochat, A., et al. (2007). Regulation of cardiomyocyte differentiation of embryonic stem cells by extracellular signalling. *Cell Mol Life Sci,* 64, 6, 704-718, 1420-682X (Print)

Friedrich Ben-Nun, I. & Benvenisty, N. (2006). Human embryonic stem cells as a cellular model for human disorders. *Mol Cell Endocrinol,* 252, 1-2, 154-159, 0303-7207 (Print)

Fu, J. D.; Jiang, P.; Rushing, S., et al. (2009). Na+/Ca2+ exchanger is a determinant of excitation-contraction coupling in human embryonic stem cell-derived ventricular cardiomyocytes. *Stem Cells Dev,* 1557-8534 (Electronic)

Fullilove, S. L. (1970). Heart induction: distribution of active factors in newt endoderm. *J Exp Zool,* 175, 3, 323-326, 0022-104X (Print)

Gallo, P.; Grimaldi, S.; Latronico, M. V., et al. (2008). A lentiviral vector with a short troponin-I promoter for tracking cardiomyocyte differentiation of human embryonic stem cells. *Gene Ther,* 15, 3, 161-170, 1476-5462 (Electronic)

Giovino, M. A.; Down, J. D.; Jackson, J. D., et al. (2002). Porcine hematopoiesis on primate stroma in long-term cultures: enhanced growth with neutralizing tumor necrosis factor-alpha and tumor growth factor-beta antibodies. *Transplantation,* 73, 5, 723-731, 0041-1337 (Print)

Graichen, R.; Xu, X.; Braam, S. R., et al. (2008). Enhanced cardiomyogenesis of human embryonic stem cells by a small molecular inhibitor of p38 MAPK. *Differentiation,* 76, 4, 357-370, 1432-0436 (Electronic)

Grepin, C.; Nemer, G. & Nemer, M. (1997). Enhanced cardiogenesis in embryonic stem cells overexpressing the GATA-4 transcription factor. *Development,* 124, 12, 2387-2395, 0950-1991 (Print)

Guhr, A.; Kurtz, A.; Friedgen, K., et al. (2006). Current state of human embryonic stem cell research: an overview of cell lines and their use in experimental work. *Stem Cells,* 24, 10, 2187-2191, 1066-5099 (Print)

He, J.; Yang, Q. & Chang, L. J. (2005). Dynamic DNA methylation and histone modifications contribute to lentiviral transgene silencing in murine embryonic carcinoma cells. *J Virol,* 79, 21, 13497-13508, 0022-538X (Print)

He, J. Q.; Ma, Y.; Lee, Y., et al. (2003). Human embryonic stem cells develop into multiple types of cardiac myocytes: action potential characterization. *Circ Res,* 93, 1, 32-39, 1524-4571 (Electronic)

Herzenberg, L. A.; Parks, D.; Sahaf, B., et al. (2002). The history and future of the fluorescence activated cell sorter and flow cytometry: a view from Stanford. *Clin Chem,* 48, 10, 1819-1827, 0009-9147 (Print)

Hiroi, Y.; Kudoh, S.; Monzen, K., et al. (2001). Tbx5 associates with Nkx2-5 and synergistically promotes cardiomyocyte differentiation. *Nat Genet,* 28, 3, 276-280, 1061-4036 (Print)

Huber, I.; Itzhaki, I.; Caspi, O., et al. (2007). Identification and selection of cardiomyocytes during human embryonic stem cell differentiation. *FASEB J,* 21, 10, 2551-2563, 1530-6860 (Electronic)

Ieda, M.; Fu, J. D.; Delgado-Olguin, P., et al. (2010). Direct reprogramming of fibroblasts into functional cardiomyocytes by defined factors. *Cell,* 142, 3, 375-386, 1097-4172 (Electronic)

Itskovitz-Eldor, J.; Schuldiner, M.; Karsenti, D., et al. (2000). Differentiation of human embryonic stem cells into embryoid bodies compromising the three embryonic germ layers. *Mol Med,* 6, 2, 88-95, 1076-1551 (Print)

Kattman, S. J.; Huber, T. L. & Keller, G. M. (2006). Multipotent flk-1+ cardiovascular progenitor cells give rise to the cardiomyocyte, endothelial, and vascular smooth muscle lineages. *Dev Cell,* 11, 5, 723-732, 1534-5807 (Print)

Kehat, I.; Kenyagin-Karsenti, D.; Snir, M., et al. (2001). Human embryonic stem cells can differentiate into myocytes with structural and functional properties of cardiomyocytes. *J Clin Invest,* 108, 3, 407-414, 0021-9738 (Print)

Kehat, I.; Khimovich, L.; Caspi, O., et al. (2004). Electromechanical integration of cardiomyocytes derived from human embryonic stem cells. *Nat Biotechnol,* 22, 10, 1282-1289, 1087-0156 (Print)

Kessler, P. D. & Byrne, B. J. (1999). Myoblast cell grafting into heart muscle: cellular biology and potential applications. *Annu Rev Physiol,* 61, 219-242, 0066-4278 (Print)

Kita-Matsuo, H.; Barcova, M.; Prigozhina, N., et al. (2009). Lentiviral vectors and protocols for creation of stable hESC lines for fluorescent tracking and drug resistance selection of cardiomyocytes. *PLoS One,* 4, 4, e5046, 1932-6203 (Electronic)

Klaus, A. & Birchmeier, W. (2009). Developmental signaling in myocardial progenitor cells: a comprehensive view of Bmp- and Wnt/beta-catenin signaling. *Pediatr Cardiol,* 30, 5, 609-616, 1432-1971 (Electronic)

Klug, M. G.; Soonpaa, M. H.; Koh, G. Y., et al. (1996). Genetically selected cardiomyocytes from differentiating embronic stem cells form stable intracardiac grafts. *J Clin Invest,* 98, 1, 216-224, 0021-9738 (Print)

Kola, I. & Landis, J. (2004). Can the pharmaceutical industry reduce attrition rates? *Nat Rev Drug Discov,* 3, 8, 711-715, 1474-1776 (Print)

Krawetz, R.; Taiani, J. T.; Liu, S., et al. (2010). Large-scale expansion of pluripotent human embryonic stem cells in stirred-suspension bioreactors. *Tissue Eng Part C Methods,* 16, 4, 573-582, 1937-3392 (Electronic)

Kubo, H.; Jaleel, N.; Kumarapeli, A., et al. (2008). Increased cardiac myocyte progenitors in failing human hearts. *Circulation,* 118, 6, 649-657, 1524-4539 (Electronic)

Laflamme, M. A.; Chen, K. Y.; Naumova, A. V., et al. (2007). Cardiomyocytes derived from human embryonic stem cells in pro-survival factors enhance function of infarcted rat hearts. *Nat Biotechnol,* 25, 9, 1015-1024, 1087-0156 (Print)

Laflamme, M. A.; Gold, J.; Xu, C., et al. (2005). Formation of human myocardium in the rat heart from human embryonic stem cells. *Am J Pathol,* 167, 3, 663-671, 0002-9440 (Print)

Laugwitz, K. L.; Moretti, A.; Lam, J., et al. (2005). Postnatal isl1+ cardioblasts enter fully differentiated cardiomyocyte lineages. *Nature,* 433, 7026, 647-653, 1476-4687 (Electronic)

Leobon, B.; Garcin, I.; Menasche, P., et al. (2003). Myoblasts transplanted into rat infarcted myocardium are functionally isolated from their host. *Proc Natl Acad Sci U S A,* 100, 13, 7808-7811, 0027-8424 (Print)

Leor, J.; Gerecht, S.; Cohen, S., et al. (2007). Human embryonic stem cell transplantation to repair the infarcted myocardium. *Heart,* 93, 10, 1278-1284, 1468-201X (Electronic)

Lev, S.; Kehat, I. & Gepstein, L. (2005). Differentiation pathways in human embryonic stem cell-derived cardiomyocytes. *Ann N Y Acad Sci*, 1047, 50-65, 0077-8923 (Print)

Li, L.; Baroja, M. L.; Majumdar, A., et al. (2004). Human embryonic stem cells possess immune-privileged properties. *Stem Cells*, 22, 4, 448-456, 1066-5099 (Print)

Liew, C. G.; Draper, J. S.; Walsh, J., et al. (2007). Transient and stable transgene expression in human embryonic stem cells. *Stem Cells*, 25, 6, 1521-1528, 1066-5099 (Print)

Lock, L. T. & Tzanakakis, E. S. (2009). Expansion and differentiation of human embryonic stem cells to endoderm progeny in a microcarrier stirred-suspension culture. *Tissue Eng Part A*, 15, 8, 2051-2063, 1937-335X (Electronic)

Ludwig, T. E.; Levenstein, M. E.; Jones, J. M., et al. (2006). Derivation of human embryonic stem cells in defined conditions. *Nat Biotechnol*, 24, 2, 185-187, 1087-0156 (Print)

Lyon, A. & Harding, S. (2007). The potential of cardiac stem cell therapy for heart failure. *Curr Opin Pharmacol*, 7, 2, 164-170, 1471-4892 (Print)

Mangi, A. A.; Noiseux, N.; Kong, D., et al. (2003). Mesenchymal stem cells modified with Akt prevent remodeling and restore performance of infarcted hearts. *Nat Med*, 9, 9, 1195-1201, 1078-8956 (Print)

Martin, C. M.; Meeson, A. P.; Robertson, S. M., et al. (2004). Persistent expression of the ATP-binding cassette transporter, Abcg2, identifies cardiac SP cells in the developing and adult heart. *Dev Biol*, 265, 1, 262-275, 0012-1606 (Print)

Marvin, M. J.; Di Rocco, G.; Gardiner, A., et al. (2001). Inhibition of Wnt activity induces heart formation from posterior mesoderm. *Genes Dev*, 15, 3, 316-327, 0890-9369 (Print)

Matin, M. M.; Walsh, J. R.; Gokhale, P. J., et al. (2004). Specific knockdown of Oct4 and beta2-microglobulin expression by RNA interference in human embryonic stem cells and embryonic carcinoma cells. *Stem Cells*, 22, 5, 659-668, 1066-5099 (Print)

Menasche, P.; Hagege, A. A.; Vilquin, J. T., et al. (2003). Autologous skeletal myoblast transplantation for severe postinfarction left ventricular dysfunction. *J Am Coll Cardiol*, 41, 7, 1078-1083, 0735-1097 (Print)

Min, J. Y.; Sullivan, M. F.; Yang, Y., et al. (2002). Significant improvement of heart function by cotransplantation of human mesenchymal stem cells and fetal cardiomyocytes in postinfarcted pigs. *Ann Thorac Surg*, 74, 5, 1568-1575, 0003-4975 (Print)

Miura, T.; Luo, Y.; Khrebtukova, I., et al. (2004). Monitoring early differentiation events in human embryonic stem cells by massively parallel signature sequencing and expressed sequence tag scan. *Stem Cells Dev*, 13, 6, 694-715, 1547-3287 (Print)

Moore, J. C.; van Laake, L. W.; Braam, S. R., et al. (2005). Human embryonic stem cells: genetic manipulation on the way to cardiac cell therapies. *Reprod Toxicol*, 20, 3, 377-391, 0890-6238 (Print)

Moretti, A.; Caron, L.; Nakano, A., et al. (2006). Multipotent embryonic isl1+ progenitor cells lead to cardiac, smooth muscle, and endothelial cell diversification. *Cell*, 127, 6, 1151-1165, 0092-8674 (Print)

Motoike, T.; Markham, D. W.; Rossant, J., et al. (2003). Evidence for novel fate of Flk1+ progenitor: contribution to muscle lineage. *Genesis*, 35, 3, 153-159, 1526-954X (Print)

Muller, M.; Fleischmann, B. K.; Selbert, S., et al. (2000). Selection of ventricular-like cardiomyocytes from ES cells in vitro. *FASEB J*, 14, 15, 2540-2548, 0892-6638 (Print)

Mummery, C.; Ward-van Oostwaard, D.; Doevendans, P., et al. (2003). Differentiation of human embryonic stem cells to cardiomyocytes: role of coculture with visceral endoderm-like cells. *Circulation,* 107, 21, 2733-2740, 1524-4539 (Electronic)
Mummery, C.; Ward, D.; van den Brink, C. E., et al. (2002). Cardiomyocyte differentiation of mouse and human embryonic stem cells. *J Anat,* 200, Pt 3, 233-242, 0021-8782 (Print)
Murry, C. E.; Wiseman, R. W.; Schwartz, S. M., et al. (1996). Skeletal myoblast transplantation for repair of myocardial necrosis. *J Clin Invest,* 98, 11, 2512-2523, 0021-9738 (Print)
Nakamura, T.; Sano, M.; Songyang, Z., et al. (2003). A Wnt- and beta -catenin-dependent pathway for mammalian cardiac myogenesis. *Proc Natl Acad Sci U S A,* 100, 10, 5834-5839, 0027-8424 (Print)
Ng, E. S.; Davis, R. P.; Azzola, L., et al. (2005). Forced aggregation of defined numbers of human embryonic stem cells into embryoid bodies fosters robust, reproducible hematopoietic differentiation. *Blood,* 106, 5, 1601-1603, 0006-4971 (Print)
Nie, Y.; Bergendahl, V.; Hei, D. J., et al. (2009). Scalable culture and cryopreservation of human embryonic stem cells on microcarriers. *Biotechnol Prog,* 25, 1, 20-31, 1520-6033 (Electronic)
Norstrom, A.; Akesson, K.; Hardarson, T., et al. (2006). Molecular and pharmacological properties of human embryonic stem cell-derived cardiomyocytes. *Exp Biol Med (Maywood),* 231, 11, 1753-1762, 1535-3702 (Print)
Nostro, M. C.; Cheng, X.; Keller, G. M., et al. (2008). Wnt, activin, and BMP signaling regulate distinct stages in the developmental pathway from embryonic stem cells to blood. *Cell Stem Cell,* 2, 1, 60-71, 1875-9777 (Electronic)
Odorico, J. S.; Kaufman, D. S. & Thomson, J. A. (2001). Multilineage differentiation from human embryonic stem cell lines. *Stem Cells,* 19, 3, 193-204, 1066-5099 (Print)
Oh, H.; Bradfute, S. B.; Gallardo, T. D., et al. (2003). Cardiac progenitor cells from adult myocardium: homing, differentiation, and fusion after infarction. *Proc Natl Acad Sci U S A,* 100, 21, 12313-12318, 0027-8424 (Print)
Oh, S. K.; Chen, A. K.; Mok, Y., et al. (2009). Long-term microcarrier suspension cultures of human embryonic stem cells. *Stem Cell Res,* 1876-7753 (Electronic)
Olson, E. N. (2004). A decade of discoveries in cardiac biology. *Nat Med,* 10, 5, 467-474, 1078-8956 (Print)
Orlic, D.; Kajstura, J.; Chimenti, S., et al. (2001). Bone marrow cells regenerate infarcted myocardium. *Nature,* 410, 6829, 701-705, 0028-0836 (Print)
Park, I. H.; Arora, N.; Huo, H., et al. (2008). Disease-specific induced pluripotent stem cells. *Cell,* 134, 5, 877-886, 1097-4172 (Electronic)
Passier, R. & Mummery, C. (2005). Cardiomyocyte differentiation from embryonic and adult stem cells. *Curr Opin Biotechnol,* 16, 5, 498-502, 0958-1669 (Print)
Passier, R.; Oostwaard, D. W.; Snapper, J., et al. (2005). Increased cardiomyocyte differentiation from human embryonic stem cells in serum-free cultures. *Stem Cells,* 23, 6, 772-780, 1066-5099 (Print)
Pera, M. F.; Andrade, J.; Houssami, S., et al. (2004). Regulation of human embryonic stem cell differentiation by BMP-2 and its antagonist noggin. *J Cell Sci,* 117, Pt 7, 1269-1280, 0021-9533 (Print)

Qyang, Y.; Martin-Puig, S.; Chiravuri, M., et al. (2007). The renewal and differentiation of Isl1+ cardiovascular progenitors are controlled by a Wnt/beta-catenin pathway. *Cell Stem Cell,* 1, 2, 165-179, 1875-9777 (Electronic)

Reinecke, H.; MacDonald, G. H.; Hauschka, S. D., et al. (2000). Electromechanical coupling between skeletal and cardiac muscle. Implications for infarct repair. *J Cell Biol,* 149, 3, 731-740, 0021-9525 (Print)

Reubinoff, B. E.; Pera, M. F.; Fong, C. Y., et al. (2000). Embryonic stem cell lines from human blastocysts: somatic differentiation in vitro. *Nat Biotechnol,* 18, 4, 399-404, 1087-0156 (Print)

Richards, M.; Fong, C. Y.; Chan, W. K., et al. (2002). Human feeders support prolonged undifferentiated growth of human inner cell masses and embryonic stem cells. *Nat Biotechnol,* 20, 9, 933-936, 1087-0156 (Print)

Robey, T. E.; Saiget, M. K.; Reinecke, H., et al. (2008). Systems approaches to preventing transplanted cell death in cardiac repair. *J Mol Cell Cardiol,* 45, 4, 567-581, 1095-8584 (Electronic)

Rubart, M.; Soonpaa, M. H.; Nakajima, H., et al. (2004). Spontaneous and evoked intracellular calcium transients in donor-derived myocytes following intracardiac myoblast transplantation. *J Clin Invest,* 114, 6, 775-783, 0021-9738 (Print)

Rust, E. M.; Westfall, M. V.; Samuelson, L. C., et al. (1997). Gene transfer into mouse embryonic stem cell-derived cardiac myocytes mediated by recombinant adenovirus. *In Vitro Cell Dev Biol Anim,* 33, 4, 270-276, 1071-2690 (Print)

Rust, W.; Balakrishnan, T. & Zweigerdt, R. (2009). Cardiomyocyte enrichment from human embryonic stem cell cultures by selection of ALCAM surface expression. *Regen Med,* 4, 2, 225-237, 1746-076X (Electronic)

Schneider, V. A. & Mercola, M. (2001). Wnt antagonism initiates cardiogenesis in Xenopus laevis. *Genes Dev,* 15, 3, 304-315, 0890-9369 (Print)

Schultheiss, T. M.; Xydas, S. & Lassar, A. B. (1995). Induction of avian cardiac myogenesis by anterior endoderm. *Development,* 121, 12, 4203-4214, 0950-1991 (Print)

Shake, J. G.; Gruber, P. J.; Baumgartner, W. A., et al. (2002). Mesenchymal stem cell implantation in a swine myocardial infarct model: engraftment and functional effects. *Ann Thorac Surg,* 73, 6, 1919-1925; discussion 1926, 0003-4975 (Print)

Siegel, D. L. (2002). Selecting antibodies to cell-surface antigens using magnetic sorting techniques. *Methods Mol Biol,* 178, 219-226, 1064-3745 (Print)

Smith-Arica, J. R.; Thomson, A. J.; Ansell, R., et al. (2003). Infection efficiency of human and mouse embryonic stem cells using adenoviral and adeno-associated viral vectors. *Cloning Stem Cells,* 5, 1, 51-62, 1536-2302 (Print)

Smith, R. R.; Barile, L.; Cho, H. C., et al. (2007). Regenerative potential of cardiosphere-derived cells expanded from percutaneous endomyocardial biopsy specimens. *Circulation,* 115, 7, 896-908, 1524-4539 (Electronic)

Snir, M.; Kehat, I.; Gepstein, A., et al. (2003). Assessment of the ultrastructural and proliferative properties of human embryonic stem cell-derived cardiomyocytes. *Am J Physiol Heart Circ Physiol,* 285, 6, H2355-2363, 0363-6135 (Print)

Steiner, D.; Khaner, H.; Cohen, M., et al. (2010). Derivation, propagation and controlled differentiation of human embryonic stem cells in suspension. *Nat Biotechnol,* 28, 4, 361-364, 1546-1696 (Electronic)

Sugi, Y. & Lough, J. (1994). Anterior endoderm is a specific effector of terminal cardiac myocyte differentiation of cells from the embryonic heart forming region. *Dev Dyn,* 200, 2, 155-162, 1058-8388 (Print)

Synnergren, J.; Adak, S.; Englund, M. C., et al. (2008). Cardiomyogenic gene expression profiling of differentiating human embryonic stem cells. *J Biotechnol,* 134, 1-2, 162-170, 0168-1656 (Print)

Synnergren, J.; Akesson, K.; Dahlenborg, K., et al. (2008). Molecular signature of cardiomyocyte clusters derived from human embryonic stem cells. *Stem Cells,* 26, 7, 1831-1840, 1549-4918 (Electronic)

Takahashi, K. & Yamanaka, S. (2006). Induction of pluripotent stem cells from mouse embryonic and adult fibroblast cultures by defined factors. *Cell,* 126, 4, 663-676, 0092-8674 (Print)

Takahashi, T.; Lord, B.; Schulze, P. C., et al. (2003). Ascorbic acid enhances differentiation of embryonic stem cells into cardiac myocytes. *Circulation,* 107, 14, 1912-1916, 1524-4539 (Electronic)

Takeuchi, J. K. & Bruneau, B. G. (2009). Directed transdifferentiation of mouse mesoderm to heart tissue by defined factors. *Nature,* 459, 7247, 708-711, 1476-4687 (Electronic)

Taylor, D. A.; Atkins, B. Z.; Hungspreugs, P., et al. (1998). Regenerating functional myocardium: improved performance after skeletal myoblast transplantation. *Nat Med,* 4, 8, 929-933, 1078-8956 (Print)

Terami, H.; Hidaka, K.; Katsumata, T., et al. (2004). Wnt11 facilitates embryonic stem cell differentiation to Nkx2.5-positive cardiomyocytes. *Biochem Biophys Res Commun,* 325, 3, 968-975, 0006-291X (Print)

Thomson, J. A.; Itskovitz-Eldor, J.; Shapiro, S. S., et al. (1998). Embryonic stem cell lines derived from human blastocysts. *Science,* 282, 5391, 1145-1147, 0036-8075 (Print)

Thomson, J. A. & Marshall, V. S. (1998). Primate embryonic stem cells. *Curr Top Dev Biol,* 38, 133-165, 0070-2153 (Print)

Toma, C.; Pittenger, M. F.; Cahill, K. S., et al. (2002). Human mesenchymal stem cells differentiate to a cardiomyocyte phenotype in the adult murine heart. *Circulation,* 105, 1, 93-98, 1524-4539 (Electronic)

Tomescot, A.; Leschik, J.; Bellamy, V., et al. (2007). Differentiation in vivo of cardiac committed human embryonic stem cells in postmyocardial infarcted rats. *Stem Cells,* 25, 9, 2200-2205, 1549-4918 (Electronic)

Tran, T. H.; Wang, X.; Browne, C., et al. (2009). Wnt3a-induced mesoderm formation and cardiomyogenesis in human embryonic stem cells. *Stem Cells,* 27, 8, 1869-1878, 1549-4918 (Electronic)

Ueno, S.; Weidinger, G.; Osugi, T., et al. (2007). Biphasic role for Wnt/beta-catenin signaling in cardiac specification in zebrafish and embryonic stem cells. *Proc Natl Acad Sci U S A,* 104, 23, 9685-9690, 0027-8424 (Print)

van Laake, L. W.; Passier, R.; Monshouwer-Kloots, J., et al. (2007). Human embryonic stem cell-derived cardiomyocytes survive and mature in the mouse heart and transiently improve function after myocardial infarction. *Stem Cell Res,* 1, 1, 9-24, 1876-7753 (Electronic) van Vliet, P.; Roccio, M.; Smits, A. M., et al. (2008). Progenitor cells isolated from the human heart: a potential cell source for regenerative therapy. *Neth Heart J,* 16, 5, 163-169, 1568-5888 (Print)

Watanabe, K.; Ueno, M.; Kamiya, D., et al. (2007). A ROCK inhibitor permits survival of dissociated human embryonic stem cells. *Nat Biotechnol,* 25, 6, 681-686, 1087-0156 (Print)

Xu, C.; He, J. Q.; Kamp, T. J., et al. (2006). Human embryonic stem cell-derived cardiomyocytes can be maintained in defined medium without serum. *Stem Cells Dev,* 15, 6, 931-941, 1547-3287 (Print)

Xu, C.; Inokuma, M. S.; Denham, J., et al. (2001). Feeder-free growth of undifferentiated human embryonic stem cells. *Nat Biotechnol,* 19, 10, 971-974, 1087-0156 (Print)

Xu, C.; Police, S.; Hassanipour, M., et al. (2006). Cardiac bodies: a novel culture method for enrichment of cardiomyocytes derived from human embryonic stem cells. *Stem Cells Dev,* 15, 5, 631-639, 1547-3287 (Print)

Xu, C.; Police, S.; Rao, N., et al. (2002). Characterization and enrichment of cardiomyocytes derived from human embryonic stem cells. *Circ Res,* 91, 6, 501-508, 1524-4571 (Electronic)

Xu, X. Q.; Graichen, R.; Soo, S. Y., et al. (2008). Chemically defined medium supporting cardiomyocyte differentiation of human embryonic stem cells. *Differentiation,* 76, 9, 958-970, 1432-0436 (Electronic)

Xu, X. Q.; Zweigerdt, R.; Soo, S. Y., et al. (2008). Highly enriched cardiomyocytes from human embryonic stem cells. *Cytotherapy,* 10, 4, 376-389, 1477-2566 (Electronic)

Xue, T.; Cho, H. C.; Akar, F. G., et al. (2005). Functional integration of electrically active cardiac derivatives from genetically engineered human embryonic stem cells with quiescent recipient ventricular cardiomyocytes: insights into the development of cell-based pacemakers. *Circulation,* 111, 1, 11-20, 1524-4539 (Electronic)

Yamashita, J. K.; Takano, M.; Hiraoka-Kanie, M., et al. (2005). Prospective identification of cardiac progenitors by a novel single cell-based cardiomyocyte induction. *FASEB J,* 19, 11, 1534-1536, 1530-6860 (Electronic)

Yang, L.; Soonpaa, M. H.; Adler, E. D., et al. (2008). Human cardiovascular progenitor cells develop from a KDR+ embryonic-stem-cell-derived population. *Nature,* 453, 7194, 524-528, 1476-4687 (Electronic)

Yao, S.; Chen, S.; Clark, J., et al. (2006). Long-term self-renewal and directed differentiation of human embryonic stem cells in chemically defined conditions. *Proc Natl Acad Sci U S A,* 103, 18, 6907-6912, 0027-8424 (Print)

Yoon, B. S.; Yoo, S. J.; Lee, J. E., et al. (2006). Enhanced differentiation of human embryonic stem cells into cardiomyocytes by combining hanging drop culture and 5-azacytidine treatment. *Differentiation,* 74, 4, 149-159, 0301-4681 (Print)

Yu, J.; Vodyanik, M. A.; Smuga-Otto, K., et al. (2007). Induced pluripotent stem cell lines derived from human somatic cells. *Science,* 318, 5858, 1917-1920, 1095-9203 (Electronic)

Zafarghandi, M. R.; Ravari, H.; Aghdami, N., et al. (2010). Safety and efficacy of granulocyte–colony-stimulating factor administration following autologous intramuscular implantation of bone marrow mononuclear cells: a randomized controlled trial in patients with advanced lower limb ischemia. *Cytotherapy,* 12, 6, 783-791, 1477-2566 (Electronic)

Zaffran, S. & Frasch, M. (2002). Early signals in cardiac development. *Circ Res,* 91, 6, 457-469, 1524-4571 (Electronic)

Zwaka, T. P. & Thomson, J. A. (2003). Homologous recombination in human embryonic stem cells. *Nat Biotechnol,* 21, 3, 319-321, 1087-0156 (Print)

24

Chemicals Regulating Cardiomyocyte Differentiation

Hideki Uosaki and Jun K Yamashita
Institute for Frontier Medical Sciences / Center for iPS Cell Research and Application
Kyoto University
Japan

1. Introduction

Differentiation toward cardiomyocytes from embryonic stem cells (ESCs) and induced pluripotent stem cells (iPSCs) has been studied widely. Many of ESC/iPSC differentiation methods have been developed using embryoid body (EB) methods, in which ESCs/iPSCs spontaneously differentiate in floating aggregates. Various marker genes (early mesoderm or mesoendoderm cells: *Brachyury*, mesoderm cells: *Mesp1*, *Flk1*, and *Pdgfra*, cardiac progenitors: *Nkx2.5*, *Islet1* and cardiomyocytes: myosin chains) have been identified as stage specific markers to recapitulate developmental process in embryo. The clues from developmental biology provided ideas to use growth factors (*Activin A/Nodal*, *Bmp4*, *Cerberus*, *Wnt3a*, and *Wnt11*) to guide cardiac differentiation. Recently these stem cell technologies have been combined with chemical biology, in which small molecules are identified and used to regulate cell fate or modulate cell reprogramming. Several chemicals were reported that they could enhance cardiomyocyte differentiation from embryonic stem cells. However, it is hard to determine how these chemicals work, which cell types they affect, and what are their molecular targets because of heterogeneity in EBs. We have established a novel systemic cardiovascular differentiation method that is based on two dimensional culture and sequential purification using cell surface markers. A mesoderm lineage population, Flk1 expressing cells derived from mouse ESC/iPSC can give rise to vascular endothelial cells, pericytes, hematopoietic cells, and cardiomyocytes. Cardiomyocytes appear after 4 days co-culture of Flk1$^+$ cells on OP9 feeder cells. A cardiac progenitor population, FCV (Flk1$^+$/CXCR4$^+$/Vascular endothelial cardherin$^-$) cells among the progeny of Flk1$^+$ mesoderm cells are observed 2 days after Flk1 purification. In our system, Flk1$^+$ and FCV cells are able to give rise to all cardiovascular cells - cardiomyocytes, endothelial cells, and pericyte even from single cell. We have identified a novel cardiogenic effect of cyclosporin-A (CsA) well-known immunosuppressant. CsA affects on mesoderm cells to drastically increase the differentiation of cardiac progenitors and cardiomyocytes up to 10 fold. In this chapter, we review recent advances of combination of stem cell technologies and chemical biology, especially chemical biology on cardiomyocyte differentiation, and our resent findings. We also describe some information to compare high throughput screening and high content screening in cell based assay.

2. Heart development *in vivo* and cardiomyocyte differentiation *in vitro*

To understand how cardiomyocytes differentiate from ESCs/iPSCs, comparison between embryo and ESC/iPSC differentiation would be helpful. In this section, we describe how a heart develops in an embryo, and how cardimyocytes differentiate from ESCs/iPSCs. (Figure 1 and 2)

2.1 Heart development *in vivo*

The heart is the first functional organ in mammals. After fertilization, a Zygote cleaves and forms blastcyst before implantation (-E3.5), which contains inner cell mass (ICM) and trophoblast. ICM cells develop toward whole body, and ESCs are established from ICM cells. ICM cells segregate to epiblast and primitive endoderm lineages in late blastcyst. In epiblast stage, Part of epiblast cells migrate to midline and form primitive streak under induction of *Nodal* secreted from node. In primitive streak, *Brachyury* is expressed. *Wnt3a* and *Wnt5* are secreted in primitive streak and regulate expansion of primitive streak (Takada et al. 1994). Cells migrating through primitive streak differentiate to mesoderm and endoderm. The outer layer cells are called endoderm that is primordia of intestines, liver, pancreas and so on. The rest cells in epiblast, which do not migrate, are called ectoderm that is primordia of neuronal system and skin. Between the layers of endoderm and ectoderm, mesoderm cells differentiate from primitive streak and loose epithelial markers such as E-cadherin. Part of mesoderm cells migrate to lateral side and are called lateral plate mesoderm (LPM). LPM expresses Flk1 and Pdgfrα on their surface and *Mesp1* that is a transcriptional factor regulating cardiac mesoderm (Kataoka et al., 1997; Saga et al., 2000). Several secreting factors are known as regulators of LPM and cardiac mesoderm formation. Spatial-temporal expression of Bmp4 regulates mesoderm fate. Anterior visceral endoderm cells secrete Noggin, Cerberus, and Dkk1, which regulate cardiac mesoderm formation thorough inhibition of BMP and Wnt signaling (Foley and Mercola, 2005). Non canonical Wnt family, Wnt11 is also known to be involved in cardiac mesoderm to heart development (Pandur et al., 2002). LPM cells migrate anteriorly and forms cardiac crescent, which expresses *Nkx2.5* and/or *Islet1*. *Nkx2.5*$^+$ population called first heart field and *Islet1*$^+$ population called second heart field. Cardiac crescent cells migrate and fuse in midline and form heart tube. Then heart tube loops rightward to form four-chambered heart. First heart field mainly contributes to left ventricle and second heart field contributes to atrium and right ventricle.

2.2 Cardiomyocyte differentiation *in vitro* – Embryoid body method

In ESC/iPSC differentiation, EB method is widely used (Figure 1). This method is rather easy to differentiate ESC/iPSC toward various cell types. However, it is hard to evaluate exact molecular mechanisms and monitor at single cell level. Using EB method in cardiac differentiation, beating EB in total EB rate is widely used for quantitative evaluation. During differentiation process, some mesoderm (*Brachyury, Mesp1, Flk1,* and *Pdgfra*) or cardiac markers (*Nkx2.5, Islet1, actinin, αMHC* and etc.) are used. These markers were evaluated using RT-PCR or transgenic line in some cases. Other than transgenic line or cell surface antigen, it is hard to evaluate in single cellular level for differentiation step. Some growth factors mentioned above increase cardiac differentiation. Wnt3a have been shown to have stage specific and bi-phasic effect (Naito et al., 2006; Ueno et al., 2007). Wnt3a treatment increased cardiac differentiation in early stage of differentiation and decreased in late stage.

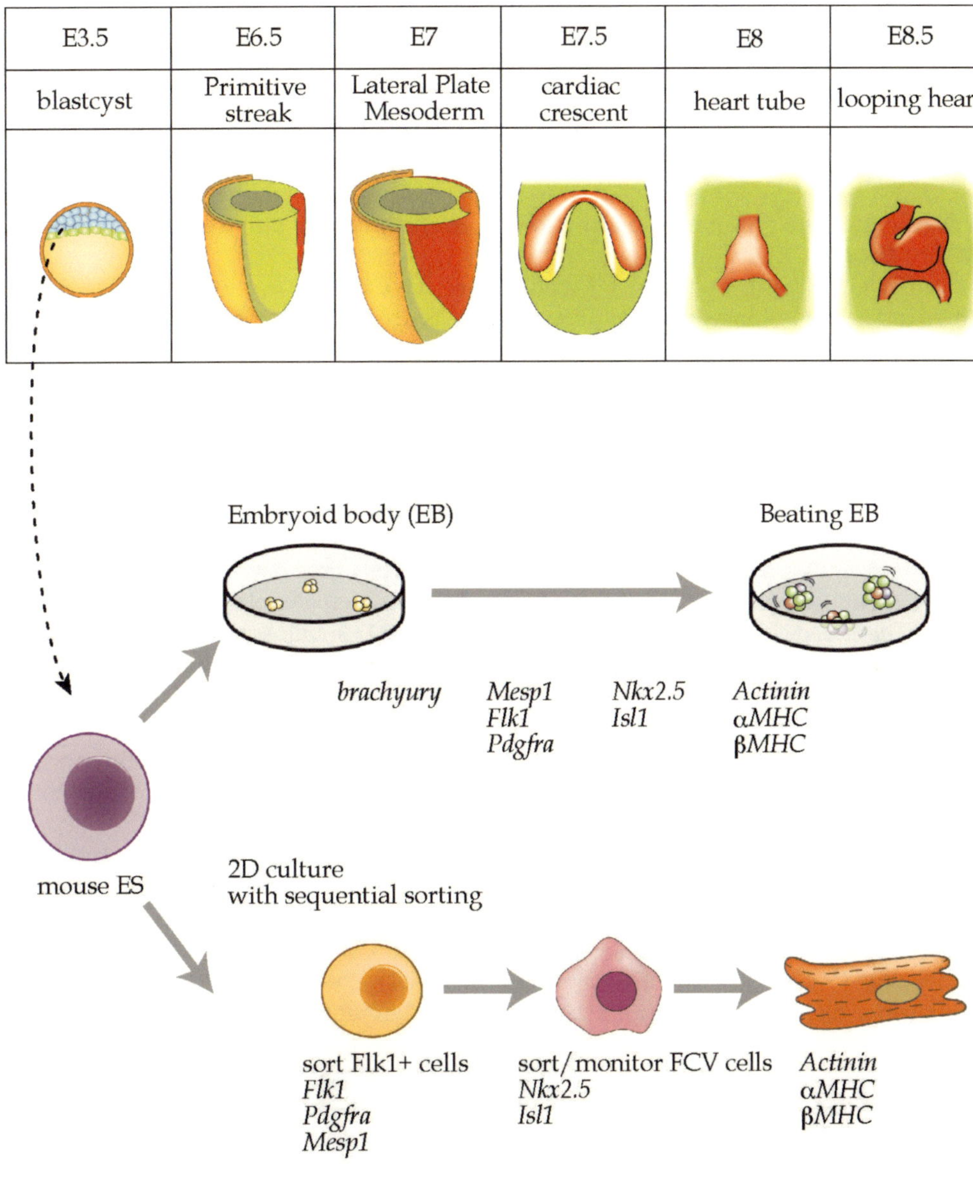

Fig. 1. Heart development in embryo and cardiac differentiation from ESCs in vitro. Cardiac lineage is mainly differentiate through primitive streak, lateral plate mesoderm, and cardiac crescent. Cardiac crescent cells migrate and fuse at midline of embryo and form heart primordial called heart tube. ESCs are established from inner cell mass of blastcyst. When ESCs are cultured in suspension, they form embryoid body, spontaneously differentiate and start beating when cardiomyocyte appear. Another differentiation method using 2D culture with sequential sorting provide us to understand stage specific prosess of differentiation.

Administration of Wnt11 in late stage of EB increase cardiomyocyte differentiation (Terami et al., 2004). Short term exposure of Noggin also directed ESCs to cardiac differentiation (Yuasa et al., 2005). Recently, directed differentiation methods using Activin A and BMP4 in EBs or high density culture of human ESCs were reported (Buehr et al., 2008; Laflamme et al., 2007). As shown here, some cardiogenic growth factors have been reported similar to embryo development. However these studies are still not clear to show the molecular mechanisms or target cells because of heterogeneity of cells in EBs or high density culture.

2.3 Cardiomyocyte differentiation *in vitro* – Two-dimensional culture and sequential differentiation

To overcome the weak points of EB method, we have developed a novel differentiation method using two-dimensional culture and FACS. In this method, we can systemically induce cardiovascular cells - cardiomyocyte, endothelial cells, pericyte, and hematopoietic cells from single Flk1 positive mesoderm cells (Figure 2). Withdrawal of LIF induces ESC/iPSC differentiation. Flk1$^+$ mesoderm cells appear during 4-4.5 days after LIF withdrawal. Flk1$^+$ cells are negative for E-cadherin, an ESC marker, and positive for Pdgfra. As previously reported (Kataoka et al., 1997), Flk1$^+$/Pdgfra$^+$ cells are observed wedge-shaped areas on proximal lateral mesoderm of mid-late streak stage embryo (E7.0) and are also observed in the anterior head fold region of head fold stage embryos (E7.5-7.75), which might consistent with cardiac crescent. Purified Flk1$^+$ cells give rise to cardiomyocyte after 3-4 days coculture on OP9 feeder cells. Endothelial cells, hematopoietic cells, and pericytes are also observed in Flk1 culture on OP9. At Flkd2, small subset of Flk1 progeny, <u>F</u>lk1$^+$/<u>C</u>XCR4$^+$/<u>V</u>E-cadherin$^-$ (FCV) cells appear and almost all cardiomyocytes are from this FCV population (Yamashita et al., 2005). Even single Flk1$^+$ and FCV cells can give rise to all these cardiovascular cells on OP9 feeder cells. Thus, it let us to know how much Flk1$^+$ and FCV cells have the differentiation potentials and to observe differentiation process under time lapse imaging (Eilken et al., 2009). Using this system, we can determine differentiation stage specific event and target cells.

In the heart, there are some subsets of cardiomyocytes, ventricular and atrial cardiomyocytes, conduction system cells, and pacemaker cells. Some molecular markers are known for these cardiomyocytes, such as myosin light chain (*Mlc2v* and *Mlc2a*), gap junction protein, and ion channels (Hyperpolarization-Activated Cyclic Nucleotide-Gated Channels, L and T type Ca channels, Na channels and K channels). Electrophysiological study clearly distinguishes them in single cellular level. In early differentiation stage (such as Flkd9.5), ESC derived cardiomyocytes show almost homogeneous phenotype, like heart tube stage. Cardiomyocytes have automaticity, higher maximum diastolic potential, and slower maximum stroke velocity (dV/dt). On the other hand, in late differentiation stage (such as Flkd23.5), cardiomyocytes show diverse character - (1) automaticity+, higher maximum diastolic potential and slower dV/dt (2) automaticity-, deep resting membrane potential, and faster dV/dt (Yanagi et al., 2007).

2.4 2D culture for EC differentiation

Flk1 positive cells are also able to differentiate to endothelial cells and pericyte on collagen IV dish with VEGF or in Collagen gel as well as on OP9 cells (Yamashita et al., 2000). Administration of cAMP with VEGF enhanced arterial endothelial cell differentiation (Yurugi-Kobayashi et al., 2006). Furthermore, this system clearly reconstituted downstream signals of cAMP. PKA enhance endothelial differentiation through enhancing VEGF signaling by

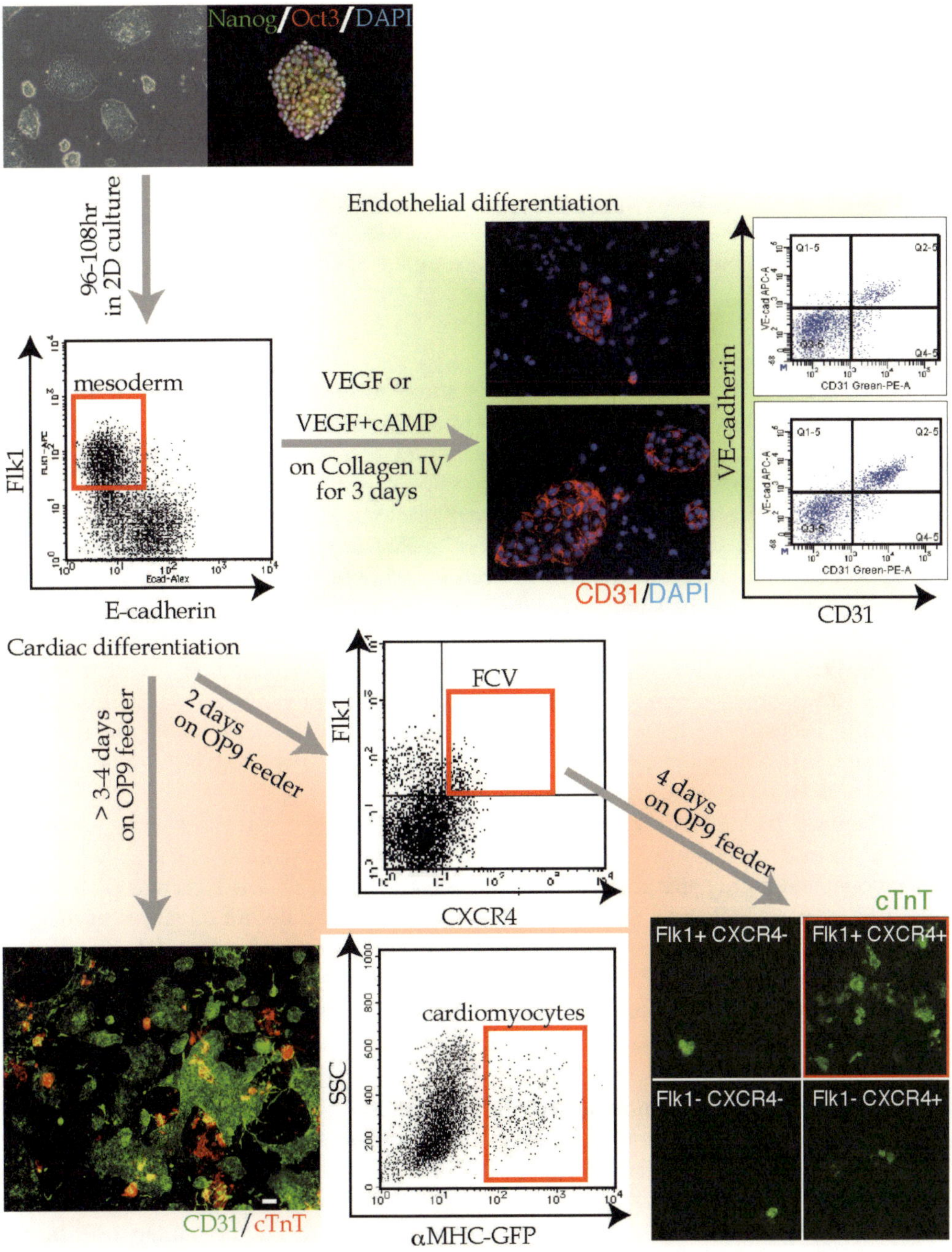

Fig. 2. Cardiac lineage differentiation from Flk1+ mesoderm cells. Cardiomyocytes are differentiated from Flk +cells on OP9 feeder cells. Self-beating cardiomyocyte appears 3-4 days after Flk1 +cells are cultured on OP9 cells (Flkd3-4). At Flkd2, cardiogenic potential is restricted in Flk1 + /CxCR4 +/VE-cadherin-(FCV) population. Endothalial cells are differentiated from same Flk1 + cells on OP9 cells and on Collagen IV coated dish with VEGF. (modified from Yamashita et al., 2005)

up-regulation of *Flk1* and *Nrp1*, which forms VEGF receptor complex (Yan et al., 2009). Addministration of cAMP also activate Notch and β-catenin, which form transcriptional complex to increase arterial endothelial cell differentiation (Yamamizu et al., 2010).

3. Chemical biology in stem cell biology

ESC Differentiation methods have been developed based on embryology as described above. Recently, combined with chemical biology, it has come to next stage. In this section, we review chemical biology and show some examples of chemical biology combined with stem cell biology

3.1 What is chemical biology?

Chemical biology or chemical genetics is the study to elucidate biological systems with small molecules that bind directly to proteins to alter protein function. There are two ways of chemical biology: forward chemical genetics and reverse chemical genetics similar to conventional genetics. "Forward" means phenotype based, so random mutation is used in conventional genetics and screenings of small molecules, which induce phenotype of interest, are used in chemical genetics. "Reverse" means target based, so that gene-targeting and transgenic animals, and RNA interference technologies are used in conventional genetics, and target-specific small molecules, such as kinase inhibitors, are used in chemical genetics. The concept itself is not so much different; however, chemical genetics have great advantages and some disadvantages compared to conventional method. Small molecules have great advantages in temporal control, rapid inhibition or activation, and regulation of functionally overlapping targets, compared to conventional method. However, they also have some disadvantages in target specificity and off-target effect. To verify or confirm the targets of small molecules are more difficult.

When chemical biology is combined with stem cell biology, some screening system is required to screen chemical libraries. High throughput screening (HTS) is one of conventional and highly efficient method to screen chemicals, in which such as plate reader is used to measure biological output (luciferase activity, etc.). Recent advances of imaging system provide a novel screening method, named high content screening (HCS), which is based on microscopes. In HCS, immunostained wells are photographed and scored using computer automatically to measure immunoreactive area or to count number of cells that show appropriate phenotypes.

3.2 Chemical biology in cardiomyocyte differentiation

In this section, we describe some example of cardiogenic small molecules.

Ascorbic acid is a firstly identified chemical that increased cardiomyocyte differentiation from ESC (Takahashi et al., 2003). Ascorbic acid was identified from "The FDA2000 Drug Library", which contains 880 of bioactive compounds approved for human use. In this screen, the authors used an ESC line carring αMHC-EGFP. The only compound found to reproducibly induce differentiation in monolayer culture was ascorbic acid. Interstingly, ascorbic acid showed no significant effect in cardiac differentiation using EB formation.

Cardiogenol was identified from 100,000 compound library using P19, embryonic carcinoma cell line, stably transfected rat atrial natiuretic factor promoter-luciferace (ANF-luc) vecter (Wu et al., 2004). In this ANF-luc reporter assay, Approximately 80 compounds were identified and 35 of 80 compounds induced MHC expression in P19CL6. 4 diaminopyrimidines, cardiogenol A-D were the most potent in inducing MHC expression. Wu et al. also tested R1, a mouse ESC line, to confirm the effects of cardiogenol. R1 cells were cultured in extremely high density (10,000cells/well in gelatin coarted 384-well plate) with cardiogenol treatment in first 3 days of 7 days culture period. Although the authors did not mention about the molecular machineries or differentiation stage specificity, R1 cells were treated in early differentiation stage indicating cardiogenol may induce cardiogenic mesoderm.

Sulfonylhydrazone (Shz-1) was identified from 147,000 compund library using P19CL6, stably transfected bacterial artificial chromosome (BAC) of Nkx2.5 locus, which luciferase is knocked-in (Nkx2.5-luc) (Sadek et al., 2008). 3,000 strong positive hits from the primary screen were selected and 1,600 compounds were still positive for the secondary screen using dose response strategy. The authors clustered these hits into 10 lists of families with chemically distinct, synthetically tractable core structural motifs. Shz was one of the strongest lead chemicals. 48 hr treatment of Szh-1 at 5μM increased Nkx2.5-luc activity and 72hr Shz1 treatment increased sarcomeric α-tropomyosin expression in P19CL6. They also tested SM1 ES cells. Because 48hr after treatment of Shz-1 endogenous myocardin and Brachyury expression level was increased, Shz-1 might also increase cardiac mesoderm differentiation. They tested various signal cascades to indentify Shz-1 target molecules but failed. They also demonstrated Shz1 increased cardiac differentiation from human M-PBMCs *in vitro*. Shz1 treated M-PBMCs rescued Cryoinjured rat heart.

Compound	structure	Assay	Screened compound number	Molecular mecahnisms
Ascoribic acid	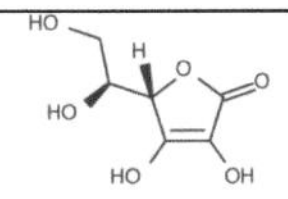	αMHC-EGFP ESC	880	Unknown
Cardiogenol C	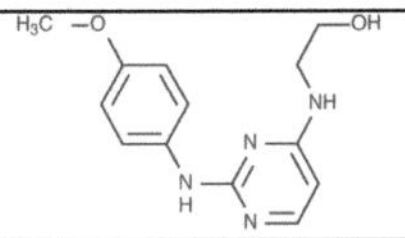	ANF-Luc P19	100,000	Unknown
Szh-1	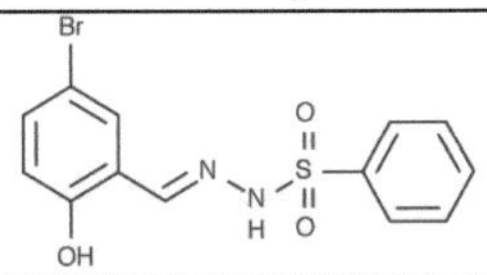	Nkx2.5-Luc P19CL6	147,000	Unknown

Table 1. Cardiogenic chemicals

3.3 Chemical biology in maintenace and generation of pluripotent stem cells

Efficient maintenance and generation method of ESCs and iPSCs are also important for cardiac differentiation and for stem cell biology. Here we describe recent advances of chemical biology in maintenance and generation of ESC/iPSC.

Mouse ESCs require LIF for their maintenance. Pluripotin was identified for the replacement of LIF from 50,000 compound library as activating Oct3/4-GFP transgene in ESC (Chen et al., 2006). Using affinity chromatography, ERK1 and RasGAP were identified as the molecular targets of pluripotin. Using similar Oct3/4-GFP transgenic mice, E-616452 a

Compound	structure	Primary Assay	Screened compound number	Molecular mechanisms
Pluripotin	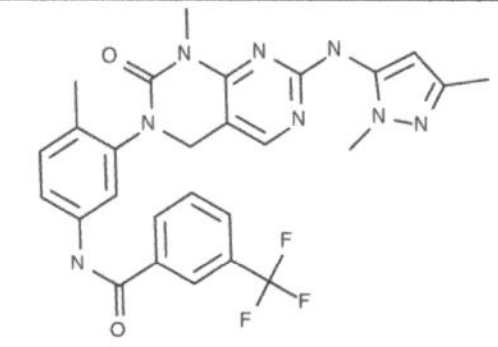	Oct3-GFP	50,000	ERK/Ras-GAP inhibition
E-616452	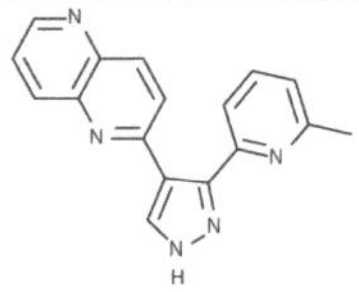	Oct3-GFP	800	Tgfbr1 inhibition
BIX01294	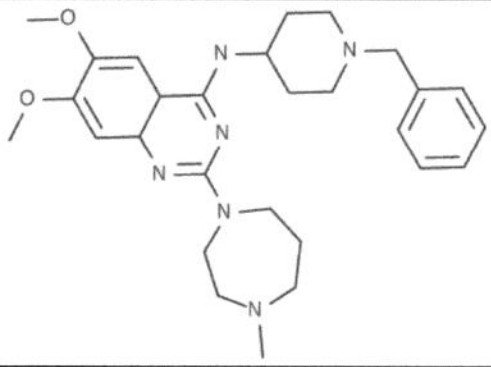	ALP positive colony number	2,000	G9a/GLP inhibition
BayK8644	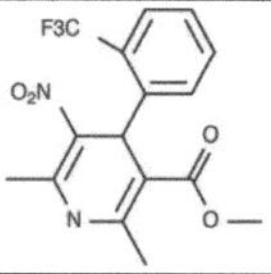			L type Ca channel agonist
RG108	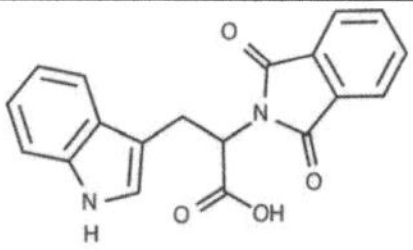			Dnmt inhibition

Table 2. Small molecules identified for mainenance of ESC/iPSC and generation of iPSC using forward chemical genetics.

Compound	structure	Primary Assay	Molecular mechanisms
5-azacytidine	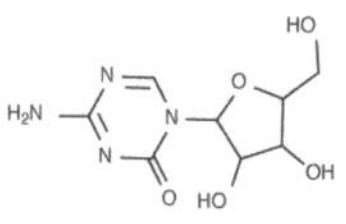	Oct3-GFP (%) from partially reprogrammed cells.	DNA methylation inhbition
Valproic acid	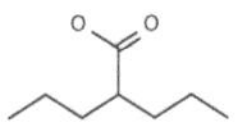	Oct3-GFP colony number	HDAC inhibition
SB431542	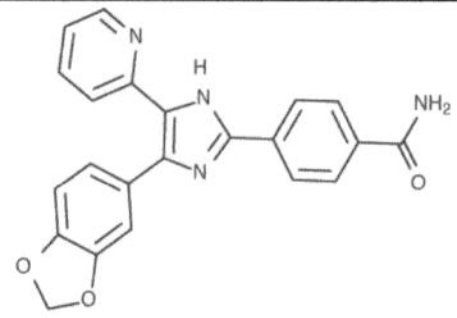	ALP positive colony number	Promoting Mesenchymal-epithelial transition through TGFβ signal inhibition

Table 3. Small molecules identified for maintenance of ESC/iPSC and generation of iPSC using reverse chemical genetics

Tgfbr1 inhibitor was identified as an alternative of Sox2 transgenes in generation of iPSCs from 800 compounds (Ichida et al., 2009). BIX01294 (G9a/GLP inhibitor), BayK8644 (L type Ca channel agonist) and RG108 (Dnmt inhibitor) were also identified that they could enhance reprogramming through screening of 2,000 bioactive molecules. These chemicals are identified through phenotype-based screening, so that forward chemical genetics was used for identification of them.

Using reverse chemical genetics, some important findings for maintenance and generation of ESCs/iPSCs were reported. As FGF4/ERK signaling is a promoter signal to induce differentiation of mouse ESCs, inhibition of FGF4/ERK using SU5402 and PD184352 suppressed ESC differentiation without LIF. In this condition, apoptosis was relatively high. To reduce apoptosis and increase ESC growth, CHIR99021 a specific GSK3 inhibitor was combined with SU5402 and PD184352. This three-inhibitor combination SU5402+PD184352+CHIR99021 (3i) support the maintenance of ESCs without Stat3 that is indispensable downstream target of LIF (Ying et al., 2008). Furthermore, 3i supported to generate rat ESCs (Buehr et al., 2008).

Generating iPSCs is robustly studied using small molecules such as 5-azacitidine (Mikkelsen et al., 2008), valproic acid (Huangfu et al., 2008) and TGFβ inhibitor (Lin et al., 2009) with reverse genetics to elucidate the machineries of reprogramming.

3.4 Chemical biology in other lineage differentiation

As like cardiogenic chemicals and small molecules for maintenance and generation of ESC/iPSC, Douglas A. Melton and his colleagues reported some small molecules induce definitive endoderm and pancreatic lineage differentiation using HCS (Borowiak et al., 2009; Chen et al., 2009). Inducers of definitive endoderm (IDE) 1 and 2 were identified from 4,000

Compound	structure	Assay	Screened compound number	Molecular targets
IDE1	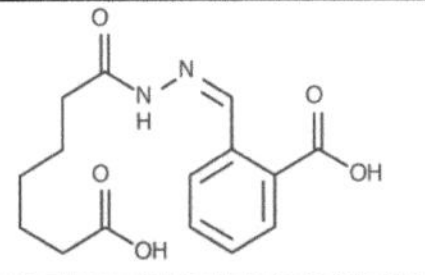	Sox17-DsRed	4,000	Trans-activate TGFβ/nodal signaling
IDE2	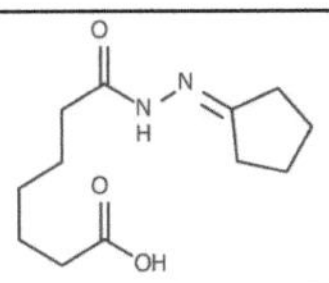			
Indolactam V	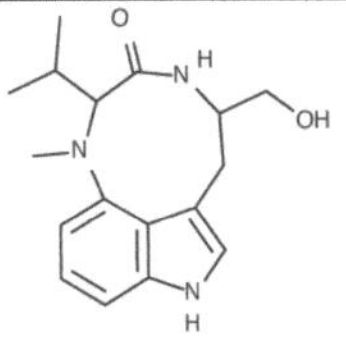	Pdx1 staining	5,000	Activate PKC signaling

Table 4. Small molecules using endodermal lineage induction from ESC/iPSC

compounds using Sox17-dTomato mESC line to replace Activin A for the initial induction of endoderm. IDE-1 and 2 transactivate TGFβ/Activin A/nodal signaling through unknown machinery. (-)-Indolactam V was identified from 5,000 compounds as inducing differentiation of $Pdx1^+$ pancreatic progenitor from definitive endoderm derived from human ESCs using high content screening. Activation of PKC signaling is the major pathway of the induction of Pdx1+ pancreatic progenitor using Indolactam V.

4. An immunosuppressant cyclosporin-A induced cardiac progenitor and cardiomyocyte differentaion

We reported cyclosporin-A, an immunosuppressant specifically and efficiently induce cardiac progenitors and cardiomyocytes from mesoderm cells (Yan et al., 2009). In this section, we summarize our results (Figure 3 and 4).

4.1 CsA induced cardiac progenitors and cardiomyocyte differentiation from mesoderm cells.

Addition of CsA (1-3μg/mL) to $Flk1^+$ cells on OP9 showed a striking effect to increase cardiomyocyte differentiation at Flkd6. Cardiac troponin T positive area was increased more than 10 times and 60% of $Flk1^+$ cell-derived cells became $\alpha MHC\text{-}GFP^+$ cardiomyocytes in maximum. To clarify stage specificity, we added CsA during the differentiation stage of undifferentiated ESC–Flkd0, Flkd0-d2, and Flkd2-d6. CsA most potently showed the cardiogenic effect during Flkd0-2. Thus, we evaluated CsA effect on FCV cardiac progenitor differentiation and demonstrated that CsA treatment potently increased cardiac progenitor differentiation more than 20 times. Interestingly, CsA treatment strongly suppressed endothelial or hematopoietic differentiation suggesting that CsA may shift the cell fate from hemangiogenic to cardiogenic.

4.2 Character, expanded cardiac progenitor

Expanded FCV progenitor cells express *Flk1, Cxcr4, Gata4, Nkx2.5,* and *Islet1* but not myosin chains nor ion channels. *Nkx2.5* and *Islet1* are the markers for the first and second heart field respectively, which were reported as cardiac progenitor markers in ESC differentiation (Moretti et al., 2006; Wu et al., 2006). In FCV cells, approximately 24% of FCV cells were positive for *Nkx2.5*, 42% were positive *Islet1*, and 14% were double positive for *Nkx2.5* and *Islet1*. This efficient expansion of the rare FCV progenitor cells allowed us to confirm their cardiogenic potential *in vivo*. We performed transplantation of CsA-induced FCV cells to rat chronic myocardial infarction model. At 2 weeks after the injection, transplanted FCV cells were successfully differentiated into cardiomyocytes and integrated in the infarct heart. This result indicates that CsA-expanded FCV cells can show highly cardiogenic potentials also *in vivo* after cell transplantation.

Cardiomyocytes induced using CsA express various cardiac markers such as *Gata4, Tbx3, Nkx2.5, Islet1* and multiple myosin chains. Cardiomyocytes also have apparent sarcomere structure, and pacemaker like potential with spontaneous beating or quiescent ventricular type potential.

4.3 Molecular mechanism of cyclosporin A

Cyclosporin A is known as inhibitor of Calcineurin-NFAT signaling. To elucidate whether NFAT cascade is important to inducing cardiomyocyte differentiation, we tested various

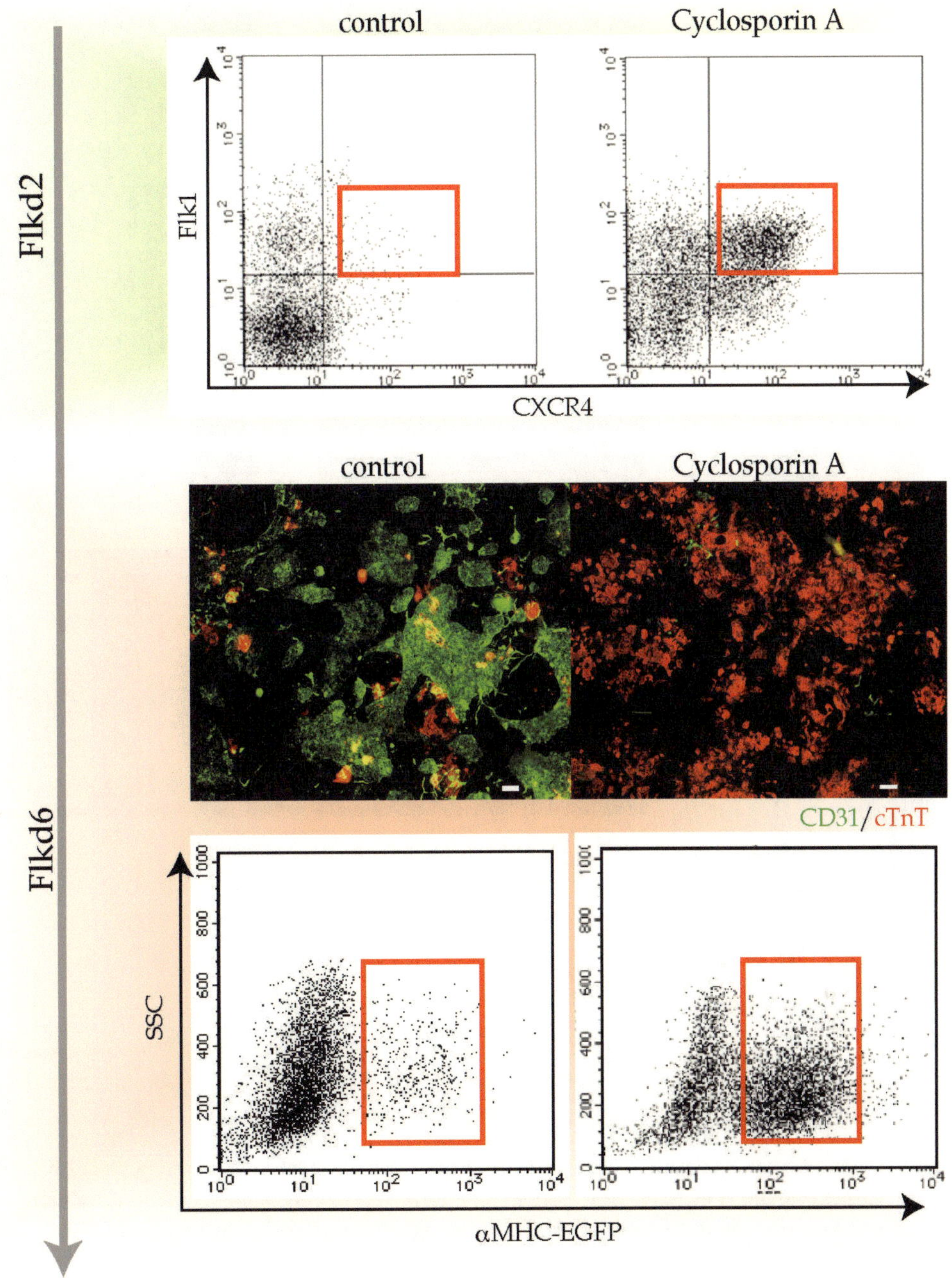

Fig. 3. An immunosuppressant cyclosporin A strikingly induce differentiation of FCV cardiac progenitor and cardiomyocytes.
Addministration of cyclosporin A increased FCV cardiac progenitor differentiation about 20 fold and cardiomyocyte differentiation about 10 fold with reciprocal inhibition of endotheleal differentiation. (modified from Yan et al., 2009)

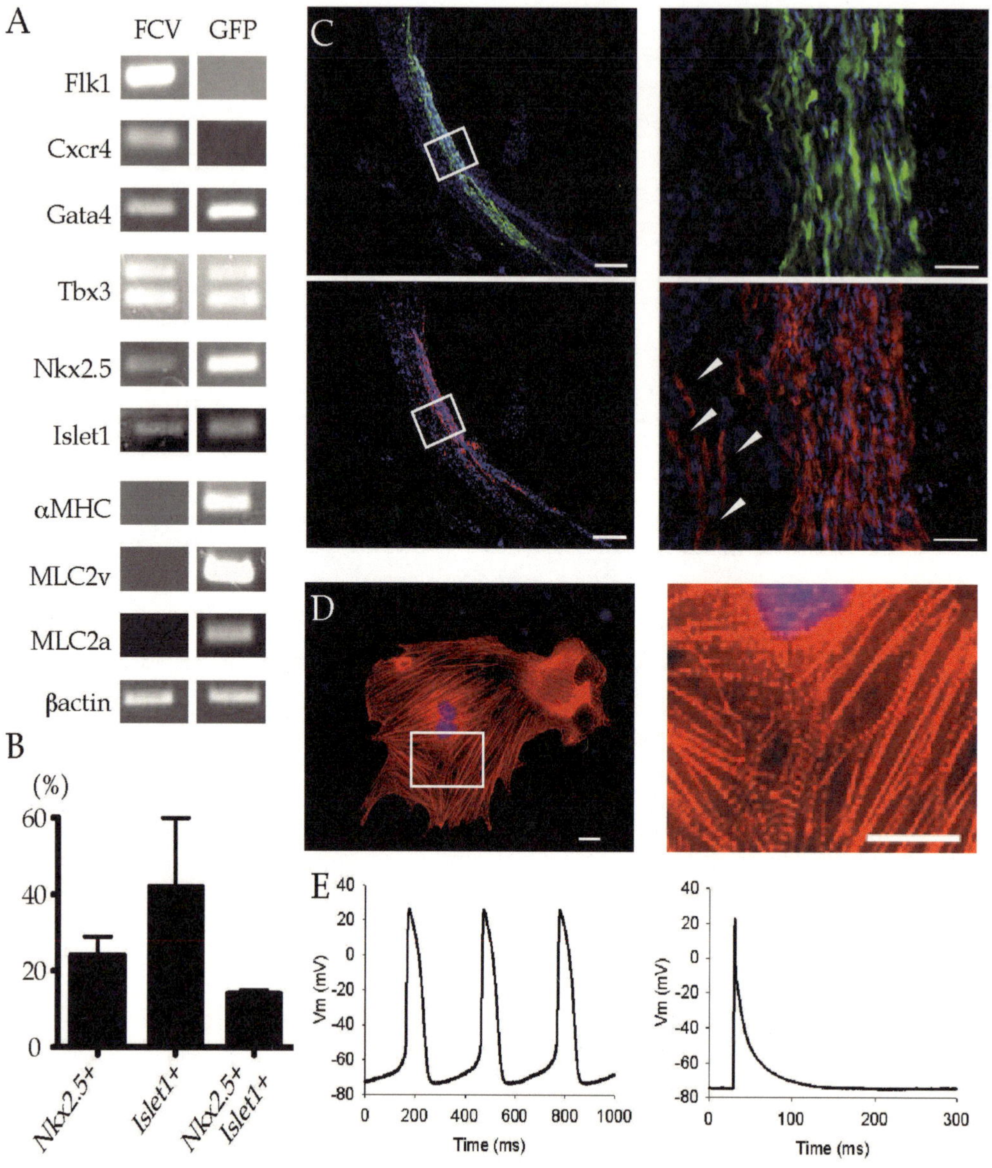

Fig. 4. Characters of expanded FCV cardiac progenitor cells and cardiomyocytes. (A) mRNA expression in purified FCV cardiac progenitor cells and aMHC-GFP+ cardiomyocytes. (B) Quantitative analysis of immunostaining for Islet1 and Nkx2.5 in purified FCV cells. (C) Cardiogenic potential of FCV cells in vivo. Purified FCV cells were injected into rat chronic myocardial infarction model. Green: aMHC-EGFP+ donor derived cardiomyocytes, Red: cardiac troponin T+ donor and recipient cardiomyocytes, Blue:DAPI (D) Appearance of isolated cardiomyocyte. Apparent sarcomere structure was observed. Red: Actinin, Blue: DAPI. (E) Action potential of isolated cardiomyocyte. Both of pacemaker like potential with spontaneous beating and quiescent ventricular type potential were observed. (modified from Yan et al., 2009)

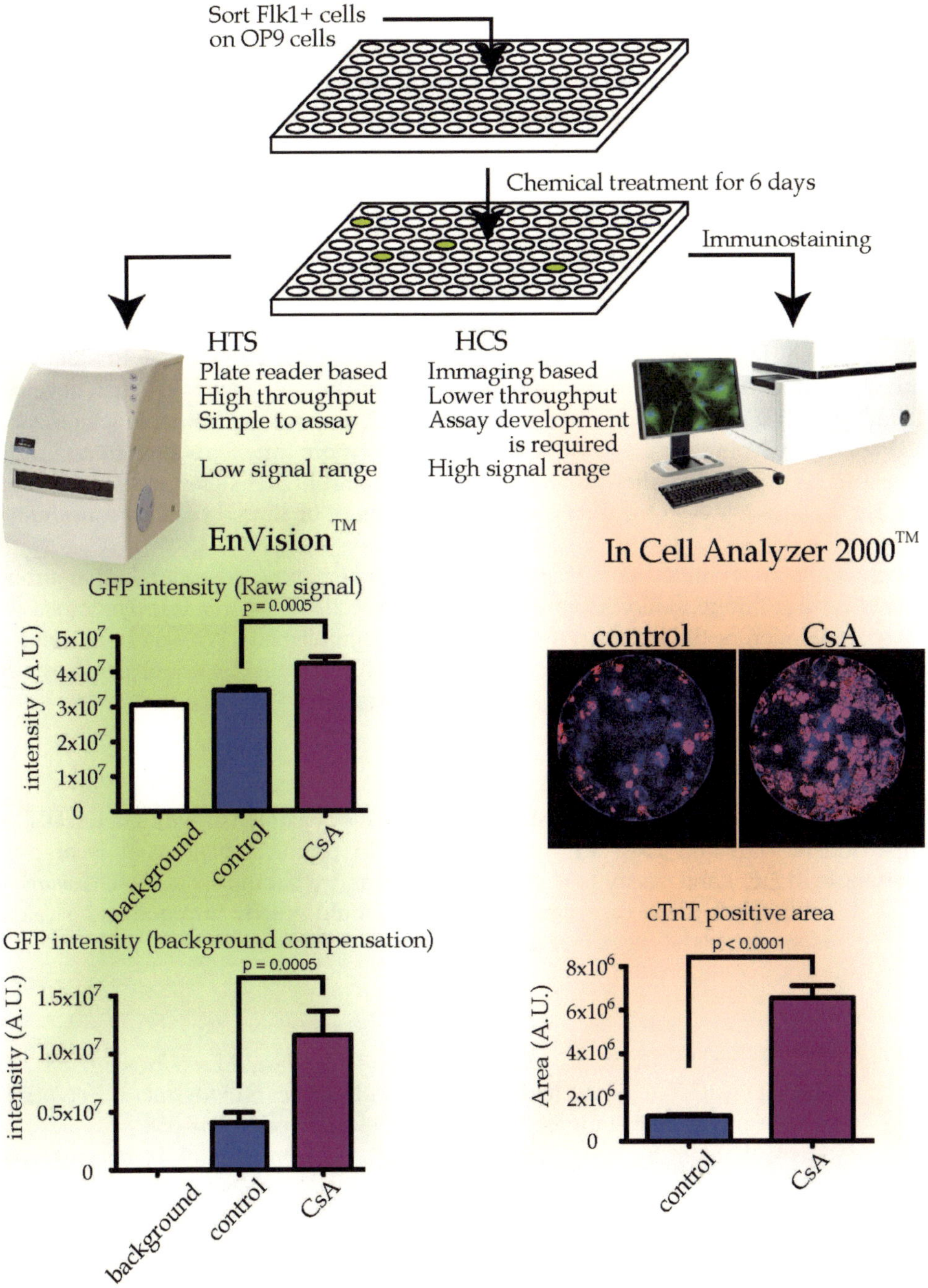

Fig. 5. Application of 2D differentiation system to HTS and HCS. Flk+1 cells are sorted directly on OP9-plated 96 well dish and treated with each chemical from chemical library. After 6 days, hight throughput screening (HTS) or high content screening (HCS) is carried out for measuring cardiomyocyte differentiation. HCS provide clear and efficient results than HTS, which show high background level.

NFAT signal inhibitors - FK506, VIVID (NFAT inhibitor) and cypermethrin. All of NFAT inhibitors failed to induce cardiomyocyte differentiation. On the other hand, NFAT inhibitors strongly inhibited endothelial differentiation. These results indicate that CsA effect on cardiomyocyte induction is through unknown pathway, and CsA effect on endothelial induction is through NFAT signaling.

5. Extension to HTS and HCS

The finding of CsA prompted us to apply our system to HTS and/or HCS (Figure 5). We first developed HTS using αMHC-GFP ESCs to determine cardiac differentiation. We found this system could work but it showed high background signals and low signals to noise ratio because of lower dynamic range of GFP (Willems et al., 2009). Thus, to validate the result we needed to reanalyze using immunostaining or flow cytometry. We then developed HCS using αMHC-GFP and cTnT staining, next. HCS required us some development for imaging and analyzing for the assay and it took much more time for staining and imaging than HTS. As expected, it showed lower background and higher signal to noise ratio than HTS. The balance between the data quality and cost and/or time should be considered to choose HTS or HCS.

Recently we have identified novel cardiogenic effects of natural compounds that strongly promote cardiac differentiation from $Flk1^+$ mesoderm cells using this HTS combined with HCS for validation of the results (Fukushima et al., unpublished). We also have identified some cardiomyocyte proliferating chemicals using HCS to directly count cardiomyocyte number and analyze cell cycle (Uosaki et al., unpublished).

6. Conclusion

Cardiac development *in vivo* and *in vitro*, chemical biology combined with stem cells, and comparison of HTS and HCS are described here. In chemical biology, it is of course important to identify and clarify how chemicals work and what molecular mechanisms are. Indeed it is difficult in some case, chemical biology would greatly improve and accelerate stem cell biology, which has been based on embryonic development.

7. References

Borowiak, M., Maehr, R., Chen, S., Chen, A.E., Tang, W., Fox, J.L., Schreiber, S.L., and Melton, D.A. (2009). Small molecules efficiently direct endodermal differentiation of mouse and human embryonic stem cells. Cell stem cell *4*, 348-358.

Buehr, M., Meek, S., Blair, K., Yang, J., Ure, J., Silva, J., McLay, R., Hall, J., Ying, Q.L., and Smith, A. (2008). Capture of authentic embryonic stem cells from rat blastocysts. Cell *135*, 1287-1298.

Chen, S., Borowiak, M., Fox, J.L., Maehr, R., Osafune, K., Davidow, L., Lam, K., Peng, L.F., Schreiber, S.L., Rubin, L.L., *et al.* (2009). A small molecule that directs differentiation of human ESCs into the pancreatic lineage. Nat Chem Biol *5*, 258-265.

Chen, S., Do, J.T., Zhang, Q., Yao, S., Yan, F., Peters, E.C., Schöler, H.R., Schultz, P.G., and Ding, S. (2006). Self-renewal of embryonic stem cells by a small molecule. Proc Natl Acad Sci U S A *103*, 17266-17271.

Eilken, H.M., Nishikawa, S., and Schroeder, T. (2009). Continuous single-cell imaging of blood generation from haemogenic endothelium. Nature *457*, 896-900.

Foley, A.C., and Mercola, M. (2005). Heart induction by Wnt antagonists depends on the homeodomain transcription factor Hex. Genes Dev *19*, 387-396.

Huangfu, D., Maehr, R., Guo, W., Eijkelenboom, A., Snitow, M., Chen, A.E., and Melton, D.A. (2008). Induction of pluripotent stem cells by defined factors is greatly improved by small-molecule compounds. Nat Biotechnol *26*, 795-797.

Ichida, J.K., Blanchard, J., Lam, K., Son, E.Y., Chung, J.E., Egli, D., Loh, K.M., Carter, A.C., Di Giorgio, F.P., Koszka, K., *et al.* (2009). A small-molecule inhibitor of tgf-Beta signaling replaces sox2 in reprogramming by inducing nanog. Cell stem cell *5*, 491-503.

Kataoka, H., Takakura, N., Nishikawa, S., Tsuchida, K., Kodama, H., Kunisada, T., Risau, W., Kita, T., and Nishikawa, S.I. (1997). Expressions of PDGF receptor alpha, c-Kit and Flk1 genes clustering in mouse chromosome 5 define distinct subsets of nascent mesodermal cells. Dev Growth Differ *39*, 729-740.

Laflamme, M.A., Chen, K.Y., Naumova, A.V., Muskheli, V., Fugate, J.A., Dupras, S.K., Reinecke, H., Xu, C., Hassanipour, M., Police, S., *et al.* (2007). Cardiomyocytes derived from human embryonic stem cells in pro-survival factors enhance function of infarcted rat hearts. Nat Biotechnol *25*, 1015-1024.

Lin, T., Ambasudhan, R., Yuan, X., Li, W., Hilcove, S., Abujarour, R., Lin, X., Hahm, H.S., Hao, E., Hayek, A., *et al.* (2009). A chemical platform for improved induction of human iPSCs. Nat Methods *6*, 805-808.

Mikkelsen, T.S., Hanna, J., Zhang, X., Ku, M., Wernig, M., Schorderet, P., Bernstein, B.E., Jaenisch, R., Lander, E.S., and Meissner, A. (2008). Dissecting direct reprogramming through integrative genomic analysis. Nature *454*, 49-55.

Moretti, A., Caron, L., Nakano, A., Lam, J.T., Bernshausen, A., Chen, Y., Qyang, Y., Bu, L., Sasaki, M., Martin-Puig, S., *et al.* (2006). Multipotent embryonic isl1+ progenitor cells lead to cardiac, smooth muscle, and endothelial cell diversification. Cell *127*, 1151-1165.

Naito, A., Shiojima, I., Akazawa, H., Hidaka, K., Morisaki, T., Kikuchi, A., and Komuro, I. (2006). Developmental stage-specific biphasic roles of Wnt/beta-catenin signaling in cardiomyogenesis and hematopoiesis. Proc Natl Acad Sci U S A *103*, 19812-19817.

Pandur, P., Läsche, M., Eisenberg, L.M., and Kühl, M. (2002). Wnt-11 activation of a non-canonical Wnt signalling pathway is required for cardiogenesis. Nature *418*, 636-641.

Sadek, H., Hannack, B., Choe, E., Wang, J., Latif, S., Garry, M.G., Garry, D.J., Longgood, J., Frantz, D.E., Olson, E.N., *et al.* (2008). Cardiogenic small molecules that enhance myocardial repair by stem cells. Proc Natl Acad Sci U S A *105*, 6063-6068.

Saga, Y., Kitajima. S,, Miyagawa-Tomita. S. (2000). Mesp1 expression is the earliest sign of cardiovascular development. Trends Cardiovasc Med. 10, 345-52.

Takahashi, T., Lord, B., Schulze, P.C., Fryer, R.M., Sarang, S.S., Gullans, S.R., and Lee, R.T. (2003). Ascorbic acid enhances differentiation of embryonic stem cells into cardiac myocytes. Circulation *107*, 1912-1916.

Terami, H., Hidaka, K., Katsumata, T., Iio, A., and Morisaki, T. (2004). Wnt11 facilitates embryonic stem cell differentiation to Nkx2.5-positive cardiomyocytes. Biochem Biophys Res Commun *325*, 968-975.

Ueno, S., Weidinger, G., Osugi, T., Kohn, A.D., Golob, J.L., Pabon, L., Reinecke, H., Moon, R.T., and Murry, C.E. (2007). Biphasic role for Wnt/beta-catenin signaling in cardiac specification in zebrafish and embryonic stem cells. Proc Natl Acad Sci U S A *104*, 9685-9690.

Willems, E., Bushway, P.J., and Mercola, M. (2009). Natural and Synthetic Regulators of Embryonic Stem Cell Cardiogenesis. Pediatric Cardiology.

Wu, S.M., Fujiwara, Y., Cibulsky, S.M., Clapham, D.E., Lien, C.L., Schultheiss, T.M., and Orkin, S.H. (2006). Developmental origin of a bipotential myocardial and smooth muscle cell precursor in the mammalian heart. Cell *127*, 1137-1150.

Wu, X., Ding, S., Ding, Q., Gray, N.S., and Schultz, P.G. (2004). Small molecules that induce cardiomyogenesis in embryonic stem cells. J Am Chem Soc *126*, 1590-1591.

Yamamizu, K., Matsunaga, T., Uosaki, H., Fukushima, H., Katayama, S., Hiraoka-Kanie, M., Mitani, K., and Yamashita, J.K. (2010). Convergence of Notch and beta-catenin signaling induces arterial fate in vascular progenitors. J Cell Biol *189*, 325-338.

Yamashita, J., Itoh, H., Hirashima, M., Ogawa, M., Nishikawa, S., Yurugi, T., Naito, M., and Nakao, K. (2000). Flk1-positive cells derived from embryonic stem cells serve as vascular progenitors. Nature *408*, 92-96.

Yamashita, J.K., Takano, M., Hiraoka-Kanie, M., Shimazu, C., Peishi, Y., Yanagi, K., Nakano, A., Inoue, E., Kita, F., and Nishikawa, S. (2005). Prospective identification of cardiac progenitors by a novel single cell-based cardiomyocyte induction. FASEB J *19*, 1534-1536.

Yan, P., Nagasawa, A., Uosaki, H., Sugimoto, A., Yamamizu, K., Teranishi, M., Matsuda, H., Matsuoka, S., Ikeda, T., Komeda, M., *et al.* (2009). Cyclosporin-A potently induces highly cardiogenic progenitors from embryonic stem cells. Biochem Biophys Res Commun *379*, 115-120.

Yanagi, K., Takano, M., Narazaki, G., Uosaki, H., Hoshino, T., Ishii, T., Misaki, T., and Yamashita, J.K. (2007). Hyperpolarization-activated cyclic nucleotide-gated channels and T-type calcium channels confer automaticity of embryonic stem cell-derived cardiomyocytes. Stem cells *25*, 2712-2719.

Ying, Q.L., Wray, J., Nichols, J., Batlle-Morera, L., Doble, B., Woodgett, J., Cohen, P., and Smith, A. (2008). The ground state of embryonic stem cell self-renewal. Nature *453*, 519-523.

Yuasa, S., Itabashi, Y., Koshimizu, U., Tanaka, T., Sugimura, K., Kinoshita, M., Hattori, F., Fukami, S., Shimazaki, T., Ogawa, S., *et al.* (2005). Transient inhibition of BMP signaling by Noggin induces cardiomyocyte differentiation of mouse embryonic stem cells. Nat Biotechnol *23*, 607-611.

Yurugi-Kobayashi, T., Itoh, H., Schroeder, T., Nakano, A., Narazaki, G., Kita, F., Yanagi, K., Hiraoka-Kanie, M., Inoue, E., Ara, T., *et al.* (2006). Adrenomedullin/cyclic AMP pathway induces Notch activation and differentiation of arterial endothelial cells from vascular progenitors. Arterioscler Thromb Vasc Biol *26*, 1977-1984.

25

Embryonic Stem Cells as a Model System to Elucidate Early Events in Cardiac Specification and Determination

Gabriella Minchiotti[1], Cristina D'Aniello[1], Roberto Ronca[2], Laura Gualandi[2,3] and Patrizia Dell'Era[2]

[1]*Stem Cell fate Laboratory, Institute of Genetics and Biophysic "A. Buzzati-Traverso," Consiglio Nazionale delle Ricerche, Naples*

[2]*Unit of General Pathology and Immunology, School of Medicine, University of Brescia*

[3]*Present address: Department of Genetics and Pathology, Rudbeck Laboratory, Uppsala University, Uppsala,*

[1,2]*Italy*

[3]*Sweden*

1. Introduction

Cardiac development is a highly ordered process that involves several steps whose features are conserved from insects to vertebrates. In the mouse, two distinct mesodermal heart fields with a common origin contribute to heart development in a temporally and spatially specific manner (Buckingham *et al.* 2005). Heart progenitor cells located in the primitive streak migrate anteriorly and spread laterally to form the two paired primary heart fields (FHFs). Mesodermal precursors for heart-forming cells express initially Brachyury T, a T-box transcription factor, and lately, at the precardiac stage, the mesoderm posterior 1 (Mesp1) marker (Solloway and Harvey 2003). Later in development, cardiac precursors coalesce into the linear heart tube and ultimately give rise to the left ventricle of the mature four-chambered mammalian heart. The secondary heart field (SHF) is instead derived from cells of the pharyngeal and the splanchnic mesoderm that will migrate into the developing heart and give rise to the right ventricle, the outflow tract, and portions of the inflow tract (Chien *et al.* 2008). FHF and SHF are characterized by the expression of specific genes: the T-box protein *tbx5* and the first wave of Nkx2.5 appear to be restricted to derivatives of the FHF (Bruneau *et al.* 2001; Zaffran *et al.* 2004; Buckingham *et al.* 2005), whereas SHF is marked by the expression of the LIM domain homeobox gene *isl1*, or the second wave of Nkx2.5 (Kelly *et al.* 2001; Cai *et al.* 2003). Finally cardiomyocyte progenitor cells that contribute to the atrial and ventricular myocardium have been identified also within the epicardium, and are marked by the expression of either Wt1 or Tbx18 (Cai *et al.* 2008; Zhou *et al.* 2008).

Instructions for cardiac commitment derive from neighboring embryonic tissues (Abu-Issa and Kirby 2007). Cell commitment occurs in two stages: specification and determination. While a cell is specified if it differentiates toward the definitive lineage in a neutral environment, it is determined if it differentiates in an antagonistic environment. Cardiac specification factors include activin or Transforming Growth Factor (TGF)-β while bone morphogenetic proteins (BMPs) and FGFs are considered as determination factors (Ladd *et al.* 1998). These inductive signals, summarized in Fig. 1, are released in a precisely timed and spatially regulated fashion (Zaffran and Frasch 2002). Generally, endoderm-derived signals act as inducers of cardiac mesoderm formation, while ectoderm secretes inhibitory factors.

Pro-cardiogenic molecules comprise specification factors, determination factors, which includes Wnt/b-catenin, TGF-ß family, Bone Morphogenetic Proteins (BMPs) and Cripto, Sonic Hedgehog, Crescent (Harvey 2002; Foley and Mercola 2004).

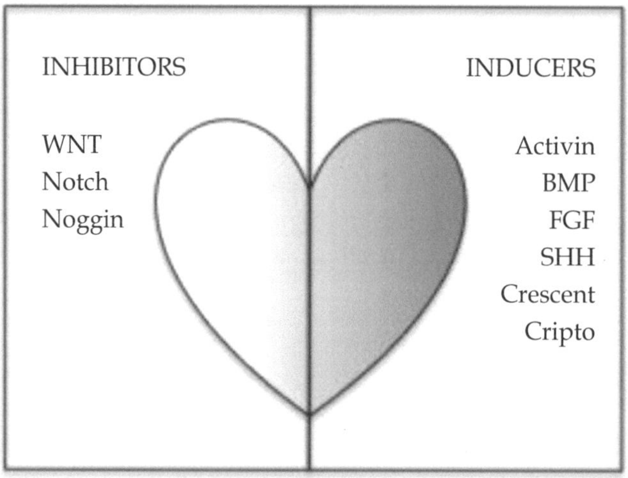

Fig. 1. Schematic diagram of cardiogenic inhibitors and inducers

At variance, myocardial differentiation is inhibited by Notch and by canonical Wnt signal through β-catenin (for a complete overview of the cardiac developmental process see (Kirby 2007). However, the scenario of the heart inducer morphogens is still incomplete, and several hints may come from *in vitro* study of the embryonic stem cell (ESC) differentiation process.

Since ESCs possess the ability to give rise to all cell lineages (Nishikawa *et al.* 2007), they represent a powerful approach to elucidate the origin and the molecular identity of the cardiovascular progenitor populations. Indeed, all the cells belonging to the cardiovascular lineage have been generated in the embryoid bodies (EBs) formed during *in vitro* ESCs differentiation protocol and gene expression analysis suggest that their development in culture recapitulates cardiogenesis in the early embryo (Boheler *et al.* 2002).

ESCs are a promising tool in cellular therapy for the repair of injured myocardium. Heart failure represents a major cause of death and hospitalization in Western countries. Because of the limited capacity of mammalian heart to regenerate, the lonely therapy for CM loss is cardiac transplantation. However, the possibility to replace damaged heart tissue with cells capable of *in situ* differentiation and of myocardial integration is very attractive. Indeed the identification, isolation, and characterization of murine ESC (mESC)-derived cardiac progenitor cells (CPCs) on the basis of Brachyury/Flk1 (Kattman *et al.* 2006), Isl1/Flk1/Nkx2-5 (Moretti *et al.* 2006), cKit/Nkx2-5 (Wu *et al.* 2006), or Nkx2-5 (Christoforou *et al.* 2008) expression, has been recently reported by several groups. These cells represent a promising source for heart repair because of their restricted capacity to differentiate into cardiac muscle, smooth muscle, and vascular endothelium (Kattman *et al.* 2006; Moretti *et al.* 2006; Wu *et al.* 2006; Christoforou *et al.* 2008). Indeed, when injected in infarcted area of murine heart, CPCs engrafted and differentiated into CMs, as well as contributed to neovascularization, thus improving the cardiac function of treated animals (Christoforou *et al.* 2010).

As stated before, the *in vitro* differentiation process of mESC mostly recapitulates the embryonic development and indeed CMs derived from mESC resemble beating cells of the embryonic heart tube (Fijnvandraat *et al.* 2003). By combining several data, the cardiac differentiation process can be traced relying on the expression of molecular markers (Fig. 2). The cardiogenic mesoderm will give rise to a cardiovascular precursor (CVP) population characterized by the expression of Flk1, Nkx2.5, Isl1, and c-kit. From these cells cardiomyocyte progenitors, as well as precursors of endothelial and smooth muscle cell lineages, will originate. Cardiomyocyte progenitors, identified by Nkx2.5, Gata4, and Mef2c expression, will afterwards differentiate into functional mature CMs, characterized by the appearance of specific structural proteins, e.g. myosin heavy chain (MHC)-α, -β, and the ventricular myosin Myl2.

Cardiomyocyte development in ESC differentiation cultures is well established and is easily detected by the appearance of areas of spontaneously contracting cells (beating foci) that display characteristics of mature CMs (Sachinidis *et al.* 2003). Beating cells are absent if the genetical manipulation of ESCs abrogates the expression of a gene involved in CM differentiation. The study of knocked-out ESCs has lead to the discovery of several genes, whose contribution to cardiogenesis could not be assessed in mice because of the early embryonic lethality. Indeed, the absence of EphB4 (Wang *et al.* 2004), Cripto-1 (Xu *et al.* 1998), Shp-2 (Qu and Feng 1998), FGFR1 (Dell'Era *et al.* 2003), Sik1 (Romito *et al.* 2010)., JSAP1 (Sato *et al.* 2005), and the overexpression of a constitutively active Rac (Puceat *et al.* 2003), severely impaired the appearance of beating foci. Due to the complexity of events, the blockade of cardiac differentiation can occur either in different step of CM differentiation or in cell lineages that secrete cardiac specification/determination peptides. Indeed, has been recently shown that the absence of CM differentiation in EphB4$^{-/-}$ mESC can be rescued by EphB4^{+}, CD31^{+} endothelial cells (Chen *et al.* 2010).

In this review we will focus on two selected membrane receptor systems, FGFR1 and Cripto, both of them involved in cardiac mesoderm formation and patterning at different levels, whose contribution to murine early cardiac development has been established by studying mESC differentiation cultures.

2. Cripto

Cripto is the original member of a family of vertebrate signaling molecules, the EGF-CFC family (Ciccodicola *et al.* 1989) which includes human, mouse and chick Cripto; human and mouse Cryptic, Xenopus FRL-1 and Zebrafish OEP (one eyed-pinhead) (Shen and Schier 2000). Initially described as secreted molecules, members of this family are extracellular membrane proteins, anchored to the lipid bilayer through a glycosilphosphatidylinositol (GPI) moiety (Minchiotti *et al.* 2000).

Early studies of *cripto* were focused on its possible role in cell transformation and tumor progression (Salomon *et al.* 1999; Persico *et al.* 2001). *Cripto* expression was first found in human and mouse embryonal carcinoma cells and male teratocarcinomas and was demonstrated to be over-expressed in breast, cervical, ovarian, gastric, lung, colon and pancreatic carcinomas, in contrast to normal tissues where *cripto* expression was invariably absent (Strizzi *et al.* 2005).

In mouse embryos, *cripto* is expressed early in the ICM and the throphoblast cells of the blastocyst (Johnson *et al.* 1994). At 6.5 dpc *cripto* is expressed in the epiblast and at the primitive streak stage in the forming mesoderm. Later on, *cripto* expression is associated with the developing heart structures; its expression is restricted to the myocardium of the developing heart tube at 8.5 dpc and in the outflow region of the heart at 9.5 dpc (Dono *et al.* 1993). This expression pattern suggests that *cripto* may play a role in the early events leading to heart morphogenesis. Mouse embryos deficient for the *cripto* gene die around day 7.5 of embryogenesis due to defects in mesoderm formation and axial organization (Ding *et al.* 1998; Liguori *et al.* 2003). Notably, mouse *cripto* mutants exhibit defects in myocardial development as evidenced by the absence of expression of terminal myocardial differentiation genes such as αMHC and MLC2v (Ding *et al.* 1998; Xu *et al.* 1999a).

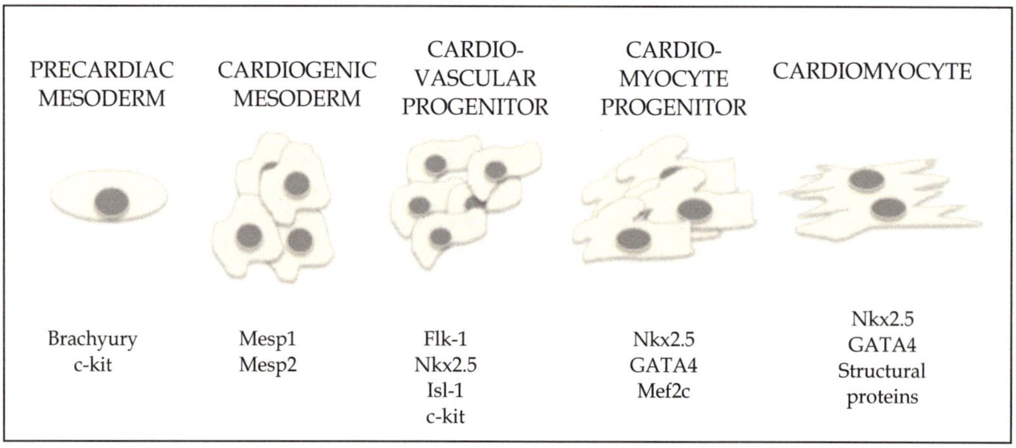

Fig. 2. Schematic illustration of cardiomyocyte development. Adapted from Chen et al., 2008

Given the early lethality of *cripto-/-* mice, ESCs have been a powerful tool to shed light into the functional role of *cripto* in mammalian cardiomiogenesis. Indeed, by using EBs derived from Cripto-/- ES cells, it has been demonstrated that genetic ablation of *cripto* block cardiac differentiation (Xu *et al.* 1998) and that forced expression of wild-type *cripto* fully rescues the cardiac phenotype, thus providing experimental evidence of a functional role of this gene in mammalian cardiomyogenesis. Interestingly, a structure function analysis using different deletion mutant derivatives of *cripto* cDNAs showed that i) a secreted form of Cripto, which lacks the C-terminus, is capable of retaining its biological activity and efficiently induces cardiogenesis and ii) the EGF-CFC domain represents the minimal functional domain, which is sufficient per sè to restore cardiac differentiation of Cripto-/- ES cells (Parisi *et al.* 2003). Worth noting, the EGF-like domain, which was previously shown to be mitogenic on mammary cell lines (Salomon *et al.* 1999), although essential, it is not sufficient per sè to promote cardiac differentiation, suggesting that there may be divergent Cripto signaling pathways depending both on different domains of the protein and/or on specific cell types, an issue that still remains undefined.

Another important issue was the timing of initiation, the strength and duration of Cripto signaling in the commitment of ESCs to a cardiac fate, which was experimentally addressed by using a recombinant secreted Cripto protein. Infact, kinetic experiments performed by adding recombinant Cripto to the culture medium of Cripto-/- ES cells indicate that Cripto is required in a precise moment during differentiation after which it fails to specify the cardiac lineage. Moreover, a transient presence of Cripto protein is inefficient and a sustained Cripto signaling is strictly required to promote cardiogenesis; thus providing evidence that timing and strength of the signaling are critical parameters for correct specification and differentiation of the cardiac lineage (Parisi *et al.* 2003).

Both the functional data and the expression profile of Cripto, which is expressed early during cardiac differentiation of ESCs (i.e., day 0-4) while is absent at stages where contracting cardiomyocites appear, indicate that Cripto is required for cardiac commitment of ESCs rather than for terminal differentiation of cardiomyocytes in culture.

Intriguingly, disruption of *cripto* leads to spontaneous neuronal differentiation of ESCs in the presence of serum, and in the absence of either specific inducers or defined culture conditions. Again, kinetic experiments indicate that the timing of Cripto signaling required for priming ESCs to cardiomyocytes resembles the competence window those cells to acquire a neural fate (Parisi *et al.* 2003). Indeed, addition of effective doses of Cripto protein to Cripto$^{-/-}$ ESCs in the 0-2 day interval of differentiation rescues the cardiac phenotype and results in a dramatic inhibition of neural differentiation. Conversely, addition of recombinant Cripto at later time points (i.e 3-6 day interval) results in progressive impairment of cardiac differentiation, and in increased competence of the cells to acquire a neural fate; thus suggesting that different timing of Cripto signaling induces different fates in ESCs.

2.1 Cripto/Smad2 signaling pathway in cardiomyogenesis

Cripto is involved in the modulation of several signaling pathways in development and tumorigenesis (Strizzi *et al.* 2005). Genetic studies and cell-based assays provide evidence for a role of Cripto and, more generally of the EGF-CFC factors, in the activation of the TGFβ-family member Nodal or related ligands GDF1 and -3 (Chen *et al.* 2006; Tanaka *et al.* 2007) through Activin type IB (ALK-4 and Alk7) and Activin type IIB serine/threonine kinase (ActRIIB) receptors (Reissmann *et al.* 2001). Upon receptor activation, the intracellular kinase domain of the type I receptor phosphorylates Smad2 and/or Smad3, which form a hexameric complex with the common Smad4, and translocate into the nucleus to regulate gene expression in conjunction with other transcription factors, such as FoxH1 (Massague and Chen 2000; Adkins *et al.* 2003; Gray *et al.* 2003; Harrison *et al.* 2005).

Notably, acute stimulation of Cripto$^{-/-}$ ESCs with recombinant Cripto protein rapidly induces Smad2 phosphorylation; thus although competent in activating Smad2, transient stimulation with Cripto is insufficient to achieve proper terminal cardiac differentiation, again highlighting the importance of Cripto signalling duration for cardiomyogenesis (Parisi *et al.* 2003).

Intracellular activation of Smad2 upon stimulation with Cripto, requires assembly of an active activin type I (ALK4) and type II receptor complex. In fact, forced overexpression of the activated forms of ALK-4 receptor are able to compensate for the lack of Cripto in cardiac differentiation. Moreover, loss-of-function experiments performed using Nodal antagonist, Cerberus-S, provide direct evidence that the TGF-β family member Nodal is required to support Cripto-regulated cardiac induction and differentiation in ES cells. Infact, addition of Cerberus Short protein, which specifically blocks Nodal by direct binding to the ligand (Piccolo *et al.* 1999) results in a strong inhibition of Cripto activity in ESCs (Parisi *et al.* 2003).

Besides the above mentioned data, several other line of evidence support the idea that temporal and spatial regulation of the Smad pathway is important for normal cardiac development from initial cardiomyocyte differentiation to terminal cardiac morphogenesis in pluripotent cells. In fact, data on P19 cells indicate that the Smad pathway is indispensable for normal cardiomyocyte differentiation (Monzen *et al.* 2001). Moreover, more recent data pointed for a key role of Nodal/Cripto for the early activation of Smad2, which was indispensable for mesendodermal induction and the subsequent cardiac differentiation of ESCs (Kitamura *et al.* 2007).

2.2 Downstream targets of Cripto signaling in cardiomyogenesis

It is now well accepted that Cripto/Smad2 is one of the key signaling pathway which regulates cardiac specification in mammals; however, little is yet known about the mechanisms of action and the identity of the factors downstream of this pathway in mammalian cardiomyogenesis. Very recently, two genes, the Angiotensin Type-I Like Receptor (AGTRL-1/APJ/*msr1*) and its ligand apelin, have been identified as previously undescribed downstream targets of Cripto-Smad2 pathway in cardiogenesis (D'Aniello *et al.* 2009).

Apj was identified and characterized in 1993 as a seven transmembrane receptor associated with G-proteins (O'Dowd *et al.* 1993); it shows a high sequence homology (30%) with angiotensin II type 1 receptor (AT-1), although APJ does not bind angiotensin II (AngII). Apj was kept "orphan" until 1998 when Tatemoto et al. identified Apelin as its selective endogenous ligand. Apelin is a prepropeptide of 77 aminoacids but its biological activity resides in the C-terminus (apelin-36 and apelin-13) (Tatemoto *et al.* 1998).

In the adult, Apelin and Apj are abundantly expressed in the heart, the central nervous system (CNS) and the lungs (Kawamata *et al.* 2001; Medhurst *et al.* 2003). The wide distribution of Apj and apelin in several organs correlates with multifunctional activities, such as the regulation of gastrointestinal and immune functions, the modulation of the hypothalamus-hypophysis axis activity and the regulation of vascular tone, cardiac contractile function and fluid balance (Kleinz and Davenport 2005).

Most remarkably, growing evidence indicate that apelin and APJ play an important role in cardiac development both in *Xenopus* (Inui *et al.* 2006; Cox *et al.* 2006) and in Zebrafish (Scott *et al.* 2007; Zeng *et al.* 2007). Indeed, under or over expression of apelin-APJ signalling results in a reduction in cardiomyocyte numbers and abnormal cardiac morphology (Scott *et al.* 2007; Zeng *et al.* 2007).

Notably, the expression of Apj and apelin i) correlates with that of *cripto* both in ESC differentiation and in gastrulating embryos and ii) is regulated by Cripto/Smad2 pathway in ESCs. Indeed, Apj and apelin expression is dramatically reduced in the absence of *cripto* (D'Aniello *et al.* 2009) both in ESC cardiac differentiation and *in vivo*. Most remarkably, APJ overexpression is capable of redirecting the neuronal fate of *cripto* knockout ESCs, restoring mesendodermal patterning and the cardiogenic program, although it fails to induce beating EBs. Finally, both apelin and Apj silencing blocks cardiac differentiation, thus pointing out a central role of APJ/Apelin in the gene regulatory cascade promoting cardiac specification and differentiation in ESCs (D'Aniello *et al.* 2009).

One of the major transduction pathways activated by Apelin depends on the interaction with a G_i-protein coupled to the Apj receptor with the subsequent interaction with the protein Kinase C (PKC) (Masri *et al.* 2004). Morever, apelin activates the phosphorylation of the intracellular kinase p70S6K through a pertussis toxin (PTX) sensitive G protein (Masri *et al.* 2002). Data on primary endothelial cells indicate that stimulation of p70S6K by Apelin depends on two mechanisms, which are either ERK or PI3K/Akt-dependent (Masri *et al.*

2004). In line with these findings, Apelin induces mammalian cardiomyogenesis via a PTX-sensitive GTP binding protein associated to Apj receptor, through the activation of ERK-dependent p70S6K signaling pathway (D'Aniello *et al.* 2009).

3. FGFR1

The FGF/FGFR system has been implicated in a variety of physiological and pathological conditions, including embryonic development, tissue growth and remodeling, inflammation, tumor growth and vascularization (Powers *et al.* 2000; Presta *et al.* 2005). FGFR1 is one of the four member of the FGFR family, whose amino acid sequence is highly conserved between members and throughout evolution (Itoh and Ornitz 2004). FGFRs differ from one another in their ligand affinities and tissue distribution. A full-length representative protein consists of an extracellular region, composed of three immunoglobulin-like domains (D1-D3), a single hydrophobic transmembrane region and an intracellular tyrosine kinase (TK) domain (Beenken and Mohammadi 2009). The ligand binding site for FGFs is located in the D2-D3 domains and the linker that connects them (Plotnikov *et al.* 2000), whereas the D1 domain is involved in receptor autoinhibition (Olsen *et al.* 2004). Ligand specificity is achieved primarily through splicing events in which the alternative exons IIIb and IIIc encode the carboxyl terminal portion of the third Ig-like loop. Indeed, alternative splicing of FGFR1 results in isoforms FGFR1-IIIb and FGFR1-IIIc with distinct FGF binding characteristics: FGFR1-IIIb binds efficiently to FGF1, FGF3 and FGF 10, whereas FGFR1-IIIc binds to FGF1, FGF2, FGF4, FGF6, FGF8 and FGF9 (Ornitz *et al.* 1996). A variety of other alternative spliced receptor molecules have been described, including the β isoforms that lack the first Ig-like domain whereas the α isoforms identify the full-length receptors (Wang *et al.* 1995).

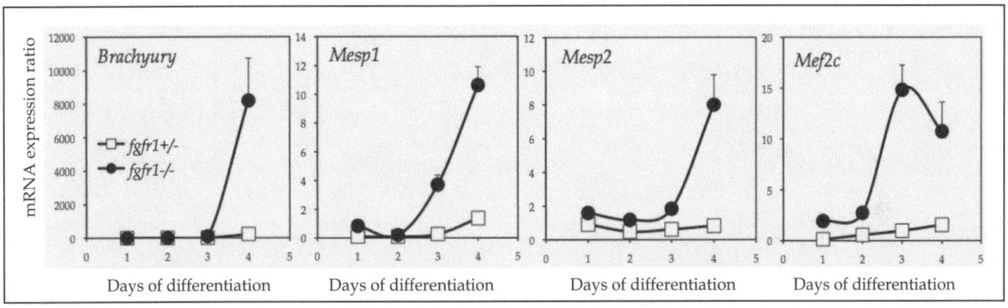

Fig. 3. Mesodermal marker expression by qPCR analysis in *fgfr1*$^{+/-}$ and *fgfr1*$^{-/-}$ EBs

FGFR1 signaling plays important functions in mesoderm formation and development (Xu *et al.* 1999b). Heterozygous animals develop normally but *fgfr1*$^{-/-}$ mice dye during gastrulation, displaying defective mesoderm patterning with reduction in the amount of paraxial mesoderm and lack of somite formation (Deng *et al.* 1994; Yamaguchi *et al.* 1994). Studies on chimeric embryos using FGFR1-deficient mESC revealed an early defect in the mesodermal and endodermal cell movement through the primitive streak, followed by deficiencies in contributing to anterior mesoderm, including heart tissue (Ciruna *et al.* 1997; Deng *et al.* 1997). We analyzed early mesodermal marker expression during the first days of mESC differentiation and indeed, *Brachyury, Mesp1, Mesp2,* and *Mef2c* genes are strongly upregulated in *fgfr1*$^{-/-}$ compared to *fgfr1*$^{+/-}$ mESC (Fig. 3). Although we still don't know the significance of this upregulation, we can confirm a bias in mesodermal lineage development also in the mESC differentiation model.

The pivotal contribution of FGF signaling in heart formation has been demonstrated in different animal models: in *C. intestinalis*, FGF signaling delineates the cardiac progenitor field (Davidson 2007); in *Drosophila*, mesoderm spreading depends upon the expression of *heartless*, homologous to vertebrate *fgfr1* (Beiman *et al.* 1996; Gisselbrecht *et al.* 1996), and *heartless* mutant is characterized by the absence of the heart (Frasch 1995; Beiman *et al.* 1996); in chicken, FGF signaling activated by FGF8 contributes to the heart-inducing properties of the endoderm (Alsan and Schultheiss 2002); in zebrafish, induction and differentiation of the heart requires FGF8 (Reifers *et al.* 2000); in mice, *Fgf8*$^{neo/-}$ mutants show complex cardiac defects (Abu-Issa *et al.* 2002).

FGFR1 has been implicated in cardiac development also during mESC differentiation. When *fgfr1*$^{-/-}$ mESC were differentiated with the "hanging drop" protocol, no beating foci were seen in EB cultures within the first fourteen days, whereas contracting areas were observed microscopically in more than 90% of the heterozygous EBs at day 8th. To verify the morphological data, total RNA was extracted at different time point of differentiation, and then subjected to retrotranscription, and semiquantitative polymerase chain reaction (PCR) for cardiac marker expression. In parallel, EBs were fixed, paraffin included, cut in 7μM sections, and analyzed by immunofluorescence for the presence of the structural protein MHC-α. Both methods confirmed the presence of cardiac markers lonely in *fgfr1*$^{+/-}$ mESC, thus indicating that the beating areas indeed correspond to CMs, whose differentiation depends upon FGFR1 expression (Dell'Era *et al.* 2003).

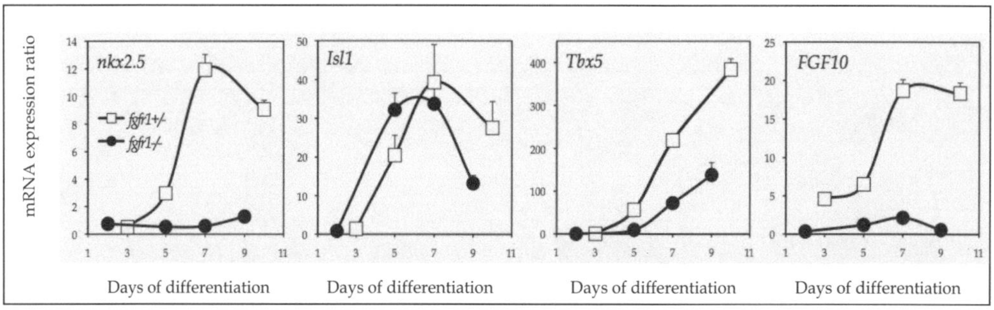

Fig. 4. Heart field marker expression by qPCR analysis in *fgfr1*$^{+/-}$ and *fgfr1*$^{-/-}$ EBs

Nkx2.5 is considered the earliest cardiac transcription factor because of its expression in cardiovascular precursors and the mRNA analysis of *fgfr1*$^{-/-}$ EBs showed that the absence of FGFR1 does not allow Nkx2.5 upregulation (Dell'Era *et al.* 2003; Ronca *et al.* 2009). This result suggest that *fgfr1*$^{-/-}$ EBs cannot make the transition to develop cardiac lineage from cardiogenic mesoderm, and indeed, Mesp1, Mesp2, and Mef2c accumulation suggests a flooding of mesodermal precursors. As mentioned, Nkx2.5^{+} CVP can give rise to CMs, endothelial, and smooth muscle cells (Chen *et al.* 2008). However, CD31 and α-smooth muscle actin immunostaining of EBs showed that differentiation of endothelial and smooth muscle cells is not affected by the lack of FGFR1 (Magnusson *et al.* 2004). Indeed, the vascular plexus in *fgfr1*$^{-/-}$ EBs is more abundant to that observed in heterozygous EBs, thus suggesting that the cells that cannot become CMs are forced toward parallel lineages, such as endothelial cells. FGFR1 was first isolated by a endothelial cell cDNA library due to its homology with the tyrosine kinase receptor Fms (Dionne *et al.* 1990), and both the prototype FGFs (FGF1 and FGF2) are considered as angiogenic growth factors (Presta *et al.* 2005). Then, it was really surprising to realize that the receptor was not really involved in endothelial development. However, as suggested by immunostaining of EBs with an antibody specific for the activated receptor, FGFR-1 is phosphorylated in a subpopulation of proliferating endothelial cells (Magnusson *et al.* 2005), thus confirming its role in endothelial cell proliferation rather than differentiation.

It has previously shown that FGFR-1 is required by epicardium-derived cells for myocardial invasion (Pennisi and Mikawa 2009), and that mature CMs expressing FGFR1 proliferate upon receptor stimulation (Seyed and Dimario 2008). Then, the open question if FGFR1 is expressed by cardiogenic precursors and/or by other mesodermal cells and why is it needed during mESC CM differentiation is still controversial. To add a little piece to the puzzle, we analyzed FHF and SHF markers in *fgfr1*$^{-/-}$ EBs and the results are reported in Fig. 4.

When compared with heterozygous EBs, both *nkx2.5* and *FGF10* genes do not show any upregulation during *fgfr1*$^{-/-}$ mESC differentiation, while a smaller increase can be seen for Tbx5; at variance, the SHF marker Isl1 seem to be comparable between the two populations, thus suggesting that the absence of FGFR1 results in a bias in primary heart field development. It should be pointed out that one of the limitation of the mESC model is that it is impossible to dissect anatomical structures of the EBs; this fact imply that cells anatomically distant can be found in close proximity, thus leading to artificial paracrine stimulations affecting the "normal" development. Taking together, our data suggest that FGFR1, probably present on cardiomyocyte precursor cell surface, may mediate cardiomyocyte differentiation by activating Nkx2.5 in Mef2c-cardiogenic mesodermal cells. Indeed, in other models FGF signaling has been shown to be sufficient to induce cardiac transcription factor expression: FGF8-soaked beads induce *nkx2.5* in chick (Alsan and Schultheiss 2002) and zebrafish (Reifers *et al.* 2000), whereas an ectopic FGF signaling results in a surplus of CMs prior to terminal differentiation (Marques *et al.* 2008).

The interaction of FGFR1 with a ligand leads to a cascade of downstream signals, ultimately influencing mitogenesis and differentiation. Seven tyrosine residues become phosphorylated in human FGFR1 (hFGFR1): Y653/654 are critical for TK activity (Mohammadi *et al.* 1996), Y463 is involved in endothelial cell proliferation by binding to Src homology (SH)2/SH3 domain-containing adaptor protein Crk (Larsson *et al.* 1999), and phosphorylated Y766 has been shown to bind phospholipase C-γ(PLC- γ) in L6 myoblasts, Shb in endothelial cells, and Grb14 in MDA-MB-231 human breast cancer cells (Mohammadi *et al.* 1991; Cross *et al.* 2002; Cailliau *et al.* 2005). Also, FGFR1 activation leads to FRS2 phosphorylation (Kouhara *et al.* 1997) followed by Grb2 and Shp-2 interactions (Hadari *et al.* 1998). Frs2, Crk, and Shb binding to FGFR1 affect the classical Ras/Raf-1/MEK/ERK/Jun proliferation pathway activated by TK receptors, while PLCγ1 activates PKC (Hug and Sarre 1993), whose role in CM differentiation has been demonstrated (Zhou *et al.* 2003).

Previous observation had shown that the FGFR1 TK inhibitor SU 5402 (Mohammadi *et al.* 1997), the $MEK_{1/2}$ inhibitor U0126 (Favata *et al.* 1998), and the classical/novel protein kinase C (PKC) inhibitor GF109203X (Kuchera *et al.* 1993) were all able to hamper beating foci formation in EBs originated by *fgfr1*$^{+/-}$ mESC (Dell'Era *et al.* 2003). In order to define the requirements for FGFR1 signaling in CM development, we transduced *fgfr1*$^{-/-}$ mESC *via* a lentiviral vector system with the IIIc isoform of either wt receptor (v-wt-hFGFR1), or hFGFR1 mutants in different tyrosine autophosphorylation sites: the tyrosine kinase defective (TK-) Y653/654F-hFGFR1 mutant, and the two TK+ Y463F-, and Y766F-h-FGFR1. Resulting cell lines were analyzed to confirm that the receptors were correctly exposed on cell surface, by ^{125}I-FGF2 binding assay and, at least for TK+ receptors, were able to upregulate the downstream signaling molecule $ERK_{1/2}$. Then, we evaluated the presence of cardiomyocyte during the differentiation process by looking microscopically to the appearance of beating foci, and by expression analysis of both early and late cardiac markers. We first observed that transduction of the human receptor molecule fully reconstitutes cardiomyocyte differentiation, while, in agreement with previous data obtained by using pharmacological inhibitors, TK- receptor does not. The analysis of Y463F and Y766F mutant EBs demonstrate that v-Y766F-hFGFR1 ES cells are able to support cardiomyocyte differentiation in a manner undistinguishable from v-wt-hFGFR1 ES cells and *fgfr1*$^{+/-}$ ES cells. In contrast, transduction of *fgfr1*$^{-/-}$ ES cells with the Y463F-hFGFR1 mutant results in a loss of rescue of cardiomyocyte formation, as assessed by the absence of beating foci and early and late cardiomyocyte markers in the corresponding EBs (Ronca *et al.* 2009). At present, the signaling cascade triggered by the autophosphorylation of Y463 in FGFR1 and its cross-talk with PKC- and ERK-mediated signaling during cardiomyocyte differentiation of mESC remains to be elucidated.

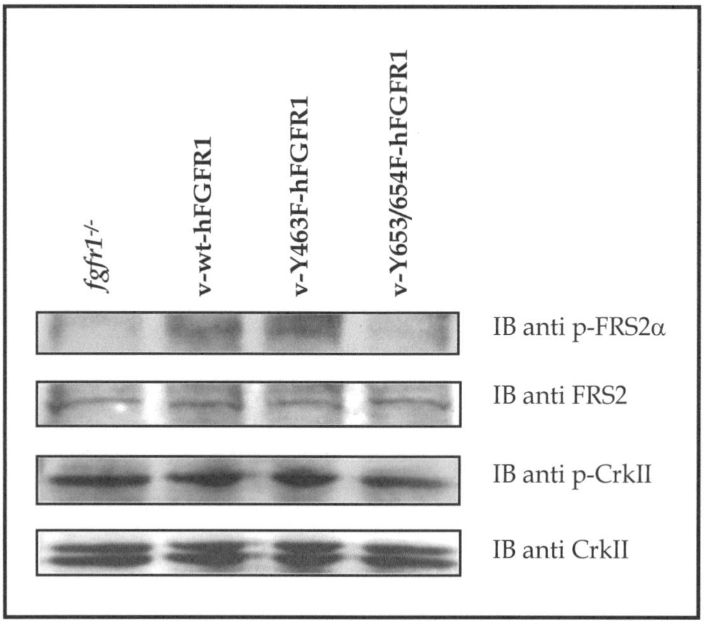

fgfr1$^{-/-}$, v-wt-hFGFR1, v-Y463F-hFGFR1, and v-Y653/654F-hFGFR1 ES cells were subjected to standard differentiation protocol. At day 9th of differentiation, cell extracts were analyzed for activated signaling molecules.

Fig. 4. Characterization of FGFR1 signalling during ES differentiation.

Indeed, Western blot analysis of total EB protein extract at day 9 of differentiation does not show any difference in CrkII phosphorylation levels among *fgfr1*$^{-/-}$, v-wt-hFGFR1, v-Y463F-hFGFR1 and v-Y653/654F-hFGFR1 EBs, thus indicating that the overall activation of these signaling molecules in differentiating EBs is not restricted to FGFR1 activity. Also, FRS2 appears to be phosphorylated in both v-Y463F-hFGFR1 and v-wt-hFGFR1 EBs but not in TK-v-Y653/654F-hFGFR1 and *fgfr1*$^{-/-}$ EBs, indicating that its activation depends on the TK activity of FGFR1 but not on Y463 phosphorylation. Again, only the isolation of the different cardiomyocyte progenitor populations from these EBs will allow the identification of the FGFR1-dependent signalling pathway(s) involved in cardiomyocyte differentiation in mESC.

4. Conclusions

Although ESCs represent a viable source of specific cellular subtypes for drug discovery and transplantation, the successful use of ES-derived donor cells would require the generation of essentially pure cultures of specific cell types. Although further analysis will be required to get insight into the regulatory networks that involve Cripto/APJ and FGFR1 in cardiac differentiation (Figure 5), these membrane receptors may represent valuable targets for the generation of pure cardiomyocyte populations from ESCs.

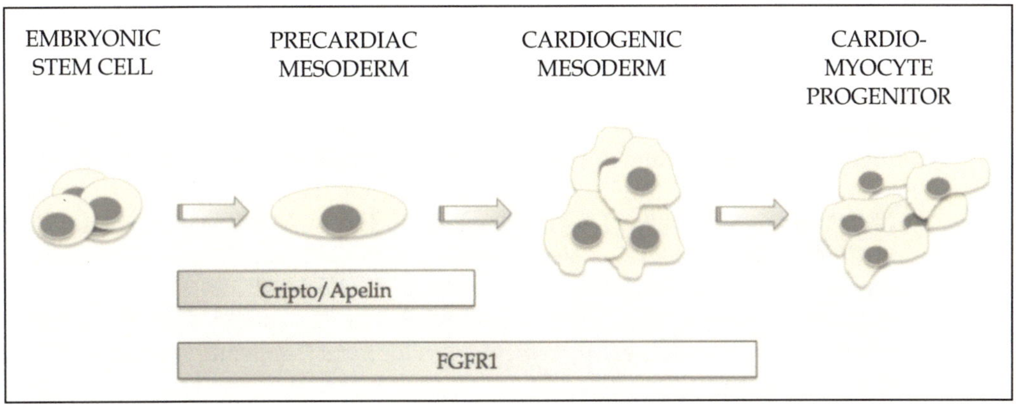

Fig. 5. Cripto/Apelin and FGFR1 influence the early phases of cardiomyocyte differentiation

5. References

Abu-Issa, R. and M. L. Kirby (2007) Heart field: from mesoderm to heart tube. *Annu Rev Cell Dev Biol.* 23, 45-68.

Abu-Issa, R., G. Smyth, I. Smoak, K. Yamamura and E. N. Meyers (2002) Fgf8 is required for pharyngeal arch and cardiovascular development in the mouse. *Development.* 129, 4613-25.

Adkins, H. B., C. Bianco, S. G. Schiffer, P. Rayhorn, M. Zafari, A. E. Cheung, O. Orozco, D. Olson, A. De Luca, L. L. Chen, K. Miatkowski, C. Benjamin, N. Normanno, K. P. Williams, M. Jarpe, D. LePage, D. Salomon and M. Sanicola (2003) Antibody blockade of the Cripto CFC domain suppresses tumor cell growth in vivo. *J Clin Invest.* 112, 575-87.

Alsan, B. H. and T. M. Schultheiss (2002) Regulation of avian cardiogenesis by Fgf8 signaling. *Development.* 129, 1935-43.

Beenken, A. and M. Mohammadi (2009) The FGF family: biology, pathophysiology and therapy. *Nat Rev Drug Discov.* 8, 235-53.

Beiman, M., B. Z. Shilo and T. Volk (1996) Heartless, a Drosophila FGF receptor homolog, is essential for cell migration and establishment of several mesodermal lineages. *Genes Dev.* 10, 2993-3002.

Boheler, K. R., J. Czyz, D. Tweedie, H. T. Yang, S. V. Anisimov and A. M. Wobus (2002) Differentiation of pluripotent embryonic stem cells into cardiomyocytes. *Circ Res.* 91, 189-201.

Bruneau, B. G., G. Nemer, J. P. Schmitt, F. Charron, L. Robitaille, S. Caron, D. A. Conner, M. Gessler, M. Nemer, C. E. Seidman and J. G. Seidman (2001) A murine model of Holt-Oram syndrome defines roles of the T-box transcription factor Tbx5 in cardiogenesis and disease. *Cell.* 106, 709-21.

Buckingham, M., S. Meilhac and S. Zaffran (2005) Building the mammalian heart from two sources of myocardial cells. *Nat Rev Genet.* 6, 826-35.

Cai, C. L., X. Liang, Y. Shi, P. H. Chu, S. L. Pfaff, J. Chen and S. Evans (2003) Isl1 identifies a cardiac progenitor population that proliferates prior to differentiation and contributes a majority of cells to the heart. *Dev Cell.* 5, 877-89.

Cai, C. L., J. C. Martin, Y. Sun, L. Cui, L. Wang, K. Ouyang, L. Yang, L. Bu, X. Liang, X. Zhang, W. B. Stallcup, C. P. Denton, A. McCulloch, J. Chen and S. M. Evans (2008) A myocardial lineage derives from Tbx18 epicardial cells. *Nature.* 454, 104-8.

Cailliau, K., D. Perdereau, A. Lescuyer, H. Chen, C. Garbay, J. P. Vilain, A. F. Burnol and E. Browaeys-Poly (2005) FGF receptor phosphotyrosine 766 is a target for Grb14 to inhibit MDA-MB-231 human breast cancer cell signaling. *Anticancer Res.* 25, 3877-82.

Chen, C., S. M. Ware, A. Sato, D. E. Houston-Hawkins, R. Habas, M. M. Matzuk, M. M. Shen and C. W. Brown (2006) The Vg1-related protein Gdf3 acts in a Nodal signaling pathway in the pre-gastrulation mouse embryo. *Development.* 133, 319-29.

Chen, K., H. Bai, M. Arzigian, Y. X. Chen, J. Bao, W. S. Wu, L. Wu and Z. Z. Wang (2010) Endothelial cells regulate cardiomyocyte development from embryonic stem cells. *J Cell Biochem.* 111, 29-39.

Chen, K., L. Wu and Z. Z. Wang (2008) Extrinsic regulation of cardiomyocyte differentiation of embryonic stem cells. *J Cell Biochem.* 104, 119-28.

Chien, K. R., I. J. Domian and K. K. Parker (2008) Cardiogenesis and the complex biology of regenerative cardiovascular medicine. *Science.* 322, 1494-7.

Christoforou, N., R. A. Miller, C. M. Hill, C. C. Jie, A. S. McCallion and J. D. Gearhart (2008) Mouse ES cell-derived cardiac precursor cells are multipotent and facilitate identification of novel cardiac genes. *J Clin Invest.* 118, 894-903.

Christoforou, N., B. N. Oskouei, P. Esteso, C. M. Hill, J. M. Zimmet, W. Bian, N. Bursac, K.W. Leong, J. M. Hare and J. D. Gearhart (2010) Implantation of mouse embryonic stem cell-derived cardiac progenitor cells preserves function of infarcted murine hearts. *PLoS One.* 5, e11536.

Ciccodicola, A., R. Dono, S. Obici, A. Simeone, M. Zollo and M. G. Persico (1989) Molecular characterization of a gene of the 'EGF family' expressed in undifferentiated human NTERA2 teratocarcinoma cells. *EMBO J.* 8, 1987-91.

Ciruna, B. G., L. Schwartz, K. Harpal, T. P. Yamaguchi and J. Rossant (1997) Chimeric analysis of fibroblast growth factor receptor-1 (Fgfr1) function: a role for FGFR1 in morphogenetic movement through the primitive streak. *Development.* 124, 2829-41.

Cox, C. M., S. L. D'Agostino, M. K. Miller, R. L. Heimark and P. A. Krieg (2006) Apelin, the ligand for the endothelial G-protein-coupled receptor, APJ, is a potent angiogenic factor required for normal vascular development of the frog embryo. *Dev Biol.* 296, 177-89.

Cross, M. J., L. Lu, P. Magnusson, D. Nyqvist, K. Holmqvist, M. Welsh and L. Claesson-Welsh (2002) The Shb adaptor protein binds to tyrosine 766 in the FGFR-1 and regulates the Ras/MEK/MAPK pathway via FRS2 phosphorylation in endothelial cells. *Mol Biol Cell.* 13, 2881-93.

D'Aniello, C., E. Lonardo, S. Iaconis, O. Guardiola, A. M. Liguoro, G. L. Liguori, M. Autiero, P. Carmeliet and G. Minchiotti (2009) G protein-coupled receptor APJ and its ligand apelin act downstream of Cripto to specify embryonic stem cells toward the cardiac lineage through extracellular signal-regulated kinase/p70S6 kinase signaling pathway. *Circ Res.* 105, 231-8.

Davidson, B. (2007) Ciona intestinalis as a model for cardiac development. *Semin Cell Dev Biol.* 18, 16-26.

Dell'Era, P., R. Ronca, L. Coco, S. Nicoli, M. Metra and M. Presta (2003) Fibroblast growth factor receptor-1 is essential for in vitro cardiomyocyte development. *Circ Res.* 93, 414-20.

Deng, C., M. Bedford, C. Li, X. Xu, X. Yang, J. Dunmore and P. Leder (1997) Fibroblast growth factor receptor-1 (FGFR-1) is essential for normal neural tube and limb development. *Dev Biol.* 185, 42-54.

Deng, C. X., A. Wynshaw-Boris, M. M. Shen, C. Daugherty, D. M. Ornitz and P. Leder (1994) Murine FGFR-1 is required for early postimplantation growth and axial organization. *Genes Dev.* 8, 3045-57.

Ding, J., L. Yang, Y. T. Yan, A. Chen, N. Desai, A. Wynshaw-Boris and M. M. Shen (1998) Cripto is required for correct orientation of the anterior-posterior axis in the mouse embryo. *Nature.* 395, 702-7.

Dionne, C. A., G. Crumley, F. Bellot, J. M. Kaplow, G. Searfoss, M. Ruta, W. H. Burgess, M. Jaye and J. Schlessinger (1990) Cloning and expression of two distinct high-affinity receptors cross-reacting with acidic and basic fibroblast growth factors. *Embo J.* 9, 2685-92.

Dono, R., L. Scalera, F. Pacifico, D. Acampora, M. G. Persico and A. Simeone (1993) The murine cripto gene: expression during mesoderm induction and early heart morphogenesis. *Development.* 118, 1157-68.

Evans, M. J. and M. H. Kaufman (1981) Establishment in culture of pluripotential cells from mouse embryos. *Nature.* 292, 154-6.

Favata, M. F., K. Y. Horiuchi, E. J. Manos, A. J. Daulerio, D. A. Stradley, W. S. Feeser, D. E. Van Dyk, W. J. Pitts, R. A. Earl, F. Hobbs, R. A. Copeland, R. L. Magolda, P. A. Scherle and J. M. Trzaskos (1998) Identification of a novel inhibitor of mitogen-activated protein kinase kinase. *J Biol Chem.* 273, 18623-32.

Fijnvandraat, A. C., A. C. van Ginneken, P. A. de Boer, J. M. Ruijter, V. M. Christoffels, A. F. Moorman and R. H. Lekanne Deprez (2003) Cardiomyocytes derived from embryonic stem cells resemble cardiomyocytes of the embryonic heart tube. *Cardiovasc Res.* 58, 399-409.

Foley, A. and M. Mercola (2004) Heart induction: embryology to cardiomyocyte regeneration. *Trends Cardiovasc Med.* 14, 121-5.

Frasch, M. (1995) Induction of visceral and cardiac mesoderm by ectodermal Dpp in the early Drosophila embryo. *Nature*. 374, 464-7.

Gisselbrecht, S., J. B. Skeath, C. Q. Doe and A. M. Michelson (1996) heartless encodes a fibroblast growth factor receptor (DFR1/DFGF-R2) involved in the directional migration of early mesodermal cells in the Drosophila embryo. *Genes Dev*. 10, 3003-17.

Gray, P. C., C. A. Harrison and W. Vale (2003) Cripto forms a complex with activin and type II activin receptors and can block activin signaling. *Proc Natl Acad Sci U S A*. 100, 5193-8.

Hadari, Y. R., H. Kouhara, I. Lax and J. Schlessinger (1998) Binding of Shp2 tyrosine phosphatase to FRS2 is essential for fibroblast growth factor-induced PC12 cell differentiation. *Mol Cell Biol*. 18, 3966-73.

Harrison, C. A., P. C. Gray, W. W. Vale and D. M. Robertson (2005) Antagonists of activin signaling: mechanisms and potential biological applications. *Trends Endocrinol Metab*. 16, 73-8.

Harvey, R. P. (2002) Patterning the vertebrate heart. *Nat Rev Genet*. 3, 544-56.

Hug, H. and T. F. Sarre (1993) Protein kinase C isoenzymes: divergence in signal transduction? *Biochem J*. 291, 329-43.

Inui, M., A. Fukui, Y. Ito and M. Asashima (2006) Xapelin and Xmsr are required for cardiovascular development in Xenopus laevis. *Dev Biol*. 298, 188-200.

Itoh, N. and D. M. Ornitz (2004) Evolution of the Fgf and Fgfr gene families. *Trends Genet*. 20, 563-9.

Johnson, S. E., J. L. Rothstein and B. B. Knowles (1994) Expression of epidermal growth factor family gene members in early mouse development. *Dev Dyn*. 201, 216-26.

Kattman, S. J., T. L. Huber and G. M. Keller (2006) Multipotent flk-1+ cardiovascular progenitor cells give rise to the cardiomyocyte, endothelial, and vascular smooth muscle lineages. *Dev Cell*. 11, 723-32.

Kawamata, Y., Y. Habata, S. Fukusumi, M. Hosoya, R. Fujii, S. Hinuma, N. Nishizawa, C. Kitada, H. Onda, O. Nishimura and M. Fujino (2001) Molecular properties of apelin: tissue distribution and receptor binding. *Biochim Biophys Acta*. 1538, 162-71.

Kelly, R. G., N. A. Brown and M. E. Buckingham (2001) The arterial pole of the mouse heart forms from Fgf10-expressing cells in pharyngeal mesoderm. *Dev Cell*. 1, 435-40.

Kirby, M. L. (2007) *Cardiac development*, Oxford University Press, Oxford ; New York

Kitamura, R., T. Takahashi, N. Nakajima, K. Isodono, S. Asada, H. Ueno, T. Ueyama, T. Yoshikawa, H. Matsubara and H. Oh (2007) Stage-specific role of endogenous Smad2 activation in cardiomyogenesis of embryonic stem cells. *Circ Res*. 101, 78-87.

Kleinz, M. J. and A. P. Davenport (2005) Emerging roles of apelin in biology and medicine. *Pharmacol Ther*. 107, 198-211.

Kouhara, H., Y. R. Hadari, T. Spivak-Kroizman, J. Schilling, D. Bar-Sagi, I. Lax and J. Schlessinger (1997) A lipid-anchored Grb2-binding protein that links FGF-receptor activation to the Ras/MAPK signaling pathway. *Cell*. 89, 693-702.

Kuchera, S., H. Barth, P. Jacobson, A. Metz, C. Schaechtele and D. Schrier (1993) Anti-inflammatory properties of the protein kinase C inhibitor, 3-[1-[3-(dimethylamino)propyl]-1H-indol-3-yl]-4-(1H-indol-3-yl)-1H- pyrrole-2,5-dione monohydrochloride (GF109203X) in the PMA-mouse ear edema model. *Agents Actions*. 39 Spec No, C169-73.

Ladd, A. N., T. A. Yatskievych and P. B. Antin (1998) Regulation of avian cardiac myogenesis by activin/TGFbeta and bone morphogenetic proteins. *Dev Biol.* 204, 407-19.

Larsson, H., P. Klint, E. Landgren and L. Claesson-Welsh (1999) Fibroblast growth factor receptor-1-mediated endothelial cell proliferation is dependent on the Src homology (SH) 2/SH3 domain-containing adaptor protein Crk. *J Biol Chem.* 274, 25726-34.

Liguori, G. L., D. Echevarria, R. Improta, M. Signore, E. Adamson, S. Martinez and M. G. Persico (2003) Anterior neural plate regionalization in cripto null mutant mouse embryos in the absence of node and primitive streak. *Dev Biol.* 264, 537-49.

Magnusson, P., C. Rolny, L. Jakobsson, C. Wikner, Y. Wu, D. J. Hicklin and L. Claesson-Welsh (2004) Deregulation of Flk-1/vascular endothelial growth factor receptor-2 in fibroblast growth factor receptor-1-deficient vascular stem cell development. *J Cell Sci.* 117, 1513-23.

Magnusson, P. U., R. Ronca, P. Dell'Era, P. Carlstedt, L. Jakobsson, J. Partanen, A. Dimberg and L. Claesson-Welsh (2005) Fibroblast growth factor receptor-1 expression is required for hematopoietic but not endothelial cell development. *Arterioscler Thromb Vasc Biol.* 25, 944-9.

Marques, S. R., Y. Lee, K. D. Poss and D. Yelon (2008) Reiterative roles for FGF signaling in the establishment of size and proportion of the zebrafish heart. *Dev Biol.* 321, 397-406.

Martin, G. R. (1981) Isolation of a pluripotent cell line from early mouse embryos cultured in medium conditioned by teratocarcinoma stem cells. *Proc Natl Acad Sci U S A.* 78, 7634-8.

Masri, B., H. Lahlou, H. Mazarguil, B. Knibiehler and Y. Audigier (2002) Apelin (65-77) activates extracellular signal-regulated kinases via a PTX-sensitive G protein. *Biochem Biophys Res Commun.* 290, 539-45.

Masri, B., N. Morin, M. Cornu, B. Knibiehler and Y. Audigier (2004) Apelin (65-77) activates p70 S6 kinase and is mitogenic for umbilical endothelial cells. *FASEB J.* 18, 1909-11.

Massague, J. and Y. G. Chen (2000) Controlling TGF-beta signaling. *Genes Dev.* 14, 627-44.

Medhurst, A. D., C. A. Jennings, M. J. Robbins, R. P. Davis, C. Ellis, K. Y. Winborn, K. W. Lawrie, G. Hervieu, G. Riley, J. E. Bolaky, N. C. Herrity, P. Murdock and J. G. Darker (2003) Pharmacological and immunohistochemical characterization of the APJ receptor and its endogenous ligand apelin. *J Neurochem.* 84, 1162-72.

Minchiotti, G., S. Parisi, G. Liguori, M. Signore, G. Lania, E. D. Adamson, C. T. Lago and M. G. Persico (2000) Membrane-anchorage of Cripto protein by glycosylphosphatidylinositol and its distribution during early mouse development. *Mech Dev.* 90, 133-42.

Mohammadi, M., I. Dikic, A. Sorokin, W. H. Burgess, M. Jaye and J. Schlessinger (1996) Identification of six novel autophosphorylation sites on fibroblast growth factor receptor 1 and elucidation of their importance in receptor activation and signal transduction. *Mol Cell Biol.* 16, 977-89.

Mohammadi, M., A. M. Honegger, D. Rotin, R. Fischer, F. Bellot, W. Li, C. A. Dionne, M. Jaye, M. Rubinstein and J. Schlessinger (1991) A tyrosine-phosphorylated carboxy-terminal peptide of the fibroblast growth factor receptor (Flg) is a binding site for the SH2 domain of phospholipase C-gamma 1. *Mol Cell Biol.* 11, 5068-78.

Mohammadi, M., G. McMahon, L. Sun, C. Tang, P. Hirth, B. K. Yeh, S. R. Hubbard and J. Schlessinger (1997) Structures of the tyrosine kinase domain of fibroblast growth factor receptor in complex with inhibitors. *Science*. 276, 955-60.

Monzen, K., Y. Hiroi, S. Kudoh, H. Akazawa, T. Oka, E. Takimoto, D. Hayashi, T. Hosoda, M. Kawabata, K. Miyazono, S. Ishii, Y. Yazaki, R. Nagai and I. Komuro (2001) Smads, TAK1, and their common target ATF-2 play a critical role in cardiomyocyte differentiation. *J Cell Biol*. 153, 687-98.

Moretti, A., L. Caron, A. Nakano, J. T. Lam, A. Bernshausen, Y. Chen, Y. Qyang, L. Bu, M. Sasaki, S. Martin-Puig, Y. Sun, S. M. Evans, K. L. Laugwitz and K. R. Chien (2006) Multipotent embryonic isl1+ progenitor cells lead to cardiac, smooth muscle, and endothelial cell diversification. *Cell*. 127, 1151-65.

Nishikawa, S., L. M. Jakt and T. Era (2007) Embryonic stem-cell culture as a tool for developmental cell biology. *Nat Rev Mol Cell Biol*. 8, 502-7.

O'Dowd, B. F., M. Heiber, A. Chan, H. H. Heng, L. C. Tsui, J. L. Kennedy, X. Shi, A. Petronis, S. R. George and T. Nguyen (1993) A human gene that shows identity with the gene encoding the angiotensin receptor is located on chromosome 11. *Gene*. 136, 355-60.

Olsen, S. K., O. A. Ibrahimi, A. Raucci, F. Zhang, A. V. Eliseenkova, A. Yayon, C. Basilico, R. J. Linhardt, J. Schlessinger and M. Mohammadi (2004) Insights into the molecular basis for fibroblast growth factor receptor autoinhibition and ligand-binding promiscuity. *Proc Natl Acad Sci U S A*. 101, 935-40.

Ornitz, D. M., J. Xu, J. S. Colvin, D. G. McEwen, C. A. MacArthur, F. Coulier, G. Gao and M. Goldfarb (1996) Receptor specificity of the fibroblast growth factor family. *J Biol Chem*. 271, 15292-7.

Parisi, S., D. D'Andrea, C. T. Lago, E. D. Adamson, M. G. Persico and G. Minchiotti (2003) Nodal-dependent Cripto signaling promotes cardiomyogenesis and redirects the neural fate of embryonic stem cells. *J Cell Biol*. 163, 303-14.

Pennisi, D. J. and T. Mikawa (2009) FGFR-1 is required by epicardium-derived cells for myocardial invasion and correct coronary vascular lineage differentiation. *Dev Biol*. 328, 148-59.

Persico, M. G., G. L. Liguori, S. Parisi, D. D'Andrea, D. S. Salomon and G. Minchiotti (2001) Cripto in tumors and embryo development. *Biochim Biophys Acta*. 1552, 87-93.

Piccolo, S., E. Agius, L. Leyns, S. Bhattacharyya, H. Grunz, T. Bouwmeester and E. M. De Robertis (1999) The head inducer Cerberus is a multifunctional antagonist of Nodal, BMP and Wnt signals. *Nature*. 397, 707-10.

Plotnikov, A. N., S. R. Hubbard, J. Schlessinger and M. Mohammadi (2000) Crystal structures of two FGF-FGFR complexes reveal the determinants of ligand-receptor specificity. *Cell*. 101, 413-424.

Powers, C. J., S. W. McLeskey and A. Wellstein (2000) Fibroblast growth factors, their receptors and signaling. *Endocr Relat Cancer*. 7, 165-97.

Presta, M., P. Dell'Era, S. Mitola, E. Moroni, R. Ronca and M. Rusnati (2005) Fibroblast growth factor/fibroblast growth factor receptor system in angiogenesis. *Cytokine Growth Factor Rev*. 16, 159-78.

Puceat, M., P. Travo, M. T. Quinn and P. Fort (2003) A dual role of the GTPase Rac in cardiac differentiation of stem cells. *Mol Biol Cell*. 14, 2781-92.

Qu, C. K. and G. S. Feng (1998) Shp-2 has a positive regulatory role in ES cell differentiation and proliferation. *Oncogene*. 17, 433-9.

Reifers, F., E. C. Walsh, S. Leger, D. Y. Stainier and M. Brand (2000) Induction and differentiation of the zebrafish heart requires fibroblast growth factor 8 (fgf8/acerebellar). *Development*. 127, 225-35.

Reissmann, E., H. Jornvall, A. Blokzijl, O. Andersson, C. Chang, G. Minchiotti, M. G. Persico, C. F. Ibanez and A. H. Brivanlou (2001) The orphan receptor ALK7 and the Activin receptor ALK4 mediate signaling by Nodal proteins during vertebrate development. *Genes Dev*. 15, 2010-22.

Romito, A., E. Lonardo, G. Roma, G. Minchiotti, A. Ballabio and G. Cobellis (2010) Lack of sik1 in mouse embryonic stem cells impairs cardiomyogenesis by down-regulating the cyclin-dependent kinase inhibitor p57kip2. *PLoS One*. 5, e9029.

Ronca, R., L. Gualandi, E. Crescini, S. Calza, M. Presta and P. Dell'Era (2009) Fibroblast growth factor receptor-1 phosphorylation requirement for cardiomyocyte differentiation in murine embryonic stem cells. *J Cell Mol Med*. 13, 1489-98.

Sachinidis, A., B. K. Fleischmann, E. Kolossov, M. Wartenberg, H. Sauer and J. Hescheler (2003) Cardiac specific differentiation of mouse embryonic stem cells. *Cardiovasc Res*. 58, 278-91.

Salomon, D. S., C. Bianco and M. De Santis (1999) Cripto: a novel epidermal growth factor (EGF)-related peptide in mammary gland development and neoplasia. *Bioessays*. 21, 61-70.

Sato, T., K. Hidaka, A. Iwanaga, M. Ito, M. Asano, Y. Nakabeppu, T. Morisaki and K. Yoshioka (2005) Impairment of cardiomyogenesis in embryonic stem cells lacking scaffold protein JSAP1. *Biochem Biophys Res Commun*. 338, 1152-7.

Scott, I. C., B. Masri, L. A. D'Amico, S. W. Jin, B. Jungblut, A. M. Wehman, H. Baier, Y. Audigier and D. Y. Stainier (2007) The g protein-coupled receptor agtrl1b regulates early development of myocardial progenitors. *Dev Cell*. 12, 403-13.

Seyed, M. and J. X. Dimario (2008) Fibroblast growth factor receptor 1 gene expression is required for cardiomyocyte proliferation and is repressed by Sp3. *J Mol Cell Cardiol*. 44, 510-9.

Shen, M. M. and A. F. Schier (2000) The EGF-CFC gene family in vertebrate development. *Trends Genet*. 16, 303-9.

Solloway, M. J. and R. P. Harvey (2003) Molecular pathways in myocardial development: a stem cell perspective. *Cardiovasc Res*. 58, 264-77.

Strizzi, L., C. Bianco, N. Normanno and D. Salomon (2005) Cripto-1: a multifunctional modulator during embryogenesis and oncogenesis. *Oncogene*. 24, 5731-41.

Tanaka, C., R. Sakuma, T. Nakamura, H. Hamada and Y. Saijoh (2007) Long-range action of Nodal requires interaction with GDF1. *Genes Dev*. 21, 3272-82.

Tatemoto, K., M. Hosoya, Y. Habata, R. Fujii, T. Kakegawa, M. X. Zou, Y. Kawamata, S. Fukusumi, S. Hinuma, C. Kitada, T. Kurokawa, H. Onda and M. Fujino (1998) Isolation and characterization of a novel endogenous peptide ligand for the human APJ receptor. *Biochem Biophys Res Commun*. 251, 471-6.

Wang, F., M. Kan, G. Yan, J. Xu and W. L. McKeehan (1995) Alternately spliced NH2-terminal immunoglobulin-like Loop I in the ectodomain of the fibroblast growth factor (FGF) receptor 1 lowers affinity for both heparin and FGF-1. *Journal of Biological Chemistry*. 270, 10231-10235.

Wang, Z., K. Cohen, Y. Shao, P. Mole, D. Dombkowski and D. T. Scadden (2004) Ephrin receptor, EphB4, regulates ES cell differentiation of primitive mammalian hemangioblasts, blood, cardiomyocytes, and blood vessels. *Blood*. 103, 100-9.

Wu, S. M., Y. Fujiwara, S. M. Cibulsky, D. E. Clapham, C. L. Lien, T. M. Schultheiss and S. H. Orkin (2006) Developmental origin of a bipotential myocardial and smooth muscle cell precursor in the mammalian heart. *Cell*. 127, 1137-50.

Xu, C., G. Liguori, E. D. Adamson and M. G. Persico (1998) Specific arrest of cardiogenesis in cultured embryonic stem cells lacking Cripto-1. *Dev Biol*. 196, 237-47.

Xu, C., G. Liguori, M. G. Persico and E. D. Adamson (1999a) Abrogation of the Cripto gene in mouse leads to failure of postgastrulation morphogenesis and lack of differentiation of cardiomyocytes. *Development*. 126, 483-94.

Xu, X., C. Li, K. Takahashi, H. C. Slavkin, L. Shum and C. X. Deng (1999b) Murine fibroblast growth factor receptor 1alpha isoforms mediate node regression and are essential for posterior mesoderm development. *Dev Biol*. 208, 293-306.

Yamaguchi, T. P., K. Harpal, M. Henkemeyer and J. Rossant (1994) fgfr-1 is required for embryonic growth and mesodermal patterning during mouse gastrulation. *Genes Dev*. 8, 3032-44.

Zaffran, S. and M. Frasch (2002) Early signals in cardiac development. *Circ Res*. 91, 457-69.

Zaffran, S., R. G. Kelly, S. M. Meilhac, M. E. Buckingham and N. A. Brown (2004) Right ventricular myocardium derives from the anterior heart field. *Circ Res*. 95, 261-8.

Zeng, X. X., T. P. Wilm, D. S. Sepich and L. Solnica-Krezel (2007) Apelin and its receptor control heart field formation during zebrafish gastrulation. *Dev Cell*. 12, 391-402.

Zhou, B., Q. Ma, S. Rajagopal, S. M. Wu, I. Domian, J. Rivera-Feliciano, D. Jiang, A. von Gise, S. Ikeda, K. R. Chien and W. T. Pu (2008) Epicardial progenitors contribute to the cardiomyocyte lineage in the developing heart. *Nature*. 454, 109-13.

Zhou, X., E. Quann and G. I. Gallicano (2003) Differentiation of nonbeating embryonic stem cells into beating cardiomyocytes is dependent on downregulation of PKC beta and zeta in concert with upregulation of PKC epsilon. *Dev Biol*. 255, 407-22.

26

Differentiation, Characterization and Applications of Human Embryonic Stem Cell – Derived Cardiomyocytes

Mari Pekkanen-Mattila[1], Erja Kerkelä[1] and Katriina Aalto-Setälä[1,2]
[1]*Institute for Regenerative Medicine, University of Tampere*
[2]*Heart Center, Tampere University Hospital, Tampere, Finland*

1. Introduction

The term "embryonic stem (ES) cell" was introduced in 1981 to distinguish embryo-derived pluripotent cells from teratocarcinoma-derived pluripotent embryonal carcinoma (EC) cells (Martin 1981). First ES cells were derived from mouse intracellular mass (ICM) in the same year (Evans and Kaufman 1981) and in 1994 Bongso and co-workers reported the successful isolation of human ICM cells and their continued culture for at least two passages in vitro (Bongso, Fong et al. 1994). The first permanent human embryonic stem cell (hESC) lines were derived more than a decade ago by Thomson and co-workers (Thomson, Itskovitz-Eldor et al. 1998) and these lines are still widely used.

hESCs are capable of proliferating extensively at undifferentiated state in vitro and have the ablity to differentiate towards all three germ layers and furthermore can, in principle, give rise to all cell types of the body. Adult human cardiomyocytes have limited capability to regenerate and the heart tissue cannot undergo extensive repair needed for example after myocardial infarction. Therefore, the rapid development of stem cell technology has raised hopes for new treatments for tissue damage of cardiac and other tissues with limited regenerative capacity. Human embryonic stem cells (hESC) have the ability to differentiate into functional cardiomyocytes by multiple differentiation methods. Traditionally hESC-derived cardiomyocytes (hESC-CM) are differentiated spontaneously in embryoid bodies (EB) or in co-culture with mouse endodermal cell like (END-2) cells (Kehat, Kenyagin-Karsenti et al. 2001; Mummery, Ward-van Oostwaard et al. 2003). In addition, more defined differentiation methods using growth factors have been developed (Laflamme, Chen et al. 2007; Yang, Soonpaa et al. 2008). However, the cardiac differentiation is still quite uncontrolled and inefficient and even though new more defined differentiation methods have been published, spontaneous differentiation in EBs and differentiation in co-culture with END-2 cells are still widely used as they are rather inexpensive and functioning for most of the hESC lines.

At the end of the year 2009, the total number of hESC lines worldwide has been estimated to be 1071 (Loser, Schirm et al. 2010). Even though the number of hESC lines has increased steadily, two lines, H1 and H9 (WiCell Research Institute), are the most used ones in stem cell research (Guhr, Kurtz et al. 2006; Scott, McCormick et al. 2009; Loser, Schirm et al. 2010).

After reporting the concept of induced pluripotent stem cells (iPS cells) in 2007 (Takahashi, Tanabe et al. 2007), these mature cell derived pluripotent stem cells have rapidly emerged in stem cell studies with hESCs or replacing them.
According to comparison studies, hESCs have been shown to be similar, but not identical according to the expression of pluripotency markers (Adewumi, Aflatoonian et al. 2007). Greater variation between cell lines has, however, been observed after cells start to differentiate - the expression of differentiation markers and differentiation propensity varies between different hESC lines (Adewumi, Aflatoonian et al. 2007; Kim, Kim et al. 2007; Osafune, Caron et al. 2008; Pekkanen-Mattila, Kerkela et al. 2009; Pekkanen-Mattila, Pelto-Huikko et al. 2010). Due to this observation, it has been suggested, that the most suitable hESC line should be chosen according to its propensity to differentiate towards lineage of interest (Osafune, Caron et al. 2008).
Pluripotent stem cell derived cardiomyocytes can be obtained in cell culture, but their therapeutic use is in its infancy. Functional cardiomyocytes can, however, be obtained with various differentiation methods and already at the moment they can be used to study cardiac differentiation, the effects of different exogenous factors on they behaviour and potentially in the near future also cardiac side effects in pharmacological and toxicological industry.

2. Spontaneous cardiac differentiation in EBs

hESCs can be differentiated spontaneously as embryoid bodies (EB). In principle, during EB formation the culture condition for stem cells is changed from two-dimension into three-dimensional. First pluripotent stem cells are either enzymatically or mechanically dissociated to small cell clumps or single cells. Secondly cells are let to form aggregates in suspension and factors needed to keep them pluripotent are removed. After few days, EBs are formed and they are plated down on matrix coated cell culture plates (Kurosawa 2007). When hESCs are removed from the environment which supports the undifferentiatied state (feeder cells and the growth factor FGF), they start to differentiate towards three germ layers in the cell aggregates (Itskovitz-Eldor, Schuldiner et al. 2000). During the early stages of suspension culture, the cell aggregate transforms into cystic body and trilayer shell composed of extra cellular proteins forms around EB (Sachlos and Auguste 2008). The paracrine and endocrine signaling determine the stem cell fate. Similarly as in embryo this signaling may lead to the formation of concentration gradient in the EBs and further influence the cell differentiation (Sachlos and Auguste 2008).
EB formation has similar characteristics as embryonal development (Keller 1995) and therefore the interplay of different germ layers and their influences into cell differentiation can be studied in EB cultures. The EB differentiation, such as cardiac differentiation, is particularly well documented with mouse ES cells (Hescheler, Fleischmann et al. 1997; Boheler, Czyz et al. 2002). However, the EB formation of hESC and the spontaneous differentiation has proven to be more difficult and more inefficient if compared to mouse counterpart (Wobus, Wallukat et al. 1991; Kehat, Kenyagin-Karsenti et al. 2001). When mouse ES cells are differentiated in EBs, beating areas appear 1 day after plating, and, within 2–10 days, 80–90% of EBs show spontaneous beating (Wobus, Wallukat et al. 1991). In the hESC differentiation beating areas are observed later and the differentiation efficiency is traditionally much lower, usually under 10% (Kehat, Kenyagin-Karsenti et al. 2001).
Cardiomyocyte differentiation from hESC and iPS cells in EBs has been described in many reports (Itskovitz-Eldor, Schuldiner et al. 2000; Kehat, Kenyagin-Karsenti et al. 2001; Zhang,

Wilson et al. 2009). EB differentiation is widely used also in production of other cell types such as neuronal cells, hematopoietic cells, adipocytes and chondrocytes (Pera and Trounson 2004). For the whole existence of hESC, EB differentiation has been widely used differentiation method for its relatively simple and inexpensive nature.

There are multiple methods for EB formation (Kurosawa 2007). Suspension culture in bacterial-grade cell culture dishes was first developed for mouse ES cells (Doetschman, Eistetter et al. 1985) and has later been used in cardiomyocyte differentiation from hESCs (Itskovitz-Eldor, Schuldiner et al. 2000; Kehat, Kenyagin-Karsenti et al. 2001). In this method enzymatically dissociated cells aggregate when cultured unattached in the culture medium. hESCs are more vulnerable to dissociation to the single cell stage (Thomson, Itskovitz-Eldor et al. 1998; Amit, Carpenter et al. 2000; Kehat, Kenyagin-Karsenti et al. 2001; Xu, Police et al. 2002) and therefore hESCs have been tradionally dissociated into small clumps of cells to retain the cell to cell contact (Amit, Carpenter et al. 2000; Pyle, Lock et al. 2006), but current technology enables also single cell stage with hESCs (Amit, Chebath et al. 2010). To scale up EB formation in suspension cultures, bioreactors and spinner flasks have also been used (Messina, De Angelis et al. 2004; Kurosawa 2007; Yirme, Amit et al. 2008).

Cardiomyocytes have also been differentiated by hanging drop-method, where single cell suspension is pipetted to small drops onto petri dish cover and cover is then inverted on top of a dish (Takahashi, Lord et al. 2003; Burridge, Anderson et al. 2007). The drop hangs because of surface tension and provides a good environment for the cells to aggregate and form the EB. Hanging drop-method is not suitable for long term EB differentiation because the medium change is impossible (Kurosawa 2007). Overall hanging drop-method is very laborious and therefore it is not suitable for large scale experiments.

Recent studies indicate that the EB size has an effect on cardiomyocyte differentiation as well as differentiation in general (Burridge, Anderson et al. 2007; Bauwens, Peerani et al. 2008; Mohr, Zhang et al. 2010). Therefore the amount of cells should be measurable in order to optimize differentiation. Hanging drop method enables to standardize the initial amount of hESC. However, Ng and coworkers developed more robust method compared to hanging drops, a forced aggregation (FA) or spinEB system for hematopoietic differentiation and it has been used also in cardiomyocyte differentiation (Ng, Davis et al. 2005; Burridge, Anderson et al. 2007). FA mimics the hanging drop method, the cells are forced to aggregate by centrifugation in a round bottomed, low-adherence 96-well plate wells. The medium change is possible to the wells and therefore longer culture times can be used and also differentiation inducing agents can be added to the culture medium (Burridge, Anderson et al. 2007). Two-dimensional cell pieces can also be produced by microprinting technique, where standard-size colonies are formed and then scraped into suspension culture (Bauwens, Peerani et al. 2008; Niebruegge, Bauwens et al. 2009). EB differentiation techiques are summarized in Table 1.

3. Cardiac differentiation in END-2 co-cultures

More directed way to differentiate cardiomyocytes from hESCs is in co-culture with mouse endodermal-like cells (END-2), particularly in a absence of serum and with ascorbic acid (Mummery, Ward-van Oostwaard et al. 2003; Passier, Oostwaard et al. 2005). The differentiation inducing factors are secreted from END-2 cells and therefore the END-2 conditioned medium can also be used in cardiomyocyte differentiation (Graichen, Xu et al. 2008). END-2 cells support the differentiation towards endodermal and mesodermal

EB differentiation techniques					
Method description	**Hanging drop**	**Forced aggregation (FA)**	**Suspension culture**	**Microprinting technique**	**Manual**
hESC colony dissociation	Enzymatic dissociation			Detachment of microprinted colonies	Manual cutting
EB formation	Single cells/small aggregates form EB in a hanging drop	Cell suspension is aggregated to EB by centrifuging in a 96-well plate	Spontaneous aggregation in suspension	One cell colony or cell colony piece forms an EB in suspension	
EB culture	Formed EBs transferred for suspension culture		Suspension culture continues		
	After suspension culture EBs are plated on a coated cell culture plate				
Advantages	Gentle EB formation in a drop because of gravity	Scalable, straightforward, cell number per EB easy to standardize	Straightforward	Cell number per EB easy to standardize	Gentle, non-enzymatic hESC colony dissociation
Disadvantages	Laborious, non-scalable	hESC colonies have to be dissociated to single cell stage	Forming EBs randomly sized	Need for microprinting technique for colony formation	Laborious, non-scalable
Reference	(Takahashi, Lord et al. 2003)	(Ng, Davis et al. 2005)	(Doetschman, Eistetter et al. 1985)	(Bauwens, Peerani et al. 2008; Niebruegge, Bauwens et al. 2009)	Pekkanen-Mattila, Peltohuikko et al., 2010)

Table 1. Summary of the EB differentiation techniques.

derivatives (Mummery, Ward-van Oostwaard et al. 2003; Passier, Oostwaard et al. 2005; Beqqali, Kloots et al. 2006) which is in accordance with embryonal development studies, which show that anterior visceral endoderm is essential in normal heart development (Lough and Sugi 2000).

The mechanism or the specific factors inducing cardiac differentiation by END-2 cells are, however, not clearly known. Systematic testing of END-2 conditioned medium revealed that END-2 cells were able to clear insulin from the medium (Xu, Graichen et al. 2008). Insulin has been shown to inhibit cardiac differentiation by suppressing endoderm and mesoderm formation and favouring ectoderm differentation (Freund, Ward-van Oostwaard et al. 2008). Insulin acts via the insulin-like growth factor-1 receptor (IGF-1R) and phosphatidylinositol 3-kinase (PI3K/Akt) pathway and has been suggested to inhibit epithelial-to mesenchymal transition by elevated levels of E-cadherin (Freund, Ward-van Oostwaard et al. 2008). However, IGF/PI3K/Akt has also been shown to have a role in proliferation of immature cardiomyocytes (McDevitt, Laflamme et al. 2005) which suggests that this pathway has a dual role in cardiomyogenesis. Additionally, END-2 cells are not the only type of cells which clear insulin from the culture media. Similar phenomenon has been observed with MES1-cells (Mummery, Feijen et al. 1986) and mouse embryonic fibroblasts (MEFs) which do not have the cardiac differentiation inducing effect (Xu, Graichen et al. 2008). Therefore insulin depletion is not likely the cardiac inducing factor of END-2 cells. A more promising cardiac differentiation inducer of END-2 cells has been suggestested to be prostaglandin I2 (PGI2). END-2 cells have been reported to secrete more PGI2 than other type of mouse cells which lack the cardiac inductive effect. In fact, addition of PGI2 to differentiation medium has been reported to result in similar levels of cardiac differentiation as END-2 conditioned medium (Xu, Graichen et al. 2008).

In addition to the PGI2, inhibition of p38 mitogen activated protein kinase (MAPK) increases cardiac differentiation rate (Graichen, Xu et al. 2008). Selective MAPK inhibitors (molecules SB203580 and SB202190) (Cuenda, Rouse et al. 1995) were found to increase the differentiation rate when added to END-2 conditioned medium. However, the inductive effect of these molecules was concentration dependent, at high concentrations (>15 μM) cardiomyocyte formation was decreased and finally inhibited (Xu, Graichen et al. 2008). The use of p38 inhibitor PD169316 also causes mouse ES cells to differentiate towards neural linage while the cardiac mesoderm formation is inhibited (Aouadi, Bost et al. 2006). Therefore the inhibition of MAPK has a partially opposite effect on mouse and human cells and its exact role in cardiac differentiation remains to be revealed.

4. Differentiation with defined growth factors

Cardiac differentiation is controlled by a complex signalling network and currently there is no single factor that would direct stem cells to differentiate effectively towards cardiac lineage. Laflamme and co-workers have used the combination of activin A and BMP-4 in cardiomyocyte differentiation (Laflamme, Chen et al. 2007). This cascade of factors enhances mesoendoderm formation, an early precursor cell lineage which gives rise to mesoderm and endoderm. Mesoderm is the origin of cardiac cells and it has been shown that cardiac differentiation inducing signals are in large extent arising from endoderm (Lough and Sugi 2000). Therefore mesoendoderm induction would yield more efficient human embryonic stem cell-derived cardiomyocyte (hESC-CM) differentiation.

Stepwise differentiation protocol was also developed by Yang and co-workers (Yang, Soonpaa et al. 2008). This protocol involves induction of primitive streak-like population, in addition to formation of cardiac mesoderm and expansion of cardiac lineages. Protocol is based on EB differentiation and is comprised of three stages. Growth factors BMP-4, FGF, activin A, vascular endothelial growth factor (VEGF) and dickkoptf homolog 1 (DKK1) were used in varying combinations.

Mesoendoderm formation has also been induced by Wnt3A, an activator of the canonical Wnt/β-catenin signalling pathway (Tran, Wang et al. 2009).

Taken together, even though the use of growth factors and other chemicals may enhance the cardiac differentiation, pure populations of cardiomyocytes can not be currently produced and enrichment methods are still needed. Due to multi-phased differentiation protocols, the high costs of growth factors, and the complexity of defining right concentrations and exposure times to different factors, the simple and functional EB differentiation method needs widely used method in cardiomyocyte production. Also the current differentiation methods produce beating aggregates, but for many purposes cardiac differentiation in monolayer would be more optimal, but this method is still to be developed further.

5. Characterization of differentiated cardiomyocytes

5.1 Structural analysis

Differentiated hESC-CM have the capacity to beat spontaneously and are, thus, easily detected in the culture as beating areas (Kehat, Kenyagin-Karsenti et al. 2001; Mummery, Ward-van Oostwaard et al. 2003). Beating cells are at early stage relatively small and round and situated in circular accumulations in the EBs. At later stages, EBs gradually develop larger and the cells turn into more elongated in shape and tend to accumulate in strands. Electron microscopy studies reveal that cardiomyocytes contain myofibrils which are first randomly and in a varying manner distributed throughout the cytoplasm. However, organised sarcomeric structures occur at later stages of differentiation with A, I, and Z bands. In the vicinity of the sarcomeres, mitochondia are also present. In addition, cells have intercalacted disks with gap junctions and desmosomes (Kehat, Kenyagin-Karsenti et al. 2001; Snir, Kehat et al. 2003; Pekkanen-Mattila, Kerkela et al. 2009).

5.2 Estimation of differentiation efficiency and expression of cardiac markers

The number of hESC-CM containing beating areas has been used in quantifying differentiation efficiency of different hESC lines (Passier, Oostwaard et al. 2005; Pekkanen-Mattila, Kerkela et al. 2009; Pekkanen-Mattila, Pelto-Huikko et al. 2010). However, Passier and co-workers reported a large variation in the number of cells in the beating areas, ranging from 1-2500 cells (Passier, Oostwaard et al. 2005). Therefore, to determine the amount of hESC-CM more accurately, cytospin analysis has been used in the estimation of percentage of cardiac marker positive cells per total amount of cells (Graichen, Xu et al. 2008; Xu, Graichen et al. 2008).
Cardiac troponin T (cTnT) is encoded by the TNNT2 gene (Thierfelder, Watkins et al. 1994) and is the tropomyosin-binding subunit of the troponin complex and therefore can be used in characterizing hESC-CM (Figure 2E). Troponin complex is located on the thin filament of striated muscles and regulates muscle contraction in response to alterations in intracellular calcium ion concentrations as revieved (Farah and Reinach 1995; Tobacman 1996). In addition to cTnT, other cardiac spesific structural proteins are used for confirming cardiac phenotype of the beating hESC-CM such as cardiac troponin I, myosins (Figure 2F) or cardiac α-actinin (Kehat, Kenyagin-Karsenti et al. 2001; Mummery, Ward-van Oostwaard et al. 2003). Even though many cardiac markers exist, there is a lack of cardiac specific surface proteins and therefore lack of antibodies for making sorting by fluorescence-activated cell sorting (FACS) possible (Mummery 2010).
Another approach to estimate the cardiac differentiation efficiency and follow the differentiation is to quantify the mRNA levels of cardiac markers and cardiac differentiation markers. hESC-CM differentiation can be predicted by the transient expression of the early mesodermal marker Brachyury T. Brachyury T expression peak is detected at day 3 in END-2 co-cultures (Beqqali, Kloots et al. 2006; Pekkanen-Mattila, Kerkela et al. 2009) and a day later in EBs (Bettiol, Sartiani et al. 2007; Pekkanen-Mattila, Pelto-Huikko et al. 2010). Brachyury T belongs to the family of transcription factors which are encoded by the T-box genes (Showell, Binder et al. 2004). Brachyury T can be nominated as a classical transcription factor, it is localized in the nucleus and is an endogenous activator of mesodermal genes (Conlon, Lyons et al. 1994; Kispert, Koschorz et al. 1995; Showell, Binder et al. 2004). In the embryo,

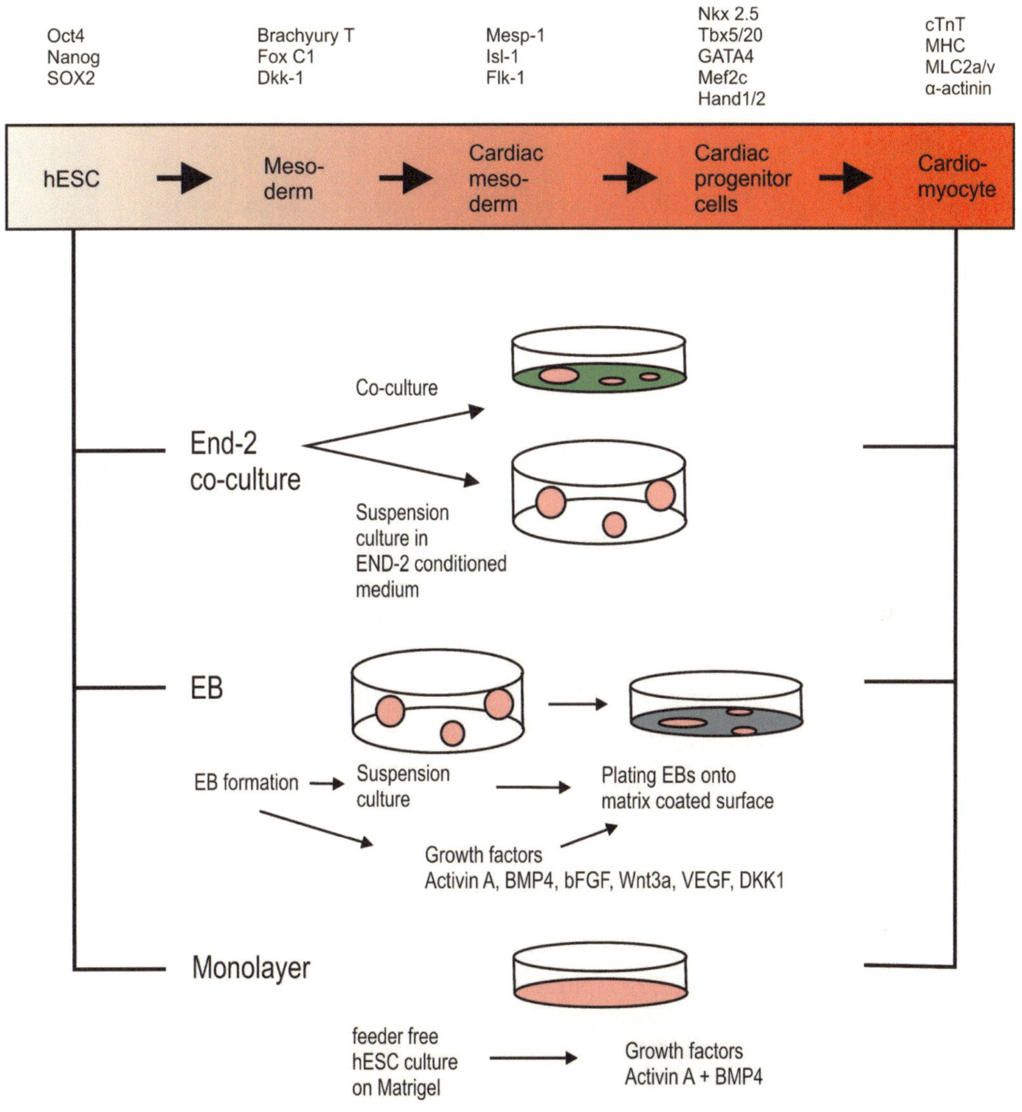

Fig. 1. ***Cardiac differentiation steps and differentiation methods.*** From the top; Markers for different stages of cardiac differentiation, steps in cardiac differentiation and schematic view of differentiation methods. **END-2** differentiation has two variables, hESC are either plated on top of END-2 cell layer or hESC are cultured as EBs in suspension in END-2 conditioned medium. **EB method**, differentiation can be performed spontaneously or with differentiation inducing growth factors. **Monolayer differentiation** is initiated with feeder free hESC cultures. Culturing of hESC and differentiation with activin A and BMP-4 is performed on top of Matrigel.

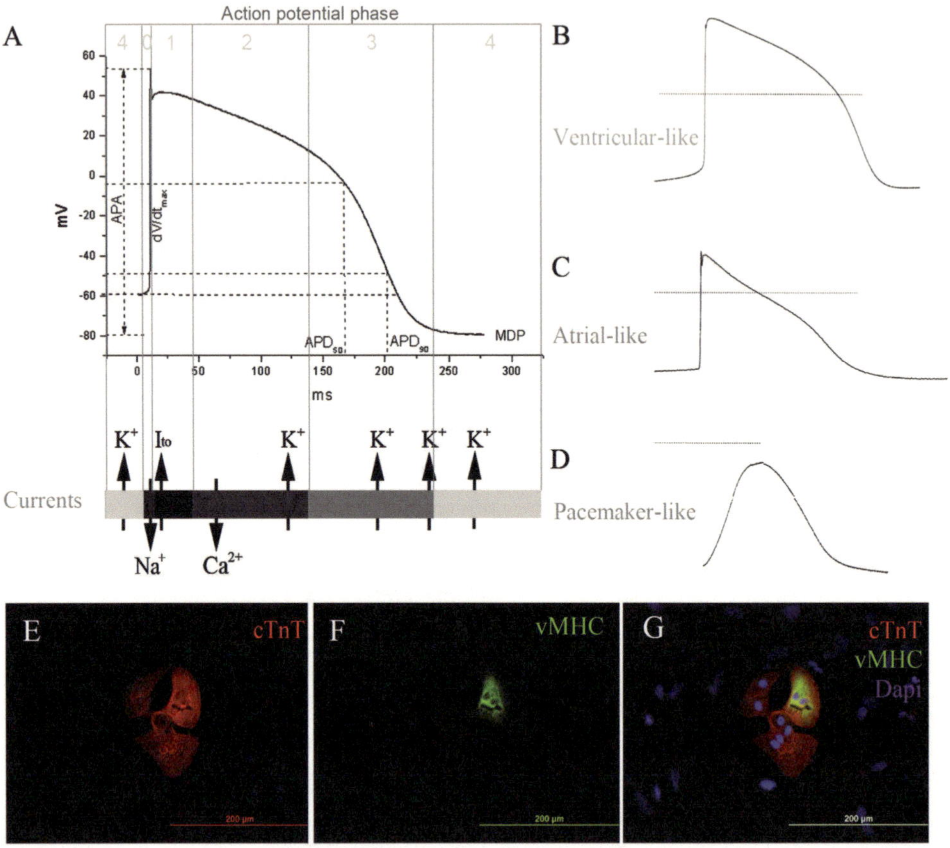

Fig. 2. ***Cardiac action potential parameters and phases, and characteristics of different cardiomyocyte subtypes*. A.** Action potential (AP) parameters: Action potential amplitude (APA), maximum rate of rise of the action potential (dV/dtmax), action potential delay (ADP) and membrane diastolic potential (MDP). AP phase 0 is a rapid depolarization phase when the sodium channels are activated and membrane permeability is increased to Na+. Rapid depolarisation is followed by rapid repolarization phase 1 and plateau phase 2, where Ca2+ ions are entered to the cell through L-type calcium channels. At phase 3, calcium channels are inactivated and repolarization is caused by outward potassium currents. Repolarization is due to the currents carried mainly by the slow Iks and rapid Ikr components of the delayed rectifier potassium channels. The Ikr current is produced by hERG channel (encoded by the human ether-à-go-go-related gene). By contrast, inward potassium current contributes to the maintenance of the resting membrane potential, phase 4. **B-D** Classification of ventricular (B), atrial (C) an pacemaker-like (D) action potentials. Ventricular action potential has a prominent plateau phase whereas atrial action potential is more triangularly shaped. Pacemaker-like cells are characterized by slower upstroke velocity and amplitude if compared to ventricular and atrial type of cells. **E-G** Immunocytochemical staining of hESC-CM by (E) cardiac Troponin-T, (F) ventricular specific myosin heavy chain, (G) Merged image with nuclei stain Dapi. Scalebar 200 µm.

Brachyury T expression is suggested to be induced by TGFβ and FGF signalling (Hemmati-Brivanlou and Melton 1992; Amaya, Stein et al. 1993). Overall, very few direct targets for T-box genes have been identified. However, embryonic FGF (eFGF) (Casey, O'Reilly et al. 1998), Brachyury-induced homeobox Bix4 (Tada, Casey et al. 1998) and XWnt11 (Tada and Smith 2000) have been suggested to be downstream targets for Brachyury T.

Differentiation cascade can be followed further by the expression of the cardiac regulatory transcription factors such as Islet-1 (Isl-1), Mesp 1, GATA-4, Nkx2.5 and Tbx6 (Graichen, Xu et al. 2008; Yang, Soonpaa et al. 2008). Markers for cardiac differentiation steps are illustrated in Figure 1.

The gene expression profiles of the hESC during cardiac differentiation (Beqqali, Kloots et al. 2006; Synnergren, Adak et al. 2008) and the differentiated hESC-CM have been studied by DNA microarray (Cao, Wagner et al. 2008; Synnergren, Akesson et al. 2008; Kita-Matsuo, Barcova et al. 2009; Xu, Soo et al. 2009). These studies reveal that the molecular signature of hESC-CM resembles the cardiomyocytes from the human heart (Vidarsson, Hyllner et al. 2010).

5.3 Electrophysiology

hESC-CM exhibit heterogenic action potential (AP) morphologies which can be divided into nodal, atrial and ventricular subtypes (Figure 2) (He, Ma et al. 2003; Zhang, Wilson et al. 2009). According to the immunocytochemisty by ventricular specific marker and action potential studies, most of the differentiated cardiomyocytes were ventricular type of cells (Pekkanen-Mattila, Chapman et al. 2010). Slight variation in the amount of ventricular type of cells has been observed in different cell lines (Moore, Fu et al. 2008) and the amount has also been observed to vary between cardiac differentiation methods (80% and 100% of END-2 and EB-derived cardiomyocytes, respectively) (Pekkanen-Mattila, Chapman et al. 2010). If compared to the human neonatal or adult atrial or ventricular cardiomyocytes, hESC-CM have relatively positive maxium diastolic potential (MDP) and slow maxium rate of rise of the AP (dV/dtmax) and therefore resemble embryonal atrial- and ventricular like cells (He, Ma et al. 2003). However, even one third of our hESC-CMs exhibited a more mature phenotype with MDPs of < -70 mV and upstroke velocities >140 V/s. The demonstration of the presence of cardiomyocytes with dV/dtmax of over 150 and MDP of close to -80 is very important and suggests that mature human cardiac cells can be produced from hESCs, but the optimal conditions are still to be defined.

Traditionally patch clamp method has been used in analyzing the AP and electrophysiological properties of cardiomyocytes. Micro-electrode array (MEA) technology provides another useful platform to study cell electrophysiology, especially for ES-derived cardiomyocytes (Hescheler, Halbach et al. 2004; Reppel, Pillekamp et al. 2004; Pekkanen-Mattila, Kerkela et al. 2009). In MEA, cells are plated on top of electrodes in a cell culture well-type platform and they can be cultured and measured repeatedly for a long period of time. In addition, MEA can be utilized in testing the effects of pharmaceutical agents on hESC-CM (Braam, Tertoolen et al. 2010).

6. Applications for hESC-CM

6.1 Human cardiac cell/tissue model

Since the establishment of the first permanent hESC line (Thomson, Itskovitz-Eldor et al. 1998) there has been a great hope for replacing damaged heart tissue by hESC derived cardiomyocytes. However, many severe problems need to be solved before hESC-CM are usable in clinics. Before clinical use becomes reality, it is likely that the hESC-CM would be applicable for drug discovery and safety pharmacological applications (Braam, Passier et al. 2009). Nevertheless, cardiac differentiation and the beating cells are already a useful tool for developmental biology and to study the pathophysiology of human cardiac diseases. In addition, iPS technology enables the production of patient specific cell lines which broadens the potential use even more.

6.2 Pathophysiology of cardiac diseases

Many cardiac diseases are caused by gene mutations or gene-enviroment interactions. Until today, these severe diseases have been studied in animal models, especially using transgenic mice. Even though mouse models can yield valuable information, differences between human and mouse physiology limit the applicability of the results, for example remarkably faster beating rate of the mouse may override the effects of arrhythmias which would be severe for human (Freund and Mummery 2009).

Cardiomyocytes derived from genetically modified hESC could be used as a disease model. To construct mutated hESC line and the disease model, the hESC line needs to be genetically manipulated. However, genetic manipulation of hESCs has proven to be more challenging if compared to mouse ES cells and only a limited amount of reports of successful gene targeting and manipulation exist (Braam, Denning et al. 2008; Giudice and Trounson 2008).

To obtain disease specific lines, the genetic manipulation step can be circumvented by deriving pluripotent stem cells from patients with genetic diseases using iPS cell technology (Park, Arora et al. 2008; Ebert, Yu et al. 2009; Freund, Davis et al. 2010). The differentiation of these model iPS-cells to desired cell type enables studying the development and the pathophysiology of the disease. In addition, the factors that affect the development and the progress of the disease can be studied (Freund, Davis et al. 2010). However, iPS-cell technology is still in its infancy and it remains to be seen if differentiated cells really manifest the disease phenotype of the mutation they carry and serve as a real disease model (Freund and Mummery 2009).

6.3 Safety pharmacology and drug discovery

The heart has been proven to be very sensitive to the side effects of pharmaceutical compounds. Severe reactions, such as syncope, arrhythmia and sudden death, related a special ventricular tachycardia, torsade de pointes (TdP), have led to the refusal of approval or the withdrawal from the market of many pharmaceutical agents (Roden 2004). In the absence of a complete understanding and direct analysis of TdP, the regulatory authorities have adopted the QT prolongation as a marker for the possible development of drug-induced TdP even though it is not a perfect marker for arrhythmogenesis (Finlayson, Witchel et al. 2004). Prolongation of the QT interval resulting from a delay in ventricular repolarization, whether drug-induced or for instance congenital arising from mutation of genes (to date LQT1-12), can be associated with TdP (Roden 2004; Zareba and Cygankiewicz 2008), though the relationship is complex (Shah and Hondeghem 2005). However, the QT interval is the cornerstone of the guidelines for the assessment of new chemical compounds in regard to proarrhythmic potential (ICH 2005; ICH 2005). Delayed rectifier potassium current (IKr) is one of the ion channels responsible for the repolarization of the action potential and the channel protein is encoded by the human ether-to-go-go-related gene (hERG) (Vandenberg, Walker et al. 2001; Pollard, Valentin et al. 2008). Inhibition of this hERG channel (KV11.1) and the following inhibition of the IKr, is the predominant basis of drug-induced QT prolongation and TdP (Redfern, Carlsson et al. 2003; Hancox, McPate et al. 2008). Currently a number of preclinical models and assays have been employed by pharmaceutical companies (Carlsson 2006; Pollard, Valentin et al. 2008). These assays include in vivo QT assays, such as ECG telemetry of conscious dogs (Miyazaki, Watanabe et al. 2005), and in vitro assays, such as of repolarization assay which detects changes in the action potential delay (APD) of cardiac tissues (isolated animal purkinje fibers, papillary muscles or cardiac myocytes) or the hERG channel assay where hERG current expressed in heterologous cell system (such as CHO- or HEK293-cells) or native IKr is characterized (Finlayson, Witchel et al. 2004; Martin, McDermott et al. 2004).

Current methods are not fully adequate (Redfern, Carlsson et al. 2003; Lu, Vlaminckx et al. 2008). In addition, they are costly and the in vivo assays are ethically doubtful because of the large number of animals is used. Therefore there is a need for an in vitro method based on human cardiac cells that would bring additional value and reliability for testing novel pharmaceutical agents.

Cardiomyocytes derived both from hESC and iPS-cells have many potential applications in pharmaceutical industry including target validation, screening and safety pharmacology. These cells would serve as an inexhaustible and reproducible human model system and preliminary report of validation of hESC-CM system already exist (Braam, Tertoolen et al. 2010). However, a lot of optimization and development remains to be done, especially because of the immature phenotype of these cells and problems due to the differentiation efficiency, heterogeneous hESC-CM populations and enrichment methods (Braam, Passier et al. 2009).

6.4 Regenerative medicine

In principle, it would be possible to restore the function of the damaged heart by transplanting differentiated hESC. However, this may be one of the most challenging tasks to put into practise. The amount of transplantable cells needed is high and they should be immunocompatible. In addition, the transplanted graft should integrate into host myocardium, receive blood flow to remain vital, electrically couple with host myocardium and contract in synchrony in response to the conduction system (Braam, Passier et al. 2009).

hESC-CM have been transplanted into healthy myocardium of rodents. The cells were reported to survive, form myocardial tissue and proliferate but they were usually separated from the rodent myocardium by a layer of fibrotic tissue (Laflamme, Gold et al. 2005; van Laake, Passier et al. 2007). When transplanted to infracted rat or mouse hearts, some beneficial effects for the function of the heart have occurred (Laflamme, Chen et al. 2007; van Laake, Passier et al. 2007). However, after longer follow-up period, positive effects were not present anymore (van Laake, Passier et al. 2007; van Laake, Passier et al. 2008; van Laake, Passier et al. 2009). It is questionable wheather these temporal benefits are due to the formed myocardium or paracrine effects, like has been proposed for adult stem cells.

Even though some information concerning transplantation can be obtained by using rodent models, studies with larger animals (pigs, goats and sheep) are warranted to give more accurate results from safety issues, electrical coupling and cardiac function. Usage of the iPS-cells or ESC from the same species would eliminate the xeno barriers (Braam, Passier et al. 2009).

In addition to the above-mentioned issues, the timing of cell therapy and the delivery methods still need to be determined. It is likely that cells need supportive material during transplantation and therefore biomaterial research is also needed before clinical studies can be properly defined (Passier, van Laake et al. 2008).

7. Future perspectives

The ultimate goal for stem cell research is to cure patients with diseases caused by the loss of functional tissue such as myocardial infarction. However, it has become clear that pluripotent stem cell derived cardiomyocytes are not ready for the clinical use in the near future and a lot of basic research on cardiac differentiation and directed differentiation of cardiomyocyte subtypes (atrial, ventricular and conduction cells) is still needed. The cardiac differentiation is still inefficient and uncontrolled. Therefore effective methods for differentiation that supply homogenous populations of cardiomyocytes of sufficient quality, reproducibility and in large quantities are prerequisite for applications in pharmaceutical industry as well as for clinical use.

The differentiated cardiomyocytes are a mixed population consisting of non-cardiac cells and cardiomyocytes with several subtypes and maturation stages. For studies of development of disease or testing new potential drug molecules the cardiomyocyte population should be of one subtype (e.g. ventricular) and they should have mature, adult-like phenotype.

It is not only a disadvantage to have non-cardiac cells present in the differentiated cell populations. If aiming at cardiac tissue model, other cell types such as fibroblasts and endothelial cells are needed for proper cardiac differentiation (Kim, Majdi et al. 2010). However, the population should, as mentioned, be standardized and composed of desired cells with right ratios. It is likely, that cells can not form three-dimensional tissue model structure by themselves and some, perhaps extra cellular matrix mimicing biomaterial is needed to give cells support, attachment surfaces and nutrition.

Even though the pluripotent stem cell derived CM are not yet ready for clinical use and many obstacles have to be overcome before their use in drug discovery, they possess already now tremendous opportunities for basic research and pharmaceutical industry.

8. References

Adewumi, O., B. Aflatoonian, et al. (2007). "Characterization of human embryonic stem cell lines by the International Stem Cell Initiative." Nat Biotechnol 25(7): 803-16.

Amaya, E., P. A. Stein, et al. (1993). "FGF signalling in the early specification of mesoderm in Xenopus." Development 118(2): 477-87.

Amit, M., M. K. Carpenter, et al. (2000). "Clonally derived human embryonic stem cell lines maintain pluripotency and proliferative potential for prolonged periods of culture." Dev Biol 227(2): 271-8.

Amit, M., J. Chebath, et al. (2010). "Suspension culture of undifferentiated human embryonic and induced pluripotent stem cells." Stem Cell Rev 6(2): 248-59.

Aouadi, M., F. Bost, et al. (2006). "p38 mitogen-activated protein kinase activity commits embryonic stem cells to either neurogenesis or cardiomyogenesis." Stem Cells 24(5): 1399-406.

Bauwens, C. L., R. Peerani, et al. (2008). "Control of human embryonic stem cell colony and aggregate size heterogeneity influences differentiation trajectories." Stem Cells 26(9): 2300-10.

Beqqali, A., J. Kloots, et al. (2006). "Genome-wide transcriptional profiling of human embryonic stem cells differentiating to cardiomyocytes." Stem Cells: 2006-0054.

Bettiol, E., L. Sartiani, et al. (2007). "Fetal bovine serum enables cardiac differentiation of human embryonic stem cells." Differentiation 75(8): 669-81.

Boheler, K. R., J. Czyz, et al. (2002). "Differentiation of Pluripotent Embryonic Stem Cells Into Cardiomyocytes." Circ Res 91(3): 189-201.

Bongso, A., C. Y. Fong, et al. (1994). "Isolation and culture of inner cell mass cells from human blastocysts." Hum Reprod 9(11): 2110-7.

Braam, S. R., C. Denning, et al. (2008). "Feeder-free culture of human embryonic stem cells in conditioned medium for efficient genetic modification." Nat Protoc 3(9): 1435-43.

Braam, S. R., R. Passier, et al. (2009). "Cardiomyocytes from human pluripotent stem cells in regenerative medicine and drug discovery." Trends Pharmacol Sci 30(10): 536-45.

Braam, S. R., L. Tertoolen, et al. (2010). "Prediction of drug-induced cardiotoxicity using human embryonic stem cell-derived cardiomyocytes." Stem Cell Res 4(2): 107-16.

Burridge, P. W., D. Anderson, et al. (2007). "Improved human embryonic stem cell embryoid body homogeneity and cardiomyocyte differentiation from a novel V-96 plate aggregation system highlights interline variability." Stem Cells 25(4): 929-38.

Cao, F., R. A. Wagner, et al. (2008). "Transcriptional and functional profiling of human embryonic stem cell-derived cardiomyocytes." PLoS ONE 3(10): e3474.

Carlsson, L. (2006). "In vitro and in vivo models for testing arrhythmogenesis in drugs." J Intern Med 259(1): 70-80.

Casey, E. S., M. A. O'Reilly, et al. (1998). "The T-box transcription factor Brachyury regulates expression of eFGF through binding to a non-palindromic response element." Development 125(19): 3887-94.

Conlon, F. L., K. M. Lyons, et al. (1994). "A primary requirement for nodal in the formation and maintenance of the primitive streak in the mouse." Development 120(7): 1919-28.

Cuenda, A., J. Rouse, et al. (1995). "SB 203580 is a specific inhibitor of a MAP kinase homologue which is stimulated by cellular stresses and interleukin-1." FEBS Lett 364(2): 229-33.

Doetschman, T. C., H. Eistetter, et al. (1985). "The in vitro development of blastocyst-derived embryonic stem cell lines: formation of visceral yolk sac, blood islands and myocardium." J Embryol Exp Morphol 87: 27-45.

Ebert, A. D., J. Yu, et al. (2009). "Induced pluripotent stem cells from a spinal muscular atrophy patient." Nature 457(7227): 277-80.

Evans, M. J. and M. H. Kaufman (1981). "Establishment in culture of pluripotential cells from mouse embryos." Nature 292(5819): 154-6.

Farah, C. S. and F. C. Reinach (1995). "The troponin complex and regulation of muscle contraction." Faseb J 9(9): 755-67.

Finlayson, K., H. J. Witchel, et al. (2004). "Acquired QT interval prolongation and HERG: implications for drug discovery and development." Eur J Pharmacol 500(1-3): 129-42.

Freund, C., R. P. Davis, et al. (2010). "The first reported generation of human induced pluripotent stem cells (iPS cells) and iPS cell-derived cardiomyocytes in the Netherlands." Neth Heart J 18(1): 51-4.

Freund, C. and C. L. Mummery (2009). "Prospects for pluripotent stem cell-derived cardiomyocytes in cardiac cell therapy and as disease models." J Cell Biochem 107(4): 592-9.

Freund, C., D. Ward-van Oostwaard, et al. (2008). "Insulin redirects differentiation from cardiogenic mesoderm and endoderm to neuroectoderm in differentiating human embryonic stem cells." Stem Cells 26(3): 724-33.

Giudice, A. and A. Trounson (2008). "Genetic modification of human embryonic stem cells for derivation of target cells." Cell Stem Cell 2(5): 422-33.

Goumans, M. J., T. P. de Boer, et al. (2007). "TGF-beta1 induces efficient differentiation of human cardiomyocyte progenitor cells into functional cardiomyocytes in vitro." Stem Cell Res 1(2): 138-49.

Graichen, R., X. Xu, et al. (2008). "Enhanced cardiomyogenesis of human embryonic stem cells by a small molecular inhibitor of p38 MAPK." Differentiation 76(4): 357-70.

Guhr, A., A. Kurtz, et al. (2006). "Current State of Human Embryonic Stem Cell Research: An Overview of Cell Lines and their Usage in Experimental Work." Stem Cells.

Hancox, J. C., M. J. McPate, et al. (2008). "The hERG potassium channel and hERG screening for drug-induced torsades de pointes." Pharmacol Ther 119(2): 118-32.

He, J.-Q., Y. Ma, et al. (2003). "Human Embryonic Stem Cells Develop Into Multiple Types of Cardiac Myocytes: Action Potential Characterization." Circ Res 93(1): 32-39.

Hemmati-Brivanlou, A. and D. A. Melton (1992). "A truncated activin receptor inhibits mesoderm induction and formation of axial structures in Xenopus embryos." Nature 359(6396): 609-14.

Hescheler, J., B. K. Fleischmann, et al. (1997). "Embryonic stem cells: a model to study structural and functional properties in cardiomyogenesis." Cardiovasc Res 36(2): 149-62.

Hescheler, J., M. Halbach, et al. (2004). "Determination of electrical properties of ES cell-derived cardiomyocytes using MEAs." Journal of Electrocardiology 37(Supplement 1): 110-116.

ICH (2005). "E14: the clinical evalution of QT/QTc interval prolongation and proarrhythmic potential for non-antiarrhythmic drugs."

ICH (2005). "S7B: The nonclinical evalution of the potential for delayed ventricular repolarization (QT interval prolongation) by human pharmaceuticals."

Itskovitz-Eldor, J., M. Schuldiner, et al. (2000). "Differentiation of human embryonic stem cells into embryoid bodies compromising the three embryonic germ layers." Mol Med 6(2): 88-95.

Kehat, I., D. Kenyagin-Karsenti, et al. (2001). "Human embryonic stem cells can differentiate into myocytes with structural and functional properties of cardiomyocytes." J Clin Invest 108(3): 407-14.

Keller, G. M. (1995). "In vitro differentiation of embryonic stem cells." Curr Opin Cell Biol 7(6): 862-9.

Kim, C., M. Majdi, et al. (2010). "Non-cardiomyocytes influence the electrophysiological maturation of human embryonic stem cell-derived cardiomyocytes during differentiation." Stem Cells Dev 19(6): 783-95.

Kim, S. E., B. K. Kim, et al. (2007). "Comparative analysis of the developmental competence of three human embryonic stem cell lines in vitro." Mol Cells 23(1): 49-56.

Kispert, A., B. Koschorz, et al. (1995). "The T protein encoded by Brachyury is a tissue-specific transcription factor." Embo J 14(19): 4763-72.

Kita-Matsuo, H., M. Barcova, et al. (2009). "Lentiviral vectors and protocols for creation of stable hESC lines for fluorescent tracking and drug resistance selection of cardiomyocytes." PLoS One 4(4): e5046.

Kurosawa, H. (2007). "Methods for inducing embryoid body formation: in vitro differentiation system of embryonic stem cells." J Biosci Bioeng 103(5): 389-98.

Laflamme, M. A., K. Y. Chen, et al. (2007). "Cardiomyocytes derived from human embryonic stem cells in pro-survival factors enhance function of infarcted rat hearts." Nat Biotechnol 25(9): 1015-24.

Laflamme, M. A., J. Gold, et al. (2005). "Formation of human myocardium in the rat heart from human embryonic stem cells." Am J Pathol 167(3): 663-71.

Loser, P., J. Schirm, et al. (2010). "Human embryonic stem cell lines and their use in international research." Stem Cells 28(2): 240-6.

Lough, J. and Y. Sugi (2000). "Endoderm and heart development." Dev Dyn 217(4): 327-42.

Lu, H. R., E. Vlaminckx, et al. (2008). "Predicting drug-induced changes in QT interval and arrhythmias: QT-shortening drugs point to gaps in the ICHS7B Guidelines." Br J Pharmacol 154(7): 1427-38.

Martin, G. R. (1981). "Isolation of a pluripotent cell line from early mouse embryos cultured in medium conditioned by teratocarcinoma stem cells." Proc Natl Acad Sci U S A 78(12): 7634-8.

Martin, R. L., J. S. McDermott, et al. (2004). "The utility of hERG and repolarization assays in evaluating delayed cardiac repolarization: influence of multi-channel block." J Cardiovasc Pharmacol 43(3): 369-79.

McDevitt, T. C., M. A. Laflamme, et al. (2005). "Proliferation of cardiomyocytes derived from human embryonic stem cells is mediated via the IGF/PI 3-kinase/Akt signaling pathway." Journal of Molecular and Cellular Cardiology 39(6): 865-873.

Messina, E., L. De Angelis, et al. (2004). "Isolation and Expansion of Adult Cardiac Stem Cells From Human and Murine Heart." Circ Res 95(9): 911-921.

Miyazaki, H., H. Watanabe, et al. (2005). "QT PRODACT: sensitivity and specificity of the canine telemetry assay for detecting drug-induced QT interval prolongation." J Pharmacol Sci 99(5): 523-9.

Mohr, J. C., J. Zhang, et al. (2010). "The microwell control of embryoid body size in order to regulate cardiac differentiation of human embryonic stem cells." Biomaterials 31(7): 1885-93.

Moore, J. C., J. Fu, et al. (2008). "Distinct cardiogenic preferences of two human embryonic stem cell (hESC) lines are imprinted in their proteomes in the pluripotent state." Biochem Biophys Res Commun 372(4): 553-8.

Mummery, C. (2010). "Sorting cardiomyocytes: a simple solution after all?" Nat Methods 7(1): 40-2.

Mummery, C., D. Ward-van Oostwaard, et al. (2003). "Differentiation of human embryonic stem cells to cardiomyocytes: role of coculture with visceral endoderm-like cells." Circulation 107(21): 2733-40.

Mummery, C. L., A. Feijen, et al. (1986). "Establishment of a differentiated mesodermal line from P19 EC cells expressing functional PDGF and EGF receptors." Exp Cell Res 165(1): 229-42.
Ng, E. S., R. P. Davis, et al. (2005). "Forced aggregation of defined numbers of human embryonic stem cells into embryoid bodies fosters robust, reproducible hematopoietic differentiation." Blood 106(5): 1601-3.
Niebruegge, S., C. L. Bauwens, et al. (2009). "Generation of human embryonic stem cell-derived mesoderm and cardiac cells using size-specified aggregates in an oxygen-controlled bioreactor." Biotechnol Bioeng 102(2): 493-507.
Oh, H., S. B. Bradfute, et al. (2003). "Cardiac progenitor cells from adult myocardium: homing, differentiation, and fusion after infarction." Proc Natl Acad Sci U S A 100(21): 12313-8.
Osafune, K., L. Caron, et al. (2008). "Marked differences in differentiation propensity among human embryonic stem cell lines." Nat Biotechnol 26(3): 313-5.
Park, I. H., N. Arora, et al. (2008). "Disease-specific induced pluripotent stem cells." Cell 134(5): 877-86.
Passier, R., D. W. Oostwaard, et al. (2005). "Increased cardiomyocyte differentiation from human embryonic stem cells in serum-free cultures." Stem Cells 23(6): 772-80.
Passier, R., L. W. van Laake, et al. (2008). "Stem-cell-based therapy and lessons from the heart." Nature 453(7193): 322-9.
Pekkanen-Mattila, M., H. Chapman, et al. (2010). "Human embryonic stem cell-derived cardiomyocytes: demonstration of a portion of cardiac cells with fairly mature electrical phenotype." Exp Biol Med (Maywood) 235(4): 522-30.
Pekkanen-Mattila, M., E. Kerkela, et al. (2009). "Substantial variation in the cardiac differentiation of human embryonic stem cell lines derived and propagated under the same conditions--a comparison of multiple cell lines." Ann Med 41(5): 360-70.
Pekkanen-Mattila, M., M. Pelto-Huikko, et al. (2010). "Spatial and temporal expression pattern of germ layer markers during human embryonic stem cell differentiation in embryoid bodies." Histochem Cell Biol 133(5): 595-606.
Pera, M. F. and A. O. Trounson (2004). "Human embryonic stem cells: prospects for development." Development 131(22): 5515-25.
Pollard, C. E., J. P. Valentin, et al. (2008). "Strategies to reduce the risk of drug-induced QT interval prolongation: a pharmaceutical company perspective." Br J Pharmacol 154(7): 1538-43.
Pyle, A. D., L. F. Lock, et al. (2006). "Neurotrophins mediate human embryonic stem cell survival." Nat Biotechnol 24(3): 344-350.
Redfern, W. S., L. Carlsson, et al. (2003). "Relationships between preclinical cardiac electrophysiology, clinical QT interval prolongation and torsade de pointes for a broad range of drugs: evidence for a provisional safety margin in drug development." Cardiovasc Res 58(1): 32-45.
Reppel, M., F. Pillekamp, et al. (2004). "Microelectrode arrays: a new tool to measure embryonic heart activity." J Electrocardiol 37 Suppl: 104-9.
Roden, D. M. (2004). "Drug-induced prolongation of the QT interval." N Engl J Med 350(10): 1013-22.

Sachlos, E. and D. T. Auguste (2008). "Embryoid body morphology influences diffusive transport of inductive biochemicals: a strategy for stem cell differentiation." Biomaterials 29(34): 4471-80.

Scott, C. T., J. B. McCormick, et al. (2009). "And then there were two: use of hESC lines." Nat Biotechnol 27(8): 696-7.

Shah, R. R. and L. M. Hondeghem (2005). "Refining detection of drug-induced proarrhythmia: QT interval and TRIaD." Heart Rhythm 2(7): 758-72.

Showell, C., O. Binder, et al. (2004). "T-box genes in early embryogenesis." Dev Dyn 229(1): 201-18.

Snir, M., I. Kehat, et al. (2003). "Assessment of the ultrastructural and proliferative properties of human embryonic stem cell-derived cardiomyocytes." Am J Physiol Heart Circ Physiol 285(6): H2355-63.

Synnergren, J., S. Adak, et al. (2008). "Cardiomyogenic gene expression profiling of differentiating human embryonic stem cells." J Biotechnol 134(1-2): 162-70.

Synnergren, J., K. Akesson, et al. (2008). "Molecular signature of cardiomyocyte clusters derived from human embryonic stem cells." Stem Cells 26(7): 1831-40.

Tada, M., E. S. Casey, et al. (1998). "Bix1, a direct target of Xenopus T-box genes, causes formation of ventral mesoderm and endoderm." Development 125(20): 3997-4006.

Tada, M. and J. C. Smith (2000). "Xwnt11 is a target of Xenopus Brachyury: regulation of gastrulation movements via Dishevelled, but not through the canonical Wnt pathway." Development 127(10): 2227-38.

Takahashi, K., K. Tanabe, et al. (2007). "Induction of pluripotent stem cells from adult human fibroblasts by defined factors." Cell 131(5): 861-72.

Takahashi, T., B. Lord, et al. (2003). "Ascorbic acid enhances differentiation of embryonic stem cells into cardiac myocytes." Circulation 107(14): 1912-6.

Thierfelder, L., H. Watkins, et al. (1994). "Alpha-tropomyosin and cardiac troponin T mutations cause familial hypertrophic cardiomyopathy: a disease of the sarcomere." Cell 77(5): 701-12.

Thomson, J. A., J. Itskovitz-Eldor, et al. (1998). "Embryonic stem cell lines derived from human blastocysts." Science 282(5391): 1145-7.

Tobacman, L. S. (1996). "Thin filament-mediated regulation of cardiac contraction." Annu Rev Physiol 58: 447-81.

Tran, T. H., X. Wang, et al. (2009). "Wnt3a-induced mesoderm formation and cardiomyogenesis in human embryonic stem cells." Stem Cells 27(8): 1869-78.

van Laake, L. W., R. Passier, et al. (2009). "Improvement of mouse cardiac function by hESC-derived cardiomyocytes correlates with vascularity but not graft size." Stem Cell Res 3(2-3): 106-12.

van Laake, L. W., R. Passier, et al. (2008). "Human embryonic stem cell-derived cardiomyocytes and cardiac repair in rodents." Circ Res 102(9): 1008-10.

van Laake, L. W., R. Passier, et al. (2007). "Human embryonic stem cell-derived cardiomyocytes survive and mature in the mouse heart and transiently improve function after myocardial infarction." Stem Cell Res 1(1): 9-24.

Vandenberg, J. I., B. D. Walker, et al. (2001). "HERG K+ channels: friend and foe." Trends Pharmacol Sci 22(5): 240-6.

Vidarsson, H., J. Hyllner, et al. (2010). "Differentiation of human embryonic stem cells to cardiomyocytes for in vitro and in vivo applications." Stem Cell Rev 6(1): 108-20.

Wobus, A. M., G. Wallukat, et al. (1991). "Pluripotent mouse embryonic stem cells are able to differentiate into cardiomyocytes expressing chronotropic responses to adrenergic and cholinergic agents and Ca2+ channel blockers." Differentiation 48(3): 173-82.

Xu, C., S. Police, et al. (2002). "Characterization and enrichment of cardiomyocytes derived from human embryonic stem cells." Circ Res 91(6): 501-8.

Xu, X. Q., R. Graichen, et al. (2008). "Chemically defined medium supporting cardiomyocyte differentiation of human embryonic stem cells." Differentiation 76(9): 958-70.

Xu, X. Q., S. Y. Soo, et al. (2009). "Global expression profile of highly enriched cardiomyocytes derived from human embryonic stem cells." Stem Cells 27(9): 2163-74.

Yang, L., M. H. Soonpaa, et al. (2008). "Human cardiovascular progenitor cells develop from a KDR+ embryonic-stem-cell-derived population." Nature 453(7194): 524-8.

Yirme, G., M. Amit, et al. (2008). "Establishing a dynamic process for the formation, propagation, and differentiation of human embryoid bodies." Stem Cells Dev 17(6): 1227-41.

Zareba, W. and I. Cygankiewicz (2008). "Long QT syndrome and short QT syndrome." Prog Cardiovasc Dis 51(3): 264-78.

Zhang, J., G. F. Wilson, et al. (2009). "Functional cardiomyocytes derived from human induced pluripotent stem cells." Circ Res 104(4): e30-41.

27

Recent Advances in Controlling Cardiacmyoctye Differentiation from Embryonic Stem Cells

Dong-Bo Ou[1], Qiang-Sun Zheng[1], Yong He, Di Zeng and Ji-Wei Teng
Department of Cardiology, Tangdu Hospital, Fourth Military Medical University, Xi' an China

1. Introduction

As a high proliferative ability cell line, embryonic stem (ES) cells can be propagated in undifferentiated state, while maintaining their pluripotency to form various kinds of adult tissue cells (Thomson et al., 1998). Under appropriate conditions, ES cells can from embryoid bodies (EBs) and subsequent differentiation into cardiacmyoctyes (CMs) that retain the function of excitability and spontaneous contractions (Wobus et al., 1991) (Maltsev et al., 1994). These spontaneously beating cells contained in EBs have almost the same ion channels that control pacemaker function in the heart (He et al., 2003) (Yanagi et al., 2007).
Although multipotent progenitors have previously been shown to give rise to CMs, smooth muscle, and endothelial cells, the mechanism governing the generation of large numbers of differentiated progeny remains poorly understood. Current research indicates that the level of differentiation is low and that few cultured EBs will ultimately convert into functional CMs (Norstrom et al., 2006). The potential use of ES cells to replace functional loss of particular tissues may depend on efficient differentiation protocols to derive tissue-specific cells. By manipulating the culture conditions in which ES cells differentiate, it has been possible to control and restrict the differentiation pathways and thereby generate cultures enriched in lineage-specific cells in vitro. Regulation of multipotent cardiac progenitor cell expansion and subsequent differentiation into CMs, smooth muscle or endothelial cells is a fundamental aspect of basic cardiovascular biology and cardiac regenerative medicine.
More recently, direct reprogramming of adult somatic cells to become pluripotent ES-like cells (a.k.a. induced pluripotent stem cells, iPS cells) has been achieved. The cardiogenic potential of iPS cells is comparable to that of ES cells and that iPS-derived CMs (iPS-CMs) possess all fundamental functional elements of a typical cardiac cell, including spontaneous beating, hormonal regulation, cardiac ion channel expression and contractility. Therefore, iPS-CMs can be regarded as a potentially valuable source of cells for in vitro studies and cellular cardiomyoplasty (Pfannkuche et al., 2009). The availability of human ES cells and iPS cells, and their successful differentiation into genuine human cardiac cells have enabled researchers to gain novel insights into the early development of the human heart as well as to pursue the revolutionary paradigm of heart regeneration.

---

[1] The first 2 authors contributed equally to this work.

2. Embryoid body formation

There are two major ways for EB from ES cells in vitro: suspension culture and hanging drop methods. In suspension culture, ES cells will differentiate into aggregates known as EBs in a manner similar to the early embryo. Suspension methods allow spontaneous aggregation of EBs but usually give rise to heterogeneous cell clusters. The heterogeneous size and shape of EBs resulting from suspension culture can influence their differentiation potential. Hanging drop methods can provide relatively uniform EB size but technically challenge by limited number of EBs for characterization. Recently, other methods such as using spinner flasks and bioreactors for stirred-suspension culture have been utilized. Enhanced growth of EBs in spinner flasks has been reported due to effective nutrient and oxygen supply. In addition to stirred-suspension culture for scalable production of EBs can provide a readily available source of CMs (Lecina et al., 2010). These strategies have been shown to be advantageous compared to conventional suspension method due to their ability to control EB-EB interactions.

Without induced differentiation, the differentiation level is low and that few cultured EBs will ultimately convert into functional CMs. Kehat et al. (Kehat et al., 2001) cultured human ES cells in suspension and plated to form aggregates termed EBs. Spontaneously contracting Cells appeared in only 8.1% of those EBs. Controlling the differentiating pathways permitting efficient generation of more mature and functional cells for basic studies or cardiac repair applications is a pressing need. As molecular mechanisms and signaling pathways leading to efficient differentiation from ES cells have not been understood and well-summarized, the potential factors that impact CM differentiation from ES cells include cell factors, drugs, hormone, culture method, and extracellular matrix.

3. Cell factors and CM differentiation

Schuldiner el at. (Schuldiner et al., 2000) examined the potential of eight growth factors including basic fibroblast growth factor (bFGF), transforming growth factor beta1 (TGF-beta1), activin-A, bone morphogenic protein 4 (BMP-4), hepatocyte growth factor (HGF), epidermal growth factor (EGF), beta nerve growth factor (βNGF), and retinoic acid to direct the differentiation of human ES-derived cells in vitro. They found that the EBs differentiate selectively into mesodermal, endodermal, or ectodermal lineages. However, these studies did not result in homogeneous differentiation of ES cells.

Many differentiating factors, including hepatocyte growth factor (HGF), transforming growth factor beta1 (TGF-β1), bone morphogenic protein (BMP)-2/4, and fibroblast growth factor (FGF) have been investigated in attempts to improve the efficiency of specific CM differentiation processes in mouse or human ES cells (Gassanov et al., 2004) (Roggia et al., 2007) (Schuldiner et al., 2000) (Ronca et al., 2009). These investigations have had some early encouraging results, but pluripotency and cellular differentiation are intricate biological processes that are coordinately regulated by a complex set of factors and epigenetic regulators, which need increasing research aiming to understand the molecular mechanisms that regulate stem cell differentiation.

Growth factor signaling is required for cellular differentiation, tissue morphogenesis, and tissue homeostasis. Wnt proteins are one important family of growth factors and its signaling is an important regulator of differentiation and morphogenesis that can also control stem cell fates. Tran et al. (Tran et al., 2009) found that mesendoderm formation and

CM differentiation were enhanced by early and transient treatment of human ES cells with Wnt3a. The induction of mesoderm and subsequent cardiac differentiation from human ES cells requires fine-tuned cross talk between activin A/BMP4 and Wnt/beta-catenin pathways (Paige et al., 2010). Cripto is one of epidermal growth factors and its signaling pathway is recently reported as essential role for cardiac myogenesis in ES cells. D'Aniello et al.(D'Aniello et al., 2009) demonstrated that APJ/apelin in the Cripto signaling pathway governs mesoderm patterning and CM differentiation via activation of mitogen-activated protein kinase/p70S6 through coupling to a Go/Gi protein. Hypoxia failed to induce differentiation of mouse ES cells into CMs in the absence of Cripto-1 expression, demonstrating that Cripto-1 is required for hypoxia to fully differentiate mouse ES cells into CMs (Bianco et al., 2009). Phosphorylation of fibroblast growth factor receptor-1 is also reported as essential process required for CM differentiation in ES cells (Ronca et al., 2009).

However, the differentiation of ES cells inevitably results in a heterogeneous mixture of cell types present in EBs. It would be more difficult to transform these into a homogenous population of CMs or to stop their differentiation at the functional CM stage. Neuregulin (NRG)-1beta/ErbB signaling regulates the ratio of nodal- to working-type cells in differentiating human ES cell-CM cultures and presumably functions similarly during early human heart development. By manipulating NRG-1beta/ErbB signaling, it will be possible to generate preparations of enriched working-type CMs (Zhu et al., 2010). Calreticulin may have a housekeeping role to play in mature CMs as well as during cardiomyogenesis. Calreticulin, an ER Ca (2+) storage protein, is a crucial regulator of cardiomyogenesis whose presence is required for controlled CM development from ES cells (Papp et al., 2009). Sik1 (salt inducible kinase 1), a serine/threonine kinase that belongs to the stress- and energy-sensing AMP-activated protein kinase family, involves in cardiac cell differentiation and/or heart development. Romito et al. (Romito et al., 2010) studied the role of sik1 in cardiac differentiation and found that sik1-mediated effects are specific for cardiomyogenesis regulating cardiomyoblast cell cycle exit toward terminal differentiation.

MicroRNAs are a newly discovered endogenous class of small noncoding RNAs that play important posttranscriptional regulatory roles by targeting messenger RNAs for cleavage or translational repression. microRNAs as important transcriptional regulators may provide a new means of manipulating stem cell fate (Sartipy et al., 2009). By modulating miR-1 and -499 expression levels, human ES cells function can be altered and differentiation directed, thereby enhancing cardiomyogenic differentiation (Sluijter et al., 2010).

4. Drugs used for the CM differentiation of embryonic stem cells

Dimethyl sulfoxide (DMSO), 5-azacytidine, retinoic Acid (RA), ascorbic acid have been investigated in attempts to improve the efficiency of CM differentiation processes in mouse or human ES cells (Paquin et al., 2002) (Yoon et al., 2006) (Drab et al., 1997) (Takahashi et al., 2003). Recently, the synergy of these drugs in inducing ES cell differentiation into matured and functional CMs is quite attractive. Chan et al. (Chan et al., 2009) have investigated Salvianolic acid B-vitamin C synergy in cardiac differentiation from ES cells and found that Salvianolic acid B (saB)- ascorbic acid synergy improves ES cell differentiation into CMs.

Recently, accumulating evidence points to reactive oxygen species (ROS) as important signaling molecules for CM differentiation from ES cells. Endogenous ROS control is important for CM formation from ES cells (Crespo et al., 2010). Nitric oxide (NO) has been shown to promote ES cell differentiation by increasing both the number and the size of beating foci in EB outgrowths (Kanno et al., 2004). Exogenous electric fields have been implied in CM differentiation of mouse ES cells and the generation of ROS, which plays a role in CM differentiation of human ES cells, through mechanisms associated with the intracellular generation of ROS (Serena et al., 2009).

5. Hormones used for the CM differentiation from embryonic stem cells

Recent researches indicate there are some hormone receptors existing in ES cells, so hormone used as inductor to promote the CM differentiation from ES cells receives increasing attention. ES cells expressed cell surface kappa opioid receptors. The prodynorphin gene and its product, dynorphin B, have been found to promote cardiogenesis in embryonic cells by inducing the expression of GATA-4 and Nkx-2.5, two transcription factor-encoding genes essential for cardiogenesis (Ventura et al., 2003). Oxytocin receptor was discovered in undifferentiated ES cells and derived differentiated cells, which plays an important role in cardiogenesis by promoting CM differentiation. Treatment with oxytocin improves the EBs with spontaneous contraction, but has no effect on ultrastructural characteristics of CMs in any stage of development (Paquin et al., 2002) (Hatami et al., 2007). Thyroid hormone is essential for normal cardiac development and function, Lee et al. (Lee et al., 2010) showed that Thyroid (3) supplementation promotes cardiac differentiation of ES cells and enhances maturation of electrophysiological, as well as calcium homeostasis, properties of ES cell-derived CMs.

6. Co-culture differentiation

Although progress has been made towards differentiating stem cells to specific cell lineages, the efficiency is often poor and the number of cells generated is not suitable for therapeutic usage. Recent studies demonstrated that controlling the stem cell microenvironment can influence differentiation. As the differentiation of ES cells closes the development of the embryonic heart (Banach et al., 2003), factors that contribute to essential functions during early embryogenesis are expected to be involved in the formation of EBs (Behfar et al., 2002). However, the precise growth factor combinations that enhance cell differentiation in ES cells are unclear (Schuldiner et al., 2000). Consequently, ES cells are co-cultured with special cells or conditioned media to promote differentiation. For example, when ES cells are co-cultured with visceral-endoderm-like (END-2) cells, the majority (>90%) of the ES -CMs have a phenotype similar to fetal ventricular cells (Mummery et al., 2003). The efficiency of cardiogenic differentiation in ES cells can be readily enhanced by a culture medium that has been conditioned by END-2 cells (Graichen et al., 2008). Similarly, when medium conditioned by mouse embryo fibroblasts is used, the homogeneity of beating EBs can be significantly improved (Burridge et al., 2007).

The niche in which stem cells reside and differentiate is a complex physicochemical microenvironment that regulates cell function. Local microenvironments are considered to be the key inducers in directing the site-specific differentiation of ES cells. Recently, it is reported that endothelial cell play an essential role in facilitating CM differentiation from pluripotent stem cells and regulate regeneration of CMs via EphB4 signaling (Chen et al., 2010). Nevertheless, the effect and the mechanism of the cardiac microenvironment on the development of CM differentiation and EB growth have not yet been systematically studied. We developed a novel protocol to generate functional CMs from ES cells by an in vitro indirect co-culture inducing strategy that looks promising for tissue engineering applications (fig.1). The propagation and CM differentiation of an ES cell line were studied in prolonged culture in this system (fig. 2). ES cells cultured on membranes by indirect co-culture with native cardiac cells were significantly enhanced ES cells to differentiate into CMs that possessed expression of cardiac specific genes and functions of excitability and

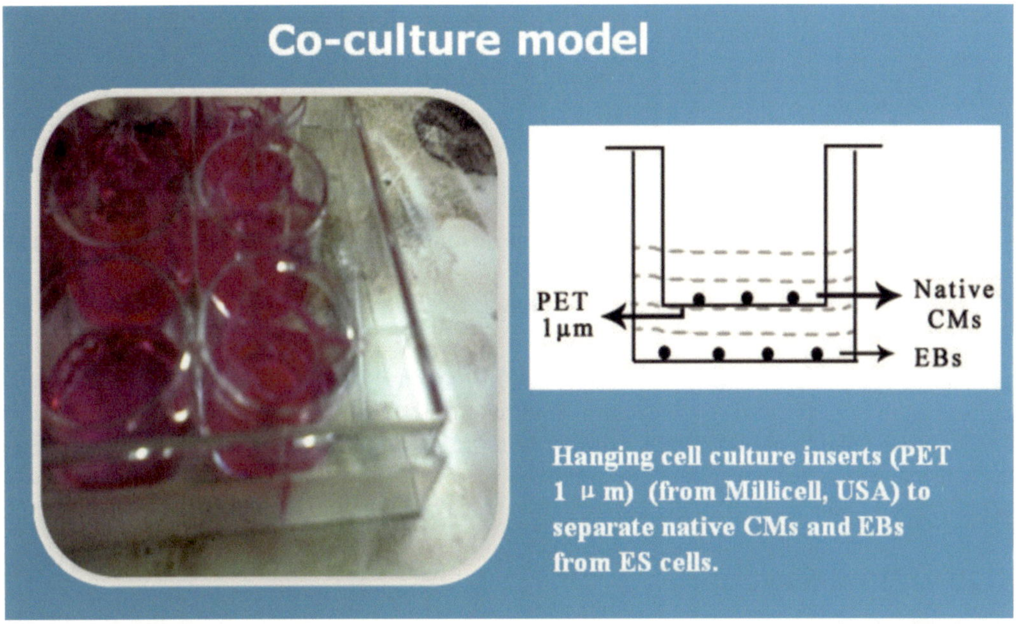

Fig. 1. The indirect co-culture system was used for CM differentiation from mouse ES cells.

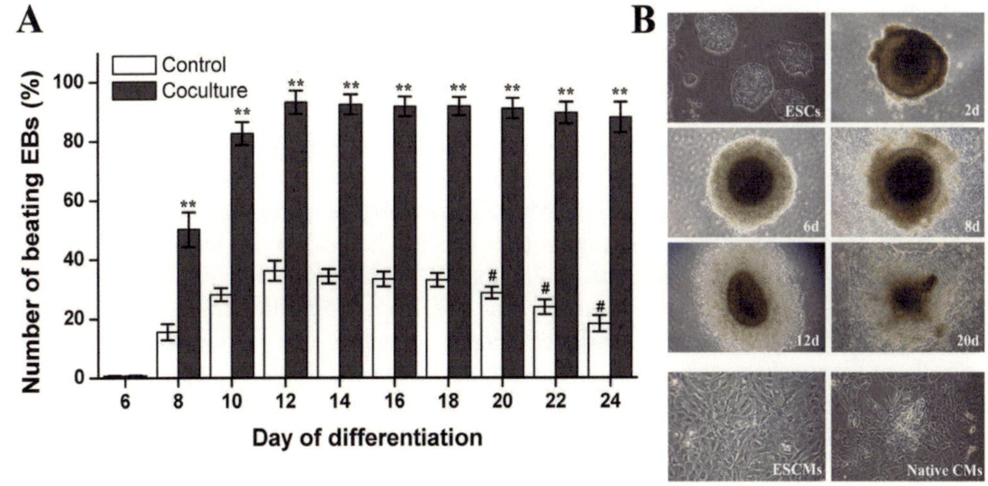

Fig. 2. The differentiation of embryonic stem cells (ES cells).
(A) The percentage of embryoid bodies (EBs) with spontaneously contracting areas in control and co-culture groups during plating culture.
(B) The in vitro differentiation process of ES cells in co-culture system.

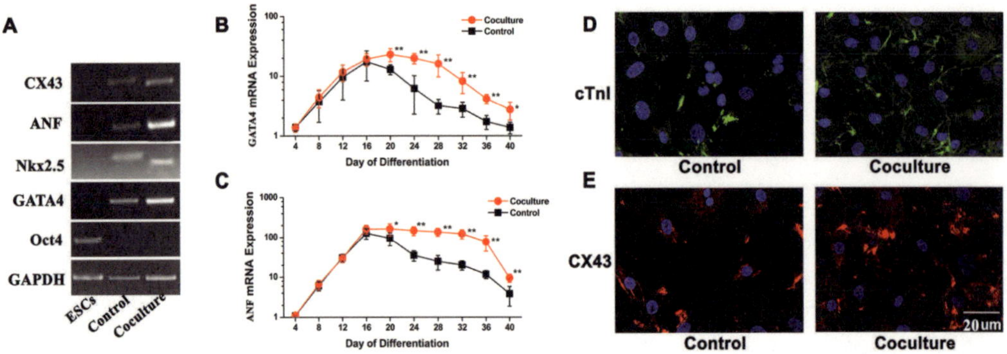

Fig. 3. The expression of cardiac markers in ES cells. (A) OCT4, GATA4, Nkx2.5, ANF and CX43 expressions were almost increased in the co-culture system. (B),(C) Real time-PCR was performed to detect mRNA expressions of GATA and ANF in prolonged time course. *: $P<0.05$, **: $P<0.01$. (D),(E) the protein expression of cardiac troponin I (cTnI) and connexin 43 (Cx43) in co-culture system.

spontaneous contractions (fig. 3). This co-culture system is a significant advance in ES cells culture methods, providing possibility to efficiently control the differentiation of ES cells and to more reliably obtain a large number of more mature and homogenous CMs for large-scale tissue engineering applications and other stem cell-based regenerative medicine (Ou et al., 2009a).

7. 2D vs. 3D differentiation

Previously the most common model in vitro for cell research has been 2D monolayer culture. Even though EB has 3D structures, the researches on the differentiation of ES cells are mainly based on 2D cultured systems. However, these 2D cultured systems don't mimic the physiological environment in vivo and may lead to low differentiated level of ES cells. Significant differences were found in the differentiation efficiency of ES cells when they were cultured in 2D and 3D environment (Liu et al., 2006).

The extracellular microenvironment for 3D differentiation plays a significant role in regulating ES cells to differentiate into special lineages, as well as in cell migration and proliferation (Philp et al., 2005) (Chen et al., 2007). The extracellular matrix is in close contact with cells and involved in their maintenance and regulation. The mechanical and chemical signals (e.g., integrin signals) elicited by the extracellular matrix regulate the activities of cytoplasmic kinases, receptors for growth factor and ion channels, impacting differentiation of stem cells. Extracellular matrix proteins independently trigger differentiation of human mesenchymal stem cells and that differentiation in this context can be guided down multiple lineages using the same Extracellular matrix protein stimulus (Santiago et al., 2009). A new scalable bioprocessing system as a scalable platform for CM production has been developed (Lecina et al., 2010). Use of a 3-D cuboidal microwell system to culture human ES cells in colonies of defined dimensions, 100-500 microm in lateral dimensions and 120 microm in depth, enabled formation of more uniform-sized EBs. The regulation of microwell-engineered EB size can be used for more efficient and reproducible formation of human ES cell-CMs (Mohr et al., 2010). In addition, proliferation of stem cell-derived CMs appears to be regulated by microtopography through tension-generation of contractility in the third-dimension (Biehl et al., 2009).

Bioreactors, recent methods to achieve CM differentiation by engineering the stem cell microenvironment, are used to control the differentiation and produce large numbers of desired cells. Based on bioreactors, it is very likely to control the differentiation of ES cells: (1) biomaterials technology, biomaterials with different growth factors or morphogenetic factors delivery systems can be utilized to mimic embryogenic process, our unpublished data demonstrates that specific factors released from biomaterials via diffusion can enhance long-term survival and differentiation of ES cells, and (2) the use of bioreactors, cultivation in rotating bioreactors can promote and maintain CMs electrophysiology and molecular properties (Bursac et al., 2003).

8. The potential applications

The availability of ES and iPS cells and their successful differentiation into genuine human cardiac cells have enabled researchers to gain novel insights into the early development of the human heart as well as to pursue the revolutionary paradigm of heart regeneration. Their potential applications for ES cells in cardiac research area include developmental biology, functional genomics, pharmacological testing, cell therapy, organ regeneration, and biological pacemaker.

8.1 Pharmacological testing

Recent withdrawals of prescription drugs from clinical use because of unexpected side effects on the heart have highlighted the need for more reliable cardiac safety pharmacology assays. Human or mouse ES cell-derived CMs serve together as a valuable model for drug safety screening and advance preclinical drug development. Assays based on human ES cell-derived CMs could complement or potentially replace some of the preclinical cardiac toxicity screening tests currently used for lead optimization and further development of new drugs (Braam et al., 2010) (Liang et al., 2010). Human ES cell-derived CMs have been used to predicate of drug-induced cardiotoxicity, e.g., to study cardioprotective effects of dexamethasone in doxorubicin cardiotoxicity (Farokhpour et al., 2009) (Braam et al., 2010).

8.2 Cell therapy

ES cells, which can be differentiated into cardiac progenitors and CMs, represent a candidate cell source for cardiac cell therapy. However, it is not known what specific cell type or condition is the most appropriate for transplantation. It is reported that transplanted mouse ES cells in the infarcted heart inhibit apoptosis, fibrosis, and hypertrophy, thereby reducing adverse remodeling, resulting in the improvement of cardiac function (Singla et al., 2007) (Singla and McDonald, 2007) (Xie et al., 2007). Several recent studies have demonstrated that human ES cell-derived CMs survive after transplantation into infarcted rodent hearts, form stable cardiac implants, and result in preserved contractile function. To investigate the ischemic environment of host myocardium affects transplanted pluripotent cells, van Laake et al. (van Laake et al., 2009) cultured mouse ES cells in medium containing ischemic myocardial interstitial fluid (iMIF) and found that the ischemic/infarcted environment is favorable to stem cell-mediated angiogenesis, but hostile to cardiac myogenesis. These findings indicated mouse ES cell-mediated improvement of cardiac function after transplantation of pluripotent cells do not reflect remuscularization. Although cell transplantation has modestly improved cardiac function, major challenges including increasing cell survival, engraftment, and functional integration with host tissue.

8.3 Cardiac muscle regeneration

Tissue-engineered cardiac muscle transplantation may have advantages over direct cell transplantation or replacement therapy for myocardial defect. The use of engineered heart tissue as a model system to accelerate development of cardiac cell therapy strategies has been demonstrated (Song et al., 2010). Human ES cell-derived CMs alone or with human endothelial cells (human umbilical vein endothelial cells) and embryonic fibroblasts (triculture constructs) were seeded onto biodegradable porous scaffolds. This tissue-engineered human vascularized cardiac muscle have been established in vitro and transplanted in vivo to form stable grafts (Lesman et al., 2010b) .

Effective engineering of viable thick complex tissue-constructs requires intense vascularization. More recently, a beating human cardiac muscle-construct containing an endothelial network was established by co-culturing human embryonic stem cell-derived-CMs, fibroblasts, and endothelial cells within biodegradable scaffolds (Lesman et al., 2010a). The success of cardiac muscle regeneration and transplantation is dependent on a combination of several factors, such as: 1) the formation of new blood vessels, 2) the release of pro-survival, pro-angiogenic and anti-inflammatory factors (paracrine effect), and 3) the functional contribution of CMs and functional integration with host tissue.

Components of the extracellular matrix are considered to be important physiological regulators and provide mechanical cues, direct differentiation and improve cell engraftment into damaged tissue. However, there are also adverse effects of scaffold materials, e.g. necrosis at the tissue core, low vascularization, and poor survival after transplantation. Scaffold-free for vascularized human cardiac muscle tissue that markedly improved viability after transplantation has been developed (Stevens et al., 2009).

8.4 Biological pacemaker

Sudden cardiac death due to abnormal heart rhythm from sick sinus syndrome kills millions of people each year, the reduction of the morbidity and mortality depends on how we can restore the generation of the cardiac impulse effectively. A conventional approach is implantation of electronic pacemakers, which are commonly used to treat cardiac impulse generation defects. However, there are many shortcomings for electronic pacemakers in clinical applications, such as the limited battery life, lack of catecholamine responsiveness, and associating with inconvenience in daily life (Rosen et al., 2004). Biological cardiac pacemaker, mimicking the sinoatrial node (SA-node), has been served as a promising way to restore the generation of the cardiac impulse for better alternative to the present routine.

ES cells can be induced to differentiate into CMs with excitability and spontaneous contraction activities (Wobus et al., 1991) (Maltsev et al., 1994), which raises the prospect of the probable therapeutic application of ES cells in cardiac pacing. Their use as biological pacemakers has also been explored (Kehat et al., 2004) (Xue et al., 2005) (Menasche, 2004 ; Yanagi et al., 2007). Kehat et al. (Kehat et al., 2004) had discussed the potential of human ES cell-derived CMs to act as a rate-responsive biological pacemaker. The transplanted human ES cell-derived CMs can functionally integrate with other quiescent ventricular cells and pace the heart of swine with complete AV block, as assessed by detailed three-dimensional electrophysiological mapping and histopathological examination. Xue et al.'s work (Xue et al., 2005) was similar with Kehat's. A functional human ES cell-derived pacemaker could be implanted in the left ventricle in vivo and successful spread of membrane depolarization was confirmed from surrounding myocardium in the site of injection.

Nevertheless, the use of ES cells in cardiac pacing is challenged by how to control the differentiation of ES cells and overcome the neoplasia, proarrhythmia or immunogenicity after transplantation. As a potential approach to solve these difficult problems, tissue engineering techniques may provide a precise control on the different cell components of multicellular aggregates and the forming construct with defined architectures and function properties (Ou et al., 2009b). For any biological pacemaker to be considered a potential medical therapy, it must functionally integrate with the heart and provide an impulse initiation enough and steady in the conducting system to ensure physiological activation of the heart. The applications of tissue engineering techniques not only provide a 3D environment for cells growth and expanding, but also provide structural support for higher order tissue organization and remodeling. The combined interactions among ES cell-derived pacemaker cells, supporting cells and matrices may result in a steady functional unit to induce rhythmic electrical and contractile activities and completely reproduce pacemaker properties.

9. The challenges in stem cell research

These investigations for ES cell differentiation have had some early encouraging results, but pluripotency and cellular differentiation are intricate biological processes that are coordinately regulated by a complex set of factors and epigenetic regulators, which need increasing research aiming to understand the molecular mechanisms that regulate stem cell differentiation. The major direction of human ES cell use is derivation of a specific differentiated progeny, which has lower proliferative potential and immune privilege, yet poses fewer risks. There is report that immaturity of the sarcoplasmic reticulum and the beta-adrenergic response can be found in CMs from ES Cells and iPS cells (Xi et al., 2010). The heterogeneity in electrical properties of the human ES cell-CMs is exist (Pekkanen-Mattila et al., 2010).

Pluripotent stem cell biology and technology is in need of further investigation and development. The development of efficient and reproducible culture systems for culturing ES cells in complete xeno-free conditions to reduce the risk of cross-transfer of pathogens without loss pluripotency is an essential pre-requisite for regenerative medicine. Synthetic, xeno-free, scalable surfaces that support the self-renewal and differentiation of ES Cells and iPS cells will be useful for both research purposes and development of cell therapies (Melkoumian et al., 2010).

Efficiently controlling the differentiation of ES Cells to produce more mature and homogenous CMs is currently the most challenging task. Thus, a future challenge will be to design strategies that eventually may allow the cells to reach a higher degree of maturation in vitro.

The recent development of reprogramming of differentiated human somatic cells to iPS cells should overcome ethics obstacles and facilitate clinical applications of stem cells. The most advantages of reprogramming is that it allows the establishment of patient- and disease-specific in vitro models of human hereditary diseases for pathophysiologic and developmental studies. However, although human ES cells and human iPS cells have been shown to share a number of similarities, there are still differences electrophysiology properties between human ES cells and human iPS cells (Jiang et al., 2010). It has been showed that foreign genes were silenced or removed after reprogramming, but those approaches have low reprogramming efficiency, and either leave residual vector sequences, or require tedious steps. Whether reprogramming methods can be improved will depend on a better understanding on the molecular cell biology of pluripotent stem cells.

10. References

Banach K., Halbach M.D., Hu P., Hescheler J. & Egert U. (2003). Development of electrical activity in cardiac myocyte aggregates derived from mouse embryonic stem cells. Am J Physiol Heart Circ Physiol, 284, H2114-2123

Behfar A., Zingman L.V., Hodgson D.M., Rauzier J.M., Kane G.C., Terzic A. & Puceat M. (2002). Stem cell differentiation requires a paracrine pathway in the heart. Faseb J, 16, 1558-1566

Bianco C., Cotten C., Lonardo E., Strizzi L., Baraty C., Mancino M., Gonzales M., Watanabe K., Nagaoka T., Berry C., Arai A.E., Minchiotti G. & Salomon D.S. (2009). Cripto-1 is required for hypoxia to induce cardiac differentiation of mouse embryonic stem cells. Am J Pathol, 175, 2146-2158

Biehl J.K., Yamanaka S., Desai T.A., Boheler K.R. & Russell B. (2009). Proliferation of mouse embryonic stem cell progeny and the spontaneous contractile activity of cardiomyocytes are affected by microtopography. Dev Dyn, 238, 1964-1973

Braam S.R., Tertoolen L., van de Stolpe A., Meyer T., Passier R. & Mummery C.L. (2010). Prediction of drug-induced cardiotoxicity using human embryonic stem cell-derived cardiomyocytes. Stem Cell Res, 4, 107-116

Burridge P.W., Anderson D., Priddle H., Barbadillo Munoz M.D., Chamberlain S., Allegrucci C., Young L.E. & Denning C. (2007). Improved human embryonic stem cell embryoid body homogeneity and cardiomyocyte differentiation from a novel V-96 plate aggregation system highlights interline variability. Stem Cells, 25, 929-938

Bursac N., Papadaki M., White J.A., Eisenberg S.R., Vunjak-Novakovic G. & Freed L.E. (2003). Cultivation in rotating bioreactors promotes maintenance of cardiac myocyte electrophysiology and molecular properties. Tissue Eng, 9, 1243-1253

Chan S.S., Chen J.H., Hwang S.M., Wang I.J., Li H.J., Lee R.T. & Hsieh P.C. (2009). Salvianolic acid B-vitamin C synergy in cardiac differentiation from embryonic stem cells. Biochem Biophys Res Commun, 387, 723-728

Chen K., Bai H., Arzigian M., Gao Y.X., Bao J., Wu W.S., Shen W.F., Wu L. & Wang Z.Z. (2010). Endothelial cells regulate cardiomyocyte development from embryonic stem cells. J Cell Biochem, 111, 29-39

Chen S.S., Fitzgerald W., Zimmerberg J., Kleinman H.K. & Margolis L. (2007). Cell-cell and cell-extracellular matrix interactions regulate embryonic stem cell differentiation. Stem Cells, 25, 553-561

Crespo F.L., Sobrado V.R., Gomez L., Cervera A.M. & McCreath K.J. (2010). Mitochondrial reactive oxygen species mediate cardiomyocyte formation from embryonic stem cells in high glucose. Stem Cells, 28, 1132-1142

D'Aniello C., Lonardo E., Iaconis S., Guardiola O., Liguoro A.M., Liguori G.L., Autiero M., Carmeliet P. & Minchiotti G. (2009). G protein-coupled receptor APJ and its ligand apelin act downstream of Cripto to specify embryonic stem cells toward the cardiac lineage through extracellular signal-regulated kinase/p70S6 kinase signaling pathway. Circ Res, 105, 231-238

Drab M., Haller H., Bychkov R., Erdmann B., Lindschau C., Haase H., Morano I., Luft F.C. & Wobus A.M. (1997). From totipotent embryonic stem cells to spontaneously contracting smooth muscle cells: a retinoic acid and db-cAMP in vitro differentiation model. Faseb J, 11, 905-915

Farokhpour M., Karbalaie K., Tanhaei S., Nematollahi M., Etebari M., Sadeghi H.M., Nasr-Esfahani M.H. & Baharvand H. (2009). Embryonic stem cell-derived cardiomyocytes as a model system to study cardioprotective effects of dexamethasone in doxorubicin cardiotoxicity. Toxicol In Vitro, 23, 1422-1428

Gassanov N., Er F., Zagidullin N. & Hoppe U.C. (2004). Endothelin induces differentiation of ANP-EGFP expressing embryonic stem cells towards a pacemaker phenotype. Faseb J, 18, 1710-1712

Graichen R., Xu X., Braam S.R., Balakrishnan T., Norfiza S., Sieh S., Soo S.Y., Tham S.C., Mummery C., Colman A., Zweigerdt R. & Davidson B.P. (2008). Enhanced cardiomyogenesis of human embryonic stem cells by a small molecular inhibitor of p38 MAPK. Differentiation, 76, 357-370

Hatami L., Valojerdi M.R. & Mowla S.J. (2007). Effects of oxytocin on cardiomyocyte differentiation from mouse embryonic stem cells. Int J Cardiol, 117, 80-89

He J.Q., Ma Y., Lee Y., Thomson J.A. & Kamp T.J. (2003). Human embryonic stem cells develop into multiple types of cardiac myocytes: action potential characterization. Circ Res, 93, 32-39

Jiang P., Rushing S.N., Kong C.W., Fu J., Lieu D.K., Chan C.W., Deng W. & Li R.A. (2010). Electrophysiological properties of human induced pluripotent stem cells. Am J Physiol Cell Physiol, 298, C486-495

Kanno S., Kim P.K., Sallam K., Lei J., Billiar T.R. & Shears L.L., 2nd. (2004). Nitric oxide facilitates cardiomyogenesis in mouse embryonic stem cells. Proc Natl Acad Sci U S A, 101, 12277-12281

Kehat I., Kenyagin-Karsenti D., Snir M., Segev H., Amit M., Gepstein A., Livne E., Binah O., Itskovitz-Eldor J. & Gepstein L. (2001). Human embryonic stem cells can differentiate into myocytes with structural and functional properties of cardiomyocytes. J Clin Invest, 108, 407-414

Kehat I., Khimovich L., Caspi O., Gepstein A., Shofti R., Arbel G., Huber I., Satin J., Itskovitz-Eldor J. & Gepstein L. (2004). Electromechanical integration of cardiomyocytes derived from human embryonic stem cells. Nat Biotechnol, 22, 1282-1289

Lecina M., Ting S., Choo A., Reuveny S. & Oh S. (2010). Scalable Platform for Human Embryonic Stem Cell Differentiation to Cardiomyocytes in Suspended Microcarrier Cultures. Tissue Eng Part C Methods, Aug 28. [Epub ahead of print]

Lee Y.K., Ng K.M., Chan Y.C., Lai W.H., Au K.W., Ho C.Y., Wong L.Y., Lau C.P., Tse H.F. & Siu C.W. (2010). Triiodothyronine Promotes Cardiac Differentiation and Maturation of Embryonic Stem Cells via the Classical Genomic Pathway. Mol Endocrinol, 24, 1728-1736

Lesman A., Gepstein L. & Levenberg S. (2010a). Vascularization shaping the heart. Ann N Y Acad Sci, 1188, 46-51

Lesman A., Habib M., Caspi O., Gepstein A., Arbel G., Levenberg S. & Gepstein L. (2010b). Transplantation of a tissue-engineered human vascularized cardiac muscle. Tissue Eng Part A, 16, 115-125

Liang H., Matzkies M., Schunkert H., Tang M., Bonnemeier H., Hescheler J. & Reppel M. (2010). Human and murine embryonic stem cell-derived cardiomyocytes serve together as a valuable model for drug safety screening. Cell Physiol Biochem, 25, 459-466

Liu H., Lin J. & Roy K. (2006). Effect of 3D scaffold and dynamic culture condition on the global gene expression profile of mouse embryonic stem cells. Biomaterials, 27, 5978-5989

Maltsev V.A., Wobus A.M., Rohwedel J., Bader M. & Hescheler J. (1994). Cardiomyocytes differentiated in vitro from embryonic stem cells developmentally express cardiac-specific genes and ionic currents. Circ Res, 75, 233-244

Melkoumian Z., Weber J.L., Weber D.M., Fadeev A.G., Zhou Y., Dolley-Sonneville P., Yang J., Qiu L., Priest C.A., Shogbon C., Martin A.W., Nelson J., West P., Beltzer J.P., Pal S. & Brandenberger R. (2010). Synthetic peptide-acrylate surfaces for long-term self-renewal and cardiomyocyte differentiation of human embryonic stem cells. Nat Biotechnol, 28, 606-610

Menasche P. (2004). Embryonic stem cells pace the heart. Nat Biotechnol, 22, 1237-1238

Mohr J.C., Zhang J., Azarin S.M., Soerens A.G., de Pablo J.J., Thomson J.A., Lyons G.E., Palecek S.P. & Kamp T.J. (2010). The microwell control of embryoid body size in order to regulate cardiac differentiation of human embryonic stem cells. Biomaterials, 31, 1885-1893

Mummery C., Ward-van Oostwaard D., Doevendans P., Spijker R., van den Brink S., Hassink R., van der Heyden M., Opthof T., Pera M., de la Riviere A.B., Passier R. & Tertoolen L. (2003). Differentiation of human embryonic stem cells to cardiomyocytes: role of coculture with visceral endoderm-like cells. Circulation, 107, 2733-2740

Norstrom A., Akesson K., Hardarson T., Hamberger L., Bjorquist P. & Sartipy P. (2006). Molecular and pharmacological properties of human embryonic stem cell-derived cardiomyocytes. Exp Biol Med (Maywood), 231, 1753-1762

Ou D.B., Chen R., Liu X.T., Guo J.J., Wang H.T. & Zheng Q.S. (2009a). Controlling the in vitro differentiation of embryonic stem cells for myocardial tissue engineering applications. Cell Biol Int, Sep 15. [Epub ahead of print]

Ou D.B., Lang H.J., Chen R., Liu X.T. & Zheng Q.S. (2009b). Using embryonic stem cells to form a biological pacemaker via tissue engineering technology. Bioessays, 31, 246-252

Paige S.L., Osugi T., Afanasiev O.K., Pabon L., Reinecke H. & Murry C.E. (2010). Endogenous Wnt/beta-catenin signaling is required for cardiac differentiation in human embryonic stem cells. PLoS One, 5, e11134

Papp S., Dziak E. & Opas M. (2009). Embryonic stem cell-derived cardiomyogenesis: a novel role for calreticulin as a regulator. Stem Cells, 27, 1507-1515

Paquin J., Danalache B.A., Jankowski M., McCann S.M. & Gutkowska J. (2002). Oxytocin induces differentiation of P19 embryonic stem cells to cardiomyocytes. Proc Natl Acad Sci U S A, 99, 9550-9555

Pekkanen-Mattila M., Chapman H., Kerkela E., Suuronen R., Skottman H., Koivisto A.P. & Aalto-Setala K. (2010). Human embryonic stem cell-derived cardiomyocytes: demonstration of a portion of cardiac cells with fairly mature electrical phenotype. Exp Biol Med (Maywood), 235, 522-530

Pfannkuche K., Liang H., Hannes T., Xi J., Fatima A., Nguemo F., Matzkies M., Wernig M., Jaenisch R., Pillekamp F., Halbach M., Schunkert H., Saric T., Hescheler J. & Reppel

M. (2009). Cardiac myocytes derived from murine reprogrammed fibroblasts: intact hormonal regulation, cardiac ion channel expression and development of contractility. Cell Physiol Biochem, 24, 73-86

Philp D., Chen S.S., Fitzgerald W., Orenstein J., Margolis L. & Kleinman H.K. (2005). Complex extracellular matrices promote tissue-specific stem cell differentiation. Stem Cells, 23, 288-296

Roggia C., Ukena C., Bohm M. & Kilter H. (2007). Hepatocyte growth factor (HGF) enhances cardiac commitment of differentiating embryonic stem cells by activating PI3 kinase. Exp Cell Res, 313, 921-930

Romito A., Lonardo E., Roma G., Minchiotti G., Ballabio A. & Cobellis G. (2010). Lack of sik1 in mouse embryonic stem cells impairs cardiomyogenesis by down-regulating the cyclin-dependent kinase inhibitor p57kip2. PLoS One, 5, e9029

Ronca R., Gualandi L., Crescini E., Calza S., Presta M. & Dell'Era P. (2009). Fibroblast growth factor receptor-1 phosphorylation requirement for cardiomyocyte differentiation in murine embryonic stem cells. J Cell Mol Med, 13, 1489-1498

Rosen M.R., Brink P.R., Cohen I.S. & Robinson R.B. (2004). Genes, stem cells and biological pacemakers. Cardiovasc Res, 64, 12-23

Santiago J.A., Pogemiller R. & Ogle B.M. (2009). Heterogeneous differentiation of human mesenchymal stem cells in response to extended culture in extracellular matrices. Tissue Eng Part A, 15, 3911-3922

Sartipy P., Olsson B., Hyllner J. & Synnergren J. (2009). Regulation of 'stemness' and stem cell differentiation by microRNAs. IDrugs, 12, 492-496

Schuldiner M., Yanuka O., Itskovitz-Eldor J., Melton D.A. & Benvenisty N. (2000). Effects of eight growth factors on the differentiation of cells derived from human embryonic stem cells. Proc Natl Acad Sci U S A, 97, 11307-11312

Serena E., Figallo E., Tandon N., Cannizzaro C., Gerecht S., Elvassore N. & Vunjak-Novakovic G. (2009). Electrical stimulation of human embryonic stem cells: cardiac differentiation and the generation of reactive oxygen species. Exp Cell Res, 315, 3611-3619

Singla D.K., Lyons G.E. & Kamp T.J. (2007). Transplanted embryonic stem cells following mouse myocardial infarction inhibit apoptosis and cardiac remodeling. Am J Physiol Heart Circ Physiol, 293, H1308-1314

Singla D.K. & McDonald D.E. (2007). Factors released from embryonic stem cells inhibit apoptosis of H9c2 cells. Am J Physiol Heart Circ Physiol, 293, H1590-1595

Sluijter J.P., van Mil A., van Vliet P., Metz C.H., Liu J., Doevendans P.A. & Goumans M.J. (2010). MicroRNA-1 and -499 regulate differentiation and proliferation in human-derived cardiomyocyte progenitor cells. Arterioscler Thromb Vasc Biol, 30, 859-868

Song H., Yoon C., Kattman S.J., Dengler J., Masse S., Thavaratnam T., Gewarges M., Nanthakumar K., Rubart M., Keller G.M., Radisic M. & Zandstra P.W. (2010). Interrogating functional integration between injected pluripotent stem cell-derived cells and surrogate cardiac tissue. Proc Natl Acad Sci U S A, 107, 3329-3334

Stevens K.R., Kreutziger K.L., Dupras S.K., Korte F.S., Regnier M., Muskheli V., Nourse M.B., Bendixen K., Reinecke H. & Murry C.E. (2009). Physiological function and transplantation of scaffold-free and vascularized human cardiac muscle tissue. Proc Natl Acad Sci U S A, 106, 16568-16573

Takahashi T., Lord B., Schulze P.C., Fryer R.M., Sarang S.S., Gullans S.R. & Lee R.T. (2003). Ascorbic acid enhances differentiation of embryonic stem cells into cardiac myocytes. Circulation, 107, 1912-1916

Thomson J.A., Itskovitz-Eldor J., Shapiro S.S., Waknitz M.A., Swiergiel J.J., Marshall V.S. & Jones J.M. (1998). Embryonic stem cell lines derived from human blastocysts. Science, 282, 1145-1147

Tran T.H., Wang X., Browne C., Zhang Y., Schinke M., Izumo S. & Burcin M. (2009). Wnt3a-induced mesoderm formation and cardiomyogenesis in human embryonic stem cells. Stem Cells, 27, 1869-1878

van Laake L.W., Passier R., den Ouden K., Schreurs C., Monshouwer-Kloots J., Ward-van Oostwaard D., van Echteld C.J., Doevendans P.A. & Mummery C.L. (2009). Improvement of mouse cardiac function by hESC-derived cardiomyocytes correlates with vascularity but not graft size. Stem Cell Res, 3, 106-112

Ventura C., Zinellu E., Maninchedda E., Fadda M. & Maioli M. (2003). Protein kinase C signaling transduces endorphin-primed cardiogenesis in GTR1 embryonic stem cells. Circ Res, 92, 617-622

Wobus A.M., Wallukat G. & Hescheler J. (1991). Pluripotent mouse embryonic stem cells are able to differentiate into cardiomyocytes expressing chronotropic responses to adrenergic and cholinergic agents and Ca2+ channel blockers. Differentiation, 48, 173-182

Xi J., Khalil M., Shishechian N., Hannes T., Pfannkuche K., Liang H., Fatima A., Haustein M., Suhr F., Bloch W., Reppel M., Saric T., Wernig M., Janisch R., Brockmeier K., Hescheler J. & Pillekamp F. (2010). Comparison of contractile behavior of native murine ventricular tissue and cardiomyocytes derived from embryonic or induced pluripotent stem cells. Faseb J, 24, 2739-2751

Xie C.Q., Zhang J., Xiao Y., Zhang L., Mou Y., Liu X., Akinbami M., Cui T. & Chen Y.E. (2007). Transplantation of human undifferentiated embryonic stem cells into a myocardial infarction rat model. Stem Cells Dev, 16, 25-29

Xue T., Cho H.C., Akar F.G., Tsang S.Y., Jones S.P., Marban E., Tomaselli G.F. & Li R.A. (2005). Functional integration of electrically active cardiac derivatives from genetically engineered human embryonic stem cells with quiescent recipient ventricular cardiomyocytes: insights into the development of cell-based pacemakers. Circulation, 111, 11-20

Yanagi K., Takano M., Narazaki G., Uosaki H., Hoshino T., Ishii T., Misaki T. & Yamashita J.K. (2007). Hyperpolarization-activated cyclic nucleotide-gated channels and T-type calcium channels confer automaticity of embryonic stem cell-derived cardiomyocytes. Stem Cells, 25, 2712-2719

Yoon B.S., Yoo S.J., Lee J.E., You S., Lee H.T. & Yoon H.S. (2006). Enhanced differentiation of human embryonic stem cells into cardiomyocytes by combining hanging drop culture and 5-azacytidine treatment. Differentiation, 74, 149-159

Zhu W.Z., Xie Y., Moyes K.W., Gold J.D., Askari B. & Laflamme M.A. (2010). Neuregulin/ErbB Signaling Regulates Cardiac Subtype Specification in Differentiating Human Embryonic Stem Cells. Circ Res,

28

Oxytocin as an Inducer of Cardiomyogenesis

Jolanta Gutkowska and Marek Jankowski
Laboratory of Cardiovascular Biochemistry, Centre de recherche, Centre hospitalier de l'Université de Montréal (CRCHUM) – Hôtel-Dieu and Department of Medicine, Université de Montréal, Montreal, Quebec
3850 St. Urbain Street, Masson Pavilion, Montreal, Quebec
Canada

1. Introduction

The first description of the uterotonic action of pituitary extracts was first described by Sir Henry Dale in 1906 (Dale, 1906). A few years later, Ott and Scott demonstrated that besides the effect on uterine activity, posterior pituitary extracts also promoted milk ejection (Ott & Scott, 1910). These are the 2 principal activities of oxytocin (OT), the structure and synthesis of which were not elucidated until 50 years later by Du Vigneaud and co-workers (Du Vigneaud, 1956). OT is mainly produced in the paraventricular nucleus and supraoptic nucleus of the hypothalamus, and released from hypothalamic nerve terminals of the posterior pituitary, where it is stored, into the bloodstreem. OT differs, by only two amino acids, from vasopressin (AVP), which is also produced in these nuclei and stored in the posterior pituitary. It was previously believed that OT as well as AVP were exclusively released from the neurohypophysis although Ott and Scott (Ott & Scott, 1910) reported that the extracts of other tissues such as the corpus luteum, pineal and thymus glands have the same milk-ejecting properties. Early studies associated OT with cardiovascular system Paton and Watson first described the blood pressure (BP) depressor response to posterior pituitary extract in avians (Paton & Watson, 1912). Hogben & Schlapp confirmed their findings and showed the effect of whole posterior pituitary extracts in amphibians and reptiles (Hogben & Schlapp, 1924). Furthermore, this effect was evoked by histamine-free extracts and was therefore a pituitary action. When separated fractions of posterior pituitary extracts obtained by fractional precipitation became available, Gaddum showed that the depressor response was attributable to the only one fraction, called oxytocin, which had oxytocic properties (Gaddum, 1928). These observations were completely overlooked.
However, recent studies have shown that OT is an ubiquitous hormone, synthesized at many locis, and a wide array of physiological activities has been attributed to this peptide. Similar numbers of oxytocinergic neurons have been found in the male and female hypothalamus, and the same stimuli induce OT release in both genders, suggesting other physiological functions that its role in female reproduction. In fact, OT receptors (OTR) are also widely expressed in diverse tissues such as the pituitary, kidney, ovary, testis, thymus, heart, vascular endothelium, osteoclasts, myoblasts, pancreatic islets, adipocytes and several types of cancer cells (Gimpl & Fahrenholz, 2001). OT elicits a variety of physiological responses such as complex sexual and maternal behavior. Indeed, OT is also involved in

cognition, tolerance and cardiovascular regulation. OT acts on one type of OTR, an integral membrane protein that is a member of the rhodopsin-type (class I) G protein-coupled receptor family, which includes AVP receptor subtypes (V1aR, V1bR and V2).

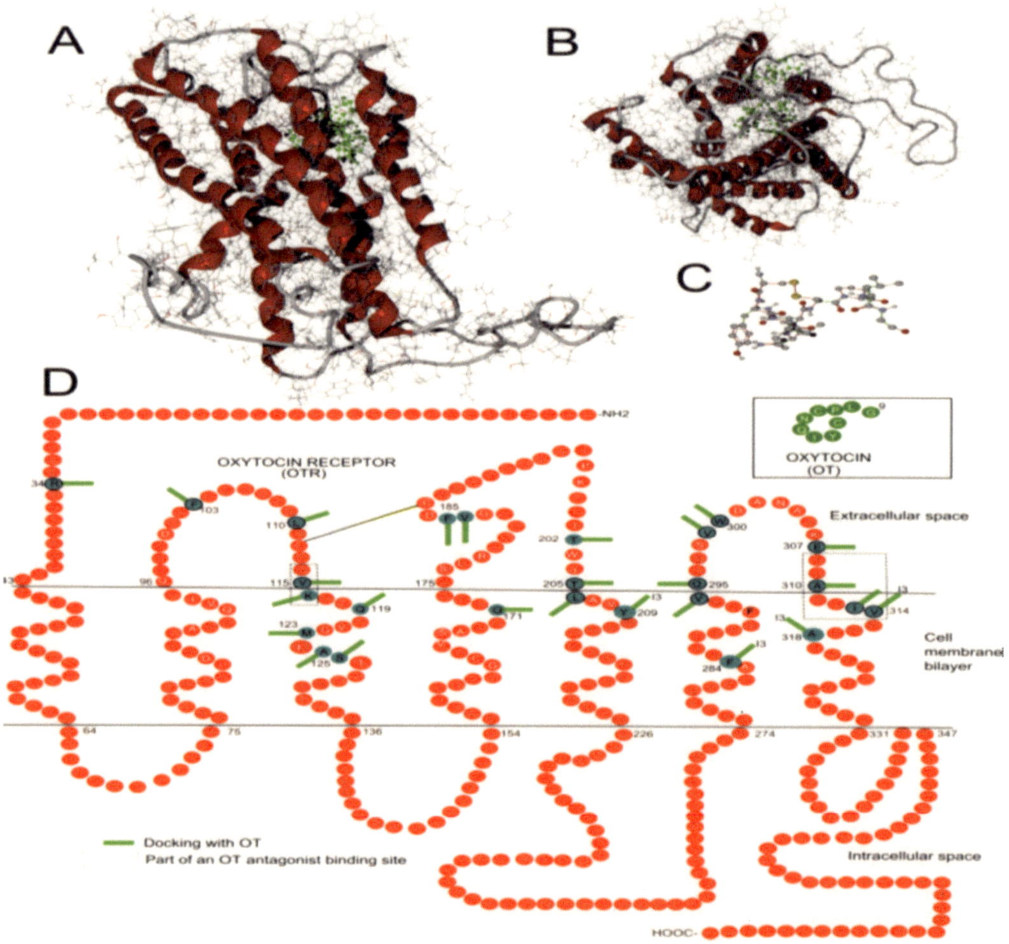

Fig. 1. Several oxytocin binding sites are present on oxytocin receptor. This figure illustrates molecular docking of the three-dimensional models of activated human oxytocin receptor with oxytocin obtained by MolDock Optimizer algorithm from Molegro Virtual Docker software. (A) The front upright view position (side view) of the receptor structure with oxytocin. (B) Panel B shows an intracellular view (i.e. rotation by 90° out of plane). (C) Conformational view of oxytocin molecule. (D) The schematic model of the human oxytocin receptor with marked amino-acid residues that are putatively involved in ligand-binding

All of these receptors have the ability to bind AVP and/or OT with different affinities. Because OT system activation was observed in fetal and newborn hearts at a stage of intense cardiac hyperplasia (Jankowski et al., 2004), we hypothesized a role for OT in cardiomyocyte (CM) differentiation. The initial experiments demonstrated that OT induces CM differentiation of the mouse embryonal carcinoma (EC) P19 cell line, a common stem cell model for studying early heart differentiation (Paquin et al., 2002). Then, several studies attempted to understand the mechanisms of cardiac differentiation by OT. Stimulation of CM differentiation can be concomitant with neovascularization because OT stimulates endothelial cells growth and angiogenesis. In pathological conditions such as cardiac ischemia and diabetes this inducer can be used to stimulate production of lost cardiac cells. The advantage of this therapy is supported by the fact that OT is endogenously produced in the organism, and does not have significant side effects when used in clinics. Moreover, it is now possible to inject (transplant) one's own stem cells after previous stimulation with OT inducers, as in the case of a heart infarct. Alternatively, direct treatment with OT molecules could promote cardiomyogenesis in situ and regeneration of a damaged heart.

2. OT induces stem cell differentiation

The manipulations of OT, during the perinatal period have become an accepted—but largely unstudied—aspect of human development. For example, synthetic OT is used to induce childbirth (Dudley, 1997). In cases of premature delivery, the OT antagonist (OTA) is used for slowing or preventing labor (Husslein, 2002). Although such complex manipulations are routine in modern obstetrics, little is known on the possible cellular and developmental consequences of these treatments. Several studies have proposed a role for OT as a growth and differentiation/maturation factor in a gestational/perinatal context. In the mother, OT is required for postpartum alveolar proliferation, and it induces differentiation and proliferation of myoepithelial cells in the mammary gland necessary for milk ejection (Gimpl & Fahrenholz, 2001). The OT/OTR system is expressed in human cumulus/luteal cells surrounding oocytes, and weak OTR gene expression is even observed in oocytes (Furuya et al., 1995). Moreover, when fertilized mouse oocytes are cultured with OT *in vitro,* they develop into the blastocyst stage at a higher rate than their unstimulated counterparts (Furuya, et al., 1995).

On the other hand, hormonal treatment may, in its simplest form, induce mammalian stem cells into a special cell type that retains the ability to self-renew (i.e. undergo cell division in an undifferentiated state) indefinitely and to differentiate into specialized cells. In this regard, it has been suggested that OT plays a role in bone homeostasis and osteoporosis based on the proliferative effects of OT on osteoblasts in vitro and the modulation of blood parameters associated with bone formation in normal rats (Petersson et al., 2002b). Recently, OT has been implicated in the regulation of the osteoblast/adipocyte balance of human multipotent adipose-derived stem cells and human bone marrow mesenchymal stromal cells (Elabd et al., 2008).

Our interest in the cardiac OT system emerged from long time investigations into the role of the brain in the control of cardio-renal homeostasis (Gutkowska & Jankowski, 2008; Gutkowska et al., 2000; McCann et al., 2002). These experiments led to the observations that OT and its receptor (OTR) are synthesized in the human and rat heart (Gutkowska et al., 1997; Jankowski et al., 1998) and that OT exerts cardio-protection either directly or via stimulation of mediators such as the natriuretic peptides (NPs) (Gutkowska et al., 1997; Gutkowska et al., 2000) and nitric oxide (NO) (Danalache et al., 2007). Since the study showing that OT induces differentiation of EC P19 cells into the functional cardiac muscle (Paquin et al., 2002), several reports have confirmed OT-stimulated cardiomyogenesis in different lines of embryonic stem cells (Fathi et al., 2009; Gassanov et al., 2008a; Hatami et al., 2007; Jankowski et al., 2004; Stefanidis et al., 2009; Uchida et al., 2007). In fact, OT can be considered as an established myogenic morphogen (Breton et al., 2002). It induces differentiation of myoepithelial cells in the mammary gland (Sapino et al., 1993). The skeletal myocytes generation has also been observed in OT-induced cardiomyogenesis (Danalache et al., 2007) as well as adipogenesis (Bouchard & Paquin, 2009). OT increases the rate of myoblast fusion and myotubule formation (Breton et al., 2001). Interestingly, most transcription factors identified so far in the heart are also present in other muscle cells, and myoblast transplantation to the injured heart improves regional systolic heart function (Thompson et al., 2003). Assuming similarities in the differentiation mechanisms of skeletal and cardiac muscles and the expression of both OT and OTR in these cells (Breton et al., 2001; Jankowski et al., 1998), we can speculate that OT plays a role in cardiac muscle regeneration.

3. OT stimulates CM differentiation in 3-dimensional cultures of stem cells

EC P19 cells are derived from a teratocarcinoma in CH3/He mice and can differentiate into all 3 germ layers (van der Heyden & Defize, 2003). Developmentally, pluripotent EC P19 cells appear to differentiate by the same mechanisms as normal embryonic stem cells (McBurney, 1993). Culture and differentiation of the cells is simple and cells remain undifferentiated without the help of feeder cells or inhibitory factors, and unlike embryonic ES cells, they do not spontaneously generate cardiomyocytes. The efficient differentiation of EC P19 cells depends on the prior formation of non-adhering cell aggregates (van der Heyden & Defize, 2003). Traditionally, cell aggregates formation in suspension culture under 0.5-1.0% dimethyl sulfoxide (DMSO) has been used to induce cardiomyocyte differentiation of EC P19 cells (McBurney, 1993). Efficiency of cardiac differentiation of ES and EC P19 cells in vitro is still not optimal in response to various agents, with yields varying between 5% and 20% of cardiomyocytes (Danalache et al., 2007; Gassanov et al., 2008a; Gassanov et al., 2008b; Gassanov et al., 2007; Paquin et al., 2002). In the EC P19 model, the order of cardiomyogenesis efficiency was OT (10^{-7} M) ≥ DMSO >retinoic acid (10^{-8}–10^{-7} M) when these agents were added to cultures during the entire period of cell aggregation (Danalache et al., 2007; Jankowski et al., 2004; Paquin et al., 2002). It is noteworthy that exposure of EC P19 cells to higher retinoic acid concentration (10^{-6} M) over the aggregation period generates neurons but not muscle cells.

The presence of OT significantly influences the shape and size of aggregated stem cells isolated from the rat heart (Gutkowska & Jankowski, 2009). This suggests that conditions inside aggregates, such as hypoxia, promote OTR and OT expression. Indeed, hypoxia can influence the functionality of OTR in cardiac cells and mechanism of natriuretic peptides secretion in response to OT treatments (Hopkins et al., 2004). Our data indicate that OT increases glucose uptake by CMs exposed to chemical hypoxia (Florian et al., 2010). Multi-cellular complex aggregate formation and exposure to various agents promotes generation of mesodermal or ectodermal lineages. In embryonic D3 stem cells, spontaneously producing beating cell colonies upon aggregation, the mesodermal derivatives formed in embryoid bodies (EB) include subtypes of cardiac cells (atrial CMs, ventricular CMs and pacemaker cells) which are potently enhanced by treatment with OT as identified by histological, molecular, and electrophysiological criteria (Gassanov et al., 2008a). However, molecular events occurring during aggregation and the necessity of their aggregation for differentiation are not entirely understood. High cell densities can at least trigger spontaneous differentiation from EC P19 cells (McBurney, 1993). Skerjanc et al. have reported that overexpression of Nkx2.5 can induce cardiomyogenesis in aggregated EC P19 cells, but not when they are maintained in monolayer cultures (Skerjanc et al., 1998). On the other hand, the addition of soluble bone morphogenic protein 4 (BMP 4) into the culture media resulted in the requirement for cell aggregation and induced cardiac differentiation being bypassed in monolayer cultures of EC P19 cells overexpressing Nkx2.5 (Jamali et al., 2001). This result demonstrated that cell aggregation is crucial for the generation of the BMP 4 signal in EC P19 cells. EC P19Cl6 cells derived from EC P19 cells seem not to be committed to a mesodermal lineage but rather represent a stage closer to differentiated cardiac muscle than the parental cell line. It was recently observed that OT does not induce cardiomyogenesis in monolayers of EC P19Cl6 cells, but does so in aggregates (Uchida et al., 2007). The same authors suggested that OT-induced cardiac differentiation is not mediated by the expression of BMP 4 signaling molecules because the induction of BMP 4 signaling cascade can bypass the requirement for prior EB formation (Fathi et al., 2009).

4. OT molecular forms of OT stimulating differentiation

OT, recognized as a female reproductive hormone, is largely produced in hypothalamic magnocellular neurons of paraventricular and supraoptic nuclei. Biochemical and recombinant DNA studies have demonstrated that it is synthesized as a non-glycosylated protein, which undergoes an initial endoproteolytic cleavage by the convertase magnolysin (EC 3.4.24.62) to OT-Gly-Lys-Arg (OT-GKR) (Brownstein et al., 1980; Burbach et al., 2001). Subsequent processing produces other OT extended molecules: OT-Gly-Lys (OT-GK) and OT-Gly (OT-G) (Burbach et al., 2001), all these forms are often referred to as OT-X (Morris et al., 1992). OT-G is converted by an α-amidating enzyme to C-amidated nonapeptide. OT is released into the bloodstream in this form. OT-X forms have been detected in the developing brain of animals and in fetal plasma. In rats, enzymatic OT-X conversion to OT is almost complete in adulthood, but not in fetuses, which accumulate OT-X in the brain (Altstein et al., 1988a; Altstein & Gainer, 1988b. Similarly, the plasma OT-X elevation reported during early fetal development in sheep (Morris et al., 1992) is reduced in late gestation, when OT begins to predominate in the cbloodstream.

We have recently shown that the OT-GKR is the main form in the fetal heart. This attracted our attention because estrogen-mediated BP reduction in humans is associated with elevated plasma OT-GKR but not OT, indicating greater bioactivity and stability of this form (Light et al., 2005). We have also found that OT-GKR even more potently than OT mediates cardiomyogenesis in EC P19 cells (Jankowski et al., 2004) and D3 stem cells (Gassanov et al., 2008a). The patch-clamp analysis indicates that OT-GKR differentiates D3 cells into CMs of the ventricular phenotype and reduces formation of pacemaker cells (Gassanov et al., 2008a). Furthermore, OT-GKR significantly increases glucose uptake in CM exposed to hypoxia (Florian et al., 2010).

To determine whether the genetic modification of stem cells also stimulates cardiomyogenesis, the OT-Gly-Lys-Arg gene was inserted into D3 stem cells. In effect, we observed stimulation of spontaneously-beating embryoid bodies and predominant stimulation of cells expressing the ventricular electrophenotype and molecular CM markers (Gassanov et al., 2008a). Interestingly, the elongated form of OT, OT-Gly-Lys-Arg, was the most potent cardiomyogen of all the OT-like molecules investigated. These findings provide a new strategy for the regeneration of diseased hearts. Transgenic cells producing OT-Gly-Lys-Arg can be injected, for cardioprotection, into the myocardial infarcted rat model. Analysis of cardiac remodeling, scar reduction, hemodynamic and echographic parameters together with histochemical and molecular analyses will provide answers as to whether these treatments can stimulate cardioprotection.

5. Mechanism in OT-induced stem cell differentiation

Some preliminary observations point to a mechanism involved in this process. Ca^{+2} mobilization in response to OT treatment has been detected in D3 ES cells differentiating into CMs (Gassanov et al., 2008a). It has also been shown that OT-induced differentiation of EC P19 stem cells into CMs is inhibited by the NOS inhibitor N(G)-nitro-L-arginine methyl ester (L-NAME).

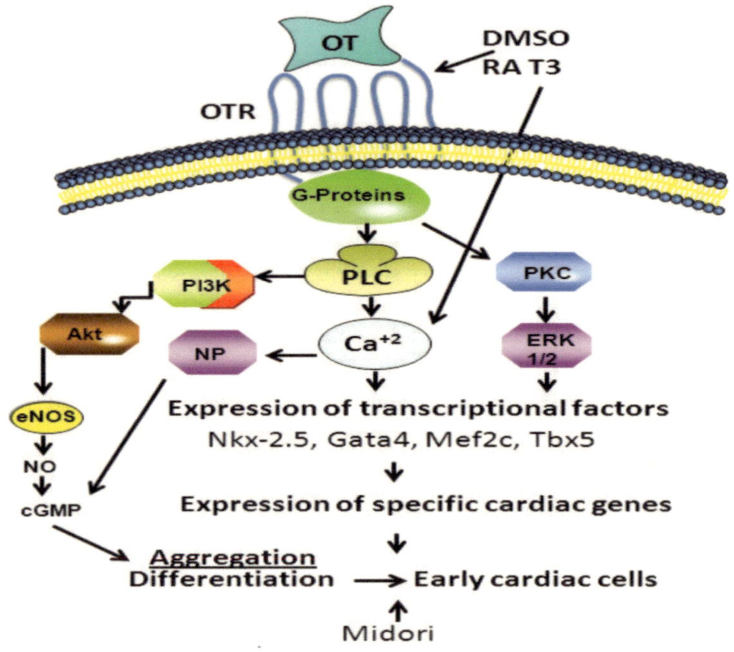

Fig. 2. Signal transduction cascades involved in oxytocin-mediated cardiac differentiation of P19 embryonal carcinoma cells. See text for details. Akt indicates protein kinase B; ERK, extracellular signal-regulated kinase; eNOS, endothelial nitric oxide synthase; midori, (myocyte induction/differentiator originator); NP, natriuretic peptides; PI3K, phosphatidylinositol 3-kinase; PLC, phospolipase C; PKC, phosphokinase C, RA, retinoic acid

The NO donor *S*-nitroso-*N*-acetylpenicillamine (SNAP) was able to reverse L-NAME-mediated inhibition of EC P19 cell differentiation into CMs (Danalache et al., 2007). This study clearly indicates a role for NO and NOS enzymes in stem cell differentiation, but it is evident that thjis may be a complex process. This complexity is highlighted by the fact that suppression of NOS activity by L-NAME has also been shown to increase the number of stem and progenitor cells differentiating into CMs (Danalache et al., 2007). Another study has reported that exposure of D3 stem cells to AVP increases the number of beating embryoid bodies and also heightens GATA-4 expression. These AVP effects on the cells were also found to be antagonized by L-NAME (Gassanov et al., 2007), again suggesting a positive role for NO in stem cell differentiation into CMs. This investigation highlighted the expression of AVP receptors in undifferentiated D3 cells, with the expression profile changing during the differentiation process (Gassanov et al., 2007). It has been observed in the EC P19 cell model that AVP not only increases spontaneously-occurring cardiomyogenesis but also initiates the process (Gassanov et al., 2007; Gutkowska et al., 2007).

The OTR-NO-cGMP pathway that is essential for OT-elicited differentiation of EC P19 stem cells into CMs is associated with elevation of GATA-4 and myocyte enhancer factor 2c (MEF2c) (Danalache et al., 2007). GATA-4 regulates the expression of genes that are critical for CM differentiation. MEF2c is a member of the MEF2 family that is involved in cardiac, skeletal, and smooth muscle development. Partial GATA-4 gene targeting in cardiac and non-cardiac cells indicates that even modest variations in GATA-4 gene level or activity can play a role in the maintenance of normal cardiac function (Bisping et al., 2006). GATA-4 has also been implicated in intercellular cross-talk by inducing hypertrophy-associated angiogenesis via vascular endothelial growth factor (VEGF) release and targeting the endothelium (Heineke et al., 2007). The upstream sequence of OTR contains putative binding sites for GATA-4 and Nkx2.5 and GATA-4 also serves as a key transcriptional regulator of numerous cardiac peptides, including ANP, BNP and OTR (Uchida et al., 2007). GATA-4 has also been identified in stem and progenitor cells of the heart in combination with OT-mediated CM differentiation (Matsuura et al., 2004; Oyama et al., 2007). A recent study has demonstrated that undifferentiated murine ES cells express BNP and its receptors, with its signaling being essential for cell survival and clonal growth (Abdelalim & Tooyama, 2009). This observation suggests possible interaction of the OT and NP systems in ES cells during cardiomyogenesis.

6. Differentiation of endogenous stem cells isolated from animal hearts

The idea that OT has cardio-regenerative capacities is supported by the observation that this hormone induces the differentiation of cultured mice (Matsuura et al., 2004) and rats (Oyama et al., 2007) resident cardiac stem cells (CSCs). In the adult heart, CSCs maintain a balance of survival, proliferation and self-renewal to replace mature cells that are lost during injury or turnover. Matsuura's group revealed the presence of a Sca-1$^+$ stem cell population in adult mouse hearts expressing telomerase reverse transcriptase, which has been associated with self-renewal potential (Matsuura et al., 2004). These cells, lacking hematopoietic markers, are easily distinguished from hematopoietic stem cells of bone marrow origin, and when treated with OT, differentiate into functional CMs. Although the cells present the early cardiac markers GATA-4 and MEF2, they do not express Nkx-2.5 or genes encoding cardiac sarcomeric proteins. When exposed to OT, a small population of Sca-1$^+$ cells manifest sarcomeric structures and form spontaneously-beating CMs. In addition, after intravenous delivery, Sca-1+ cardiac stem cells can be recruited to the myocardium injured by ischemia/reperfusion and can functionally differentiate *in situ* (Matsuura et al., 2004). Importantly, these cells had positive ionotropic responses to isoproterenol via β1-adrenergic receptor signaling. Given the apparently small number of CMs generated in vitro by OT stimulation, this raises the questionof whether or not OT-mediated cardiomyogenesis is a default pathway for CSCs. Accordingly, Matsuura et al.

reported that OT induces about 0.5-1% of Sca-1^{POS} ckit^{NEG} CD45^{NEG} cells from the adult murine heart to differentiate into functional, spontaneously-beating immature CMs (Matsuura, et al., 2004). In this regard, the cardiac differentiation of Sca-1+ cells does not require cell aggregation for the process to proceed (Matsuura, et al., 2004). On the other hand, a study by the same group in another (CSC) type isolated from the rat heart (Oyama, et al., 2007) disclosed that OT treatment resulted in the generation of 5% CMs. These cells, termed cardiac side population cells (CSPs), but not to corresponding side population cells isolated from bone marrow, differentiated into CMs in response to OT treatment. Therefore, OT possesses more powerful cardiogenic activity against CSCs than previously reported. CSPs have the ability to efflux Hoechst dye, a process dependent on ABC transporters. CSCs, especially Abcg2-dependent CSPs, have been associated with stem/progenitor cells. These cells are positive for Abcg2, Sca-1, ckit (low), CD34 (low), CD45 (low) and negative for CD31 (Mouquet et al., 2005; Pfister et al., 2005). A possible role for CSCs in heart healing is indicated by increased numbers of Abcg2-expressing cells in the border zone adjacent to myocardial infarcts (Pfister, et al., 2005). Stimulation of CM differentiation could be concomitant with neovascularization because OT stimulates endothelial cell growth (Thibonnier et al., 1999) and angiogenesis (Cattaneo et al., 2008).

7. OT mediates cardioprotection

The absence of either OT or its receptors in knockout mice, however, has not been reported to produce cardiac insufficiencies (Nishimori et al., 2008; Takayanagi et al., 2008). Although OT knockout mice have a normal heart structure, experiments have shown augmented intrinsic heart rates in these animals, indicating that an intracardiac OT system controls cardiac electrical activity (Bernatova et al., 2003).

Although the pathophysiological role of OT is beginning to be understood, accumulating evidence indicates multiple beneficial effects in the heart and vasculature. To date, OT's cardiovascular properties include: i. natriuresis (Soares et al., 1999) and decreased blood pressure (BP), possibly secondary to atrial natriuretic peptide (ANP) release (Gutkowska et al., 1997; Petersson, 2002a) ii. negative inotropic and chronotropic effects (Favaretto et al., 1997; Ondrejcakova et al., 2009) and parasympathetic neuromodulation (Mukaddam-Daher et al., 2001); iii. vasodilatation via the OTR-induced NO pathway; iv. endothelial cell growth and possible vessel generation (Cattaneo et al., 2008; Cattaneo et al., 2009; Thibonnier et al., 1999); and v. modulation of insulin release (Sirotkin et al., 2003) and anti-diabetic actions (Florian, et al., 2010). At the cellular level, protective OT: i. has antioxidant properties (Iseri et al., 2005a, 2005b) ii. has anti-inflammatory actions (Jankowski et al., 2010a; Szeto et al., 2008), iii. potentiates glucose uptake in neonatal and adult CMs exposed to hypoxia and conditions of insulin resistance mimicked by the presence of ketone bodies (Florian, et al., 2010) iv. stimulates endothelial markers in mesenchymal cells (Kim et al., 2010) and stem cells isolated from the heart as a side population (Oyama et al., 2007).

Central, intraventricular infusion of OT is accompanied by an increase in blood pressure; this effect is probably associated with the stimulation of substance P forebrain receptors by OT (McCann et al., 2002). OT's ability to raise blood pressure is caused not only by its vasoconstrictory activity but also by antidiuretic activity. Peripheral administration of OT, on the contrary, lowers the average arterial pressure in rats and does not affect heart rate (Petersson, 2002a). On the other hand, in the absence of a central regulatory influence, OT can bring down the heart rate and reduce the strength of contractions of isolated atria during perfusion of rat hearts (Favaretto et al., 1997; Mukaddam-Daher et al., 2001). In addition, intracardiac OT stimulating the release of ANP may control cardiovascular homeostasis and the body's internal environment (Favaretto et al., 1997; Gutkowska et al., 1997).

OT's negative chronotropic action was recently associated with attenuation of cardiac damage evinced by ischemia-reperfusion (Ondrejcakova et al., 2009). Positive cardiac effects can also be attributed to the fact that OT stimulates ANP release (Gutkowska et al., 1997) by improving hydromineral homeostasis as well as cardiac hypertrophy and reducing pro-inflammatory mediators (Jankowski et al., 2010a). ANP, a member of the NPs family that includes BNP, C type natriuretic peptide and urodilatin, is released into the circulation after volume expansion, atrial stretch (Dietz, 2005), hypoxia (Toth et al., 1994) and in response to various hormones and neurotransmitters (Antunes-Rodrigues et al., 1997; McCann et al., 2002). ANP causes BP to decline with a concomitant increment of diuresis, natriuresis and decrease plasma volume (Christensen, 1993; Ruskoaho, 1992). NPs also inhibit the sympathetic nervous system and hormones involved in cardiac hypertrophy, such as angiotensin II, endothelin and AVP (Gerstberger et al., 1992; Jankowski, 2009; Kaneko et al., 1988; Mukaddam-Daher et al., 2009; Neuser et al., 993). NPs signaling via functional receptors (NPR-A and NPR-B) prevents pathological hypertrophy (Oliver et al., 1997) and cardiac fibrosis (Calderone et al., 1998) by attenuating both DNA and collagen synthesis in cardiac fibroblasts, oxidative stress (Baldini et al., 2005; De Vito et al., 2003) and inflammation (Kiemer & Vollmar, 2001). Recent reports indicate that BNP and ANP activity is associated with lipolysis and postprandial lipid oxidation (Birkenfeld et al., 2008). Both hormones modulate fatty acid trafficking and prevent triglyceride accumulation in CMs via cGMP signaling (Khairallah et al., 2008).

The different cardioprotective actions of OT were recently demonstrated in animal models of myocardial infarction (MI). In rat and rabbit models of ischemic heart disease, OT treatment significantly reduced infarct size and improved parameters of heart function (Alizadeh et al., 2010; Houshmand et al., 2009; Jankowski et al., 2010a; Kim et al., 2010; Kobayashi et al., 2009; Ondrejcakova et al., 2009).

8. OT Signaling in cardiovascular system

In cardiac cells, several signaling pathways have also been postulated in conjunction with specific functions in the heart. Figure 1 illustrates the hypothetical pathways in the heart that are associated with OT-mediated cardioprotection, such as the prevention of apoptosis, CM hypertrophy and fibrosis, with stimulation of glucose uptake, cell proliferation and differentiation.

OT signaling depends on coupling to specific G-proteins, cell type and localization on the cell membrane surface. As a result, OTR stimulates different second messengers, which consequently exerts various physiological effects (Reversi et al., 2005). Due to its organ- and tissue-specific expression patterns, it is believed that OTR is regulated largely at the gene transcription level (Devost et al., 2008; Zingg & Laporte, 2003). In the cardiovascular system, OTR is associated with the ANP-cGMP and NO-cGMP pathways, which reduce the force and rate of contraction and increase vasodilatation.

Our recent report shows that OT increases glucose uptake in CMs via phosphoinositide-3-kinase (PI3K) and potentiates the glucose uptake effect of 2,4-dinitrophenol, an uncoupler of oxidative phosphorylation targeting the mitochondria (Florian, et al., 2010). PI3K pathways are considered beneficial during myocardial injuries (Cantley, 2002; Fujio et al., 2000; Miki et al., 2007). The calcium-calmodulin kinase kinase (Ca-CAMKK) and AMP-activated protein kinase (AMPK) pathways are also involved in OT-mediated glucose uptake in CMs (Florian et al., 2010). AMPK activation in the heart after ischemia and reperfusion is recognized as cardioprotective because AMPK limits both apoptosis and cell damage because both limits apoptosis and cell damage (Lee et al., 2008; Miki et al., 2007; Russell et al., 2004). We should also consider p38 MAPK and extracellular signal-regulated kinase 1/2 (ERK 1/2) phosphorylation which may contribute to OT's proliferative activity (Devost et al., 2008). More recently, in a rabbit model of myocardial ischemia-reperfusion, OT treatment induced ERK1/2, AKT and eNOS phosphorylation in cardiac tissues (Kobayashi et al., 2009). Therefore, OT, like other G-protein-coupled ligands, can act by PI3K/AKT activation and projection onto downstream kinases. Recent studies have demonstrated that the cardioprotective effects of OT are mediated through opening the mitochondrial ATP-dependent potassium (mitoKATP) channels in the rat heart (Alizadeh et al., 2010).

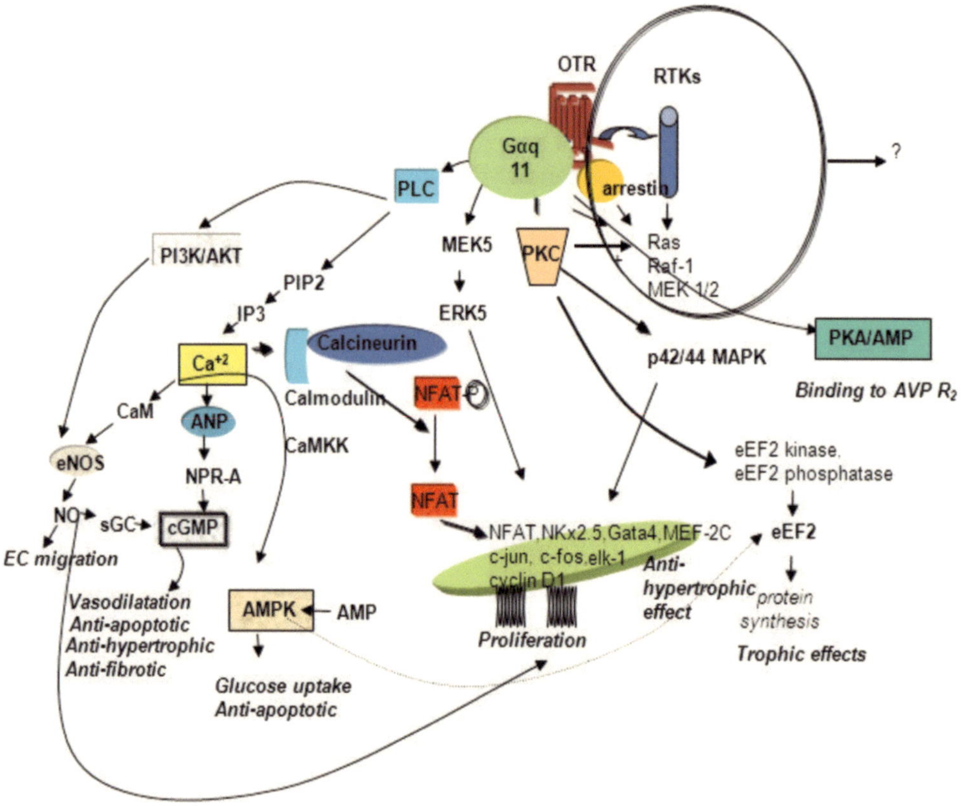

Fig. 3. Schematic diagram of potential protective pathways of oxytocin in the cardiomyocytes. AMPK, AMP activated protein kinase; ANP, atrial natriuretic peptide; CaMKK - Ca^{+2} calmodulin-dependent protein kinase; $AVPR_2$ -Vasopressin receptor R2; EC, endothelial cells; eEF2, eukaryotic translation elongation factor 2; MEK, protein kinase, mitogen activated kinase; NFAT, nuclear factor of activated T-cells

OT promotes the migration of human dermal EC, breast-derived EC (Cassoni et al., 2006) and human umbilical vein EC (HUVEC) (Cattaneo et al., 2008; Cattaneo et al., 2009). The pro-migratory effect of OT requires OTR activation of the phosphatidylinositol-3-kinase (PI3K)/AKT/eNOS pathway (Cattaneo et al., 2009). Moreover, OT increases proliferation of EC and alters gene expression for adhesion molecules and matrix metalloproteinases (MMPs), contributing to improved cell motility and growth (Cassoni et al., 2006). Angiogenic and anti-apoptotic OT effect was indicated by increased $CD31^+$ microvessels (Jankowski et al., 2010b). In this way, OT can control blood flow to the heart.

9. OT in stem cells therapy

Cardiovascular disease is one of the leading causes of death throughout the world (Jain et al., 2005). Following myocardial infarct (MI), endogenous repair mechanisms are insufficient for meaningful regeneration, therefore, muscle lost is replaced by non-contractile fibrotic scar (Laflamme et al., 2007). Because cardiomyocytes are unable to regenerate in the adult heart, cell-based therapy of transplantation provides a potential alternative approach to replace damaged myocardial tissue and restore cardiac function. A major roadblock toward this goal is the lack of donor cells, therefore, it is urgent to identify the cardiovascular cells that are necessary for achieving cardiac muscle regeneration (Dowell et al., 2003), to treat heart failure (Dowell et al., 2003; Raeburn et al., 2002) and restore function. (Hassink et al., 2003; Orlic et al., 2002). Several candidates have been investigated: fetal (Li et al., 1997) and neonatal (Watanabe et al., 1998) CM, embryonic stem cells (Min et al., 2002; Min et al., 2003), cardiac resident stem cells, and skeletal myoblasts (Dimmeler, Zeiher, & Schneider, 2005; Leor et al., 1996; Menasche et al., 2001; Menasche et al., 2003; Murry et al., 1996; Taylor et al., 1998). Alternative source are the bone marrow-derived stem cells (Dimmeler, et al., 2005). However, recent studies have questioned the ability of implanted, untreated stem cells to generate new CMs (Laflamme et al., 2007; Murry et al., 2004; Noiseux et al., 2006). Most cells die within hours of transplantation due to the interplay of ischemia, inflammation, and apoptosis (Menasche, 2009; Rosenzweig, 2006).

Among the cells used, the MSCs are more suitable for the cell therapy because of easy isolation, high expansion potential giving an unlimited pool of transplantable cells, low immunogenicity, amenability to *ex vivo* genetic modification and multipotency. It has been shown that injection of MSC either, directly into infarcted hearts (Gnecchi & Melo, 2009) by intramuscular (Shabbir et al., 2009) or intraperitoneal injections (Takahashi et al., 2006) improves myocardial function and repair. Following exposure to hypoxia and serum starvation, conditions that mimic MI, MSC are stimulated to secrete several growth factors and cytokines (Gnecchi & Melo, 2009). The exact mechanisms underlying MSC therapeutic effects require further investigations and challenges remain in optimizing the culture-expansion conditions, the MSC capacity for growth factor production and conditions to direct stem cells differentiation into CMs or endothelial cells. Recent studies have shown that direct injection with OT-treated MSCs into the rat heart after ischemia-reperfusion injury improves the engraftment rate and results in an enhanced cardio-protective effect via: increased transmigration activity, the upregulation of matrix metalloproteinase-2 mRNA, the integration of MSCs into the myocardium as well as the anti-fibrotic and anti-inflammatory effects (Kim et al., 2010).

We propose that MSCs and other types of the stem cells should be treated with OT before their engraftment *in vivo*. Based on our preliminary data, we expect that preconditioning with OT will enhance the capacity of these cells to repair infarcted myocardium due to reduced cell death after implantation, increased angiogenic potential, and by enhanced secretion of paracrine factors. Alternatively, the therapeutical potential of these cells can be improved by the introduction of the OT-expressing construct as recently demonstrated (Gassanov et al., 2008a).

10. Summary and conclusions

Our research lead to the observation that OT and OTR are synthesized in CMs, and we have identified OT as a potent, naturally-occurring cardiomyogenic factor, which by OTR up-regulation promotes the differentiation of embryonic and somatic stem cells residing in the heart to mature and functional CMs.

All these OT actions have physiological relevance, particularly on glucose uptake in CMs, since it is reduced in hearts from insulin-resistant diabetic mice, a disturbance that culminates in cardiac dysfunction.

Understanding the mechanisms of cardiac differentiation by OT can provide therapeutic approaches to the management of heart diseases. Currently, it is still extremely difficult to obtain new cardiac cells *in vitro* using stem cells isolated from the heart, as the only method that has provided satisfactory results is limited to co-culture with mature CMs. OT and some OT agonists, such as OT-Gly-Lys-Arg, that do not interfere with other physiological processes in the body (for example, without renal and hemodynamic effects), can successfully stimulate the differentiation of stem cell residing in the heart. In pathological conditions, such as cardiac ischemia and diabetes, this inducer can serve to stimulate the production of cardiac cells lost during the development of these pathologies. The advantage of such a therapy is supported by the fact that OT is produced endogenously in the organism, and does not have significant side effects when administered clinically. Moreover, it is now possible to inject (transplant) stem cells after previous stimulation with OT inducers, as in the case of heart attacks. Alternatively, direct treatment with OT molecules could promote cardiomyogenesis *in situ* and the regeneration of damaged hearts.

11. Acknowledgements

This work was supported by the Canadian Institutes of Health Research [MOP-53217 to J.G. and M.J.]; and the Canadian Heart and Stroke Foundation [NET SRD-63193 to J.G. and M.J.].

12. References

Abdelalim, E. M., & Tooyama, I. (2009). BNP signaling is crucial for embryonic stem cell proliferation. *PLoS One, 4*(4), e5341.

Alizadeh, A. M., Faghihi, M., Sadeghipour, H. R., Mohammadghasemi, F., Imani, A., Houshmand, F., et al. (2010). Oxytocin protects rat heart against ischemia-reperfusion injury via pathway involving mitochondrial ATP-dependent potassium channel. *Peptides, 31*(7), 1341-1345.

Altstein, M., Whitnall, M. H., House, S., Key, S., & Gainer, H. (1988a). An immunochemical analysis of oxytocin and vasopressin prohormone processing in vivo. *Peptides, 9*(1), 87-105.

Altstein, M., & Gainer, H. (1988b). Differential biosynthesis and posttranslational processing of vasopressin and oxytocin in rat brain during embryonic and postnatal development. *J Neurosci, 8*(11), 3967-3977.

Antunes-Rodrigues, J., Favaretto, A. L., Gutkowska, J., & McCann, S. M. (1997). The neuroendocrine control of atrial natriuretic peptide release. *Mol Psychiatry, 2*(5), 359-367.

Baldini, P. M., De Vito, P., Vismara, D., Bagni, C., Zalfa, F., Minieri, M., et al. (2005). Atrial natriuretic peptide effects on intracellular pH changes and ROS production in HEPG2 cells: role of p38 MAPK and phospholipase D. *Cell Physiol Biochem, 15*(1-4), 77-88.

Bernatova, I., Rigatto, K. V., Key, M. P., & Morris, M. (2004). Stress-induced pressor and corticosterone responses in oxytocin-deficient mice. *Exp Physiol, 89*(5), 549-557.

Birkenfeld, A. L., Budziarek, P., Boschmann, M., Moro, C., Adams, F., Franke, G., et al. (2008). Atrial natriuretic peptide induces postprandial lipid oxidation in humans. *Diabetes, 57*(12), 3199-3204.

Bisping, E., Ikeda, S., Kong, S. W., Tarnavski, O., Bodyak, N., McMullen, J. R., et al. (2006). Gata4 is required for maintenance of postnatal cardiac function and protection from pressure overload-induced heart failure. *Proc Natl Acad Sci U S A, 103*(39), 14471-14476.

Bouchard, F., & Paquin, J. (2009). Skeletal and cardiac myogenesis accompany adipogenesis in P19 embryonal stem cells. *Stem Cells Dev, 18*(7), 1023-1032.

Breton, C., Di Scala-Guenot, D., & Zingg, H. H. (2001). Oxytocin receptor gene expression in rat mammary gland: structural characterization and regulation. *J Mol Endocrinol, 27*(2), 175-189.

Breton, C., Haenggeli, C., Barberis, C., Heitz, F., Bader, C. R., Bernheim, L., et al. (2002). Presence of functional oxytocin receptors in cultured human myoblasts. *J Clin Endocrinol Metab, 87*(3), 1415-1418.

Brownstein, M. J., Russell, J. T., & Gainer, H. (1980). Synthesis, transport, and release of posterior pituitary hormones. *Science, 207*(4429), 373-378.

Burbach, J. P., Luckman, S. M., Murphy, D., & Gainer, H. (2001). Gene regulation in the magnocellular hypothalamo-neurohypophysial system. *Physiol Rev, 81*(3), 1197-1267.

Calderone, A., Thaik, C. M., Takahashi, N., Chang, D. L., & Colucci, W. S. (1998). Nitric oxide, atrial natriuretic peptide, and cyclic GMP inhibit the growth-promoting effects of norepinephrine in cardiac myocytes and fibroblasts. *J Clin Invest, 101*(4), 812-818.

Cantley, L. C. (2002). The phosphoinositide 3-kinase pathway. *Science, 296*(5573), 1655-1657.

Cassoni, P., Marrocco, T., Bussolati, B., Allia, E., Munaron, L., Sapino, A., et al. (2006). Oxytocin induces proliferation and migration in immortalized human dermal microvascular endothelial cells and human breast tumor-derived endothelial cells. *Mol Cancer Res, 4*(6), 351-359.

Cattaneo, M. G., Chini, B., & Vicentini, L. M. (2008). Oxytocin stimulates migration and invasion in human endothelial cells. *Br J Pharmacol, 153*(4), 728-736.

Cattaneo, M. G., Lucci, G., & Vicentini, L. M. (2009). Oxytocin stimulates in vitro angiogenesis via a Pyk-2/Src-dependent mechanism. *Exp Cell Res, 315*(18), 3210-3219.

Christensen, G. (1993). Release of atrial natriuretic factor. *Scand J Clin Lab Invest, 53*(2), 91-100.

Dale, H. H. (1906). On some physiological actions of ergot. *Journal of Physiology, 34*, 163-206.

Danalache, B. A., Paquin, J., Donghao, W., Grygorczyk, R., Moore, J. C., Mummery, C. L., et al. (2007). Nitric oxide signaling in oxytocin-mediated cardiomyogenesis. *Stem Cells, 25*(3), 679-688.

De Vito, P., Di Nardo, P., Palmery, M., Peluso, I., Luly, P., & Baldini, P. M. (2003). Oxidant-induced pHi/Ca2+ changes in rat aortic smooth muscle cells. The role of atrial natriuretic peptide. *Mol Cell Biochem, 252*(1-2), 353-362.

Devost, D., Wrzal, P., & Zingg, H. H. (2008). Oxytocin receptor signalling. *Prog Brain Res, 170*, 167-176.

Dietz, J. R. (2005). Mechanisms of atrial natriuretic peptide secretion from the atrium. *Cardiovasc Res, 68*(1), 8-17.

Dimmeler, S., Zeiher, A. M., & Schneider, M. D. (2005). Unchain my heart: the scientific foundations of cardiac repair. *J Clin Invest, 115*(3), 572-583.

Dowell, J. D., Rubart, M., Pasumarthi, K. B., Soonpaa, M. H., & Field, L. J. (2003). Myocyte and myogenic stem cell transplantation in the heart. *Cardiovasc Res, 58*(2), 336-350.

Du Vigneaud, V. (1956). Hormones of the posterior pituitary gland. Oxytocin and vasopressin. *Harvey Lectures, 51*, 1-26.

Dudley, D. J. (1997). Oxytocin: use and abuse, science and art. *Clin Obstet Gynecol, 40*(3), 516-524.

Elabd, C., Basillais, A., Beaupied, H., Breuil, V., Wagner, N., Scheideler, M., et al. (2008). Oxytocin controls differentiation of human mesenchymal stem cells and reverses osteoporosis. *Stem Cells, 26*(9), 2399-2407.

Fathi, F., Murasawa, S., Hasegawa, S., Asahara, T., Kermani, A. J., & Mowla, S. J. (2009). Cardiac differentiation of P19CL6 cells by oxytocin. *Int J Cardiol, 134*(1), 75-81.

Favaretto, A. L., Ballejo, G. O., Albuquerque-Araujo, W. I., Gutkowska, J., Antunes-Rodrigues, J., & McCann, S. M. (1997). Oxytocin releases atrial natriuretic peptide from rat atria in vitro that exerts negative inotropic and chronotropic action. *Peptides, 18*(9), 1377-1381.

Florian, M., Jankowski, M., & Gutkowska, J. (2010). Oxytocin increases glucose uptake in neonatal rat cardiomyocytes. *Endocrinology, 151*(2), 482-491.

Fujio, Y., Nguyen, T., Wencker, D., Kitsis, R. N., & Walsh, K. (2000). Akt promotes survival of cardiomyocytes in vitro and protects against ischemia-reperfusion injury in mouse heart. *Circulation, 101*(6), 660-667.

Furuya, K., Mizumoto, Y., Makimura, N., Mitsui, C., Murakami, M., Tokuoka, S., et al. (1995). A novel biological aspect of ovarian oxytocin: gene expression of oxytocin and oxytocin receptor in cumulus/luteal cells and the effect of oxytocin on embryogenesis in fertilized oocytes. *Adv Exp Med Biol, 395*, 523-528.

Gaddum, J. H. (1928). Some properties of the separated active principles of the pituitary (posterior lobe). *Journal of Physiology, 65*, 434-440.

Gassanov, N., Jankowski, M., Danalache, B., Wang, D., Grygorczyk, R., Hoppe, U. C., et al. (2007). Arginine vasopressin-mediated cardiac differentiation: insights into the role of its receptors and nitric oxide signaling. *J Biol Chem, 282*(15), 11255-11265.

Gassanov, N., Devost, D., Danalache, B., Noiseux, N., Jankowski, M., Zingg, H. H., et al. (2008a). Functional activity of the carboxyl-terminally extended oxytocin precursor

Peptide during cardiac differentiation of embryonic stem cells. *Stem Cells, 26*(1), 45-54.

Gassanov, N., Er, F., Zagidullin, N., Jankowski, M., Gutkowska, J., & Hoppe, U. C. (2008b). Retinoid acid-induced effects on atrial and pacemaker cell differentiation and expression of cardiac ion channels. *Differentiation, 76*(9), 971-980.

Gerstberger, R., Schutz, H., Luther-Dyroff, D., Keil, R., & Simon, E. (1992). Inhibition of vasopressin and aldosterone release by atrial natriuretic peptide in conscious rabbits. *Exp Physiol, 77*(4), 587-600.

Gimpl, G., & Fahrenholz, F. (2001). The oxytocin receptor system: structure, function, and regulation. *Physiol Rev, 81*(2), 629-683.

Gnecchi, M., & Melo, L. G. (2009). Bone marrow-derived mesenchymal stem cells: isolation, expansion, characterization, viral transduction, and production of conditioned medium. *Methods Mol Biol, 482*, 281-294.

Gutkowska, J., Jankowski, M., Lambert, C., Mukaddam-Daher, S., Zingg, H. H., & McCann, S. M. (1997). Oxytocin releases atrial natriuretic peptide by combining with oxytocin receptors in the heart. *Proc Natl Acad Sci U S A, 94*(21), 11704-11709.

Gutkowska, J., Jankowski, M., Mukaddam-Daher, S., & McCann, S. M. (2000). Oxytocin is a cardiovascular hormone. *Braz J Med Biol Res, 33*(6), 625-633.

Gutkowska, J., Miszkurka, M., Danalache, B., Gassanov, N., Wang, D., & Jankowski, M. (2007). Functional arginine vasopressin system in early heart maturation. *Am J Physiol Heart Circ Physiol, 293*(4), H2262-2270.

Gutkowska, J., & Jankowski, M. (2008). Oxytocin revisited: It is also a cardiovascular hormone. *J Am Soc Hypertens, 2*(5), 318-325.

Gutkowska, J., & Jankowski, M. (2009). Oxytocin: Old Hormone, New Drug. *Pharmaceuticals, 2*(3), 168-183.

Hassink, R. J., Brutel de la Riviere, A., Mummery, C. L., & Doevendans, P. A. (2003). Transplantation of cells for cardiac repair. *J Am Coll Cardiol, 41*(5), 711-717.

Hatami, L., Valojerdi, M. R., & Mowla, S. J. (2007). Effects of oxytocin on cardiomyocyte differentiation from mouse embryonic stem cells. *Int J Cardiol, 117*(1), 80-89.

Heineke, J., Auger-Messier, M., Xu, J., Oka, T., Sargent, M. A., York, A., et al. (2007). Cardiomyocyte GATA4 functions as a stress-responsive regulator of angiogenesis in the murine heart. *J Clin Invest, 117*(11), 3198-3210.

Hopkins, W. E., Chen, Z., Fukagawa, N. K., Hall, C., Knot, H. J., & LeWinter, M. M. (2004). Increased atrial and brain natriuretic peptides in adults with cyanotic congenital heart disease: enhanced understanding of the relationship between hypoxia and natriuretic peptide secretion. *Circulation, 109*(23), 2872-2877.

Houshmand, F., Faghihi, M., & Zahediasl, S. (2009). Biphasic protective effect of oxytocin on cardiac ischemia/reperfusion injury in anaesthetized rats. *Peptides, 30*(12), 2301-2308.

Husslein, P. (2002). Development and clinical experience with the new evidence-based tocolytic atosiban. *Acta Obstet Gynecol Scand, 81*(7), 633-641.

Iseri, S. O., Sener, G., Saglam, B., Gedik, N., Ercan, F., & Yegen, B. C. (2005a). Oxytocin ameliorates oxidative colonic inflammation by a neutrophil-dependent mechanism. *Peptides, 26*(3), 483-491.

Iseri, S. O., Sener, G., Saglam, B., Gedik, N., Ercan, F., & Yegen, B. C. (2005b). Oxytocin protects against sepsis-induced multiple organ damage: role of neutrophils. *J Surg Res, 126*(1), 73-81.

Jain, M., Pfister, O., Hajjar, R. J., & Liao, R. (2005). Mesenchymal stem cells in the infarcted heart. *Coron Artery Dis, 16*(2), 93-97.

Jamali, M., Karamboulas, C., Rogerson, P. J., & Skerjanc, I. S. (2001). BMP signaling regulates Nkx2-5 activity during cardiomyogenesis. *FEBS Lett, 509*(1), 126-130.

Jankowski, M., Hajjar, F., Kawas, S. A., Mukaddam-Daher, S., Hoffman, G., McCann, S. M., et al. (1998). Rat heart: a site of oxytocin production and action. *Proc Natl Acad Sci U S A, 95*(24), 14558-14563.

Jankowski, M., Danalache, B., Wang, D., Bhat, P., Hajjar, F., Marcinkiewicz, M., et al. (2004). Oxytocin in cardiac ontogeny. *Proc Natl Acad Sci U S A, 101*(35), 13074-13079.

Jankowski, M. (2009). GATA4, a new regulator of cardiac fibroblasts, is sensitive to natriuretic peptides. *Cardiovasc Res, 84*(2), 176-177.

Jankowski, M., Bissonauth, V., Gao, L., Gangal, M., Wang, D., Danalache, B., et al. (2010a). Anti-inflammatory effect of oxytocin in rat myocardial infarction. *Basic Res Cardiol, 105*(2), 205-218.

Jankowski, M., Wang, D., Danalache, B., Gangal, M., & Gutkowska, J. (2010b). Cardiac oxytocin receptor blockade stimulates adverse cardiac remodeling in ovariectomized spontaneously hypertensive rats. *Am J Physiol Heart Circ Physiol, 299*(2), H265-274.

Kaneko, K., Fukuda, S., Ishikawa, S., & Saito, T. (1988). Comparative biological activities of alpha-rat and alpha-human atrial natriuretic peptide in the inhibition of arginine vasopressin release in conscious rats. *Endocrinol Jpn, 35*(6), 809-817.

Khairallah, R. J., Khairallah, M., Gelinas, R., Bouchard, B., Young, M. E., Allen, B. G., et al. (2008). Cyclic GMP signaling in cardiomyocytes modulates fatty acid trafficking and prevents triglyceride accumulation. *J Mol Cell Cardiol, 45*(2), 230-239.

Kiemer, A. K., & Vollmar, A. M. (2001). The atrial natriuretic peptide regulates the production of inflammatory mediators in macrophages. *Ann Rheum Dis, 60 Suppl 3*, iii68-70.

Kim, Y. S., Kwon, J. S., Hong, M. H., Kim, J., Song, C. H., Jeong, M. H., et al. (2010). Promigratory activity of oxytocin on umbilical cord blood-derived mesenchymal stem cells. *Artif Organs, 34*(6), 453-461.

Kobayashi, H., Yasuda, S., Bao, N., Iwasa, M., Kawamura, I., Yamada, Y., et al. (2009). Postinfarct treatment with oxytocin improves cardiac function and remodeling via activating cell-survival signals and angiogenesis. *J Cardiovasc Pharmacol, 54*(6), 510-519.

Laflamme, M. A., Zbinden, S., Epstein, S. E., & Murry, C. E. (2007). Cell-based therapy for myocardial ischemia and infarction: pathophysiological mechanisms. *Annu Rev Pathol, 2*, 307-339.

Lee, E. S., Uhm, K. O., Lee, Y. M., Kwon, J., Park, S. H., & Soo, K. H. (2008). Oxytocin stimulates glucose uptake in skeletal muscle cells through the calcium-CaMKK-AMPK pathway. *Regul Pept, 151*(1-3), 71-74.

Leor, J., Patterson, M., Quinones, M. J., Kedes, L. H., & Kloner, R. A. (1996). Transplantation of fetal myocardial tissue into the infarcted myocardium of rat. A potential method for repair of infarcted myocardium? *Circulation, 94*(9 Suppl), II332-336.

Li, R. K., Mickle, D. A., Weisel, R. D., Mohabeer, M. K., Zhang, J., Rao, V., et al. (1997). Natural history of fetal rat cardiomyocytes transplanted into adult rat myocardial scar tissue. *Circulation, 96*(9 Suppl), II-179-186; discussion 186-177.

Light, K. C., Grewen, K. M., Amico, J. A., Brownley, K. A., West, S. G., Hinderliter, A. L., et al. (2005). Oxytocinergic activity is linked to lower blood pressure and vascular resistance during stress in postmenopausal women on estrogen replacement. *Horm Behav, 47*(5), 540-548.

Matsuura, K., Nagai, T., Nishigaki, N., Oyama, T., Nishi, J., Wada, H., et al. (2004). Adult cardiac Sca-1-positive cells differentiate into beating cardiomyocytes. *J Biol Chem, 279*(12), 11384-11391.

McBurney, M. W. (1993). P19 embryonal carcinoma cells. *Int J Dev Biol, 37*(1), 135-140.

McCann, S. M., Antunes-Rodrigues, J., Jankowski, M., & Gutkowska, J. (2002). Oxytocin, vasopressin and atrial natriuretic peptide control body fluid homeostasis by action on their receptors in brain, cardiovascular system and kidney. *Prog Brain Res, 139*, 309-328.

Menasche, P., Hagege, A. A., Scorsin, M., Pouzet, B., Desnos, M., Duboc, D., et al. (2001). Myoblast transplantation for heart failure. *Lancet, 357*(9252), 279-280.

Menasche, P., Hagege, A. A., Vilquin, J. T., Desnos, M., Abergel, E., Pouzet, B., et al. (2003). Autologous skeletal myoblast transplantation for severe postinfarction left ventricular dysfunction. *J Am Coll Cardiol, 41*(7), 1078-1083.

Menasche, P. (2009). Cell-based therapy for heart disease: a clinically oriented perspective. *Mol Ther, 17*(5), 758-766.

Michelini, L. C., Marcelo, M. C., Amico, J., & Morris, M. (2003). Oxytocinergic regulation of cardiovascular function: studies in oxytocin-deficient mice. *Am J Physiol Heart Circ Physiol, 284*(6), H2269-2276.

Miki, T., Miura, T., Tanno, M., Nishihara, M., Naitoh, K., Sato, T., et al. (2007). Impairment of cardioprotective PI3K-Akt signaling by post-infarct ventricular remodeling is compensated by an ERK-mediated pathway. *Basic Res Cardiol, 102*(2), 163-170.

Min, J. Y., Yang, Y., Converso, K. L., Liu, L., Huang, Q., Morgan, J. P., et al. (2002). Transplantation of embryonic stem cells improves cardiac function in postinfarcted rats. *J Appl Physiol, 92*(1), 288-296.

Min, J. Y., Yang, Y., Sullivan, M. F., Ke, Q., Converso, K. L., Chen, Y., et al. (2003). Long-term improvement of cardiac function in rats after infarction by transplantation of embryonic stem cells. *J Thorac Cardiovasc Surg, 125*(2), 361-369.

Morris, M., Castro, M., & Rose, J. C. (1992). Alterations in oxytocin prohormone processing during early development in the fetal sheep. *Am J Physiol, 263*(3 Pt 2), R738-740.

Mouquet, F., Pfister, O., Jain, M., Oikonomopoulos, A., Ngoy, S., Summer, R., et al. (2005). Restoration of cardiac progenitor cells after myocardial infarction by self-proliferation and selective homing of bone marrow-derived stem cells. *Circ Res, 97*(11), 1090-1092.

Mukaddam-Daher, S., Yin, Y. L., Roy, J., Gutkowska, J., & Cardinal, R. (2001). Negative inotropic and chronotropic effects of oxytocin. *Hypertension, 38*(2), 292-296.

Mukaddam-Daher, S., Menaouar, A., Paquette, P. A., Jankowski, M., Gutkowska, J., Gillis, M. A., et al. (2009). Hemodynamic and cardiac effects of chronic eprosartan and moxonidine therapy in stroke-prone spontaneously hypertensive rats. *Hypertension, 53*(5), 775-781.

Murry, C. E., Wiseman, R. W., Schwartz, S. M., & Hauschka, S. D. (1996). Skeletal myoblast transplantation for repair of myocardial necrosis. *J Clin Invest, 98*(11), 2512-2523.

Murry, C. E., Soonpaa, M. H., Reinecke, H., Nakajima, H., Nakajima, H. O., Rubart, M., et al. (2004). Haematopoietic stem cells do not transdifferentiate into cardiac myocytes in myocardial infarcts. *Nature, 428*(6983), 664-668.

Neuser, D., Stasch, J. P., Knorr, A., & Kazda, S. (1993). Inhibition by atrial natriuretic peptide of endothelin-1-stimulated proliferation of vascular smooth-muscle cells. *J Cardiovasc Pharmacol, 22 Suppl 8*, S257-261.

Nishimori, K., Takayanagi, Y., Yoshida, M., Kasahara, Y., Young, L. J., & Kawamata, M. (2008). New aspects of oxytocin receptor function revealed by knockout mice: sociosexual behaviour and control of energy balance. *Prog Brain Res, 170*, 79-90.

Noiseux, N., Gnecchi, M., Lopez-Ilasaca, M., Zhang, L., Solomon, S. D., Deb, A., et al. (2006). Mesenchymal stem cells overexpressing Akt dramatically repair infarcted myocardium and improve cardiac function despite infrequent cellular fusion or differentiation. *Mol Ther, 14*(6), 840-850.

Oliver, P. M., Fox, J. E., Kim, R., Rockman, H. A., Kim, H. S., Reddick, R. L., et al. (1997). Hypertension, cardiac hypertrophy, and sudden death in mice lacking natriuretic peptide receptor A. *Proc Natl Acad Sci U S A, 94*(26), 14730-14735.

Ondrejcakova, M., Ravingerova, T., Bakos, J., Pancza, D., & Jezova, D. (2009). Oxytocin exerts protective effects on in vitro myocardial injury induced by ischemia and reperfusion. *Can J Physiol Pharmacol, 87*(2), 137-142.

Orlic, D., Hill, J. M., & Arai, A. E. (2002). Stem cells for myocardial regeneration. *Circ Res, 91*(12), 1092-1102.

Ott, I., & Scott, J. C. (1910). The action of infundibulin upon the mammary secretion. *Proceedings of the Society for Experimental Biology, 8*, 48-49.

Oyama, T., Nagai, T., Wada, H., Naito, A. T., Matsuura, K., Iwanaga, K., et al. (2007). Cardiac side population cells have a potential to migrate and differentiate into cardiomyocytes in vitro and in vivo. *J Cell Biol, 176*(3), 329-341.

Paquin, J., Danalache, B. A., Jankowski, M., McCann, S. M., & Gutkowska, J. (2002). Oxytocin induces differentiation of P19 embryonic stem cells to cardiomyocytes. *Proc Natl Acad Sci U S A, 99*(14), 9550-9555.

Paton, D. N., & Watson, A. (1912). The actions of pituitrin, adrenalin and barium on the circulation of the bird. *Journal of Physiology, 44*(5-6), 413-424.

Petersson, M. (2002a). Cardiovascular effects of oxytocin. *Prog Brain Res, 139*, 281-288.

Petersson, M., Lagumdzija, A., Stark, A., & Bucht, E. (2002b). Oxytocin stimulates proliferation of human osteoblast-like cells. *Peptides, 23*(6), 1121-1126.

Pfister, O., Mouquet, F., Jain, M., Summer, R., Helmes, M., Fine, A., et al. (2005). CD31- but Not CD31+ cardiac side population cells exhibit functional cardiomyogenic differentiation. *Circ Res, 97*(1), 52-61.

Raeburn, C. D., Zimmerman, M. A., Arya, J., Banerjee, A., & Harken, A. H. (2002). Stem cells and myocardial repair. *J Am Coll Surg, 195*(5), 686-693.

Reversi, A., Cassoni, P., & Chini, B. (2005). Oxytocin receptor signaling in myoepithelial and cancer cells. *J Mammary Gland Biol Neoplasia, 10*(3), 221-229.

Rosenzweig, A. (2006). Cardiac cell therapy--mixed results from mixed cells. *N Engl J Med, 355*(12), 1274-1277.

Ruskoaho, H. (1992). Atrial natriuretic peptide: synthesis, release, and metabolism. *Pharmacol Rev, 44*(4), 479-602.

Russell, R. R., 3rd, Li, J., Coven, D. L., Pypaert, M., Zechner, C., Palmeri, M., et al. (2004). AMP-activated protein kinase mediates ischemic glucose uptake and prevents postischemic cardiac dysfunction, apoptosis, and injury. *J Clin Invest, 114*(4), 495-503.

Sapino, A., Macri, L., Tonda, L., & Bussolati, G. (1993). Oxytocin enhances myoepithelial cell differentiation and proliferation in the mouse mammary gland. *Endocrinology, 133*(2), 838-842.

Shabbir, A., Zisa, D., Suzuki, G., & Lee, T. (2009). Heart failure therapy mediated by the trophic activities of bone marrow mesenchymal stem cells: a noninvasive therapeutic regimen. *Am J Physiol Heart Circ Physiol, 296*(6), H1888-1897.

Sirotkin, A. V., Florkovicova, I., Makarevich, A. V., Schaeffer, H. J., Kotwica, J., Marnet, P. G., et al. (2003). Oxytocin mediates some effects of insulin-like growth factor-I on porcine ovarian follicles. *J Reprod Dev, 49*(2), 141-149.

Skerjanc, I. S., Petropoulos, H., Ridgeway, A. G., & Wilton, S. (1998). Myocyte enhancer factor 2C and Nkx2-5 up-regulate each other's expression and initiate cardiomyogenesis in P19 cells. *J Biol Chem, 273*(52), 34904-34910.

Soares, T. J., Coimbra, T. M., Martins, A. R., Pereira, A. G., Carnio, E. C., Branco, L. G., et al. (1999). Atrial natriuretic peptide and oxytocin induce natriuresis by release of cGMP. *Proc Natl Acad Sci U S A, 96*(1), 278-283.

Stefanidis, K., Loutradis, D., Anastasiadou, V., Beretsos, P., Bletsa, R., Dinopoulou, V., et al. (2009). Embryoid bodies from mouse stem cells express oxytocin receptor, Oct-4 and DAZL. *Biosystems, 98*(2), 122-126.

Szeto, A., Nation, D. A., Mendez, A. J., Dominguez-Bendala, J., Brooks, L. G., Schneiderman, N., et al. (2008). Oxytocin attenuates NADPH-dependent superoxide activity and IL-6 secretion in macrophages and vascular cells. *Am J Physiol Endocrinol Metab, 295*(6), E1495-1501.

Takahashi, M., Li, T. S., Suzuki, R., Kobayashi, T., Ito, H., Ikeda, Y., et al. (2006). Cytokines produced by bone marrow cells can contribute to functional improvement of the infarcted heart by protecting cardiomyocytes from ischemic injury. *Am J Physiol Heart Circ Physiol, 291*(2), H886-893.

Takayanagi, Y., Kasahara, Y., Onaka, T., Takahashi, N., Kawada, T., & Nishimori, K. (2008). Oxytocin receptor-deficient mice developed late-onset obesity. *Neuroreport, 19*(9), 951-955.

Taylor, D. A., Atkins, B. Z., Hungspreugs, P., Jones, T. R., Reedy, M. C., Hutcheson, K. A., et al. (1998). Regenerating functional myocardium: improved performance after skeletal myoblast transplantation. *Nat Med, 4*(8), 929-933.

Thibonnier, M., Conarty, D. M., Preston, J. A., Plesnicher, C. L., Dweik, R. A., & Erzurum, S. C. (1999). Human vascular endothelial cells express oxytocin receptors. *Endocrinology, 140*(3), 1301-1309.

Thompson, R. B., Emani, S. M., Davis, B. H., van den Bos, E. J., Morimoto, Y., Craig, D., et al. (2003). Comparison of intracardiac cell transplantation: autologous skeletal myoblasts versus bone marrow cells. *Circulation, 108 Suppl 1*, II264-271.

Toth, M., Vuorinen, K. H., Vuolteenaho, O., Hassinen, I. E., Uusimaa, P. A., Leppaluoto, J., et al. (1994). Hypoxia stimulates release of ANP and BNP from perfused rat ventricular myocardium. *Am J Physiol, 266*(4 Pt 2), H1572-1580.

Uchida, S., Fuke, S., & Tsukahara, T. (2007). Upregulations of Gata4 and oxytocin receptor are important in cardiomyocyte differentiation processes of P19CL6 cells. *J Cell Biochem, 100*(3), 629-641.

van der Heyden, M. A., & Defize, L. H. (2003). Twenty one years of P19 cells: what an embryonal carcinoma cell line taught us about cardiomyocyte differentiation. *Cardiovasc Res, 58*(2), 292-302.

Watanabe, E., Smith, D. M., Jr., Delcarpio, J. B., Sun, J., Smart, F. W., Van Meter, C. H., Jr., et al. (1998). Cardiomyocyte transplantation in a porcine myocardial infarction model. *Cell Transplant, 7*(3), 239-246.

Zingg, H. H., & Laporte, S. A. (2003). The oxytocin receptor. *Trends Endocrinol Metab, 14*(5), 222-227.

29

Impact of Reactive Oxygen and Reactive Nitrogen Species for Stem Cell Mobilization, Function and Cardiovascular Differentiation

Heinrich Sauer[1] and Maria Wartenberg[2]
[1]*Department of Physiology, Justus Liebig University Giessen,*
[2]*Department of Internal Medicine I, Cardiology Division,*
Friedrich Schiller University Jena,
Germany

1. Introduction

Reactive oxygen species (ROS), i.e. substances like hydrogen peroxide (H_2O_2), the superoxide anion (O_2^-) or the highly reactive hydroxyl ion (HO^-) as well as reactive nitrogen species (RNS) with nitric oxide (NO) as its most important member are ideally suited to serve as signaling molecules since they are locally generated, are highly and rapidly diffusible and can be neutralized by a bulk of anti-oxidative agents organized in the cellular anti-oxidative defense system. So far the signaling pathways regulating organ-specific differentiation of stem cells are largely unknown. Differentiation processes of stem cells embedded in tissues and organs are tightly regulated by the cellular microenvironment which is critically determined by the availability of nutrients and oxygen as well as by the balance of ROS and NO generation. Already during early embryogenesis NADPH oxidases and NO synthases are expressed in the growing embryo, suggesting that gradients of ROS and NO may exist in the developing organs and may be involved in proper functioning of commitment programs. During pathophysiological insults, e.g. during hypertension, atherosclerosis and cardiac infarction high levels of ROS and NO are generated, thus creating an inflammatory microenvironment which on one side contributes to cell damage, apoptosis and remodeling, however, on the other side may activate repair processes that involve recruitment and differentiation of stem cells of the cardiovascular cell lineage. During recent years emerging evidence suggests that ROS and RNS are involved in cardiovascular differentiation of embryonic (ES) and adult stem cells. Comparable effects may occur during differentiation processes of resident cardiac stem cells. A pivotal role for NADPH oxidases and NO synthases in cardiomyogenesis and vasculogenesis of ES cells has been recently outlined. In this chapter the current knowledge on activation, recruitment and differentiation of different cardiovascular stem cell populations by ROS and NO and the involved signal transduction cascade is reviewed. Furthermore the specific microenvironmental requirements for proper stem cell engraftment and maintenance are outlined.

2. Hypoxia and the anti-oxidative defense of stem cells

Life on earth requires oxygen for generating energy by metabolic processes. Paradoxically, life is on the same time threatened by oxygen, since highly reactive ROS may destruct cellular components such as DNA, proteins and lipids by oxidation. Organisms have acquired a complex system of antioxidant metabolites and antioxidative enzymes to prevent oxidative damage to cellular components. This system comprises of divergent agents such as ascorbic acid (Vitamin C), polyphenols, flavonoids, tocopherols, uric acid, glutathione and thioredoxins as well as antioxidant enzymes e.g. peroxiredoxins, superoxide dismutase and catalase. These antioxidant molecules may be supplemented by the antioxidative function of other enzymes that are not directly related to the anti-oxidative defense, like the sirtuins which are a phylogenetically conserved NAD+-dependent protein deacetylase/ADP-ribosyltransferase family implicated in diverse biological processes. Treatment of mouse preimplantation embryos with sirtuin inhibitors resulted in increased intracellular ROS levels and decreased blastocyst formation. These effects were recapitulated by siRNA-induced knockdown of Sirt3, which is involved in mitochondrial energy metabolism, and in Sirt3-/- embryos (Kawamura et al., 2010). Furthermore Prdm16, a transcription factor that regulates leukaemogenesis, palatogenesis and brown-fat development has been demonstrated to be involved in the maintenance of stem cell function by modulating the intracellular redox state. In stem cells of the haematopoietic and nervous systems, Prdm16 deficiency led to increase in ROS levels, depletion of stem cells, increased cell death and altered cell-cycle distribution. This could be rescued in the presence of the anti-oxidant, N-acetyl-cysteine (Chuikov et al., 2010). Previously it was assumed that antioxidant systems either prevent reactive species from being formed, or remove them before they can damage vital components of the cell. However, research of the last decade has unraveled a decisive role of ROS and RNS in cellular signaling where reactive species are involved in growth factor, hormone and cytokine function. From this time on it was speculated that the function of the antioxidative system is to keep ROS and RNS at an optimum homeostatic level that allows the regulation of specific signaling pathways. Apparently physiological levels of ROS are necessary to maintain genome stability in ES cells and cardiac stem cells. Interestingly antioxidants suppressed DNA damage at low concentrations, but potentiated such damage at higher concentrations. High-dose antioxidants decreased cellular levels of ATM (ataxia-telangiectasia mutated) and other DNA repair enzymes (Li & Marban, 2010). How the redox state inside cells is balanced is yet not known. Current methods of cellular analysis are apparently not sensitive enough to discriminate the time and spatial distribution of antioxidants within cells. It may be speculated that distinct combinations of antioxidant molecules are spatially distributed within cells and/or may be bound to specific cell structures thus allowing the generation of gradients of the anti-oxidative defense within cells. These gradients may be further diversified by the different anti-oxidative strength of members of the anti-oxidative capacity as well as different time durations of the chemical reactions that lead to the neutralization of the respective reactive species and termination of the respective signal transduction pathway.

During embryogenesis the early embryo is extremely sensitive towards oxidative stress. Extensive research on in vitro fertilization and embryo transfer has demonstrated that during embryo culture cell culture media have to be supplemented with distinct cocktails of anti-oxidants to avoid serious tissue damage (Agarwal et al., 2008). In the morula state the early embryo passing through the ovary duct as well as the blastocyst prior to implantation in the uterus is subjected to severe hypoxia. Since hypoxia via hypoxia-activated transcription factors can regulate gene expression it has been speculated that the low oxygen pressure present in early embryo may direct the earliest possible differentiation steps (Simon & Keith, 2008). Also during early organogenesis, the embryo is in a state of relative hypoxia associated with a major decrease in terminal electron transport system activity and a marked increase in anaerobic glycolysis (Shepard et al., 2000). Hypoxia can be harmful to the embryo which requires the development of protection mechanisms. As recently described one of those may involve the adenosine A1 receptor (A1R) since A1R$^{-/-}$ embryos were more sensitive towards ischemic stress then normoxic controls and just the heart was identified as the site of A1R-mediated embryo protection (Wendler et al., 2007; Wendler et al., 2010). The susceptibility of the very early embryo towards oxidative stress may imply that also stem cells need either a low oxygen and/or high anti-oxidative status for stem cell maintenance. Indeed it has been demonstrated that human ES cells remained in an undifferentiated state when cultivated under low oxygen conditions (Ezashi et al., 2005). Furthermore it has been shown that bovine blastosysts produced under reduced O_2 (< 2% O_2) tensions displayed significantly more inner cell mass cells displaying pluripotency (Harvey et al., 2004). In the adult one of the major sources for stem cells is the bone marrow. The bone marrow microenvironment displays a lower oxygen concentration than other tissues, and stem cells are localized within the hypoxic regions thus suggesting that hypoxia may be crucial for stem cell maintenance. In hypoxic human ES cells hypoxia inducible factor-1α (HIF-1α), a principal mediator of hypoxic adaptations, modulates Wnt/β-catenin signalling by enhancing β-catenin activation and expression of the downstream effectors LEF-1 and TCF-1. O_2 availability, therefore, may have a direct role in stem cell regulation through HIF-1α modulation of Wnt/β-catenin signalling (Mazumdar et al., 2010). In a further study it has been recently shown that external addition of anti-oxidant and low oxygen tension results in reprogramming of human adipose stromal cells towards a more primitive stem cell type (Jee et al., 2010). HIF-2α but not HIF-1α is a direct upstream regulator of the stemness gene Oct-4 since HIF-2α was shown to be capable of binding hypoxic regulatory elements in the murine *Oct-4* promotor (Covello et al., 2006).

3. ROS generation during embryogenesis

During later stages of embryogenesis, when the utero-placental circulation is established the embryo is more capable to defend against oxidative stress which is due to a stronger anti-oxidative stress response and at least partially due to the metabolic switch from glycolysis to oxidative phosphorylation which occurs at times where the embryonic heart starts to contract, thus requesting more energy for heart performance. Despite the sensitivity of the early embryo towards oxidative stress few studies demonstrated that ROS at very low concentrations are already actively generated during the blastocyste state in rabbits and in postimplantation mouse embryos harvested on day 8 of pregnancy; here ROS generation is localized to the trophoblast cell layers (Gagioti et al., 1995). Moreover, placental NADPH oxidase-mediated ROS generation occurs in women during early pregnancy and may contribute to elevated ROS levels in embryos (Raijmakers et al., 2006). These data suggest that very low but physiologically relevant concentrations of ROS may already be involved in developmental processes during organogenesis and differentiation of stem cells from the inner cell mass. The meaning of ROS during later stages of organ maturation and morphogenesis is not well defined but may at least be involved in neuronal, cardiac and vascular growth, where ROS have been shown in several studies to be involved in growth factor and cytokine-mediated signaling pathways such as the vascular endothelial growth factor/flk-1 (VEGF/flk-1) (Roy et al., 2008), platelet-derived growth factor BB (PDGF-BB) (Lange et al., 2009), cardiotrophin-1 (CT-1) (Sauer et al., 2004) and nerve growth factor (NGF)-mediated signaling pathways (Suzukawa et al., 2000). These pathways are associated to vasculogenesis, angiogenesis as well as the development of the central and peripheral nerve system, where ROS may be involved in the regulation of axon guidance through semaphorin 3A (Schwamborn et al., 2004). Furthermore, high levels of ROS have been implicated in site-specific cell death in inter-digital regions of the developing limb (Schnabel et al., 2006), where peroxidase activity and glutathione peroxidase-4 gene (Gpx4) expression were restricted to the non-apoptotic tissue (e.g., digits) of the developing autopod, thus suggesting that differential tissue growth may be regulated by redox gradients which are determined by distinct expression patterns of anti-oxidant molecules.

4. Oxidative stress during myocardial infarction – a potential stimulus for stem cell activation

During cardiovascular repair processes embryonic genes are activated, suggesting that comparable signaling pathways are involved in embryonic development of the cardiovascular system and in cardiac repair during adult live. During hypertension and hypertrophic cardiac growth (Akki et al., 2009; Anilkumar et al., 2009) but also in acute myocardial infarction (Hori & Nishida, 2009; Di Lisa et al., 2007; Webster et al., 2006), ROS are generated in the ischemic myocardium especially after reperfusion. ROS in high concentrations directly injure the cell membrane and cause cell death. However, ROS in low concentrations also stimulate signal transduction to elaborate inflammatory cytokines, e.g. tumour necrosis factor-α (TNF-α), interleukin (IL)-1β and -6, in the ischemic region and surrounding myocardium as a host reaction. These inflammatory cytokines regulate cell survival and cell death in the chain reaction with ROS (Frangogiannis, 2008). Other cytokines like transforming growth factor-β (TGF-β) are upregulated upon inflammation (Czarkowska et al., 2004), suggesting that TGF-β signaling may be crucial for repression of inflammatory gene synthesis in healing infarcts, presumably by mediating resolution of the inflammatory infiltrate. Furthermore TGF-β may play an important role in modulating fibroblast phenotype and gene expression, promoting extracellular matrix deposition in the infarct by upregulating collagen and fibronectin synthesis and by decreasing matrix degradation through induction of protease inhibitors (Frangogiannis, 2008). TGF-β is also a key mediator in the pathogenesis of hypertrophic and dilative ventricular remodeling by stimulating cardiomyocyte growth and by inducing interstitial fibrosis (Ellmers et al., 2008). Furthermore TGF-β has been demonstrated to enhance cardiomyogenesis of mouse ES cells, thus suggesting that stem cell differentiation requires a paracrine pathway within the heart (Behfar et al., 2002).

5. Stem cells within the heart and potential redox-regulated signaling pathway involved in stem cell proliferation and specification

Cardiac repair following myocardial injury is restricted due to the limited proliferative potential of adult cardiomyocytes. The ability of mammalian cardiomyocytes to proliferate is lost shortly after birth as cardiomyocytes withdraw from the cell cycle and differentiate. However, recent research using integration of carbon-14, generated by nuclear bomb tests during the Cold War, into DNA to establish the age of cardiomyocytes in humans revealed that cardiomyocytes indeed renew, with a gradual decrease from 1% turning over annually at the age of 25 to 0.45% at the age of 75. Fewer than 50% of cardiomyocytes are exchanged during a normal life span (Bergmann et al., 2009). In contrast Hsieh et al did not find significant cardiac repopulation to occur during normal aging in mice, However, they found cardiomyocyte repopulation, albeit modest, by endogenous progenitors following injury, e.g. during cardiac infarction (Hsieh et al., 2007), thus suggesting that cardiac repair and renewal processes may occur through stem cell-mediated cell replacement.

The cellular basis for the exchange of cardiomyocytes during human life is not yet known but could be comparable to mice due to mobilization of bone marrow-derived stem cells (BMSC) and/or the activation of resident stem cells in the heart. Several studies on patients have shown that myocardial infarction results in mobilization of various populations of BMSCs which may be involved in cardiac repair processes (Leone et al., 2005; Leone et al., 2005; Leone et al., 2006; Wojakowski et al., 2009; Wojakowski et al., 2006) Besides BMSCs and circulating multipotent progenitor cells (Cesselli et al., 2009) several populations of resident cardiac stem cells have been described during recent years. In the early embryo, progenitor cells in pharyngeal mesoderm contribute to the rapid growth of the heart tube during looping morphogenesis. These progenitor cells constitute the second heart field and were first identified in 2001 (Rochais et al., 2009). Side population (SP) cells residing within the adult heart and comprising about 1 % of all cells were identified in 2002 by Hierlihy et al. who used the Hoechst 33342 dye exclusion procedure which was previously used to isolate stem cell populations expressing ATP-binding cassette (ABC) membrane transporters, e.g. P-glycoprotein, which confers multidrug resistance in cancer disease (Hierlihy et al., 2002). Upon coculture of SP cells from GFP^+ mice with adult cardiac cells from wild type mice this cell population gained positive α-actinin immunoreactivity, suggesting that a cardiac phenotype was attained (Martin et al., 2004). A subpopulation of SP cells comprising approximately 10% of the total SP cells expressing the stem cell marker Sca-1 was identified by Pfister et al. in 2005. This cell population was negative for the endothelial cell marker CD31, expressed Nkx2.5 and GATA-4, but not α-actinin or α-MHC. The cells could be differentiated into a more mature cardiac phenotype upon coculture with ventricular cardiomyocytes (Pfister et al., 2005). Upon cardiac infarction the CD31 negative cell population in the heart was depleted both within the infarct and non-infarct areas. SP pools were subsequently reconstituted to baseline levels within 7 days after myocardial infarction, through both proliferation of resident SP cells, as well as through homing of BMSCs to specific areas of myocardial injury and immunophenotypic conversion of BMCs to adopt a SP phenotype (Mouquet et al., 2005). Besides the SP cell population $Sca\text{-}1^+$ $c\text{-}Kit^-$ cells have been reported to be present in the mouse heart (Tallini et al., 2009), and so-called cardiospheres were isolated by mild enzymatic digestion of mouse and human heart tissues (Messina et al., 2004). A further resident stem cell population within the heart are $Isl1^+$ cells which express the islet-1 (Isl1) LIM homeodomain transcription factor (Laugwitz et al., 2005). $Isl1^+$ cells give rise to cardiomyocyte, endothelial, and smooth muscle lineages *in vitro* and may be involved in embryonic development of the coronary artery tree and for coronary artery growth. Previously it was shown that $Isl1^+$ cells with the transcriptional signature of $Isl1^+/Nkx2.5^+/flk1^+$ defines a multipotent cardiovascular progenitor which is not only capable to differentiate into cardiac cells but also to smooth muscle and endothelial cells which may participate in coronary artery formation (Moretti et al., 2006). During embryonic development Isl1 is expressed by progenitor cells of the second heart field that gives rise to the formation of the outflow tract, the atria and the right ventricle and is required for proliferation, survival, and migration of these progenitors into the forming heart (Cai et al., 2003). Isl1 also marks cardiac progenitors found within postnatal hearts of rodents and humans (Laugwitz et al., 2005). Recently it has been shown that β-catenin directly regulates *Isl1* expression in cardiovascular progenitors and is required for multiple aspects of cardiogenesis (Lin et al., 2007). β-catenin is also required upstream of a number of genes required for pharyngeal arch, outflow tract, and/or atrial septal morphogenesis, including *Tbx2, Tbx3, Wnt11, Shh,* and *Pitx2* (Lin et al., 2007).

The signaling pathways that regulate differentiation of BMSCs and resident cardiac stem cells and/or stimulate proliferation of cardiac progenitor cells are just emerging. Potentially inflammation and elevation of ROS levels following cardiac infarction are involved in the initiation of signaling pathways that activate quiescent resident cardiac stem cells and BMSCs. A beneficial effect of pro-inflammatory signals during bone marrow stem cell therapy has been recently outlined (Sun et al., 2009). In the latter study transplanted BMSCs increased heart tissue inflammation, and elevated TNF-α, TGF-β and fibroblast growth factor-2 (FGF-2) levels which resulted in improved heart function and capillary density in the border zone of the myocardial infarct (Sun et al., 2009). Recently it has been shown that the redox effector protein-1 (Ref-1) which plays an essential role in DNA repair and redox regulation of several transcription factors is involved in the maintenance of cardiac stem cells, since Ref-1 inhibition in the presence of exogenous hydrogen peroxide resulted in the initiation of cardiac differentiation programs thus suggesting that Ref-1 plays an important role in maintaining the redox status of cardiac stem cells and protects them from oxidative injury-mediated cell death and differentiation (Gurusamy et al., 2009).

Many answers on signaling pathways involved in stem cell activation can be given from lessons in cardiac embryology where several signaling pathways that are involved in the development of the first heart field and the second heart field have been recently deciphered (Rochais et al., 2009). One of the main features of second heart field is the control of cardiac progenitor cell proliferation. The latter has been recently shown to be regulated by β-catenin, the intracellular mediator of the canonical Wnt pathway, which is likewise known to be involved in the regulation of several stem cell populations (Reya & Clevers, 2005). Wnt signaling displays positive as well as negative effects on early mesoderm commitment and cardiac specification depending on the developmental stage of the embryo (Rochais et al., 2009). In ES cells Wnt signals are required for early mesoderm differentiation (Lin et al., 2007) whereas during later stages of cardiomyogenesis Wnt signaling restricts cardiac differentiation to the lateral splanchnic mesoderm (Tzahor & Lassar, 2001; Nakamura et al., 2003). Recently it was shown that the Wnt/β-catenin pathway is essential for cardiac myogenesis to occur in ES cells, acting at a gastrulation-like stage, and mediating mesoderm formation and patterning. Among genes associated temporally with this step was Sox17, encoding an endodermal HMG-box transcription factor (Liu et al., 2007). β-catenin interacts with TCF/LEF-1 transcription factors to activate the expression of Wnt target genes. In the absence of Wnt signaling, β-catenin function is blocked by a destruction complex consisting of Axin, APC and the kinases GSK3ß and CK1α, which targets β-catenin for destruction by the proteasome. Binding of Wnt to its receptor Frizzled and LRP leads to inhibition of the destruction complex and allows β-catenin signaling. The cytoplasmic protein Dishevelled (Dvl) is involved in this process by binding to the redox-sensitive protein nucleoredoxin

(NRX) which belongs to the thioredoxin protein family known to be involved in the regulation of a variety of ROS mediated signaling pathways (Korswagen, 2006). ROS are presumably involved in a variety of signaling pathways that are crucial for heart development. Recently it was shown that ROS can modulate signaling by the Wnt/β-catenin pathway (Funato et al., 2006). Oxidative stress inhibits the interaction between NRX and Dvl thus stabilizing β-catenin and leading to an increase in the expression of endogenous Wnt target genes. Further studies have demonstrated that ROS can also inhibit Wnt/β-catenin signaling (Shin et al., 2004) which suggests that a specific time frame and concentration of ROS may be necessary for redox-mediated modulation of the Wnt/β-catenin signaling pathway. Another important pathway known to be crucial for cardiac mesoderm specification and differentiation is bone morphogenic protein (BMP) pathway. *BMP-4* overexpression promotes a cardiac cell lineage in the cranial mesoderm (Tirosh-Finkel et al., 2006). BMP-4 is known to be regulated by Wnt/β-catenin and FGF signaling and is involved in outflow tract septation which includes smooth muscle and endocardial cushion development (Liu et al., 2004). Furthermore BMP-2, another member of the BMP family is essential for cardiac cushion epithelial-mesenchymal transition and myocardial patterning (Ma et al., 2005). Proinflammatory cytokine TNF-α and H_2O_2 significantly increased endothelial expression of BMP-2 but not BMP-4 and induced a proinflammatory endothelial phenotype (Csiszar et al., 2006). In further studies the same group demonstrated that BMP-4 exerts prooxidant, prohypertensive, and proinflammatory effects but only in the systemic circulation, whereas pulmonary arteries are protected from these adverse effects of BMP-4 (Csiszar et al., 2008). BMP-4 by itself may increase ROS generation which has been shown in endothelial cells where oscillatory shear stress elevates BMP-4 and induces monocyte adhesion by stimulating ROS production from a Nox-1-based NADPH oxidase (Sorescu et al., 2004). In malformed embryos from diabetic rats which exert elevated levels of systemic ROS sonic hedgehog homolog (Shh) expression was decreased, and BMP-4 was increased, thus pointing to a redox sensitive regulation of the Shh/BMP-4 pathway. Recently it has been shown that Shh, which is secreted by stem cells in the amphibian intestine, induces BMP-4 in subepithelial fibroblasts suggesting that both Shh and BMP-4 are involved in the development of the cell-renewable epithelium (Ishizuya-Oka & Hasebe, 2008).

6. Impact of redox-regulated pro-angiogenic signals during cardiac infarction

During cardiac insults not only growth factors and cytokines which are involved in the proliferation and differentiation of resident cardiac stem cells towards cardiac cells are upregulated. The healing of infarction is also grossly dependent on proper revascularization. Healing may be dependent on redox-mediated expression/release of pro-angiogenic growth factors like FGF-2 (Detillieux et al., 2003), VEGF (Wojakowski et al., 2004) and PDGF (Zymek et al., 2006) which has been demonstrated to occur after cardiac infarction. Pro-angiogenic factors are also released by monocytes and neutrophils (Lambert et al., 2008) which are migrating to the area of infarction where they induce formation of granulation tissue, containing myofibroblasts and neovessels (Nahrendorf et al., 2007). Increasing angiogenic growth factors in the infarcted hearts has therefore been recently used for cardioprotection and/or to improve cardiac healing (Lahteenvuo et al., 2009; Zhang et al., 2009; Harada et al., 1994; Hsieh et al., 2006b; Hsieh et al., 2006a; House et al., 2003). Conversely, inhibition of pro-angiogenic signaling, e.g. PDGF-signalling in infarcted hearts of mice resulted in impaired maturation of the infarct vasculature, enhanced capillary density, and formation of dilated uncoated vessels. Defective vascular maturation in antibody-treated mice was associated with increased and prolonged extravasation of red blood cells and monocyte/macrophages (Zymek et al., 2006). VEGF is critical for stem cell-mediated cardioprotection which was shown in experiments where VEGF was downregulated in mesenchymal stem cells by si-RNA approaches. When these cells were infused in the coronary circulation the increase in postischemic myocardial recovery after ischemia reperfusion injury was significantly impaired (Markel et al., 2008). Recently suicide genes under the control of endothelium (endothelial nitric oxide synthase)-, smooth muscle (SM22α)-, and cardiomyocyte (α-MHC)-specific promoters, were used to selectively deplete the individual cell lineage acquired by transplanted undifferentiated bone marrow-derived cells into an acute myocardial infarction model. It was demonstrated that elimination of transplanted endothelium-committed or SM22α-expressing cells, but not cardiac-committed cells, induced a significant deterioration of ejection fraction. Moreover, elimination of endothelial NO synthase-expressing cells 2 weeks after injection reduced capillary and arteriole density (Yoon et al., 2010). The angiogenic factors VEGF, PDGF-BB and FGF-2 are all upregulated by exogenous ROS (Sen et al., 2002; Eyries et al., 2004) and exert cardioprotective effects under conditions of ischemia-reperfusion injury (Hsieh et al., 2006c; Iwai-Kanai et al., 2002). Bone marrow mesenchymal stem cells by themselves release VEGF as a potentially beneficial paracrine response which is enhanced by TGF-α and TNF-α (Wang et al., 2008). Furthermore VEGF upregulation has been observed under tissue stress conditions associated with ROS generation, e.g. physical exercise (Roy et al., 2008) and cardiac infarction, where not only the *VEGF* gene but also the VEGF receptors flt-1 and flk-1 were upregulated (Li et al., 1996). Exogenous FGF-2 increased endogenous FGF-2 promoter activity and protein levels in ovine pulmonary arterial smooth muscle cells (PASMC). These increases in FGF-2 expression were mediated by elevations in superoxide levels via NADPH oxidase activation. In addition, FGF-2-mediated increases in FGF-2 expression and pulmonary arterial smooth muscle cell (PASMC) proliferation were attenuated by inhibition of phosphatidylinositol 3-kinase, Akt, and NADPH oxidase (Black et al., 2008). Comparably exogenous ROS increased VEGF and VEGFR expression (Gonzalez-Pacheco et al., 2006; Chua et al., 1998), stimulated endothelial cell proliferation and migration (Luczak et al., 2004) as well as cytoskeletal reorganization (Vepa et al., 1999) and tubular morphogenesis (Shono et al., 1996) which all utilize ROS within their signal transduction pathways. Addition of PDGF-BB, FGF-2 and VEGF to non-phagocytic cells has been shown to rapidly increase ROS generation (Thannickal et al., 2000) which may likewise occur in stem cells thus stimulating cardiovascular differentiation.

7. Redox-regulated pathways involved in mobilization of stem cells from the bone marrow

Stem cells and progenitor cells are mobilized from the bone marrow in response to inflammation, tissue injury and cytokines (Aicher et al., 2005). A cytokine playing a prominent role in stem cell mobilization, endothelial cell differentiation and vascular repair is stromal cell-derived factor-1α (SDF-1α), a CXC chemokine known to play a critical role in the trafficking of hematopoietic, lymphopoietic cells as well as stem cell progenitors, and in maintaining hematopoietic stem cell niches in bone marrow (Kucia et al., 2004). The high SDF-1α content in the bone marrow creates a concentration gradient, which retains hematopoietic stem cells within the stem cell niche. Disruption of this SDF-1α gradient results in mobilization of stem cells into the circulation. This degradation occurs after upregulation of G-CSF levels during systemic stress or injury. Under these conditions elastase is secreted from neutrophils which cleaves membrane-bound SDF-1/CXCR4 complexes on the surface of bone marrow stem cells in the marrow (Heissig et al., 2002; Jin et al., 2008). SDF-1 is released by stromal cells and binds to its CXCR4 receptor on stem and progenitor cells. The signaling cascade following interaction between SDF-1 and CXCR4 may involve the generation of ROS. This has been recently evidenced in studies on B-lymphocytes where ROS were involved in CXCR4-induced Akt activation (Lee et al., 2007). If high concentration gradients of circulating SDF-1 exist, CXCR4-positive cells are leaving the bone marrow to be directed to sites of tissue injury. During tissue damage, ischemia and inflammation plasma and tissue levels of SDF-1α are upregulated (Schober, 2008). Consequently SDF-1α expression is significantly upregulated in experimental rat and mouse models of infarction (Pillarisetti & Gupta, 2001), and in plasma and cardiac tissue of patients with myocardial infarction (Yamani et al., 2005). Furthermore, SDF-1α expression has been shown to increase under hypoxic conditions (Ceradini et al., 2004), and thus may serve to attract stem cells to sites of tissue injury and ischemia. Recently it has been shown that expression of SDF-1α on circulating platelets is increased in patients with acute coronary syndrome and correlates with the number of $CD34^{+}$ progenitor cells (Stellos et al., 2009). Expression of SDF-1α appears to be correlated to the expression of eNOS in the heart since eNOS $^{-/-}$ mice displayed reduced SDF-1α levels in isolated cardiomyocytes. eNOS in the host myocardium promoted mesenchymal stem cell migration to the ischemic myocardium and improved cardiac function through cGMP-dependent increases in SDF-1α expression (Li et al., 2009). The local inflammatory response implying adhesion molecule expression and eNOS-dependent signaling was required for SDF-1α-induced adhesion of c-kit^{+} cells to the vascular endothelium (Kaminski et al., 2008). Furthermore, oxidative stress from lactate metabolism by circulating stem/progenitor cells accelerated further stem cell recruitment and differentiation through thioredoxin-1 (Trx1)-mediated elevations in HIF-1 levels and the subsequent synthesis of HIF-1-dependent growth factors including VEGF and SDF-1α (Milovanova et al., 2008). Taken together these data suggest a model whereby, in response to tissue injury and inflammation, stem cells within the bone marrow are expanded and primed through G-CSF which then results in mobilization of stem cells via degradation of SDF-1α in the marrow and recruitment of the stem cells to sites of elevated SDF-1α levels within the injured, inflamed or ischemic tissues. Mobilization is then terminated when the increased SDF-1α gradient in the marrow is re-established and retains newly formed or non-mobilized stem cells as a reserve for future emergency signals (Mays et al., 2007). Interestingly G-CSF stimulation induced ROS generation in bone marrow neutrophils correlating with activation of Lyn, PI3-kinase and Akt, whereas the anti-oxidant N-acetyl cysteine diminished G-CSF-induced ROS production and cell proliferation (Zhu et al., 2006). Further research on the priming function of G-CSF on ROS generation by neutrophils revealed that mitogen-activated protein kinase (MAPK) pathways are involved in phosphorylation of Ser345 of p47phox, a cytosolic component of NADPH oxidase in human neutrophils (Dang et al., 2006). Previously it was shown that several hematopoietic growth factors including G-CSF signal through the formation of ROS (Sattler et al., 1999) which has been associated to a stimulation of cell proliferation of hematopoetic stem cells upon treatment with G-CSF (Pyatt et al., 1996). Furthermore the blood oxidative status was found to be significantly increased in healthy hematopoetic stem cell donors receiving G-CSF, which indicated that during stem cell mobilization a transient inflammatory status is generated (Cella et al., 2006) which may facilitate further stem cell mobilization. ROS mediated stem cell mobilization and recruitment may be used in therapeutic angiogenesis approaches. In this respect hyperbaric oxygen has been shown to stimulate recruitment and differentiation of circulating stem/progenitor cells in subcutaneous Matrigel which was inhibited by antagonists of NADPH oxidase and free radical scavengers (Milovanova et al., 2009).

Mostly ROS elicited upon growth factor and cytokine signaling are only acting within a narrow time window. Recently the interesting concept of the redox window of coronary collateral growth was formulated. This concept suggests that the redox window constitutes a range in the redox state of cells, which not only is permissive for the actions of growth factors but amplifies their actions. Initial changes in cellular redox are arising from different events, e.g. from the oxidative burst during reperfusion following ischemia, to recruitment of various types of inflammatory cells capable of producing ROS. Any event that upsets the normal redox equilibrium is capable of amplifying growth. However, extremes of the redox window, oxidative and reductive stresses, are associated with diminished growth factor signaling and reduced activation of redox-dependent kinases (Yun et al., 2009). Previously the same group had demonstrated that ROS are involved in human coronary artery endothelial cell (HCAEC) tube formation, coronary collateral growth in vivo, and signaling (p38 MAP kinase) by which ROS may stimulate vascular growth (Yun et al., 2009).

8. ROS and NO generation in bone marrow-derived stem cells

Besides the role of ROS and NO generated during states of tissue inflammation, ischemia and injury, stem cells per se are generating ROS as well as NO which may be involved in proliferation and differentiation processes. ROS and NO generation in stem cells could occur in response to transient changes in systemic redox balance and could initiate a feedforward cycle of ROS/NO generation and elaboration of a balanced anti-oxidative response system that may be the basis of stem cell proliferation, migration and differentiation. An increasing number of studies reported on the crucial role of ROS/NO on mesenchymal stem cell differentiation. It was shown that neuronal differentiation of mesenchymal stem cells involved upregulation of NADPH oxidase and increased ROS generation (Wang et al., 2007). Furthermore physical shockwave treatment was shown to increase osteogenic activity of human umbilical cord blood (HUCB) mesenchymal progenitor cells through superoxide-mediated TGF-β1 induction (Wang et al., 2004). ROS generation through the activity of the Nox-2 and Nox-4 isoform of NADPH oxidase has been demonstrated in human $CD34^+$ cells which may contribute to the activation of intracellular signaling pathways leading to mitochondriogenesis, cell survival, and differentiation in hematopoietic stem cells (Piccoli et al., 2007a). In the latter study the authors suggested that the coordinated activity of the Nox isoforms in hematopoietic stem cells functions as environmental oxygen sensor and generates low level of ROS, which likely serve as second messengers. The pro-oxidant setting, entering into play when hematopoietic stem and progenitor cells leave the hypoxic bone marrow niche, would enable them to be more responsive to proliferative/differentiative stimuli. Moreover it is suggested that enhanced ROS elicit mitochondrial "differentiation" in a pre-commitment phase needed to match the bioenergetic request in the oncoming proliferation/differentiation process (Piccoli et al., 2007b). Mesenchymal stem cells from the bone marrow have been shown to express iNOS (Sato et al., 2007) as well as eNOS (Klinz et al., 2005). Recently it was shown that hematopoietic stem cell development is dependent on blood flow and and is closely associated to NO generation since intrauterine NOS inhibition or embryonic eNOS deficiency resulted in a reduction of hematopoietic clusters and transplantable murine hematopoietic stem cells (North et al., 2009). Generation of NO by eNOS has been reported in mouse endothelial progenitor cells (EPCs) and was utilized to identify the EPC population (Loomans et al., 2006). Administration of Angiotensin II (Ang II) significantly promoted NO release, inhibited EPC apoptosis and enhanced EPC adhesion potential (Yin et al., 2008). In a recent study it was shown that two NO agents (SNAP and DEA/NO), able to activate both cGMP-dependent and -independent pathways, were increasing the cardiomyogenic potential of bone marrow-derived mesenchymal stem cells and adipose tissue-derived stem cells (ADSCs) (Rebelatto et al., 2009). Furthermore it was recently shown that the nitric oxide-soluble guanylyl cyclase pathway is involved in the oxytocin-mediated differentiation of porcine bone marrow stem cells into cardiomyocytes (Ybarra et al., 2010).

9. ROS and NO in cardiovascular differentiation of ES cells

Most evidence about the role of NO and ROS in cardiovascular differentiation has been obtained in mouse ES cells. It was shown that undifferentiated self-renewing stem cells are devoid of endogenous ROS generation and expression of NADPH oxidase. Undifferentiated ES cells were demonstrated to be equipped with highly efficient mechanisms to defend themselves against various stresses and to prevent or repair DNA damage. One of these mechanisms is high activity of a verapamil-sensitive multidrug efflux pump. During the differentiation process anti-oxidative genes are downregulated which should result in increased ROS generation (Saretzki et al., 2004). Consequently, during the differentiation process the gp91-phox homologues Nox-1, Nox-2 and Nox-4 are upregulated in a distinct time frame, starting with Nox-1 and followed by Nox-4 (Buggisch et al., 2007), whereas Nox-2 is closely correlated to the differentiation of phagocytic cells from ES cells which occurs subsequent to cardiovascular differentiation (Hannig et al., 2010). During the early stages of ES cell differentiation, i.e. between day 4 and day 10 of cell culture ROS generation is elevated and downregulated during later stages. The stages of active ROS generation are just those where cardiovascular differentiation occurs, i.e. between day 4 and day 9 of cell culture. Any approaches to increase intracellular ROS, e.g. by addition of nanomolar concentrations of H_2O_2 to differentiating embryoid bodies (Buggisch et al., 2007; Sauer et al., 1999; Sauer et al., 2000), treatment with direct current electrical fields (Sauer et al., 2005; Sauer et al., 1999; Serena et al., 2009), application of mechanical strain (Schmelter et al., 2006), treatment with cardiotrophin-1 (CT-1) (Sauer et al., 2004), PDGF-BB (Lange et al., 2009) or peroxisome proliferator-activated receptor α (PPARα) (Sharifpanah et al., 2008) resulted in prominent stimulation of cardiovascular differentiation of ES cells. Interestingly elevation of intracellular ROS by exogenous stimulators resulted in upregulation of Nox-1 and Nox-4 thus initiating a feed-forward stimulation of prolonged ROS generation (Schmelter et al., 2006; Buggisch et al., 2007). Consequently, si-RNA inactivation of Nox-4 resulted in complete inhibition of embryonic stem cell-derived cardiomyogenesis (Li et al., 2006). Stimulation of ROS generation by different means resulted in activation of the MAPK pathways ERK1,2, p38 and JNK. Furthermore stimulation of embryoid bodies by ROS resulted in activation of the cardiogenic transcription factors BMP-10, MEF2C, GATA-4, DTEF-1 and Nkx-2.5 (Buggisch et al., 2007). Interestingly vasculogenesis required activation of ERK1,2 and JNK whereas p38 activation was dispensable. Cardiomyogenesis, however, required the activation of all three pathways since pharmacological inhibition of either pathway abolished cardiac cell differentiation (Schmelter et al., 2006). Recently it was demonstrated that also smooth muscle differentiation from ES cells is under the control of Nox-4. Overexpression of Nox-4 specifically resulted in increased smooth muscle cell marker production, whereas knockdown of Nox-4 induced a decrease. Furthermore, smooth muscle cell-specific transcription factors, including serum response factor (SRF) and myocardin were activated by Nox4 gene expression (Xiao et al., 2009). When cardiomyogenesis was stimulated with CT-1, additionally activation of NF-κB and the JAK/STAT signaling pathway in a redox-sensitive manner was observed (Sauer et al., 2004). CT-1 has been previously shown to exert cardioprotective effects which may be related to the activation of anti-apoptotic signaling pathways (Calabro et al., 2009). CT-1 is expressed in post-myocardial infarct heart, and may play an important role in infarct scar formation and ongoing remodeling of the scar (Freed et al., 2003). Furthermore, CT-1 is a cytokine that induces hypertrophy and has been shown to be increased in hypertensive patients (Lopez et al., 2009). An additional role of CT-1 may involve the activation and differentiation of resident cardiac stem cells. In this respect is has been recently shown that CT-1 signaling through glycoprotein 130 (gp130) regulates the endothelial differentiation of cardiac stem cells (Mohri et al., 2009). Recently CT-1 in combination of 5-azacytidine which is an inhibitor of DNA methylation was shown to induce cardiac gene expression in mesenchymal stem cells (Xinyun et al., 2009). In human ES cells telomere maintenance, oxidative stress generation, and genes involved in antioxidant defense and DNA repair were investigated during spontaneous differentiation of two human embryonic stem cell lines. Telomerase activity was quickly downregulated during differentiation, probably due to deacetylation of histones H3 and H4 at the hTERT promoter and deacetylation of histone H3 at hTR promoter. Telomere length decreased accordingly. Mitochondrial superoxide production and cellular levels of ROS increased as result of stimulated mitochondrial biogenesis. The expression of major antioxidant genes was downregulated despite this increased oxidative stress. DNA damage levels increased during differentiation, whereas expression of genes involved in different types of DNA repair decreased (Saretzki et al., 2008). Besides the evident role of ROS for cardiovascular differentiation a prominent involvement of NO in cardiomyogenesis of ES cells has been evidenced. In murine undifferentiated ES cells, NOS-1, NOS-3, and sGCβ(1) were detected while NOS-2, sGCα(1), and PKG were very low or undetectable. When ES cells were subjected to differentiation, NOS-1 abruptly decreased within one day, NOS-2 mRNA became detectable after several days, and NOS-3 increased after 7-10 days (Mujoo et al., 2006). Components of NO signaling were likewise expressed in human ES cells (Mujoo et al., 2006). Nkx2.5 and myosin light chain (MLC2) mRNA expression was increased on exposure of mouse and human ES cells to NO donors and a decrease in mRNA expression of both cardiac-specific genes was observed with nonspecific NOS inhibitor (Mujoo et al., 2008). Recently it was shown that NO acts as a repressor of the stemness gene *Nanog* in mouse and human ES cells. The suppressive action of NO on *Nanog* gene depended on the activation of p53 repressor protein by covalent modifications, such as pSer15, pSer315, pSer392 and acetyl Lys 379. NO-induced repression of *Nanog* was also associated with binding of trimethylated histone H3 and pSer315 p53 to its promoter region (Mora-Castilla et al., 2010). In several studies it was reported that NO is acting as signaling molecule during cardiomyogenesis of ES cells (Mujoo et al., 2008; Kanno et al., 2004; Bloch et al., 1999). Furthermore NO donors stimulated vasculogenesis of ES cells (Milosevic et al., 2010). Studies on NO generating agents revealed that sGC activators alone exhibited an increase in mRNA expression of cardiac genes (MLC2 and Nkx2.5). Robust inductions of mRNA and protein expression of marker genes were observed when NO donors and sGC activators were combined. Measurement of NO metabolites demonstrated an increase in the nitrite levels in the conditioned media and cell lysates on exposure of cells to the different concentrations of NO donors. cGMP analysis in undifferentiated stem cells revealed a lack of stimulation with NO donors. Differentiated cells however, acquired the ability to be stimulated by NO donors (Mujoo et al., 2008). Generation of NO is apparently also the mediator of cardiomyogenesis of mouse ES cells achieved with the hormone oxytocin (Muller et al., 2008) and arginine vasopressin (Gassanov et al., 2007). Hence these data suggest that the interplay of ROS and NO is required to direct undifferentiated ES cells into the cardiovascular cell lineage.

10. Conclusions

Emerging evidence of recent years outlines a decisive role of ROS and RNS in the initiation of differentiation programs of stem cells. How this can be achieved is currently difficult to investigate due to technical limitations of in vivo detection of short lived radical species and subcellular analysis of the distribution of diffusible intracellular anti-oxidants. More than 20 years ago the free radical theory of development was established by Allen and Balin (Allen & Balin, 1989). According to this theory metabolic gradients exist in developing organisms and are believed to influence development. Metabolically generated oxidants that are counteracted by the antioxidative defense may be one decisive factor that directs the initiation of certain developmental events (Allen & Balin, 1989). The literature reviewed in the current article outlines that ROS as well as RNS are involved in stem cell mobilization, function and differentiation in a very complex way. To understand how this can function in a tissue where many different cell types exist in close proximity remains to be unraveled. Barriers of antioxidants have to be postulated that hinder the diffusibility of oxidants in one and pave their way in other cellular locations thus achieving distinct redox microenvironments and differentiation patterns. The prooxidant role of NO is often attributed to RNS intermediates rather than NO itself. NO may react with superoxide which - in contrast to H_2O_2 - is not diffusible across cell membranes. The resulting peroxynitrite ($ONOO^-$) can diffuse freely within and out of the cell, and react with lipids, proteins and DNA. The balance between NO and superoxide may help to define the prooxidant action of NO versus ROS in biological tissues. This interplay may discriminate between pathways resulting in oxidative stress and induction of apoptotic pathway versus signaling cascades resulting in stem cell maintenance and/or stem cell differentiation.

11. References

Agarwal A, Gupta S, Sekhon L, and Shah R. (2008). Redox considerations in female reproductive function and assisted reproduction: from molecular mechanisms to health implications. *Antioxid Redox Signal* 10, 1375-1403.

Aicher A, Zeiher AM, and Dimmeler S. (2005). Mobilizing endothelial progenitor cells. *Hypertension* 45, 321-325.

Akki A, Zhang M, Murdoch C, Brewer A, and Shah AM. (2009). NADPH oxidase signaling and cardiac myocyte function. *J Mol Cell Cardiol* 47, 15-22.

Allen RG and Balin AK. (1989). Oxidative influence on development and differentiation: an overview of a free radical theory of development. *Free Radic Biol Med* 6, 631-661.

Anilkumar N, Sirker A, and Shah AM. (2009). Redox sensitive signaling pathways in cardiac remodeling, hypertrophy and failure. *Front Biosci* 14, 3168-3187.

Behfar A, Zingman LV, Hodgson DM, Rauzier JM, Kane GC, Terzic A, and Puceat M. (2002). Stem cell differentiation requires a paracrine pathway in the heart. *FASEB J* 16, 1558-1566.

Bergmann O, Bhardwaj RD, Bernard S, Zdunek S, Barnabe-Heider F, Walsh S, Zupicich J, Alkass K, Buchholz BA, Druid H, Jovinge S, and Frisen J. (2009). Evidence for cardiomyocyte renewal in humans. *Science* 324, 98-102.

Black SM, DeVol JM, and Wedgwood S. (2008). Regulation of fibroblast growth factor-2 expression in pulmonary arterial smooth muscle cells involves increased reactive oxygen species generation. *Am J Physiol Cell Physiol* 294, C345-C354.

Bloch W, Fleischmann BK, Lorke DE, Andressen C, Hops B, Hescheler J, and Addicks K. (1999). Nitric oxide synthase expression and role during cardiomyogenesis. *Cardiovasc Res* 43, 675-684.

Buggisch M, Ateghang B, Ruhe C, Strobel C, Lange S, Wartenberg M, and Sauer H. (2007). Stimulation of ES-cell-derived cardiomyogenesis and neonatal cardiac cell proliferation by reactive oxygen species and NADPH oxidase. *J Cell Sci* 120, 885-894.

Cai CL, Liang X, Shi Y, Chu PH, Pfaff SL, Chen J, and Evans S. (2003). Isl1 identifies a cardiac progenitor population that proliferates prior to differentiation and contributes a majority of cells to the heart. *Dev Cell* 5, 877-889.

Calabro P, Limongelli G, Riegler L, Maddaloni V, Palmieri R, Golia E, Roselli T, Masarone D, Pacileo G, Golino P, and Calabro R. (2009). Novel insights into the role of cardiotrophin-1 in cardiovascular diseases. *J Mol Cell Cardiol* 46, 142-148.

Cella G, Marchetti M, Vignoli A, Randi ML, Saggiorato G, Pasetto L, Pagnan A, Barbui T, and Falanga A. (2006). Blood oxidative status and selectins plasma levels in healthy donors receiving granulocyte-colony stimulating factor. *Leukemia* 20, 1430-1434.

Ceradini DJ, Kulkarni AR, Callaghan MJ, Tepper OM, Bastidas N, Kleinman ME, Capla JM, Galiano RD, Levine JP, and Gurtner GC. (2004). Progenitor cell trafficking is regulated by hypoxic gradients through HIF-1 induction of SDF-1. *Nat Med* 10, 858-864.

Cesselli D, Beltrami AP, Rigo S, Bergamin N, D'Aurizio F, Verardo R, Piazza S, Klaric E, Fanin R, Toffoletto B, Marzinotto S, Mariuzzi L, Finato N, Pandolfi M, Leri A, Schneider C, Beltrami CA, and Anversa P. (2009). Multipotent progenitor cells are present in human peripheral blood. *Circ Res* 104, 1225-1234.

Chua CC, Hamdy RC, and Chua BH. (1998). Upregulation of vascular endothelial growth factor by H2O2 in rat heart endothelial cells. *Free Radic Biol Med* 25, 891-897.

Chuikov S, Levi BP, Smith ML, and Morrison SJ. (2010). Prdm16 promotes stem cell maintenance in multiple tissues, partly by regulating oxidative stress. *Nat Cell Biol* 12, 999-1006.

Covello KL, Kehler J, Yu H, Gordan JD, Arsham AM, Hu CJ, Labosky PA, Simon MC, and Keith B. (2006). HIF-2alpha regulates Oct-4: effects of hypoxia on stem cell function, embryonic development, and tumor growth. *Genes Dev* 20, 557-570.

Csiszar A, Ahmad M, Smith KE, Labinskyy N, Gao Q, Kaley G, Edwards JG, Wolin MS, and Ungvari Z. (2006). Bone morphogenetic protein-2 induces proinflammatory endothelial phenotype. *Am J Pathol* 168, 629-638.

Csiszar A, Labinskyy N, Jo H, Ballabh P, and Ungvari Z. (2008). Differential proinflammatory and prooxidant effects of bone morphogenetic protein-4 in coronary and pulmonary arterial endothelial cells. *Am J Physiol Heart Circ Physiol* 295, H569-H577.

Czarkowska P, Przybylski J, Marciniak A, Pawlowska M, Juskowa J, Foroncewicz B, and Mucha K. (2004). Proteolytic enzymes activities in patients after myocardial infarction correlate with serum concentration of TGF-beta. *Inflammation* 28, 279-284.

Dang PM, Stensballe A, Boussetta T, Raad H, Dewas C, Kroviarski Y, Hayem G, Jensen ON, Gougerot-Pocidalo MA, and El Benna J. (2006). A specific p47phox -serine phosphorylated by convergent MAPKs mediates neutrophil NADPH oxidase priming at inflammatory sites. *J Clin Invest* 116, 2033-2043.

Detillieux KA, Sheikh F, Kardami E, and Cattini PA. (2003). Biological activities of fibroblast growth factor-2 in the adult myocardium. *Cardiovasc Res* 57, 8-19.

Di Lisa F, Canton M, Menabo R, Kaludercic N, and Bernardi P. (2007). Mitochondria and cardioprotection. *Heart Fail Rev* 12, 249-260.

Ellmers LJ, Scott NJ, Medicherla S, Pilbrow AP, Bridgman PG, Yandle TG, Richards AM, Protter AA, and Cameron VA. (2008). Transforming growth factor-beta blockade down-regulates the renin-angiotensin system and modifies cardiac remodeling after myocardial infarction. *Endocrinology* 149, 5828-5834.

Eyries M, Collins T, and Khachigian LM. (2004). Modulation of growth factor gene expression in vascular cells by oxidative stress. *Endothelium* 11, 133-139.

Ezashi T, Das P, and Roberts RM. (2005). Low O2 tensions and the prevention of differentiation of hES cells. *Proc Natl Acad Sci U S A* 102, 4783-4788.

Frangogiannis NG. (2008). The immune system and cardiac repair. *Pharmacol Res* 58, 88-111.

Freed DH, Moon MC, Borowiec AM, Jones SC, Zahradka P, and Dixon IM. (2003). Cardiotrophin-1: expression in experimental myocardial infarction and potential role in post-MI wound healing. *Mol Cell Biochem* 254, 247-256.

Funato Y, Michiue T, Asashima M, and Miki H. (2006). The thioredoxin-related redox-regulating protein nucleoredoxin inhibits Wnt-beta-catenin signalling through dishevelled. *Nat Cell Biol* 8, 501-508.

Gagioti S, Colepicolo P, and Bevilacqua E. (1995). Post-implantation mouse embryos have the capability to generate and release reactive oxygen species. *Reprod Fertil Dev* 7, 1111-1116.

Gassanov N, Jankowski M, Danalache B, Wang D, Grygorczyk R, Hoppe UC, and Gutkowska J. (2007). Arginine vasopressin-mediated cardiac differentiation: insights into the role of its receptors and nitric oxide signaling. *J Biol Chem* 282, 11255-11265.

Gonzalez-Pacheco FR, Deudero JJ, Castellanos MC, Castilla MA, Alvarez-Arroyo MV, Yague S, and Caramelo C. (2006). Mechanisms of endothelial response to oxidative aggression: protective role of autologous VEGF and induction of VEGFR2 by H2O2. *Am J Physiol Heart Circ Physiol* 291, H1395-H1401.

Gurusamy N, Mukherjee S, Lekli I, Bearzi C, Bardelli S, and Das DK. (2009). Inhibition of ref-1 stimulates the production of reactive oxygen species and induces differentiation in adult cardiac stem cells. *Antioxid Redox Signal* 11, 589-600.

Hannig M, Figulla HR, Sauer H, and Wartenberg M. (2010). Control of leukocyte differentiation from embryonic stem cells upon vasculogenesis and confrontation with tumour tissue. *J Cell Mol Med*303-312.

Harada K, Grossman W, Friedman M, Edelman ER, Prasad PV, Keighley CS, Manning WJ, Sellke FW, and Simons M. (1994). Basic fibroblast growth factor improves myocardial function in chronically ischemic porcine hearts. *J Clin Invest* 94, 623-630.

Harvey AJ, Kind KL, Pantaleon M, Armstrong DT, and Thompson JG. (2004). Oxygen-regulated gene expression in bovine blastocysts. *Biol Reprod* 71, 1108-1119.

Heissig B, Hattori K, Dias S, Friedrich M, Ferris B, Hackett NR, Crystal RG, Besmer P, Lyden D, Moore MA, Werb Z, and Rafii S. (2002). Recruitment of stem and progenitor cells from the bone marrow niche requires MMP-9 mediated release of kit-ligand. *Cell* 109, 625-637.

Hierlihy AM, Seale P, Lobe CG, Rudnicki MA, and Megeney LA. (2002). The post-natal heart contains a myocardial stem cell population. *FEBS Lett* 530, 239-243.

Hori M and Nishida K. (2009). Oxidative stress and left ventricular remodelling after myocardial infarction. *Cardiovasc Res* 81, 457-464.

House SL, Bolte C, Zhou M, Doetschman T, Klevitsky R, Newman G, and Schultz JJ. (2003). Cardiac-specific overexpression of fibroblast growth factor-2 protects against myocardial dysfunction and infarction in a murine model of low-flow ischemia. *Circulation* 108, 3140-3148.

Hsieh PC, Davis ME, Gannon J, MacGillivray C, and Lee RT. (2006a). Controlled delivery of PDGF-BB for myocardial protection using injectable self-assembling peptide nanofibers. *J Clin Invest* 116, 237-248.

Hsieh PC, MacGillivray C, Gannon J, Cruz FU, and Lee RT. (2006b). Local controlled intramyocardial delivery of platelet-derived growth factor improves postinfarction ventricular function without pulmonary toxicity. *Circulation* 114, 637-644.

Hsieh PC, MacGillivray C, Gannon J, Cruz FU, and Lee RT. (2006c). Local controlled intramyocardial delivery of platelet-derived growth factor improves postinfarction ventricular function without pulmonary toxicity. *Circulation* 114, 637-644.

Hsieh PC, Segers VF, Davis ME, MacGillivray C, Gannon J, Molkentin JD, Robbins J, and Lee RT. (2007). Evidence from a genetic fate-mapping study that stem cells refresh adult mammalian cardiomyocytes after injury. *Nat Med* 13, 970-974.

Ishizuya-Oka A and Hasebe T. (2008). Sonic hedgehog and bone morphogenetic protein-4 signaling pathway involved in epithelial cell renewal along the radial axis of the intestine. *Digestion* 77 Suppl 1, 42-47.

Iwai-Kanai E, Hasegawa K, Fujita M, Araki M, Yanazume T, Adachi S, and Sasayama S. (2002). Basic fibroblast growth factor protects cardiac myocytes from iNOS-mediated apoptosis. *J Cell Physiol* 190, 54-62.

Jee MK, Kim JH, Han YM, Jung SJ, Kang KS, Kim DW, and Kang SK. (2010). DHP-derivative and low oxygen tension effectively induces human adipose stromal cell reprogramming. *PLoS One* 5, e9026.

Jin F, Zhai Q, Qiu L, Meng H, Zou D, Wang Y, Li Q, Yu Z, Han J, Li Q, and Zhou B. (2008). Degradation of BM SDF-1 by MMP-9: the role in G-CSF-induced hematopoietic stem/progenitor cell mobilization. *Bone Marrow Transplant* 42, 581-588.

Kaminski A, Ma N, Donndorf P, Lindenblatt N, Feldmeier G, Ong LL, Furlani D, Skrabal CA, Liebold A, Vollmar B, and Steinhoff G. (2008). Endothelial NOS is required for SDF-1alpha/CXCR4-mediated peripheral endothelial adhesion of c-kit+ bone marrow stem cells. *Lab Invest* 88, 58-69.

Kanno S, Kim PK, Sallam K, Lei J, Billiar TR, and Shears LL. (2004). Nitric oxide facilitates cardiomyogenesis in mouse embryonic stem cells. *Proc Natl Acad Sci U S A* 101, 12277-12281.

Kawamura Y, Uchijima Y, Horike N, Tonami K, Nishiyama K, Amano T, Asano T, Kurihara Y, and Kurihara H. (2010). Sirt3 protects in vitro-fertilized mouse preimplantation embryos against oxidative stress-induced p53-mediated developmental arrest. *J Clin Invest* 120, 2817-2828.

Klinz FJ, Schmidt A, Schinkothe T, Arnhold S, Desai B, Popken F, Brixius K, Schwinger R, Mehlhorn U, Staib P, Addicks K, and Bloch W. (2005). Phospho-eNOS Ser-114 in human mesenchymal stem cells: constitutive phosphorylation, nuclear localization and upregulation during mitosis. *Eur J Cell Biol* 84, 809-818.

Korswagen HC. (2006). Regulation of the Wnt/beta-catenin pathway by redox signaling. *Dev Cell* 10, 687-688.

Kucia M, Jankowski K, Reca R, Wysoczynski M, Bandura L, Allendorf DJ, Zhang J, Ratajczak J, and Ratajczak MZ. (2004). CXCR4-SDF-1 signalling, locomotion, chemotaxis and adhesion. *J Mol Histol* 35, 233-245.

Lahteenvuo JE, Lahteenvuo MT, Kivela A, Rosenlew C, Falkevall A, Klar J, Heikura T, Rissanen TT, Vahakangas E, Korpisalo P, Enholm B, Carmeliet P, Alitalo K, Eriksson U, and Yla-Herttuala S. (2009). Vascular endothelial growth factor-B induces myocardium-specific angiogenesis and arteriogenesis via vascular endothelial growth factor receptor-1- and neuropilin receptor-1-dependent mechanisms. *Circulation* 119, 845-856.

Lambert JM, Lopez EF, and Lindsey ML. (2008). Macrophage roles following myocardial infarction. *Int J Cardiol* 130, 147-158.

Lange S, Heger J, Euler G, Wartenberg M, Piper HM, and Sauer H. (2009). Platelet-derived growth factor BB stimulates vasculogenesis of embryonic stem cell-derived endothelial cells by calcium-mediated generation of reactive oxygen species. *Cardiovasc Res* 81, 159-168.

Laugwitz KL, Moretti A, Lam J, Gruber P, Chen Y, Woodard S, Lin LZ, Cai CL, Lu MM, Reth M, Platoshyn O, Yuan JX, Evans S, and Chien KR. (2005). Postnatal isl1+ cardioblasts enter fully differentiated cardiomyocyte lineages. *Nature* 433, 647-653.

Lee RL, Westendorf J, and Gold MR. (2007). Differential role of reactive oxygen species in the activation of mitogen-activated protein kinases and Akt by key receptors on B-lymphocytes: CD40, the B cell antigen receptor, and CXCR4. *J Cell Commun Signal* 1, 33-43.

Leone AM, Rutella S, Bonanno G, Abbate A, Rebuzzi AG, Giovannini S, Lombardi M, Galiuto L, Liuzzo G, Andreotti F, Lanza GA, Contemi AM, Leone G, and Crea F. (2005). Mobilization of bone marrow-derived stem cells after myocardial infarction and left ventricular function. *Eur Heart J* 26, 1196-1204.

Leone AM, Rutella S, Bonanno G, Contemi AM, de Ritis DG, Giannico MB, Rebuzzi AG, Leone G, and Crea F. (2006). Endogenous G-CSF and CD34+ cell mobilization after acute myocardial infarction. *Int J Cardiol* 111, 202-208.

Li J, Brown LF, Hibberd MG, Grossman JD, Morgan JP, and Simons M. (1996). VEGF, flk-1, and flt-1 expression in a rat myocardial infarction model of angiogenesis. *Am J Physiol* 270, H1803-H1811.

Li J, Stouffs M, Serrander L, Banfi B, Bettiol E, Charnay Y, Steger K, Krause KH, and Jaconi ME. (2006). The NADPH oxidase NOX4 drives cardiac differentiation: Role in regulating cardiac transcription factors and MAP kinase activation. *Mol Biol Cell* 17, 3978-3988.

Li N, Lu X, Zhao X, Xiang FL, Xenocostas A, Karmazyn M, and Feng Q. (2009). Endothelial nitric oxide synthase promotes bone marrow stromal cell migration to the ischemic myocardium via upregulation of stromal cell-derived factor-1alpha. *Stem Cells* 27, 961-970.

Li TS and Marban E. (2010). Physiological levels of reactive oxygen species are required to maintain genomic stability in stem cells. *Stem Cells* 28, 1178-1185.

Lin L, Cui L, Zhou W, Dufort D, Zhang X, Cai CL, Bu L, Yang L, Martin J, Kemler R, Rosenfeld MG, Chen J, and Evans SM. (2007). Beta-catenin directly regulates Islet1 expression in cardiovascular progenitors and is required for multiple aspects of cardiogenesis. *Proc Natl Acad Sci U S A* 104, 9313-9318.

Liu W, Selever J, Wang D, Lu MF, Moses KA, Schwartz RJ, and Martin JF. (2004). Bmp4 signaling is required for outflow-tract septation and branchial-arch artery remodeling. *Proc Natl Acad Sci U S A* 101, 4489-4494.

Liu Y, Asakura M, Inoue H, Nakamura T, Sano M, Niu Z, Chen M, Schwartz RJ, and Schneider MD. (2007). Sox17 is essential for the specification of cardiac mesoderm in embryonic stem cells. *Proc Natl Acad Sci U S A* 104, 3859-3864.

Loomans CJ, Wan H, de Crom R, van Haperen R, de Boer HC, Leenen PJ, Drexhage HA, Rabelink TJ, van Zonneveld AJ, and Staal FJ. (2006). Angiogenic murine endothelial progenitor cells are derived from a myeloid bone marrow fraction and can be identified by endothelial NO synthase expression. *Arterioscler Thromb Vasc Biol* 26, 1760-1767.

Lopez B, Gonzalez A, Querejeta R, Barba J, and Diez J. (2009). Association of plasma cardiotrophin-1 with stage C heart failure in hypertensive patients: potential diagnostic implications. *J Hypertens* 27, 418-424.

Luczak K, Balcerczyk A, Soszynski M, and Bartosz G. (2004). Low concentration of oxidant and nitric oxide donors stimulate proliferation of human endothelial cells in vitro. *Cell Biol Int* 28, 483-486.

Ma L, Lu MF, Schwartz RJ, and Martin JF. (2005). Bmp2 is essential for cardiac cushion epithelial-mesenchymal transition and myocardial patterning. *Development* 132, 5601-5611.

Markel TA, Wang Y, Herrmann JL, Crisostomo PR, Wang M, Novotny NM, Herring CM, Tan J, Lahm T, and Meldrum DR. (2008). VEGF is critical for stem cell-mediated cardioprotection and a crucial paracrine factor for defining the age threshold in adult and neonatal stem cell function. *Am J Physiol Heart Circ Physiol* 295, H2308-H2314.

Martin CM, Meeson AP, Robertson SM, Hawke TJ, Richardson JA, Bates S, Goetsch SC, Gallardo TD, and Garry DJ. (2004). Persistent expression of the ATP-binding cassette transporter, Abcg2, identifies cardiac SP cells in the developing and adult heart. *Dev Biol* 265, 262-275.

Mays RW, van't Hof W, Ting AE, Perry R, and Deans R. (2007). Development of adult pluripotent stem cell therapies for ischemic injury and disease. *Expert Opin Biol Ther* 7, 173-184.

Mazumdar J, O'Brien WT, Johnson RS, LaManna JC, Chavez JC, Klein PS, and Simon MC. (2010). O2 regulates stem cells through Wnt/beta-catenin signalling. *Nat Cell Biol* 12, 1007-1013.

Messina E, De Angelis L, Frati G, Morrone S, Chimenti S, Fiordaliso F, Salio M, Battaglia M, Latronico MV, Coletta M, Vivarelli E, Frati L, Cossu G, and Giacomello A. (2004). Isolation and expansion of adult cardiac stem cells from human and murine heart. *Circ Res* 95, 911-921.

Milosevic N, Bekhite MM, Sharifpanah F, Ruhe C, Wartenberg M, and Sauer H. (2010). Redox Stimulation of Cardiomyogenesis Versus Inhibition of Vasculogenesis Upon Treatment of Mouse Embryonic Stem Cells with Thalidomide. *Antioxid Redox Signal.*, in press.

Milovanova TN, Bhopale VM, Sorokina EM, Moore JS, Hunt TK, Hauer-Jensen M, Velazquez OC, and Thom SR. (2008). Lactate stimulates vasculogenic stem cells via the thioredoxin system and engages an autocrine activation loop involving hypoxia-inducible factor 1. *Mol Cell Biol* 28, 6248-6261.

Milovanova TN, Bhopale VM, Sorokina EM, Moore JS, Hunt TK, Hauer-Jensen M, Velazquez OC, and Thom SR. (2009). Hyperbaric oxygen stimulates vasculogenic stem cell growth and differentiation in vivo. *J Appl Physiol* 106, 711-728.

Mohri T, Fujio Y, Obana M, Iwakura T, Matsuda K, Maeda M, and Azuma J. (2009). Signals through glycoprotein 130 regulate the endothelial differentiation of cardiac stem cells. *Arterioscler Thromb Vasc Biol* 29, 754-760.

Mora-Castilla S, Tejedo JR, Hmadcha A, Cahuana GM, Martin F, Soria B, and Bedoya FJ. (2010). Nitric oxide repression of Nanog promotes mouse embryonic stem cell differentiation. *Cell Death Differ* 17, 1025-1033.

Moretti A, Caron L, Nakano A, Lam JT, Bernshausen A, Chen Y, Qyang Y, Bu L, Sasaki M, Martin-Puig S, Sun Y, Evans SM, Laugwitz KL, and Chien KR. (2006). Multipotent embryonic isl1+ progenitor cells lead to cardiac, smooth muscle, and endothelial cell diversification. *Cell* 127, 1151-1165.

Mouquet F, Pfister O, Jain M, Oikonomopoulos A, Ngoy S, Summer R, Fine A, and Liao R. (2005). Restoration of cardiac progenitor cells after myocardial infarction by self-proliferation and selective homing of bone marrow-derived stem cells. *Circ Res* 97, 1090-1092.

Mujoo K, Krumenacker JS, Wada Y, and Murad F. (2006). Differential expression of nitric oxide signaling components in undifferentiated and differentiated human embryonic stem cells. *Stem Cells Dev* 15, 779-787.

Mujoo K, Sharin VG, Bryan NS, Krumenacker JS, Sloan C, Parveen S, Nikonoff LE, Kots AY, and Murad F. (2008). Role of nitric oxide signaling components in differentiation of embryonic stem cells into myocardial cells. *Proc Natl Acad Sci U S A* 105, 18924-18929.

Muller P, Kazakov A, Semenov A, Bohm M, and Laufs U. (2008). Pressure-induced cardiac overload induces upregulation of endothelial and myocardial progenitor cells. *Cardiovasc Res* 77, 151-159.

Nahrendorf M, Swirski FK, Aikawa E, Stangenberg L, Wurdinger T, Figueiredo JL, Libby P, Weissleder R, and Pittet MJ. (2007). The healing myocardium sequentially mobilizes two monocyte subsets with divergent and complementary functions. *J Exp Med* 204, 3037-3047.

Nakamura T, Sano M, Songyang Z, and Schneider MD. (2003). A Wnt- and beta -catenin-dependent pathway for mammalian cardiac myogenesis. *Proc Natl Acad Sci U S A* 100, 5834-5839.

North TE, Goessling W, Peeters M, Li P, Ceol C, Lord AM, Weber GJ, Harris J, Cutting CC, Huang P, Dzierzak E, and Zon LI. (2009). Hematopoietic stem cell development is dependent on blood flow. *Cell* 137, 736-748.

Pfister O, Mouquet F, Jain M, Summer R, Helmes M, Fine A, Colucci WS, and Liao R. (2005). CD31- but not CD31+ cardiac sid population cells exhibit functional cardiomyogenic differentiation. *Circ Res* 97, 52-61.

Piccoli C, D'Aprile A, Ripoli M, Scrima R, Lecce L, Boffoli D, Tabilio A, and Capitanio N. (2007a). Bone-marrow derived hematopoietic stem/progenitor cells express multiple isoforms of NADPH oxidase and produce constitutively reactive oxygen species. *Biochem Biophys Res Commun* 353, 965-972.

Piccoli C, D'Aprile A, Scrima R, Ripoli M, Boffoli D, Tabilio A, and Capitanio N. (2007b). Role of reactive oxygen species as signal molecules in the pre-commitment phase of adult stem cells. *Ital J Biochem* 56, 295-301.

Pillarisetti K and Gupta SK. (2001). Cloning and relative expression analysis of rat stromal cell derived factor-1 (SDF-1)1: SDF-1 alpha mRNA is selectively induced in rat model of myocardial infarctio. *Inflammation* 25, 293-300.

Pyatt DW, Stillman WS, and Irons RD. (1996). Reactive oxygen species mediate stem cell factor synergy with granulocyte/macrophage colony-stimulating factor in a subpopulation of primitive murine hematopoietic progenitor cells. *Mol Pharmacol* 49, 1097-1103.

Raijmakers MT, Burton GJ, Jauniaux E, Seed PT, Peters WH, Steegers EA, and Poston L. (2006). Placental NAD(P)H oxidase mediated superoxide generation in early pregnancy. *Placenta* 27, 158-163.

Rebelatto CK, Aguiar AM, Senegaglia AC, Aita CM, Hansen P, Barchiki F, Kuligovski C, Olandoski M, Moutinho JA, Dallagiovanna B, Goldenberg S, Brofman PS, Nakao LS, and Correa A. (2009). Expression of cardiac function genes in adult stem cells is increased by treatment with nitric oxide agents. *Biochem Biophys Res Commun* 378, 456-461.

Reya T and Clevers H. (2005). Wnt signalling in stem cells and cancer. *Nature* 434, 843-850.

Rochais F, Mesbah K, and Kelly RG. (2009). Signaling pathways controlling second heart field development. *Circ Res* 104, 933-942.

Roy S, Khanna S, and Sen CK. (2008). Redox regulation of the VEGF signaling path and tissue vascularization: Hydrogen peroxide, the common link between physical exercise and cutaneous wound healing. *Free Radic Biol Med* 44, 180-192.

Saretzki G, Armstrong L, Leake A, Lako M, and von Zglinicki T. (2004). Stress defense in murine embryonic stem cells is superior to that of various differentiated murine cells. *Stem Cells* 22, 962-971.

Saretzki G, Walter T, Atkinson S, Passos JF, Bareth B, Keith WN, Stewart R, Hoare S, Stojkovic M, Armstrong L, von Zglinicki T, and Lako M. (2008). Downregulation of multiple stress defense mechanisms during differentiation of human embryonic stem cells. *Stem Cells* 26, 455-464.

Sato K, Ozaki K, Oh I, Meguro A, Hatanaka K, Nagai T, Muroi K, and Ozawa K. (2007). Nitric oxide plays a critical role in suppression of T-cell proliferation by mesenchymal stem cells. *Blood* 109, 228-234.

Sattler M, Winkler T, Verma S, Byrne CH, Shrikhande G, Salgia R, and Griffin JD. (1999). Hematopoietic growth factors signal through the formation of reactive oxygen species. *Blood* 93, 2928-2935.

Sauer H, Bekhite MM, Hescheler J, and Wartenberg M. (2005). Redox control of angiogenic factors and CD31-positive vessel-like structures in mouse embryonic stem cells after direct current electrical field stimulation. *Exp Cell Res* 304, 380-390.

Sauer H, Neukirchen W, Rahimi G, Grunheck F, Hescheler J, and Wartenberg M. (2004). Involvement of reactive oxygen species in cardiotrophin-1-induced proliferation of cardiomyocytes differentiated from murine embryonic stem cells. *Exp Cell Res* 294, 313-324.

Sauer H, Rahimi G, Hescheler J, and Wartenberg M. (1999). Effects of electrical fields on cardiomyocyte differentiation of embryonic stem cells. *J Cell Biochem* 75, 710-723.

Sauer H, Rahimi G, Hescheler J, and Wartenberg M. (2000). Role of reactive oxygen species and phosphatidylinositol 3-kinase in cardiomyocyte differentiation of embryonic stem cells. *FEBS Lett* 476, 218-223.

Schmelter M, Ateghang B, Helmig S, Wartenberg M, and Sauer H. (2006). Embryonic stem cells utilize reactive oxygen species as transducers of mechanical strain-induced cardiovascular differentiation. *FASEB J* 20, 1182-1184.

Schnabel D, Salas-Vidal E, Narvaez V, Sanchez-Carbente MR, Hernandez-Garcia D, Cuervo R, and Covarrubias L. (2006). Expression and regulation of antioxidant enzymes in the developing limb support a function of ROS in interdigital cell death. *Dev Biol* 291, 291-299.

Schober A. (2008). Chemokines in vascular dysfunction and remodeling. *Arterioscler Thromb Vasc Biol* 28, 1950-1959.

Schwamborn JC, Fiore R, Bagnard D, Kappler J, Kaltschmidt C, and Puschel AW. (2004). Semaphorin 3A stimulates neurite extension and regulates gene expression in PC12 cells. *J Biol Chem* 279, 30923-30926.

Sen CK, Khanna S, Babior BM, Hunt TK, Ellison EC, and Roy S. (2002). Oxidant-induced vascular endothelial growth factor expression in human keratinocytes and cutaneous wound healing. *J Biol Chem* 277, 33284-33290.

Serena E, Figallo E, Tandon N, Cannizzaro C, Gerecht S, Elvassore N, and Vunjak-Novakovic G. (2009). Electrical stimulation of human embryonic stem cells: cardiac differentiation and the generation of reactive oxygen species. *Exp Cell Res* 315, 3611-3619.

Sharifpanah F, Wartenberg M, Hannig M, Piper HM, and Sauer H. (2008). Peroxisome proliferator-activated receptor alpha agonists enhance cardiomyogenesis of mouse

ES cells by utilization of a reactive oxygen species-dependent mechanism. *Stem Cells* 26, 64-71.

Shepard TH, Muffley LA, and Smith LT. (2000). Mitochondrial ultrastructure in embryos after implantation. *Hum Reprod* 15 Suppl 2, 218-228.

Shin SY, Kim CG, Jho EH, Rho MS, Kim YS, Kim YH, and Lee YH. (2004). Hydrogen peroxide negatively modulates Wnt signaling through downregulation of beta-catenin. *Cancer Lett* 212, 225-231.

Shono T, Ono M, Izumi H, Jimi SI, Matsushima K, Okamoto T, Kohno K, and Kuwano M. (1996). Involvement of the transcription factor NF-kappaB in tubular morphogenesis of human microvascular endothelial cells by oxidative stress. *Mol Cell Biol* 16, 4231-4239.

Simon MC and Keith B. (2008). The role of oxygen availability in embryonic development and stem cell function. *Nat Rev Mol Cell Biol* 9, 285-296.

Sorescu GP, Song H, Tressel SL, Hwang J, Dikalov S, Smith DA, Boyd NL, Platt MO, Lassegue B, Griendling KK, and Jo H. (2004). Bone morphogenic protein 4 produced in endothelial cells by oscillatory shear stress induces monocyte adhesion by stimulating reactive oxygen species production from a nox1-based NADPH oxidase. *Circ Res* 95, 773-779.

Stellos K, Bigalke B, Langer H, Geisler T, Schad A, Kogel A, Pfaff F, Stakos D, Seizer P, Muller I, Htun P, Lindemann S, and Gawaz M. (2009). Expression of stromal-cell-derived factor-1 on circulating platelets is increased in patients with acute coronary syndrome and correlates with the number of CD34+ progenitor cells. *Eur Heart J* 30, 584-593.

Sun J, Li SH, Liu SM, Wu J, Weisel RD, Zhuo YF, Yau TM, Li RK, and Fazel SS. (2009). Improvement in cardiac function after bone marrow cell thearpy is associated with an increase in myocardial inflammation. *Am J Physiol Heart Circ Physiol* 296, H43-H50.

Suzukawa K, Miura K, Mitsushita J, Resau J, Hirose K, Crystal R, and Kamata T. (2000). Nerve growth factor-induced neuronal differentiation requires generation of Rac1-regulated reactive oxygen species. *J Biol Chem* 275, 13175-13178.

Tallini YN, Greene KS, Craven M, Spealman A, Breitbach M, Smith J, Fisher PJ, Steffey M, Hesse M, Doran RM, Woods A, Singh B, Yen A, Fleischmann BK, and Kotlikoff MI. (2009). c-kit expression identifies cardiovascular precursors in the neonatal heart. *Proc Natl Acad Sci U S A* 106, 1808-1813.

Thannickal VJ, Day RM, Klinz SG, Bastien MC, Larios JM, and Fanburg BL. (2000). Ras-dependent and -independent regulation of reactive oxygen species by mitogenic growth factors and TGF-beta1. *FASEB J* 14, 1741-1748.

Tirosh-Finkel L, Elhanany H, Rinon A, and Tzahor E. (2006). Mesoderm progenitor cells of common origin contribute to the head musculature and the cardiac outflow tract. *Development* 133, 1943-1953.

Tzahor E and Lassar AB. (2001). Wnt signals from the neural tube block ectopic cardiogenesis. *Genes Dev* 15, 255-260.

Vepa S, Scribner WM, Parinandi NL, English D, Garcia JG, and Natarajan V. (1999). Hydrogen peroxide stimulates tyrosine phosphorylation of focal adhesion kinase in vascular endothelial cells. *Am J Physiol* 277, L150-L158.

Wang FS, Yang KD, Wang CJ, Huang HC, Chio CC, Hsu TY, and Ou CY. (2004). Shockwave stimulates oxygen radical-mediated osteogenesis of the mesenchymal cells from human umbilical cord blood. *J Bone Miner Res* 19, 973-982.

Wang N, Xie K, Huo S, Zhao J, Zhang S, and Miao J. (2007). Suppressing phosphatidylcholine-specific phospholipase C and elevating ROS level, NADPH oxidase activity and Rb level induced neuronal differentiation in mesenchymal stem cells. *J Cell Biochem* 100, 1548-1557.

Wang Y, Crisostomo PR, Wang M, Markel TA, Novotny NM, and Meldrum DR. (2008). TGF-alpha increases human mesenchymal stem cell-secreted VEGF by MEK- and PI3-K- but not JNK- or ERK-dependent mechanisms. *Am J Physiol Regul Integr Comp Physiol* 295, R1115-R1123.

Webster KA, Graham RM, Thompson JW, Spiga MG, Frazier DP, Wilson A, and Bishopric NH. (2006). Redox stress and the contributions of BH3-only proteins to infarction. *Antioxid Redox Signal* 8, 1667-1676.

Wendler CC, Amatya S, McClaskey C, Ghatpande S, Fredholm BB, and Rivkees SA. (2007). A1 adenosine receptors play an essential role in protecting the embryo against hypoxia. *Proc Natl Acad Sci U S A* 104, 9697-9702.

Wendler CC, Poulsen RR, Ghatpande S, Greene RW, and Rivkees SA. (2010). Identification of the heart as the critical site of adenosine mediated embryo protection. *BMC Dev Biol* 10, 57.

Wojakowski W, Maslankiewicz K, Ochala A, Wyderka R, Zuk-Popiolek I, Flak Z, Mroz I, and Tendera M. (2004). The pro- and anti-inflammatory markers in patients with acute myocardial infarction and chronic stable angina. *Int J Mol Med* 14, 317-322.

Wojakowski W, Tendera M, Kucia M, Zuba-Surma E, Paczkowska E, Ciosek J, Halasa M, Krol M, Kazmierski M, Buszman P, Ochala A, Ratajczak J, Machalinski B, and Ratajczak MZ. (2009). Mobilization of bone marrow-derived Oct-4+ SSEA-4+ very small embryonic-like stem cells in patients with acute myocardial infarction. *J Am Coll Cardiol* 53, 1-9.

Wojakowski W, Tendera M, Zebzda A, Michalowska A, Majka M, Kucia M, Maslankiewicz K, Wyderka R, Krol M, Ochala A, Kozakiewicz K, and Ratajczak MZ. (2006). Mobilization of CD34(+), CD117(+), CXCR4(+), c-met(+) stem cells is correlated with left ventricular ejection fraction and plasma NT-proBNP levels in patients with acute myocardial infarction. *Eur Heart J* 27, 283-289.

Xiao Q, Luo Z, Pepe AE, Margariti A, Zeng L, and Xu Q. (2009). Embryonic stem cell differentiation into smooth muscle cells is mediated by Nox4-produced H2O2. *Am J Physiol Cell Physiol* 296, C711-C723.

Xinyun C, Zhi Z, Bin Z, Li R, Yucheng C, Yafei Y, Tingjie Z, and Shengfu L. (2009). Effects of cardiotrophin-1 on differentiation and maturation of rat bone marrow mesenchymal stem cells induced with 5-azacytidine in vitro. *Int J Cardiol.* 143, 171-177

Yamani MH, Ratliff NB, Cook DJ, Tuzcu EM, Yu Y, Hobbs R, Rincon G, Bott-Silverman C, Young JB, Smedira N, and Starling RC. (2005). Peritransplant ischemic injury is associated with up-regulation of stromal cell-derived factor-1. *J Am Coll Cardiol* 46, 1029-1035.

Ybarra N, Castillo JR, and Troncy E. (2010). Involvement of the nitric oxide-soluble guanylyl cyclase pathway in the oxytocin-mediated differentiation of porcine bone marrow stem cells into cardiomyocytes. *Nitric Oxide*, in press.

Yin T, Ma X, Zhao L, Cheng K, and Wang H. (2008). Angiotensin II promotes NO production, inhibits apoptosis and enhances adhesion potential of bone marrow-derived endothelial progenitor cells. *Cell Res* 18, 792-799.

Yoon CH, Koyanagi M, Iekushi K, Seeger F, Urbich C, Zeiher AM, and Dimmeler S. (2010). Mechanism of improved cardiac function after bone marrow mononuclear cell therapy: role of cardiovascular lineage commitment. *Circulation* 121, 2001-2011.

Yun J, Rocic P, Pung YF, Belmadani S, Carrao AC, Ohanyan V, and Chilian W. (2009). Redox-Dependent Mechanisms in Coronary Collateral Growth The Redox Window Hypothesis. *Antioxid Redox Signal* 11, 1961-1974.

Zhang J, Ding L, Zhao Y, Sun W, Chen B, Lin H, Wang X, Zhang L, Xu B, and Dai J. (2009). Collagen-targeting vascular endothelial growth factor improves cardiac performance after myocardial infarction. *Circulation* 119, 1776-1784.

Zhu QS, Xia L, Mills GB, Lowell CA, Touw IP, and Corey SJ. (2006). G-CSF induced reactive oxygen species involves Lyn-PI3-kinase-Akt and contributes to myeloid cell growth. *Blood* 107, 1847-1856.

Zymek P, Bujak M, Chatila K, Cieslak A, Thakker G, Entman ML, and Frangogiannis NG. (2006). The role of platelet-derived growth factor signaling in healing myocardial infarcts. *J Am Coll Cardiol* 48, 2315-2323.

30

ES Cell Differentiation as a Model to Study Cell Biological Regulation of Vascular Development

Kiyomi Tsuji-Tamura, Hiroshi Sakamoto and Minetaro Ogawa
Institute of Molecular Embryology and Genetics, Kumamoto University
Japan

1. Introduction

The organized vascular system is established through three successive steps, vasculogenesis, angiogenesis and vascular remodeling. Vasculogenesis is the initiation of nascent embryonic vessel formation by the differentiation of mesoderm-derived angioblasts into endothelial cells. The nascent primitive vascular plexus then expands and reorganizes into hierarchical vascular structures by the process of angiogenesis, which involves sprouting, bridging, and intussusceptive division of preexisting vessels. Vascular smooth muscle cells are then recruited to the vessels and the highly organized mature vascular system is established after vascular remodeling. By using gene disruption strategies, a number of molecules have been identified as being involved in the development of the vascular system. These identified molecules include transcription factors, cell adhesion molecules, secretory molecules and their receptors, of which deprivation lead to failure of angiogenesis and/or vascular remodeling in most cases. Despite the substantial expansion of the list of molecules that are essential for angiogenesis, we do not yet understand sufficiently how these molecules regulate the discrete processes of angiogenesis and vascular remodeling. Indeed, we know very little how normal endothelial cells behave in response to various angiogenic stimuli and cooperate to generate blood vessels during embryogenesis. It is critically important to establish an experimental model to study the functions of key molecules at the cellular level by monitoring the behavior of normal endothelial cells involving in angiogenesis.

In vitro differentiation systems of embryonic stem (ES) cells have been serving as an excellent experimental maneuver of developmental biology since Doetschman *et al.* reported that hematopoietic cells were generated *in vitro* from ES cells (Doetschman *et al.*, 1985). The ES cell differentiation systems have three important features. First, as differentiation of cells and tissues takes place in a culture dish instead of uterus, the process of differentiation can be directly monitored. Second, intermediate stages of differentiation can be identified and isolated as a cell population. Third, cell differentiation and proliferation can be controlled by either adding exogenous signaling molecules, introducing exogenous genes or modifying endogenous genes. Following the pioneering report of angiogenic differentiation from mouse ES cells engrafted onto quail chorioallantoic membrane (Risau *et al.*, 1988), *in vitro* derivation of vascular endothelial cells from ES cells has a history of almost two decades (Bloch *et al.*, 1997; Feraud *et al.*, 2001; Vittet *et al.*, 1996; Wang *et al.*, 1992). These earlier studies relied upon formation of cell aggregates termed cystic embryoid bodies from ES cells, whereby real-time monitoring of cell differentiation with a high resolution was hampered. Since Nishikawa's group developed a culture system in which ES cells differentiate into hematopoietic and endothelial cell lineages on type IV collagen-coated substrates or OP9 stromal cell layers (Hirashima *et al.*, 1999; Yamashita *et al.*, 2000), the simplicity of the culture system allowed detailed analysis of developmental steps essential in the formation of vascular and hematopoietic systems (Eilken *et al.*, 2009; Kondo *et al.*, 2009). This system thus provides a useful model to study cell biological regulation of endothelial cells in the process of angiogenesis and vascular remodeling. The aim of this chapter is to review our experimental models for the direct examination of differentiation, cell-cell adhesion, migration and morphological regulation of ES cell-derived endothelial cells, and discuss the cellular mechanisms underlying the process of vascular development.

2. Endothelial cell-specific gene promoters and enhancers

2.1 Endothelial differentiation from murine ES cells

It has been demonstrated that murine ES cells differentiate into Flk-1-expressing lateral mesodermal cells when cultured on a layer of OP9 stromal cell line (Endoh *et al.*, 2002; Hashimoto *et al.*, 2007; Kataoka *et al.*, 1997; Nakano *et al.*, 1994). In this culture system, undifferentiated ES cells are simply added to a culture flask pre-seeded with OP9 cells and incubated for 4 days in α-MEM supplemented with 2-mercaptoethanol and fetal calf serum. Cultured cells are dissociated by incubating in a buffer containing chelating reagent, and stained with a monoclonal antibody against Flk-1. The cells expressing Flk-1, an equivalent population of proximal lateral mesoderm in the embryo (Kataoka *et al.*, 1997), can be purified by fluorescence-activated cell sorting (FACS). When the purified Flk-1$^+$ mesodermal cells are further cultured on a freshly prepared OP9 cell layer, they continue to differentiate into several cell lineages including primitive and definitive hematopoietic cells, cardiomyocytes, smooth muscle cells and endothelial cells. Vascular endothelial (VE)-cadherin can be used as a specific marker to identify the endothelial cells. VE-cadherin$^+$ (and also Flk-1$^+$ CD31$^+$) endothelial cells are finally isolated by FACS with specific monoclonal antibodies. Alternatively, when undifferentiated ES cells are allowed to differentiate on the OP9 cell layer for more than 5 days, endothelial cells can be isolated in a single step from the culture by FACS. In this case, Flk-1$^+$ VE-cadherin$^-$ mesodermal cells become detectable on the 3rd day of differentiation, and Flk-1$^+$ VE-cadherin$^+$ endothelial cells on the 5th day of differentiation (Endoh *et al.*, 2002; Hashimoto *et al.*, 2007). Consequently, this culture system is a potent and reproducible way to induce ES cell differentiation into the lateral mesoderm and its derivatives.

While the lineage-specific markers such as Flk-1 and VE-cadherin can be detected by using specific monoclonal antibodies thereby providing means to isolate desired cell populations, the induced expression of reporter genes under the control of lineage-specific gene promoter/enhancer is another useful method especially for real-time monitoring of the process of cell differentiation. For example, Hirai *et al.* reported that a combination of 5′ flanking region and 3′ portion of the first intron of the *Kdr* (Flk-1) gene drove the expression of enhanced green fluorescent protein (EGFP) reporter gene shortly after generation of VE-cadherin$^+$ endothelial cells during the differentiation of ES cell-derived Flk-1$^+$ mesodermal cells (Hirai *et al.*, 2003). As the *Kdr* promoter/enhancer discriminates mature committed endothelial cells from immature endothelial cells that still possess a hematopoietic potential, it should be of use for monitoring the maturation process of endothelial cells. By contrast, a combination of 5′ flanking region and 5′ half of the first intron of the *Cdh5* (VE-cadherin) gene was reported to be active exclusively in all of the VE-cadherin⁺ endothelial cells during the course of ES cell differentiation, suggesting that the activity of the *Cdh5* promoter/enhancer reflects that of the endogenous *Cdh5* gene (Hisatsune *et al.*, 2005). Therefore, it provides a valuable tool for the endothelial cell-specific induction of a transgene expression regardless of differentiation stages. In the next section, we show another example of an endothelial-related enhancer element that might be useful to monitor endothelial cell differentiation from their precursor stages.

2.2 *Mef2c* enhancer element and endothelial cell commitment

De Val *et al.* reported that a 44bp element (F10-44) of an endothelial-specific transcriptional enhancer of the *Mef2c* gene directed expression of *lacZ* reporter specifically to the developing vascular endothelium of the mouse embryo (De Val *et al.*, 2008). The element consists of a FOX:ETS motif that is present in many known endothelial-specific enhancers. The F10-44-*lacZ* reporter mice demonstrated that this element was sufficient to direct endothelial cell-specific expression from the blood island stage at embryonic day (E) 7.5 through angiogenesis and remodeling at E9.5, and extinguished after E10.5 (De Val *et al.*, 2008). The activity of F10-44 as early as E7.5, when the expression of VE-cadherin also becomes detectable (Breier *et al.*, 1996; Nishikawa *et al.*, 1998), prompted us to determine exactly when F10-44 is activated upon commitment of mesodermal cells to endothelial cells.

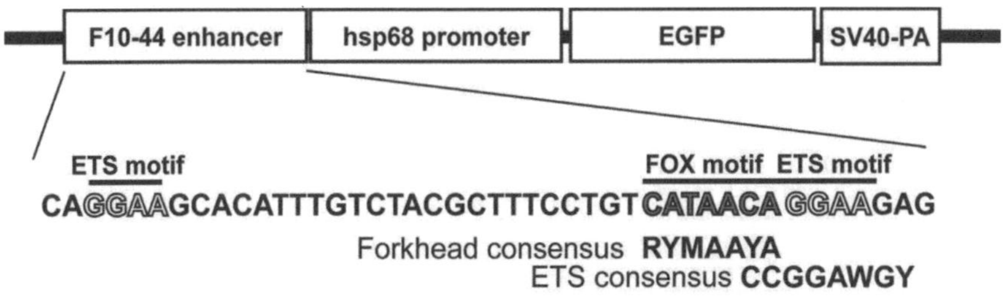

Fig. 1. Construction of F10-44-EGFP expression vector.
The EGFP reporter gene was connected to downstream of the F10-44 enhancer element and *Hsp68* minimal promoter. Two ETS motifs and a FOX motif in the F10-44 element are indicated together with consensus sequences.

We transfected wild-type ES cells with a plasmid vector harboring EGFP reporter gene that is connected to downstream of F10-44 enhancer element and *Hsp68* minimal promoter (Figure 1). ES cell clones were screened for proper controllability of EGFP expression by heat shock and sufficient capability of differentiating into endothelial cell lineage. We obtained three independent clones that showed essentially the same results. The F10-44-EGFP ES cells were cultured on OP9 stromal cell layer for 6 days to induce endothelial cell differentiation as described above. Flow cytometry analyses revealed that essentially all the VE-cadherin⁺ CD31⁺ endothelial cells activated the F10-44 enhancer as judged by expression of EGFP, indicating that the ES cell differentiation system recapitulated the activity of F10-44 in developing vascular endothelium of the mouse embryos (Figure 2). Expression of EGFP also distributed in the VE-cadherin⁻ cell population, and the majority of them expressed Flk-1 yet.

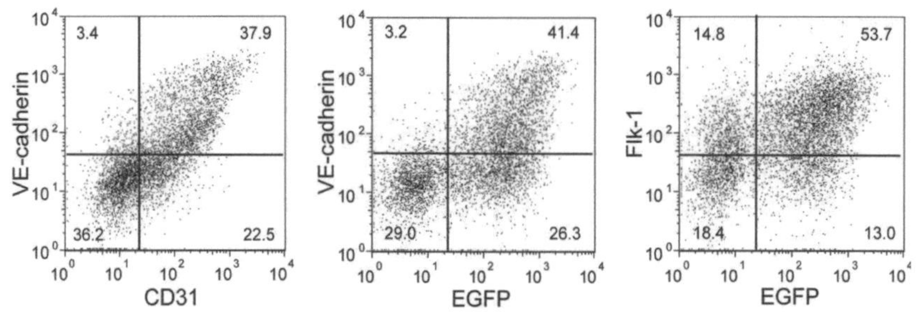

Fig. 2. Activity of F10-44 in ES cell-derived endothelial cells
F10-44-EGFP ES cells were co-cultured with OP9 cells for 6 days. Differentiating cells were dissociated and analyzed for expression of endothelial cell markers and EGFP by flow cytometry. Numbers indicate the percentage of the cells in the quartered areas.

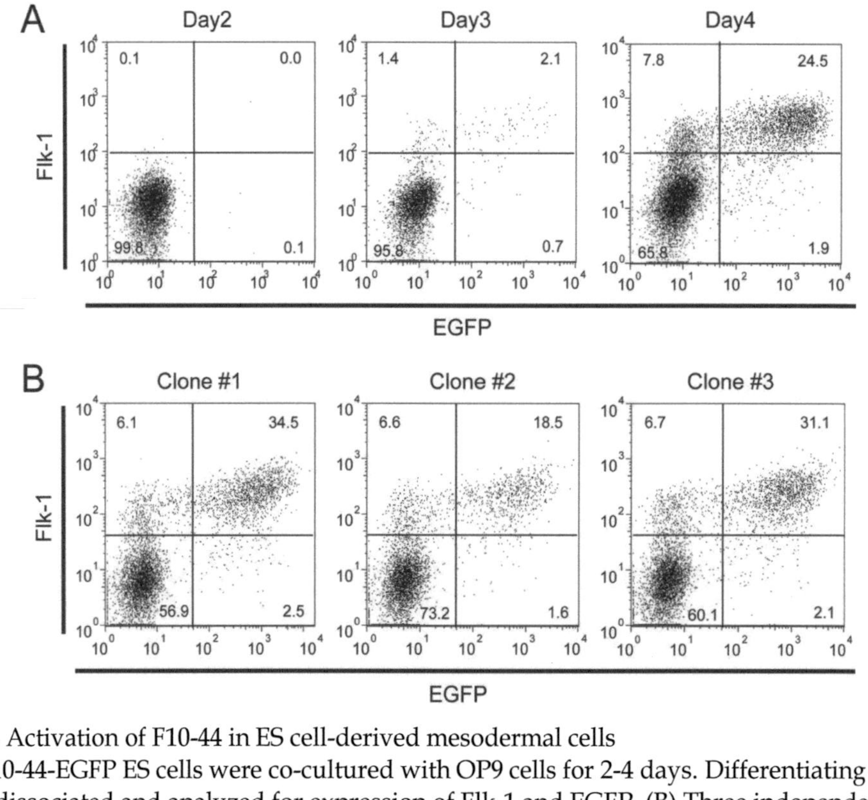

Fig. 3. Activation of F10-44 in ES cell-derived mesodermal cells
(A) F10-44-EGFP ES cells were co-cultured with OP9 cells for 2-4 days. Differentiating cells were dissociated and analyzed for expression of Flk-1 and EGFP. (B) Three independent clones of F10-44-EGFP ES cells were co-cultured with OP9 cells for 4 days and analyzed in the same way as (A). Numbers indicate the percentage of the cells in the quartered areas.

In order to determine when the expression of EGFP begins during the course of ES cell differentiation, we examined the kinetics of mesodermal differentiation and EGFP expression. When F10-44-EGFP ES cells were allowed to differentiate on OP9 cell layer, an EGFP-expressing cell population emerged exclusively as a major subset of Flk-1^{+} mesodermal cells (Figure 3). As three independent ES clones showed the same pattern of

EGFP expression, the activation of F10-44 in mesodermal cells is not a peculiar case of a selected clone, although we cannot exclude the possibility that it is specific to *in vitro* differentiation of ES cells.

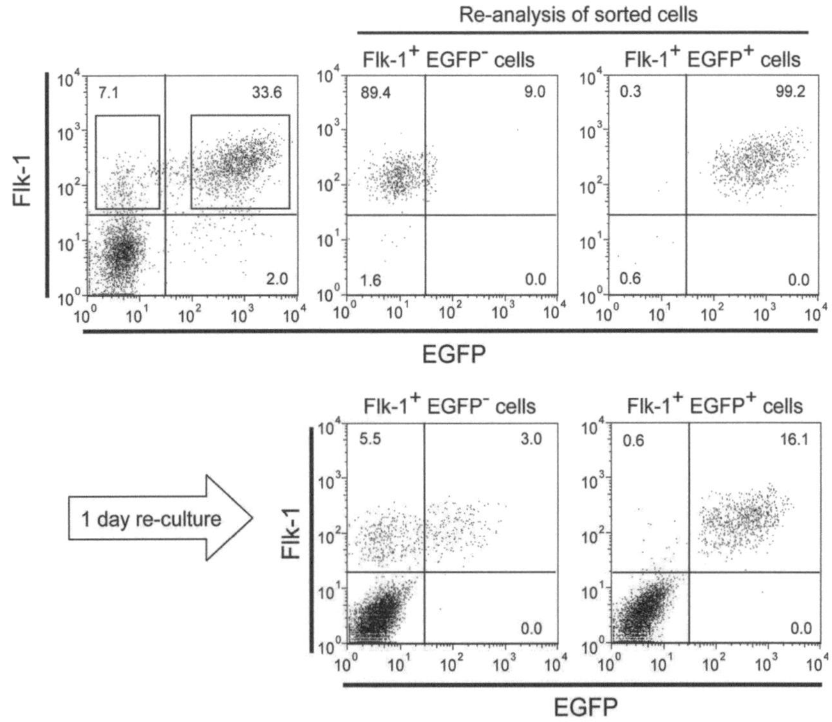

Fig. 4. Differentiation of Flk-1+ EGFP- cells to Flk-1+ EGFP+ cells
F10-44-EGFP ES cells were co-cultured with OP9 cells for 4 days. Flk-1+ EGFP- cells and Flk-1+ EGFP+ cells (indicated by the rectangles) were purified by FACS and re-cultured with OP9 cells. After 24 hours, cells were harvested and analyzed for expression of Flk-1 and EGFP. Numbers indicate the percentage of the cells in the quartered areas.

We next tested whether the Flk-1+ EGFP- cells differentiate to the Flk-1+ EGFP+ cells, and vice versa, by separating these cell populations with FACS and re-culturing them for 24 hours on a freshly prepared layer of OP9 cells. Figure 4 shows that the Flk-1+ EGFP- cells gave rise to the Flk-1+ EGFP+ cells within the culture period, while the Flk-1+ EGFP+ cells did not produce the Flk-1+ EGFP- cells. This observation suggests that the Flk-1+ EGFP- subset is more immature than the Flk-1+ EGFP+ subset. With the aim of characterizing the two subsets of mesodermal cells in terms of multi-lineage differentiation capacity, we performed single cell sorting of these populations and examined for capabilities of each single cells to differentiate into endothelial cells, smooth muscle cells and cardiomyocytes, the three major derivatives of lateral mesoderm. Single cells were cultured separately on OP9 stromal cell layers for 7 days, followed by immunofluorescence staining of VE-cadherin (endothelial cells), desmin (smooth muscle cells) and troponin I (cardiomyocytes). Out of 48 single cells isolated from each of the Flk-1+ EGFP- and Flk-1+ EGFP+ cell subsets by FACS, thirty-six cells of each subsets showed the differentiation of at least one of the three cell lineages (Figure 5).

Among of them, eighty percent of the Flk-1+ EGFP+ cells produced unipotential colonies containing only endothelial cells, although the rest of them retained a potential to differentiate into smooth muscle cells. By contrast, the Flk-1+ EGFP- cells produced a bipotential and a tripotential colony that contained cardiomyocytes, in addition to colonies containing endothelial cells and/or smooth muscle cells. This observation supports the above assumption that the Flk-1+ EGFP- subset represents earlier differentiation stage of lateral mesoderm than the Flk-1+ EGFP+ subset.

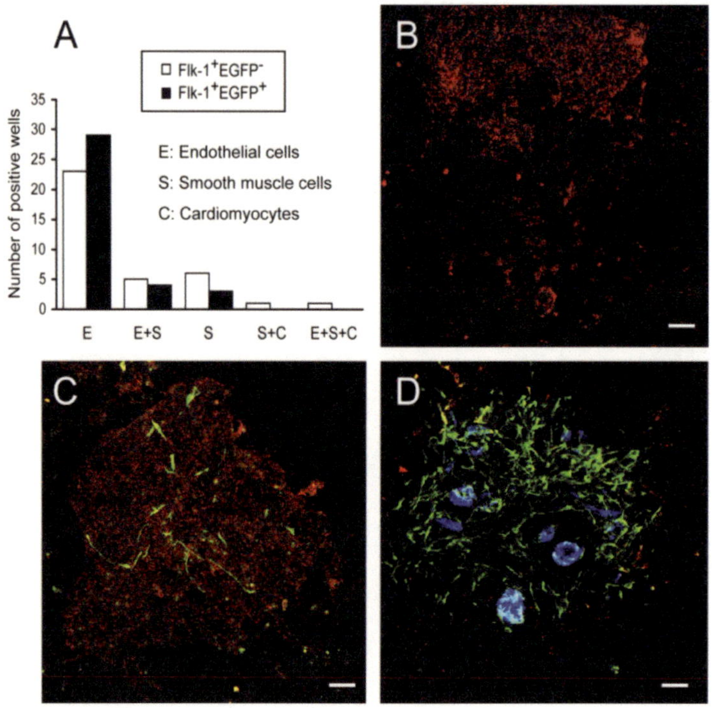

Fig. 5. Multi-lineage differentiation potential of mesodermal cells
F10-44-EGFP ES cells were co-cultured with OP9 cells for 4 days. Forty-eight single cells of Flk-1+ EGFP- and Flk-1+ EGFP+ subsets were separately deposited into wells pre-seeded with OP9 cells. After 7 days cultivation, wells were stained with antibodies against VE-cadherin (red), desmin (green) and troponin I (blue). (A) Number of wells that were positive for any of the three cell lineages. (B) A colony containing only endothelial cells. (C) A colony containing endothelial cells and smooth muscle cells. (D) A colony containing all the three cell lineages. Scale bars indicate 100 μm.

Our results suggest that the F10-44 enhancer element is first activated in a subset of lateral mesoderm, of which differentiation potential becomes restricted to the endothelial and smooth muscle cell lineages. As smooth muscle cells derived from Flk-1+ EGFP+ cells did not retain the expression of EGFP (data not shown), the F10-44 activity persists specifically in the endothelial cell lineage. It remains to be elucidated whether the early mesodermal activation of F10-44 also takes place in the mouse embryos, and whether the initiation of the F10-44 activity in the mesoderm and its maintenance in the endothelium are regulated by the same trans-activators.

2.3 Monitoring endothelial differentiation of mesodermal cells

As mentioned above, F10-44 directed EGFP expression from endothelial precursors through differentiation of mature endothelial cells in ES cell cultures. Taking advantage of high visibility of EGFP fluorescence in living cells, the developmental process of endothelial precursors can be directly monitored by time-lapse analysis of differentiating ES cells. In order to detect the initiation of the F10-44 activity in mesodermal cells, we seeded F10-44-EGFP ES cells onto OP9 stromal cell layer and performed time-lapse imaging analyses of differentiating cells.

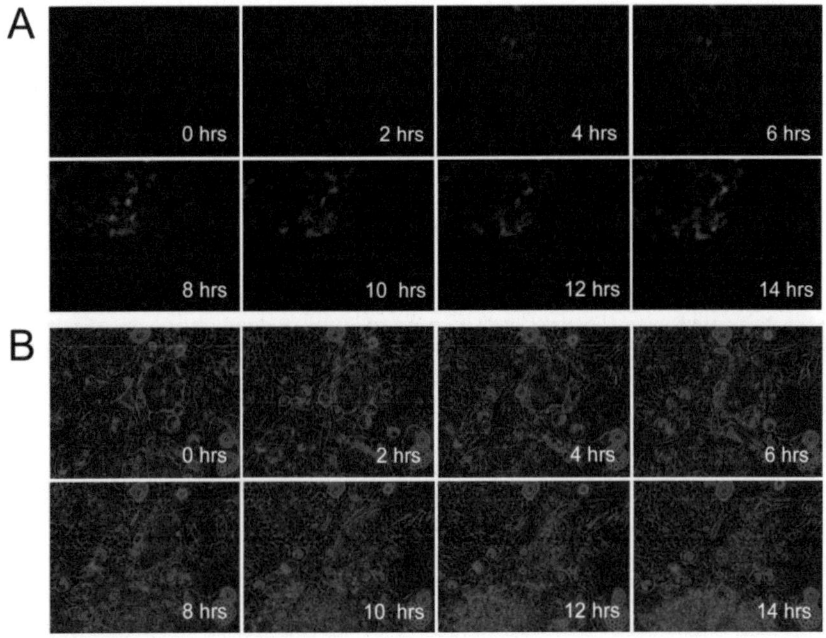

Fig. 6. Time-lapse analysis of differentiating mesodermal cells
F10-44-EGFP ES cells were seeded on the OP9 cell layer to induce the mesodermal differentiation. After 3 days, the cultures were subjected to time-lapse analyses under a fluorescence microscope. Numbers indicate elapsed time counted by hours. (A) EGFP fluorescence images. (B) Phase contrast images.

As shown in Figure 6, progression of mesodermal cell differentiation as revealed by the emergence of EGFP fluorescence occurred within a group of cells in a relatively synchronous manner, rather than a clonal expansion of a single ancestral EGFP$^+$ cells. This observation suggests that a group of mesodermal cells has determined their fate toward the endothelial cell lineage at a certain point, and then a combination of trans-activators activate F10-44 enhancer element synchronously in this population. The same sequence of events may also take place independently in other groups of mesoderm cells in culture.

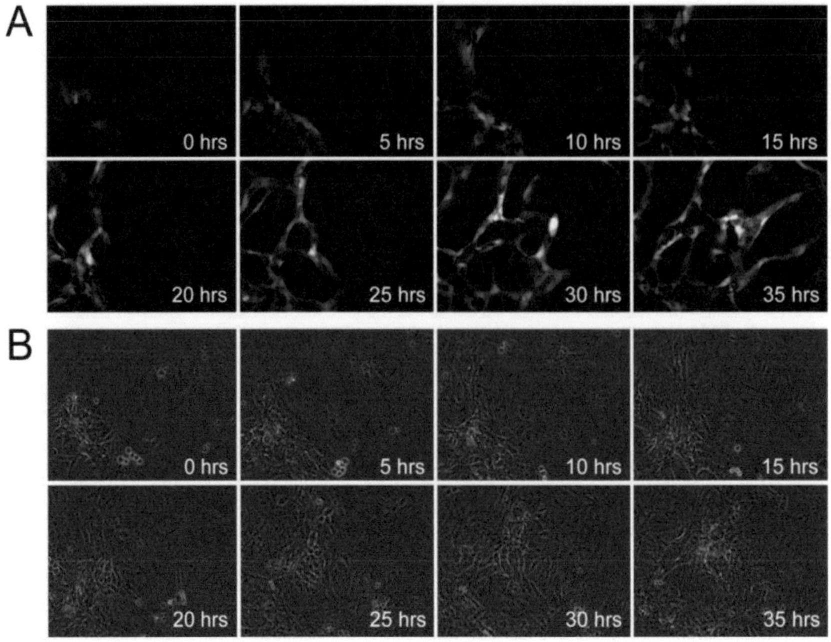

Fig. 7. Time-lapse analysis of differentiating endothelial cells
F10-44 EGFP ES cells were co-cultured with OP9 cells for 4 days, and Flk-1$^+$ EGFP$^+$ cells were purified by FACS. Cells were seeded on OP9 cell layer, and cultured in the presence of VEGF-A. After 36 hours, the cultures were subjected to time-lapse analyses under a fluorescence microscope. Numbers indicate elapsed time counted by hours. (A) EGFP fluorescence images. (B) Phase contrast images.

The F10-44 enhancer element also provides means of tracking how endothelial precursors are organized into endothelial structures in culture. F10-44-EGFP ES cells were co-cultured with OP9 stromal cells for 4 days. FACS-purified Flk-1$^+$ EGFP$^+$ cells were re-cultured on OP9 cell layer in the presence of 50 ng/mL vascular endothelial growth factor A (VEGF-A), and subjected to time-lapse analyses (Figure 7). The EGFP$^+$ endothelial precursors first generated small groups of endothelial cells, and then the cells were rearranged to form large elongated clusters. Endothelial cells were further organized into a network of cord-like structures through splitting of elongated clusters, sprouting of endothelial cells with lamellipodial protrusions, migration and interconnection of elongated endothelial cells. These processes are reminiscent of those found in vasculogenesis and angiogenesis in the yolk sac vasculatures. Therefore, by combining the endothelial cell differentiation on OP9 stromal cell layer with the expression of fluorescent molecular probes, ES cell differentiation systems provide powerful experimental models to investigate the behavior of normal endothelial cells in response to angiogenic stimuli. We will discuss this point a little bit more in the next section.

3. Cell-cell adhesion and motility of endothelial cells

3.1 Cell-cell junction as a regulating factor of endothelial cells

Endothelial cells adhere to each other through tight junctions and adherens junctions. The tight junction regulates paracellular permeability and maintains cell polarity (Tsukita *et al.*,

2001). The tight junction is mediated by occludin, claudins and other transmembrane proteins associated with intracellular components such as ZO-1. The loss of claudin-5, a major adhesion molecule involved in the tight junction of endothelium, caused size-selective loosening of the blood-brain barrier (Nitta *et al.*, 2003). Meanwhile, the adherens junction regulates the integrity of endothelium, paracellular permeability to solute, and transmigration of leukocytes across the endothelium (Bazzoni and Dejana, 2004). The adherens junction in endothelium is mediated by the homotypic binding of VE-cadherin and its anchorage to actin microfilaments through intracellular catenins. Disruption of the *Cdh5* gene that encodes VE-cadherin resulted in impaired vascular remodeling and vessel collapse thereby led to embryonic lethality at E9.5 (Carmeliet *et al.*, 1999; Gory-Faure *et al.*, 1999).

Endothelial cells growing as sparse cells in culture are generally deemed to mimic the behavior of cells undergoing angiogenesis. They are stimulated to proliferate by growth factors and behave like fibroblasts, exhibiting elongated spindle-like shape and high motility. By contrast, confluent endothelial cells in culture are generally considered to be a model of resting endothelium. They develop organized adherens junctions and tight junctions, lose the ability to proliferate in response to growth factors, exhibit phenotypes of epithelioid cells, and remain static (Dejana, 2004). Indeed, in the absence of VE-cadherin, endothelial cells fail to display contact inhibition in culture and grow to higher density than cells expressing VE-cadherin (Grazia Lampugnani *et al.*, 2003). Endothelial cells lacking VE-cadherin also fail to respond to VEGF-A that protects them from apoptosis (Carmeliet *et al.*, 1999). VE-cadherin and β-catenin form a complex with Flk-1/VEGFR-2 and modulate its downstream signaling by influencing the phospholipase Cγ and phosphatidylinositol 3-kinase pathways, which regulates cell proliferation and survival, respectively. Therefore, the adherens junction appears to regulate not only the integrity of endothelium but also the proliferation and survival of endothelial cells. It is also reported that oligomerized angiopoietin-1 bridges Tie2 receptor at cell-cell contacts of confluent endothelial cells, resulting in formation of trans-association of Tie2 and activation of the Akt-Foxo1 and Akt-endothelial NO synthase signaling pathways, which may enhance endothelial survival and integrity (Fukuhara *et al.*, 2008).

Although many features of endothelial cells in the activated or resting state, especially cell proliferation and survival, are modeled by cultured endothelial cells, the supposed incompatibility between endothelial cell junctions and motility remains uncertain. Furthermore, it is also obscure whether the cell biological behaviors characterized by using cultures of mature endothelial cells such as human umbilical vein endothelial cells (HUVECs) are applicable to that of nascent endothelial cells involved in vascular development in the embryos. In this context, endothelial differentiation of ES cells may provide a means to investigate the behavior of nascent endothelial cells derived directly from progenitor cells.

3.2 Endothelial cell movement compatible with junctional integrity

We had developed a culture system in which endothelial cells derived from ES cells grow to form sheet-like colonies on a layer of OP9 stromal cells (Hirashima *et al.*, 1999; Matsumura *et al.*, 2003). In this culture system, differentiation of VE-cadherin$^+$ CD31$^+$ endothelial cells is induced first from ES cells by co-cultivating with OP9 stromal cells. Endothelial cells are then purified by FACS and re-cultured on OP9 cell layers to allow colonies to form. In the absence of exogenous factors except for those secreted by OP9 cells or already contained in the serum, most of the colonies are round clusters of flat polygonal endothelial cells that are connected to each other via adherens junctions and tight junctions, which are revealed by VE-cadherin and claudin-5 immunostaining, respectively (Figure 8 and Guo *et al.*, 2007).

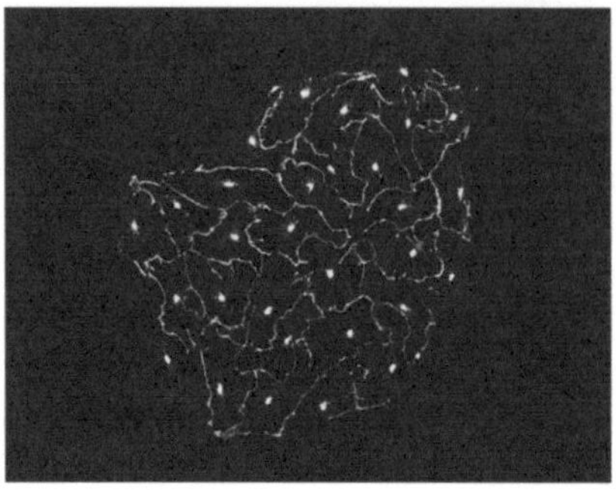

Fig. 8. Morphology of ES cell-derived endothelial cell colony
ES cells were co-cultured with OP9 cells for 5 days. VE-cadherin$^+$ CD31$^+$ endothelial cells were purified by FACS, and seeded on OP9 cell layer. After 3 days, cultures were fluorescently stained with anti-VE-cadherin antibody to reveal endothelial cell colonies.

Interestingly, time-lapse analyses under a phase contrast microscope showed that endothelial cells moved continuously within the colonies. The average rate of cell movement was about 15 μm per hour. Moving endothelial cells frequently displayed protrusive expansion of cell membrane in the direction of movement. To examine whether the endothelial cell movement involves active cell locomotion, we investigated subcellular dynamics of actin and Arp2/3 complex by using ES cell clones that constitutively express cytoplasmic β-actin or p41-Arc, a subunit of the Arp2/3 complex, both fused to EGFP. Time-lapse analyses of EGFP-β-actin- or EGFP-p41Arc-expressing endothelial cell colonies under a fluorescence microscope revealed that the protrusive expansion of cell membrane was accompanied by dynamic accumulation of actin and Arp2/3 complex, which resembled the process of lamellipodium formation seen at the leading edge of a moving cell (Guo *et al.*, 2007). These observations indicate an involvement of lamellipodium formation and active cell locomotion in the movement of endothelial cells within the colonies, even though the endothelial cells stay connected via adherens junctions and tight junctions.

In order to visualize the adherens junction and the tight junction on living endothelial cells, we established ES cell clones that express VE-cadherin or claudin-5, both tagged with Venus (a derivative of yellow fluorescent protein), under the control of the endothelial specific promoter/enhancer of the *Cdh5* (VE-cadherin) gene (Guo *et al.*, 2007). When endothelial differentiation on the OP9 cell layer was induced from these ES cell clones, the VE-cadherin-Venus or Venus-claudin-5 proteins were selectively localized at cell-cell boundaries of endothelial cell colonies. Time-lapse analyses of endothelial cell colonies under a fluorescence microscope demonstrated that moving endothelial cells changed shape continuously to maintain the integrity of adherens junctions and tight junctions (Figure 9 and Guo *et al.*, 2007). Intercellular junctions underwent dynamic remodeling at the leading edge of moving endothelial cells, especially when a cell protrudes membrane and moves toward a lateral side of a contiguous cell.

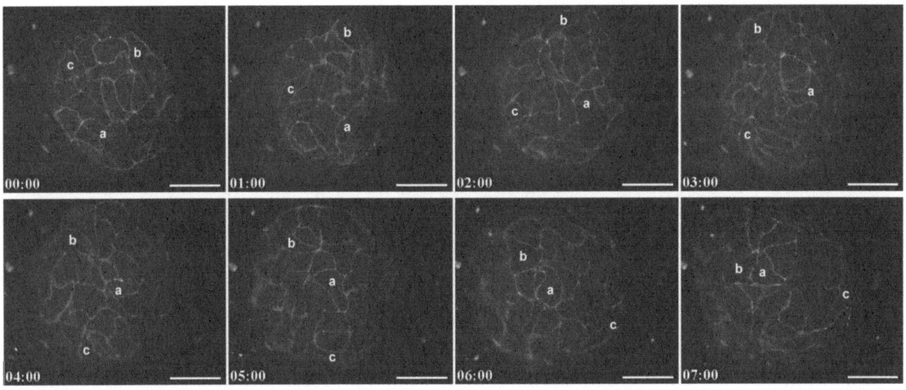

Fig. 9. Time-lapse imaging of adherens junctions on endothelial cells
An ES cell clone that express VE-cadherin-Venus under the control of endothelial-specific promoter/enhancer of the *Cdh5* gene was induced to differentiate into endothelial cells. FACS-purified endothelial cells were cultured on OP9 cell layer to form colonies. Colonies were subjected to time-lapse analysis under a fluorescence microscope to reveal dynamics of VE-cadherin-mediated adherens junctions. Numbers indicate elapsed time counted by hours. Three cells are individually labeled with alphabets to show the cell movement. Scale bars indicate 50 μm.

These results clearly indicate that the integrity of the adherens junction and the tight junction can be maintained as an endothelial cell migrates. It was reported that a monoclonal antibody against VE-cadherin disrupted adhesion of endothelial cells in growing vessels and prevented angiogenesis, suggesting that continuity of homotypic interaction of VE-cadherin is necessary for the process of angiogenesis (Corada *et al.*, 2002). The compatibility of cell motility with junctional integrity should be an important characteristic of endothelial cells participating in angiogenesis. ES cell-derived endothelial cell colony formation serves as a useful model to study cell biological regulation of vascular development as further emphasized in the next section.

4. Regulation of endothelial cell morphology during angiogenesis

4.1 Essential role of Foxo1 in vascular development

Vascular development involves morphogenetic processes such as vasculogenesis, angiogenesis, and vascular remodeling. Mutant mouse models have implicated several key molecules in vascular development, including VEGF, platelet-derived growth factor, transforming growth factor-β (TGF-β), angiopoietin, Notch, and ephrin/Eph, the loss of which result in abnormal vascular development. However, little is known about how these molecules regulate the behavior of vascular components, indicating the importance of research at the cellular level that connects function of molecules and phenotype of mutant animals. A study on the regulation of angiogenesis by Foxo1 would be an example of such research.

Foxo1 is a member of the Foxo subfamily of forkhead box transcription factors that promote cell cycle arrest, repair of damaged DNA, oxidative stress resistance, apoptosis and gluconeogenesis by regulating specific genes (van der Horst and Burgering, 2007). A role for Foxo1 in the developing vasculature was demonstrated by the observation that *Foxo1*(-/-)

mice die around E11 due to defects in the development of branchial arches and malformation in major vessels of the embryo and the yolk sac (Furuyama *et al.*, 2004; Hosaka *et al.*, 2004). *Foxo1*(-/-) embryos developed a small first branchial arch, but no second branchial arch, and often exhibited marked pericardial swelling. The dorsal aorta was severely underdeveloped and irregularly formed. Hypoplasia of aortic arch artery was also observed. *Foxo1*(-/-) yolk sacs showed primitive vascular plexus similar to that of wild-type yolk sacs at E8.75, but failed to develop a normal vasculature at E9.5. These observations suggested that vasculogenesis, but not angiogenesis and remodeling, proceeds without Foxo1. Expression of the *Foxo1* gene was detected in developing vasculature of normal mouse embryos including the dorsal aorta, the intersomitic vessels, the vitelline and umbilical vessels and others (Furuyama *et al.*, 2004; Hosaka *et al.*, 2004). Although Foxo1 has been considered as a key regulator of energy metabolism and lifespan, the role of this transcription factor in the vascular development can hardly be understood from the well-known target genes. There was no significant difference between wild-type and *Foxo1*(-/-) yolk sac in the expression levels of genes involved in vascular development including VEGF-A, VEGFR-1, VEGFR-2, angiopoietin-1, angiopoietin-2, Tie-1, Tie-2 and EphB4. Still, *Foxo1*(-/-) yolk sac vasculature appeared to had lost an arterial property as indicated by reduced expression of arterial markers such as ephrinB2 and connexin-40 (Furuyama *et al.*, 2004).

In order to delineate the role of Foxo1 in angiogenesis, we employed an *in vitro* differentiation system of ES cells for investigating the expression and function of Foxo1 in endothelial cells. *Foxo1* transcripts were detected in the VE-cadherin$^+$ CD31$^+$ endothelial cells derived from wild-type ES cells differentiating in the co-culture with OP9 stromal cells (Furuyama *et al.*, 2004). Comparable number of endothelial cells were able to be obtained from the cultures initiated from wild-type and *Foxo1*(-/-) ES cells. Endothelial cells of the two genotypes also gave rise to comparable number and size of monostratal endothelial colonies when purified by FACS and re-seeded on the OP9 cell layer, suggesting that the loss of Foxo1 did not influence the proliferation and survival of endothelial cells. This notion may contrast with a previous report that Foxo1 induced apoptosis in HUVECs, which was inhibited by angiopoietin-1 signaling via activation of Akt pathway (Daly *et al.*, 2004). This discrepancy could reflect the difference in the differentiation stage of endothelial cells, namely, nascent immature endothelial cells derived from ES cells and mature endothelial cells isolated from established endothelium. The former may resemble the cells that undertake vascular development in the embryos.

We took advantage of the ES cell-derived endothelial cell colony formation assay to compare behavior of wild-type and *Foxo1*(-/-) endothelial cells in response to various angiogenic stimuli. In the absence of exogenously added factors, endothelial cells of both genotypes formed monostratal colonies composed of rough-edged flat cells as described above. When VEGF-A was added in the medium during colony formation, however, wild-type and *Foxo1*(-/-) endothelial cells exhibited quite different morphological responses. VEGF-A induced elongation of wild-type endothelial cells, thereby led to form colonies with corded structure composed of long spindle-shaped endothelial cells entwining with each other. In contrast to the response of wild-type endothelial cells, *Foxo1*(-/-) endothelial cells remained flatten in the presence of exogenous VEGF-A. Yet they changed morphology from rough-edged to polygonal shape with rather straight adherens junctions and partially overlapped to the neighboring cells (Figure 10). (Endothelial cell overlapping as a stable condition is rarely observed in unstimulated monostratal colonies.) Induced expression of

Foxo1 cDNA in the *Foxo1*(-/-) endothelial cells by using the endothelial-specific promoter/enhancer of the *Cdh5* (VE-cadherin) gene restored cell elongation in response to VEGF-A, implying a cell-autonomous function of Foxo1 in the morphological response of endothelial cells (Furuyama *et al.*, 2004). Taken together, these results suggest that Foxo1 regulates endothelial cell elongation in response to VEGF-A signaling, and the failure of proper morphological response of endothelial cells to angiogenic stimuli likely accounts for the compromised angiogenesis in the *Foxo1*(-/-) embryos.

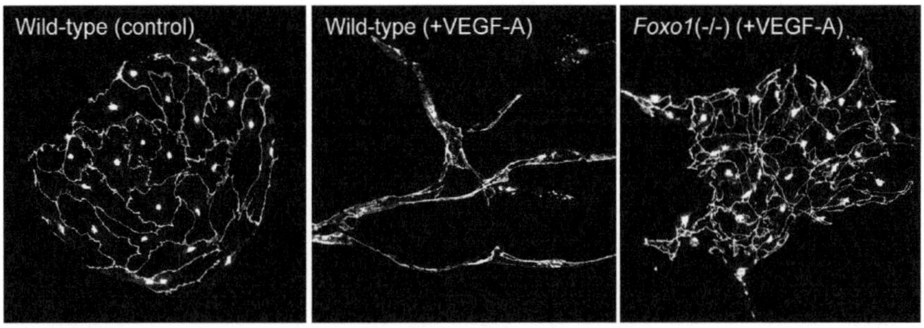

Fig. 10. Morphological response of endothelial cells to VEGF-A
Wild-type and *Foxo1*(-/-) ES cells were co-cultured with OP9 cells for 5 days. VE-cadherin+ CD31+ endothelial cells were purified by FACS and seeded onto OP9 cell layer. Cells were cultured for 3 days in the presence or absence of exogenously added VEGF-A. Endothelial cell colonies were revealed by immunofluorescent staining of VE-cadherin.

4.2 Endothelial cell elongation induced by angiogenic factors

Severe dilation of the yolk sac blood vessels in *Foxo1*(-/-) mice are reminiscent of mice lacking components of the TGF-β signaling pathway including Alk5, Alk1, endoglin and Smad5 (Larsson *et al.*, 2001; Li *et al.*, 1999; Oh *et al.*, 2000; Yang *et al.*, 1999). It is well known that endothelial cells provide TGF-β signaling to neighboring mesenchymal cells and promote their differentiation into smooth muscle cells which support and stabilize the vessels (ten Dijke and Arthur, 2007). However, endothelial cell-specific disruption of the *Tgfbr2* or *Alk5* gene resulted in the phenotype indistinguishable from that of the *Tgfbr2*-null or *Alk5*-null embryos, suggesting an endothelial cell-intrinsic role of TGF-β signaling in the regulation of angiogenesis (Carvalho *et al.*, 2004; Jiao *et al.*, 2006). In order to examine whether Foxo1 and TGF-β are involved in a common process of angiogenesis, we tested the effect of TGF-β on the morphology of ES cell-derived endothelial cells. VE-cadherin+ CD31+ endothelial cells derived from wild-type and *Foxo1*(-/-) ES cells differentiating in the co-culture with OP9 cells were purified by FACS and re-seeded on the OP9 cell layer in the presence of TGF-β. Wild-type endothelial cells produced cord-like colonies composed of spindle-shaped cells in response to TGF-β. By contrast, *Foxo1*(-/-) endothelial cells failed to produce cord-like colonies, but they exhibited partial overlapping as in the case of VEGF-A treatment (Matsukawa *et al.*, 2009). These observations indicate that TGF-β induces elongation of endothelial cells in a *Foxo1*-dependent manner.

Sequestration of a low concentration of endogenous VEGF-A pre-existing in the culture (mainly secreted by OP9 cells) by addition of Flt1-Fc chimeric protein diminished TGF-β-induced elongation of wild-type endothelial cells, indicating that endothelial cell elongation induced by TGF-β depends upon VEGF-A signaling (Matsukawa *et al.*, 2009). As the amount of *Vegfa* transcripts in OP9 cells was not influenced by the stimulation with TGF-β, two possible mechanisms might be implied. First, TGF-β may synergize with VEGF-A to regulate endothelial cell morphology, and Foxo1 is possibly involved in the TGF-β signaling. It is indeed reported that Foxo proteins interacted with Smad3 and Smad4 to form a *p21Cip1* transactivation complex in the regulation of neuroepithelial and glioblastoma cell proliferation by TGF-β (Seoane *et al.*, 2004). Alternatively, TGF-β may reduce the threshold of VEGF-A-responsiveness of endothelial cells, thereby facilitates cell elongation under the presence of a limited amount of VEGF-A, which is dependent upon Foxo1. Nevertheless, the observations that TGF-β induced endothelial cell elongation in a Foxo1-dependent manner and that disruption of Foxo1 and TGF-β signaling resulted in a similar phenotype in vascular development suggest important roles of Foxo1-dependent endothelial cell elongation in the process of angiogenesis.

4.3 Distinct functions of Foxo subfamily members

The Foxo subfamily of forkhead box transcription factors consists of four members (Foxo1, Foxo3, Foxo4 and Foxo6). Somatic deletion of the *Foxo1, Foxo3* and *Foxo4* genes in adult mice revealed that Foxo1 is the major regulator of endothelial stability which suppresses endothelial growth and hemangioma development, yet a certain degree of functional redundancy appears to exist among the Foxo members (Paik *et al.*, 2007). Indeed, both Foxo1 and Foxo3 were reported to induce apoptosis in HUVECs (Daly *et al.*, 2004; Kim *et al.*, 2005). Foxo1 and Foxo3, but not Foxo4, were also shown to comparably inhibit endothelial cell migration and tube formation *in vitro* (Potente *et al.*, 2005). However, Foxo1 appears to have nonredundant functions in vascular development during embryogenesis. In contrast to the defect of angiogenesis in *Foxo1*(-/-) embryos, both *Foxo3*(-/-) and *Foxo4*(-/-) mice are born alive with no detectable abnormality of vascular development (Castrillon *et al.*, 2003; Hosaka *et al.*, 2004). Dispensability of these two genes for vascular development possibly due to differential tissue distribution of the Foxo members in developing embryos, since the expression of *Foxo1, Foxo3* and *Foxo4* mRNA in the embryos were complementary to each other (Furuyama *et al.*, 2000). Alternatively, molecular functions of the Foxo members in the endothelium are possibly diversified. For instance, expression of several genes that are related to inflammation and angiogenesis (e.g. angiopoietin-2 and interleukin-8) are differentially regulated by Foxo1 and Foxo3 in HUVECs (Potente *et al.*, 2005).

We examined expression of *Foxo3* during the course of ES cell differentiation into endothelial cell lineage. While *Foxo1* was expressed constantly by undifferentiated ES cells, Flk-1$^+$ mesodermal cells and various stages of VE-cadherin$^+$ CD31$^+$ endothelial cells, expression of *Foxo3* was not detected in any of these cell populations (Matsukawa *et al.*, 2009). Therefore, Foxo1 appears to be a primary and nonredundant factor that regulates morphological response of developing endothelial cells to angiogenic stimulations. In order to gain insight into the molecular mechanisms underlying the cell morphological regulation by Foxo1, we investigated whether or not Foxo3 is able to functionally replace Foxo1 when expressed as a transgene in *Foxo1*(-/-) endothelial cells (Matsukawa *et al.*, 2009). We recruited a tetracycline-regulated gene expression system to induce the expression of either Foxo1 or Foxo3 protein in *Foxo1*(-/-) endothelial cells. In this system, a tetracycline responsive promoter drives expression of either the *Foxo1* or *Foxo3* transgene when activated by a tetracycline transactivator in the absence of tetracycline, or suppresses it in the presence of tetracycline. The expression system was introduced into *Foxo1*(-/-) ES cells, and tightly controllable clones were selected. The Foxo1- or Foxo3-inducible *Foxo1*(-/-) ES cell clones were allowed to differentiate into endothelial cells by co-culturing with OP9 stromal cells for 5 days in the presence of tetracycline (to suppress the transgene expression), followed by FACS purification of endothelial cells. The sorted endothelial cells were seeded onto OP9 cell layer in the presence of VEGF-A, and the transgene was activated by withdrawal of tetracycline. Morphological examination of endothelial cell colonies revealed that induction of Foxo1 in *Foxo1*(-/-) endothelial cells led to cell elongation that is almost comparable to wild-type endothelial cells. In contrast to the effect of Foxo1, induction of Foxo3 in *Foxo1*(-/-) endothelial cells failed to restore the VEGF-A-dependent elongation, suggesting that Foxo3 is not able to exert the same function as does Foxo1 in promoting endothelial cell elongation in response to VEGF-A signaling (Matsukawa *et al.*, 2009).

ES cell-derived endothelial cell cultures represent ongoing processes of endothelial cell differentiation such as the upregulation of CD34 and Flt1 expression and activation of the promoter/enhancer element of the *Flk-1* gene (Hirai *et al.*, 2003; Hirashima *et al.*, 1999; Hirashima *et al.*, 2003). Interestingly, when endothelial cells were isolated from 6 days culture of differentiating Foxo3-inducible ES cell clones (i.e. 24 hours later than the above-described experiments) and examined for morphological response to VEGF-A, induction of Foxo3 restored the VEGF-A-dependent endothelial cell elongation (Matsukawa *et al.*, 2009). It is thus suggested that endothelial cells serially activate two different mechanisms of morphological regulation during the differentiation process. Foxo1 is involved in both mechanisms, while Foxo3 is able to participate only in the late-acting mechanism. Taken together, we hypothesize that distinct functions as well as distribution of Foxo1 and Foxo3 in endothelial cells may account for the difference of the phenotypes between *Foxo1*(-/-) and *Foxo3*(-/-) embryos.

The functional disparity of Foxo members in endothelial cells should provide valuable clue to clarify the molecular mechanisms underlying the regulation of endothelial cell morphology by Foxo1. Identification of Foxo1 target genes that are responsible for the morphological regulation is of critical importance. We performed DNA microarray analysis to compare gene expression profiles of the endothelial cells derived from wild-type and *Foxo1*(-/-) ES cells by using the 3D-Gene Mouse Oligo chip 24 (Toray Industries Inc.). Among a total of 25,392 genes, we identified 207 genes that were differentially expressed in the wild-type and *Foxo1*(-/-) endothelial cells. From the 207 genes, we further excluded genes of which expression was influenced in the same direction (increased or decreased) by the induction of Foxo1 or Foxo3 in *Foxo1*(-/-) endothelial cells. Thirty-one genes were finally selected as candidate genes, and *in vitro* functional screening is currently underway. Once a target gene of Foxo1 that is responsible for the regulation of endothelial cell morphology is identified, expression of the candidate gene should be genetically manipulated in the *Foxo1*(-/-) embryos to examine whether or not it restores the abnormal angiogenesis caused by *Foxo1*-deficiency, which will finally testify to the importance of *Foxo1*-dependent morphological regulation of endothelial cells in vascular development.

4.4 Vessel-like structure formation from ES cells

In order to understand the role of Foxo1 in the regulation of endothelial cell morphology, it is important to investigate how the cytoskeletal organization of endothelial cells is influenced in the absence of functional Foxo1. However, the OP9 cell-dependent two-dimensional culture described in the preceding sections is not suitable for microscopic observation of cytoskeletal structures of endothelial cells due to interference of the robust cytoskeleton of OP9 cells. In addition, endothelial cell behavior is induced not only by exogenously added factors but might also be influenced by unknown molecules secreted by OP9 cells. It is thus necessary to establish an experimental model in which endothelial cell behavior can be examined in the absence of stromal cells. Yamashita *et al.* previously reported that ES cell-derived Flk-1^{+} mesodermal cells produced vessel-like structures consisting of endothelial tubes supported by smooth muscle cells in a type I collagen gel culture with VEGF-A (Yamashita *et al.*, 2000). This three-dimensional culture system is able to circumvent the need of OP9 feeder cells and provide an *in vitro* model for studying the morphological change of endothelia cells into a capillary-like structure and its association with smooth muscle cells. We employed and further improved the culture system to investigate the impact of *Foxo1*-deficiency on the cytoskeletal organization of endothelial cells and the formation of vessel-like structures (Park *et al.*, 2009).

In the culture system, ES cells were cultured with OP9 cells for 4 days to induce differentiation of Flk-1^{+} mesodermal cells. Flk-1^{+} cells were purified by FACS and allowed to aggregate in suspension culture. Flk-1^{+} cell aggregates were embedded in type I collagen gel and cultured for 4 days in the presence of VEGF-A to form vessel-like structures. Dome-like collagen gels were finally flattened by liquid absorption for fixation and immunostaining. Gel flattening greatly facilitated immunofluorescence staining and confocal microscopic observation of the vessel-like structures deeply embedded in the gels (Park *et al.*, 2009). Wild-type endothelial cells showed spindle-shaped elongation as revealed by rectilinear staining of VE-cadherin along the long axis of the cells, and organized into long cord-like structures. On the other hand, *Foxo1*(-/-) endothelial cells failed to elongate and produced only short bundles with irregularly kinked adherens junctions, demonstrating that Foxo1 is essential for elongation of endothelial cells that are organized into vessel-like structures *in vitro* (Figure 11, 12 and Park *et al.*, 2009).

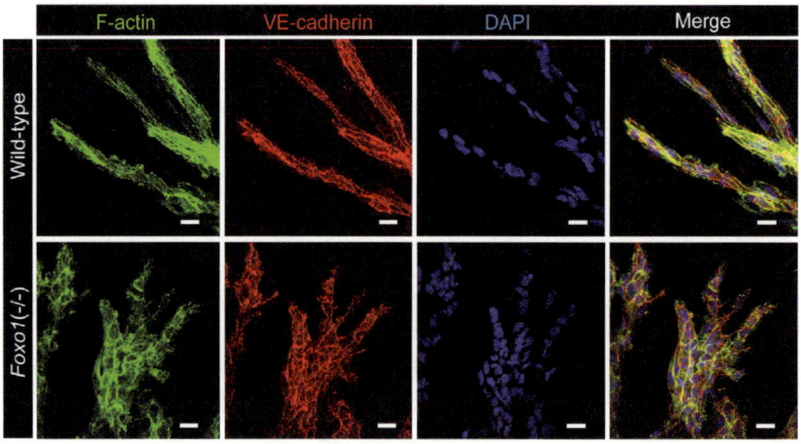

Fig. 11. Abnormal vessel-like structure formation from *Foxo1*(-/-) mesodermal cells
Wild-type and *Foxo1*(-/-) ES cells were co-cultured with OP9 cells for 4 days. FACS-purified mesodermal cells were cultured in type I collagen gel in the presence of VEGF-A for 4 days. Vessel-like structures were fluorescently stained with phalloidin, anti-VE-cadherin antibody and DAPI. Scale bars indicate 10μm.

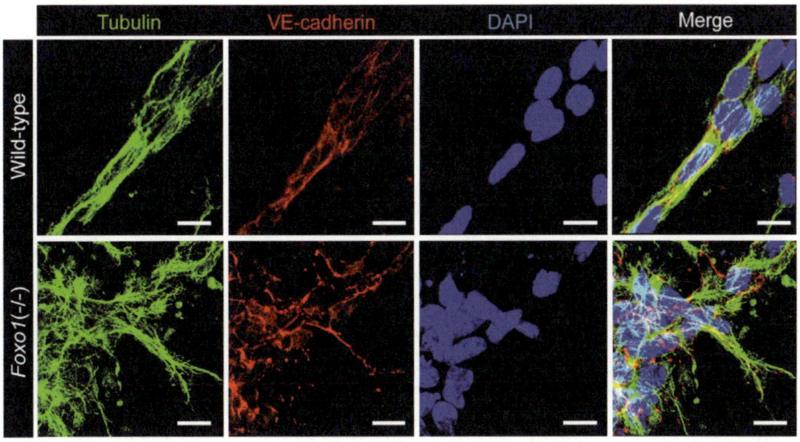

Fig. 12. Microtubular organization of endothelial cells in the vessel-like structures
Wild-type and *Foxo1*(-/-) ES cells were co-cultured with OP9 cells for 4 days. FACS-purified mesodermal cells were cultured in type I collagen gel in the presence of VEGF-A for 4 days. Vessel-like structures were fluorescently stained with antibodies against tubulin and VE-cadherin together with DAPI. Scale bars indicate 10μm.

Phalloidin staining revealed that prominent filamentous cortical actin which co-localized with VE-cadherin at the adherens junction was present in wild-type endothelial cells. By contrast, F-actin was not only accumulated to the adherens junctions but also scattered as punctate structures in *Foxo1*(-/-) endothelial cells. The punctate accumulation of F-actin is unlikely to represent the focal adhesion complex, because only vinculin, but not paxillin nor focal adhesion kinase, was co-localized with the punctate structures. Indeed, the stress fiber formation was not observed in both wild-type and *Foxo1*(-/-) endothelial cells (Park *et al.*, 2009). These results suggested that *Foxo1*-deficiency led to abnormal organization of actin cytoskeleton, which might influence the adhesion and migration of endothelial cells.

Cell elongation requires the reorganization of microtubules as well as actin cytoskeleton. While long filamentous microtubules in a mesh-like network were observed in wild-type endothelial cells, *Foxo1*(-/-) endothelial cells developed thick circumferential accumulation of microtubules with small spikes at the tip of cells (Park *et al.*, 2009). It is thus suggested that *Foxo1*-deficiency leads to disorganization of microtubular system in endothelial cells, which may resemble a condition resulted from hyper stabilization of microtubules (Gloushankova *et al.*, 1994). It was proposed that cell length is controlled by the equilibrium of two antagonistic forces: elongation exerted by integrated microtubular system and contraction exerted by actin-myosin system (Kharitonova and Vasiliev, 2008). We hypothesize that hyper stabilization of microtubules as well as circumferential accumulation of actin microfilaments may contribute to the failure of elongation of *Foxo1*(-/-) endothelial cells.

Pharmacological approaches by using chemicals that influence the cytoskeletal organization would be useful to investigate the molecular basis of impaired elongation of *Foxo1*(-/-) endothelial cells. One of the advantages of the three-dimensional culture system is that it circumvents the need of OP9 stromal cell layer, which might otherwise be affected by such chemicals and secondarily influence the morphology of endothelial cells. Taking advantage of the culture system, we tested effects of Y27632, a compound that inhibits Rho kinase (ROCK), on the formation of vessel-like structures (Figure 13 and 14).

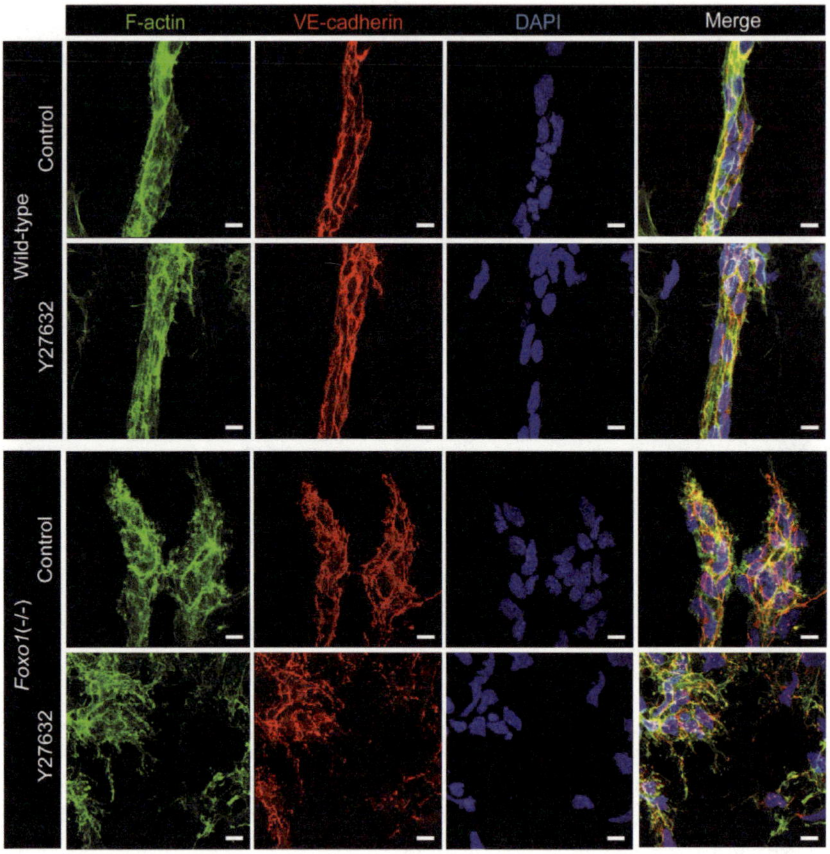

Fig. 13. Aberrant dependency on RhoA/ROCK pathway in *Foxo1*(-/-) endothelial cells
Wild-type and *Foxo1*(-/-) ES cells were co-cultured with OP9 cells for 4 days. FACS-purified mesodermal cells were cultured in type I collagen gel in the presence of VEGF-A and presence or absence of Y27632. Vessel-like structures were fluorescently stained with phalloidin, anti-VE-cadherin antibody and DAPI. Scale bars indicate 10μm.

Flk-1^{+} cells isolated from differentiating wild-type or Foxo1(-/-) ES cells were allowed to aggregate and embedded in type I collagen gel and cultured in the presence of 50 ng/mL VEGF-A and presence or absence of 10 μM Y27632. VE-cadherin immunofluorescence staining of flattened gels revealed that wild-type Flk-1^{+} cells generated capillary-like structures consisting of spindle-shaped endothelial cells regardless of whether Y27632 was present or absent in the culture. Actin microfilaments and microtubule networks were not also affected by Y27632 treatment. These observations indicate that ROCK inhibition does not affect the elongation of wild-type endothelial cells and the formation of capillary-like structures *in vitro*. By contrast, treatment with Y27632 resulted in extreme disintegration of cell organization in the cultures of *Foxo1*(-/-) endothelial cells. Adherens junctions revealed by VE-cadherin and F-actin staining became fragmentary and distributed in a disorderly manner. Microtubules also exhibited short fluffiness with random distribution. Consequently, *Foxo1*(-/-) endothelial cells no longer maintained even short bundles in the presence of Y27632. *Foxo1*(-/-) endothelial cells thus appear to be abnormally dependent on the RhoA/ROCK pathway.

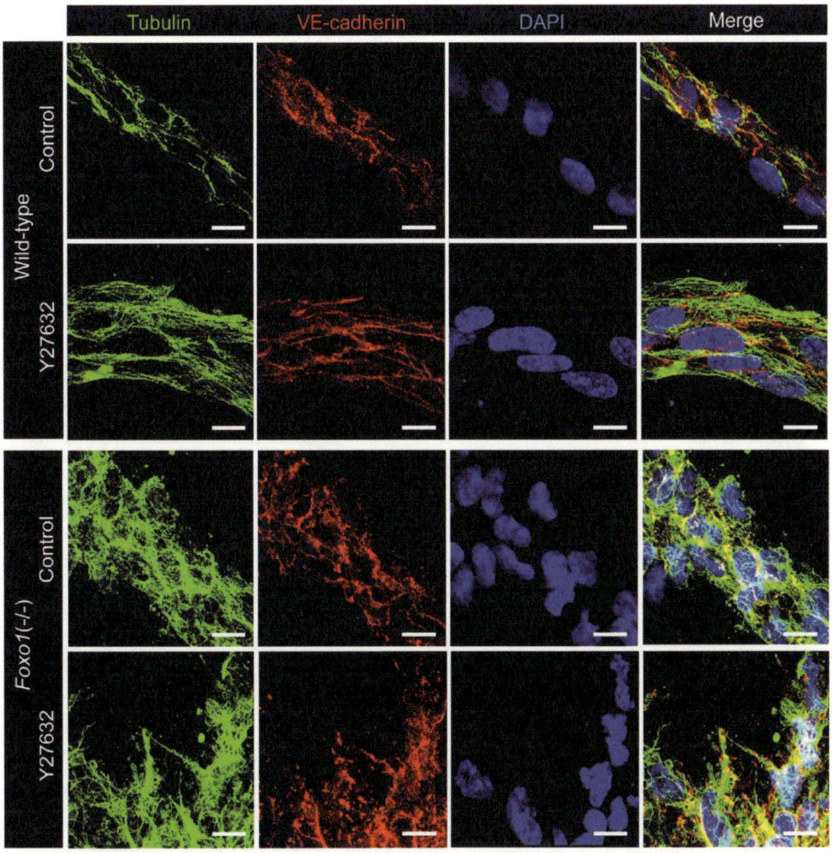

Fig. 14. Microtubular organization of ROCK-inhibited endothelial cells
Wild-type and *Foxo1*(-/-) ES cells were co-cultured with OP9 cells for 4 days. FACS-purified mesodermal cells were cultured in type I collagen gel in the presence of VEGF-A and presence or absence of Y27632. Vessel-like structures were fluorescently stained with antibodies against tubulin and VE-cadherin together with DAPI. Scale bars indicate 10μm.

Our results suggest that the RhoA/ROCK pathway is not involved in the regulation of endothelial cell morphology or there might be some compensatory pathways in the wild-type genetic background. On the other hand, *Foxo1*(-/-) endothelial cells are suggested to harbor an abnormal enhancement of the RhoA/ROCK pathway or lack the putative compensatory pathways. However, the enhancement of ROCK-dependency itself might be the effect, rather than the cause, of abnormal morphological response of *Foxo1*(-/-) endothelial cells to VEGF-A, because the inhibition of ROCK activity did not restore the abnormal phenotype. Nevertheless, the aberrant requirement of the RhoA/ROCK pathway for the maintenance of cell organization provides clue to how a loss of Foxo1 compromises cytoskeletal organization and morphological response of endothelial cells during angiogenesis.

4.5 Association between endothelial cells and smooth muscle cells

Correct association and interaction between endothelial cells and smooth muscle cells is known to be essential for vascular maturation. The capillary-like structure generated *in vitro* from ES cell-derived Flk-1$^+$ mesodermal cells consists of endothelial tubes associated with mural cells that express α-smooth muscle actin (αSMA). A massive investment of mural cells is usually observed along the vascular-like structures. Endothelial cells and mural cells exhibit patchy formation of desmosome-like junctions and some collagen fibers are evident adjacent to the mural cells, indicating that these cells interact with each other to form mature vessel-like structures *in vitro* (Yamashita *et al.*, 2000).

Interestingly, while the capillary-like structures generated from wild-type mesodermal cell aggregates were thoroughly covered by αSMA$^+$ mural cells, no such coverage of αSMA$^+$ cells was observed along the short endothelial cell bundles generated from *Foxo1*(-/-) mesodermal cell aggregates (Figure 15 and Park *et al.*, 2009). Yet, αSMA$^+$ cells were abundantly detected in the core region of *Foxo1*(-/-) cell aggregates and occasionally apart from cell aggregates, and *Foxo1*(-/-) αSMA$^+$ cells are morphologically indistinguishable from wild-type αSMA$^+$ cells (Figure 16). Therefore, differentiation of smooth muscle cells from Flk-1$^+$ cells was not affected by the absence of Foxo1. These results suggest that *Foxo1* is essential for either the migration of mural cells or the physical interaction between mural cells and endothelial cells, although it is obscure whether functional Foxo1 is required autonomously in mural cells or heteronomously in endothelial cells.

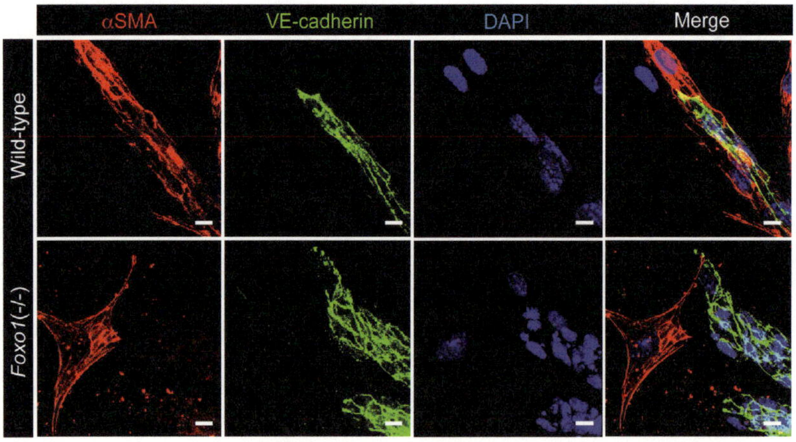

Fig. 15. Impaired association of mural cells with endothelial cells in *Foxo1*(-/-) cultures Wild-type and *Foxo1*(-/-) ES cells were co-cultured with OP9 cells for 4 days. FACS-purified mesodermal cells were cultured in type I collagen gel in the presence of VEGF-A for 4 days. Vessel-like structures were fluorescently stained with antibodies against αSMA and VE-cadherin together with DAPI. Scale bars indicate 10μm.

We hypothesize from the above observations that Foxo1 is involved in the recruitment of mural cells to nascent blood vessels. However, it has not been documented whether or not the interaction of endothelial cells and mural cells is maintained in the blood vessels of *Foxo1*(-/-) embryos (Furuyama *et al.*, 2004; Hosaka *et al.*, 2004). Thus, further phenotypic characterization of the knockout embryos is clearly important. The yolk sac vasculature of *Foxo1*(-/-) embryos had dilated and lost the arterial identity as mentioned above. This might be an indicative of a loss of supportive force exerted by mural cells to maintain the hemodynamic property that is necessary to induce arterial differentiation (Heil *et al.*, 2006). The functional necessity of *Foxo1* in vascular smooth muscle cells in the embryo is an important subject for future studies.

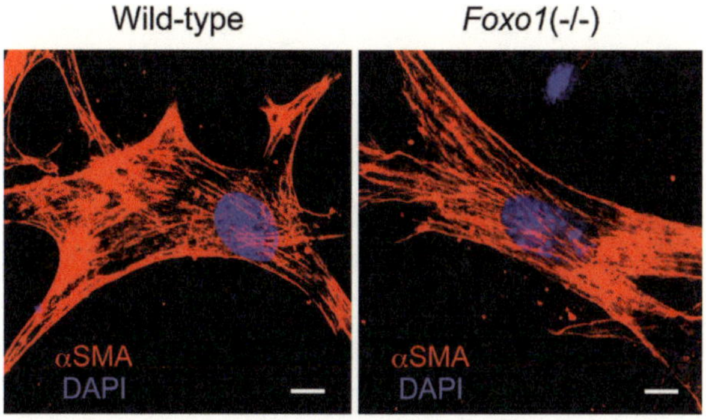

Fig. 16. Morphology of ES cell-derived smooth muscle cells
Wild-type and *Foxo1*(-/-) ES cells were co-cultured with OP9 cells for 4 days. FACS-purified mesodermal cells were cultured in type I collagen gel in the presence of VEGF-A. Smooth muscle cells were revealed by immunofluorescent staining with anti-αSMA antibody. Scale bars indicate 10μm.

5. Conclusion

Differentiation, migration, cell-cell adhesion and morphological regulation of vascular endothelial cells are the important cell biological processes that drive vasculogenesis and angiogenesis. *In vitro* differentiation of ES cells provides useful means to elucidate the mechanisms underlying cell biological regulation of vascular development. Further improvement will, however, be necessary for simplification of the system to raise the controllability of cell biological processes. Such an improvement should encompass the replacement of the OP9 stromal cell layer with a defined extracellular matrix with variable elasticity. Establishing the cell biology of normal endothelial cells and understanding the cell biological control of vascular development will contribute to not only developmental biology but also clinical approaches for regulation of physiological and pathological neoangiogenesis in ischemic tissues and tumors.

6. Acknowledgements

We wish to thank the members, past and present, of the Department of Cell Differentiation, Institute of Molecular Embryology and Genetics, Kumamoto University. This work was supported by grants from the Ministry of Education, Culture, Sports, Sciences and Technology (MEXT) of Japan and Japan Society for the Promotion of Science. This work was supported in part by Global COE Program (Cell Fate Regulation Research and Education Unit), MEXT, Japan.

7. References

Bazzoni, G. & Dejana, E. (2004). Endothelial cell-to-cell junctions: molecular organization and role in vascular homeostasis. *Physiol Rev*, Vol. 84, No. 3, 869-901.

Bloch, W., Forsberg, E., Lentini, S., Brakebusch, C., Martin, K., Krell, H.W., Weidle, U.H., Addicks, K. & Fassler, R. (1997). Beta 1 integrin is essential for teratoma growth and angiogenesis. *J Cell Biol*, Vol. 139, No. 1, 265-278.

Breier, G., Breviario, F., Caveda, L., Berthier, R., Schnurch, H., Gotsch, U., Vestweber, D., Risau, W. & Dejana, E. (1996). Molecular cloning and expression of murine vascular endothelial-cadherin in early stage development of cardiovascular system. *Blood*, Vol. 87, No. 2, 630-641.

Carmeliet, P., Lampugnani, M.G., Moons, L., Breviario, F., Compernolle, V., Bono, F., Balconi, G., Spagnuolo, R., Oostuyse, B., Dewerchin, M., *et al.* (1999). Targeted deficiency or cytosolic truncation of the VE-cadherin gene in mice impairs VEGF-mediated endothelial survival and angiogenesis. *Cell*, Vol. 98, No. 2, 147-157.

Carvalho, R.L., Jonker, L., Goumans, M.J., Larsson, J., Bouwman, P., Karlsson, S., Dijke, P.T., Arthur, H.M. & Mummery, C.L. (2004). Defective paracrine signalling by TGFbeta in yolk sac vasculature of endoglin mutant mice: a paradigm for hereditary haemorrhagic telangiectasia. *Development*, Vol. 131, No. 24, 6237-6247.

Castrillon, D.H., Miao, L., Kollipara, R., Horner, J.W. & DePinho, R.A. (2003). Suppression of ovarian follicle activation in mice by the transcription factor Foxo3a. *Science*, Vol. 301, No. 5630, 215-218.

Corada, M., Zanetta, L., Orsenigo, F., Breviario, F., Lampugnani, M.G., Bernasconi, S., Liao, F., Hicklin, D.J., Bohlen, P. & Dejana, E. (2002). A monoclonal antibody to vascular endothelial-cadherin inhibits tumor angiogenesis without side effects on endothelial permeability. *Blood*, Vol. 100, No. 3, 905-911.

Daly, C., Wong, V., Burova, E., Wei, Y., Zabski, S., Griffiths, J., Lai, K.M., Lin, H.C., Ioffe, E., Yancopoulos, G.D., *et al.* (2004). Angiopoietin-1 modulates endothelial cell function and gene expression via the transcription factor FKHR (FOXO1). *Genes Dev*, Vol. 18, No. 9, 1060-1071.

De Val, S., Chi, N.C., Meadows, S.M., Minovitsky, S., Anderson, J.P., Harris, I.S., Ehlers, M.L., Agarwal, P., Visel, A., Xu, S.M., *et al.* (2008). Combinatorial regulation of endothelial gene expression by ets and forkhead transcription factors. *Cell*, Vol. 135, No. 6, 1053-1064.

Dejana, E. (2004). Endothelial cell-cell junctions: happy together. *Nat Rev Mol Cell Biol*, Vol. 5, No. 4, 261-270.

Doetschman, T.C., Eistetter, H., Katz, M., Schmidt, W. & Kemler, R. (1985). The in vitro development of blastocyst-derived embryonic stem cell lines: formation of visceral yolk sac, blood islands and myocardium. *J. Embryol. Exp. Morphol.*, Vol. 87, 27-45.

Eilken, H.M., Nishikawa, S. & Schroeder, T. (2009). Continuous single-cell imaging of blood generation from haemogenic endothelium. *Nature*, Vol. 457, No. 7231, 896-900.

Endoh, M., Ogawa, M., Orkin, S. & Nishikawa, S. (2002). SCL/tal-1-dependent process determines a competence to select the definitive hematopoietic lineage prior to endothelial differentiation. *Embo J*, Vol. 21, No. 24, 6700-6708.

Feraud, O., Cao, Y. & Vittet, D. (2001). Embryonic stem cell-derived embryoid bodies development in collagen gels recapitulates sprouting angiogenesis. *Lab Invest*, Vol. 81, No. 12, 1669-1681.

Fukuhara, S., Sako, K., Minami, T., Noda, K., Kim, H.Z., Kodama, T., Shibuya, M., Takakura, N., Koh, G.Y. & Mochizuki, N. (2008). Differential function of Tie2 at cell-cell contacts and cell-substratum contacts regulated by angiopoietin-1. *Nat Cell Biol*, Vol. 10, No. 5, 513-526.

Furuyama, T., Kitayama, K., Shimoda, Y., Ogawa, M., Sone, K., Yoshida-Araki, K., Hisatsune, H., Nishikawa, S., Nakayama, K., Nakayama, K., *et al.* (2004). Abnormal angiogenesis in Foxo1 (Fkhr)-deficient mice. *J Biol Chem*, Vol. 279, No. 33, 34741-34749.

Furuyama, T., Nakazawa, T., Nakano, I. & Mori, N. (2000). Identification of the differential distribution patterns of mRNAs and consensus binding sequences for mouse DAF-16 homologues. *Biochem J*, Vol. 349, No. Pt 2, 629-634.

Gloushankova, N.A., Lyubimova, A.V., Tint, I.S., Feder, H.H., Vasiliev, J.M. & Gelfand, I.M. (1994). Role of the microtubular system in morphological organization of normal and oncogene-transfected epithelial cells. *Proc Natl Acad Sci U S A*, Vol. 91, No. 18, 8597-8601.

Gory-Faure, S., Prandini, M.H., Pointu, H., Roullot, V., Pignot-Paintrand, I., Vernet, M. & Huber, P. (1999). Role of vascular endothelial-cadherin in vascular morphogenesis. *Development*, Vol. 126, No. 10, 2093-2102.

Grazia Lampugnani, M., Zanetti, A., Corada, M., Takahashi, T., Balconi, G., Breviario, F., Orsenigo, F., Cattelino, A., Kemler, R., Daniel, T.O., *et al.* (2003). Contact inhibition of VEGF-induced proliferation requires vascular endothelial cadherin, beta-catenin, and the phosphatase DEP-1/CD148. *J Cell Biol*, Vol. 161, No. 4, 793-804.

Guo, R., Sakamoto, H., Sugiura, S. & Ogawa, M. (2007). Endothelial cell motility is compatible with junctional integrity. *J Cell Physiol*, Vol. 211, No. 2, 327-335.

Hashimoto, K., Fujimoto, T., Shimoda, Y., Huang, X., Sakamoto, H. & Ogawa, M. (2007). Distinct hemogenic potential of endothelial cells and CD41(+) cells in mouse embryos. *Dev Growth Differ*, Vol. 49, No. 4, 287-300.

Heil, M., Eitenmuller, I., Schmitz-Rixen, T. & Schaper, W. (2006). Arteriogenesis versus angiogenesis: similarities and differences. *Journal of cellular and molecular medicine*, Vol. 10, No. 1, 45-55.

Hirai, H., Ogawa, M., Suzuki, N., Yamamoto, M., Breier, G., Mazda, O., Imanishi, J. & Nishikawa, S. (2003). Hemogenic and nonhemogenic endothelium can be distinguished by the activity of fetal liver kinase (Flk)-1 promoter/enhancer during mouse embryogenesis. *Blood*, Vol. 101, No. 3, 886-893.

Hirashima, M., Kataoka, H., Nishikawa, S., Matsuyoshi, N. & Nishikawa, S. (1999). Maturation of embryonic stem cells into endothelial cells in an in vitro model of vasculogenesis. *Blood*, Vol. 93, No. 4, 1253-1263.

Hirashima, M., Ogawa, M., Nishikawa, S., Matsumura, K., Kawasaki, K., Shibuya, M. & Nishikawa, S. (2003). A chemically defined culture of VEGFR2+ cells derived from embryonic stem cells reveals the role of VEGFR1 in tuning the threshold for VEGF in developing endothelial cells. *Blood*, Vol. 101, No. 6, 2261-2267.

Hisatsune, H., Matsumura, K., Ogawa, M., Uemura, A., Kondo, N., Yamashita, J.K., Katsuta, H., Nishikawa, S. & Chiba, T. (2005). High level of endothelial cell-specific gene expression by a combination of the 5' flanking region and the 5' half of the first intron of the VE-cadherin gene. *Blood*, Vol. 105, No. 12, 4657-4663.

Hosaka, T., Biggs, W.H., 3rd, Tieu, D., Boyer, A.D., Varki, N.M., Cavenee, W.K. & Arden, K.C. (2004). Disruption of forkhead transcription factor (FOXO) family members in mice reveals their functional diversification. *Proc Natl Acad Sci U S A*, Vol. 101, No. 9, 2975-2980.

Jiao, K., Langworthy, M., Batts, L., Brown, C.B., Moses, H.L. & Baldwin, H.S. (2006). Tgfbeta signaling is required for atrioventricular cushion mesenchyme remodeling during in vivo cardiac development. *Development*, Vol. 133, No. 22, 4585-4593.

Kataoka, H., Takakura, N., Nishikawa, S., Tsuchida, K., Kodama, H., Kunisada, T., Risau, W., Kita, T. & Nishikawa, S.I. (1997). Expressions of PDGF receptor alpha, c-Kit and Flk1 genes clustering in mouse chromosome 5 define distinct subsets of nascent mesodermal cells. *Dev. Growth. Differ.*, Vol. 39, No. 6, 729-740.

Kharitonova, M.A. & Vasiliev, J.M. (2008). Controlling cell length. *Seminars in cell & developmental biology*, Vol. 19, No. 6, 480-484.

Kim, H.S., Skurk, C., Maatz, H., Shiojima, I., Ivashchenko, Y., Yoon, S.W., Park, Y.B. & Walsh, K. (2005). Akt/FOXO3a signaling modulates the endothelial stress response through regulation of heat shock protein 70 expression. *Faseb J*, Vol. 19, No. 8, 1042-1044.

Kondo, N., Ogawa, M., Wada, H. & Nishikawa, S. (2009). Thrombin induces rapid disassembly of claudin-5 from the tight junction of endothelial cells. *Exp Cell Res*, Vol. 315, No. 17, 2879-2887.

Larsson, J., Goumans, M.J., Sjostrand, L.J., van Rooijen, M.A., Ward, D., Leveen, P., Xu, X., ten Dijke, P., Mummery, C.L. & Karlsson, S. (2001). Abnormal angiogenesis but intact hematopoietic potential in TGF-beta type I receptor-deficient mice. *Embo J*, Vol. 20, No. 7, 1663-1673.

Li, D.Y., Sorensen, L.K., Brooke, B.S., Urness, L.D., Davis, E.C., Taylor, D.G., Boak, B.B. & Wendel, D.P. (1999). Defective angiogenesis in mice lacking endoglin. *Science*, Vol. 284, No. 5419, 1534-1537.

Matsukawa, M., Sakamoto, H., Kawasuji, M., Furuyama, T. & Ogawa, M. (2009). Different roles of Foxo1 and Foxo3 in the control of endothelial cell morphology. *Genes Cells*, Vol. 14, No. 10, 1167-1181.

Matsumura, K., Hirashima, M., Ogawa, M., Kubo, H., Hisatsune, H., Kondo, N., Nishikawa, S., Chiba, T. & Nishikawa, S. (2003). Modulation of VEGFR-2-mediated endothelial-cell activity by VEGF-C/VEGFR-3. *Blood*, Vol. 101, No. 4, 1367-1374.

Nakano, T., Kodama, H. & Honjo, T. (1994). Generation of lymphohematopoietic cells from embryonic stem cells in culture. *Science*, Vol. 265, No. 5175, 1098-1101.

Nishikawa, S.I., Nishikawa, S., Kawamoto, H., Yoshida, H., Kizumoto, M., Kataoka, H. & Katsura, Y. (1998). In vitro generation of lymphohematopoietic cells from endothelial cells purified from murine embryos. *Immunity*, Vol. 8, 761-769.

Nitta, T., Hata, M., Gotoh, S., Seo, Y., Sasaki, H., Hashimoto, N., Furuse, M. & Tsukita, S. (2003). Size-selective loosening of the blood-brain barrier in claudin-5-deficient mice. *J Cell Biol*, Vol. 161, No. 3, 653-660.

Oh, S.P., Seki, T., Goss, K.A., Imamura, T., Yi, Y., Donahoe, P.K., Li, L., Miyazono, K., ten Dijke, P., Kim, S., *et al.* (2000). Activin receptor-like kinase 1 modulates transforming growth factor-beta 1 signaling in the regulation of angiogenesis. *Proc Natl Acad Sci U S A*, Vol. 97, No. 6, 2626-2631.

Paik, J.H., Kollipara, R., Chu, G., Ji, H., Xiao, Y., Ding, Z., Miao, L., Tothova, Z., Horner, J.W., Carrasco, D.R., *et al.* (2007). FoxOs are lineage-restricted redundant tumor suppressors and regulate endothelial cell homeostasis. *Cell*, Vol. 128, No. 2, 309-323.

Park, S.H., Sakamoto, H., Tsuji-Tamura, K., Furuyama, T. & Ogawa, M. (2009). Foxo1 is essential for in vitro vascular formation from embryonic stem cells. *Biochem Biophys Res Commun*, Vol. 390, No. 3, 861-866.

Potente, M., Urbich, C., Sasaki, K., Hofmann, W.K., Heeschen, C., Aicher, A., Kollipara, R., DePinho, R.A., Zeiher, A.M. & Dimmeler, S. (2005). Involvement of Foxo transcription factors in angiogenesis and postnatal neovascularization. *J Clin Invest*, Vol. 115, No. 9, 2382-2392.

Risau, W., Sariola, H., Zerwes, H.G., Sasse, J., Ekblom, P., Kemler, R. & Doetschman, T. (1988). Vasculogenesis and angiogenesis in embryonic-stem-cell-derived embryoid bodies. *Development*, Vol. 102, No. 3, 471-478.

Seoane, J., Le, H.V., Shen, L., Anderson, S.A. & Massague, J. (2004). Integration of Smad and forkhead pathways in the control of neuroepithelial and glioblastoma cell proliferation. *Cell*, Vol. 117, No. 2, 211-223.

ten Dijke, P. & Arthur, H.M. (2007). Extracellular control of TGFbeta signalling in vascular development and disease. *Nat Rev Mol Cell Biol*, Vol. 8, No. 11, 857-869.

Tsukita, S., Furuse, M. & Itoh, M. (2001). Multifunctional strands in tight junctions. *Nat Rev Mol Cell Biol*, Vol. 2, No. 4, 285-293.

van der Horst, A. & Burgering, B.M. (2007). Stressing the role of FoxO proteins in lifespan and disease. *Nat Rev Mol Cell Biol*, Vol. 8, No. 6, 440-450.

Vittet, D., Prandini, M.H., Berthier, R., Schweitzer, A., Martin Sisteron, H., Uzan, G. & Dejana, E. (1996). Embryonic stem cells differentiate in vitro to endothelial cells through successive maturation steps. *Blood*, Vol. 88, No. 9, 3424-3431.

Wang, R., Clark, R. & Bautch, V.L. (1992). Embryonic stem cell-derived cystic embryoid bodies form vascular channels: an in vitro model of blood vessel development. *Development*, Vol. 114, No. 2, 303-316.

Yamashita, J., Itoh, H., Hirashima, M., Ogawa, M., Nishikawa, S., Yurugi, T., Naito, M., Nakao, K. & Nishikawa, S. (2000). Flk1-positive cells derived from embryonic stem cells serve as vascular progenitors. *Nature*, Vol. 408, No. 6808, 92-96.

Yang, X., Castilla, L.H., Xu, X., Li, C., Gotay, J., Weinstein, M., Liu, P.P. & Deng, C.X. (1999). Angiogenesis defects and mesenchymal apoptosis in mice lacking SMAD5. *Development*, Vol. 126, No. 8, 1571-1580.

31

Direct Differentiation of Human Embryonic Stem Cells toward Osteoblasts and Chondrocytes through an Intermediate Mesenchyme Progenitor Lineage

Basem M. Abdallah[1], Linda Harkness[1],
Amer Mahmood[1,2] and Moustapha Kassem[1,2]
[1]Molecular Endocrinology Laboratory (KMEB), Medical Biotechnology Centre and Department of Endocrinology,
University Hospital of Odense and University of South Denmark, Odense,
[2]Stem Cell Unit, Department of Anatomy, King Saud University, Riyadh,
[1]Denmark
[2]Saudi Arabia

1. Introduction

The recent establishment of in vitro cultures of human embryonic stem cells (hESC) from preimplantation embryos has provided a biologically relevant in vitro model for studying early human embryonic development and the signals involved in early stages of cellular commitment to different lineages. In addition, hESC represent a good source of an unlimited supply of cells that can generate clinical-grade, transplantable, lineage-specific cells to treat a large number of skeletal diseases including bone and cartilage diseases (e.g. localized bone and cartilage defects, non-healing fractures and systemic age-related degenerative conditions such as osteoporosis and osteoarthritis) (for review, [1], [2]). Typically, the basic in vitro methods for differentiating hESC into specific cell lineages are based on three procedures; 1) direct differentiation as a monolayer on extracellular matrix substrates, 2) differentiation in co-culture with primary tissue derived cells, and 3) the formation of 3-D spherical structures in suspension culture, termed embryoid bodies (EBs) [3-6]. However, developing clinically relevant protocols for directing the differentiation of hESC into definitive, homogenous populations of osteogenic or chondrogenic cells still faces several challenges due to the complexity of pathways that are conserved between embryonic and hESC differentiation for all cell types of the three germ layers. Our group, and others, have employed several strategies to first, generate mesenchymal progenitor cells (termed MSC) from hESCs demonstrating the characteristic phenotype of adult bone marrow-derived mesenchymal stem cells (BM-MSC), and second differentiate these cells into homogenous populations of osteoblast or chondrocytes by using protocols established for adult MSC. Thus, our objective in this chapter is to provide an overview on the majority of published protocols for derivation of osteoblasts and chondrocytes from hESC through an intermediate mesenchymal progenitor lineage.

2. Differentiation of hESC into mesenchymal progenitor cells

Multipotential mesenchymal stem/stromal cells (MSCs) have been isolated from diverse tissues including bone marrow (BM), adipose, muscle, periodontal ligament, umbilical cord blood and other connective tissues. MSCs are typically characterized by their adherance to plastic culture plates, expression of Stro-1, CD29, CD73, CD90, CD105, CD106, CD166, CD146, and CD44, lack of hematopoietic markers and MHC class II expression, and ability to differentiate into cells of mesenchymal origin such as bone, cartilage and adipose tissue [7]. MSC have documented evidence for their potential use in cell-based therapy for the treatment of bone and cartilage defects [8].

Derivation of MSC-like cells from hESC has previously been reported using a number of different approaches including direct differentiation employing withdrawal of feeders, addition of PDGF, or isolation and culture of spontaneous differentiation [9-13]; through an intermediate EB formation [14-16]; or by co-culturing hESC with primary bone derived cells [17], OP9 cells [18] or periodontal ligament fibroblasts [19]. In addition, methods were utilised for isolation of mesenchymal-like populations, based on cell sorting, employing a number of CD marker combinations: CD13+/-, CD71+, CD105+ [20] and CD105+, CD24- [21].

Culturing hESC as hEBs activates well-conserved cascades of genes that govern the earliest events during gastrulation and germ layer formation [5] that may be important for subsequent differentiation. Previous publications from mouse [22] and human [23] ESCs have shown that early mesoderm markers are expressed during hEB formation.

EB formation of hESC is initiated by enzymatic dissociation of hESC colonies into small clumps and differentiation is initiated by removal of bFGF and 2-mercaptoethanol, as well as by culturing the dissociated cells in ultra low adhesion culture dishes [24-26]. The length of time of EB culture is dependent upon the ultimate target cell type and EB differentiation appears to correlate well with post-implantation development of embryos [5].

Our profiling of these cells during 20 days of EB differentiation has confirmed up-regulation of several mesoderm specific markers *MEOX1, MIXL1, ALX4, TBX6* and *FOXF1* in association with increasing percentages of cells expressing MSC markers including CD29, CD44, CD63, CD73 and CD166 (fig. 1) [27]. Thus, differentiation of hESC via EB formation represents a promising method for directing the hESC not only to the mesoderm lineage but also to mesenchymal progenitor cells.

Since, TGFβ/BMP signaling pathways have been shown to play a role in mesoderm induction in vertebrates [28-30], we studied the effect of several members of the TGFβ/BMP family on mesoderm induction and derivation of MSC in the hESC-EBs model. We succeeded in developing clinically relevant protocols for generating MSC from hESC by blocking TGF-β signaling during hEB formation using SB-431542 (an inhibitor of activin receptor-like kinase (ALK)5, 4 and 7 (the TGF-β type I receptor) [27]. This led to selective up-regulation of several markers involved in mesoderm induction and myogenic differentiation including, *TBX6* and *Myf5* genes, additionally, these findings were confirmed by microarray analysis. In another study, we showed that the treatment of hESC-derived EBs with 50ng/ml Activin B (member of TGFb superfamily) for 10 days was sufficient to up-regulate several mesoderm specific markers (*FOXF1, VE-cadherin, KDR*) [31]. Further enrichment for MSC population was performed by establishing monolayer outgrowth culture from hEBs induced either by Activin B- or SB-induced. Our data showed that this strategy is very efficient in obtaining a homogenous MSC population with high percentage of cells expressing the characteristic surface markers for bone marrow derived-MSC including CD44, CD146, CD166, CD63, CD166 and CD73 [31].

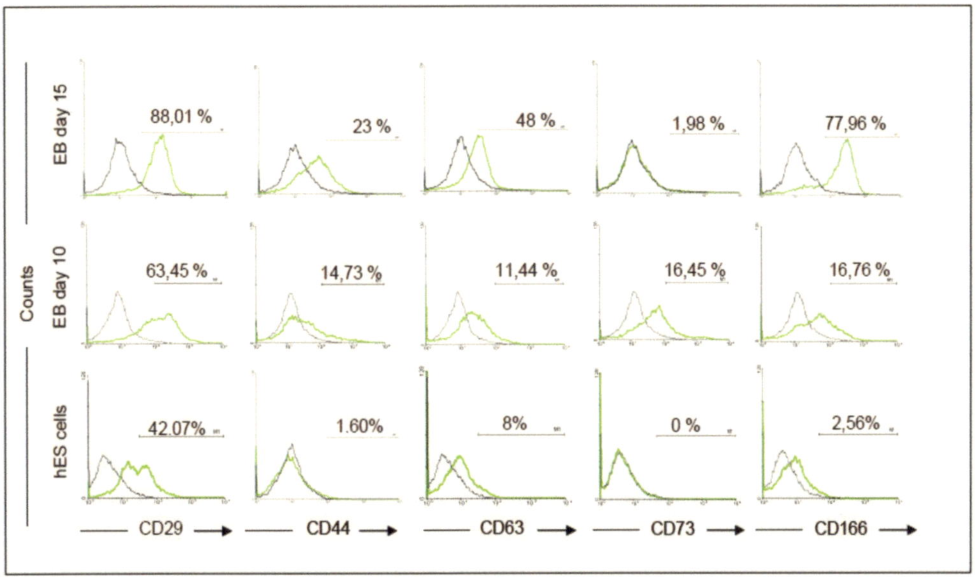

Fig. 1. FACS analysis of the MSC characteristic CD marker expression in hESC during their spontaneous differentiation as hEBs at day 10 and 15. The percentage of positive cells is indicated for each graph

3. Differentiation of hESC into chondrogenic lineages

Embryonic chondrogenesis is a complex biological process that plays essential roles in endochondral ossification and skeletal patterning. It involves migration of committed mesenchymal cells from the cranial neural crest, paraxial mesoderm and lateral plate mesoderm to the site of skeletogenesis to form cell mass condensations that define the morphology of the future bone [32, 33]. Inside these condensations, mesenchymal cells are differentiated into chondrocytes that proliferate and secrete an increasing amount of cartilage matrix macromolecules. Chondrogenesis is terminated through chondrocyte hypertrophy and apoptosis and is followed by blood vessel invasion and bone matrix formation by osteoblasts.

Current protocols for differentiating hESC into the chondrogenic lineage are principally based on differentiating hESC into the mesoderm/mesenchyme lineage via hEBs formation. This is followed by induction towards chondrogenesis using pellet micromass culture in combination with:TGFβ1, TGFβ3 BMP-2 and -4 [18, 31, 34], bovine chondrocyte conditioned medium [14] or via encapsulation in hydrogels [35], or arginine-glycine-aspartate (RGD)-modified PEGDA hydrogels [36]. Other published strategies include, co-culture of hESC-derived MSC with primary chondroprogenitor cells [37] and co-culture of enriched mesodermal cell populations ($CD73^{+}$) with murine OP9 cells [18]. However, most of these protocols failed to generate clinically compliant homogenous populations of chondrocytes that could be utilized for cartilage repair due either to their complexity or use of animal products such as fetal calf serum. Table 1 provides an overview on published chondrogenic differentiation protocols for hESC.

Initial Differentiation	Chondrogenic Differentiation	In vitro Analysis	In Vivo experiments	Ref
hESC were cultured as EBs for 10days followed by outgrowth culture on fibronectin coated plates in DMEM+15%KSR+ 50ng/ml Activin B for mesoderm induction. MSC-like cells were FACS sorted based of expressing Dlk1/FA1^{+}.	Pellet micromass culture of dlk1/FA1^{+}cells induced in CDM medium supplemented with 100 nM Dex+50ug/ml VitC+1%ITS+ 10 ng/ml TGFβ1.	-FACS analysis for MSC markers. -Alcian blue staining. -Real-time PCR	n/a	[31]
hESC were cultured as EBs for 10days followed by outgrowth culture on gelatine coated plates and subculturing at high density in DMEM+10% FBS to obtain MSC-like cells.	hESC-MSC induced in Argininge-glycine-aspartic acid (RGD)-modified PEG-based hydrogels in bovine chondrocyte conditioned media: DMEM 10% FBS + 1μm dex, 200μm Indometh, 10μg/ml insulin, 0.5mM IBMX	-Immune-staining for Col type 2. -Real-time PCR	-hESC-MSC encapsulated in PEGDA hydrogels and implanted s.c. in athymic mice. -Cartilage repair capacity of hESC-MSC pellets was evaluated in rat knee defect model.	[14]
hESC were plated on mouse stromal OP9 cells in αMEM with 20% FBS, followed by FACS sorting of CD73^{+}cells to obtain MSC-like cells.	Pellet micromass culture induced in DMEM with 10ng/ml TGFβ3+200μm Vit C in +10% FBS for 28 days.	-Alcian Blue staining. -Real-time PCR	n/a	[18]
hESC were cultured as EB for 5 days in DMEM/F12 +10% FBS, trypsinised and plated for outgrowth culture.	Cells induced in co-culture with human primary chondrocytes in DMEM + 10%FBS, +1mM L-Glut + 50μg/ml Vit C for 28days.	-Alcian Blue staining. -Immune-staining for Col type 2 and SOX9	Co-culture hESC seeded onto PDLLA foam and implanted s.c. in SCID mice for 35 days.	[16]

hESC were cultured as EB for 10 days and plated on gelatine coated plates in DMEM 10% FBS.	Pellet micromass cultures induced in DMEM +1mM L-glut, 100nm Dex+ 50μg/ml Vit C+ 1mM Na pyruvate + 1% ITS with 10ng/ml TGFβ1 or 25ng/ml BMP2 for 2 weeks .	-Safranin o - staining . -PCR	n/a	[36]
Undifferentiated hESC cells were co-cultured with irradiated neonatal or adult articular chondrocytes in high-density pellet mass cultures for 14 days, followed by monolayer expansion.	-Pellet micromass cultures of co cultured hESC were induced in DMEM with 10%FCS and 10ng/ml TGFβ3. -Co-cultures induced in hyaluronan-based Hyaff-11 scaffolds	-Safranin o -and Alcian Blue staining	n/a	[44]
hESC were cultured as EBs for 5 days in DMEM +10% FBS. Single cells were obtained either from the hESC aggregates or EBs day5.	hESC or 5-day EB-derived cells were resuspended at a high density in DMEM + ITS, 1.25mg/ml BSA+ 5.35μg/ml linoleic acid, 1% KSR+ 40μg/ml L-proline, 50μg/ml Vit C, 1% NEAA, 100nM Dex+100ng/ml BMP2 and cultured in micromass on gelatin coated plates.	-Immune-staining for Col type 2.	n/a	[45]
Dissociated single human ESCs were cultured and passaged on gelatin-coated plates in DMEM + 10% FBS, 5ng/ml FGF2.	Pellet micromass culture of hESC aggregates induced in DMEM + 0.1μM dex+ 50μM Vit C+ 40μg/ml proline, 1mM Na pyruvate+ 1% ITS+ premix, and/or 10ng/ml TGFβ1 or 300ng/ml BMP7 for 14days.	-Toluidine blue staining, -Col type 2 staining -RT-PCR	n/a	[46]

Used the same protocol as [45]	Used the same protocol as [45] except used 10ng/ml of TGFβ1 or 1 μM SB431542 instead of BMP2 in chondrogenic media to study the effect of TGFβ signaling pathway.	Real-time PCR Col2 immune-staining	n/a	[47]
hESC were cultured on Matrigel and differentiated as EBs in chondrogenic media consisting of DMEM with 100nM Dex+1% ITS+ 40 μg/ml L-proline, 50 μg/ml VitC+ 100 μg/ml sodium pyruvate and 1 ng/ml TGFβ1. EBs were cultured in this medium for 3 weeks.		col1, 2 and GAG staining. Mechanical testing using unconfined creep cytocompression	n/a	[48]

Abbreviations used, EB: embryoid bodies; KSR:knockout serum replacement; FBS :fetal bovine serum; CDM: chemically defined medium; Dex: dexamethasone; VitC: ascorbic acid/vitamin C; RGD: arginine-glycine-aspartic acid; PEG: poly ethylene glycol; Indometh: indomethacin; IBMX: 3-isobutyl-1-methylxanthine; NEAA: non essential amino acids; PEGDA: Poly (Ethylene Glycol)-Diacrylate; s.c: subcutaneously; PDLLA: Poly-DL-lactic acid

Table 1. hESC chondrogenic differentiation protocols

We have recently identified dlk1/FA1 as a surface marker for chondrogenesis that indicates transition of immature proliferating to pre-mature hypertrophic chondrocytes during mouse embryonic limb development as well as in vivo hESC-derived teratoma formation. Our data showed that Dlk1 is a dynamic surface marker and responds to some external stimuli controlling chondrogenesis. dlk1/FA1 (delta-like 1 protein, fetal antigen 1, also named preadipocyte factor 1 [Pref-1]) is a membrane-associated protein belonging to Notch/Serrata/Delta family [38]. Dlk1/FA1 plays an important role in controlling the cell fate decisions during embryogenesis [39] and extends its function to regulate many mesoderm differentiation processes in the post-natal organism including adipogenesis [40], myogenesis [41] and osteoblastogenesis [42].

We have designed a clinically relevant protocol for directing the differentiation of hESC into the definitive early chondrogenic lineage based on tracking the expression of dlk1/FA1 as a mesoderm/chondroprogenitor surface marker. In this protocol, treatment of hESC-EB cultures with Activin B is shown to markedly up-regulate dlk1/FA1 expression in association with increasing several mesoderm induction markers (i.e. *FOXF1, KDR* and *MEOX1*) [31]. By establishing an hEB-outgrowth culture we were able to enrich for cells expressing Dlk1/FA1 and obtain a homogenous population of mesenchymal/chondroprogenitor cells that could be further differentiated as micromass cultures into chondrocytes in serum free medium containing TGFβ1 [31]. Thus, there is a need to identify novel surface markers that define early stages of hESC commitment to chondrocytes. Tracking these markers during the progression of hESC-EBs formation will be valuable for developing well-defined and efficient protocols for directing hESC differentiation into the chondrogenic lineage in vitro.

4. Differentiation of hESC into osteogenic lineages

Osteoblasts originate from mesenchymal stem cells in bone marrow through a differentiation process that is controlled by numerous hormones and growth factors. Osteoblasts are characterized by expressing various phenotypic markers such as high alkaline phosphatase (ALP) activity and synthesizing collagenous and noncollagenous bone matrix proteins including osteocalcin. The most important function of osteoblasts is to form mineralized bone [43].

Initial Differentiation	Osteogenic Induction	In vitro assays	In Vivo assays	Ref
hESC were cultured as EBs for 10days followed by outgrowth culture on fibronectin coated plates in DMEM+15%KSR+ 10 µM SB421543. hEBs-outgrowth were passaged and maintained in CDM medium for enrichment for MSC.	hESC-MSC induced in αMEM medium with 10%FCS+10mM β-glyc+ 100µg/ml Vit C for 20 days.	-FACS analysis for MSC surface markers. -Alzarin Red and ALP staining. -Real-time PCR.	hESC-MSC were mixed with HA/TCP s.c. into Nod/SCID mice for 8 weeks.	[27]
hESC were cultured as EBs for 10 days and plated for outgrowth culture on gelatine coated plates in DMEM+10% FBS+2mM L-Glut to obtain MSC-like cells.	hESC-MSC induced in DMEM media with 50µM Vit C + 10mM β-glyc + 100nM Dex in for 14 d	-Alzarin Red and ALP staining. -Real-time PCR.	hESC-MSC were seeded onto the polymer scaffold of PLLA/PLGA, cultured for 10 days and implanted s.c. in athymic mice for 8 weeks.	[14]
hESC were plated on mouse stromal OP9 cells in αMEM with 20% FBS, followed by FACS sorting of $CD73^+$cells.	$CD73^+$cells induced in αMEM medium with 10% FBS+10mM β-glyc+ 0.1µM Dex+ 200µM Vit C for 3 to 4weeks.	-Alzarin Red and -Von Kossa staining. -Real-time PCR.	n/a	[18]
hESC were cultured as EBs for 3 to 4 days, trypsinised and plated onto gelatin plates in KODMEM + 10% FBS, L-Glut, NEAA, BME	Cells induced in KODMEM + 10% FBS+ BME +50µM Vit C, 10mM β-glyc +100nM Dex	-Von Kossa staining	Cells induced for 4d, injected into diffusion chamber and implanted into nude mice for 7 weeks.	[15]

hESC were cultured as EBs for 7 days, and plated onto gelatin coated plates in αMEM with 10% FBS+ 200mM L-glut+10mM NEAA+4ng/ml FGF for up to 2 weeks.	hESC-MSC were induced as described [18]	-Von Kossa staining	n/a	[49]
hEBs day3 were plated with inactivated human primary bone derived cells (hPBDs).	The co-culture of hEBs-derived cells with hPBDs was continued for 2 weeks.	-Real Time PCR	Cells seeded on PLGA/HA scaffolds implanted with BMP2 s.c. into SCID mice for 4 & 8 weeks.	[17]
Removal of MSC-like cells from long term culture of hESC colonies in feeder free system.	hESC-MSC induced in αMEM with 20% FBS +10nM Dex+ 0.2 mM VitC+ 10 mM +β-glyc for 3 weeks	-Alzarin Red staining.	n/a	[20]
hESC aggregates were removed from the MEF and placed onto a tissue culture dish for 24 hours, followed by single-cell suspensions and subsequent plating.	Cells were induced in aMEM with FBS was supplemented with 10nM DEX+ 50μg/ml VitC + 5 mM β-glyc	-Von Kossa and ALP staining .	n/a	[50]
hESC single cell suspension plated on gelatin coated plate in DMEM-HG, L-Glut+ 10% FBS+ 10ng/ml FGF. Continuous passaging to obtain hESC-MSC.	hESC-MSC induced in DMEM + 1% PEST, L-glutamine (2mM), 10% FCS+ 45 nM VitC+ 100nM dex+ 20mM β-glyc for 6 weeks.	-Von Kossa staining	n/a	[51]
hESC were cultured as EBs for 10 days, and plated onto gelatin plates in aMEM + 20% FBS, NEAA, L-Glut, 1ng/ml FGF. hESC-MSC outgrowth cells taken to p4	hESC-MSC induced in DMEM, 10%FBS, 50 μg/mL Vit C, 10mM β-glyc +100nM Dex or 50ng/ml BMP7 or Dex+BMP7 for 4 weeks. Cells induced also on Scaffolds (6 weeks), matrices or film cultured with Dex+BMP7.	-Von Kossa and ALP staining	n/a	[52]

hESC were cultured as EBs for 5 days in DMEM/F12, 10% KSR, NEAA, BME. EBs were placed on Matrigel coated plated .	Attached EBs were induced in medium with 0.1 mM VitC+ 10mM β-glyc, 0.1 mM dex +/- 5μM LY294002+ 0.5μM AKT inhibitor,+1nM Rapamycin or 10nM FK506 (PI3K/AKT/mTOR inhibitors) for 3 weeks.	-Von Kossa and ALP staining. -Real time-PCR.	n/a	[53]
Prolonged in vitro cultivation and starvation for hESC mesenchymal differentiation based on method used by [20].	Medias: #1 KODMEM + 20% KSR, with/without FBS (hi or non hi) or 5% hPL + 40u/ml heparin; + 1% ITS+, and/or 10^{-8} M Dex, and 10^{-4} M L-ascorbic acid phosphate magnesium salt n-hydrate (AscP). 32-45days #2 aMEM +10% FBS +/- Dex/AscP 45-69days #3 KODMEDM + 20 KSR, +/- FGF or media #2, 53 days	-Real time-PCR.	Differentiated hESC using different cocktails were mixed with HA/TCP and implanted s.c. in immunodeficient mice.	[54]

Abbreviations used, β-glyc: beta glycerophosphate; KODMEM: Knockout DMEM; BME: BME amino acids; HA/TCP: Hydroxyapatite tri calcium phosphate; hi: heat inactivated

Table 2. hESC osteogenic differentiation protocols

A number of cell culture conditions have been used to induce differentiation of hESCs into the osteogenic lineage with and without EB formation. Osteoblastic differentiation of hESCs has principally been achieved by differentiating hESC into MSC-like cells followed by induction of the hESC-MSC to the osteogenic lineage using a well known osteogenic cocktail containing ascorbic acid, β-glycerophosphate, and dexamethasone, or through co-culture with primary bone derived cells (more detailed protocols are shown in table 2). However, the majority of these protocols still show some limitations in providing homogeneous cell populations which are efficient in differentiating into osteoblasts and could be used in tissue engineering for skeletal regeneration. These limitations include; 1) insufficient data comparing the phenotype of hESC-derived osteoblasts with well-established osteoblastic cells; 2) using in vitro assays as the main criteria for evaluation of the osteogenic differentiation protocols; 3) small numbers of published in vivo studies examining the capacity of hESC-derived osteoblasts for bone regeneration; 4) little or no available data on the reproducibility of a specific protocol on different hESC cell lines.

Recently, we have reported an efficient protocol for directing hESC into mesenchymal osteo-progenitors by inhibiting TGF-β signaling with SB-431542 (SB) during hEBs formation.

Detailed cellular and molecular analysis revealed the differentiation of SB-treated hEBs into muscle progenitor cells (MPC)[27]. Successive outgrowth cultures of the MCP, in the presence of 10% fetal bovine serum (FBS), demonstrated a step-wise progression from a heterogeneous EB outgrowth cell population, through an MCP population, to a homogeneous population of mesenchymal progenitors that expressed CD markers characteristic of mesenchymal stem cells (MSCs): CD44 (100%), CD73 (98%), CD146 (96%), and CD166 (88%). Lastly these mesenchymal progenitors verified their ability to differentiate into osteoblasts in vitro and could form ectopic bone upon subcutaneous implantation with hydroxyl-apatite/tricalcium phosphate ceramic powder (HA/TCP) in immune deficient mice [27].

5. Conclusions

Due to the unlimited self-renewal and differentiation capacity of hESC-MSC over adult MSC the derivation of MSC from hESC for skeletal regeneration demonstrates a promising alternative to the use of adult stem cells. Despite the increasing number of published protocols for differentiating hESC into definitive osteogenic and chondrogenic lineages, the use of hESC in tissue engineering for bone/cartilage repair in pre-clinical studies is elusive due to either the complexity of available protocols or lack of data on testing the functional activity of cells in vivo. More work is needed, to design robust xeno-free protocols that can provide large numbers of homogenous and efficiently differentiated osteoblastic /chondrocytic cells that work on multiple hESC cell lines. To achieve this goal, new strategies need to be developed to improve the in vitro and in vivo assessment assays, to develop a predictive molecular signature for each particular cell type at different differentiation stages, and to identify new and dynamic surface markers that recognize early chondro-progenitor and osteo-progenitor cells during hESC differentiation.

6. Acknowledgments

This work was supported by grants from the Lundbeck foundation, the Danish Medical Research Council and the Novo Nordisk Foundation.

7. Reference

[1] A.Trounson, M.Pera, Human embryonic stem cells Fertil.*Steril.* 76, (2001) 660-661.

[2] M.X.Doss, C.I.Koehler, C.Gissel, J.Hescheler, A.Sachinidis, Embryonic stem cells: a promising tool for cell replacement therapy, *J.Cell Mol.Med. 8*, (2004) 465-473.

[3] C.M.Cameron, W.S.Hu, D.S.Kaufman, Improved development of human embryonic stem cell-derived embryoid bodies by stirred vessel cultivation, *Biotechnol.Bioeng.* 94, (2006) 938-948.

[4] J.Itskovitz-Eldor, M.Schuldiner, D.Karsenti, A.Eden, O.Yanuka, M.Amit, H.Soreq, N.Benvenisty, Differentiation of human embryonic stem cells into embryoid bodies comprising the three embryonic germ layers, *Mol.Med.* 6, (2000) 88-95.

[5] G.Keller, Embryonic stem cell differentiation: emergence of a new era in biology and medicine *Genes & Development* 19, (2005) 1129-1155.

[6] A.Kubo, K.Shinozaki, J.M.Shannon, V.Kouskoff, M.Kennedy, S.Woo, H.J.Fehling, G.Keller, Development of definitive endoderm from embryonic stem cells in culture, *Development* 131, (2004) 1651-1662.

[7] B.M.Abdallah, M.Kassem, Human mesenchymal stem cells: from basic biology to clinical applications, *Gene Ther.* 15, (2008) 109-116.

[8] B.M.Abdallah, M.Kassem, The use of mesenchymal (skeletal) stem cells for treatment of degenerative diseases: current status and future perspectives, *J.Cell Physiol* 218, (2009) 9-12.

[9] N.L.Boyd, K.R.Robbins, S.K.Dhara, F.D.West, S.L.Stice, Human Embryonic Stem Cell-Derived Mesoderm-like Epithelium Transitions to Mesenchymal Progenitor Cells Tissue, *Engineering* Part A 15, (2009) 1897-1907.

[10] B.C.Heng, W.S.Toh, B.P.Pereira, B.L.Tan, X.Fu, H.Liu, K.Lu, J.F.Yeo, T.Cao, An autologous cell lysate extract from human embryonic stem cell (hESC) derived osteoblasts can enhance osteogenesis of hESC, *Tissue and Cell* 40, (2008) 219-228.

[11] C.Karlsson, K.Emanuelsson, F.Wessberg, K.Kajic, M.Z.Axell, P.S.Eriksson, A.Lindahl, J.Hyllner, R.Strehl, Human embryonic stem cell-derived mesenchymal progenitors--Potential in regenerative medicine, *Stem Cell Research* 3, (2009) 39-50.

[12] I.Mateizel, A.De Becker, H.Van de Velde, M.De Rycke, A.Van Steirteghem, R.Cornelissen, J.Van der Elst, I.Liebaers, I.Van Riet, K.Sermon, Efficient differentiation of human embryonic stem cells into a homogeneous population of osteoprogenitor-like cells, *Reproductive BioMedicine Online* 16, (2008) 741-753.

[13] P.Trivedi, P.Hematti, Derivation and immunological characterization of mesenchymal stromal cells from human embryonic stem cells, *Experimental Hematology* 36, (2008) 350-359.

[14] N.S.Hwang, S.Varghese, H.J.Lee, Z.Zhang, Z.Ye, J.Bae, L.Cheng, J.Elisseeff, In vivo commitment and functional tissue regeneration using human embryonic stem cell-derived mesenchymal cells, *Proc.Natl.Acad.Sci.U.S.A* 105, (2008) 20641-20646.

[15] J.L.Tremoleda, N.R.Forsyth, N.S.Khan, D.Wojtacha, I.Christodoulou, B.J.Tye, S.N.Racey, S.Collishaw, V.Sottile, A.J.Thomson, A.H.Simpson, B.S.Noble, J.McWhir, Bone tissue formation from human embryonic stem cells in vivo, *Cloning Stem Cells* 10, (2008) 119-132.

[16] A.Vats, R.C.Bielby, N.Tolley, S.C.Dickinson, A.R.Boccaccini, A.P.Hollander, A.E.Bishop, J.M.Polak, Chondrogenic differentiation of human embryonic stem cells: the effect of the micro-environment, *Tissue Eng* 12, (2006) 1687-1697.

[17] S.Kim, S.S.Kim, S.H.Lee, A.S.Eun, S.J.Gwak, J.H.Song, B.S.Kim, H.M.Chung, In vivo bone formation from human embryonic stem cell-derived osteogenic cells in poly(d,l-lactic-co-glycolic acid)/hydroxyapatite composite scaffolds, *Biomaterials* 29, (2008) 1043-1053.

[18] T.Barberi, L.M.Willis, N.D.Socci, L.Studer, Derivation of multipotent mesenchymal precursors from human embryonic stem cells, *PLoS.Med.* 2, (2005) e161.

[19] B.Inanç, A.E.Elçin, Y.M.Elçin, Effect of Osteogenic Induction on the in Vitro Differentiation of Human Embryonic Stem Cells Cocultured With Periodontal Ligament Fibroblasts, *Artificial Organs* 31, (2007) 792-800.

[20] E.N.Olivier, A.C.Rybicki, E.E.Bouhassira, Differentiation of human embryonic stem cells into bipotent mesenchymal stem cells, *Stem Cells* 24, (2006) 1914-1922.

[21] Q.Lian, E.Lye, K.Suan Yeo, E.Khia Way Tan, M.Salto-Tellez, T.M.Liu, N.Palanisamy, R.M.El Oakley, E.H.Lee, B.Lim, S.K.Lim, Derivation of Clinically Compliant MSCs from CD105+, CD24- Differentiated Human ESCs, *Stem Cells* 25, (2007) 425-436.

[22] H.J.Fehling, G.Lacaud, A.Kubo, M.Kennedy, S.Robertson, G.Keller, V.Kouskoff, Tracking mesoderm induction and its specification to the hemangioblast during embryonic stem cell differentiation, *Development* 130, (2003) 4217-4227.

[23] V.Sottile, A.Thomson, J.McWhir, In vitro osteogenic differentiation of human ES cells, *Cloning Stem Cells* 5, (2003) 149-155.

[24] U.Frandsen, A.D.Porneki, C.Floridon, B.M.Abdallah, M.Kassem, Activin B mediated induction of Pdx1 in human embryonic stem cell derived embryoid bodies, *Biochem Biophys Res Commun* 362, (2007) 568-574.

[25] G.M.Keller, In vitro differentiation of embryonic stem cells, *Curr Opin.Cell Biol* 7, (1995) 862-869.

[26] A.G.Smith, Embryo-derived stem cells: of mice and men, *Annu.Rev Cell Dev Biol* 17, (2001) 435-462.

[27] A.Mahmood, L.Harkness, H.D.Schroder, B.M.Abdallah, M.Kassem, Enhanced differentiation of human embryonic stem cells to mesenchymal progenitors by inhibition of TGF-beta/activin/nodal signaling using SB-431542, *J.Bone Miner.Res.* 25, (2010) 1216-1233.

[28] T.Era, N.Izumi, M.Hayashi, S.Tada, S.Nishikawa, S.I.Nishikawa, Multiple Mesoderm Subsets Give Rise to Endothelial Cells Whereas Hematopoietic Cells are Differentiated only From a Restricted Subset in ES Cell Differentiation Culture, *Stem Cells* (2007).

[29] P.Gadue, T.L.Huber, P.J.Paddison, G.M.Keller, Wnt and TGF-beta signaling are required for the induction of an in vitro model of primitive streak formation using embryonic stem cells, *Proc.Natl.Acad.Sci.U.S.A* 103, (2006) 16806-16811.

[30] D.Kimelman, Mesoderm induction: from caps to chips, *Nat.Rev.Genet.* 7, (2006) 360-372.

[31] L.Harkness, H.Taipaleenmaki, A.Mahmood, U.Frandsen, A.M.Saamanen, M.Kassem, B.M.Abdallah, Isolation and Differentiation of Chondrocytic Cells Derived from Human Embryonic Stem Cells Using dlk1/FA1 as a Novel Surface Marker, *Stem Cell Rev.Rep.*(2009).

[32] R.Cancedda, P.Castagnola, F.D.Cancedda, B.Dozin, R.Quarto, Developmental control of chondrogenesis and osteogenesis, *Int.J.Dev.Biol.* 44, (2000) 707-714.

[33] V.Lefebvre, P.Smits, Transcriptional control of chondrocyte fate and differentiation, *Birth Defects Res.C.Embryo.Today* 75, (2005) 200-212.

[34] J.Kramer, C.Hegert, K.Guan, A.M.Wobus, P.K.Muller, J.Rohwedel, Embryonic stem cell-derived chondrogenic differentiation in vitro: activation by BMP-2 and BMP-4, *Mech.Dev.* 92, (2000) 193-205.

[35] N.S.Hwang, S.Varghese, P.Theprungsirikul, A.Canver, J.Elisseeff, Enhanced chondrogenic differentiation of murine embryonic stem cells in hydrogels with glucosamine, *Biomaterials* 27, (2006) 6015-6023.

[36] N.S.Hwang, S.Varghese, Z.Zhang, J.Elisseeff, Chondrogenic differentiation of human embryonic stem cell-derived cells in arginine-glycine-aspartate-modified hydrogels, *Tissue Eng* 12, (2006) 2695-2706.

[37] Y.Sui, T.Clarke, J.S.Khillan, Limb bud progenitor cells induce differentiation of pluripotent embryonic stem cells into chondrogenic lineage, *Differentiation* 71, (2003) 578-585.

[38] S.J.Bray, S.Takada, E.Harrison, S.C.Shen, A.Ferguson-Smith, The atypical mammalian ligand Delta-like1 (Dlk1) can regulate Notch signalling in Drosophila, *BMC.Dev.Biol.* 8, (2008) 11.

[39] J.Laborda, The role of the epidermal growth factor-like protein dlk in cell differentiation 3, *Histol.Histopathol.* 15, (2000) 119-129.

[40] C.M.Smas, H.S.Sul, Molecular mechanisms of adipocyte differentiation and inhibitory action of pref-1, *Crit Rev.Eukaryot.Gene Expr.* 7, (1997) 281-298.

[41] R.M.Crameri, H.Langberg, P.Magnusson, C.H.Jensen, H.D.Schroder, J.L.Olesen, C.Suetta, B.Teisner, M.Kjaer, Changes in satellite cells in human skeletal muscle after a single bout of high intensity exercise, *Journal of Physiology-London* 558, (2004) 333-340.

[42] B.M.Abdallah, C.H.Jensen, G.Gutierrez, R.G.Leslie, T.G.Jensen, M.Kassem, Regulation of human skeletal stem cells differentiation by Dlk1/Pref-1, *J.Bone Miner.Res.* 19, (2004) 841-852.

[43] M.Kassem, B.M.Abdallah, H.Saeed, Osteoblastic cells: differentiation and trans-differentiation, *Arch.Biochem.Biophys.* 473, (2008) 183-187.

[44] N.Bigdeli, C.Karlsson, R.Strehl, S.Concaro, J.Hyllner, A.Lindahl, Coculture of human embryonic stem cells and human articular chondrocytes results in significantly altered phenotype and improved chondrogenic differentiation, *Stem Cells* 27, (2009) 1812-1821.

[45] W.S.Toh, Z.Yang, H.Liu, B.C.Heng, E.H.Lee, T.Cao, Effects of culture conditions and bone morphogenetic protein 2 on extent of chondrogenesis from human embryonic stem cells, *Stem Cells* 25, (2007) 950-960.

[46] T.Nakagawa, S.Y.Lee, A.H.Reddi, Induction of chondrogenesis from human embryonic stem cells without embryoid body formation by bone morphogenetic protein 7 and transforming growth factor beta1, *Arthritis Rheum.* 60, (2009) 3686-3692.

[47] Z.Yang, L.Sui, W.S.Toh, E.H.Lee, T.Cao, Stage-Dependent Effect of TGFβ1 on Chondrogenic Differentiation of Human Embryonic Stem Cells, *Stem Cells and Development* 18, (2009) 929-940.

[48] G.Ofek, V.P.Willard, E.J.Koay, J.C.Hu, P.Lin, K.A.Athanasiou, Mechanical characterization of differentiated human embryonic stem cells, *J.Biomech.Eng* 131, (2009) 061011.

[49] S.E.Brown, W.Tong, P.H.Krebsbach, The Derivation of Mesenchymal Stem Cells from Human Embryonic Stem Cells, *Cells Tissues Organs* 189, (2009) 256-260.

[50] J.M.Karp, L.S.Ferreira, A.Khademhosseini, A.H.Kwon, J.Yeh, R.S.Langer, Cultivation of human embryonic stem cells without the embryoid body step enhances osteogenesis in vitro, *Stem Cells* 24, (2006) 835-843.

[51] G.M.de Peppo, P.Sjovall, M.Lenneras, R.Strehl, J.Hyllner, P.Thomsen, C.Karlsson, Osteogenic Potential of Human Mesenchymal Stem Cells and Human Embryonic Stem Cell-Derived Mesodermal Progenitors: A Tissue Engineering Perspective, *Tissue Eng Part A* 16 (2010) 3413-3426.

[52] J.Hu, L.A.Smith, K.Feng, X.Liu, H.Sun, P.X.Ma, Response of Human Embryonic Stem Cells Derived Mesenchymal Stem Cells to Osteogenic Factors and Architectures of Materials during in vitro, *Osteogenesis Tissue Eng Part A* 16 (2010) 3507-3514.

[53] K.W.Lee, J.Y.Yook, M.Y.Son, M.J.Kim, D.B.Koo, Y.M.Han, Y.S.Cho, Rapamycin Promotes the Osteoblastic Differentiation of Human Embryonic Stem Cells by Blocking the mTOR Pathway and Stimulating the BMP/Smad Pathway, *Stem Cells and Development* 19, (2010) 557-568.

[54] S.Kuznetsov, N.Cherman, R.P.Gehron, In Vivo Bone Formation by Progeny of Human Embryonic Stem Cells, *Stem Cells Dev.*(2010).

32

From Pluripotent Stem Cells to Lineage-Specific Chondrocytes: Essential Signalling and Cellular Intermediates

Naoki Nakayama and Katsutsugu Umeda
Brown Foundation Institute of Molecular Medicine,
The University of Texas Health Science Center at Houston
USA

1. Introduction

Cartilage is a type of connective tissue found throughout the body. The three major types of cartilage are distinguished by their physical properties, which reflect their relative contents of the extracellular matrices: proteoglycan, elastin fiber and collagen. Hyaline cartilage, which makes smooth joint surfaces at the ends of bone, is hard and translucent. It is rich in type II collagen and hydrophilic proteoglycans, and highly resistance to pressure and shear forces. Elastic cartilage, found in the epiglottis and the Eustachian tube, contains large amounts of elastin fibers throughout the matrix, making it stiff but elastic. Fibrocartilage is often found in areas subjected to frequent compressive stress, such as intervertebral discs and menisci. It contains more collagen (especially type I collagen) and less proteoglycan than hyaline cartilage, enabling fibrocartilage to resist pressure effectively.

Among the many orthopedic conditions, osteoarthritis is one of the most common degenerative joint disorders. Osteoarthritis progresses slowly but steadily to the stage where articular cartilage has to be entirely regenerated or replaced with an artificial joint. Traditional treatments for osteoarthritis deal with the symptoms. While current treatments might offer transient relief of pain, no regenerative therapy has yet been established. Instead, the best available treatment for patients with advanced osteoarthritic disease is surgery to provide an artificial joint. Joint articular cartilage lacks blood vessels and nerves, so damaged cartilage is not spontaneously healed by resident, (if they exist), or nearby (e.g. synovial) chondroprogenitors. In the current climate, chondroprogenitor transplantation is considered a promising therapy for the regeneration of articular cartilage.

Focal damage to articular cartilage, such as that caused by sports injury, can be repaired by regenerative surgery such as the microfracture method, which has been performed for many years. Injured cartilage is removed, subchondral bone is exposed and several tiny holes (microfractures) are created in the bone to cause bone marrow cells and blood to clot and cover the cleared area, thereby promoting healing and tissue repair.

Osteochondral grafting is also performed, in which full-thickness cylindrical plugs of articular cartilage attached to the subchondral bone are directly inserted into matching holes drilled in the chondral defect. Recently, cell-based therapies such as autologous chondrocyte implantation (ACI) have been developed for use in various clinical settings (Brittberg et al.,

1994; Gikas et al., 2009; Moriya et al., 2007; Roberts et al., 2009). ACI involves clearing the damaged site and covering it with a periosteal patch or a collagen membrane, under which expanded articular chondrocytes are placed. ACI therefore requires the sampling of chondrocytes from normal area of the patient's articular cartilage and their expansion in culture to obtain enough cells for treatment. The drawback of the procedure is that *in vitro* expansion causes de-differentiation of the cells (i.e. loss of chondrocytic phenotypes). Most of the clinical trials of these methods published to date have reported unsatisfactory results. The endogenous marrow cells or implanted de-differentiated chondrocytes used for the microfracture method and ACI, respectively, have shown limited capacity for proliferation and re-differentiation, and tended to produce fibrocartilaginous tissue that fills the cleared, damaged sites or the spaces around the plugs, but is poorly integrated into the surrounding hyaline cartilage. The osteochondral grafting method suffers from a similar integration problem. As a result, the repaired site is often destabilized by tensile and shear forces and lost as a result, leading to the need for subsequent surgery (Steinert et al., 2007).

2. Stem cells for articular cartilage regeneration

Chondrocytes developed from endogenous marrow cells or from those once dedifferentiated *in vitro* apparently fail to make hyaline cartilage matrix *in vivo,* even after a long-period of time. It is unclear whether the cells of either origin are intrinsically incapable of generating hyaline cartilage, or the environment of diseased/damaged sites is not conducive to the formation of hyaline cartilage. To date, no attempt to overcome the problem has been successful. Furthermore, ACI has the additional drawbacks of 1) requiring the isolation of chondrocytes from the patient's normal cartilage, thereby introducing the long-term risk of the procedure itself inducing a degenerative cartilage disorder, and 2) requiring an effective method, not yet available, for the expansion/dedifferentiation of chondrocytes before transplantation that maintains their redifferentiation capacity. A better source of chondrocytes for joint repair and better knowledge of the critical environmental cues for the development of hyaline cartilage by chondrocytes are clearly needed (Bianco et al., 2008).

2.1 Adult stem cells

Mesenchymal stem cells (MSCs), originally isolated from bone marrow (Pittenger et al., 1999), are defined *in vitro* based on their tri-lineage (i.e. bone, cartilage and fat) differentiation potential and capacity to grow in a clonal fashion (i.e. MSC represents colony-forming unit-fibroblast or CFU-F). Similar cells are known to reside in other adult tissues such as adipose tissue, skeletal muscle, periosteum, synovial membrane, and the dental pulp of deciduous baby teeth (Asakura et al., 2001; De Bari et al., 2001a; De Bari et al., 2001b; Gronthos et al., 2000; Zuk et al., 2002). Since MSCs are capable of proliferating and differentiating into chondrocytes *in vitro* and *in vivo,* they were expected to contribute to regenerative therapies for cartilage. However, as for other adult stem cells, the major drawbacks of MSCs for clinical purposes are 1) MSCs are present at low incidence in adult tissues, necessitating *in vitro* expansion before transplantation, and 2) MSCs have a strong proliferative capacity that diminishes over time; however during proliferation their chondrogenic potential is not stable. As for chondrocytes, these limitations make it difficult to prepare sufficient fully functional MSCs for transplantation. Furthermore, the yield, proliferation capacity, and differentiation potential of MSCs decrease with advancing age, limiting the usefulness of clinical application of autologous MSCs in the patient group that would most benefit, namely aged adults (Murphy et al., 2002; Tokalov et al., 2007).

Despite the limitations described, MSCs have been used instead of dedifferentiated chondrocytes for ACI. However, to date, the reported results show that the modified method is no better than conventional ACI in the longer term. The MSC treatment, in combination with growth/differentiation factor-treatment, demonstrates a benefit at 1-2 months, after which the MSC-derived chondrocytes are lost or have matured into hypertrophic chondrocytes that form osteophytes (Steinert et al., 2007).

2.2 Pluripotent stem cells: embryonic stem cells and induced pluripotent stem cells

Embryonic stem (ES) cells were first established from the inner cell mass of mouse blastocysts (Evans & Kaufman, 1981; Martin, 1981). They are described as 'pluripotent' because when injected back into blastocysts, ES cells contribute to all somatic tissues formed during subsequent embryogenesis. Eighteen years later, ES-like cells from human blastocysts were established (Thomson et al., 1998), although their pluripotency has only been inferred based on their teratoma forming activity in immunocompromised mice. Human ES cells have two advantages over adult stem cells for transplantation therapy. They proliferate in culture without the loss of differentiation capacity over a prolonged period, allowing the production of a large number of stem cells. Second, they are theoretically capable of differentiating into any somatic cell types in culture, given the right growth conditions, i.e. those that mimic stages of embryogenesis (reviewed in Gadue et al., 2005; Keller, 2005; Nishikawa et al., 2007).

With the exception of growth plate chondrogenesis, which continues into early adolescence, chondrogenesis is largely completed before birth. Robust chondrogenesis during embryogenesis is a multi-step process involving development of chondrogenic mesenchyme, precartilage condensation of the mesenchyme, the development of chondrocytes, and their maturation into different types. The chondrogenic mesenchymal cells migrate, proliferate and condense through interaction with the adjacent epithelium. In precartilage condensation, cells continue to proliferate and start to differentiate toward chondroblasts, while depositing extracellular matrix containing type II, IX, X and XI collagens and aggrecan. The chondroblasts then proliferate and differentiate into chondrocytes. Finally, growth plate chondrocytes mature into the larger "hypertrophic" chondrocytes that are destined for mineralization. These processes are known to be regulated by hormones and growth factors, such as transforming growth factor (TGF) β, Wnts, fibroblast growth factors (FGFs), bone morphogenetic proteins (BMPs), insulin-like growth factors (IGFs), sonic/indian hedgehog (Shh/Ihh), and parathyroid hormone-related peptide (PTHrP) (reviewed in Goldring et al., 2006).

Embryonic chondroprogenitors thus have potential as a promising alternative for cartilage cell therapy. Due to the large number of cells required for clinical purposes, embryonic tissues are not a practical source of chondroprogenitors. For humans, progeny derived from human ES cells seem to be the only source of embryonic cells. Major impediments to the clinical use of human ES cell progeny are the risk that contamination of undifferentiated ES cells could lead to teratoma formation in patients, and the limited immunological diversity of available human ES cell lines. The technical and ethical issues involved in producing more such lines have created the need for allogeneic transplantation, which carries the risk of immunological complications.

There is now a solution to the problem of too few human ES cell lines. In the past couple of years, a nuclear reprogramming technology has been established that enables the induction of ES cell-like pluripotent stem (PS) cells from mouse and human somatic cells (Park et al., 2008; Takahashi & Yamanaka, 2006; Takahashi et al., 2007; Yu et al., 2007). The technology is based on viral vector-mediated gene transfer into the target somatic cells in culture. Forced expression of pluripotency genes, such as Pou5f1 (Oct3/4), Sox2, Klf4, Nanog and/or Lin28, and the myc oncogene alters the pattern of epigenetic modifications that accumulate on chromatin, to re-establish the transcriptional profile of ES cells, and gives rise to colonies of cells morphologically similar to ES cells. The reprogrammed cells, induced pluripotent stem (iPS) cells, display self-renewal and differentiation capacities equivalent to those of ES cells. Since somatic cells from individual patients can give rise to iPS cells, this technology has opened a practical path toward establishing individualized human PS cells that avoid the immunological problems associated with allografting. Use of the technology also avoids the controversies surrounding the establishment of human ES cells. However, the current iPS technology depends on stable integration of pluripotency gene cDNAs into the chromatin of the recipient, a manipulation that could introduce harmful genetic alterations that in the long term manifest, for example, as tumors, as observed in cases of gene therapy (Nienhuis et al., 2006). In this respect, novel non-integrative methods using a transient-transfection protocol (Okita et al., 2008), EBV-based episomal vector (Yu et al., 2009), an adenoviral vector (Stadtfeld et al., 2008), a Sendai virus vector (Fusaki et al., 2009), small molecules (Shi et al., 2008), or the piggyBac method (Kaji et al., 2009; Woltjen et al., 2009; Yusa et al., 2009) have all been reported. Such methods are likely to lead to the creation in the near future of safer, transgene-free human iPS cell lines suitable for clinical applications.

3. Chondrogenesis from mouse and human PS cells.

The ability of ES cells to form cartilage was first seen as cartilage nodule formation in teratomas. Chondrogenic differentiation of mouse and human ES cells has been demonstrated. Here we summarize methods for the initial differentiation of ES cells and the subsequent derivation of chondrocytes.

3.1 Chondrogenesis in the embryoid body and the outgrowth.

Embryoid bodies (EBs), the floating cellular aggregates spontaneously generated from PS cells *in vitro,* allow stem cells to differentiate into a number of cell types derived from each of the three germ layers (endoderm, mesoderm, and ectoderm) in a process that can recapitulate early embryogenesis. As a result, the method of EB formation is widely used for directing ES cell differentiation toward chondrocytic lineages.

For the induction of chondrogenesis, mouse ES cells are differentiated for 5 days in the form of EBs in a serum-containing medium (Fig. 1). EBs are then attached to plastic surface to allows outgrowth of mesenchymal cells, so-called 'outgrowth culture', Fig. 1-1). When outgrowth culture is performed in a chondrogenic medium, some of the mesenchymal cells condense to form cartilage nodules (Kramer et al., 2000). Periodic stimulation of chondrogenic gene expression and electron-microscopic analysis of the nodules (Kramer et al., 2005b) have indicated that the process of *in vivo* chondrogenesis, from mesenchymal cells through chondroprogenitor cells to mature chondrocytes, and finally to hypertrophic chondrocytes, can be reproduced in culture. However, the efficiency of cartilage nodule formation varies significantly among mouse ES cell lines (Kramer et al., 2005a).

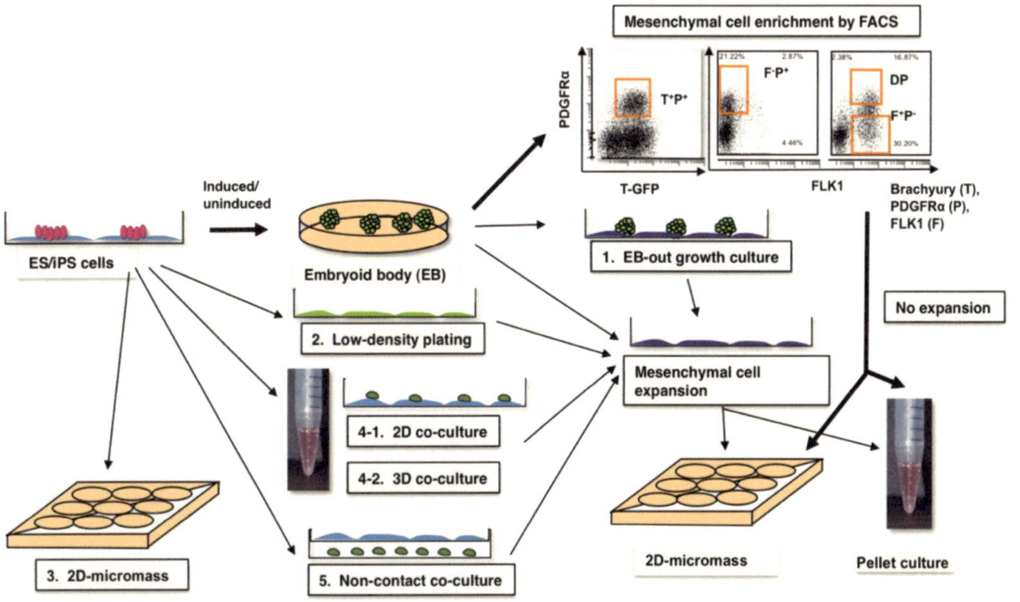

Fig. 1. General strategies for differentiating pluripotent stem cells to chondrocytes

Various growth factors have been tested for their ability to enhance cartilage formation in the outgrowth area. TGFβ-superfamily proteins, such as TGFβ and BMP, are commonly used to stimulate chondrogenesis from chick embryonic limb mesenchyme (Duprez et al., 1996; Kulyk et al., 1989). According to Kramer's group, addition of BMP2 (2-10 ng/ml) or BMP4 (10 ng/ml) increases, but TGFβ1 (2 ng/ml) slightly reduces, the number of cartilaginous nodules (Kramer et al., 2000). Treatment with BMP2 from day 2 to 5 of differentiation (EB formation stage) is sufficient to produce the stimulatory effect. However, these observations differ from ours in three respects. 1) We found that the first 5-day treatment with 2 ng/ml BMP4 induced mesoderm, including hemogenic angioblasts, in EBs (Nakayama et al., 2000; Nakayama et al., 2003). 2) Treatment of the isolated mesoderm with BMP4 alone inhibited *in vitro* chondrogenesis. 3) TGFβ3 (10 ng/ml) and BMP4 (10-50 ng/ml), or TGFβ3 followed by BMP4, synergistically enhanced chondrogenesis (Nakayama et al., 2003). Consistent with our observations, Kawaguchi et al. (2005) have reported that exogenous TGFβ3 (10 ng/ml), but not BMP4 (100 ng/ml), added from the stage of outgrowth culture, enhances chondrogenesis in the outgrowth, and zur Nieden et al. (2005) have demonstrated that addition of BMP2 (10 ng/ml) and TGFβ1 (2 ng/ml) between days 3 and 5 of differentiation (i.e. EB culture stage) and addition of BMP2 alone during the subsequent outgrowth culture results in pronounced up-regulation of cartilage-specific gene expression.

In contrast, Toh et al. (2007) have demonstrated that mesenchymal cells growing out of human day 5 EBs generated in the absence of the above-mentioned factors form few, if any, cartilage nodules during 21-day outgrowth culture in a serum-containing chondrogenesis medium. However, addition of BMP2 (100 ng/ml) during the outgrowth culture enhanced cartilage nodule formation (Type II collagen positive) in the outgrowth area, a finding similar to that seen for mouse EB outgrowth culture, although sulfated proteoglycan levels was not changed significantly.

Thus, differences in timing/duration and magnitude of TGFβ and BMP signaling lead to different effects of the two factors on chondrogenic differentiation of ES cells.

3.2 Chondrogenesis from ES cells during 2D-differentiation culture.

Although the standard differentiation protocol for ES cells involves a 3-dimensional (3D) culture leading to EB formation, ES cells can also differentiate as a monolayer during 2-dimensional (2D) culture. The 2D culture method had not been widely used for the generation of mesodermal cell lineages from mouse ES cells, primarily because fewer functional cell types are induced and sustained with this method compared with EB formation culture, especially in an undefined serum-containing medium. However, the advantage of 2D monolayer differentiation is the ease of visualizing and quantitating intermediate cell types. Dr. Nishikawa et al. (1998) have elegantly demonstrated that differentiation of mouse ES cells on a type IV collagen-coated surface, combined with a serial cell sorting using E-cadherin, platelet derived growth factor receptor-α (PDGFRα), vascular endothelial growth factor receptor-2 (FLK1/KDR) and vascular endothelial (VE)-cadherin markers, not only led to the effective generation of hemogenic angioblasts, but also allowed the determination of the intermediate cell types in the differentiation pathway (Low-density plating, Fig. 1-2). As described below, a similar 2D-differentiation method has been applied to human ES cells to generate MSC-like mesenchymal cells (Barberi et al., 2007; Nakagawa et al., 2009).

While there have been no reports describing application of the 2D monolayer differentiation protocol directly to the development of chondrocytes from human ES cells, one recent publication provided interesting data, though limited, on 2D-micromass culture using chondrogenic medium with BMP2 (100 ng/ml) added on day 2-3 of culture for the differentiation of human ES cells (Gong et al., 2010) (Fig. 1-3). In contrast to the regular EB formation method (using KSR-medium with no added protein factors), the method resulted in a cell mass that stained strongly with Alcian blue at 14 days. A noteworthy feature of differentiation under the conditions described was its unusual rapidity. The chondrocyte marker gene Sox9 is already expressed by the time BMP2 is added on day 3 and induces expression of the mesendoderm marker gene Brachyury on day 4. Both transcripts disappear by day 7, when aggrecan and Col2a1 transcripts are maximal. Since Sox9 is also functionally important for cranial neural crest formation, it was not clear what type of chondrocyte was preferentially produced under the culture conditions.

3.3 Chondrogenesis from ES cells using co-culture methods.

Chondrogenesis is controlled by multiple environmental cues, including signals provided through cell-to-cell interaction (e.g. mesenchymal condensation) and secreted morphogenetic factors. Co-culture of ES cells with cells that provide the necessary environmental cues offers an alternative for chondrogenic differentiation of ES cells. 2D differentiation culture of mouse and human ES cells on a layer of supportive cells has been used for the development of lymphohematopoietic cells and endothelial cells (Fig. 1-4-1). One of the earliest and most successful examples is the demonstration of pre-B cell genesis and two-phase (from primitive to definitive) erythropoiesis from mouse ES cells on the op/op mouse bone-derived stromal cell line, OP9 (Nakano et al., 1994, 1996). Since the mid-90s, OP9 has become the co-culture partner of choice for developing mesodermal cell types from mouse and human PS cells. As for chondrogenesis, Sui et al. (2003) co-cultured drug-resistant mouse ES cells with primary limb bud mesenchymes in 2D micromass culture (Fig. 1-3/1-4-1). After 4 days, the corresponding drug is added to remove progeny from limb bud cells and enrich those from the drug-resistant ES cells. Chondrocytes were detected among the surviving cells. Limb bud mesenchymes proliferate, condense and differentiate into bone and cartilage cells *in vivo*, and thus are potentially a suitable co-culture partner for chondrogenesis. However, as they are isolated from embryos, mesenchymes are not appropriate for use in humans. Interestingly, the co-culture partner for chondrogenesis can be derived from tissues unrelated to skeleton: e.g. hepatic cells (Lee et al., 2008).

Probably one of the most successful co-culture methods for directing differentiation of human ES cells to chondrocytes is the use of chondrocytes themselves as co-culture partner. Primary human (Vats et al., 2006) or bovine (Hwang et al., 2008a) chondrocytes are placed in a well insert either directly, or after brief expansion, respectively, and human ES cells are 2D cultured underneath for 3-4 weeks in a serum-containing MSC medium in the presence or absence of ascorbate, respectively (Fig. 1-5). The resultant human ES cell progeny express 2.5-3.5 times more type I and type II collagen (the ratio is not changed) and sulfated glycosaminoglycan (sGAG) than those generated without chondrocytes, and form a cartilage-like area *in vivo* when transplanted encapsulated in a polymer-based scaffold (Vats et al., 2006). Such chondrogenic progeny also grow in the MSC medium, which causes dedifferentiation; however, the cells expanded and dedifferentiated for 3 weeks retain significant chondrogenic activity when subjected to 3D-pellet culture or hydrogel (arginine-glycine-aspartate [RDG]-coated polyethyleneglycol [PEG]-based hydrogel, Hwang et al., 2006b) culture in the presence of TGFβ1 (Hwang et al., 2008a). A 3D coculture of human ES cells with irradiated human chondrocytes in a serum-free chondrogenic medium including TGFβ1, that resembles the pellet culture for cartilage particle formation, facilitates mesenchymal progeny formation from human ES cells. Such mesenchymal cells form cartilage particles that stain uniformly with Alcian Blue after expansion in a human serum-containing medium (Fig. 1-4-2) (Bigdeli et al., 2009). Identification of the critical factors produced by the primary chondrocytes may thus provide a route for more efficient chondrogenic differentiation of mouse and human PS cells.

3.4 Cooperation with tissue-engineering technology.

Simple implantation of mesenchymal cells in combination with a periosteal patch as in ACI, necessitates additional surgery to isolate periosteum from the patient, and also fails to faithfully reconstitute the architecture of host cartilage within the damaged or diseased area. However, conventional cartilage-formation culture results in a small-size cartilaginous particle that is structurally disorganized and mechanically weak, and therefore is not suitable for clinical-scale cartilage repair. Therefore, attempts have been made to generate large engineered cartilage using artificial or bioartificial scaffold with proper chondroprogenitor cells. In general, the engineered cartilage is a construct consisting of either chondrocytes matured *in vitro* before implantation, or chondrogenic mesenchymal cells, such as expanded chondrocytes and immature chondroprogenitor cells, which are expected to (re)differentiate or mature *in vivo* after implantation. In either case, it is intended that the seeded chondroprogenitor cells will survive, grow, differentiate, and reform cartilage structure in the scaffold *in vitro* and *in vivo*. Many attempts, especially using biodegradable polymer scaffolds and hydrogel have been described (reviewed in Jukes et al., 2010). Cartilage tissue engineering using MSCs and both types of scaffold has been reported (Fukumoto et al., 2003; Richardson et al., 2006).

Porous biodegradable polymer scaffolds are based on polylactic acid (PLA), polyglycolic acid (PGA), polyvinyl alcohol (PVA) or polyurethanes. Such solid-type scaffolds are rigid enough to withstand continuous loading, whereas invasive surgical incision is required for implantation. As for application to mouse ES cells, 4 day-old EBs or cells isolated from such EBs (formed in a serum-containing medium) were further cultured on a polyethyleneoxide terephthalate (PEOT)/polybutylene terephthalate (PBT) scaffold in a serum-free chondrogenic medium (Jukes et al., 2008). However, the seeding efficiency of EB cells was extremely low and retention of EBs and seeded EB cells within the scaffold during chondrogenic differentiation was also poor, leading to inefficient, non-homogeneous cartilaginous constructs. In another example, a scaffold made from polycaprolactone (PCL) was used. EBs were formed in a serum-containing medium including retinoic acid, then outgrowth culture was performed in a serum-containing chondrogenic medium, and the outgrowth cells were seeded onto the PCL scaffold (Fecek et al., 2008). Implantation of the resulting constructs yielded a poor result, with only a few chondrocyte clusters inside.

More refined tissue-engineering technology has been used for chondrogenic differentiation from human ES cells. The first report demonstrated that polylactic-co-glycolic acid (PLGA)/PLA scaffold in combination with Matrigel or fibronectin improved the seeding efficiency of human EB-derived cells, leading to cartilaginous area formation *in vitro* in the presence of TGFβ1 (Levenberg et al., 2003). Furthermore, a poly-D,L-lactide (PDLLA) scaffold seeded with human ES cells differentiated using the chondrocyte-coculture method and implanted in SCID mice, resulted in areas within the scaffold that expressed type II collagen at 35 days (Vats et al., 2006), although signs of proteoglycan accumulation were weak.

Hydrogels are water-insoluble but highly absorbent materials, and include a variety of gels based on collagen, GAG, agarose, gelatin, alginate, and hyaluronic acid. Hydrogel lacks structural integrity but have the major benefits of supporting the homogeneous distribution of cells at the desired density within the scaffold, and being associated with the less invasive method of transcutaneous injection for implantation. Mouse EB cells encapsulated and cultured in natural hydrogels such as agarose and Matrigel, showed homogenous seeding; however the viability of the seeded cells was very poor and the cells did not form cartilage (Jukes et al., 2008). In contrast, when EBs were directly encapsulated, cells within the EBs had greater viability and chondrogenic differentiation was observed *in vitro*. Implantation into immunodeficient mice, however, resulted in teratoma formation and a failure to detect cartilage nodules. As found with synthetic hydrogels, Dr. Elisseeff's group has reported that mouse EBs encapsulated in a PEG-based hydrogel formed cartilaginous tissue structures during *in vitro* culture in the presence of TGFβ1 (Hwang et al., 2006a). As noted above, they have further modified the PEG-based hydrogels to promote cell adhesion and spreading by the addition of RGD-containing peptides. In the modified gel, MSC-like cells (Hwang et al., 2008b) and chondrogenically committed cells (Hwang et al., 2008a), derived from human ES cells using the EB-outgrowth culture method and the chondrocyte co-culture method, respectively, survived and gave rise to many chondrocytes *in vitro*. Furthermore, when the MSC-like cells were pre-treated with chondrocyte-conditioned medium, implantation of the PEG-hydrogel construct resulted in cartilage-like tissue formation in damaged articular cartilage by 8 weeks (Hwang et al., 2008b).

In summary, polymer scaffolds and hydrogels work to some extent but are not ideal for promoting generation of cartilage from ES cell-derived mesenchymal cells *in vitro* or *in vivo*.

Although expansion in a chondrocyte-conditioned medium improves the chondrogenicity of the mesenchymal cells, there is considerable scope for improvements to scaffolds to enhance chondrogenesis. In this sense, novel biomaterials, such as hybrid gel/scaffolds (Richardson et al., 2008), nanofibrous scaffold (Hu et al., 2009; Li et al., 2005) and bioartificial matrixes are of great interest.

3.5 Chondrogenesis enhancement by gene modification

Directed differentiation of ES cells can be controlled by intracellular proteins such as cell-type specific transcription factors. Controlled expression of key differentiation regulators via gene transfer is therefore an alternative strategy for the achievement of lineage-specific differentiation of PS cells: e.g. hepatocytes (Ishizaka et al., 2002) and skeletal muscles (Darabi et al., 2008). The most likely gene candidates for directing chondrogenesis are the high-mobility group (HMG) box protein Sox9 and paired-box-gene Pax1 (Bi et al., 1999; Wallin et al., 1994). Constitutive overexpression of Sox9 or transfection of Sox9, in combination with Sox5 and Sox6, can induce the expression of cartilage marker genes in the absence of specific growth factors or culture conditions (Ikeda et al., 2004; Kim et al., 2005). However, such gene transfer-based facilitation of chondrogenesis is unlikely to be of use in the clinical setting.

4. Identification of chondroprogenitor cells derived from pluripotent stem cells: undefined mesenchymal cells and MSC-like cells

Homogenous populations of chondroprogenitor cells are unlikely to be obtained from PS cells solely by directed differentiation culture. However, either for cell transplantation or for engineered cartilage formation before transplantation, purified chondroprogenitor cells are needed, especially because the least contamination with teratoma-forming, undifferentiated ES cells will be harmful. Therefore, a convenient yet effective method for purifying the chondroprogenitors is required. Surface molecules specifically expressed on chondroprogenitor cells have been used to isolate them by fluorescence-activated cell sorting (FACS). We published the first attempt to induce, identify, purify and characterize chondroprogenitors from differentiating mouse ES cells (Nakayama et al., 2003). In particular, the research described focused on 1) the genesis/induction of $FLK1^{-}PDGFR\alpha^{+}$ progeny, which showed a transcriptional profile of lateral plate/extraembryonic mesoderm and posterior paraxial mesoderm, and the $FLK1^{+}PDGFR\alpha^{-}$ progeny, which showed lymphohemogenic activity, and the corresponding transcriptome in a serum-free medium, and 2) the control of chondrogenic activity in these fractions of progeny. We demonstrated using 2D and 3D chondrogenesis assays that both mesenchymal progeny have full chondrogenic activity: from mesenchymal condensation to hypertrophic chondrocyte formation. Similarly, Sakurai et al. (2006) demonstrated, as part of their attempt to characterize what they claim to be primitive mesoderm/mesendoderm ($FLK1^{+}PDGFR\alpha^{+}$ progeny), paraxial mesoderm ($FLK1^{-}PDGFR\alpha^{+}$ progeny) and lateral plate mesoderm ($FLK1^{+}PDGFR\alpha^{-}$ progeny), generated simultaneously from mouse ES cells using Nishikawa's 2D-differentiation method in a serum-containing medium, that the primitive mesoderm and paraxial mesoderm were able to form cartilage nodules during 2D micromass culture. Chondroprogenitor cells from human ES cells have not yet been purified and characterized to the extent of those derived from murine ES cells.

4.1 Undefined mesenchymal cells from EBs and EB-outgrowth.

Chondrogenic activity in human ES cell-derived progeny was first reported by Levenberg et al. (2003) from a study to demonstrate the effect of a 3D polymer scaffold on differentiation of human ES cell progeny. The progeny were EB cells prepared from 8 to 9-day-old EBs formed in KSR-based serum-free medium in the absence of exogenously added protein factors, and used directly without expansion or purification. Toh et al. (2007) have also demonstrated that uncultured, unfractionated EB cells isolated from human day 5 EBs formed in the absence of exogenous protein factors give rise to type II collagen-positive cartilage nodules when subjected to 2D-micromass culture in a serum-containing medium. The effect of BMP2 (100 ng/ml) added during micromass culture was, however, minimal. Dr. Athanasiou and his colleagues have published a series of reports on chondrogenesis from human ES cells using an EB cell-re-aggregation method (Hoben et al., 2008; Koay et al., 2007; Koay & Athanasiou, 2008, 2009). ES cells are differentiated as a form of EB for 3 weeks in various media based on a 0-20% serum-containing chondrogenic medium with or without the addition of combinations of growth factors such as TGFβ, PDGF, IGF, sonic hedgehog and/or BMP2/4/6. EBs are then disrupted, and the cells are resuspended without expansion or purification at a high density in a small volume (e.g. 20 μl) of 0-1% serum-containing chondrogenic medium with or without TGFβ1 or IGF1, and transferred into a 3 mm, 2% agarose well. Although a macroscopic particle forms in 4 weeks, there is only a weak indication that cartilage particles form, as assessed by the production of ECM rich in type II collagen and GAG. Interestingly, TGFβ+BMP2/4 treatment during EB culture in the serum-containing chondrogenic medium is most effective at stimulating the accumulation of GAG and type VI collagen in 3 weeks, whereas EB culture in an agarose-coated Transwell, conditions under which chondrocytes are co-cultured as a monolayer, strongly stimulates the accumulation of type II collagen within EBs (Hoben et al., 2009). In summary, crude EB cells without expansion or sub-fractionation failed to show robust chondrogenic activity in any of the *in vitro* culture methods used. The addition of protein factors at various steps of differentiation had no dramatic effect.

Attempts have also been made to generate morphologically uniform mesenchymal progeny from crude EB cells, before assaying chondrogenesis. Human ES cells are differentiated in the form of EB for 4 to 10 days in the absence of exogenous protein factors in a serum-containing medium (Toh et al., 2009; Xu et al., 2004), or the standard KSR-based serum-free medium (Hwang et al., 2006b; Hwang et al., 2008b; Terraciano et al., 2007). The resultant EBs are either attached to a plate and cultured in the corresponding factor-free EB medium to allow the outgrowth of mesenchymal cells (Fig. 1-1), or subjected to 2D micromass culture, from which mesenchymal cells are isolated (Toh et al., 2009). Multiple passaging of the outgrowth cells in the serum-containing MSC medium results in a morphologically homogeneous mesenchymal cell population (Fig. 1-6). However, such cells are not chondrogenic in standard 3D pellet culture in the presence of TGFβ. Weak chondrogenic activity was detected only when the cells were encapsulated in a special scaffold (e.g. RDG-PEG hydrogel) and stimulated by periodic pressures (Terraciano et al., 2007). Interestingly, expansion of the mesenchymal cells from 2D micromass culture in the presence of growth factors to some extent preserved the ability of the cells to accumulate Alcian blue-positive matrix in the following 3D pellet assay (Toh et al., 2009). The significant level of type I collagen in the resulting particles suggested the formation of fibrocartilage.

As described above, chondrocyte-coculture during EB formation improved the "expansion" strategy for obtaining morphologically uniform chondroprogenitors. For example, human

ES cell differentiation with irradiated chondrocytes using the 3D pellet culture-like method in a serum-free chondrogenic medium with TGFβ1 resulted at day 14 in a mixture of cells that included chondrocytes. Although the chondrocytes bore similarities to co-cultured irradiated chondrocytes, their expansion in a human serum-containing medium led to the accumulation of mesenchymal cells with the capacity to reform a cartilage pellet that stained uniformly with Alcian Blue (Bigdeli et al., 2009). While such cells were not adipogenic and only weakly osteogenic, the chondrocyte co-culture method is suitable for specifying chondroprogenitor cells from human ES cells.

4.2 ES cell-derived MSC-like cells.

MSCs are *in vitro*-defined chondroprogenitor cells present in various adult tissues. Therefore, ES cell-derived chondroprogenitor cells may possess MSC-like properties. The first demonstration of the genesis of MSC-like cells from human ES cells was that of Barberi et al. (2005), who used a 2D co-culture method for differentiation. A long-term (40 day) human ES cell differentiation on the OP9 stromal cell line in a serum-containing medium was followed by $CD73^+$ cell isolation and further culture in a serum-containing MSC medium without OP9. The method was later modified to include a stroma-free 2D differentiation culture, in which human ES cells are cultured at low cell density on a gelatin-coated plate in serum-free medium without protein factors for 20 days (Barberi et al., 2007) to produces mesenchymal cells. $CD73^+$ cells emerged at 14 days of expansion of the mesenchymal cells in serum-containing MSC medium. The $CD73^+$ cells, sorted then maintained in the same medium, showed bone-cartilage-fat "tri-lineage" potential. Nakagawa et al. (2009) used a similar method to produce mesenchymal cells with *in vitro* tri-lineage potential. They plated human ES cells directly onto gelatin-coated plates and maintained them in serum-containing medium in the presence of FGF2. Interestingly, however, unlike other studies, TGFβ did not stimulate the accumulation of GAG and type II collagen during 3D-pellet chondrogenesis culture.

The expanded EB-outgrowth cells in serum-containing MSC medium were osteogenic and adipogenic, but poorly chondrogenic (Hwang et al., 2008b), and lacked the expression of the MSC marker CD73 (Hwang et al., 2006b). Interestingly, expansion of the outgrowth cells in a chondrocyte-conditioned medium significantly increased the chondrogenic ability of the cells and was associated with a better outcome after transplantation of the cells into a critical size cartilage defect in the femoropatellar groove of nude rats. Probably the most extensive analysis of MSC-genesis through EB-outgrowth cells was the report by Mahmood et al. (2010), which describes the differentiation of human ES cells in the absence of exogenous protein factors, but the presence of a high concentration (10 μM) of the small molecule inhibitor of Nodal/Activin/TGFβ receptor (ALK7/4/5) kinase, SB431542, for 10 days in a KSR-based serum-free medium. Although SB431542 is inhibitory for mesoderm specification, even at a lower concentration (Gadue et al., 2006), and induces neuronal differentiation (Smith et al., 2008) from mouse and human ES cells, transcripts representing myogenic cells (as well as neural cells) were induced at day 10. EB-outgrowth culture and subsequent maintenance culture for the mesenchymal cells was performed in a chemically defined serum-free medium in the presence of 1 μM SB431542. Cells expressing myocytic and neuronal proteins persisted after cell expansion. To generate $CD73^+$ MSCs, such cell populations must be cultured in a serum-containing medium for 20 days, during which signs of myocytes and neuronal cells disappear. The chondrogenic activity of the MSC-like cells developed by this method was demonstrated *in vivo* by studies in nude mice.

Overall, the chondrogenic activity of the human ES cell progeny reported thus far is weak, and not highly susceptible to enhancement by the protein factors tested. Interestingly, for the genesis of MSC, expansion culture of the progeny in a serum-containing MSC medium appears to be essential. Therefore, the chondrogenic activity as part of the MSC activity can be generated/enhanced during such culture. However, expansion culture of human ES cell progeny does not always accumulate MSC-like chondroprogenitor cells, possibly reflecting the low yield of MSCs within the progeny population. It is worth noting that, again, chondrocyte-coculture, either during EB formation or during the expansion culture phase, improved the chondrogenic activity of the human ES cell progeny, even after expansion.

5. Lineage–specific chondroprogenitors from pluripotent stem cells

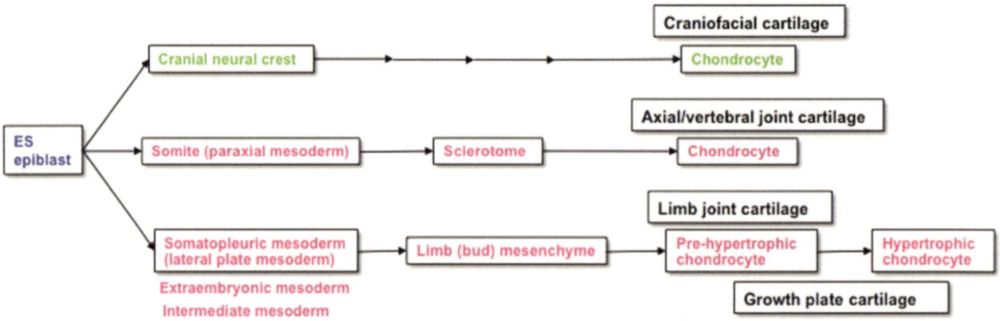

Fig. 2. Key developmental intermediates from pluripotent stem cells to chondrocytes

Embryonic chondrocytes are responsible for the robust chondrogenesis that occurs during embryogenesis. The three sources of embryonic chondrocytes and their precursor cells (chondrogenic mesenchymes and chondroprogenitors) are lateral plate mesoderm for limb joint and growth plate chondrocytes, paraxial mesoderm for vertebral and rib chondrocytes, and cranial neural crest for craniofacial chondrocytes (Fig. 2). The mesodermal chondrocytes give rise to hyaline cartilage at joint surfaces and fibrocartilage in the patella, meniscus, and vertebral disk. In contrast, cranial neural crest-derived chondrocytes make craniofacial cartilage that is mostly fibrocartilage or elastic cartilage. Therefore, embryonic chondrocytes from lateral plate mesoderm, paraxial mesoderm or cranial neural crest might preferentially generate different types of cartilage in a given environment. Recent observations that bone marrow MSCs arise from the neural crest in young mice (Morikawa et al., 2009; Takashima et al., 2007) suggest that the chondrogenic potential of adult MSCs and mesodermal chondroprogenitor cells might also be different.

In vitro differentiation of mouse and human ES/iPS cells is thought to represent the early stages of embryogenesis. In fact, progeny displaying characteristics of a particular germ-layer can be induced in an exogenous protein factor-dependent manner from mouse ES (reviewed in Gadue et al., 2005; Keller, 2005; Nishikawa et al., 2007). The factors used are developmental factors, defined by genetic and embryological studies, which provide essential signaling between pluripotent cells (inner cell mass and epiblast) and the differentiated cell type of interest. However, to date, only a small number of reports describe such a "factor biology" approach to the generation of chondrocytes with distinct germ-layer specificity (Kawaguchi et al., 2005; Nakayama et al., 2003; Sakurai et al., 2006; Tanaka et al., 2009), and these have all dealt with mouse ES cells. There have been no reports of controlled differentiation of human ES/iPS cells to lineage-specific chondrogenic progeny. A few reports have addressed the effects of TGFβ1, BMP2, and Activin (or its inhibitor SB431542) on human ES cell differentiation. Although the origin of the developed chondroprogenitors has not been explored, the expression of mesodermal genes in the progenitor population after extensive expansion has been reported (Harkness et al., 2009).

6. Mesodermal chondroprogenitors: differential specification and prospective isolation from differentiating pluripotent stem cells

Applying the "factor biology" strategy, progeny displaying characteristics of different types of mesoderm are induced from mouse ES cells in a defined medium (Gadue et al., 2006; Nostro et al., 2008; Tanaka et al., 2009). A similar principle has been applied to human ES/iPS cells to develop various mesodermal cell types (reviewed in Murry & Keller, 2008; Nishikawa et al., 2008). Mesoderm is developed through gastrulation *in vivo*, a time when posterior primitive ectoderm (or epiblast) undergoes epithelial-to-mesenchymal transition to give rise to extraembryonic mesoderm, cardiac and lateral plate mesoderm, and subsequently to paraxial mesoderm and axial mesendoderm. The TGFβ-superfamily (TGFβ, Nodal/Activin, and BMP), Wnt, and FGF are involved in these processes, and are effective in different doses and combinations in determining the differential specification of mesodermal cells from ES cells (reviewed in Gadue et al., 2005; Keller, 2005; Nishikawa et al., 2007). For transplantation in the clinic, it will be necessary to purify the generated mesodermal chondroprogenitor cells from unwanted contaminants. FLK1/KDR and PDGFRα are useful surface markers with which to monitor the early stages of mesoderm development in the mouse. Flk1 expression is detected in extraembryonic and lateral plate mesoderm (Yamaguchi et al., 1993), which gives rise to somatopleuric mesoderm, the origin of limb bud chondrogenic mesenchymal cells. In contrast, Pdgfra expression is detected in most early mesodermal cells, but surface expression of PDGFRα is restricted in paraxial mesoderm derivatives (Takakura et al., 1997). Dr. Nishikawa and his colleagues (1998) have started using these markers for the identification and isolation by FACS of mesodermal progeny from mouse ES cells. Our group has combined factor biology and FACS purification to selectively generate, isolate and characterize chondroprogenitors derived from mouse and human PS cells.

6.1 Signaling code for differential specification of ES cell progeny.

In applying the combined factor biology/FACS purification approach, mouse ES cells are differentiated in a serum-free medium and the resulting EBs are harvested at an appropriate time. Single-cell preparations from EBs are immunostained for FACS using antibodies to E-cadherin (E), N-cadherin (N), PDGFRα (P), and FLK1 (F). The E^- EB cells, free of undifferentiated ES cells and endoderm, are further fractionated with N, F and P, and subjected to bioassays and *in vivo* assays. Brachyury$(T)^+$ mesendodermal/early mesodermal progeny are generated in 3 days in the presence of exogenous WNT3a and endogenous Nodal-like activity. Although exogenous BMP4 induces T, it does so through induction of endogenous WNT and Nodal-like activities (Tanaka et al., 2009). The P^+ progeny, the first sign of mesenchymal cell genesis, become obvious by day 4. When Activin A is provided exogenously with WNT3a or BMP4, F^+ cells start to accumulate, as was also reported by the Keller group (Fehling et al., 2003). Approximately 90% of the F^+ and/or P^+ EB cells developed by day 5 are derived from T^+ progeny and are also N^+, implying that the cells include mesodermal derivatives (Fig. 3).

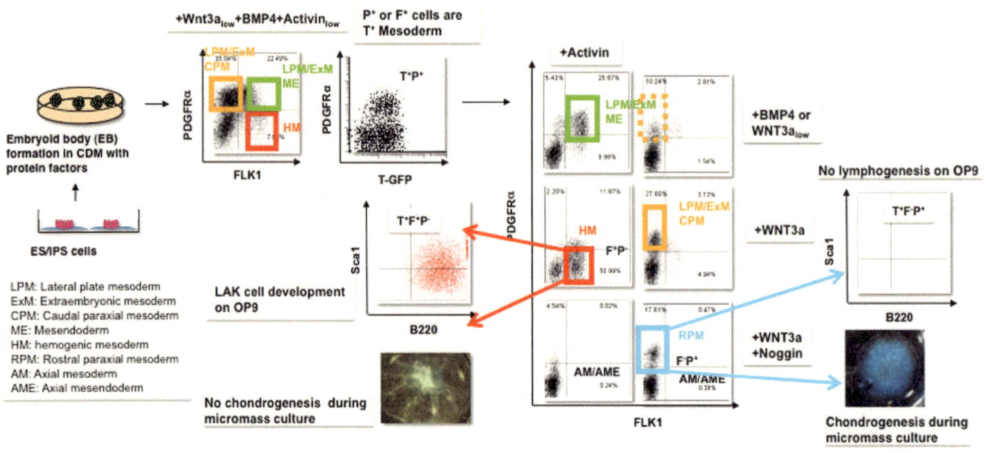

Fig. 3. Chondrogenic paraxial mesoderm from mouse pluripotent stem cells

Global transcriptional profiling has shown that the expression of genes associated with pluripotent stem cells (Oct3/4/Pou5f1 and Nanog) and epiblasts (Fgf5) is confined to the $T^-N^-F^-P^-$ fraction (which is mostly E^+). Markers for rostral paraxial mesoderm such as Tcf15/Paraxis, Uncx, Meox1, Tbx18 and Mesp2, are confined to the $T^+N^+F^-P^+$ cells generated when WNT signaling is activated and BMP signaling is inactivated. In contrast, Foxf1 and Hand2, expressed in lateral plate and extraembryonic mesoderm, are detected mostly in the same $T^+N^+F^-P^+$ cell fraction, but generated only under active WNT and BMP signaling. The genes associated with hemogenic mesoderm/endothelium, such as Scl/Tal1, Hhex, Ikaros, Chd5, Tie2/Tek and Cd34, are exclusively expressed in the $T^+N^+F^+P^-$ fraction generated efficiently when WNT, BMP and Nodal/Activin signaling are activated via exogenous factors (Tanaka et al., 2009).

Consistent with the notion that BMP signaling is inhibitory to ES cell-derived neurogenesis, and that head induction depends on both BMP/Nodal-inhibition and WNT inhibition *in vivo*, $T^-N^+F^-P^-$ neural precursors, expressing early neural markers such as Sox1, Nestin, Pax6, Neurod and Neurog, are differentiated from ES cells when none of the three signaling mechanisms is activated (Fig. 5). Interestingly though, when BMP signaling was activated in the presence of WNT inhibitors, non-mesodermal $T^-N^{dim}F^-P^+$ mesenchymal cells were generated at low levels, as discussed in detail below.

Lateral plate/extraembryonic mesoderm specification from mouse ES cells, as judged by gene expression profiling, is achieved by exogenous BMP4 or WNT3a, although both WNT and BMP signaling are activated under either condition. In contrast, rostral paraxial mesoderm specification requires BMP inhibition when an optimized level of WNT3a is provided. These results imply that activation of either BMP or WNT signaling results in the activation of the other during ES cell differentiation. In support of this idea, exogenous BMP4 upregulated the levels of Wnt mRNAs (Wnt5a, Wnt4 and, weakly, Wnt2, Wnt5b and Wnt11) and suppressed the level of the Wnt inhibitor gene (Sfrp1 and Sfrp2) mRNAs. On the other hand, exogenous WNT3a elevated expression of Bmp and Nodal transcripts.

6.2 Prospective isolation of mesodermal chondroprogenitors from mouse ES cells.

We were the first to report the use of the cell surface markers FLK1 (F) and PDGFRα (P) to enrich by FACS a population of mesenchymal progeny that demonstrated chondrogenic activity *in vitro* from a crude mixture of differentiated ES cells (Nakayama et al., 2003). The conditions for ES cells differentiation produced lateral plate mesoderm and extraembryonic mesoderm, but no rostral paraxial mesoderm (WNT3a[or BMP4]±Activin A, Fig. 3). The isolated F^-P^+ progeny contained lateral plate/extraembryonic mesoderm and posterior paraxial mesoderm. The F^+P^- progeny contained hemoangiogenic (lateral plate/extraembryonic) mesoderm. Both cell fractions reproducibly showed chondrogenic activity *in vitro*. Rostral paraxial mesoderm transforms itself into somite, an epithelial structure that later gives rise to progenitors for skeletomyogenesis and axial skeletogenesis. We have also recently reported that under conditions in which no lateral plate/extraembryonic mesoderm is specified but rostral paraxial mesoderm is selectively induced from ES cells (WNT3a+Noggin, Fig. 3), the F^-P^+ progeny represents the rostral paraxial mesoderm and is also rich in chondroprogenitors (Tanaka et al., 2009). Unfortunately, F^-P^+ is not a characteristic of either type of chondrogenic mesoderm. The availability of a reporter cell line using a mesoderm type-specific transcription factor gene (e.g. Meox1) would be a useful tool for studying the cellular developmental pathway and further purification. However, considering future clinical application, cell surface markers specific for each mesoderm-type and the chondroprogenitors will be needed.

6.3 Prospective isolation of mesodermal chondroprogenitors from human ES cells.

The prospective isolation of lineage-specific chondroprogenitor cells from human PS cells using specific cell surface markers without extensive expansion has not been reported. However, an interesting report published recently describes how the delta-like 1/fetal antigen 1 (dlk1/FA1), a transmembrane protein belonging to the Notch/Delta/Serrate family, is useful in some cases (Harkness et al., 2009). We reported a serum-free differentiation method for human ES and iPS cells that favors the generation of hematopoietic progenitor cells in an exogenous WNT3a-dependent manner (Wang & Nakayama, 2009; Wang et al., 2010) (Fig. 4, right panel), which we have now adapted to promote paraxial mesoderm formation. Specification of paraxial mesoderm from human ES and iPS cells was achieved using conditions similar to those described for mouse ES cells. In addition, FACS-purified human paraxial mesoderm was found to have chondrogenic activity when measured by the conventional 2D micromass and 3D pellet chondrogenesis assays (under the same condition

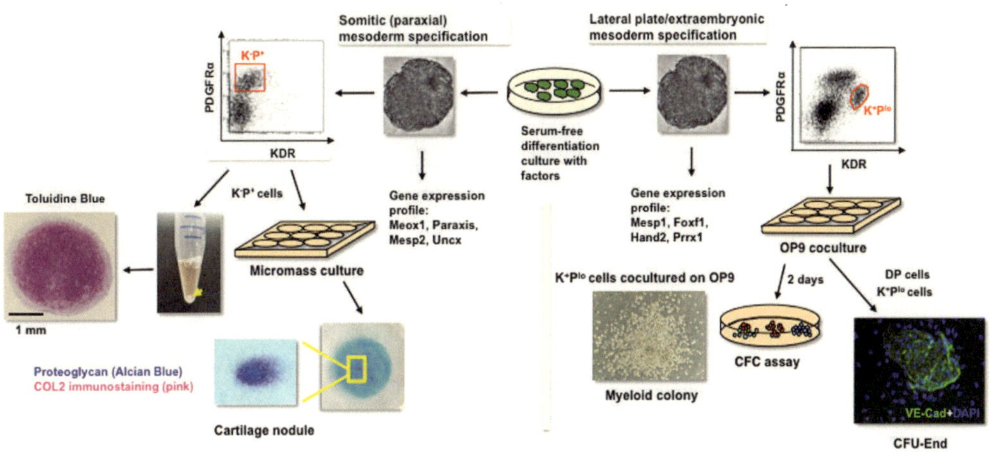

Fig. 4. Chondrogenic paraxial mesoderm from human pluripotent stem cells

established for the mouse ES cell-derived rostral paraxial mesoderm) (Umeda, et al. unpublished observations; Fig. 4, left panel). Chondrogenic activity of the isolated paraxial mesoderm is robust, so that RDG-hydrogel or another scaffold is not necessary for *in vitro* assays.

7. Directed specification of MSC-like chondroprogenitors from pluripotent stem cells: neural or potentially other non-mesodermal lineages

Cranial neural crest cells arise from the forebrain (to form the frontonasal skeleton), the midbrain (to form the facial bone and cartilage and jaw bone) and the hindbrain (to form the middle ear and neck bones), in distinct populations. They differ from trunk neural crest in their potential to differentiate into cartilage, bone and connective tissues even *in vitro* (Abzhanov et al., 2003). Cranial neural crest cells are mesenchymal cells, generated from the neuroepithelium/surface ectoderm border of the fore-, mid- and hindbrain via epithelial-to-mesenchymal transition. They migrate anteriorly or ventrally to colonize distinct regions (the frontonasal process or pharyngeal arch 1 to 3) and form region-specific skeletal elements. Wnt, BMP, FGF, and Notch pathways are involved in neural crest cell specification, although the timing and order of the signals required is not fully understood (Steventon et al., 2005).

7.1 MSC-like cells from mouse ES cell: non-mesodermal origin.

Neural crest derivatives such as peripheral nerve cells were first generated from mouse ES cells using undefined stromal cell-derived activity (which is thought to contain Nodal-inhibitor and WNT-inhibitor) to induce neural commitment, followed by BMP4 treatment (Mizuseki et al., 2003). In contrast, while a brief treatment with retinoic acid during EB culture in a serum-containing medium failed to induce T, MSC-like tri-lineage potential cells developed during the subsequent EB-outgrowth culture. A sign of neural crest cell formation was also detected (Kawaguchi et al., 2005), consistent with the finding from

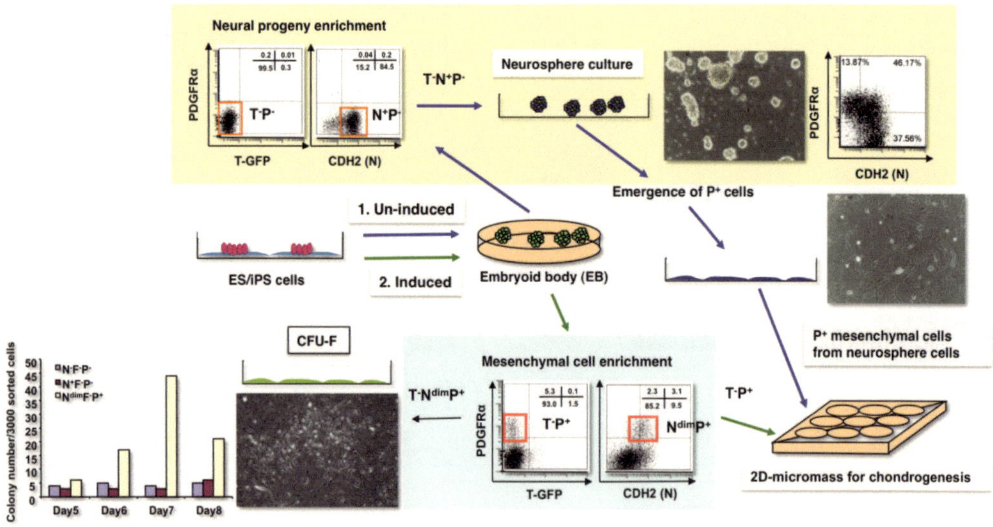

Fig. 5. Non-mesodermal chondroprogenitor from mouse pluripotent stem cells

lineage-tracing analysis that MSCs arise from the neural crest during mouse embryonic development (Morikawa et al., 2009; Takashima et al., 2007). We have also produced non-mesodermal (as judged by the lack of T expression) mesenchymal cells, i.e. $T^-N^{dim}F^-P^+$ cells (hereafter called T^-P^+ cells) from the isolated $T^-N^+F^-P^-$ neural progeny (hereafter called T^-P^- cells) that are generated from mouse ES cells without retinoic acid or any particular protein factors (Fig. 5, Un-induced). When the isolated T^-P^- cells were subjected to neurosphere culture in the neural crest stem cell culture medium (Wong et al., 2006), T^-P^+ cells emerged in 4 days. MSC-like spindle-shaped mesenchymal cells, which are strongly chondrogenic in 2D micromass culture (data not shown), can be generated from such T^-P^+ cells after brief expansion in serum-containing MSC medium. These mesenchymal cells are probably of neural origin but their relevance to neural crest has not been established.

Direct differentiation of mouse ES cells to similar non-mesodermal T^-P^+ mesenchymal cells has also been achieved by the addition of BMP4 or WNT3a from day 0 of differentiation in a serum-free medium (Fig. 5, Induced). The T^-P^+ cell population appearing from day 4 of differentiation is enriched in chondrogenic activity that is detectable from day 6. In addition, CFU-F activity also emerges from day 6, and is confined in the T^-P^+ fraction. The mesenchymal/fibroblastic cells derived from the resulting colonies spontaneously gave rise to Oil Red O^+ adipocytes on reaching confluence (data not shown). Mineral deposition was readily detected within 20 days of culture in osteogenic medium (data not shown). Thus, the chondrogenic T^-P^+ progeny possess adult MSC-like (tri-lineage potential plus CFU-F) activity, although the lineage relationship of this population of cells with neural crest has not been demonstrated.

7.2 Non-mesodermal MSC-like cells from human ES cells.

Mesenchymal cells with MSC-like phenotypes can also be isolated during directed neural crest cell differentiation from human PS cells (Lee et al., 2007; Mahmood et al., 2010; Zhou & Snead, 2008). Although a morphologically homogenous cell population can be obtained simply by subsequent passaging (Bigdeli et al., 2009; Hwang et al., 2006b; Mahmood et al., 2010), purification of particular mesenchymal cells by mechanical dissection based on morphological criteria (Hwang et al., 2008a) or by FACS using MSC-specific cell surface markers (Barberi et al., 2005; Lee et al., 2007) helps to enrich functional MSC-like cells and demonstrate their multi-lineage differentiation into fat, cartilage, bone, and muscle cells. The development of MSCs often requires long-term culture (sometimes for more than 2 months) in a serum-containing medium. A challenge for the future will be to determine how extensive expansion culture is associated with the genesis of MSCs from ES cell progeny.

8. Optimization of the factor environment for chondrogenesis from PS cell-derived chondroprogenitor cells

The 2D micromass or 3D pellet cultures, in which cells are maintained at a very high density, can mimic precartilage condensation as observed during the early stage of chondrogenesis. Both culture systems have been extensively used to differentiate embryonic limb mesenchymal cells into chondrocytes, and to induce chondrogenic differentiation of mesenchymal cell lines (Denker et al., 1995) and MSCs (Johnstone et al., 1998; Mackay et al., 1998; Yoo et al., 1998). Therefore, these assays have been used for verifying chondrogenic activity of the ES cell progeny. Cartilaginous particles formed by the 3D pellet culture allow histological and immunohistochemical analyses, which in turn provide semi-quantitative information on the type of cartilage formed. Chondrogenesis does not occur spontaneously, particularly in a serum-free environment, but must be stimulated by the addition of protein factors. According to the reports, factors thus far tested for crude, expanded or purified ES cell progeny are largely resemble those used for chondrogenic induction from human MSCs, namely TGFβ, BMP2/4, BMP6/7. However, many other signalling molecules are known to be essential for chondrogenesis (Fig. 6A). There is a possibility that such non-TGFβ-BMP signaling mechanisms may functionally distinguish the chondroprogenitor cells of different origin, which would support the hypothesis that the different lineage-specific chondroprogenitor cells would behave differently in a given environment *in vivo*.

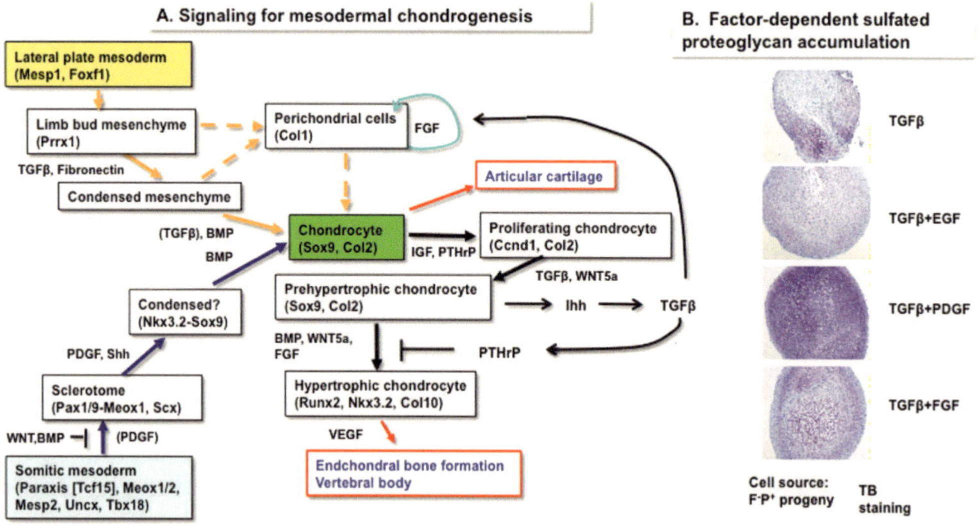

Fig. 6. Growth factor signaling requirement for mesodermal chondrogenesis *in vivo* and *in vitro*

To demonstrate that the mesodermal progeny generated from mouse ES cells are capable of forming a macroscopic cartilage particle dependent on added protein factors, we employed the 3D pellet assay method based on a chemically-defined serum-free medium, originally established for adult MSCs (Johnstone et al., 1998; Mackay et al., 1998; Yoo et al., 1998), and developed a 2D micromass culture method using the same serum-free medium (Nakayama et al., 2003). Combined with the serum-free EB formation method (Johansson & Wiles, 1995; Nakayama et al., 2000; Wiles & Johansson, 1997), we have established an entirely serum-free culture system that supports differentiation from ES cells to mesodermal progeny, and on to mesodermal chondrocytes (Nakayama et al., 2003; Tanaka et al., 2009). The resulting ES cell-derived mesodermal progeny has proven sensitive to environmental cues such as PDGF, TGFβ and BMP4 in the serum-free pellet culture and responds to them by producing different types of cartilage particle: e.g. proteoglycan-rich, type I collagen-poor hyaline-like cartilage, or proteoglycan-poor, type I collagen-containing fibrocartilage (Nakayama et al., 2003). TGFβ alone facilitates stable particle formation but cartilage matrix deposition is weak, whereas TGFβ+PDGF allows uniform deposition of cartilage matrix in the particle (Fig. 6B). TGFβ+epidermal growth factor (EGF) forms a large particle consisting mainly of undifferentiated mesenchymal cells that do not deposit cartilage matrix. TGFβ+FGF results in a large particle comprising a thick perichondrial cell layer on the surface and hypertrophic-like chondrocytes inside, which is associated with weak deposition of cartilage matrix.

9. Future perspective

Since first reported by Kremer, et al. in 2000, research on chondrogenesis from PS cells has made rapid progress. The question of whether ES cell-derived chondroprogenitor cells are MSCs has been addressed primarily with human ES cells. A consistent observation in all reported methods is that the genesis of MSCs from ES cells requires expansion of precursor (mesenchymal) cells for various times in a serum-containing MSC medium. What is the critical role of serum in the process? Furthermore, another interesting question relates to the identity of chondrocyte-derived secreted signals that best support the genesis and maintenance of the chondroprogenitor and MSC activities from human ES cells. Defining both signaling components and mechanisms will shed light on the fundamental biological mechanism of the genesis of MSC activity *in vivo*. On the other hand, lineage origin of the developed chondrocytes or their precursor (mesenchymal) cells has never been addressed. There are, however, good reasons to hypothesize that the ES/iPS cell-derived embryonic chondrocytes differ from adult MSCs in their ability to regenerate hyaline cartilage, and ES/iPS cell-derived embryonic chondrocytes also differ, depending on their lineage origin, in their efficiency and preference for forming particular types of cartilage. To address these suggestions, greater control of the differentiation of PS cells is needed to allow production of sufficient lineage-specific chondroprogenitors for comparative preclinical (or clinical) studies with adult MSCs. For example, methods need to be established for facilitating maturation of rostral paraxial mesoderm to Pax1-positive sclerotome, producing Prrx1-expressing limb mesenchyme from lateral plate mesoderm, and improving differentiation and isolation methods for cranial neural crest and their osteochondrogenic progeny from human PS cells. In addition, finding the critical environmental cue or cues that direct the type of cartilage formed from human PS cell-derived chondrocytes of a particular developmental origin will help to generate that type of cartilage *in vitro* (in scaffold) and *in vivo*. Such investigations will facilitate the establishment of a new cell-based therapy for the treatment of cartilage defects. In this context, it is worth noting that a human ES cell-derived fibrocartilagenous particle transplanted to a size-matched osteochondral defect in rat knee showed multi-step changes in the matrix and type of component cells (Toh et al., 2010). In addition during preparation of this review, one new paper reporting that mesodermal differentiation of human ES cells in a 2D culture using altered growth factor treatments in a chemically defined medium: i.e. WNT3a and Activin A, then BMP4 and FGF2, and then BMP4 is replaced gradually with GDF5, which resulted in Sox9 and type II collagen expressing chondrocytes, was published (Oldershaw et al., 2010).

10. References

Abzhanov, A.; Tzahor, E.; Lassar, A.B. & Tabin, C.J. (2003). Dissimilar regulation of cell differentiation in mesencephalic (cranial) and sacral (trunk) neural crest cells in vitro. *Development*, Vol. 130, No. 19, 4567-4579. 0950-1991

Asakura, A.; Komaki, M. & Rudnicki, M. (2001). Muscle satellite cells are multipotential stem cells that exhibit myogenic, osteogenic, and adipogenic differentiation. *Differentiation*, Vol. 68, No. 4-5, 245-253. 0301-4681

Barberi, T.; Willis, L.M.; Socci, N.D. & Studer, L. (2005). Derivation of multipotent mesenchymal precursors from human embryonic stem cells. *PLoS Med*, Vol. 2, No. 6, e161. 1549-1676

Barberi, T.; Bradbury, M.; Dincer, Z.; Panagiotakos, G.; Socci, N.D. & Studer, L. (2007). Derivation of engraftable skeletal myoblasts from human embryonic stem cells. *Nat Med*, Vol. 13, No. 5, 642-648. 1078-8956

Bi, W.; Deng, J.M.; Zhang, Z.; Behringer, R.R. & de Crombrugghe, B. (1999). Sox9 is required for cartilage formation. *Nat Genet*, Vol. 22, No. 1, 85-89. 1061-4036

Bianco, P.; Robey, P.G. & Simmons, P.J. (2008). Mesenchymal stem cells: revisiting history, concepts, and assays. *Cell Stem Cell*, Vol. 2, No. 4, 313-319. 1875-9777

Bigdeli, N.; Karlsson, C.; Strehl, R.; Concaro, S.; Hyllner, J. & Lindahl, A. (2009). Coculture of human embryonic stem cells and human articular chondrocytes results in significantly altered phenotype and improved chondrogenic differentiation. *Stem Cells*, Vol. 27, No. 8, 1812-1821. 1549-4918

Brittberg, M.; Lindahl, A.; Nilsson, A.; Ohlsson, C.; Isaksson, O. & Peterson, L. (1994). Treatment of deep cartilage defects in the knee with autologous chondrocyte transplantation. *N Engl J Med*, Vol. 331, No. 14, 889-895. 0028-4793

Darabi, R.; Gehlbach, K.; Bachoo, R.M.; Kamath, S.; Osawa, M.; Kamm, K.E.; Kyba, M. & Perlingeiro, R.C. (2008). Functional skeletal muscle regeneration from differentiating embryonic stem cells. *Nat Med*, Vol. 14, No. 2, 134-143. 1546-170X

De Bari, C.; Dell'Accio, F. & Luyten, F.P. (2001a). Human periosteum-derived cells maintain phenotypic stability and chondrogenic potential throughout expansion regardless of donor age. *Arthritis Rheum*, Vol. 44, No. 1, 85-95. 0004-3591

De Bari, C.; Dell'Accio, F.; Tylzanowski, P. & Luyten, F.P. (2001b). Multipotent mesenchymal stem cells from adult human synovial membrane. *Arthritis Rheum*, Vol. 44, No. 8, 1928-1942. 0004-3591

Denker, A.E.; Nicoll, S.B. & Tuan, R.S. (1995). Formation of cartilage-like spheroids by micromass cultures of murine C3H10T1/2 cells upon treatment with transforming growth factor-beta 1. *Differentiation*, Vol. 59, No. 1, 25-34. 0301-4681

Duprez, D.M.; Coltey, M.; Amthor, H.; Brickell, P.M. & Tickle, C. (1996). Bone morphogenetic protein-2 (BMP-2) inhibits muscle development and promotes cartilage formation in chick limb bud cultures. *Dev Biol*, Vol. 174, No. 2, 448-452.

Evans, M.J. & Kaufman, M.H. (1981). Establishment in culture of pluripotential cells from mouse embryos. *Nature*, Vol. 292, No. 5819, 154-156. 0028-0836

Fecek, C.; Yao, D.; Kacorri, A.; Vasquez, A.; Iqbal, S.; Sheikh, H.; Svinarich, D.M.; Perez-Cruet, M. & Chaudhry, G.R. (2008). Chondrogenic derivatives of embryonic stem cells seeded into 3D polycaprolactone scaffolds generated cartilage tissue in vivo. *Tissue engineering Part A*, Vol. 14, No. 8, 1403-1413. 1937-3341

Fehling, H.J.; Lacaud, G.; Kubo, A.; Kennedy, M.; Robertson, S.; Keller, G. & Kouskoff, V. (2003). Tracking mesoderm induction and its specification to the hemangioblast during embryonic stem cell differentiation. *Development*, Vol. 130, No. 17, 4217-4227.

Fukumoto, T.; Sperling, J.W.; Sanyal, A.; Fitzsimmons, J.S.; Reinholz, G.G.; Conover, C.A. & O'Driscoll, S.W. (2003). Combined effects of insulin-like growth factor-1 and transforming growth factor-beta1 on periosteal mesenchymal cells during chondrogenesis in vitro. *Osteoarthritis Cartilage*, Vol. 11, No. 1, 55-64. 1063-4584

Fusaki, N.; Ban, H.; Nishiyama, A.; Saeki, K. & Hasegawa, M. (2009). Efficient induction of transgene-free human pluripotent stem cells using a vector based on Sendai virus, an RNA virus that does not integrate into the host genome. *Proc Jpn Acad Ser B Phys Biol Sci*, Vol. 85, No. 8, 348-362. 1349-2896

Gadue, P.; Huber, T.L.; Nostro, M.C.; Kattman, S. & Keller, G.M. (2005). Germ layer induction from embryonic stem cells. *Exp Hematol*, Vol. 33, No. 9, 955-964.

Gadue, P.; Huber, T.L.; Paddison, P.J. & Keller, G.M. (2006). Wnt and TGF-beta signaling are required for the induction of an in vitro model of primitive streak formation using embryonic stem cells. *Proc Natl Acad Sci U S A*, Vol. 103, No. 45, 16806-16811. 0027-8424

Gikas, P.D.; Bayliss, L.; Bentley, G. & Briggs, T.W. (2009). An overview of autologous chondrocyte implantation. *J Bone Joint Surg Br*, Vol. 91, No. 8, 997-1006. 0301-620X

Goldring, M.B.; Tsuchimochi, K. & Ijiri, K. (2006). The control of chondrogenesis. *J Cell Biochem*, Vol. 97, No. 1, 33-44. 0730-2312

Gong, G.; Ferrari, D.; Dealy, C.N. & Kosher, R.A. (2010). Direct and progressive differentiation of human embryonic stem cells into the chondrogenic lineage. *J Cell Physiol*, Vol. 224, No. 3, 664-671. 1097-4652

Gronthos, S.; Mankani, M.; Brahim, J.; Robey, P.G. & Shi, S. (2000). Postnatal human dental pulp stem cells (DPSCs) in vitro and in vivo. *Proc Natl Acad Sci U S A*, Vol. 97, No. 25, 13625-13630. 0027-8424

Harkness, L.; Taipaleenmaki, H.; Mahmood, A.; Frandsen, U.; Saamanen, A.M.; Kassem, M. & Abdallah, B.M. (2009). Isolation and differentiation of chondrocytic cells derived from human embryonic stem cells using dlk1/FA1 as a novel surface marker. *Stem Cell Rev*, Vol. 5, No. 4, 353-368. 1558-6804

Hoben, G.M.; Koay, E.J. & Athanasiou, K.A. (2008). Fibrochondrogenesis in two embryonic stem cell lines: effects of differentiation timelines. *Stem Cells*, Vol. 26, No. 2, 422-430. 1549-4918

Hoben, G.M.; Willard, V.P. & Athanasiou, K.A. (2009). Fibrochondrogenesis of hESCs: growth factor combinations and cocultures. *Stem Cells Dev*, Vol. 18, No. 2, 283-292. 1557-8534

Hu, J.; Feng, K.; Liu, X. & Ma, P.X. (2009). Chondrogenic and osteogenic differentiations of human bone marrow-derived mesenchymal stem cells on a nanofibrous scaffold with designed pore network. *Biomaterials*, Vol. 30, No. 28, 5061-5067. 1878-5905

Hwang, N.S.; Kim, M.S.; Sampattavanich, S.; Baek, J.H.; Zhang, Z. & Elisseeff, J. (2006a). Effects of three-dimensional culture and growth factors on the chondrogenic differentiation of murine embryonic stem cells. *Stem Cells*, Vol. 24, No. 2, 284-291. 1066-5099

Hwang, N.S.; Varghese, S.; Zhang, Z. & Elisseeff, J. (2006b). Chondrogenic differentiation of human embryonic stem cell-derived cells in arginine-glycine-aspartate-modified hydrogels. *Tissue Eng*, Vol. 12, No. 9, 2695-2706. 1076-3279

Hwang, N.S.; Varghese, S. & Elisseeff, J. (2008a). Derivation of chondrogenically-committed cells from human embryonic cells for cartilage tissue regeneration. *PLoS ONE*, Vol. 3, No. 6, e2498. 1932-6203

Hwang, N.S.; Varghese, S.; Lee, H.J.; Zhang, Z.; Ye, Z.; Bae, J.; Cheng, L. & Elisseeff, J. (2008b). In vivo commitment and functional tissue regeneration using human embryonic stem cell-derived mesenchymal cells. *Proc Natl Acad Sci U S A*, Vol. 105, No. 52, 20641-20646. 1091-6490

Ikeda, T.; Kamekura, S.; Mabuchi, A.; Kou, I.; Seki, S.; Takato, T.; Nakamura, K.; Kawaguchi, H.; Ikegawa, S. & Chung, U.I. (2004). The combination of SOX5, SOX6, and SOX9 (the SOX trio) provides signals sufficient for induction of permanent cartilage. *Arthritis Rheum*, Vol. 50, No. 11, 3561-3573. 0004-3591

Ishizaka, S.; Shiroi, A.; Kanda, S.; Yoshikawa, M.; Tsujinoue, H.; Kuriyama, S.; Hasuma, T.; Nakatani, K. & Takahashi, K. (2002). Development of hepatocytes from ES cells after transfection with the HNF-3beta gene. *Faseb J*, Vol. 16, No. 11, 1444-1446. 1530-6860

Johansson, B.M. & Wiles, M.V. (1995). Evidence for involvement of activin A and bone morphogenetic protein 4 in mammalian mesoderm and hematopoietic development. *Mol Cell Biol*, Vol. 15., No. 1., 141-151. 0270-7306

Johnstone, B.; Hering, T.M.; Caplan, A.I.; Goldberg, V.M. & Yoo, J.U. (1998). In vitro chondrogenesis of bone marrow-derived mesenchymal progenitor cells. *Exp Cell Res*, Vol. 238, No. 1, 265-272.

Jukes, J.M.; Moroni, L.; van Blitterswijk, C.A. & de Boer, J. (2008). Critical Steps toward a tissue-engineered cartilage implant using embryonic stem cells. *Tissue engineering Part A*, Vol. 14, No. 1, 135-147. 1937-3341

Jukes, J.M.; van Blitterswijk, C.A. & de Boer, J. (2010). Skeletal tissue engineering using embryonic stem cells. *J Tissue Eng Regen Med*, Vol. 4, No. 3, 165-180. 1932-7005

Kaji, K.; Norrby, K.; Paca, A.; Mileikovsky, M.; Mohseni, P. & Woltjen, K. (2009). Virus-free induction of pluripotency and subsequent excision of reprogramming factors. *Nature*, Vol. 458, No. 7239, 771-775. 1476-4687

Kawaguchi, J.; Mee, P.J. & Smith, A.G. (2005). Osteogenic and chondrogenic differentiation of embryonic stem cells in response to specific growth factors. *Bone*, Vol. 36, No. 5, 758-769. 8756-3282

Keller, G. (2005). Embryonic stem cell differentiation: emergence of a new era in biology and medicine. *Genes Dev*, Vol. 19, No. 10, 1129-1155

Kim, J.H.; Do, H.J.; Yang, H.M.; Oh, J.H.; Choi, S.J.; Kim, D.K.; Cha, K.Y. & Chung, H.M. (2005). Overexpression of SOX9 in mouse embryonic stem cells directs the immediate chondrogenic commitment. *Exp Mol Med*, Vol. 37, No. 4, 261-268. 1226-3613

Koay, E.J.; Hoben, G.M. & Athanasiou, K.A. (2007). Tissue engineering with chondrogenically differentiated human embryonic stem cells. *Stem Cells*, Vol. 25, No. 9, 2183-2190. 1549-4918

Koay, E.J. & Athanasiou, K.A. (2008). Hypoxic chondrogenic differentiation of human embryonic stem cells enhances cartilage protein synthesis and biomechanical functionality. *Osteoarthritis Cartilage*. 1063-4584

Koay, E.J. & Athanasiou, K.A. (2009). Development of Serum-Free, Chemically Defined Conditions for Human Embryonic Stem Cell-Derived Fibrochondrogenesis. *Tissue Eng Part A*. 1937-3341

Kramer, J.; Hegert, C.; Guan, K.; Wobus, A.M.; Mueller, P.K. & Rohwedel, J. (2000). Embryonic stem cell-derived chondrogenic differentiation in vitro: activation by BMP-2 and BMP-4. *Mech Dev*, Vol. 92, No. 2, 193-205. 0925-4773

Kramer, J.; Hegert, C.; Hargus, G. & Rohwedel, J. (2005a). Mouse ES cell lines show a variable degree of chondrogenic differentiation in vitro. *Cell Biol Int*, Vol. 29, No. 2, 139-146. 1065-6995

Kramer, J.; Klinger, M.; Kruse, C.; Faza, M.; Hargus, G. & Rohwedel, J. (2005b). Ultrastructural analysis of mouse embryonic stem cell-derived chondrocytes. *Anat Embryol*, Vol. 210, No. 3, 175-185. 0340-2061

Kulyk, W.M.; Rodgers, B.J.; Greer, K. & Kosher, R.A. (1989). Promotion of embryonic chick limb cartilage differentiation by transforming growth factor-beta. *Dev Biol*, Vol. 135, No. 2, 424-430.

Lee, G.; Kim, H.; Elkabetz, Y.; Al Shamy, G.; Panagiotakos, G.; Barberi, T.; Tabar, V. & Studer, L. (2007). Isolation and directed differentiation of neural crest stem cells derived from human embryonic stem cells. *Nat Biotechnol*, Vol. 25, No. 12, 1468-1475. 1546-1696

Lee, H.J.; Yu, C.; Chansakul, T.; Varghese, S.; Hwang, N.S. & Elisseeff, J.H. (2008). Enhanced chondrogenic differentiation of embryonic stem cells by coculture with hepatic cells. *Stem Cells Dev*, Vol. 17, No. 3, 555-563. 1547-3287

Levenberg, S.; Huang, N.F.; Lavik, E.; Rogers, A.B.; Itskovitz-Eldor, J. & Langer, R. (2003). Differentiation of human embryonic stem cells on three-dimensional polymer scaffolds. *Proc Natl Acad Sci U S A*, Vol. 100, No. 22, 12741-12746. 0027-8424

Li, W.J.; Tuli, R.; Huang, X.; Laquerriere, P. & Tuan, R.S. (2005). Multilineage differentiation of human mesenchymal stem cells in a three-dimensional nanofibrous scaffold. *Biomaterials*, Vol. 26, No. 25, 5158-5166. 0142-9612

Mackay, A.M.; Beck, S.C.; Murphy, J.M.; Barry, F.P.; Chichester, C.O. & Pittenger, M.F. (1998). Chondrogenic differentiation of cultured human mesenchymal stem cells from marrow. *Tissue Eng*, Vol. 4, No. 4., 415-428.

Mahmood, A.; Harkness, L.; Schroder, H.D.; Abdallah, B.M. & Kassem, M. (2010). Enhanced differentiation of human embryonic stem cells to mesenchymal progenitors by inhibition of TGF-beta/activin/nodal signaling using SB-431542. *J Bone Miner Res*, Vol. 25, No. 6, 1216-1233. 1523-4681

Martin, G.R. (1981). Isolation of a pluripotent cell line from early mouse embryos cultured in medium conditioned by teratocarcinoma stem cells. *Proc Natl Acad Sci U S A*, Vol. 78, No. 12, 7634-7638. 0027-8424

Mizuseki, K.; Sakamoto, T.; Watanabe, K.; Muguruma, K.; Ikeya, M.; Nishiyama, A.; Arakawa, A.; Suemori, H.; Nakatsuji, N.; Kawasaki, H.; Murakami, F. & Sasai, Y. (2003). Generation of neural crest-derived peripheral neurons and floor plate cells from mouse and primate embryonic stem cells. *Proc Natl Acad Sci U S A*, Vol. 100, No. 10, 5828-5833.

Morikawa, S.; Mabuchi, Y.; Niibe, K.; Suzuki, S.; Nagoshi, N.; Sunabori, T.; Shimmura, S.; Nagai, Y.; Nakagawa, T.; Okano, H. & Matsuzaki, Y. (2009). Development of mesenchymal stem cells partially originate from the neural crest. *Biochem Biophys Res Commun*, Vol. 379, No. 4, 1114-1119. 1090-2104

Moriya, T.; Wada, Y.; Watanabe, A.; Sasho, T.; Nakagawa, K.; Mainil-Varlet, P. & Moriya, H. (2007). Evaluation of reparative cartilage after autologous chondrocyte implantation for osteochondritis dissecans: histology, biochemistry, and MR imaging. *J Orthop Sci*, Vol. 12, No. 3, 265-273. 0949-2658

Murphy, J.M.; Dixon, K.; Beck, S.; Fabian, D.; Feldman, A. & Barry, F. (2002). Reduced chondrogenic and adipogenic activity of mesenchymal stem cells from patients with advanced osteoarthritis. *Arthritis Rheum*, Vol. 46, No. 3, 704-713. 0004-3591

Murry, C.E. & Keller, G. (2008). Differentiation of embryonic stem cells to clinically relevant populations: lessons from embryonic development. *Cell*, Vol. 132, No. 4, 661-680

Nakagawa, T.; Lee, S.Y. & Reddi, A.H. (2009). Induction of chondrogenesis from human embryonic stem cells without embryoid body formation by bone morphogenetic protein 7 and transforming growth factor beta1. *Arthritis Rheum*, Vol. 60, No. 12, 3686-3692. 0004-3591

Nakano, T.; Kodama, H. & Honjo, T. (1994). Generation of lymphohematopoietic cells from embryonic stem cells in culture. *Science*, Vol. 265, No. 5175, 1098-1101.

Nakano, T.; Kodama, H. & Honjo, T. (1996). In vitro development of primitive and definitive erythrocytes from different precursors. *Science*, Vol. 272, No. 5262, 722-724.

Nakayama, N.; Lee, J. & Chiu, L. (2000). Vascular endothelial growth factor synergistically enhances bone morphogenetic protein-4-dependent lymphohematopoietic cell generation from embryonic stem cells in vitro. *Blood*, Vol. 95, No. 7, 2275-2283. 0006-4971

Nakayama, N.; Duryea, D.; Manoukian, R.; Chow, G. & Han, C.Y. (2003). Macroscopic cartilage formation with embryonic stem-cell-derived mesodermal progenitor cells. *J Cell Sci*, Vol. 116, No. Pt 10, 2015-2028. 0021-9533

Nienhuis, A.W.; Dunbar, C.E. & Sorrentino, B.P. (2006). Genotoxicity of retroviral integration in hematopoietic cells. *Molecular therapy : the journal of the American Society of Gene Therapy*, Vol. 13, No. 6, 1031-1049. 1525-0016

Nishikawa, S.; Goldstein, R.A. & Nierras, C.R. (2008). The promise of human induced pluripotent stem cells for research and therapy. *Nat Rev Mol Cell Biol*, Vol. 9, No. 9, 725-729. 1471-0080

Nishikawa, S.-I.; Jakt, L.M. & Era, T. (2007). Embryonic stem-cell culture as a tool for developmental cell biology. *Nat Rev Mol Cell Biol*, Vol. 8, No. 6, 502-507. 1471-0072

Nishikawa, S.I.; Nishikawa, S.; Hirashima, M.; Matsuyoshi, N. & Kodama, H. (1998). Progressive lineage analysis by cell sorting and culture identifies FLK1+VE-cadherin+ cells at a diverging point of endothelial and hemopoietic lineages. *Development*, Vol. 125, No. 9, 1747-1757.

Nostro, M.C.; Cheng, X.; Keller, G.M. & Gadue, P. (2008). Wnt, activin, and BMP signaling regulate distinct stages in the developmental pathway from embryonic stem cells to blood. *Cell Stem Cell*, Vol. 2, No. 1, 60-71. 1934-5909

Okita, K.; Nakagawa, M.; Hyenjong, H.; Ichisaka, T. & Yamanaka, S. (2008). Generation of mouse induced pluripotent stem cells without viral vectors. *Science*, Vol. 322, No. 5903, 949-953. 1095-9203

Oldershaw, R.A.; Baxter, M.A.; Lowe, E.T.; Bates, N.; Grady, L.M.; Soncin, F.; Brison, D.R.; Hardingham, T.E. & Kimber, S.J. (2010). Directed differentiation of human embryonic stem cells toward chondrocytes. *Nat Biotechnol*, Vol. 28, No. 11, 1187-1194. 1546-1696

Park, I.H.; Zhao, R.; West, J.A.; Yabuuchi, A.; Huo, H.; Ince, T.A.; Lerou, P.H.; Lensch, M.W. & Daley, G.Q. (2008). Reprogramming of human somatic cells to pluripotency with defined factors. *Nature*, Vol. 451, No. 7175, 141-146. 1476-4687

Pittenger, M.F.; Mackay, A.M.; Beck, S.C.; Jaiswal, R.K.; Douglas, R.; Mosca, J.D.; Moorman, M.A.; Simonetti, D.W.; Craig, S. & Marshak, D.R. (1999). Multilineage potential of adult human mesenchymal stem cells. *Science*, Vol. 284, No. 5411, 143-147.

Richardson, S.M.; Curran, J.M.; Chen, R.; Vaughan-Thomas, A.; Hunt, J.A.; Freemont, A.J. & Hoyland, J.A. (2006). The differentiation of bone marrow mesenchymal stem cells into chondrocyte-like cells on poly-L-lactic acid (PLLA) scaffolds. *Biomaterials*, Vol. 27, No. 22, 4069-4078. 0142-9612

Richardson, S.M.; Hughes, N.; Hunt, J.A.; Freemont, A.J. & Hoyland, J.A. (2008). Human mesenchymal stem cell differentiation to NP-like cells in chitosan-glycerophosphate hydrogels. *Biomaterials*, Vol. 29, No. 1, 85-93. 0142-9612

Roberts, S.; Menage, J.; Sandell, L.J.; Evans, E.H. & Richardson, J.B. (2009). Immunohistochemical study of collagen types I and II and procollagen IIA in human cartilage repair tissue following autologous chondrocyte implantation. *Knee*, Vol. 16, No. 5, 398-404. 1873-5800

Sakurai, H.; Era, T.; Jakt, L.M.; Okada, M.; Nakai, S.; Nishikawa, S. & Nishikawa, S.I. (2006). In vitro modeling of paraxial and lateral mesoderm differentiation reveals early reversibility. *Stem Cells*, 575-586. 1066-5099

Shi, Y.; Desponts, C.; Do, J.T.; Hahm, H.S.; Scholer, H.R. & Ding, S. (2008). Induction of pluripotent stem cells from mouse embryonic fibroblasts by Oct4 and Klf4 with small-molecule compounds. *Cell Stem Cell*, Vol. 3, No. 5, 568-574. 1875-9777

Smith, J.R.; Vallier, L.; Lupo, G.; Alexander, M.; Harris, W.A. & Pedersen, R.A. (2008). Inhibition of Activin/Nodal signaling promotes specification of human embryonic stem cells into neuroectoderm. *Dev Biol*, Vol. 313, No. 1, 107-117. 1095-564X

Stadtfeld, M.; Nagaya, M.; Utikal, J.; Weir, G. & Hochedlinger, K. (2008). Induced pluripotent stem cells generated without viral integration. *Science*, Vol. 322, No. 5903, 945-949. 1095-9203

Steinert, A.F.; Ghivizzani, S.C.; Rethwilm, A.; Tuan, R.S.; Evans, C.H. & Noth, U. (2007). Major biological obstacles for persistent cell-based regeneration of articular cartilage. *Arthritis Res Ther*, Vol. 9, No. 3, 213. 1478-6362

Steventon, B.; Carmona-Fontaine, C. & Mayor, R. (2005). Genetic network during neural crest induction: from cell specification to cell survival. *Semin Cell Dev Biol*, Vol. 16, No. 6, 647-654. 1084-9521

Sui, Y.; Clarke, T. & Khillan, J.S. (2003). Limb bud progenitor cells induce differentiation of pluripotent embryonic stem cells into chondrogenic lineage. *Differentiation*, Vol. 71, No. 9-10, 578-585. 0301-4681

Takahashi, K. & Yamanaka, S. (2006). Induction of pluripotent stem cells from mouse embryonic and adult fibroblast cultures by defined factors. *Cell*, Vol. 126, No. 4, 663-676. 0092-8674

Takahashi, K.; Tanabe, K.; Ohnuki, M.; Narita, M.; Ichisaka, T.; Tomoda, K. & Yamanaka, S. (2007). Induction of pluripotent stem cells from adult human fibroblasts by defined factors. *Cell*, Vol. 131, No. 5, 861-872. 0092-8674

Takakura, N.; Yoshida, H.; Ogura, Y.; Kataoka, H. & Nishikawa, S. (1997). PDGFR alpha expression during mouse embryogenesis: immunolocalization analyzed by whole-mount immunohistostaining using the monoclonal anti-mouse PDGFR alpha antibody APA5. *J Histochem Cytochem*, Vol. 45, No. 6, 883-893.

Takashima, Y.; Era, T.; Nakao, K.; Kondo, S.; Kasuga, M.; Smith, A.G. & Nishikawa, S. (2007). Neuroepithelial cells supply an initial transient wave of MSC differentiation. *Cell*, Vol. 129, No. 7, 1377-1388. 0092-8674

Tanaka, M.; Jokubaitis, V.; Wood, C.; Wang, Y.; Brouard, N.; Pera, M.; Hearn, M.; Simmons, P. & Nakayama, N. (2009). BMP inhibition stimulates WNT-dependent generation of chondrogenic mesoderm from embryonic stem cells. *Stem Cell Res*, Vol. 3, No. 2-3, 126-141. 1876-7753

Terraciano, V.; Hwang, N.; Moroni, L.; Park, H.B.; Zhang, Z.; Mizrahi, J.; Seliktar, D. & Elisseeff, J. (2007). Differential response of adult and embryonic mesenchymal progenitor cells to mechanical compression in hydrogels. *Stem Cells*, Vol. 25, No. 11, 2730-2738. 1549-4918

Thomson, J.A.; Itskovitz-Eldor, J.; Shapiro, S.S.; Waknitz, M.A.; Swiergiel, J.J.; Marshall, V.S. & Jones, J.M. (1998). Embryonic stem cell lines derived from human blastocysts. *Science*, Vol. 282, No. 5391, 1145-1147. 0036-8075

Toh, W.S.; Yang, Z.; Liu, H.; Heng, B.C.; Lee, E.H. & Cao, T. (2007). Effects of culture conditions and bone morphogenetic protein 2 on extent of chondrogenesis from human embryonic stem cells. *Stem Cells*, Vol. 25, No. 4, 950-960. 1066-5099

Toh, W.S.; Guo, X.M.; Choo, A.B.; Lu, K.; Lee, E.H. & Cao, T. (2009). Differentiation and enrichment of expandable chondrogenic cells from human embryonic stem cells in vitro. *J Cell Mol Med*, Vol. 13, No. 9B, 3570-3590. 1582-4934

Toh, W.S.; Lee, E.H.; Guo, X.M.; Chan, J.K.; Yeow, C.H.; Choo, A.B. & Cao, T. (2010). Cartilage repair using hyaluronan hydrogel-encapsulated human embryonic stem cell-derived chondrogenic cells. *Biomaterials*, Vol. 31, No. 27, 6968-6980. 1878-5905

Tokalov, S.V.; Gruner, S.; Schindler, S.; Wolf, G.; Baumann, M. & Abolmaali, N. (2007). Age-related changes in the frequency of mesenchymal stem cells in the bone marrow of rats. *Stem Cells Dev*, Vol. 16, No. 3, 439-446. 1547-3287

Vats, A.; Bielby, R.C.; Tolley, N.; Dickinson, S.C.; Boccaccini, A.R.; Hollander, A.P.; Bishop, A.E. & Polak, J.M. (2006). Chondrogenic differentiation of human embryonic stem cells: the effect of the micro-environment. *Tissue Eng*, Vol. 12, No. 6, 1687-1697. 1076-3279

Wallin, J.; Wilting, J.; Koseki, H.; Fritsch, R.; Christ, B. & Balling, R. (1994). The role of Pax-1 in axial skeleton development. *Development*, Vol. 120, No. 5, 1109-1121. 0950-1991

Wang, Y. & Nakayama, N. (2009). WNT and BMP signaling are both required for hematopoietic cell development from human ES cells. *Stem Cell Res*, Vol. 3, No. 2-3, 113-125. 1876-7753

Wang, Y.; Umeda, K. & Nakayama, N. (2010). Collaboration between WNT and BMP signaling promotes hemoangiogenic cell development from human fibroblast-derived iPS cells. *Stem Cell Res*, Vol. 4, No. 3, 223-231. 1876-7753

Wiles, M.V. & Johansson, B.M. (1997). Analysis of factors controlling primary germ layer formation and early hematopoiesis using embryonic stem cell in vitro differentiation. *Leukemia*, Vol. 11 Suppl 3, 454-456.

Woltjen, K.; Michael, I.P.; Mohseni, P.; Desai, R.; Mileikovsky, M.; Hamalainen, R.; Cowling, R.; Wang, W.; Liu, P.; Gertsenstein, M.; Kaji, K.; Sung, H.K. & Nagy, A. (2009). piggyBac transposition reprograms fibroblasts to induced pluripotent stem cells. *Nature*, Vol. 458, No. 7239, 766-770. 1476-4687

Wong, C.E.; Paratore, C.; Dours-Zimmermann, M.T.; Rochat, A.; Pietri, T.; Suter, U.; Zimmermann, D.R.; Dufour, S.; Thiery, J.P.; Meijer, D.; Beermann, F.; Barrandon, Y. & Sommer, L. (2006). Neural crest-derived cells with stem cell features can be traced back to multiple lineages in the adult skin. *J Cell Biol*, Vol. 175, No. 6, 1005-1015. 0021-9525

Xu, C.; Jiang, J.; Sottile, V.; McWhir, J.; Lebkowski, J. & Carpenter, M.K. (2004). Immortalized fibroblast-like cells derived from human embryonic stem cells support undifferentiated cell growth. *Stem Cells*, Vol. 22, No. 6, 972-980. 1066-5099

Yamaguchi, T.P.; Dumont, D.J.; Conlon, R.A.; Breitman, M.L. & Rossant, J. (1993). flk-1, an flt-related receptor tyrosine kinase is an early marker for endothelial cell precursors. *Development*, Vol. 118, No. 2, 489-498.

Yoo, J.U.; Barthel, T.S.; Nishimura, K.; Solchaga, L.; Caplan, A.I.; Goldberg, V.M. & Johnstone, B. (1998). The chondrogenic potential of human bone-marrow-derived mesenchymal progenitor cells. *J Bone Joint Surg Am*, Vol. 80, No. 12, 1745-1757. 0021-9355

Yu, J.; Vodyanik, M.A.; Smuga-Otto, K.; Antosiewicz-Bourget, J.; Frane, J.L.; Tian, S.; Nie, J.; Jonsdottir, G.A.; Ruotti, V.; Stewart, R.; Slukvin, II & Thomson, J.A. (2007). Induced pluripotent stem cell lines derived from human somatic cells. *Science*, Vol. 318, No. 5858, 1917-1920. 1095-9203

Yu, J.; Hu, K.; Smuga-Otto, K.; Tian, S.; Stewart, R.; Slukvin, II & Thomson, J.A. (2009). Human induced pluripotent stem cells free of vector and transgene sequences. *Science*, Vol. 324, No. 5928, 797-801. 1095-9203

Yusa, K.; Rad, R.; Takeda, J. & Bradley, A. (2009). Generation of transgene-free induced pluripotent mouse stem cells by the piggyBac transposon. *Nat methods*, Vol. 6, No. 5, 363-369. 1548-7105

Zhou, Y. & Snead, M.L. (2008). Derivation of cranial neural crest-like cells from human embryonic stem cells. *Biochem Biophys Res Commun*, Vol. 376, No. 3, 542-547. 1090-2104

Zuk, P.A.; Zhu, M.; Ashjian, P.; De Ugarte, D.A.; Huang, J.I.; Mizuno, H.; Alfonso, Z.C.; Fraser, J.K.; Benhaim, P. & Hedrick, M.H. (2002). Human adipose tissue is a source of multipotent stem cells. *Molecular biology of the cell*, Vol. 13, No. 12, 4279-4295. 1059-1524 zur Nieden, N.I.; Kempka, G.; Rancourt, D.E. & Ahr, H.J. (2005). Induction of chondro-, osteo- and adipogenesis in embryonic stem cells by bone morphogenetic protein-2: effect of cofactors on differentiating lineages. *BMC Dev Biol*, Vol. 5, 1. 1471-213X

33

Derivation of Mesenchymal Stem Cells from Human Embryonic Stem Cells

Shelley E. Brown[1,2] and Paul H. Krebsbach[1,2]
[1]*Biologic and Materials Sciences Department, University of Michigan School of Dentistry*
[2]*Biomedical Engineering Department, University of Michigan College of Engineering*
USA

1. Introduction

In the early to mid 1970's, hallmark studies were published demonstrating the ability to isolate and derive adult fibroblast cell colonies from the bone marrow stroma and the spleen (Friedenstein *et al.*, 1970; Friedenstein and Kuralesova, 1971). These fibroblast-like cells, later termed bone marrow stromal cells (BMSCs) or bone marrow-derived mesenchymal stem cells (MSCs), were shown to proliferate in culture, to continually grow upon passaging while maintaining stable karyotypic characteristics, and were comprised of cells that had multipotent potential to differentiate along multiple mesenchymal cell lineages such as bone, cartilage, fat and could support hematopoietic stem cell (HSC) differentiation (Bab *et al.*, 1984; Bab *et al.*, 1986; Friedenstein *et al.*, 1970; Friedenstein *et al.*, 1974b). Numerous studies spurred from these findings, which also led researchers in this area to explore the functions of these cells in vitro and in their normal microenvironment. Bone marrow stromal cells were transplanted *in vivo* to determine if they had the ability to re-establish the marrow microenvironment, and it was reported that the *ex vivo* expanded stromal cells did indeed restore the hematopoietic niche within the bone marrow (Friedenstein *et al.*, 1974a). These experiments further developed the hypothesis that within the bone marrow stroma resided a heterogeneous mixture of cells that function as a repository of progenitors, known as MSCs, that may migrate out of their stem cell niche in response to disease, injury, and aging. Therefore, extensive investigation into the identification of MSCs and their utility for cell-replacement therapies were the basis for a new emerging field known as tissue engineering (Ashton *et al.*, 1980; Bab *et al.*, 1986; Owen and Friedenstein, 1988; Beresford, 1989; Jaiswal *et al.*, 1997; Krebsbach *et al.*, 1999).

In the 1980's and 1990's, many groups further demonstrated that culture-adherent MSCs present in the marrow stroma were capable of differentiation into bone, cartilage, muscle, tendon, and fat for multiple species such as canine, chicken, rabbit, rat, and mouse (Jaiswal *et al.*, 1997). Using the expertise gained from these culture systems, MSCs were then isolated and propagated from human adult bone marrow (hMSCs) (Bab *et al.*, 1988); (Krebsbach *et al.*, 1997). Human MSCs were then used with site-specific delivery vehicles to repair bone, cartilage, and other connective tissues (Haynesworth *et al.*, 1992a; Haynesworth *et al.*, 1992b). Additionally, a series of monoclonal antibodies were developed to identify characteristic surface markers on hMSCs, which would prove to be beneficial to researchers interested in not only identifying MSCs, but also subpopulations of osteoprogenitor cells

(Haynesworth *et al.*, 1992a; Haynesworth *et al.*, 1992b); (Gronthos *et al.*, 1999). Simultaneously, Caplan et al. used the embryonic chick limb bud mesechymal cell culture system as an assay for the purification of inductive factors in bone to further develop the technology for isolating, expanding, and preserving the stem cell capacity of adult human bone marrow-derived mesenchymal stem cells (Caplan, 2005). With this newly acquired knowledge and the emerging technologies in biomedical engineering, hMSCs became the principle cell source for cell-based pre-clinical bone tissue engineering studies.

Currently, substantial advances have been made to address clinical needs for regeneration of damaged or diseased tissues. The three main approaches of cell-based clinical therapies that employ the use of hMSCs are: 1) from a tissue engineering standpoint where cells are incorporated into 3D biomaterial scaffolds for the replacement of tissue *in vivo*, 2) from a cell replacement therapy standpoint where allogeneic donor cells are used to replace ablated tumors and diseased cells; and 3) from a inductive standpoint where cells provide cytokine and growth factor cues that stimulate host reparative events and inhibit degenerative events (Caplan, 2005). Thus, clinical protocols were developed to establish that autologous hMSCs could be safely implanted back in order to reconstitute the marrow microenvironment for breast cancer and osteogenesis imperfecta (OI) patients following chemotherapy treatment (Koc *et al.*, 2000; Horwitz *et al.*, 2002). Additionally, hMSCs have been shown to have immunomodulatory effects and could induce immune suppression in patients (Le Blanc and Pittenger, 2005); (Aggarwal and Pittenger, 2005). Although the use of hMSCs has been successfully used in some cases, there are challenges that scientists and clinicians must overcome before the transplantation of these cells is incorporated into routine clinical practice. Specifically, the classic method to isolate MSCs from bone marrow relies on their capacity to adhere to plastic, their resistance to trypsinization during passaging, and proliferation in growth medium containing serum (Olivier *et al.*, 2006). However, cell availability is greatly limited with this method because MSCs are present at low concentrations in the marrow, occurring at less than 1 in 100,000-500,000 nucleated cells (Caplan, 2005). Also, the availability of tissues for their isolation remains limiting and requires invasive procedures that may cause severe donor site morbidity.

Therefore, an alternative source for generating MSCs can be found in human embryonic stem cells (hESCs) (Thomson *et al.*, 1998). Human ESCs are an alternative source for generating MSCs due to the fact that they can theoretically be expanded infinitely and also because using these cells would eliminate the need for invasive cell harvesting techniques. Host immune rejection could also be circumvented by the use of autologous hESCs generated from nuclear transfer or from immune compatible allogeneic hESCs. Derivation of mesenchymal stem cells from human ES cells will further the understanding of the differentiation pathways and important cellular events that occur during early human development and could also have useful clinical applications. Because of the therapeutic potential, particularly in the areas of cell therapy and regenerative medicine, derivation of MSCs from hESCs (hESC-MSCs) has specific advantages over the current "gold standard" use of autologous and allogeneic adult hMSCs for bone tissue engineering. (Olivier *et al.*, 2006).

2. Human embryonic stem cells

The major advancements in the area of stem cell culture, derivation, propagation, and differentiation paved the way for a pivotal discovery that was reported in a 1998 study from the University of Wisconsin. Thomson et. al. described the first successful isolation and long term sustained culture of a small cluster cells from the inner cell mass of four-day old embryos (Thomson *et al.*, 1998). These cells, known as human embryonic stem cells (hESCs), represent a robust biologic tool and model system through which the scientific and medical communities will better understand human development, disease pathophysiology, organogenesis, and mechanisms for cellular differentiation; all of which will help develop and improve the field of regenerative medicine. These embryonic stem cells are derived by the selection and expansion of individual colonies rather than clonal expansion of a single cell. Human ESCs are pluripotent cells that are presumed to have virtually unlimited proliferation capacity in vitro, maintain normal karyotypic characteristics, sustain high levels of telomerase activity, and retain uniform undifferentiated morphology in prolonged culture (Thomson *et al.*, 1998). In addition, hESCs have the ability to differentiate along the three embryonic germ layers *in vivo* as evidenced by teratoma formation after injection into severe combined immunodeficient (SCID) mice. The teratomas can contain gut epithelium (endoderm); cartilage, bone, smooth muscle, and striated muscle (mesoderm); and neural epithelium, embryonic ganglia, and stratified squamous epithelium (ectoderm). They have also been shown to express certain cell surface markers that are widely used to confirm pluripotency, such as stage-specific embryonic antigen SSEA-4, TRA-1-60, TRA-1-81, and alkaline phosphatase. Oct-4, a transcription factor, has been identified as another key indicator of the undifferentiated state.

To maintain their self renewal capacity, hESCs were originally cultured on mouse embryonic fibroblast (MEF) feeder layers and grown under serum-free conditions using serum replacement (SR) with supplements of basic fibroblast growth factor (bFGF). Under these culture conditions, hESCs have been passaged continuously and maintained pluripotency as well as a normal karyotype. However, it has been reported that hESCs have been successfully cultured with feeder cells of human origin, such as human bone marrow stromal cells (hBMSCs), human placental fibroblasts, human foreskin fibrobalsts (hFFs), feeders derived from hESCs, and on polymeric substrates in feeder free conditions (Cheng *et al.*, 2003; Genbacev *et al.*, 2005; Wang *et al.*, 2005; Stojkovic *et al.*, 2005; Hovatta *et al.*, 2003; Villa-Diaz *et al.*, 2010). In order to safely use hESCs in a clinical setting, it is imperative that that feeder-free and animal product-free culture conditions are explored further to overcome the risks of cross-transfer of pathogens from xenogeneic sources.

The ability of hESCs to maintain an undifferentiated state indefinitely in culture and to differentiate into all cell types and tissues within the human body has created a high demand for research. Although the cells are of great scientific interest, progression of this type of research has been met with great controversy and resistance due to the ethical concern of destroying early human embryos for derivation of hESC lines (Knowles, 2004; Baschetti, 2005; Gruen and Grabel, 2006). Nevertheless, once the ethical concerns are abated through placement of the appropriate guidelines and policies on research, the hESC field will not only evolve, but will continue to rapidly progress toward monumental medical and scientific breakthroughs.

3. Human embryonic stem cell derived mesenchymal stem cells

The current major goal for hESC research in regenerative medicine is the controlled differentiation into specific progenitor cells for the purpose of replacing or regenerating damaged tissue. Therefore, the ability to obtain large quantities of multipotent cells from hESCs represents a challenge for cell based therapy and tissue engineering strategies that currently rely on human bone marrow stromal cells (hMSCs). Within the diverse population

of hMSCs, there exist early progenitor mesenchymal stem cells capable of self-renewal and multi-lineage differentiation into cell types such as osteoblasts, chondrocytes, and adipocytes (Bianco *et al.*, 2003; Wagers and Weissman, 2004). While hMSCs make a useful source of osteoprogenitor cells for tissue engineering strategies, they have limited proliferation and differentiation capacity. In contrast, hESCs which are able to proliferate indefinitely in vitro, represent a potentially unlimited source of mesenchymal stem cells.
Recent studies demonstrate that the derivation of hESC-MSCS, mesenchymal precursors derived from hESCs, has been achieved via various isolation methods, and the generation of osteoblasts has been achieved in co-culture with primary bone derived cells (PBDs), in the presence of known osteogenic supplements, and in transwell co-culture with hBMSCs (Ahn *et al.*, 2006; Cao *et al.*, 2005; Duplomb, 2007; Karner *et al.*, 2007; Karp *et al.*, 2006; Sotille *et al.*, 2003; Tong *et al.*, 2007). Although the identification and characterization of hESC-MSCs has been reported, the data are quite vast and varied in terms of the derivation method, cell culture conditions, the mechanism of differentiation (epithelial-mesenchymal transition vs neural crest stem cell-mesenchymal differentiation), multilineage differentiation potential, and surface markers used to select for a pure mesenchymal stem cell subpopulation. As the field continues to evolve, careful attention should be placed on standardizing these parameters along clinical-grade good manufacturing practice (GMP guidelines). Through the isolation and identification of hESC-MSCs and the ability to produce a large supply of progenitor cells that can be genetically modified, the field hESC-MSC based tissue engineering and regenerative medicine strategies holds great promise.

3.1 Derivation methods

Thorough and extensive investigation into the definition, differentiation, and identification of mesenchymal stem cells has occurred over the last three decades. However, there are fundamental mechanistic and developmental concepts that remain poorly understood. The foundation laid by pioneers in the MSC field has provided current researchers with a breadth of knowledge to draw upon because the same fundamental questions are being investigated to identify the true "MSC" from differentiating hESCs. Many investigators state that although MSCs isolated from the adult bone marrow have been shown to differentiate *in vitro* and *in vivo*, as well as have been successfully used in a clinical setting to repopulate the marrow environment in cancer patients, harvesting and utilizing adult hMSCs has disadvantages such as tissue availability, donor site morbidity, and host immune rejection (Caplan, 2005; Horwitz *et al.*, 2002; Karp *et al.*, 2006). Therefore, hESCs have been the topic of great discussion and interest as a potential repository of cells that can provide an unlimited number of specialized mesenchymal stem cells known as hESC-MSCs.
Numerous isolation protocols have been reported describing successful derivation and differentiation of hESC-MSCs (Arpornmaeklong *et al.*, 2009; Barberi *et al.*, 2005; Brown *et al.*, 2009; de Peppo *et al.*, 2010a; de Peppo *et al.*, 2010b; Evseenko *et al.*, 2010; Karlsson *et al.*, 2009; Karp *et al.*, 2006; Kopher *et al.*, 2010; Kuznetsov *et al.*, ; Lian *et al.*, 2007; Olivier *et al.*, 2006; Smith *et al.*, 2009; Trivedi and Hematti, 2007; Xu *et al.*, 2004). One of the first reports of the derivation of a MSC-like progenitor population was in 2004, where fibroblast-like hESC derivatives were infected with a human telomerase reverse transcriptase (hTERT) retrovirus, as a result showed extended proliferative capacity, supported undifferentiated growth of hESCs as a feeder layer, and differentiated into osteoblasts (Xu *et al.*, 2004). Following that study, another group reported the successful production of hESC-MSCs when cultured on murine OP9 stromal cells in the presence of heat-inactivated FBS, and

indicated that the hESC-MSCs had a similar immunophenotype to hMSCs after flow cytometry was performed to purify the hESC-MSC population from the stromal cell feeder (Barberi *et al.*, 2005). Another method for hESC-MSC production involved the use of spontaneously differentiated hESC colonies. The cells obtained became morphologically fibroblastic and homogenous after multiple passages, possessed a characteristic MSC immunophenotype, and supported hESC and hematopoietic progenitor cell growth (Olivier *et al.*, 2006). Of particular importance, two reports showed the ability to reproducibly derive clinically compliant hESC-MSCs in a xeno-free environment where all contaminating animal-derived components were replaced with human-derived or recombinant components. Thus, they cultivated a hESC-MSC line suitable for clinical use ((Karlsson *et al.*, 2009; Lian *et al.*, 2007). Other groups described similar findings, demonstrating that hESCs had the ability to reproducibly proliferate, differentiate, and commit to the mesodermal lineage in various cell culture conditions (both in monolayer and 3D) while retaining their multilineage differentiation potential and self renewal capacity, further demonstrating their high potential for tissue engineering applications (Arpornmaeklong *et al.*, 2009; Brown *et al.*, 2009; de Peppo *et al.*, 2010a; de Peppo *et al.*, 2010b; Evseenko *et al.*, 2010; Kopher *et al.*, 2010; Lian *et al.*, 2007; Smith *et al.*, 2009; Trivedi and Hematti, 2007).

In summary, multiple approaches have attempted to achieve the most direct and efficient derivation of hESC-MSCs. A variety of studies have compared using the embryoid body (EB) step versus omitting this step, using multiple media formulations with and without serum, and using feeder-free cultures versus co-culture. These reports greatly contributed to the field, however, a concensus on the most appropriate method of isolation and culture is absolutely necessary to make hESC-MSC based therapies in a clinical setting a reality.

3.2 Osteoprogenitor cell differentiation from hESCs

Currently, there are major gaps in the knowledge about the growth factors and three-dimensional milieu that influence and direct osteoblast differentiation. The generation of osteoprogenitors from hESC-MSCs has been shown to be successful as evidenced by osteogenic gene expression of runt-related transcription factor 2 (Runx2), collagen type 1A (Col1A1), bone-specific alkaline phosphatase (ALP), and osteocalcin (OCN); mineralized matrix confirmed by von Kossa and Alizarin Red staining; bone nodule formation *in vitro;* and bone formation *in vivo* in diffusion chambers and transplants to orthotopic sites (Duplomb, 2007). One of the first differentiation studies used cultured hESCs in the presence of defined osteogenic supplements for 21 days, and was able to demonstrate mineralization and induction of osteoblastic marker expression (Sotille *et al.*, 2003). Human ESCs have been co-cultured with primary bone derived cells (PBDs) to induce osteoblast differentiation without the addition of exogenous factors, and cultured *in vitro* in the presence of known osteogenic factors without the embryoid body (EB) formation step - both studies confirming that hESCs have the capacity to differentiate into osteoblasts (Ahn *et al.*, 2006; Karp *et al.*, 2006). Whereas, other findings suggest that 12 day EB-derived hESC-MSCs are equally capable of undergoing multilineage differentiation *in vitro* (Cao *et al.*, 2005). It has also been shown that hESC-MSCs can not only differentiate into functional osteoblasts and adipocytes and express markers characteristic of hMSCs, but they can also be successfullly transduced with an osteogenic lineage specific Col2.3-GFP lentivirus in order to track and isolate cells as they underwent differentiation. The transgene construct used has been shown to be a useful tool for studying hBMSC differentiation (Brown *et al.*, 2009). When the hESC-MSCs began as pre-osteoblasts there was low GFP expression, however, increased GFP expression was detected after 28 days culture in osteogenic medium, suggesting that hES-MSCs differentiated into mature osteoblasts. The ability to track differentiation allowed the isolation of osteoprogenitor cells from the derived hESC-MSC population. These studies suggest that in particular, the osteoprogenitor populations derived from hESCs have tremendous potential, and can serve as a tool through which we can characterize early bone development and cellular behavior on bone-related biomaterials.

3.3 Gene transcription and proteomic array analyses

The therapeutic capacity of hESC-MSCs to treat a variety of diseases lies within their capability to different into numerous cell phenotypes to repair or regenerate tissues and organs. However, it remains to be determined if transplanted MSCs, whether of hESC or adult stem cell origin, contribute to and integrate within the majority of newly formed tissue, or perhaps via paracrine action mediate and stimulate host repair and regeneration. To that end, investigation into the therapeutic potential of the hESC-MSC paracrine proteome has been conducted. Within the study, defined serum-free culture medium was conditioned by hESC-MSCs and subsequently analyzed via multidimensional protein identification and cytokine antibody array analysis (Sze *et al.*, 2007). The array data revealed over 200 unique gene products that play a role in biological processes such as metabolism, defense, response, and tissue differentiation including vascularization, hematopoiesis, and skeletal development. These processes and pathways are associated with numerous cellular processes that are activated to participate in injury, repair, and regeneration, as well as to facilitate immune cell migration to the site of injury, ECM remodeling, and increases in cellular metabolism (Sze *et al.*, 2007). The identification of a large number of MSC secretory products that can act as paracrine modulators provides insight into the potential mechanism of action by which hESC-MSCs may participate in tissue repair and disease treatment.

Another study investigated the gene expression profile of differentiating hESC-MSCs and reported that during derivation major transcriptional changes occurred, resulting in an expression profile very similar to that of hMSCs (de Peppo *et al.*, 2010b). The major questions addressed were how the transcriptome may be affected by the hESC-MSC derivation process and whether hESCs and their MSC derivatives were distinct or equivalent to one another. The findings in the hESC-MSC population revealed a down-regulation in pluripotency genes such as the *OCT* family of genes, *NANOG, TDGF1, LIN28, GDF3,* and *ZIC3,* down regulation in tumor development *p53*-associated genes *LTBP2* and *TFAP2A*, up-regulation of mesodermal lineage commitment genes such as *RUNX2, TGBR2, BMPR2,* and *TFAP2A,* and up-regulation of genes supportive of craniofacial development and osteogenesis such as *DLX1, DLX2,* and *MSX1.* Lastly, and importantly, the immunological profile of hESC-MSCs displayed lower expression than hMSCs of HLA-ABC and HLA-DR, two markers characteristic of the inflammatory immune response. These findings suggest that the hESC-MSCs may be more immuno-privileged than hMSCs, thus another piece of evidence supporting the notion that hESC-MSCs represent a suitable alternative for cell transplantation therapies (Romieu-Mourez *et al.*, 2007; de Peppo *et al.*, 2010a).

3.4 Epithelial-mesenchymal transition

Cells within the body are derived from a single cell, with variations of cell phenotypes resulting from expression of a specific and defined transcriptome, thus further imparting diversity in cellular signaling and function. Epithelia are considered to be highly plastic during embryogenesis and have the ability to shuttle back and forth between mesenchyme and epithelia through the process know as epithelial-mesenchymal transition (EMT). It is one mechanism that gives rise to mesenchymal-like behavior to cells in numerous different settings (Kalluri, 2009). Historically, it has been proposed that epithelial cells have to be terminally differentiated to in order to perform defined functions involved in organ development. However, experimental evidence has suggested that epithelial cells can alter their phenotype based on the influence of microenvironment (Boyer *et al.*, 2000). Therefore, EMT has been accepted as a mechanism by which fibroblasts and mesenchymal cells are formed in injured tissues. In the adult, the process of EMT occurs during tissue regeneration and wound healing by facilitating mesenchymal cell migration to invade surrounding tissues. This was described as one of the three EMT subtypes that occurs, and is also suggested to be an underlying mechanism for derivation of hESC-MSCs (Zeisberg and Neilson, 2009; Ullmann *et al.*, 2007).

It has been reported that hESCs grown in monolayer in feeder-free conditions, without MEFs or other supporting cells, form uniform sheets of epithelial cells after removal from standard feeder culture systems (Boyd *et al.*, 2009; Ullmann *et al.*, 2007). The uniform epithelial sheets exhibit characteristic mesodermal gene expression patterns that appear to undergo EMT that results in a highly proliferative population of cells that over time become uniformly homogenous with a mesenchymal stem cell morphology. It is in fact these homogenous cells that many researchers identify as hESC-MSCs, which have the ability to differentiate along multiple mesenchymal cell lineages *in vitro*. More specifically, these studies find that the hESCs that underwent mesenchymal differentation in monolayer culture were over 80% positive for E-cadherin, a characteristic epithelium marker, and maintained expression while cell morphology changed. Additionally, the cells that were undergoing apparent EMT were positive for the characteristic markers such as CD73, CD90, CD 105 and CD166, and negative for CD31, CD34, CD45, CD133 and CD146, further confirming the formation of a mesenchymal progenitor cell population (Boyd *et al.*, 2009). The key significance of these studies is the finding that hESCs are behaving in culture in a manner similar to that of normal embryogenesis, thus underscoring the importance of using hESCs as a tool for better understanding overall human development.

3.5 Tissue engineering strategies for human clinical applications

A major challenge for using stem cells in a clinical setting is the need to identify an ideal stem cell candidate that is multipotent while retaining its self-renewal capacity. Although hMSCs make a useful source of progenitor cells for tissue engineering strategies, as evidenced by their multipotent potential and immunosuppressive characteristics, their limited proliferative and differentiation capacity represent an obstacle for therapeutic application. In contrast, hESCs with their ability to proliferate indefinitely *in vitro* and multi lineage differentiation capacity represent an unlimited source of progenitor cells, specifically, mesenchymal progenitor cells. Therefore, it is necessary to establish clinical-grade GMP protocols for the derivation, identification, and isolation of hES-MSCs, to produce large quantities of genotypically homogenous progenitor cells that can be modified, and to fully characterize these cells for tissue regeneration strategies. Tissue engineering is an emerging field of research aimed at regenerating functional tissues by combining cells with a supporting substrate or biomaterial that possesses design characteristics that deliver progenitor cells and important signalling molecules in a spatially and temporally controlled manner, while promoting vascularization and tissue invastion into the interior of the scaffold. Ideally, biomimetic scaffolds designed for hESC-MSC based tissue engineering strategies would contain inductive signaling cues for proliferation and differentiation, possess composite material properties that conferred the ability to generate multi-layered hybrid tissues, and have tunable three-dimensional geometrical architecture that appropriately restores form and function to anatomical defects or diseased tissues.

Within the hESC field, the use of 3D scaffolds has been employed in only a few reports (Arpornmaeklong *et al.*, 2009; Ferreira *et al.*, 2007; Kaufman *et al.*, 2010; Levenberg *et al.*, 2003; Kim *et al.*, 2007; Kuznetsov *et al.*, 2010; Smith *et al.*, 2009). Investigators have used collagen scaffolds for hepatocye differentiation, and porous polylactic/polyglycolic biomaterial sponges to direct neural, chondrogenic, or hepatocytic lineages (Levenberg et al., 2003). While other studies have shown that 3D porous alginate scaffolds to provide a conducive environment for generation of well-vascularized embryoid body derived hESCs (Ferreira et al., 2007). Within the bone tissue engineering field, the use of architecturally designed scaffolds with hESC-MSCs is seen even less frequently. It has been reported that hESC-MSCs were capable of forming bone tissue *in vivo* when implanted subcutaneously after 8 weeks in the presence of BMP-2 (Kim *et al.*, 2007). In 2009, Arpornmaeklong et al. reported the influence of composite collagen scaffolds on the osteogenic differentiation of hESC-MSCs *in vitro* as indicated by osteogenic gene induction, increased ALP activity, and the presence of mature bone ECM proteins; all of which are characteristic of the osteoblast phenotype. From an *in vivo* standpoint, enriched osteoprogenitor cells were encaspulated in fibrin gels mixed with ceramic particles and implanted in a rat calvarial defect model. After six weeks, the identification of transplanted hESC-MSCs in newly formed bone verified the role that MSCs derived from hESCs played in the bone regeneration process (Arpornmaeklong *et al.*, 2009). Another study demonstrated that hESC-MSCs can form mineralized tissue *in vitro* when cultured on 3D nanofibrous polylactic acid (PLLA) in the presence of BMP-7, illustrating the capability of hESC-MSCs to differentiate in 3D culture for bone regeneration purposes (Smith *et al.*, 2009). Most recently, a comprehensive study investigated multiple media formulations and cell culture conditions for efficient derivation of a homogenous hESC-MSC population. To determine their *in vivo* osteogenic potential, cells were implanted up to 16 wks with biphasic ceramic particles and histology revealed cells of human origin were embedded with the bone, including broad areas of multiple intertwining trabeculae (Kuznetsov *et al.*, 2010).

It is hypothesized that the hESC-MSCs not only require a 3D biomaterial, but also inductive cues. This suggests that for tissue formation, hESCs may require additional biological cues such as pro-osteogenic factors for attachment, proliferation, and directed differentiation on biomaterials. For bone formation specifically, hESCs may require an osteoconductive biomaterial with not only the appropriate scaffold architecture, but one that also can associate cellular and molecular elements to increase cellular response to the biomaterial.

4. Conclusion

Human ESC research is a rapidly developing field, and has the potential to impact the medical and scientific community immensely. It is vitally important that we continue to explore hESC biology in order to realize the potential of hESCs to cure diseases. The derivation of mesenchymal stem cells from human embryonic stem cells is an area of active investigation in that hESC-MSCs potentially offer insight into embryonic mesodermal development events, as well as provide information about underlying differentiation mechanisms and signaling pathways that have been unclear heretofor. In addition to elucidating the mechanisms by which hESC-MSCs differentiate, it is equally important to better understand how the 3D biomaterial microenvironment can be manipulated to direct and control this process. In general, stem cell research advances the knowledge and understanding of how an organism develops and how progenitor cells migrate from the stem cell niche to the site of damaged or diseased tissue. To improve upon the overall quality of human health, scientists must continue to work collaboratively with clinicians to drive translational "bench-to-bedside" research. To this end, extensive investigation into the xeno-free derivation, robustness, and non-tumorigenic safety of hESC-MSCs will be absolutely necessary as the field progresses toward the realization of clinical tissue engineering and regenerative medicine therapies.

5. References

Aggarwal S, Pittenger MF (2005). Human mesenchymal stem cells modulate allogeneic immune cell responses. *Blood* 105(4):1815-22.

Ahn S, Kim S, Park K, Moon S, Lee H, Kim G, Lee Y, Park K, Cha K, Chung H (2006). Primary bone-derived cells induce osteogenic differentiation without exogenous factors in human embryonic stem cells. *Biochem Biophys Res Commun* 340(2):403-8.

Arpornmaeklong P, Brown SE, Wang Z, Krebsbach PH (2009). Phenotypic characterization, osteoblastic differentiation, and bone regeneration capacity of human embryonic stem cell-derived mesenchymal stem cells. *Stem Cells Dev* 18(7):955-68.

Ashton BA, Allen TD, Howlett CR, Eaglesom CC, Hattori A, Owen M (1980). Formation of bone and cartilage by marrow stromal cells in diffusion chambers in vivo. *Clin Orthop Relat Res* (151):294-307.

Bab I, Howlett CR, Ashton BA, Owen ME (1984). Ultrastructure of bone and cartilage formed in vivo in diffusion chambers. *Clin Orthop Relat Res* (187):243-54.

Bab I, Ashton BA, Gazit D, Marx G, Williamson MC, Owen ME (1986). Kinetics and differentiation of marrow stromal cells in diffusion chambers in vivo. *J Cell Sci* 84:139-51.

Bab I, Passi-Even L, Gazit D, Sekeles E, Ashton BA, Peylan-Ramu N, Ziv I, Ulmansky M (1988). Osteogenesis in in vivo diffusion chamber cultures of human marrow cells. *Bone Miner* 4(4):373-86.

Barberi T, Willis LM, Socci ND, Studer L (2005). Derivation of Multipotent Mesenchymal Precursors from Human Embryonic Stem Cells. *PLoS Medicine* 2(6):e161.

Baschetti R (2005). Evolutionary, biological origins of morality: implications for research with human embryonic stem cells. *Stem Cells Dev* 14(3):239-47.

Beresford JN (1989). Osteogenic stem cells and the stromal system of bone and marrow. *Clin Orthop Relat Res* 240:270-80.

Bianco P, Riminucci M, Gronthos S, Robey P (2003). Bone Marrow Stromal Stem Cells: Nature, Biology, and Potential Applications. *Stem Cells* 19(3):180-92.

Boyd NL, Robbins KR, Dhara SK, West FD, Stice SL (2009). Human embryonic stem cell-derived mesoderm-like epithelium transitions to mesenchymal progenitor cells. *Tissue Eng Part A* 15(8):1897-907.

Boyer B, Valles AM, Edme N (2000). Induction and regulation of epithelial-mesenchymal transitions. *Biochem Pharmacol* 60(8):1091-9.

Brown SE, Tong W, Krebsbach PH (2009). The derivation of mesenchymal stem cells from human embryonic stem cells. *Cells Tissues Organs* 189(1-4):256-60.

Cao T, Heng B, Ye C, Liu H, Toh W, Robson P, Li P, Hong Y, Stanton L (2005). Osteogenic differentiation within intact human embryoid bodies result in a marked increase in osteocalcin secretion after 12 days of in vitro culture, and formation of morphologically distinct nodule-like structures. *Tissue and Cell* 37(4):325-34.

Caplan AI (2005). Review: mesenchymal stem cells: cell-based reconstructive therapy in orthopedics. *Tissue Eng* 11(7-8):1198-211.

Cheng L, Hammond H, Ye Z, Zhan X, Dravid G (2003). Human adult marrow cells support prolonged expansion of human embryonic stem cells in culture. *Stem Cells* 21(2):131-42.

de Peppo GM, Sjovall P, Lenneras M, Strehl R, Hyllner J, Thomsen P, Karlsson C (2010a). Osteogenic Potential of Human Mesenchymal Stem Cells and Human Embryonic Stem Cell-Derived Mesodermal Progenitors: A Tissue Engineering Perspective. *Tissue Eng Part A*.

de Peppo GM, Svensson S, Lenneras M, Synnergren J, Stenberg J, Strehl R, Hyllner J, Thomsen P, Karlsson C (2010b). Human embryonic mesodermal progenitors highly resemble human mesenchymal stem cells and display high potential for tissue engineering applications. *Tissue Eng Part A* 16(7):2161-82.

Duplomb L, Dagouassat, M, Jourdon, P, Heymann, D (2007). Concise Review: Embryonic Stem Cells: A New Tool to Study Osteoblast and Osteoclast Differentiation. *Stem Cells* 25(3):544-52.

Evseenko D, Zhu Y, Schenke-Layland K, Kuo J, Latour B, Ge S, Scholes J, Dravid G, Li X, MacLellan WR, Crooks GM (2010). Mapping the first stages of mesoderm commitment during differentiation of human embryonic stem cells. *Proc Natl Acad Sci U S A* 107(31):13742-7.

Friedenstein A, Kuralesova AI (1971). Osteogenic precursor cells of bone marrow in radiation chimeras. *Transplantation* 12(2):99-108.

Friedenstein AJ, Chailakhjan RK, Lalykina KS (1970). The development of fibroblast colonies in monolayer cultures of guinea-pig bone marrow and spleen cells. *Cell Tissue Kinet* 3(4):393-403.

Friedenstein AJ, Chailakhyan RK, Latsinik NV, Panasyuk AF, Keiliss-Borok IV (1974a). Stromal cells responsible for transferring the microenvironment of the hemopoietic tissues. Cloning in vitro and retransplantation in vivo. *Transplantation* 17(4):331-40.

Friedenstein AJ, Deriglasova UF, Kulagina NN, Panasuk AF, Rudakowa SF, Luria EA, Ruadkow IA (1974b). Precursors for fibroblasts in different populations of hematopoietic cells as detected by the in vitro colony assay method. *Exp Hematol* 2(2):83-92.

Genbacev O, Krtolica A, Zdravkovic T, Brunette E, Powell S, Nath A, Caceres E, McMaster M, McDonagh S, Li Y, Mandalam R, Lebkowski J, Fisher SJ (2005). Serum-free derivation of human embryonic stem cell lines on human placental fibroblast feeders. *Fertil Steril* 83(5):1517-29.

Gronthos S, Graves SE, Ohta S, Simmons PJ (1994). The STRO-1+ Fraction of Adult Human Bone Marrow Contains the Osteogenic Precursors. *Blood* 84(12):4164-73.

Gronthos S, Zannettino AC, Graves SE, Ohta S, Hay SJ, Simmons PJ (1999). Differential cell surface expression of the STRO-1 and alkaline phosphatase antigens on discrete developmental stages in primary cultures of human bone cells. *J Bone Miner Res* 14(1):47-56.

Gruen L, Grabel L (2006). Concise review: scientific and ethical roadblocks to human embryonic stem cell therapy. *Stem Cells* 24(10):2162-9.

Haynesworth SE, Baber MA, Caplan AI (1992a). Cell surface antigens on human marrow-derived mesenchymal cells are detected by monoclonal antibodies. *Bone* 13(1):69-80.

Haynesworth SE, Goshima J, Goldberg VM, Caplan AI (1992b). Characterization of cells with osteogenic potential from human marrow. *Bone* 13(1):81-8.

Horwitz EM, Gordon PL, Koo WK, Marx JC, Neel MD, McNall RY, Muul L, Hofmann T (2002). Isolated allogeneic bone marrow-derived mesenchymal cells engraft and stimulate growth in children with osteogenesis imperfecta: Implications for cell therapy of bone. *Proc Natl Acad Sci U S A* 99(13):8932-7.

Hovatta O, Mikkola M, Gertow K, Stromberg AM, Inzunza J, Hreinsson J, Rozell B, Blennow E, Andang M, Ahrlund-Richter L (2003). A culture system using human foreskin fibroblasts as feeder cells allows production of human embryonic stem cells. *Hum Reprod* 18(7):1404-9.

Jaiswal N, Haynesworth SE, Caplan AI, Bruder SP (1997). Osteogenic differentiation of purified, culture-expanded human mesenchymal stem cells in vitro. *J Cell Biochem* 64(2):295-312.

Kalajzic I, Kalajzic Z, Clark S, Lichtler A, Rowe D (2002). Use of Col1a1GFP transgenes to identify subpopulations of cells at different stages of the osteoblast lineage. *Journal of Bone and Mineral Research* 17(1):15-25.

Kalajzic I, Staal A, Yang W, Wu Y, Johnson S, Feyen J, Krueger W, Maye P, Yu F, Zhao Y, Kuo L, Gupta R, Achenie L, Wang H, Shin D, Rowe D (2005). Expression profile of osteoblastic lineage at defined stages of differentiation. *J Biol Chem* 280(26):24618-26.

Kalluri R (2009). EMT: when epithelial cells decide to become mesenchymal-like cells. *J Clin Invest* 119(6):1417-9.

Karlsson C, Emanuelsson K, Wessberg F, Kajic K, Axell MZ, Eriksson PS, Lindahl A, Hyllner J, Strehl R (2009). Human embryonic stem cell-derived mesenchymal progenitors-Potential in regenerative medicine. *Stem Cell Res.*

Karner E, Unger C, Sloan AJ, Ahrlund-Richter L, Sugars RV, Wendel M (2007). Bone matrix formation in osteogenic cultures derived from human embryonic stem cells in vitro. *Stem Cells Dev* 16(1):39-52.

Karp J, Ferreira L, Khademhosseini A, Kwon A, Yeh J, Langer R (2006). Cultivation of human embryonic stem cells without the embryoid body step enhances osteogenesis in vitro. *Stem Cells* 24(4):835-43.

Knowles LP (2004). A regulatory patchwork--human ES cell research oversight. *Nat Biotechnol* 22(2):157-63.

Koc ON, Gerson SL, Cooper BW, Dyhouse SM, Haynesworth SE, Caplan AI, Lazarus HM (2000). Rapid hematopoietic recovery after coinfusion of autologous-blood stem cells and culture-expanded marrow mesenchymal stem cells in advanced breast cancer patients receiving high-dose chemotherapy. *J Clin Oncol* 18(2):307-16.

Kopher RA, Penchev VR, Islam MS, Hill KL, Khosla S, Kaufman DS (2010). Human embryonic stem cell-derived CD34(+) cells function as MSC progenitor cells. *Bone.*

Krebsbach PH, Kuznetsov SA, Satomura K, Emmons RV, Rowe DW, Robey PG (1997). Bone formation in vivo: comparison of osteogenesis by transplanted mouse and human marrow stromal fibroblasts. *Transplantation* 63(8):1059-69.

Krebsbach PH, Kuznetsov SA, Bianco P, Robey PG (1999). Bone marrow stromal cells: characterization and clinical application. *Crit Rev Oral Biol Med* 10(2):165-81.

Kuznetsov S, Cherman N, Gehron Robey P In Vivo Bone Formation by Progeny of Human Embryonic Stem Cells. *Stem Cells Dev.*

Le Blanc K, Pittenger M (2005). Mesenchymal stem cells: progress toward promise. *Cytotherapy* 7(1):36-45.

Lian Q, Lye E, Suan Yeo K, Khia Way Tan E, Salto-Tellez M, Liu TM, Palanisamy N, El Oakley RM, Lee EH, Lim B, Lim SK (2007). Derivation of clinically compliant MSCs from CD105+, CD24- differentiated human ESCs. *Stem Cells* 25(2):425-36.

Olivier EN, Rybicki AC, Bouhassira EE (2006). Differentiation of Human Embryonic Stem Cells in Bipotent Mesenchymal Stem Cells. *Stem Cells* 24(8):1914-22.

Owen M, Friedenstein AJ (1988). Stromal stem cells: marrow-derived osteogenic precursors. *Ciba Found Symp* 136:42-60.

Pittenger M, Mackay A, Beck S (1996). Human mesenchymal stem cells can be directed into chondrocytes, adipocytes or osteocytes. *Mol Biol Cell* 7:305a.

Romieu-Mourez R, Francois M, Boivin MN, Stagg J, Galipeau J (2007). Regulation of MHC class II expression and antigen processing in murine and human mesenchymal stromal cells by IFN-gamma, TGF-beta, and cell density. *J Immunol* 179(3):1549-58.

Smith LA, Liu X, Hu J, Wang P, Ma PX (2009). Enhancing osteogenic differentiation of mouse embryonic stem cells by nanofibers. *Tissue Eng Part A* 15(7):1855-64.

Sotille V, Thomson A, McWhir J (2003). In vitro osteogenic differentiation of human ES cells. *Cloning and Stem Cells* 5(2):149-55.

Stewart K, Walsh S, Screen J, Jefferiss CM, Chainey J, Jordan G, Beresford JN (1999). Further Characterization of Cells Expressing STRO-1 in Cultures of Adult Human Bone Marrow Stromal Cells. *Journal of Bone and Mineral Research* 14(8):1345-1356.

Stojkovic P, Lako M, Stewart R, Przyborski S, Armstrong L, Evans J, Murdoch A, Strachan T, Stojkovic M (2005). An autogeneic feeder cell system that efficiently supports growth of undifferentiated human embryonic stem cells. *Stem Cells* 23(3):306-14.

Sze SK, de Kleijn DP, Lai RC, Khia Way Tan E, Zhao H, Yeo KS, Low TY, Lian Q, Lee CN, Mitchell W, El Oakley RM, Lim SK (2007). Elucidating the secretion proteome of human embryonic stem cell-derived mesenchymal stem cells. *Mol Cell Proteomics* 6(10):1680-9.

Thomson J, Itskovitz-Eldor J, Shapiro S, Waknitz M, Swiergiel J, Marshall V, Jones J (1998). Embryonic stem cell lines derived from human blastocysts. *Science* 282(5391):1145-47.

Tong W, Brown SE, Krebsbach PK (2007). Human Embryonic Stem Cells Undergo Osteogenic Differentiation in Human Bone Marrow Stromal Cell Microenvironments. *Journal of Stem Cells* 2(3):139-47.

Tremoleda JL, Forsyth NR, Khan NS, Wojtacha D, Christodoulou I, Tye BJ, Racey SN, Collishaw S, Sottile V, Thomson AJ, Simpson AHWR, Noble BS, McWhir J (2008). Bone Tissue Formation from Human Embryonic Stem Cells In Vivo. *Cloning and Stem Cells* 1(10):119-131.

Trivedi P, Hematti P (2007). Simultaneous generation of CD34+ primitive hematopoietic cells and CD73+ mesenchymal stem cells from human embryonic stem cells cocultured with murine OP9 stromal cells. *Exp Hematol* 35(1):146-54.

Ullmann U, In't Veld P, Gilles C, Sermon K, De Rycke M, Van de Velde H, Van Steirteghem A, Liebaers I (2007). Epithelial-mesenchymal transition process in human embryonic stem cells cultured in feeder-free conditions. *Mol Hum Reprod* 13(1):21-32.

Villa-Diaz LG, Nandivada H, Ding J, Nogueira-de-Souza NC, Krebsbach PH, O'Shea KS, Lahann J, Smith GD (2010). Synthetic polymer coatings for long-term growth of human embryonic stem cells. *Nat Biotechnol* 28(6):581-3.

Wagers A, Weissman I (2004). Plasticity of adult stem cells. *Cells* 116(5):639-48.

Wang Q, Fang ZF, Jin F, Lu Y, Gai H, Sheng HZ (2005). Derivation and growing human embryonic stem cells on feeders derived from themselves. *Stem Cells* 23(9):1221-7.

Xu C, Jiang J, Sottile V, McWhir J, Lebkowski J, Carpenter MK (2004). Immortalized fibroblast-like cells derived from human embryonic stem cells support undifferentiated cell growth. *Stem Cells* 22(6):972-80.

Zeisberg M, Neilson EG (2009). Biomarkers for epithelial-mesenchymal transitions. *J Clin Invest* 119(6):1429-37.

Part 5

Hormonal Signals that Regulate the Differentiation of Endodermal Cells - Thymogenesis

34

Generation of Thyrocytes from Embryonic Stem Cells

Reigh-Yi Lin
Department of Medicine, Division of Endocrinology, Diabetes and Bone Disease
Mount Sinai School of Medicine, New York, New York 10029
USA

1. Introduction

The thyroid gland is one of the largest endocrine glands in the body. It is found on the anterior of the neck and is composed of two connected, cone-like lobes. The thyroid inherits its name from the Greek word for "shield", which the related thyroid cartilage resembles. It is the only endocrine organ in the body that can absorb iodine and synthesize the thyroid hormones triiodothyronine (T_3) and thyroxine (T_4), which regulate homeostasis and metabolism and affect the growth and function of the entire body. The thyroid gland also produces another hormone called calcitonin, which regulates calcium homeostasis. The thyroid hormones and calcitonin are produced by two distinct cell types in the thyroid gland. Thyroid follicular cells (also known as thyrocytes) produce the thyroid hormones. They are derived from the endoderm of the pharyngeal floor. Parafollicular cells, derived from neural crest cells within the ultimobranchial bodies of the fourth pharyngeal arch, synthesize calcitonin. Almost 70% of the thyroid gland is composed of thyrocytes arranged in follicles; parafollicular cells are scattered throughout the interfollicular space (Di Lauro, 2003; Di Lauro et al., 1995).

Many thyroid diseases are caused by the abnormal proliferation and differentiation of thyroid cells. These abnormalities can lead to developmental defects and benign and malignant tumor formation. Most commonly, the thyroid gland is either overactive—a condition called hyperthyroidism—or underactive—a condition called hypothyroidism. Congenital hypothyroidism occurs in about one of every 4,000 births, making it the most frequent endocrine disorder in newborns. 85 percent of cases are caused by a disturbance in the thyroid gland's organogenesis that causes the gland to be absent, hypoplastic or located in an unusual position. In the remaining 15 percent of cases, congenital hypothyroidism is caused by defects in thyroid hormone synthesis (Kopp, 2002; LaFranchi, 1999; Van Vliet, 2003).

The function of the thyroid gland is controlled by the hypothalamus and the pituitary gland (Di Lauro, 2003). Although most scientists agree on the causes of disorders involving aberrations of the hypothalamus-pituitary-thyroid axis, the cellular and molecular mechanisms leading to thyroid developmental abnormalities remain largely elusive. Many of the early studies documenting the complex process of thyroid gland development were performed in animals with genetically engineered mutations (Mansouri et al., 1999). Although these sophisticated mouse models have furthered our understanding of the genetic and cellular events that lead to thyroid development *in vivo* (De Felice et al., 1998; Macchia et al., 1998; Parlato et al., 2004; Postiglione et al., 2002; Zannini et al., 1997), their utility is limited by the fact that many important mutations cause embryonic lethality. As a result, cellular models have been generated to screen for genes involved in various aspects of thyroid cell physiology. Important studies have been carried out with established thyroid cell lines (FRTL-5 and PC C13) (Ambesi-Impiombato and Coon, 1979; Zimmermann-Belsing et al., 1998) and primary cultures from fetal or adult thyroids. However, despite their experimental utility, primary and immortalized thyroid cells are themselves not without limitations. Primary thyroid cultures are often contaminated with other cell types. They are also unstable and have a limited life span in culture. And although the widely used FRTL-5 cell line, which originated from a normal rat thyroid, can be propagated indefinitely and retain most of the features of differentiated thyrocytes, FRTL-5 cells are incapable of forming follicles in culture. Furthermore, some FRTL-5 cell clones develop into tumors when injected into nude mice (Ossendorp et al., 1990). Finally, these cell lines are limited in their ability to model human development and provide few clues about the developmental aspects of thyroid disease.

It is only recently that embryonic stem (ES) cells have offered the possibility to establish a robust cell-based model for examining the genetic and epigenetic mechanisms of human disorders from a developmental biology perspective. Originally isolated from the inner cell mass of a blastocyst-stage embryo, ES cells can replicate for long periods of time *in vitro*. They are pluripotent, which means that they can differentiate into all derivatives of the three primary germ layers: ectoderm, mesoderm, and endoderm (Evans and Kaufman, 1981; Martin, 1981; Thomson et al., 1998). Under defined culture conditions, ES cells can be instructed to form cellular aggregates known as embryoid bodies. The differentiation of these embryoid bodies, to a limited extent, recapitulates embryonic development (Keller, 1995): the cellular aggregates first appear as hollow balls, called cystic embryoid bodies, and next form internal structures such as a yolk sac and cardiomyocytes. They are also able to differentiate into the more than 220 cell types found in the human body.

The availability of human ES cells and the ability to access cell populations representing the earliest events of embryogenesis offer an unprecedented opportunity to model human development. Analysis of human ES cells derived from patients with specific genetic disorders will provide novel insights into the disease process and should prove useful for identifying new therapeutic targets. Ultimately, the ability to genetically engineer stem cells may allow clinicians to test the effects of new drugs and to develop clinically relevant screening assays that would not otherwise be possible.

2. Embryonic thyroid gland development

Some of the first insights into thyroid organogenesis came from histological analyses of the developing thyroid glands. The thyroid gland develops from the foramen cecum, the endoderm located in midline of the floor of the pharynx between the first and second pharyngeal pouches. The morphogenesis of the thyroid gland, like that of many endoderm-derived organs, begins with the recruitment of a specific group of cells to the thyroid fate. At human embryonic day 20 (E20), and mouse E8, the median thyroid bud appears as a thickening in the floor of the pharynx. One of the earliest markers defining the thyrocyte lineage is the transcription factor paired box gene 8 (Pax8). Mutations in this gene have been associated with thyroid dysgenesis, thyroid follicular carcinoma and thyroid adenoma

(Macchia et al., 1998). Thyroid transcription factor-1 (TTF-1; also known as Nkx2.1, T/EBP (thyroid-specific-enhancer-binding protein) or TITF1) and thyroid transcription factor-2 (TTF-2; or FoxE1) are also important (Di Lauro, 2003; Di Lauro et al., 1995; Kopp, 2002; Zhang et al., 2006). Expression of these transcription factors in the thyroid gland begins with thyroid morphogenesis and is essential for the development of thyrocytes.

At about E50 in humans and E13.5 in mice, the thyroid diverticulum starts its migration from the pharyngeal floor to its definitive pretracheal position. By E60 (E14 in mouse), the thyrocyte precursors express thyroid stimulating hormone receptor (TSHR). By E70 (E15.5 in mouse), thyroid follicular organization appears with the expression of multiple thyroid-specific proteins essential for thyroid hormone biosynthesis, including sodium-iodide symporter (which transports iodide into thyroid cells), thyroglobulin (the precursor of thyroid hormones), and thyroid peroxidase (the enzyme responsible for thyroglobulin iodination). The organogenesis of the thyroid gland is often disturbed in humans, which leads to a variety of clinical conditions including agenesis, ectopy and hypoplasia. All of these conditions are collectively called thyroid dysgenesis. Currently, one of the biggest challenges in thyroid research is the identification of the key molecules that regulate thyroid development. Dissecting the genetic and signaling pathways involved might facilitate the understanding of the crucial roles of various molecules in these developmental processes.

3. Differentiation of mouse ES cells to thyrocyte lineage

Previous biological tools used to study thyroid development were limited to human thyroid cell lines, mouse xenograft models and organ cultures. However, the recent availability of a range of ES cell lines has allowed us to begin to define key molecular and cellular characteristics of thyrocyte-fated populations. Unique molecular markers including early-stage transcription factors, structural proteins and metabolic regulators facilitate the distinction between undifferentiated ES cells and those that have committed to the thyrocyte lineage.

These molecular markers are particularly important because the thyrocyte lineage is one of the least recognizable cell types in a culture of differentiating ES cells. Unlike cardiovascular lineage cells, which exhibit characteristic, spontaneous, and rhythmic contractions, identification of the thyrocyte lineage by conventional screening is very challenging. One thyrocyte identification strategy relies on an ES cell reporter line in which enhanced green fluorescent protein (GFP) cDNA has been targeted to a specific cell surface protein. This approach has allowed us to monitor thyrocytes in differentiating cultures and to isolate cellular representatives of early developmental stages. Another strategy mimics embryonic development by exposing ES cells and their derivatives to growth factors and hormones that they would normally encounter *in vivo*. One such factor is thyroid stimulating hormone (TSH or thyrotropin). TSH is synthesized and secreted by the pituitary gland to regulate the development and function of the thyroid gland. Previous studies have shown that the binding of TSH to TSHR is an important trophic stimulus of the thyroid gland (Davies et al., 2005). Furthermore, the thyroid gland is much smaller than normal in mutant mouse lines with nonfunctional TSH or TSHR (e.g. $Tshr^{hyt}Tshr^{hyt}$, *TSHR-KO*, $pit^{dw}pit^{dw}$) (Marians et al., 2002; Pichurin et al., 2004; Postiglione et al., 2002). Histological analysis has indicated that thyroids from these mice have fewer follicles and more non-follicle-associated interstitial cells than do wild-type thyroids.

In 2003, my laboratory developed a protocol to convert mouse CCE ES cells into thyrocyte-like cells (Lin et al., 2003). In our protocol, ES cells are first grown in standard media for six days as embryoid bodies and then for an additional 11 days in the presence of TSH. We first observed thyrocyte-like cells after 17 days of differentiation. Detailed analysis of these cells reveals that they have some of the properties of thyrocytes: They express the thyroid markers Pax8, sodium-iodide symporter, thyroglobulin, and thyroid peroxidase as well as functional TSHR (Lin and Davies, 2006; Lin et al., 2003). However, this strategy produced only variable and transient thyrocyte-like cells, which were neither pure nor present in sufficient quantities for additional functional studies.

3.1 TSH-dependent differentiation of thyrocytes from mouse ES cells

We next used a mouse ES cell line with a GFP cDNA targeted to the *TSHR* locus to investigate the role of TSHR in thyroid development *in vitro*. We found that the appearance of GFP-positive cells was dependent on the formation of embryoid bodies from undifferentiated ES cells. In our protocol, we first treat ES cell-derived embryoid bodies with TSH for three days under serum-free conditions to direct the differentiation of the embryoid bodies into the pre-thyrocyte stage. Next, the embryoid bodies are dissociated and a fluorescence-activated cell sorter is used to generate a pure sample of TSHR-expressing cells (based on their GFP expression). Upon reaggregation on Matrigel, GFP-positive cells can be further induced by TSH to differentiate into mature thyrocytes after a period of 21 days (Arufe et al., 2006).

Mouse ES cell-derived thyrocytes arising from this approach are phenotypically similar to thyroid follicles derived from primary tissue cultures (Arufe et al., 2006). The thyrocytes are organized into one or more neofollicle-like clusters that express follicular thyroid markers and they have a functional sodium-iodide symporter (Arufe et al., 2006) as depicted in Fig. 1 and Fig. 2. This observation is of particular interest because cells from immortalized thyroid cell lines such as FRTL-5 cannot form follicles when grown as monolayer in culture. These studies demonstrate that the early development of a mouse thyrocyte is clearly regulated by TSH.

Similarily, Jiang *et al* have demonstrated that, under the influence of TSH and insulin, cells of the mouse E14 ES cell line begin to co-express TSHR, sodium-iodide symporter, thyroid peroxidase and thyroglobulin mRNA (Jiang et al., 2010). Although the ultrastructural features of these cells (identified with electron microscopy) are similar to those of adult thyrocytes, the cells had no signs of secretory vesicles. This may explain why these thyrocytes fail to produce thyroid hormones and suggests that conditions must be further optimized to allow for *in vitro* differentiation of ES cells into functional thyrocytes.

3.2 TSH-independent induction of thyrocytes from mouse ES cells

Recently, the Davies laboratory demonstrated the TSH-independent induction of thyroid endoderm from ES cells by activin A. Activin A is a member of the TGFβ superfamily, which has been shown to be critical to the regulation of endoderm formation *in vitro* and *in vivo* (D'Amour et al., 2005; Kubo et al., 2004; Mfopou et al., 2007; Ninomiya et al., 1999; Rippon et al., 2006). Adding activin A to cultures of embryoid bodies markedly increased endodermal markers, including Gata4, CXCR4 and Foxa2 and α-fetoprotein. In fact, a small population of cells expressing Pax8, TSHR and sodim-iodide symporter can be generated with activin A alone (Ma et al., 2009). Although these findings parallel those of several other reports suggesting that thyroid cell development can occur in the absence of TSH (Marians et al., 2002; Valentine et al., 1994), this strategy failed to sustain a stably differntiated thyrocyte phenotype.

3.3 Insulin and insulin-like growth factor-1 are critical to mouse thyrocyte maturation

Our discovery of the importance of step-wise exposure of ES cells to a synchronized sequence of factors has allowed us to learn more about thyroid differentiation *in vitro.* We have recently used the step-wise administration of activin A, TSH, insulin and insulin-like growth factor-1 to successfully generate mature, thyroglobulin-expressing thyrocytes from pluripotent ES cells. First, activin A and TSH were used to induce the differentiation of definitive endoderm and thyrocyte progenitor cells expressing *Sox17, Foxa2,* and *TSHR.* These progenitor cells were then converted into cellular aggregates that, in the presence of insulin and insulin-like growth factor-1, further differentiated into thyroglobulin-expressing thyrocytes (Arufe et al., 2009). These data show that the addition of insulin and insulin-like growth factor-1 to the late stages of embryoid body culture enable the long-term propagation and differentiation of mature thyrocytes.

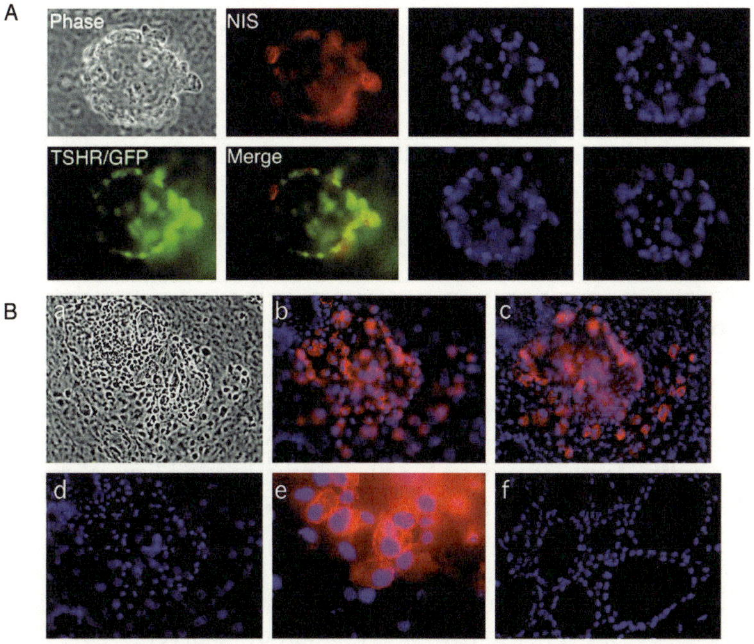

Fig. 1. Thyroid follicle-like clusters derived from mouse *TSHR*$^{+/-}$ ES cells after 21 days of differentiation visualized microscopically (*phase*) or after staining with an antibody to sodium-iodide symporter (*red*). A. TSHR expression is indicated by the green GFP signal. An overlaid image shows the colocalization of sodium-iodide symporter with TSHR (*yellow*). B. Immunofluorescent images of several thyroid follicle-like clusters. a. Phase contrast exposure. b and c, Immunofluorescent staining demonstrating the expression of sodium iodide symporter. Note that a rim of cells several layers thick adhered closely to the sodium iodide symporter-positive cells. d. Nuclear DAPI staining. e. High magnification of anti-sodium iodide symporter immunofluorescence, demonstrating the expression of sodium iodide symporter protein in the plasma membrane. f. Nuclear DAPI staining of native mouse thyroid tissue (from Arufe et al. (2006))

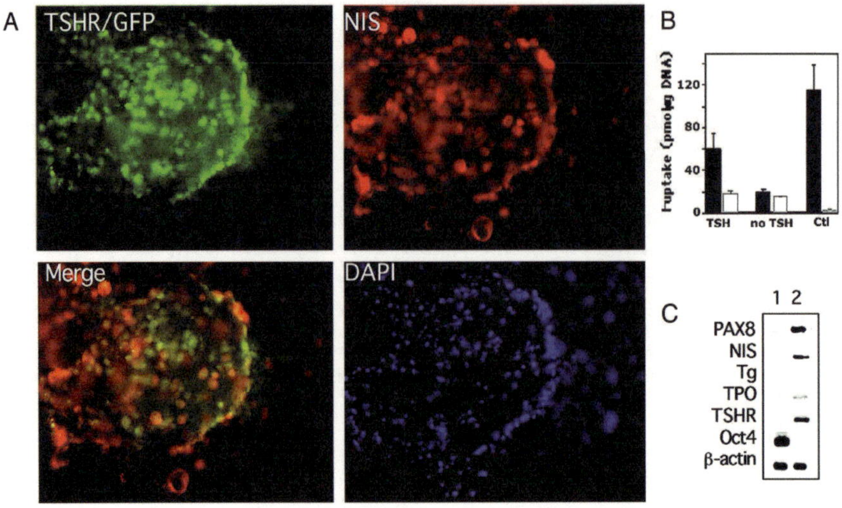

Fig. 2. Thyroid potential of mouse ES cell-derived thyrocytes. A. Several clusters of GFP+ sodium iodide symporter+ cells derived from *TSHR*$^{+/-}$ ES cells after staining with an antibody to sodium iodide symporter (*red*). In merging, some areas showed overlay (*yellow*). *Blue* indicates nuclear DAPI staining. B. Cells treated with TSH showed I^- uptake activity whereas cells maintained in the absence of TSH did not. C. Gene expression analysis by RT-PCR shows differentiation of thyrocytes from ES cells. RNA was isolated from undifferentiated ES cells (lane 1) and thyrocytes grown for 21 days (lane 2) (from Arufe et al. (2006))

4. Generation of thyrocyte progenitors from human ES cells

We are currently using the HES2 human ES cell line to develop a human model of thyroid cell differentiation. Undifferentiated HES2 cells stain strongly for ES-alkaline phosphatase, a non-tissue-specific isozyme precursor expressed at high levels in undifferentiated mouse and human ES cells. They also express the pluripotent stem markers *Nanog, Rex1* and *Oct4.* Together, the expression profiles for these human ES cells contribute to their "stemness" phenotype.

We used combinations of growth factors as well as enrichment through cell surface receptors to design a step-wise differentiation protocol for human ES cells similar to that used in mouse ES cells. This included the formation of embryoid bodies, the induction of endoderm, and the specification of thyroid endoderm. We cultured day 4 embryoid bodies, which express TSHR, in the presence of various combinations of TSH and activin A. RT-PCR analysis confirmed that the endoderm markers *Foxa2* and *Sox17* were strongly expressed by day 4 in cell cultures treated with activin A, but were not expressed in the non-treated cultures. By day 7, *Foxa2* and *Sox17* genes, indicative of endoderm development, were up-regulated in cell cultures treated with activin A. In contrast, treatment with activin A did not significantly alter the expression of the mesoderm marker *c-fms* – the receptor for macrophage colony-stimulating factor – or *Gata1*, a zinc finger transcription factor necessary for hemopoisis (Bang and Goulding, 1996; Simon, 1993; Simon et al., 1992; Yamane et al., 1997). These data suggest that activin A up-regulates endodermal gene expression in the embryoid bodies without inducing mesoderm development. After seven total days of embryoid body development, we found that the expression of *TSHR* and *TTF-1* was also significantly up-regulated in TSH-treated cell cultures (Thomas & Lin unpublished observations). Since *TSHR* and *TTF-1* are nearly undetectable in the absence of TSH, the high levels of expression in these cells indicate that the endoderm has undergone specification to a thyrocyte fate. We are currently optimzing this protocol in an attempt to generate mature thyrocytes from these human ES cells.

5. Concluding remarks

The first ES cell lines were isolated from mouse blastocysts in 1981. Now, almost three decades later, the world's first human clinical trial of ES cell-based therapy in patients with acute spinal cord injury has been approved by the U.S. Food and Drug Administration. Stem cell technology has revolutionized modern biology and medicine and provides us with unique opportunities to explore the molecular mechanisms that control basic biological and disease processes. The studies described in this review represent an important step in the establishment of a new model to investigate key regulators and events of thyrocyte development. They also outline how to generate a viable thyrocyte from mouse ES cells. Although the molecular triggers for directing thyrocyte progenitors toward mature human thyrocytes remain unknown, our investigation of human ES cell differentiation will no doubt provide an attractive model with which to study human thyroid disease. Such cell-based models of disease are important. They could forward our understanding of the molecular basis of thyroid development and aid in the elucidation of the underlying pathology of thyroid disease and the screening for potential therapeutic agents.

Human ES cell lines were first reported in 1998. Since that time, various laboratories have derived a spectrum of genetically diverse human ES cell lines. Although most of the lines expressed similar profiles of genes and surface marker antigens characteristics of human ES cells, there were intrinsic genetic variations. Furthermore, the slow growth of human ES cells may allow the accumulation of genetic changes during prolonged passage. It is therefore important to clearly identify the characteristics of specific human ES cell lines – including clonability, chromosome stability, the expression of a variety of molecular and biochemical markers, and the ability to differentiate into various types of cells and tissues and to form teratomas in immunocompromised animals – before using these lines to screen new drug compounds and develop innovative cell therapies for diseases.

6. Acknowledgements

This work has been partially supported by the National Institutes of Health Grant R01DK068057.

7. References

Ambesi-Impiombato, F. S., and Coon, H. G. (1979). Thyroid cells in culture. *Int Rev Cytol Suppl*, 163-172.

Arufe, M. C., Lu, M., Kubo, A., Keller, G., Davies, T. F., and Lin, R. Y. (2006). Directed differentiation of mouse embryonic stem cells into thyroid follicular cells. *Endocrinology 147*, 3007-3015.

Arufe, M. C., Lu, M., and Lin, R. Y. (2009). Differentiation of murine embryonic stem cells to thyrocytes requires insulin and insulin-like growth factor-1. *Biochem Biophys Res Commun 381,* 264-270.

Bang, A. G., and Goulding, M. D. (1996). Regulation of vertebrate neural cell fate by transcription factors. *Curr Opin Neurobiol 6,* 25-32.

D'Amour, K. A., Agulnick, A. D., Eliazer, S., Kelly, O. G., Kroon, E., and Baetge, E. E. (2005). Efficient differentiation of human embryonic stem cells to definitive endoderm. *Nat Biotechnol 23,* 1534-1541.

Davies, T. F., Ando, T., Lin, R. Y., Tomer, Y., and Latif, R. (2005). Thyrotropin receptor-associated diseases: from adenomata to Graves disease. *J Clin Invest 115,* 1972-1983.

De Felice, M., Ovitt, C., Biffali, E., Rodriguez-Mallon, A., Arra, C., Anastassiadis, K., Macchia, P. E., Mattei, M. G., Mariano, A., Scholer, H., *et al.* (1998). A mouse model for hereditary thyroid dysgenesis and cleft palate. *Nat Genet 19,* 395-398.

Di Lauro, R. (2003). Molecular abnormalities of organogenesis and differentiation of the thyroid gland. *Ann Endocrinol* (Paris) *64,* 53.

Di Lauro, R., Damante, G., De Felice, M., Arnone, M. I., Sato, K., Lonigro, R., and Zannini, M. (1995). Molecular events in the differentiation of the thyroid gland. *J Endocrinol Invest 18,* 117-119.

Evans, M. J., and Kaufman, M. H. (1981). Establishment in culture of pluripotential cells from mouse embryos. *Nature 292,* 154-156.

Jiang, N., Hu, Y., Liu, X., Wu, Y., Zhang, H., Chen, G., Liang, J., Lu, X., and Liu, S. (2010). Differentiation of E14 mouse embryonic stem cells into thyrocytes in vitro. *Thyroid 20,* 77-84.

Keller, G. M. (1995). In vitro differentiation of embryonic stem cells. *Curr Opin Cell Biol 7,* 862-869.

Kopp, P. (2002). Perspective: genetic defects in the etiology of congenital hypothyroidism. *Endocrinology 143,* 2019-2024.

Kubo, A., Shinozaki, K., Shannon, J. M., Kouskoff, V., Kennedy, M., Woo, S., Fehling, H. J., and Keller, G. (2004). Development of definitive endoderm from embryonic stem cells in culture. *Development 131,* 1651-1662.

LaFranchi, S. (1999). Congenital hypothyroidism: etiologies, diagnosis, and management. *Thyroid 9,* 735-740.

Lin, R. Y., and Davies, T. F. (2006). Derivation and characterization of thyrocyte-like cells from embryonic stem cells in vitro. *Methods Mol Biol 330,* 249-261.

Lin, R. Y., Kubo, A., Keller, G. M., and Davies, T. F. (2003). Committing embryonic stem cells to differentiate into thyrocyte-like cells in vitro. *Endocrinology 144,* 2644-2649.

Ma, R., Latif, R., and Davies, T. F. (2009). Thyrotropin-independent induction of thyroid endoderm from embryonic stem cells by activin A. *Endocrinology 150,* 1970-1975.

Macchia, P. E., Lapi, P., Krude, H., Pirro, M. T., Missero, C., Chiovato, L., Souabni, A., Baserga, M., Tassi, V., Pinchera, A., *et al.* (1998). PAX8 mutations associated with congenital hypothyroidism caused by thyroid dysgenesis. *Nat Genet 19,* 83-86.

Mansouri, A., St-Onge, L., and Gruss, P. (1999). Role of Genes in Endoderm-derived Organs. *Trends Endocrinol Metab 10,* 164-167.

Marians, R. C., Ng, L., Blair, H. C., Unger, P., Graves, P. N., and Davies, T. F. (2002). Defining thyrotropin-dependent and -independent steps of thyroid hormone synthesis by using thyrotropin receptor-null mice. *Proc Natl Acad Sci U S A 99,* 15776-15781.

Martin, G. R. (1981). Isolation of a pluripotent cell line from early mouse embryos cultured in medium conditioned by teratocarcinoma stem cells. *Proc Natl Acad Sci U S A 78,* 7634-7638.

Mfopou, J. K., De Groote, V., Xu, X., Heimberg, H., and Bouwens, L. (2007). Sonic Hedgehog and other soluble factors from differentiating embryoid bodies inhibit pancreas development. *Stem Cells 25,* 1156-1165.

Ninomiya, H., Takahashi, S., Tanegashima, K., Yokota, C., and Asashima, M. (1999). Endoderm differentiation and inductive effect of activin-treated ectoderm in Xenopus. *Dev Growth Differ 41,* 391-400.

Ossendorp, F. A., Bruning, P. F., Schuuring, E. M., Van Den Brink, J. A., van der Heide, D., De Vijlder, J. J., and De Bruin, T. W. (1990). Thyrotropin dependent and independent thyroid cell lines selected from FRTL-5 derived tumors grown in nude mice. *Endocrinology 127,* 419-430.

Parlato, R., Rosica, A., Rodriguez-Mallon, A., Affuso, A., Postiglione, M. P., Arra, C., Mansouri, A., Kimura, S., Di Lauro, R., and De Felice, M. (2004). An integrated regulatory network controlling survival and migration in thyroid organogenesis. *Dev Biol 276,* 464-475.

Pichurin, P. N., Pichurina, O., Marians, R. C., Chen, C. R., Davies, T. F., Rapoport, B., and McLachlan, S. M. (2004). Thyrotropin receptor knockout mice: studies on immunological tolerance to a major thyroid autoantigen. *Endocrinology 145,* 1294-1301.

Postiglione, M. P., Parlato, R., Rodriguez-Mallon, A., Rosica, A., Mithbaokar, P., Maresca, M., Marians, R. C., Davies, T. F., Zannini, M. S., De Felice, M., and Di Lauro, R. (2002). Role of the thyroid-stimulating hormone receptor signaling in development and differentiation of the thyroid gland. *Proc Natl Acad Sci U S A 99,* 15462-15467.

Rippon, H. J., Polak, J. M., Qin, M., and Bishop, A. E. (2006). Derivation of distal lung epithelial progenitors from murine embryonic stem cells using a novel three-step differentiation protocol. *Stem Cells 24,* 1389-1398.

Simon, M. C. (1993). Transcription factor GATA-1 and erythroid development. *Proc Soc Exp Biol Med 202,* 115-121.

Simon, M. C., Pevny, L., Wiles, M. V., Keller, G., Costantini, F., and Orkin, S. H. (1992). Rescue of erythroid development in gene targeted GATA-1- mouse embryonic stem cells. *Nat Genet 1,* 92-98.

Thomson, J. A., Itskovitz-Eldor, J., Shapiro, S. S., Waknitz, M. A., Swiergiel, J. J., Marshall, V. S., and Jones, J. M. (1998). Embryonic stem cell lines derived from human blastocysts. *Science 282,* 1145-1147.

Valentine, M., Martin, A., Unger, P., Katz, N., Shultz, L. D., and Davies, T. F. (1994). Preservation of functioning human thyroid "organoids" in the severe combined immunodeficient mouse. III. Thyrotropin independence of thyroid follicle formation. *Endocrinology 134,* 1225-1230.

Van Vliet, G. (2003). Development of the thyroid gland: lessons from congenitally hypothyroid mice and men. *Clin Genet 63,* 445-455.

Yamane, T., Kunisada, T., Yamazaki, H., Era, T., Nakano, T., and Hayashi, S. I. (1997). Development of osteoclasts from embryonic stem cells through a pathway that is c-fms but not c-kit dependent. *Blood 90*, 3516-3523.

Zannini, M., Avantaggiato, V., Biffali, E., Arnone, M. I., Sato, K., Pischetola, M., Taylor, B. A., Phillips, S. J., Simeone, A., and Di Lauro, R. (1997). TTF-2, a new forkhead protein, shows a temporal expression in the developing thyroid which is consistent with a role in controlling the onset of differentiation. *Embo J 16*, 3185-3197.

Zhang, P., Zuo, H., Nakamura, Y., Nakamura, M., Wakasa, T., and Kakudo, K. (2006). Immunohistochemical analysis of thyroid-specific transcription factors in thyroid tumors. *Pathol Int 56*, 240-245.

Zimmermann-Belsing, T., Rasmussen, A. K., and Feldt-Rasmussen, U. (1998). Lack of thyroglobulin synthesis as an indicator of early random dedifferentiation of the Fischer rat thyroid cell line FRTL-5. *Scand J Clin Lab Invest 58*, 529-535.